OXYGEN TRANSPORT
TO TISSUE XIV

ADVANCES IN EXPERIMENTAL MEDICINE AND BIOLOGY

Recent Volumes in this Series

Volume 310
IMMUNOLOGY OF MILK AND THE NEONATE
Edited by Jiri Mestecky, Claudia Blair, and Pearay L. Ogra

Volume 311
EXCITATION–CONTRACTION COUPLING IN SKELETAL,
CARDIAC, AND SMOOTH MUSCLE
Edited by George B. Frank, C. Paul Bianchi, and Henk E. D. J. ter Keurs

Volume 312
INNOVATIONS IN ANTIVIRAL DEVELOPMENT AND THE DETECTION OF
VIRUS INFECTIONS
Edited by Timothy M. Block, Donald Jungkind, Richard L. Crowell,
Mark Denison, and Lori R. Walsh

Volume 313
HEPARIN AND RELATED POLYSACCHARIDES
Edited by David A. Lane, Ingemar Björk, and Ulf Lindahl

Volume 314
CELL–CELL INTERACTIONS IN THE RELEASE OF
INFLAMMATORY MEDIATORS: Eicosanoids, Cytokines, and Adhesion
Edited by Patrick Y-K Wong and Charles N. Serhan

Volume 315
TAURINE: Nutritional Value and Mechanisms of Action
Edited by John B. Lombardini, Stephen W. Schaffer, and Junichi Azuma

Volume 316
OXYGEN TRANSPORT TO TISSUE XIII
Edited by Thomas K. Goldstick, Michael McCabe, and David J. Maguire

Volume 317
OXYGEN TRANSPORT TO TISSUE XIV
Edited by Wilhelm Erdmann and Duane F. Bruley

Volume 318
NEUROBIOLOGY OF ESSENTIAL FATTY ACIDS
Edited by Nicolas G. Bazan, Mary G. Murphy, and Gino Toffano

A Continuation Order Plan is available for this series. A continuation order will bring delivery of each new volume immediately upon publication. Volumes are billed only upon actual shipment. For further information please contact the publisher.

OXYGEN TRANSPORT TO TISSUE XIV

Edited by

Wilhelm Erdmann

Erasmus University
Rotterdam, The Netherlands

and

Duane F. Bruley

University of Maryland Baltimore County
Baltimore, Maryland

SPRINGER SCIENCE+BUSINESS MEDIA, LLC

Library of Congress Cataloging-in-Publication Data

Oxygen transport to tissue XIV / edited by Wilhelm Erdmann and Duane
  F. Bruley.
      p.   cm. -- (Advances in experimental medicine and biology ; v.
  317)
      "Proceedings of the Nineteenth Annual Meeting of the International
  Society on Oxygen Transport to Tissue, held August 24-30, 1991, in
  Willemstad, Curaçao"--T.p. verso.
    Includes bibliographical references and indexes.
      ISBN 978-1-4613-6516-7   ISBN 978-1-4615-3428-0 (eBook)
      DOI 10.1007/978-1-4615-3428-0
      1. Tissue respiration--Congresses. 2. Oxygen--Physiological
  transport--Congresses.   I. Erdmann, W. (Wilhelm), 1940-   .
  II. Bruley, Duane F. III. International Society on Oxygen Transport
  to Tissue. Meeting (19th : 1991 : Willemstad, Curaçao) IV. Title:
  Oxygen transport to tissue 14. V. Title: Oxygen transport to tissue
  fourteen. VI. Series.
      [DNLM: 1. Biological Transport--congresses. 2. Oxygen--blood-
  -congresses. 3. Oxygen Consumption--physiology--congresses.
  4. Spectrophotometry, Infrared--congresses.   W1 AD559 v.317 / WF
  110 098 1991]
  QP121.O88 1992
  612.2'2--dc20
  DNLM/DLC
  for Library of Congress   92-16987
                                                                  CIP

Proceedings of the nineteenth annual meeting of the International Society
on Oxygen Transport to Tissue, held August 24-30, 1991,
in Willemstad, Curaçao

© 1992 Springer Science+Business Media New York
Originally published by Plenum Press, New York in 1992

INTERNATIONAL SOCIETY ON OXYGEN TRANSPORT TO TISSUE
1990—1991

Officers

President:	W. Erdmann, The Netherlands/U.S.A.
President-Elect:	P. Vaupel, Germany
Past President:	M. McCabe, Australia
Secretary:	N.S. Faithfull, U.S.A.
Treasurer:	S.N. Cain, U.S.A.

Executive Committee

D.F. Bruley, U.S.A.	D.T. Delpy, United Kingdom
K. Groebe, Germany	A. Hudetz, Hungary
I.S. Longmuir, U.S.A.	H. Metzger, Germany
N. Sato, Japan	Z. Turek, The Netherlands

CURAÇAO MEETING, AUGUST 24–30, 1991

General Organizing Committee

W. Erdmann (chairman)	D. Haas (general secretary)
D. Bruley (vice-chairman)	L. Visser (scientific secretary)

Local Organizing Committee

D. Kibbelaar	G. Fabian	V. de Man
F. Pinedo	G. Werleman	J. de Windt

American Coordinator	**South American Coordinator**
A. Grogono, U.S.A.	F.J. Rua, Colombia

International Scientific Committee
(in addition to ISOTT officers and ISOTT Executive Committee)

SPONSORS

We are most grateful for the financial support for the 1991 ISOTT Meeting received from the following:

Boehringer Ingelheim
Hewlett Packard
KLM
Lilly Nederland
Medical System Corp.
Ortomed
Physio
Stöpler
Tonometrics
Viggo Spectramed

Sponsoring Organizations in Curaçao

ABN
Biomedical
Curaçaosche Zuurstoffabriek
Curoil N.V.
St. Elisabeth Hospital
Refineria Isla Curaçao
Visser Trading Curaçao N.V.

PREFACE

The International Society on Oxygen Transport to Tissue (ISOTT) was founded in 1973 as a scientific society providing a forum for bioengineers, basic scientists (physiologists, biochemists and physicists) and clinicians (including anesthesiologists, intensive care specialists, pediatricians, neonatologists, internists, surgeons and other specialists) to facilitate the exchange of scientific information among those interested in any aspect of the transport and/or utilization of oxygen in tissues. From the ranks of its members, many fundamental discoveries and inventions have been made involving the many aspects of oxygen transport and utilization by biological tissues. The ISOTT proceedings, now in its 14th edition, has become a standard work in the field as witnessed by the inclusion in the Science Citation Index of all volumes published so far.

The 19th ISOTT Meeting was held in Curaçao from August 24th through August 30th, 1991. The Curaçao Meeting attracted 145 registrants and 45 accompanying persons. The format originated by Dr. Ian Longmuir in 1985, consisting of posters accompanied by an abbreviated oral summary, was again successfully handled with slight modifications. The meeting was introduced by 6 review lectures covering the whole field of oxygen transport from bioengineering, the problem of diffusion in lung, blood and tissue through pathology of oxygen uptake in the lung, oxygen supply dependency of the critically ill to artificial oxygen carriers. A special session dealt with oxygen supply under ambiant pressure changes. Furthermore, each special session was introduced by one or several introductory reviews and main lectures on topics of current interest. Technical developments for oxygen measurement in tissue, pulmonary gas exchange registration, acid base analysis and the evaluation of tissue metabolism and whole body metabolism were further basic topics discussed. A second topic of great clinical implication was pharmacological regulation and augmentation of tissue oxygen supply, pulmonary gas exchange (with main emphasis on pulmonary surfactant replacement) and, further, tissue protection from oxygen deficit damage and artificial oxygen transporting agents. In accordance with the overall concept of the congress, demonstrations with personal discussions and thereafter, plenary reports and discussions took place.

The editors and organizers would like to thank the staff of the Curaçao St. Elisabeth Teaching Hospital under the coordination of Mr D. Kibbelaar, for their enormous input, excellent local organization of the meeting, the warm and friendly atmosphere, care of the participants and the attractive social program giving scientists the chance to meet each other personally and continue discussion of their interests. Special thanks are due to the Hospital Board, with Dr. K. de Monte as President, for the great support to make all this possible, and the Medical Direction under Dr. E. Martis. The Governor of Curaçao offered to personally open the meeting, and gave a most stimulating introduction into local problems concerning adequate oxygen supply in relation to environmental pollution.

All manuscripts were reviewed by members of the society and experts in the concerned special field. The editors are most grateful for the generous support and willingness of the many ISOTT members who offered to undertake this difficult task and to Mrs. Laraine Visser and Denise Haas, of the Congress Secretariat, for their excellent coordination of the activities related to the proceedings book. Extensive revisions were made in 22% and modest revisions in another 48% of the manuscripts. Papers not in an appropriate format were processed by scanning and thereafter revised with the generous help of the secretarial staff of the Anesthesia Department at the Erasmus University Rotterdam. All of the original camera-ready manuscripts in this volume were prepared by the authors themselves and we greatly appreciate their cooperation, especially in meeting the very short deadlines.

The editors congratulate Dr. Paul Okunieff for the honour of being selected as the 1991 Melvin Knisely Award Winner for his outstanding contributions in the field of research on oxygen transport to tissue.

We look forward to the continued expansion of ISOTT and to the next meeting to be held in summer 1992 at the Johannes Gutenberg University in Mainz, Germany.

For the Editors

Wilhelm Erdmann

January, 1992

CONTENTS

GENERAL INTRODUCTION

Bioengineering: The Fifth Traditional Engineering
    Discipline? 3
 D.F. Bruley

The Oxygen Molecule and its Course from Air to Cell 7
 W. Erdmann

SPECIAL INTRODUCTION

Basic Mechanisms of Diffusive and Diffusion-Related Oxygen
    Transport in Biological Systems. A Review 21
 K. Groebe and G. Thews

Oxygen Supply Dependency in the Critically Ill -
    A Continuing Conundrum 35
 S.M. Cain

Oxygen Uptake in the Lungs under Pathological Conditions
    and its Therapeutic Effects 47
 D. Gommers, G.J. van Daal, and B. Lachmann

Artificial Oxygen Carrying Blood Substitutes 55
 N.S. Faithfull

AMBIANT PRESSURE CHANGES AND OXYGENATION

Adaptation of O_2 Transport and Utilization at Altitude in Man 75
 P.D. Wagner

Hyperbaric Oxygen Therapy: Past, Present and Future Indications 95
 D.J. Bakker

Rat Brain Adaptation to Chronic Hypobaric Hypoxia 107
 J.C. LaManna

Whole Lung Lavage under Hyperbaric Conditions:
  1. The Monitoring of Circulation During
     Treatment of PAP 115
  J.D. Biervliet, J.A. Peper, C.M. Roos, A.J. van der Kleij,
     D.J. Bakker, and H.M. Jansen

Whole Lung Lavage under Hyperbaric Conditions:
  2. Monitoring Tissue Oxygenation 121
  A.J. van der Kleij, J.A. Peper, J.D. Biervliet,
     D.J. Bakker, C.M. Roos, and H.M. Jansen

Skeletal Muscle PO_2 in Anaerobic Soft Tissue Infections
     during Hyperbaric Oxygen Therapy 125
  A.J. van der Kleij, D.J. Bakker, M. Lubbers,
     and Ch. P. Henny

Effect of Hyperbaric Oxygen Treatment and Perfluorochemical
     Administration on Glutathione Status of the Lung 131
  E. Purucker and J. Lutz

TUMOR PHYSIOLOGY AND OXYGENATION

Blood Flow and Tissue Oxygenation of Human Tumors:
     An Update 139
  P. Vaupel, K. Schlenger, and M. Hoeckel

Computerized Histographic Oxygen Tension Measurements of
     Murine Tumors 153
  D.J. Terris, A.I. Minchinton, E.P. Dunphy, and J.M. Brown

Oxygenation and Bioenergetic Status of Murine Fibrosarcomas 161
  C. Schäefer, P. Okunieff, and P. Vaupel

Measurement of Human Tumor Blood Flow: A Positron Technique
     Using an Artifact of High Energy Radiation Therapy 169
  P. Okunieff, J. Lee, and P. Vaupel

Improving the Effectiveness of the Bioreductive Antitumor
     Agent SR 4233 by Induced Hypoxia 177
  A.I. Minchinton and J.M. Brown

A Computer Simulation of Oxygen Partial Pressure and Temperature
     Profiles during Hyperthermia 183
  K.A. Kang, S.A. Afuwape, and D.F. Bruley

ADVANCEMENT IN TECHNOLOGY AND INSTRUMENTATION

Oxygen Dependent Quenching of Phosphorescence:
     A Perspective 195
  D.F. Wilson

Reflection Spectrometry 203
    M. Kessler, K. Frank, J. Höper, D. Tauschek, and J. Zündorf

Can the Flow Dependency of the Polarographic PO_2 Electrode be
    Used to Measure Arterial PO_2 and Local Capillary Flow
    Transcutaneously? 213
    A. Grundmann and D.W. Lübbers

The Use of EPR for the Measurement of the Concentration
    of Oxygen In Vivo in Tissues under Physiologically
    Pertinent Conditions and Concentrations 221
    H.M. Swartz, S. Boyer, D. Brown, K. Chang, P. Gast,
    J.F. Glockner, H. Hu, K.J. Liu, M. Moussavi,
    M. Nilges, S.W. Norby, A. Smirnov, N. Vahidi,
    T. Walczak, M. Wu, and R.B. Clarkson

In Vivo EPR Oximetry Using Two Novel Probes: Fusinite and
    Lithium Phthalocyanine 229
    J.F. Glockner and H.M. Swartz

Measurement of Cerebral Blood Flow in Adult Humans Using Near
    Infrared Spectroscopy - Methodology and Possible Errors 235
    C.E. Elwell, M. Cope, A.D. Edwards, J.S. Wyatt,
    E.O.R. Reynolds, and D.T. Delpy

Design and Evaluation of a Reflectance Oxygen Sensor in
    Critically Ill Patients 247
    S. Takatani, G.P. Noon, Y. Nose, and M.E. DeBakey

Biologically Active Cyanine Dyes as Probes for the Identification
    of Active Oxygen Species 255
    H. Hori, Y. Nakagawa, H. Ojima, T. Niijima, and H. Terada

Analysis of Multiple Multipole Scattering by Time-Resolved
    Spectroscopy and Angular Dependent Spectrometry 261
    K. Frank, M. Kessler, J. Wiesner, and A. Wokaun

In Vivo NADH and Pd-Porphyrin Video Fluori-/Phosphorimetry 267
    C. Ince, J.F. Ashruf, E.A. Sanderse, E.G.J.M. Pierik,
    J.M.C.C. Coremans, and H.A. Bruining

In Vivo NADH Fluorescence 277
    C. Ince, J.M.C.C. Coremans, and H.A. Bruining

Quantitation of Tissue Optical Characteristics and Hemoglobin
    Desaturation by Time and Frequency Resolved Multi-Wavelength
    Spectrophotometry 297
    B. Chance, N.G. Wang, M. Maris, S. Nioka, and E. Sevick

Evaluation of the Algorithm Used in Near Infrared
    Spectrophotometry 305
    W.N.J.M. Colier, B.E.M. Ringnalda, J.A.M. Evers, and B. Oeseburg

THE LUNG, ENTRANCE PORT OF OXYGEN

Informative Imaging of Oxygen Supply Parameters
in Clinical Practice 315
A.W. Grogono, A.P.K. Verkaaik, and W. Erdmann

Diffusion Limitation of Oxygen in Heterogeneous Lung
and Tissue Models 319
J. Piiper

Gas Exchange in the Lung, Computer Feed Back Controlled
Physiological Matching of Artificial Ventilation 325
A.P.K. Verkaaik, G. van Dijk, B. Westerkamp,
and W. Erdmann

Non-Invasive, On-Line Measurement of Oxygen Consumption during
Anesthesia 331
A.P.K. Verkaaik, J.W. Kroon, H.G.M. van den Broek, and W. Erdmann

Oxygen Transport Through a Model Lung Surfactant Surface Layer
Influence of the Film Compression on the Kinetics 343
E. Ladanyi, R.C. Ahuja, D. Möbius, and K. Stalder

Different Surfactant Treatment Strategies for Respiratory Failure
Induced by Hydrochloric Acid Aspiration in Rats 349
E.P. Eijking, D. Gommers, J. Kullander, E. de Buijzer,
R.W. Beukenholdt, and B. Lachmann

Gas Exchange Uniformity Within Individual Lung Lobes 357
M.J. Emery, M.E. Middaugh, T. Tran, and M.P. Hlastala

Effect of Artificial Ventilation on Pulmonary Antioxidant
Enzyme Activities in a Congenital Diaphragmatic
Hernia Rat Model 363
R. Tenbrinck, W. Sluiter, F. Silveri, A.P. Bos, E.C. Scheffers,
A.T.J.I. Go, J.A.H. Bos, D. Tibboel, and B. Lachmann

Comparison of Pressure Support Ventilation (PSV) and Intermittent
Mandatory Ventilation (IMV) during Weaning in Patients
with Acute Respiratory Failure 371
F. Esen, T. Denkel, L. Telci, J. Kesecioglu, A.S. Tütüncü,
K. Akpir, and B. Lachmann

Alterations of the Respiratory Minute Volume Cause Changes
in the Muscle Oxygenation of Rats 377
M. Günderoth

Optical Measurements of Oxygen and Electrical Measurements of
Oxygen Chemoreception in the Cat Carotid Body 387
W.L. Rumsey, S. Lahiri, R. Iturriaga, A. Mokashi,
D. Spergel, and D.F. Wilson

Dose Dependent Improvement of Gas Exchange by Intratracheal
Perflubron (Perfluorooctylbromide) Instillation in
Adult Animals with Acute Respiratory Failure 397
A. S. Tütüncü, B. Lachmann, N.S. Faithfull, and W. Erdmann

Gas Exchange and Lung Mechanics during Long-Term Mechanical
Ventilation with Intratracheal Perfluorocarbon
Administration in Respiratory Distress Syndrome 401
A.S. Tütüncü, K. Akpir, P. Mulder, N.S. Faithfull,
W. Erdmann, and B. Lachmann

Perflubron (Perfluorooctylbromide) Instillation Combined with
Mechanical Ventilation: An Alternative Treatment of
Acute Respiratory Failure in Adult Animals 409
B. Lachmann, A.S. Tütüncü, J.A.H. Bos, N.S. Faithfull,
and W. Erdmann

Clinical Use of Oxygen Stores: Pre-oxygenation and Apneic
Oxygenation 413
R. Zander and F. Mertzlufft

A New Device for the Oxygenation of Patients:
The NasOral-System 421
F. Mertzlufft and R. Zander

OXYGEN TRANSPORT IN BLOOD

Hemodilution and Oxygen Transport 431
A. Trouwborst, E.C.S.M. van Woerkens, and R. Tenbrinck

Second Generation Fluorocarbons 441
N.S. Faithfull

Properties of Chemically Cross-Linked Hemoglobin Solutions
Designed as Temporary Oxygen Carriers 453
P.E. Keipert

Elaboration of Fluorocarbon Emulsions with Improved Oxygen-
Carrying Capabilities 465
J.G. Riess and M.P. Krafft

The Respiratory Potential of Oxygen: A New Quantity to
Characterize State, Effects and Bio-Availability
of the Gas in Organism 473
W.K.R. Barnikol

Facilitation of Oxygen Transfer by Perflubron in
Hemodiluted Dogs 479
S.M. Cain, S.E. Curtis, and W.E. Bradley

Monitoring of Intracapillary HbO_2 in Foetal Scalp
during Delivery 485
J. Höper, M. Kessler, K. Frank, D. Tauschek,
J. Zündorf, N. Lang, and E. Mauch

O_2 Transport during Exercise After Cardiac Transplantation. 491
M. Meyer, P. Cerretelli, C. Cabrol, and J. Piiper

Effect of Low Dose Oxygen[TM] Added to Blood on
Muscle VO_2MAX 497
M.C. Hogan, D. Willford, N.S. Faithfull, and P.D. Wagner

Fetal Oxygenation in Chronic Maternal Hypoxia:
What's Critical? 499
B. Oeseburg, B.E.M. Ringnalda, J. Crevels, H.W. Jongsma,
P. Mannheimer, J. Menssen, and J.G. Nijhuis

The Relation of Oxygen Delivery to Utilization During Liver
Transplantation: Is There a Critical Value? 503
H. Steltzer, M. Hiesmayr, G. Tüchy, and M. Zimpfer

Comparative Study of the Accuracy of Two Fiberoptic Mixed Venous
Saturation Catheters (Spectracath[R] vs Opticath[R]) during Acute
Changes in Hematocrit and Cardiac Output in Humans 509
E.C.S.M. van Woerkens, A. Trouwborst, L. Snel,
A. van Dorp van Vliet, and R. Tenbrinck

Erythropoietin Induction by Hypoxia. A Comparison of In Vitro
and In Vivo Experiments 515
H. Pagel, A. Engel, and W. Jelkmann

Altered Concentrations of Aldosterone in Neonatal Calves During
Chronic Hypoxia and the Subsequent Recovery Period 521
H. Tyler and H. Ramsey

OXYGENATION OF THE HEART, THE MOVING MOTOR

Myocardial Oxygen Supply under Critical Conditions, the Effects
of Hemodilution and Fluorocarbons 527
M. Fennema, W. Erdmann, and N.S. Faithfull

Regional Cardiac Hemodynamics and Oxygenation during Isovolemic
Hemodilution in Anesthetized Pigs. 545
E.C.S.M. van Woerkens, A. Trouwborst,
D.J.G.M. Duncker, and P.D. Verdouw

Gradients of Capillarization in the Subendocardium of
Rat Heart Septum 553
S. Batra, P. Veprek, and K. Rakusan

Oxygen Pressure Histograms Calculated in a Block of
      Rat Heart Tissue 561
    L. Hoofd and Z. Turek

The Effect of Realistic Geometry of Capillary Networks on Tissue PO_2
      in Hypertrophied Rat Heart 567
    Z. Turek, L. Hoofd, S. Batra, and K. Rakusan

Contractile Dysfunction of "Reperfused" Neonatal Rat Heart Cells:
      A Model for Studying "Myocardial Stunning" at the Cellular
      Level? 573
    P. Boekstegers, A. Pfeifer, W. Peter, and K. Werdan

Monitoring of Redox-State of Respiratory Enzymes and Myoglobin
      Oxygenation in the Working Rat Heart in Normoxia and
      Oxygen Deficiency 583
    J. Zündorf, D. Tauschek, K. Frank, K. Ito, S. Nioka,
    M. Kessler, and B. Chance

Spatial Distribution of Oxygen Supply Units in Heart and
      Skeletal Muscle and their Regulatory Significance 593
    M. Kessler, and J. Höper

Effects of CPPV, PC-IRV, and LFPPV-ECCO$_2$R on Right Ventricular
      Functions in Pigs with ARDS 599
    L. Telci, J. Kesecioglu, F. Esen, T. Denkel, A.S. Tütüncü,
    K. Akpir, and B. Lachmann

OXYGEN SUPPLY OF THE TISSUE

Factors that Determine the Oxygen Supply of the Cell and
      their Possible Disruption 607
    W. Erdmann, M. Fennema, and R. van Kesteren

Oxygen Supply by Perfusion and Diffusion in Heterogeneous
      Tissue Models 623
    J. Piiper

In Situ Determination of Convection and Diffusion Profiles
      in Heterogeneous Media 629
    N.A. Busch and M.L. Yarmush

Tissue Alterations By the Penetration of a PO_2 Sensing
      Needle Probe 639
    K.F. Wagner, W. Bossen, and U. Schramm

A New Program to Evaluate Data of PO_2 Histograph on Personal
      Computers under MS-DOS 645
    K.F. Wagner and B.G. Steppan

A Placental Perfusion pO_2 Logger 649
    D.J. Maguire, R. Blums, R. Morgan, J. Collie, and G.R. Cannell

A Method for Simultaneous Recording of Tissue PO_2 and EP in the
      Brain Cortex of a Test Animal with a Single Electrode 653
    H. Vermariën, K. van Rossem, R.T. Altan, and K. Decuyper

Computer Simulation of Erythrocyte Transit in the Cerebrocortical
      Capillary Network 659
    A.G. Hudetz

Instrumentation and Technology for Multiparametric Mapping of
      Intraparenchymal Circulation in the Brain Cortex 671
    A. Eke

THE BRAIN

Classification of Oxygen Transport to Tissue with
      Neural Networks 681
    W. Babel, N. Hetterich, and T. Müller

Oxygenation of the Cortex of the Brain of Cats during Occlusion
      of the Middle Cerebral Artery and Reperfusion 689
    D.F. Wilson, S. Gomi, A. Pastuszko, and J.H. Greenberg

Active and Basal Whole Brain Blood Flow, Oxygen and Glucose
      Metabolism in Monkeys 695
    E.M. Nemoto, L. Yao, H. Yonas, and J. Darby

The Regional Cerebral Blood Flow Response to Cortical Microelectrode
      Insertion is Neutrophil Dependent 701
    M.W. Uhl, P.M. Kochanek, J.K. Schiding, J.A. Melick,
      and E.M. Nemoto

Cerebral Blood Flow and Brain Mitochondrial Redox State Responses
      to Various Perturbations in Gerbils 707
    A. Mayevsky

Local Tissue PO_2 during and after Focal Brain Cortical Infarction
      in Rabbits 717
    K. van Rossem, H. Vermariën, K. Decuyper, J. van Reempts,
    M. Laureys, and R. Bourgain

Brain Surface PO_2 and rCBF in Rabbits with a Focal Cerebral Lesion
      and Pulmonary Hypoxia under Fentanyl-, Isoflurane or
      Thiopental-Anesthesia 723
    S. Berger, R. Murr, L. Schürer, R. Enzenbach,
    K. Peter, and A. Baethmann

Treatment of Hemorrhagic Hypotension with Hypertonic Saline/Dextran:
      Effects on Brain Surface Oxygen Tension in Experimentally
      Traumatized Brain 731
    C. Dautermann, L. Schürer, R. Härtl, F. Röhrich,
      A. Baethmann, and K. Messmer

Monitoring of Cortical Intracapillary Hemoglobin Oxygenation in
       Patients during Brain Surgery - First Results 737
    D. Tauschek, J. Höper, M.R. Gaab, and M. Kessler

MUSCLE

Inert Gas Washout Measurement of Muscle Blood Flow Distribution -
       Roles of Hypoxia and Diffusion Limitation 745
    M. P. Hlastala, G.M. Malvin, C. Quartararo, and J. Grønlund

Effects of Endotoxin on Canine Skeletal Muscle Oxygen Delivery-Uptake
       Relations During Progressive Hypoxic Hypoxia 751
    S.E. Curtis, W.E. Bradley, and S.M. Cain

Skeletal Muscle Capillary Flow and Oxygenation in Hypoxic Hypoxia:
       Effect of a $5-HT_2$ Receptor Antagonist 759
    U. Gustafsson, D.H. Lewis, and P. Thorborg

Distribution Pattern of Capillary and Venular Red Blood Cell Velocity
       Following Ischemia-Reperfusion in Striated Muscle. 765
    M.D. Menger, G. Feifel, and K. Messmer

Oxygen Consumption of Human Skeletal Muscle by Near Infrared
       Spectroscopy during Tourniquet-Induced Ischemia in Maximal
       Voluntary Contraction 771
    R.A. De Blasi, M. Cope, and M. Ferrari

Tissue Oxygenation Measurement: A Directly Applied Clark-Type
       Electrode in Muscle Tissue 779
    S.O.P. Hofer, A.J. van der Kleij, and K.E. Bos

OTHER TISSUES

Oxygen Tension and Blood Flow in the Retina of Normal
       and Diabetic Rats 787
    S. Cringle, D.Y. Yu, V. Alder, and E.N. Su

Arteriolar Spasm and Ischemia in the Ocular Fundus of NaCl-Loaded
       Salt Sensitive Dahl Rats. Vascular Protection by Long-term
       Treatment with the Calcium Antagonist Nitrendipine 793
    F. Thimm, M. Frey, K. Spitzmüller, W. Hofgärtner,
    and G. Fleckenstein-Grün

Influence of Cryopreservation on Viability and Nutritional
       Microcirculation of Islets of Langerhans 799
    M.D. Menger, J. Pattenier, B. Wolf, S. Jäger, and G. Feifel

O₂ SUPPLY: CLINICAL PROBLEMS

Role of Arachidonic Acid Metabolites in Pulmonary
Oxygen Toxicity 807
J. Klein, A. Trouwborst, and W. Erdmann

The Relationships between Oxygen Delivery and Consumption
and Continuous Mixed Venous Oximetry are Predictive
Parameters in Septic Shock 813
F. Giunta, L.S. Brandi, T. Mazzanti, M. Oleggini,
G. Tulli, and A.M.R. Cuttano

Oxygen Delivery and Postoperative Mortality 825
H. van der Zee and R.C. Evans

Is Oxygen Consumption Measurement during Anesthesia for Liver
Transplantation Valuable for a Rapid Assessment of
Adequate Function of the Graft? 835
T.H.N. Groenland, C.G.O.T. Bouman, A. Trouwborst,
and W. Erdmann

Assessment of Cerebral Oxygenation and Hemodynamics by Near
Infrared Spectrophotometry during Induction of ECMO:
Preliminary Results 841
K.D. Liem, J.C.W. Hopman, L.A.A. Kollée, and B. Oeseburg

Oxygen Uptake and Static Lung Compliance during Automatic
Ventilation 847
Y.A. Ruetsch, C.P. Naumann, A.P.K. Verkaaik,
W. Erdmann, and G.A. Zäch

Oxygen Uptake/Supply Dependency in Human Sepsis: Does it
Increase the Risk of Multisystem Organ Failure? 855
F. Esen, L. Telci, K. Akpir, J. Kesecioglu,
T. Denkel, and K. Pembeci

Intra-Anesthetic On-Line Monitoring of Oxygen Consumption Using
a Closed Circuit System 863
H. Wauer, M. Schädlich, A.P.K. Verkaaik, and W. Erdmann

PO₂ Profiles in Human Muscle Tissue as Indicator of Therapeutical Effect
in Septic Shock Patients 869
C.P. Naumann, Y. Ruetsch, W. Fleckenstein,
M. Fennema, W. Erdmann, and G.A. Zäch

Preoperative Gastrocnemius Muscle PO₂ as Predictor of Healing after
Below Knee Amputation 879
K.F. Wagner, E.M. Noah, R. Perner, F.W. Busse, and H.P. Bruch

Changes of Tissue PO₂ in the Lower Leg Muscles after Vascular
Surgery 885
K. Wagner, U. Krüger, R. Schäfer, M. Albrecht, and G. Hohlbach

Comparison of Different Modes of Artificial Ventilation with
        Extracorporeal CO_2 Elimination on Gas Exchange in
        an Animal Model of Acute Respiratory Failure 893
    J. Kesecioglu, L. Telci, T. Denkel, A. Tütüncü, F. Esen,
        K. Akpir, and B. Lachmann

Evaluation of Oxygenation with Different Modes of Ventilation
        in Patients with Adult Respiratory Distress Syndrome 901
    J. Kesecioglu, L. Telci, F. Esen, T. Denkel, K. Akpir,
        A. Tütüncü, W. Erdmann, and B. Lachmann

GROUP PHOTOGRAPH 907

AUTHOR INDEX 909

SUBJECT INDEX 913

GENERAL INTRODUCTION

BIOENGINEERING: THE FIFTH TRADITIONAL ENGINEERING DISCIPLINE?

Duane F. Bruley

University of Maryland Baltimore Country

Baltimore, Maryland 21228

To begin, I will provide a general definition of BIOENGINEERING; "The Application of Engineering Principles and Fundamentals to Engineering Problems that Require A Basic Understanding of Biology and/or Living Systems", which will be the basis for the remainder of this paper. After all, if engineering problem analysis or solution does not require a knowledge of biology or living systems then the existing traditional disciplines that deal specifically with non-living systems would be appropriate.

As Charlie Brown would say, "Good Grief Not Another Prophet", I will go out on the limb and predict that Bioengineering will emerge as the fifth traditional discipline of engineering. Of course, the quote by Neils Bohr is probably appropriate at this point, "Prediction is Always Difficult, Especially About the Future".

Figure 1 illustrates the engineering discipline progression. Civil Engineering evolved from Military Engineering and Mechanical and Electrical Engineering followed as recognized "traditional" disciplines in engineering. All three are based primarily on mathematics and physics. As the need for chemical processing became important Chemical Engineering became recognized as the fourth traditional discipline in engineering. This result was heavily influenced by the need for mathematics, physics and a second basic science, chemistry for the development of the Chemical Process Industries. It seems obvious that of all the remaining engineering subdisciplines that Bioengineering would be the most appropriate choice as the fifth traditional discipline of engineering. This conclusion is supported by the fact that bioengineering is dependent upon mathematics, physics, chemistry and the third basic science, biology.

Bioengineering, or the engineering of living systems, can be broken into two subsets. Figure 2 illustrates some topical areas in Bioprocess Engineering (primarily dealing with cells and cell cultures) and in Biomedical Engineering (primarily addressing the health related needs of human beings). Some dotted lines have been included in the figure to show that there are overlaps between the two subdivisions.

For instance, it would be difficult to categorize the growth of transplantable tissue in a bioreactor as specifically bioprocess engineering or biomedical engineering. An example of such an effort is the recently announced NASA achievement of growing small intestine tissue in vitro.

Figure 3 more vividly depicts the interfaces and overlaps of biomedical and bioprocess engineering in the context of the total field for engineering. The illustration shows the cross disciplinary nature of bioengineering where biomedical engineering relies more upon formal training in physiology and anatomy while bioprocess engineering relies more on formal training in biochemistry, industrial microbiology, and molecular and cellular biology.

```
• CIVIL ENGINEERING

• MECHANICAL ENGINEERING          MATHEMATICS/PHYSICS

• ELECTRICAL ENGINEERING
------------------------------------------------
• CHEMICAL ENGINEERING         MATHEMATICS/PHYSICS/CHEMISTRY
================================================
• BIO ENGINEERING    MATHEMATICS/PHYSICS/CHEMISTRY/LIFE SCIENCES
================================================
```

Fig. 1   The engineering discipline progression.

A basic appreciation for molecular and cellular biology is necessary for all Bioengineers, since all problems with living systems eventually get back to a need of that level of understanding. For example, in the case of prostheses design, the major consideration is the interface between the prosthesis and the living tissue, and its cellular response to the device. Also, the control of a bioreactor is ultimately determined by the metabolic processes of single cells.

The comparison of bioreactor design and oxygen transport to tissue in the microcirculation again illustrates the overlap between biomedical engineering and bioprocess engineering. Both cases are concerned about the environment of single cells and the transport processes carrying nutrients to the cells and those removing waste products from the system. Bioengineers can quantify these processes via mathematics and good experimental data to gain a detailed understanding of the rate limiting processes in the various systems. Quantification of the microcirculation has been an important component of every ISOTT meeting since the beginning of the society. Simulation techniques have been implemented to study heat, mass and momentum transfer and chemical reaction in specific vascular beds and in single cells. Theoretical calculations to predict energy generation (ATP) at the cellular level is fundamental to the success of metabolic engineering efforts to understand and enhance cellular expression of important proteins and other bioproducts. A deliberate effort is being made to develop realistic, simplistic simulation strategies to examine transport phenomena in three dimensional, heterogeneous, non-linear, time-dependent, convection-diffusion and reaction systems. The ultimate objective is to study system disturbances and their impact on cell functioning, which then translates directly into human health considerations.

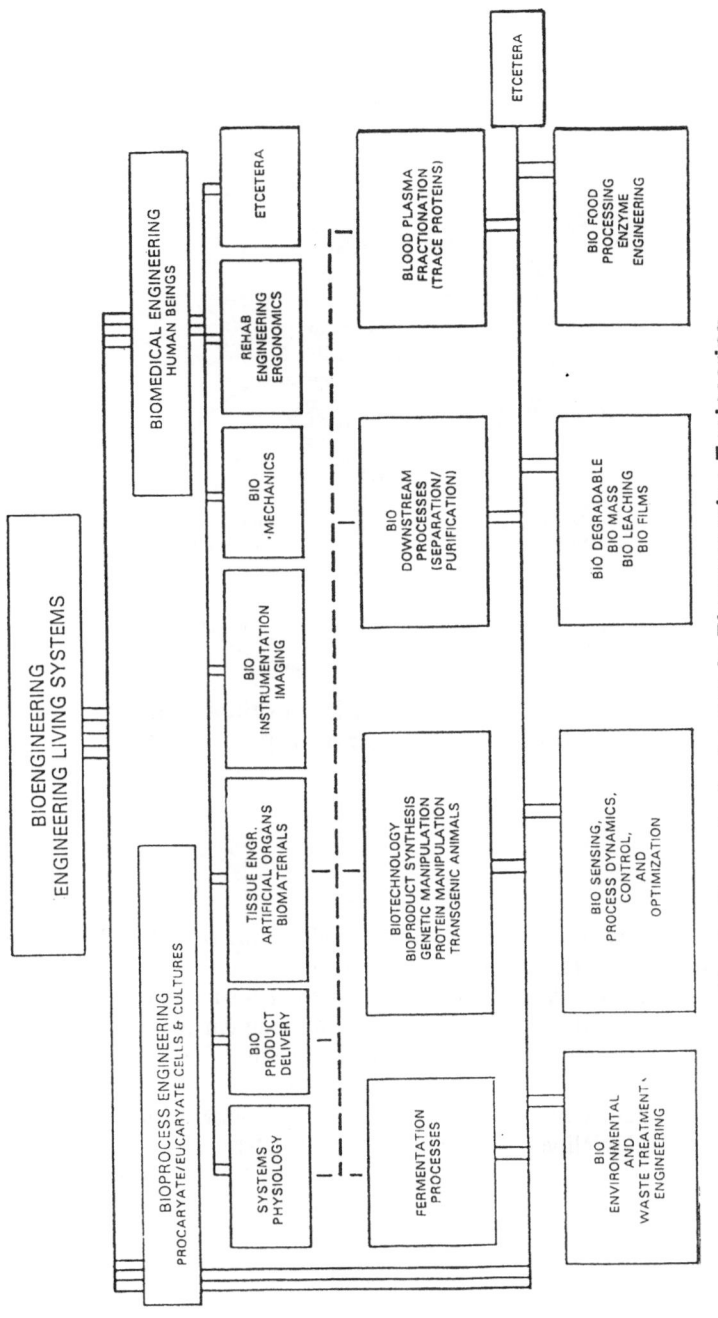

Fig. 2 Some topical areas in Bioprocessing Engineering and in Biomedical Engineering.

Bioengineering is a discipline built upon mathematics, physics, chemistry and the life and/or medical sciences. Projections by many experts indicate that both the bioprocess engineering and biomedical engineering industries will grow rapidly during the next decade and continue to expand after year 2000. This development is so significant that many knowledgeable engineers and scientists are predicting that

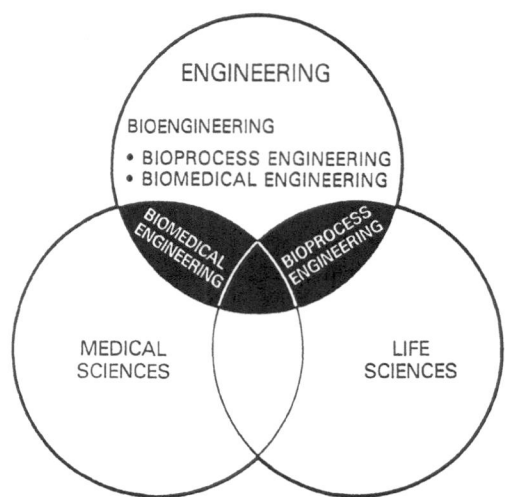

Fig. 3   The interfaces of Biomedical and Bioprocess Engineering

the twenty-first century will be the century of biotechnology. The bioengineering industries will require the talents of highly trained bioengineers (both bioprocess and biomedical) cross linked with many disciplines in engineering and the sciences. This involvement will require that the traditional engineering curriculum be modified to provide a background in the engineering of living systems.

Only time will tell if bioengineering will take its place with the present four accepted traditional disciplines. WHAT DO YOU THINK?

THE OXYGEN MOLECULE AND ITS COURSE FROM AIR TO CELL

W. Erdmann

Department of Anesthesiology, Erasmus University Rotterdam,
The Netherlands

ORIGIN OF OXYGEN

An understanding of the role of oxygen requires that we should go back to the origin of oxygen in the atmosphere of this planet.

Oxygen is a very reactive gas and it is by no means obvious why it should be present in our atmosphere at all. The formation of the earth occurred $\pm$ 4.5 billion years ago. The generally favored theory of cold accretion envisages the increasing gravational attraction of the mass raising the pressure in the center of the earth to a level which, whether we like it or not, initiated a thermonuclear reaction, rocks were decomposed and emission of various gaseous and volatile constituents occurred to form the primitive atmosphere (predominantly water vapour, nitrogen, carbon dioxide, ammonia, sulphur dioxide and hydrogen sulphide) no oxygen being present yet.

Abiogenetic synthesis, in the surface water, of quite complex organic molecules from outgassed material, the energy deriving from ultra-violet light and ionising radiation, was the next step followed by aggregation of these organic compounds within primitive membranes. Photosynthesis, reproduction capacity and ensymbiosis in which they engulfed chloroplasts including own DNA followed and release of oxygen as a waste product started.

Excess oxygen escaped into the atmosphere where it blocked the solar radiation and made it possible for living organisms to leave the protective shielding of water. The third fate of photosynthetically-derived oxygen was oxidative metabolism, not only in animals, but also in plants providing a potent increase in the liberation of biological energy from substrates (Fig.1).

Oxygen Transport to Tissue XIV, Edited by W. Erdmann and
D.F. Bruley, Plenum Press, New York, 1992

OXIDATIVE PHOSPHORYLATION

Indeed, anaerobic conditions, before the atmosphere contained molecular O_2, were important for the formation of nucleotides by the "first spark" - a process that probably was the origin of life since it led to the development of DNA and RNA, the carriers of genetic information. Energy was generated by anaerobic oxidation thus simple transfer of electrons to some suitable electron acceptor. The essential part in this process of "combustion" is the liberation of energy when electrons are transferred from an "electron donor" to an "electron acceptor" of higher electron affinity; all this can happen without

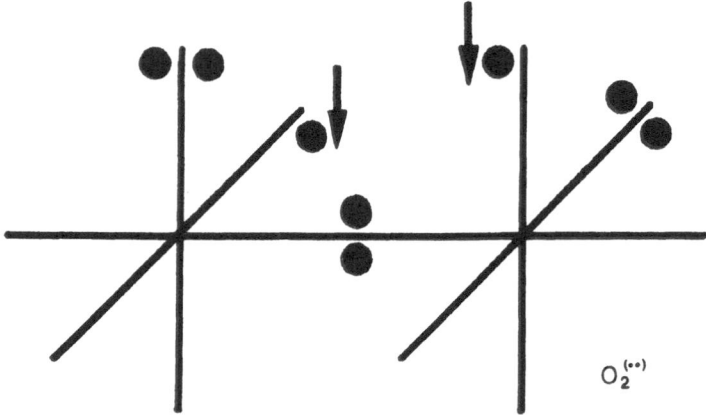

Fig. 1.   The outer shell of electrons of an oxygen molecule.
There are two impaired electrons (shown by arrows) in separate orbits on each of the two oxygen atoms. These unpaired electrons confer the property of paramagnetism, but in addition oxygen may be considered as a double free radical.

O_2, but when O_2 became available in the earth's atmosphere it became the ideal electron acceptor because of its very high electron affinity, and "combustion" could now be carried to the end with true "oxidation" liberating the full amount of energy the substrates can give off resulting in the formation of water. However, our system is not equipped to utilize the energy contained in heat and thus it was necessary to capture this energy in high-energy bonds between phosphate groups (attached to nucleotides) in the form of adenosine triphosphate, ATP. The energy can then be given to the energy-requiring processes of cells. The most fundamental step in the evolution of the cells was the development of a complex set of enzymes (catalysts) that could transfer the energy liberated by the various steps of oxidation to ATP, the oxidative phosphorylation.

As everyone in ISOTT knows, we are now totally dependent upon the access of oxygen to our tissues and interruption of supply for more than a few minutes has disastrous results.

OXYGEN: DIFFUSION AND CONVECTION

In eukaryotic cells, the enzymes of oxidative phosphorylation, are housed in specialized organelles, the mitochondria, and thus the ATP generation is separated from the sites of ATP-utilization. The mitochondria thus are the cell furnaces where molecular O_2 is converted to water, the endproduct of combustion. This "fire of life" can only be kept burning if there is a continuous inflow of O_2 from the cell's environment. In protozoa this environment is the aqueous medium in which the cell lives and oxygen is derived from air. In higher organisms, made of a large number of specialized cells each equipped with mitochondria, this environment is the intercellular fluid. The problem is now how to ensure an adequate flow of O_2 from the environmental air to all the cells in the body. The mechanisms by which a flow of O_2 into the tissues can be maintained are:

1. diffusion or molecular movement of O_2 through air and fluid;

2. convection or mass transport of O_2 by moving the O_2 containing medium.

A pathway for diffusive flow of O_2 in a given direction needs compartments with different PO_2's to be established and to be separated by some structural barrier. The content of the barrier is considered static, only O_2 moves by displacement of molecules in the direction of the concentration gradient, and the O_2 provides its own driving force, the partial pressure difference. Diffusive flow of O_2 is the main supply force from the carrying medium to the cell and from the environment into the carrying medium.

Convective flow is different. Here we assume that O_2 molecules are dissolved in a fluid or bound to some carrier (e.g. blood with hemoglobin) which is moved by outside forces (e.g. the heart). The amount of O_2 that can be conveyed this way depends essentially on the quantity of O_2 required to raise the PO_2 by one unit, the concentration of the oxygen dissolved in the medium proportional to partial pressure, the amount of oxygen bound to the carrier at a certain partial pressure and the volume flow rate of the carrier.

OXYGEN TRANSPORT AND THE CARRYING MEDIUM

Water itself has a very low oxygen transport capacity with a solubility coefficient of only 0.00003 ml O_2/ml of water for each mm Hg, or in other words one fiftieth as compared to air at sea level. Thus, nature has chosen two basically different approaches to solve the problem of getting adequate amounts of O_2 to the cells in the depths of higher organisms:

1. using air as a carrier for O_2 flow by convection and diffusion through a system of channels (insects), and:

2. developing an intermediate O_2 carrier in fluid (e.g. blood) which connects the cells to an external gas exchanger.

The body of insects is pervaded by a system of air-filled tubes, called tracheae. They open to the outside air via lateral openings, called spiracles, and branch into the tissues, finally penetrating into the muscle cells where the terminal tracheoles come to lie in the immediate vicinity of the mitochondria. The tracheae usually show some thin-walled dilatations and convection of air is produced by ventilation of these sacs through pumping movement of the abdomen (as pumping motor), but also by accordeon-like shortening and lengthening of the trachea through contraction of the muscles. However, the tracheolae are too thin for adequate convection and oxygen mainly diffuses into and through these to the cells from the larger tracheae. As diffusion of O_2 through air is about 300.000 times faster than through water adequate O_2 supply to the cells can indeed be assured (Fig.2).

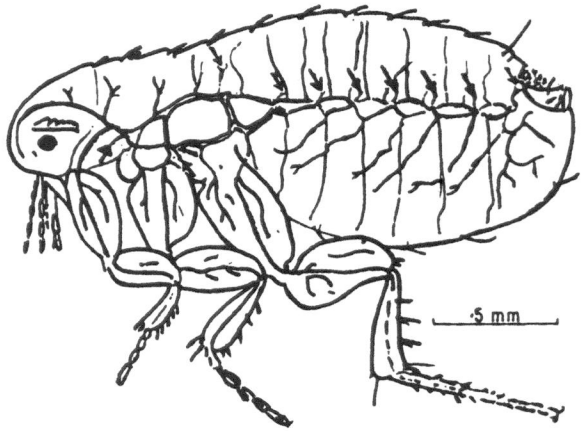

Fig. 2.   Gas transport in insects is achieved by a system of air-filled tracheae through which the air is convected actively entering from the outside through openings called spiracles (arrows). From the tracheae the air penetrates into the tissue through tracheoles, mainly by diffusion.

For most other higher animals the second possibility was chosen by nature and a variety of O_2 carriers have evolved. They all share one common feature: they are metal-containing proteins that are colored and often called respiratory pigments (e.g. hemocyanin which contains copper - mollucs and arthropodes; or hemoglobin as most common carrier).

THE EXTERNAL GAS EXCHANGE

An O_2 carrying intermediate in the oxygen transport chain, however, asks for a sophisticated external gas exchanger that has evolved in higher organisms - fish gills, bird lungs and alveolar lungs of amphibia, reptiles and mammals.

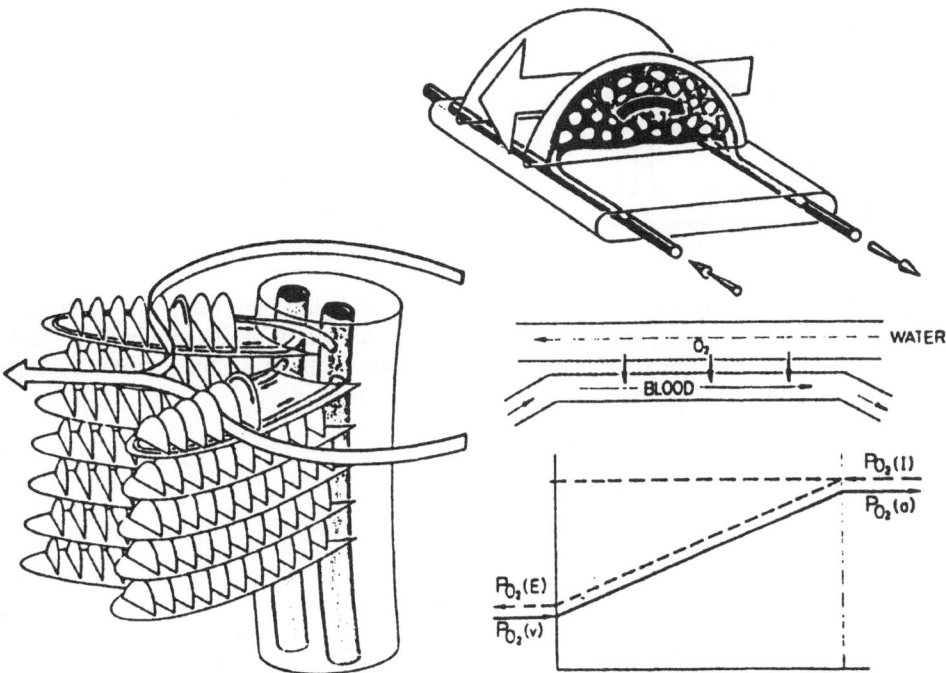

Fig. 3A.   Fish gills: evaginated large surface for gas exchange. A complex structure with filaments mounted in paired rows on the gill arches, gas exchange occurs at the surface of numerous thin lamellae which are mounted on the filaments. Flow of water between the lamellae is in the opposite direction to blood flow. This countercurrent exchanger allows the PO_2 in arterial blood to approach that in inspired water.

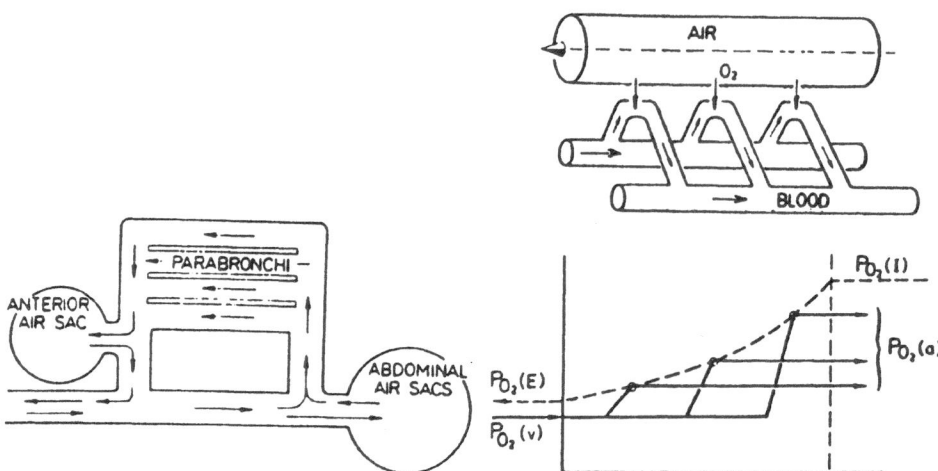

Fig. 3B. Bird lung: air flows from rear to front through parabronchi, with abdominal air sacs serving as bellows. The gas exchanger is arranged along the parabronchi and is perfused with blood according to a cross-current system. Arterial PO_2 may be higher than that in expired air.

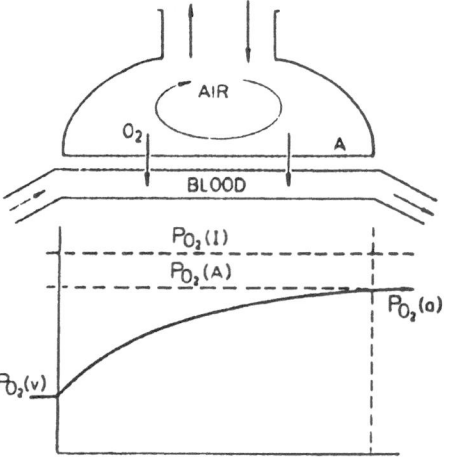

Fig. 3C. Alveolar lung: blood flows by an air containing alveolar pool that is ventilated through the airways. Blood PO_2 becomes equilibrated to alveolar PO_2 which is, however, lower than that of inspired air.

The feature in common, irrespective of the different principles that determine their function in detail, is that the flow of O_2 across the barrier separating blood from air or water occurs entirely by diffusion. Hence, it depends on the existence of a PO_2 gradient as the driving force for O_2 flow, and O_2 flow decreases as the barrier gets thicker. Thus gas exchange requires that a very thin barrier is established which confronts the blood and the environmental O_2 store over a very large surface.

Nature has developed them accordingly either in the form of gills, the exchange surface evaginated (turned outward) or in the form of lungs with the exchange surface invaginated (turned inward). Taking the mammalian lung as an example this fits quite well into the overall concept. Air is mainly moved into the gas exchanging alveolus with large surface and thin barrier to the blood by convection. Although diffusion (similar to that of insect tracheolae) plays a small role. This is clinically applied in high frequency ventilation (or jet ventilation) where diffusion of O_2 through the bronchiolae into the alveoli becomes the main cause of oxygen flow (Figs. 3A-C).

THE OXYGEN FLOW CASCADE

The pathway for O_2 transport from the O_2 store in the air or water, continuously replenished through photosynthesis, to the O_2 consuming furnace in the mitochondria, where it disappears in the process of oxidation is a cascade. According to the steps in this cascade model the O_2 molecules flow by convection and diffusion from air (or water) through the lungs (or gills) as entrance port of oxygen into the blood, the oxygen carrying medium, where they are transported by convection with the heart as moving motor to the tissue and by diffusion to the consuming mitochondria in the cells. The PO_2 falls step by step from environmental values to eventually zero in the furnace.

Development of technical possibilities to clarify the difficult processes involved at the various levels of the cascade are of utmost importance to increase understanding and to ultimately enable selective correction of disturbances. The advanced methods of bioengineering and biomathematic modelling to prospect further details based on the somewhat limited measurement that are currently available help in this process (Fig. 4).

THE REACTIVE OXYGEN MOLECULE

Having thus defined the course of the oxygen molecule from air to cell it is necessary to look to some drawbacks that have to be taken into account considering the energy source, the reactive oxygen molecule. Our dependence on oxygen is specially precarious because of the reactive nature of oxygen. Certain alterations to the basic molecular structure (two unpaired electrons in separate orbits on each of the two oxygen atoms) cause a great increase in reactivity. A molecule of oxygen may receive a single electron from a variety of different sources and this electron will pair with one of the unpaired electrons. This configuration is termed the superoxide free radical, a basis of oxygen toxicity, and mainly influential in many disease states. Thus, the two impaired electrons confer the property of paramagnetism, but in addition oxygen may be considered as a double free radical. However, the two unpaired electrons do in some sense neutralize each other's activity and, in the absence of other compounds with which it can react, oxygen itself is stable with an indefinite half-life, while the extremely reactive oxygen free radicals have a very short half-life. Besides formation of oxygen free radicals the direction of rotation

13

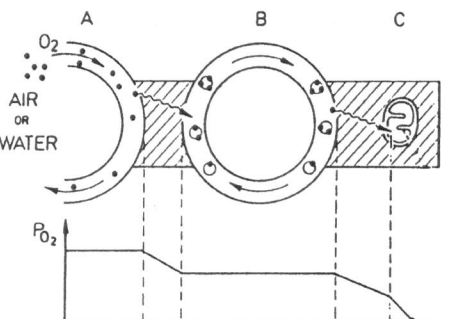

Fig. 4.   Cascade model of oxygen flow from air to the cell: O_2 flows by diffusion and convection from air or water (A) through blood (B) to the mitochondria in the cells (C) where it eventually disappears in the metabolic process. The PO_2 falls step by step from environmental values to eventually zero in the "furnace" (C), with O_2 partial pressure gradients as driving force at all diffusion barriers.

of one of the unpaired electrons might reverse, this so-called singlet oxygen is also highly reactive with a very short half-life. A third toxicity problem might be created by the fact that the two unpaired electrons may each share an electron with a hydrogen atom to give hydrogen peroxide.

The oxygen consumed by the cells is fully reduced in the terminal cytochrome of the electron transport chain. Four electrons are passed along the cytochrome chain and, with four hydrogen ions, there is a single-stage reduction of one molecule of oxygen to two molecules of water.

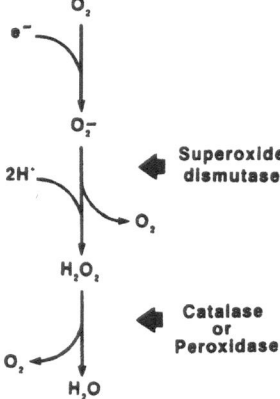

Fig. 5.   Intermediate stages of oxygen reduction: Superoxide free radical - dismutation by superoxide dismutase and hydrogen peroxide - catalisation by catalase or peroxidase.

14

However, under certain conditions the reduction of oxygen passes through two intermediate stages, the acquisition of a single electron to form superoxide free radical. Two molecules of the superoxide radical then undergo a dismutation (under the influence of superoxide dismutase) in which one donates and the other receives an electron to form one molecule of oxygen and one molecule of an intermediate, which then combines with two hydrogen ions to form hydrogen peroxide, which itself is broken down into water and oxygen under the influence of catalase or one of the various peroxidases (Fig.5).

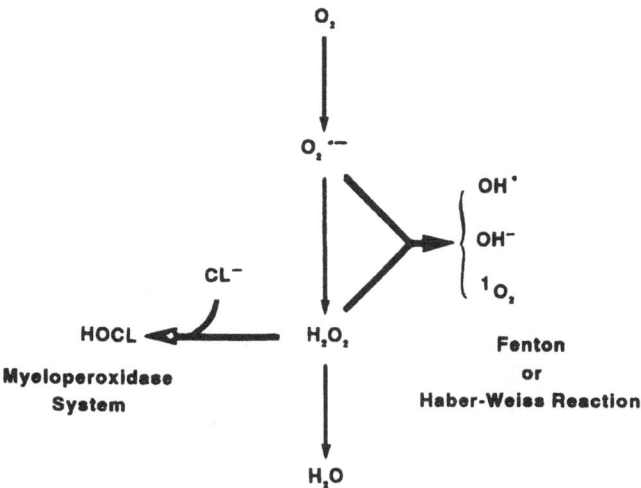

Fig. 6.   Superimposed reaction between superoxide free radical and hydrogen peroxide: Reaction occurs according to the Haber-Weiss reaction or the modified Fenton reaction. The products include: hydroxyl free radical, hydroxyl ion and singlet oxygen, furthermore hypochlorous acid. All substances are extremely reactive and/or toxic.

However, during the described reduction of oxygen to water, a reaction between superoxide free radical and hydrogen peroxide might be superimposed. They may react directly according to the Haber-Weiss reaction (described in 1934), or indirectly in a modified form, the Fenton reaction which requires the presence of an oxidised form of a trace metal (mostly iron) by changing from the ferrous to the ferric state). The products of both reactions are the same and include the extremely reactive hydroxyl free radical, the inoccuous hydroxyl ion and lastly the highly reactive singlet oxygen ($O_2^{( )}$). Hydrogen peroxide may also react with chloride ions to form hypochlorous acid, a very toxic substance formed by the enzyme myeloperoxidase (Fig.6).

Leaving this short detour we conclude that nature in having chosen for oxygen as energy source had not only to develop means to supply the cells with adequate oxygen at all times but also to develop a whole spectrum of enzymes and neutralizing reactions to detoxify oxygen free radicals as normal side products of metabolism.

These protective factors are:

1. superoxide dismutase, 2. catalase and peroxidase, 3. introduction of these enzymes by e.g.; a. increasing O_2 concentration; b. endotoxin (better in young than elderly), 4. antioxidants and free radical scavengers; a. glutathione peroxidase; b. ascorbic acid; c. vitamin E; d. dimethyl thiourea, 5. corticosteroids?

There are two main conditions under which the production of oxygen free radicals may overwhelm the natural defence:
1.   increased oxidative metabolism (e.g. sepsis)
2.   high concentrations and partial pressures of oxygen (luxury O_2 supply).

Thus returning back to our main subject we conclude that not only the maintenance of adequate or sufficient oxygen supply is of importance but also avoidance of luxury supply to any organ. The damage produced by high oxygen concentrations or better partial pressures is closely related to the factor time of exposure and can be best illustrated by an example: cultures of Chinese hamster lung fibroblasts are exposed to various concentrations of oxygen and the chromosomes, stained with Orcein, are examined (Fig. 7).

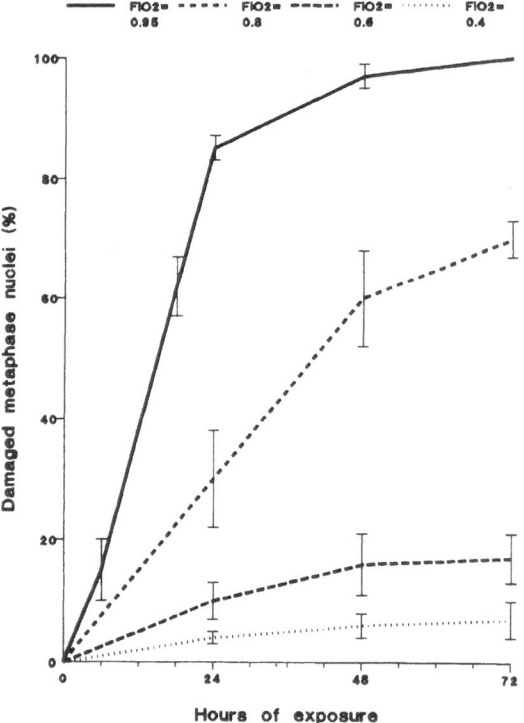

Fig. 7.   Cultures of Chinese hamster lung exposed to varying concentrations of oxygen and times of exposure (a model to investigate oxygen toxicity to the lung): the degree of damage is plotted in percent of damaged cells (metaphase nuclei). Low concentrations of oxygen between 40-60% are well tolerated, higher concentrations for more than 24 hours lead to extensive tissue damage (Nunn, 1988).

The percentage of cells that show nuclei in metaphase with chromosome breaks is counted and plotted against time of exposure. It clearly shows that the defense mechanisms, primarily peroxide dismutase, catalase or the various peroxidase systems can handle the formation of free radicals only to a certain level (0.4 FiO_2). Another point emerges from these studies: oxygen concentrations of 60-95% also induce mutations and it is known that most mutagens are also carcinogenics.

In conclusion, the necessity to guarantee sufficient oxygen supply to each cell at any moment does not exclude that oxygen itself might be the basis for severe tissue damage under certain conditions.

REFERENCES

This overview was based on:

Nunn, J. F., 1988, The origin of free oxygen, lecture to the department of anesthesia, Erasmus University Rotterdam.

Weibel, E. R., 1984, The pathway for oxygen: structure and function in the mammalian respiratory system. Harvard University Press, Cambridge, Massachusetts, U.S.A. and London, England.

SPECIAL INTRODUCTION

BASIC MECHANISMS OF DIFFUSIVE AND DIFFUSION-RELATED OXYGEN TRANSPORT IN BIOLOGICAL SYSTEMS: A REVIEW

K. Groebe and G. Thews

Institut für Physiologie und Pathophysiologie
Johannes Gutenberg-Universität Mainz
Saarstr. 21, D-6500 Mainz, Germany

In mammals, energy metabolism of active tissues requires permanent availability of oxygen. Because cessation of O_2 supply results in loss of organ function within seconds or minutes, continual feed of adequate amounts of O_2 to tissue is the most vital task for living organisms. For many years, it therefore has been one of the greatest challenges to physiologists to understand the mechanisms provided by nature to satisfy this need for oxygen.

Oxygen is taken up from ambient air into the lungs mainly by gas bulk flow, and is transported from there to the O_2 consuming tissues by circulation of blood. These convective transport steps are linked to each other and to mitochondrial O_2 consumption in tissues by diffusion and diffusion-related mechanisms which effect O_2 transport from alveolar gas to lung capillary blood and from tissue capillary blood to tissue cell mitochondria. In the first section of this article, these mechanisms are discussed in some detail. As an illustration of the interactions between the basic mechanisms, results on O_2 exchange of capillary contained blood are briefly summarized in a second section. Blood O_2 exchange has been chosen because this process takes place in a similar fashion during lung O_2 uptake and during tissue O_2 release.

1 Physical laws of O_2 diffusion with special regard to biological systems

Substance diffusion is defined to represent an equilibration process in which — according to the second fundamental law of thermodynamics — substance molecules are transferred from loci of higher concentration to ones of lower concentration. The energy required in this process originates from the thermokinetic energy of the molecules which undergo collisions during their undirected Brownian motions. Random movements in a field of variable concentration result in a net substance flux[1] along the direction of

---

[1]In the physical and physiological literature, use of the term "substance flux" is ambiguous. Some authors define "substance flux" to denote the total amount of substance transported per unit of time between two specified locations. Others would term this same quantity

decreasing concentration. In the absence of other processes, the net flux continues to exist until concentrations have equalized.

There are two fundamentally different approaches to the mathematical analysis of this kind of equilibration process: On the one hand, it may be assessed on a molecular level within the framework of thermodynamic statistics. On the other hand, its regularities may be described phenomenologically on a macroscopic level[2]. This latter view which represents the classical understanding of diffusion is suitable for assessing most physiological problems. However, in some recent theoretical studies on diffusional O_2 transport the former approach has been adopted (*cf.* [6]).

1.1 Fick's first law of diffusion. The physical laws governing substance diffusion in solvents have first been established quantitatively by the physiologist ADOLF FICK in 1855 [11]. Specifically for the case of oxygen, FICK's first law of diffusion states that a *partial pressure gradient* (*i.e.* the drop in partial pressure per unit length of diffusion path) acts as the driving force for diffusive particle flux. Furthermore, this law gives the relation between gradient and flux for the case of stationary diffusion: It quantifies $\dot{M}_{O_2}$, the amount of dissolved oxygen transported per unit of time across a plane layer of solvent, to be proportional to layer cross sectional area S and to the difference in O_2 partial pressure P_{O_2} at both sides of the layer (ΔP) divided by layer thickness (Δx).

$$\dot{M}_{O_2} = -K \cdot \frac{\Delta P}{\Delta x} \cdot S \tag{1}$$

A negative sign has been inserted because solute moves in the direction of decreasing concentration. The constant of proportionality K which, *e.g.*, may be given in units of $(ml\ O_2) \cdot cm^{-1} \cdot s^{-1} \cdot mm\ Hg^{-1}$ is referred to as gas "conductivity" or "KROGH's diffusion coefficient".

There is a second quantity characterizing O_2 diffusion which is the O_2 "diffusion coefficient" or "diffusivity" D. D applies in place of K if equation (1) is expressed for O_2 concentration $[O_2]$ instead of O_2 partial pressure P. Diffusion coefficients depend upon the characteristics of solute and solvent as well as on temperature, and are usually specified in units of cm^2/s. From HENRY-DALTON's law it follows that conductivity K is related to diffusivity D as

$$K = \alpha \cdot D, \tag{2}$$

where α is the solubility (specified, *e.g.*, in units of $(ml\ \text{gas}) \cdot (cm^3\ \text{tissue})^{-1} \cdot mm\ Hg^{-1}$).

1.2 Fick's second law of diffusion. Satisfactory descriptions of diffusive O_2 transport in physiological systems mostly cannot be obtained from FICK's first law of diffusion because it is applicable to a very special, *stationary* situation with constant

---

"substance flow" and define "substance flux" to be the substance flow per unit of cross sectional area occupied by the flow. In this article the former notation is used, and flux/flow per unit of cross sectional area is termed more specifically "substance flux density".

[2]The term "macroscopic level" is used here with respect to the molecular level. In physiological applications of the laws of diffusion, the lengths of typical diffusion paths are in the range of about 1 μm to 100 μm, which is clearly "microscopic" in a physiological sense.

P_{O_2} boundary conditions only.[3] In contrast, conditions for tissue O_2 supply quite commonly include more complicated boundary conditions, P_{O_2} variations with time, and presence of phenomena like O_2 carriers, convective O_2 fluxes, or O_2 sources and sinks which need to be accounted for. Mathematical treatment of such complexities first of all requires reformulating FICK'S first law of diffusion, Eq. (1), for infinitesimally small differential volumes:

$$\vec{j} = -K \cdot \nabla P, \tag{3}$$

where ∇ is the gradient operator and $\vec{j}$ is the O_2 flux density vector (which is oriented parallel to O_2 flux, and the length of which equals the amount of O_2 transported per unit of time across a unit area of the normal plane to $\vec{j}$). This equation states that flux density be proportional to and of opposite direction as the gradient in P_{O_2}. The denotation "(diffusive) O_2 conductivity" for K, introduced above, originates in an analogy of Eq. (3) to the differential version of OHM'S law in electrodynamics, that relates current density to the electric field (which is the gradient of electrostatic potential).

Eq. (3) can be used to consider *transient* conditions, *i.e.* temporal changes in the amount of O_2 present at a given tissue location. The time dependent change in the amount of O_2 contained in a small volume of tissue clearly is equal to the difference between inward and outward flux which may be expressed using FICK'S first law, Eq. (3), and which depends on the spatial change in P_{O_2} gradient. By passage to the limit of differential volumes, the resulting balance equation may be transformed to yield a relation between local concentration change with time and spatial change in P_{O_2} gradient:

$$\alpha \frac{\partial P}{\partial t} = K \cdot \nabla^2 P, \tag{4}$$

where α is the O_2 solubility, $\partial P / \partial t$ is the change in P_{O_2} with time, and ∇^2 is the Laplace operator. (According to HENRY-DALTON'S law $\alpha \partial P / \partial t$ is the concentration change with time.) This equation is known as (a slight generalization of) FICK'S second law of diffusion.

1.3 The O_2 transport equation.

It was mentioned before that other mechanisms like presence of O_2 sources or sinks also exert influence on local O_2 concentration. As detailed below, their effects may be readily implemented in FICK'S second law by adding appropriate terms which describe the change in O_2 concentration brought about by the respective mechanism. The resulting modification of FICK'S second law of diffusion is given in Eq. (5), first equation.

(i) The most important physiological O_2 sink is tissue O_2 consumption occurring at a rate $\dot{V}_{O_2}$ (= amount of O_2 consumed per unit volume of tissue and unit of time) which may vary with local P_{O_2} and location (due to variations in mitochondrial density).

(ii) Secondly, there may be an O_2-carrier c present, which is hemoglobin (Hb) within erythrocytes and myoglobin (Mb) in red muscle cells. This carrier may release or bind oxygen. The degree of O_2 binding to a carrier c generally is expressed as its saturation $S = [c\text{-}O_2]/[c]_T$ where $[c\text{-}O_2]$ is the concentration of the carrier-O_2 compound and $[c]_T$ is the total carrier concentration. Carrier-bound O_2 represents a second species of oxygen,

---

[3]*I.e.*, FICK'S first law quantifies the diffusive flux between two "reservoirs" held at a given P_{O_2} and separated by a tissue layer of known dimensions and diffusivity. The requirement that P_{O_2} at the boundaries of the tissue layer is to be equal to a given constant value is called a (constant P_{O_2}) boundary condition.

the concentration of which needs to be considered in a separate equation, formulated, *e.g.*, in terms of carrier O_2 saturation (Eq. (5), second equation). Exchange between carrier-bound and free O_2 is quantified via the (net) O_2 release rate r which constitutes the link between both equations, in that r has to be added to the one equation and subtracted from the other.

(iii) Hemoglobin and myoglobin not only can reversibly bind oxygen. Rather, compounds of these molecules and O_2 are small enough to diffuse at a considerable rate, by that enhancing free O_2 diffusive transport (*cf.* [21,23]). This mechanism has been termed "carrier-facilitated O_2 diffusion". In analogy to FICK's second law for free O_2 diffusion the above mentioned equation for carrier-bound O_2 (Eq. (5), second equation) relates the temporal change in c-O_2 concentration, $[c]_T \cdot \partial S / \partial t$, carrier release rate r, and spatial variation in c-O_2 concentration gradient, $[c]_T \cdot \nabla^2 S$ (where the constant of proportionality, D_C, is the carrier diffusion coefficient).

(iv) Finally, local convection displaces free and carrier bound O_2 at a rate depending on the velocity vector $\vec{v}$. The effect of this displacement on local O_2 and c-O_2 concentrations is proportional to the magnitudes of $\vec{v}$ and of the concentration gradients, and depends on their mutual orientation, so it may be expressed by the scalar products $\vec{v} \cdot \alpha \nabla P$ and $\vec{v} \cdot [c]_T \nabla S$, respectively.

With these additions, the O_2 transport equation becomes:

$$
\underbrace{\alpha \frac{\partial P}{\partial t}}_{\substack{\text{change} \\ \text{in } [O_2]}} + \underbrace{\dot{V}_{O_2}}_{\substack{O_2 \text{ con-} \\ \text{sumption}}} = \underbrace{K \cdot \nabla^2 P}_{\substack{\text{free } O_2 \\ \text{diffusion}}} - \underbrace{\vec{v} \cdot \alpha \nabla P}_{\substack{\text{convective} \\ O_2 \text{ transport}}} + \underbrace{r}_{\substack{\text{carrier} \\ O_2 \text{ release}}}
$$

$$
\underbrace{[c]_T \frac{\partial S}{\partial t}}_{\substack{\text{change} \\ \text{in } [c\text{-}O_2]}} = \underbrace{D_C [c]_T \nabla^2 S}_{\substack{\text{facilitated} \\ O_2 \text{ diffusion}}} - \underbrace{\vec{v} \cdot [c]_T \nabla S}_{\substack{\text{convective} \\ c\text{-}O_2 \text{ transport}}} - \underbrace{r}_{\substack{\text{carrier} \\ O_2 \text{ release}}}
$$

(5)

The reactions between carrier and O_2 molecules occur extremely fast, fast enough so they may be assumed to reside in equilibrium under most physiological conditions (see [21] and section **2** below). As a consequence, carrier O_2 saturation S solely depends on partial pressure P (*i.e.*, $S = S(P)$) which eliminates the need of solving the system of coupled differential equations (5) for P and S. The function $S(P)$ is referred to as carrier "O_2 dissociation curve" (ODC). $S(P)$ depends on a number of parameters, the most important being temperature, pH, and concentrations of carbon dioxide (CO_2) and 2,3-diphosphoglycerate (2,3-DPG). pH, CO_2, and 2,3-DPG impacts on $S(P)$ are mediated via *allosteric* mechanisms. In the case of chemical equilibrium, the temporal change in $[c\text{-}O_2]$ is given by $[c]_T \cdot \partial S(P) / \partial t = [c]_T \cdot S'(P) \cdot \partial P / \partial t$ (S' is the derivative of S), and the two equations in (5) may be added to eliminate carrier O_2 release rate r:

$$
\underbrace{\alpha \frac{\partial P}{\partial t}}_{\substack{\text{change} \\ \text{in } [O_2]}} + \underbrace{[c]_T S'(P) \frac{\partial P}{\partial t}}_{\substack{\text{change} \\ \text{in } [c\text{-}O_2]}} + \underbrace{\dot{V}_{O_2}}_{\substack{O_2 \text{ con-} \\ \text{sumption}}} =
$$

$$
\underbrace{K \cdot \nabla^2 P}_{\substack{\text{free } O_2 \\ \text{diffusion}}} + \underbrace{D_C [c]_T \nabla^2 S(P)}_{\substack{\text{facilitated} \\ O_2 \text{ diffusion}}} - \underbrace{\vec{v}(\alpha \nabla P + [c]_T \nabla S(P))}_{\substack{\text{convective } O_2 \text{ and} \\ c\text{-}O_2 \text{ transport}}}
$$

(6)

Depending on the geometry inherent to the actual problem of interest, Eqs. (5) and (6) may also be expressed in coordinate systems other than the Cartesian system, *e.g.*, in polar or cylindrical coordinates.

1.4   Heterogeneous tissue domains, interface conditions, boundary conditions, solution of the O₂ transport equation.   There is one more peculiarity of O_2 diffusion in biological systems, complicating its mathematical treatment: Systems typically are composed of a number of different tissues (*e.g.*, capillary blood, interstitial space, muscle fibers) exhibiting distinctly different conditions for O_2 diffusion (*e.g.*, differences in conductivity, presence of O_2 carriers, *etc.*). This leads to "discontinuities" in transport characteristics at the interfaces between the various tissue regions and requires description of O_2 supply by an individual differential equation in each region. In solving differential equations from adjacent regions, their solutions need to be matched such that P_{O_2} and O_2 flux densities are continuous across the common interface. In other words, solutions need to satisfy so-called interface conditions.

Along with interface and (appropriate) boundary conditions (and an initial condition in the transient case), the O_2 transport equation uniquely defines P_{O_2} in a given biological system. Actual P_{O_2} distributions throughout the system may be obtained by solving this so-called boundary value problem. Solutions cannot be given as closed-form expressions for the general case. Rather, they depend on the particular geometry, on actual conditions for O_2 diffusion, as well as on boundary and interface conditions. If analytical solutions exist, their derivation frequently involves application of advanced mathematical techniques. In other cases, approximations may be obtained by use of appropriate numerical algorithms on digital computers.

P_{O_2} distributions represent the integrated end-product of the assembly of all O_2 transport processes. After P_{O_2} distributions have been found, they may be employed to evaluate the significance of any subprocess involved in O_2 transport. Up to this point prerequisites and individual steps necessary for obtaining these P_{O_2} distributions have been presented. In the following subsections, some special aspects of the physics of O_2 transport in biological systems will be elaborated on, and some useful tools in evaluating P_{O_2} profiles will be discussed.

1.5   The physiological O₂ carriers hemoglobin and myoglobin.   In this subsection, the general statements made on carrier O_2 binding in subsection **1.3** *(ii)* are going to be specialized to the cases of hemoglobin (Hb) and myoglobin (Mb).

1.5.1   General remarks on hemoglobin. Hb is present within erythrocytes at a concentration of 0.02 *mol/l* (Hb monomer), and O_2 binding by Hb accounts for more than 98 % of the 0.2 *ml* of oxygen which are contained in 1 *ml* of normal arterial blood. The hemoglobin molecule is composed of four Hb monomers each of which can bind one O_2 molecule. Within the tetramer — which commonly is abbreviated as Hb_4 —, Hb reactions with O_2 are cooperative, *i.e.*, reaction rate coefficients of free binding sites depend on the number of O_2 molecules already bound to the tetramer.

1.5.2   Four-step reaction scheme and O₂ dissociation curve.   According to ADAIR'S intermediate compound reaction scheme [1], reaction between Hb and O_2

may be described to proceed in four steps ($i = 1 \ldots 4$) each of which is characterized by individual rate constants for O_2 binding (k_i') and O_2 release (k_i):

$$Hb_4(O_2)_{i-1} + O_2 \quad \overset{k_i'}{\underset{k_i}{\rightleftharpoons}} \quad Hb_4(O_2)_i, \qquad i = 1 \ldots 4 \qquad (7)$$

In chemical equilibrium, at any O_2 concentration (or P_{O_2}), concentration of each species, $[Hb_4(O_2)_i]$, $i = 0 \ldots 4$, is uniquely defined by the on and off reaction rate coefficients k_i' and k_i, respectively. From these species equilibrium concentrations, O_2 saturation $S = \left(\sum_{i=1}^{4} i[Hb_4(O_2)_i] \right) \cdot \left(4 \cdot \sum_{i=0}^{4} [Hb_4(O_2)_i] \right)^{-1}$ for any given O_2 concentration or P_{O_2} may be calculated, which results in the typical S-shaped Hb-O_2-dissociation curve (ODC), Fig. 1, solid curve labelled "Hb" (data from [2]). As opposed to ODCs directly measured in chemical equilibrium, we are going to refer to the ones that are determined by a given reaction scheme as "implicit".

1.5.3 Transport equations for four-step reaction scheme. As a consequence of ADAIR'S intermediate compound reaction scheme [1], for each species of Hb-bound oxygen, $Hb_4(O_2)_i$, $i = 1 \ldots 4$, a separate differential equation similar to the second equation (carrier equation) of (5) exists. These equations are coupled to each other via net O_2 release rates $r_1 \ldots r_4$. In other words, the carrier equation in the transport equation actually represents a system of four coupled differential equations — one for each species — which, in turn, is coupled to the free O_2 transport equation (first equation in Eq. (5)) via the sum of the four net O_2 release rates. Thus, from this approach a system of five coupled partial differential equations results.

1.5.4 Approximations to hemoglobin ODC. Empirical equations for the Hb-ODC have been developed, the most popular of which is HILL'S equation [18] (dashed curve in Fig. 1).

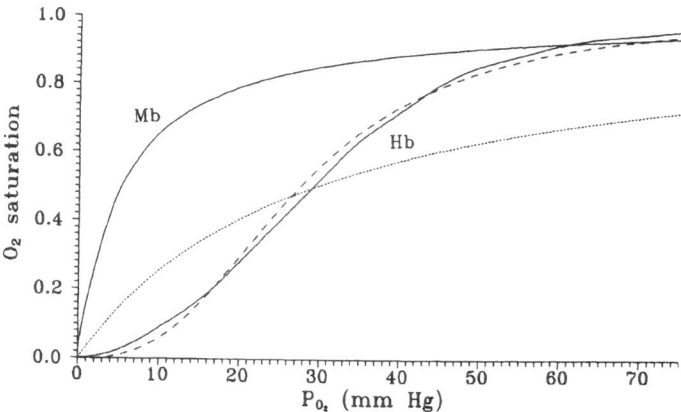

Figure 1. Oxygen dissociation curves (ODC) of hemoglobin (solid curve labelled "Hb") and myoglobin (solid curve labelled "Mb"). Dashed curve is the HILL approximation to the Hb-ODC. Dotted curve is a best fit of a hyperbolic ODC to the Hb-ODC which obviously represents but a poor approximation.

$$S(P) = \frac{(P/P_{50})^n}{1 + (P/P_{50})^n} \qquad \text{with its inverse} \qquad P(S) = P_{50}\left(\frac{S}{1-S}\right)^{1/n} \qquad (8)$$

where P_{50} is the Hb half-saturation P_{O_2} of about 26 $mm\,Hg$ and $n \approx 2.7$. BUERK and BRIDGES [7] have given a more precise and also invertible approximation to the Hb-ODC which accounts for its dependence on pH, P_{CO_2}, temperature, and [2,3-DPG].

1.5.5 Simplification: Variable rate coefficient one-step reaction scheme.

In order to keep the mathematical problem of solving the O_2 transport equation tractable, it is desirable not to deal with the system of five coupled partial differential equations resulting from the four-step reaction scheme Eq. (7) (see **1.5.3**). If Eq. (7) is replaced by a one-step reaction of Hb monomer with O_2 (MICHAELIS-MENTEN kinetics),

$$Hb + O_2 \underset{k}{\overset{k'}{\rightleftharpoons}} HbO_2, \qquad (9)$$

this reduces the system of carrier equations to the one given in Eq. (5). The ODC resulting from this simplification is

$$S(P) = \frac{k'[O_2]}{k + k'[O_2]} = \frac{P}{P_{50} + P}, \qquad (10)$$

where P_{50} is the half-saturation P_{O_2} of Hb. Thus, the function $S(P)$ is of hyperbolic shape and even a best fit of its parameter P_{50} can achieve but a poor approximation to the true Hb-ODC (dotted curve in Fig. 1).

This problem has been circumvented in the following way (*cf.* [22]): Given an approximation $S(P)$ to the Hb-ODC and an (average) off reaction rate coefficient k, a variable on reaction rate coefficient k' may be defined which is consistent with the ODC. *E.g.*, for

$$k'(P) = \frac{kP^{n-1}}{\alpha P_{50}^{\,n}} \qquad (11)$$

the HILL approximation to the Hb ODC (dashed curve in Fig. 1) would be reproduced. In this case, the net O_2 release rate for actual P_{O_2}, P, and actual saturation, S, may be obtained in the following way:

$$r(P,S) = k[HbO_2] - k'[O_2][Hb] = k[Hb]_T\left(S - \frac{P^n}{P_{50}^{\,n}}(1-S)\right), \qquad (12)$$

where $[Hb]_T$ is total Hb concentration. Note that $r(P,S)$ vanishes if S is the equilibrium saturation at oxygen partial pressure P.

1.5.6 The physiological O_2 carrier myoglobin.

Myoglobin is present in oxidative skeletal muscles and in heart at concentrations of up to about 0.5 $mmol/l$, *i.e.*, concentrations are at least 40 times smaller than Hb concentrations in RBC's. O_2 storage capacity of myoglobin in heart at maximum exercise is slightly greater than that of capillary-contained erythrocytes and suffices to maintain O_2 consumption for only about 1 s. Mb molecules are non-cooperative when binding to O_2. Hence, reaction kinetics are well described by the MICHAELIS-MENTEN approach, Eqs. (9) and (10), and Mb ODC is hyperbolic (curve labelled "Mb" in Fig. 1). Myoglobin P_{50} is 5.3 $mm\,Hg$ and thus much lower than that of Hb.

1.6 Free and facilitated O_2 diffusion. Due to its great physiological importance particularly in Mb containing muscles, some further considerations on facilitated diffusion appear to be appropriate. In Eq. (3), free O_2 flux density was given as the product of conductivity times P_{O_2} gradient. In the presence of an O_2 carrier, total (free and facilitated) O_2 flux density is described by an analogous relation:

$$\vec{j} = -K \cdot \nabla P - [c]_T \cdot D_C \cdot \nabla S(P)$$

($[c]_T$ and D_C are carrier concentration and diffusivity, respectively), which may be transformed to yield

$$\vec{j} = -\underbrace{K\left(1 + \frac{[c]_T \cdot D_C}{\alpha \cdot D} \cdot S'(P)\right)}_{K_{\text{eff}}} \nabla P. \tag{13}$$

The term $K \cdot (1 + \ldots)$ may be interpreted to represent an effective conductivity $K_{\text{eff}} = K \cdot (1 + \ldots)$ that replaces K in FICK's first law, Eq. (3). In Fig. 2, K_{eff} is shown for conditions in erythrocytes ("$K_{\text{eff E}}$") and in muscle fibers ("$K_{\text{eff M}}$") where carriers are Hb or Mb, respectively. Horizontal lines are the respective K's for free O_2 diffusion. K_{eff} is always greater than or equal to K and depends on the ratio $([c]_T \cdot D_C)/(\alpha \cdot D)$ (that relates characteristic quantities of carrier and free O_2 diffusion and has been termed "facilitation pressure" [21]) and on the slope $S'(P)$ of carrier ODC at actual P_{O_2}. For small carrier concentrations $[c]_T$ or diffusivities D_C, $K_{\text{eff}} \approx K$. Moreover, $K_{\text{eff}} \approx K$ in locations in which carrier ODC is essentially flat (which is true for large values of P; see Fig. 1). K_{eff} becomes maximal at the O_2 partial pressure at which carrier ODC is steepest (which is about half-saturation pressure of $26\ mm\ Hg$ for Hb, and $0\ mm\ Hg$ for Mb) and may be as large as 7 times K in erythrocytes and 6 times K in red muscle.

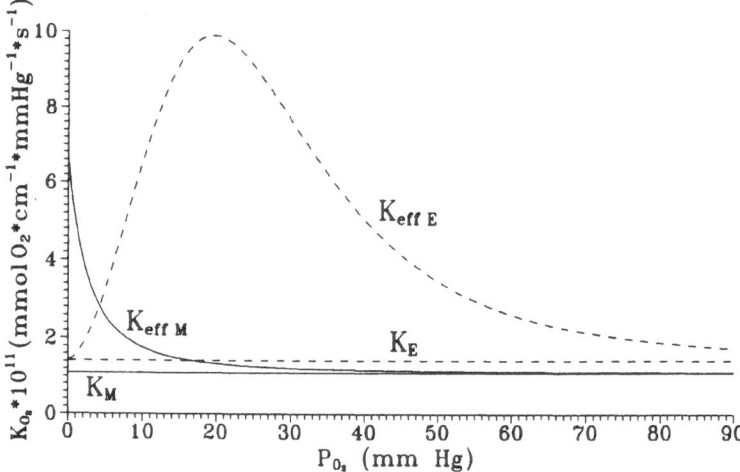

Figure 2. Effective conductivity, K_{eff}, for conditions in erythrocytes ("$K_{\text{eff E}}$") and in muscle fibers ("$K_{\text{eff M}}$") where carriers are Hb or Mb, respectively. Horizontal lines are the conductivities for free O_2 diffusion in RBC's ("K_E") and in muscle cells ("K_M").

1.7 Transport resistances, conductances, and diffusing capacity. The process of O_2 transport in tissue — similar considerations are true for O_2 uptake in the lungs — encompasses transport in different compartments (e.g., in RBC, plasma, capillary endothelium, etc.) which are linked to each other by interface conditions (cf. **1.4**). In each individual compartment, O_2 transport involves a number of interrelated subprocesses taking place simultaneously (e.g., free and Hb-facilitated O_2 diffusion and O_2 release from Hb in RBC's). These subprocesses are described in the O_2 transport equation, Eqs. (5) and (6). Thus, the overall O_2 transport process consists of a long chain of series or parallel coupled events. When analyzing the organism's O_2 exchange apparatus it is of interest to know to which extent any one of the various steps in the process of O_2 exchange is rate limiting for O_2 flux.

Conceptually, limitation of organ O_2 flux may be due to two distinct causes (for which, however, in a given situation identification of the principle cause is not easily possible): Restrictions in convective flow of Hb-bound oxygen (O_2 delivery) or limited O_2 release from capillary contained blood, the latter one of which we shall primarily be concerned with in this article. P_{O_2} acts as the driving force not only for free O_2 diffusion but also, *cum grano salis*, for any other of the non-convective steps involved in capillary O_2 release (i.e. in reaction with carriers, facilitated O_2 diffusion, O_2 consumption). Thus, the most obvious quantity characterizing the hindrance offered by any one of the various steps would be the amount of driving force "consumed" in it, i.e., the P_{O_2} drop across the respective step. This is illustrated in Fig. 3 which shows a P_{O_2} profile in heavily working muscle (solid curve). P_{O_2} drops across RBC ("$\Delta P_{O_2\,RBC}$"), capillary wall ("$\Delta P_{O_2\,CAP}$"), and muscle fiber ("$\Delta P_{O_2\,MUS}$") represent hindrances to O_2 transport offered by the respective structures. As can be seen from the P_{O_2} profile for low muscle performance (dotted curve), however, these P_{O_2} drops strongly rise and fall with the magnitude of actual O_2 flux. Evaluating hindrances via comparing P_{O_2} drops is therefore feasible only if O_2 fluxes are the same.

To meet actual tissue energy requirements by maintaining adequate O_2 fluxes is the ultimate goal of the organism's O_2 excha ĝe apparatus. Thus, when approaching

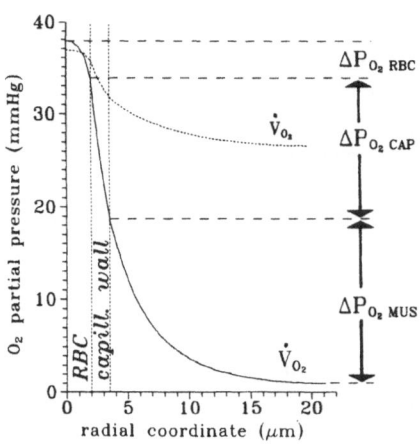

Figure 3. P_{O_2} profile in heavily working muscle (solid curve labelled $\dot{V}_{O_2}$). Labels "RBC" and *"capill. wall"* indicate positions of red blood cell and capillary wall, respectively. P_{O_2} drops across RBC ("$\Delta P_{O_2\,RBC}$"), capillary wall ("$\Delta P_{O_2\,CAP}$"), and muscle fiber ("$\Delta P_{O_2\,MUS}$") represent hindrances to O_2 transport offered by the respective structures. P_{O_2} profile for low muscle performance (dotted curve labelled $\dot{V}_{O_2}$), demonstrates that these P_{O_2} drops strongly depend on the magnitude of actual O_2 flux.

the limits of performance, the magnitude of O_2 flux is the decisive quantity that needs to be maximized, and capillary P_{O_2} represents the driving force that is available for (non-convective) O_2 exchange and is limiting its magnitude. Therefore, a measure of the efficiency of this exchange process (or of the hindrance offered by its individual steps) is the mean P_{O_2} drop ΔP_{O_2} (total or across an individual step), divided by the O_2 flux $\dot{V}_{O_2}$ that can be driven by ΔP_{O_2}. ($\dot{V}_{O_2}$ usually is given as total body or organ O_2 flux or as flux per unit of body mass, of tissue mass, or of tissue volume (as defined in **1.3** (i)), but other normalizations are not uncommon either.) In analogy to OHM's law in electrodynamics, this measure has been denoted "(transport) resistance" R, and its inverse k has been termed "conductance" or "mass transfer coefficient":

$$R = \frac{\Delta P_{O_2}}{\dot{V}_{O_2}} \qquad\qquad k = \frac{\dot{V}_{O_2}}{\Delta P_{O_2}} \qquad\qquad (14)$$

Depending on the exact definition of $\dot{V}_{O_2}$, R and k represent resistance/conductance (total or across an individual transport step) of an organ, of a unit of body or tissue mass, *etc.* The total conductance of an entire organ is frequently called its "diffusing capacity" (which notion was first introduced for the lung by BOHR [5]).

It should be noted, though, that — unless any P_{O_2} dependent processes and heterogeneities are absent — "resistances" defined in this way by no means are material constants of a given tissue region (or parts thereof), like resistances of electrical resistors are. Rather, R may depend very sensitively upon actual P_{O_2} distribution which, in turn is a function of a number of parameters as discussed in subsection **1.3** when presenting the O_2 transport equation. Nevertheless, the notions of resistance and conductance are useful, particularly for studying their changes with variations in physiological parameters like hematocrit, O_2 consumption rate, *etc.* Since resistances of serial subprocesses and conductances of parallel ones are additive in reasonably good approximation, they may also be employed for studying contributions of individual mechanisms in a complex system to overall system resistance/conductance.

2 Diffusive O_2 transport in capillary blood

The first breakthrough in kinetics of blood O_2 exchange is owed to HARTRIDGE and ROUGHTON who as early as 1923 introduced the first experimental setup for monitoring extremely rapid chemical reactions [16]. Their work laid the foundations for recognizing blood resistance to O_2 transport and for developing the concept of blood O_2 diffusing capacity. In later years, experimental findings have been complemented by theoretical investigations.

For a complete description of capillary O_2 exchange, the following circumstances and mechanisms of possible relevance for O_2 transport need to be accounted for. These mechanisms are schematically graphed in Fig. 4.

1. *Free O_2 diffusion in plasma and erythrocytes.*

2. *O_2 storage by hemoglobin* according to actual Hb-O_2-dissociation curve.

3. *Realistic ranges of capillary Hb-O_2-saturation and transcapillary O_2 flux.*

4. *Hemoglobin-O_2-reaction kinetics.* May be important in a thin boundary layer in which Hb and O_2 are in disequilibrium.

5. *Hemoglobin-facilitated O_2 diffusion in RBC's.*

30

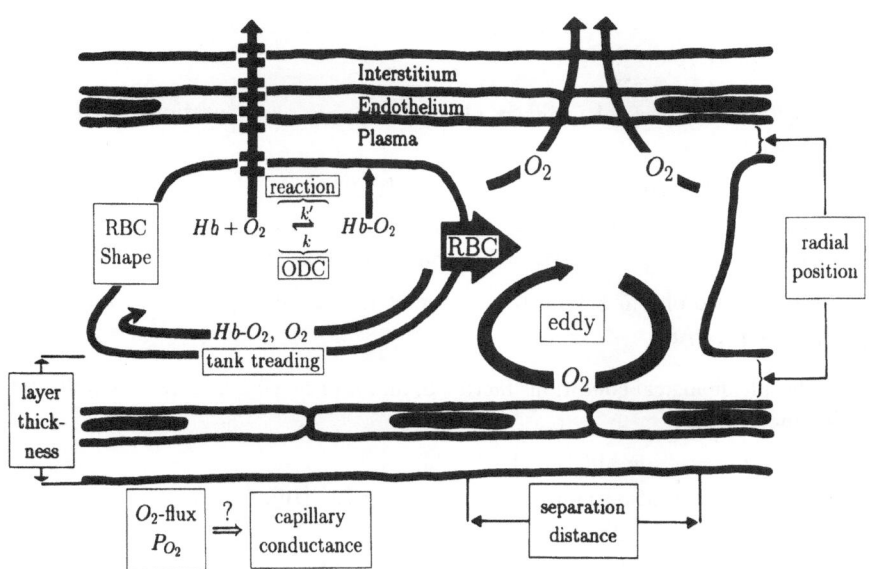

Figure 4. Schematic presentation of circumstances and mechanisms of possible relevance for capillary O_2 exchange. For more explanations see text.

6. *Realistic geometrical arrangement of red blood cells and plasma within capillaries* (including precise red cell shape and radial position). In narrow capillaries, RBC's are droplet- or slipper-shaped and travel in single file flow separated by plasma gaps.

7. *Penetration of O_2 through red cell and other membranes* in the diffusion path.

8. *Motion of capillary blood relative to adjacent tissue* resulting in loading and unloading of adjacent tissue with oxygen when RBC's or plasma gaps pass by.

9. *Small scale convection in plasma and RBC's.* Effects of eddy like convection in plasma gaps and tank-treading motion of RBC membranes on O_2 transport.

Space does not allow to discuss the numerous studies investigating into the above questions and to develop the conclusions that can be drawn from their results in any detail. The following listing briefly summarizes an evaluation of the importance of factors affecting capillary O_2 exchange, which has been extracted from the presently available literature, and cites some of the pertinent investigations on which the evaluation was based.[4]

1. Most important determinants of capillary O_2 exchange are

   (i) free O_2 diffusion in RBC and plasma [12,19],

   (ii) thickness of layer surrounding RBC's (adding extra diffusion resistance) [10, 12,14],

   (iii) cell separation distance (reducing surface effective in O_2 exchange) [10,13, 17], and

---

[4]Obviously, these conclusions can only be correct to a degree to which the underlying — and widely accepted — experimental data are correct. It should be noted that particularly regarding conductivities and reaction rates some controversies about their true *in vivo* values remain.

(iv) correction of a hyperbolic Hb-ODC (if 1-step Hb-O_2-reaction kinetics are employed) [4,22].

2. Important modifiers of capillary O_2 exchange are augmentation of red cell O_2 conductance by

   (i) Hb-facilitated O_2 diffusion in RBC's [12,22],
   (ii) diffusion of O_2 through plasma gaps [10,13], and
   (iii) motion of RBC's relative to surrounding tissue [13],

   the latter two of which tend to counteract the loss of O_2 exchange area by cell separation 1 *(iii)*.

3. Under physiological conditions factors of minor or no significance for capillary O_2 conductance are

   (i) Hb-O_2-reaction kinetics [4,9,12,15],
   (ii) Hb-O_2-saturation and transcapillary O_2-flux [10],
   (iii) resistance of cell membranes [20],
   (iv) convection in plasma and RBC [3,8,19], and
   (v) precise RBC shape and radial position within capillary [10,13].

From this evaluation it follows that O_2 transport resistance in capillaries may be described by an empirical "conductance function", in the derivation of which only the mechanisms listed under "essential" or "important" have to be accounted for. Accordingly, the function would not need to depend on any parameters but thickness of layer surrounding RBC's, cell separation distance, and cell velocity (and on invariant species-specific blood parameters). This equation may then be plugged into any model for tissue O_2 transport — coupled by continuity conditions for P_{O_2} and O_2 flux density at the interfaces — to correctly describe O_2 release from capillaries in a larger environment.

References

[1] G.S. ADAIR, The hemoglobin system, *J.Biol.Chem.* 63:529 (1925)

[2] P.L. ALTMAN, D.S. DITTMER, "Biology Data Book", Federation of American Societies for Experimental Biology, Bethesda, 1972

[3] J. AROESTY, J.F. GROSS, Convection and diffusion in the microcirculation, *Microvasc.Res.* 2:247 (1970)

[4] P.T. BAXLEY, J.D. HELLUMS, A simple model for simulation of oxygen transport in the microcirculation, *Ann.Biomed.Eng.* 11:401 (1983)

[5] C. BOHR, Über die spezifische Tätigkeit der Lungen bei der respiratorischen Gasaufnahme, *Scand.Arch.Physiol.* 22:221 (1909)

[6] D.F. BRULEY, Probabilistic solutions and models: Oxygen transport in the brain microcirculation, p. 133 in: "Mathematics of Microcirculation Phenomena", J.F. Gross, A. Popel (eds.), Raven, New York, 1980

[7] D.G. BUERK, E.W. BRIDGES, A simplified algorithm for computing the variation in oxyhemoglobin saturation with pH, PCO_2, T, and DPG, *Chem.Eng.Commun.* 47:113 (1986)

[8] A. CLARK, JR., G.R. COKELET, W.J. FEDERSPIEL, Erythrocyte motion and oxygen transport, *Bibl.Anat.* 20:385 (1981)

[9] A. CLARK, P.A.A. CLARK, The end-points of the oxygen path: Transport resistance in red cells and mitochondria, *Adv.Exp.Med.Biol.* 200:43 (1986)

[10] W.J. FEDERSPIEL, A.S. POPEL, A theoretical analysis of the effect of the particulate nature of blood on oxygen release in capillaries, *Microvasc.Res.* 32:164 (1986)

[11] A. FICK, Über Diffusion, *Pogg.Ann.* 94:59 (1855)

[12] K. GROEBE, G. THEWS, Theoretical analysis of oxygen supply to contracted skeletal muscle, *Adv.Exp.Med.Biol.* 200:495 (1986)

[13] K. GROEBE, G. THEWS, Effects of red cell spacing and red cell movement upon oxygen release under conditions of maximally working skeletal muscle, *Adv.Exp.Med.Biol.* 248:175 (1989)

[14] K. GROEBE, G. THEWS, Calculated intra- and extracellular P_{O_2} gradients in heavily working red muscle, *Am.J.Physiol.* 259:H84 (1990)

[15] G. GUTIERREZ, The rate of oxygen release and its effect on capillary O_2 tension: A mathematical analysis, *Respir.Phsiol.* 63:79 (1986)

[16] H. HARTRIDGE, F.J.W. ROUGHTON, A method for measuring the velocity of very rapid chemical reactions, *Proc.Roy.Soc. A* 104:376 (1923)

[17] J.D. HELLUMS, The resistance to oxygen transport in the capillaries relative to that in the surrounding tissue, *Microvasc.Res.* 13:131 (1977)

[18] A.V. HILL, The possible effects of the aggregation of the molecules of haemoglobin on its dissociation curve, *J.Physiol. (London)* 41:iv (1910)

[19] C. HOOK, K. YAMAGUCHI, P. SCHEID, J. PIIPER, Oxygen transfer of red blood cells: experimental data and model analysis, *Respir.Phsiol.* 72:65 (1988)

[20] F. KREUZER, W.Z. YAHR, Influence of red cell membrane on diffusion of oxygen, *J.Appl.Physiol.* 15:1117 (1960)

[21] F. KREUZER, L. HOOFD, Facilitated diffusion of oxygen and carbon dioxide, in: "Handbook of Physiology, Sect. 3: The Respiratory System, Vol. IV: Gas Exchange", L.E. FAHRI, S.M. TENNEY, eds., American Physiological Society, Bethesda, 1987

[22] W. MOLL, The influence of hemoglobin diffusion on oxygen uptake and release by red cells, *Respir.Physiol.* 6:1 (1968/69)

[23] J.B. WITTENBERG, Myoglobin-facilitated oxygen diffusion: Role of myoglobin in oxygen entry into muscle, *Physiol.Rev.* 50:559 (1970)

OXYGEN SUPPLY DEPENDENCY IN THE CRITICALLY ILL -

A CONTINUING CONUNDRUM

Stephen M. Cain

Department of Physiology and Biophysics
University of Alabama at Birmingham
Birmingham, AL 35294-0005, U.S.A.

INTRODUCTION

Although they were not the first to observe it, a seminal paper by Danek et al. (1980) ushered in a decade of concern for oxygen supply dependency in critically ill patients. That concern continues today with some urgency because of the uncertainty of what it means. The observation made by Danek et al. (1980) was that O_2 uptake (VO_2) varied directly with O_2 delivery (DO_2) even at very high delivery rates in patients that were critically ill with adult respiratory distress syndrome (ARDS). A similar linear relationship between VO_2 and DO_2 was not seen in another group of critically ill ventilator patients who had not been diagnosed with ARDS. In this and in the earlier studies, it appeared that O_2 demand was higher than expected and was not satisfied in ARDS patients even if DO_2 were raised to well above normal resting levels. The authors suggested that this was due to a possible disturbance of normal regulatory control of blood flow in the periphery as well as a possible disruption of cellular structure with mitochondrial dysfunction.

In a subsequent review article, I (1984) pointed out that defective tissue O_2 extraction appeared to be one of the hallmarks in conjunction with an increased O2 demand in patients with ARDS and that a similar state had been identified in patients with sepsis. I termed this "pathological O_2 supply dependency" to distinguish it from the more normal response to an increase in O_2 demand/supply ratio; i.e., to increase O_2 extraction ratio (O_2ER) and thus to utilize more of the O_2 reserves in blood as a first resort. The normal response was termed "physiological O_2 supply dependency." Increased O_2ER is seen in normal animals made hypoxic (Cain, 1978) and as an early compensation to an increase in O_2 demand initiated by exercise in humans (Rowell, 1986).

The clinical significance of pathological O_2 supply dependency was clearly indicated in a prospective study by Bihari et al. (1987). They increased DO_2 with a potent vasodilator, prostacyclin, in 27 critically ill patients. They also gave the same treatment to 7 normal volunteers. They found that 13 patients increased VO_2 when DO_2 increased whereas 14 did not nor did the 7 volunteers. All 13 of the responders did not survive their hospital stay whereas all of the nonresponders did. The interpretation they made of these results was that there was a substantial O_2 debt in the patients who subsequently died and that this may have contributed to irreversible multiorgan failure.

Other investigators subscribed to the idea that a pervasive occult tissue hypoxia can be present in septic patients even with the elevated cardiac outputs associated with "warm" or hyperdynamic sepsis. Tuchschmidt et al. (1991) recently reviewed the VO_2/DO_2 status of septic patients. They found O_2 supply dependency below a DO_2 of 15 ml/kg min, which was approximately twice the critical DO_2 found by others in nonseptic patients who were undergoing coronary bypass surgery. Below that critical level of DO_2, arterial lactates were elevated. The conviction that this signified widespread tissue hypoxia was reinforced by several studies that were reviewed in another recent article (Cain and Curtis, 1991a). In all of these studies, elevated lactates were associated with O_2 supply dependency and they decreased as VO_2 increased with an increase in DO_2. The association of elevated lactates with pathological O_2 supply dependency is so strong, in fact, that some studies have shown one to predict the other (Kruse et al., 1990; Fenwick et al., 1990) and that lactate is a superior predictor of outcome in human septic shock (Bakker et al., 1991).

There was no lack of indirect evidence to support the idea that a defect in peripheral O_2 extraction can lead to a pervasive tissue hypoxia which in turn may underlie the high mortality rate in patients with ARDS or sepsis. Conversely, there is relatively little direct evidence that these are causative factors or indeed are even present. Dantzker et al. (1991) challenged the basic idea of pathological O_2 supply dependency in humans as unproven. They suggested that DO_2 in critically ill patients is simply following increased O_2 demands in a physiological fashion and there is no abnormal manifestation of impaired O_2 extraction. Others such as Ronco et al. (1991) claim that the finding of pathological O_2 supply dependency in severe ARDS is a methodological error that arises from "coupling error" as described by Archie (1981). It arises when two variables to be correlated share the same measurement error. The most common way to obtain DO_2 and VO_2 at the bedside, for example, is to use thermodilution to measure cardiac output and then to multiply that by arterial O_2 content and the arteriovenous O_2 content difference to arrive at DO_2 and VO_2, respectively. Multiplication of the measurement error insures that some positive correlation will be found when the regression line is calculated whether or not any correlation truly exists.

The information that can be gotten at the bedside of critically ill patients may be influenced by the demands of treatment and will be limited by the invasiveness of any additional procedure. This leads to an examination of the information that is currently available from animal experiments where better control and more invasive procedures can be used. The most common model utilizes lipopolysaccharide endotoxin derivative to emulate the septic state in patients. A logical first question is whether this constitutes an adequate model of human sepsis. Cain and Curtis (1991a) recently examined the available evidence relevant to this question and concluded that endotoxin did serve this purpose very well. Clearly, endotoxin liberates many of the same cytokines and initiates the immune cascade similar to sepsis. The next question is whether endotoxin treatment can produce an "O_2 extraction defect." If the extraction defect is found, is it indicative of pervasive tissue hypoxia and the reason for elevated lactate levels? The ultimate question, and one that won't be resolved here, is whether debilitating tissue hypoxia can be present even when blood flow is elevated such as in hyperdynamic sepsis.

O_2 EXTRACTION DEFECT IN ANIMAL MODELS

In three separate studies, anesthetized dogs were injected with either live bacteria cultured from a septicemic patient or with endotoxin (5-8 mg/kg). The animals were then hemorrhaged in steps to lower cardiac output. In this manner, sufficient measurements of VO_2 and DO_2 were made to identify the critical DO_2 below which VO_2 became linearly dependent upon DO_2. In addition to the systemic measurements, two regional circulations were isolated for study in skeletal muscle and gut. The results are shown in Table 1.

Clearly the endotoxin or bacteremia raised the critical DO_2 and lowered the critical O_2ER in the whole body. Muscle appeared to be resistant to the effects of endotoxin whereas gut was particularly susceptible. These results supported the idea that injections of endotoxin into anesthetized dogs could reproduce some of the postulated events in septic patients, in this case the O_2 extraction defect.

Table 1. Critical values obtained in hemorrhaged dogs.

	Control	Endotoxin	Ref.
Critical DO_2 (ml/kg/min)	7.4 $\pm$1.2	11.4$\pm$2.2**	[1]Nelson et al. (1987)
Critical O_2ER	0.71$\pm$0.10	0.51$\pm$0.009**	Nelson et al. (1988)
Critical DO_2 (ml/kg/min)	6.8 $\pm$1.2	12.8$\pm$2.0**	
Critical O_2ER			
systemic	0.78$\pm$0.04	0.54$\pm$0.11**	
gut	0.69$\pm$0.06	0.47$\pm$0.10**	
Critical DO_2 (ml/kg/min)	8.0 $\pm$0.7	11.4$\pm$2.7*	Samsel et al. (1988)
Critical O_2ER			
systemic	0.70$\pm$0.07	0.61$\pm$0.11*	
gut	0.67	0.69 (ns)	

[1]These results were obtained with injection of live bacteria rather than endotoxin.
*$P<0.05$; **$P<0.01$

Endotoxin is a potent stimulus to activate macrophages and monocytes and to release tumor necrosis factor (TNF, cachectin) (Messmer et al., 1989). A profoundly disturbed microcirculation is one consequence of these actions and the one that is of most concern for O_2 delivery to tissues. Leukocyte plugging, increased microvascular permeability with resultant pericapillary edema, endothelial swelling, and dysfunctional endothelial mediated microvascular controls can all accrue to a severe defect in O_2 extraction. Each of these events is tantamount to lengthening the effective diffusion distance for O_2 One solution to overcome such added obstacles to O_2 transfer is to increase the PO_2 gradient.

Bredle et al. (1989) examined the effect of increasing arterial PO_2 on the critical O_2 delivery and extraction in hindlimb muscle of dogs with and without injections of endotoxin. In their experiments, the arterial inflow to the limb muscles was controlled by a pump so that O_2 delivery could be selectively lowered to that region in three steps while the remainder of the animal was normally perfused. In addition, a membrane oxygenator was inserted into the perfusion circuit so that that PO_2 could be set at either 60 or 200 torr. Contrary to the results of Samsel et al. (1988), endotoxin injections had a significant effect on the critical O_2 extraction which was decreased significantly from 0.73 in control animals to 0.60 in the treated animals. No difference was seen in either the control or the endotoxin treated animals as a result of raising PO_2 from 60 to 200 torr.

One possible explanation for the lack of any benefit from raising arterial PO_2 is that the postulated interference with O_2 transfer was not present. In that there was a significant difference in O_2 extraction between the control and endotoxin treated groups, that patently was not the case. A more likely explanation can be found in Figure 1 showing some of the results of Bredle et al. (1989). When limb O_2 uptake was graphed against limb venous PO_2, two significantly different lines resulted with the high PO_2 line lying to the right of the low PO_2 line in both groups with no difference between the two groups. Furthermore, both sets of lines had a positive intercept on the abscissa. The extrapolation would signify a positive venous PO_2 even as O_2 uptake fell to zero. An alternative and more likely explanation for these results was that precapillary O_2 diffusion between adjoining artery and vein prevented any marked rise in capillary PO_2. Diffusional shunting of this type would have been favored as flow was lowered and fits with the rightward displacement of the high PO_2 group in the figure. Consequently, raising arterial PO_2 was ineffective in increasing the ability of muscle to extract O_2 in endotoxic dogs. The extent to which these results can be applied to critically ill patients is open to question but the results do seem to indicate that little purpose would be served

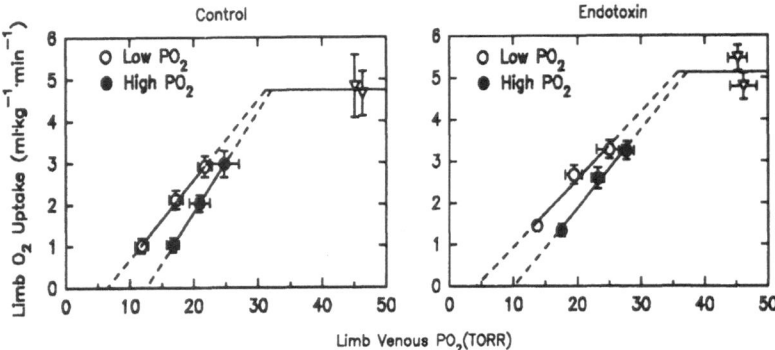

Fig. 1. Limb O_2 uptake and venous PO_2 for normal and three levels of reduced flow at high and low PO_2 in control and endotoxin treated dogs (redrawn from Bredle et al., 1989).

by extraordinary measures to raise arterial PO_2.

Another approach to answer the question of whether there is a diffusional limitation to the transfer Of O_2 in endotoxic dogs was taken by Curtis et al. (this volume). In experiments similar to those of Bredle et al. (1989), they progressively lowered PO_2 with a membrane oxygenator while pump-perfusing hindlimb muscles of anesthetized dogs at a constant flow. Their rationale was that any diffusional limitation should be discernable by a worse critical 0_2ER if O_2 delivery is decreased by lowering PO_2 rather than flow. As in the earlier study, comparisons were made between control animals and those given endotoxin with sufficient donor red blood cells and dextran to maintain cardiac output at normal or higher levels.

Their pooled results for the limb muscles are shown in Fig. 2 taken from their paper (Curtis et al., this volume). The dual linear regression lines for the pooled data accurately depict the information obtained by fitting lines for individual dogs. The critical DO_2 was elevated in the endotoxin group but that was primarily because O_2 demand was significantly increased. The critical 0_2ERs were not different at $66\pm4\%$ for controls and $72\pm8\%$ for the endotoxin group. In other words, ability of the limb muscles to increase O_2 extraction was not affected by lowering the PO_2 diffusion gradient which would apparently rule against endotoxin creating an extraordinary diffusion limitation. The effect of the endotoxin to create an apparent O_2 supply dependency at deliveries above critical is also evident in this figure. VO_2 increased with increasing DO_2 to produce a picture that is very similar to the type of information presented for critically ill patients

(Danek et al., 1980). According to Curtis et al. (this volume), the effect in muscle was enough to account for the increase they found in whole body VO_2.

Once again the results are beclouded by additional factors that must be considered. The goal in the experiments just described was to demonstrate an increased diffusion limitation as a result of endotoxin treatment by lowering PO_2 to a regional circulation to produce a purely local hypoxic hypoxia.

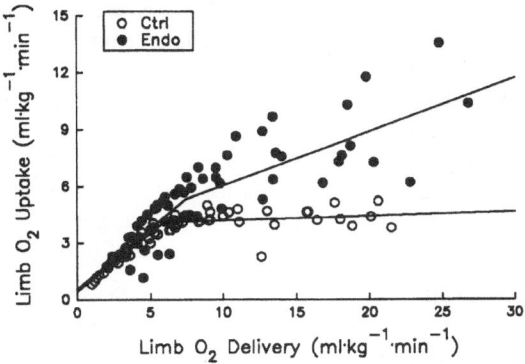

Fig. 2. Limb O_2 uptake and delivery for control (Ctrl) and endotoxin treated (Endo) groups as O_2 delivery was decreased by lowering PO_2 (from Curtis et al., this volume).

Low PO_2, however, has a direct vasodilatory effect on the vasculature which, in this case, would not be countered by any increase in sympathetic tone, as usually occurs with systemic hypoxic hypoxia. In other experiments similar to these but in which no endotoxin was given, we used a pump-membrane oxygenator system to lower DO_2 to the limb muscles either by lowering flow with normoxic PO_2 or by lowering PO_2 at normal flow while the rest of the animal was kept normoxic. The critical O_2ER was significantly lower for hypoxic hypoxia than for ischemic hypoxia (unpublished data). The perfusion heterogeneity and effective increase in diffusion distances that have been ascribed to the endotoxin animals may have been replicated by the direct effect of low PO_2 on the vasculature. This is still under investigation so there are no final answers as yet.

To summarize the information to this point, an O_2 extraction defect can be produced by giving endotoxin to experimental animals. The evidence for that statment was the increase in critical DO_2 and decrease in critical O_2ER. Although there is much circumstantial evidence to suggest that it is caused by dysfunctions induced in the microvasculature and a resultant diffusion limitation, we were unable to generate any direct support for such a limitation by manipulating PO_2. In Fig. 2, there was a definite biphasic relationship between DO_2 and VO_2 but even at DO_2 above critical, there was still a positive slope. Similar findings in patients have been interpreted as representing unsatisfied O_2 demand. That does not necessarily mean that there was widespread tissue hypoxia.

LACTATE AS A SIGN OF TISSUE HYPOXIA

When hypoxia was deliberately induced in experimental animals, arterial lactate concentration increased in fairly exact proportion to the measured O_2 deficit that was accumulated (Cain, 1967). This association of hypoxia and lactate has long been recognized to the point where any increase in lactate is sometimes accepted as evidence that tissue hypoxia must have been present. Elevated blood lactate levels have been found in critically ill patients with ARDS or sepsis who have also shown pathological O_2 supply dependency (see Cain and Curtis, 1991a for citations). In most cases, blood lactate levels decreased as DO_2 and VO_2 increased. In all cases, elevated blood lactate levels were only seen in patients who also showed O_2 supply dependency. The assumption that pervasive tissue hypoxia must have also been present was either tacit or explicit in the discussion of each of these papers and the elevated lactate levels were seen as confirmatory.

Examination of the schematic diagram of cellular energy production in Figure 3 provides some clues to the origin of elevated lactates whether hypoxia is the root cause or not. The hatched bar labelled "hypoxia" at the lower right corner of the figure shows the entry of O_2 into the electron transport chain at Cytochrome aa_3. Without a continuing supply Of O_2, oxidative phosphorylation is slowed or stopped with a consequent inability to handle the hydrogen ions being generated in the citric acid and glycolytic cycles. This forces pyruvate to become the primary hydrogen ion acceptor and mass action causes formation of lactate as a blind pocket for the combination of pyruvate and hydrogen ion.

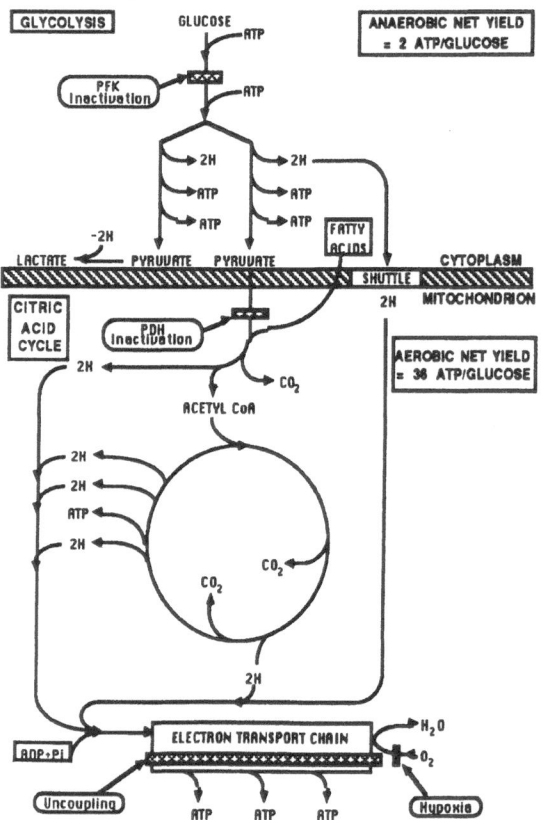

Fig. 3.   Schematic representation of glycolysis and the citric
             acid cycle (from Cain and Curtis, 1991a).

The lowermost position of lactate on the electrochemical gradient dictates this direction for the reactions. Increased blood lactate levels caused by hypoxia, therefore, will be marked by an increased L/P ratio. The other event with inadequate oxygenation is, of course, an energy deficit as ATP production fails to keep up with demand. The decrease in energy potential removes any inhibition of phosphofructokinase so that glycolysis is promoted. The much less efficient utilization of glucose in the anaerobic energy cycle can only partially correct the energy deficit engendered by hypoxia.

A similar chain of events can occur if oxidation is uncoupled from phosphorylation in the mitochondrion. The hatched bar labelled "uncoupling" at the bottom of Figure 3 represents this event. Treatment with an uncoupling agent such as dinitrophenol will cause a relative energy deficit with a consequent increase in lactate production. This will occur with increased VO_2 together with increased glycolytic flux rates. A hyperdynamic circulatory state will ensue and this has been used as a chemical simulation of muscular exercise because the effects are most marked in skeletal muscle (Cain and Chapler, 1985). Although uncoupling is not hypoxia, its net effect to create an energy deficit will cause both lactate and L/P ratio to rise.

Another key control point is at phosphofructokinase (PFK). If this were inactivated, however, as shown by the hatched bar near the top of Figure 3, it would prevent the entry of glucose into the energy generation cycle so that the most likely effect on lactate levels would be a decrease.

One other control point is particularly germane to this discussion, pyruvate dehydrogenase (PDH). PDH is a protein kinase which exists as a complex of active and inactive forms on the inner membrane of the mitochondrion. It is responsible for the entry of pyruvate into the citric acid cycle. If it were inactivated (as represented by the hatched bar at the center of Figure 3), pyruvate would accumulate and spill over into lactate so that lactate increased but the L/P ratio would be relatively unchanged. What makes these events germane to a discussion of the presence of tissue hypoxia in critically ill patients is the fact that endotoxin will inactivate PDH (Vary et al., 1986). The conversion of PDH to the inactive form by the action of endotoxin causes blood lactate levels to be elevated whether hypoxia is present or not. In the case of sepsis, there is a reduction in the energy derived from glucose and an increase in the oxidation of 2-carbon fragments from fat and amino acids in the citric acid cycle (Barton and Cerra, 1989).

We (Curtis and Cain, 1992) examined regional and systemic VO_2 and DO_2 together with lactate flux in two groups of endotoxin treated (2 mg/kg infused over 1 hr) dogs of which one group was also given dichloroacetate (DCA). Volumes of donor red cells and dextran were given as necessary to keep cardiac output at or above the pre-endotoxin level. DCA is an activator of PDH and can reverse the inactivation caused by endotoxin (Stacpoole, 1989). Our objective was to see how much of the lactate elevation with endotoxin was reversible by DCA. By measuring DO_2 at the same time and with foreknowledge of the critical DO_2 in endotoxin treated dogs, we were able to make some judgment about whether the lactate signified tissue hypoxia. As a further challenge to tissue oxygenation, we also ventilated the animals for 30 min with an hypoxic gas mixture (F_IO_2 12%). In all cases, sufficient donor red cells and dextran were given to resuscitate the animals from the hypotension caused by endotoxin infusion and to convert them to a hyperdynamic circulatory state.

Figure 4 shows DO_2 for whole body, muscle, and gut in the two groups of animals. In all cases, DO_2 was safely above the critical levels reported in Table 1 for anesthetized dogs that were bacteremic or injected with endotoxin. There were some significant differences in whole body DO_2 but even the lower values in the untreated group were higher than the critical whole body DO_2 of 12.8 ml/kg/min reported by Nelson et al. (1988). The hypoxic challenge had the unexpected result of unchanged or even increased DO_2 by virtue of an increase in cardiac output, particularly in the untreated group. This was attributed to the increased sympathetic stimulation caused by hypoxia acting through the peripheral chemoreceptors.

The action of the DCA, presumably its reactivation of PDH, was very apparent in Figure 5 which shows arterial lactate concentration and the regional lactate flux rates in the two groups of dogs. DCA lowered arterial lactate to below starting levels whereas the untreated endotoxic dogs had a continuing increase to above 6 mmol. In contrast, the DCA did not decrease the two regional flux rates. The endotoxin caused gut, in particular, to become a notable lactate producer. Because DCA was so efficacious in lowering arterial lactate and because DO_2 was always above critical levels in both groups of animals, we concluded that most of the lactate that appeared in arterial blood was the result of PDH inactivation.

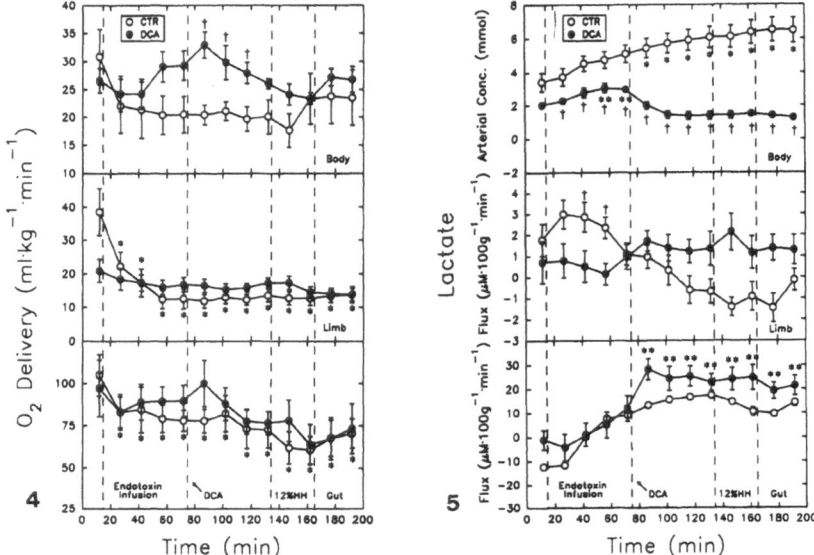

Fig. 4. DO_2 to whole body, muscle, and gut of dogs infused with endotoxin and either treated with dichloroacetate (DCA) or not (CTR). Asterisks denote significant differences from initial value; daggers are significant differences between groups. (From Curtis and Cain, 1992).

Fig. 5.   Arterial lactate concentration and lactate flux across muscle and gut of CTR and DCA dogs. Asterisks denote significant differences from initial value; daggers are significant differences between groups. (From Curtis and Cain, 1992).

This leaves tissue hypoxia as a questionable cause of lactacidosis in sepsis.

Arterial lactate levels reflect the balance between production and disposal and give no information on the origin of the lactate. The regional information that we obtained in the DCA study indicated that gut might be a significant contributor to arterial lactate levels. Furthermore, lactate production there was unaffected by DCA so it may have been more related to hypoxia in the gut. A similar study that we recently completed on anesthetized dogs tends to bear out that conclusion (Cain and Curtis, 1991b).

The protocol was almost the same as in the DCA study but in this case we were studying the possible beneficial effects of dopexamine, a potent β_2-adrenoceptor and dopaminergic agonist. Once again the animals were infused with 2 mg/kg of endotoxin over a 1-hr period while cardiac output was supported by volumes of donor red cells and dextran. We expected dopexamine to increase DO_2 as we had found before (Bredle and Cain, 1991) but were interested to see if the increase in DO_2 went more to gut than to muscle. The increased DO_2 (see Fig. 6) with dopexamine did not persist into the second hour after starting endotoxin infusion. All values of DO_2 were above critical as before. There were no differences between the two groups during the second hour after starting endotoxin.

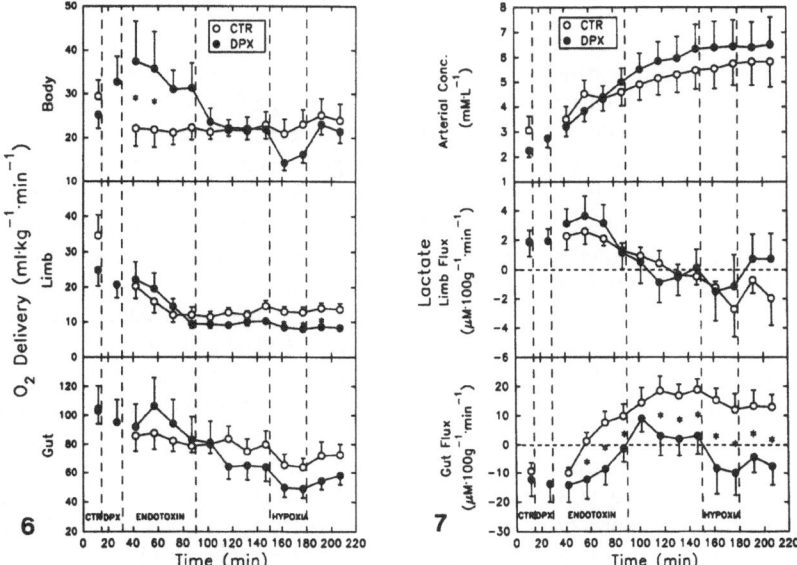

Fig. 6. DO_2 to whole body, limb, and gut of two groups of endotoxic dogs that were either continuously infused with dopexamine (DPX, 12 μg/kg/min) or not (CTR) (from Cain and Curtis, 1991b). Asterisks denote significant differences between groups.

Fig. 7.   Arterial lactate concentration and lactate flux rates across limb and gut of control (CTR) and dopexamine treated (DPX) endotoxic dogs (from Cain and Curtis, 1991b). Asterisks denote significant differences between groups.

The arterial lactate concentrations and regional lactate fluxes in the dopexamine experiments are shown in Figure 7. As the arterial lactate reached high levels in both groups, limb muscles converted from producing lactate to taking it up. Conversely, gut switched from taking up lactate as the endotoxin infusion was started to becoming again a strong lactate producer in the control group. The dopexamine appeared to prevent that because the lactate flux in DPX never became significantly positive. The lesser production of lactate with dopexamine occurred despite the fact that neither gut DO_2 nor VO_2 (not shown) differed between the groups.

We (Cain and Curtis, 1991b), interpreted these results as a sparing action of dopexamine on the gut which was attributable primarily to its strong ß$_2$-adrenergic agonist property. The gut was less hypoxic because of the effect of dopexamine to redistribute blood flow away from muscularis to the mucosa where more of the energy demand resides. Other information was cited to support the fact that such redistribution within the gut wall had been previously documented with other ß-agonists such as isoproterenol. We also observed that the lumenal surface of the gut segment in the dopexamine treated dogs was pink and healthy appearing whereas in the control dogs, the lumenal surface was hemorrhagic with grossly distorted rugae. These observations supported our suggestion that the mucosa was better oxygenated in the dopexamine group. In these experiments, a regional lactate flux may have been more useful than an overall measurement such as arterial lactate to indicate that tissue hypoxia may have differed markedly from region to region.

SUMMARY

There was little dispute that endotoxin treatment of experimental animals could recreate the O_2 extraction defect that had been observed in critically ill patients. The remaining question was whether or not this necessarily signified pervasive tissue hypoxia. Some limitation to O_2 diffusion in the tissues had been postulated because of known effects of endotoxin that ultimately result in damage to endothelium. We were unable to alter the critical DO_2 or O_2ER in endotoxic dogs by manipulating the arterial PO_2. This tended to rule against there being a diffusion limitation created by the endotoxin as a result of endothelial disruption or microvascular dysfunction. The results of the DCA and dopexamine experiments served to remind us that arterial lactate measurements may or may not indicate widespread tissue hypoxia. Sepsis, as emulated by endotoxin infusions, is also a metabolic disease that can cause inactivation of PDH and thus cause lactacidosis without tissue hypoxia. Regional measurements of lactate flux indicated that gut was hypoxic in spite of DO_2 above critical because of maldistribution of blood flow between muscularis and mucosa.

The questions persist of how much tissue hypoxia is caused by sepsis or endotoxin when DO_2 is supported at supposedly adequate levels and whether there are marked regional differences. Such questions still await answers. Newer technological advances that permit assessment of tissue oxygenation by noninvasive methods, such as near infrared spectrophotometry or nuclear magnetic resonance measurement of tissue energy potential, may soon be feasible in critically ill patients. This kind of information will be of vast importance in designing the most effective therapeutic regimen.

ACKNOWLEDGMENTS

This research was supported in part by Grant #HL 269207 from National Heart, Lung, and Blood Institute and by funds from Fisons Pharmaceuticals.

REFERENCES

Archie, J.P., 1981, Mathematic coupling of data - a common source of error. Ann. Surg. 193:296.

Bakker, J., Coffernils, M., Leon, M., Gris, P., Vincent, J.-L., 1991, Blood lactate levels are superior to oxygen-derived variables in predicting outcome in human septic shock. Chest 99:956.

Barton, R. and Cerra, F.B., 1989, The hypermetabolism multiple organ failure syndrome. Chest 96:1153.

Bihari, D., Smithies, M., Gimson, A., and Tinker, J., 1987, The effects of vasodilation with prostacyclin on oxygen delivery and uptake in critically ill patients. N. Engl. J. Med. 317:397.

Bredle, D.L., Samsel, R.W., Schumacker, P.T., and Cain, S.M., 1989, Critical O_2 delivery to skeletal muscle at high and low PO_2 in endotoxemic dogs. J. Appl. Physiol. 66:2553.

Bredle, D.L. and Cain, S.M., 1991, Systemic and muscle O_2 uptake/delivery after dopexamine infusion in endotoxic dogs. Crit. Care Med. 19:198.

Cain, S.M., 1967, Oxygen deficit incurred during hypoxia and its relation to lactate and excess lactate. Am. J. Physiol. 213:57.

Cain, S.M., 1978, Effects of time and vasoconstrictor tone on O_2 extraction during hypoxic hypoxia. J. Appl. Physiol. 45:219.

Cain, S.M., 1984, Supply dependency of oxygen uptake in ARDS: Myth or reality? Am J. Med. Sci. 288:119.

Cain, S.M. and Chapler, C.K., 1985, Circulatory responses to 2,4dinitrophenol in dog limb during normoxia and hypoxia. J. Appl. Physiol. 59:698.

Cain, S.M. and Curtis, S.E., 1991a, Experimental models of pathological oxygen supply dependency. Crit. Care Med. 19:603.

Cain, S.M. and Curtis, S.E., 1991b, Systemic and regional O_2 uptake/delivery and lactate flux in endotoxic dogs infused with dopexamine. Crit. Care Med. 19:1552.

Curtis, S.E., Bradley, W.E., and Cain, S.M., (this volume), Effects of endotoxin on canine skeletal muscle oxygen delivery-uptake relations during progressive hypoxic hypoxia. Adv. Exp. Med. Biol.

Curtis, S.E. and Cain, S.M., 1992, Regional and systemic oxygen delivery/uptake relations and lactate flux in hyperdynamic, endotoxin-treated dogs. Am. Rev. Resp. Dis. 145:348.

Danek, S.J., Lynch, J.P., Weg, J.G., and Dantzker, D.R., 1980, The dependence of oxygen uptake on oxygen delivery in the adult respiratory distress syndrome. Am. Rev. Respir. Dis. 122:387.

Dantzker, D.R., Foresman, B., and Gutierrez, G., 1991, Oxygen supply and utilization relationships. Am. Rev. Respir. Dis. 143:675.

Fenwick, J.C., Dodek, P.M., Ronco, J.J., Phang, P.T., Wiggs, B. and Russell, J.A., 1990, Increased concentrations of plasma lactate predict pathologic dependence of oxygen consumption on oxygen delivery in patients with adult respiratory distress syndrome. J. Crit. Care 5:81.

Kruse, J.A., Haupt, M.T., Puri, V.K., and Carlson, R.W., 1990, Lactate levels as predictors of the relationship between oxygen delivery and consumption in ARDS. Chest 98:959.

Messmer, K., Kreimeier, U., and Hammersen, F., 1989, Changes in the microcirculation in sepsis and septic shock. In: "Sepsis," Reinhart, K. and Eyrich, K.(eds.), Springer-Verlag, Berlin, pp 148.

Nelson, D.P., Beyer, C., Samsel, R.W., Wood, L.D.H., and Schumacker, P.T., 1987, Pathological supply dependence Of O_2 uptake during bacteremia in dogs. J. Appl. Physiol. 63:1487.

Nelson, D.P., Samsel, R.W., Wood, L.D.H., and Schumacker, P.T., 1988, Pathological supply dependence of systemic and intestinal O_2 uptake during endotoxemia. J. Appl. Physiol. 64:2410.

Ronco, J.J., Phang, P.T., Walley, K.R., Wiggs, B., Fenwick, J.C., and Russell, J.A., 1991, Oxygen consumption is independent of changes in oxygen delivery in severe adult respiratory distress syndrome. Am. Rev. Respir. Dis. 143:1267.

Rowell, L.B., 1986, "Human Circulation," Oxford Univ. Press, New York, pp 236.

Samsel, R.W., Nelson, D.P., Sanders, W.M., Wood, L.D.H., and Schumacker, P.T., 1988, Effect of endotoxin on systemic and skeletal muscle O_2 extraction. J. Appl. Physiol. 65:1377.

Stacpoole, P.W., 1989, The pharmacology of dichloroacetate. Metabolism 38:1124.

Tuchschmidt. J., Oblitas, D., and Fried, J.C., 1991, Oxygen consumption in sepsis and shock. Crit. Care Med. 19:664.

Vary, T.C., Siegel, J.H., Nakatani, T., Sato, T., and Aoyama, H., 1986, Effect of sepsis on activity of pyruvate dehydrogenase complex in skeletal muscle and liver. Am. J. Physiol. 250:E634.

OXYGEN UPTAKE IN THE LUNGS UNDER PATHOLOGICAL CONDITIONS AND

ITS THERAPEUTIC EFFORTS

D. Gommers, G.J. van Daal and B. Lachmann

Dept. of Anesthesiology, Erasmus University
Rotterdam, The Netherlands

INTRODUCTION

One of the most important clinical syndromes in which failure of oxygen uptake in the lung leads to severe hypoxia, with a mortality rate of 40-70%, is the so-called adult respiratory distress syndrome (ARDS).

ARDS is a complex of clinical signs and symptoms which occur following diverse pulmonary or systemic insults including sepsis, shock, pneumonia, trauma, liquid aspiration, hematologic disorders, smoke inhalation, and many others[1,2]. Despite diverse etiologies, the common pathological hallmark is increased capillary permeability associated with damage to the alveolar epithelium[3,4]. Regardless of the etiology of the lung injury, ARDS is characterized by: severe arterial hypoxemia refractory to oxygen therapy, increased intrapulmonary shunting, decreased lung compliance, decreased lung volumes, and the absence of indicators of left ventricular failure[5].

PATHOGENESIS OF ARDS

Alveolar-capillary permeability

Although acute lung injury may be the result of a diversity of predisposing conditions, the end-result is always remarkably similar[3]. Direct or indirect mechanisms lead to increased capillary permeability, often associated with damage to the alveolar membrane[4,6], resulting in pulmonary edema containing serum proteins of all classes and increased numbers of inflammatory cells[7]. Bronchoalveolar lavage (BAL) fluid from ARDS patients often contains over 90% polymorphonuclear neutrophils (PMNs), which normally comprise less than 2% of BAL fluid cells[8].

Direct injury to the alveolar membrane may result from several situations, including aspiration of liquids, inhalation of gases or smoke, and generation of inflammatory mediators or microbial (endo)toxins during pneumonia. Increased permeability resulting from indirect, distant causes, is not always understood. Both cellular and humoral mechanisms have been investigated extensively and in the etiology of ARDS active roles have been attributed to neutrophils, basophils, macrophages, platelets, arachidonic acid metabolites, oxygen-derived free radicals, complement, proteases, interleukins, serotonin, platelet activating factor (PAF), tumor necrosis factor (TNF), surfactant inhibiting proteins, drugs, and many other substances[2]. It is, therefore, conceivable that different mechanisms operate in different predisposing conditions of

Oxygen Transport to Tissue XIV, Edited by W. Erdmann and
D.F. Bruley, Plenum Press, New York, 1992

Surfactant deficiency

Increased capillary permeability, associated with damage to the alveolar epithelium, can lead to wash-out of alveolar phospholipids into the blood[9]; inactivation of surfactant by plasma components [10]; surfactant depletion by foaming[11] or ventilation with large tidal volumes[12]; and disturbed synthesis, storage, or release of surfactant secondary to direct injury to type II cells[13], leading to functional impairment of the surfactant system.

Functional impairment of the surfactant system has far-reaching consequences for the functioning of the lung. Independent of the cause, decreased surfactant function will directly or indirectly lead to[14] (for normal surfactant function see[15]):

- decreased pulmonary compliance
- decreased functional residual capacity of the lung
- atelectasis
- pulmonary edema with decreased gas exchange and respiratory acidosis
- hypoxemia with anaerobic metabolism and metabolic acidosis
- further inactivation of surfactant by plasma constituents
- enlargement of the functional right-to-left shunt of the lung

THERAPEUTIC MEASURES

ARDS requires treatment with artificial ventilation, a high inspiratory oxygen concentration and infusion of buffer solutions[16]. This management aims at correcting hypoxemia, as well as metabolic and respiratory acidosis. In a few highly specialized intensive care units, patients not responding to this therapeutic regimen have been treated with extracorporeal membrane oxygenation or with extracorporeal elemination of CO_2[17]. Lung transplantation as a therapeutic measure is not clinically relevant at present, at least not for patients with ARDS. But despite all these therapeutic efforts, the mortality in ARDS has remained consistently high over the last 20 years[4].
Current therapeutic endeavors in ARDS are now focused on three main areas:
1) blockade of various biochemical pathways following an alveolotoxic insult with the aim of preventing the development of ARDS. Examples of this biochemical approach include control of blood coagulation and complement release, inhibition of the arachidonic acid cascade, inhibition of proteases, and the use of free radical scavengers, heavy metal chelators etc.
2) optimal ventilator settings which compensate for the high retractive forces due to a damaged surfactant system
3) re-establishment of the functional integrity of the damaged surfactant system in ARDS lungs by pharmacological stimulation of surfactant production and release; and/or by tracheal instillation of surfactant.

SURFACTANT REPLACEMENT THERAPY IN ARDS

Animal studies

Surfactants are now being used worldwide in clinical trials for treatment, or prevention, of the neonatal respiratory distress syndrome and have demonstrated improvement of lung function, reduction in the rate of pneumothrax and decreased numbers of deaths[18,19]. The neonatal respiratory distress syndrome is characterized by surfactant deficiency due to immature enzyme systems which also leads to increased membrane permeability with leakage of protein-rich fluid into the alveoli, formation of hyaline membranes and interstitial edema. Adult and neonatal respiratory distress syndromes have many features in common.

Very few reports have been published concerning surfactant replacement in ARDS. Clinical research focused on surfactant replacement therapy is up to now almost impossible due to the fact that there is not enough surfactant available. ARDS models have been applied in studies concerning the pathogenesis of lung lesions and functional disturbances in ARDS, as well as for evaluation of therapeutic regimens including various ventilator settings, pharmacological intervention and surfactant therapy. Currently many different animal models are used; in vivo lung lavage, neurogenic ARDS, oxygen toxicity, oxydant producing enzymes, hydrochloric acid instillation, anti-lung serum, monoclonal antibodies, different pulmonary infections, N-nitroso-N-methylurethane, oleic acid infusion, paraquat intoxication, soft tissue trauma, hemorrhagic shock and tumor necrosis factor[14]. The next part describes three selected ARDS models with surfactant replacement therapy:

In vivo lung lavage; Experiments on postmortem lung specimens have revealed that a considerable portion of the alveolar phospholipids can be recovered by repeated washings of the airspaces[20]. We have used guinea pigs[21], rabbits[22], dogs[23] to develop ARDS models, in which alveolar surfactant is removed by in vivo lavage.

In these lavage models, severe respiratory insufficiency is defined as a fall in P_aO_2 below 60 mm Hg during artificial ventilation with 100% oxygen. Already the first lavage results in significant reduction of thorax-lung compliance and repeated lavage results in a derangement of lung function persisting for at least 8 h. These disturbances, which include a 35% decrease in functional residual capacity (FRC), are almost certainly secondary to increased surface tension in the alveolar lining, especially as lung lavage with saline does not alter the elastic properties of the pulmonary parenchyma[24]. However, removal of alveolar surfactant by in vivo lung lavage significantly increases permeability of the alveolo-capillary membrane, which was demonstrated by Wollmer and coworkers in their studies on ^{99m}Tc-DTPA clearance[25]. Histologic examination of lungs from animals 5 min after the last lavage shows atelectasis, desquamation of bronchial and bronchiolar epithelium, and incipient formation of hyaline membranes[21].

Tracheal instillation of exogenous surfactant in surfactant depleted animals results in dramatic improvement of gas exchange, even if the treatment is given 2 h after the lavage procedure[26,27]. Improved blood gases are stable for at least 5 h[28], whereas P_aO_2 in control animals remains low despite ventilation with PEEP and pure oxygen. Histologic sections from surfactant-treated animals showed a uniform pattern of well-aerated alveoli with only minimal intra-alveolar edema and hyaline membranes, whereas control animals, ventilated with the same ventilator settings had extensive atelectasis and prominent hyaline membranes[28].

The lung lavage models are useful for a variety of experimental purposes, including testing of alternative surfactant preparations[27], evaluation of pharmacologic agents stimulating discharge of alveolar phospholipids[28] and studies on the significance of various ventilator settings[22,23].

Hydrochloric acid instillation; The pulmonary effects of acid aspiration have been documented extensively in animal models. Data from these studies show that in the pH range below 2.5, the degree of lung injury is proportional to the hydrogen ion concentration of the aspirated material. Toung and coworkers[29,30] studied the effects of acid injury on excised, perfused, ventilated canine lung lobes. Three hours after intrabronchial instillation of 1 ml 0.1 M HCl, the weight of the lobe tripled, lobar shunt increased from about 6-53% and hemorrhagic edema appeared in the alveolar spaces. These findings reflect an acute increase in pulmonary capillary permeability after acid instillation[31]; in the injured lobes, perfusion decreases while right-to-left shunting increases[32]. Shortly after acid instillation, type II pneumocytes show cytoplasmic swelling[33] and the surface properties of alveolar lavage fluid and lung extracts are characterized by increased minimal surface tension and decreased hysteresis[34,35].

Recently, the results of the first studies on surfactant replacement therapy during HCl aspiration pneumonitis in rabbits were published. Lamm and coworkers[36] demonstrated that with marked pulmonary edema caused by HCl aspiration, surfactant replacement with a natural modified surfactant improved pulmonary mechanics, whereas it did not improve gas exchange. However, Kobayashi and coworkers[37,38] demonstrated that treatment with natural porcine surfactant with proir lung lavage with saline resulted in significantly improved gas exchange, whereas surfactant treatment alone did not. Because surfactant treatment alone does not suffice to improve pulmonary function it was concluded that lung lavage prior to surfactant instillation removes or dilutes the inhibitory proteins present in the alveolar edema fluid.

These studies indicate that it is possible to treat respiratory failure due to aspiration of gastric content by instillation of surfactant after removal or dilution of inhibitory proteins by lung lavage.

Oxygen toxicity; Severe lung damage and progressive deterioration of pulmonary function, resembling ARDS, can be induced in animals by prolonged exposure to pure oxygen [for review see[39]]. The sequence of morphologic changes that occur in the lungs secondary to prolonged O_2 exposure is similar in different animal species[39], interstitial edema appears within 24-48 h, prolonged experimental exposure leads to destruction of alveolar capillary endothelium and intra-alveolar edema. Recordings of lung mechanics reveal reduced lung compliance, reflecting decreased surfactant activity[40] and impaired synthesis of surfactant phospholipids in degenerating type II cells[41].

Several authors [for review see[39]] have suggested that the lung lesions in ARDS result from mechanisms similar to those operating in oxygen toxicity[42]. These mechanisms involve generation of unstable O_2 derivates such as O_2^-, H_2O_2, hydroxyl radicals, singlet oxygen, and other free radicals. Reactive and unstable O_2 metabolites can damage the lung tissue directly by oxidative and peroxidative decomposition of thiol groups and essential membrane-associated fatty acids. The result is altered permeability of endothelial and epithelial cell membranes, with intracellular edema, swelling of organelles, and transudation of fluid into the alveolar airspaces.

Matalon, Holm and coworkers[43-45] demonstrated, in a chain of studies in rabbits, the devastating effects of prolonged exposure to pure oxygen on the surfactant system and pulmonary function and the efficacy of treatment with natural surfactant during respiratory failure due to oxygen toxicity[46]. However, in another study on surfactant replacement during respiratory distress due to oxygen toxicity in guinea pigs, surfactant instillation failed to improve blood gases and pulmonary mechanics[47], probably due to inhibition of surfactant by pulmonary edema constituents.

However, it was shown[48] that prophylactic instillation of exogenous natural surfactant can decrease mortality and minimize the development of most of the symptoms of oxygen toxicity after prolonged exposure to pure oxygen in the rabbit model. Furthermore, it was shown in the same model that rabbits with higher levels of endogenous surfactant survive considerably longer in hyperoxia than normal rabbits[49].

Presently, there are only preliminary reports about the efficacy of exogenous surfactant instillation in the management of patients with ARDS. Richman et al.[50] found a transient increase in PaO_2 in two patients with ARDS after administration of a single dose of 4 g exogenous surfactant. These authors suggested that multiple instillations are needed for a sustained beneficial effect. Lachmann[51] reported that tracheal instillation of natural surfactant (300 mg/kg BW) in a patient with sepsis and severe ARDS led to a significant improvement of gas exchange. Nosoka et al.[52] demonstrated some improvement in PaO_2 and X-ray after multiple instillations of surfactant (each time 240 mg) in two adult patients with ARDS. Joka et al.[53] treated an 18-year-old victim of a motorcycle accident with a posttraumatic ARDS. They observed an improvement in gas exchange and pulmonary hemodynamics, and a decrease of alveolar protein-leakage after surfactant substitution (50 mg/kg BW). In all of these cases the surfactant was dissolved in saline and given via the trachea as a bolus. The surfactant preparations were all natural surfactants, prepared from animal lungs.

These few reports indicate that instillation of exogenous surfactant in patients with ARDS is feasible and has at least a temporary beneficial effect. Clearly, additional studies are needed to document whether surfactant instillation improves the morbidity and mortality of patients with ARDS.

CONCLUDING REMARKS

If it becomes possible in the near future to control mechanisms of blood coagulation and complement release as well as to prevent or delay toxic influence of high oxygen concentration and free oxygen radicals by specific agents and if, on the other hand, early damage in the surfactant system of the lung can be prevented by surfactant replacement combined with "non-traumatic" artificial ventilation, the high mortality rate associated with the respiratory distress syndrome can probably be reduced.

REFERENCES

1. T. L. Petty, and D. G. Ashbaugh, The adult respiratory distress syndrome. Clinical features, factors influencing prognosis and principles of management. Chest, 60: 233-239 (1971).
2. R. G. Spragg, and R. M. Smith, Biology of acute lung injury, in: The Lung: Scientific Foundations, R.G. Crystal et al., ed, Raven Press Ltd., New York, 1991, pp. 2003-2017 (1991).
3. J. H. Stevens, and T. A. Raffin, Adult respiratory distress syndrome - I. Aetiology and mechanisms, Postgrad. Med. J., 60: 505-513 (1984).
4. J. F. Murray, Mechanism of acute respiratory failure, Am. Rev. Respir. Dis. 115: 1071-1078 (1977).
5. B. A. Holm, and S. Matalon, Role of pulmonary surfactant in the development and treatment of adult respiratory distress syndrome, Anesth Analg., 69: 805-818 (1989).
6. J. E. Rinaldo, and R. M. Rogers, Adult respiratory distress syndrome: changing concepts of lung injury and repair, New Engl. J. Med., 306: 900-909 (1982).
7. J. F. Holter, J. E. Weiland, E. R. Pacht, J. E. Gadek, and W. B. Davis, Protein

permeability in the adult respiratory distress syndrome: loss of size selectivity of the alveolar epithelium, J. Clin. Invest., 78 (6):1513-1522 (1986).

8. H. Y. Reynolds, Bronchoalveolar lavage - state of the art, Am. Rev. Respir. Dis., 135: 250-263 (1987).

9. R. J. Fallat, and M. Lamy, Pathophysiologic correlates in adult respiratory distress syndrome, in: Acute respiratory failure, O. Mayrhofer-Krammel, G. Schlag, and H. Stoeckel, ed., Thiem, Stuttgart, pp 180-188 (1979).

10. B. A. Holm, G. E. Enhorning, and R. H. Notter, A biophysical mechanism by which plasma proteins inhibit surfactant activity, Chem. Phys. Lipids, 49: 49-55 (1988).

11. W. R. Harlan, Jr., S. I. Said, and C. M. Banerjee, Metabolism of pulmonary phospholipids in normal lung and during acute pulmonary edema, Am. Rev. Respir. Dis., 94: 938-947 (1966).

12. I. Wyszogrodski, K. Kyei-Aboagye, H. W. Taeusch, Jr., and M. E. Avery, Surfactant inactivation by hyperventilation: conservation by end-expiratory pressure, J. Appl. Physiol., 38: 461-466 (1975).

13. E. Malmquist, G. Grossmann, B. Ivemark, and B. Robertson, Pulmonary phospholipids and surface properties of alveolar wash in experimental paraquat poisoning, Scand. J. Resp. Dis., 54: 206-214 (1973).

14. B. Lachmann, and E. Danzmann, Adult respiratory distress syndrome, In: B. Robertson, L. M. G. Van Golde, J. J. Batenburg, eds., Pulmonary Surfactant, Amsterdam, Elsevier, pp. 505-548 (1984).

15. L. M. G. Van Golde, J. J. Batenburg, and B. Robertson, The pulmonary surfactant system: Biochemical aspects and functional significance, Physiological Rev., 68: 374 (1988).

16. H. Pontoppidan, R. S. Wilson, M. A. Rie, and R. C. Schneider, Respiratory intensive care, Anesthesiology, 47: 96-116 (1977).

17. L. Gattinoni, A. Presenti, G. P. Possi, S. Vesconi, U. Fox, T. Kolobow, A. Agostini, A. Pelizzola, M. Langer, L. Uziel, F. Longoni, and G. Damia, Treatment of acute respiratory failure with low frequency positive-pressure ventilation and extracorporeal removal of CO_2, Lancet, 2: 292-294 (1980).

18. T. A. Merritt, M. Hallman, R. Spragg, G. P. Heldt, and N. Gillard, Exogenous surfactant treatments for neonatal respiratory distress syndrome and their potential role in the adult respiratory distress syndrome, Drugs, 38 (4): 591-611 (1989).

19. H. Segerer, and M. Obladen, Surfactant substitution treatment of neonatal respiratory distress syndrome, Pediatric Rev. Commun., 5: 67-82 (1990).

20. T. Fujiwara, F. H. Adams, A. El-Salawy, and S. Sipos, "Alveolar" and whole lung phospolipids of newborn lambs, Proc. Soc. Exp. Biol. Med.(N.Y.), 127: 962-969 (1968).

21. Lachmann, B., Robertson, B. and J. Vogel, In vivo lung lavage as an experimental model of the respiratory distress syndrome, Acta. Anaesthesiol. Scand., 24: 231-236 (1980).

22. B. Lachmann, B. Jonson, M. Lindroth, and B. Robertson, Modes of artificial ventilation in severe respiratory distress syndrome. Lung function and morphology studied in rabbits after washout of alveolar surfactant, Crit. Care Med., 10: 724-732 (1982).

23. B. Lachmann, E. Danzmann, B. Haendly, and B. Jonson, Ventilator setting and gas exchange in respiratory distress syndrome, in: Applied Physiology in Clinical

Respiratory Care, O.Prakash, ed., Martinus Nijhoff, The Hague, pp. 141-176 (1982).

24. G. L. Huber, L. H. Edmunds, Jr., and T. N. Finley, Effects of experimental saline lavage on pulmonary mechanics and morphology, Am. Rev. Respir. Dis., 104: 337-347 (1971).

25. P. Wollmer, E. Evander, B. Jonson, and B. Lachmann, Pulmonary clearance of inhaled ^{99m}Tc-DTPA: effect of surfactant depletion in rabbits, Clin. Physiol., 6: 85-89 (1986).

26. T. Kobayashi, H. Kataoka, T. Ueda, S. Murakami, Y. Takada, and M. Kokubo, Effect of surfactant supplement and end-expiratory pressure in lung-lavaged rabbits, J. Appl. Physiol., 57: 995-1001 (1984).

27. B. Lachmann, T. Fujiwara, S. Chida, T. Morita, M. Konishi, K. Nakamura, and H. Maeta, Improved gas exchange after tracheal instillation of surfactant in the experimental adult respiratory distress syndrome, Crit. Care Med., 9: 158 (1981).

28. B. Lachmann, T. Fujiwara, S. Chida, T. Morita, M. Konishi, K. Nakamura, and H. Maeta, Surfactant replacement therapy in the experimental adult respiratory distress syndrome (ARDS), in: Pulmonary Surfactant System, E.V. Cosmi and E.M. Scarpelli, ed., Elsevier, Amsterdam, pp. 231-235 (1983).

29. T. J. Toung, K. D. Bordos, D. W. Benson, D. Carter, G. D. Zuidema, S. Permutt, and J. L. Cameron, Aspiration pneumonia: experimental evaluation of albumin and steroid therapy, Ann. Surg., 183: 179-184 (1976).

30. T. J. Toung, P. Saharia, S. Permutt, G. D. Zuidema, and J. L. Cameron, Aspiration pneumonia: beneficial and harmful effects of positive end-expiratory pressure, Surgery, 82: 279-283 (1977).

31. F. A. Grimbert, J. C. Parker, and A. E. Taylor, Increased pulmonary vascular permeability following acid aspiration, J. Appl. Physiol., 51: 335-345 (1981).

32. C. J. Fisher, and L. D. H. Wood, Effect of lobar acid injury on pulmonary perfusion and gas exchange in dogs, J. Appl. Physiol., 49: 150-156 (1980).

33. L. J. Greenfield, R. P. Singleton, D. R. McCaffree, and J. J. Coalson, Pulmonary effects of experimental graded aspiration of hypochloric acid, Ann. Surg., 170: 74-86 (1969).

34. K. F. Baum, and D. L. Beckman, Aspiration pneumonitis and pulmonary phospholipids, J. Trauma, 16: 782-787 (1967).

35. E. S. Brown, Aspiration and lung surfactant, Anesth. Analg. Curr. Res., 46: 665-672 (1967).

36. W. J. E. Lamm, and R. K. Albert, Surfactant replacement improves recoil in rabbit lungs after acid aspiration, Am. Rev. Respir. Dis., 142: 1279-1283 (1990).

37. T. Kobayashi, M. Ganzuka, T. Ueda, J. Taniguchi, and S. Marakami, Surfactant replacement in respiratory failure induced by aspiration of hydrochloric acids in rabbits, in: B. Lachmann, ed., Surfactant Replacement Therapy in Neonatal and Adult Respiratory Distress Syndrome, Berlin: Springer-Verlag, pp. 245-257 (1988).

38. T. Kobayashi, M. Ganzuka, J. Taniguchi, K. Nitta, and S. Murakami, Lung lavage and surfactant replacement for hydrochloric acid aspiration in rabbits, Acta Anaesthesiol. Scand., 34: 216-221 (1990).

39. J. Klein, Normobaric pulmonary oxygen toxicity, Anesth. Analg., 70: 195-207 (1990).

40. D. L. Beckman, and H. W. Weiss, Hyperoxia compared to surfactant washout

on pulmonary compliance in rats, J. Appl. Physiol., 26: 700-709 (1969).

41. T. Fujiwara, F. H. Adams, and K. Seto, Lipids and surface tension of extracts of normal and oxygen treated guinea pigs, J. Pediat., 65: 45-52 (1964).

42. G. C. Cochrane, R. G. Spragg, and S. D. Revak, Pathogenesis of the adult respiratory distress syndrome. Evidence of oxidant activity in bronchoalveolar lavage fluid, J. Clin. Invest., 71: 754-761 (1983).

43. S. Matalon, and E. A. Egan, Effects of 100% oxygen breathing on permeability of alveolar epithelium to solute, J. Appl. Physiol., 50: 859-863 (1981).

44. S. Matalon, and E. A. Egan, Interstitial fluid volumes and albumin spaces in pulmonary oxygen toxicity, J. Appl. Physiol., 57: 1767-1772 (1984).

45. B. A. Holm, R. H. Notter, J. Seigle, and S. Matalon, Pulmonary physiological and surfactant changes during injury and recovery from hyperoxia, J. Appl. Physiol., 59: 1402-1409 (1985).

46. S. Matalon, B. A. Holm, and R. H. Notter, Mitigation of pulmonary hyperoxic injury by administration of oxygenous surfactant, J. Appl. Physiol., 62: 756-761 (1987).

47. J. J. Ennema, T. Kobayshi, B. Robertson, and T. Curstedt, Inactivation of exogenous surfactant in experimental respiratory failure induced by hyperoxia, Acta Anaesthesiol. Scand., 32: 665-671 (1988).

48. G. M. Loewen, B. A. Holm, I. Milanowski, L. M. Wild, and S. Matalon, Alveolar hyperoxic injury in rabbits receiving exogenous surfactant, J. Appl. Physiol., 66: 1087-1092 (1988).

49. R. R. Baker, B. A. Holm, P. C. Panus, and S. Matalon, Development of O_2 tolerance in rabbits without increase in antioxydant enzymes, J. Appl. Physiol., 66: 1679-1684 (1989).

50. P. S. Richmann, R. G. Spragg, B. Robertson, T. A. Merritt, and T. Curstedt, The adult respiratory distress syndrome: first trials with surfactant replacement, Eur Respir J., 2: Suppl. 3, 109s-111s (1989).

51. B. Lachmann, The role of pulmonary surfactant in the pathogenesis and therapy of ARDS, in: J. L. Vincent, ed., Update in Intensive Care and Emergency Medicine, Springer, Berlin, pp 123-134 (1987).

52. S. Nosoka, T. Sakai, M. Yonekura, and K. Yoshikawa, Surfactants for adults with respiratory failure, Lancet, 336: 947-948 (1991).

53. Th. Joka, and U. Obertacke, Neue medikamentose Behandlung in ARDS: Effekt einer intrabronchialen xenogenen Surfactantapplikation, Z. Herz, Thorax, Gefasschir., 3: Suppl. 1, 21-24 (1989).

ARTIFICIAL OXYGEN CARRYING BLOOD SUBSTITUTES

N. Simon Faithfull

Anesthesiologist and Vice President Medical Research, Alliance

Pharmaceutical Corp., 3040 Science Park Rd, San Diego, Ca, USA

INTRODUCTION

Oxygen (O2) is essential for animal life on this planet. In lower single celled organisms it is obtained by simple diffusion from the surrounding aqueous mileau, but higher animals have systems designed for transporting O2 to the tissues needing it. In the case of insects a complex system of branching tubes (tracheae) has been developed that ducts air to the tissues. These tracheae end in the vicinity of mitochondria and allow efficient O2 transport in animals below a critical body mass. Higher organisms have a system in which an oxygen transporting fluid is pumped to the tissues. This fluid (blood) usually contains a metal containing pigment of high O2 affinity. In mollusks and some arthropods this pigment is the copper containing hemocyanin. In almost all higher animals this is an iron-porphyrin ring containing protein (hemoglobin) packaged in cells dedicated to its transport in blood. Interestingly enough under very cold conditions the antarctic icefish, Chaenocephalus aceratus, makes use of high solubility for O2 in the plasma and exists without the need for hemoglobin (Hb) in its blood.

The dream of developing a blood substitute capable of carrying and delivering significant quantities of oxygen began as long ago as 1868 when Nauyn injected a hemoglobin solution into dogs [1]. By 1937 Amberson [2] was able to review a considerable body of knowledge of the properties of the crude Hb solutions then available, and recognized a number of side effects that still concern researchers today. Since then the effects of unmodified and modified hemoglobin solutions have been extensively studied and some of the main details will be given below.

Fluorocarbons, more correctly known as perfluorocarbons or perfluorochemicals (PFCs) were first synthesized just before World War II where they were used in development of the atom bomb. They have subsequently been developed and are extensively used in industry as coolants, electrical insulating agents, and as non-flammable lubricants and polymers [3]. They are inert and not metabolized. One of their most important properties, from a medical and biological point of view, is their very high solubilities for gases, including oxygen and carbon dioxide. A summary of work performed in the O2 transport field will be given below together with brief notes on other less well known and medically utilizable properties.

Oxygen Transport to Tissue XIV, Edited by W. Erdmann and
D.F. Bruley, Plenum Press, New York, 1992

The first injections of substantial quantities of Hb in a clinical setting was in 1916 [4]. Further human trials were performed by Amberson et al [5] in which they were confronted by the first indications of vasoconstrictive action of the product - an effect that had not been predicted by animal work up to that time.

Since then a large quantity of work has been performed using many different preparations of modified or unmodified Hb. A lot of confusing and sometimes contradictory data has been generated and the need exists for large scale production of purified hemoglobin with consistent lot to lot characteristics. Till this is achieved no thorough comparisons can be made and no universally accepted protocols can be designed.

Types of Hemoglobin Solutions

Unmodified hemoglobin solutions prepared by lysis and stroma removal are readily available from outdated blood [6,7]. They are essentially all prepared by washing red cells and lysing them in distilled water [8] or other hypotonic media including phosphate buffers [9,10,11]. The stroma is removed in a variety of ways including centrifugation [8], filtration [9] or extraction by toluene or chloroform [12]. Alternatively lysate that is filtered to remove coarse particles can be dialyzed against hypertonic phosphate buffer causing hemoglobin to crystallize on the inside of the dialysis tubing [10,13].

These solutions offer a number of advantages. They can be stored for considerable lengths of time - 6.5 months at 4°C and 10 months at -80°C [14]. In the crystallized state they can be kept at -20°C for 2 years or one year at 4°C [15]. Blood type specificity is not present when these solutions are devoid of stroma; and nor is there need for crossmatching - Hb from one species can be given to a second. The agents are probably not allergenic [16] though higher antibody levels are obtained after heterologous compared to homologous Hb infusions [17]. Highly polymerized hemoglobin preparations may be immunogenic [18,19].

Unmodified Hb solutions have significant intrinsic oncotic pressure [20] and have lower viscosity than that of blood; they are almost Newtonian fluids. They can have obvious rheological advantages and hemodilution with Hb solutions can block the hyperviscosity state that develops early in myocardial ischemia [21]. The solutions neither induce nor inhibit production of platelet aggregation [10].

Simple unmodified hemoglobin solutions also have a number of important disadvantages. The main problem is often considered to be the position of the oxyhemoglobin dissociation curve. Normal standardized human blood has a P50 of 26.5 mm Hg [22]. Simple unmodified hemoglobin solutions have a far greater affinity for O2 and this results in a much lower P50 - somewhere in the region of 12 - 14 mm Hg [23], as compared to 26.5 mm Hg for human blood [22]. Figure 1 shows the oxyhemoglobin dissociation curves for normal blood and stroma free hemoglobin (SFH).

The lower P50 of free hemoglobin implies a lower tissue PO2 for the same extraction of oxygen from the blood passing through that tissue and a lower mixed venous oxygen tension (PvO2) for the same cardiac output, oxyhemoglobin saturation and hemoglobin concentration. These conclusions were confirmed when baboons were hemodiluted with pyridoxalated Hb (P50 of 20-25 mm Hg). At low hematocrits these animals had a mean PvO2 of 13.8 mm Hg; control hemodiluted animals had a PvO2 of 31.3 mm Hg [24].

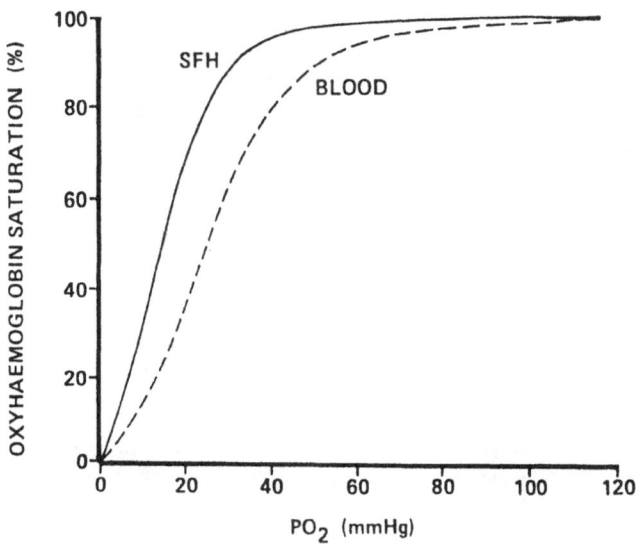

Fig.1. Oxyhemoglobin Dissociation Curves for Normal Human Blood and Stroma Free Hemoglobin Solution.

Simple Hb solutions often have an intravascular half life of 4 hours or less [11, 25, 26]. The Hb tetramer splits into 32 kilodalton dimers, which are readily filtered through the renal glomeruli. This can cause renal toxicity in both animals [27] and in man [28]. Renal abnormalities have been an almost universal finding when both unmodified and modified hemoglobin solutions are administered and still form a substantial barrier to the introduction of Hb solutions into clinical practice. Toxicity is often considered mild and reversible [25, 29]; and some workers did not detect it [30]. Unfortunately no animal models have yet been developed that can reliably predict renal dysfunction in man [31].

Disseminated intravascular coagulation has been seen following Hb solutions [32] even when stroma was apparently absent [33], while other studies demonstrated anticoagulant activity with Factor VIII deficiency [34]. Hemoglobin can also activate complement via the alternate pathway; this is not seen with erythrocyte stroma [35]. Though many of the side effects attributed to stroma may be solved by the availability of pure hemoglobin produced by recombinant technology [36], problems may occur due to contaminants of bacterial or yeast origin below current detection ability.

The oncotic pressure of unmodified hemoglobin solutions is about 3 times that of blood [37]. To achieve an oncotic pressure approaching the normal value of 20 mm Hg, the Hb concentration must be reduced to about 7 gm per 100 ml - this might be considered rather "anemic" and not suitable for resuscitation purposes. Modifications of the hemoglobin molecule are therefore designed to overcome some of the above problems.

The P50 of Hb can be increased by both intramolecular and intermolecular crosslinkage. Pyridoxal 5'-phosphate is an organic phosphate analogue of 2,3-diphosphoglycerate, which increases P50 [37]. Pyridoxalated hemoglobin with P50 values of 22-26 mm Hg have been produced [38, 39]. Intravascular persistence varied from about 1 to 4 hours depending on the species and the dose administered. Beta chains of the Hb dimer can be covalently crosslinked using 2-nor-2 formyl-pyridoxal 5'-phosphate [40]. This results in a P50 of about 45 mm Hg and a plasma half life of about 3-5 hours in rodents.

Molecular polymerization of the Hb molecule causes increased "molecular" size and decrease in glomerular filtration, hence increasing its circulation half life. Unfortunately, this polymerization, which is most often performed using glutaraldehyde [39], causes a decrease in the P50 and a variable degree of loss of cooperativity of the molecule. The reaction is random and batch reproducibility is difficult. A further advantage of polymerization is that it decreases the particle concentration and reduces oncotic pressure, allowing the use off higher Hb concentrations in an iso-oncotic solution [23].

Pyridoxalated polymerized hemoglobin has been produced with Hb concentration between 14 and 15 gm per 100 ml, normal oncotic pressure and P50 of 14-15 mm Hg; half life in baboons was about 46 hours [41]. Commercialization produced a product with a half life of 38 hours, P50 of 16-20 mm Hg and (Hb) of 14-16 gm/dl. A similar product recently completed Phase I clinical trials in the USA; early Phase II trials were stopped in the USA by the Food and Drug Administration (FDA) due to safety concerns.

Intramolecular crosslinking of Hb alpha chains by 3,5-bis-dibromosalicyl fumarate can, when performed under deoxygenated conditions, both decrease oxygen affinity and prolong retention. This has produced a P50 of 18-28 mm Hg and a half life of about 20 hours [42]. Other approaches to Hb modification include linking the molecule to synthetic high molecular weight compounds such as dextran [43] or polyethylene glycol [44]. These products have high inherent stability and have long retention times in the circulation.

Bovine hemoglobin has a P50 of about 28.5 mm Hg in free solution [39]. This is decreased to about 24 mm Hg when it is polymerized and hence remains in a suitable range [45]. It does not appear to cause antibody formation when injected into non-bovine species. A commercial preparation of polymerized bovine hemoglobin has recently been in Phase I clinical trials in the USA. Trials have recently been suspended due to side effects of unknown aetiology involving smooth muscle spasms.

Hemoglobin filled liposomes, or neohemocytes as they are sometimes termed [46] may be uni- or multi-lamellar and can have a normal O2 transport capacity [47]; P50 values between 26 and 28 mm Hg have been given [48, 49]. Circulation half life is about 6 hours [49]. The liposomes are cleared by a mixture of tissue binding, breakdown and uptake by the reticuloendothelial system.. Short term hemolysis is not a problem and they can be frozen for at least 6 weeks [48].

Medical Applications of Hemoglobin Solutions

For use as blood substitutes Hb solutions must have O2 dissociation characteristics approximating that of normal blood with correct cooperativity. They should have an intravascular retention time of 24 - 48 hours and have normal oncotic pressures at Hb contents of about 14 Gm/dl. Hyperoncotic solutions may be suitable for initial resuscitation following trauma. Obvious applications would be for intraoperative and postoperative blood replacement and for trauma resuscitation. "Clean" preparations would not transfer infections such as hepatitis and AIDS; meticulous quality control would be necessary to ensure this. Less obvious opportunities for use of Hb are conferred by the low viscosity of the solutions.

Low viscosity implies high penetrability into areas from which red cells are excluded. This confers applications in treating ischemic conditions such as cerebral or myocardial infarctions [21], peripheral vascular disease with incipient gangrene or during sickle cell crises. Other uses would be in Percutaneous Transluminal Coronary Angioplasty (PTCA) or as a radiosensitizer in radiotherapy.

FLUOROCARBONS

The first dramatic demonstration of the biological usefulness of PFCs was the demonstration of the ability of small mammals to survive breathing oxygenated liquid PFC [50]. The first emulsions of PFCs were made using albumin as the emulsifier [51]. Subsequently pluronic emulsifiers were utilized in work in which all blood was removed from small animals [52].

FLUOSOL®

Fluosol® contains 14 gm/dl perfluoro-decalin (FDC) and 6 gm/dl perfluoro-tripropylamine (FTPA). The emulsion, which is stabilized with Pluronic F68 and egg yolk phospholipids (EYP), is manufactured by The Green Cross Corporation of Osaka, Japan. FDC has a relatively short whole body half life of about 7 days [53], but needs the presence of FTPA to ensure emulsion stability. The latter has an undesirably long life of about 65 days [53]. Even so, Fluosol® is not very stable and the emulsion must be stored in a frozen state.

Pluronic F68 is thought to cause many of the side effects seen with Fluosol® including complement activation [54], inhibition of chemotaxis and neutrophil phagocytosis [55], microvascular flow inhomogeneity [56], decreased foetal growth rates [57] and severe hypotensive reactions in dogs [58]. Some side effects may be advantageous such as reduction of reperfusion consequent upon reducing leucocyte plugging and decreased free radical formation [59, 60].

Fluosol® was first used in Japan and the USA in Jehovah's Witness patients refusing blood transfusion on religious grounds [61, 62]. These patients were given up to a maximum dose of 5gm/kg PFC. The FDA rejected the oxygen transporting blood substitute claim even though safety and the ability of PFCs to transport O2 were not in question [63]. Late in 1989 Fluosol® was finally approved for use during (PTCA) in high risk patients [64, 65, 66]. It is also licensed for use in PTCA in Europe and is used in Jehovahs Witness patients on a named patient basis.

Since 1985 Fluosol® has been in trials as a radiosensitiser in therapy of tumors of the head and neck [67], brain [68] and lung [69]. The encouraging preclinical efficacy results [70, 71, 72, 73, 74] have not been seen in clinical trials, due probably to its never being given in sufficiently high doses. In 1990 trials began in the USA using Fluosol® as an adjuvant to thrombolytic therapy in acute myocardial infarction. The use of Fluosol® in sickle cell crisis will start shortly.

OXYPHEROL

Oxypherol (FC-43) contains 20 gm/dl perfluoro-tributylamine (FTBA) emulsified with Pluronic F68 to produce an emulsion with excellent room temperature stability, but with unacceptably long RES half life for clinical use [75]. The emulsion has been used extensively as an artificial O2 carrier in animal work.

Other first generation PFC emulsions include Emulsion No II, which is made in china and is of similar composition to Fluosol® [76]; it is reported to have been used in military surgery [77]. Ftorosan is made in the Soviet Union [76] and has been extensively used under clinical conditions.

PERFLUBRON

Perflubron is the generic name for perfluorooctyl bromide, a second generation PFC in development by Alliance Pharmaceutical Corp. A single bromide atom confers radiopacity to the compound, which is emulsified with EYP to produce highly concentrated emulsions [78]. EYPs do not induce complement activation [79] and have been used for many years in parenteral fat emulsions. Room temperature stable Perflubron emulsions as concentrated as 125 gm/dl can be produced [80].

Oxygen Transport by Fluorocarbons

The O2 content of PFCs is proportional to the O2 tension; this is not so for Hb which has a sigmoid oxygen dissociation curve (Fig. 2).

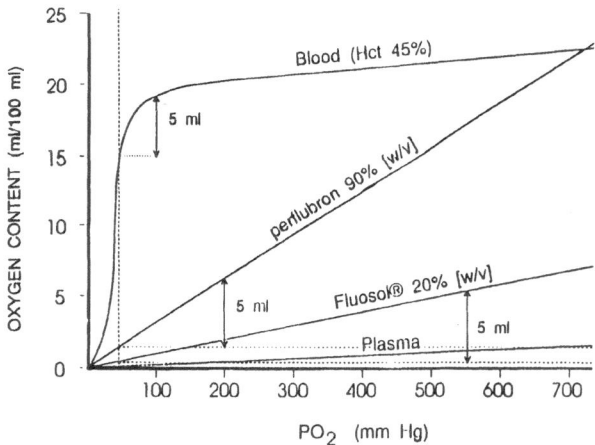

Fig.2. Relationship between Oxygen Content and PO2 for Whole Blood, 90% w/v Perflubron emulsion, Fluosol® and Plasma.

Between about 100 mm Hg (normal arterial oxygen tension - PaO2) and 45 mm Hg (normal PvO2) blood with an hematocrit of 45% will release about 5 ml of O2. Assuming that the circulation contains only Fluosol®, PaO2 must be increased to 550 mm Hg to allow "normal " delivery of O2; with 90% Perflubron emulsion in the circulation, the same O2 delivery can be obtained if the PaO2 is 200 mm Hg.

It should be noted that, at high PO2s PFC emulsions, due to their straight line "dissociation curve", have a much greater oxygen extraction coefficient for oxygen than Hb. When the circulation contains a mixture of blood and PFC emulsion, this results in the contribution of PFC to O2 consumption being many times greater than the PFC contribution to O2 delivery to the tissues. In an early clinical case report,for instance, Fluosol® contributed 27 per cent to total oxygen delivery while at the same time contributing more than 50 per cent of the oxygen consumed [81].

Metabolism and Excretion of Fluorocarbons

PFCs are considered to be biologically inert due to the very high carbon-fluorine bond strengths in the molecule [82]. One study published in the chemical literature [83] has suggested that metabolism of PFCs can occur in vivo. The reactions reported require an aprotic solvent while biological media are protic (water-based). The reactants used do not exist in nature and the report is totally irrelevant to biological conditions. Early work [53] demonstrated no new gas chromatograph peaks in the liver, indicating no metabolic processing of the compounds. Additionally no increased fluoride concentrations were seen in either animals [53] or man [84]. No evidence has been advanced to show that fluorocarbons themselves when chosen with due regard to their vapor pressure and molecular weights have any more toxicity than polytetrafluoroethylene - teflon™ [85].

PFCs are taken up in the reticuloendothelial system (RES), a normal mechanism for clearance of particulate materials such as fat emulsions, liposomes and PFC emulsions. They remain there and slowly leave in the vapor phase to be redistributed in very small quantities to other organs depending on their vapor pressure and organ solubility. They are finally eliminated via the expired air and possibly very slightly through the skin [86, 87, 88, 89, 90]. Excretion in urine and feces is practically nil apart from FTBA; during the first 14 days after administration 25% of recovered FTBA is found in the feces, presumably via the bile [91]. The half life of a number of PFCs mentioned in this article are: FTBA: 895 days [91]; FTPA: 65 days [53]; FDC: 7 days [53]; Perflubron: 4 days. The very advantageous excretion rate of the latter is attributed to the lipophilic character of its terminal bromine atom [85].

Administration of all fluorocarbon emulsions have been followed by increases in levels of hepatic enzymes, particularly SGOT and SGPT, in both animals and man [86, 92, 93, 94, 95, 96]. Though of some concern it should be noted that all changes rapidly resolve without progression to any degree of hepatic failure. The reason for these changes is unclear but would seem to be related to rapid engorgement of Kupffer cells with PFC and consequent compression of hepatic tissue [97].

Medical Applications of Fluorocarbons

Apart from the obvious applications of PFC emulsions for replacement of perioperative blood loss and in trauma resuscitation many other potential applications have been explored. Though many of these applications are theoretical or only demonstrated in animals, there is no doubt that they will become widely used as second and third generations of PFC emulsions enter clinical practice.

PFC emulsions have shown a remarkable ability to support ischemic microcirculation and to limit infarct damage. This was once thought to be partially due to low viscosity of the emulsions, though recent work using Perflubron in myocardial infarction indicates that this may not be as important as previously thought [97].

The mean particle diameter in PFC emulsions is 0.1 - 0.25 microns. These may be able to perfuse border areas of ischemic tissue through available collaterals [98] and bypass obstructions in the circulation caused by endothelial blebs and sludged red cells [99]. They may thus be able to penetrate deep into hypoxic tissue beds bypassing sludged cells (8 microns in diameter) and reoxygenating them [47, 100], hence reversing the stiffening of the red cell membrane that occurs under conditions of hypoxia and tissue acidosis [101, 102].

PFCs may also improve ischemic hypoxia by their actions on leukocytes because some PFC emulsions can inhibit neutrophil phagocytosis, chemotaxis and superoxide release [55, 103, 104]. This may explain the decreased inflammatory response and increased myocardial salvage observed following early post infarction hemodilution with PFCs [105] and their protection against the reperfusion syndrome [59, 104, 106]. Following myocardial ischemia, PFC emulsions can increase local oxygenation [107, 108], improve microcirculatory flow [98, 107] and increase intramyocardial O2 tensions [109, 110, 111]. The volume of infarcted tissue is reduced [98, 105, 108, 112, 113].

Cerebral damage can be reduced by treatment with PFC emulsions [114, 115] particularly when used together with oxygen radical scavengers [116, 117]. Oxygenated PFC nutrient solution improves severe global cerebral ischaemia [118, 119] and spinal cord injury [120]. PFC emulsions may be of benefit in peripheral vascular disease [121, 122] and during intestinal ischemia [123].

Neat (unemulsified) Perflubron has been used clinically for bronchography and alveolography [124] and as an MRI contrast agent of the gastrointestinal tract [125]. F-19 MRI imaging can be used to detect tissue oxygenation [126, 127] and to estimate tumor blood flow [128, 129]. Perflubron emulsion is a computer tomographic (CT) blood pool imaging agent and an excellent agent for clinical identification of secondary malignant tumors in the liver [96]. PFCs are selectively taken up by certain tumors outside the RES including cerebral neoplasms [130]. When Perflubron is used high radiographic definition can be obtained.

Owing to their compressibility, PFCs are good ultrasound contrast agents. Perflubron can be used to evaluate cardiac and vascular flow, detect tumors [131] and abscesses, assess myocardial infarction volume [132] and to assess renal function [133, 134]. When Perflubron emulsions are injected subcutaneously the PFC particles are picked up in the lymphatics and transported to lymph nodes. These can then be imaged using CT [135].

A number of different organs have been successfully perfused with oxygenated PFC emulsions including kidney [136], brain [137], heart [138], liver [139, 140] and isolated limbs [141, 142]. Clinically, successful renal transplantation has ben performed [143, 144] and traumatically amputated human extremities have been successfully reimplantated [47].

Total liquid ventilation of preterm infants with unemulsified PFCs have resulted in improvement of oxygenation and decrease in pulmonary compliance[145]. Recent work on partial liquid ventilation with PFCs using conventional gas ventilators point the way to new therapies for respiratory failure and the possible use of Perflubron as replacement therapy for pulmonary surfactant.

REFERENCES

1. Nauyn B, Untersuchungen uber blutgerinnungen lebenden thiere und ihre folgen , Archiv Pathol Pharmakol 1 (1873): 1-17.

2. Amberson WR, Blood Substitutes, Biol Rev 12 (1937): 48-86.

3. Slinn DSL, and Green SW, "Fluorocarbon fluids for use in the electronics industry ," Preparation, properties and industrial applications of organofluoride compounds., Ed. Banks RE (Chichister: Ellis Horwood, 1982) 45-82.

4. Sellards AW, and Minot GR, Injection of Hemoglobin in man and its relation to blood destruction with special reference to the anemias , J Med Res 34 (1916): 469-475.

5. Amberson WR, Jennings JJ, and Rhode CM, Clinical Experience with hemoglobin-saline solutions , J Appl Physiology 1 (1949): 469-489.

6. Biro GP, Current status of erythrocyte substitutes, Can Med Assoc J 129 (1983): 237-244.

7. Baldwin JE, and Gill B, Approaches to the preparation of oxygen carriers for use as blood substitutes, Medical Lab Sciences 39 (1982): 45-51.

8. Christensen SM, Medina F, Winslow RW, Snell SM, Zegna A, and Marini MA, Preparation of human hemoglobin Ao for possible use as a blood substitute, J Biochem Biophys Methods 17 (1988): 143-154.

9. Hsia JC, Hronowski LJ, Persaud K, and Ansari MR, ATP-hemoglobin purification by ATP-agarose affinity chromatography, J Chromatography 381 (1986): 153-157.

10. De Venuto F, Zuck TF, Zegna AI, and Moores WY, Characteristics of stroma-free hemoglobin prepared by crystallization, J Lab Clin Med 89.3 (1977): 509-516.

11. Rabiner SF, Helbert JR, Lopas H, and Friedman LH, Evaluation of a stroma-free hemoglobin solution for use as a plasma expander, J Exp Med 126 (1968): 1127-1142.

12. Wilson BG, and Odling-Smee W, Blood Substitutes: effect of haemoglobin solution on neutrophil killing , Irish J Med Sci 155 (1986): 334.

13. Drabkin DL, A simplified technique for large scale crystallization of human oxyhemoglobin. Isomorphous transformations of hemoglobin and myoglobin in the crystalline state , Arch Biochem 21 (1949): 224-232.

14. Greenburg AG, Ginsburg K, and Peskin GW, Preservation of stroma-free hemoglobin solution, Surgical Forum 29 (1977): 5-9.

15. De Venuto F, Stability of crystalline hemoglobin solution during extended storage , J Lab Clin Med 92 (1978): 976.

16. Cochin A, Das Gupta TK, De Woskin R, and Moss GS, Immunogenic properties of stroma vs stroma-free hemoglobin solution, Surgical Forum (1972): 19-22.

17. Bruzzese FJ, Dix JA, Rava RP, and Cerny LC, (Abstract) Resonance raman spectra of potential blood substitutes, Biomater Artif Cells Artif Organs 15.2 (1987): 350.

18. Winslow RM, Blood substitutes - Minireview, <u>The Red Cell:</u> <u>Seventh Ann Arbor Conference</u>, Alan R Liss, 1989) 305-323.

19. Hertzman CM, Keipert PE, and Chang TMS, Serum antibody titers in rats receiving repeated small subcutaneous injections of hemoglobin or polyhemoglobin: a preliminary report, <u>Int J Artif Organs</u> 9.3 (1986): 179-182.

20. Kaplan HR, and Murthy VS, Hemoglobin solution: a potential oxygen transporting plasma volume expander, <u>Fed Proceed</u> 34 (1975): 1461-1465.

21. Biro GP, Beresford-Kroeger D, and Hendry P, Early deleterious hemorheologic changes following acute experimental coronary occlusion and salutary antihyperviscosity effect of hemodilution with stroma-free hemoglobin, <u>Am Heart J</u> 103 (1982): 870-878.

22. Nunn JF, <u>Applied Respiratory Physiology</u>, 2nd ed. (London and Boston: Butterworths, 1977).

23. Gould SA, Sehgal LR, Rosen AL, Sehgal HL, and Moss GS, "(Chapter 6) Artificial oxygen carriers," <u>Modern Transfusion Therapy</u>, Ed. Janice P Dutcher (Boca Raton: CRC Press, 1990) I: 107-123.

24. Gould SA, Rosen A, Sehgal L, Noud G, Sehgal H, DeWoskin R, Levine H, Kerstein M, Rice C, and Moss GS, The effect of altered hemoglobin-oxygen affinity on oxygen transport by hemoglobin solution, <u>J Surg Res</u> 28 (1980): 246-251.

25. Savitsky JP, Doczi J, Black J, and Arnold JD, A clinical safety trial of stroma-free hemoglobin, <u>Clin Pharmacol Ther</u> (1978): 73-80.

26. Bonhard K, Acute oxygen supply by infusion of hemoglobin solutions, <u>Fed Proceed</u> 34.6 (1975): 1466-1467.

27. Hamilton PB, Hiller A, and Van Slyke DD, Renal effects of hemoglobin infusions in dogs in hemorrhagic shock, <u>J Exp Med</u> 86 (1947): 477-487.

28. Brandt JL, Frank NR, and Lichtman HC, The effects of hemoglobin solutions on renal functions in man, <u>Blood</u> 6 (1951): 1152-1158.

29. Lee R, Atsumi N, Jacobs EE, Austen WG, and Vlahakes GJ, Ultrapure, stroma-free, polymerized bovine hemoglobin solution: Evaluation of renal toxicity, <u>J Surg Res</u> 47 (1989): 407-411.

30. Relihan M, and Litwin MS, Clearance rate and renal effects of stroma-free hemoglobin on acidotic dogs, <u>Surgery, Gynecology and Obstetrics</u> 137 (1973): 73-79.

31. Moss GS, Gould SA, Rosen AL, Sehgal L, and Sehgal HL, Animal model for nephrotoxicity of haemoglobin tetramer, <u>The Lancet</u> May24 (1986): 1219.

32. Birndorf NI, Lopas H, and Robboy SJ, Disseminated intravascular coagulation and renal failure: Production in the monkey with autologous red blood cell stroma, <u>Laboratory Investigation</u> 25.4 (1971): 314-319.

33. Bolin R, Smith D, Moore G, Boswell G, and De Venuto F, "Hematologic effects of hemoglobin solutions in animals," <u>Advances in Blood Substitute Research</u>, Ed. R. G. a. G. N. RB Bolin (New York: Alan R Liss, 1982) 117-126.

34. Moss GS, DeWoskin R, and Cochin A, Stroma-free hemoglobin. I. Preparation and observations on in vitro changes in coagulation, Surgery 74.2 (1973): 198-203.

35. Wilson WA, and Thomas EJ, Activation of the alternative pathway of human complement by haemoglobin, Clin Exp Immunol 36 (1979): 140-144.

36. FDC Reports, Somatogen lipid-encapsulated recombinant hemoglobin blood substitute under Navy contract, FDC Reports: "The Pink Sheet" 52.45 (1990): 8.

37. Moss GS, Gould SA, Sehgal LR, and Sehgal HL, Hemoglobin solution-From tetramer to polymer, Surgery 95.3 (1984): 249-255.

38. Sehgal LR, Rosen AL, Noud G, Sehgal HL, Gould SA, DeWoskin R, '. Rice CL, and Moss GS, Large-volume preparation of pyridoxylated hemoglobin with high P50, J Surg Res 30 (1981): 14-20.

39. De Venuto F, and Zegna A, Preparation and evaluation of pyridoxalated-polymerized human hemoglobin, J Surg Res 34 (1983): 205-212.

40. Bakker JC, Bleeker WK, and van der Plas J, Properties of hemoglobin interdimerically cross-linked with NFPLP, Transfusion Med: Proceedings of the XVII Annual Scientific Symposium of the American Red Cross (1986): 49-55.

41. Sehgal LR, Gould SA, Rosen AL, Sehgal HL, and Moss GS, Polymerized pyridoxylated hemoglobin: A red cell substitute with normal oxygen capacity, Surgery 95.4 (1984): 433-438.

42. Tye RW, (US Patent) Preparation of stroma-free non-heme protein-free hemoglobin, US Patent 4,473,494 (1984):

43. Tam S-C, Blumenstein J, and Wong T-F, Soluble dextran-hemoglobin complex as a potential blood substitute, Proc Natl Acad Sci USA 73 (1976): 2128-2131.

44. Ajisaka K, and Iwashita Y, Modification of human hemoglobin with polyethylene glycol: A new candidate for blood substitute, Biochem and Biophysical Res Commun 97.3 (1980): 1076-1081

45. Feola M, Gonzalez H, and Canizaro PC, Development of a bovine stroma-free hemoglobin solution as a blood substitute, Surgery Gynecology and Obstetrics 157.5 (1983): 399-408.

46. Hunt CA, and Burnette RR, "Neohemocytes," Advances in blood Substitute Research, Ed. R. g. a. G. N. RB Bolin (New York: Alan R Liss, 1982) 59-70.

47. Smith AR, Van Alphen W, Faithfull NS, and Fennema M, Limb preservation in replantation surgery, J Plast Reconstr Surg 75.2 (1985): 227-237.

48. Djordjevich L, and Ivankovich AD, Half-life of synthetic erythrocytes. in-vivo, Anesthesiology 65.3A (1986): A94.

49. Hunt CA, Burnette RR, MacGregor RD, Strubbe AE, Lau DT, Taylor N, and Kawada H, Synthesis and evaluation of a prototypal artificial red cell, Science 230 (1985): 1165-1168.

50. Clark LC, and Gollan F, Survival of mammals breathing organic liquids equilibrated with oxygen at atmospheric pressure, Science 152 (1966): 1755-1756.

51. Sloviter HA, "Perfusion of the brain and other isolated organs with dispersed perfluoro compounds," <u>Blood Substitutes and Plasma Expanders</u>, (New York: Alan R Liss, 1978) 28-39.

52. Geyer RP, Monroe RG, and Taylor K, "Survival of rats totally perfused with a fluorocarbon-detergent preparation," <u>Organ Perfusion and Preservation</u>, Ed. JC Norman 1968) 85-96.

53. Yokoyama K, Yamanouchi K, Ohyanagi H, and Mitsuno T, Fate of perfluorochemicals in animals after intravenous injection of hemodilution with their emulsions, <u>Chem Pharm Bull</u> 26.3 (1978): 956-966.

54. Vercellotti GM, Hammerschmidt DE, Craddock PR, and Jacob HS, Activation of plasma complement by perfluorocarbon artificial blood: Probable mechanism of adverse pulmonary reactions in treated patients and rationale for corticosteroid prophylaxis, <u>Blood</u> 59.6 (1982): 1299-1304.

55. Virmani R, Fink LM, Gunter K, and English D, Effect of perfluorochemical blood substitutes on human neutrophil function, <u>Transfusion</u> 24 (1984): 343-347.

56. Faithfull NS, King CE, and Cain SM, Peripheral vascular responses to fluorocarbon administration, <u>Microvasc Res</u> 33.2 (1987): 183-193.

57. Faithfull NS, and Marshall HW, The effect of fluorocarbon emulsion on placental insufficiency, <u>Adv Exp Med Biol</u> 248 (1989): 357-369.

58. Faithfull NS, and Cain SM, Cardiorespiratory consequences of fluorocarbon reactions in dogs, <u>Biomater Artif Cells Artif Organs</u> 16.1-3 (1988): 463-472.

59. Forman MB, Bingham S, Kopeman HA, Wehr C, Sandler MP, Kolodgie F, Vaughn WK, Friesinger GC, and Virmani R, Reduction of infarct size with intracoronary perfluorochemical in a canine preparation of reperfusion, <u>Circulation</u> 71.5 (1985): 1060-1068.

60. Forman MB, Puett DW, Wilson BH, Vaughn WK, Friesinger GC, and Virmani R, Beneficial long-term effect of intracoronary perfluorochemical on infarct size and ventricular function in a canine reperfusion model, <u>JACC</u> 9.5 (1987): 1082-1090.

61. Tremper KK, Friedman AE, Levine EM, Lapin R, and Camarillo D, The preoperative treatment of severely anemic patients witha perfluorochemical oxygen-transport fluid, Fluosol-DA, <u>New England J Med</u> 307 (1982): 277-283.

62. Gould SA, Rosen AL, Sehgal LR, Sehgal HL, Langdale LA, Krause LM, Rice CL, Chamberlain WH, and Moss GS, Fluosol-DA as a red-cell substitute in acute anemia, <u>New England J Med</u> 314.26 (1986): 1653-1656.

63. Marwick C, FDA committee questions Fluosol efficacy; US approval not imminent, <u>JAMA</u> 250 (1983): 2585-2586.

64. Anderson HV, Leimgruber PP, Roubin GS, Nelson DL, and Gruentzig AR, Distal coronary artery perfusion during percutaneous transluminal coronary angioplasty, <u>Am Heart J</u> 110.4 (1985): 720-726.

65. Jaffe CC, Wohlgelernter D, Cabin H, Bowman L, Deckelbaum L, Remetz M, and Cleman M, Preservation of left ventricular ejection fraction during percutaneous transluminal coronary angioplasty by distal transcatheter coronary perfusion of oxygenaed Fluosol DA 20%, <u>Am Heart J</u> 115.6 (1988): 1156-1164.

66. Kent KM, Cleman MW, Cowley MJ, Forman M, Jafft CC, Kaplan M, King SB, Kurcoff MW, Lassar T, McAuley B, Smith R, Wisdom C, and Wohlgelernter D, Reduction of myocardial ischemia during percutaneous transluminal coronary angioplasty with oxygenated Fluosol, Am J Cardiology 66 (1990): 279-284.

67. Lustig R, McIntosh-Lowe N, Rose C, Haas J, Krasnow s, Spaulding M, and Prosnitz L, Phase I/II study of Fluosol-DA and 100% oxygen as an adjuvant to radiation in the treatment of advanced squamous cell tumors of the head and neck, Int J Radiation Oncology Biol Phys 16 (1989): 1587-1593.

68. Evans RG, Kimler BF, Morantz RA, Vats TS, Gemer LS, Liston V, and Lowe N, A phase I/II study of the use of Fluosol as an adjuvant to radiation therapy in the treatment of primary high-grade brain tumors, Int J Radiation Oncology Biol Phys 19 (1990): 415-420.

69. Lustig R, Lowe N, Prosnitz L, Spaulding M, Cohen M, Stitt J, and Brannon R, Fluosol and oxygen breathing as an adjuvant to radiation therapy in the treatment of locally advanced non-small cell carcinoma of the lung: Results of a phase I/II study, Int J Radiation Oncology Biol Phys 19 (1990): 97-102.

70. Martin DF, Porter E, Fischer JJ, and Rockwell S, Effect of a perfluorochemical emulsion on the radiation response of BA1112 Rhabdomyosarcomas, Radiation Research 112 (1987): 45-53.

71. Rockwell S, Use of a perfluorochemical emulsion to improve oxygenation in a solid tumor, Int J Radiation Oncology Biol Phys 11 (1985): 97-103.

72. Rockwell S, Irvin CG, and Kelley M, Preclinical studies of a perfluorochemical emulsion as an adjunct to radiotherapy, Int J Radiation Oncology Biol Phys 15 (1988): 913-920.

73. Teicher BA, and Rose CM, Oxygen-carrying perfluorochemical emulsion as a adjuvant to radiation therapy in mice, Cancer Research 44 (1984): 4285-4288.

74. Teicher BA, Herman TS, and Rose CM, Effect of Fluosol-DA on the response of intracranial 9L tumors to x rays and BCNU, Int J Radiation Oncology Biol Phys 15 (1988): 1187-1192.

75. Rosenblum WI, Hadfield F, Martinez AJ, and Schatzki P, Alterations of liver and spleen following intravenous infusion of fluorocarbon emulsions, Arch Pathol Lab Med 100 (1976): 213-217.

76. Chen HS, and Yang ZH, Abstract: Perfluorocarbon as blood substitute in clinical applications and in war casualties, Biomater Artif Cells Artif Org 16.1-3 (1988): 403-409.

77. Lowe KC, Perfluorochemical: Blood Substitutes and Beyond, Adv Mater 3.2 (1991): 87-93.

78. Arlen C, Follana R, Le Blanc M, Long C, Long D, Riess JG, and Valla A, Formulation of highly concentrated fluorocarbon emulsions and assessment by near-total exchange perfusion of the conscious rat, Biomater Artif Cells Artif Org 16.1-3 (1988): 455-457.

79. Hammerschmidt DE, and Vercellotti GM, Limitation of complement activation by perfluorocarbon emulsions: Superiority of lecithin-emulsified preparations, Biomater Artif Cells Artif Org 16.1-3 (1988): 431-438.

80. Long DM, Long DC, Mattrey RF, Long RA, Burgan AR, Herrick WC, and Shellhamer DF, An overview of perfluoroctylbromide - Application as a synthetic oxcygen carrier and imaging agent for X-ray, ultrasound, and nuclear magnetic resonance, Biomater Artif Cells Artif Organs 16.1-3 (1988): 411-420.

81. Tremper KK, Lapin R, Levine E, Friedman A, and Shoemaker WC, Hemodynamic and oxygen transport effects of a perfluorochemical blood substitute, Fluosol-DA (20%), Crit Care Med 8.12 (1980): 738-741.

82. Sargent JW, and Sefel RJ, Properties of perfluorinated liquids, Fed Proceed 1970 29.5 (1970): 1699-1703.

83. MacNicol DD, and Robertson CD, New and unexpected reactivity of saturated fluorocarbons, Nature 332 (1988): 59-61.

84. Ackerman NB, and Hechmer PA, Effects of pharmacological agents on the microcirculation of tumors implanted in the liver, 9th European Conference on Microcirculation, (Antwerp: 1976) 301-303.

85. Riess JG, Blood substitutes: where do we stand with the fluorocarbon approach?, Current Surgery 45.5 (1988): 365-370.

86. Naito R, and Yokoyama K, Perfluorochemical Blood Substitutes Fluosol-43, Fluosol-DA, 20%, and 35% for preclinical studies as a candidate for erythrocyte substitution, Green Cross Corporation, 1978).

87. Biro GP, Blais P, and Rosen A, Perfluorocarbon Blood Substitutes, CRC Critical Reviews in Oncology/Hematology 6.4 (1987): 311-374.

88. Lowe KC, Perfluorocarbons as oxygen-transport fluids, Comp Biochem Physiol 87/a.4 (1987): 825-838.

89. Faithfull NS, Fluorocarbons, Anaesthesia 42 (1978): 234-242.

90. Pillai R, Bando K, Schueler S, Zebley M, Reitz BA, and Baumgartner WA, Leukocyte depletion results in excellent heart-lung function after 12 hours of storage, Ann Thorac Surg 50 (1990): 211-214.

91. Yokoyama K, Yamanouchi K, and Murashima R, Excretion of perfluorochemicals after intravenous injection of their emulsion, Chem Pharm Bull 23.6 (1975): 1368-1373.

92. Oyama T, Matsuki A, Wakayama S, Tanioka F, Kudo T, and Noguchi T, "Effects of Fluosol-DA 20% infusion on circulatory and endocrine function in surgical patients," Oxygen Carrying Colloidal Blood Substitutes (Mainz, march 1981), Ed. H. B. a. K. S. R Frey (Munchen : Zuckschwerdt, 1982) 187-192.

93. Honda K, Hoshino S, Shoji M, Usuba A, Motoki R, Tsuboi M, Inoue H, and Iwaya F, Letter to Editor: Clinical use of a blood substitute, New England J Med 303.7 (1980): 391-392.

94. Mitsuno T, Ohyanagi H, and Naito R, Clinical studies of a perfluorochemical whole blood substitute (Fluosol-DA), Ann Surg 195.1 (1982): 60-69.

95. Hirlinger WK, Grunert A, Herrmann M, Petutschnigk D, and Langer K, Auswirkungen eines teilweisen Blutaustausches mit Fluosol DA 20% auf den intakten Organismus des Schweines, Anaesthesist 31 (1982): 660-666.

96. Bruneton JN, Falewee MN, Francois E, Cambon P, Philip C, Riess JG, Balu-Maestro C, and Rogopoulos A, Liver, spleen, and vessels: Preliminary clinical results of CT with perfluoroctylbromide, Radiology 170 (1989): 179-183.

97. Mitsuno T, Ohyanagi H, and Yokoyama K, Development of a perfluorochemical emulsion as a blood gas carrier, Artif Organs 8.1 (1983): 25-33.

98. Biro GP, Effect of hemodilution with Dextran, Stroma-Free Hemoglobin Solution and Fluosol-DA on experimental myocardial ischemia in the dog, Biblthca haemat 47 (1981): 54-69.

99. Kloner RA, and Glogar DH, "Overview of the used of perfluorochemical for myocardial ischemic rescue," Perfluorochemical Oxygen Transport, Ed. Tremper KK (Boston: Little, Brown & Co., 1985) 23: 115-130.

100. Faithfull NS, Fennema M, Erdmann W, Lapin r, Smith AR, Van Alphen W, Essed CE, and Trouwborst A, Tissue oxygenation by fluorocarbons, Adv Exp Med Biol 180 (1984): 569-580.

101. Schmid-Schonbein H, Weiss J, and Ludwig H, A simple method for measuring red cell deformability in models of the microcirculation, Blut XXVI (1973): 369-379.

102. Barnikol WKR, and Burkhard O, Die abhangigkeit der erythrozyten-deformierbarkeit von der osmolaritat, dem pH-Wert, der temperatur und der proteinkonzentration, Funkt Biol Med 4 (1985): 55-60.

103. Virmani R, Warren D, Rees R, Fink LM, and English D, Effects of perfluorochemical on phagocytic function of leukocytes, Transfusion 23 (1983): 512-515.

104. Virmani R, Osmialowski AF, Kolodgie FD, and Forman MB, The effect of perfluorochemical Fluosol-DA (20%) on myocardial infarct healing in the rabbit, Am J Cardio Path 3.1 (1990): 69-80.

105. Kolodgie FD, Dawson AK, Forman MB, and Viramani R, Effect of perfluorochemical (Fluosol-DA) on infarct morphology in dogs with permanent coronary artery occlusion, Virchow Arch B 50 (1985): 119-134.

106. Parrish MD, Olson RD, Mushlin PS, Artman M, and Boucek RJ, Treatment of postischemic reperfusion cardiac injury with a perfluorochemical solution, J Cardiovasc Pharmacol 6.1 (1984): 159-164.

107. Rude RE, Glogar DH, Khure S, Karaffa S, Kloner RA, Clark LC, Muller Je, and Braunwald E, (Abstract) Beneficial effects of fluorocarbons (synthetic oxygen-carrying compounds) on intramyocardial pO_2 during acute myocardial ischemia, Clinical Research 28 (1980): 617A.

108. Biro GP, Fluorocarbon and Dextran hemodilution in myocardial ischemia, Canadian J Surgery 26.2 (1983): 163-168.

109. Faithfull NS, Fennema M, Erdmann W, Dhasmana MK, and Eilers G, Effects of acute ischaemia on intramyocardial oxygen tensions, Adv Exp Med Biol 200 (1986): 339-348.

110. Faithfull NS, Erdmann W, Fennema M, and Kok A, Effects of haemodilution with fluorocarbons or dextran on oxygen tensions in the acutely ischaemic myocardium, Br J Anaesth 58.9 (1986): 1031-1040.

111. Rahamatbulla P, Watanabe K, Ashraft M, and Millard RW, Myocardial function and morphology in rat hearts perfused with Krebs-Henseleit solution and perfluorocarbon emulsion, _Physiologist_ (1984).

112. Nunn GR, Dance G, Peters J, and Cohn LH, Effect of fluorocarbon exchange transfusion on myocardial infarction size in dogs, _Am J Cardiology_ 52 (1983): 203-205.

113. Glogar DH, Kloner RA, Muller J, DeBoer WV, Braunwald E, and Clark LC, Fluorocarbons reduce myocardial ischemic damage after coronary occlusion, _Science_ 211 (1981): 1439-1441.

114. Peerless SJ, Ishikawa R, Hunter IG, and Peerless MJ, Protective effect of Fluosol-DA in acute cerebral ischemia, _Stroke_ 12.5 (1981): 558-563.

115. Peerless SJ, Nakamura R, Rodriquez-Salazar A, and Hunter IG, Modification of cerebral ischemia with fluosol, _Stroke_ 16.1 (1985): 38-43.

116. Mizoi K, Yoshimoto T, and Suzuki J, Experimental study of new cerebral protective substances - functional recovery of severe, incomplete ischaemic brain lesions pretreated with mannitol and fluorocarbon emulsion, _Acta Neurochirurgica_ 56 (1981): 157-166.

117. Kagawa S, Koshu K, Yoshimoto T, and Suzuki J, The protective effect of mannitol and perfluorochemicals on hemorrhagic infarction: an experimental study, _Surgical Neurology_ 17.1 (1982): 66-70.

118. Osterholm JL, Alderman JB, Triolo AJ, D'Amore BR, Williams HD, and Frazer G, Severe cerebral ischemia treatment by ventriculosubarachnoid perfusion with an oxygenated fluorocarbon emulsion, _Neurosurgery_ 13.4 (1983): 381-387.

119. Bose B, Osterholm J, Payne JB, and Chambers K, Preservation of neuronal function during prolonged focal cerebral ischemia by ventriculocisternal perfusion with oxygenated fluorocarbon emulsion, _Neurosurgery_ 18.3 (1986): 270-276.

120. Osterholm JL, Alerman JB, Triolo AJ, D'Amore BR, and Williams H, Oxygenated fluorocarbon nutrient solution in the treatment of experimental spinal cord injury, _Neurosurgery_ 15.3 (1984): 373-380.

121. Geyer RP, Potential uses of artificial blood substitutes, _Fed Proceed_ 34.6 (1975): 1525-1528.

122. Long DM, Higgins CB, Mattrey RF, Mitten RM, and Multer FK, "Is there a time and place for radiopaque fluorocarbons?," _Preparation, Properties, and Industrial Applications of Organofluorine Compounds_, Ed. RE Banks (New York, Brisbane, Chichester, Toronto: Ellis Horwood Limited, 1982) 139-156.

123. Ricci JL, Sloviter HA, and Ziegler MM, Intestinal ischemia: Reduction of mortality utilizing intraluminal perfluorochemical, _Am J Surgery_ 149 (1985): 84-90.

124. Liu MS, and Long DM, Biological disposition of perfluoroctylbromide: Tracheal administration in alveolography and bronchography, _Investigative Radiology_ 11 (1976): 479-485.

125. Mattrey RF, Trambert MA, Brown J, Bruneton JN, Young S, Kortman K, Wesby G, and Schooley G, Summary of phase III clinical trials of imagent GI (PFOB) as a negative oral MR contrast agent, _RSNA_, (Chicago, IL: 1990).

126. Delpuech J-J, Hamza MA, and Serratrice G, Determination of oxygen by a nuclear magnetic resonance method, Journal of Magnetic Resonance 36 (1979): 173-179.

127. Reid RS, Koch CJ, Castro ME, Lunt JA, Treiber EO, Boisvert DJP, and Allen PS, The influence of oxygenation of the 19F spin-lattice relaxation rates of Fluosol-DA, Phys Med Biol 30.7 (1985): 677-686.

128. Authier B, Reactive hyperemia monitored on rat muscle using perflurorocarbons and 19F NMR, Magnetic Resonance in Medicine 8 (1988): 80-83.

129. Ceckler TL, Gibson SL, Hilf R, and Bryant R, In situ assessment of tumor vascularity using fluorine NMR imaging, Magnetic Resonance in Medicine 13 (1990): 416-433.

130. Patronas N, Miller DI, and Girton M, Experimental comparison of EOE-13 and perfluoroctylbromide for the CT detection of hepatic metastases, Invest Radiol 19 (1984): 570-573.

131. Mattrey RF, Scheible FW, Gosink BB, Leopold GR, Long DM, and Higgins CB, Perfluoroctylbromide: A liver/spleen-specific and tumor-imaging ultrasound contrast material, Radiology 145 (1982): 759-762.

132. Mattrey RF, and Andre MP, Ultrasonic enhancement of myocardial infarction with perfluorocarbon compounds in dogs, Am J Cardiol 54 (1984): 206-210.

133. Coley BD, Mattrey RF, Roberts A, and Keane S, Potential role of PFOB enhanced sonography of the kidney. II. Detection of partial infarction., Kidney International 39 (1991): 740-745.

134. Munzing D, Mattrey RF, Reznik VM, Mitten RM, and Peterson T, Potential role of PFOB enhanced sonography of the kidney. I. Detection of renal function and dacute tubular necrosis., Kidney Inter 39 (1991): 733-739.

135. Wolf GL, Long D, and Reiss J, (Abstract) Percutaneous lymphography with PFOB emulsions, RSNA, (Chicago: 1990).

136. Berkowitz HD, McCombs P, Sheety S, Miller LD, and Sloviter H, Fluorochemical perfusates for renal preservation, J Surg Res 20 (1976): 595-600.

137. Sloviter HA, and Kamimoto T, Erythrocyte substitute for perfusion of brain, Nature 216 (1967): 458-460.

138. Toyohira H, Taira A, Arikawa K, Hamada Y, Ohzono H, and Akita H, Isolated heart perfusion with FC-43: an experimental study, Proceedings of the IVth International Symposium Perfluorochemical Blood Substitutes, (Kyoto: 1978) 161-172.

139. Novakova V, Birke G, Plantin LO, and Wretlind A, Studies on isolated rat liver perfused by perfluoro-compound emulsion, Fed Proceed 34.6 (1975): 1488-1492.

140. Skibba JL, Sonsalla J, Petroff RJ, and Denor P, Canine liver isolation-perfusion at normo- and hyperthermic temperatures with perfluorochemical emulsion (Fluosol-43), Eur Surg Res 17 (1985): 301-309.

141. Tauber A, Wendt P, Mittlmeier T, Stamatopoulos C, Besibarth H, Maurer P, and Blumel G, "Initial perfusion of extremities with Fluosol-43 prior to replantation-metabolic and hemodynamic investigations in rabbits," Oxygen Carrying Colloidal Blood Substitutes (Mainz, March 1981), Ed. H. B. a. K. S. R Frey (Munchen: Zuckschwerdt, 1982) 245-254.

142. Schindler H-G, Pennig D, Schlake W, Schonleben K, and Brug E, "Preliminarty results: The use of Fluosol-43 for intermediary perfusion in amputated limbs," <u>Oxygen Carrying Colloidal Blood Substitutes</u>, Ed. H. B. a. K. S. R Frey (Munchen: Zuckschwerdt, 1981) 255-260.

143. Honda K, Motoki R, Hoshino S, Inoue H, Usuba A, Hamada O, Iwaya F, and Ando M, Use of perfluorochemical artificial blood (Fluosol-DA) for perfusion of cadaveric kidneys for transplantation, <u>Current Therapeutic Research</u> 28.3 (1980): 309-318.

144. Fuchinoue S, Takahashi K, Teraoka S, Toma H, Ashishi T, and Ota K, Clinical experience in kidney preservation with a new fluorocarbon emulsion perfusate, <u>Transplantation Proceed</u> XVIII.3 (1986): 566-570.

145. Greenspan JS, Wolfson MR, Rubenstein D, and Shaffer TH, Liquid ventilation of human preterm neonates, <u>J Pediatr</u> 117 (1990): 106-111.

AMBIANT PRESSURE CHANGES AND OXYGENATION

ADAPTATION OF O_2 TRANSPORT AND UTILIZATION AT ALTITUDE IN MAN

Peter D. Wagner

Department of Medicine, 0623, University of California, San Diego
9500 Gilman Drive
La Jolla, CA 92093-0623, U.S.A.

INTRODUCTION

The mammalian O_2 transport system is remarkable in at least two major respects. First, several organs must work together in an integrated manner to achieve maximal rates of transfer of O_2 from inspired air to muscle mitochondria during exercise. Too little or too much functional ability of one organ compared to another produces inefficiency. For example, a high cardiac output clearly enhances the bulk transport of O_2 between lungs and muscle, but if too high, both diffusive O_2 loading in the lung and diffusive O_2 unloading in the tissues are compromised by rapid red cell transit times. Thus, the pulmonary system (using convective and diffusive modes of O_2 transport) is linked to the cardiovascular system, which in turn delivers O_2 to the muscle mitochondria at rates dependent on blood transport ability (Hb concentration, shape and position of the O_2 Hb dissociation curve) and muscle O_2 transport capacity by diffusion between the red cell and the mitochondrion. The second remarkable aspect of O_2 transport is that at maximal exercise, virtually all of the O_2 that can be transferred into arterial blood is able to be extracted and used by the exercising muscles. In trained normal subjects, O_2 extraction can exceed 90%.

Exposure to the hypoxia of altitude has substantial effects on the O_2 transport system, affecting essentially every step in the O_2 pathway. This short review will highlight the ways in which altitude produces these changes in O_2 transport, to the extent known, and in the setting of maximal or near maximal exercise which is where the system is obviously most stressed. The focus is on short-term effort lasting only a few minutes, not on prolonged lower intensity exercise.

THE PATHWAY FOR O_2

The important steps in the pathway for O_2 are well-established (Weibel, 1984), mostly non-controversial, and require little elaboration in this review. Briefly, the first step is inspiration of ambient air, the PO_2 of which is imposed upon the subject beyond his control. Convective ventilation, accomplished by inspiratory muscle contraction, delivers this O_2 to the alveolar region. Mixing of inspired gas with resident alveolar gas occurs very rapidly by diffusion interacting (Piiper and Scheid, 1987) with convection, such that inefficiency of O_2 transport due to impaired gas phase mixing is considered to be small (Hlastala et al., 1981) under normal circumstances (even in exercise when the time available for mixing is reduced by the approximately fourfold increase in respiratory rate from rest). However, even in the normal lung there is some degree of ventilation-perfusion inequality (West, 1988), such that alveolar PO_2 is not everywhere the same. This contributes a small amount to inefficiency in O_2 transport, as will be discussed. The next step is diffusion of O_2 across the alveolar membrane into pulmonary capillary blood, a process which occurs to completion at sea level except during extremely heavy exercise ($VO_2 > 3-4$ L/min) in trained subjects (Wagner et al., 1986). The cardiac pump then distributes all of the O_2 loaded into the blood to the organs in need of O_2, mostly the muscles during heavy exercise, wherein O_2 leaves the capillary red cell by dissociating from Hb, diffusing through the red cell and plasma space to and through the capillary wall and interstitial space to the myocyte cell membrane. Thereafter, diffusion to the mitochondria continues, facilitated both by the presence of myoglobin in the cytoplasm and the propensity for mitochondria to cluster nearer the myocyte cell wall than at the center of the cell (Hoppeler et al., 1981).

RECENT DEVELOPMENTS IN THE UNDERSTANDING OF O_2 TRANSPORT, WITH SPECIAL REFERENCE TO ALTITUDE

A. The Lungs

Of the several organs (lungs, heart, blood, blood vessels and muscles) necessary for O_2 transport, the notion is emerging that the lungs are the weakest link in the chain, especially in well-trained human (Dempsey, 1986) or animal (Wagner et al., 1989) athletes, and even more so at altitude. This is attested to by several pieces of evidence:

1. The alveolar arterial PO_2 difference uniformly increases (Fig. 1) with increasing exercise, such that in elite athletes, values of 40 torr or more are not uncommon (Dempsey et al., 1984). When arterial PO_2 remains on the flat part of the O_2 Hb dissociation curve as happens at sea level, the consequences for O_2 transport are minor. However, at altitude when resting arterial PO_2 is already on the steep portion of the O_2 Hb dissociation curve, even a small decrease in PO_2 can have a considerable effect on systemic O_2 delivery. Thus, the first consequence of altitude exposure to O_2 transport during exercise is due to the reduced PIO_2 and persistence of gas exchange defects causing a high alveolar-arterial PO_2 difference (A-a PO_2). While the well-known adaptive response of hyperventilation at altitude is very helpful in preserving alveolar PO_2 values, this does not prevent the fall in arterial PO_2 during exercise that is reflected in the increased A-a PO_2. What is the physiological basis of exercise-induced gas exchange defect? There is no doubt that alveolar-capillary diffusion limitation is a major reason for the high A-a PO_2 under these conditions (Wagner et al., 1987). As pointed out many years ago (Wagner, 1977), gas exchange occurring on the steep portion of the O_2 Hb dissociation curve stresses the ability of the lungs to transfer O_2 into the blood by diffusion. Such diffusion limitation appears closely linked to total pulmonary blood flow

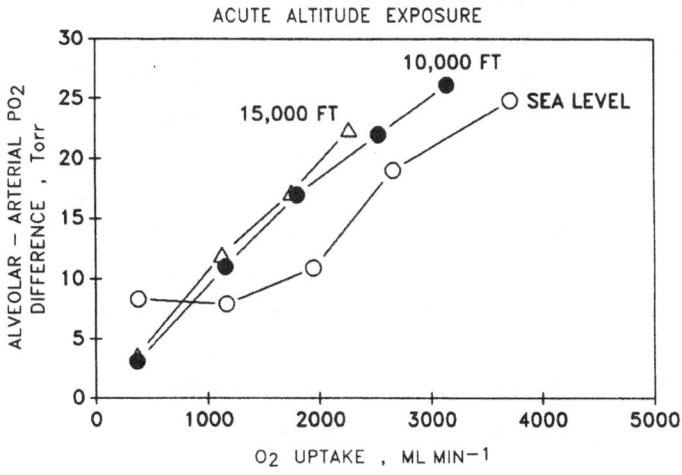

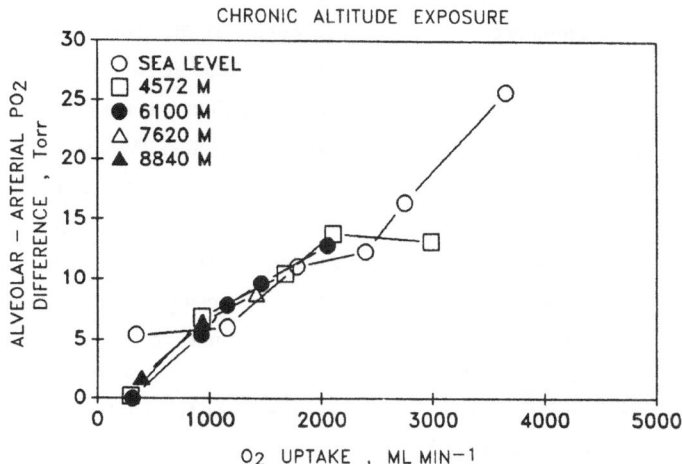

Figure 1. Increase in alveolar-arterial PO_2 difference with increasing exercise at sea level and altitude. Top panel shows data with acute altitude exposure, bottom panel with chronic altitude exposure. Note that A-a PO_2 is higher with acute than than chronic hypoxia compared to sea level values.

rates, being a greater problem in acute than chronic hypoxia (at the same PIO_2 and work rate) because of the higher cardiac output in acute hypoxia (Bebout et al., 1989). Diffusion limitation is difficult to separate from other causes of gas exchange defect (shunting and ventilation/blood flow mismatch in particular) but this can be done with the multiple inert gas elimination technique (Wagner et al., 1974; Evans and Wagner, 1977; Wagner and West, 1980). This method directly determines that portion of the A-a PO_2 due to shunt and $\dot{V}_A/\dot{Q}$ mismatch, any residual being due to diffusion limitation. It should be pointed out that from measurement of morphometric pulmonary diffusing capacity, Karas et al. (1987) believe that there is sufficient diffusive capability that diffusion limitation should not be seen, at least at sea level. Their analysis does not take into account the indisputable fact that not all red cells will traverse the gas exchange pathway across alveoli in the same time. Those cells undergoing more rapid transit than average may well experience diffusion limitation. Furthermore, it is not established that morphological estimates of diffusing capacity represent functional diffusing capacity during life -- they may overestimate this variable if not all of the alveolar structure is utilized for gas exchange, or if the alveolar membrane is transiently thicker during exercise (than at rest) due to the high flux of water out of the capillaries (Coates et al., 1984) into the interstitial space. Data from the MIGET however reflect all of these possibilities by determining the net overall effect of exercise on O_2 transport.

In addition to diffusion limitation, the lungs experience an increase in $\dot{V}_A/\dot{Q}$ inequality in about 50% of subjects during heavy exercise (Wagner et al., 1986). This occurs despite the well-known improvement in topographical $\dot{V}_A/\dot{Q}$ distribution (Bake et al., 1968) due to increased pulmonary artery pressure. Such $\dot{V}_A/\dot{Q}$ inequality also contributes to the increase in A-a DO_2 on exercise but the effects are not as marked as for diffusion limitation (Wagner et al., 1987). There is good evidence that worsening $\dot{V}_A/\dot{Q}$ relationships: a) are exacerbated at altitude, either with acute or chronic hypoxia (Wagner et al., 1986a; Wagner et al., 1987), b) relate quantitatively to the degree to which pulmonary artery pressure is raised by combination of altitude and exercise level (Wagner et al., 1986a), and c) persist for some 20 minutes after cessation of exercise (Schaffartzik et al., 1990b). In trained normal pigs, exercise produces perivascular edema (Schaffartzik et al., 1990a) and in some species there is even evidence of loss of pulmonary microvascular integrity on maximum effort (Pascoe et al., 1981). All of these observations are consistent if circumstantial pointers to accumulation of extravascular fluid as the likely basis for the increased $\dot{V}_A/\dot{Q}$ mismatch of exercise, but this hypothesis remains to be directly evaluated in man. Alternate hypotheses related to the airway side of the alveolar membrane are less tenable. There is no evidence of bronchoconstriction under these circumstances (Wagner et al., 1986b) or impaired airway gas mixing efficiency (Wagner et al., 1986a) (although the reduced dead space/tidal volume ratio could reveal more $\dot{V}_A/\dot{Q}$ mismatch than seen at rest (Tsukimoto et al., 1990)). It should be pointed out that at extreme altitude (and thus with chronic, not acute, altitude exposure: above 20,000 feet or 6000 meters), occasionally dramatic degrees of $\dot{V}_A/\dot{Q}$ mismatch are seen on exercise (Wagner et al., 1987). Under these conditions, there may be clinical signs of acute pulmonary edema and one is probably seeing the development of H.A.P.E. Even under such extreme conditions, however, the contribution of $\dot{V}_A/\dot{Q}$ mismatch to O_2 transport reduction is much less than that due to O_2 diffusion limitation as presented above (fig. 2).

2. <u>Pulmonary hypertension develops</u>, the more so the higher the altitude and the higher the level of exercise. Clearly, at altitude, the develoment of hypoxic vasoconstriction acts in concert with the metabolic demand for high cardiac output to raise pulmonary artery pressures. In acute hypoxia at equivalent altitudes of 15,000 feet (P_B 430 torr), pulmonary artery mean pressure regularly reaches 40 torr. In normal subjects exercising at the equivalent of the Mt. Everest summit, these pressures reach 50-60 torr (Groves et al., 1987). Even more remarkable is the observation of near systemic pulmonary artery mean pressures in the heavily exercising horse (at sea level) of 80-100 torr (Erickson et al., 1990). It is not known what would happen to the horse at altitude! These high vascular pressures, which may also be contributed to by high left ventricular filling pressures (Wagner et al., 1986a; Reeves et al., 1987) may contribute to H.A.P.E. through Starling's pressure balance equation and even through capillary disruption (West et al., 1991). In any event, the tripling of pulmonary artery pressures between rest and maximal exercise far exceeds the minimal relative increase in mean systemic pressures under the same conditions, and indicates that the pulmonary vascular bed is marginally sufficient to accommodate the high flows of maximum exercise. These conclusions, based only on hemodynamic considerations, fit with similar conclusions based on gas exchange data: the gas exchange defects seen -- diffusion limitation, and possibly interstitial edema causing $\dot{V}_A/\dot{Q}$ mismatch -- probably reflect inadequate circulatory structure for the level of function required.

3. <u>CO_2 retention may develop</u> during exercise. While the absolute value of PCO_2 in arterial blood does not rise (from resting levels) during exercise at altitude, there is some evidence that in elite athletes both human (Dempsey, 1986) and equine (Wagner et al., 1989), $PaCO_2$ does not fall in the manner seen for less elite athletes, at least at sea level. This may represent the reaching of maximum sustainable ventilation in such individuals. A real increase in $PaCO_2$ is frequently seen in patients with obstructive airways disease during exercise where it clearly represents ventilatory limitation. That normal lungs are close to their ventilatory limit during maximal exercise is further suggested by the observation that respiratory flow-volume loops reach the maximum flow-volume envelope during heavy exercise (Leaver and Pride, 1971) during a portion of expiration.

In summary, there is now a wealth of evidence that the lungs sustain temporary functional defects during very heavy exercise, mostly related to the pulmonary circulation. Those related to gas exchange and to pulmonary hemodynamics are clearly even more problematic at altitude, whether this be either acutely or chronically imposed. Most of these defects have little consequence for O_2 transport at sea level, but this is definitely not so at altitude where frank pulmonary edema may occur. Even if this does not happen, significant reduction in systemic O_2 transport occurs from diffusion limitation of O_2 transfer in the lungs.

B. The Heart

In contrast to the lungs, current evidence suggests that cardiac function is well-preserved during maximal exercise at sea level (and altitude). It is of interest that while there is evidence of limited pulmonary function (discussed above) which impacts little on O_2 transport at sea level, the

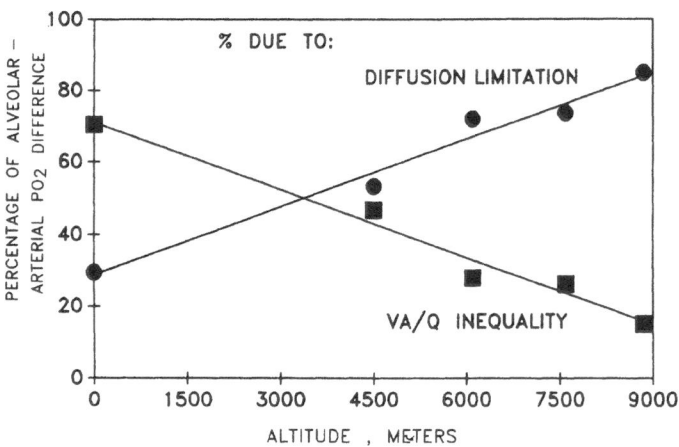

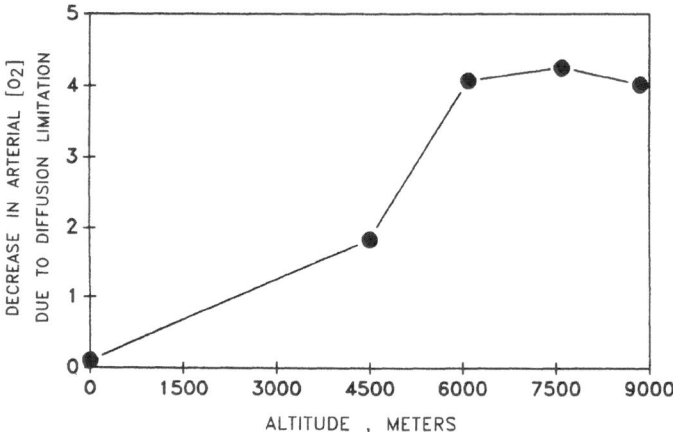

Figure 2. Upper panel: Percentage contribution of ventilation-perfusion inequality and diffusion limitation to the alveolar-arterial PO_2 difference. With increasing altitude, diffusion limitation becomes far more important than ventilation-perfusion inequality. Lower panel: Effect of incomplete alveolar-capillary diffusion equilibration on arterial oxygen concentration, showing the marked effect of altitude.

general consensus is that cardiac function is a major component of the limits to maximum O_2 transport (yet one sees normal function of the heart even at maximum exercise). What appears to be happening is that cardiac output reaches its maximum values because heart rate is so high that diastolic filling reaches its limit. Thus, further increases in heart rate would necessarily be accompanied by reduction in stroke volume. But, in terms of classical functional indices, there is no decrement of myocardial function *per se*. Thus, at $\dot{V}O_{2max}$: 1) normal subjects do not experience anginal chest pain, 2) EKG does not show ischemic changes, 3) cardiac filling pressures are not inappropriately high (Reeves et al., 1987) for the cardiac output mounted, 4) cardiac output remains essentially proportional to work load and $\dot{V}O_2$ all the way to $\dot{V}O_{2max}$, and 5) ventricular ejection remains normal by Echocardiographical assessment. Occasional reports of reduced contractility have been published following ascent to altitude (Tucker et al., 1976), but in the situation of short-term high intensity exercise addressed by the current review, this does not appear to be the case.

Against this backdrop of functional response of the heart to exercise at sea level, the effects of altitude on cardiac function are of considerable interest and pose perplexing, unsolved questions. The effects of acute hypoxia are entirely different from, and in some respect opposite to, those of chronic hypoxia. When one plots cardiac output against either work rate or VO_2 in subjects exercised in acute hypoxia, one finds a higher cardiac output at any given submaximal work rate (Fig. 3) (Wagner et al., 1986a). When one dissects cardiac output into its two components, heart rate and stroke volume, one sees that the higher submaximal cardiac output is achieved by higher submaximal heart rates as well as greater stroke volumes. This is consistent with the widely held notion that hypoxia acutely stimulates sympathetic activity as reflected by higher circulating catecholamine levels than at the same work load at sea level. Maximum cardiac output at altitude, however, equals that at sea level. The same plots obtained at altitude after acclimatization show a different response: the cardiac output-$\dot{V}O_2$ relationship overlies that at sea level, but does not extend to as high levels (Fig. 4). Thus, submaximal cardiac output is no different at altitude than at sea level at the same $\dot{V}O_2$, but maximal cardiac output is lower than at sea level in precise proportion to the lower maximal $\dot{V}O_2$. Figure 4 also shows that the reduced maximal cardiac output at altitude is due to a lower maximal heart rate than at sea level. We still do not know the reason for this fundamental difference between acute and chronic hypoxia in cardiac output. Logically, it does not appear to be attributable to greater hypoxia in chronic altitude exposure since the above effects are seen at constant altitude when acute and chronic altitude measurements are compared (Bebout et al., 1989). It is also not attributable to decrement of intrinsic myocardial function (Reeves et al., 1987) since the classical physiological indices and markers of function remain normal (Figs. 5,6). It appears not to be due to dehydration causing loss of vascular volume based on general markers of hydration and blood volume. The probably significant clue is the associated lower maximal heart rate, suggesting that sympathetic down regulation may occur with chronic hypoxia. This postulate has not been verified nor excluded at this time, and how this fits with the higher submaximal heart rates at a given $\dot{V}O_2$ (Fig. 4) is not clear.

What is clear from the behavior of cardiac output in acute hypoxia is that maximal O_2 transport is not reduced because of cardiac pump abnormalities, quite the reverse since cardiac output is higher than (or similar to) sea level values. However, a negative consequence of the higher submaximal cardiac output in acute hypoxia is the more rapid pulmonary red cell capillary transit, resulting in more diffusion limitation and thus larger A-a PO_2 than would otherwise be the case.

In chronic hypoxia, the lower maximal cardiac output must be a major factor reducing systemic O_2 transport at $\dot{V}O_{2max}$. However, one positive consequence of the lower maixmal cardiac output is the longer time red cells spend in both pulmonary and muscle capillaries which can only benefit the diffusive transfer of O_2 between the environment and the mitochondria.

C. The Blood

In man, short-term exercise at sea level produces little change in the blood during the few minutes of activity. Unlike the horse (Persson, 1967), the spleen does not act as a red cell reservoir to be emptied into the circulation during exercise. There is some movement of fluid out of the blood vessels, such that [Hb] rises slightly, by about 1 gm/deciliter between rest and maximum effort. Consequently, virtually all of the increase in O_2 delivery to the muscles during exercise must be accomplished by the associated increase in muscle blood flow, since arterial $[O_2]$ is minimally different from rest.

The horse behaves entirely differently than man. By virtue of splenic contraction that is sympathetically controlled and closely related to the level of exercise (Carlson, 1986), circulating [Hb] essentially doubles from rest to exercise. Thus, the normal resting situation of [Hb] ~ 10 gm/deciliter and hematocrit ~ 30% is transformed into one of [Hb] ~ 18-20 gm/deciliter and hematocrit ~ 55-60% at full gallop (Fig. 7). The time course for this is remarkably rapid -- [Hb] is doubled in 90 seconds.

While the horse's enormous increase in [Hb] must greatly increase systemic O_2 delivery, the increase in blood viscosity creates a greater rise in pulmonary artery pressure than if the Hb response were ablated by prior splenectomy (Davis and Manohar, 1988). It is not yet known whether maximal cardiac output is affected by the increase in [Hb], nor what the implications are for diffusive loading of O_2 in the lungs or unloading of O_2 at the tissues. This must await controlled studies before and after splenectomy.

Exercise during acute exposure to altitude has no apparent additional effects on [Hb] than does exercise at sea level, but chronic altitude exposure has the very well-known effect of increasing [Hb] via stimulation of the marrow by erythropoietin released from the kidney in response to hypoxia (Fisher and Langston, 1967). The amount by which [Hb] increases is clearly dependent on two major factors -- the altitude level itself and the time spent at altitude, as nicely summarized in (Ward, 1989). The increase in [Hb] is also contributed to by a reduction in plasma volume that occurs rapidly on ascent and reflect a diuresis and natiuresis (Myhre et al., 1970; Pugh, 1964). Environmental factors affecting state of hydration and thus peripheral vascular tone combined with fluid intake and exertional activity are additional modulators of fluid volume -- a complex situation

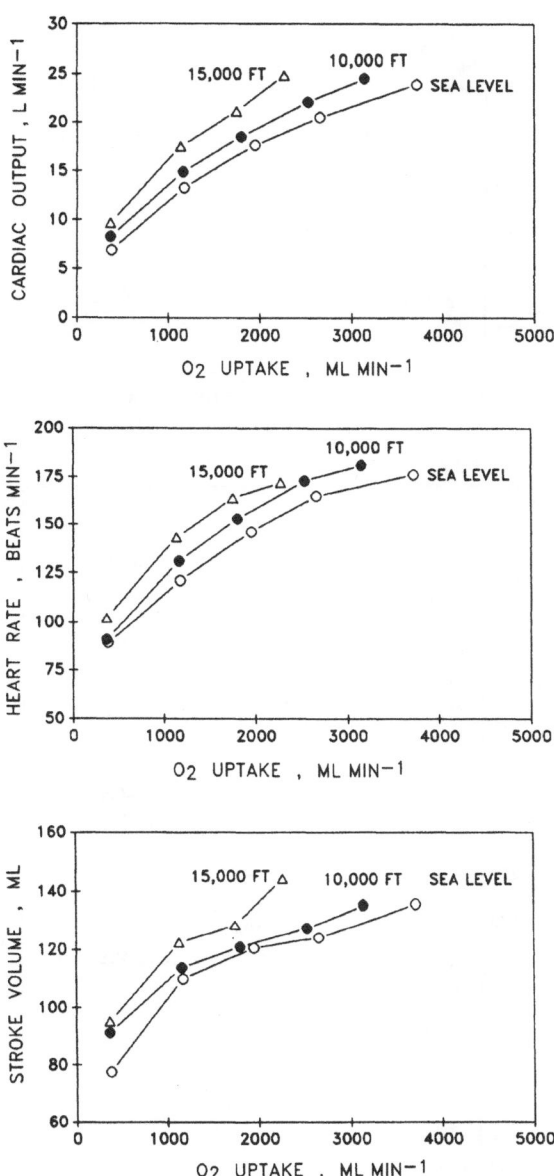

Figure 3. Cardiac output, heart rate and stroke volume as functions of oxygen uptake during exercise at sea level and altitude following acute exposure. All variables are higher at a given submaximal $\dot{V}O_2$ in hypoxia, but maximal values are the same at all altitudes.

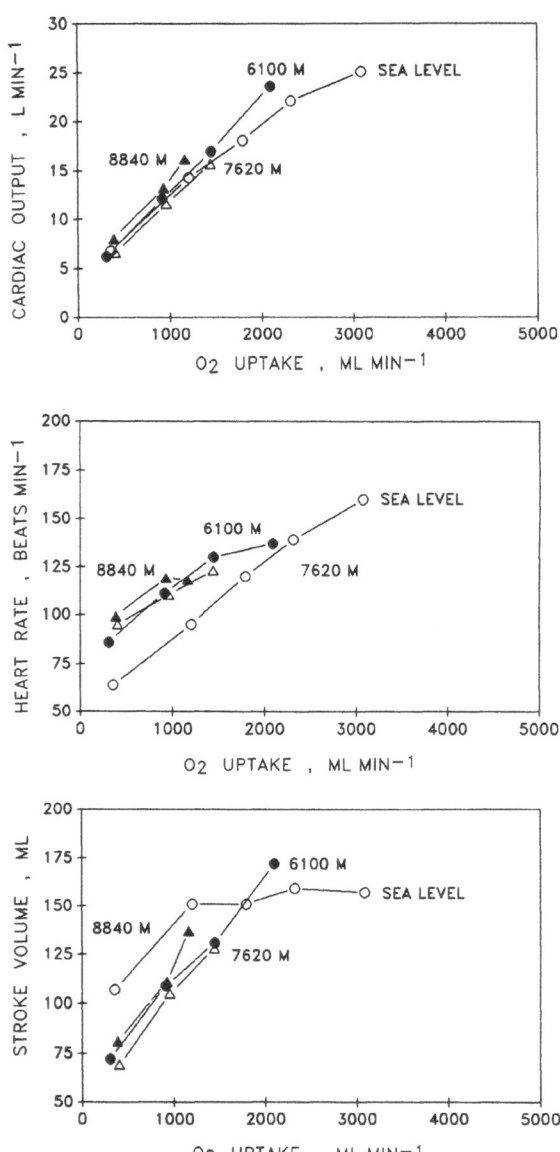

Figure 4. Cardiac output, heart rate and stroke volume as functions of oxygen uptake during exercise after chronic altitude acclimatization. Maximal cardiac output is reduced with altitude. However, the cardiac output-$\dot{V}O_2$ relationship is similar to that at sea level. This is produced by a somewhat higher submaximal heart rate and lower stroke volume. The lower maximal cardiac output is due primarily to the lower maximal heart rate.

ACUTE ALTITUDE EXPOSURE

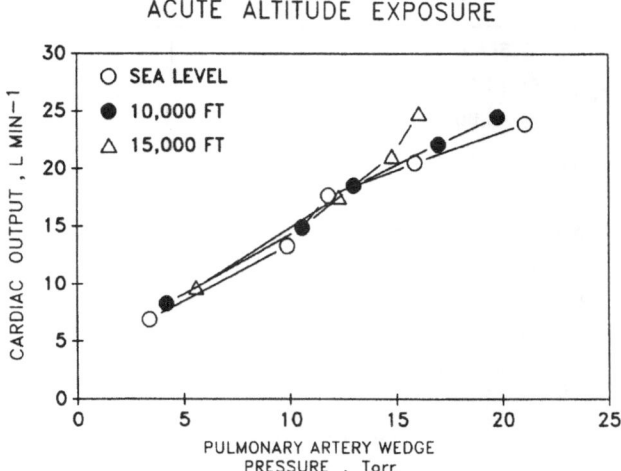

CHRONIC ALTITUDE EXPOSURE

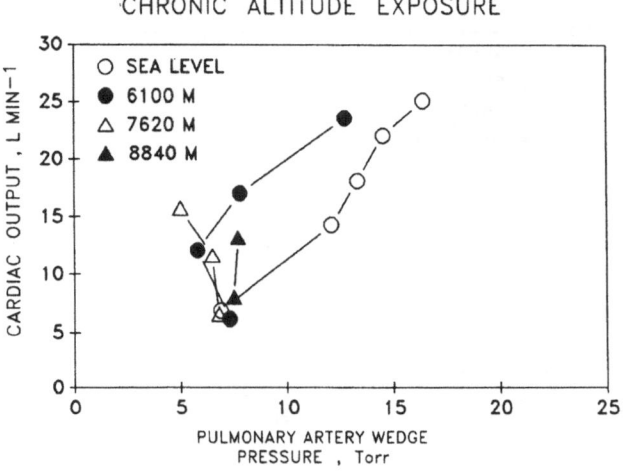

Figure 5. Relationship between cardiac output and left ventricular filling pressures estimated by pulmonary artery wedge pressure with both acute and chronic altitude exposure. Acute hypoxia does not alter the relationship (upper panel), and in chronic hypoxia, cardiac output is, if anything, increased for a given wedge pressure at altitude.

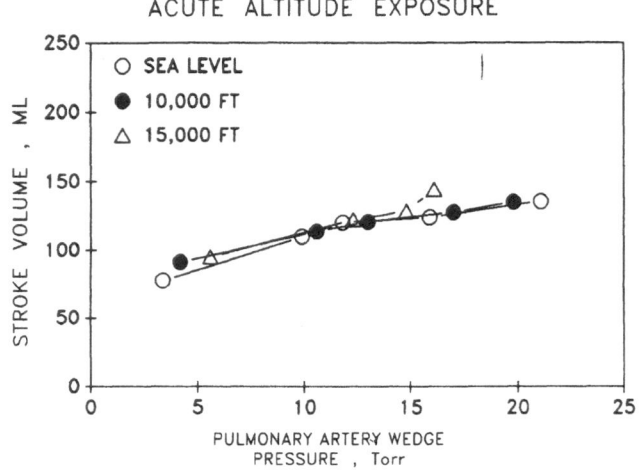

ACUTE ALTITUDE EXPOSURE

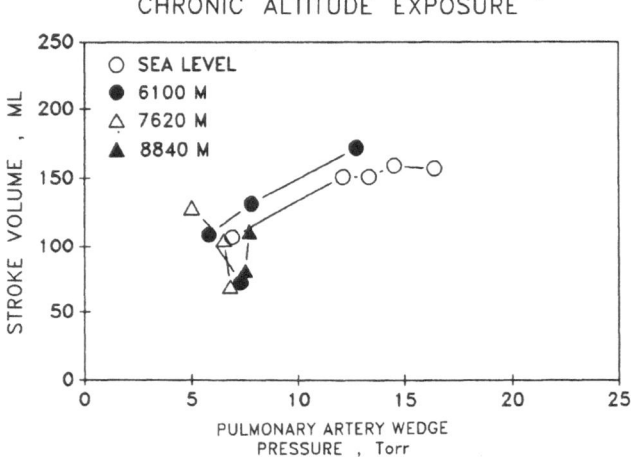

CHRONIC ALTITUDE EXPOSURE

Figure 6. Stroke volume as a function of pulmonary artery wedge pressure following acute and chronic altitude exposure. The relationship is unaffected by acute altitude exposure (upper panel) and also chronic exposure (lower panel). Figures 5 and 6 together suggest no impairment of cardiac pump function with acute or chronic altitude exposure, even to extreme hypoxia.

with many possibilities. Over the longer term, dietary factors may be yet additional modulators of [Hb]. Suffice it to say that [Hb] rises uniformly with altitude to values commonly around 20 gm/deciliter at Mt. Everest altitude. As a footnote, these levels were not reached in Operation Everest II because of: a) the relatively short time spent at altitude (40 days from sea level to the summit), and b) substantial blood sampling for physiological research throughout the study.

Physiologically, systemic O_2 delivery is somewhat preserved at altitude due to the increase in [Hb] especially because, as presented above, cardiac output at sea level, and after chronic altitude exposure are similar at any given $\dot{V}O_2$. Whether there are significant hemodynamic consequences of the high [Hb] are difficult to know in man, but there are some data to suggest that at a given altitude, and at the same exercise level, the increase in arterial $[O_2]$ with chronic altitude is accompanied by a reduction in muscle blood flow compared to acute hypoxia between acute and chronic situations (Bender et al., 1988).

In terms of gas exchange, O_2 diffusing capacity should be increased in the lung if [Hb] rises since there are more O_2 binding sites per ml of blood when [Hb] is increased. The lower cardiac output at any $\dot{V}O_2$ (compared to acute hypoxia, not sea level conditions) should prolong capillary red cell transit time and further assist O_2 diffusion (as has been found (Bebout et al., 1989)). Studies of the effect of [Hb] on muscle O_2 diffusion are fewer, but suggest similar directional effects as in the lungs: Hogan et al. (1991) found that about half of the total resistance to O_2 diffusion in the muscle is related to [Hb], such that diffusion of O_2 is enhanced by the higher [Hb]. This finding has been corroborated by similar observations in man (Barton et al., 1990), which necessarily encompassed a much narrower range of [Hb] levels.

As shown by Piiper et al. (1984), the degree of diffusion equilibration in a capillary bed (lungs or muscles) is determined by the ratio of diffusive to perfusive conductance for O_2 in that bed. This ratio, call it R, is given by the ratio of diffusing capacity (DO_2) to the product of blood flow (Q) and slope of the O_2 Hb dissociation curve (β):

$$R = DO_2 / (\beta Q) \tag{1}$$

Because β is proportional to [Hb] and DO_2 is only in part affected by [Hb], a fall in [Hb] will reduce β more than DO_2, so that R will increase (at constant Q). This increase in R increases the degree of diffusion equilibration. The opposite occurs when [Hb] rises.

In summary, the effects of chronic altitude exposure are to raise [Hb]. As a result, the following changes occur: 1) the increase in [Hb] increases O_2 diffusing capacity of both lungs and muscles. 2) However, through equation (1), the degree of diffusion equilibration will be reduced - - a negative effect. 3) If an increase in [Hb] causes a fall in muscle blood flow, transit time is prolonged to act to increase the degree of diffusion equilibrium. 4) The net effect of increased [Hb] and (possibly) decreased blood flow modulate systemic convective O_2 delivery in addition to the effects on diffusion (1,2 and 3 above). These multiple factors seem to combine so as to always increase O_2 supply as [Hb] is increased and decrease it as [Hb] is reduced.

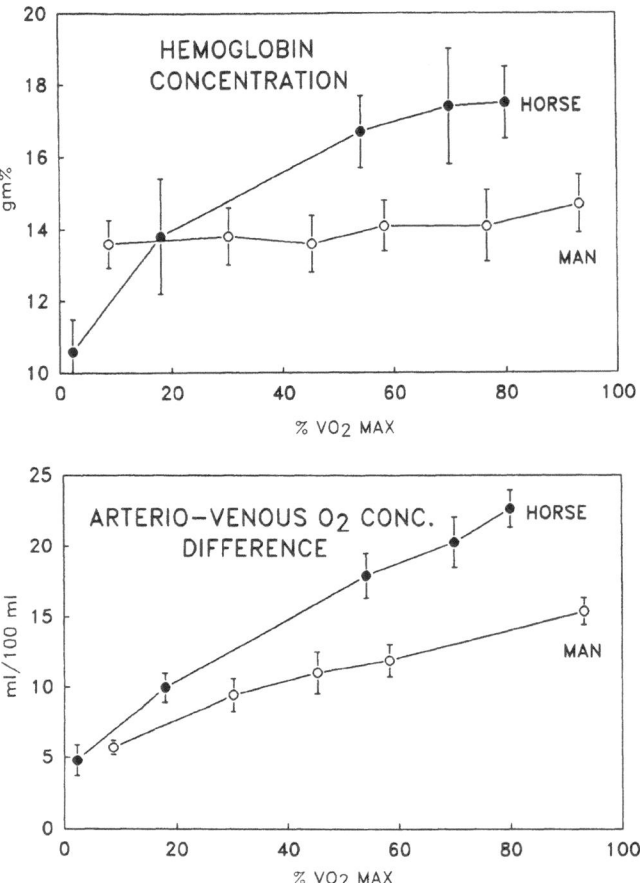

Figure 7. Effect of exercise on arterial hemoglobin concentration and arterio-venous oxygen concentration difference, comparing man and horse. The difference is explained by the large splenic reservoir of red cells in the horse, which is released to the circulation in proportion to the imposed work load.

D. The Muscles

Of all steps in the O_2 transport pathway, effects of altitude on O_2 transport in muscle are the least well-studied. Yet, they may be of considerable significance to performance at altitude.

At sea level, progressive increases in exercise level result in proportional increases in muscle blood flow and thus in O_2 delivery. At the same time, muscle venous PO_2 (PvO_2) falls, but never to zero. PvO_2 reaches some asymptotic value varying widely amongst individuals (Roca et al., 1989). Generally, PvO_2 is lower in the fittest individuals and higher in those less fit (Roca et al., 1989).

Acute exposure to altitude reduces maximum $\dot{V}O_2$, has little effect on maximal cardiac output or muscle blood flow during cycle or treadmill exercise, and reduces muscle venous PO_2 (Wagner et al., 1987). $\dot{V}O_{2max}$ is found to be reduced in direct proportion to each of the following variables: 1) muscle O_2 delivery, 2) muscle venous PO_2, and 3) calculated mean muscle capillary PO_2. These observations are readily explained by the interaction between convective supply of O_2 into the muscle capillary and subsequent diffusive movement of O_2 from the capillary to the mitochondria (Wagner, 1988a; Wagner, 1988b; Wagner, 1991). Structurally, acute hypoxia produces no changes, although the functional response of hypoxia-induced vasodilatation is well-documented. At constant blood flow, this could act to prolong capillary transit time, which would enhance diffusive unloading of O_2. There is suggestive but not conclusive evidence of such an effect in isolated muscle studies (Hogan et al., 1988).

Chronic altitude exposure may produce structural changes in muscle, although it appears that this happens unequivocally only at extreme altitudes. At moderate altitudes (4-5000 m), there are conflicting observations (Mathieu-Costello et al., 1989). Recent data suggest that of the possible structural factors that could modulate diffusion of O_2 in the muscle (capillary number per fiber and fiber diameter in particular), the former is important to O_2 diffusion, while the latter is not (Honig et al., 1984). This conclusion is counter to the classical Kroghian viewpoint (Krogh, 1919) which assumes a uniform diffusing medium between the red cell and the mitochondria -- Krogh would have argued that intercapillary distance was the key factor and that this is indeed determined in part by fiber diameter. However, the importance of capillary rather than fiber geometry fits closely with the ideas of Honig et al. (1984). It supports the notion that due to facilitated transport of O_2 within the myocyte (itself due to myoglobin), fiber diameter is unimportant, and that the capillary wall area is the key variable.

Given this background, one may ask what are the muscle structural consequences of chronic exposure to altitude? While chronic altitude exposure has little effect at moderate altitudes such as White Mountain or Pike's Peak at extreme altitudes, the obvious overall muscle wasting is accompanied by microscopic evidence of reduced muscle fiber diameter (Green et al., 1989). On the other hand, capillary number per fiber appears unaffected, such that the increase in capillary density that occurs is fully explained by decreases in fiber size. This would, on the basis of the foregoing, suggest no effect of altitude on muscle O_2 diffusional characteristics despite a reduction in mean intercapillary distance, but this hypothesis has not been definitively tested.

To approach this issue from another point of view, the dependence of muscle O_2 diffusing capacity on capillary number per fiber (as opposed to dependence on fiber diameter) may help at least in part to explain how it is possible for some individuals to reach the summit of Mt. Everest without supplemental O_2. Oelz et al. (1986) studied such elite climbers but could not discern any functional respiratory differences from normal subjects (although pulmonary diffusing capacity was not measured). Those authors did take muscle biopsies and found that capillary number per fiber was increased 20% compared to controls. Theoretical calculations (Wagner, 1989) point to the importance of muscle O_2 diffusing capacity as a major determinant of O_2 supply to the mitochondria at extreme altitude (at sea level it is blood flow rather than diffusing capacity that is the biggest single determinant of O_2 supply, by similar calculations). It is therefore possible that (in addition to psychological factors), muscle O_2 diffusing capacity was a significant factor assisting the unaided ascent of Mt. Everest, and was above average in the elite climbers due to a greater number of capillaries/fiber (another hypothesis remaining to be tested).

SUMMARY

All stages of the O_2 transport pathway are affected in some way by both acute and chronic altitude exposure. At any one stage, the effects are multiple, sometimes subtle, and frequently opposing. Clear-cut differences in responses to acute and to chronic altitude exposure are detectable but not in every case explainable, leaving important and perplexing problems still to be solved. Perhaps the most interesting of these relate to control of cardiac output and to determinants of O_2 diffusion from muscle capillary red cells to the muscle mitochondria.

ACKNOWLEDGEMENTS

It is a pleasure to acknowledge the contributions of a large number of collaborators too numerous to mention individually who were involved in collecting some of the data summarized in this paper. These include the investigators assocated with Operation Everest II, and my collaborators at Kansas State University where the equine data were collected. The data pertaining to acute altitude exposure come from two studies done in collaboration with Dr. Herbert A. Saltzman et al. at Duke University, North Carolina.

This research was supported by NIH grant HL 17731, TRRP 1RT-227 and Alliance Pharmaceutical Corp. (R-05-07).

The author also wishes to thank Tania Davisson for preparation of this paper.

REFERENCES

Bake, B., Bjure, J., and Widimsky, J., 1968, The effect of sitting and graded exercise on the distribution of pulmonary blood flow in healthy subjects studied with the [133]xenon technique, Scan. J. Clin. Lab. Invest., 22:99-106.

Bebout, D. E., Hogan, M. C., and Wagner, P. D., 1990, The effects of exercise training and immobilization on $\dot{V}O_{2max}$ and estimated diffusing capacity (DO_2) in canine gastrocnemius muscle in situ, FASEB J., Part II, 4(4):5494, A1212.

Bebout, D. E., Story, D., Roca, J., Hogan, M. C., Poole, D. C., Gonzalez-Camerena, R., Ueno, O., Haab, P. and Wagner, P. D., 1989, Effects of altitude acclimatization on pulmonary gas exchange during exercise, J. Appl. Physiol., 67(6):2286-2295.

Barton, E. D., Schaffartzik, W., Poole, D. C., Hogan, M. C., Tsukimoto, K., Bebout, D. E., and Wagner, P. D., 1990, The effect of altered hemoglobin concentration on O_2 diffusion from blood to muscle at maximal exercise, FASEB J., Pt. II, 4(4):3449, A861.

Bender, P. R., Groves, B. M., McCullough, R. E., McCullough, R. G., Huang, S-Y, Hamilton, A. J., Wagner, P. D., Cymerman, A., and Reeves, J. T., 1988, Oxygen transport to exercising leg in chronic hypoxia, J. Appl. Physiol., 65(6):2592-2597.

Carlson, G. P., 1986, Hematology and body fluids in the equine athlete: A review, in: "Equine Exercise Physiology 2," J. R. Gillespie, and N. E. Robinson, eds., ICEEP Publications, Davis.

Coates, G., O'Brodovich, H., Jefferies, A. L., and Gray, G. W., 1984, Effects of exercise on lung lymph flow in sheep and goats. J. Clin. Invest., 74:133-141.

Davis, J., and Manohar, M., 1988, Effect of splenectomy on exercise-induced pulmonary and systemic hypertension in ponies, Am. J. Vet. Res., 49:1169-1172.

Dempsey, J. A., 1986, Is the lung built for exerise? J.B. Wolffe Memorial Lecture, Med. Sci. Sports & Exercise, 18(2):143-155.

Dempsey, J. A., Hanson, P. G., and Henderson, K. S., 1984, Exercise-induced arterial hypoxaemia in healthy human subjects at sea level, J. Physiol. Lond., 355:161-175.

Erickson, B. K., Erickson, H. H., and Coffman, J. R., 1990, Pulmonary artery, aortic and esophageal pressure changes during high-intensity treadmill exercise in the horse: A possible relation to exercise-induced pulmonary hemorrhage, Eq. Vet. Jnl., (Exercise Physiology Suppl.), in press.

Evans, J. W., and Wagner, P. D., 1977, Limits on $\dot{V}_A/Q$ distributions from analysis of experimental inert gas elimination, J. Appl. Physiol., 42:889-898.

Fisher, J. W., and Langston, J. W., 1967, The influence of hypoxia and cobalt on erythropoietin production in the isolated perfused dog kidney, Blood, 29, 114-115.

Green, H. J., Sutton, J. R., Cymerman, A., Young, P. M., and Houston, C. S., 1989, Operation Everest II: adaptations in human skeletal muscle, J. Appl. Physiol., 66(5):2454-2461.

Groves, B. M., Reeves, J. T., Sutton, J. R., Wagner, P. D., Cymerman, A., Malconian, M. K., Rock, P. B., Young, P. M., and Houston, C. S., 1987, Operation Everest II: elevated high-altitude pulmonary resistance unresponsive to oxygen, J. Appl. Physiol., 63(2):521-530.

Hlastala, M. P., Scheid, P., and Piiper, J., 1981, Interpretation of inert gas retention and excretion in the presence of stratified inhomogeneity, Respir. Physiol., 46:247-259.

Hogan, M. C., Bebout, D. E., and Wagner, P. D., 1991, Effect of hemoglobin concentration on maximal O_2 uptake in canine gastrocnemius muscle in situ, J. Appl. Physiol., 70(3):1105-1112.

Hogan, M. C., Roca, J., Wagner, P. D., and West, J. B., 1988, Limitation of maximal O_2 uptake and performance by acute hypoxia in dog muscle *in situ*, J. Appl. Physiol., 65:815-821.

Honig, C. R., Gayeski, T. E. J., Federspiel, W., Clark Jr., A., Clark, P., 1984, Muscle O_2 gradients from hemoglobin to cytochrome: new concepts, new complexities, Adv. Exp. Med. Biol., 169:23-38.

Hoppeler, H., Mathieu, O., Krauer, R., Claassen, H., Armstrong, R. B., and Weibel, E. R., 1981, Design of the mammalian respiratory system. VI. Distribution of mitochondria and capillaries in various muscles, Respir. Physiol., 44:87-111.

Karas, R. H., Taylor, C. R., Jones, J. J., Lindstedt, S. L., Reeves, R. B., and Weibel, E. R., 1987, Adaptive variation in the mammalian respiratory system in relation to energetic demand: VII. Flow of oxygen across the pulmonary gas exchanger, Respir. Physiol., 69(1):101-115.

Krogh, A., 1919, The number and distribution of capillaries in muscle with calculations of the pressure head necessary for supplying the tissue, J. Physiol (Lond.), 52:409-415.

Leaver, D. G., and Pride, N. B., 1971, Flow volume curves and expiratory pressures during exercise in patients with chronic airways obstruction, Scand. J. Resp. Dis., Suppl. 77:23-77.

Mathieu-Costello, O., Poole, D. C., and Logemann, R. B., 1989, Muscle fiber size and chronic exposure to hypoxia, Adv. Exp. Med. Biol., 248:305-311.

Myhre, L. G., Dill, D. B., Hall, F. G., and Brown, K. K., 1970, Blood volume changes during three-week residence at high altitude, Clin. Chem., 16:7-14.

Oelz, O., Howald, H., Di Prampero, P. E., Hoppeler, H., Glaassen, H., Jenni, R., Bühlmann, A., Ferretti, G., Brückner, J-C., Veicsteinas, A., Gussoni, M., and Cerretelli, P., 1986, Physiological profile of world-class high-altitude climbers, J. Appl. Physiol., 60(5):1734-1742.

Pascoe, J. R., Ferrato, G. L., Cannon, H. H., Arthur, R. M., and Wheat, J. D., 1981, Exercise-induced pulmonary hemorrhage in racing Thoroughbreds: A preliminary study, Am. J. Vet. Res., 42:703-707.

Persson, S. G. B., 1967, On blood volume and working capacity in horses, Acta Vet. Scand., (Suppl.), 19:1-189.

Piiper, J., Meyer, M., and Scheid, P., 1984, Dual role of diffusion in tissue gas exchange: Blood-tissue equilibration and diffusion shunt, Respir. Physiol., 56:131-144.

Piiper, J., and Scheid, P., 1987, Diffusion and convection in intrapulmonary gas mixing, chapt. 4, in: "Handbook of Physiology, Section 3: The Respiratory System," vol. IV, A. P. Fishman, L. E. Farhi, S. M. Tenney, and S. R. Geiger, eds., pp. 51-69.

Pugh, L. G. C. E., 1964, Blood volume and haemoglobin concentration at altitudes above 18,000 ft (5500 m), J. Physiol., 170, 344-354.

Reeves, J. T., Groves, B. M., Sutton, J. R., Wagner, P. D., Cymerman, A., Malconian, M. K., Rock, P. B., Young, P. M., and Houston, C. S., 1987, Operation Everest II: preservation of cardiac function at extreme altitude, J. Appl. Physiol., 63(2):531-539.

Roca, J., Hogan, M. C., Story, D., Bebout, D. E., Haab, P., Gonzalez, R., Ueno, O., and Wagner, P. D., 1989, Evidence for tissue diffusion limitation of $\dot{V}O_{2max}$ in normal humans, J. Appl. Physiol., 67(1):291-299.

Schaffartzik, W., Arcos, J., Tsukimoto, K., Mathieu-Costello, O., and Wagner, P. D., 1990a, Interstitial pulmonary edema in pigs after exercise, FASEB J., Pt. I., 4(3):900, A422.

Schaffartzik, W., Poole, D. C., Derion, T., Tsukimoto, K., Hogan, M. C., Arcos, J., Bebout, D. E., and Wagner, P.D., 1990b, $\dot{V}_A/\dot{Q}$ mismatch, blood gas and acid-base status during and after exercise in hypoxia, The Physiologist, 33:A-74.

Tsukimoto, K., Arcos, J. P., Schaffartzik, W., Wagner, P. D., and West, J. B., 1990, Effect of common dead space on $\dot{V}_A/\dot{Q}$ distribution in the dog, J. Appl. Physiol., 68:2488-2493.

Tucker, C. E., James, W. E., Berry, M. A., Johnston, C. L., and Grover, R. F., 1976, Depressed myocardial function in the goat at high altitude, J. Appl. Physiol., 41:356-361.

Wagner, P. D., 1989, Algebraic analysis of the determinants of maximum oxygen uptake, 1989, in: "Proceedings of the Meeting of the International Society on Oxygen Transport to Tissue", (Abstract), Göttingen, July 21-24.

Wagner, P. D., 1988b, An integrated view of the determinants of maximum oxygen uptake, in: "Oxygen Transfer from Atmosphere to Tissues", vol. 227, N. C. Gonzalez, and M. R. Fedde, eds., Plenum Press, New York, pp. 245-256.

Wagner, P. D., 1991, Central and peripheral aspects of oxygen transport and adaptations with exercise, Sports Medicine, 11(3):133-142.

Wagner, P. D., 1977, Diffusion and chemical reaction in pulmonary gas exchange, Physiol. Rev., 57:257-312.

Wagner, P. D., 1988a, The determinants of VO_{2max}, Annals of Sports Med., 4(4):196-212.

Wagner, P. D., Gale, G. E., Moon, R. E., Torre-Bueno, J. R., Stolp, B. W., and Saltzman, H. A., 1986a, Pulmonary gas exchange in humans exercising at sea level and simulated altitude, J. Appl. Physiol., 61(1):260-270.

Wagner, P. D., Gillespie, J. R., Landgren, G. L., Fedde, M. F., Jones, B. W., DeBowes, R. M., Pieschl, R. L., and Erickson, H. H., 1989, Mechanism of exercise-induced hypoxemia in horses, J. Appl. Physiol., 66(3):1227-1233.

Wagner, P. D., Naumann, P. F., and Laravuso, R. B., 1974, Simultaneous measurement of eight foreign gases in blood by gas chromatography, J. Appl. Physiol., 36:600-605.

Wagner, P. D., Sutton, J. R., Malconian, M. K., Cymerman, A., Groves, B. M., and Reeves, J. T., 1986b, Lung volumes and flow rates in man during a simulated ascent of Mt. Everest, Am. Rev. Respir. Dis., 133(4):A76.

Wagner, P. D., Sutton, J. R., Reeves, J. T., Cymerman, A., Groves, B. M., and Malconian, M. D., 1987, Operation Everest II: pulmonary gas exchange during a simulated ascent of Mt. Everest, J. Appl. Physiol., 63(6):2348-2359.

Wagner, P. D., and West, J. B., 1980, Ventilation-perfusion relationships, in: "Pulmonary Gas Exchange", vol. 1, chapt. 7, J. B. West, ed., Academic Press, New York, pp. 263-306.

Ward, M., 1989, in: "High Altitude Medicine and Physiology," M. P. Ward, J. S. Milledge, and J. B. West, eds., Univ. of Pennsylvania Press, Philadelphia.

Weibel, E. R., ed., 1984, in: "The Pathway for Oxygen," Harvard Univ. Press, Cambridge/Massachusetts/London.

West, J. B., Tsukimoto, K., Mathieu-Costello, O., and Prediletto, R., 1991, Stress failure in pulmonary capillaries, J. Appl. Physiol., 70:1731-1742.

West, J. B., 1988, Ventilation, bloodflow and gas exchange, in: "Textbook of Respiratory Medicine", J. F. Murray, and J. A. Nadel, eds., Saunders, Philadelphia.

HYPERBARIC OXYGEN THERAPY: PAST, PRESENT AND FUTURE INDICATIONS

D.J.Bakker

Academic Medical Center, University of Amsterdam

Meibergdreef 9, 1105 AZ   Amsterdam, The Netherlands

INTRODUCTION

Hyperbaric medicine is concerned with the medical and physiological problems and therapeutic applications of barometric pressure higher than at
sea level.
Hyperbaric oxygen therapy is defined as a mode of medical treatment in
which the patient is entirely enclosed in a pressure chamber, breathing
oxygen at a pressure greater than one atmosphere.
Treatment may be carried out either in a monoplace chamber, pressurized
with pure oxygen or in a large multiplace chamber pressurized with air,
in which case the patient receives pure oxygen by mask, head-tent or
endotracheal tube.
Breathing pure oxygen at a pressure of one atmosphere, or applying oxygen
topically to parts of the body without the use of a pressurized chamber
which encloses the patient completely, is not considered as hyperbaric
oxygen therapy (8).

Oxygen is transported by the blood in two different ways: (1) Chemically
bound to hemoglobin in the erythrocytes, and (2) physically dissolved in
plasma according to Henry's Law, stating that the degree to which a gas
enters into physical solution in body fluids is directly proportional to
the partial pressure of the gas to which the fluid is exposed. Since
hemoglobin is 97 % saturated with oxygen at 1 ATA breathing air, the total
oxygen content of the blood can not be raised very much by saturating
hemoglobin to 100%. When the pressure rises to 3 ATA while breathing 100 %
O_2, the ideal dissolved oxygen content in the plasma becomes 6,80 vol%. The
total circulating oxygen content of the blood is then 20,1 + 6,80 = 26,9
vol% O_2. Consequently under these circumstances the total oxygen requirement of the body, for which approximately 6,0 vol% O_2 is necessary, can
already be fulfilled only by the dissolved oxygen in the plasma, so that
hemoglobin is not necessary (4).
This amount of oxygen in arterial blood can never be reached due to possible venous admixture caused by (1) alveolar blood barrier preventing a complete equilibrium; (2) shunting via bronchial, thebesian and pleural vessels; (3) some physiological shunting in the lungs by differences between
ventilation and perfusion; and (4) possibly some atelectases in the lung.
Calculated dissolved O_2 according to the alveolar gas equation for O_2 is
5,6-5,8 ml/dl. Because of the high tissue-oxygen gradient, arterial oxygen tensions in the tissues will only rise moderately.

In normal tissues, the primary action of increased arterial pO_2 seems to be generalized vasoconstriction and although this effect is greater in the kidney and the brain, it is not restricted to these organs. This vasoconstrictive side effect can be beneficial when edema and swelling are major complications (for example in burns, gas embolism, decompression sickness, peripheral trauma, etc.). Another effect of hyperbaric oxygenation is a 10-20 % reduction in cardiac output, which is primarily due to bradycardia rather than to reduction in stroke volume (12).

THE PAST

Soon after the invention of the air pump, experiments in the use of compressed air in medicine started. The British physician Henshaw was the first to use compressed air in a specially equipped room called a "Domicilium", in 1664. Henshaw hoped that "a person might receive the benefit of removal to another climate at any season of the year while remaining in his own city without neglecting his employment". His results are unknown and obviously his method was not followed by others.
The Dutch Society of Sciences in Haarlem awarded a prize in 1783 for the best apparatus for experiments with compressed air in order to study its influence on plants and animals. There were however, no contenders not even after the offer was repeated in three subsequent years.
An apparatus for variable air pressure was described by Junod in 1843, but the real promotor of the so-called "pneumatic therapy" was Tabarié, who made his proposals to the "Académie des Sciences" in 1832. He was the first to open an official "pneumatic center" in Montpellier in 1840. Pravaz opened a center in Lyon.
Treatment with air under increased pressure, "le bain d'air comprimé", was supposed to have a beneficial effect on a wide range of disorders, mainly pulmonary. Cholera, chest deformities and rickets were also treated and the pressures used ranged from 2-4 ATA. From 1860 on, pneumatic centers spread all over Europe: Stockholm, London, Altona in Germany, Johannesburg, Copenhagen, Munich, Dresden, Stuttgart, Odessa, Moscow, Vienna etc. The first hyperbaric chamber on the North American continent was built and used in 1860 in Ashawa, Canada.
Very large chambers were constructed by Cunningham in Kansas City (1927) and Cleveland, Ohio (1927), the latter having a diameter of sixty-four feet. All these pneumatic centers however disappeared. Arntzenius warned already in 1887 that the confidence the treatment, in his opinion, deserved might be lost by overemphasizing the value. All types of diseases were treated; even singers went in for intermittent treatment of their voices.
In assessing the value of hyperbaric treatment in the last century, there is serious doubt whether many of the effects of the treatment were more than a pure psychological effect. This is especially true where pressures of no more than 1/3 or 1/4 atmosphere overpressure were used; the luxurious environment inside the chambers will have been helpful from this point of view. In the Netherlands pneumatic chambers were build by Pyan in Haarlem in 1882; in Amsterdam by Arntzenius in 1885 and in Hilversum by Tresling in 1887.

Oxygen was discovered by Priestley in 1775 and by Scheele in 1777. Its toxic effects were first noted by Sequin and Lavoisier in 1789. Beddoes reported the first medicinal use of oxygen in 1794. Paul Bert (1878) considered pure oxygen as particularly dangerous, even at normal pressure. The only therapeutic indication, according to Bert was carbon monoxide poisoning (Claude Bernard reported in 1857 on the chemical affinity of CO and hemoglobin). Bert used enriched air, containing 60% oxygen and not pure oxygen. He also demonstrated that inhalation of pure oxygen by skylarks produced oxygen seizures. The so-called Lorrain-Smith effect (inhalation of pure oxygen at 1 ATA causes pulmonary damage) is known since the early

1900s. Mosso performed experiments with monkeys in 1900 and used oxygen at
2 ATA.

The first article of using oxygen under pressure was written by Fontaine in
1879.
It is not exactly known how much oxygen was given in the pneumatic centers
but certainly not pure oxygen (1,6).
An upsurge in hyperbaric oxygen treatment followed in 1956 in Amsterdam,
when Boerema and his co-workers started using it in cardiac surgery. The
aim of the preceding experimental work, carried out in the chambers of the
Royal Navy in Den Helder, was to determine the value of increased atmospheric pressure in cardiac surgery. The idea of using oxygen under these circumstances was closely related to the use of hypothermia. By lowering the
body temperature to $28°C$, oxygen consumption was reduced to about half its
level at $37°C$. First surface cooling was used but later on Meyne developed
the extra-corporeal blood cooling. Complete circulatory arrest for 7-8
minutes was well tolerated and it was possible to perform small intracardi-
ac repairs in this way. Since the time required for ventricular surgery was
still longer Boerema looked for methods that permitted a longer cardiac arrest without the necessity of deeper cooling. (Temperatures below $28°C$ caused irreversible cardiac arrhythmia's). The idea was to ventilate the patient with pure oxygen under increased pressure "to fill every cell of the
body with a large amount of oxygen before circulatory arrest". After completion of the large hyperbaric chamber in our hospital in 1959, heart
surgery was done under pressure in this chamber. From 1960-1977, 187 open
heart operations were performed; palliative procedures in cyanotic heart
disease but also complete corrections in children with cyanotic tetralo-
gies, severe pulmonary stenosis with atrial septal defects or patent fora-
men ovale and corrections of complete transpositions of the great vessels.
The large hyperbaric chamber was pressurized with air and the patient was
ventilated with 100% O_2 under a pressure of 3 ATA. As soon as the actual
correction was finished, when the surgeon started closing the thorax, the
chamber was decompressed (12).
With the development and perfection however, of the heart-lung machine,
this first and at that time only indication for the use of hyperbaric oxygen disappeared.

In 1960 Boerema and Brummelkamp demonstrated the value of hyperbaric oxygen
in the treatment of gas gangrene, caused by the anaerobic micro-organism
Clostridium Welchii or perfringens. From 1960-1990, 588 patients suspected
of having gas gangrene were admitted to our department. The diagnosis gas
gangrene could be confirmed both clinically and bacteriologically in 439
patients or 74,4%.
Our results show very clearly that a combination of hyperbaric oxygen,
antibiotics and surgery gives the lowest mortality and morbidity in this
disease (5,2).
From then on the value of hyperbaric oxygen was studied in the treatment of
necrotizing soft tissue infections. A new classification was proposed and
published, together with studies on the etiology and bacteriology of these
infections (3).
In 1963 the first International Congress on Clinical Application of
Hyperbaric Oxygen was held in Amsterdam, followed by many others; Glasgow
1964, Durham NC 1966, Sapporo 1969, Vancouver 1974, Aberdeen 1978, Moscow
1981, Long Beach Ca. 1984, Sydney 1987 and Amsterdam again in 1990.
Over the years there have been many indications for hyperbaric oxygen therapy; some very serious, some however only "wishful thinking", or even
anecdotal. A serious attempt to re-evaluate and categorize all known indications started in October 1975 at San Francisco Medical School under
chairmenship of J.C.Davis and T.K.Hunt. This resulted in publication of the
book "Hyperbaric Oxygen Therapy" by the Undersea and Hyperbaric Medical Society (8). In November 1976 the Hyperbaric Oxygen Therapy Committee was
established. This Committee prepares and publishes a report presenting the

best medical opinion of the moment on the efficacy of hyperbaric oxygen in a great variety of disorders. This report is reviewed and updated yearly. Accepted conditions are diseases where sufficient evidence of the usefulness of hyperbaric oxygen exists. This evidence must be proven by publications in the international literature on experimental and clinical studies. The task of the Committee is to weigh the evidence and accordingly categorize the diseases (6).

THE PRESENT

From the 28 indications for which third party reimbursement was recommended in the 1976 and 1979 reports, the number of approved indications has been refined to 12 in the 1989 report. Evidence accepted by the Committee includes sound physiologic rationale, in vivo and/or in vitro studies that demonstrate effectiveness, controlled animal studies, prospective randomized controlled clinical studies, and extensive clinical experience from multiple, recognized hyperbaric medicine centers. Accepted indications are:
1. Air or gas embolism
2. Carbon monoxide poisoning and smoke inhalation
   CO poisoning complicated by cyanide poisoning
3. Clostridial myonecrosis (gas gangrene)
4. Crush injury, compartment syndrome, and other acute traumatic ischemias
5. Decompression sickness
6. Enhancement of healing in selected problem wounds
7. Exceptional blood loss (anemia)
8. Necrotizing soft tissue infections (subcutaneous tissue, muscle, fascia)
9. Osteomyelitis (refractory)
10. Radiation tissue damage (osteoradionecrosis)
11. Skin grafts and flaps (compromised)
12. Thermal burns

The list of research indications shows at present 21 disorders. These include:
1. Brain abscess
2. Carbon tetrachloride poisoning
3. Acute CVA
4. Cerebral edema
5. Head injury
6. Fracture healing and bone grafting
7. Hydrogen sulfide poisoning
8. Lepromatous leprosy
9. Meningitis
10. Pyoderma gangrenosum
11. Multiple sclerosis
12. Pseudomonas colitis
13. Spinal cord injury
14. Radiation enteritis and proctitis
15. Radiation myelitis
16. Acute central retinal artery insufficiency
17. Selected refractory mycosis-mucormycosis, aspergillosis
18. Sepsis
19. Intra-abdominal abscess
20. Sickle cell crisis
21. Brown recluse spider bite

Mechanisms: Two different events can be distinguished when patients are treated with hyperbaric oxygen: 1. Elevation of pressure and 2. Elevation of PaO_2.
1. Elevation of pressure: The physiological effects of increased pressure comprehend the compression of gas in closed spaces such as the middle ear, intestinal gas, paranasal sinuses etc. These changes follow Boyle's Law: PV = constant. A doubling of the environmental pressure will cause the volume of gas inside the body to decrease by one half. This effect underlies one of the major beneficial effects of hyperbaric oxygen treatment upon the pathological presence of gas, such as:
- arterial and venous gas embolism
- decompression sickness
- pneumatosis cystoides intestinalis
- iatrogenic gas embolism.

Elevation of the PaO_2 in the blood has an added beneficial effect on ischemic tissues, when the obstruction of small blood vessels by gas bubbles is abolished by the pressure effect.

2. **Increased PaO_2:** Increasing the PaO_2 has three major pharmacological effects: a. Increased blood O_2 content
   b. Vasoconstriction
   c. Antibacterial action.

a. <u>Increased blood O_2 content:</u> The increased arterial O_2 content underlies the use of hyperbaric oxygen for the treatment of ischemic conditions, for example non-healing wounds, "problem wounds". Problem wounds are wounds that fail to heal after adequate surgical and antibiotic treatment. This usually results from systemic or local host factors or environmental insults. Diabetes mellitus, rheumatoid vasculitis, trauma, immunosuppressive drugs, burns or former therapeutic irradiation are well known examples, leading from limb-threatening diabetic foot ulcers to massive hemorrhage after carotid erosion in radiation necrosis of the neck. A common denominator in problem wounds is tissue hypoxia. Not all hypoxic wounds are ischemic. A wound with adequate perfusion may be hypoxic because infection raises the oxygen consumption. All ischemic wounds are hypoxic. Whether the oxygen tension in hypoxic wounds is elevated by raising the dose of inspired oxygen is dependent on the status of the regional perfusion. Oxygen cannot diffuse more than a few microns from functioning capillaries, even at the high arterial oxygen tensions that can be reached with hyperbaric oxygen. Data from recent literature support however the following statements:
- Hypoxia in partially ischemic and infected wounds or in irradiated tissues can often be corrected by high-dose oxygen inhalation. The dose of inspired oxygen that is required, varies from wound to wound and ranges from 40% at sea level to 100% at 2-3 ATA.
- Intermittent correction of problem wound hypoxia increases fibroblast replication and production of collagen to support capillary proliferation; neovascularisation. When the environment of the fibroblast has an oxygen tension of 10 mm Hg or less, the cell can no longer divide, synthesize collagen or migrate. The collagen produced forms a matrix in which new capillaries grow.
- Healing is significantly improved following surgery in previously irradiated tissue if it is preceded by 20 - 30 days of intermittent high dose oxygen inhalation (hyperbaric oxygen treatment) (13).
- Also osteoblasts and osteoclasts need an adequate tissue oxygen tension to be able to function normally. This is the rationale for the use of hyperbaric oxygen in chronic refractory osteomyelitis. It has been shown that oxygen tensions in chronic osteomyelitic bone are far too low to permit normal healing (10). These experimental data have been confirmed in large clinical series (13).

b. <u>Vasoconstriction:</u> This effect is the rationale for the treatment of edema with hyperbaric oxygen. This edema can be present in burns, cerebral and spinal cord edema after trauma and/or surgery and in peripheral compartment syndromes. As we mentioned before, the elevation of arterial pO_2 which occurs while breathing 100% O_2 at 3 ATA results in a significant reduction in cardiac output and heart rate, and in an increase in total peripheral resistance. There is also a slight increase in mean arterial pressure.

c. <u>Antibacterial action:</u> In infections, tissue oxygen tensions have been found to be markedly reduced. Hyperbaric oxygen treatment can increase these tissue oxygen tensions to normal or even above normal. Increasing oxygen tension has a direct lethal effect on strict anaerobic microorganisms and some microaerophilic aerobes. An increase in oxygen tension leads to an increased concentration of the oxygen free radical superoxide, both intracellularly and extracellularly. These increased superoxide levels lead to an increased production of hydrogen peroxide and other toxic oxygen radicals. Most anaerobic micro-organisms lack the defense mechanisms against these toxic substances; superoxide dismutase and catalase. In this way, hyperbaric oxygen has a direct bactericidal effect on most Clostridium species. Furthermore the production of the lethal alpha-toxin is stopped. The effect on other strict anaerobic organisms has been less well documented although there are several beneficial clinical reports. This is called the **direct effect**.

There is also an **indirect effect**. Most organisms that are termed aerobic bacteria are, in fact, facultative anaerobes. These facultative anaerobes and strict aerobes will be termed aerobic organisms. Hyperbaric oxygen raises again oxygen tensions which lead to an increased production of superoxide, hydrogen peroxide and other toxic oxygen free radicals. This induces further production of superoxide dismutase by aerobic organisms. They are able to detoxify superoxide, even under hyperbaric conditions. Thus, there is no direct effect of hyperbaric oxygen on aerobic organisms. However, the oxygen dependent intracellular killing mechanisms of the polymorphnuclear leukocytes require oxygen for adequate killing of many aerobic micro-organisms. In infected areas, tissue oxygen tension has been found to be markedly reduced. By the use of hyperbaric oxygen, the required oxygen is made available so that the polymorphnuclear leukocyte killing of aerobic organisms returns to normal. It is likely that also the monocyte-macrophage returns to a normal function when using hyperbaric oxygen in infected areas.

EXPERIENCES: RESULTS

The hyperbaric medical department as subdivision of the surgical clinic of the Academic Medical Center at the University of Amsterdam exists since the large hyperbaric chamber was installed in the Wilhelmina Gasthuis in 1959. This chamber was moved to the new hospital in 1984.
We will discuss our large experience in the use of adjunctive hyperbaric oxygen in the following selected diseases:
 - Gas gangrene
 - Selected aerobic and anaerobic soft tissue infections
 - Osteoradionecrosis of the jaw
 - Radiation cystitis
 - Vascular surgery
Our methods and results in tissue oxygen monitoring in various diseases where we use hyperbaric oxygen is discussed elsewhere in this book.

GAS GANGRENE

Gas gangrene is an acute, rapidly progressive, non-pyogenic, gasforming and necrotizing infection of muscles, skin and subcutaneous tissues. The infection is caused by anaerobic sporeforming bacteria of the genus Clostridium, primarily Clostridium Welchii or perfringens. Gas gangrene is characterized by profound toxemia, extensive edema, massive death of tissue and a variable degree of gas production. For the induction of gas gangrene two conditions have to be fulfilled: the presence of clostridial spores and an area of lowered oxidation-reduction potential (Eh), caused by circulatory failure in a local area or by extensive soft tissue damage and necrotic muscle tissue. This results in an area with a low oxygen tension where clostridial spores can develop into the vegetative form. Clostridial bacteria surround themselves with produced toxins; local host defense mechanisms are abolished when toxin concentration is sufficiently high, followed by ever increasing tissue destruction and further clostridial growth. Clinically a very rapid spreading wound infection is seen. The most dangerous toxin is the hemolytic, tissue necrotizing and lethal alpha-toxin.
Modern treatment of gas gangrene includes surgical intervention, antibiotics, general resuscitative and ancillary measures and, most important, hyperbaric oxygen.

The action of hyperbaric oxygen on clostridia is based on the formation of oxygen free radicals in the absence of free radical scavengers. In this way hyperbaric oxygen is bacteriostatic and bactericidal on clostridia. A tissue pO_2 of 250 mm Hg is necessary to stop alpha toxin production completely. Since the progressive nature of gas gangrene depends on the continuous production of alpha toxin, hyperbaric oxygen is the quickest way initially to break that vicious circle.

This treatment is **lifesaving** because less heroic surgery needs to be performed in very ill patients, and the cessation of alpha toxin production is rapid; our mortality is 11,8% in the whole series of 439 patients, with a mortality of 6,8% in the posttraumatic group, compared to a mortality of 30-50%, even up to 100% before hyperbaric oxygen was used.

The treatment is **limb- and tissuesaving** because no major amputations or excisions are done in advance, and when demarcation becomes clear, far less tissue appears to be lost than initially thought; our amputation percentage because of gas gangrene is 7,1% or 24 patients, compared to 50% when HBO was used late, and even higher when HBO was not used at all.

HBO **clarifies the demarcation** and allows a more rapid distinction between dead and viable tissue, generally within 24-30 hours after the first treatment. Timing of surgery has always been a point of discussion. Our approach has always been very conservative with first hyperbaric oxygen and late surgery. This approach went not uncriticised, however our results prove the validity of our opinion.

The combination of hyperbaric oxygen, local debridement and antibiotics leads to less mortality and morbidity than any of these treatment modalities alone (2).

SELECTED AEROBIC AND ANAEROBIC SOFT TISSUE INFECTIONS

These infections are classified as progressive bacterial gangrene, necrotizing fasciitis and the above mentioned myositis and myonecrosis. In these infections anaerobic micro-organisms are often found in combination with aerobic Gram-negative organisms. In traumatically, surgically or medically compromised patients, local tissue hypoxia and a decreased oxidation-reduction potential (Eh) is usually present, thus promoting the growth of more anaerobic micro-organisms. The vast majority of these necrotizing soft tissue infections has an endogenous anaerobic component. Hypoxic conditions also allow the proliferation of facultative aerobic organisms since polymorphnuclear leucocytes function poorly under decreased oxygen tensions. The growth of aerobic micro-organisms further lower the Eh; more fastidious anaerobes become established and the disease process can rapidly accelerate.

Clinically the most important signs of these infections are tissue necrosis, a putrid discharge, gas production, the tendency of the process to burrow through fascial planes and in many cases the absence of the classical signs of tissue inflammation.

Treatment of these infections is a combination of surgical debridement (timely, limited or aggressive), appropriate antibiotics, good nutritional support and optimal oxygenation of the infected tissues. When ambient oxygen is insufficient, hyperbaric oxygen can be used. Therapy must be monitored by transcutaneous or, even better, by direct intraphlegmonous and\or intramuscular pO_2 measurements.

The main goals are: (a) improvement of tissue pO_2, necessary for normal wound healing, (b) improvement of phagocytic function by stimulating the oxygen-dependent killing mechanisms, either direct or indirect, and (c) the diminishing of edema and improvement of the circulation in the affected areas. This can be roughly summarized as stimulation of the host defense and repair mechanisms.

We recommend the adjunctive use of hyperbaric oxygen in progressive bacterial gangrene of skin and subcutaneous tissues in cases with serious underlying systemic diseases and symptoms of general toxicity and in other immune-compromised patients. In necrotizing fasciitis of the deep fascia surgical treatment is of the utmost importance. However the best results can only be obtained with a combination of surgery, antibiotics and hyperbaric oxygen.

OSTEORADIONECROSIS OF THE JAW

Osteoradionecrosis of the mandible is a serious complication of radiotherapy in the treatment of head and neck neoplasms. Although the exact pathogenesis is not completely understood, most of the authors agree that three factors are involved: Radiation, trauma and infection. It is not primarily infection of irradiated bone but a complex metabolic and tissue homeostatic deficiency created by radiation induced cellular injury and characterized by the sequence: radiation, formation of a hypoxic, hypovascular and hypocellular tissue, followed by tissue breakdown, resulting in a chronic, non-healing wound. More recently Bras et al showed that osteoradionecrosis of the mandible is an ischemic necrosis, due to a radiation induced obliteration of the inferior alveolar artery and branches of this artery. Revascularisation, as normally occurs from branches of the facial artery, has been disturbed by radiation induced vascular disease and periosteal damage. We performed controlled prospective randomized trials in the use of hyperbaric oxygen in osteoradionecrosis of the mandible. In many patients resection of large parts of the mandible is necessary. When hyperbaric oxygen was used as pretreatment, followed by aggressive surgery and again a course of postoperative hyperbaric oxygen treatment, results are good (11).

RADIATION CYSTITIS

In 1985, the possible beneficial effects of hyperbaric oxygen therapy in radiation-induced hemorrhagic cystitis in three patients were published (16). Treatment of this disease had, so far, been a very frustrating experience. There is no known cure and all palliative measures, like intravesical instillation of formalin, alum or silver-nitrate, treatment with corticosteroids, cyclokapron or antibiotics together with hydrostatic dilatation etc., frequently fail. Total cystectomy with urinary diversion is then the only solution.
We developed a research protocol for patients with serious complaints of macroscopic hematuria, necessitating regular bloodtransfusions and abacterial cystitis caused by previous pelvic radiation for urologic, gynecologic or any other malignancies in that region and where all other known treatment modalities had failed.
Any other therapy was discontinued and only hyperbaric oxygen treatment was given. In the first 10 patients hematuria stopped completely in 5 and greatly diminished in the other 5. No more blood transfusions were needed nor bladder irrigations, previously frequently necessary. In all patients where the hematuria did not disappear completely, residual tumor was present. This was unnnoticed before, because of the hematuria. There seemed to occur some kind of demarcation between healthy mucosa and tumor so that in three patients a transurethral resection of tumor could be done. The mechanism of action of HBO in this disease is not known. Bladder biopsies before and after HBO and the macroscopic appearance of the bladder mucosa showed a quick return to normal during treatment. We assume that increasing the oxygen tension has a beneficial effect on vascular compromised tissues. It promotes healing by stimulating neovascularisation and formation of granulation tissue (14).

HYPERBARIC OXYGEN IN VASCULAR SURGERY

There are several indications for HBO in vascular surgery. The most important indications are those where there is a temporary need for hyperbaric oxygen; that is in reversible diseases. HBO can be used as presurgical therapy of poorly vascularized limbs with edema. HBO treatment makes it possible to supply the ischemic peripheral tissues with sufficient oxygen.

The extensive edema caused by the vascular insufficiency can be markedly reduced. The oxygen supply together with the local fibrinolytic activity gives a possibility for restoration of the peripheral flow. When reconstructive vascular surgery is performed, the succes rate can thus be improved. Patient selection for this treatment, however is very difficult.
When the arteriosclerotic or diabetic vascular insufficiency is so severe that amputation of an extremity becomes necessary, intermittent hyperbaric oxygen therapy can postpone amputation or can change the necessary above-knee amputation in a below-knee amputation. With transcutaneous tissue oxygen measurements the exact level of amputation can be determined very precisely.
Other indications are carotid artery surgery in high risk patients; the use of HBO in chronic ischemia of the limbs without the possibility of vascular reconstruction. Our experience is that good results can be obtained even in chronic vascular insufficiency with ulcerations. Amputation can be postponed, the ulcers heal and the serious pain by ischemic neuritis disappear after 2-3 weeks of daily treatments.
Other indications are found in patients with functional vasospastic diseases. In this group we find M.Raynaud, Raynaud phenomenon, livedo reticularis, erythrocyanosis and artery spasms. This artery spasm can be caused by trauma, artery puncture or ergotamin intoxication.
The symptoms of acute ergot-poisoning are due to extreme vasoconstriction caused by stimulation of the alpha-receptors in the vessel wall. All forms of treatment are directed to the relief of the intense vasospasm, but this can be so severe that no vasodilator drugs can reach the affected parts, mainly the lower extremities. Frequently amputations are necessary.
We have treated 14 patients with impending gangrene due to this form of vasospasm. The very small circulation left, carried enough oxygen to keep the peripheral tissues alive during the acute period. A mean treatment period of three days was sufficient to prevent peripheral gangrene and amputation in all cases.
Also patients with Morbus Buerger are chronically treated with hyperbaric oxygen in our center.

THE FUTURE

The field of hyperbaric oxygen therapy is changing. Many new indications have come into focus, while others have become obsolete.
Originally a mode of treatment largely based on clinical experience, more and more indications are now accepted only on the basis of sound experimental evidence and prospective randomized clinical trials. For some indications like gas gangrene and carbon monoxide poisoning, clinical experience is so overwhelming that further clinical trials are considered unethical.
Other indications are relatively seldom and so variable in their presentation that prospective randomized trials are virtually impossible, such as chronic osteomyelitis and necrotizing fasciitis. Other indications like multiple sclerosis show very conflicting results in different, both open and controlled trials so that the debate goes on and will probably never end.
Although this may be true, we have a strong obligation in hyperbaric medicine to design controlled clinical trials, especially in the group of research indications, in order to prove the efficacy of hyperbaric oxygen and have them accepted. On the other hand we have to strengthen the available evidence also in the accepted category. The Hyperbaric Oxygen Therapy Committee of the Undersea and Hyperbaric Medical Society considers it her duty to design multicenter trials and many are under way already. Sometimes it is difficult to use the same protocols and use the same study design, because science is practised differently in different parts of the world. The interpretation of "good clinical practice" is not the same in many, even neighbouring, countries.
Other questions that need an urgent answer are more basic and therefore

conferences like these are extremely useful for us, simple clinicians. We
need the basic sciences to answer questions on oxygen transport to tissues.
We know that we start with oxygen in the lungs and we finally end in cells
in peripheral tissues. We know however very little about the way oxygen is
delivered into those tissues. Can the optimum dose of oxygen be determined
by accurate measurement of wound oxygen tension? What is a dose of oxygen
and what factors influence this? Oxygen to tissue transport depends on many
factors that are not fully understood. Even the way from capillary to cell
remains unclear in many aspects. The role of oxygen free radicals in many
disease processes has to be determined. In these questions we need the
investigations of physiologists, pharmacologist, cell biologists etc. etc.
New fields that are explored in our center in Amsterdam are:
- Different ways of monitoring and measuring tissue oxygen tensions.
- Hyperbaric lung lavage in pulmonary alveolar proteinosis (9).
- Prospects of zetotherapy in the combination of hyperbaric oxygen and 131
  I-Metaiodobenzylguanidine (131 I-MIBG) in patients with neuroblastoma
  (15).
- The influence of hyperbaric oxygen on cultivated neuroblastoma cell li-
  nes in vitro.
- The influence of hyperbaric oxygen on organ preservation for transplan-
  tation purposes (7).
- The influence of hyperbaric oxygen on mycosis and yeasts both in vitro
  and in vivo (Candidiasis and Mucormycosis etc.).

I sincerely hope that this 19th ISOTT-conference marks the beginning of a
long and fruitful cooperation of basic and clinical sciences in all aspects
of oxygen transport and utilization for the benefit of many future patients.

REFERENCES

1. Bakker DJ. The use of hyperbaric oxygen in the treatment of certain
    infectious diseases especially gas gangrene and acute dermal gangre-
    ne. Drukkerij Veenman BV., Wageningen. University of Amsterdam, 1984.

2. Bakker DJ. Clostridial myonecrosis. In: Davis JC, Hunt TK, (eds).
    Problem wounds: the role of oxygen. New York:Elsevier, 153-172, 1988.

3. Bakker DJ, Kox C. Classification and therapy of necrotizing soft
    tissue infections: The role of surgery, antibiotics and hyperbaric
    oxygen. In: Current problems in general surgery 5 (4): 489-500, 1988.

4. Boerema I, Meijne NG, Brummelkamp WH, Bouma S, Mensch MH, Kamermans
    F, Hanf S, van Aalderen A. Life without blood. J Cardiovasc Surg 182:
    133-146, 1960.

5. Brummelkamp WH, Hogendijk J, Boerema I. Treatment of anaerobic
    infections (clostridial myositis) by drenching the tissues with
    oxygen under high atmospheric pressure. Surgery 49: 299-302, 1961.

6. Committee Report on Hyperbaric Oxygen Therapy. Undersea and Hyperba-
    ric Medical Society Inc., Bethesda, Maryland USA (publ. no. 30CR(HBO)
    1989), revised edition 1989.

7. van Gulik TM, Hullett DA, Boudjema K, Landry AS, Southard JH, Sollin-
    ger HW, Belzer FO. Prolonged survival of murine thyroid allografts
    after 7 days' hyperbaric oxygen culture in the UW preservation
    solution at hypothermia. Transplantation 49: 971-975, 1990.

8. Hyperbaric Oxygen Therapy. Davis JC, Hunt TK, (eds). Undersea Medical
    Society Inc., Bethesda, Maryland USA, 1977.

9. Jansen HM, Zuurmond WAA, Roos CM, Schreuder JJ, Bakker DJ. Whole-lung lavage under hyperbaric oxygen conditions for alveolar proteinosis with respiratory failure. Chest 91, 6: 829-832, 1987.

10. Mader JT, Brown GL, Guckian JC, Wells CH, Reinarz JA. A mechanism for the amelioration by hyperbaric oxygen of experimental staphylococcal osteomyelitis in rabbits. J Infect Dis 142: 915-922, 1980.

11. van Merkesteyn JPR, Bakker DJ, Dellemijn HL. The use of hyperbaric oxygen in the treatment of osteomyelitis and osteoradionecrosis of the jaw. In: Proc 2nd Swiss Symposium on Hyperb Med. Bakker DJ, Schmutz J, (eds). Foundation for Hyp Med, Basel, 151-154, 1990.

12. Meijne NG. Hyperbaric oxygen and its clinical value (with special emphasis on biochemical and cardiovascular aspects). Charles C Thomas, Springfield Ill, 1970.

13. Problem Wounds: The role of oxygen. Davis JC, Hunt TK, (eds). New York: Elsevier, 1988.

14. Rijkmans BG, Bakker DJ, Dabhoiwala NF, Kurth KH. Successful treatment of radiation cystitis with hyperbaric oxygen. Eur Urol 16: 354-356, 1989.

15. Voute PA, Hoefnagel CA, de Kraker J, Valdes Olmos R, Bakker DJ, van der Kleij AJ. Results of treatment with 131 I-Metaiodobenzylguanidine (131 I-MIBG) in patients with neuroblastoma. Future prospects of Zetotherapy. In: Advances in Neuroblastoma Research 3, Evans AE, D'Angio GJ, Knudson Jr AG, Seeger RC (eds). Wiley-Liss, Inc. 439-445, 1991.

16. Weiss JP, Boland FP, Mori H, et al: Treatment of radiation-induced cystitis with hyperbaric oxygen. J Urol 134: 352-354, 1985.

RAT BRAIN ADAPTATION TO CHRONIC HYPOBARIC HYPOXIA

Joseph C. LaManna

Depts of Neurology, Physiology & Biophysics, and Neurosciences
Case Western Reserve University School of Medicine
Cleveland, Ohio 44106, USA

INTRODUCTION

The normal function of the brain is entirely dependent on an adequate supply of oxygen for the provision of ATP via oxidative energy metabolism. But, the presence of oxygen in excess of metabolic needs can be toxic, with especially drastic consequences for an organ such as the brain in which neurons cannot be regenerated once lost. Thus, there are mechanisms that have evolved, apparently in response to these opposing requirements, which control oxygen delivery to the brain so that it is tightly coupled to tissue oxygen consumption, such that energy demand is satisfied but brain oxygen exposure is minimized. This principle of just sufficient oxygen implies that under normal ambient oxygen pressures and normal physiological conditions, regional brain blood flow is relatively low and brain tissue oxygen tension is also low (Sick et al., 1982). Increases in functional energy demand would have to be accompanied by at least transient increases in local oxygen delivery (LaManna et al., 1987).

Exposure to hypoxic conditions necessitates some form of compensatory responses to restore the balance between oxygen delivery and consumption (LaManna et al., 1984). The primary systemic compensatory mechanism is increased ventilatory rate while the initial response within the brain is increased blood flow. These early reactions seem capable of supporting normal function; but, they cannot be optimal because prolonged hypoxic exposure results in further functional and structural refinements (Dempsey and Forster, 1982). Changes in environmental oxygen availability are encountered in normal physiology, for example by deep burrowing animals or changes in altitude of habitation. Therefore, hypoxic adaptations are part of the normal acclimatization mechanisms.

In considering hypoxia, there are two main factors which must always be defined: severity and duration. Severity can be qualitatively defined as mild, moderate or severe. The significance of these qualitative categories can be conveniently understood by reference to physiological responses to altitude. "High" altitude is considered to begin at about 3000m where the first signs of acclimatization are observed. The oxygen pressure at this altitude is the boundary between mild and moderate hypoxia. At the other end, successful adaptation may be possible up to, at most, 6000m and this defines the upper limit for moderate hypoxia. The ranges thus specified are mild hypoxia, sea level to 3000m; moderate hypoxia, 3000 - 6000m; and severe hypoxia, >6000m. In terms of duration, three arbitrary divisions can be defined. Acute hypoxic exposure lasts for minutes to hours; short term exposure is from days to weeks; and long term exposure is months to years. A fourth division might be added to consider generations of exposure.

The potential structural and functional adaptations to hypoxia can be considered at 5 different levels (Schmidt-Nielsen, 1983): 1) Lung ventilation; 2) O_2 diffusion from lung to blood; 3) O_2 transport in blood; 4) O_2 diffusion from blood to tissue; 5) O_2 utilization by tissues. This paper discusses the adaptations to acute and short term moderate hypoxia in the rat.

METHODS

Male Wistar rats (250-350 g), 6 to 12 at a time, were kept for 3 weeks in hypobaric chambers kept constantly at a pressure of 380 torr except for up to 1 hr daily when the pressure was returned over 10 minutes to atmospheric for cage cleaning, water and food. Littermate rats were kept outside the chambers but in the same location and were watered and fed with the same schedule. Rats were studied under 6 conditions:

Group I   normoxic controls, kept for 3 weeks under normobaric, normoxic
                 conditions;
Group II   15 minute hypoxic, kept for 3 weeks under normobaric, normoxic
                 conditions then exposed for 15 minutes to 10% O_2 (bal. N_2)
Group III   3 hour hypoxic, the same as group II, but exposed to 3 hours of the
                 hypoxic gas mixture;
Group IV   3 week hypoxic, kept for 3 weeks at 0.5 ATM then, after surgical
                 preparation, exposed for 3 hours to normobaric 10% oxygen;
Group V   15 minute hypoxic hypercapnic, the same as group II, but the hypoxic
                 gas mixture also contains 5% carbon dioxide;
Group VI   4 hour recovery after 3 weeks hypoxic, kept for 3 weeks at 0.5 ATM
                 then, after surgical preparation, exposed for 4 hours to
                 normobaric normoxia.

Surgical Preparation

Surgical procedures on all rats were done under conditions of normobaric normoxia. On the day of the experimental procedure, rats were anesthetized with chloral hydrate (400

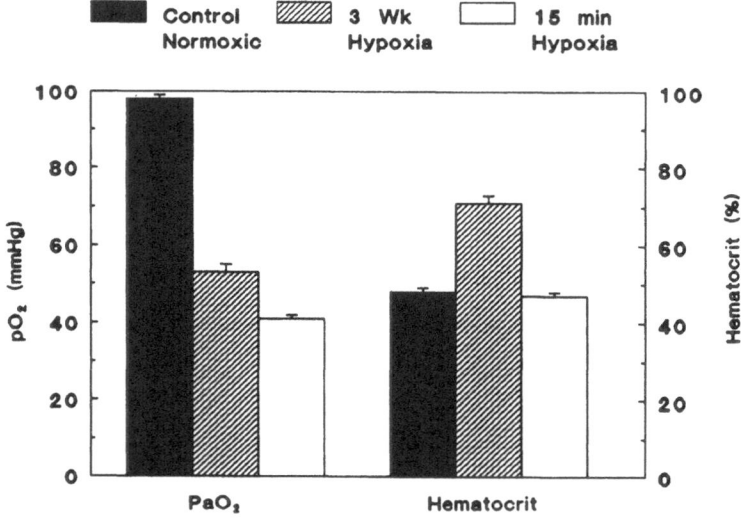

Figure 1. Arterial oxygen tension and hematocrit in acute and chronic hypoxia

mg/kg, i.p.). A cannula was placed in the tail artery for monitoring systemic arterial pressure, and another placed in the right atrium via the external jugular for intravenous drug and radioactive tracer administration. Arterial blood gases and pH were determined using a Radiometer ABL2 Blood Gas Analyzer, which also provided temperature corrected estimates of bicarbonate and hemoglobin (Hb) levels (g/dl). Plasma glucose (mM) was determined in separate samples using a Beckman Glucose Analyzer. All skin incisions were infiltrated with a long-lasting local anesthetic, and sutured. As the rats awakened from anesthesia, they were immobilized in shoulder-to-hip plaster casts. The time elapsed from anesthesia to immobilization was about 1 hour, and an additional 3 hours was allowed for recovery from anesthesia. Rats studied under conditions that required altered ventilation gas mixtures were placed in a plexiglass chamber which was flushed constantly with the appropriate humidified gas mixture. Body temperature was maintained near 37°C with a feedback regulated heat lamp and a rectal probe.

Capillary Density

Rats were anesthetized with sodium pentobarbital (75 mg/kg, i.p.), and FITC-dextran (60,000 MW, 200mg/kg) was administered through an intravenous cannula and allowed to circulate for a few minutes. The rat was then decapitated, the brain was removed from the skull, placed in a solution of 2-methylbutane cooled by liquid nitrogen for about 30 sec, for 1-2, and then placed in Liquid N_2. The brains were stored at -80° C until sectioned. Five μm thick coronal sections of the frozen brains were cut by a cryostat microtome at -25° C. Sections were transferred to slides and allowed to dry for 2 hrs at room temperature. Fluorescence photomicrographs were taken at 100x total magnification. Alternatively, some sections were stained for alkaline phosphatase activity before light photomicrography.

Statistical Analyses

Except where noted, all data were analyzed and all six groups compared for statistically significant differences ($p < 0.05$) using the analysis of variance function ONEWAY, corrected by Tukey's test for multiple comparisons, in SPSS/PC+.

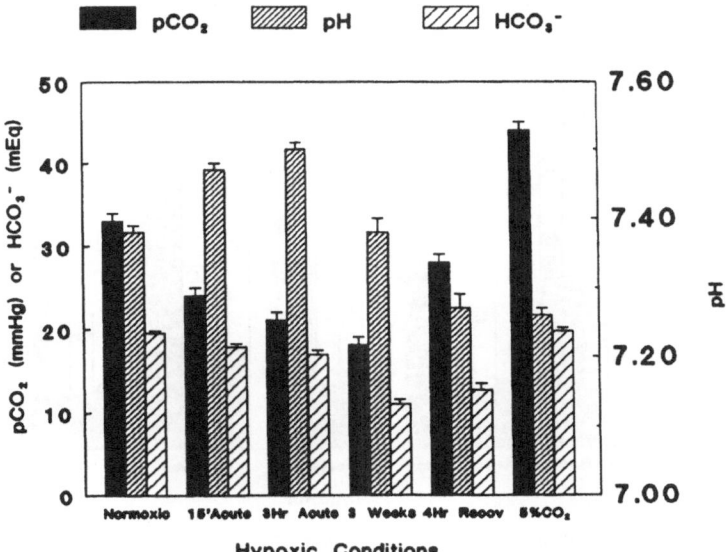

Figure 2. Blood acid-base changes in acute and chronic hypoxia

RESULTS AND DISCUSSION

Brain tissue oxygen tension falls almost immediately when the partial pressure of inspired oxygen is reduced even moderately (Kreisman et al., 1981; Sick et al., 1982; Metzger et al., 1971; Leniger-Follert et al., 1975). In mild hypoxia, brain capillary hemoglobin becomes more disoxygenated (Rosenthal et al., 1976) despite increased blood volume (Shockley and LaManna, 1988), increased blood flow (LaManna et al., 1992; Beck and Krieglstein, 1987; Dahlgren, 1990), faster capillary mean transit times (Shockley and LaManna, 1988), increased red cell velocity (Ivanov et al., 1985) and capillary recruitment (Francois-Dainville et al., 1986; Kissen and Weiss, 1989). However, these compensatory mechanisms that are activated by mild hypoxia are not sufficient to avoid electrophysiological consequences (LaManna et al., 1984; Speckmann et al., 1988). In awake rats the hypoxia-triggered ventilatory increase (Dempsey and Forster, 1982) exacerbates the compensatory inadequacy because hyperventilation blunts the cerebrovascular responses by decreasing $PaCO_2$ (Kreisman et al., 1981; Rosenthal et al., 1976; Dahlgren, 1990).

Rats continuously exposed to hypoxia respond with longer term adaptive changes that supersede the acute responses. Although the systemic mechanisms elicited by continuous hypoxic exposure differ by species, the rat responses appear to be very similar to those of the human adapting to high altitude (Olson and Dempsey, 1978; Dempsey and Forster, 1982). As in humans (Boyer and Blume, 1984), rats exposed to long term hypoxia fail to gain weight (Fregly and Waters, 1966; LaManna et al., 1992). This is considered part of the successful adaptation to hypoxia at altitudes up to 5000m (Krzywicki et al., 1971; Blume et al., 1984). Adaptation also includes the well known increase in blood hemoglobin concentration and hematocrit (Figure 1). Another change that occurs in most species is a right shift of the hemoglobin dissociation curve. The ratio of arterial blood hemoglobin to hematocrit (Hct, %) measured in 37 normoxic control rats (LaManna et al., 1992) was 0.42 ± 0.02 (mean ± sd). The Hb/Hct ratio should be used to calculate Hb content in hypoxic rats (this ratio is unchanged with prolonged hypoxic exposure (Fregly and MacArthur, 1990)), as the ABL2 estimate

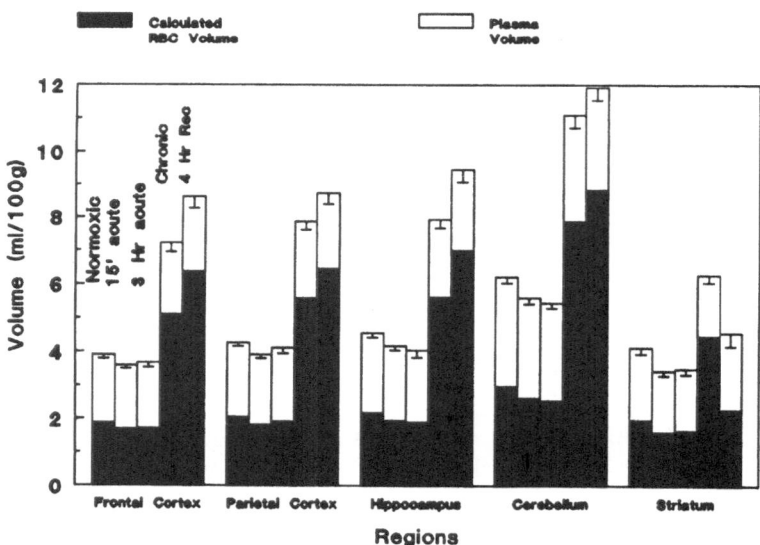

Figure 3. Regional brain blood volume in hypoxia. Open bars indicate plasma volume measured by [^{14}C]sucrose space. Filled bars are RBC space calculated based on the arterial hematocrit.

becomes less accurate with high hemoglobin values (Sjöberg et al., 1990). Oxygen saturation (SaO$_2$) was calculated from the PaO$_2$ by the Hill equation using a P$_{50}$ of 38.7 and a Bohr effect constant of -0.603 (Dhindsa et al., 1981). No correction in the P$_{50}$ was made for the 3 week hypoxic rats because, although there appears to be a short-lived increase in P$_{50}$ to 47 (David, 1975), there is no report of a hypoxic-induced right shift in the hemoglobin saturation curves in rats, and since there is apparently no change in the 2,3-DPG/Hemoglobin ratio with prolonged hypoxic exposure in rats (Dux et al., 1984).

An indication that the rats in this study achieved successful acclimatization to hypobaric hypoxia over the three weeks is the preservation of brain wet weight/dry weight ratios and constancy of brain ionic composition (LaManna et al., 1992).

Increases in lung ventilation frequency and tidal volume are early responses to hypoxic exposure. This response is demonstrated by a hyperventilation induced decrease in PaCO$_2$ observed in acute hypoxia that persists while hypoxia is maintained (Figure 2). Arterial oxygen partial pressure increases by about the same amount as the fall in PaCO$_2$. Hyperventilation leads to an acute respiratory alkalosis which resolves due to increased renal HCO$_3^-$. O$_2$ diffusion from lung to blood may be increased after long term adaptation to hypoxia through increased lung size.

Hypoxia-induced increased regional cerebral blood flow has been consistently reported for a variety of mammalian species including human (Siesjö, 1978; Krasney et al., 1990; Edelman et al., 1984). Cerebral blood flow increases despite decreased PaCO$_2$ (Kissen and Weiss, 1989; Beck and Krieglstein, 1987; Dahlgren, 1990). After continued hypoxic exposure there is "resetting" of the relationship between PaCO$_2$ and CBF (Severinghaus et al., 1966), so that after 3 weeks regional cerebral blood flow is returned to control levels in spite of continued hyperventilation (LaManna et al., 1992). The increased blood flow is associated with an increased blood volume, mainly through an increase in RBC volume since plasma volume appears to remain stable (Figure 3). Although PaCO$_2$ remained low, blood pH returned to normal due to renal excretion of HCO$_3^-$ (Dempsey and Forster, 1982). After acute normalization of atmospheric oxygen atmosphere, the rats went through a period of rising PaCO$_2$ resulting in acidosis (Figure 2), but without an increase in regional cerebral blood flow. This suggests that control of cerebral blood flow by CO$_2$ under conditions of elevated

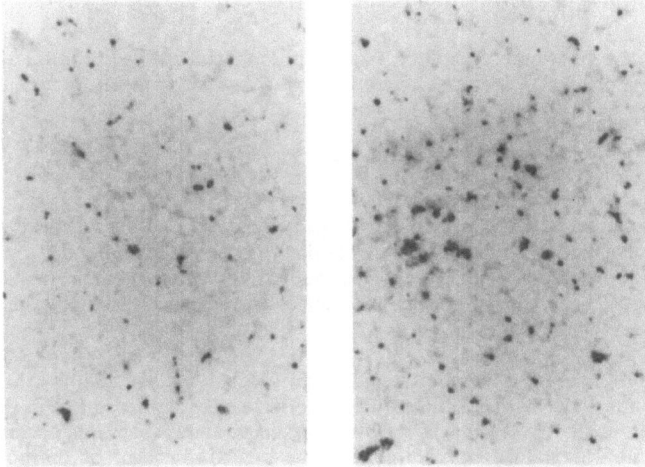

Figure 4. Reversed FITC-dextran fluorescence micrographs from the parietal cortex of normoxic control rats (left), and rats after 3 weeks of hypobaric hypoxia (right).

hematocrit is blunted, and implies that oxygen delivery may be the controlling variable for CBF (Brown et al., 1985), although direct mechanical effects from the increased viscosity accompanying elevated HCT cannot be ignored (Harrison, 1989).

Elevation of blood hemoglobin concentration was sufficient to maintain oxygen delivery when CBF returned to baseline during prolonged hypoxic exposure. The maintenance of oxygen delivery by itself cannot guarantee that the brain tissue oxygen tension returns to normal. This is because the diffusional driving force is the capillary PO_2 and the flux is determined (at constant use) primarily by the intercapillary distance. Thus, the most efficient way to return tissue oxygen tension to normal is to increase the capillary density, thus decreasing intercapillary distance. And indeed, increased capillary density (Figures 4 and 5) is found after prolonged hypoxic exposure (Opitz, 1951; Diemer and Henn, 1965; Miller and Hale, 1970). Increasing oxygen diffusion at the tissue level is, therefore, accomplished mainly through capillary recruitment which increases the diffusional surface area. After 3 weeks of

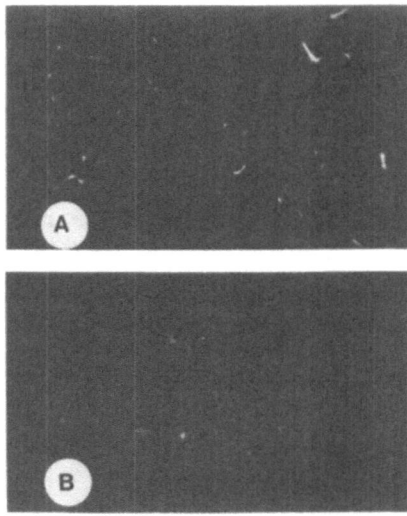

Figure 5. Negative photomicrograph of alkaline phosphatase stained brain sections from the parietal cortex of hypoxic-adapted (A) and normoxic control (B) rats.

hypoxia the dramatic increase in capillary density suggests that angiogenesis is initiated by hypoxia (Figures 4 and 5). The increase in capillary density is compounded by an increase in the density of the stereo-specific D-glucose facilitative transport carrier at the blood-brain barrier (Harik et al., 1991), suggesting a markedly increased capacity for glucose influx into the brain of hypoxic-adapted rats.

In addition to the systemic responses and local vascular adaptations to prolonged hypoxia, there must also be mechanisms which activate adaptive responses in the tissue itself. Chronic (Musch et al., 1983) changes in cerebral glucose metabolism that accompany hypoxia are not completely explained by continued tissue hypoxia because of the normalization of oxygen delivery (Musch et al., 1983). The reason for the increased brain lactate production that continues during prolonged hypoxia remains to be determined.

CONCLUSIONS

The results indicate that the brain response to hypoxia follows a pattern consisting of 3 main mechanisms which attempt to maintain constant oxygen delivery. The acute response is an increase in regional blood flow. Blood flow returns to control levels when the oxygen carrying capacity of the blood is increased by erythropoiesis. Finally, there is a decrease in intercapillary distance as a result of capillary angiogenesis.

ACKNOWLEDGEMENTS

These studies were supported in part by PHS grants HL42215 and NS22077.

REFERENCES

Beck, T. and Krieglstein, J., 1987, Cerebral circulation, metabolism, and blood-brain barrier in rats in hypocapnic hypoxia, Am. J. Physiol., 252: H504-H512.

Blume, F.D., Boyer, S.J., Braverman, L.E., Cohen, A., Dirkse, J., and Mordes, J.P., 1984, Impaired osmoregulation at high altitude, JAMA, 252: 524-526.

Boyer, S.J. and Blume, F.D., 1984, Weight loss and changes in body composition at high altitude, J. Appl. Physiol., 57: 1580-1585.

Brown, M.M., Wade, J.P.H., and Marshall, J., 1985, Fundamental importance of arterial oxygen content in the regulation of cerebral blood flow in man, Brain, 108: 81-93.

Dahlgren, N., 1990, Local cerebral blood flow in spontaneously breathing rats subjected to graded isobaric hypoxia, Acta Anesthesiol. Scand., 34: 463-467.

David, J.N., 1975, Adaptation of brain monoamine synthesis to hypoxia in the rat, J. Appl. Physiol., 39: 215-220.

Dempsey, J.A. and Forster, H.V., 1982, Mediation of ventilatory adaptations, Physiol. Rev., 62: 262-346.

Dhindsa, D.S., Metcalfe, J., Blackmore, D.W., and Koler, R.D., 1981, Postnatal changes in oxygen affinity of rat blood, Comp. Biochem. Physiol., 69A: 279-283.

Diemer, K. and Henn, R., 1965, Kapillarvermehrung in der hirnrinde der ratte unter chronischem sauerstoffmangel, Die Natur., 52: 135-136.

Dux, E., Temesvári, P., Joó, F., Adám, G., Clements, F., Dux, L., Hideg, J., and Hossmann, K.-A., 1984, The blood-brain barrier in hypoxia: ultrastructural aspects and adenylate cyclase activity of brain capillaries, Neurosci., 12: 951-958.

Edelman, N.H., Santiago, T.V., and Neubauer, J.A., 1984, Hypoxia and brain blood flow, in: "High Altitude and Man," J.B. West and S. Lahiri, eds., pp. 101-113, American Physiological Society, Bethesda, Maryland.

Francois-Dainville, E., Buchweitz, E., and Weiss, H.R., 1986, Effect of hypoxia on percent of arteriolar and capillary beds perfused in the rat brain, J. Appl. Physiol., 60: 280-288.

Fregly, M.J. and MacArthur, S.A., 1990, Some characteristics of post-hypoxic-induced drinking in rats, Aviat. Space Environ. Med., 61: 1116-1120.

Fregly, M.J. and Waters, I.W., 1966, Posthypoxic drinking response of rats, Fed. Proc., 25: 1220-1226.

Harik, S.I., Behmand, R.A., and LaManna, J.C., 1991, Chronic hypobaric hypoxia increases the density of cerebral capillaries and their glucose transporter protein, J. Cereb. Blood Flow Metab., 11 (Suppl): S496.(Abstract)

Harrison, M.J.G., 1989, Influence of hematocrit on the cerebral circulation, Cerebrovasc. Brain Metab. Rev., 1: 55-67.

Ivanov, K.P., Kalinina, M.K., and Levkovich, Yu.I., 1985, Microcirculation velocity changes under hypoxia in brain, muscles, liver, and their physiological significance, Microvasc. Res., 30: 10-18.

Kissen, I. and Weiss, H.R., 1989, Cervical sympathectomy and cerebral microvascular and blood flow responses to hypocapnic hypoxia, Am. J. Physiol., 256: H460-H467.

Krasney, J.A., Jensen, J.B., and Lassen, N.A., 1990, Cerebral blood flow does not adapt to sustained hypoxia, J. Cereb. Blood Flow Metab., 10: 759-764.

Kreisman, N.R., Sick, T.J., LaManna, J.C., and Rosenthal, M., 1981, Local tissue oxygen tension - cytochrome a,a_3 redox relationships in rat cerebral cortex in vivo, Br. Res., 218: 161-174.

Krzywicki, H.J., Consolazio, C.F., Johnson, H.L., Nielsen, W.C.Jr., and Barnhart, R.A., 1971, Water metabolism in humans during acute high-altitude exposure (4,300 m), J. Appl. Physiol., 30: 806-809.

LaManna, J.C., Light, A.I., Peretsman, S.J., and Rosenthal, M., 1984, Oxygen insufficiency during hypoxic hypoxia in rat brain cortex, Br. Res., 293: 313-318.

LaManna, J.C., Sick, T.J., Pikarsky, S.M., and Rosenthal, M., 1987, Detection of an oxidizable fraction of cytochrome oxidase in intact rat brain, Am. J. Physiol., 253: C477-C483.

LaManna, J.C., Vendel, L.M., and Farrell, R.M., 1992, Brain adaptation to chronic hypobaric hypoxia in rats, J. Appl. Physiol., (submitted):

Leniger-Follert, E., Lübbers, D.W., and Wrabetz, W., 1975, Regulation of local tissue P_{O2} of the brain cortex at different arterial O_2 pressures, Pflüg. Arch., 359: 81-95.

Metzger, H., Erdmann, W., and Thews, G., 1971, Effect of short periods of hypoxia, hyperoxia, and hypercapnia on brain O_2 supply, J. Appl. Physiol., 31: 751-759.

Miller, A.T.,Jr. and Hale, D.M., 1970, Increased vascularity of brain, heart, and skeletal muscle of polycythemic rats, Am. J. Physiol., 219: 702-704.

Musch, T.I., Dempsey, J.A., Smith, C.A., Mitchell, G.S., and Bateman, N.T., 1983, Metabolic acids and [H^+] regulation in brain tissue during acclimatization to chronic hypoxia, J. Appl. Physiol., 55: 1486-1495.

Olson, E.B. and Dempsey, J.A., 1978, Rat as a model for humanlike ventilatory adaptation to chronic hypoxia, J. Appl. Physiol., 44: 763-769.

Opitz, E., 1951, Increased vascularization of the tissue due to acclimatization to high altitude and its significance for the oxygen transport, Exp. Med. Surg., 9: 389-403.

Rosenthal, M., LaManna, J.C., Jöbsis, F.F., Levasseur, J.E., Kontos, H.A., and Patterson, J.L.Jr., 1976, Effects of respiratory gases on cytochrome a in intact cerebral cortex: is there a critical PO_2, Br. Res., 108: 143-154.

Schmidt-Nielsen, K., 1983, "Animal Physiology: Adaptation and environment," 3rd edn., Cambridge University Press, New York.

Severinghaus, J.W., Chiodi, H., Eger, E.I., Brandstater, B., and Hornbein, T.F., 1966, Cerebral blood flow in man at high altitude, Circ. Res., 19: 274-282.

Shockley, R.P. and LaManna, J.C., 1988, Determination of rat cerebral cortical blood volume changes by capillary mean transit time analysis during hypoxia, hypercapnia, and hyperventilation, Br. Res., 454: 170-178.

Sick, T.J., Lutz, P.L., LaManna, J.C., and Rosenthal, M., 1982, Comparative brain oxygenation and mitochondrial redox activity in turtles and rats, J. Appl. Physiol., 53: 1354-1359.

Siesjö, B.K., 1978, "Brain Energy Metabolism," John Wiley & Sons, Chichester.

Sjöberg, F., Vegfors, M., and Lennmarken, C., 1990, The clinical value of the hemoglobin concentration analysis (g/L) and the calculation of oxyhemoglobin saturation (%) in an automatic blood gas analyzer (ABL2, Radiometer A/S, Denmark), Scand. J. Clin. Lab. Invest., 50 (Suppl 203): 241-245.

Speckmann, E.-J., Bingmann, D., Lehmenkühler, A., and Lipinski, H.G., 1988, Changes of the bioelectric activity and extracellular micromilieu in the central nervous system during variations of local oxygen pressure, in: "Oxygen Sensing in Tissues," H. Acker, ed., pp. 179-191, Springer-Verlag, Berlin.

WHOLE LUNG LAVAGE UNDER HYPERBARIC CONDITIONS:

1. THE MONITORING

JD Biervliet, JA Peper, CM Roos, AJ vd Kleij, DJ Bakker,
HM Jansen

Dpts of Anesthesiology, Pulmonology and Hyperbaric Medicine
Academic Medical Center
Amsterdam, The Netherlands

INTRODUCTION

This chapter, part 1 of 2, focuses on monitoring of circulation and
ventilation during the treatment of Pulmonary Alveolar Proteinosis (PAP).

The disease is characterized by accumulation of lipid- and protein-
rich substances in alveoli and small airways. Affected individuals suffer
from progressive hypoxaemia at rest, requiring continuous oxygen support.
Massive unilateral bronchoalveolar lavage (BPL) is the most effective
treatment; large quantities of infusion fluid are instilled in one lung
under conditions of general anesthesia and intubation with a double lumen
tube to ventilate the other lung. Sessions will last for 3 to 4 hours.
However, during this treatment tissue oxygenation is in jeopardy by the
combined detrimental effects of the disease itself, the lavage and the one-
lung-ventilation.

Hyperbaric oxygen (HBO) might prevent the deterioration of hypoxemia:
a higher ambient pressure at the alveolar site enhances O_2 gradient across
the alveolar membrane, resulting in an elevation of arterial oxygen
tension (PaO_2). Arterial oxygen content (CaO_2) benefits from a higher
fraction of both oxyhemoglobin and of physically dissolved oxygen.

Application of HBO, however, is not without risks. Apart from O_2
toxicity, hyperbaric pressure will redistribute bloodflow and interferes
with ventilator settings, thus partially neutralizing the objective on
tissue oxygenation. The problems are related to the altered physical
conditions and though avoidable, they will at least create
anesthesiological pitfalls.

An extensive monitoring was set up for the direct assessment of
ventilation and circulation during these lavaging procedures under HBO. It
allowed fast interpretation of rapidly changing vital parameters, thus
optimizing the treatment within safety limits.

METHODS

In a multiplace hyperpressure walk-in tank 11 sessions of massive whole-lung lavages were performed in 6 patients. The main chamber has a content of 98 m^3 and an inner floor surface of 30 m^2; it is constructed to manage a pressure of 5 atm. The tank serves a regional function in the treatment of disorders as diving decompression syndrome, gas gangrene, carbon monoxide intoxication; it is equipped with all facilities of an operating theatre.

Five males and one female, ages ranging from 29 to 50 years, were treated in a severely disabling stage of a diagnosed PAP; all required continuous oxygen support for an adequate arterial oxygenation. Diffusion capacities, as measured by a carbon monoxide diffusion test, ranged from 20 to 30% of the predicted normal. Right-to-left shunting was calculated more than 20% of cardiac output. Apart from their pulmonary disease, all patients were in a good physical condition.

After an oral premedication with diazepam, 10 mg, anesthesia was induced and maintained with continuous infusions of propofol (2 $mg.kg^{-1}$, followed by 6 $mg.kg^{-1}.hr^{-1}$) and fentanyl (200 μg; 3 $\mu g.kg^{-1}.hr^{-1}$). Pancuronium was administered every hour. All patients received low doses heparin in a continuous infusion and a perioperative regimen of corticosteroids.

The largest possible left dual lumen tube was placed in the trachea, using a flexible bronchoscope to confirm the optimal lung separation; cuffs were filled with saline to avoid leakage during the pressure build-up and to protect the airways against hyperinflated cuffs during decompression. Controlled ventilation was started at a $FiO_2 = 1.0$ using a Servo 900B ventilator. Tidal volumes and ventilatory rates were adjusted to maintain a stable acid-base balance and $PaCO_2$ level, monitored in end-tidal (ET) PCO_2 and confirmed by repeated blood gas analysis.

Pulseoximetry was used for the continuous monitoring of arterial oxygenation (SpO_2). After introduction of arterial and venous lines, a pulmonary artery catheter, equipped with a fiberoptic channel for measurement of mixed venous O_2 saturation (SvO_2) was inserted.

Circulatory parameters were kept within physiological limits; 500 ml of haemaccel was administered iv, followed by Ringer lactate 5 $ml.kg^{-1}.hr^{-1}$ to compensate for the insensible loss throughout the procedure; reflex bradycardia was prevented with 0.5 mg atropine intravenously.

The following parameters were monitored:
- central circulation: ECG, heart rate, systemic arterial blood pressure, pulmonary artery pressure, right atrial pressure, pulmonary capillary wedge pressure,
- peripheral circulation: urine production, body temperature, SvO_2 and SpO_2.
- ventilation: respiratory rate and pressure, FiO_2, ET CO_2 concentration, fluid filling pressure.
ECG, systemic arterial blood pressure, pulmonary artery pressure, right atrial pressure, SvO_2, Presp, and $ETCO_2$ were continuously and digitally recorded (EDR8000, United Kingdom, see fig.2).

Blood gas sampling was performed at regular intervals. Arterial and mixed venous blood samples were withdrawn simultaneously in glass syringes, stored on ice, depressurized and analyzed within 5 minutes (ABL3). Continuously monitored SvO_2 values (Oximetrix3, Abbott) were compared to saturation values from samples, taken from the pulmonary artery.

After bi-aural paracentesis patients were placed in lateral decubitus position with the lung to be lavaged dependent. In 15 minutes the chamber was pressurized to 2.2 bar, respiratory minute volume was increased with 10% to avoid alveolar hypoventilation. The lavaging procedure was then started. The airway of the dependent lung was blocked, while ventilation of the upper lung was continued at a FiO_2 of 1.0. Hereafter a strategy was instituted towards a controlled reduction of FiO_2 without endangering tissue oxygenation, as monitored by the pertinent variables. The blocked lung was degassed passively and infused with a volume of saline at 37 C. In the process of filling a hydrostatic pressure of 50 cm H_2O was applied; filling was continued until a sharp decline in any of the observed parameters indicated that a maximum was reached; by then 1.0 to 1.5 L was instilled. Five minutes were allowed to reach an equilibrium in protein diffusion, drainage was achieved by applying negative pressure.

The lavages continued until the drainage was more or less clear; in total between 15 and 20 L was instilled. At the end of the procedure the "wet" lung was ventilated with 100% oxygen and expanded by applying an end expiratory pressure of 10 cm H_2O, while the "dry" and more compliant lung was blocked away from ventilation, protecting it against distension during this procedure. After a stabilization period both lungs were ventilated with decreasing oxygen fractions and the chamber was depressurized. The patients were extubated and transferred to the recovery room.

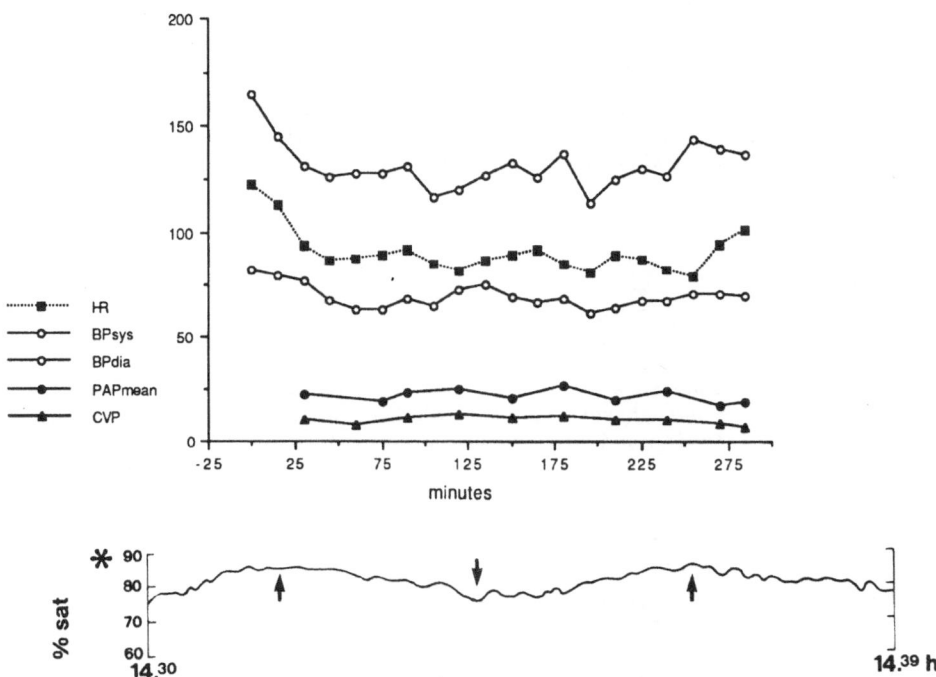

Fig. 1. Upper panel: circulatory parameters in stable periods of sessions. Lower panel: variations in on-line SvO_2 during instillation.

RESULTS

Throughout the procedures, in periods between the instillation of fluid, the monitored parameters of circulation (fig 1, upper panel) and ventilation remained stable. We did not observe any consistent increase in pulmonary arterial and right atrial pressure nor did a change in regularly measured hematocrit and in urine production suggest significant fluid absorption from the lavaged lung into the circulation.

After instituting controlled ventilation, SpO_2 increased in all patients from values <85% to values between 96 and 100%, to remain so throughout the session, pressurizing and lavaging included. Moreover, $PaCO_2$ values at 1 bar and 2 bar were 40.9 ($\pm$9.0) and 41.6 ($\pm$7.9) respectively, as an indication of adequate ventilation.

During pressurizing and two lung ventilation both PaO_2 and PvO_2 increased:

	1.0 bar	2.2 bar	
PaO_2	160 $\pm$ 135	485 $\pm$ 270	mmHg
PvO_2	49 $\pm$ 9	81 $\pm$ 54	mmHg

At an FiO_2 of 1.0 and comparing values of an ambient pressure of 2.2 bar vs. 1 bar, a significant rise was observed in $-O_2$ tension: PaO_2 ($p<0.0005$) and PvO_2 ($p<0.05$), and in -saturation: SaO_2 ($p<0.005$), SvO_2 ($p<0.05$). Calculated O_2 transport improved, demonstrated in an increase of CaO_2 from 17.0 $\pm$ 2.5 to 18.8 $\pm$ 2.4 ml.dl^{-1} ($p<0.05$). $C(a-v)O_2$ did not change significantly, as an indication of a more or less stable oxygen extraction and cardiac output.

On-line obtained SvO_2 and SpO_2 values matched SO_2 values from samples. The correlation coefficient for 46 comparisons between SvO_2 and in vitro analyzed saturation, was 0.8. During all lavages, SvO_2 increased in the process of filling the lung (fig 1, lower panel) to return to baseline values after draining. The mean increase of PaO_2 from the empty lung to the filled lung, measured in 7 sessions at 12 occasions, was 133% with a median of 104%. The simultaneous mean increase of PvO_2 was 7% with a median of 6%.

DISCUSSION

Broncho-alveolar lavage was successfully advocated in the treatment of alveolar proteinosis. All patients recovered well from this rather aggressive approach, which for safety reasons, was performed under hyperbaric conditions in a hyperpression tank. To our opinion the starting point -low PaO_2 values at rest, a considerable intra-pulmonary shunt and diminished CO diffusion-, justifies the treatment.

Pure oxygen ventilation under hyperbaric conditions will give rise to a very high alveolar oxygen tension, enlarging O_2 gradient across the alveolar membrane. Beside an elevation of saturated oxyhaemoglobin to its maximum, a high arterial oxygen tension (PaO_2) also enhances the amount of physically dissolved oxygen according to Henry's law, stating that the concentration of a dissolved gas equals its (partial) pressure times its solubility coefficient in a fluid. As the overall result the arterial oxygen content (CaO_2) improves.

Objective in monitoring was the early detection of circulatory and ventilatory impairment. Extensive monitoring was installed, including on-line peripheral arterial and mixed venous saturation measurement, which were controlled by repeated off-line biochemical analysis. Pulse-oximetry is a well established, reliable, non-invasive monitor of oxygenation. Since ambient pressure doesn't interfere with the assessment of saturation by this technique, it was to be expected that the high SpO_2 values under circumstances of hyperpression matched the off-line measured SaO_2 values.

The good correlation (R=0.8) between directly measured SvO_2 and in vitro values, indicates reliability of this parameter; the on-line response to a potentially rapid change in SvO_2 further emphasized its usefulness. It should be noticed that the range of measured values during the sessions is located on the steep part of the oxy-haemoglobuline dissociation curve. In that range a minor decrease in SvO_2 corresponds with a substantial decrease in PvO_2. Assuming a stable O_2 consumption under the circumstances of general anesthesia and complete muscle relaxation, a decreasing PvO_2 might indicate:
-a decrease in PaO_2 due to intercurrent ventilatory problems, such as leakage to or compression on the ventilated lung,
-a diminished cardiac output to the tissues, to a level that O_2 consumption becomes transport-dependent and O_2 extraction increases.

Another solid argument for the feasibility of SvO_2 monitoring results from the fact that assessment of cardiac output by thermodilution is highly unreliable during these procedures; the instillation of fluid, a subsequent change both in intrathoracic temperature and in intrapulmonary blood flow are contributing factors to the inaccuracy of the measurement.

A lavage procedure is expected to be most effective when instilled volumes are maximal or at least sub-maximal. Leakage into the ventilated lung was prevented by the upper position of this lung and by filling both cuffs with saline. Furthermore, the lavaging was continued until one of the monitored parameters showed a change, (unequivocally) indicative for an impaired circulatory or ventilatory system. A further infusion might endanger the patient, although we are lacking evidence to that respect, never having reached such a volume. The importance of hyperpression and monitoring is best illustrated in the recording shown in fig.2. During instillation the following unexpected changes were observed: respiratory pressure increased and ET CO_2 decreased; simultaneously a rise in pulmonary artery and right atrial pressure and a decrease of systemic arterial pressure was recorded. SvO_2 did not show its usual increase. It was decided to interrupt the filling process. Since 500 ml lavage fluid was not recovered, we assumed an accidental leak into the ventilated lung; suction was applied to this lung, hereafter all variables returned to baseline values and the procedure was continued. Out-chamber assessment of oxygenation cannot supply this vital information on rapidly changing circumstances. In the same manner, the on-line information allowed us to reduce FiO_2 gradually from 1 to 0.5 without jeopardizing oxygenation. In doing so, we diminished the potential risk of O_2 toxicity to lung tissue.

The observed increase in PaO_2 and in PvO_2 during instillation is a well known phenomenon in literature; it is explained by a gradual hydrostatic obstruction of pulmonary capillary flow by the instilled fluid, deviating the blood flow away from this non-ventilated lung and diminishing intra-pulmonary shunt. A higher PaO_2, once the filling is completed, reflects the improvement in ventilation/perfusion mismatch.

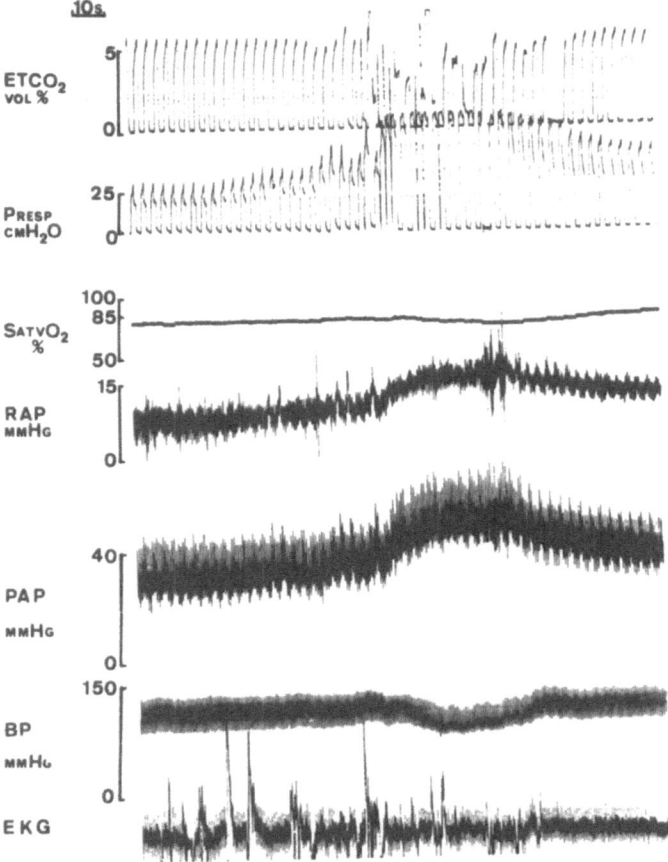

Fig. 2. Recording of monitored parameters during event, described in text.

CONCLUSION

Bronchoalveolar lavage improved the course of pulmonary alveolar proteinosis; all patients were able to resume a normal life. Performed under hyperbaric circumstances with an important increase in tissue oxygenation, it is the treatment of choice in case of severe respiratory failure. An intensive monitoring will create additional safety margins that will permit a more effective lavage. Combined, on-line monitoring of arterial and mixed venous saturation was considered the standard of care.

REFERENCES

Jansen, H.M., Zuurmond, W.W.A. and Roos, C.M., 1987, Whole-lung lavage under hyperbaric oxygen conditions for Alveolar proteinosis with respiratory failure. Chest, 91,6:829-832.
Cohen, E. and Eisenkraft, J.B., 1990, Bronchopulmonary lavage: effects on oxygenation and hemodynamics, J Cardiothorac Anesth, 4:609-615.

WHOLE LUNG LAVAGE UNDER HYPERBARIC CONDITIONS: 2. MONITORING TISSUE OXYGENATION

AJ van der Kleij[1], JAK Peper[2], JD Biervliet[2],
DJ Bakker[1], CM Roos[3], HM Jansen[3]

Departments of Hyperbaric Medicine[1],
Anesthesiology[2], Pulmonology[3],
Academic Medical Center, Meibergdreef 9,
1105 AZ, Amsterdam, The Netherlands

INTRODUCTION

Pulmonary Alveolar Proteinosis (PAP) is characterized by dyspnea and progressive detoriation of arterial oxygenation, caused by accumulation of lipid- and protein-rich insoluble material in the alveoli. Bronchoalveolar lavage is the basic principle of treatment and can be performed while ventilating one lung. However due to severe pulmonary insufficiency, ventilation of one lung lessens arterial PO_2 to unacceptable levels. In order to prevent arterial and tissue hypoxemia we performed bronchoalveolar lavage under hyperbaric conditions as previously described[1]. During these lavage procedures hemodynamic and central circulatory oxygen parameters were continuously monitored[2]. Oxygenation of peripheral tissue was monitored by multiple assessments of skeletal muscle PO_2 (quadriceps musculature) with a polarographic pO_2 needle electrode[3] and by transcutaneous pO_2 measurements. The aim of this study was to study alterations in muscle oxygenation during and after bronchoalveolar lavage.

MATERIAL AND METHODS

Four pulmonary lavage procedures were performed in two patients. The right and the left lung of both patients (a male and a female) were lavaged during separate sessions with an interval of 6 to 12 weeks. During normobaric conditions the patients were intubated with a dual lumen tube (Bronchocath, Mallinckrodt). After the radial artery and the pulmonary artery were catheterized the patient was turned on the side which had to be lavaged. Before the multiplace chamber was pressurized to 2.2 ATA the position of the dual lumen tube was checked with a fiberoptic-bonchoscope. Before, during and after the chamber was pressurized arterial as well as mixed venous blood gas analyses (ABL3, Radiometer) were determined at regular intervals.

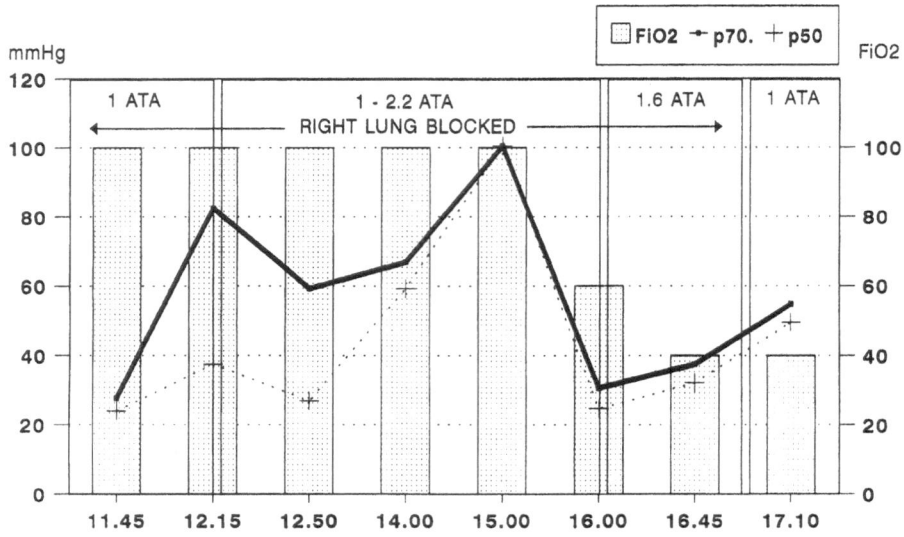

MALE, 44 YEARS

Figure 1. Course of skeletal muscle PO$_2$ values (p50 and p70) during bronchoalveolar lavage under hyperbaric conditions in a male patient.

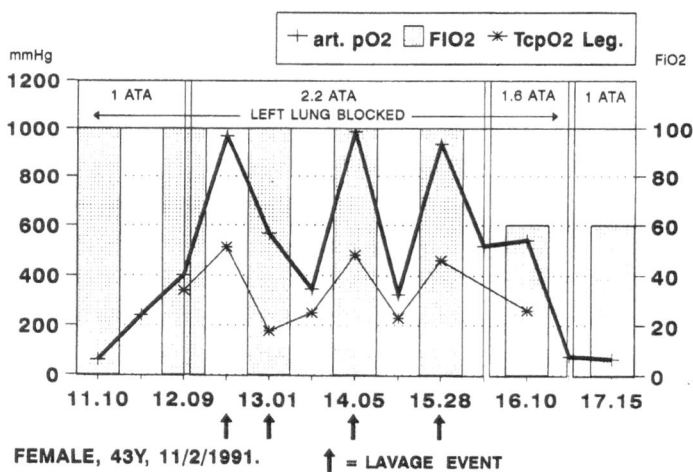

FEMALE, 43Y, 11/2/1991.   ↑ = LAVAGE EVENT

Figure 2. Arterial PO$_2$ and TcPO$_2$ during bronchoalveolar lavage at different ambient pressures in a female patient.

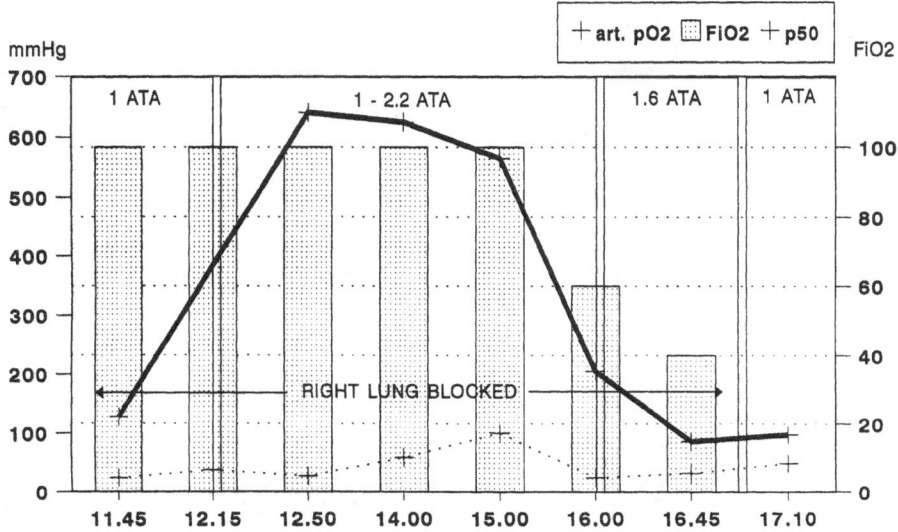

MALE, 44 YEARS

Figure 3. Arterial PO$_2$ and skeletal muscle PO$_2$ (p50 value)during bronchoalveolar lavage under hyperbaric conditions.

Skeletal muscle pO$_2$ was measured in the quadriceps muscululature with a polarographic pO$_2$ needle electrode. The p50 and p70 values of the cumulative distribution curve (n=100) were used to quantify muscle oxygenation. Transcutaneous pO$_2$ was monitored infraclavicularly. Bronchoalveolar lavage was performed by instilling an amount of NaCl 0.9% fluid into the blocked tube until a drop in CvO$_2$ was observed or a drop in mean arterial blood pressure. The total lavage amount ranged between 15 and 20 liters of NaCl 0.9%.

RESULTS AND DISCUSSION

Both patients sustained the hyperbaric conditions as well as the lavage procedures without complications and recovered from their PAP.

Increasing the ambient pressure from 1 ATA to 2,2 ATA (FiO$_2$ of 100%) increased p50 and p70 values of skeletal muscle pO$_2$ (Fig.1, 12.15). Instilling lavage fluid resulted in an immediate drop in the p50 and p70 values (Fig. 1, 12.50). After the total lavage procedure improved oxygen availability within skeletal muscle tissue was reflected by higher p50 and p70 values even with a lower FiO$_2$ (Fig.1, 17.10).

Instilling lavage fluid and ventilating one diseased lung revealed a sharp decline in arterial PO$_2$ (Fig. 2) compared to ventilating a lung which had already been lavaged in a previous session. The level course of the p50 and p70 values was in concordance with arterial pO$_2$ values (Fig. 3).

Transcutaneous pO$_2$ values were 4 to 5 times higher than skeletal muscle PO$_2$ values. Thusfar this difference in tissue oxygenation has not been elucidated. One explanation could be that chronic hypoxia affects autoregulatory mechanisms of skeletal muscle flow earlier than that of the skin.

CONCLUSION

During pulmonary lavage procedures under normo- and hyperbaric conditions assessment of tissue oxygenation with a polarographic pO_2 needle electrode and $TcPO_2$ electrodes provides information about lavage efficacy, restored pulmonary diffusion capacity and peripheral oxygenation.

REFERENCES

1. Jansen H. M., Zuurmond W. A., Roos C. M., Schreuder J. J., Bakker D.J. Whole-lung lavage under hyperbaric oxygen con ditions for alveolar proteinosis with respiratory failure. Chest 91, 6: 829-832 (1987).

2. Biervliet J. D., Peper J. A. K., Roos C. M., Kleij van der A. J., Bakker D.J. Whole Lung Lavage Under Hyperbaric Con ditions: 1. The Monitoring. This volume.

3. Kleij van der A. J., Koning de J., Beerthuizen J., Goris R. J. A., Kreuzer F., Kimmich H. P., 1983. Early detection of hemorrhagic hypovolemia by muscle oxygen pressure assessment: Preliminary report, Surgery. 93:518-524.

SKELETAL MUSCLE PO$_2$ IN ANAEROBIC SOFT TISSUE INFECTIONS DURING HYPERBARIC OXYGEN THERAPY

AJ van der Kleij[1], DJ Bakker[1], M Lubbers[1], Ch.P. Henny[2]

Departments of Surgery[1] and Anesthesiology[2],
Academic Medical Center, Meibergdreef 9,
1105 AZ, Amsterdam, The Netherlands

INTRODUCTION

Gas gangrene is caused by anaerobic, spore-forming gram positive encapsulated bacilli of the genus clostridium and is characterized by a rapidly progressive, painful infection of the skin, subcutis, fascia and muscle. Since the introduction of Hyperbaric Oxygen (HBO) therapy for gas gangrene in 1960 by Boerema and Brummelkamp[1] this therapy has been widely accepted. HBO therapy has reduced the extent of surgical treatment resulting in more limb and tissue salvage. Additionally HBO therapy is nowadays used as adjunctive therapy for mixed anaerobic and aerobic soft tissue infections, progressive bacterial synergistic gangrene, necrotizing fasciitis and nonclostridial myonecrosis[2 3]. Since hyperbaric oxygen therapy increases the amount of free molecular oxygen in the tissues, we were interested in the skeletal muscle PO$_2$ levels of the infected tissue or the tissues adjacent to the initial wound, before, during and after HBO treatment. In four patients with an anaerobic soft tissue infection we were able to measure skeletal muscle PO$_2$ with a polarographic needle PO$_2$ electrode under different ambient pressures.

PATIENTS AND METHODS

Two patients with gas gangrene and two patients with an anaerobic soft tissue infection were treated with HBO therapy in a multiplace chamber. The first patient had developed gas gangrene over the region of the right scapula after a car accident. The second patient developed gas gangrene in the right upper leg after drainage of an intraabdominal abscess. The third patient, with an amputation below the knee, developed an anaerobic soft tissue infection just above the amputation level. The fourth patient, with a diabetic foot ulcer, developed an ascending necrotizing fasciitis of the left leg. The first two patients were treated according to the Amsterdam Therapeutic Regimen (8 liter oxygen/min. by mask during 90 minutes at 3 ATA, 7 sessions in three days). The other two pa-

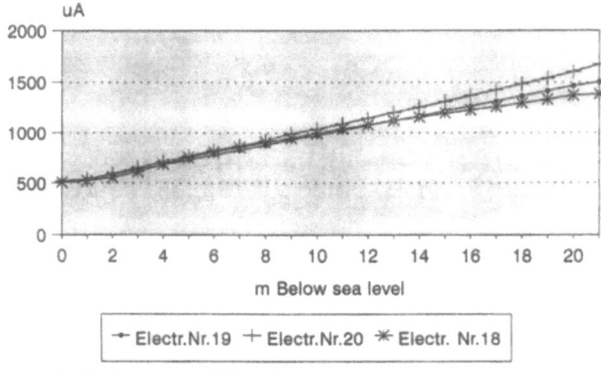

PO2 ELECTRODE LINEARITY TESTS
DURING DIFFERENT AMBIENT PRESSURES

m Below sea level

Electr.Nr.19   Electr.Nr.20   Electr. Nr.18

Date: 16-10-90, Vpol: -654 , 5.6 VOL% O2
0.5 h Polarization. Gain:0.53 Zero:0.79

Figure 1. Linearity tests of three polarographic PO_2 needle electrodes.

tients were treated according to the gas gangrene protocol the first day only, the following days they received HBO treatment twice a day. Skeletal muscle PO_2 measurements were performed with a polarographic needle electrode as previously described by van der Kleij et al [4]. During the investigation, the linearity of the needle electrodes were tested in a calibration unit at different ambient pressures (Fig. 1). In clinical practice calibration of the PO_2 needle electrode preceded each tissue PO_2 measurement under normobaric as well as under hyperbaric conditions. The p50 value of the cumulative frequency distribution curve (n=100) was used to express the state of oxygenation within the tissues.

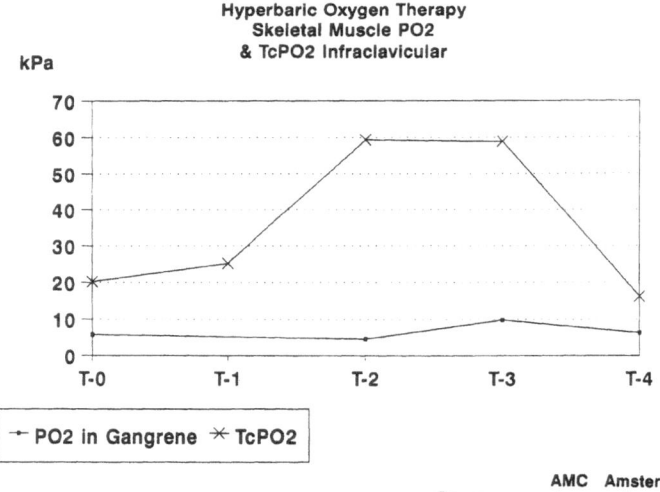

Hyperbaric Oxygen Therapy
Skeletal Muscle PO2
& TcPO2 Infraclavicular

PO2 in Gangrene   TcPO2

AMC   Amsterdam
Gas gangrene in scapula region

Figure 2. Infraclavicular transcutaneous PO_2 values and p50 values of skeletal muscle PO_2 measurements(n=100) measured in a gas gangrene patient.

Skeletal muscle PO_2 measurements were performed at 5 intervals. Before the chamber was pressurized (T-0) to an ambient air-pressure of 3 ATA and subsequently at 5 minutes after the ambient air pressure reached 3 ATA (T-1), thirty and sixty minutes at 3 ATA (T-2 and T-3, respectively), and after decompression to 1 ATA (T-4). Patients who were artificially ventilated received a FiO_2 of 80% whereas spontaneously breathing patients received 8 liters of oxygen per minute by mask. To objectivate the effect of hyperoxygenation, infraclavicular $TcPO_2$ measurements (TCM1, Radiometer) were performed (Fig. 2).

RESULTS

The linearity tests of the polarographic PO_2 needle electrodes revealed the expected proportional change in current at different ambient pressures (Fig. 1). The electrodes showing optimal linearity were used for clinical application.

All patients tolerated HBO treatment and remained hemodynamically stable. The patient with an ascending necrotizing fasciitis of the left leg did not survive and died two weeks after HBO treatment due to Multiple Organ Failure. The patient with gas gangrene over the right scapula survived, which is very rare for this type of localization. This might be due to the fact that he was a young healthy male (25 years of age). The infraclavicular $TcPO_2$ measurements parallelled arterial PO_2 partial pressures and revealed much higher values than the p50 (median) values measured within the infected region (Fig. 2).

In patients with mixed anaerobic soft tissue infections (Fig. 3) it was found that skeletal muscle PO_2 values were lower than in patients with gas gangrene (Fig. 3).

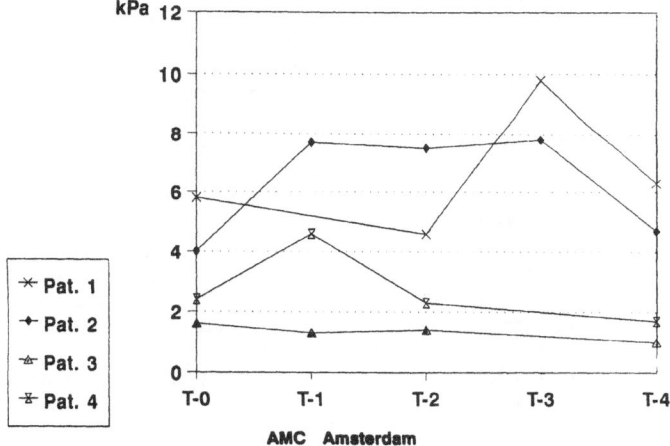

Figure 3. Change in p50 values of skeletal muscle PO_2 measurements before (T-0), during (T-1, T-2, T-3) and after (T-4) HBO treatment. Patients 1 & 2 are gas gangrene, patients 3 & 4 are anaerobic soft tissue infection.

DISCUSSION

Skeletal muscle blood flow is regulated by neural, hormonal, and autoregulatory control mechanisms. Additionally tissue PO_2 values are dependent on a number of variables such as, pulmonary and cardiac function, oxygen-carrying capacity of the blood, intravascular volume, local oxygen consumption, and the type of tissue[5]. Skeletal muscle PO_2 measurements reflect the overall balance of all above mentioned variables and the interaction of mechanisms controlling blood flow within the muscle. As oxygenation within the tissues has a heterogeneous distribution, it is generally accepted to express the skeletal muscle PO_2 measurements as median or mean of multiple single measurements[6][7]. As micro-electrodes are not suitable for clinical application, polarographic needles and multiwire surface electrodes are generally used[6][7][8].

Using an implanted polarographic PO_2 electrode in non-healing, chronic wounds Sheffield[9] found PO_2 values ranging from 5 to 20 mm Hg, whereas his control values ranged between 30-50 mmHg. The underlying cause of a non-healing wound is often a inadequate localized blood supply, reflected by low tissue PO_2 values. The two patients suffering anaerobic soft tissue infections revealed low tissue PO_2 values, even though the infected region was hyperaemic and did not show any sign of inadequate blood supply. Necrotizing fasciitis is an anaerobic infection of the fascia which subsequently induces the cellular metabolic rate of the underlying muscle tissue to increase. According to Silver[10], oxygen tension adjacent to cells is influenced by the utilization of oxygen within that cell. High oxygen consumption in the cell reduces oxygen concentration outside the cell. This means that tissues with a high cellular oxygen consumption will show an increased number of steep oxygen gradients resulting in lower p50 (median) values. This may explain why patients with an anaerobic soft tissue infection showed decreased median PO_2 values. Skeletal muscle PO_2 measurements performed in septic patients were found to have a high mean of 48.8 ± 8.5 mm Hg. The authors suggested that these findings are problably caused by a reduction in oxygen consumption by the tissues and not by a reduction in oxygen transport capacity[7]. Similar values were also found in the present study in patients with gas gangrene before hyperbaric oxygen therapy was initiated. Sixty minutes after therapy with IPPV with a FiO_2 of 80% under hyperbaric conditions of 3 ATA mean PO_2 values between 60 and 75 mm Hg were found.

Finally the infraclavicular $TcPO_2$ values measured revealed a significant increase in PO_2 after the initiation of hyperbaric oxygen therapy. This reflects the increase of oxygen availability within the infraclavicular region. This may in turn reflect that hyperbaric oxygen therapy increases the systemic oxygen availability.

SUMMARY

In the present study skeletal muscle PO_2 measurements were performed in patients with gas gangrene and anaerobic soft tissue infections before, during and after hyperbaric oxygen therapy. Polarographic PO_2 needle electrodes appeared to be suitable for application during different ambient pressures. We found that patients with gas gangrene revealed higher skeletal muscle PO_2 values than patients with an anaerobic soft tissue infection. This may be explained by a higher metabolic rate within the anaerobically infected soft tissues. The higher PO_2 values in gas gangrene may be caused by alpha toxins, affecting cellular and intracellular membranes thus destroying PO_2 diffusion barriers.

REFERENCES

1. Brummelkamp W. H., Hogendijk J., Boerema I. Treatment of anearobic infections (clostridium myositis) by drenching the tissue with oxygen under high atmospheric pressure. Surgery. 49:229-302, (1961).
2. Mader J. T., Mixed Anaerobic and Aerobic Soft Tissue Infections, In: Problem wounds: The role of oxygen, J. C. Davis, T. K. Hunt, eds., Elsevier, New York, (1988).
3. Bakker D. J.. The use of hyperbaric oxygen in the treatment of certain infectious diseases especially gas gangrene and acute dermal gangrene. Thesis, University of Amsterdam (1984).
4. Kleij van der A. J., Koning de J., Beerthuizen J., Goris R. J. A., Kreuzer F., Kimmich H. P., 1983. Early detection of hemorrhagic hypovolemia by muscle oxygen pressure assessment: Preliminary report, Surgery. 93:518-524.
5. Kleij van der A. J., in: "Skeletal muscle PO_2 in shock," Thesis, Nijmegen, The Netherlands (1984).
6. Beerthuizen G. I. J. M., Goris R.J. A., Bredel J. J., Mashhour Y. A., Kimmich H. P., Kleij van der A.J., Kreuzer F. Muscle oxygen tension, hemodynamics, and oxygen transport after extracorporal circulation. Crit. Care Med. 16:748-750 (1988).
7. Wiederhöfer St, Boekstegers P., Pilz G., Werdan K.. Tissue oxygen partial pressure within skeletal muscle is high in septic patients with multiple organ failure. Circulatory Shock. 34:1, 87 (1991).
8. Haus J., Schönleben K., Spiegel H. U.. Therapiekontrolle durch Überwachung des Gewebe-pO_2. In: Aktuelle Probleme in der Angiologie:41. Hans Huber, Bern. (1982).
9. Sheffield P. J. Tissue oxygen measurements with respect to soft-tissue wound healing with normobaric and hyperbaric oxygen. Hyperb Oxyg Rev. 6(1):18-46, (1984).
10. Silver I. A.. Heterogeneity in tissue oxygenation; systemic and local factors. In: Advances in Physiological Sciences. Vol 25: Oxygen Transport to Tissue, Eds., A. G. B. Kovach, E. Dora, M. Kessler, I. A. Silver, Pergamon Press, Oxford, England, 67-76. (1981).

EFFECT OF HYPERBARIC OXYGEN TREATMENT AND PERFLUOROCHEMICAL ADMINISTRATION ON GLUTATHIONE STATUS OF THE LUNG

E. Purucker and J. Lutz

Physiologisches Institut, Universität Würzburg, Röntgenring 9
D/W 8700 Würzburg, Germany

INTRODUCTION

Application of hyperbaric oxygen, either in form of sessions with mild hyperoxia (1-2.5 ATA $\approx$ 100-250 kPa) or under pressures (up to 4-6.8 ATA) is used for different causes (enhancing cancer radiosensitity, treatment of anaerobic infections, decompression sickness, consequences of cerebral ischemia a.o.). However, the danger of an increase of oxygen free radicals is given, whereas somatic adaptations will also occur.

From previous studies using normobaric hyperoxia over 2-14 days a distinct stimulation in the cellular defence system has been evaluated (1-4). Thus, reduced glutathione (GSH), glucose-6-phosphate dehydrogenase (G6PDH), glutathione reductase (GR), glutathione peroxidase (GPX) and superoxide dismutase (SOD) have been shown to increase in response to hyperoxia. These increments represent an adaptive stimulation of the cellular antioxidant systems that protect lung cells from damage mediated by O_2^- and oxygen free radicals (5,6). The importance of the glutathione mediated defence is underlined by the finding of a drastic rise in mortality in glutathione depleted mice exposed to hyperoxia (7).

As all these findings were derived from experiments using long lasting normobaric hyperoxia, we designed the following study to evaluate, whether toxic effects must be taken into consideration, when intermittent mild hyperbaric (2.5 ATA) hyperoxia is used for therapeutic reasons. Thus, we tested the status of reduced (GSH) and oxidized (GSSG) glutathione as one of the organism's most effective systems to reduce oxidative stress.

The study was expanded by groups receiving perfluorochemicals (PFC) used as artificial oxygen carriers in addition to HBO. To test the effect of a single exposure to hyperbaric hyperoxia we exposed rats to a pressure of 7 ATA for 60 min.

METHODS

Experiments were performed using male Wistar rats of 230-260 g body weight. They were housed in groups of four animals in macrolon cages, fed a standard diet (Altromin,

Lage, Germany), and had access to food and water *ad libidum*. Besides a control group the animals were exposed to hyperbaric oxygen (HBO) in random groups of four in a cylindrical pressure chamber (30 cm length, 20 cm internal diameter) that was closed with a 2 cm thick steel-framed acrylic glass. Draegersorb 600 (Draeger, Lübeck, Germany) was placed on a dish at the bottom of the chamber to prevent accumulation of carbon dioxide. Compressed oxygen from a commercial cylinder was used to perform pressurization. In the first experiment the animals underwent daily sessions for 8 or 14 days with 2.5 ATA for 90 min between 9-12 a.m.. The pressure was increased and at the end decreased by 2 ATA/min. The PFC-group received Fluosol-DA (Green Cross Corp., Osaka, Japan) in a dose of 2g PFC/kg b.w. on the first day and thereafter every second day 15 min before the HBO session. The last PFC dose was given 48h and the last HBO session was performed 24 h before killing.

In the second experiment animals underwent a single session with 7 ATA for 60 min. The pressure was increased beginning with a slow initial phase of 1 ATA/30 s and then doubled every 30 s according to a previously validated schedule (8). PFC-emulsion was given in a dose of 8 g/kg. An additional group received allopurinol in a dose of 50 mg/kg b.w. together with PFC as a scavenger. The animals were investigated immediately after decompression. They were killed in ether anesthesia by exsanguination, lungs were perfused with ice-cold NaCl 0.9%, excised, and homogenized in ice-cold 5% 5-sulfosalicylic acid. The glutathione content (GSH and GSSG) was determined enzymatically (9).

Statistics: The values denote the GSH or GSSG content of the whole lung related to kg body weight as mean plus/minus the standard error of the mean. Analysis of variance was used to calculate differences between the sample means of the groups. When relevant, statistical testing between the groups were done using the Scheffé-test. A probability value $p < 0.05$ was considered to signify a true difference in mean value.

RESULTS

Under *repeated mild oxygen treatment* for 8 or 14 days a significant increase in the content of lung tissue GSH occurred: + 19% and +16%, resp. ($p<0.05$) compared to the controls value of 6.43 ± 0.39 μM/lung·kg b.w. (x±SEM). This increase was more pronounced by treatment with PFC: +47% and +39% after 8 an 14 days ($p<0.01$) [Fig.1].

GSSG tented to increase in the HBO group after 8 days. It exhibited 28% higher levels ($p<0.05$) under PFC as compared to the control value of 411 ± 20 nM/lung·kg. After 14 days of treatment the GSSG increase accounted for 57% in the HBO group and for 118% in the HBO+PFC group [Fig. 2].

The ratio GSSG/GSH (control value: 0.052 ± 0.003) tended to lower values in both groups after 8 days of treatment, but increased after 14 days by 39% and 57% [Fig. 3].

After *acute 7-ATA hyperbaric oxygen treatment* GSH increased by 40% ($p<0.05$). No increase occurred under additional PFC-treatment [Fig. 4]. However, GSSG increased by 241% and 163% ($p<0.05$) in HBO and HBO+PFC group [Fig. 4], thereby the ratio GSSG/GSH was also elevated by + 175 and 203%, respectively ($p<0.01$). By means of allopurinol, given together with PFC in one group, the increase of GSSG as well as that of the ratio GSSG/GSH could be nearly totally suppressed.

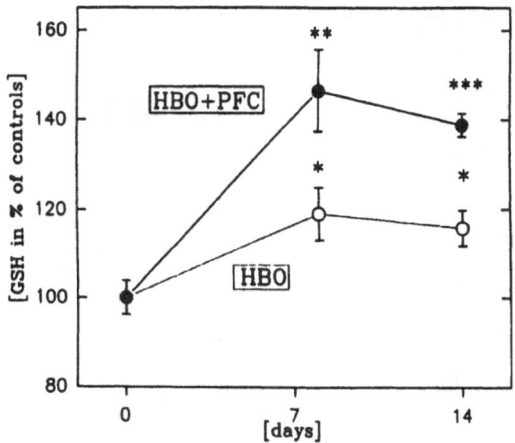

Fig. 1. GSH changes after 8 and 14 days of repeated HBO-treatment (open circles) and HBO+PFC treatment (filled circles) [x±SEM, p<0.05 *, <0.01 **, <0.001 ***]

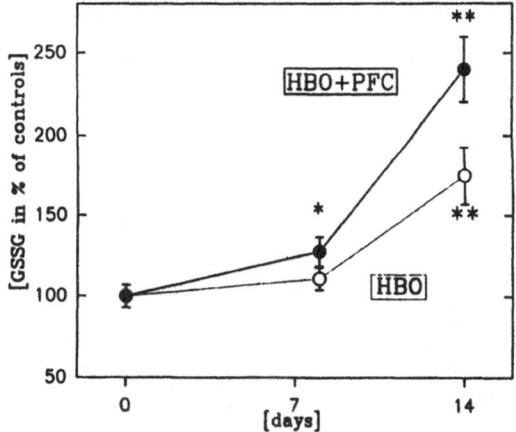

Fig. 2. GSSG changes after 8 and 14 days of repeated HBO-treatment (open circles) and HBO+PFC treatment (filled circles) [x±SEM, p<0.05 *, <0.01 **]

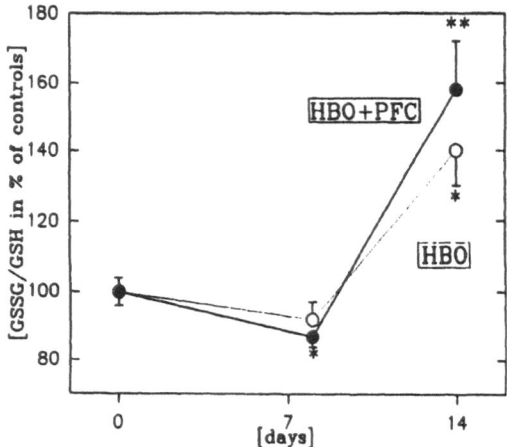

Fig. 3. GSSG/GSH changes after 8 and 14 days of repeated HBO-treatment (open circles) and HBO+PFC treatment (filled circles) [x±SEM, p<0.05 *, <0.01 **]

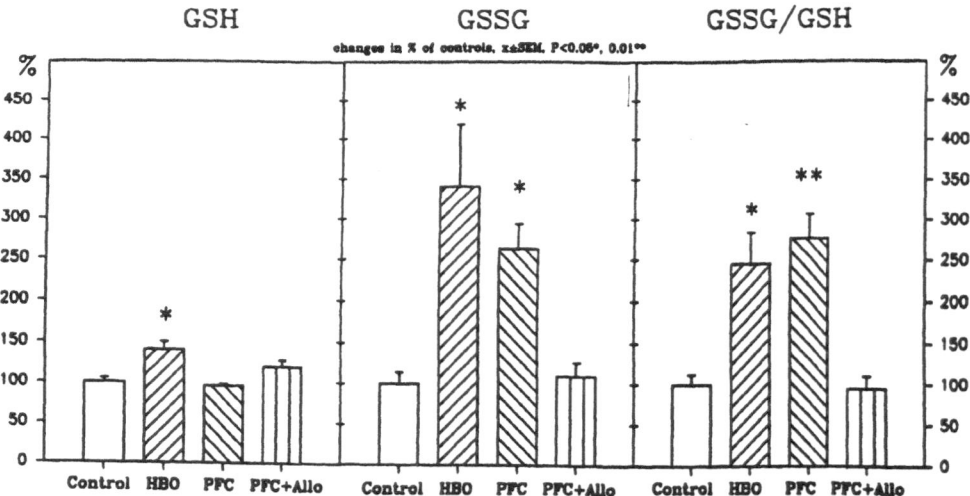

Fig. 4. Changes in the GSH status of the lung after a 60 min treatment with oxygen at 7.5 ATA (x±SEM, p< 0.05 *, <0.01 **)

DISCUSSION

Increased levels of reduced glutathione in the lung after exposure to hyperoxia reflect an adaptive mechanism to counteract an enhanced oxidative stress produced by high oxygen tensions (5,10). The primary effect, however, is an augmented production of intracellular oxidized glutathione that resulted from the hydrogen peroxide and lipid peroxides degrading action of GSH-peroxidase (11). High intracellular concentrations of GSSG increase the GSSG/GSH ratio, a change of which adversely alters enzyme functions by oxidation of protein-sulfhydryl groups (12). To counteract an increased GSSG/GSH ratio, GSSG is either reduced by glutathione reductase, which catalyses the NADPH-dependent reduction to GSH, or actively transported out of the cell (13), which causes an irreversible GSH-loss. A third mechanism is the stimulation of GSH synthesis that has been documented in previous publications using normobaric hyperoxia for 1 day to 2 weeks (1-3,14).

The difference in the extent of GSH and GSSG responses between HBO and HBO+PFC was smaller than expected, especially in the acute experiment. The larger effects produced by the additional PFC treatment resulted in an increased GSSG production. Thus, taking GSSG as a measure for oxidative stress, a 10% elevation after 8 days in the HBO group caused a 20% elevation of GSH, while a 37% GSSG elevation in the PFC+HBO group induced a 47% GSH elevation. The GSH response seems either to be limited, as no further increase occurred after 14 days, when GSSG increased by 56% and 118 %, resp., or to be overwhelmed by a GSSG translocation to the extracellular space.

Thus, the higher GSH after 8 days was even able to lower the GSSG/GSH ratio, but as no further rise had occurred after 14 days the ratio dramatically increased. Since we did not perform morphologic evaluations, we are not able to correlate the biochemical changes with an exudative or proliferative phase of O_2-toxicity (15).

The short and strong hyperbaric hyperoxia (60 min 7 ATA) revealed a 40% GSH response that, however, could not compensate the 240% GSSG increase. Thus, the GSH-response seems to be a rapid defence mechanism as it is present as soon as after 1 h of hyperbaric hyperoxia. The exceeding elevation of the GSSG/GSH ratio is equal pronounced in the HBO and HBO+PFC group, although no GSH response could be observed in the latter. Probably, the GSH response in the PFC group is overcome by the rapid formation of GSSG, that is transported out of the cell. Allopurinol was able to compensate the effects of acute HBO. The dose used acts not only as inhibitor of the conversion of xanthine dehydrogenase to an oxidase, but has been shown to be a very potent scavenger (16).

In conclusion adaptive and overloading processes must be considered under treatment with increased oxygen pressures, especially in the case of administration of PFC.

SUMMARY

Exposure to hyperoxia, especially under hyperbaric conditions, causes an enhanced oxidative stress particularly in lung tissue. To test the potential hazardous effect of either a single or repeated hyperbaric oxygen treatment (HBO) on the cellular defence system the glutathione status of lung tissue from rats exposed to HBO was investigated. When daily exposed to 2.5 ATA of >95% O_2 for 90 min over 8 or 14 days the content of reduced glutathione in lung tissue (GSH) increased by 16-19%. Oxidized glutathione (GSSG) tended to increase after 8 days and was 56% higher after 14 days. While the GSSG/GSH ratio was unchanged after 8 days, it increased by 39% after 14 days. Thus, the GSH increase after 8 days can be understood as a adaptive process to protect the lung from oxidative stress. The distinct increment of the cellular GSSG that lead to an increase of the GSSG/GSH ratio after 14 days reflects a situation, in which the cellular defence system is overwhelmed by oxidative stress. The additional pretreatment with perfluorochemicals in a dose of 2g/kg every second day aggravated the observed changes (GSH +39-47%, GSSG +118%).

In a second experiment rats were exposed to a single session with 7 ATA of O_2 for 60 min. GSH in the lungs increased for 40%, it was not elevated by PFC. However, GSSG increased to a much higher degree in untreated as well as in PFC-treated animals (+240%, +163%), elevating the ratio GSSG/GSH markedly (+145%, +176%). Allopurinol given as radical scavenger in a dose of 50 mg/kg was able to suppress the increased oxidative stress widely.

Thus adaptive and overloading processes are involved under the treatment with increased oxygen pressures. As the administration of PFC aggravates the observed changes, a still increased blood oxygen offer must be considered as the causative agent. A radical scavenger is capable to suppress the increased oxidative stress widely.

REFERENCES

1) Yam J, Frank L, Roberts RJ (1978) Oxygen toxicity: comparison of lung biochemical response in neonatal and adult rats. Pediat Res 12, 115-119
2) Yam J, Roberts RJ (1979) Pharmacological alteration of oxygen-induced lung toxicity. Toxicol Appl Pharmacol 47, 367-375
3) Kimball RE, Reddy K, Peirce TH, Schwartz LW, Mustafa MG, Cross CE (1976) Oxygen toxicity: augmentation of antioxidant defence mechanisms in rat lung. Am J Physiol 230, 1425-1431
4) Crouch LS, Prough RA, Kennedy KA, Snyder JB, Warshaw JB (1988) Rat lung antioxidant enzyme activities and their specific proteins during hyperoxia. J Appl Physiol 65, 797-804
5) Fridovich I, Freeman B (1986) Antioxidant defences in the lung. Ann Rev Physiol 48, 693-702
6) Jamieson D, Chance B, Cadenas E, Boveris A (1986) The relation of free radical production to hyperoxia. Ann Rev Physiol 48, 703-719
7) Smith LJ, Anderson J, Hamsuddin M, Hsueh W (1990) Effect of fasting on hyperoxic lung injury in mice. Am Rev Respir Dis 141, 141-149
8) Lutz J, Stark M (1988) Administration of perfluorochemicals under hyperbaric oxygen pressure and treatment with free oxygen radical scavengers. Biomat Art Cells Art Org 16, 395-402
9) Griffith OW (1980) Determination of glutathione and glutathione disulfide using glutathione reductase and 2-vinylpyridine. Anal Biochem 106, 207-212
10) Deneke SM, Fanburg BL (1989) Regulation of cellular glutathione. Am J Physiol 257, L163-L173
11) Sies H, Cadenas E (1983) Biological basis of detoxication of oxygen free radicals. In: Caldwell J, Jakoby WB (eds.) Biological Basis of Detoxication. Academic Press, New York, 181-211
12) Ziegler DM (1985) The role of reversible oxidation-reduction of enzyme thiol-disulfide in metabolic regulation. Ann Rev Biochem 54, 305-329
13) Nishiki K, Jamieson D, Chance B (1976) Oxygen toxicity in the perfused rat liver and lung under hyperbaric conditions. Biochm J 160, 343-355
14) Deneke SM, Gershoff SN, Fanburg BL (1983) Potentiation of oxygen toxicity in rats by dietary protein or amino acid deficiency. J Appl Physiol 4, 147-151
15) Crapo JD (1986) Morphologic changes in pulmonary oxygen toxicity. Ann Rev Physiol 48, 721-731
16) Moorehouse PC, Grootveld M, Halliwell JG, Quinlan G, Gutteridge JMC (1987) Allopurinol and oxypurinol are hydroxyl radical scavengers. FEBS Lett 213, 23-28

TUMOR PHYSIOLOGY AND OXYGENATION

BLOOD FLOW AND TISSUE OXYGENATION OF HUMAN TUMORS: AN UPDATE

P. Vaupel[1], K. Schlenger[1], and M. Hoeckel[2]

[1]Institute of Physiology and Pathophysiology
[2]Department of Obstetrics and Gynecology
University of Mainz, D-6500 Mainz, Germany

INTRODUCTION

It is generally accepted that tumor microcirculation, blood flow, oxygen and nutrient supply, tissue pH distribution, and the bioenergetic status (factors which are usually closely linked and which define the so-called cellular microenvironment) can markedly influence the therapeutic response of malignant tumors. Tumor blood flow is the major determinant for intra-tumor pharmacokinetics and (through modulation of the cellular microenvironment) of pharmacodynamics. The oxygen supply greatly determines the radiosensitivity of the tumors to be treated. The oxygen enhancement ratio, i.e., the ratio of doses with and without oxygen to produce the same biological effect is 2.7 to 3.0. O_2 partial pressures (O_2 tensions) of 3 to 4 mmHg (i.e., 0.5 to 0.6% O_2) result in a sensitivity halfway between radiobiological hypoxia and full oxygenation (Vaupel, 1992).

The influence of O_2 on the cytotoxicity resulting from chemotherapeutic agents is distinctly more complicated than with ionizing radiation and more difficult to understand.

This may explain why published data are somewhat contradictory. Some agents, such as bleomycin, vincristine, cyclophosphamide, actinomycin D, adriamycin, 5-fluorouracil (5-FU), carmustine (BCNU), and carboplatin seem to be more toxic to well oxygenated cells than to chronically hypoxic cells. By contrast, agents such as mitomycin C, misonidazole and etamidazole are substantially more toxic to hypoxic than to normoxic cells in vivo (Teicher et al., 1990).

In addition to the effects already discussed, oxygen has other, more indirect influences on the efficacy of chemo- or radiotherapy. Oxygen-deprived cells have been found to be slowly proliferating. Since most anticancer drugs and radiation are more active against rapidly dividing cells, a certain "resistance" has to be expected in areas distant from tumor blood vessels, because these areas are considered to contain non-cycling, but probably clonogenic cells. Further-

Oxygen Transport to Tissue XIV, Edited by W. Erdmann and
D.F. Bruley, Plenum Press, New York, 1992

more, O_2 deprivation in tumors necessitates anaerobic glyco- lysis with production of lactic acid which is inefficiently removed in low-flow tumors. Tissue acidosis is the mandatory result of this metabolic condition. Low pH values in turn have been shown to inhibit cell proliferation, to decrease the efficacy of anticancer drugs (e.g., adriamycin, bleomy- cin) and radiation sensitivity (for further details see Wike-Hooley et al., 1984).

Despite the apparent importance of tumor blood supply and tissue oxygenation for early tumor response to treatment and (probably) for prediction of longterm outcome, data on human tumors in situ are scarce although the number of clini- cal investigations dealing with this subject is increasing.

This synopsis attempts (i) to summarize the currently available information on tumor blood supply and tissue oxygen distribution, (ii) to demonstrate that an inadequate and heterogeneous blood supply and oxygenation of many solid tumors can be a causative factor for drug resistance and radioresistance in oncology, and (iii) to show that a knowl- edge of tumor oxygenation might be most beneficial for de- signing specifically tailored treatment protocols for indi- vidual subjects.

BLOOD FLOW OF HUMAN TUMORS

A number of studies on blood flow in human tumors have been reported. Many of them are more or less anecdotal re- ports rather than systematic investigations and, therefore, definite conclusions cannot be drawn. Flow studies have been performed on primary and metastatic brain tumors (Beaney, 1984; Beaney et al., 1985; Brooks et al., 1986; Cronqvist et al., 1966; Ito et al., 1982; Lammertsma, 1987; Lammertsma et al., 1983, 1984, 1985; Olesen and Paulson, 1971), lymphomas (Mäntylä et al., 1982), primary and metastatic liver tumors (Plengvanit et al., 1972; Taylor et al, 1979; Wartnaby et al., 1963), breast cancers (Beaney et al., 1984; Johnson, 1976; Waterman et al., 1991; Wilson et al., 1991), bronchial carcinomas (Rowell et al., 1990), squamous cell carcinomas of the head and neck region (Wheeler et al., 1986), and uterus tumors (Nyström et al., 1969).

Considering the presently available data on human tumors in situ, the following (preliminary) conclusions can be drawn if flow data for different tumor types are pooled (see Fig. 1):
a) Blood flow and thus O_2 supply can vary considerably de- spite similar histological classification and primary site. The O_2 availability is calculated to be 1.2 - 136.0 $\mu l\ O_2 \cdot g^{-1} \cdot min^{-1}$ in bronchial carcinomas, and 22 - 140 $\mu l\ O_2 \cdot g^{-1} \cdot min^{-1}$ in breast cancers (see Fig. 2; calculations based on measured blood flow data and assuming an arterial O_2 concentration of 0.2 ml O_2/ml blood);
b) tumors can have flow rates (O_2 availabilities) which are similar to those measured in organs with a high metabolic rate such as liver, heart or brain;
c) some tumors exhibit flow rates (O_2 availabilities) which are even lower than those of tissues with a low metabolic rate such as skin, resting muscle or adipose tissue;

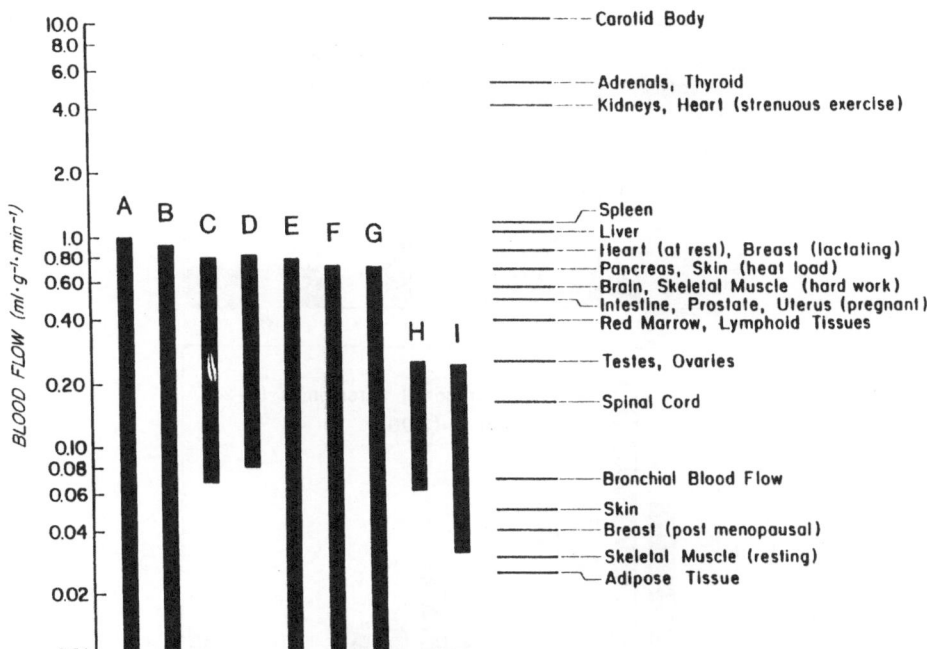

Fig.1. Variability of blood flow in human malignancies (vertical, black bars) and mean flow values of normal tissues (horizontal lines). Pooled data are given for the various tumors. A: primary brain tumors, B: lymphomas, C: metastatic liver tumors, D: breast cancers, E: metastatic brain tumors, F: uterus tumors, G: bronchial carcinomas, H: squamous cell carcinomas of the head and neck region, I: primary liver tumors (updated from Vaupel et al., 1990).

d) blood flow (O_2 availability) in human tumors can be higher or lower than that of the tissue of origin, depending on the functional state of the latter tissue (e.g., average blood flow in breast cancers is substantially higher than that of postmenopausal breast and significantly lower than flow data obtained in the lactating, parenchymal breast);

e) the average perfusion rate (O_2 availability) of carcinomas does not deviate substantially from that of sarcomas;

f) metastatic lesions exhibit a blood supply (O_2 availability) which is comparable to that of the primary tumors (Wilson et al., 1991) with the exception of those secondary lesions in the liver which are preferentially supplied by the portal system;

g) in some tumor entities blood flow (O_2 availability) in the tumor periphery is distinctly higher than in the center (Rowell et al., 1990) whereas in others blood flow is significantly higher at the tumor center compared with tumor edge (Zografos et al., 1990);

h) flow data (O_2 availabilities) from multiple sites of measurement show marked heterogeneity within individual tumors (up to 55-fold differences). However, these differences are smaller than flow data between different tumors (> 100-fold differences, i.e., tumor-to-tumor was more pronounced than intra-tumor heterogeneity (Acker et al., 1990), and

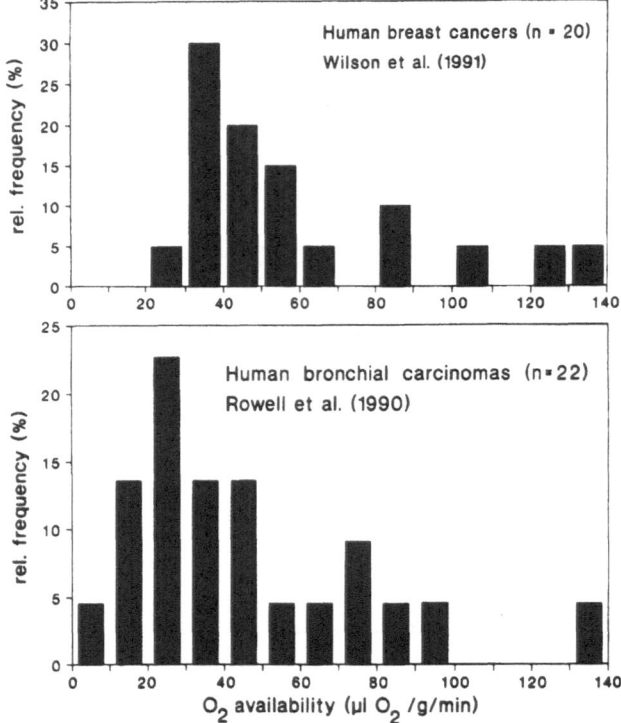

Fig.2. Histograms of calculated O_2 availabilities in human breast cancers (upper panel) and in bronchial carcinomas (calculated data based on measured blood flow values, n = number of tumors investigated).

i) there is no association between tumor size (as determined by TNM staging) and blood flow (T1-2 tumors: 0.26 $ml \cdot ml^{-1} \cdot min^{-1}$; T3-4 tumors: 0.29 $ml \cdot ml^{-1} \cdot min^{-1}$; Wilson et al., 1991).

From recent studies there is convincing evidence that a significant correlation between the perfusion rate of soft tissue sarcomas (Nishizawa et al., 1991) or the density of microvessels in invasive breast carcinomas (Weidner et al., 1991) and the occurrence of metastases exists. Assessment of tumor vascular density and/or blood flow may therefore prove valuable in identifying patients who will ultimately develop distant metastases and in selecting for aggressive therapy.

OXYGENATION OF SOLID TUMORS

Besides mathematical evaluations of the pO_2 distribution in tumor tissue (e.g., Groebe et al., 1988; Vaupel et al., 1988), polarographic and cryospectrophotometric microtechniques have been used to gain an insight into the oxygenation of microareas in human tumors.

Oxyhemoglobin saturation of single red blood cells in tumor microvessels

Characterization of the oxygen status of human tumors is possible using a cryospectrophotometric ex vivo microtechnique which allows for the measurements of the oxyhemoglobin (HbO_2) saturation of individual red blood cells in tumor microvessels (O < 12 μm; Mueller-Klieser et al., 1981; Wendling et al., 1984). This technique has been applied to both isotransplanted fast-growing rat tumors and primary human tumors (for a review see Vaupel, 1992). A modification of this technique has also been applied to isotransplanted mouse tumors and human tumor xenografts using tumor tissue surfaces prepared from rapidly frozen tumors (Rofstad et al., 1988; Mueller-Klieser et al., 1990).

The oxyhemoglobin saturation (HbO_2) frequency distribution for differentiated adenocarcinomas of the rectum and for the normal rectal mucosa is shown in Fig. 3 (Wendling et al., 1984; Vaupel and Kallinowski, 1987). As a rule, the mean HbO_2 saturation values observed in the tumors are distinctly lower than those found in the normal tissue at the site of tumor growth. The same holds for squamous cell carcinomas of the oral cavity (Mueller-Klieser et al., 1981). The medians of the HbO_2 frequency distributions of the normal oral mucosa and of the tumors decreased from 80 to 49 sat.% and correlated with changes in vascular density. Here, in various malignant tumors, considerable inter- and intra-individual differences were observed, even when tumors of the same clinical stage and histological grade were investigated. Substantial

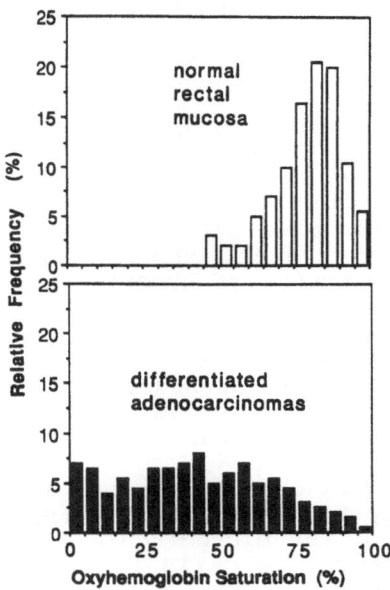

Fig.3.
Frequency distributions of measured oxyhemoglobin (HbO_2) saturation values of individual red blood cells within microvessels of the normal rectal mucosa (top), and of differentiated adenocarcinomas of the rectum (bottom; adapted from Vaupel et al., 1990).

intra-individual differences were also found in a metastatic lesion of a lung cancer (Vaupel et al., 1989). In these measurements of the oxyhemoglobin saturation of individual red blood cells in tumor microvessels, radiobiological hypoxia is to be expected in central portions of the intercapillary space if intravascular HbO_2 values fall below 30 sat.%.

Oxygen partial pressure distribution

The peculiarities of tumor tissue oxygenation can be attributed mainly to characteristic structural and functional abnormalities of the tumor microcirculation. A restriction and uneven distribution of nutritive perfusion during tumor growth leads to a "convection-dependent" reduction of the O_2 delivery to the cancer cells. The deterioration of the diffusion geometry in malignancies (e.g., by enlargement of the intercapillary distances) results in a "diffusion-dependent" restriction in the O_2 supply. The significance of these limitations has been evaluated systematically by measurements of tumor tissue oxygen partial pressures (tensions) using O_2-sensitive microelectrodes.

Frequency distributions of measured pO_2 values for various normal human tissues have been described in detail (for reviews see Vaupel 1990, 1992; Vaupel et al., 1989, 1991a). As expected, there is a scattering of the pO_2 values in normal tissues between 1 mmHg and values typical for arterial blood (80 to 100 mmHg). Whereas in normal tissues the median pO_2 values usually range from 13 to 66 mmHg, the respective values in the malignancies are 10 - 35 mmHg. If frequently occurring tumors are considered, median pO_2 values > 20 mmHg have only been found in the periphery of squamous cell carcinomas of the head and neck region (Fleckenstein et al., 1988) and in breast cancers (Vaupel et al., 1991b) to date. Comparing the pO_2 histograms of normal tissues with those of squamous cell carcinomas or breast cancers there is clear evidence that in tumors
a) there is a distinct shift of the distribution curve to the left,
b) on average, mean pO_2 values are lower in malignancies than in the surrounding normal tissues,
c) there is an accumulation of pO_2 values in the lower pO_2 classes indicating tissue hypoxia and, thus, reduced radiosensitivity in some tumors,
d) the oxygenation patterns in some tumor entities and the occurrence of hypoxia and/or anoxia do not correlate with either the pathological stages or the histological grades (Vaupel et al., 1991b; Hoeckel et al., 1991),
e) tumor-to-tumor variability in the oxygenation pattern is often more pronounced than intra-tumor heterogeneity, and
f) intratumor pO_2 variations cannot be predicted, e.g., as a function of the measuring site (center vs. periphery).

Data obtained so far are indicative of an inadequate O_2 supply to many solid tumors, most probably due to a restriction of the microcirculation and thus of the O_2 availability to the cancer cells in vivo, and a progressive expansion of the intercapillary space, i.e., an increase in intercapillary diffusion distances.

In most studies performed to date large needle electrodes with diameters > 0.7 mm were used. There is therefore
an urgent need for systematic investigations of human tumors
employing invasive hypodermic needle pO_2 probes with smaller
diameters. In a current study, direct pO_2 readings in cancers
of the breast and of the uterine cervix are being obtained
using a technique first described by Fleckenstein and Weiss
(1984). pO_2 distributions are assessed in conscious patients
by means of steel-sheated, polarographic needle electrodes
(outer tip diameter: 200 to 300 μm). The sharply ground tips
of the probes contain a membranized, recessed microelectrode
(diameter: 12 μm; pO_2 histograph, Eppendorf, Hamburg, Germany).

Fig.4.
Cumulative pO_2
histograms for nor-
mal breast (cir-
cles), fibrocystic
disease (squares)
and for breast
cancers (dots: T1
and T2 tumors, n =
12; triangles: T3
and T4 malignancies,
n = 6). The broken
line indicates the
pO_2 value for half-
maximum radiosensi-
tivity (top) and the
frequency of pO_2
readings in the
respective tissues
with less than
half-maximum radio-
sensitivity.

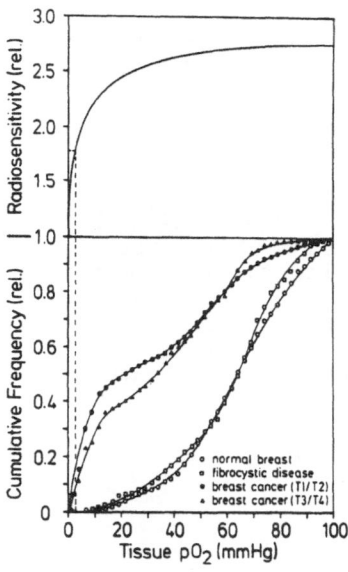

pO_2 values have been obtained in breast and cervix
cancers of different pathological stages and histological
grades (Vaupel et al., 1991b; Hoeckel et al., 1991). As a
rule, the mean (and median) pO_2 values are distinctly lower
in the malignancies than in the normal tissues. The pO_2
histograms obtained for tumors are usually shifted to lower
O_2 tensions whereas in the respective normal tissues, the pO_2
distribution is usually Gaussian (see Fig. 4). O_2 tensions
measured in the normal breast of 16 patients (1009 pO_2 readings) revealed a mean (median) pO_2 value of 65 mmHg (see Fig.
4, circles), whereas in 18 cancers of the breast (stages pT
1-4, 1218 pO_2 readings) the median pO_2 was 28 mmHg. Thus far,
6 out of 18 breast cancers exhibited pO_2 values between zero

and 2.5 mmHg, i.e., tissue areas with less than half-maximum radiosensitivity, whereas in the normal breast pO_2 values $\leq$ 12.5 mmHg could not be detected. Since 33% of the tumors investigated contained hypoxic areas and the remaining tumors were normoxic and comparable to normal breast tissue, the pO_2 distribution curve is clearly bimodal. The proportion of pO_2 readings between zero and 2.5 mmHg ranged from 4% (in a T4 breast cancer) to 64% (in a T3 tumor).

Pooled data for all breast cancers of pathological stages T1 and T2 (dots) and of T3 and T4 tumors are shown in Fig. 4. This compilation provides clear evidence that there are no statistically significant differences between the two groups (median pO_2 in T1/2 tumors: 26 mmHg; median pO_2 in T3/4 tumors: 30 mmHg) and implies that the oxygenation pattern in breast cancers and the occurrence of hypoxia and/or anoxia do not correlate with the pathological stage. This is in agreement with blood flow data of breast cancers, because there was no association between tumor size, as determined by TNM staging, and blood flow (Wilson et al., 1991). Furthermore, there is substantial evidence that the oxygenation patterns do not correlate with the histological grades, the menopausal status, the tumor histology (ductal vs. lobular), the extent of necrosis or fibrosis and with a series of other clinically relevant parameters (Fig. 5).

Statistical analysis of O_2 tensions measured in the normal cervix of nulliparous women resulted in oxygenation patterns which are characteristic for normal, adequately supplied tissues (median pO_2: 48 mmHg) with approximately 1% of the pO_2 values grouped between zero and 2.5 mmHg (see Fig. 6, circles). As a rule, the mean (and median) pO_2 values were distinctly lower in the normal cervix of parous women (most probably due to scar formation following vaginal delivery).

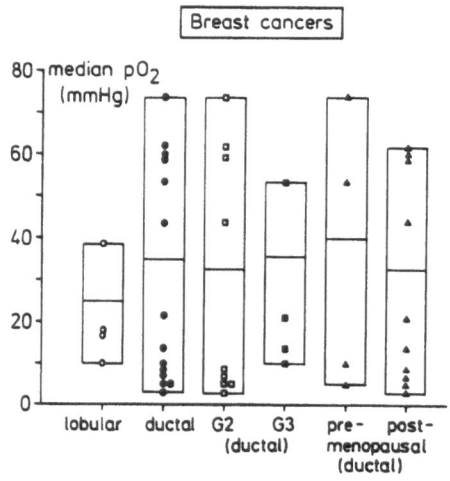

Fig.5.
Median pO_2 values in breast cancers as a function of tumor histologies (ductal vs. lobular), histological grades (G2 vs. G3), and menopausal status (pre- vs. postmenopausal tumors). The boxes indicate the variability of the median pO_2 values measured in the individual tumors, the horizontal line within the boxes denotes the mean pO_2 values of the respective tumors investigated.

Here, the median pO_2 value was 13 mmHg (with approximately 14% of the pO_2 readings in the lowest class; squares in Fig. 6). In 29 cancers of the cervix (stages FIGO I-IV, 3071 pO_2 readings) the median pO_2 was 12 mmHg. To date, 10 out of 29 cervix cancers exhibited pO_2 values between zero and 2.5 mmHg. In the case of cervix cancers the relative number of pO_2 readings between zero and 2.5 mmHg ranged from 1% (in a FIGO IV cancer) to 82% (in a FIGO III tumor).

Fig.6.
Cumulative pO_2 histograms for the normal cervix of nulliparous (circles) or parous women (squares), and for carcinomas of the cervix (dots: FIGO I and II tumors, n = 17; triangles: FIGO III and IV malignancies, n = 12). For further explanations see legend.

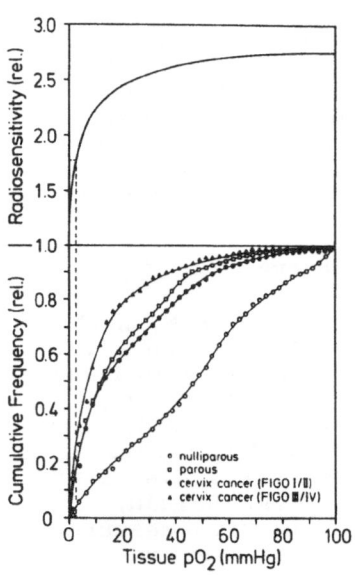

The distribution of the O_2 tensions measured in FIGO I/II tumors is presented in Fig. 6 (dots). In this group the median pO_2 is 12 mmHg with approximately 5% of the readings in the lowest pO_2 class. In FIGO III/IV tumors the median pO_2 is also 12 mmHg and 6% of the pO_2 values are grouped in the lowest class (Fig. 6, triangles).

As was the case with the breast cancers, the oxygenation pattern in cervix cancers and the occurrence of hypoxia and/or anoxia did not correlate with any of the above mentioned relevant parameters. From our clinical studies there is, therefore, clear indication that the oxygenation status of individual tumors before therapy cannot be predicted on the basis of staging and/or grading. Evaluation of the tissue oxygenation of individual tumors may help in the design of specifically tailored treatment protocols for individual patients in order to improve tumor response to treatment using "O_2-dependent" anticancer drugs and/or standard irradiation.

The lack of predictability is predominantly caused by pronounced tumor-to-tumor variabilities even if tumors of the same pathological stage and histological grade are compared. Tumor-to-tumor variability in the oxygenation pattern is often more pronounced than intra-tumor heterogeneity although in many tumors marked variability in tissue pO_2 values could be observed. For this reason, even pO_2 variations within a tumor cannot be predicted, e.g., as a function of the measuring site (tumor center vs. periphery).

On the basis of distributions such as those presented in Figs. 4 and 6 the fraction of tumor cells having reduced sensitivity towards "O_2-dependent" anticancer drugs (and standard radiation) as a result of their poor oxygenation status may be estimated. Specifically, cells in microenvironments with pO_2 values $\leq$ 2.5 mmHg will have a significant reduction in sensitivity compared to well-oxygenated tissue.

REFERENCES

Acker, J.C., M.W. Dewhirst, G.M. Honore, T.V. Samulski, J.A. Tucker, and J.R. Oleson, 1990, Blood perfusion measurements in human tumours: evaluation of laser Doppler methods, Int. J. Hyperthermia, 6:287.

Beaney, R.P., 1984, Positron emission tomography in the study of human tumors, Sem. Nucl. Med., 14:324.

Beaney, R.P., A.A. Lammertsma, T. Jones, C.G. McKenzie, and K.E. Halnan, 1984, Positron emission tomography for in-vivo measurements of regional blood flow, oxygen utilisation, and blood volume in patients with breast carcinoma, Lancet, 1:131.

Beaney, R.P., D.J. Brooks, K.L. Lenders, D.G.T. Thomas, T. Jones, and K. Halnan, 1985, Blood flow and oxygen utilisation in the contralateral cerebral cortex of patients with untreated intracranial tumours as studied by positron emission tomography, with observations on the effects of decompressive surgery, J. Neurol. Neurosurg. Psychol., 48:310.

Brooks, D.J., R.P.Beaney, A.A.Lammertsma, D.R. Turton, J. Marshall, D.T.G. Thomas, and T. Jones, 1986, Studies on regional cerebral haematocrit and blood flow in patients with cerebral tumours using positron emission tomography, Microvasc. Res., 31:267.

Cronqvist, S., D.H. Ingvar, and N.A. Lassen, 1966, Quantitative measurements of regional cerebral blood flow related to neuroradiological findings, Acta Radiol. Diagn., 5:760.

Fleckenstein, W., and C.A. Weiss, 1984, A comparison of pO_2 histograms from rabbit hind-limb muscles obtained by simultaneous measurements with hypodermic needle electrodes and with surface electrodes. Adv. Exp. Med. Biol., 169:447.

Fleckenstein, W., J.R. Jungblut, M. Suckfüll, W. Hoppe, and C. Weiss, 1988, Sauerstoffdruckverteilungen in Zentrum und Peripherie maligner Kopf-Hals-Tumoren. Dt. Z. Mund-Kiefer-Gesichtschir., 12:205.

Groebe, K., and P. Vaupel, 1988, Evaluation of oxygen diffusion distances in human breast cancer xenografts using tumor-specific in vivo-data: Role of various mechanisms in the development of tumor hypoxia. Int. J. Radiat. Oncol. Biol. Phys., 15:691.

Hoeckel, M., K. Schlenger, C. Knoop, and P. Vaupel, 1991, Oxygenation of carcinomas of the uterine cervix: Evaluation by computerized O_2 tension measurements. Cancer Res., (submitted).

Ito, M., A.A. Lammertsma, R.J.S. Wise, S. Bernardi, R.S.J. Frackowiak, J.D. Heather, C.G. McKenzie, D.G.T. Thomas, and T. Jones, 1982, Measurement of regional cerebral blood flow and oxygen utilisation in patients with cerebral tumours using ^{15}O and positron emission tomography: analytical techniques and preliminary results. Neuroradiology, 23:63.

Johnson, R., 1976, A thermodynamic method for investigation of radiation induced changes in the microcirculation of human tumors, Int. J. Radiat. Oncol. Biol. Phys., 1:659.

Lammertsma, A.A., 1987, Positron emission tomography and in vivo measurements of tumour perfusion and oxygen utilisation. Cancer Metastasis Rev., 6:521.

Lammertsma, A.A., R.J.S. Wise, and T. Jones, 1983, In vivo measurements of regional cerebral blood flow and blood volume in patients with brain tumours using positron emission tomography. Acta Neurochir., 69:5.

Lammertsma, A.A., R.J.S. Wise, and T. Jones, 1984, Regional cerebral blood flow and oxygen utilisation in edema associated with cerebral tumors, in: Recent progress in the study and therapy of brain edema, K.G. Go, and A. Baethmann, eds., pp. 331, Plenum Publ. Corp., New York.

Lammertsma, A.A., R.J.S. Wise, T.C.S. Cox, D.G.T. Thomas, and T. Jones, 1985, Measurement of blood flow, oxygen utilization, oxygen extraction ratio, and fractional blood volume in human brain tumours and surrounding oedematous tissue, Br. J. Radiol., 58:725.

Mäntylä, M.J., J.T. Toivanen, M.A. Pitkänen, and A.H. Rekonen, 1982, Radiation-induced changes in regional blood flow in human tumors, Int. J. Radiat. Oncol. Biol. Phys., 8:1711.

Mueller-Klieser, W., P. Vaupel, R. Manz, and R. Schmidseder, 1981, Intracapillary oxyhemoglobin saturation of malignant tumors in humans, Int. J. Radiat. Oncol. Biol. Phys., 7:1397.

Mueller-Klieser, W., C. Schaefer, S. Walenta, E.K. Rofstad, B.M. Fenton, and R.M. Sutherland RM, 1990, Assessment of tumor energy and oxygenation status by bioluminescence, nuclear magnetic resonance spectroscopy, and cryospectrophotometry, Cancer Res., 50:1681.

Nishizawa, K., P. Okunieff, D. Elmaleh, K.A. McKusick, H.W. Strauss, and H.D. Suit, 1991, Blood flow of human soft tissue sarcomas measured by thallium-201 scanning: Prediction of tumor response to radiation, Int. J. Radiat. Oncol. Biol. Phys., 20:593.

Nyström, C., L. Forssman, and B. Roos, 1969, Myometrical blood flow studies in carcinoma of the corpus uteri, Acta Radiol. Ther., 8:193.

Olesen, J., and O.B. Paulson, 1971, The effect of intra-arterial papaverine on the regional cerebral blood flow in patients with stroke or intracranial tumor, Stroke, 2:148.

Plengvanit, K., R. Suwanak, O. Chearanai, S. Intrasupt, S. Sutayavanich, C. Kalayasir, and V. Viranuvatti, 1972, Regional hepatic blood flow studied by intrahepatic injection of [133]xenon in normals and in patients with primary carcinoma of the liver, with particular reference to the effect of hepatic artery ligation, Aust. NZ J. Med., 1:44.

Rofstad, E.K., D. DeMuth, B.M. Fenton, and R.M. Sutherland, 1988, ^{31}P nuclear magnetic resonance spectroscopy studies of tumor energy metabolism and its relationship to intracapillary oxyhemoglobin saturation status and tumor hypoxia, Cancer Res., 48:5440.

Rowell, N.P., M.A. Flower, V.R. McCready, B. Cronin, and A. Horwich, 1990, The effects of single dose oral hydralazine on blood flow through human lung tumours, Radiother. Oncol., 18:283.

Taylor, I., R. Bennett, and S. Sherriff, The blood supply of colorectal liver metastases, Br. J. Cancer, 39:749.

Teicher, B.A., S.A. Holden, A. Al-Achi, and T.S. Herman, 1990, Classification of antineoplastic treatments by their differential toxicity toward putative oxygenated and hypoxic tumor subpopulations in vivo in the FSaIIC murine fibrosarcoma, Cancer Res., 50:3339.

Vaupel, P., 1990, Oxygenation of human tumors. Strahlenther. Onkol., 166:377.

Vaupel, P., 1992, Oxygenation of solid tumors, in: Drug resistance in oncology, B.A. Teicher, ed., Marcel Dekker Inc., New York.

Vaupel, P., and F. Kallinowski, 1987, Tissue oxygenation of primary and xenotransplanted human tumours, in: Radiation Research, Vol. 2, E.M. Fielden, J.F. Fowler, J.H. Hendry, and D. Scott, eds., pp. 707, Taylor & Francis, London, New York, Philadelphia.

Vaupel, P., F. Kallinowski, and K. Groebe, 1988, Evaluation of oxygen diffusion distances in human breast cancer using cell line-specific in vivo data: Role of various pathogenetic mechanisms in the development of tumor hypoxia, Adv. Exp. Med. Biol., 222:719.

Vaupel, P., F. Kallinowski, and P. Okunieff, 1989, Blood flow, oxygen and nutrient supply, and metabolic microenvironment of human tumors. A review. Cancer Res., 49:6449.

Vaupel, P., F. Kallinowski, and P. Okunieff, 1990, Blood flow, oxygen consumption, and tissue oxygenation of human tumors. Adv. Exp. Med. Biol., 277:895.

Vaupel, P., K. Schlenger, and M. Höckel, 1991a, Blood flow and oxygenation of human tumors, in: Tumor Blood Supply and Metabolic Microenvironment, P.Vaupel, and R.K. Jain, eds., pp. 165, Fischer, Stuttgart, New York.

Vaupel, P., K. Schlenger, C. Knoop, and M. Hoeckel, 1991b, Oxygenation of human tumors: Evaluation of tissue oxygen distribution in breast cancers by computerized O_2 tension measurements, Cancer Res., 51:3316.

Wartnaby, K.M., I.A.D. Bouchier, C.E. Pope, and S. Sherlock, 1963, Hepatic blood flow in patients with tumors of the liver, Gastroenterology, 44:733.

Waterman, F.M., L. Tupchong, R. Nerlinger, and J. Matthews, 1991, Blood flow in human tumors during local hyperthermia, Int. J. Radiat. Oncol. Biol. Phys., 20:1255.

Weidner, N., J.P. Semple, W.R. Welch, and J. Folkman, 1991, Tumor angiogenesis and metastasis - Correlation in invasive breast carcinoma, New Engl. J. Med., 324:1.

Wendling, P., R. Manz, G. Thews, and P. Vaupel, 1984, Inhomogeneous oxygenation of rectal carcinomas in humans. A critical parameter for preoperative irradiation? Adv. Exp. Med. Biol., 180:293.

Wheeler, R.H., H.A. Ziessman, B.R. Medvec, J.E. Juni, J.H. Thrall, J.W. Keyes, S.R. Pitt, and S.R. Baker, 1986, Tumor blood flow and systemic shunting in patients receiving intraarterial chemotherapy for head and neck cancer, Cancer Res., 46:4200.

Wike-Hooley, J.L., J. Haveman, and H.S. Reinhold, 1984, The relevance of tumour pH to the treatment of malignant disease, Radiother. Oncol., 2:343.

Wilson, C.B.J.H., A.A. Lammertsma, C.G. McKenzie, K. Sikora, and T. Jones, 1991, Measurements of blood flow and exchanging water space in breast tumors using positron emission tomography: A rapid and non invasive dynamic method, Cancer Res., 51: in press.

Zografos, G.C., S.Y. Iftikhar, J. Harrison, and D.L. Morris, 1990, Evaluation of blood flow in human rectal tumours using a laser Doppler flowmeter, Europ. J. Surg. Oncol. 16:497.

COMPUTERIZED HISTOGRAPHIC OXYGEN TENSION MEASUREMENTS

OF MURINE TUMORS

D.J. Terris, A.I. Minchinton†, E.P. Dunphy†, and J.M. Brown†

Division of Otolaryngology/Head and Neck Surgery and
†Department of Radiation Oncology
Stanford University Medical Center
Stanford, California, U.S.A.

ABSTRACT

As further work on tumor oxygenation results in the development of agents capable of modulating hypoxic cell radiosensitivity, and chemotherapeutic agents capable of targeting hypoxic cells, knowledge of relative tumor oxygenation takes on greater importance.

The pO_2 Histograph was used to characterize the oxygen tension of a murine sarcoma (RIF1) and a murine carcinoma (SCCVII), each with different hypoxic fractions. These tumors were studied sequentially to assess the suitability of the Histograph both as a research tool and, ultimately, as a clinical monitor of tumor hypoxia.

INTRODUCTION

The importance of tumor hypoxia and its relevance in radioresistance has been previously investigated.[1] Gray and his colleagues were instrumental in focusing research efforts toward quantifying the degree of hypoxia in various animal and human tumors. This resulted in indirect methods which have most recently included ^{14}C-misonidasole binding, ^{31}P nuclear magnetic resonance spectroscopy and laser Doppler flow, among others.[2-4]

Intuitively, a more direct measurement is that of tumor oxygen tension.[5-7] Although even this method does not directly measure those cells which are hypoxic, it does give a representation of microenvironments within tissues that may be hypoxic. This concept has been thoroughly discussed by Vaupel, Kallinowski, and Mueller-Klieser.[8-10] Previous oxygen tension measurement has required large custom-fashioned probes which obtain a limited number of measurements of variable reliability.[11] Gatenby has reported on perhaps the most reproducible arrangement, in which he measured human tumors using computerized tomographic guidance.[12] Recently, the pO_2 Histograph has been developed, which is capable of taking a large number of highly reproducible measurements in a short period of time, using a fine needle probe.[13]

We have used this instrument to characterize the oxygen tension of two murine cancers, a sarcoma and a carcinoma. Although histographic measurements of animal tumors have been reported,[14] this is the first report of sequential measurements in two different tumor lines. In addition to establishing the progression of tumor hypoxia over time, this study further evaluates the utility of the Histograph as a research tool.

METHODS AND MATERIALS

Tumor Model

C3H mice were implanted intradermally with 2×10^5 RIF1 sarcoma cells (N=8) and 2×10^5 SCCVII carcinoma cells (N=9). After two weeks (when tumors had reached a diameter of approximately of 10 mm), serial oxygen tension measurements were taken of both the tumors and areas of normal subcutaneous tissue (SQ) in anaesthetized animals (50 mg/kg of i.p. pentobarbital). Approximately 40-60 measurements were taken at each site. These measurements were repeated at 48 hour intervals.

Prior to each measurement, animals were weighed, and tumor dimensions recorded. Rectal temperatures, ambient temperature, and barometric pressure were documented. Animals were given standard laboratory chow ad libitum. They were sacrificed by cervical dislocation after the final measurements.

Oxygen Tension Measurements

All measurements were performed using the Sigma-Eppendorf pO_2 Histograph Kimoc 6650 (Hamburg, FRG). The tissue oxygen tension is measured polarographically using a fine needle O_2 electrode. The 12 micron glass-insulated gold microcathode is covered with a Teflon membrane and is recessed in a jacket tube of spring steel with a diameter of 0.3 millimeters. The gold cathode is biased with -700 mV toward a Ag/AgCl anode, with a resulting current of 0.01 to 3.0 nA. This current is proportional to the oxygen tension in the connecting elecrolyte (represented by test tissue).

All measurements were immediately preceded and followed by calibration and recalibration, respectively, in room air (20.9% O_2) and pure nitrogen (0.0% O_2). Following calibration, a 22 gauge angiocath was used to pierce the dermis overlying either the tumor or subcutis of the prepared animal.

Measurements were then performed with the assistance of the microprocessor-controlled manipulator, by which the needle probe is moved through the tissue in a series of movements termed a "pilgrim step" process (small forward motion, followed immediately by a much smaller, backward step). This helps to eliminate so-called pressure artifact (localized tissue compression leading to spuriously low readings). After an equilibration period of approximately one second, oxygen tension measurements were then automatically recorded. The results are presented in the form of histograms; these may be pooled to include all animals in a single group.

RESULTS

RIF1 Tumor and SQ - The pooled histograms for the RIF1 tumor measurements obtained over time are pictured in Figure 1. For comparison, the normal SQ histogram is shown in Figure 2. The measurements were taken every 48 hours from the 20th day to the 26th day following tumor implant.

The mean values of the histograms were calculated, standard errors computed, and plotted over time in Figure 3. The mean SQ measurement was 54.3 ± 11.1 mmHg. The mean tumor measurement fell from 19.9 ± 6.2 mmHg on the 20th post-implant day to 10.6 ± 4.8 mmHg on the 26th post-implant day.

Tumor volume measurements increased from 796.5 ± 66.9 mm³ on post-implant day 20 to 1445.9 ± 137.8 mm³ on post-implant day 26. These values are pictured in Figure 4.

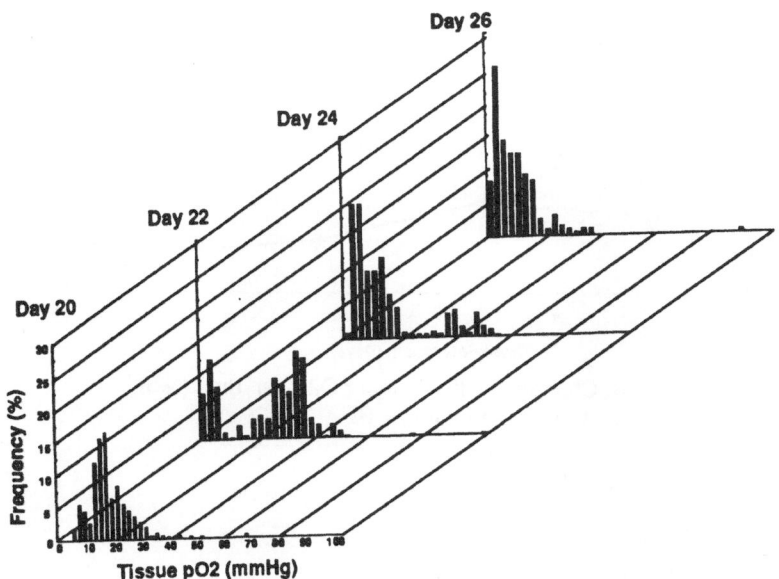

Figure 1: Change in pO2 distribution of RIF1 tumors over 6 days (N=8).

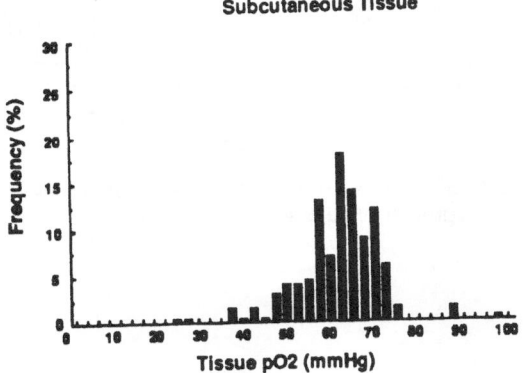

Figure 2: Distribution of oxygen tension of SQ. Mean ± SE=54.3 ± 11.1 mmHg.

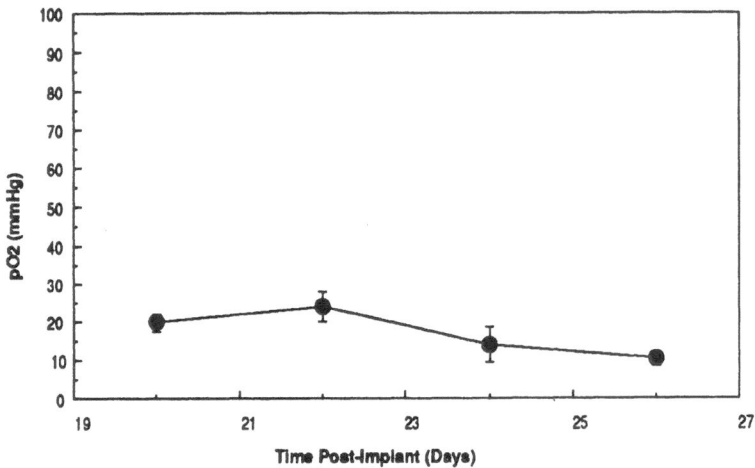

Figure 3: Change in RIF1 Tumor pO2 over Time (N=8).
Values are means ± SE.

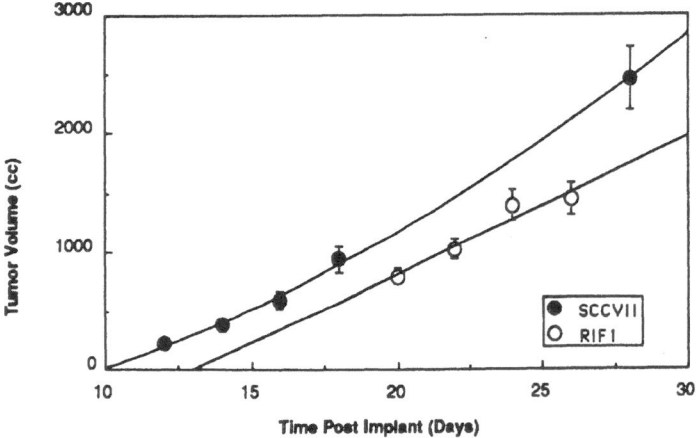

Figure 4: SCCVII (N=9) and RIF1 (N=8) tumor volume
growth. Values are means ± SE.

SCCVII Tumor and SQ   -   The pooled histograms for the SCCVII tumor obtained over time are depicted in Figure 5. The mean values and standard errors were computed for both the SCCVII tumor and SQ measurements, and are plotted in Figure 6. The mean tumor measurement fell from 36.3 ± 14.0 mmHg on post-implant day 12 to 8.4 ± 2.0 mmHg on post-implant day 28. The tumor volume measurements increased from 222.6 ± 27.7 mm³ on post-implant day 20 to 2463.9 ± 271.1 mm³ on day 26, and are plotted in Figure 4.

SCCVII and RIF1 - A quantitative comparison between RIF1 and SCCVII tumors was made by plotting both sets of measurements on the same graph describing oxygen tension as it relates to tumor volume (Figure 7). It can be seen that at comparable volumes, the oxygen tension of the SCCVII tumors was lower than that of the RIF1 tumors.

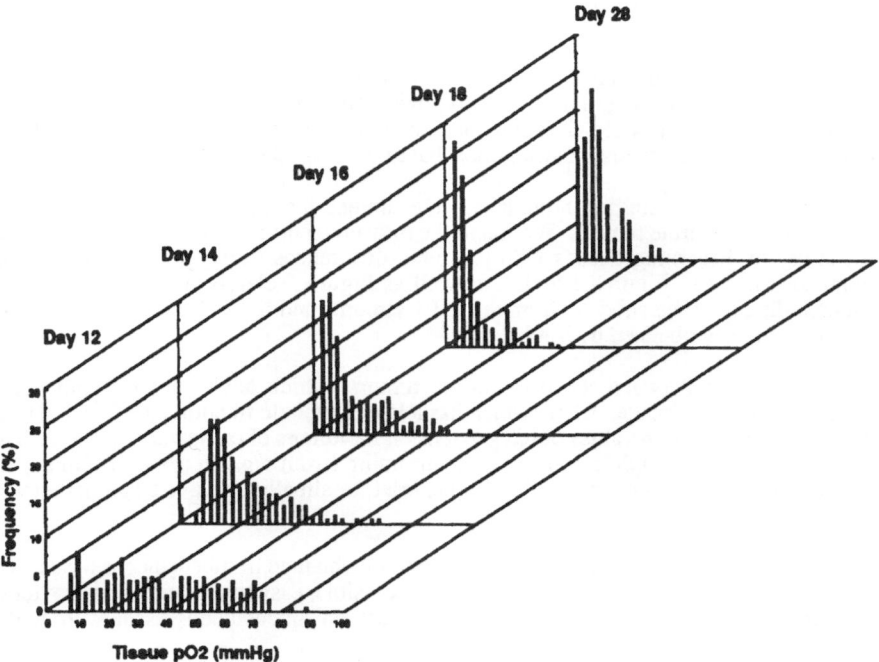

Figure 5: Change in pO2 of SCCVII tumors over 16 days (N=9).

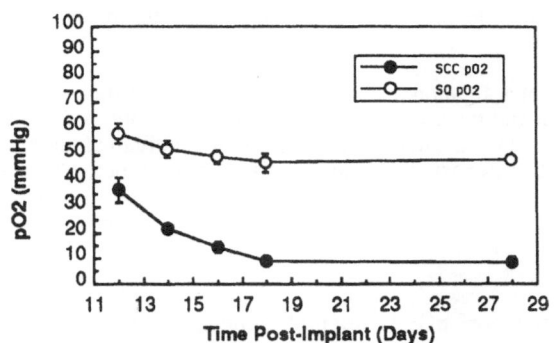

Figure 6: Change in SCCVII Tumor and SQ mean pO2 over time (N=9).

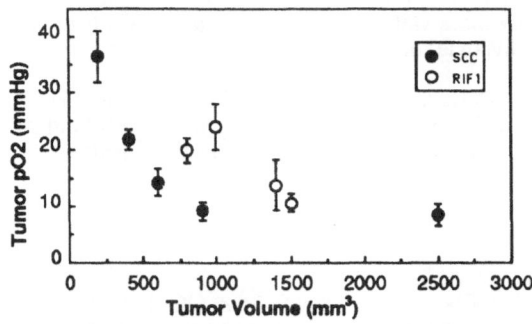

Figure 7: Change in RIF1 and SCCVII mean pO2 with volume (N=8).

DISCUSSION

There is continued interest in methods for overcoming hypoxic cell radioresistance in radiation therapy of solid tumors.[15] As hypoxic cell sensitizers such as SR 2508 and bioreductive agents such as SR 4233 demonstrate increasing promise for clinical use,[16-17] it becomes more important to distinguish tumors based on their relative hypoxia.

As a prelude to studies in humans, we examined the utility of the Eppendorf pO_2 Histograph in two murine tumors. We found the measurements to be reliable, and there was no difficulty in discerning tissues with profound differences in oxygenation (tumor and subcutaneous tissue). Figures 1 and 2, as well as Figure 5 clearly demonstrate that the Histograph distinguished subcutis from the RIF1 sarcoma and SCCVII carcinoma. Similar data were previously obtained by Vaupel et al.[18]

Comparison between the two murine tumors, which have known hypoxic cell fractions,[19,20] was also made. RIF1 (which has a lower hypoxic fraction than SCCVII) was found to have higher mean values of oxygen tension for tumors of comparable volume. This suggests that the Histograph is capable of determining (from a given array of tumors) the relative tumor hypoxia on the basis of a simple, brief, minimally invasive examination with minimal risk.

We chose to compare the tumor values to those obtained in subcutaneous tissue. The mouse and human subcutaneous tissue is highly accessible, easy to measure, and provides a consistent reference tissue. All oxygen measurements in subcutaneous tissue were higher than those in either of the two tumors.

The main limitation of the pO_2-Histograph in assessing tumor oxygenation is that it yields a representation of a "microenvironment,"[8-10] rather than quantifying the number of hypoxic cells. This may be sufficient as a predictor of radioresistance, but has yet to be demonstrated.

Our ongoing investigation includes examining the contribution of radiation therapy to tumor reoxygenation and the ability of the Histograph to monitor these changes. Potential clinical uses would include the determination of which individual tumors would be suitable for treatment with either hypoxic cell radiosensitizers or bioreductive agents.

CONCLUSIONS

(1) The Eppendorf pO_2 Histograph appears to be a useful instrument for the characterization of relative hypoxia of murine tumors.

(2) Substantial differences in tissue oxygen tension (tumor versus subcutaneous) can be discerned in the same individual, although the resolution is as yet undefined.

(3) The RIF1 tumor (which is known to have a relatively low hypoxic fraction) appears to be generally more well-oxygenated than the SCCVII tumor (which is known to have a relatively high hypoxic fraction).

(4) Further studies (both in animals and in humans) using the Sigma-Eppendorf pO_2 Histograph are warranted as it may ultimately become a useful indicator of which tumors would benefit from hypoxic cell sensitizers or bioreductive agents.

REFERENCES

1. Gray LH, Conger AD, Ebert M, Hornsey S, Scott OC, 1953, Concentration of oxygen dissolved in tissues at the time of irradiation as a factor in radiotherapy. Brit J Radiol, 26:638-648.

2. Garrecht BM, Chapman JD, 1983, The labelling of EMT-6 tumours in BALB/C mice with ^{14}C-misonidazole. Brit J Radiol, 56:745-753.
3. Okunieff P, Vaupel P, Sedlacek R, Neuringer LJ, 1989, Evaluation of tumor energy metabolism and microvascular blood flow after glucose or mannitol administration using ^{31}P nuclear magnetic resonance spectroscopy and laser Doppler flowmetry. Int J Rad Oncol Biol Phys, 16:1493-1500.
4. Vaupel P, Kluge M, Ambroz MC, 1988, Laser doppler flowmetry in subepidermal tumours and in normal skin of rats during localized ultrasound hyperthermia. Int J Hyperthermia, 3:307-321.
5. Urbach F, 1956, Pathophysiology of malignancy: I. Tissue oxygen tension of benign and malignant tumors of the skin. Proc Soc Exp Biol Med, 92:644-649.
6. Cater DB, Silver IA, 1960, Quantitative measurements of oxygen tension in normal tissues and in the tumours of patients before and after radiotherapy. Acta Radiol, 53:233-256.
7. Badib AO, Webster JH, 1969, Changes in tumor oxygen tension during radiation therapy. Acta Rad Ther Phys Biol, 8:247-257.
8. Vaupel P, Kallinowski F, Okunieff P, 1989, Blood flow, oxygen and nutrient supply, and metabolic microenvironment of human tumors: a review. Cancer Res, 49:6449-6465.
9. Kallinowski F, Schlenger KH, Runkel S, Kloes M, Stohrer M, Okunieff P, Vaupel P, 1989, Blood flow, metabolism, cellular microenvironment, and growth rate of human tumor xenografts. Cancer Res, 49:3759-3764.
10. Mueller-Klieser WF, Walenta SM, Kallinowski F, Vaupel PW, 1989, Tumor physiology and cellular microenvironments, in: "Prediction of tumor treatment response," JD Chapman, LJ Peters, HR Withers, eds., Pergamon Press, Elmsford, NY.
11. Davies PW, Brink F, 1942, Microelectrodes for measuring local oxygen tension in animal tissues. Rev Sci Instrum, 13:524-533.
12. Gatenby RA, Kessler HB, Rosenblum JS, Coia LR, Moldofsky PJ, Hartz WH, Broder GJ, 1988, Oxygen distribution in squamous cell carcinoma metastases and its relationship to outcome of radiation therapy. Int J Rad Oncol Biol Phys, 14:831-838.
13. Fleckenstein W, Weiss C, Heinrich R, Schomerus H, Kersting T, 1984, A new method for the bed-side recording of tissue pO_2-Histograms. Verh Dtsch Ges Inn Med, 90:439-443.
14. Kallinowski F, Zander R, Hoeckel M, Vaupel P, 1990, Tumor tissue oxygenation as evaluated by computerized-pO_2-histography. Int J Rad Oncol Biol Phys, 19:953-961.
15. Dische S, 1985, Chemical sensitizers for hypoxic cells: A decade of experience in clinical radiotherapy. Radiotherapy Oncol, 3:97-115.
16. Brown JM, 1982, Clinical perspectives for the use of new hypoxic cell sensitizers. Int J Rad Oncol Biol Phys, 8:1491-1497.
17. Zeman EM, Brown JM, Lemmon MJ, Kirst VK, Lee WW, 1986, SR-4233: a new bioreductive agent with high selective toxicity for hypoxic mammalian cells. Int J Rad Oncol Biol Phys, 12:1239-1242.
18. Vaupel P, Okunieff P, Kallinowski F, Neuringer LJ, 1989, Correlations between 31P-NMR spectroscopy and tissue O2 tension measurements in a murine fibrosarcoma. Radiat Res, 120:477-493.
19. Kitakabu Y, Shibamoto Y, Sasai K, Ono K, Abe M, 1991, Variations of the hypoxic fraction in the SCCVII tumors after single dose and during fractionated radiation therapy: assessment without anesthesia or physical restraint of mice. Int J Rad Oncol Bio Phys, 20:709-714.
20. Moulder JE, Rockwell S, 1984, Hypoxic fractions of solid tumors: experimental techniques, methods of analysis, and a survey of existing data. Int J Rad Oncol Biol Phys, 10:695-612.

OXYGENATION AND BIOENERGETIC STATUS OF MURINE FIBROSARCOMAS

C. Schaefer[1], P. Okunieff[2], and P. Vaupel[1]

[1]Institute of Physiology & Pathophysiology,
University of Mainz, D-6500 Mainz, Germany
[2]Department of Radiation Oncology, Mass. General
Hospital, Harvard Med. School, Boston, MA 02114,
USA

INTRODUCTION

The heterogeneity of cellular response to therapy is a major problem in non-surgical cancer therapy. This heterogeneity is influenced by both the genetic variability between different tumor cells and by epigenetic, physiological factors, such as the local metabolic milieu. A restriction of tumor microcirculation concomitant with regional hypoxia, nutrient depletion, accumulation of lactate, and an intensified tumor acidosis becomes evident during growth of many solid tumors[1]. These critical factors can greatly influence the efficiency of various non-surgical tumor therapies.

In an earlier study, size-dependent changes of some therapeutically relevant parameters such as tissue oxygenation and bioenergetic status of a murine fibrosarcoma (FSaII) were characterized[2]. Tumor bioenergetics were determined by ^{31}P-NMR spectroscopy. pO_2 values, NTP/P_i and PCr/P_i ratios decreased as the tumors increased in size. The key question of the present study was to clarify whether changes in the NTP and PCr concentrations on the one hand or P_i levels on the other caused these reductions in the NTP/P_i or PCr/P_i ratios. To answer this question, size-dependent changes in tumor oxygenation, energy status and metabolic parameters of FSaII tumors were studied using pO_2 histography, HPLC techniques and quantitative bioluminescence. The new data obtained were related to the earlier results derived from ^{31}P-NMR measurements.

MATERIAL AND METHODS

Animals and tumors

10- to 12-week old C3Hf/Sed mice were used as experimental animals. FSaII tumors were transplanted s.c. in the hind foot dorsum. The tumors grew rapidly with a volume doubling time of 2 days at a volume of 100 - 200 mm^3. Tumors were studied when they reached volumes between 44 and 680 mm^3.

Tumor oxygenation

Tumor oxygenation was measured with O_2-sensitive needle electrodes (recessed gold in glass electrode; shaft diameter: 300 μm, diameter of the cathode: 12 μm; pO_2 Histograph, Eppendorf, Hamburg, Germany; for details see Vaupel et al.[2]).

^{31}P-NMR Spectroscopy

Spectra were obtained from unanaesthetized animals at 8.5 Tesla on a homemade spectrometer as described by Vaupel et al.[2]. The following ratios were determined: nucleoside triphosphate/inorganic phosphate (NTP/P_i), phosphocreatine/inorganic phosphate (PCr/P_i). pH_{NMR} was estimated using the PCr to P_i chemical shift.

High performance liquid chromatography (HPLC)

The tumor bearing hind foot of anaesthetized animals was rapidly frozen in liquid nitrogen. Tumor samples were prepared under liquid nitrogen, ground to a fine powder and lyophilized. After extraction with 0.6 M perchloric acid, centrifugation and neutralisation with 2 M KOH the concentrations of adenylate phosphates (ATP, ADP, and AMP) were determined as described earlier[3].
Concentrations of PCr and P_i, energy charge (ECh) and the phosphorylation potential were calculated using the NMR ratios and the absolute concentrations of ATP, ADP and AMP determined by HPLC.

$$ECh = \frac{[ATP] + 0.5[ADP]}{([ATP] + [ADP] + [AMP])} \qquad PPot. = \frac{[ATP]}{[ADP] \cdot [P_i]}$$

Metabolic status

Tumor glucose and lactate concentrations were determined using standard enzymatic assays for tissue extracts as described earlier[3].

Metabolic imaging with quantitative bioluminescence

Quantitative bioluminescence was used to measure microregional concentrations of ATP, glucose, and lactate by linking the substrate of interest to the luminescence of luciferase through the respective enzymes as described by Mueller-Klieser et al.[4,5].

RESULTS

Bioenergetic and oxygenation status

^{31}P-NMR spectroscopy demonstrates a progressive drop of PCr/P_i and NTP/P_i ratios as the tumors increase in size (Fig. 1). The median tissue pO_2 progressively declines from 33 mmHg in small fibrosarcomas to 8 mmHg in larger tumors. During tumor growth, a steady increase in the incidence of O_2 par-

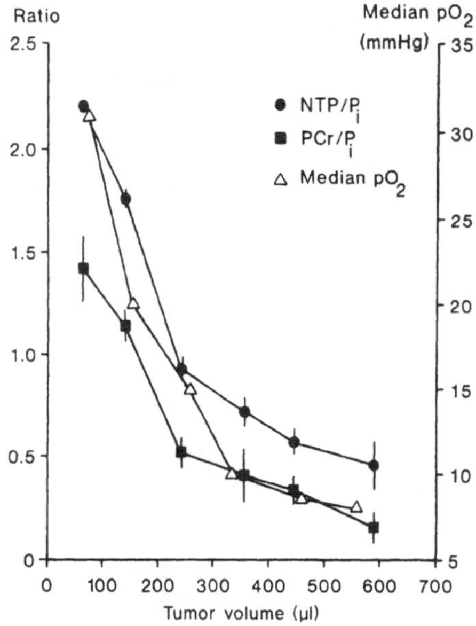

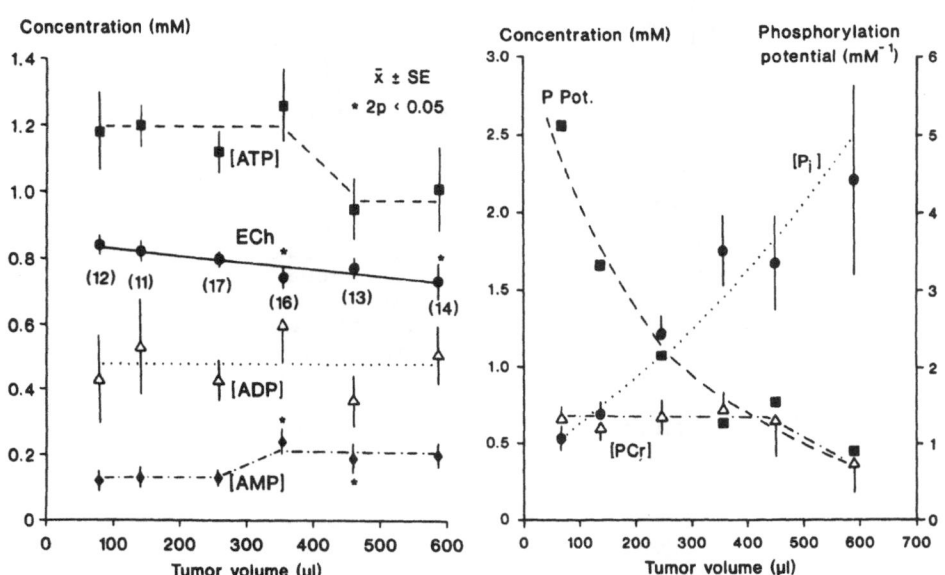

Fig. 1.
NTP/P_i, PCr/P_i ratios (means ± SE) as determined by ^{31}P-NMR spectroscopy and median pO_2 values measured with polarographic needle electrodes as a function of tumor volume. SE of tumor volumes are within the symbol sizes.

Fig. 2.   Left panel: ATP, ADP and AMP concentrations and energy charge (ECh) of FSaII tumors as a function of tumor volume. Numbers in parentheses indicate the numbers of tumors investigated. SE of tumor volumes are within the symbol sizes.
Right panel: Calculated PCr and P_i concentrations and phosphorylation potentials (P Pot.) as a function of tumor volume. Values for PCr and P_i are means ± SE. SE of tumor volumes are within the symbol size.

tial pressures in the range between 0 and 2.5 mmHg (areas with less than half maximum radiosensitivity) is found.
Tumor ATP concentrations determined with HPLC techniques and the calculated PCr levels decrease marginally only at tumor sizes ≥ 450 mm^3, while the calculated P_i levels show a steady increase during tumor growth from 0.54 to 2.20 mM (Fig. 2). Since there are only slight changes of ATP and PCr levels, drops in the NTP/P_i and PCr/P_i ratios are predominantly due

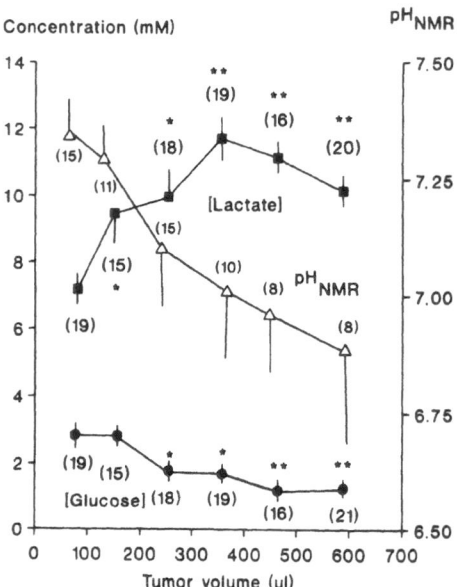

Fig. 3.
Tumor pH as a function of tumor volume determined by ^{31}P-MRS (pH$_{NMR}$), and tissue glucose and lactate concentrations as a function of tumor volume. Numbers in parentheses indicate the number of tumors investigated. Values are means $\pm$ SE. SE of tumor volumes are within the symbol sizes. Significance levels: * 2p < 0.05; ** 2p < 0.0001.

to a drastic increase in the P_i resonances. Consequently, the phosphorylation potential (P Pot.) steadily decreases from 5.1 to 0.9 mM^{-1} with increasing tumor volume. ATP synthesis is being driven by the dysequilibrium of the ATP synthetase reaction induced by increased intracellular P_i concentrations. This suggests that cells regulate both the ATP concentration and to a lesser extent the PCr concentration by adjusting their P_i concentration to the stress of the micromilieu. In accordance, energy charge significantly declines only at tumor sizes ≥ 350 mm^3 from 0.84 to 0.73 (Fig. 2).

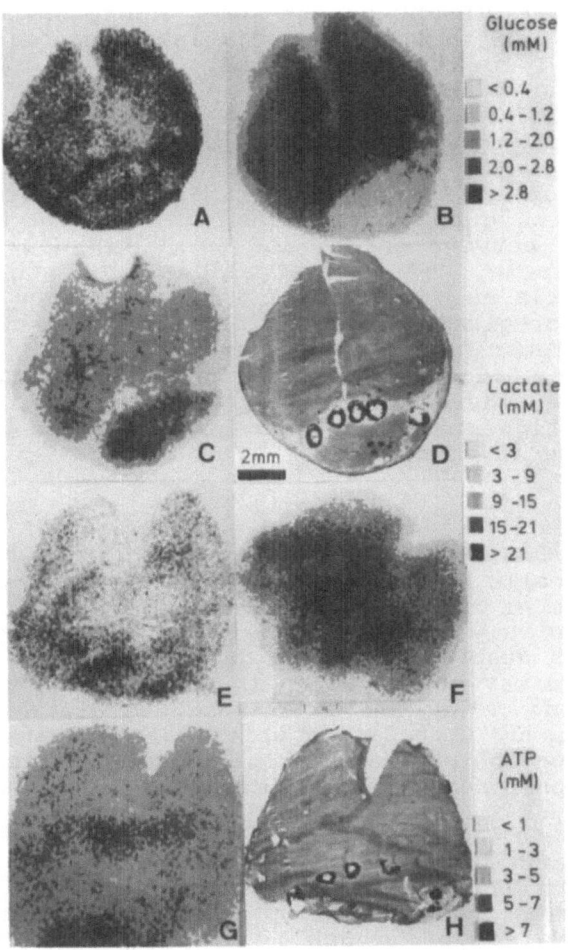

Fig. 4. Regional distributions of ATP, glucose and lactate
concentrations in serial cryostat sections and
respective histological sections (HE stained) of
tumor-bearing hind feet of mice. Skeletal muscle
(below metatarsal bones) and tumor tissue (above
bones) can be clearly distinguished.
(A) regional glucose, (B) lactate, and (C) ATP
distribution of a 73 mm^3 FSaII tumor; (D) histological section.
(E) regional glucose, (F) lactate, and (G) ATP
distribution of a 169 mm^3 FSaII tumor; (H) histological section. (For Fig. 4 see next page.)

Metabolic status

Global glucose concentrations drop from 2.8 mM in small FSaII tumors to 1.2 mM in larger malignancies, whereas the mean lactate levels increase from 7 to 11 mM. Tumor pH as determined by ^{31}P-NMR spectroscopy decreases during tumor growth from 7.35 to 6.88 (Fig. 3). Tumor acidosis (pH < 7.00) was found only in tumors $\geq$ 450 mm^3.

Metabolic imaging with quantitative bioluminescence

The regional ATP, glucose and lactate distributions measured with quantitative bioluminescence show drastic differences between normal and tumor tissues (Fig. 4). In skeletal muscle, ATP concentrations are much higher, and lactate levels much lower than in tumor tissue. Differences in the microregional glucose distributions between skeletal muscle and tumor become evident only in larger malignancies. Tumor ATP and glucose concentrations decrease with increasing tumor volume, whereas lactate levels increase, as described above. The highest lactate concentrations are usually found in the centre of tumors.

CONCLUSIONS

Living cells do maintain a "homeostatically" controlled steady state of ATP concentration[6,7]. In FSaII tumors < 450 mm^3 a relatively constant ATP level is maintained during tumor growth despite a drastic deterioration in tumor oxygenation. This may be due to an increasing proportion of non-proliferating cells, and an intensified cleavage of glucose to lactic acid, a condition which is accompanied by a substantial drop in intracellular pH. ATP levels decline only in larger fibrosarcomas, probably due to a decreased activity of glyceraldehyde-3-P-dehydrogenase following a reduced availability of NAD$^+$. Under normoxic conditions NAD$^+$ is provided by the transfer of electrons from NADH + H$^+$ to O$_2$ through the electron-transport chain; under hypoxic conditions NAD$^+$ is regained by the reduction of pyruvate to lactate. In larger tumors, the lack of oxygen leads to an inhibition of the oxidative pathway and NAD$^+$ is regenerated only by the lactate-dehydrogenase reaction. Since the activity of pyruvate kinase is low in tumor areas with reduced glucose supply, insufficient pyruvate is available to regenerate NAD$^+$[8]. The glyceraldehyde-3-P-dehydrogenase reaction is thus inhibited[8] and the fixation of P$_i$ in energy rich metabolites is blocked. In consequence the mean tumor lactate levels reached a plateau at tumor sizes $\geq$ 450 mm^3. ATP and PCr concentrations were not major parameters responsible for the pronounced changes in the ATP/P$_i$ and PCr/P$_i$ ratios at sizes $\leq$ 450 mm^3. In accordance with very small changes in ATP and AMP, there was only a small but significant drop in adenylate energy charge. Consequently it seems that the major parameter responsible for the drop in the NTP/P$_i$ and PCr/P$_i$ ratios was the change in P$_i$. Hence the adenylate phosphates and probably PCr are all regulated homeostatically by cellular enzymes, apparently by adjusting the P$_i$ concentrations and thus the phosphorylation potential. Based on these observations the

ECh has little value as an estimator of cellular bioenergetic status since it will only change significantly under conditions where homeostasis is failing and cell death is likely. Future studies of tumor bioenergetics should therefore focus on P_i levels, PPot., or the Δ G of ATP hydrolysis as estimators of cellular metabolic state rather than ATP concentration or ECh.

REFERENCES

1. P. Vaupel, S. Frinak and H.I. Bicher, Heterogeneous oxygen partial pressure and pH distribution in C3H mouse-mammary adenocarcinoma, Cancer Res. 41: 2008 (1981).
2. P. Vaupel, P. Okunieff, F. Kallinowski, and L. Neuringer, Correlation between ^{31}P-NMR spectroscopy and tissue O_2 tension measurements in a murine fibrosarcoma, Radiat. Res. 120:477 (1989).
3. W. Krueger, W.-K. Mayer, C. Schaefer, M. Stohrer, and P. Vaupel, Acute changes of systemic parameters in tumour-bearing rats, and of tumour glucose, lactate, and ATP levels upon local hyperthermia and/or hyperglycaemia, J. Cancer Res. Clin. Oncol. 117: in press (1991).
4. W. Mueller-Klieser, S. Walenta, W. Paschen, F. Kallinowski, and P. Vaupel, Metabolic imaging in microregions of tumors and normal tissues with bioluminescence and photon counting, J. Natl. Cancer Inst. 80:842 (1988).
5. W. Mueller-Klieser, C. Schaefer, S. Walenta, E.K. Rofstad, B.M. Fenton, and R.M. Sutherland, Assessment of tumor energy and oxygenation status by bioluminescence, NMR spectroscopy, and cryospectrophotometry, Cancer Res. 50: 1681 (1990).
6. Atkinson D.E., Cellular Energy Metabolism and Its Regulation, Academic Press, New York (1977).
7. M. Stubbs, L.M. Rodrigues and J.R. Griffiths, Growth studies of subcutaneous rat tumours: comparison of ^{31}P-NMR spectroscopy, acid extracts and histology, Br. J. Cancer 60:701 (1989).
8. E. Eigenbrodt, P. Fister, and M. Reinacher, New perspectives on carbohydrate metabolism in tumor cells, in: Regulation of Carbohydrate Metabolism (Vol. II), R. Breitner, CRC Press, Boca Raton (1985).

MEASUREMENT OF HUMAN TUMOR BLOOD FLOW: A POSITRON TECHNIQUE USING

AN ARTIFACT OF HIGH ENERGY RADIATION THERAPY*

P. Okunieff[1], J. Lee[1], and P. Vaupel[2]

[1]Dept. Radiation Oncology, Massachusetts General
Hospital, Boston, MA 02114, USA
[2]Institute of Physiology and Pathophysiology,
Pathophysiology Division, University of Mainz, 6500
Mainz, Germany

INTRODUCTION

For at least three decades (1-6) there has been an interest
in measuring tumor blood flow (TBF) and in the determination of
its relation to the response of human tumors to radiation, drug
therapies, and to the probability of development of distant
metastases. The proton activation method which will be described
below allows daily measurements of blood flow, in only 7 minutes,
in patients being irradiated by photons of ≥ 20 MV, or by protons
and other heavy particles.

Patients treated for tumors using radiation beams of greater
than 10 MV become minimally but measurably radioactive due to ^{15}O
production. In theory this artifact of treatment can
conveniently be used to measure blood flow in irradiated tissues
in much the same way that ^{15}O is used to measure blood flow using
standard positron emission tomography (PET) scans (7). This
technique however is a wash-out rather than wash-in technique and
thus requires no arterial line for quantitative analysis.
Furthermore, the patient receives only his usual radiation
treatment, hence unlike conventional PET there is no additional
whole body radiation dose and no radioactive gasses to manage.

A positron camera was built and tested on patients undergoing
radiation treatment at the Harvard Cyclotron Laboratory, where a
160 MV proton beam is routinely employed. Preliminary
measurements of human tumor blood flow, calibration measurements
in phantoms, and Monte Carlo modelling studies to estimate
sensitivity and accuracy are presented.

---

*This work supported in part by NIH grants CA48096, RR00995,
and CA13311, and by the American Cancer Society Career
Development Award.

MATERIALS AND METHODS

Radiation Therapy

Tumors treated using x-ray beams of greater than 10 MV become minimally but measurably radioactive due to ^{15}O production after neutron expulsion from ^{16}O. The latter is naturally present in many biological molecules, including water. In normal tissues water makes up approximately 70% of tissue weight, in many tumors more than 80% of the mass is water, hence sufficiently high concentrations of ^{16}O can be produced with just 2 Gy of radiation. This artifact of treatment, while posing no danger to the patient or personnel, can be conveniently used to measure Fick type blood flow, often termed nutritive blood flow. The "injection" is non-invasive and is perfectly homogeneous, thus avoiding the usual pitfalls of washout techniques. To test the feasibility of the above methodology patients treated with curative intent at the Harvard cyclotron were studied. A 160 MV proton beam was routinely employed. No modification of the patient's planned treatment was needed or would be permitted.

Phantom

A glass cylindrical phantom fitted with an inflow jet and outflow nozzle was constructed. Flow was adjusted using a peristaltic pump, and measured by collecting the outflow in a graduated cylinder. The cylinder had a volume of 23 ml and a mass of 17 grams. The silicon dioxide of glass is expected to model the immobile oxygen of tissue while the water motion models nutritive blood flow. Studies with dyed water indicated that the input jet did not achieve homogeneous mixing even at high flow rates. This was corrected by attaching a vortex by rubber band to the phantom. The vibrations from the vortex achieved sufficient homogeneous mixing. To confirm that reflow of radioactive blood does not appreciably affect the flow measurements, a separate series of flow measurements were made where the water source was supplied from and returned to a 1 liter flask. In the case of humans, the reflow will have a volume of dilution of total body water (approx 50 liters), but the volume of radiation will be determined by the tumor location and size. To a first approximation therefore this reflow model is sufficient to account for irradiation of tumor volumes in patients $50 * (23 + 17)$ g = 2 kg. A mass likely to be greater than most tumors.

Equipment Used

The detectors used were two inch NaI detectors operated at 800 V using a high voltage source and preamplifiers. Energy was gated for optimal coincidence detection and minimum noise. A coincidence discrimination period of 110 microseconds was used. Detectors were mounted in cylindrical lead collimators and shielded to reduce any false coincidence. Patients were treated using a horizontal proton beam. Using lasers and computerized treatment plans the tumor was centered between the two NaI detectors, which were thus positioned symmetrically on either side of the patient. The detectors were typically 0.5 to 0.9 meters apart. The output of the coincidence detectors was sent to an amplifier which feeds a 4096 channel multi-channel analyzer operated by a IBM PC XT computer. Coincidences per 100 milliseconds are then stored by the computer software for later analysis. A total of 409.6 seconds of data was collected beginning immediately after the proton beam treatment was completed. Studies of phantoms involve an identical setup except that the phantom replaces the patient.

Flow Model

The flow model is shown schematically in Figure 1. Blood is assumed to enter the tumor free of radioactivity. This water then homogeneously mixes with the radioactive tissue water, diluting it, and then exiting the tumor. The equation describing this model is:

$$A = A_O \{m*\exp[(-1/\tau - F/V)t]+(1-m)*\exp[(-1/\tau)t]\}$$

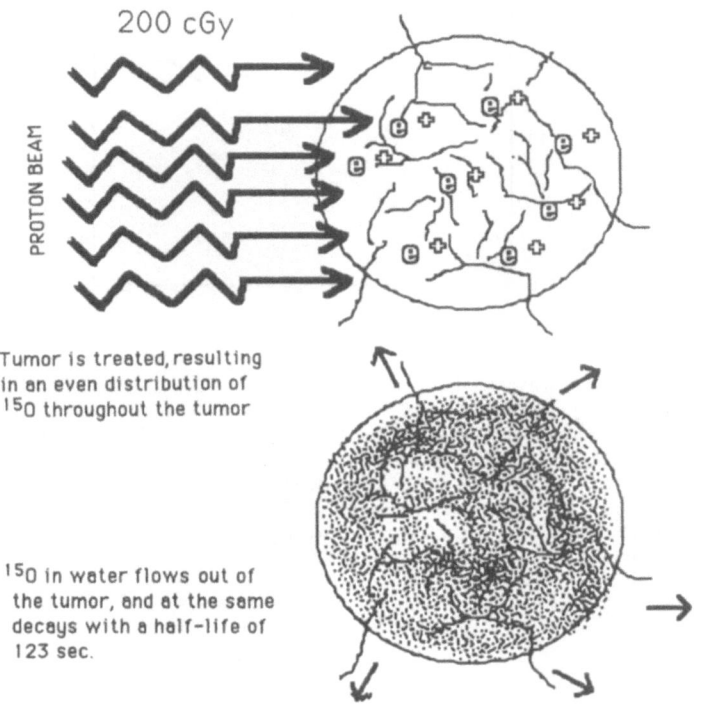

200 cGy

PROTON BEAM

Tumor is treated, resulting in an even distribution of ^{15}O throughout the tumor

^{15}O in water flows out of the tumor, and at the same decays with a half-life of 123 sec.

Figure 1. Schematic blood flow model.

Where A is the activity of ^{15}O in the tumor at time t, A_O is the initial activity, m is the fraction of ^{15}O that is mobile, F/V is the flow per unit fluid volume, and $(1/\tau)$ is the natural decay rate of ^{15}O (half-life is 123 sec). Whole blood is approximately 85% water (depending on hematocrit) and tumor is about 75-80% water, so that the nutritive blood flow is closer to 1.10*F/V.

RESULTS

Figure 2 shows the results of phantom flow measurements. The
fitting program estimated both m and [F/V] though for the case of
an ideal phantom the exact value of m was known. The fit to m
was accurate ± 5%. In Figure 2, true flow refers to flow as
measured in a graduated cylinder collecting the outflow of the
phantom. The correlation between flow measurements was excellent
with a slope of near unity and an intercept of approximately
zero.

PRELIMINARY DATA USING IDEAL PHANTOM

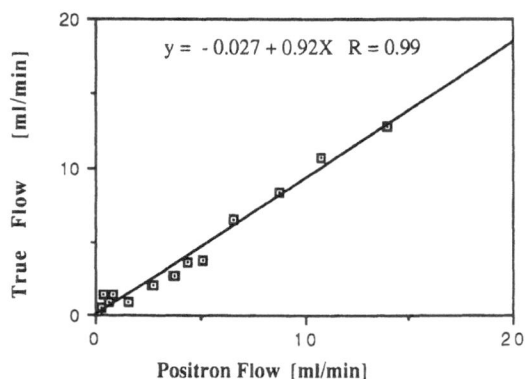

Fig. 2. Flow rates were measured through an idealized
phantom using the positron activation technique. The
experiments were performed at the Harvard Cyclotron
Laboratory in Cambridge. The radiation dose used for each
flow measurement was 250 cGy, which is similar to the daily
dose used in cancer treatment. The correlation between the
true flow and the flow measured by the positron technique,
was extremely high (R=0.99).

Figure 3 shows a typical positron decay curve measured from a
patient being treated for a clivus chordoma. A typical treatment
produces an Ao of between 100 and 200 counts/sec. Since the
initial activity is critical to the degree of accuracy possible
in making flow measurements, Monte Carlo experiments were
performed in order to evaluate the importance of this factor and
thus estimate the possible imaging resolution which might be
obtained (Table 1).

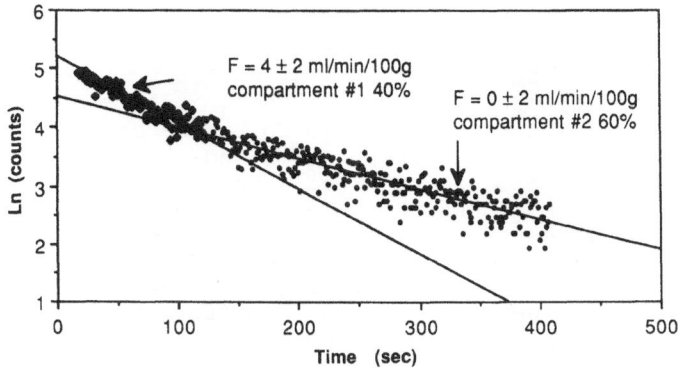

Fig. 3. Flow measurement made in a female patient with a clivus chordoma treated at the Harvard cyclotron. Flow measurements begin immediately following radiation therapy, using just the radioactivity normally induced by the 180 cGy treatment. The results suggest that approx. 40% of the tumor had a flow rate of 4 ml/min/100g. Since the tumor had a very low flow rate, this could explain the classical insensitivity of these tumors to radiation. Sixty percent of the oxygen in this cartilaginous tumor was non-diffusible, or had a flow rate undistinguishable from zero. A meningioma, unresectable due to hypervascularity, was also studied (not shown). Preliminary analysis demonstrated a flow rate of 172 ml/min/100g (approx. triple that of normal brain).

Table 1

EXPECTED ACCURACY OF POSITRON FLOW MEASUREMENTS

True Flow Rate ml/min/100g	Initial Activity (counts/sec)			
	25	75	300	1000
2	±2*	±2	±0.5	±0.5
5	±2	±1	±1	±0.2
10	±5	±1	±1	±1.0
50	±4	±2	±1	±0.5
150	±12	±3	±4	±2.0

* Errors are standard deviations. N=10. Values have units of ml/min/100g.

For Monte Carlo experiments, data sets were created assuming that ^{15}O is the only nucleus activated that produces positrons, that 30% of the ^{15}O is bound in non-diffusible form, and that the initial count rate is on average that shown on the top row of

Table 1. Noise was modeled using the large sample Gaussian approximation to the Poisson distribution, with a standard deviation equal to the square root of the count rate convolved with a uniform (0,6 cps) background noise. The simulated count rate data was analyzed assuming ^{15}O to be the only nucleus, but without knowledge of the fraction of bound ^{15}O. Ten modeled data sets were analyzed and are reported. The degree of accuracy possible, as determined by this modeling, suggests that even at very slow flow rates, and even when the initial count rate is very low, the accuracy of flow measurement will rarely be in error by more than a few ml/min/100 g. Estimation of unbound (immobile) ^{15}O was never more than 5% different than the correct value of 30%, at any initial count rate.

DISCUSSION

   Perhaps the most important determinants of the efficacy of non-surgical cancer treatment are the metabolic and physiological state of the tumor; and among these general categories of prognostic factors, tumor blood flow (TBF) is the most important since it is a primary determinant of the others(8). For example, chemotherapeutic efficacy is highly dependent on drug delivery, cellular pH, and oxygenation status. The incidence of distant metastasis is modified by tumor blood flow and angiogenesis(9,10). Likewise, in most studies performed over the past three decades, the curability by radiation of tumors of human or animal origin (grown in experimental animals) is predominantly dependent on the fraction of cells with low pO_2 values(11,12). The efficacy of many new treatment modalities including toxin and radiolabelled antibodies, photodynamic therapy and hyperthermia is also directly or indirectly affected by blood flow. For these and other reasons the measurement of TBF is an important area of medical research(13-15).
   Towards the goal of routine measurement of tumor blood flow, several investigators have measured human TBF, but most of these studies have technical difficulties in performance or in analysis of the resultant data. Probably the best techniques have employed PET (16-18). Nevertheless, the various techniques under clinical investigation are distinctly sub-optimal as they are either invasive, time consuming, expensive, or logistically difficult(13,19), and therefore have not become routine.
   Reliable and quantitative routine measurements of blood flow, would constitute an important advance in the ability to characterize tumors physiologically and radiobiologically(8,13,19), and should have immediate clinical applications.
   The purpose of the project described is to develop and implement a clinical system for non-invasive measurement of tumor blood flow(TBF). The system uses similar assumptions to that for standard positron emission tomography(PET), but takes advantage of the endogenous creation of positron emitters in patients undergoing radiation therapy. Hence, a relatively crude quality positron image, with approximately 10 cc volume resolution is possible in only 5 to 7 minutes immediately following proton beam radiation treatment. We do not yet know if the proposed system will be feasible on high energy linear accelerators routinely used for patient treatment such as the Siemens KD-2 23MV, or the 24MV Clinac 2500 linear accelerators.

Though this preliminary data is optimistic, we must have better calibration. For example, in addition to ^{15}O, small quantities of ^{11}C, ^{13}N, and ^{14}O will be produced in small quantities, and have been ignored in this analysis. Including these isotopes in the model will reduce errors and improve sensitivity. To handle this we will augment the mathematical modelling of the distribution of rates of blood flow by computer implementation of statistical equations like the following:

$E(C) = A\Sigma p_i * e^{-\lambda_i t}$ where C is the counts with a Poisson distribution and convolved with a uniform distribution due to background radiation. p_i is the portion of the brain in the ith compartment (or i^{th} positron emitting source). λ_i is the decay rate in the ith compartment and A is the initial activity (dependent on the voxel size and radiation dose). Using the technique of maximum likelihood (23) we will estimate (p_i and λ_i) and their standard errors, giving a method which jointly estimates the separate flow rates and compartment sizes. Estimating the number of compartments is important and methodologically challenging, however using methods developed by Tiago de Oliveria (In Classical and Contagious Discrete Distributions pp379-84, Pergamon, NY) on tests for component of mixtures, we will be able to perform statistical tests which discriminate between models with different numbers of flow compartments (or types of positron decaying nuclei).

Though in theory one could perform daily high resolution PET scans on patients, the respiration (or systemic administration) of radioactive tracers and the associated whole body radiation doses, the expense, the technical difficulty and need of arterial line placement are all avoided when the flow measurement is incidentally made by the non-invasive and local "injection" of ^{15}O tracer by the treatment beam. Based on our experience in the field of tumor physiology we believe these advantages easily offset the greater spatial resolution of standard PET, and that ultimately the proposed system will become a routine part of patient management.

REFERENCES

1. Moulder JE, Rockwell S: Hypoxic fractions of solid tumors: experimental techniques, methods of analysis and a survey of existing data. Int J Radiat Oncol Biol Phys 10:695-712, 1984.
2. Thomlinson RH: Changes of oxygenation in tumors in relation to irradiation. Front Radiat Ther Oncol 3: 109-121,1968.
3. Chaplin DJ: Postirradiation modification of tumor blood flow: a method to increase the effectiveness of chemical radiosentizers. Rad Res 115:292-302, 1988.
4. Okunieff P, Walsh CS, Vaupel P, Kallinowski F, Hitzig BM, Neuringer LJ, and Suit HD: Effects of hydralazine on in vivo tumor energy metabolism, hematopoietic radiation sensitivity, and cardiovascular parameters. Int J Radiat Biol Oncol Phys 16:1145-1148, 1989.
5. Okunieff P, Kallinowski F, Vaupel P, Neuringer LJ: Effect of hydralazine-induced vasodilation on the energy metabolism of murine tumors studied by in vivo ^{31}P-nuclear magnetic resonance spectroscopy. J Natl Cancer Inst 80:745-750, 1988.
6. Okunieff P, Vaupel P, Sedlacek R, Neuringer LJ: Evaluation of tumor energy metabolism and microvascular blood flow after

glucose or mannitol administration using ^{31}P magnetic resonance spectroscopy and laser Doppler flowmetry. Int J Radiat Oncol Biol Phys 16:1493-1500, 1989.

7. Hughes WL, Nussbaum GH, Connolly R, Emami B, Reilly P: Tissue perfusion rate determined from the decay of oxygen-15 activity after photon activation in situ. Science 204:1215-7, 1979.

8. Kallinowski F, Schlenger KH, Runkel S, Kloes M, Stohrer M, Okunieff P, Vaupel P: Blood flow, metabolic functions, cellular microenvironment and growth rate of human tumor xenografts. Cancer Res 49:3759-3764, 1989.

9. Folkman J: Tumor angiogenesis: therapeutic implications. N Engl J Med 285:1182-6, 1971.

10. Folkman J: What is the role of angiogenesis in metastasis from cutaneous melanoma? Eur J Cancer Clin Oncol 23:361-3, 1987.

11. Denekamp F: Cytotoxicity and radiosensitization in mouse and man. Br J Radiol 51:636-637, 1978.

12. Denekamp J, Fowler JF, Dische S: The proportion of hypoxic cells in a human tumor. Int J Radiat Oncol Biol Phys 2:1227-1228, 1977.

13. Vaupel P, Kallinowski F, Okunieff P: Blood flow, oxygen and nutrient supply, and metabolic microenvironment of human tumors:a review. Cancer Res 49:6449-6465, 1989.

14. Sutherland RM: Cell and environment interactions in tumor microregions: the multicell spheroid model. Science 240:177-84, 1988.

15. Auerbach R, Auerbach W: Regional differences in the growth of normal and neoplastic cells. Science 215:127-34, 1982.

16. Hawkins RA, Phelps ME: PET in clinical oncology. Cancer Met Rev 7:119-142, 1988.

17. Beaney RP: Positron emission tomography in the study of human tumors. Seminars Nucl Med 14:324-341, 1984.

18. McEwan AJB: Positron-emission tomography and predicting tumor treatment response. In: Chapman JD, Peters LJ, Withers HR (eds) Prediction of tumor treatment response. Pergamon, New York, 1989.

19. Okunieff, P.: Relationship of ^{31}P NMR measurements to tumor biology. In: Evelhoch, J., Negendank, W., Valeriote, F., Baker, L. (eds.) Magnetic Resonance in Experimental and Clinical Oncology, Kluwer Academic Publishers, Boston, pp 23-58, 1990.

20. Okunieff PG, Koutcher JA, Gerweck L, McFarland E, Urano M, Hitzig B, Neuringer L, and Suit HD: Tumor size dependent metabolic changes in a murine fibrosarcoma: Use of Fourier transform ^{31}P NMR to evaluate tumor energy metabolism. Int J Radiat Oncol Biol Phys 12:793-799, 1986.

21. Vaupel P, Okunieff P, Kallinowski F, Neuringer LJ: Correlations between ^{31}P-NMR spectroscopy and tissue O_2 tension measurements in a murine fibrosarcoma. Radiat Res 120:477-493, 1989.

22. Vaupel P, Okunieff P, Neuringer LJ: Blood flow, tissue oxygenation, pH distribution, and energy metabolism of murine mammary adenocarcinomas during growth. Adv Exptl Med Biol 248:835-846, 1989.

23. Titterington DM, Smith AFM, Makov UE: Statistical analysis of finite mixture distributions. § 4.3, p. 89, Wiley, NY, 1985.

IMPROVING THE EFFECTIVENESS OF THE BIOREDUCTIVE ANTITUMOR AGENT

SR 4233 BY INDUCED HYPOXIA

Andrew I. Minchinton and J. Martin Brown

Stanford University Medical Center
Division of Radiation Biology
Stanford University
Stanford, CA 94305-5105, USA

Introduction

The benzotriazine N-oxide, SR 4233 (1,2,4-benzotriazine-3-amine 1,4-di-N-oxide) is selectively toxic to hypoxic cells in vitro[1,2,3] and shows significant anti-tumor activity against a selection of experimental tumors in vivo.[4,5] The probable mechanism underlying the cytotoxicity of this compound is thought to involve the 1-electron reduction of the parent molecule to a free radical intermediate which results in DNA double strand breaks and subsequent cell death.[6,7,8,9] Under oxygenated conditions this cytotoxicity is ameliorated by back oxidation to the parent molecule and a second oxygen dependent pathway involving 2-electron reduction to a non-cytotoxic metabolite. Therefore, in order for this agent to be effective when combined with radiotherapy, tumors must contain cells at subnormal oxygen tension. While many previous modalities aimed at improving cancer therapy have focused on reducing the population of hypoxic cells within tumors, this therapy should work more effectively on tumors containing a high proportion of hypoxic cells. An aim of this study is to evaluate the effectiveness of SR 4233 when the proportion of hypoxic cells within the tumor has been artificially increased.

Hypoxic cells comprise a small, but crucial population of cells within tumors. Since they are known to be refractory to treatment with radiation[10,11] and may also be resistant to certain chemical agents[12] their eradication represents an important goal of research to improve the control of solid tumors.[13] While there is no doubt that hypoxic cells are present in many tumors at the beginning of therapy, their persistence throughout a course of treatment is less well understood. Hypoxic cells are presently thought to arise through two separate mechanisms termed diffusion dependent hypoxia and perfusion dependent hypoxia. Diffusion dependent hypoxia is thought to arise because the vasculature within a tumor cannot supply all the cells in the tumor with sufficient oxygen.[14] Continual cellular proliferation within the tumor forces a proportion of cells away from their nutritive blood vessels to such an extent that they eventually reside beyond the oxygen diffusion distance. Such cells are doomed if this state persists, but these cells can be 'rescued' by therapies which preferentially kill oxygenated cells. These hypoxic cells could then become 'reoxygenated' and then act as the focus for new clonogenic growth.

The existence of perfusion dependent hypoxia was originally hypothesized to explain certain experimental observations[15] and is thought to result from intermittent blood flow. In this model, regions of tumor tissue can become temporarily starved of oxygen and therefore rendered radioresistant. Convincing experimental evidence for the presence of perfusion dependent hypoxia has come from animal experiments using flow and static cytometry, both of which exploit the perivascular staining properties of Hoechst 33342 following intravenous injection.[16,17,18,19,20] These transiently hypoxic cells would also be refractory to therapies targeted primarily towards oxygenated cells and the survival of these cells could also lead to clonogenic regrowth.

Materials and Methods

Male and female C3H/km mice were bred under defined flora conditions and allowed access to food and water *ad libitum* except during the treatment time and for approximately one hour subsequently. SCCVII tumors were treated when they reached a geometric mean diameter of 8.24±1.2 mm about 17 days after intradermal implantation of $2x10^5$ cells into the lower sacral region of the mice. Each group consisted of eight mice: four of each sex. No sex difference in the response to the treatments was observed. Mice were immobilized in lead jigs which allowed the tumor and a minimal amount of normal tissue to be irradiated. X-rays (250 kVp) were delivered with a Phillips RT250 unit operating at 12.5 mA at a dose rate of approximately 1.8 Gy min^{-1} filtered with 0.35 mm Cu (half-value layer 1.3 mm Cu).

Two treatment protocols were used: the first comprised one administration of SR 4233 following a single irradiation and subsequent hypoxic breathing (see below); the second employed a fractionated treatment comprising eight fractions of radiation followed on each occasion by SR 4233 administration, and one hour hypoxic breathing. In the single treatment protocol, mice receiving radiation were given 20 Gy and then immediately injected with SR 4233 (0.3 mmol kg^{-1} i.p.) and then transferred to a plastic box of dimensions 0.14 x 0.14 x 0.3 m purged with 10% oxygen 90% nitrogen at a flow rate of 5 l min^{-1} (hypoxic breathing). The mice remained in this box for one hour and were then returned to their cages under normal atmospheric conditions. The fractionated studies consisted of eight irradiations of 2.5 Gy given twice daily at 8:30 am and 5:00 pm for four consecutive days. Immediately after each irradiation the mice were administered SR 4233 i.p. at a dose of 0.08 mmol kg^{-1} and some groups breathed 10% oxygen 90% nitrogen for one hour before being returned to their cages.

Mice were killed by cervical dislocation the day after the last treatment. The tumors were excised, weighed, minced and further dissociated using a mixture of pronase, DNAse and collagenase at a concentration of 0.6, 0.2 and 0.2 mg ml^{-1} in 10 mls Hanks buffered saline solution. This mixture was agitated while being maintained at 37 °C for 30 min. The dissociated cells were then filtered through a steel mesh and centrifuged at 450 x G for 10 min prior to being resuspended in 10 mls Waymouth's media containing 15% fetal calf serum. Cells excluding trypan blue were counted and plated after appropriate dilution and incubated in a humidified atmosphere at 37°C for 13 days in air containing 5% CO_2. Clonogenic survival was deemed as the development of a colony of not less than approximately 50 cells, and scoring was facilitated by staining the washed plates with 2.5% crystal violet dissolved in 95% ethanol/water.

The results are expressed in terms of 'relative clonogenic cells per tumor' (RCC/T), 'surviving fraction' (SF) and 'fractional yield' (FY). RCC/T is the product of the surviving fraction and the fractional yield where 'fractional yield' (FY) is equal to the total number of cells recovered from the treated tumor divided by the total number of cells recovered from control tumors. The surviving fraction is the proportion of cells excluding trypan blue which subsequently form colonies relative to control tumors.

Results and discussion

Single treatment

Panel A of Figure 1 shows the effect of various treatments on the tumor, expressed in terms of relative clonogenic cells per tumor (RCC/T). Irradiation of the tumor in air breathing mice resulted in a reduction of the RCC/T to approximately 10^{-3} compared to control unirradiated tumors. No significant difference was noted when mice breathed the 10% oxygen mixture for one hour following the irradiation. Unirradiated tumors from mice administered SR 4233 and subsequently breasthing normal air showed no effect compared to mice not receiving the drug. However, tumors from the irradiated group showed a decrease in RCC/T of at least a further logarithm compared to the tumors of mice not administered SR 4233. Irradiated tumors from mice receiving SR 4233 and subsequently exposed to the hypoxic breathing conditions had a RCC/T similar to that of mice not breathing the reduced oxygen atmosphere, but the unirradiated group showed a decreased RCC/T indicating enhanced cytotoxicity of SR 4233. Panel B shows that this decrease in the RCC/T of the unirradiated group of animals receiving SR 4233 and hypoxic breathing arose from a decrease in the yield of cells from the tumors. The surviving fraction of cells extracted from the tumor and excluding trypan blue are shown in panel C. The unirradiated groups all show a similar level of survival while the irradiated groups show that treatment with SR 4233 results in a decrease in surviving fraction of about one logarithm.

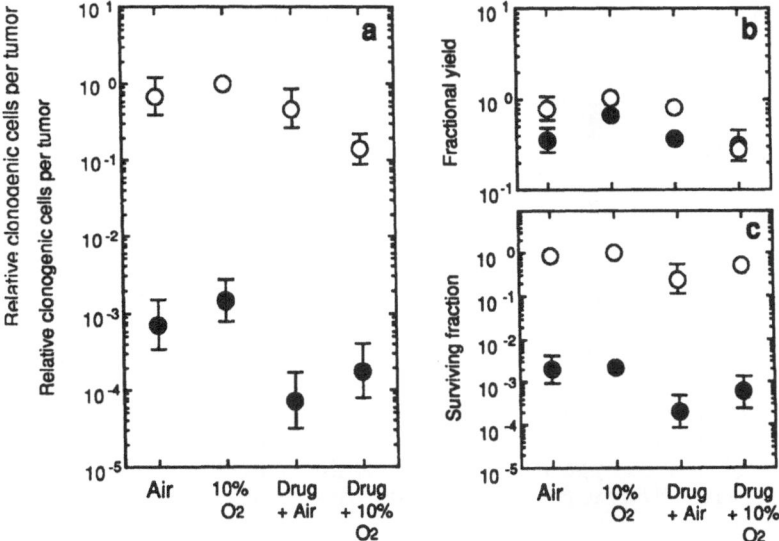

Figure 1 *Effect of breathing 10% oxygen - 90% nitrogen ('hypoxic breathing') on the antitumor effectiveness of SR 4233 in combination with a single dose of 20 Gy X-rays. SR 4233 was administered immediately after the radiotherapy. Hypoxic breathing immediately followed the SR 4233/radiation treatment and lasted one hour. Each point represents the geometric mean and standard error of 8 mice (4 females and 4 males). Open symbols represent cells from unirradiated tumors and closed symbols represent cells from irradiated tumors.*

Fractionated treatment

In the fractionated studies the profile of responses to the various treatments is different from the single dose studies. Panel A of Figure 2 shows the overall antitumor effect of the treatments. In the normal air-breathing groups, irradiation resulted in a reduction of the RCC/T to about 2.5×10^{-3} compared to the unirradiated group. Hypoxic breathing after irradiation did not alter the response of the irradiated (or unirradiated) groups. Treatment with SR 4233 resulted in a modest decrease in the RCC/T of the unirradiated group, but caused a larger decrease in the RCC/T of the irradiated group by about one logarithm. The effect of hypoxic breathing after administration of SR 4233 was profound. It reduced the RCC/T of both the unirradiated group and the irradiated groups by about a logarithm. Panels B and C of Figure 2 illustrate the effects the different treatments have on the fractional yield and the surviving fraction. They demonstrate that the combination of SR 4233 and hypoxic breathing after the treatment reduces both the surviving fraction and the yield of cells from the tumor.

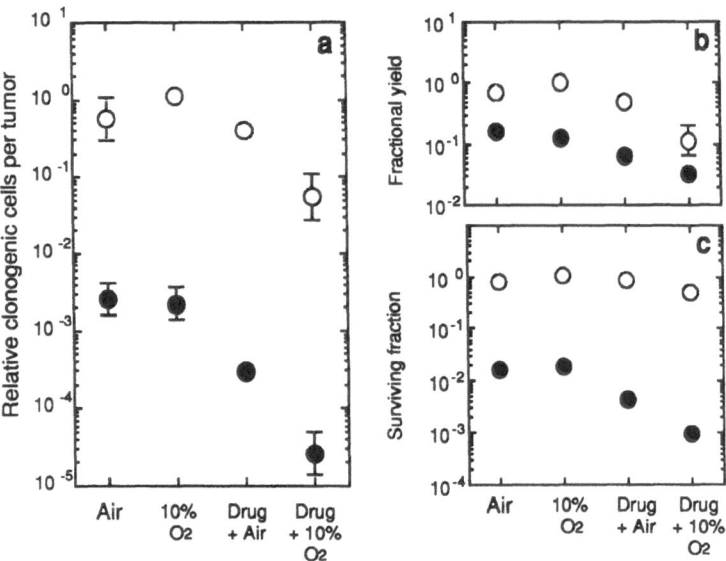

Figure 2. *The effect of hypoxic breathing on the antitumor effectiveness of SR 4233 in a fractionated protocol involving 8 treatments over 4 days, comprising twice daily irradiations of 2.5 Gy per fraction and administration of SR 4233 (0.08 mmol kg[-1]) followed by one hour of hypoxic breathing. Each point represents the geometric mean and standard error of 8 mice (4 females and 4 males). Open symbols represent cells from unirradiated tumors and closed symbols represent cells from irradiated tumors.*

Conclusions

SR 4233 combined with radiation therapy shows activity against a variety of experimental tumors grown in mice, but with differential potency. It is likely that several factors determine the activity of SR 4233 *in vivo* such as the complement of enzymes necessary for its reductive metabolism as well as pharmacological factors, but it is very likely that the oxygenation of the tumor is a most crucial factor. The experimental results in this article confirm that inducing hypoxia artificially by means of breathing a reduced content of oxygen can enhance the antitumor potency of SR 4233 though this is observed only in fractionated treatment.

Acknowledgements

The technical assistance of Doug Menke and Nixy Zutshi is gratefully acknowledged. This investigation was supported by Grant 25990 from the National Cancer Institute, Department of Health and Human Services.

References

1. Zeman, E.M., Brown, J.M., Lemmon, M.J., Hirst, V.K. & Lee, W.W. SR-4233: A bioreductive agent with high selective toxicity for hypoxic cells *Int. J. Radiat. Biol. Phys.* **12**: 1239-1242 (1986).

2. Baker, M.A., Zeman, E.M., Hirst, V.K. & Brown, J.M. Metabolism of SR-4233 by Chinese hamster ovary cells: Basis of selective cytotoxicity *Cancer Res.* **48**: 5947-5952 (1988).

3. Zeman, E.M., Hirst, V.K., Lemmon, M.J. & Brown, J.M. Enhancement of radiation-induced tumor cell killing by the hypoxic cell toxin SR-4233 *Radiother. Oncol.* **12**: 209-218 (1988).

4. Brown, J.M. & Lemmon, M.J. Potentiation by the hypoxic cytotoxin SR 4233 of cell killing produced by fractionated irradiation of mouse tumors *Cancer Res.* **50**: 7745-7749 (1990).

5. Brown, J.M. & Lemmon, M.J. SR 4233 - A tumor specific radiosensitizer active in fractionated radiation regimes *Radiother Oncol* **20**: 151-156 (1991).

6. Biedermann, K.A., Wang, J., Graham, R.P. & Brown, J.M. SR 4233 cytotoxicity and metabolism in DNA repsir-competent and repair-deficient cell cultures *Br. J. Cancer* **63**: 358-362 (1991).

7. Laderoute, K.R. & Rauth, A.M. Identification of two major reduction products of the hypoxic cell toxin 3-amino1,2,4-benzotriazine-1,4-dioxide *Biochem. Pharmacol.* **35**: 3417-3420 (1986).

8. Laderoute, K., Wardman, P. & Rauth, A.M. Molecular mechanisms for the hypoxia dependent activation of 3-amino-1,2,4-benzotriazine-1,4-dioxide (SR 4233) *Biochem. Pharmacol.* **37**: 1487-1495 (1988).

9. Zeman, E.M. & Brown, J.M. Pre- and post-irradiation radiosensitization by SR 4233 *Int. J. Radiat. Oncol. Biol. Phys.* **16**: 967-971 (1989).

10. Bush, R.S., Jenkin, R. D. T., Allt, W. E. C., Beale, F. A., Bean, H., Dembo, A. J., Pringle, J. F. Definitive evidence for hypoxic cells influencing cure in cancer therapy *Br. J. Cancer* **37**:302-306 (1978).

11. Dische, S., Anderson, P.J., Sealy, R. & Watson, E.R. Carcinoma of the cervix-anaemia, radiotherapy and hyperbaric oxygen supports the importance of hypoxia in radiotherapy *Br. J. Radiol.* **56**: 251-255 (1983).

12. Tannock, I. & Guttman, P. Response of chinese hamster ovary cells to anti-cancer drugs under aerobic and hypoxic conditions *Br. J. Cancer* **43**: 245-248 (1981).

13. Fowler, J.F. La Ronde - Radiation sciences and medical radiology *Radiother. Oncol.* **1**: 1-22 (1983).

14. Thomlinson, R.H. & Gray, L.H. The histological structure of some human lung cancers and the possible implications for radiotherapy *Br. J. Cancer* **9**: 539-549 (1955).

15. Brown, J.M. Evidence for acutely hypoxic cells in mouse tumours and a possible mechanism of reoxygenation *Br. J. Radiol.* **52**: 650-656 (1979).

16. Chaplin, D.J., Durand, R.E. & Olive, P.L. Cell selection from a murine tumour using the fluorescent probe Hoechst 33342 *Br. J. Cancer* **51**: 569-572 (1985).

17. Chaplin, D.J., Durand, R.E. & Olive, P.L. Acute hypoxia in tumour: Implications for modifiers of radiation effects *Int. J. Radiat. Oncol. Biol. Phys.* **12**: 1279-1282 (1986).

18. Chaplin, D.J., Olive, P.L. & Durand, R.E. Intermittent blood flow in a murine tumour: Radiobiological effects *Cancer Res.* **47**: 597-601 (1987).

19. Trotter, M.J., Chaplin, D.J., Durand, R.E. & Olive, P.L. The use of fluorescent probes to identify regions of transient perfusion in murine tumours *Int. J. Radiat. Oncol. Biol. Phys.* **16**: 931-934 (1989).

20. Minchinton, A.I., Durand, R.E. & Chaplin, D.J. Intermittent blood flow in the KHT sarcoma - flow cytometry studies usinf Hoechst 33342 *Br. J. Cancer* **62**: 195-200 (1990).

A COMPUTER SIMULATION OF OXYGEN PARTIAL PRESSURE
AND TEMPERATURE PROFILES DURING HYPERTHERMIA

Kyung A. Kang*[1], Samuel A. Afuwape*[2], and Duane F. Bruley*[3]

*[1] Dept. of Biochem. & Biophy., Univ. of Penn., Philaldelphia, PA 19104
*[2] Dept. of Biomed. Engr, Univ. of Southern Cal., LA, CA 90089
*[3] College of Engineering, UMBC, Baltimore, MD 21228

INTRODUCTION

Hyperthermia is a cancer therapy utilizing the difference in heat transfer characteristics between the tumor and normal tissue. In many cases, preferential heating of the tumor takes place because the perfusion rate is less in the tumor tissue relative to the normal tissue. This therapy has the advantage of fewer side effects during and after the treatment compared with other currently used cancer therapies. Hyperthermia can be used not only by itself and also in conjunction with other cancer therapies, such as chemo-therapy (Haranaka et al., 1987), x-irradiation (George et al., 1989; Fujiwara, et al., 1990), or both (Howard and Bleehen, 1988), to enhance tumor eradication or regression.

During the hyperthermia treatment optimum energy deposition can be defined as the minimum amount of heat necessary to destroy the tumor with minimum damage to normal cells. Therefore, estimating the optimum energy applied to the normal and tumor tissue requires a knowledge of the oxygen and temperature distributions which are a result of the diffusion (for the oxygen transport), conduction (for heat transport), convection, and oxygen consumption characteristics of the system.

Mathematical models and modern computer technology have been used to better predict system conditions (Strohbehn and Roemer, 1984; Busch, Bruley, and Bicher, 1982; Kang, Bruley, and Bicher, 1988). However, it is still very difficult to estimate temperature and oxygen profiles when the system has an irregular geometry and/or includes heterogeneous properties and convection. The B-W-K technique is proving to be a powerful probabalistic-numerical method for solving three dimensional, time dependent, conduction, diffusion, convection, and reaction problems (Williford, Bruley, and Artigue, 1974).

Oxygen partial pressure and temperature profiles in a cube of normal tissue, with a tumor containing necrotic tissue and regions of perfusion, are calculated considering locally applied microwave energy. The finite element method was used to compute the energy deposition in tissue and, therefore, the energy source term in the dielectric medium. This term allowed coupling the energy source with the heat and oxygen transport problem using the B-W-K technique for computation.

SYSTEM SIMULATED

A cube of tissue (3 x 3 x 3 cm) with a cube of tumor (1 x 1 x 1 cm) were chosen as the system of simulation (Figure 1). Necrotic tissue (0.4 x 0.4 x 0.4 cm) was located at the center of the tumor. Microwave energy was applied at the top of the normal tissue (y = 3 cm).

Oxygen Transport to Tissue XIV, Edited by W. Erdmann and
D.F. Bruley, Plenum Press, New York, 1992

In order to simplify the simulated system, microwave energy was assumed to be placed directly on the top of the normal tissue (i.e. no skin, fat, or bone) The blood flow was assumed only in the z direction with the flow rate in the tumor assumed to be one half of the normal tissue (for normal tissue, 6 ml/100g/min).

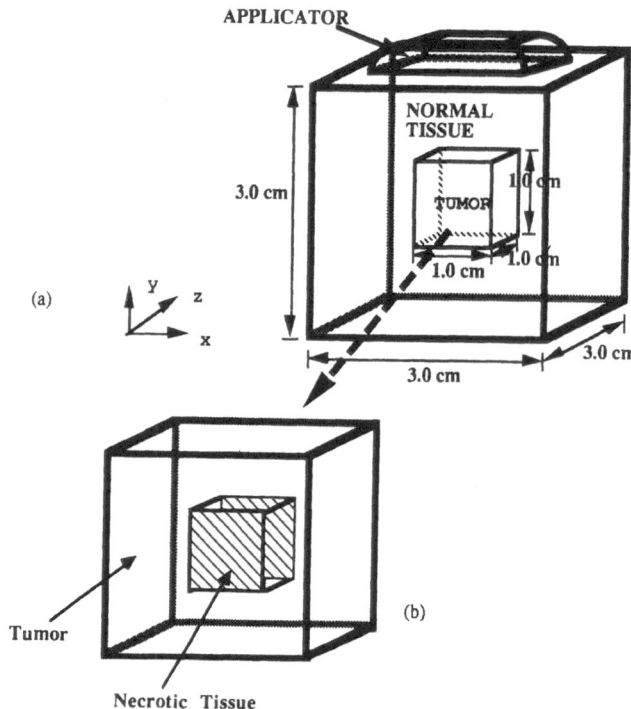

Figure 1. The system simulated. A tumor (1 x 1 x 1 cm) is imbedded in a normal tissue (3 x 3 x 3 cm). At the center of the tumor, necrotic tissue (0.4 x 0.4 x 0.4 cm) is located.

COMPUTATIONAL PROCEDURE

(1) The simulation system is discretized: Node distribution was determined (31 x 31 x 31 nodes) after studying the geometry and important system parameters and then the optimum time step size was determined using the spatial step size and system parameters.

(2) Microwave heat source characteristics are chosen for the application.

(3) The heat generation term at each nodal point was computed by the finite element method (Afuwape and Bruley, 1989; 1990).

(4) The temperature at each nodal point, after a single time step, for three different phases (normal tissue, tumor, and necrotic tissue) was computed using the heat generation term, by the B-W-K method.

(5) Blood flow (convective velocity) with changing temperature in normal tissue was computed using established experimental results (Song et al., 1984).

(6) Oxygen partial pressure values, after a single time step, for three phases (normal tissue, tumor, and necrotic tissue) were computed with changing blood flow by the B-W-K method.

(7) Steps (4), (5), and (6) were repeated until the final time step is reached (summation of the individual time steps yields the total time of the process).

Since this simulation represents an initial attempt to couple four major variables in the system (heat generation, heat transfer, blood flow change due to the temperature change, and oxygen transfer), much of the detailed correlation of physiological change has been simplified. However, the computer code is designed such that any known physiological information for each nodal point can be integrated into the problem solution.

MATHEMATICAL MODEL

I. Heat Generation by Microwaves

Governing Equations

A theoretical model of electromagnetic wave propagation in biological media was developed from fundamental curl and divergent Maxwell's equations, which describe the macroscopic interactions of nonionizing electromagnetic waves with living matter. The system of equations is asymmetrical in the time domain, they were transformed into the complex phasor form and numerically solved in a discrete element of the medium of microwave propagation using the finite element method (FEM). This hyperbolic Helmholtz equation is the wave equation that governs the behavior of the electromagnetic field in biological media and is stated as follows:

$$\nabla^2 E + \gamma^2 \, \nabla E = 0 \tag{1}$$

with

$$\gamma^2 = \omega^2 \, \mu \, \varepsilon + j \, \omega \, \mu \, \sigma \tag{2}$$

where γ is the square of the medium constitutive complex propagation constant (neper/m) that mediates interaction with the microwave excitation, E is the distributed vectorial field (V/m), ω is the angular frequency (radian/s), μ is permeability of the biomedium(H/m), ε is the permissivity of the biomedium (F/m), and σ is the conductivity of the biomedium (siemens/m).

For an excited case, a nonhomogeneous system is given as follows:

$$\nabla^2 E + \gamma^2 \, \nabla E = f \, (\omega, \, z) \tag{3}$$

where the directional microwave excitation (z-axis); f is volt/meter. Eq. (3) gives the Helmholtz equation in its general form of the wave equation.

The implementation of the model system (Eq. (3)) in a discretized three dimensional finite element formulation of the medium led to the functional (Silvester, 1983) that is developed into a tetrahedron finite element routine for an arbitrary biological tissue divisible into a cubical geometry. The simplification of the stationary functional in the form of a surface integral (Silvester, 1983) is given as follows:

$$F(E) = \frac{\varepsilon}{2} \int_V \nabla E \ \nabla E \ dV - \frac{\gamma^2}{2} \int_V \nabla E \ dV - \int_S E \ f \ dS \qquad (4)$$

with the assumption of microwave propagation into the z-axis and the imposition of Dirichlet and natural boundary conditions. The EM energy $f(\omega,z)$ is impinge on the Dirichlet surface. F(E) is the variational energy functional in an arbitrary volume.

Initial Condition

$$E \ (x,y,z) = 0.0 \qquad (5)$$

Boundary Condition

$$\frac{\partial E}{\partial x} \ (x=0.0, \ x=3.0) = \frac{\partial E}{\partial y} \ (y=0.0, \ y=3.0) = 0.0 \qquad (6)$$

where x, y, and z are spatial coordinates.

II. Oxygen Transfer

Typical Governing Partial Differential Equation

$$\frac{\partial P}{\partial t} = D \ (\frac{\partial^2 P}{\partial x^2} + \frac{\partial^2 P}{\partial y^2} + \frac{\partial^2 P}{\partial z^2}) - v_z \ (\frac{\partial P}{\partial z}) - R_x \qquad (7)$$

where P is oxygen partial pressure (mmHg), D is the diffusion coefficient of oxygen (cm^2/s), v_z is the superficial velocity of the blood (cm/s), R_x is zeroth order oxygen consumption rate (mmHg/s), and t is time (s). Parameters, D, v_z, and R_x are a function of location in the location of the simulated system (i.e. values for the normal tissue are different from those of tumor or necrotic tissue) and the mass transfer between adjacent regions is conserved.

Blood Flow Rate Change and Perfusion Area

According to experimental measurements (Song, et al., 1984) the blood flow in normal tissue increases very rapidly with temperature rise while, in may cases, the blood flow in tumor does not change significantly. The actual increase of blood flow as tissue temperature increase varies from a type of tissue to another. In this study, the blood flow rate at various tissue temperature between 36 °C and 45 °C was curve-fitted and used for the computation. Blood flow rate in the tumor was assumed to be independent of the temperature increase.

In this simulation, the concept of convective velocity is treated in a non-conventional way. Previously, it has been emphasized that the heterogeneity of the blood vessels is important for heat and oxygen transport calculations. However, in practice, it is impossible to see the heterogeneous structure of the vasculature before or during the treatment. Therefore, in this simulation the convective region is treated as a perfused region instead of a single blood vessels and tissue. This is accomplished by distributing the convective transport among the node points of tumor and normal tissues, except in the necrotic region.

186

Oxygen Consumption Rate

Oxygen consumption is assumed to be zeroth order and the oxygen consumption rate for tumor was assumed to be one half of the normal tissue, except for the necrotic tissue.

Initial Conditions

Initial conditions for oxygen transfer were arbitrarily chosen. Since the oxygen partial pressure profiles before hyperthermia were not available for the simulated system it was assumed that the partial pressure at the inlet of the system, P (z=0.0 cm), was 50 mmHg, and the outlet of the system, P (z=3.0 cm) was 30 mmHg and between z = 0.0 cm and z = 3.0 cm, a linear PO_2 profile was assumed.

$$P(z) = 50 + (30-50) \cdot z/L \tag{8}$$

where L is the total system length in the z direction (3 cm).

The initial conditions chosen for this simulation may not accurately represent an actual physiological condition. It is possible that the oxygen partial pressure at the center of the tumor would be very low and at positions farther from the center of the tumor the oxygen partial pressure would be higher. However, since this is a trial simulation and actual oxygen partial pressure values for the system were not available a simple linear pressure profile was assumed.

Boundary Conditions

$$P(z=0.0 \text{ cm}) = 50 \text{ mmHg} \tag{9}$$

$$\frac{\partial P}{\partial x}(x=0.0, \ x=3.0) = \frac{\partial P}{\partial y}(y=0.0, \ y=3.0) = \frac{\partial P}{\partial z}(z=3.0)=0.0 \tag{10}$$

III. Heat Transfer

Typical Governing Partial Differential Equation

$$\frac{\partial T}{\partial t} = D_T\left(\frac{\partial^2 T}{\partial x^2}+\frac{\partial^2 T}{\partial y^2}+\frac{\partial^2 T}{\partial z^2}\right) - v_z\left(\frac{\partial T}{\partial z}\right) + H_G \tag{11}$$

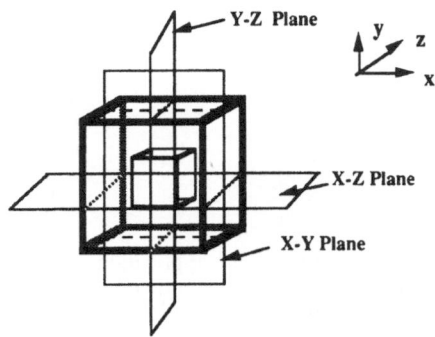

Figure 2. Planes that oxygen partial pressure and temperature profiles are shown.

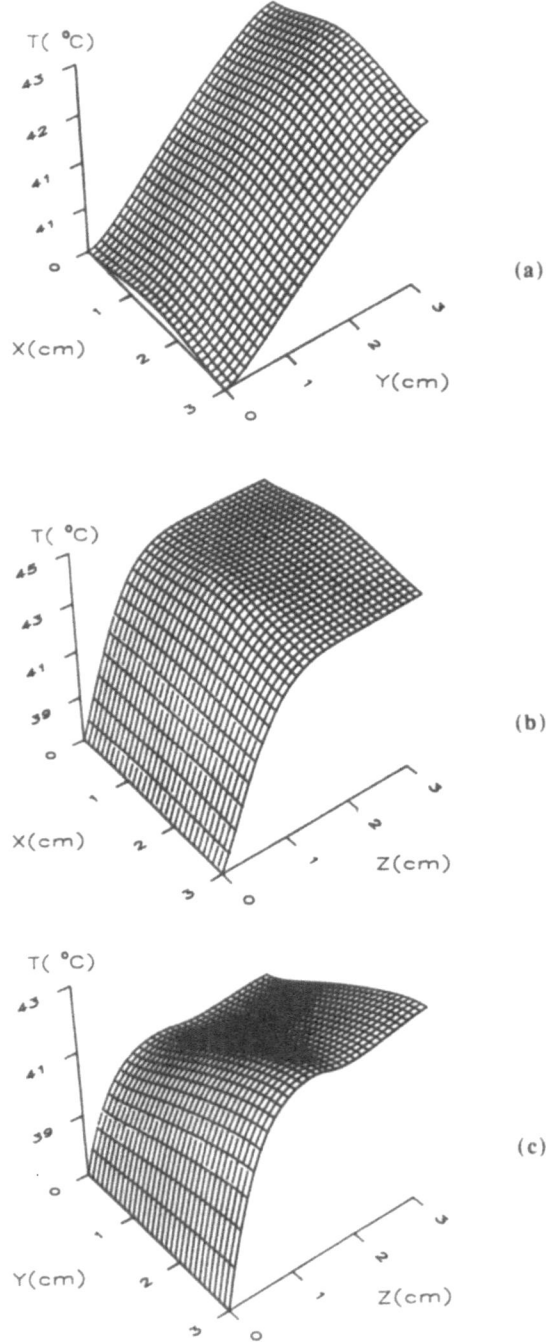

Figure 3. Temperature profiles
    (a) Profiles on x-y plane after two minutes of microwave application.
    (b) Profiles on y-z plane after two minutes of microwave application.
    (c) Profiles on y-z plane after two minutes of microwave application.

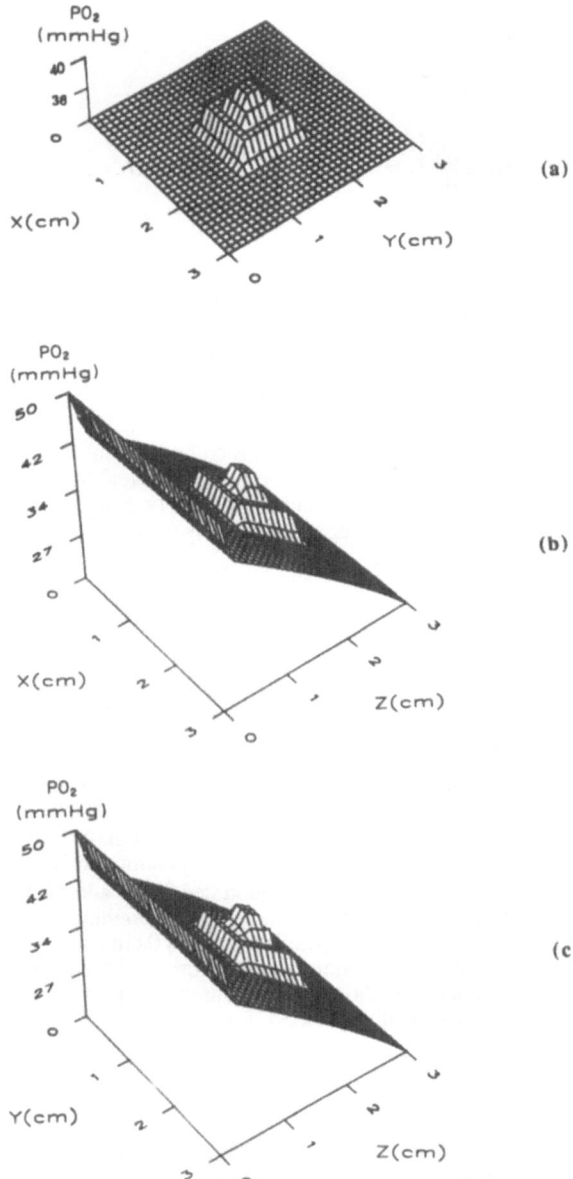

Figure 4. Oxygen concentration profiles.
   (a) Profiles on x-y plane after two minutes of microwave application.
   (b) Profiles on x-z plane after two minutes of microwave application.
   (c) Profiles on y-z plane after two minutes of microwave application.

where T is Temperature (^{0}C), D_T is thermal diffusivity (thermal conductivity/specific gravity/specific heat), and H_G is heat generation (W/cm^3) by microwave.

Initial Condition

$$T = 37\,^{0}C \tag{12}$$

Boundary Conditions

$$\frac{\partial T}{\partial x}\,(x=0.0,\ x=3.0) = \frac{\partial T}{\partial y}\,(y=0.0,\ y=3.0) = \frac{\partial T}{\partial z}\,(z=3.0) = 0.0 \tag{13}$$

$$T\,(z=0.0) = 37\ ^{\circ}C. \tag{14}$$

The B-W-K technique was used for the computation of oxygen partial pressure and temperature profiles on the Cray Y-MP computer located in San Diego, California, USA.

RESULTS

Oxygen partial pressure and temperature at each nodal point (system is divided into 31 x 31 x 31 nodal points) were computed for two minutes of real time. Three planes [x = 1.5 cm (y-z plane), y = 1.5 cm (x-y plane), and z = 1.5 cm (x-z plane); figure 2] were selected to illustrated the computed results and the results on the planes were shown in figures 3 and 4.

DISCUSSION

I. Temperature Profiles

Figure 3 shows the temperature profile after two minutes of microwave application. From figure 3 (a) it can be shown that the temperature rise near to the surface is much greater than that inside of the tissue. This results because the heat generation via microwave energy deposition at the upper surface is very high and as the penetration depth increases the heat generation decreases rapidly. Figure 3 (b) shows the temperature profiles of the x-z plane at y = 1.5 cm. Since the temperature at z = 0 cm is assumed 37 $^{\circ}$C and the blood flow is in the z direction, in the other regions the temperature increases due to the heat generated by microwave deposition and the temperature gradient at z = 0.0 cm changes rapidly. In figure 3 (c), the effect of different blood flows in the three regions is also shown at the center of the tissue. Over the time period of computation, temperature in the necrotic area tends to be higher than the other areas because of zero blood flow.

II. Oxygen Profiles

The oxygen partial pressure values for the three regions are distinctly different. In the profiles for all three planes (Figure 4), the oxygen pressure in the necrotic region is the highest and the normal tissue has the lowest values. This appears to be opposite to expected physiological phenomena. This is thought to be due to improper initial conditions for the oxygen transfer computation. It would be expected that when the tumor is rapidly growing, oxygen consumption in the tumor is very high. The blood vasculature in the tumor is not well developed, therefore, convective oxygen transfer in the tumor is much lower than that of normal tissue. Therefore, it is possible that the oxygen partial pressure at the center of the tumor becomes low and the necrotic region develops (Gatenby, et al., 1988). Since the initial condition for oxygen transport was assumed to be linear, oxygen consumption in the tumor is only half that of the normal tissue and there was no oxygen consumption in the necrotic region, the oxygen partial pressure becomes highest in the necrotic region.

In the figures 4 (b) and (c), the convection effect by blood flow is shown. Because the high oxygen partial pressure at z = 0.0 is transported by the blood flow and the oxygen is consumed in the other regions the profiles in the z direction show steep gradient around z = 0.0 cm.

CONCLUSIONS AND FURTHER STUDY

This simulation presents the coupling of electromagnetic wave energy deposition, in biological tissue, with heat and mass transfer consideration to produce theoretical temperature and oxygen profiles in three dimensions and time in a heterogeneous system. Calculations of this type, with the appropriate initial and boundary conditions will be useful for the design of microwave applicators and their positioning and power optimization during clinical hyperthermia.

To insure validity the simulation will have to be compared with experimental data from phantom models or in *in vivo* animal testing. The simulation, probably, may not be refined to the point that they will give exact predictions, however, they will be useful for best estimates of the energy transport processes for local heating with microwaves.

More realistic simulations including skin, fat, and bone, with irregular geometries will be important in the future.

REFERENCES

Afuwape, S. A. and Bruley, D. F. ," Simulation of Electrical Field Distribution and Penetration of Therapeutic Microwave Energy Deposited in Biological Tissue." AIChE Symposium series #269, 85, pp. 374-382, 1989.

Afuwape, S. A. and Bruley, D. F. ,"Coupling of EM Microwave Energy Deposition with Distributed Perfusion for Therapeutic Thermal Distribution", ASME Bioheat Transfer Application in Hyperthermia Emerging Horisons in Instrumentation and Modeling, HTD-126, BED-12 (Roemer, R. B., Mcgrath, J. J., and Bowman, H. F., eds.), pp. 67-74, 1990.

Andersen, B. J., "EM Heating - a Review", pp. 113-128, in Proceedings of the 4th International Symposium on Hyperthermia Oncology, (Aarhus ed.), 2, Taylor and Francis, London, 1984.

Busch, N., Bruley, D. F., and Bicher, H. I., "Identification of Viable Regions in *in vitro* Spheroidal Tumors: A Mathematical Investigation", pp. 185-189, in Hyperthermia (Bicher, H. I. and Bruley, D. F., eds.), Plenum Press, New York, 1982.

Dickinson, R. J., "An Ultrasound System for Local Hyperthermia using Scanned Focused Transducers", IEEE Transactions on Biomedical Engineering, BME-31 (1), pp. 120-125, 1984.

George, K. C., Streffer, C., and Pelzer, T., "Combined Effects of X Rays, Ro 03-8799, and Hyperthermia on Growth, Necrosis, and Cell Proliferation in a Mounse Tumor", Int. J. Radiation Oncology Biol. Phys., 16, pp. 1119-1122, 1989.

Fujiwara, K. and Watanabe, T., "Effects of Hyperthermia, Radiotherapy and Thermoradiotherapy on Tumor Microvasculature Permiability", Japanese Society of Pathology, 40(2), pp. 79-84, 1990.

Gatenby, R. A., Kessler, H. B., Rosenblum, J. S., and Coia, L. R., "Oxygen Distribution in Squamous Cell Carcinoma Metastases and its Relationship to outcome of Radiation Therapy", I.J. Radiation Oncology, Biol. Phys., 10, pp.831-834, 1988.

Haranaka, K., Sakurai, A., and Satomi, N., "Antitumor Activity of Recombinant Human Tumor Necrosis Factor in Conbination with Hyperthermia, Chemotherapy, of Immunotherapy", Journal of Biological Modifiers, 6, pp. 379-391, 1987.

Howard, G. C. W. and Bleehen N. M., "Clinical Experience in the Conbination of Hyperthermia with Chemotherapy or Radiotherapy", Recent Result in Cancer Research, 107, pp.214-248, 1988.

Kang, K. A., Bruley, D. F., and Bicher, H., "A Computer Simulation of Simultaneous Heat and Oxygen Transport during Heterogeneous Three Dimensional Tumor Hyperthermia.", pp. 747-756, Advances in Experimental Medicine and Biology, (Mochizuki, M., Honig, C., Koyama, T., Goldstick, T., and Bruley, D. F., eds.), Plenum Press, New York, 1988.

Silvester, P. P. and Ferrari, R. L., "Finite Element for Electrical Engineers", Cambridge University Press, Cambridge, 2nd edition, 1990.

Song, C. W., Lokshina, A., Rhee, J. G., Patten, M., and Levitt, S. H., "Implication of Blood Flow in Hyperthermic Treatment of Tumors", IEEE Transactions on Biomedical Engineering, BME-31 (1), pp. 9-16, 1984.

Strohbehn, J. W. and Roemer, R. B., "A Survey of Computer Simulations of Hyperthermia Treatments", BME-31 (1), pp. 136-149, 1984.

Williford, Jr., C., Bruley, D., and Artique, R., "Probabilistic Modelling of Oxygen Tranpost in Brain Tissue", NeuroResearch, 2, pp. 153-170, 1974.

ADVANCEMENT IN TECHNOLOGY AND INSTRUMENTATION

OXYGEN DEPENDENT QUENCHING OF PHOSPHORESCENCE:

A PERSPECTIVE

D.F. Wilson

Department of Biochemistry and Biophysics, Medical School
University of Pennsylvania Philadelphia, PA 19104 U.S.A.

SUMMARY

Oxygen quenches phosphorescence by energy transfer from the phosphor when oxygen molecules collide with molecules of the phosphor in the excited triplet state. Thus increasing oxygen pressure causes an increase in the rate of decay of phosphorescence (shorter lifetimes) and a decrease in total phosphorescence intensity. Phosphors have been selected which decay with a single exponential and for which the relationship between phosphorescence lifetime and oxygen pressure is quantitatively described by the Stern-Volmer equation. The use of phosphorescence lifetime as the measure of oxygen pressure makes the method insensitive to the absorbance changes of other chromophores in the system. This method has permitted quantitative, rapid (less than 10 msec) and sensitive (to less than 10^{-8} Torr) measurements of oxygen pressure in suspensions of cells or subcellular organells.

In tissues, oxygen pressure has been evaluated by measuring phosphorescence using an intensified CCD camera. Maps of oxygen pressure in the vasculature of the cortex of the brain and of other tissues demonstrate the method is limited only by the optics of the system and resolutions of a few microns are readily attained.

INTRODUCTION

The oxygen dependent quenching of phosphorescence has been shown to accurately measure oxygen consumption by suspensions of mitochondria and cells from air saturation to less than 10^{-8} torr (see for example Vanderkooi et al, 1987; Wilson et al, 1988; Robiolio et al, 1989, Rumsey et al, 1990; Wilson et al, 1991). The oxygen pressure dependence of the rate of oxygen consumption was determined for suspensions of isolated mitochondria (Wilson et al, 1988) and cells (Robiolio et al, 1989; Rumsey et al, 1990). Video imaging of phosphorescence has been shown to be a valid method for obtaining two dimensional maps of oxygen pressure in tissues (see Rumsey et al, 1988; 1992; Wilson et al, 1991; 1992A; 1992B; Shonat et al, 1992).

METHODS AND MATERIALS

Oxygen dependent quenching of phosphorescence provides a very sensitive measure of oxygen pressure in the environment of phosphorescent molecules. The Pd-porphyrins are particularly useful because intersystem crossing is rapid enough to quantitatively convert the excited singlet state to the triplet state. As a result, these compounds show little or no fluorescence and the phosphorescence quantum efficiency is high. Thus the phosphorescence lifetime can be readily measured not only using conventional photomultipliers but also using intensified video cameras. Photomultipliers are single detectors and all the light collected from the phosphorescent sample is concentrated on one detector. Intensified video cameras function as arrays of detectors, in the case of the camera used in the present studies this was a 512 x 480 array, and each of these must receive sufficient light energy to make a reliable measurement of light intensity. If the detectors are of equal sensitivity, the video camera requires 245,000 times (512 x 480) as much light as the photomultiplier for measurements of equal accuracy. Phosphorimeters appropriate for measuring phosphorescence lifetimes, and thereby oxygen pressure, in aqueous media including suspensions of biological material, as well as in local regions of tissue *in vivo* have been described in some detail by Green et al (1988) and Pawlowski and Wilson (1992). The present paper will, therefore, focus on the technology used to obtain two dimensional images of oxygen distribution in tissue.

Oxygen dependent quenching of phosphorescence of selected compounds can be quantitatively described by the Stern-Volmer relationship:

$$I°/I = T°/T = 1 + k_Q T° \ PO_2 \qquad (1)$$

For the Pd complex of tetra-(4-carboxyphenyl) porphine bound to bovine serum albumin and at physiological pH and 38°C, for example, k_Q has a value of 325 $Torr^{-1}$ sec^{-1} and T° is 600 usec (see Wilson et al, 1991; Pawlowski and Wilson, 1992). I° and I are the phosphorescence intensities and T° and T the phosphorescence lifetimes at zero oxygen pressure and at an oxygen pressure PO_2 respectively. Phosphorescence lifetime is a more accurate measure of oxygen pressure because the lifetime measurements are independent of probe concentration and of illumination light intensity. Moreover, lifetime measurements are essentially unaffected by changes in the absorbance of other chromophores in tissues, such as hemoglobin, myoglobin and the cytochromes. The measurements are therefore independent of the degree of oxygenation of hemoglobin and myoglobin as well as the state of reduction of cytochromes.

Calibration of the oxygen probes

If the values of the phosphorescence lifetime are determined both at a known oxygen pressure and at zero oxygen pressure, the quenching constant, k_Q can be calculated. In order to obtain the necessary measurements, the phosphorescence lifetimes have been measured in the samples either equilibrated with air or treated with glucose oxidase and glucose to remove the oxygen and sealed in glass chambers. In each case, the measurements have been made at temperatures from 21°C to 38°C (see Wilson et al, 1991; Pawlowski and Wilson, 1992).

Effect of temperature and pH on phosphorescence lifetime and quenching constant of selected oxygen probes

The effects of temperature and pH on different probes have been reported (Wilson et al, 1991; Pawlowski and Wilson, 1992). The phosphorescence lifetime for Pd-meso-tetra (4-carboxyphenyl) porphine in the absence of oxygen was almost temperature independent (less than 0.5%/degree), but increased significantly with increasing pH. There was little effect on $T°$ at pH values more alkaline than 7.2 (about 6.7% increase from 7.2 to 7.75) but was somewhat larger at pH values more acidic than 7.2 (about 17% decrease from 6.2 to 6.8). The quenching constant (k_Q) was independent of pH between pH 7.2 and 7.75 but increased on the acidic side of pH 7.2 (about 8% increase from pH 6.8 to pH 6.2).

For most experimental conditions the value of pH and temperature can be measured and the correct values of $T°$ and k_Q used. Where this is not possible, it is useful to note that the pH induced changes in $T°$ have little effect on the calculated oxygen pressure when the latter is above approximately 5 Torr (see Wilson et al, 1991). At a measured phosphorescence lifetime of 50 usec, for example, and using the value of $T°$ for pH 7.2 the calculated oxygen pressures at pH 6.4, 6.8, 7.2 and 7.75 are 47 Torr, 50.9 Torr, 56.5 Torr and 56.5 Torr respectively. Thus, at these relatively high oxygen pressures the effect of an uncertainity in $T°$ due to pH uncertainty is minimal. At a lifetime of 200 usec, the respective oxygen pressures are 7 Torr, 8.5 Torr, 10.1 Torr and 10.4 Torr. The pH effect increases with decreasing oxygen pressure but from pH 7.2 to 7.75 is very small until values are attained which *in vivo* would indicate severe hypoxia.

Other probes, such as Pd-coproporphyrin and Pd-mesoporphyrin, have very little dependence on pH compared to Pd-meso-tetra (4-carboxyphenyl) porphine (see Wilson et al, 1991; Pawlowski and Wilson, 1992) and their use can avoid any effect of changing pH in the physiological range of values. The Pd-coproporphyrin is, however, substantially more expensive than the Pd-meso-tetra (4-carboxyphenyl) porphine and this can be a consideration. Pd-mesoporphyrin, while more reasonably priced, has not yet been extensively tested in *in vivo* experiments.

Use of phosphorescent oxygen probes for *in vivo* experiments

For measurements in tissue *in vivo* the Pd complex of meso tetra-(4-carboxyphenyl) porphine (Porphyrin Products, Logan, Utah; approx. 20 mg/kg) has been infused through an artery or vein as a complex with bovine serum albumin (Fraction V, ICN ImmunoBiologicals, Costa Mesa, CA; 60-70 mg/ml) dissolved in physiological saline at pH 7.4. When the Pd-porphyrins are injected into the blood, there is no measurable decrease in phosphorescence over the course of several hours. Thus the probe is not removed from the blood at a significant rate and a single bolus injection is sufficient for experiments extending over periods of several hours. There have been no adverse reactions to the probes in our experience and the properties of the Pd-porphyrin probes suggest they should not have significant toxicity for animals.

Method for obtaining maps of phosphorescence lifetime and oxygen pressure

In our laboratory, observations are generally made using a Wild Macrozoom microscope with an epifluorescence attachment. The phosphorescence is imaged using a

Xybion intensified CCD camera (Xybion Electronics Systems Corp., San Diego, CA). The camera is focused on the surface of the tissue prior to injection of the Pd-porphyrin. The phosphorescence can be measured using excitation at either near 400 nm (the Soret band) or in the visible region near 530 nm (alpha and beta bands). The depth of tissue sampled is dependent on absorbance of the excitation light, and the blue (about 400 nm) light penetrates only 0.1-0.2 mm whereas the green light penetrates about 1 mm. Phosphorescence emission is generally at wavelengths greater than 630 nm where there is little absorption of light by the tissue. For the tissues examined to date, there has been no detectable phosphorescence before the probe was injected.

The illuminating light for the epifluorescence attachment was a EG&G 45 watt xenon flashlamp (EG&G, Salem, MA) with a flash duration of less than 5 usec mounted in a Leitz lamp housing. The flash lamp was controlled by a 16 MHz 80386 microcomputer (Spear Technology, Northbrook, IL) which determined the timing of the flashes and the gating of the video camera intensifier. A typical protocol was as follows: Number of flashes averaged for each delay time, 8; Delay times after the flash, 20 usec, 40 usec, 80 usec, 160 usec, 300 usec, 600 usec, and 2,500 usec; gate width in all cases, 2,500 usec. The image processor averaged the 8 frames for each delay time and this averaged image was displayed and recorded on the hard disk. Between 1 and 1.5 seconds was used to acquire the image for each delay time and each set of 7 images was acquired in total of about 40 seconds, including saving the images to the hard disk. It was assumed during analysis of the data that the brain did not move significantly during collection of a set of images. Direct comparison of the positions of prominent features of the images in the sequence confirmed that this was a reasonable assumption.

Analysis of the phosphorescence data

Two different methods have been used to analyze the data:

1. The regions of interest (small rectangular areas), are identified and the average intensity in that region determined for each image. These are plotted as the logarithm of phosphorescence intensity against delay time. The fit to a straight line is determined and where the decay is biphasic (usually indicating a fit to a straight line with a correlation coefficient of less than 0.97) fits to two or more straight lines can be used. In most cases the fit to a straight line is satisfactory (decay follows a single exponential). The measured decay constants (T) are substituted into the Stern-Volmer equation (equation 1), and the oxygen pressure in each region of interest is calculated from known values of k_Q and $T°$. This approach is useful for following oxygen pressure in discrete anatomical features, particularly when there is reason to suspect the oxygen pressures may not be homogeneous (the phosphorescence decay will not follow a single exponential).

2. The images of phosphorescence are digitized as 512 x 480 pixel arrays of data. These data arrays are filtered and used to calculate a phosphorescence lifetime for each pixel of the image set. The computation involved passing a filter over each image and then subtracting the background (the image taken with a 2,500 usec delay) from each of the other images of the set. The decrease in intensity at each pixel of the array which occurs with increasing delay time is then fitted to a single exponential decay curve. The result is two dimensional maps of the distribution of phosphorescence lifetimes in the observed area of the tissue.

Albumin as an aid in using Pd-porphyrins as oxygen probes

It is important to use the probes in a homogeneous environment, since the phosphorescence lifetime at zero oxygen and access of the excited triplet state to oxygen can be influenced by the local environment. For most physiological conditions it is very convenient to have the probes in a medium with an excess of bovine serum albumin. The latter binds porphyrins and thus provides:

1. The same environment for all the molecules of probe, assuring that the population of probes have the same lifetimes in the absence of oxygen and the same quenching constant.

2. Self-quenching, which occurs when molecules of probe in the excited triplet state collide with and transfer energy to molecules of probe in the ground state, is suppressed. This is due to a combination of shielding of the probe by the surrounding albumin and thereby decreasing the efficiency of energy transfer and the fact that the diffusion constants of the probe-albumin complexes are smaller than those for the probe alone (this decreases collisional frequency).

3. The environment provided by the albumin binding site includes restricted access to oxygen, decreasing the quenching constant for oxygen by approximately a factor of 10 (see Vanderkooi et al, 1987; Wilson et al, 1991). The result is a quenching constant optimally suited for measurements of oxygen *in vivo* i.e. for Pd-meso-tetra (4-carboxyphenyl) porphine the phosphorescence lifetime at air saturation is approximately 30 usec and that at zero oxygen is approximately 600 usec.

4. Last, but not least, the currently used Pd-porphyrins have very limited water solubility and albumin is required to keep them from precipitating in physiological media or at least binding to the cells in the blood and to vessel walls.

When the Pd-porphyrins are used as solutions in the presence of albumin and the oxygen pressure is homogeneous, the phosphorescence decay curves have been found to follow a single exponential (correlation coefficient of greater than 0.99).

Representative data obtained by imaging of phosphorescence of oxygen probes in the blood *in vivo*

The images of phosphorescence intensity of the brain cortex (Wilson et al, 1991; 1992A,B) show that under control conditions the veins as relatively bright, well defined vessels on a background of lower phosphorescence intensity. The oxygen probe is dissolved in the blood and phosphorescence intensity is dependent primarily on:

1. The amount of probe exposed to the excitation light (this is proportional to the concentration of blood in the approximately 1 mm thickness of surface tissue which is penetrated by the excitation light).

2. The oxygen pressure in the blood in the region of observation (phosphorescence intensity increases with decreasing oxygen pressure as described by the Stern-Volmer equation).

3. The intensity of the illuminating light (phosphorescence intensity increases in proportion to the illuminating light intensity). To the extent that the illuminating light flash provides uneven illumination, the phosphorescence intensity is uneven. These factors combine to provide an image in which the brightest regions are the veins because of their high blood concentration and low oxygen pressure. The capillary beds have only 10% or less of their volume as blood, the rest being cells and interstitial space. Therefore the capillary beds have a lower phosphorescence than veins although the oxygen pressure is similar to that in the veins. Arteriols are not seen without using experimental conditions emphasizing short phosphorescence lifetimes. Although the concentration of blood, and therefore of probe, in the arteriols is high, the vessels are small in diameter and the oxygen pressure is high. The latter quenches the

phosphorescence emission of the blood in the arteriols to below that of the capillary bed. The excitation light penetrates to the capillary bed below the arteriol, and the greater phosphorescence from the capillaries dominates the images and the calculated phosphorescence lifetimes.

The phosphorescence lifetimes are independent of the concentration of phosphorescent probe (amount of blood in the observed tissue). The oxygen pressures in the veins and capillary beds are normally not very different, and therefore the phosphorescence lifetimes are not very different. The two dimensional maps of phosphorescence lifetime and oxygen pressure thus often do not show the pattern of vessels in the surface of the brain and these morphological features are lost. The phosphorescence characteristics associated with any observable structure, including veins, can, however, be determined by reading the values of the pixels at the corresponding positions of the initial intensity, phosphorescence lifetime, and oxygen pressure maps.

Oxygen dependent quenching of phosphorescence is generally applicable to study of the distribution of oxygen in living tissue. The types of tissue that have been examined and the questions addressed is rapidly expanding. Initial measurements were made of isolated perfused rat liver (Rumsey et al, 1989; Wilson et al, 1989) but more recent studies have included the cortex of the brain of newborn piglets (Wilson et al, 1991; 1992B) and adult cats (Wilson et al, 1992A), the carotid body of the cat *in vitro* (Rumsey et al, 1991) and *in vivo* (Rumsey et al, 1992), the retina of the cat eye *in vivo* (Shonat et al, 1992), rat heart *in vivo* (Ince et al, 1992), and subcutaneous tumors in the rat *in vivo* (Wilson and Cerniglia, 1992). In each case the method has been able to provide new and valuable data on the delivery and utilization of oxygen *in situ,* suggesting it will make a major contribution to research in this important area of biochemistry and physiology. Although initial indications are that it will also become a powerful new tool for the diagnosis and treatment of medical problems such as tumors, wound healing, vascular disease and eye disease, there remain many barriers to overcome before this future can be realized.

Acknowledgements: This work was supported in part by a grant NS-10939 from the National Institutes of Health and N00014-89-J-1243 from the Office of Naval Research.

BIBLIOGRAPHY

Green, T.J., Wilson, D.F., Vanderkooi, J.M., and DeFeo, S.P. (1989) Phosphorimeters for analysis of decay profiles and real time monitoring of exponential decay and oxygen concentrations. Analy. Biochem. 174: 73-79.
Ince, C., Ashruf, J., Sanderse, E.A. Pierik, E.G.J.M., Coremans, J.M.C.C., and Bruining, H.A. (1992) *In vivo* NADH and Pd-porphyrin video fluori-/phosphorimetry. Adv. Exptl. Med. Biol. these proceedings.
Pawlowski, M. and Wilson, D.F. (1992) Monitoring of the oxygen pressure in the blood of live animals using the oxygen dependent quenching of phosphorescence. Adv. Exptl. Med. Biol. in press.
Robiolio, M., Rumsey, W.L., and Wilson, D.F. (1989) Oxygen diffusion and mitochondrial respiration in neuroblastoma cells. Amer. J. Physiol. 256: C1207-C1213.

Rumsey, W.L., Iturriaga, R., Spergel, D., Lahiri, S., and Wilson, D.F. (1991) Optical measurements of the dependence of chemoreception on oxygen pressure in the cat carotid body. Amer. J. Physiol. 261: in press.

Rumsey, W.L., Lahiri, S., Iturriaga, R., Mokashi, A., Spergel, D., and Wilson, D.F. (1992) Optical measurements of oxygen and electrical measurements of oxygen chemoreception in the cat carotid body. Adv. Exptl. Med. Biol. These proceedings.

Rumsey, W.L., Robiolio, M., and Wilson, D.F. (1989) Contribution of diffusion to the oxygen dependence of energy metabolism in human neuroblastoma cells. Adv. Exptl. Med. Biol. 248: 829-833.

Rumsey, W.L., Schlosser, C., Nuutinen, E.M., Robiolio, M., and Wilson, D.F. (1990) Cellular energetics and the oxygen dependence of respiration in cardiac myocytes isolated from adult rats. J. Biol. Chem. 265: 15392-15399.

Rumsey, W.L., Vanderkooi, J.M., and Wilson, D.F. (1989) Imaging of phosphorescence: a novel method for measuring oxygen distribution in perfused tissue. Science, Wash. DC 241: 16491-1651.

Shonat, R.D., Wilson, D.F., Riva, C.E., and Pawlowski, M. (1992) Oxygen distribution in the retinal and choroidal vessels of the cat as measured by a new phosphorescence imaging method. Applied Optics in press.

Vanderkooi, J.M., Maniara, G, Green, T.J., and Wilson, D.F. (1987) An optical method for measurement of dioxygen concentration based on quenching of phosphorescence. J. Biol. Chem. 262: 5476-5482.

Vanderkooi, J.M., and Wilson, D.F. (1986) A new method for measuring oxygen in biological systems. Adv. Exptl. Med. Biol. 200: 189-193.

Vanderkooi, J.M., Wright, W.W., and Erecinska, M. (1991) Oxygen gradients in mitochondria examined with delayed luminescence from excited-state triplet probes. Biochemistry 29: 5332-5338.

Wilson, D.F. and Cerniglia, G.J. (1992) Localization of tumors and evaluation of their state of oxygenation by phosphorescence imaging. J. Cancer Res., in press.

Wilson, D.F., Gomi, S., Pastuszko, A., and Greenberg, J.H. (1992A) Oxygenation of the cortex of the brain of cats during occlusion of the middle cerebral artery and reperfusion. Adv. Exptl. Med. Biol. These proceedings.

Wilson, D.F., Pastuszko, A., DiGiacomo, J.E., Pawlowski, M., Schneiderman, R., Delivoria-Papadopoulos, M. (1991) Effect of hyperventilation on oxygenation of the brain cortex of newborn piglets. J. Appl. Physiol. 70(6): 2691-2696.

Wilson, D.F., Pastuszko, A., Schneiderman, R., DiGiacomo, J.E., Pawlowski, M. and Delivoria-Papadopoulos, M. (1992B) Effect of hyperventilation on the oxygenation of the brain cortex of neonates. Adv. Exptl. Med. Biol. In press.

Wilson, D.F., Rumsey, W.L., Green, T.J., and Vanderkooi, J.M. (1988) The oxygen dependence of mitochondrial oxidative phosphorylation measured by a new optical method for measuring oxygen concentration. J. Biol. Chem. 263: 2712-2718.

Wilson, D.F., Rumsey, W.L., and Vanderkooi, J.M. (1989) Oxygen distribution in isolated perfused liver observed by phosphorescence imaging. Adv. Exptl. Med. Biol. 248: 109-115.

Wilson, D.F., Vanderkooi, J.M., Green, T.J., Maniara, G., DeFeo, S.P., and Bloomgarden, D.C. (1987) A versatile and sensitive method for measuring oxygen. Adv. Exptl. Med. Biol. 215: 71-77.

REFLECTION SPECTROMETRY

M. Kessler, K. Frank, J. Höper, D. Tauschek, J. Zündorf

Institut für Physiologie und Kardiologie der Universität Erlangen-Nürnberg, Waldstraße 6, D-8520 Erlangen

1. General aspects of light scattering

Light waves irradiated into tissues are altered by absorption and scattering. Both physical phenomena decrease the intensity of incident light.

Most of the light is scattered in the forward direction while a smaller part is directed backward into the direction of the incident light.

Before we go into details of light scattering phenomena in the tissues the basic physical mechanisms of light scattering should be discussed.

What is light scattering?

Principally, all wave based scattering phenomena follow rather similar laws. The nature of interactions between waves and tissue structures can be illustrated in quite an intelligible way using a floating cork model in order to describe classic wave optics.

When a plane wave propagates in water and hits a floating cork an excitation of the cork is induced. As a result, the cork acts as a small antenna and causes circular waves which propagate on the surface of the water (Fig. 1A). At a certain time and at a certain distance a distinct amplitude can be measured independent of the direction of observation. Close to the floating cork interference phenomena are produced between the incoming plane wave and the excited circular wave.

The amplitude of the radiated wave depends upon the inertia of the floating cork. When the inertia is zero the floating cork will oscillate in phase with the incident wave and thus will not produce any radiation wave. Increasing values of inertia will produce increasing phase shift between the incident waves and the radiation waves of the floating cork with increasing wave amplitudes. The higher the amplitude the more energy is absorbed transiently by the floating cork from the incident wave. This phenomenon is called elastic scattering.

Light absorption (inelastic scattering) can be defined as an energy uptake by the floating cork in such a way that the energy of the incident wave is either transferred into heat or is stored "wave mechanically". When applied to light waves this means that the electrons are raised to a higher energy level and when they return to their initial energy level fluorescence waves are emitted.

Single scattering occurs when the scattering particles are located at such a distance that the radiated waves do not interfere with each other. In practical terms this can only happen when the transit time of waves radiated by two neighbouring floating corks is longer than the life time of the primary wave which successively hits the two corks.

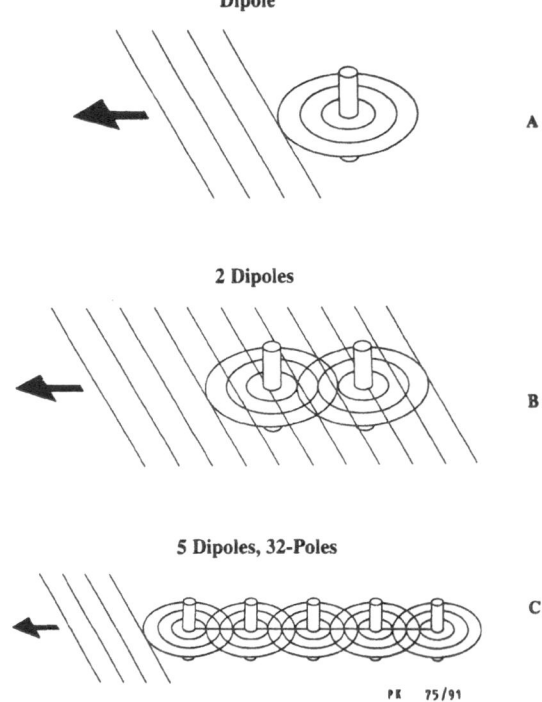

Dipole

2 Dipoles

5 Dipoles, 32-Poles

PK 75/91

Figure 1.
A : A floating cork exited by plane waves to harmonic oscillations radiates circular waves.
B : A plane wave interacts with two non-coupled floating corks. The radiated circular waves of each floating cork interfere with each other (multiple scattering).
C : Due to the interference phenomena at a chain of coupled floating corks, induced by a incident plane wave, the scattered waves are oriented in the scattering plane. They form a reflected and refracted wave.

The term "multiple scattering" describes a situation in which the waves radiated by two noncoupled floating corks interact with each other in such a way that their radiated waves show a phase shift, and thus are able to cause interference phenomena which lead to the multiple formation of new radiation centres (Fig 1B.).
The wave optics described above apply to floating corks smaller in size than the wavelength of the incident plane waves. However, when molecular interconnections between the floating corks form chains (Fig. 1C), rings or spheres a wave optical phenomenon called multipole scattering is induced. Due to such coupling the radiation states produced by the oscillations of the different partners give rise to well defined phase relationships.

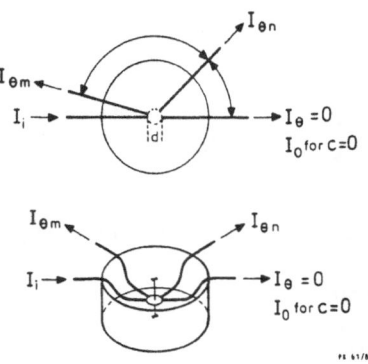

Figure 2. Schematic drawing of the scattering chamber. The volume under observation is illuminated by a micro-lightguide. The bundle of detecting micro-lightguides can be moved around the centre of the scattering chamber detecting the angular intensity distribution.

I_0 : Irradiated intensity.

$I\Theta$: Scattered intensity at $0°$.

$I\Theta'$, $I\Theta''$: Angular dependent scattered intensity.

d   : distance between illuminating and detecting micro-lightguide.

2. Scattering Chamber

Scattering diagrams obtained from measurements in scattering chambers (figure 2) can be analysed using the theories formulated by Rayleigh and Mie.

Such a scattering chamber is an old physical tool to investigate scattering phenomena in molecules and particles in which the angular distribution of scattered light can be measured.

An overall theory of wave optics applicable for multiple dipole and multipole scattering processes in living tissues does not exist as yet and there is little evidence that the very complex problems of light scattering can be solved primarily by use of mathematical models.

Nevertheless, there is no doubt that a detailed theoretical analysis of the interactions between electromagnetic waves and the tissue will be possible based upon valid experimental data. Finally the formulation of a coherent mathematical model based upon experimental data, able to describe theoretically the scattering of electromagnetic waves in living tissues might be possible.

3. Rayleigh Scattering

One of the important discoveries among Lord Rayleigh's extensive investigations in the fields of heat radiation, acoustics and optics was the Rayleigh law (1871). He explained mathematically the blueness of the sky as wavelength dependent light scattering phenomenon at the molecules of the atmosphere (Fig 3).

$$\frac{I_\theta}{I_0} = \frac{4\pi^2 (n-1)^2}{N\lambda^4 R^2} \frac{1+\cos^2\theta_S}{2}$$

Figure 3: Rayleigh's law.

I_θ	:	Scattered intensity in the direction S.
λ	:	Vacuum wavelength.
I_0	:	Reference intensity
R	:	Distance of the detector from the scattering centre.
n	:	Refractive index.
θ_S	:	Scattering angle.
N	:	Number of scattering centres per volume.

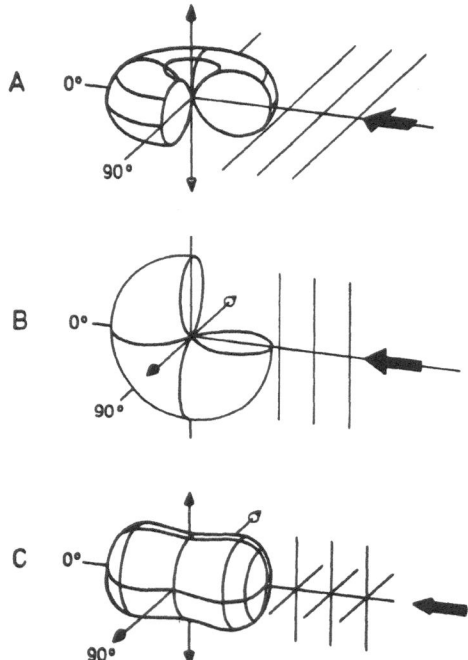

Figure 4. Calculated angular intensity distribution of three dipoles in a threedimensional drawing for different polarisation states of irradiated light.

A : Perpendicular polarised light.
B : Parallel polarised light.
C : Incident unpolarised light.

Based upon, Rayleigh's algorism (Fig. 3) the angular intensity of three dipoles was calculated (Fig. 4). The bodies A and B of figure 4 are obtained when perpendicular and parallel polarized light hits a scattering center while C shows an elipsoid which is produced when non-polarized light hits the scattering center.

4. Mie Scattering

One of the important discoveries of Mie (1908) was the realisation that when the diameter of a sphere exceeds about 0.001 of the wavelength of the irradiated light used for scattering measurements, an increased number of dipole radiation centres emerge and the resulting interference between these dipoles can be evaluated in terms of additive contributions of the series of electrical and magnetic multipoles starting with a dipole and progressing to the final multipole of the particle to be analyzed.

In a theoretical study and based upon Mie's algorism (Fig. 5) Frank calculated angular scattering diagrams in the forward and backward direction for single Mie scattering.

$$\frac{I_\theta}{I_0} = \frac{\lambda^2}{8\pi^2 R^2}\left[\Sigma \frac{2n+1}{n(n+1)} (a_n\pi_n(\cos\theta_S)+b_n\tau_n(\cos\theta_S))\right]^2$$
$$+ \frac{\lambda^2}{8\pi^2 R^2}\left[\Sigma \frac{2n+1}{n(n+1)} (b_n\pi_n(\cos\theta_S)+a_n\tau_n(\cos\theta_S))\right]^2$$

Figure 5: Mie's law:
I_θ : Scattered intensity.
I_0 : Reference intensity.
λ : Vacuum wavelength
R : Radius of the sphere.
π_n, τ_n : Angular dependent functions.
a_n, b_n : Scattering coefficients.
Θ_S : Scattering angle.

For his calculation he chose 5 spherical (Fig. 6) particles of sizes between 0.1 - 2.0 μm. As can be seen, the scattering in forward and backward direction increases drastically with increasing particle size. However, this yields only for single Mie scattering as it happens for example in the atmosphere of the sky.

5. Multiple multipole scattering

In figure 7A wavelength and concentration dependent scattering diagrams of isolated liver mitochondria measured in the conventional scattering are shown. The lower part of the diagrams indicates scattering in backward direction.

As can be seen in fig. 7A the wavelength dependent light absorption as induced by multiple scattering decreases with increasing wavelength.

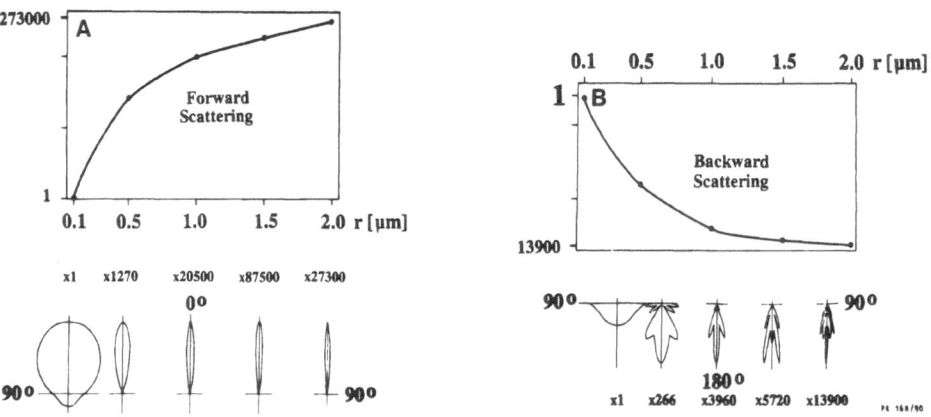

Figure 6. Forward (A) and backward (B) scattered intensities and patterns as a function of size of multipole particles.

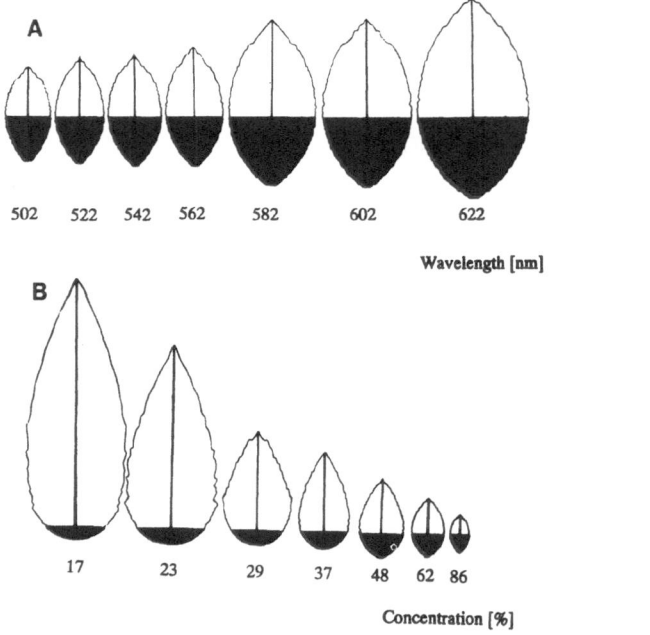

Figure 7. Wavelength dependent angular scattering diagrams as a function of wavelenghts (A) and concentrations (B). The black areas correspond to backscattered light.

As shown in figure 7B increase in multiple multipole scattering as induced by an increasing concentration of mitochondria increases the light absorption by light scattering. However with increasing concentration the scattering in backward direction is augmented.

6. Light intensity profiles in suspensions of mitochondria

The conventional angular dependent scattering diagrams are very informative but do not give any information about the light intensity distribution within the suspension of the scattering elements.

In order to determine the intensity distribution within scattering samples a scanning technique which enables a systematic measurement of the true intensity profiles within suspensions or tissue slides was developed (Frank et al., this vol.).

In the figure 8 the intensity profiles determined quantitatively in suspensions of liver mitochondria are depicted.

Figure 8A shows light profiles for mitochondria of a size of > 1 μm while in figure 8B the profiles of mitochondria < 1 μm can be seen.

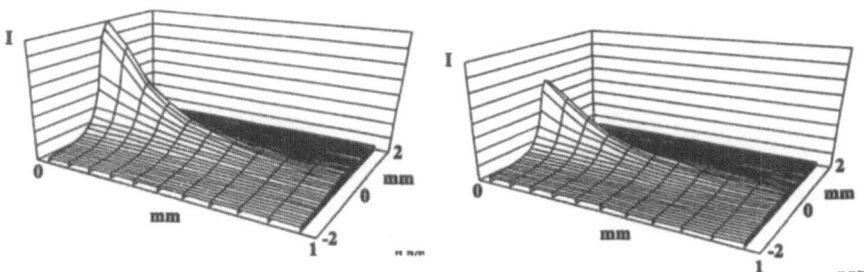

Figure 8. Intensity distributions of mitochondrial suspensions. A: >1 μm, B: < 1 μm .

As a matter of fact we can conclude that the dependency from the particle size found by Mie for single Mie scattering can also be reproduced in multiple Mie scattering.

Hitherto, the knowledge of intensity distribution of light irradiated into tissue was mainly based upon calculations using algorisms derived from diffusion equations. The aim in this context should be to develop new algorisms on the basis of such experimental work and to reconstruct mathematically the wave optics of multiple multipole scattering.

An important parameter in this context will be the exact determination of the pathlength of irradiated light which is a precise function of the number of scattering events. The techniques which have to be applied are the time resolved spectroscopy (TRS) or the phase modulated spectroscopy (PMS). Both techniques have been promoted by B. Chance most successfully.

Remission spectrometry in tissues is only possible because of the fact that light irradiated into tissue is scattered in backward direction.

The Bessel function of light which has entered the tissue is given by the intensity profiles shown in the figures. Of course, the intensity functions of back scattered light measured with a receiving micro-lightguide depends directly on such intensity profiles which are produced in tissues in a similar way. Based upon these profiles the catchment volume of the applied light fibres can be defined as well.

7. Optical multicomponent system of tissue

Signals obtained in biological systems by remission spectrometry are the result of physical phenomena described above. When light is irradiated they are induced in the optical multicomponent system formed by tissue. The following light absorbing parameters determine this multicomponent system:

1. Tissue pigments:
   a) Haemoglobin.
   b) Myoglobin.
   c) Cytochromes aa3, b, c.
   d) Flavoproteins.
   e) Pyridinnucleotides.
   f) Melanin.
2. Rayleigh scattering (multiple dipole).
3. Mie scattering (multiple multipole).

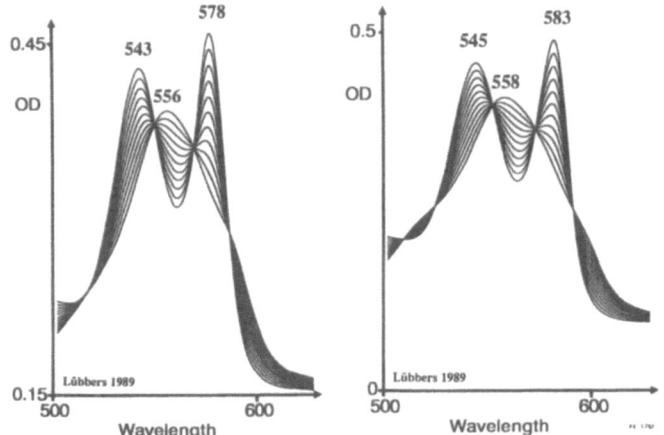

Figure 9. Cuvette spectra of hemoglobin (A) and myoglobin (B).

What happens with spectra of these tissue pigments when they are mixed with a fantastic number of different light scattering and light absorbing elements?

Based upon experimental results it has beeome possible to answer this question rather precisely.

For the analysis, a synthetic procedure can be applied in such a way that in a first step the spectra of isolated and purified pigments are analyzed in the cuvette (Figures 9 and 10).

In order to avoid a too extended discussion of the optical interactions occurring in optical multicomponent system represented by tissue we will focus on only three pigments: hemoglobin, myoglobin and cytochrome aa3. The spectra of the three

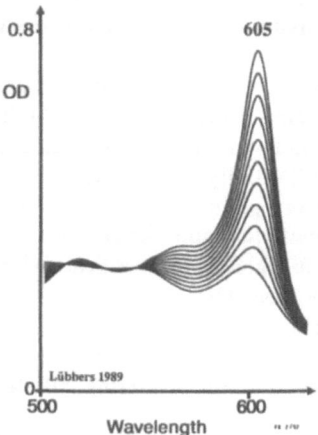

Figure 10. Cuvette spectra of cytochrome aa3.

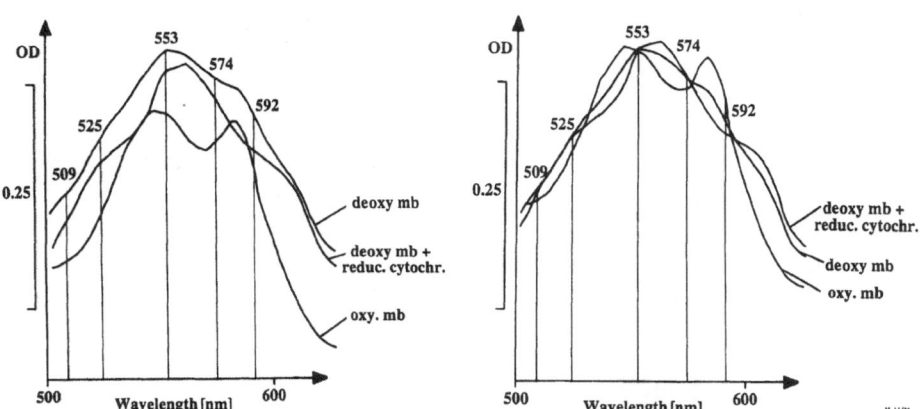

Figure 11. Spectra of hemoglobin free working myocardium in three oxygenation states.
A: 1. Oxygenated myoglobin (70%), 2. Partly oxygenated myoglobin (20%), 3. Fully de-oxygenated myoglobin and reduced cytochromes.
B: Spectra of left figure linked up at the isosbestic wavelength of 553 nm.

pigments lying in the wavelength range of visible light have several peaks which show distinct changes with varying oxygenation (Hb, Mb) and oxidation (aa3) states. The three spectra also reveal several precise isosbestic wavelengths at which OD changes do not occur when oxygenation or redox states are modified.

In the tissues the natural pigments listed above are located in "micro cuvettes" represented by capillaries and living cells which are filled with subcellular particles and macromolecules.

The light scattering induced by scattering elements induces a pronounced distortion of the spectra of tissue pigments. This is depicted in Fig. 11 in which myoglobin spectra of the working myocardium of hemoglobin free perfused rat heart are plotted. Three defined states were induced: Partially oxygenated (70%) and deoxygenated (20%) myoglobin, fully deoxygenated myoglobin and reduced cytochrome aa3.

The effect of different oxygenation and redox states on light absorption is most impressive. Deoxygenation of myoglobin to a critical value of 20% induces a dramatic shift towards higher light absorption (OD).

When total anoxia is caused in the myocardium OD decreases in a pronounced way.

When all 3 spectra are linked up at the isosbestic wavelength of 553 nm it becomes apparent that by the influence of light scattering all definite isosbestic points have disappeared.

This observation as well as the fact that the slopes of the base lines which connect the wavelengths of 500 and 630 nm of each spectrum can be altered dramatically by changes in tissue oxygenation clearly demonstrate that the effects of light scattering must be measured precisely when any kind of quantitative spectrometry is supposed to be performed. The disappearance of defined isosbestic wavelengths further proves that calculations for these wavelengths are not valid.

Summarizing the results presented in this review we can conclude that the realization of a quantitative tissue spectrometry based upon a physical analysis of the optical multicomponent system existing in tissues has become possible.

CAN THE FLOW DEPENDENCY OF THE POLAROGRAPHIC PO_2 ELECTRODE BE USED TO MEASURE ARTERIAL PO_2 AND LOCAL CAPILLARY FLOW TRANSCUTANEOUSLY?

A. Grundmann and Dietrich W. Lübbers

Max-Planck-Institut für Systemphysiologie

Rheinlanddamm 201, 4600 Dortmund, Germany

INTRODUCTION

The parameters which determine the skin surface pO_2 ($sspO_2$) are given by parameters which determine the oxygen supply of the tissue as well as describe the influence of the pO_2 measurement. During steady state conditions oxygen delivery to the tissue is equal to its oxygen consumption, i.e. the difference between arterial and venous oxygen concentration ($C_{a,O2}$ - $C_{v,O2}$) times blood flow, BF is equal to tissue oxygen consumption, M_{O2}

or

$$(C_{a,O2} - C_{v,O2}) \cdot BF = M_{O2} \qquad (1)$$

$$C_{v,O2} = C_{a,O2} (1 - \frac{M_{O2}}{C_{a,O2} \cdot BF}) \qquad (2)$$

$$\frac{C_{v,O2}}{C_{a,O2}} = f = (1 - \frac{M_{O2}}{C_{a,O2} \cdot BF}) \qquad (3)$$

The state of tissue oxygen supply can be characterized by its relative oxygen extraction $f = C_{v,O2}/C_{a,O2}$ (eq. 3). f approaches unity in a hyperbolic way if the oxygen offer (denominator: $C_{a,O2} \cdot BF$) becomes large as compared to the oxygen consumption (numerator: M_{O2}). If gas exchange through the skin surface is excluded skin surface pO_2 depends similarly on the convective O_2 supply as venous oxygen concentration (eq. 1), but it is additionally influenced by the O_2 transport within the tissue which follows the laws of diffusion. Convective and diffusive O_2 transport determine the pO_2 profiles within the tissue, the pO_2 field of the tissue. However, if $sspO_2$ is measured polarographically by a membrane covered platinum electrode the O_2 consumption of the electrode influences the pO_2 field within the tissue. It could be experimentally shown that electrodes with cathodes of different sizes and with different membranes measure different skin surface pO_2 values although the oxygen offer by the blood and M_{O2} did not change[1,2,3]. Therefore, we investigated in a theoretical analysis whether it should be possible to determine transcutaneously arterial pO_2 as well as local blood flow by simultaneous application of two polarographic pO_2 electrodes of different measuring properties.

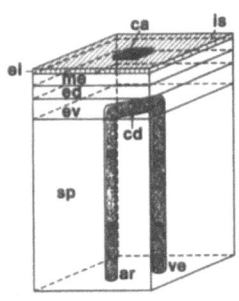

Fig. 1.   Schematic view of a microcirculatory unit of the upper skin (MU, capillary loop model) covered by a polarographic pO_2 electrode. Side length: 140 μm; height (sp+ev+ed) = 240 μm. sp: stratum papillare with the capillary loop (ar: arterial part, cd: capillary dome, ve: venous part); ev: viable tissue (str. germinativum, str. spinosum and partly str. granulosum); ed: dead tissue (str. corneum); me: membrane of the electrode; el: electrode with cathode (ca) and gas impermeable insulation (is).

METHOD

Since the upper skin has a rather regular structure it can be simulated by a network of similar microcirculatory units, MU's (fig. 1, capillary loop model[4,5]). The dimensions of the MU are taken from literature. It is a square piece with a side length of 140 μm (ca 50 cap/mm^2) and a height of 240 μm. In the diagonal of the MU the capillary loop (ar, cd, ve) is situated which is surrounded by connective tissue (sp, str. papillare) and covered by a viable (ev, str. germinativum, str. spinosum and partly str. granulosum) and a dead layer (ed, str. corneum). Blood is simulated by a hemoglobin solution with a variable p_{50} and the O_2 dissociation curve by the Adair approximation. The CO_2 binding curve is simulated by a straight line. With this capillary loop model the pO_2 distribution within the microcirculatory unit is calculated for steady state conditions as well as for transients. It allows also to take into account the pCO_2 and heat distribution. The MU is covered by a membrane covered polarographic electrode (el) with cathode (ca) and gas impermeable insulation (is).

To demonstrate the influence of the polarographic pO_2 electrode on the pO_2 field the pO_2 distribution within the microcirculatory unit has been calculated using the capillary loop model (fig. 2). In fig. 2a an electrode with a cathode of a diameter of 15 μm and a teflon membrane of 15 μm thickness has been applied, in fig. 2b an electrode with the same membrane is used but with a cathode diameter of 200 μm covering the whole MU. In the case of the small electrode small local changes occur in front of the cathode mainly in the membrane, but the whole pO_2 field depends on the properties of the tissue and not the influence of the electrode. The large electrode, however, influences the whole pO_2 field: the pO_2 at the level of the capillary dome decreases from 35.5 Torr (a) to 26.8 Torr (b) and at the skin surface from 34.6 Torr to 7.7 Torr. Fig. 2 demonstrates also that the influence of the cathode diameter depends on the size of the surface of the MU. Since the largest effect is observed if the cathode covers the MU, larger cathodes covering several MU's give similar $sspO_2$ values. During polarographic measurement the pO_2 at the surface of the cathode amounts to 0 Torr.

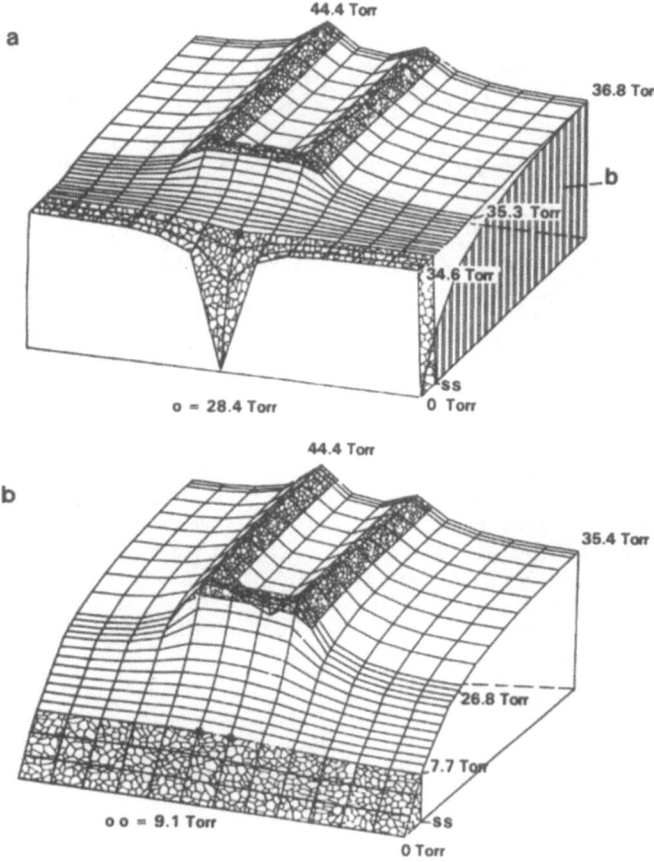

44.4 Torr

a

36.8 Torr

b

35.3 Torr

34.6 Torr

o = 28.4 Torr   0 Torr

ss

44.4 Torr

b

35.4 Torr

26.8 Torr

7.7 Torr

o o = 9.1 Torr   ss

0 Torr

Fig. 2. Simulated pO_2 field of the diagonal section of the microcirculatory unit of the upper skin calculated for 2 different membrane
covered polarographic pO_2 electrodes. pO_2 electrodes: diameter of
the pt-wire: a) 15 μm and b) 200 μm (covering the whole surface of
the MU); membrane: 15 μm thick teflon membrane. p_aO_2: 44.4 Torr
(43°C); BF: 100 ml·100g^{-1}·min^{-1}; Hb = 8.0 g·dl^{-1}; M_{O2}: 0.4
ml·100g^{-1}·min^{-1}. - The influence of the small electrode is practically restricted to the electrode membrane, whereas the large
electrode (b) disturbs the whole pO_2 field of the MU (skin surface
(ss) is marked by dots).

RESULTS AND DISCUSSION

Simulation was carried out for two electrodes, one with a cathode
diameter of 30 μm (el 1), the other cathode covered the surface of the MU
(diameter of 200 μm). With the larger cathode the thickness of the teflon
membranes was varied between 15 (el 2), 30 (3) and 40 μm (4). Flow and p_aO_2
was varied systematically and the oxygen current towards the cathode calculated. The O_2 current (z-axis) can be drawn as a 3 dimensional plane above
the p_aO_2 (x-axis)-blood flow (y-axis) plane. Fig. 3 shows this plane calculated for a cathode diameter of 200 μm and a 15 μm thick teflon membrane
(el 2). Similar O_2 current planes were calculated for the other electrodes.
To compare the O_2 current values of these planes, iso-O_2-current curves were
constructed (see fig. 3) and projected on the p_aO_2-BF plane. Fig. 4 demonstrates that there are different iso-O_2 current curves for the different

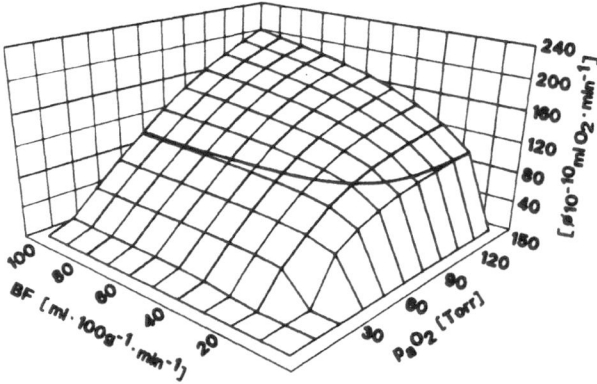

Fig. 3. Oxygen current of the MU towards the polarographic cathode in dependence on arterial pO_2 and blood flow.
x-axis: p_aO_2, arterial pO_2; y-axis: BF, blood flow; z-axis: O_2 current towards the cathode. pO_2 electrode: el(2) with a diameter of the cathode of 200 μm and a 15 μm thick teflon membrane. M_{O2}: 0.4 ml·100g^{-1}·min^{-1}; Hb: 16 g·dl^{-1}; pCO_2: 40 Torr; pH: 7.4. The calculation was carried out for t = 43°C. - As an example the iso-O_2-current curve for an oxygen current of $100 \cdot 10^{-10}$ ml O_2·min^{-1} is drawn.

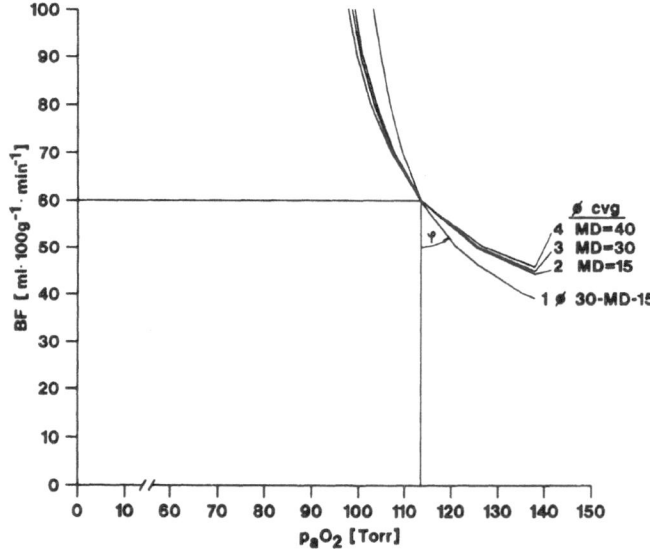

Fig. 4. Intersecting point of iso-O_2-current curves obtained by different pO_2 electrodes at a defined combination of arterial pO_2 and blood flow.
pO_2 electrodes: el(1) with a cathode diameter of 30 μm ($\emptyset$) and a 15 μm thick teflon membrane (MD-15) and el(2) with a cathode diameter of 200 μm (covering the MU: $\emptyset$ cvg) and a teflon membrane of a thickness of 15 μm (el2), 30 μm (3) and 40 μm (4). There is a defined intersecting point which allows to extract from the measurement - e.g. with el(1) and el(2) - the p_aO_2 as well as the blood flow: p_aO_2 = 114 Torr, BF = 60 ml·100g^{-1}·min^{-1}.

electrodes. These iso-O_2-current curves intersect at a defined intersecting point. This shows that, in principle, two simultaneous sspO_2 measurements using two sufficiently different polarographic pO_2 electrodes can be used to measure the arterial pO_2 as well as the local capillary flow. Under our conditions the variation of the cathode diameter produces a larger change of the signal than the changes of membrane thickness.

The precision of the measurements depends on the accuracy by which the intersection point can be determined. For example, assuming that the reduction current of the pO_2 electrode can be determined with an error of ±0.05nA the corresponding change of the O_2 current was calculated and the 2 iso-O_2 current curves drawn in fig. 5. By this error the intersecting point at BF= 60ml$\cdot100$g$^{-1}\cdot$min^{-1} and $p_aO_2=$ 114.5 Torr can change between BF-values of 66.5 and 56.5ml$\cdot100$g$^{-1}\cdot$min^{-1} and p_aO_2-values of 112.0 and 118.0 Torr.

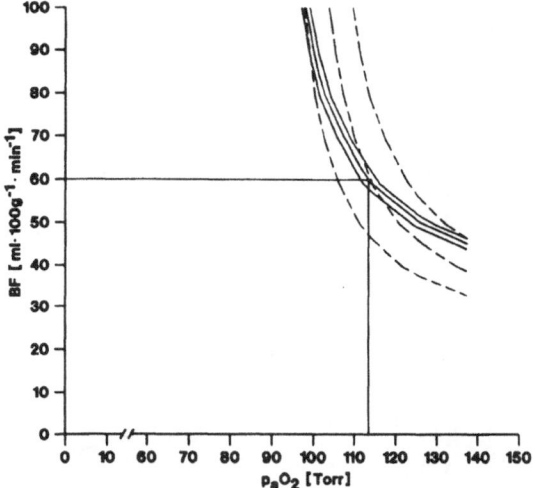

Fig. 5.   The effect of a measuring error of the pO_2 electrodes on the precision of the intersection point of the iso-O_2-current curves. The change of the O_2-current produced by a measuring error of ±0.05 nA is calculated and marked by 2 iso-O_2 current curves. The intersection point at BF = 60 ml$\cdot100$ g$^{-1}\cdot$min^{-1} and p_aO_2 = 114.5 Torr varies from BF = 66.5 ml$\cdot100$ g$^{-1}\cdot$min^{-1} and p_aO2 = 112.0 Torr to BF = 56.5 ml$\cdot100$ g$^{-1}\cdot$min^{-1} and p_aO_2 - 118 Torr.

In fig. 6 the iso-O_2-current curves in dependence on the paO_2 and blood flow are shown. At low values of flow and paO_2 the curves are rather in parallel to the x- or y-axis, respectively. Only at higher values the form of the curves changes in such a way that sufficiently well defined intersection points (fig. 6, circles) occur, i.e. in a range in which is p_aO_2 larger than 70 Torr and blood flow larger than 40 ml$\cdot100$g$^{-1}\cdot$min^{-1}. In the network of lines exist additional intersection points (marked by triangles), but the calculations easily show that these points can be excluded on theoretical reasons.

CONCLUSIONS

The O_2 current towards the cathode of a polarographic pO_2 electrode measuring skin surface pO_2 is determined
1) by the properties of the pO_2 electrode: a) diameter of the cathode in relation to the size of the surface of the MU (number of capillaries/min), b) properties of the electrode membrane and c) O_2 consumption of the electrode as well as:

2) by the properties of the microcirculatory unit (MU): a) its convective and diffusive O_2 supply and b) its O_2 consumption.

The analysis shows that the different behaviour of electrodes can be described by a set of iso-O_2 current curves projected on the p_aO_2-blood flow plane (fig. 4 and 6). In a certain measuring range the iso-O_2 current curves of different electrodes have distinct intersection points characterizing the actual p_aO_2 and blood flow (fig. 4). The analysis shows that, in principle, the influence on the $sspO_2$ exerted by two sufficiently different polarographic pO_2 electrodes can be used to measure simultaneously arterial pO_2 and local blood flow. In practice, however, the application will be probably restricted since
1) because of the almost hyperbolic form of the iso-O_2 current curves distinct intersection points are only obtained in a range which is clinically probably not very important (fig. 6, circles) and since
2) because of the relative small angle at the intersection point of the intersecting curves pO_2 measurements of a very high precision and accuracy are needed.

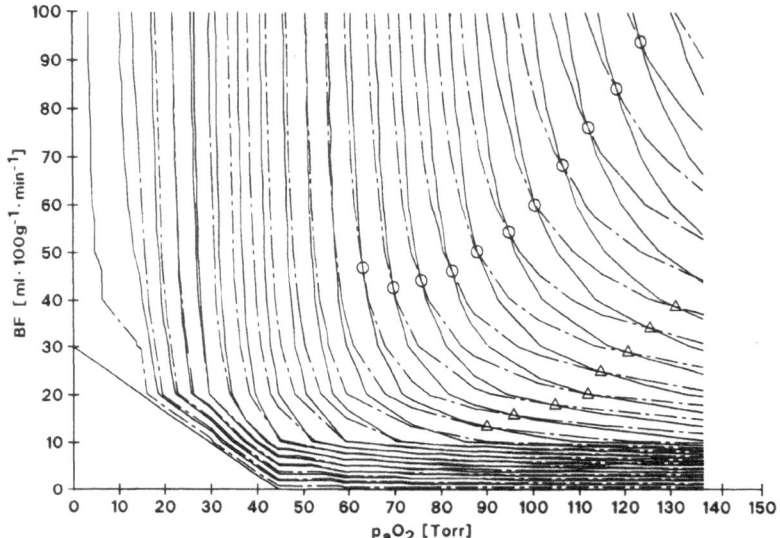

Fig. 6.   Intersecting points of iso-O_2-current curves obtained by 2 different electrodes in dependence of arterial pO_2 and blood flow. solid lines: el(1): cathode diameter 30 μm, thickness of the teflon membrane 15 μm; iso-O_2-current curves:
1) 2 ml $O_2 \cdot 10^{-10} \cdot min^{-1}$, 2) 2+2 = 4 ml $O_2 \cdot 10^{-10} \cdot min^{-1}$...
broken lines: el(2): cathode diameter 200 μm, thickness of the teflon membrane 15 μm; iso-O_2-current curves:
1) 4 ml $O_2 \cdot 10^{-10} \cdot min^{-1}$, 2) 4+11 = 15 $\cdot 10^{-10}$ ml $O_2 \cdot min^{-1}$... . - Defined intersection points (circles) are only obtained with p_aO_2 values above 70 Torr and blood flow values above 45 ml $\cdot 100$ g$^{-1} \cdot$min^{-1}. Other intersection points (triangles) can be excluded on theoretical reasons.

REFERENCES

1.   R. Huch, A. Huch and D.W. Lübbers, "Transcutaneous pO_2," Thieme-Stratton Inc. New York, Thieme-Verlag, Stuttgart-New York (1981).
2.   D.W. Lübbers, Possibilities and limitations of the transcutaneous measuring technique: A theoretical analysis, in: "Continuous Transcutaneous Monitoring," A. Huch, R. Huch, G. Rooth, eds., Plenum Press, New York-London. Adv. Exp. Med. Biol. 220:9-17 (1987).

3. D.W. Lübbers, Theory and development of transcutaneous oxygen pressure measurement, in: "Advances in Oxygen Monitoring," K.K. Tremper, S.J. Barker, eds., Little, Brown & Company, Boston (1987), pp. 31-65.

4. U. Grossmann, P. Winkler and D.W. Lübbers, Coupled transport of O_2 and CO_2 within the upper skin simulated by the capillary loop model, in: "Oxygen Transport to Tissue-V," Adv. Exp. Med. & Biol. 169:125-132, D.W. Lübbers, H. Acker, E. Leniger-Follert, eds., Plenum Press, New York-London (1984).

5. U. Grossmann and P. Winkler, Transients of gas exchange processes in the upper skin calculated by the capillary loop model, in: "Oxygen Transport to Tissue-VI,", Adv. Exp. Med. & Biol. 180:35-41, D. Bruley, H.I. Bicher, D. Reneau, eds., Plenum Press, New York-London (1984).

THE USE OF EPR FOR THE MEASUREMENT OF THE CONCENTRATION OF

OXYGEN IN VIVO IN TISSUES UNDER PHYSIOLOGICALLY PERTINENT

CONDITIONS AND CONCENTRATIONS

H. M. Swartz, S. Boyer, D. Brown, K. Chang, P. Gast, J. F. Glockner, H. Hu, K. J. Liu, M. Moussavi, M. Nilges, S. W. Norby, A. Smirnov, N. Vahidi, T. Walczak, M. Wu, and R. B. Clarkson

University of Illinois College of Medicine at Urbana-Champaign
506 South Mathews
Urbana, IL 61801   USA

INTRODUCTION

The aim of this paper is to describe an emerging methodology which appears to offer new capabilities for the measurement of the concentration of oxygen ($[O_2]$) in complex biological systems in vitro and in vivo with the sensitivity and accuracy needed to make these measurements under conditions that are pertinent for physiological and pathological processes. The method uses electron paramagnetic resonance (EPR or, equivalently, ESR); the approach is often termed EPR oximetry. In particular EPR oximetry can make measurements selectively in the intracellular compartment and in tissues in vivo and can detect $[O_2]$ as low as 0.1 micromolar. Our use of EPR to measure intracellular $[O_2]$ with nitroxides has been described previously (1-2) and therefore this paper will concentrate on the methodology to make measurements in vivo and also, at very low $[O_2]$. We will review the basis of the methodology briefly but will emphasize results in tissues which illustrate the capabilities of this new technique. Because of the newness of this approach the results obtained so far have been aimed at determining and illustrating the potential of this technique, rather than providing detailed information on a specific biological problem. A companion paper illustrates a more detailed application of the new technology, the measurement of $[O_2]$ in skeletal muscle (3).

EPR spectroscopy is specifically and uniquely sensitive to the presence of unpaired electrons (paramagnetic molecules). The principles are similar to other spectroscopies such as UV-visible light spectroscopy, but because EPR is based on the presence of unpaired electrons, it requires a magnetic field to define their energy states. The use of EPR in biological systems now usually requires electromagnetic radiation in the microwave range (500 to 10,000 MHz) to excite the unpaired electrons and a corresponding magnetic field of 180 to 3600 Gauss. The principal naturally occurring paramagnetic species are free radicals and certain valence states of metal ions; none of these are present in sufficient concentrations to use for the measurement of $[O_2]$ and their concentration is too low to interfere with the measurements, which rely on the presence of stable paramagnetic molecules which are introduced into the system.

EPR oximetry is based on the sensitivity of EPR spectra to $[O_2]$ (4,5). Although in principle any paramagnetic species could be used, until the last two years EPR oximetry has usually been based on nitroxides. Recently several new types of paramagnetic

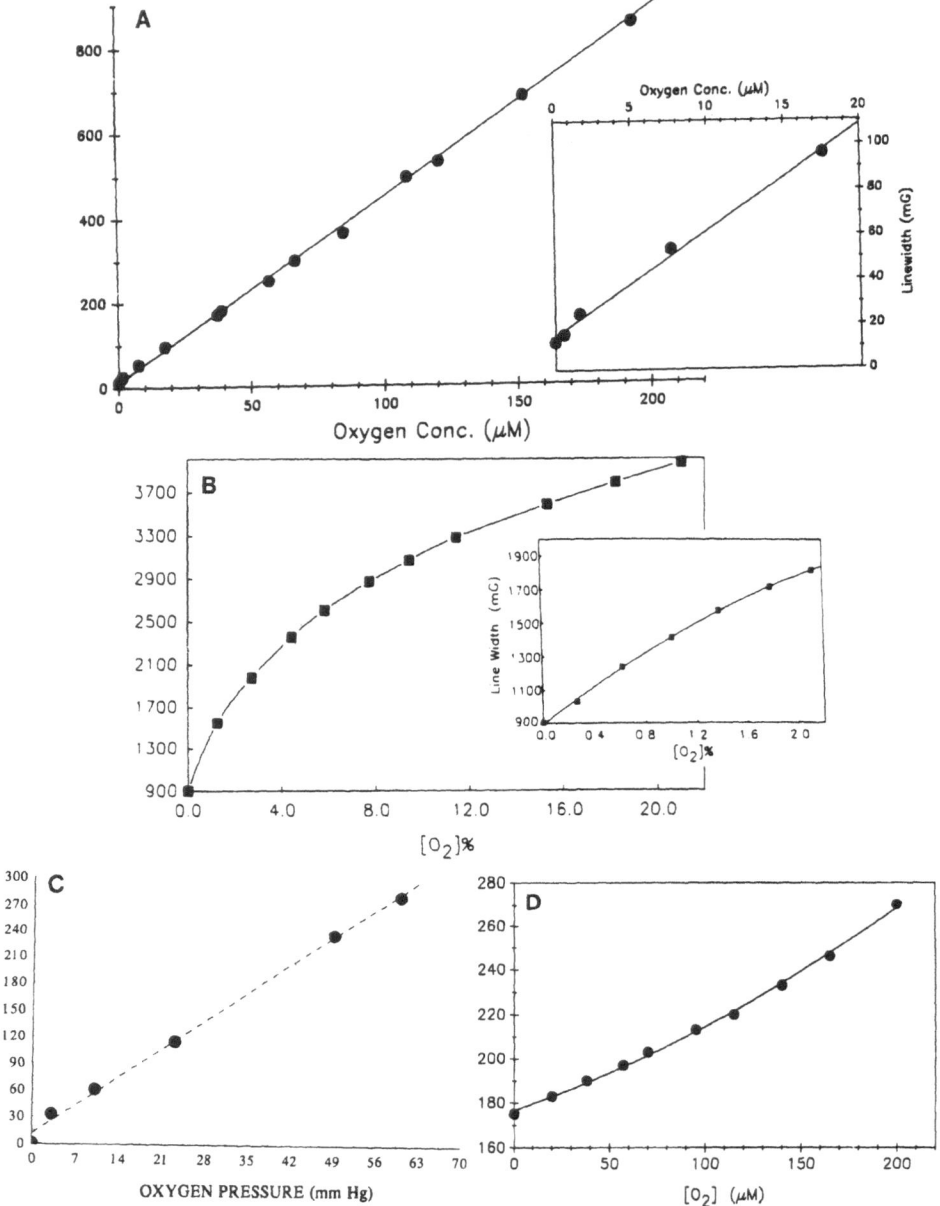

Figure 1. Calibration curves of linewidth vs $[O_2]$: linewidth in milligauss (except C., in gauss); $[O_2]$ in Torr or equivalent μM in solution. The spectra were obtained at 9 GHz with modulation amplitudes and microwave power levels selected to avoid effects on linewidth [Figure 1B reproduced with permission, from (6).]

A. Lithium phthalocyanine crystals, in gaseous oxygen
B. Acid-washed fusinite particles, in gaseous oxygen
C. Sucrose char formed at 620°C, in gaseous oxygen
D. 0.1 mM perdeuterated Tempone (4-one-2,2,6,6,tetramethylpiperidine-d$_{16}$-1-oxyl)

	Nitroxides	Fusinite	LiPc$^+$	Sucrose Char
spins/gm	*	1.0×10^{19}	9.2×10^{19} $(2.0 \times 10^{20})^+$	8×10^{19}
sensitivity of line to O$_2$ (mG/μM)	0.57	18.0	4.2 $(33.1)^+$	2,900
pH Sensitivity	Usually small, depends on structure of nitroxide	none	below pH = 2	none
resistance to oxidation	high	very high	very high	X
resistance to reduction	low	very high	very high	very high
stability in tissues	becomes bioreduced in seconds to hours	stable for at least 150 days	stable for at least 150 days	X
toxicity	minimal for concentrations < 1 mM	none observed**	none observed**	X
can use line heights instead of line widths, to increase sensitivity	only in absence of bioreduction	yes	yes	yes
spin-spin self broadening	yes	no	no	no
broadening by other spin systems	yes	no	no	no
nature of dependency of effect of oxygen	product of concentration and diffusion	partial pressure of O$_2$	partial pressure of O$_2$	partial pressure of O$_2$
physical nature	usually soluble	insoluble particles of any size	insoluble crystals 5 mm to .5 micron	insoluble particles of any size

---

*due to concentration dependent spin-spin broadening in nitroxides, it usually is necessary to keep the local concentration below 1 mM.

**exposure of cultured cells to large amounts of < 1 micron diameter particles for more than 24 hours leads to loss of colony forming ability--apparently due to physical engorgement.

+LiPc = lithium phthalocyanine. Values for (CH$_3$O)$_8$ derivative of LiPc indicated in parenthesis, other properties probably same as LiPc

X=to be measured

molecules have been introduced (6) which provide enhanced sensitivity to $[O_2]$ (figure 1) and also high chemical stability and very low toxicity (Table 1). The various types of stable paramagnetic species that are now available provide a versatile set of approaches to solve different experimental needs. The nitroxides respond instantly to changes in the concentration of oxygen and reflect the actual $[O_2]$ in the environment. The particulates, lithium phthalocyanine, fusinite, and carbohydrate chars reflect partial pressure of oxygen. The particulates vary in their rate of response to changes in $[O_2]$, with lithium phthalocyanine responding very quickly while the others have response times which vary from seconds to minutes, depending in part on the size of the particles. The nitroxides may be inactivated in functional biological systems by bioreduction to the hydroxylamines, but appropriate selection of the type of nitroxide can minimize this problem (7). All of the particulates appear to be completely resistant to bioreduction and other types of metabolic transformations. The magnitude of the responses to $[O_2]$ of these paramagnetic probes varied from a charge of less than one milligauss/Torr in perdeuterated Tempone to 4,670 milligauss/Torr for the sucrose char.

There also have been significant developments in EPR instrumentation which now allow the use of this technique in living animals (8). The most important advances include: development of low frequency EPR spectrometers which enables the technique to be used for samples containing large amounts of aqueous materials, including tissues; the use of loop gap resonators and other resonant structures designed for use in biological samples as surface detectors or detectors shaped to go around or into the tissue of interest; the development of coupled circuits which have permitted the development of much more flexible detector configurations; and the development of gradient techniques which permit simultaneous measurements of $[O_2]$ in several sites.

ILLUSTRATIVE APPLICATIONS OF THE METHODOLOGY

A. Measurement of the deoxygenation of an isolated rat heart. Figure 2 indicates the capability of the technique to measure very small changes in $[O_2]$ in a rapid manner. The heart was removed intact from an anesthetized rat, a crystal of lithium phthalocyanine placed in the myocardium of the left ventricle, and the heart which continued to beat actively was then placed on the surface of the EPR detector. Under these conditions the heart rapidly utilized the available oxygen, as shown by the EPR spectra which were taken at the indicated intervals. As $[O_2]$ decreased there was a corresponding decrease in the line width of the EPR signal. Because the total number of the unpaired spins remained constant, the change in line width could be followed very sensitively by simply measuring the peak-peak height of the EPR signal (note that, as is usually the case in EPR spectroscopy, the EPR spectrum is a first derivative of the absorption because of the use of modulation of the magnetic field and phase sensitive detection). The corresponding changes in $[O_2]$ were determined from a calibration curve which was obtained in a separate experiment by measuring linewidth in a system in the presence of known $[O_2]$. As indicated in the figure, $[O_2]$ values could be readily resolved down to at least 0.2 micromolar.

B. Effect of anesthesia on $[O_2]$ in the brain. Figure 3 illustrates the type of results which were obtained when $[O_2]$ was measured in the brain of a living mouse. The animal was anesthetized and then under sterile conditions, a skin flap was made to expose the skull and a small hole was drilled in the cranium. A crystal of lithium phthalocyanine was introduced into the brain through a 25 gauge needle and the surgical site was closed. Measurements of $[O_2]$ were made by placing an external loop on the head of the animal and measuring the linewidth of the EPR spectra. The figure shows the results of measurements which were obtained seven days after the surgery, first with the animal awake but restrained and then after the administration of nembutal intraperitoneally. The figure indicates the depressing effect of the anesthetic on $[O_2]$ in the brain. This experiment also indicates the feasibility of using this technique for repeated measurements of $[O_2]$ in the brain; there was no apparent chemical toxicity associated with the presence of the crystal of lithium phthalocyanine in the substance of the brain.

C. Measurements of $[O_2]$ in the myocardium in vivo. Some of the most interesting and potentially important applications of methods to measure $[O_2]$ in tissues are in the heart. Consequently many of our initial explorations of the capabilities of the new methodology have involved the heart, usually using lithium phthalocyanine crystals which have been inserted into the myocardium of rats whose chest is open and who are maintained on a ventilator.

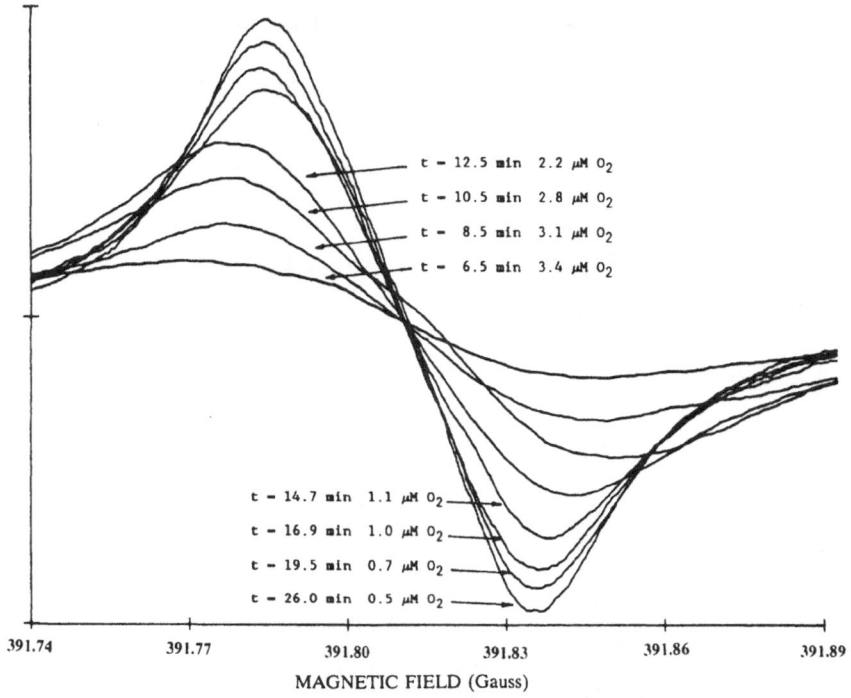

t – 12.5 min 2.2 μM O₂

t – 10.5 min 2.8 μM O₂

t – 8.5 min 3.1 μM O₂

t – 6.5 min 3.4 μM O₂

t – 14.7 min 1.1 μM O₂

t – 16.9 min 1.0 μM O₂

t – 19.5 min 0.7 μM O₂

t – 26.0 min 0.5 μM O₂

391.74   391.77   391.80   391.83   391.86   391.89

MAGNETIC FIELD (Gauss)

Figure 2. Use of lithium phthalocyanine to measure very low concentrations of oxygen in tissues. EPR spectra and the corresponding values of [O₂] in a freshly isolated heart of a rat in which a small crystal of LiPc was placed in the wall of the heart and then repeated measurements were made with the heart placed on the surface of the detector of a 1.1-GHz EPR spectrometer at room temperature. The linewidths and estimated [O₂] at intervals after removal of the heart are indicated. The lower range values of [O₂] were calculated by a standard curve extrapolated from calibration points at higher concentrations, assuming a linear relationship. The differences between [O₂] for different experimental points are considered to be accurate but the absolute values of [O₂] may have a systematic error. Reprinted with permission, from (6).

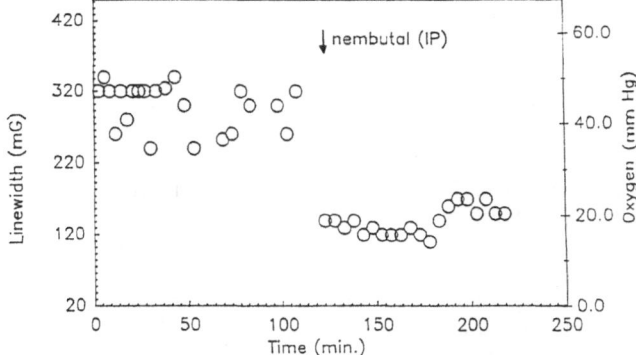

Figure 3. Measurement of [O₂] in the brain of a living mouse. The measurements were made 7 days after the insertion of a small crystal of lithium phthalocyanine into the brain.

Figure 4 demonstrates the changes observed when the left anterior descending branch of the coronary artery was ligated. The figure also illustrates the rapid and sensitive responses that have been obtained, even at the relatively early state of development of this technology. The changes are, perhaps, less than might have been anticipated. This may reflect the extensive complexing networks of vessels in this species. This also may reflect the limitations of using macroscopic crystals in small organs; this problem should diminish as the techniques are applied to larger animals.

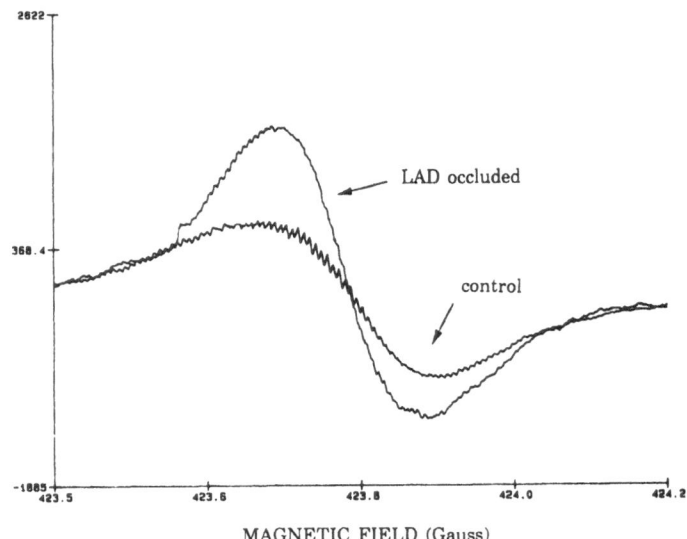

MAGNETIC FIELD (Gauss)

Figure 4. Measurement of $[O_2]$ by a crystal of lithium phthalocyanine in the myocardium of a ventilated rat before and after ligation of the left anterior descending branch of the coronary artery. The corresponding $[O_2]$ were 45 μM and 36 μM. The measurement time for each spectrum was 60 seconds, using a loop coupled to a loop gap cavity of an EPR spectrometer operating at 1100 MHz.

Figure 5 demonstrates an important extension of the capabilities of the method: the ability to measure $[O_2]$ in several areas simultaneously. This figure also illustrates the rapid and sensitive responses that have been obtained, even at the relatively early state of development of this technology. This experiment was performed in a rat which had been made hypertensive by ligation of the aorta between the renal arteries. As a result of this treatment fibrotic areas formed in the heart; the purpose of the experiment was to determine whether there was a difference in $[O_2]$ in these areas and how they respond to changes in arterial $[O_2]$. A modest magnetic field gradient was superimposed on the main magnetic field so that in the direction of the additional magnetic field the value of the main field required for resonance became a function of position in regard to the gradient, thereby altering the position in the spectrum where the resonances for each crystal occurred. The resulting spectra seen with two crystals are shown in Figure 5, along with the time dependent changes in the spectra which occurred as the amount of oxygen in the breathing gas was varied. The data indicate that the fibrotic areas had a lower $[O_2]$ and also responded more slowly to changes in $[O_2]$ in the blood.

D. <u>Measurement of $[O_2]$ in the kidney in vivo</u>. Figure 6 summarizes the results of an experiment in which a crystal of lithium phthalocyanine was placed in the cortex of the kidney of a rat and the $[O_2]$ followed when the aorta above the kidney was ligated. As indicated in the figure, after an initial drop in $[O_2]$ to essentially zero, $[O_2]$ gradually increased, presumably via retrograde flow in the aorta below the ligature. As is also indicated in the figure, the lithium phthalocyanine crystal could be left in the kidney and $[O_2]$ remeasured several days later.

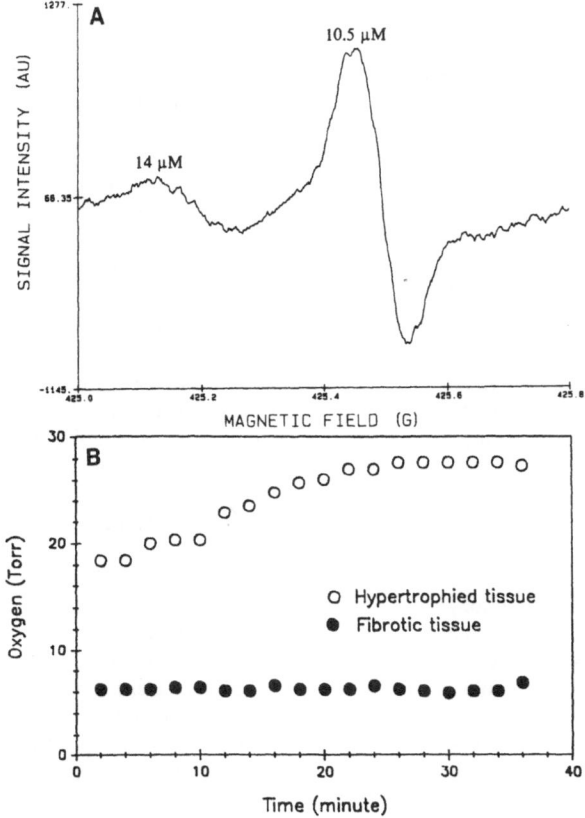

Figure 5. Measurement of [O$_2$] in two different areas of the heart. A rat was made hypertensive by ligation of the aorta between the renal arteries. When the chest was opened (respiration was maintained by a ventilator) fibrotic and normal appearing areas were found and small crystals of lithium phthalocyanine placed into representative areas of both types. A small magnetic field gradient was applied to separate the EPR spectra of the two crystals. Figure 5A shows the spectra that were observed and the [O$_2$] which corresponded to the observed linewidths. Figure 5B shows the response of the two areas to an increase in [O$_2$] in the breathing gas; [O$_2$] in the fibrotic area did not change.

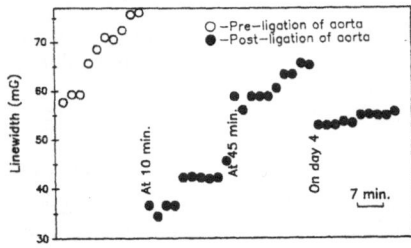

Figure 6. Measurement of [O$_2$] in kidney in vivo. A small crystal of lithium phthalocyanine was placed in the cortex of the kidney of a rat and baseline measurement of [O$_2$] made from the linewidths of the EPR spectra. After [O$_2$] stabilized, the aorta above the kidney was ligated and measurements were repeated. The [O$_2$] is shown 10 minutes, 45 minutes, and 4 days after the ligation; the crystal was left in the kidney and the measurements at 4 days were obtained at a second operation. The EPR measurements were made with a coupled loop placed on the surface of the kidney.

FUTURE DIRECTIONS AND POTENTIAL ROLE OF EPR OXIMETRY IN

The results described above indicate the types of measurements of [O$_2$] possible with the existing paramagnetic materials and instrumentation. It is feasible to measure [O$_2$] at essentially all levels of biological interest ranging from <0.1 µM to hyperbaric concentrations. For at least 24 hours and, in some organs, for considerably longer times lithium phthalocyanine can be used as the sensitive paramagnetic material and therefore very rapid changes in [O$_2$] (< 1 second) can be followed. In other experiments which are not described in this paper we have found that in organs such as skeletal muscle, lithium phthalocyanine may lose its response to [O$_2$] after 24 hours. Materials such as fusinite, however, remain fully active and responsive for >200 days; because of their slower responses to changes in [O$_2$] these materials are best suited for measuring steady state [O$_2$].

These techniques appear to be suitable for a wide variety of biological systems ranging from cell suspensions to tissues in vivo. It also appears that this approach might readily be adapted to develop a new type of oxygen meter for general use, which would be very sensitive and not subject to some of the limitations of Clark electrodes such as the consumption of oxygen and a lack of sensitivity to very low [O$_2$]. There do not appear to be any substantial barriers to the early adoption of these techniques for use in a wide variety of experimental situations.

Although at the present time it should be considered speculative, it appears feasible to use this methodology clinically. The new paramagnetic materials appear to be remarkably nontoxic and nonreactive in tissues, as might be expected on the basis of their chemical inertness and the experience from the exposure of biological materials to charcoal and other similar substances. The most readily apparent clinical use is to measure [O$_2$] in various regions of tumors and use this information as guide for individualizing radiation and chemotherapy. The techniques also might be readily adapted for making intraoperative measurements such as myocardial oxygenation during cardiac surgery. Eventually these techniques might be used for a wider range of clinical measurements as experience is gained in their safe and effective use.

ACKNOWLEDGEMENTS

This work was supported by NIH grant GM 34250, and used the facilities of the ESR Center at the University of Illinois, which is supported by NIH grant RR 01811. J. F. G. and N. V. received support from NIH training grant CA 09067.

REFERENCES

1. H. M. Swartz and M. A. Pals, in Handbook of Biomedicine of Free Radicals and Antioxidants, Jaime Miquel, Hans Weber, and Alex Quintanilha, (Eds.), CRC Press Inc., Boca Raton, FL, 3:141-151 (1989).
2. J. Glockner, H. M. Swartz, and M. Pals, J. Cell. Physiol., 140:505-511 (1989).
3. J. Glockner and H. Swartz, this volume
4. J. Hyde and W. K. Subczynski, Biol. Magn. Reson. 8:399 (1989).
5. H. M. Swartz and J. F. Glockner, in Advanced EPR in Biology and Biochemistry, Arnold J. Hoff, ed., Elsevier Science Publishers, Amsterdam, The Netherlands, 753-782 (1989).
6. H. M. Swartz, S. Boyer, P. Gast, J. F. Glockner, H. Hu, K. J. Liu, M. Moussavi, S. W. Norby, T. Walczak, N. Vahidi, M. Wu, and R. B. Clarkson, Magn. Reson. in Med., 20:333-339 (1991).
7. H. M. Swartz, Pure & Appl. Chem., 62:235-239, 1990.
8. H. M. Swartz and J. F. Glockner, in EPR Imaging and In Vivo EPR, G. R. Eaton, S. S. Eaton, and K. Ohno, (Eds.), CRC Press, Inc., Boca Raton, FL, 261-290 (1991).

IN VIVO EPR OXIMETRY USING TWO NOVEL PROBES:

FUSINITE AND LITHIUM PHTHALOCYANINE

James F. Glockner and Harold M. Swartz

College of Medicine, Depts. of Medicine and Biophysics
University of Illinois
Urbana, IL 61801

INTRODUCTION

The concentration of oxygen is an important variable in many physiological and pathological processes; it is not surprising, therefore, that a variety of methods have been developed to measure oxygen tensions in biological systems. Many techniques provide accurate and reliable determinations of in vitro oxygen tensions; in vivo measurements are more difficult, however, and the currently available methodologies have significant limitations. In this study we employ two novel EPR oximetric probes, fusinite and lithium phthalocyanine (LiPc), along with a low frequency (1.1 GHz) electron paramagnetic resonance (EPR) spectrometer and surface probe to measure intramuscular oxygen pressures in anesthetized mice. Principles of in vivo EPR oximetry are discussed in a preceding article by Swartz.

MATERIALS AND METHODS

Materials. Lithium phthalocyanine was a kind gift of Dr. Mehdi Moussavi and the LETI Corporation, Grenoble, France. Fusinite was generously provided by Dr. Robert Clarkson, University of Illinois, and particles smaller than 15 μ in diameter were used in all experiments.

Animals. ICR female mice were anesthetized with intramuscular ketamine and injected in leg muscle with either fusinite or LiPc. The animals were secured on top of a plastic holder which was then positioned in the spectrometer so that the tissue of interest rested directly above the surface probe and the EPR signal intensity was maximal. Temperature was monitored with a rectal probe and maintained at 32 $\pm$ 1°C by blowing warm air around the animal. The composition of the breathing gas was controlled by means of a plastic hood placed over the animal's head.

Electron Paramagnetic Resonance. EPR measurements at 1 GHz were recorded on a spectrometer consisting of a home built L-band bridge integrated with a Varian spectrometer. The surface coil used with this system has a resonant frequency of 1.1 GHz and an unloaded Q of approximately 800. The external B_1 field is localized in a

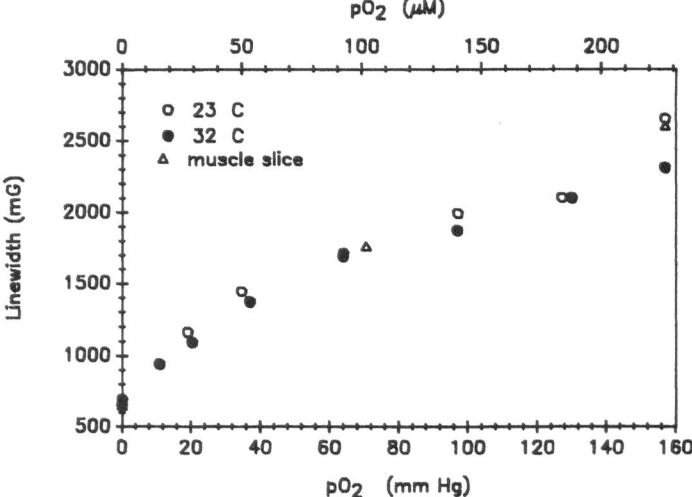

Fig. 1. Calibration of linewidth vs. oxygen pressure for a suspension
of fusinite in PBS. See text for a description of the experiment.

hemispherical region above the surface probe with a radius of about 8 mm.

<u>Preparation and Injection of Probes</u>. Particles of fusinite less than 15 μ in diameter
were suspended in phosphate buffered saline (PBS) and approximately 0.05 ml injected
into leg muscle using a 22 gauge needle. LiPc crystals were individually loaded into a
22 gauge needle (3-4 crystals per injection) and then injected into the leg muscle by
means of a wire plunger. The site of injection was verified in roughly 20% of the
animals following the experiment by visualizing the probe with the aid of a dissecting
microscope. All samples were appropriately located within the skeletal muscle of the
leg.

<u>Calibration</u>. Calibration of linewidth as a function of oxygen pressure for a sample of
fusinite contained in gas-permeable Teflon tubing is shown in Fig. 1. Warm gas was
blown across the sample to heat it to 32 $\pm$ 0.5°C, and the oxygen composition of the
gas was monitored with a Clark electrode. A second calibration was performed at room
temperature, approximately 23°C, indicating that within the sensitivity of our linewidth
measurements, moderate changes in temperature have relatively little effect.

The third set of points on the calibration figure represents an attempt to ensure
that injection into tissue does not change in some way the oxygen-dependent spectral
properties of fusinite. In this experiment, a mouse was sacrificed and slices of leg
muscle removed. Approximately 0.05 ml of the fusinite suspension was injected into
each slice, and the slices were then placed into a beaker containing PBS and NaCN (to
inhibit tissue oxygen consumption) and equilibrated at 32°C with an appropriate gas
mixture (equilibration times were typically 2 h). The slices were removed from the
beaker and the spectra recorded quickly to minimize changes in temperature. The 3
point tissue calibration agrees fairly well with the other data, indicating that it is
reasonable to assume that the calibration curve is valid for samples of muscle in vivo.
The 32°C calibration curve was fit to a second order regression equation, and the
regression equation was then used to calculate values of pO_2 from linewidths measured
in intact animals.

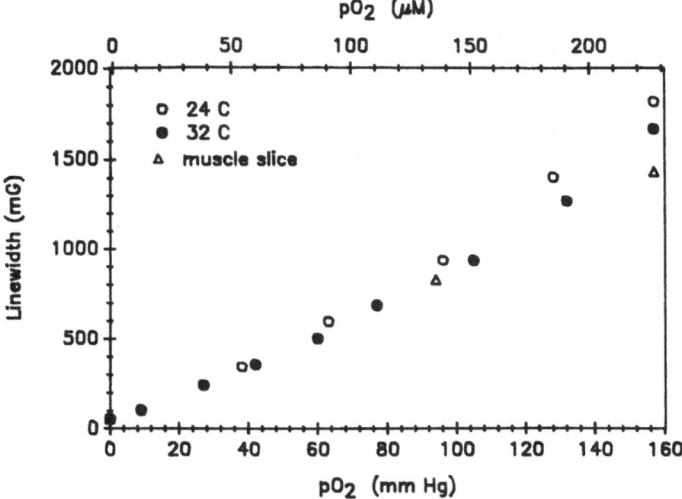

Fig. 2. Calibration of linewidth vs. oxygen pressure for a sample consisting of 3-4 crystals of LiPc in Teflon tubing. See text for experimental details.

Figure 2 shows similar calibrations for a sample of LiPc. 3-4 dry crystals were placed in Teflon tubing and equilibrated with appropriate gas mixtures. The LiPc calibration shows relatively little temperature effect, and the 3 point muscle slice calibration is reasonably consistent with the rest of the data. pO_2 values were calculated in the same manner as described for fusinite: the calibration curve was fit to a second order regression equation, and the equation was then used to convert measured linewidths into oxygen pressures.

RESULTS

Following anesthesia, injection of the probe, and placement of the mouse over the surface probe of the spectrometer, the animal breathed air for 20 minutes followed by 20 minutes of 85% oxygen/15% CO_2 and was then asphyxiated with 100% CO_2 and monitored for 10 minutes. Spectra were recorded at 5 minute intervals throughout each experiment, and approximately 40 mice were used for each probe.

Figure 3 shows representative data obtained following IM injection of fusinite and LiPc. Fig. 3a shows the variation in linewidth over time for a fusinite injection, and Fig. 3b shows the corresponding data obtained following injection of LiPc. Control experiments were also performed in which a probe was injected and the mouse breathed air for 50 minutes - significant changes in linewidth were not observed in the air breathing animals.

Figures 4 and 5 are tissue pO_2 histograms showing the percentage of measurements falling within a given oxygen pressure range for air-breathing and 85% oxygen-breathing mice. Both the fusinite and LiPc show a fairly compact distribution of air-breathing pO_2's, with average values of 23 ± 3 torr for the fusinite experiments and 18 ± 3 torr for LiPc. The oxygen breathing histograms are much more widely scattered, and the average values do not agree as well (69 ± 5 torr for fusinite and 83 ± 6 torr for LiPc).

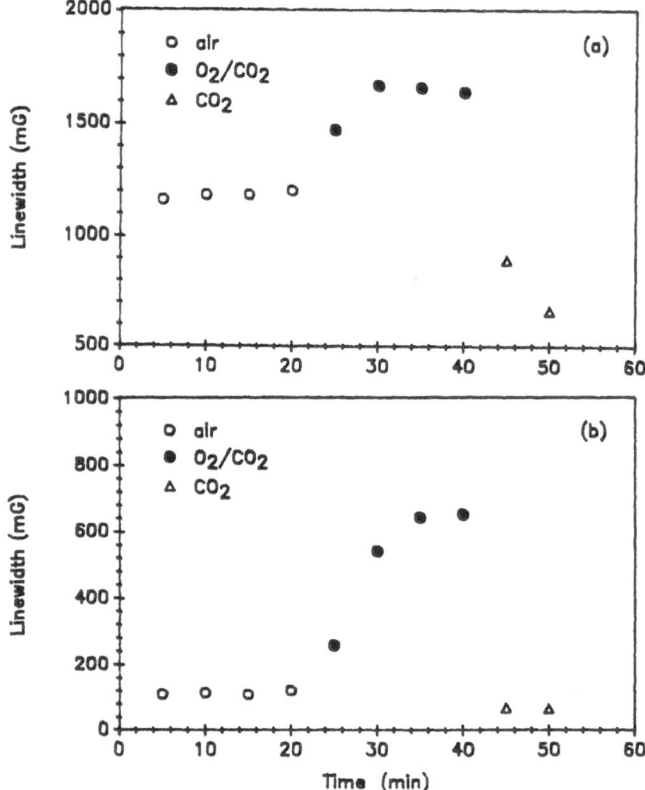

Fig. 3. (a): Variation in linewidth vs. time following IM injection
of 0.05 ml fusinite suspension. The mouse breathed air for 20 minutes,
85% O_2/15% CO_2 for 20 minutes, and then was asphyxiated with CO_2 and
monitored for 10 minutes. Spectra were recorded at 5 minute intervals,
and animal temperature was maintained at $32 \pm 1°C$. (b): The same
experiment following IM injection of LiPc crystals.

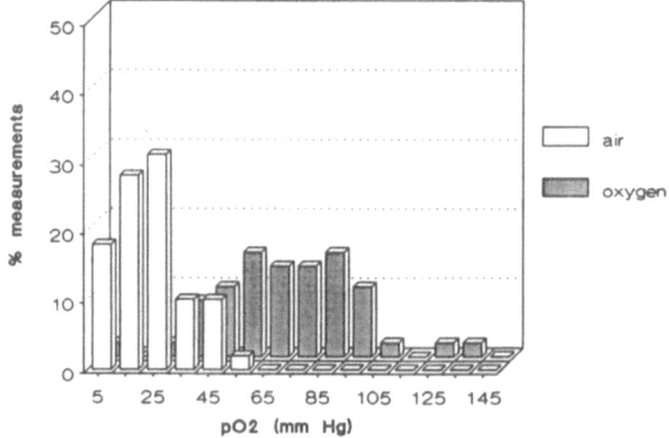

Fig. 4. pO_2 frequency histogram for air breathing and oxygen breathing
mice following injection of fusinite (n=38). Air and oxygen breathing
values were obtained by averaging the two pO_2's measured in the final
10 minutes of each period.

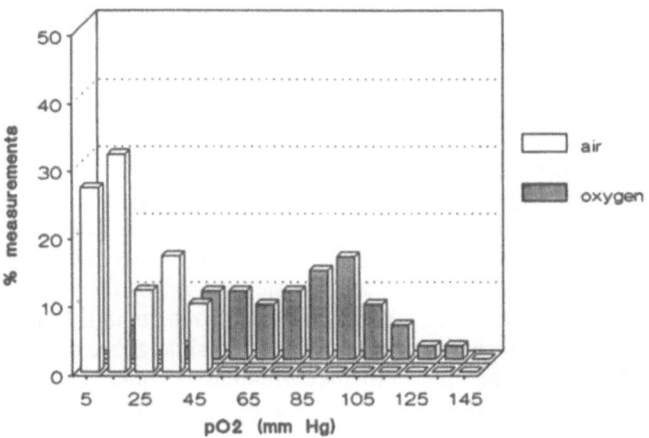

Fig. 5. pO_2 frequency histogram for air-breathing and oxygen-breathing mice following IM injection of LiPc (n=40).

DISCUSSION

The data obtained with fusinite and LiPc are fairly consistent for air-breathing mice, with a narrow distribution of pO_2's and average values which fall reasonably close together. Both probes show large increases in average tissue pO_2 in the oxygen breathing mouse, and both histograms were much more widely scattered. The average values, however, do not agree as well. This may simply be a reflection of the fact that relatively few measurements were used to construct the histogram (100 is a more typical number), it may be caused by systemic errors in calibration or application of one of the techniques, or it may be the result of genuine oxygen concentration differences - that is, the two methods may sample slightly different environments.

The average air breathing values measured with fusinite and LiPc (23 torr and 18 torr respectively) are well within the range of previous measurements of skeletal muscle oxygen tensions. The large response of tissue pO_2 to a change in the FI_{O2} from 21% to 85% is surprising, however, since hemoglobin accounts for most of the blood's oxygen carrying capacity, and should be almost 100% saturated in healthy air breathing animals. There are a number of possible explanations for these results. Perhaps most likely is the possibility that the anesthetic and relative hypothermia of the animal compromised respiration and/or ventilation - i.e., the air breathing anesthetized mouse did not have a normal hemoglobin saturation. Other possibilities include a direct effect of the anesthetic or hypothermia on oxygen delivery to tissues. A similar increase in tissue pO_2 in response to an elevation in FI_{O2} has also been noted in a few other reports which used polarographic methods to obtain tissue oxygen measurements in vivo[1-3]. The widely scattered O_2 histogram pattern has also been noted by others[4]. This issue should be fairly easy to resolve in future experiments by monitoring either arterial pO_2 or hemoglobin saturation throughout the course of the experiment.

A final aspect of this study which merits discussion is the oxygen sensitivity, accuracy, and precision of the two probes. The precision of the oxygen measurements depends on the precision of the spectral linewidth measurements, and the uncertainty in these depends on the signal to noise ratio. Signal strength depends on the injected volume, the depth of injection, and the degree of success in placing the site of injection directly over the center of the surface probe. In addition, the signal to noise ratio depends on the tissue oxygen concentration - at very low oxygen tensions the linewidth is narrow, so that the signal to noise ratio is high and the linewidth can be determined very accurately. (This is so because the concentration of oxygen does not affect the total number of spins, and therefore the double integral of the first derivative EPR spectrum remains constant - thus a narrow linewidth results in a large signal height.) At higher oxygen levels, linewidths are broad and signal heights correspondingly lower, so that S/N diminishes and linewidth measurement becomes less precise. This is particularly true with LiPc, which has a very narrow intrinsic linewidth.

We estimate that the average error in linewidth determination is $\pm$ 3 mG for both fusinite and LiPc, which corresponds to an error in oxygen pressure of $\pm$ 0.3 torr for fusinite and $\pm$ 0.4 torr for LiPc. There are additional factors that could affect accuracy and precision: inhomogeneities in the external magnetic field; any variability in the oxygen response of different LiPc crystals or fusinite particles, so that the experimental probes behaved slightly differently from the calibration samples; and power saturation effects either during calibration or the experiment. It should be relatively easy to control these latter problems, however, and thus it is not unreasonable to postulate an oxygen measurement that is reasonably accurate and has a precision of < 1 torr; this precision should improve at low oxygen tensions.

In summary, we have demonstrated the feasibility of using a low frequency EPR spectrometer in conjunction with fusinite and LiPc to obtain in vivo tissue pO_2 measurements. The oxygen sensitivity of both probes compares favorably to that of the more commonly used nitroxides, and in addition their high spin density, biologic inertness, and lack of spectral dependence on probe concentration makes these substances very attractive alternatives for future in vivo EPR oximetry.

ACKNOWLEDGMENTS

This work was supported by NIH grant GM 34250, and used the facilities of the EPR Center at the University of Illinois, which is supported by NIH grant RR 01811. J. F. Glockner received support from NIH training grant CA 09067.

REFERENCES

1. D. K. Harrison, J. Hoper, H. Gunther, H. Vogel, K. H. Fronk, M. Brunner, R. Ellerman, and M. Kessler, Microcirculation and pO_2 in Skeletal Muscle During Respiratory Hypoxia and Stimulation, Adv. Exptl. Med. Biol. 169:477 (1984).
2. W. Fleckenstein, R. Heinrich, W. Grauer, H. Schomerus, W. Dolle, and C. Weiss, Fast Local Regulations of Muscle pO_2 Fields in Patients Suffering from Cirrhosis of the Liver, Adv. Exptl. Med. Biol. 180:687 (1984).
3. W. Fleckenstein and C. Weiss, A Comparison of pO_2 Histograms From Rabbit Hind Limb Muscles Obtained by Simultaneous Measurements with Hypodermic Needle Electrodes and with Surface Electrodes, Adv. Exptl. Med. Biol. 169:447 (1984).
4. N. Lund, Skeletal Muscle Surface Oxygen Pressure Fields in Humans, Adv. Physiol. Sci. 25:251 (1981).

MEASUREMENT OF CEREBRAL BLOOD FLOW IN ADULT HUMANS USING NEAR INFRARED

SPECTROSCOPY - METHODOLOGY AND POSSIBLE ERRORS

C.E. Elwell, M. Cope, A.D. Edwards[*], J.S. Wyatt[*],
E.O.R. Reynolds[*], D.T. Delpy

Departments of Medical Physics and Paediatrics[*]
University College London, U.K.

INTRODUCTION

The principle of near infrared spectroscopy (NIRS) was first clearly defined by Jobsis in 1977, and recent technical and methodological advances have made it possible to use NIRS to make non-invasive, quantitative measurements of tissue oxygenation and haemodynamics (Reynolds et al 1988). The NIRS technique has been applied predominantly to the measurement of neonatal cerebral haemodynamics (Brazy et al 1985, 1986, Ferrari et al 1986, Wyatt et al 1986, 1991); in particular the quantification of cerebral blood volume (Wyatt et al 1990) and blood flow (CBF)(Edwards et al 1988a).

Measurement of blood flow by NIRS is derived from the Fick principle and uses a rapid change in arterial oxyhaemoglobin concentration as an intravascular tracer. The technique has been successfully validated by studies directly comparing NIRS measurements of CBF in sick newborn infants with measurements made using [133]Xenon (Skov et al 1991, Bucher et al 1991), and in the adult forearm by comparison with venous occlusion plethysmography.(Edwards et al 1988b). These studies show that the technique is accurate for measurement of flows up to approximately 40 ml.100g^{-1}.min^{-1}.

For this technique to be applied to the measurement of adult CBF, the problems associated with measuring higher flow rates must be identified and addressed. The purpose of this report is to highlight the errors which can occur when measuring adult cerebral blood flow using NIRS, and to indicate the relative contribution that different methods of data collection, averaging and analysis can make to these errors. The results of CBF measurements on 10 normal adults are presented.

THEORY

NIRS

The details of near infrared spectroscopy have already been described elsewhere (Cope et al 1988). Briefly the technique depends upon absorption by the chromophores oxyhaemoglobin (HbO$_2$)and deoxyhaemoglobin (Hb) of near infrared light transmitted through the organ of interest. The changes in concentration of these chromophores can be quantified using a modified Beer-Lambert law which describes optical attenuation in a highly scattering medium (Delpy et al 1988). This can be expressed as:

$$Attenuation(OD) = \log\frac{I_o}{I} = \alpha cLB + G \qquad (1)$$

where OD is optical densities, I_0 the incident light intensity, I the detected light intensity, α the absorption coefficient of the chromophore (mM^{-1}.cm^{-1}), c the concentration of chromophore (mM), L the physical distance between the points where light enters and leaves the tissue (cm), B a 'pathlength factor' which takes into account the scattering of light in the tissue and G a factor related to the geometry of the tissue. If measurements are only made of the changes in attenuation, then L, B and G can be assumed to remain constant and changes in chromophore concentration can be derived from the expression:

$$\delta c = \frac{\delta OD}{\alpha LB} \tag{2}$$

The absorption coefficients of HbO_2 and Hb are known (Wray et al 1988) and B has been measured in adult head to have a mean value of $5.93 \pm$ S.D. 0.42 (van der Zee et al 1991).

Measurement of blood flow

The Fick principle states that the rate of accumulation of a tracer in an organ (dQ/dt) is equivalent to the difference between its rate of arrival (flow (F) x arterial concentration (P_a)) and the rate of its departure (F x venous concentration (P_v)). If measurements of the rate of accumulation of the tracer are made within the transit time of that tracer through the organ, T_t, then the venous outflow ($F.P_v$) will be zero. The flow can then be calculated from the ratio of the quantity of tracer accumulated to the quantity of tracer introduced during time t :

$$F = \frac{Q(t)}{\int_0^t P_a(t)\ dt} \qquad where\ t < T_t \tag{3}$$

When a sudden increase is induced in arterial oxygen saturation (ΔSaO_2), the resulting initial increase in cerebral HbO_2 concentration represents the accumulation of tracer Q. The actual quantity of the tracer introduced (disregarding changes in dissolved oxygen content) is given by the product of the integral of ΔSaO_2 with respect to time and the arterial total haemoglobin concentration [tHb]. CBF in $ml.100g^{-1}.min-1$. can then be calculated using the following equation (Edwards et al 1991):

$$CBF(ml.100g^{-1}.min^{-1}) = \frac{\Delta(HbO_2)}{K \cdot \int_0^t \Delta SaO_2\ dt} \tag{4}$$

where K is a constant which reflects the molecular weight of haemoglobin, cerebral tissue density and the concentration of haemoglobin in the subject's blood which is determined from a blood sample. To obtain data from which to calculate CBF, simultaneous oximetry and NIRS measurements are made while the subject's FiO_2 is slowly reduced and then rapidly increased to produce the sudden increase in SaO_2.

When tissue haemoglobin concentration is constant it may be assumed that the changes in HbO_2 and Hb concentration are equal and opposite. Under these conditions the signal representing the difference between the change in HbO_2 and Hb concentrations is twice the amplitude of the corresponding signal representing the change in HbO_2 concentration alone. To improve the signal to noise ratio of the NIRS data, equation 4 can therefore be modified to :

$$CBF(ml.100g^{-1}.min^{-1}) = \frac{\cdot\Delta(HbO_2) - \Delta(Hb)}{2 \cdot K \cdot \int_0^t \Delta SaO_2\ dt} = \frac{\Delta Hbdiff}{2 \cdot K \cdot \int_0^t \Delta SaO_2\ dt} \tag{5}$$

In the subsequent explanation of the analysis procedure the term Hbdiff will be used to represent the difference between the HbO_2 and Hb signals and the term Hbsum to represent the sum of the two signals.

SOURCES OF ERROR

The method requires accurate measurement of the rapid change in arterial HbO_2 concentration in the organ by the NIR spectrometer and of the change in arterial SaO_2 by a pulse oximeter. The limitations in the performance of both these monitors which can contribute errors to the calculation of blood flow will be considered in turn:

1) NIRS measures the increase in cerebral HbO_2 concentration and thus quantifies the amount of tracer accumulation Q. The measured changes in HbO_2 concentration do not however differentiate between arterial inflow and venous outflow of tracer. In order to justify the assumption that the changes in cerebral HbO_2

concentration are due to arterial accumulation of the tracer in the brain, all measurements must be made within the minimum transit time of blood through the brain. Cerebral blood flow and vascular transit time will vary according to the relative amounts of grey and white matter in the field of view. If a mean value of 5.5 seconds is taken (Leenders et al 1990) with perhaps a range of ± 2 seconds, then it is clear that the NIRS data must be collected over short time periods in order to obtain a sufficient number of data points before the transit time is reached.

2) With the relatively short adult cerebral vascular transit time, the success of the NIRS flow measuring technique depends upon the sudden input of the tracer, oxyhaemoglobin, into the brain. In many pulse oximeters some degree of signal averaging is performed which will damp the monitored SaO_2 input function and thus lead to an underestimate of the actual quantity of tracer introduced to the brain during time, t. To obtain SaO_2 values which will accurately monitor the step change, a pulse oximeter which performs minimal signal averaging is needed. This can be achieved by measuring SaO_2 on every heart beat.

3) HbO_2 and SaO_2 are measured by NIRS and pulse oximetry respectively at two different sites on the body. Thus, the induced change in oxyhaemoglobin concentration may be seen first by the NIRS or by the pulse oximeter depending upon the time taken for the oxygenated blood to reach the brain and the pulse oximetry site

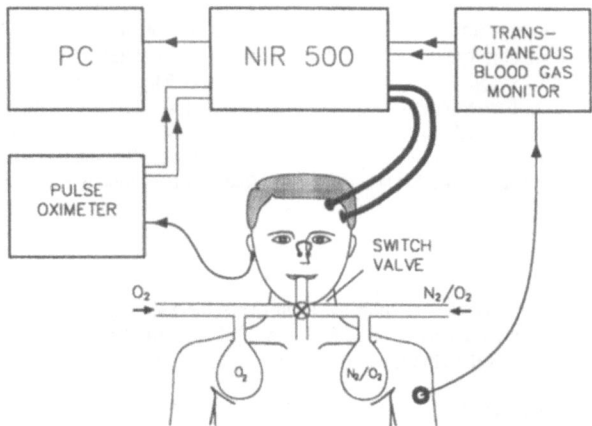

Figure 1. Experimental Setup

respectively. To minimise this delay the ear is usually chosen as the monitoring site for SaO_2, but the delay will be exaggerated by the choice of a more distal site such as the finger for the oximetry measurements. The Fick principle involves calculating the ratio of the tracer accumulated to the tracer introduced during a given time period. Any delay between the detection of the signals must therefore be accurately compensated for since temporal mismatch of the signals will introduce errors in the calculation.

DATA COLLECTION

Five measurements of CBF were made on each of 10 male subjects (age range 24-36 years, median 27.5) who had no known respiratory or circulatory disorders.

Instrumentation and Procedure

Near infrared light was carried to and from the NIR spectrometer (NIR 500, Hamamatsu Photonics

KK, Japan) through fibre optic bundles. The optodes were positioned high on the left side of the forehead 4 - 5 cm apart. The exact position of the optodes was dependent upon the level of the hairline in each subject, but did not vary more than (±1 cm) for all the subjects (Figure 1). The optodes were held in position with double sided adhesive rings and self adhesive tape and the head was then wrapped in black cloth to reduce background light. The prototype NIR 500 used in these studies used pulsed laser diodes at three wavelengths (775, 829, 909 nm) as its light source and a photomultiplier tube for detection. Data was collected every 0.5 second and the changes in concentration of HbO_2 and Hb were calculated using a previously established algorithm (Wray et al 1988).

Arterial blood oxygen saturation and heart rate were monitored using a pulse oximeter (Novametrix 500, USA) modified to measure in beat to beat mode via a probe positioned on the right ear. A transcutaneous blood gas electrode (Novametrix 850, USA) was placed on a fleshy part of the upper left arm and was used to continuously monitor transcutaneous oxygen ($TcPO_2$) and carbon dioxide tension ($TcPCO_2$). The analogue outputs of both the oximeter and transcutaneous monitor were linked directly to the spectrometer for real time display and storage along with the NIRS data.

A modified anaesthetics trolley was used to supply a controlled mixture of nitrogen and oxygen to the subject via an anaesthetics bag and mouthpiece. A separate circuit supplied 100% oxygen to a second anaesthetics bag. Both bags were connected to the mouthpiece by means of a two way switch valve which allowed the gas the subject was breathing to be changed from an hypoxic mixture to 100% O_2 virtually instantaneously. The subject wore a nose clip to ensure the inspired oxygen concentration was well controlled and a flap valve vented expired gas to atmosphere to prevent rebreathing.

All measurements were made with the subjects lying flat on a couch. The subject was made comfortable and started breathing through the mouthpiece from the first anaesthetics bag. FiO_2 in this circuit was gradually reduced until a baseline SaO_2 of 85-90% was achieved. This baseline was maintained for a few minutes and then at the end of an expiration the valve was switched to allow the subject to inspire 100% O_2 on the next breath which returned SaO_2 to 98-100% within 1.5 - 2 seconds. After one or two breaths of 100% O_2 the nose clip was removed to allow the subject to breathe room air. This method allowed the subject to maintain a constant breathing rate while SaO_2 was both lowered and then rapidly increased.

DATA ANALYSIS

Figure 2(a) shows one set of data collected from one subject during a CBF measurement. The step change in SaO_2 can be seen on an expanded scale in Figure 2(b), with the time delay between the initial rise of the SaO_2 and the initial rise of the Hbdiff clearly evident, as is the cycling of SaO_2 due to respiration.

To calculate CBF both the integral of the SaO_2 rise and the change in Hbdiff must be evaluated over a given time t. To do this, the start point for the rise in SaO_2 was defined by inspection of the data and a baseline value calculated from the mean value over the previous three respiratory cycles. The cumulative SaO_2 integral at each point over time t was estimated by Simpsons's rule, and was then plotted against the equivalent Hbdiff data for the same time period (Figure 3(a)). A fourth order polynomial was fitted to this data and the first order differential of this polynomial used to calculate a set of values for CBF over the time period, t (Figure 3(b)).

To estimate the true value of the time difference between the ear (SaO_2) and the brain (Hbdiff), a range of CBF calculations were made. The cumulative SaO_2 integral over time t was fixed as above and then plotted against the measured Hbdiff for the same time duration, but beginning 0.5 second away from the point of the initial rise in SaO_2. This procedure was then repeated changing the offset point by 0.5 second each time until CBF had been calculated on the data up to 5 seconds before and 10 seconds after the initial rise in SaO_2. This produced a set of calculated CBF values for every 0.5 second shift within this 15 second period. The point at which the change in Hbdiff is correctly temporally matched with the SaO_2 step change was defined by the CBF - time curve which had the maximum initial CBF value. Five CBF measurements were performed on each subject, and the CBF - time curves calculated from the analysis of these measurements were averaged to produce a mean curve for each subject.

The time over which the integral of the SaO_2 rise and the corresponding change in Hbdiff are measured will depend upon the transit time of the organ of interest. To investigate this, the analysis procedure described above was initially performed over a period of 3 seconds and then repeated on the same data for periods of 4, 5 & 6 seconds. To quantify the effects of signal averaging of the oximetry data, a rolling average over 1.5, 2.5 and 3.5 seconds was applied in software to the non averaged 'beat to beat' SaO_2 data. The analyses described

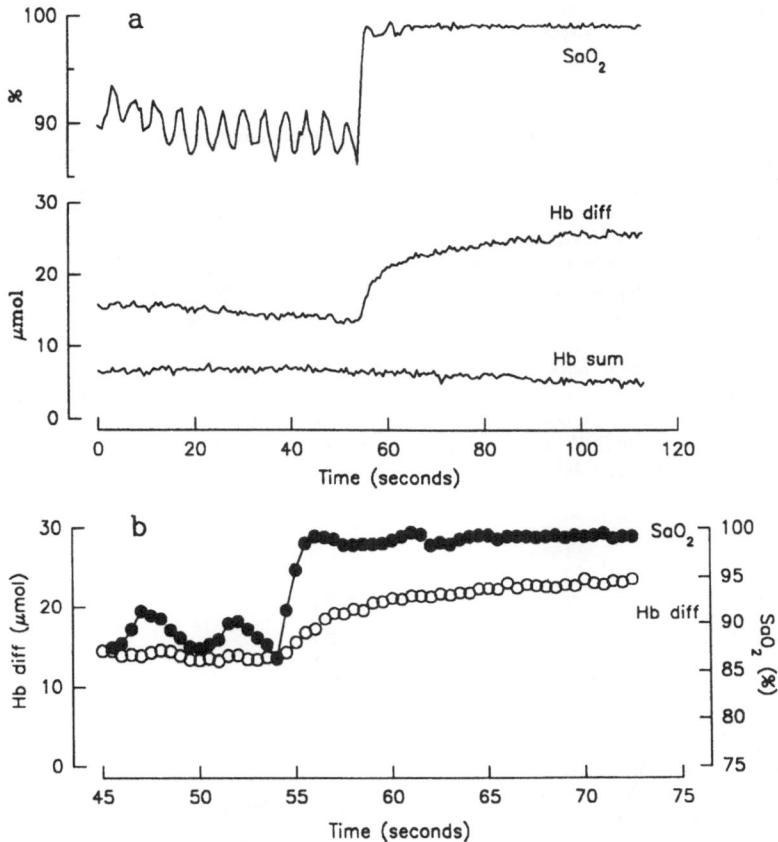

Figure 2 (a) Data collected from one subject during a CBF measurement and (b) the same data expanded to show the SaO₂ step change.

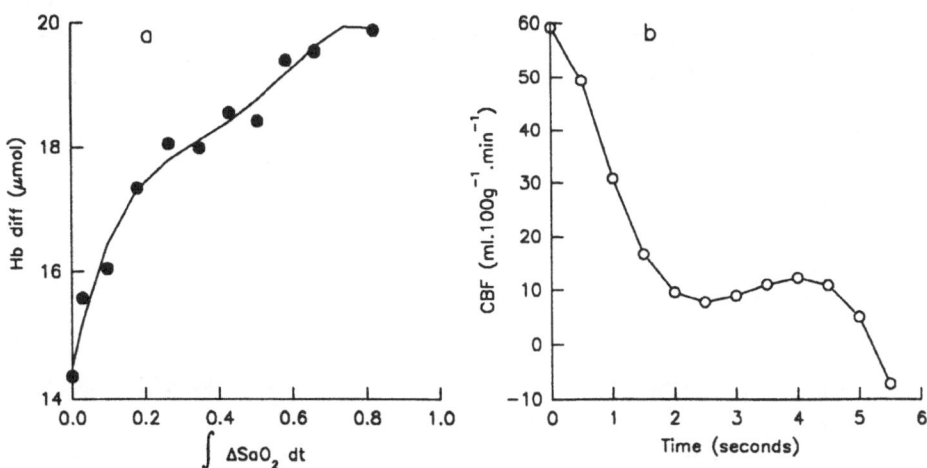

Figure 3 (a) Cumulative SaO₂ integral plotted against the equivalent change in Hbdiff and (b) the calculated CBF - time curve for this data.

above were then repeated for each degree of averaging. The effect of a deliberate temporal mismatch of the Hbdiff rise and SaO_2 step was investigated by comparing the CBF - time curves obtained 1 and 2 seconds before and after the curve which produced the maximum initial CBF.

RESULTS

Figure 4 shows the mean CBF - time curves, calculated over periods of 3, 4, 5, and 6 seconds, for each subject, and the overall mean for all ten subjects for each time period. The negative flow values of the final point of each curve are an artifact due to the effects of the polynomial fit, and are included only to demonstrate the high and low flow component trend seen on each measurement (see Discussion). The intra subject coefficient of variation for the maximum CBF measurements ranged from 4% to 57% (mean 20%), while the inter subject variations ranged from 30% to 34% (mean 32%). The absolute values of $TcPCO_2$ did not vary by more than 0.53 kPa during any of the CBF measurements. The change in total cerebral haemoglobin concentration (Hbsum) as a percentage of the change in Hbdiff had a mean value of 23%.

The relative effects of SaO_2 signal averaging on the mean CBF - time curve for all subjects over 6 seconds is shown in Figure 5. Table I summarises the effects of SaO_2 averaging upon the calculated maximum flow. Figure 6 shows a plot, for each time period, of the maximum CBF against temporal offset from the previously defined maximum CBF. The percentage decrease in calculated maximum CBF due to +2, +1, -1, and -2 second offset are shown in Table II.

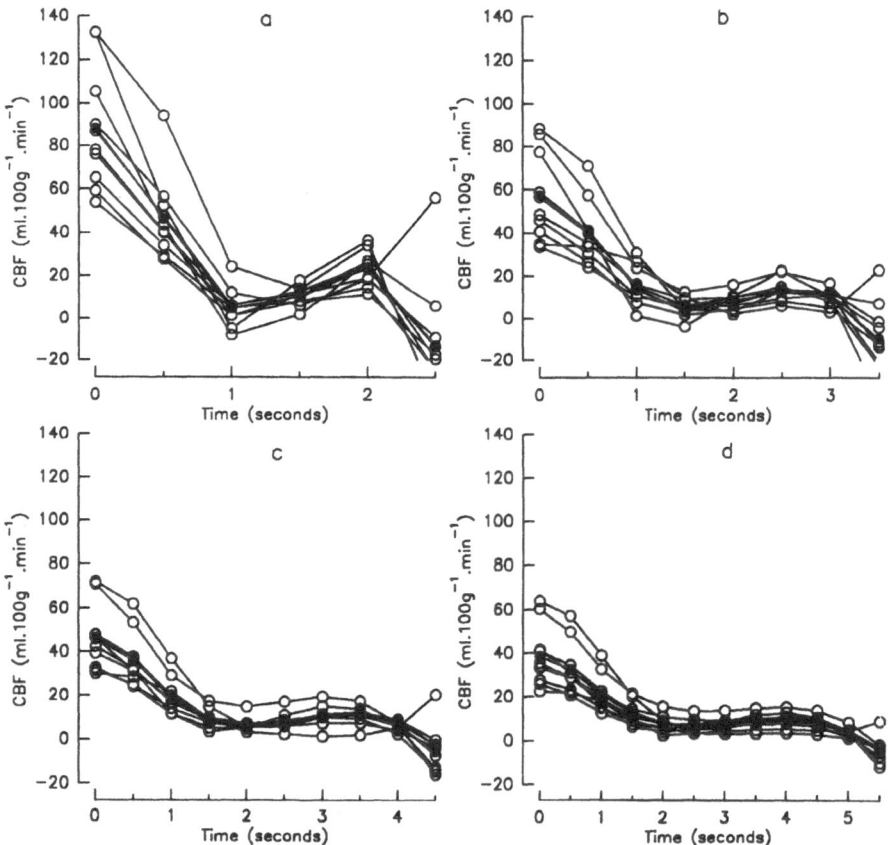

Figure 4. Mean CBF time curves for each subject for (a) t = 3 seconds, (b) t = 4 seconds, (c) t = 5 seconds and (d) t = 6 seconds with the overall mean for all ten subjects in each time period shown as filled circles.

Table I Percentage increase in calculated maximum CBF due to 1.5, 2.5, and 3.5 second averaging of the SaO$_2$ data.

	t = 3 seconds	t = 4 seconds	t = 5 seconds	t = 6 seconds
1.5 second SaO$_2$ averaging	45.8%	34.9%	26.7%	23.3%
2.5 second SaO$_2$ averaging	95.7%	64.0%	51.1%	46.3%
3.5 second SaO$_2$ averaging	122.4%	91.3%	75.7%	72.8%

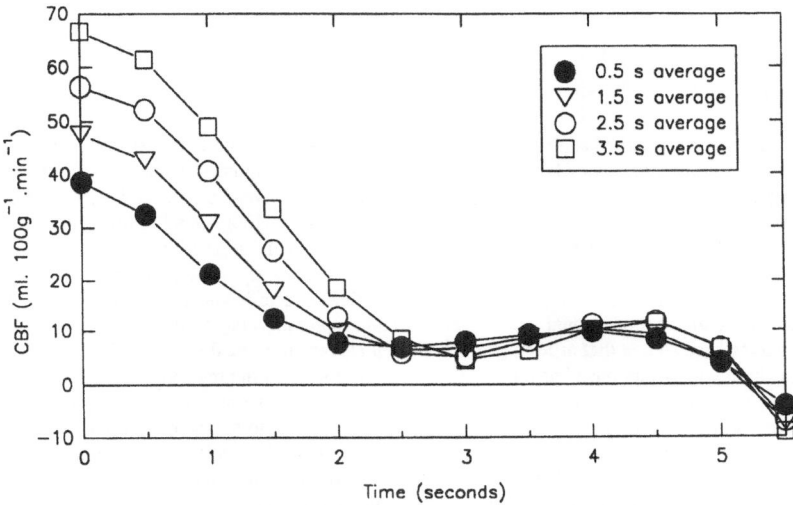

Figure 5. The effect of SaO$_2$ signal averaging on the mean CBF - time curves for all subjects for t = 6 seconds.

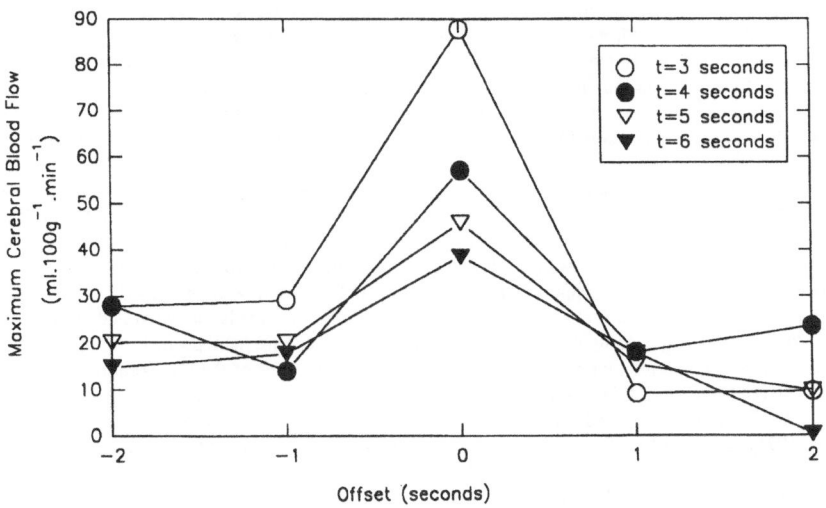

Figure 6. The effects of temporal offset on the calculated maximum CBF.

Table II Percentage decrease in the calculated maximum CBF due to +2, +1, -1, and -2 second offset from the correctly temporally matched CBF calculations.

	t = 3 seconds	t = 4 seconds	t = 5 seconds	t = 6 seconds
+ 2 second offset	89.0%	58.7%	78.7%	99.2%
+ 1 second offset	89.6%	68.4%	66.8%	53.9%
- 1 second offset	66.9%	75.3%	55.8%	54.0%
- 2 second offset	68.4%	50.8%	55.9%	61.2%

DISCUSSION

These measurements of adult CBF by NIRS show a consistent pattern of variation, with the value of CBF depending on the period over which the measurement is made. On all measurements there is an immediate fall from a peak value, and then a slight rise to a second peak. The average of the flows measured on normal subjects where t = 3 seconds show an initial maximum flow of 88 ml.100g^{-1}.min^{-1} falling within 1.5 seconds to a lower mean flow of 13 ml.100g^{-1}.min^{-1}. The most consistent explanation of this phenomenon is that the distribution of cerebral vascular transit times is bimodal, and that the early measurements detect a rapid transit compartment while the later values represent slower transit times, tracer having already reached a steady state in tissue where the transit time is rapid. There is evidence to support this conjecture. Measurements of ^{133}Xenon clearance curves have shown that at least two exponential curves are needed to fit the data (Ginsberg 1986), and this has been widely interpreted as representing different flow rates in grey and white matter. If these data are combined with measurements of cerebral blood volume by positron emission tomography (Leenders et al 1990), mean transit times for grey and white matter can be calculated to be about 3.5 seconds and 9 seconds respectively. Measurements of regional cerebral transit time using radionuclide cerebral angiography (Carlsen et al 1987) give a mean value over all regions of 3.2 ±0.67 seconds. As the minimum transit time must be less than this, these values are consistent with the hypothesis that the variation in NIRS measurements is due to a bimodal distribution of transit times which may represent the perfusion of cerebral grey and white matter. The alternative explanation, that the minimum cerebral transit time is less than one second is unlikely, and is not supported by studies in animals (Jones et al 1991).

If the shortest cerebral transit times is around 1.5 - 2 seconds, and transit times were normally distributed then the first two or three values for calculated CBF would be expected to be similar. However, a different pattern was observed, with an immediate fall from the maximum calculated value. The likely reason for this is that the NIRS measurement is made over a relatively large volume of tissue in which there is not only a distribution of transit times, but also a distribution of arrival times of the HbO$_2$ tracer. This latter distribution is also likely to vary by up to ±1 second. The measured data represents a convolution of these two distributions, and hence the calculated flow is unlikely to show a constant value if the data sampling period is 0.5 second.

There are several possible sources of error in the NIRS method of CBF measurement. Some uncertainty exists about the absolute quantification of NIRS data because of difficulties in determining the geometry and pathlength of light propagated through tissue. The light entering the head from the NIR spectrometer will pass through a proportion of skin, fat, muscle and bone as well as brain tissue before being detected. In neonates the contribution to scattering and absorption by these tissues is not thought to be significant (Cope 1991). In adults however the percentage of non brain tissue through which the light travels is increased and as such will have an effect on the pathlength of light in the head/region of interest. This increase in pathlength is not as yet quantified, but will at a minimum be equal to the anatomical thicknesses of the non brain tissues between the optodes. The contribution of bone blood flow to CBF measurement is unknown but must be considered. Given the short period over which the NIRS measurements are made, and the relatively low values of bone blood flow (Brookes 1971), it is possible that the rise in HbO$_2$ in the bone blood will not have commenced and thus bone blood flow will not be detected over the period chosen for CBF calculation. A prolonged 'low flow' component can be seen in the NIRS data (Figure 2(a)), and it may be possible to attempt a separate calculation for this region. The error due to system noise is small and has been discussed elsewhere (Cope 1991). More significant are the uncertainties due to methods of data collection and analysis, and violations of the theoretical basis of the technique. These will be addressed in turn.

Data Collection and Analysis

In order to make the measurement of CBF, SaO_2 must have both a stable baseline and a rapid change. If SaO_2 is reduced to about 90%, oscillations occur in arterial saturation due to respiration. These are only partially transmitted to cerebral tissue, presumably because of homeostatic mechanisms, and because of this damping we have been able to take a mean value as our baseline for measuring the change in SaO_2.

The technique requires that the change in arterial saturation is determined continuously, and we have approximated this by measuring arterial saturation on every heart beat by pulse oximetry. Any systematic error due to the pulse oximeter, will compromise the accuracy of the CBF measurement particularly if the data is processed to produce a rolling average over more than one heart beat. Many commercial instruments average over several seconds to improve the stability of the calculated saturation measurement. The current data show that averaging causes an artificial increase in the calculated flow, and demonstrates that a non-averaging oximeter is essential for adult CBF measurements. Fortunately we have found that shielding the oximeter probes from changes in ambient light, and minimising sensor movement allowed sufficiently stable measures of SaO_2 to be obtained without averaging between heartbeats.

This study also found that slight temporal mismatching of the SaO_2 and Hbdiff signals caused a significant error in the calculated flow. If no time delay existed between the measured SaO_2 and Hbdiff signals, the calculated SaO_2 integral could simply be matched directly with the equivalent Hbdiff rise over the same time period. However, time delays in the order of seconds are seen even when the oximeter probe is sited on the ear. Several calculations of CBF with different time offsets are then necessary to determine the correct CBF, and at high flow rates, a temporal mismatch of even one second can cause a gross underestimation of the true value. This means that all data must be sampled at intervals of 0.5 second or less.

Theoretical Considerations

The two theoretical violations of the basis of the technique that are possible are firstly that cerebral transit time is too short for measurements to be possible and secondly that changes in SaO_2 cause alterations in the CBF, cerebral blood volume or cerebral oxygen consumption. The first problem has been discussed above, but the second requires further consideration.

Experimental data from adults and animals has shown that small changes in blood oxygenation close to the normal range has little effect on CBF and cerebral oxygen extraction (Siesjo 1978). However we have observed that the sudden increase in SaO_2 during CBF measurement is associated in most subjects with a small fall in total cerebral haemoglobin concentration and thus in cerebral blood volume. The fall is small and thus will only introduce a small error into the measurement. It may represent a small change in CBF caused by the measurement procedure. It is likely that this problem can be ameliorated by preventing PaO_2 from rising above 14 kPa (Siesjo 1978), and this possibility is under investigation.

CONCLUSION

NIRS provides a practical, rapid, noninvasive method of measuring CBF. There are some methodological problems associated with both the collection of the data and its analysis. An experimental system which can collect data at time intervals of 0.5 seconds or less is required, and the step change in SaO_2 must occur and be accurately monitored over a similarly short period. The average of the flows measured on normal subjects show a bimodal distribution with fast component and slow component values which compare well with the previously established data for grey and white matter. The slow flow component is relatively insensitive to both errors in the data collection and analysis. However, the errors induced in the higher flow rates are far more significant and as such must be taken into account in the measurement and analysis procedures.

ACKNOWLEDGEMENTS

This work has been supported by grants from the Wellcome Trust, the M.R.C., the S.E.R.C., the Wolfson Foundation and Hamamatsu Photonics KK.

REFERENCES

Brazy, J.E., Lewis, D.V., Mitnick, M.H., Jöbsis, F.F., 1985, Noninvasive monitoring of cerebral oxygenation in preterm infants: preliminary observations. Pediatrics, 75, 217-225.

Brazy, J.E., Lewis, D.V., 1986, Changes in cerebral blood volume and cytochrome aa3 during hypertensive peaks in preterm infants. Pediatrics, 108, 983-987.

Brookes, M. 1971. The blood supply to bone. Butterworth. London., 236.

Bucher, H.V., Edwards, A.D., Lipp, A.E., Duc, G. 1991. Comparison between [133]Xenon clearance and near infrared spectroscopy for estimation of cerebral blood flow. Paed. Res. (in press).

Carlsen, O., Hedegaard, O. 1987, Evaluation of regional cerebral circulation based on absolute mean transit times in radionuclide cerebral angiography. Phys. Med. Biol., 32, 11, 1457-1467.

Cope, M., Delpy, D.T. 1988. A system for long term measurement of cerebral blood and tissue oxygenation in newborn infants by near infrared transillumination. Med. Biol. Eng. & Comp., 26, 3, 289-294.

Cope, M. 1991. The development of a near infrared spectroscopy system and its application for non invasive monitoring of cerebral blood and tissue oxygenation in the newborn infant. PhD Thesis, University of London.

Delpy, D.T., Cope, M., van der Zee, P., Arridge, S.R., Wray, S., Wyatt, J.S. 1988. Estimation of optical pathlength through tissue from direct time of flight measurement. Phys. Med. & Biol., 33, 12, 1433-1442.

Edwards, A.D., Wyatt, J.S., Richardson, C.E., Delpy, D.T., Cope, M., Reynolds, E.O.R. 1988a. Cotside measurement of cerebral blood flow in ill newborn infants by near infrared spectroscopy. Lancet, ii, 770-771.

Edwards, A.D., Reynolds, E.O.R., Richardson, C.E., Wyatt, J.S. 1988b. Estimation of blood flow in man using near infrared spectroscopy (NIRS). J. Physiol., 410, 50P.

Edwards, A.D., Richardson, C., van der Zee, P., Elwell, C., Wyatt, J.S., Cope, M., Delpy, D.T., Reynolds, E.O.R. 1991. Measurement of haemoglobin flow and blood flow by near infrared spectroscopy. J. Appl. Physiol. (submitted).

Ferrari, M., De Marchis., Giannini., Nicola, A., Agostino, R., Nodari, S., Bucci, G., 1986, Cerebral blood volume and haemoglobin oxygen saturation monitoring in neonatal brain by near infrared spectroscopy. Adv. Exp. Med. Biol., 200, 203-212.

Ginsberg M.D., 1986. Cerebral circulation and its regulation and pharmacology, in : 'Diseases of the nervous system' Asbury A.K., McKhann G.M., McDonald W.I, ed., W. B. Saunders, Philadelphia.

Jöbsis, F.F. 1977. Noninvasive, infrared monitoring of cerebral and myocardial oxygen sufficiency and circulatory parameters, Science, 198, pp. 1264-1267.

Jones, S.C., Korfali, E., Marshall, S.A. 1991. Cerebral blood flow with the indicator fractionation of [^{14}C] Iodoantipyrine: Effect of $PaCO_2$ on cerebral venous appearance time. J. Cereb. Blood Flow & Met., 11, 236-241.

Leenders, K.L., Perani, D., Lammertsma, A.A., Heathers, J.D., Buckingham, P., Healy, M.J.R., Gibbs, J.M., Wise, R.J.S. et al. 1990, Cerebral blood flow, blood volume and oxygen utilisation. Normal values and effect of age. Brain, 113, 27-47.

Reynolds, E.O.R., Wyatt, J.S., Azzopardi, D., Delpy, D.T., Cady, E.B., Cope, M., Wray, S. 1988. New non-invasive methods for assessing brain oxygenation and haemodynamics. Brit. Med. Bull., 44, 4, 1052-1075.

Siesjo, B.K. 1978. Brain energy metabolism. J. Wiley & Sons, N.Y.

Skov, L., Pryds, O., Grieisen, G. 1991. Estimating cerebral blood flow in newborn infants: comparison of near infrared spectroscopy and [133]Xenon clearance. Paed. Res. (in press).

van der zee, P., Cope, M., Arridge, S.R., Essenpries, M., Potter, L.A., Edwards, A.D., Wyatt, J.S., McCormick, D.C., Roth, S.C., Reynolds, E.O.R., Delpy, D.T. 1991. Experimentally measured optical pathlengths for the adult head, calf and forearm and the head of the newborn infant as a function of interoptode spacing. Adv. Exp. Med. & Biol. (in press).

Wray, S., Cope, M., Delpy, D.T., Wyatt, J.S., Reynolds, E.O.R. 1988. Characterisation of the near infrared absorption spectra of cytochrome aa$_3$ and haemoglobin for the non invasive monitoring of cerebral oxygenation. Biochim. Biophys. Acta, 933, 184-192.

Wyatt, J.S., Cope, M., Delpy, D.T., Wray, S., Reynolds, E.O.R. 1986. Quantification of cerebral oxygenation and haemodynamics in sick newborn infants by near infrared spectrophotometry. Lancet, 2, 1063-1066.

Wyatt, J.S., Cope, M., Delpy, D.T., Richardson, C.E., Edwards, A.D., Wray, S.C., Reynolds, E.O.R. 1990. Quantitation of cerebral blood volume in newborn infants by near infrared spectroscopy. J. Appl. Physiol. 68, 3, 1086-1091.

Wyatt, J.S., Edwards, A.D., Cope, M., Delpy, D.T., McCormick, D.C., Potter, A., Reynolds, E.O.R. 1991. Response of cerebral blood volume to changes in arterial carbon dioxide tension in preterm and term newborn infants. Paed. Res. 29, 553-557.

DESIGN AND EVALUATION OF A REFLECTANCE OXYGEN SENSOR IN CRITICALLY ILL PATIENTS

Setsuo Takatani, George P. Noon, Yukihiko Nose, and Michael E. DeBakey

Department of Surgery, Baylor College of Medicine
Houston, TX 77030, USA

INTRODUCTION

The optical pulse oximeter has gained broad clinical applications in rapidly and noninvasively measuring arterial hemoglobin oxygen saturation (S_aO_2)(Yelderman and New, 1983; Severinghaus, 1987; Pologe, 1987). However, major shortcoming of the current transmission oximeters is the requirement of a cuvette through which light signals must be transmitted, thus limiting its application to fingertip or earlobe. More broad applications can be achieved with the reflectance mode.

The reflectance pulse oximeter, as reported by Mendelson et al (1983), was the first to measure S_aO_2 from the fingertip. Since then, several reflectance pulse oximeters have been reported, but no clinical application has of yet been reported (Mendelson et al, 1988; Mendelson and Ochs, 1988). The major difficulty of reflectance pulse oximeters is small pulsatile signal level, particularly at red wavelength, in comparison to the transmission mode. Also, sensor characteristics, and physiological factors such as arterial-to-venous blood volume distribution, local venous saturation, blood flow, tissue inhomogeneity, optical properties of tissue, etc, all affect the accuracy of the measurement. For proper design of the reflectance sensor, the effects of these variables on the S_pO_2 computation methodology must be well understood. In this study, theoretical modeling of reflectance pulse oximetry was undertaken using the three-dimensional photon diffusion theory whose applicability to oximetry in whole blood and tissue has been well established (Takatani and Graham, 1979; Reynolds et al, 1976; Cohen and Longini, 1971). The reflectance pulse oximeter sensor was then designed, tested in animals, and used in the critically ill patients who were undergoing open heart and lung surgery.

REFLECTANCE PULSE OXIMETRY THEORETICAL MODELING

In modeling the reflectance pulse oximetry, the three-dimensional photon diffusion theory was used. For the sensor geometry of Fig. 1, the diffuse reflectance equation was

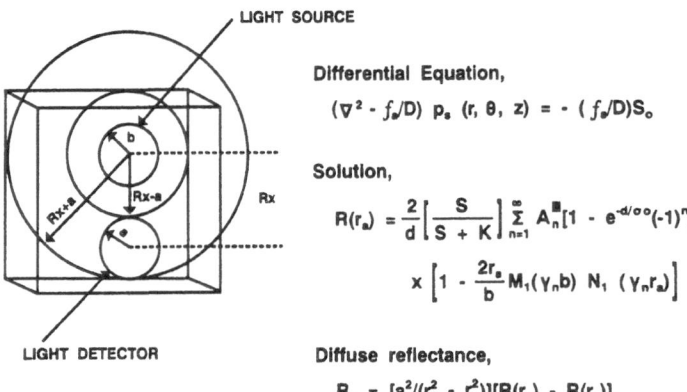

LIGHT SOURCE

Differential Equation,

$$(\nabla^2 - f_a/D)\, p_s\,(r,\,\theta,\,z) = - (f_a/D)S_o$$

Solution,

$$R(r_s) = \frac{2}{d}\left[\frac{S}{S+K}\right] \sum_{n=1}^{\infty} A_n^{\blacksquare}[1 - e^{-d/\infty}(-1)^n]$$

$$\times \left[1 - \frac{2r_s}{b} M_1(\gamma_n b)\, N_1\,(\gamma_n r_s)\right]$$

LIGHT DETECTOR

Diffuse reflectance,

$$R_d = [a^2/(r_2^2 - r_1^2)][R(r_2) - R(r_1)].$$

Fig. 1. Sensor model and diffuse reflectance equation.

described earlier by Takatani (1989). Equations 1 through 6 from Takatani were used to derive diffuse reflectance from tissue in terms of S_aO_2, S_vO_2, arterial and venous hemoglobin content ($[Hb]_a$, $[Hb]_v$). In addition, venous hemoglobin oxygen saturation (S_vO_2) was expressed as a function of S_aO_2, total hemoglobin content ($[Hb]$), blood flow (Q) and tissue metabolism (V_{O2}) by the following equation;

$$S_vO_2 = S_aO_2 - V_{O2} / \{ 1.32cc/g \times [Hb] \times Q \} \qquad (1)$$

Thus, the Equation (1) allows to include the effect of the local blood flow and metabolism on the S_aO_2 measurement. The effects of sensor design including wavelength selection and sensor geometry, physiological parameters such as arterial-to-venous blood volume distribution, venous saturation, blood flow, tissue metabolism, tissue optical characteristics were studied to optimize the sensor design and S_aO_2 computation algorithm.

The theoretical results revealed that dual wavelengths, 665 and 820 nm, in combination with the following computation algorithm will yield the best results (Fig. 2);

$$S_aO_2 = A \times Ratio + B, \qquad (2)$$

where A and B are the constants that depend on the sensor geometry and physiological parameters, and Ratio is given by;

$$Ratio = (AC/DC)_{665nm} / (AC/DC)_{820nm}$$

with AC and DC being defined as the pulsatile amplitude and average level of the signals at each wavelength. This methodology can also minimize the effects of physiological variables such as arterial-to-venous distribution, venous saturation, tissue optical characteristics.

REFLECTANCE PULSE OXIMETER SYSTEM

Optical Sensor
Fig. 3a and 3b show the prototype optical sensor and its

schematics. The optical sensor consists of four light emitting diode (LED) chips for each wavelength spaced equally around the photodiode chip to sample tissue spectra. Four chips for each wavelength were employed to enhance signal-to-noise ratio and to average out inhomogeneous effect of tissue. An optical barrier was placed between the LED and photodiode to prevent direct coupling effect and control the detection depth in tissue. The current sensor can detect the light returning from 0.5 mm or deeper in the tissue. This design, thus, minimizes the multiple scattering effect of the shallow layer,

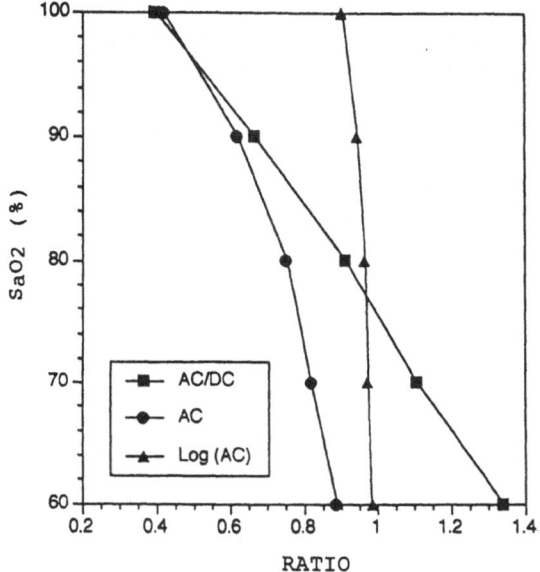

Fig. 2. Comparison of three S_aO_2 computation algorithms. AC/DC model gives the best linear relation with respect to S_aO_2.

particularly at the vicinity of the light source and detector and measures diffusely scattering light in the deeper region. In addition to measuring wavelengths of 665 and 820 nm, near-infrared wavelength (940 nm) LEDs were also included in the sensor; these are used to warm the tissue in case the tissue-sensor interface temperature decreases. A thermocouple was mounted at the sensor surface to continuously monitor the sensor-tissue interface temperature. The dimension of the sensor is approximately 1 cm square and it is placed at the center of a plexi-glass mount (3 cm diameter) whose main purpose is to prevent heat loss to the ambient.

Oximeter System
    Fig. 4 shows the block diagram of the oximeter system. The LEDs are excited sequentially with a narrow width pulse (10 micro-second) at 1 kHz. The measured reflectance from tissue is

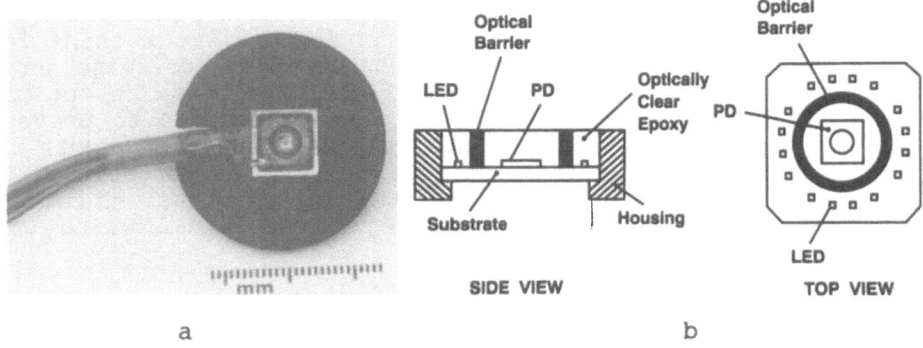

Fig. 3. Prototype reflectance pulse oximeter sensor (a) and
schematic diagram of the sensor (b).

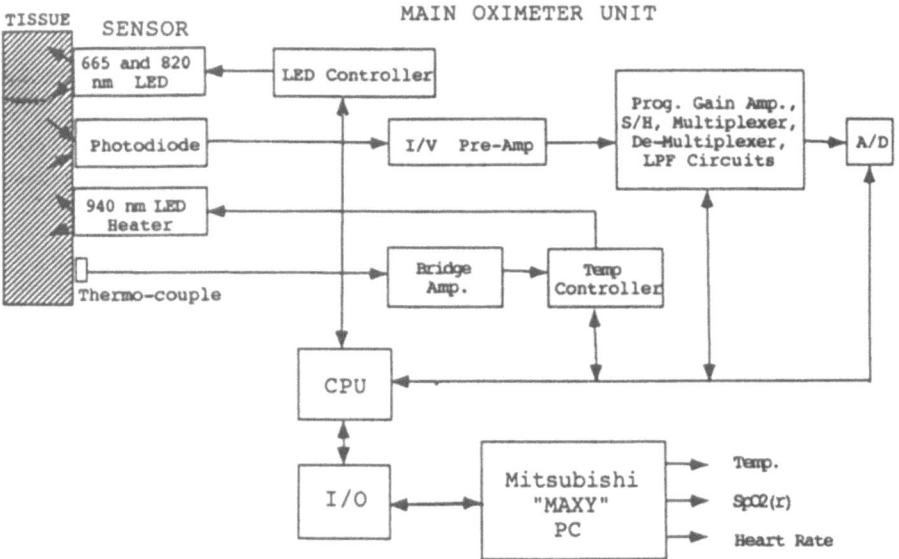

Fig. 4. Block diagram of the reflectance pulse oximeter control
and data acquisition system.

amplified, smapled, and AC and DC components of each wavelength
are computed. A personal computer, Maxy (Mitsubishi), is used
to compute the S_aO_2 on-line using the Equation (2) with pre-
determined A and B constants.

RESULTS AND DISCUSSION

Prior to clinical applications, the prototype sensor was
evaluated in animals. The mongrel dogs with weight ranging from
10 to 20 Kg were anesthetized and their respiration was
controlled. The femoral artery was cannulated to obtain blood
samples and to measure the S_aO_2 and [Hb] using an IL-282

bench-top CO-Oximeter. A commercially available transmission pulse oximeter was placed on the earlobe to obtain a reference S_aO_2. The oxygen content of the gas ventillating the dogs was varied to alter S_aO_2. Fig. 5 shows the correlation plot between the S_aO_2 measured by the reflectance pulse oximeter vs. those by the IL-282 CO-Oximeter.

For clinical evaluation of the sensor, the critically ill patients who might have low S_aO_2 such as lung and heart diseases were selected. After obtaining the consent from the patients, the sensor was placed on the forehead or cheek with a double-sided tape where the best pulsatile signals were obtainable.

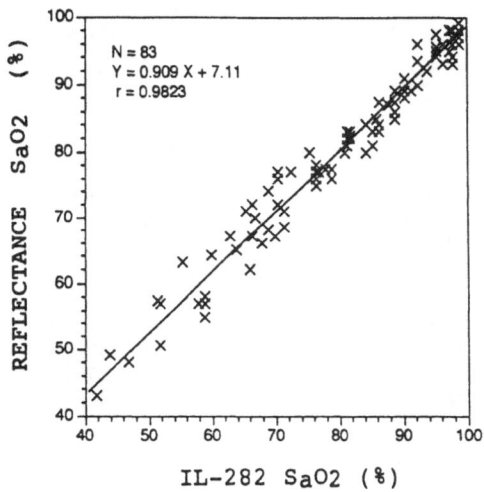

Fig. 5. Correlation plot between the S_aO_2 by the reflectance pulse oximeter and those by the IL-282 CO-Oximeter in dogs.

A total of 18 patients were studied in the operating room over the duration of 4-5 hours; of the patients studied, two were black and one was a 4 day old infant. Fig. 6 shows the correlation plot between the S_aO_2 measured by the reflectance pulse oximeter vs. those analyzed by the IL-282 CO-Oximeter. The errors may have developed during blood sampling, because the patient's S_aO_2 were unstable. Although the better accuracy was obtained in dog experiments, dog experiment was well controlled and S_aO_2 was stable during blood sampling. Concerning the application site, cheek showed better signal level and stability over 4-5 hour monitoring. As for the sensor-skin interface temperature and signal level, when the interface temperature was maintained at or above 35 C, usually there was no problem in detection of signal from the cheek. Since the head is exposed to the anesthesiologist during surgery, application of the sensor is easier in comparison to the fingertip. Also, since the circulation to the cheek or forehead is related to the circulation to the head, the

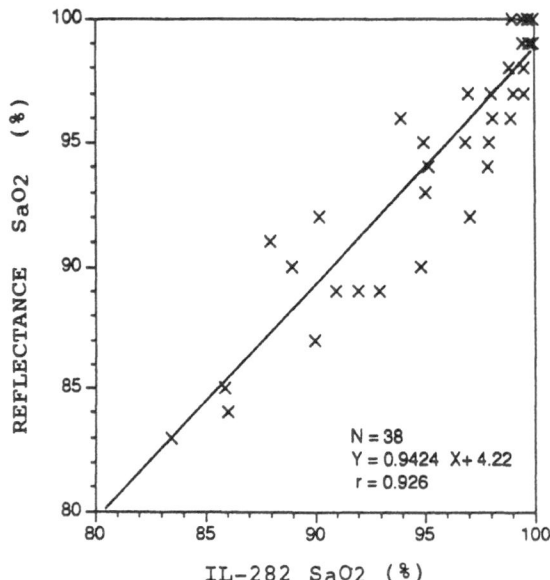

Fig. 6. Correlation plot between the S_aO_2 by the reflectance pulse oximeter and those by the IL-282 CO-Oximeter in critically ill patients.

monitoring from the head area may be indicative of the circulation to the vital organ. The reflectance pulse oximeter can be a powerful alternative to the transmission oximeter in the operating room and intensive care unit.

ACKNOWLEDGEMENT

This research was partially supported by a grant in aid from the Colin Electronics Inc., Komaki, Japan.

REFERENCES

Cohen, A. and Longini, R.L., 1971, Theoretical determination of the blood's relative oxygen saturation in vivo. Med Biol Eng 9:61-69.

Mendelson, Y., Cheung, P.W., Neuman, M.R., Fleming, D.G. and Cahn, S.D., 1983, Spectrophotometric investigation of pulsatile blood flow for transcutaneous reflectance oximetry. Oxygen Transport to Tissue VI:93-102.

Mendelson, Y., Kent, J.C., Yocum, B.L. and Birle, J., 1988, Design and evaluation of a new reflectance pulse oximeter sensor. Medical Instrumentation 22(4):167-173.

Mendelson, Y. and Ochs, B.D., 1988, Noninvasive pulse oximetry utilizing skin reflectance photoplethysmography. IEEE Trans Biomed Engr 35(10):798-805.

Pologe, J.A., 1987, Pulse oximetry; Technical aspects of machine design. Int Anesthesiol Clin 25:137-153.

Severinghaus, J.W. and Naifeh, K.H., 1987, Accuracy of response of six pulse oximeters to profound hypoxia. Anesthesiology 67:551-558.

Reynolds, I.O., Johnson, C.C., and Ishimaru, A., 1976, Diffuse reflectance from a finite blood medium; Application to the modeling of fiber optic catheters. Appl Opt 15:2059-2067.

Takatani, S. and Graham, M.D., 1979, Theoretical analysis of diffuse reflectance from a two-layer tissue model. IEEE Trans Biomed Engr 26(12):656-664.

Takatani, S., 1989, Toward absolute reflectance oximetry: I. Theoretical consideration for noninvasive tissue reflectance oximetry. Oxygen Transport to Tissue XI:91-102.

Yelderman, M. and New W., 1983, Evaluation of pulse oximetry. Anesthesiology 59:349-352.

BIOLOGICALLY ACTIVE CYANINE DYES AS PROBES FOR THE

IDENTIFICATION OF ACTIVE OXYGEN SPECIES

Hitoshi Hori,[1]   Yoshinori Nakagawa,[2]   Hiroshi Ojima,[3]
Takehiro Niijima,[4] Hiroshi Terada[5]

[1]Department of Biological Science and Technology, Faculty of
 Engineering, University of Tokushima, Tokushima 770,
 Japan
[2]Nippon Kankoh-Shikiso Kenkyusho Co.,LTD,Okayama 700, Japan.
[3]Pearl Co., LTD, Tokyo, 135, Japan
[4]Harajuku Immunity Center, Tokyo, Japan
[5]Faculty of Pharmaceutical Sciences, University of
 Tokushima, Tokushima 770, Japan

INTRODUCTION

Active oxygen species are involved in both normal cellular processes
and in a wide spectrum of pathologies. Hydroxyl radicals ($\cdot$OH) are highly
reactive oxygen species that are likely to cause cellular damage.
Superoxide($\cdot O_2^-$) and hydrogen peroxide(H_2O_2) are frequently formed in
biological systems and participates in reactions that produce $\cdot$OH, if iron
ions are present. Cyanine dyes are known to have biologically acitve
properties, the mechanism of which have been suggested to involve oxygen-
derived species.[1-3] To estimate the possibility of using cyanine dyes as
probes to identify these active oxygen species, especially $\cdot$OH and $\cdot O_2^-$,
we studied the bleaching of the dyes in a controlled Fenton reaction
[Fe(II) and H_2O_2]. The Fenton reaction was originally proposed as a
chemical source of $\cdot$OH and also as a biological source of hydroxyl
radicals and a likely cause of free radical-induced tissue damage.[4,5]

MATERIALS AND METHODS

The cyanine dyes (NK-2, 13, 19, and 24) shown in Fig. 1 were
synthesized in our laboratory as described previously.[1] Other compounds
used were obtained from commercial sources.

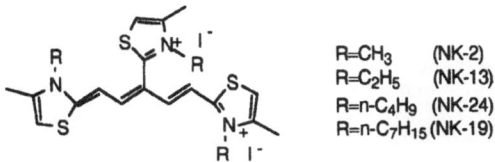

R=CH$_3$ (NK-2)
R=C$_2$H$_5$ (NK-13)
R=n-C$_4$H$_9$ (NK-24)
R=n-C$_7$H$_{15}$ (NK-19)

Fig.1. Chemical structures of cyanine dyes.

Oxygen Transport to Tissue XIV, Edited by W. Erdmann and
D.F. Bruley, Plenum Press, New York, 1992

Bleaching of dyes by irons and H_2O_2 (Fenton reagnet): All experiments were carried out at 25°C in the dark.

Bleaching of dyes by Fe(II) and H_2O_2: The reaction mixtures contained, in a total volume of 2.22ml, 4.95μM dye dissolved in methanol, 0.1ml of various concentrations of Fe(II) in sodium citrate (4.5mM, pH 3.5), 0.1ml of various concentrations (in excess to [Fe(II)]) of H_2O_2 and 2.0ml of water. The reaction was initiated by adding H_2O_2 and dye bleaching was measured as change of absorbance at 590nm . The methanol in the reaction mixture didn't have any afffect on the bleaching of dyes. *Bleaching of dye by the secondary Fenton reaction products (Run A):* The reaction mixtures contained, in a total volume of 2.22ml, 4.95μM dye in methanol, 0.45mM Fe(II) in sodium citrate (4.5mM, pH 3.5), 4.5mM H_2O_2 and 2.0ml water. After 2-min incubation of the mixture of Fe(II) and H_2O_2, the dye was added and its bleaching was measured as described above.

Bleaching of dye by Fe(III) and H_2O_2 (Run B): The reaction mixtures contained, in a total volume of 2.22ml, 4.95μM dye, 0.45mM Fe(II) in sodium citrate (4.5mM, pH 3.5), 4.5mM H_2O_2 and 2.0ml of water. The dye was incubated with Fe(II) for 30 sec., the reaction was initiated by adding H_2O_2 and dye bleaching was measured at 590nm.

Bleaching of dye by the xanthine-xanthine oxidase system[6]: The reaction mixtures contained, in total volume of 2.12ml, 5.19μM dye in phosphate buffer (0.47mM, pH 7.8), 94μM xanthine in phosphate buffer (0.47mM, pH 7.8) and 0.16 units of xanthine oxidase solution. The reaction was initiated by adding the xanthine oxidase and the absorbance change of the dye at 590nm was monitored. The effect of superoxide dismutase(SOD)(85 units/ml) to xanthine-xanthine oxidase reaction of the dye was measured.

Detection of ·OH by measuring hyaluronic acid degradation[7]: The reaction mixtures contained , in a total volume of 8.21ml, 2.4mM H_2O_2, either 12μM Fe(II) or 24μM Fe(III) solution in sodium citrate (0.12mM, pH 3.5) and 0.09% hyaluronic acid in saline at room temperature. The viscosity was measured in an Ostwold viscosimeter and determined from the time (in sec) required for passage of 4ml of reaction mixture.

RESULTS

Bleaching of dyes by Fenton's reagent. Studies were made on the bleaching of dyes by the Fenton's reagent [Fe(II) and H_2O_2] , which generated hydroxyl radicals quantitatively by Fenton reaction (reation 1).

$$Fe^{2+} + H_2O_2 \rightarrow Fe^{3+} + \cdot OH + OH^- \qquad [1]$$

When 2.25mM H_2O_2 was added to a mixture of 4.95μM dye NK-19 and 22.5mM Fe(II) with at pH 3.5 in the dark, bleaching was instantaneous and complete, as shown in Fig.2. Decrease in the concentrations of either component of Fenton's reagent, Fe(II) or H_2O_2, in the presence of a large excess of [H_2O_2] over [Fe(II)] resulted in two phase-bleaching, first rapid bleaching (normal Fenton reaction) and the gradual bleaching dependent on secondary Fenton reaction products (Fig.2). The bleaching of other dyes were similar to that of NK-2.

For examination of the second gradual phase of bleaching, the dye was added after mixing of Fe(II) and H_2O_2 (run A). In this condition, only the gradual bleaching occurred, as shown in Fig.3.

The bleaching of dyes with shorter side-chain was faster, the order of the rate of bleaching being NK-2 =13 >24 >19. This run A-bleaching was probably caused by reactive radicals formed through the further reaction of Fe(III) and H_2O_2. When the dye was added to the Fenton reaction mixture of Fe(III) and H_2O_2 at pH 3.5 (run B), results were similar to those in run A, but the side-chain effect was greater.

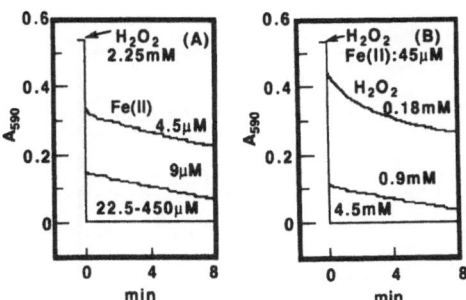

Fig.2. Effects of the concentrations of Fe(II) (A) and H_2O_2 (B) on the
    bleaching of dye NK-19 by the Fenton reaction. (A) See *Bleaching
    of dyes by Fe(II) and H_2O_2*' in 'Methods' for experimental
    details.

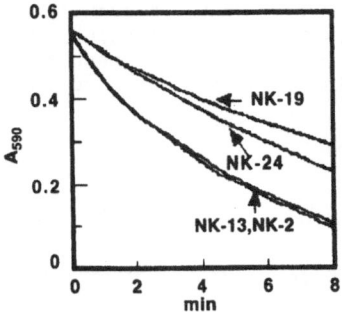

Fig.3. Bleaching of dyes by secondary Fenton reaction products of Fe(II)
    and H_2O_2 (Run A). See 'Methods' for experimental details.

Bleaching reaction of dye by superoxide. To determine whether or not superoxides were involved in runs A and B, we checked the bleaching of dyes in xanthine-xanthine oxidase system which generates superoxide(pH 7.8). Fig. 4 shows that addition of xanthine oxidase to a mixture of xanthine and NK-19 caused decrease in the absorbance at 590nm to a constant level after 1 min and SOD inhibited this bleaching. It suggests the involvement of superoixdes in the bleaching. The bleaching of NK-2 with a shorter side chain was more rapid than that of the NK-19 with a longer side chain.

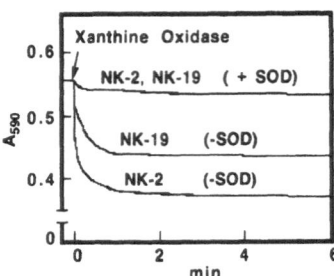

Fig.4. Bleaching of dye by superoxide generated by a xanthine-xanthine oxidase system. See 'Methods' for experimental details.

Identification of the hydroxyl radical in Fenton reaction mixtures

We used hyaluronic acid to examine whether ·OH was involved in runs A and B. Decrease in viscosity was observed in the Fenton reaction mixture containing H_2O_2 and Fe(II) (Fig.5a,b), but not in that containing H_2O_2 and Fe(III) (Fig.5c, d). These results indicate that hydroxyl radicals were produced in run A, but not in run B.

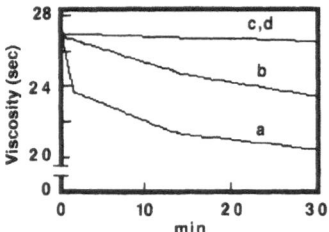

Fig.5 Effects of Fenton reaction products on hyaluronic acid degradation. (a) Fe(II) + hyaluronic acid, then H_2O_2; (b) Fe(II) + H_2O_2 , then hyaluronic acid; (c) Fe(II) + H_2O_2 + dipyridyl, then hyaluronic acid; (d) Fe(III) + hyaluronic acid, then H_2O_2. See 'Methods' for experimental details.

DISCUSSION

Iron ions are concerned to the generation of active oxygen species such as $\cdot OH$, $\cdot O_2^-$, and $^1O_2$ in biological systems. We examined the possibility of using cyanine dyes as probes to identify the active oxygen species generated by the Fenton reaction [Fe(II) or Fe(III) / H_2O_2]. The hydroxyl radical ($\cdot OH$) was produced by the Fe(II) catalyzed reaction of H_2O_2 (reaction 1) (i.e. the normal Fenton's reaction). Fig.2 shows that the instant bleaching of cyanine dyes occurred in two phase, first rapidly and then gradually. The first rapid phase was due to scavenging $\cdot OH$, and occurred at the same rate with the cyanine dyes examined. The latter phase alone was observed by controlling the Fenton reaction condition (run A) as shown in Fig.3, and in this phase dyes with shorter side chains were bleached faster than those with longer ones.

As the second phase of bleaching of dyes by Fe(III) and H_2O_2 (run B) took place continuously and gradually in the same manner as in run A, this gradual bleaching was probably due to secondary Fenton reaction products derived from $\cdot OH$, Fe(III) and H_2O_2. To ascertain this assumption we checked the bleaching of dyes by Fe(III) and H_2O_2 (run A). The bleaching of dyes in the presence of a xanthine-xanthine oxidase system was similar to that in the controlled Fenton reaction (i.e., run A) and this bleaching was inhibited by SOD (Fig.4). These findings indicate that the cyanine dyes scavenged $\cdot O_2^-$ as well as $\cdot OH$.

As an alternative method for detection of the hydroxyl radical we studied the degradation of hyaluronic acid by monitoring decrease in viscosity. Fig.5 shows that the hydroxyl radical was generated in run A (line b) as well as in the normal Fenton reaction (line a) but was not generated in run B (line d). The finding that addition of dipyridyl to the reaction mixture in run A presented hyaluronic acid degradation also indicates that the presence of the hydroxyl radical was formed in the reaction mixture. Moreover the above results suggest that superoxide was generated in run B.

The bleaching of cyanine dyes by the controlled Fenton reaction can be summarized as follow. In run A, the hydroxyl radical generated by reaction 1 reacts with H_2O_2 to produce $\cdot O_2^-$ (reaction 2), and then $\cdot O_2^-$ reduces Fe(III) to yield Fe(II) (reaction 3).

$$\cdot OH \quad + \quad H_2O_2 \quad \rightarrow \quad \cdot O_2^- \quad + \quad H_3O^+ \qquad [2]$$

$$\cdot O_2^- \quad + \quad Fe^{3+} \quad \rightarrow \quad O_2 \quad + \quad Fe^{2+} \qquad [3]$$

This Fe(II) then recycles to generate $\cdot OH$ by the Fenton reaction (reaction 1). Thus, the combined reactions ([1] + [3]) can be presented by reaction 4, which is sometimes called as iron catalyzed Haber-Weiss reaction.[8]

$$\cdot O_2^- \quad + \quad H_2O_2 \quad \rightarrow \quad O_2 \quad + \quad \cdot OH \quad + \quad OH^- \qquad [4]$$

We conclude that $\cdot O_2^-$ and $\cdot OH$ are the reactive oxygen species in run A. Run B may generate the superoxide, as we obtained no evidence of formation of Fe(II) or $\cdot OH$ in this condition. Generally Fe(III) reacts with H_2O_2 to form the superoxide according to the following reaction (reaction 5).[4]

$$Fe^{3+} + H_2O_2 \quad \rightarrow \quad Fe^{2+} + \cdot O_2^- + 2H^+ \qquad [5]$$

However, the rate constant for this reaction is known to be very slow ($\approx 10^{-3}$ M^{-1} s^{-1})[4] , and this reaction may be represented as reactions 3 and 6.

$$H_2O_2 \quad + \quad O_2 \quad \rightarrow \quad 2 \cdot O_2^- \quad + \quad 2H^+ \quad\quad [6]$$

In the dark, we consider that the complex $[Fe^{3+}-O_2^-]$ was formed from traces of $\cdot O_2^-$ and large amounts of Fe(III) (reaction 7) rather than that of Fe(II) was generated by reaction 3 to promote reaction 6.

$$\cdot O^{2-} \quad + \quad Fe^{3+} \quad \rightleftharpoons \quad Fe^{3+}-O^{2-} \quad\quad [7]$$

In fact we found by trapping with dipyridyl that Fe(II) ions were generated in this reaction of Fe(III) and H_2O_2 in the light(data were not shown here). Thus reaction 5 can take place in normal light conditions, although it probably does not occur in the dark as reaction 3 is unlikely to occur.

Our results showed that the bleaching profile of cyanine dyes could be changed by varying the reaction conditions. This finding indicates the possibility of using these cyanine dyes as probes to identify the active oxygens species involved in pathological and normal biological phenomena such as lung injury,[5] carcinogenesis,[9] and aging.[10]

REFERENCES

1. T.Murofushi, Fundamental studies on antimicrobial activities of photosensitizing dyes having the characteristics of food preservatives. Part 1. Studies on antimicrobial activities of 148 sorts of photosensitizing dyes against various microorganisms in relation to their chemical structures. Med. Res. Photosensitizing Dyes 58:1-25 (1959).
2. K.Fukuzumi and N.Ikeda The effect of photosensitizing dyes as antioxidants on autoxidation of methyl linoleate. J. Am. Oil Chemists. Soc. 48:384-386 (1972).
3. I.Yamamoto, S.Ohkuma, and M.Naitoh, Platonin, a photosensitive cyanine dye as immunomodulator. Int. J. of Immuno-pharmacology 4:312 (1982).
4. C. Walling, Fenton's reagent revisited. Acc. Chem. Res. 8:125-131 (1975).
5. B. Halliwell and J.M. Gutteridge, Oxygen Free Radical and Iron in Relation to Biology and Medicine: some Problems and Concepts. Arch. Biochem. Biophys. 246: 501-504 (1986).
6. McCord and I.Fridovich, Superoxide dismutase. An enzyme function for erythrocuprein (hemocuprein). J. Biol. Chem. 244 : 6049-6055 (1969).
7. J. M. McCord, Free radicals and Inflammation: Protection of synovial fluid by superoxide dismutase. Science 185: 529-531(1974) .
8. M. McCord and E. D. Jr. Day, Superoxide-dependent production of hydroxyl radical catalyzed by iron-EDTA complex. FEBS Lett. 86: 139-142 (1978).
9. P.A. Cerutti, Prooxidation States and Tumor Promotion. Science 227: 375-381(1985).
10. R. Adelman, R.L.Saul, and B.N.Ames, Oxidative Damage to DNA: Relation to species metabolic rate and life span. Proc. Natl. Acad. Sci. USA 85: 2706-2708 (1988).

ANALYSIS OF MULTIPLE MULTIPOLE SCATTERING BY TIME-RESOLVED
SPECTROSCOPY AND ANGULAR DEPENDENT SPECTROMETRY

Frank K, Kessler M, Wiesner J[1], Wokaun A[1]

Institut für Physiologie und Kardiologie, Universität Erlangen-Nürnberg
[1]Institut für Physikalische Chemie II, Universität Bayreuth

1. INTRODUCTION

Molecules, subcellular particles and cells dissolved in aqueous solutions scatter light. The number of scattering events in such suspensions depends upon the number of particles per volume as well as the size, shape, and refractive index of these elements.

The high density of subcellular particles within tissues of humans and mammals should cause predominantly multiple multipole scattering. The resulting physical effects strongly influence the angular dependent scattering characteristics (Frank et al. 1989, 1991) and the depth of penetration of light (Delpy et al 1988, Chance et al. 1988, 1990).

During the past few years the activity in the development of optical systems for tissue measurements increased. The result was a number of instruments, based upon contineous wave and time resolved techniques, applicable to measurements in micro environments of capillaries and cells (Frank et al. 1989) and in macro volumes of whole organs (for review see: Chance et al. 1988).

To understand the influence of the optical properties modulating the light in tissues experiments were performed using red cells, microspheres (Frank et al. 1989), mitochondria suspensions (Frank et al. 1991), yeast and milk (Chance et al. 1989) as well as intralipid suspensions (Jacques and Flock 1991).

In parallel mathematical methods like Monte Carlo simulation (Delpy et al. 1988, van de Zee and Delpy 1988) and photon diffusion theory (Bonner et al. 1987) were applied. Based upon the photon diffusion theory Patterson et al. (1989) derived a formula, which enabled the calculation of absorption and scattering coefficients for macro volume measurements.

Due to the fact, that these calculations are only valid for a large number of scattering events, only an integral information about the interaction of light with tissues is gained. No information about the environment of few and single cells can be extracted. Therefore we initiated measurements in mitochondrial suspensions using contineous wave spectrometry and time resolved spectroscopy. Our aim was to analyze the events of optical scattering found in tissues by systematic investigation of the basic physical phenomena.

In the present study the influence of size and concentration changes in mitochondrial suspensions were investigated. Effects of other subcellular particles and structures as well as the influence of changes of the refractive index were not taken into consideration.

2. METHODS

2.1 SCATTERING CHAMBER

A windowless scattering chamber (Frank et al.1989) combined with the EMPHO (Frank et al. 1989) was used for the determination of the scattering diagrams. The light from a Xenon high pressure arc lamp illuminated the mitochondrial suspension via a lightguide fiber (diameter 250 µm). A collecting fiber (diameter 250 µm) received the scattered light and transfered it to the detection unit of the EMPHO. Measurements were carried out from backward (178°) to forward (0°) direction by moving the detecting lightguide in steps of 2° around the center of the scattering chamber. The distance between the two lightguides in the forward direction was 1 mm. The lightguide fibers were dipped directly into the suspension.

2.2 INTENSITY FIELDS

An xy-scanning device in combination with the EMPHO was used for the measurements of the light intensity fields in the mitochondrial suspensions. The probe was illuminated by a 250 µm lightguide. By use of a 50 µm fiber, which was mounted in an angle of 90° with respect to the illuminating fiber, the intensity profiles were mapped. The detecting fiber was moved in steps of 50 µm in x respectivly y direction. During the measurements the cuvette was cooled to 4^oC.

The calibration of the measured light intensities was performed by use of a reference light source. The intensity of the light source at 600 nm was set to one intensity unit [IU].

2.3 TIME-RESOLVED TECHNIQUE

For the time-resolved measurements a modified version of the instrument described before (Frank et al. 1991) was used. In order to increase the measuring volume of the system the laser light was coupled into a lightguide fiber with a diameter of 250 µm, which served for illumination of the probe. A second identical fiber faced the illuminating fiber in a variable distance and was used for detection. The total length of the two lightguides was 2.5 m. The reference arm of the instrument was compensated for the physical light path with a similar lightguide fiber (length 2.5 m). The time resolution of the combined system corresponded to 200 ps.

The scattered radiation was detected by degenerated frequency upconversion with a delayed probe pulse.

2.4 MITOCHONDRIA PREPARATION

The mitochondrial suspensions were prepared as described before (Frank et al. 1991). To separate mitochondrial fractions of different size the homogenate was centrifuged at 1000 g. The supernatant was treated by differential centrifugation at 3000 g and 10000 g. This procedure resulted in two fractions containing mitochondria greater 1 µm and smaller 1 µm, respectively.

After the centrifugation steps the respective sediments obtained were set to 100 % and thus corresponded to the initial concentration of mitochondria. The weight per volume was determined before filling the isolated mitochondria into the scattering cuvettes. This latter value is proportional to the ratio of liver weight/liver volume.

The size of the particles was microscopically controlled by comparing them with LATEX beads (LB6, LB8, and LB30, SIGMA).

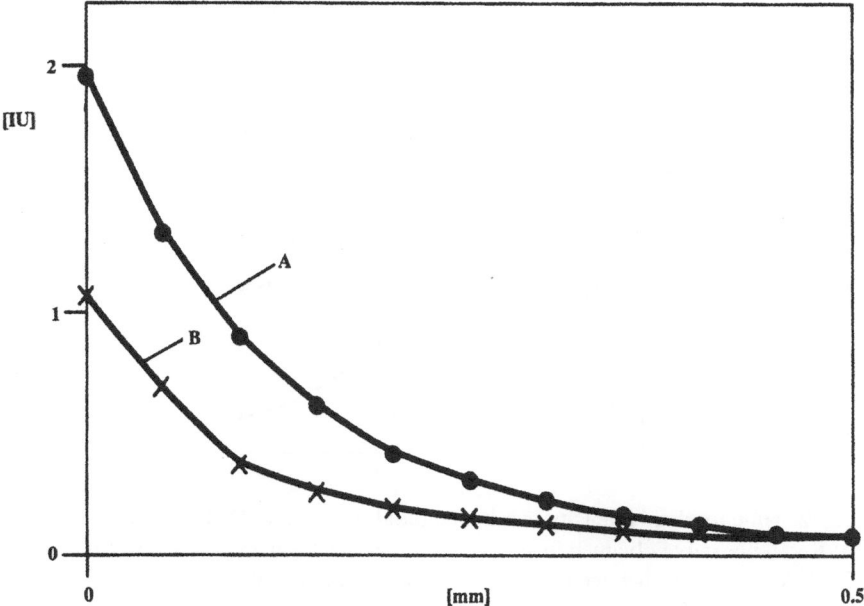

Fig. 1. Distance dependent intensity curves obtained in mitochondrial suspensions. The wavelength used was 600 nm.
A: Diameter > 1 µm, c=83%.
B: Diameter < 1 µm, c=88%.

3. RESULTS

3.1 INTENSITY FIELDS

The results of the measurements performed in mitochondrial suspensions are shown in figure 1. The calibrated light intensity unit [IU] at 600 nm is plotted as a function of the distance of the illuminating and detecting micro-lightguides. Curve A shows the result obtained in a suspension containing 83 % of mitochondria greater than 1 µm. An increase of the distance of the two lightguides leads to a decrease of the light intensity. When the measurements are performed with mitochondria, containing 88 % of particles smaller than 1 µm, curve B results.

3.2 TIME-RESOLVED MEASUREMENTS

Figure 2 shows the results of a time-resolved measurement in a suspension containing mitochondria with a diameter greater than 1 μm. Curve A represents the reference curve obtained in pure sucrose solution. The difference between the mitochondrial suspension (c=40 %) and the reference curve is plotted in curve B. The time delay between the reference peak and the peak in the difference curve is 102 ps.

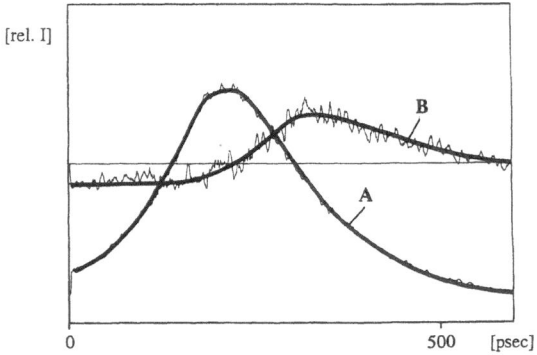

Fig. 2: Time resolved intensity in mitochondrial suspension. Distance 200 μm.
A: Reference curve obtained in sucrose solution. The wavelength used was 600 nm.
B: Difference curve between the time curve obtained in a mitochondrial suspension containing particle > 1 μm (c=40 %) and curve A.

4. DISCUSSION

4.1 INTENSITY FIELDS

Our recent experiments performed in highly concentrated mitochonderial suspensions show that pronounced effects are induced when the particle concentration or the particle size are changed. For the interpretation of these phenomena the basic physical interaction between light waves and the particles must be considered. Calculations based upon Mie's theory for single scattering predicted a tremendeous increase in forward and backward directed intensities and a narrowing of the radiation pattern as function of particle size. The results of our experiments indicate that none of the effects predicted from single scattering calculations are found.

These unexpected results can be explained as follows:

Within the concentration range of mitochondria used in this study the density of the subcellular particles is such that they are more or less in contact with each other. In practical terms this means that the propagating wave hits a series of particles almost simultaneously. This results in an angular radiation with high intensities in the lateral and backward direction as compared to the single multipole scattering defined by Mie. A small increase in the size of mitochondria causes a significant increase in light intensity.

4.2 TIME-RESOLVED MEASUREMENTS

In our experiments a time delay of 100 ps was observed. This indicates, that the pathlength of the light was increased by a factor of 160. This macroscopic effect is not explainable by use of theoretical considerations for the pathlength increase due to scattering. *In vivo* measurements in human skeletal muscle (Chance et a 1. 1988) and rat brain (Delpy et al. 1988) resulted in an increase by factor 5 to 10.

A possible explanation for the effect found in our micro volume measurements could be that adsorbed mitochondria form an optically active cladding at the surface of the lightguide fibers acting as a semi transparent mirror. The light, which leaves the illuminating fiber perpendicular to the plain of the surface might be multiply reflected at the surfaces of the fibers. This might be responsible for the large increase in pathlength. The system would then act like a multiple-path cell, which might be a convenient tool for the investigation of light particle interactions in small measuring volumes.

5. REFERENCES

Chance B, Leight J S, Miyake H, Smith D S, Nioka S, Greenfeld R, Finnander M, Kaufmann K, Levy W, Young M, Cohen P, Yoshioka H, Brotsky R, 1988, Comparison of time resolved and unresolved measurements of deoxyhemoglobin in brain, Proc.Natl.Acad.Sci.USA, 85:4971-4975.

Chance B, Nioka S, Kent J, McCully K, Fountain M, Greenfeld R, Holtom G, 1988, Time-resolved spectroscopy of hemoglobin and myoglobin in resting and ischemic muscle, Analytical Biochemistry 174:698-707.

Chance B, Smith D S, Nioka S, Hiyake H, Holtom G, Maris M, 1989, Photon migration in muscle and brain, in Photon Migration in Tissues, ed. B Chance, 121-135.

Chance B, Maris M, Sorge J, Zhang M Z, 1990, A phase modulation system for dual wavelength difference spectroscopy of hemoglobin deoxygenation in tissues, SPIE, 1204.

Delpy D T, Arridge S R, Cope M, Edwards D, Reynolds E O R, Richardson C E, Wray E, Wyatt J, van de Zee P, 1988, Quantitation of pathlength in optical spectroscopy, Adv.Exp.Med.Biol, 222:191-198.

Delpy D T, Cope M, van de Zee P, Arridge S, Wray S, Wyatt J, 1988, Estimation of optical pathlength through tissue from direct time of flight measurements, Phys.Med.Biol., 33:1433-1442.

Frank K H, Kessler M, Appelbaum K, Dümmler W, 1989, The Erlangen micro-lightguide spectrophotometer EMPHO I, Phys.Med.Biol., 34:1885-1900.

Frank K H, Kessler M, Appelbaum K, Albrecht H P, Mauch E D, 1989, Measurements of angular distribution of Rayleigh and Mie scattering events in biological models, Phys.Med.Biol, 34:1901-1916.

Frank K, Höper J, Zündorf J, Tauschek D, Kessler M, Wiesner J, Wokaun A, 1991, Analysis of multiple multipole scattering by time-resolved spectroscopy and spectrometry, SPIE, 1431:2-11.

Jacques S L, Flock S T, 1991, Effect of surface boundary on time-resolved reflectance: measurements with prototype endoscopic catheder, SPIE, 1431:12-20.

Maris M, Mayevsky A, Evick E, Chance B, 1991, Frequency domain measurements of changes of optical pathlength during spreading depression in a rodent brain model, SPIE, 1431:136-148.

van de Zee P, Delpy D T, 1988, Computed point spread function for light in tissue using a measured volume scattering function, Adv.Exp.Med.Biol., 222:191-198.

IN VIVO NADH AND Pd-PORPHYRIN VIDEO FLUORI-/PHOSPHORIMETRY

C. Ince, J.F. Ashruf, E.A. Sanderse, E.G.J.M. Pierik,
J.M.C.C. Coremans and H.A. Bruining

Department of Surgery
Erasmus University Rotterdam
Dr. Molewaterplein 40
3015 GD Rotterdam, The Netherlands

INTRODUCTION

Optical spectroscopy can be used to visualize the extent of tissue hypoxia and of blood oxygenation, using either endogenously present chromophores or artificial dyes added to the circulation. Reduced nicotinamide adenine dinucleotide (NADH) is a naturally occurring intracellular fluorophore which plays a key role in the transfer of reducing equivalents from the tricarboxylic acid cycle to the respiratory chain in the mitochondria. Inhibition of the respiratory chain due to inadequate oxygen supply is reflected by increased NADH levels. Upon illumination of tissue with ultraviolet light NADH (and not NAD^+) fluoresces in the blue. In cardiac tissue the blue fluorescence emitted from the organ surface mainly originates from the mitochondrial NADH pool. In this way NADH fluorescence provides direct information about the mitochondrial redox state. Introduced by Chance [Chance et al. 1962, Chance 1976], NADH fluorescence has been applied to a wide range of organs both in vivo and in vitro to investigate the oxygen dependence of mitochondrial oxidative phosphorylation in situ (for review of technique see [Ince et al., this volume]).

Wilson and co-workers developed the phosphorescent properties of Pd-porphyrin compounds as indicators of oxygen concentrations (for recent review see [Wilson et al., this volume]). Not only the phosphorescence intensity of these Pd-porphyrin compounds is oxygen dependent, also the time constant of the quenching of phosphorescence is directly related to the oxygen concentration as described by the Stern-Volmer relation. Hence after excitation of a Pd-porphyrin compound, measurement of the lifetimes of phosphorescence extinction enables quantitative determination of the O_2 concentration [Vanderkooi et al. 1987, Wilson et al. 1991]. When complexes of Pd-porphyrin and bovine serum albumin are injected into the blood, the oxygen probe is confined to the circulation and in vivo measurements of the microcirculatory O_2 concentration can be made [Wilson et al. 1991]. Images of the distribution of O_2 in the vasculature of intact organs can be obtained through continuous registration of the phosphorescence intensity of the Pd-porphyrin dye [Rumsey et al. 1988].

Taking the above into consideration it would be expected that combination of NADH fluorescence (oxygen in tissue) and Pd-porphyrin phosphorescence measurements (oxygen in microcirculation) in vivo would provide additional information about the factors effecting mitochondrial respiration in a continuous and non-invasive way. In this study we present a video fluorimeter based on a second generation charge-coupled-device (CCD) video camera which is suitable for in vivo observation of NADH fluorescence and Pd-porphyrin phosphorescence. Consequently, this instrument allows investigation of both intracellular O_2 demand (as reflected by the mitochondrial redox state) and O_2 distribution within the vascular system of the gut and heart of mechanically ventilated rats during anoxic and ischemic hypoxia created by N_2-ventilation and vascular ligations, respectively.

MATERIALS AND METHODS

The video fluori-/phosphorimeter (Fig. 1A) consists of an image-intensified CCD video camera with storage and processing features, coupled to an optical unit which contains the illumination source and filter combinations for selection of suitable wavelengths of excitation and emission light. The apparatus is attached to an operation microscope stand, allowing flexible positioning of the instrument above an experimental animal. Images are taken with a second generation CCD video camera (MXRi 5051 camera, HCS Vision Technology, Eindhoven, The Netherlands) in which the existing photocathode is replaced with a cathode tube with optimal sensitivity in the ultraviolet/blue

A

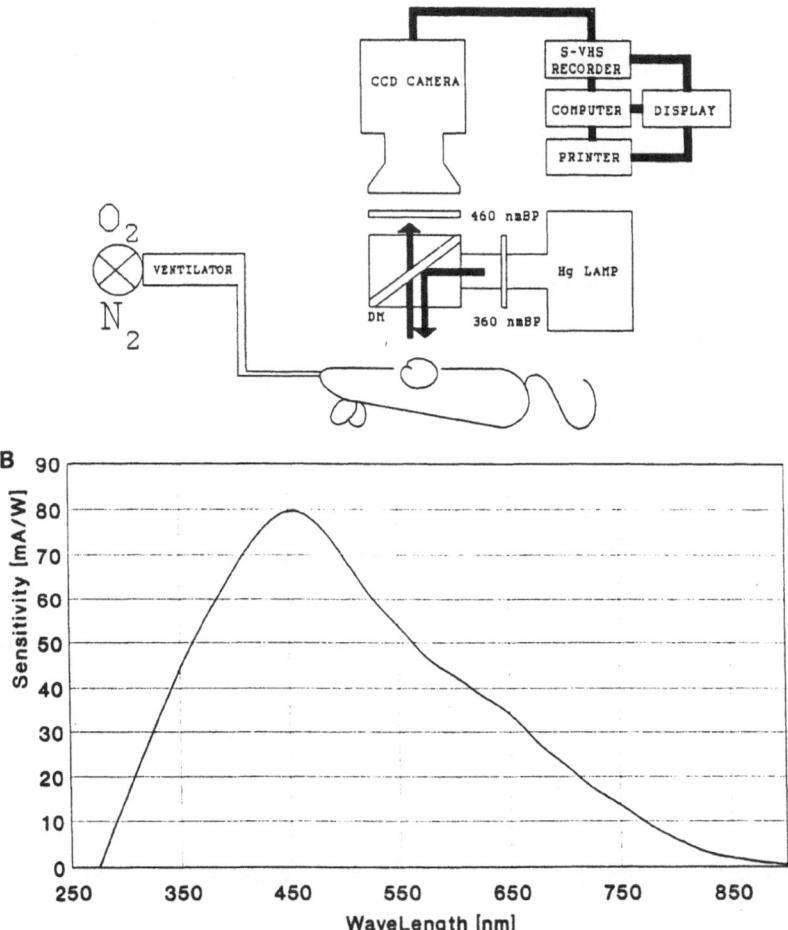

Figure 1.   A. Block diagram of the optical system of the video fluori-,phosphorimeter. The illumination unit consists of a 100 W mercury arc lamp. The filter unit consists of two sets of filters to select excitation and emission light for NADH fluorescence measurements (360 nm and 460 nm band pass (BP) filters, respectively) and for Pd-porphyrin phosphorescence measurements (545 nm band pass and 600 nm long pass filters, respectively). Excitation and emission light are separated by a dichroic mirror (DM). A second generation CCD camera fitted with a cathode tube with optimal sensitivity in the ultraviolet/blue spectral region is able to detect in vivo organ surface NADH fluorescence and Pd-porphyrin phosphorescence. Images are displayed on a monitor and stored on a S-VHS video recorder. Hard copies of the fluorescence and phosphorescence images can be obtained with a video printer and a computer can be used for image processing and quantification. B. Spectral characteristics of the image-intensifier used in the second generation CCD video camera.

spectral region (S20 Philips; Fig. 1B). The camera is fitted with a C-mount to which a 105 mm Micro-Nikkor macrolens is attached. The use of this highly sensitive second generation CCD video camera was a prerequisite for detection of the low intensity fluorescence and phosphorescence images in vivo. Images were continuously recorded on a S-VHS video recorder (Panasonic Type AG 7330) and printed on a Mitsubishi video copy processor. Images could also be computer analysed off-line for quantification of the intensity of fluorescence.

The camera is connected to a B2-RFA optical unit of an Olympus BH-2 fluorescence microscope composed of an illumination unit and a filter unit. The illumination unit consists of a 100 W mercury arc lamp which provides the excitation light for both NADH fluorescence (360 nm) and Pd-porphyrin phosphorescence (545 nm). The filter unit is composed of two switchable filter blocks; one for NADH fluorescence measurements and one for Pd-porphyrin phosphorescence measurements. Each filter block contains a band pass filter to select the desired excitation light, a dichroic mirror to separate excitation and emission light and a filter to select the emission light from the organ surface (i.e. a 460 nm band pass filter for NADH fluorescence and a 600 nm long pass filter for Pd-porphyrin phosphorescence).

Measurements were made of heart and gut surfaces of thoraco-laparotomized Wistar rats. The animals were mechanically ventilated with 98% O_2 and 2% of the anaesthetic ethrane. In order to create anoxic hypoxia, N_2 was substituted for O_2 in the respiratory gas. Ischemic hypoxia in the heart was created by ligation of the coronary artery as described by Barlow et al. (1976). NADH fluorescence measurements were performed as described above. The lumiphore Pd-meso-tetra (4-carboxyphenyl) porphine (Porphyrin Products, Logan, Utah, USA) was chosen as artificial oxygen probe. This phosphorescent dye was coupled to bovine serum albumin dissolved in physiological saline at pH 7.4 [Wilson et al. 1991], and injected intravenously to a final concentration of about 30 μM.

RESULTS

The applicability of the developed video fluorimeter for detection of NADH fluorescence and Pd-porphyrin phosphorescence images of organ surfaces in vivo was examined in heart and gut of mechanically ventilated rats (Figs. 2,3). NADH fluorescence images of the organ surfaces during O_2-ventilation were characterised by dull fluorescence of the tissue, interrupted by the dark contours of the strongly absorbing blood vessels (Figs. 2A,2C). Due to the limited depth of penetration of the excitation light (estimated to range from 300-600 nm) crisp images of the coronary vasculature could be observed enabling vessel diameter changes to be measured by gray level profile analysis. Substitution of N_2 for O_2 in the respiratory gas resulted in an almost two-fold enhancement of the NADH fluorescence intensity of the myocardial tissue and in enlargement of the coronary blood vessels associated with hypoxia induced vasodilation (Figs. 2A,B). Measurements of the gut in the same animal also revealed enhanced NADH fluorescence during N_2-ventilation (Figs. 2C,2D) but here constriction of the vasculature

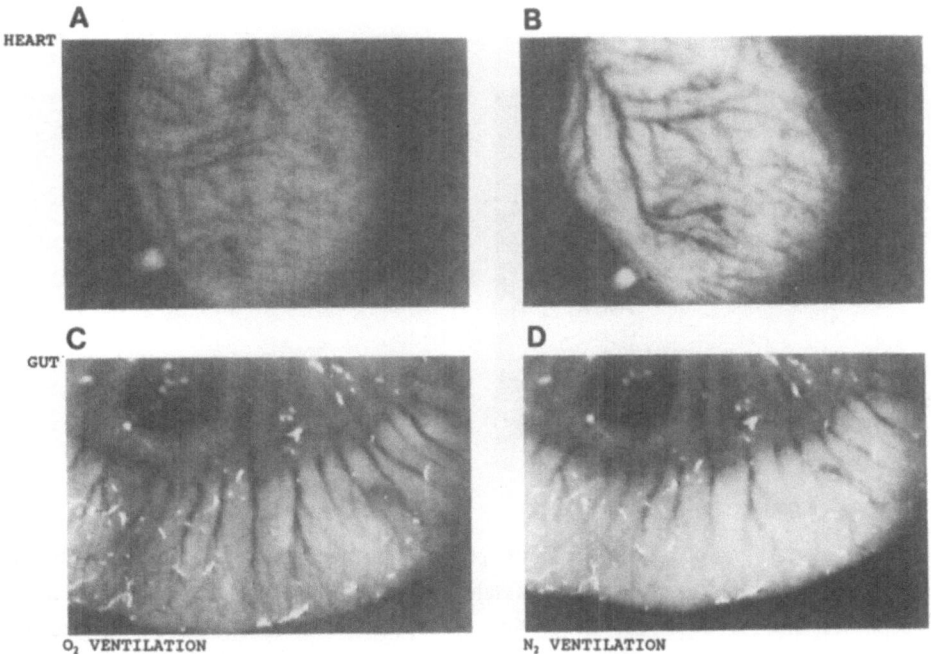

Figure 2: NADH fluorescence recordings of the heart and gut in the same mechanically ventilated rat before and during anoxic hypoxia created by N_2-ventilation. A. Normoxic heart tissue shows a dull fluorescence indicative of low NADH concentrations in the tissue. Due to the absorption of light by blood, blood vessels appear as dark lines. B. Substitution of N_2 for O_2 in the respiratory gas results in higher tissue NADH fluorescence, accompanied by hypoxia induced coronary vasodilation. C. NADH fluorescence image of the normoxic gut also shows a dull fluorescence. D. During anoxic hypoxia enhanced NADH fluorescence can also be observed in the gut tissues. In contrast to the heart, however, the gut vasculature shows constriction during N_2-ventilation. Stray rat hairs in the abdominal cavity cause the bright fluorescent spots in panels C and D.

is observed due to opposite autoregulation mechanisms of gut and heart vasculature in response to anoxic hypoxia. Hypovolemic hypoxia induced by bleeding resulted in a slight further increase in NADH fluorescence, demonstrating attenuation of the fluorescence signal due to the absorption of excitation and emission light by blood (not shown).

To monitor O_2 concentration changes in the coronary microcirculation, Pd-porphyrin bound to bovine serum albumin was injected intravenously into the mechanically ventilated rat. During O_2-ventilation a dull phosphorescence is emitted from the well-oxygenated heart surface (Fig. 3A). Anoxic hypoxia created by N_2-ventilation, however, resulted in a marked increase in phorphorescence intensity, indicative of a decrease in the microcirculatory oxygen concentration (Fig. 3B). These effects did not occur when the dye was excluded from the circulation.

A   B

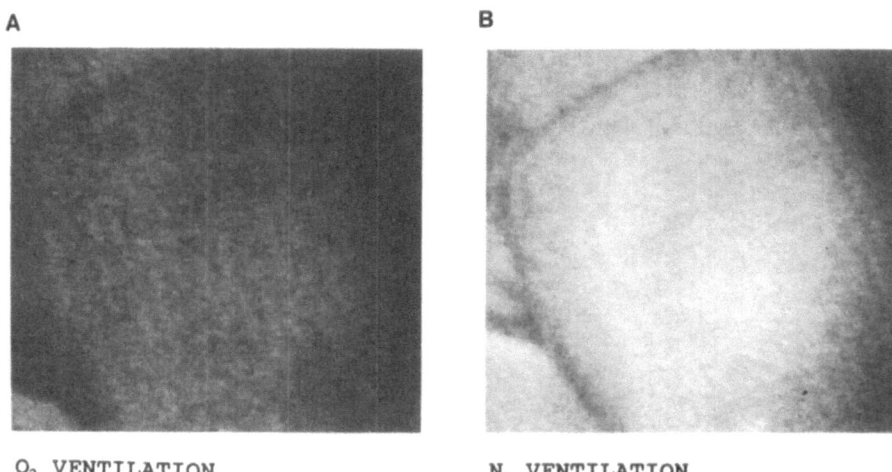

O_2 VENTILATION N_2 VENTILATION

Figure 3.   Alteration in Pd-porphyrin phosphorescence associated with microcirculation hypoxia in the mechancially ventilated rat heart. A. During O_2-ventilation low phosphorescence is observed. B. During N_2-ventilation, however, enhanced phosphorescence is observed due to decrease of the microcirculatory O_2 concentration. This effect is not seen in the absence of dye in the circulation.

To investigate whether both the fluorescence and phosphorescence technique could also be applied for detection of ischemic hypoxia in vivo, videofluorimetric measurements were made of the rat heart surface during ligations of coronary arteries. Under normoxic conditions the NADH fluorescence (Fig. 4A) as well the Pd-porphyrin phosphorescence (Fig. 4C) were low in intensity, reflecting high oxygenation in both tissue and microcirculation. Ligation of the coronary artery, however, resulted in an ischemic area in the

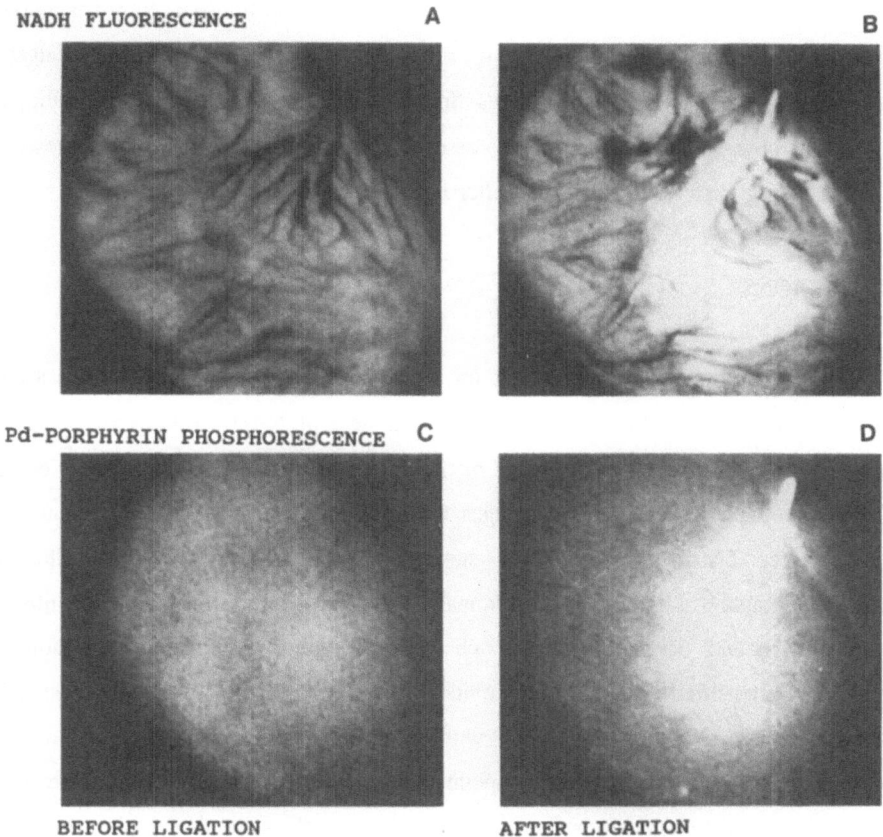

NADH FLUORESCENCE   A   B

Pd-PORPHYRIN PHOSPHORESCENCE   C   D

BEFORE LIGATION   AFTER LIGATION

Figure 4.   The effect of ischemic hypoxia on NADH fluorescence and Pd-porphyrin phosphorescence in rat heart as a result of coronary ligation in the left ventricle. Prior to the experiment albumin-coupled Pd-porphyrin compound was injected into the blood of the rat (final concentration 30 μM). A. NADH fluorescence image of the rat heart surface prior to ligation. B. After coronary ligation a sharp region of increased NADH fluorescence is seen, indicative of tissue hypoxia. C. Pd-porhyrin phosphorescence image of the same heart as in (A), prior to ligation. D. After coronary ligation enhanced phosphorescence is observed in the area distal to the ligation, demonstrating decreased O_2 concentrations in the microcirculation. The hypoxic areas as indicated by increased NADH fluorescence (B) and Pd-porphyrin phosphorescence (D) correspond well with each other.

cardiac tissue with high NADH fluorescence intensity, clearly and abruptly demarkated from the surounding normoxic tissue (Fig. 4B). Enhanced Pd-porphyrin phosphorescence distal to the ligation also revealed impaired oxygen supply to a specific zone of the myocardium (Fig. 4D). As can be seen by comparison of Fig. 4B with Fig. 4D the hypoxic areas indicated by the fluorescent and phosphorescent probes correspond well with each other. In experiments where no Pd-porphyrin dye was injected prior to ligation of coronary arteries, alteration of NADH fluorescence could still be observed following ligation, but observations at wavelengths associated with Pd-porphyrine phosphorescence did not show any difference before and after ligation.

CONCLUSIONS

In this communication we report on a new video fluorimeter based on a second generation CCD camera for simultaneous measurement of NADH fluorescence and Pd-porphyrin phosphorescence in vivo. Use of the described second generation CCD camera provided a number of technical advantages for photometric imaging of fluorescence and phophorescence distributions of organ surfaces. CCD cameras are well suited for quantitative imaging as they provide minimal geometric distortion and the light intensity of each pixel is well defined. Moreover, choice of an ultraviolet/blue sensitive phototube in the image-intensified CCD camera provides the extra sensitivity required for detection of the low light intensities associated with fluorimetry and phosphorimetry in vivo. Mounting the whole instrument on an operation microscope stand provides the flexibility to select a convenient optical configuration; e.g. horizontal positioning of the camera for spectrophotometric measurements of saline-perfused organs in vitro [Ashruf et al. 1990] or vertical set-up of the camera for study of organ surfaces in vivo [Ince et al. 1991].

To our knowledge we present the first combined NADH fluorescence and Pd-porphyrin phosphorescence measurements in the rat heart in vivo. Application of these optical techniques enable identification of both anoxic as well as ischemic hypoxia in intact organs. It is expected that the ability to image the distribution of the mitochondrial redox state of tissue as well as the distribution of oxygen concentration in the microcirculation (as described in this study), will provide new insights into the relation between local oxygen supply and demand in cells, tissues and organs in vivo.

ACKNOWLEDGEMENTS

This study was in part supported by grants from the Netherlands Heart Foundation and by the Rotterdam Blood Bank.

REFERENCES

Ashruf, J.F., Avontuur, J.A.M., Van Bavel, E., Ince, C., and Bruining, H.A. (1990), Local ischemia created by microspheres in the rat heart induces heterogeneous epicardial NADH fluorescence. Pflügers Arch. 416,S2.

Barlow, C.H., and Chance, B. (1976), Ischemic areas in perfused rat hearts: Measurement by NADH fluorescence photography. Science 193, 909-910.

Chance, B., Cohen, P., Jöbsis, F., Schoener, B. (1962), Intracellular oxidation-reduction states in vivo. Science 137, 499-508.

Chance, B. (1976), Pyridine nucleotide as an indicator of the oxygen requirements for energy-linked functions of mitochondria. Circ. Res. 38 (Suppl. I), 31-38.

Ince, C., and Bruining, H.A. (1991), Optical Spectroscopy for the measurement of tissue hypoxia. In: Update in Intensive Care and Emergency Medicine: Update 1991 (Vincent, J.L., ed.), pp. 161-171, Springer-Verlag, New York.

Ince, C., Coremans, J.M.C.C., and Bruining, H.A. (1992), In vivo NADH fluorescence. Adv. Exp. Med. Biol., this volume.

Rumsey, W.L., Vanderkooi, J.M., and Wilson, D.F. (1988), Imaging of phosphorescence: A novel method for measuring oxygen distribution in perfused tissue. Science Wash. DC 241, 1649-1651.

Vanderkooi, J.M., Maniare, G., Green, T.J., and Wilson, D.F. (1987), An optical method for measurement of dioxygen concentration based upon quenching of phosphorescence. J. Biol. Chem. 262, 5476-5482.

Wilson, D.F. (1992), Oxygen dependent quenching of phosphorescence: A perspective. Adv. Exp. Med. Biol., this volume.

Wilson, D.F., Pastuszko, A., DiGiacomo, J.E., Pawlowski, M., Schneiderman, R., and Delivoria-Papadopoulos, M. (1991), Effect of hyperventilation on oxygenation of the brain cortex of newborn piglets. J. Appl. Physiol. 70(6), 2691-2696.

IN VIVO NADH FLUORESCENCE

C. Ince, J.M.C.C. Coremans, H.A. Bruining

Department of Surgery
Erasmus University Rotterdam
Dr. Molewaterplein 40
3015 GD Rotterdam, The Netherlands

INTRODUCTION

Reduced nicotinamide adenine dinucleotide (NADH) offers one of the main means to transfer energy from the tricarboxylic acid cycle to the respiratory chain in the mitochondria. NADH is situated at the high-energy side of the respiratory chain and during tissue hypoxia accumulates in concentration because less NADH is oxidized to NAD^+. The optical properties of NADH and NAD^+ clearly differ. The absorption spectrum of a NADH solution shows two maxima at the ultraviolet end of the light spectrum, one at 250 nm and the other at about 340 nm. NAD^+, on the other hand has an absorption maximum at 250 nm and almost does not absorb light above 300 nm [Renault et al. 1982]. Upon excitation with UV-light NADH, unlike NAD^+, fluoresces in the blue (broad-band emission centered around 460 nm). Chance and co-workers pioneered this fluorescence property of NADH as an indicator of the intramitochondrial redox state and, in the presence of sufficient substrate and phosphates, as an indicator of cellular oxygen requirements [Chance et al. 1962, 1976]. In order to gain insight into the metabolic properties of biological material from various sources, NADH fluorescence measurements have now been applied to single cells, tissue-slices and intact organs (both saline-perfused and in vivo). As absorption and scatter of light by chromophores and corpuscles in tissue affect measurement of NADH fluorescence, several compensation methods have been introduced. Especially when studying NADH fluorescence in vivo, the effect of blood has to be taken into consideration for correct interpretation of the fluorescence measurements. The purpose of this brief review is to summarize the work that has been done on the application of NADH fluorescence measurements to the study of the metabolic state of tissue in vivo.

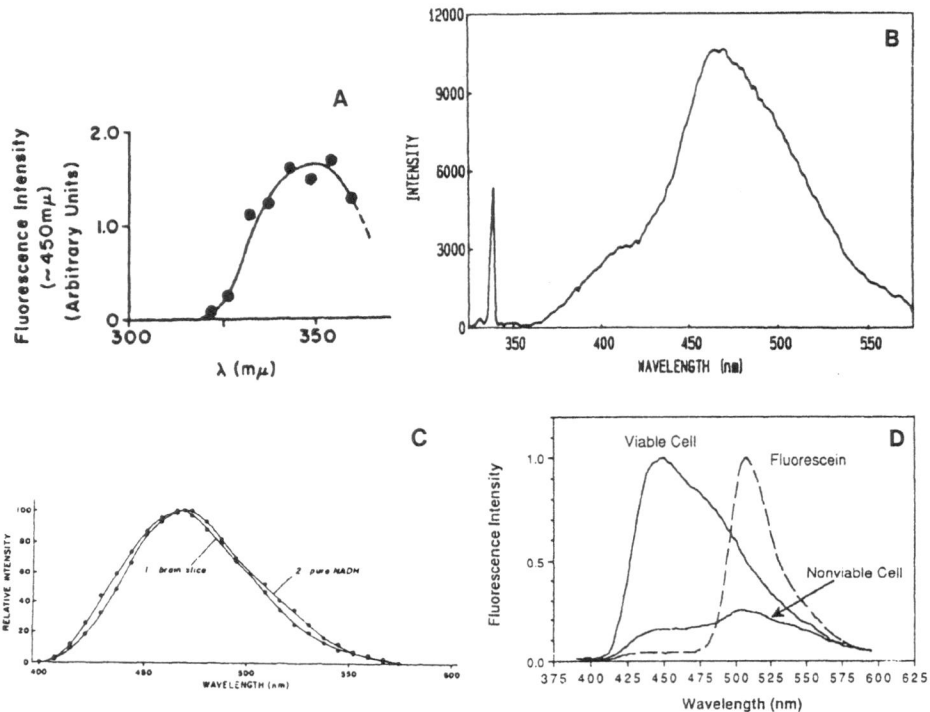

Figure 1.   Comparison of excitation and emission spectra of UV-light induced blue fluorescence from various biological preparations with that of a pure NADH solution. A. Excitation spectrum of perfused rat heart in the carbon-monoxide inhibited state. From Chance et al. (1965) with permission. B. Emission spectrum from a perfused rat heart surface with some contribution of reflected excitation light (337 nm). From Koretsky et al. (1987) with permission. C. Emission spectra from (1) a brain cortex slice examined just after death and (2) from a solution of β-NADH (2 mg/ml), 366 nm excitation. From Harbig et al. (1976) with permission. D. Emission spectra from viable and nonviable single rat cardiac myocytes under 366 nm excitation. The fluorescence emission spectrum of a fluorescein solution (used as an external standard) does not overlap with the blue fluorescence maximum. From <u>Biophysical Journal</u> (1989), 55, 621-630, by copyright permission of the Biophysical Society.

When excited by UV-light, NADH, NADPH and flavins can all contribute to the fluorescence signal of different cell types. The blue-green flavin emission is readily separated from the blue fluorescence of the reduced pyridine nucleotides (NADH + NADPH). To ascertain, that the UV-light induced blue fluorescence originates from reduced pyridine nucleotides, excitation and emission spectra of the biological sources have been compared with those of pure NADH solutions. The excitation spectrum of a saline-perfused rat heart, shown in Fig. 1A [Chance et al. 1965], corresponds to the 340 nm absorption band of pure NADH [Renault et al. 1982]. The absorption band of NADH around 250 nm was not reflected in the excitation spectrum of the heart surface, as increased absorption of proteins and tissue pigments in this wavelength region attenuate the excitation light below detection levels. Fluorescence emission spectra from different sources, such as from saline-perfused rat heart (Fig. 1B; [Koretsky et al. 1987b]), cat brain cortex slice (Fig. 1C; [Harbig et al. 1976]) and from a single rat cardiac myocyte (Fig. 1D; [Eng et al. 1989]) all show broad bands with maxima around 460 nm similar to the emission spectrum of NADH in solution (Fig. 1C).

Biochemical determinations in the isolated perfused rat heart have shown that the blue fluorescence of the intact tissue originates primarily from NADH and NADPH in the mitochondria, with insignificant contributions of the cytoplasmic fractions [Jöbsis and Duffield 1967, Nuutinen 1984]. This selective detection of mitochondrial NAD(P)H is due to the formation of specific protein-nucleotide complexes in the mitochondria, in which the nucleotide fluorescence is enhanced [Avi-Dor et al. 1962, Estabrook 1962, Jöbsis and Duffield 1967]. Recently, detailed fluorescence images of single rat cardiac myocytes have provided additional evidence for the mitochondrial origin of the blue fluorescence [Eng et al. 1989]. Since in mitochondria of muscle and heart, NADH is the predominant reduced pyridine nucleotide [Glock and McLean 1955, Klingenberg et al. 1959] and the fluorescence yield of NADH is substantially higher (2- to 4-fold) than that of NADPH [Avi-Dor et al. 1962, Estabrook 1962], the contribution of NADPH to the measured blue fluorescence in these tissues is negligible. Furthermore, comparison of analytical determinations of reduced pyridine nucleotides with fluorescence changes in perfused rat heart has revealed a linear relationship between tissue NADH concentration and blue fluorescence intensity, while no evident correlation between intracellular NADPH and the fluorescence measurement was observed [Chance et al. 1965].

As NADH plays a central role in the transfer of reducing equivalents from the tricarboxylic acid cycle to the respiratory chain, mitochondrial NADH levels are dependent on the metabolic state of the tissue. By far the most widespread application of NADH fluorescence measurements in vivo have involved its use as an indicator of the oxygen requirement of mitochondria. Fig. 2A shows the relationship between oxygen concentration and the reduction of pyridine nucleotides in actively respiring pigeon mitochondria [Chance 1976] and illustrates how mitochondrial NADH levels can be successfully used to measure mitochondrial oxygen requirements. Whereas, oxygen (the ultimate electron acceptor of the respiratory chain) can modify the NADH redox state, also alterations in extracellular substrate concentrations, e.g. in the delivery of reducing equivalents to the respiratory chain, can affect the mitochondrial NADH concentrations as well as the rate of oxidative phosphorylation (Fig. 2B) [Jöbsis and Duffield 1967, Hassinen and Hiltunen 1975, Chance 1976, Nuutinen 1984, Balaban and Mandel 1988]. Apart from oxygen and substrate, any potential regulatory site of cellular energy metabolism can control mitochondrial NADH levels. A discussion for the possible role of the NADH redox state in the regulation of oxidative phosphorylation has been discussed in a number of excellent reviews [Erecinska and Wilson 1982, Tamura et al. 1989, Balaban 1990, Heineman and Balaban 1990]. From the above considerations it is clear that one must be aware of the metabolic state of the tissue under investigation when interpreting the meaning of blue fluorescence signals and changes therein. In this light, it may be important to note that anesthetics have been shown to affect NADH fluorescence in saline-perfused hearts [Kissin et al. 1983, Renault et al. 1987].

INSTRUMENTATION

Since the introduction of NADH fluorimetry for nondestructive investigation of organ metabolism in vivo in 1959 [Chance and Thorell 1959], the advancing technology led to continuous adaptation of the instrumentation.

Illuminators: Illumination sources have included Xenon lamps, tungsten-halogen lamps [Chance et al. 1975, Hassinen and Jämsä 1982], water- or air-cooled mercury arc lamps [e.g. Chance and Thorell 1959, Kobayashi et al. 1971a,b, Eke et al. 1979] and nitrogen lasers [Renault et al. 1982]. High-powered Xenon lamps have been used for NADH excitation in flash-photography of saline-perfused hearts [Barlow et al. 1976, 1977

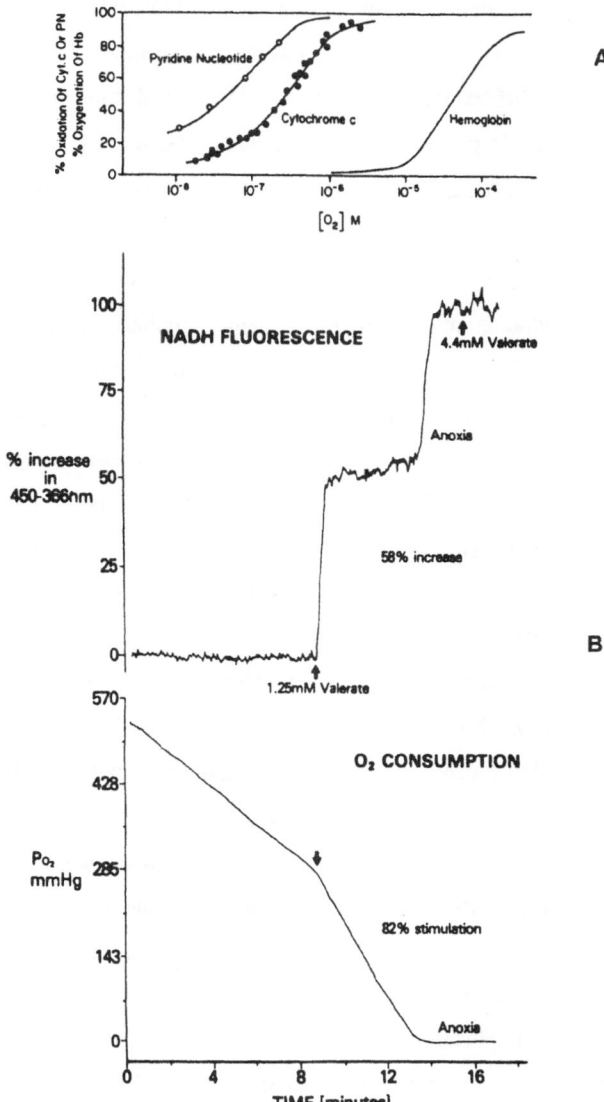

Figure 2. Effect of oxygen and substrate on reduced pyridine nucleotide levels. A. Pyridine nucleotide fluorescence as indicator of mitochondrial oxygen requirement together with the oxygen requirement for cytochrome c oxidation (log of [oxygen] against percent oxidation of cytochrome c) in the actively metabolizing state 3 mitochondria and, for comparison, a sketch of the oxygen requirement for hemoglobin. Reproduced from Chance, B. (1976) by permission of the American Heart Association, Inc., B. Effects of valerate on NADH fluorescence and oxygen consumption of a cortical tubule suspension from rabbit kidney. ↓ ↑, time of injection of a bolus of valerate. A bolus of valerate was added after reaching anoxia to test for nonspecific effects of substrate on 450-nm fluorescence. From Balaban et al. (1988) with permission.

Steenbergen et al. 1977]. For continuous illumination of biological samples, mercury arcs with increased UV emission are preferred. From the two spectral lines of the mercury arc that fall within the NADH excitation spectrum (334 nm and 366 nm), the more intense 366 nm line is selected. To provide sufficient excitation power, light from 50 W to 1000 W mercury arcs has been used. The excitation power provided by pulsed nitrogen lasers (wavelength 337 nm) exceeds that of mercury arcs but the actual radiation flux intensity that is delivered to the tissue depends on attenuation of the UV-light in the optical system. Since intense UV irradiation may cause tissue damage [Chance et al. 1978, Mayevsky 1984, Tamura et al. 1989], the compromise between maximalisation of the excitation energy and prevention of tissue damage is critically balanced. Excitation light that is reflected by the tissue is often used as reference light for compensation of the fluorescence light for changes in the optical properties of tissue not related to the NADH/NAD$^+$ redox state. Tunable dye lasers, transversely pumped by nitrogen lasers, have been used as source of reference light with a wavelength in the visible or NIR region [Renault et al. 1982, 1984].

Detectors: Initially blue fluorescence measurements, both in vitro and in vivo, were made with photomultipliers. The high sensitivity of photomultipliers enables detection of the weak fluorescence signals. Images of the distribution of blue fluorescence in vivo have, for the first time, been achieved by computer reconstruction of measurements made by flying-spot fluorimeters equipped with photomultiplier detection systems [Stuart and Chance 1974, Chance et al. 1978, Paddle 1988]. Photocameras with sensitive film were also used to image the distribution of blue fluorescence in saline perfused organ surfaces [Barlow and Chance 1976, Steenbergen et al. 1977, Ince et al. 1988, 1990b]. In vivo application of this type of fluorescence photography has been limited to the brain since the relative immobility of the brain allowed for the long exposure times needed for photography [Ji et al. 1977]. The development of sensitive video cameras has led to the introduction of videofluorimetry. A television detection system (image intensifier + video camera tube) combined with a video densitometer was first applied by Schuette et al. 1974 for monitoring blue fluorescence images of the brain cortex of patients undergoing brain surgery. For investigation of the spectral content of the blue fluorescence from a perfused heart surface a monochromator was coupled to a silicon-intensified target video camera [Koretsky et al. 1987b]. A similar rapid-scan videofluorimeter was used to image the blue fluorescence of single rat cardiac myocytes [Eng et al. 1989]. Single-cell fluorescence observations have also been made on human macrophages using a photon counting CCD camera [Ince et al. 1990a]. Recently we have developed a videofluorimeter based on a second generation CCD video camera with which it is possible to image NADH fluorescence distribution in organ surfaces in vivo [Ince et al. 1991, 1992].

To select the desired band of excitation light from the illumination source and to protect the detector from reflected excitation light, a wide variety of optical filters have been used. In some experimental set-ups, multi-wavelength photometry was combined with analog electronics to correct for the effect of chromophores other than NADH on the measured blue fluorescence [Kobayashi et al. 1971b, Hassinen and Jämsä 1982, Harbig et al. 1976]. For simultaneous measurement of multiple wavelengths, spatial separation of the individual spectral components has been employed [Chance et al. 1972, Schuette et al. 1974, Vern et al. 1975, Harbig et al. 1976] and time-sharing multichannel photometers have been developed using filterwheels with variable filter combinations [Chance et al. 1975, Hassinen and Jämsä 1982].

Calibration: Due to noise and drift present in the optical and electronic equipment regular calibration procedures need to be carried out. Examples of fluorescence standards providing constant blue fluorescence, when excited by UV-light, have included a piece of paper [Jöbsis and Duffield 1967], fluorescent glasses: Uranyl [Ploem 1970] and Corning type 360 [Chance et al. 1962], aqueous fluorescein solutions [Eng et al. 1989] and solutions of β-NADH [Harbig et al. 1976, Ji et al. 1977, Renault et al. 1987, Duboc et al. 1990]. Calibration values for reflection (backward scattering) of the excitation light have been obtained from paper [Eke et al. 1979], surgical tape [Ji et al. 1979], magnesium oxide powder [Harbig et al. 1976] and latex suspension [Renault et al. 1987]. Absolute values of fluorescence and reflection intensities have been measured with neutral Kodak test cards [Renault et al. 1984].

Optical configurations: For measurement of NADH organ surface fluorescence a variety of optical configurations have been developed. Initially fluorimeters operated in a fixed geometry of instrumentation and studied surface (Fig. 3A,3B). In case fluorescence excitation is directed perpendicularly to the organ surface and fluorescence emission is recorded along the same path, the fluorescence emission light and the reflected excitation light can be separated by a dichroic mirror (Fig. 3A [Ashruf et al. 1990]). When diffusely reflected excitation light is used to compensate for possible artifacts in the measured blue fluorescence, excitation and detection systems are positioned at convenient angles to prevent contribution of the specularly reflected excitation light to the measured

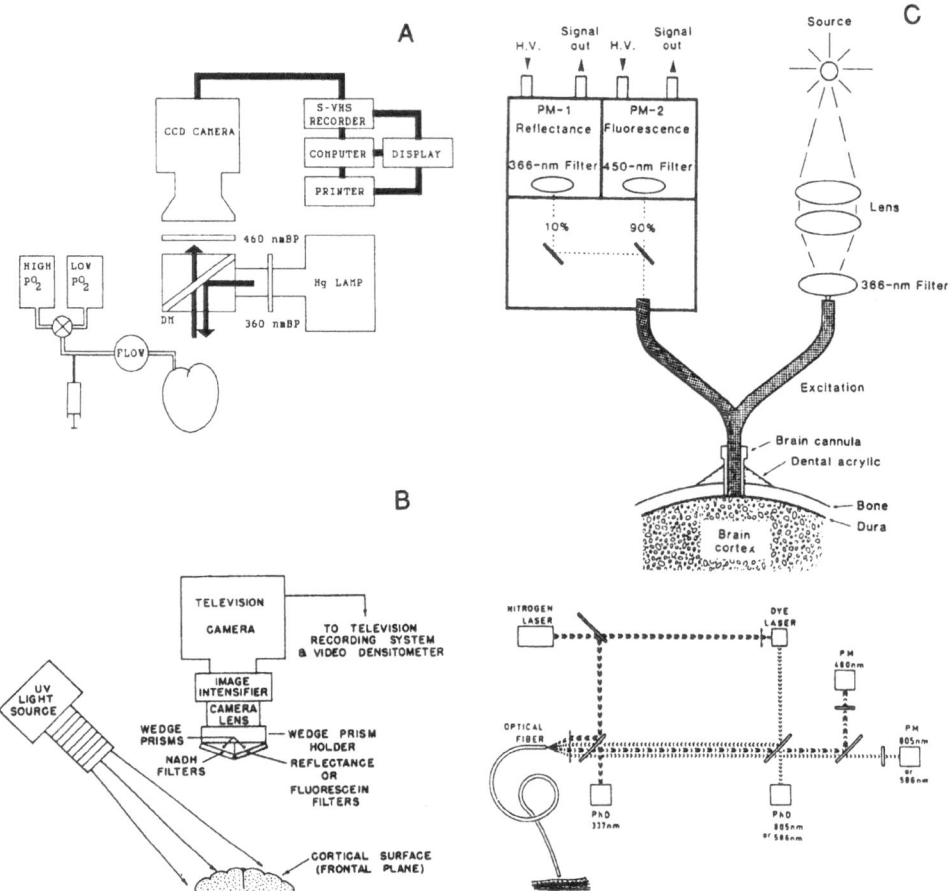

Figure 3.   Schematic representation of devices with distinct optical configurations for measurement of NADH organ surface fluorescence: A. An example of fluorescence excitation and emission light separated by a dichroic mirror. B. Angular arrangement of the UV-light source and television camera in relation to the brain cortex. From Vern et al. (1975) with permission. C. Y-shaped light guide with separate bundles of excitation and emission fibers (H.V., high-voltage input; PM-1 and PM-2, photomultipliers). From Mayevsky et al. (1982) with permission. D. Laser fluorimeter with a single optical fiber for transmittance of both fluorescence excitation and emission light. Beside photoreceivers (PhD, photodiode; PM, photomultipliers) are indicated the selected wavelengths: UV excitation light, 337 nm; fluorescence of NADH, 480 nm; reference light, 586 or 805 nm. From Renault et al. (1984) with permission.

diffuse reflection (Fig. 3B) [Chance et al. 1972, Schuette et al. 1974, Vern et al. 1975]. In these "rigid" configurations the optical components have to meet specific requirements to maintain focusation of a moving organ surface.

In the seventies optical fibers have been introduced into surface fluorimetry, allowing a flexible geometry of the fluorimeter and the organ surface [Mayevsky and Chance 1973]. When light-guides are used the main optical configurations involve either separate bundles of excitation and emission fibers arranged in a Y-shaped light guide (Fig. 3C) [e.g. Mayevsky and Chance 1982, Ji et al. 1979], or a single fiber for transmittance of both fluorescence excitation and emission light (Fig. 3D) [Renault et al. 1983, 1984]. With the fiber optic technique movement artifacts can be largely eliminated and due to the well defined geometry of the fiber tissue interface the contribution of specular reflection can be canceled [Renault et al. 1983, 1985].

TISSUE PROPERTIES AFFECTING ORGAN SURFACE NADH FLUORIMETRY

The intensity of the blue fluorescence emitted from organ surfaces when excited by UV-light is not wholly determined by the mitochondrial NADH concentration. The properties of tissues which affect organ surface fluorescence can be classified into three main categories: (i) movement, (ii) hemodynamic, and (iii) oximetric effects. To what extent these effects interfere with measurement of NADH fluorescence in vivo is strongly dependent on the organ studied and on the optical configuration used. Movement is most prominent in heart, but also in brain respiration and hemodynamic changes cause movement of the organ surface [Harbig et al. 1976]. In addition, the changing geometry of the organ surface and the optical system can give rise to fluctuations in the intensity of the specular reflected excitation light. The hemodynamic effect may interfere with fluorescence measurements when metabolic changes are accompanied by appreciable changes in blood volume. While the spectral characteristics of the blood pigments make absorption of both ultraviolet excitation light and blue fluorescence dependent on perfusion of the organ surface, varying numbers of blood corpuscles in the irradiated area alter the relative intensity of scattered light and diffusely reflected excitation light. Therefore, the hemodynamic effect is in fact composed of absorption, and light-scattering effects. Finally, the possible presence of an oximetric effect should be considered. Apart from adjustment of the redox-balance of the cytochromes to fluctuating tissue oxygen tensions, variations in the oxygenation level of hemoglobin and myoglobin also result in changing absorption spectra of these chromophores.

Despite these drawbacks, however, in vivo organ surface fluorimetry clearly reveals the blue NADH fluorescence from mitochondria. In the heart of mechanically ventilated rats, for instance, an increase in emitted fluorescence is readily observed upon induction of hypervolemic hypoxia by N_2-ventilation (Fig. 4) [Renault et al. 1984, Pierik et al. 1991, Ince et al., 1991, this volume]. This finding is significant since anoxia in the heart is associated with increases in blood volume which would be expected to lower and not raise fluorescence levels if the signal were dominated by a hemodynamic effect. For accurate assessment of the NADH-fluorescence from the measured blue fluorescence signal workers in the field have employed different compensation methods (Table I). The extent to which these compensation methods take account for the earlier mentioned effects varies.

The majority of compensation methods make use of variations in the intensity of reflected excitation light in order to correct for hemodynamic and oximetric effects. In the simplest case, where the oximetric effect is regarded of minor importance, the corrected fluorescence is expressed as a linear relationship between the measured fluorescence and the reflected excitation light [e.g. Jöbsis et al. 1971, Mayevsky and Chance 1974, 1982, Harbig et al. 1976, Ji et al. 1977, Dóra et al. 1984]. When the oxygenation level of hemoglobin is found to contribute seriously to the optical density of the tissue, higher order polynomials and exponential expressions are used for computation of the corrected fluorescence [Kobayashi et al. 1971a, Ji et al. 1979, Renault et al. 1984]. Additional reduction of the oximetric effect has been achieved by registration of the blue fluorescence at an isosbestic wavelength of hemoglobin (e.g. 445 nm [Harbig et al. 1976], 448 nm [Kramer and Pearlstein 1979]). Most correction methods apply fiber optics to follow organ surface movements. Through physical contact of the fiber tip with the tissue surface, the adverse effect of specularly reflected excitation light on detection optics can be abolished. Without use of optical fibers motion artifacts can be conveniently reduced by careful selection of the angle of incidence of the excitation light and appropriate positioning of the detection optics with respect to the organ. Fluorescence recordings from the beating heart may require additional synchronization of the detection system to the beat frequency [Schuette and Simon 1968].

Other investigators have developed compensation methods which are exclusively based on fluorescence measurements [Vern et al. 1975, Kramer and Pearlstein 1979,

TABLE I. Overview of compensation methods employed for the assessment of NADH fluorescence from the measured blue fluorescence. $F^{\lambda}_{...}$ and $R^{\lambda}_{...}$, fluorescence and reflectance intensities at wavelength λ. CICF: 5(6)-carboxy-2',7'-dichlorofluorescein, k, k_i : constants.

CORRECTED FLUORESCENCE	ORGAN	Reference
$F^{445} - R^{366}$	brain (cortex)	Harbig '76
$F^{450} - R^{366}$	brain (cortex) liver, kidney, testis	Mayevsky '74 Mayevsky '82
$F^{450} - k \cdot R^{366}$	brain (cortex)	Dóra '84
$F^{=460} - k \cdot R^{366}$	brain (cortex)	Jöbsis '71
$\left(F^{450}_{anoxic} - F^{450}_{normoxic}\right) - \left(R^{360}_{anoxic} - R^{360}_{normoxic}\right)$	brain (cortex)	Ji '77
$F^{450} - k_1 \cdot (\% \Delta R^{350} + 100)^{k_2} + k_3$	heart, liver	Ji '79
$F^{460}_{no\ blood} \cdot e^{-k \cdot \left(\frac{R^{720}_{no\ blood}}{R^{720}} - 1\right)}$	heart, liver, kidney	Kobayashi '71a
$F^{480}_{no\ blood} \cdot \left(\frac{R^{586}}{R^{586}_{no\ blood}}\right)^{k}$	heart	Renault '84
$\dfrac{F^{448}_{NADH}}{F^{549}_{tissue}}$	brain (cortex)	Kramer '79
$\dfrac{F^{426}_{NADH}}{F^{592}_{CICF}}$	heart	Koretsky '87b
$F^{450} - F^{520}_{FLUORESCEIN}$	brain (cortex)	Vern '75

Koretsky et al. 1987b]. Internal fluorescent standards were obtained from artificial fluorochromes (e.g. rhodamine B, fluorescein and CICF: 5(6)-carboxy-2',7'-dichlorofluorescein) [Vern et al. 1975, Kramer and Pearlstein 1979, Katz et al. 1987, Koretsky et al. 1987a, 1987b], or from the tissue itself [Kramer and Pearlstein 1979]. Fluorescence

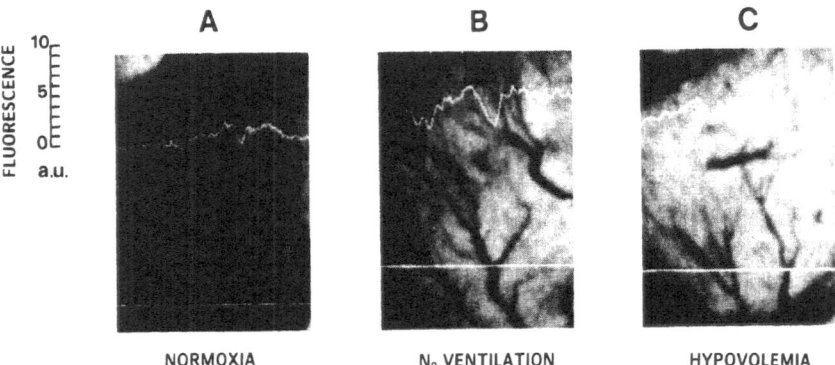

Figure 4.   NADH video fluorimetry of a rat heart in vivo. A. NADH fluorescence image of normoxic heart tissue in a mechanically ventilated rat (ventilation gas: 97% O_2 and 3% ethrane). Due to the absorption of light by blood, blood vessels appear as dark lines. B. Substitution of N_2 for O_2 in the ventilation gas induced hypervolemic hypoxia as reflected by enhanced blue fluorescence and increased blood vessel diameters. C. Further enhancement of the blue fluorescence signal in hypovolemic hypoxia illustrates the attenuation of the fluorescence signal by absorptive effects of blood. Images were taken with a sensitive CCD video camera as described elsewhere in this volume and processed by a computer to provide the grey-level profile along the horizontal (chosen at the same bifurcation in each image), shown across the image of the heart. Fluorescence on the vertical axis is expressed in arbitrary units with 0 corresponding to the background fluorescence level.

intensity fluctuations of the internal standards readily compensate for motion artifacts. To minimize possible oximetric and hemodynamic effects on the fluorescence measurement, these methods also make use of isosbestic wavelengths in either the tissue absorption spectrum (426, 520 and 592 nm) [Koretsky et al. 1987b] or the hemoglobin spectrum (448 and 549 nm) [Kramer and Pearlstein 1979]. An important benefit of fluorescence compensation compared to the reflection approach is that the fluorescence correction is not complicated by questions concerning the relative contributions of the unpredictable intensities of specularly reflected light and/or diffusely reflected light to the recorded reflection signal, thereby releasing some constraints on the geometry of the optical system.

The great diversity in compensation methods for accurate determination of the NADH fluorescence not only reflects the tissue-specific predominance of either movement, hemodynamic or oximetric effects; it also indicates that the correction procedures are strongly dependent on the design of the fluorimeter and its spatial organization with respect to the organ surface.

Despite all attention for methodological considerations, the NADH concentration in tissue cannot be determined from the corrected blue fluorescence signal because of the lack of suitable calibration procedures. Absolute calibration of NADH fluorescence in vivo is not feasible due to deviations from Beer-Lambert's law in biological samples and change of the fluorescence emission spectrum of NADH upon interaction with cytosolic and mitochondrial proteins. As a consequence most compensation methods still express fluorescence and reflection intensities proportionally, with the intensities of normoxic and anoxic tissue (blood-perfused or saline-perfused) as references. Hence, the corrected blue fluorescence yields a qualitative and not a quantitative impression of the NADH redox state in living tissue.

BLUE FLUORESCENCE AS A MEASURE OF THE MITOCHONDRIAL NADH REDOX STATE IN VIVO

Although caution needs to be excercised when interpreting in vivo blue fluorescence measurements, the potential importance of information supplied by knowledge of mitochondrial NADH levels has led to extensive application of in vivo NADH fluorimetry for study of metabolic processes in organ surfaces under different (patho)physiological conditions. A complete survey of this literature would go beyond the scope of this paper, although a brief survey is useful in gaining an impression as to the wide scope of application of NADH fluorimetry.

Organs in which NADH fluorescence has been studied in vivo have included heart [e.g. Chance et al. 1965, Mills et al. 1977, Renault et al. 1984, Ince et al. 1991, this volume], brain [e.g. Chance et al. 1962, Jöbsis et al. 1971, O'Connor et al. 1972, Vern et al. 1975, Sundt et al. 1975, Harbig et al. 1976, Ji et al. 1977, Eke et al. 1979, Kramer and Pearlstein 1979, Mayevsky and Chance 1982, Dóra et al. 1984, 1986], liver [e.g.

Mayevsky and Chance 1982], kidney [e.g. Chance et al. 1962, Kobayashi et al. 1971a, Mayevsky and Chance 1982], gut [Avontuur et al. 1990, Ince et al. this volume], testis [Mayevsky and Chance 1982], muscle [e.g. Duboc et al 1986a, Paddle 1988], skin [e.g. Pappajohn et al. 1972] and eye [Tsubota et al. 1989]. Although NADH fluorimetry has been most widely used for investigations into the (patho)physiology of ischemia, other functions affecting mitochondrial redox state have been studied. These have included allograft transplantation rejection (Fig. 5A) [Duboc et al. 1990], altered rates of respiration in tumors [Galeotti et al. 1970, Gosalvez et al. 1972], epilepsy [O'Connor et al. 1972, Mayevsky and Chance 1975], sepsis [Avontuur et al. 1990] and changes in cardiac respiration following changes in work load (Fig. 5B) [Katz et al. 1987]. Although most NADH fluorimetry studies have been carried out in animal experiments, NADH fluorimetry has also gained some clinical applications. A double-beam laser fluorimeter [Renault et al. 1984] has been used in single point optical fiber measurements in patients suffering from ischemic heart [Duboc et al. 1986b, Toussaint et al. 1987] and muscle disease [Duboc et al.1986a].

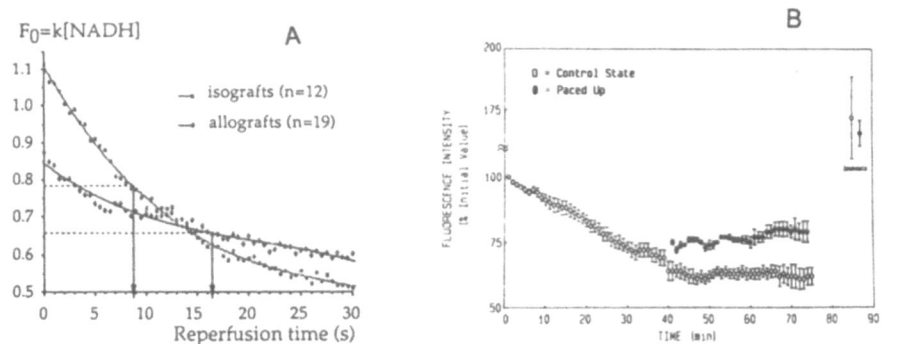

Figure 5.   Alterations in the mitochondrial NADH redox state during cardiac allograft rejection (A) and during changes in cardiac work output (B). A. Representation of half-recovery time (t½) determination for both allografts (♦) and isografts (*) demonstrate a slower rate of postischemic NADH fluorescence decrease in rejecting hearts than in nonrejecting ones. Dotted horizontal lines are median levels between maximum and baseline F_o levels. From Duboc et al. (1990) with permission. B. Effects of increased heart rate on NADH fluorescence. Time course of fluorescence changes (NADH/CICF) for hearts perfused with glucose. For the first 40 min hearts were paced at 250 bpm and then either pacing was continued at this rate (O, controls) or paced up to 590 bpm (●, paced up). From Katz et al. (1987) with permission.

It is expected that the advances in technology will enhance the applicability of NADH fluorimetry in the study of tissue metabolism in vivo. The particular advantages that NADH fluorimetry holds, has sustained interest in this technique since its introduction by Chance. NADH fluorescence provides direct information about the activity of mitochondrial respiration and enables detailed imaging of its spatial distribution in organ surfaces in vivo. It is expected that advancements in imaging technologies and combination of NADH fluorescence with other optical indicators sensitive to, for example, oxygen [Vanderkooi et al. 1987, Wilson et al. 1988, Ince et al. this volume], Ca^{2+} [Uematsu et al. 1989, Fralix et al. 1990] and pH [Tsien 1989] will enable construction of detailed maps of the distribution of the metabolic activity of tissue in vivo under various (patho)physiological conditions.

ACKNOWLEDGEMENTS

This study was supported in part by the Netherlands Heart Foundation (grant NHS 99001).

REFERENCES

Ashruf, J., Avontuur, J.A.M., Van Bavel, E., Ince, C., and Bruining, H.A. (1990), Local ischemia created by microspheres in the rat heart induces heterogeneous epicardial NADH fluorescence. Pflügers Arch. 416, S2.

Avi-Dor, Y., Olson, J.M., Doherty, M.D., and Kaplan, N.O. (1962), Fluorescence of pyridine nucleotides in mitochondria. J. Biol. Chem. 237, 2377-2383.

Avontuur, J.A.M., Ashruf, J., Yzermans, J.N.M., Scheringa, M., Ince, C., and Bruining, H.A. (1990), NADH videofluorimetry reveals early gut ischemia caused by tumor necrosis factor. Pflügers Arch. 418, R143.

Balaban, R.S., and Mandel, L.J. (1988), Metabolic substrate utilization by rabbit proximal tubule. An NADH fluorescence study. Am. J. Physiol. 254, F407-F416.

Balaban, R.S. (1990), Regulation of oxidative phosphorylation in the mammalian cell. Am. J. Physiol. 258, C377-C389.

Barlow, C., and Chance, B. (1976), Ischemic areas in perfused rat hearts: measured by NADH fluorescence photography. Science Wash. DC 193, 909-910.

Barlow, C.H., Harken, A.H., and Chance, B. (1977), Evaluation of cardiac ischemia by NADH fluorescence photography. Ann. Surg. 186, 737-740.

Chance, B., and Thorell, B. (1959), Localization and kinetics of reduced pyridine nucleotides in living cells by microfluorimetry. J. Biol. Chem. 234, 3044-3050.

Chance, B., Cohen, P., Jöbsis, F., and Schoener, B. (1962), Intracellular oxidation-reduction states in vivo. The microfluorometry of pyridine nucleotide gives a continuous measurement of the oxidation state. Science 137, 499-508.

Chance, B., Williamson, J.R., Jamieson, D., and Schoener, B. (1965), Properties and kinetics of reduced pyridine nucleotide fluorescence of the isolated and in vivo rat heart. Biochem. Zeitschrift 341, 357-377.

Chance, B., Salkovitz, I.A. and Kovach, A.G.B. (1972), Kinetics of mitochondrial flavoprotein and pyridine nucleotide in perfused heart. Am. J. Physiol. 213(1), 207-218.

Chance, B., Legallais, V., Sorge, J., and Graham, N. (1975), A versatile time-sharing multichannel spectrophotometer, reflectometer, and fluorometer. Anal. Biochem. 66, 498-514.

Chance, B. (1976), Pyridine nucleotide as an indicator of the oxygen requirements for energy-linked functions of mitochondria. Supp. 1 Circ. Res. 38, I31-I38.

Chance, B., Barlow, C., Nakase, Y., Takeda, H. Mayevsky, A., Fischetti, R., Graham, N., and Sorge, J. (1978), Heterogeneity of oxygen delivery in normoxic and anoxic states: a fluorimeter study. Am. J. Physiol. 235(6), H809-H820.

Dóra, E., Gyulai, L., and Kovách, A.G.B. (1984), Determinants of brain activation-induced cortical NAD/NADH responses in vivo. Brain Research 299, 61-72.

Dóra, E., Tanaka, K., Greenberg, J.H., Gonatas, N.H., and Reivich, M. (1986), Kinetics of microcirculation, NAD/NADH, and electrocorticographic changes in cat brain cortex during ischemia and recirculation. Ann. Neurol. 19, 536-544.

Duboc, D., Renault, G., Polianski, J., Muffat-Joly, M., Toussaint, M., Guerin, F., Pocidalo, J., and Fardeau, M. (1986a), NADH measured by laser fluorimetry in skeletal muscle in McArdle's disease. New Eng. J. Med. 316, 1664-1665.

Duboc, D., Toussaint, M., Donsez, D., Weber, S., Guerin, F., Degeorges, M., Renault, G., Polianski, J., and Pocidalo, J.-J. (1986b), Detection of regional myocardial ischemia by NADH laser fluorimetry during human left heart catheterization. The Lancet, Aug. 30, 522.

Duboc, D., Abastado, P., Muffat-Joly, M., Perrier, P., Toussaint, M., Marsac, C., Francois, D., Lavergne, T., Pocidalo, J.-J., Guerin, F., and Carpentier, A. (1990), Evidence of mitochondrial impairment during cardiac allograft rejection. Transplantation 50, 751-755.

Eke, A., Hutiray, G., and Kovách, A.G.B. (1979), Induced hemodilution detected by reflectometry for measuring microregional blood flow and blood volume in cat brain cortex. Am. J. Physiol. 236(5), H759-H768.

Eng, J., Lynch, R.M., and Balaban, R.S. (1989), Nicotinamide adenine dinucleotide fluorescence spectroscopy and imaging of isolated cardiac myocytes. Biophys. J. 55, 621-630.

Erecińska, M., and Wilson, D.F. (1982), Regulation of cellular energy metabolism. J. Membrane Biol. 70, 1-14.

Estabrook, R.W. (1962), Fluorometric measurement of reduced pyridine nucleotide in cellular and subcellular particles. Anal. Biochem. 4, 231-245.

Fralix, T.A., Heineman, F.W., and Balaban, R.S. (1990), Effects of tissue absorbance on NAD(P)H and Indo-1 fluorescence from perfused rabbit hearts. FEBS Lett. 262, 287-292.

Galeotti, T., van Rossum, G.D.V., Mayer, D.H., and Chance, B. (1970), On the fluorescence of NAD(P)H in whole cell preparations of tumours and normal tissues. Eur. J. Biochem. 17, 485-496.

Glock, G.E., and McLean, P. (1955), Levels of oxidized and reduced diphosphopyridine nucleotide and triphosphopyridine nucleotide in animal tissues. Biochem. J. 61, 388-390.

Gosalvez, M., Thurman, R.G., Chance, B., and Reinhold, H.S. (1972), Indication of hypoxic areas in tumors from in vivo NADH fluorescence. Eur. J. Cancer 8, 267-269.

Harbig, K., Chance, B., Kovách, A.G.B. and Reivich, M. (1976), In vivo measurement of pyridine nucleotide fluorescence from cat brain cortex. J. Appl. Physiol. 41, 480-488.

Hassinen, I.E., and Hiltunen, K. (1975), Respiratory control in isolated perfused rat heart. Role of the equilibrium relations between the mitochondrial electron carriers and adenylate system. Biochim. Biophys. Acta 408, 319-330.

Hassinen, I. and Jämsä, T. (1982), A reflectance spectrophotometer-surface fluorometer suitable for monitoring changes in hemoprotein spectra and fluorescence of flavins and nicotinamide nucleotides in intact tissues. Anal. Biochem. 120, 365-372.

Ince, C., Wieringa, P.A., van der Laarse, A., Nederlof, P.M., Tanke, H.J., van Marle, J., van Weeren-Kramer, J., and Spaan, J.A.E. (1988), A microscopic video-enhanced fluorescence technique for measurement of [NADH] in the myocardium. Pflügers Arch. 412, 72.

Ince, C., Beekman, R.E., and Verschragen, G. (1990a), A micro-perfusion chamber for single-cell fluorescence measurements. J. Immunological Methods, 128, 227-234.

Ince, C., Vink, H., Wieringa, P.A., Giezeman, M., and Spaan, J.A.E. (1990b), Heterogeneous NADH fluorescence during postanoxic reactive hyperemia in saline perfused rat heart. In: Oxygen Transport to Tissue (Piiper, J., Goldstick, T.K., and Meyer, M., eds.), Adv. Exp. Med. Biol. 277, pp. 477-482, Plenum Press, New York.

Ince, C., and Bruining, H.A. (1991), Optical spectroscopy for identification of tissue hypoxia. In: Update in intensive care and emergency medicine: Update 1991. (Vincent, J.L., ed.), pp. 161-171, Springer-Verlag, New York.

Ince C., Ashruf, J.F., Pierik, E.G.J.M., Coremans, J.M.C.C., and Bruining, H.A. (1992), NADH fluorescence and Pd-porphyrin phosphorescence in the rat heart in vivo. Adv. Exp. Med. Biol., this volume.

Ji, S., Chance, B., Stuart, B.H., and Nathan, R. (1977), Two-dimensional analysis of the redox state of the rat cerebral cortex in vivo by NADH fluorescence photography. Brain Res. 119, 357-373.

Ji, S., Chance, B., Nishiki, K., Smith, T., and Rich, T. (1979), Micro-light guides: a new method for measuring tissue fluorescence and reflectance. Am. J. Physiol. 236(3), C144-C156.

Jöbsis, F.F., and Duffield, J.C. (1967), Oxidative and glycolytic recovery metabolism in muscle. J. Gen. Physiol. 50, 1009-1047.

Jöbsis, F.F., O'Connor, M., Vitale, A. and Vreman, H. (1971), Intracellular redox changes in functioning cerebral cortex. I. Metabolic effects of epileptiform activity. J. Neurophysiol. 34, 735-749.

Katz, L.A., Koretsky, A.P., and Balaban, R.S. (1987), Respiratory control in the glucose perfused heart. A ^{31}P NMR and NADH fluorescence study. FEBS Lett. 221, 270-276.

Kissin, I., Aultmann, D.F., Smith, L.R. (1983), Effects of volatile anesthetics on myocardial oxidation-reduction status assessed by NADH fluorometry. Anesthesiology 59, 447-452.

Klingenberg, M., Slenczka, W., and Ritt, E. (1959), Vergleichende Biochemie der Pyridinnucleotid-systeme in Mitochondrien verschiedener Organe. Biochem. Z. 332, 47-66.

Kobayashi, S., Nishiki, K., Kaede, K. and Ogata, E. (1971a), Optical consequences of blood substitution on tissue oxidation-reduction state fluorometry. J. Appl. Physiol. 31(1), 93-96.

Kobayashi, S., Kaede, K., Nishiki, K., and Ogata, E. (1971b), Microfluorimetry of oxidation-reduction state of the rat kidney in situ. J. Appl. Physiol. 31(5), 693-696.

Koretsky, A.P., and Balaban, R.S. (1987a), Changes in pyridine nucleotide levels alter oxygen consumption and extra-mitochondrial phosphates in isolated mitochondria: a ^{31}P-NMR and NAD(P)H fluorescence study. Biochim. Biophys. Acta 893, 398-408.

Koretsky, A.P., Katz, L.A. and Balaban, R.S. (1987b), Determination of pyridine nucleotide fluorescence from the perfused heart using an internal standard. Am. J. Physiol. 253, H856-H862.

Kramer, R.S., and Pearlstein, R.D. (1979), Cerebral cortical microfluorometry at isosbestic wavelengths for correction of vascular artifact. Science 205, 693-696.

Mayevsky, A., and Chance, B. (1973), A new long-term method for the measurement of NADH fluorescence in intact rat brain with implanted cannula. In: Oxygen Transport to Tissue (Bicher, H.J., and Bruley, D.F., eds.), Adv. Exp. Med. Biol., 37A, pp. 239-244, Plenum Press, New York.

Mayevsky, A., and Chance, B. (1974), Repetitive patterns of metabolic changes during cortical spreading depression of the awake rat. Brain Res. 65, 529-533.

Mayevsky, A., and Chance, B. (1975), Metabolic responses of the awake cerebral cortex to anoxia hypoxia spreading depression and epileptiform activity. Brain Res. 98, 149-165.

Mayevsky, A., and Chance, B. (1982), Intracellular oxidation-reduction state measured in situ by a multichannel fiber-optic surface fluorometer. Science 217, 537-540.

Mayevsky, A. (1984), Brain NADH redox state monitored in vivo by fiber optic surface fluorometry. Brain Res. Rev. 7, 49-68.

Mills, S.A., Jöbsis, F.F., and Seaber, A.V. (1977), A fluorometric study of oxidative metabolism in the in vivo canine heart during acute ischemia and hypoxia. Ann. Surg. 186, 193-200.

Nuutinen, E.M. (1984), Subcellular origin of the surface fluorescence of reduced nicotinamide nucleotides in the isolated perfused rat heart. Basic Res. Cardiol. 79, 49-58.

O'Connor, M.J., Herman, C.J., Rosenthal, M., and Jöbsis, F.F. (1972), Intracellular redox changes preceding the onset of epileptiform activity in intact cat hippocampus. J. Neurophysiol. 35, 471-483.

Paddle, B.M. (1988), A scanning fluorometer for imaging ischaemic areas in traumatized muscle. Suppl. 1 J. Trauma 28, S189-S193.

Pappajohn, D.J., Penneys, R., and Chance, B. (1972), NADH spectrofluorometry of rat skin. J. Appl. Physiol. 33, 684-687.

Pierik, E.G.J.M., Ince, C., Avontuur, J.A.M., Ashruf, J., and Bruining, H.A. (1991), The application of NADH fluorescence to identify noninvasively tissue hypoxia in vivo. European Surgical Research 23, 12-13.

Ploem, J.S. (1970), In: Standardization in immunofluorescence (E.J. Holborow ed.), pp. 137-153, Blackwell, Oxford.

Renault, G., Raynal, E., Sinet, M., Berthier, J.P., Godard, B., and Cornillault, J. (1982), A laser fluorimeter for direct cardiac metabolism investigation. Optics and Laser Techn. 14, 143-148.

Renault, G., Raynal, E., and Cornillault, J. (1983), Cancelling of Fresnel reflections in in situ, double beam laser, fluorimetry using a single optical fibre. J. Biomed. Eng. 5, 243-247.

Renault, G., Raynal, E., Sinet, M., Muffat-Joly, M., Berthier, J.-P., Cornillault, J., Godard, B., and Pocidalo, J.-J. (1984), In situ double-beam NADH laser fluorimetry: a choice of a reference wavelength. Am. J. Physiol. 246, H491-H499.

Renault, G., Sinet, M., Muffat-Joly, M., Cornillault, J., and Pocidalo, J.J. (1985), In situ monitoring of myocardial metabolism by laser fluorimetry: relevance of a test of local ischemia. Lasers in Surgery and Medicine 5, 111-122.

Renault, G., Muffat-Joly, M., Polianski, J., Hardy, R.I., Boutineau, J.-L., Duvent, J.-L., and Pocidalo, J.-J. (1987), NADH in situ laser fluorimetry: effect of pentobarbital on continuously monitored myocardial redox state. Lasers in Surgery and Medicine 7, 339-346.

Schuette, W.H., and Simon, A.L. (1968), A new device for recording cardiac motion. Med. Res. Eng. 7, 25-27.

Schuette, W.H., Whitehouse, W.C., Lewis, D.V., O'Connor, M., and Van Buren, J.M. (1974), A television fluorometer for monitoring oxidative metabolism in intact tissue. Med. Instrum. 8, 331-333.

Steenbergen, C., DeLeeuw, G., Barlow, C., Chance, B., and Williamson, J.R. (1977), Heterogeneity of the hypoxic state in perfused rat heart. Circ. Res. 41, 606-615.

Stuart, B.H., and Chance, B. (1974), NADH brain surface scanning and 3-D computer display. Brain Res. 76, 473-479.

Sundt, T.M., and Anderson, R.E. (1975), Reduced nicotinamide adenine dinucleotide fluorescence and cortical blood flow in ischemic and nonischemic squirrel monkey cortex. 1. Animal preparation, instrumentation, and validity of model. Stroke 6, 270-278.

Tamura, M., Hazeki, O., Nioka, S., and Chance, B. (1989), In vivo study of tissue oxygen metabolism using optical and nuclear magnetic resonance spectroscopies. Ann. Rev. Physiol. 51, 813-834.

Toussaint, M., Duboc, D., Renault, G., Polianski, J., Schved, M., Donsez, D., Weber, S., Dessault, O., Pocidalo, J.J., Guérin, F., and Degeorges, M. (1987), Exploration du métabolisme myocardique par fluorimétrie laser du NADH au cours du cathétérisme cardiaque. Arch. Mal. Couer 9, 1341-1349.

Tsien, R.Y. (1989), Fluorescence indicators of ion concentrations. Meth. Cell. Biol. 30, 127-153.

Tsubota, K., Krauss, J.M., Kenyon, K.R., Laing, R.A., Miglior, S., and Cheng, H.-M. (1989), Lens redox fluorometry: pyridine nucleotide fluorescence and analysis of diabetic lens. Exp. Eye Res. 49, 321-334.

Uematsu, D., Greenberg, J.H., Reivich, M., and Karp, A. (1989), Cytosolic free calcium and NAD/NADH redox state in the cat cortex during in vivo activation of NMDA receptors. Brain Res. 482, 129-135.

Vanderkooi, J.M., Maniare, G., green, T.J., and Wilson, D.F. (1987), An optical method for measurement of dioxygen concentration based upon quenching of phosphorescence. J. Biol. Chem. 262, 5476-5482.

Vern, B., Whitehouse, W.C., and Schuette, W.H. (1975), Sodium fluorescein: a new reference for NADH fluorometry. Brain Res. 98, 405-409.

Wilson, D.F., Rumsey, W.L., and Vanderkooi, J.M. (1989), Oxygen distribution in isolated perfused liver observed by phosphorescence imaging. In: Oxygen Transport to Tissue (Rakusan, K., Biro, G.P., Goldstick, T.K., and Turek, Z., eds.), Adv. Exp. Med. Biol. 248, 109-118.

QUANTITATION OF TISSUE OPTICAL CHARACTERISTICS AND HEMOGLOBIN DESATURATION BY TIME- AND FREQUENCY-RESOLVED MULTI-WAVELENGTH SPECTROPHOTOMETRY

B. Chance[1], N.G. Wang[1], M. Maris[1], S. Nioka[1], and E. Sevick[2]

[1]Dept. of Biochemistry/Biophysics, University of Pennsylvania
Philadelphia, PA, 19104; [2]Department of Chemical Engineering
Vanderbilt University, Nashville, TN, USA

ABSTRACT

Photon migration in highly scattering tissues such as muscle and brain gives optical pathlengths that are dependent upon absorption and scattering parameters, μ_a, μ_s'. Determination of these parameters gives the correct concentration of principal absorber such as hemoglobin in the red region of the spectrum. Determinations of scatter factor in functioning and pathological tissues are made.

INTRODUCTION

The rapid development of the pulse time technique, particularly the availability of laser diode light sources that can be pulsed in the picosecond region modulated at hundreds of megahertz, has made a cumbersome and unwieldy technique available at the clinical bedside (1-3). The concept of pulse time spectroscopy of the body organs originated with the idea of pulse radar in the late 1930s, and its applicability to cornea was demonstrated by Alfano (4), to rat head by Tamura (5), to animal brain and human subject by Chance, et al. (1,6) and Delpy et al. (7,8). The idea that the scattering coefficient, μ_s', and the absorption coefficient, μ_a, could be independently determined from the photon kinetics was shown by Patterson's analysis of Chance's data (3,9,10). This showed a good fit to the arm and the head data by the diffusion equation in which the parameters, μ_a, μ_s', are separately calculated. The correction of data obtained from existing systems using continuous light by a differential path length factor was developed by Delpy et al. (7) and has been used by them to quantitate data obtained from existing near infrared (NIR) spectrophotometers, as applied to animal models. Delpy has, at the same time, put forward at this meeting a masterful review of the potentialities of time and frequency domain methods (8). Concurrent development in laser power generation and gated or modulated light amplifiers have led to considerable steps forward in time-resolved spectroscopy and imaging (11,12).

This communication describes currently available technology for time and frequency domain studies of human organs, as well as for spectroscopy and imaging of hemoglobin desaturation in human head, limbs and tumors.

METHOD

The principle of the pulse time method is very similar to time-resolved fluorometry as developed for various fluorochromes of chemical and biological interest. However, those methods are cumbersome and unreliable, since the flash lamps used to generate the excitation are limited in light output, often with erratic triggering, and are limited to times in the nanosecond region. A giant step forward has been afforded by pulsed laser diodes that shift the time domain into the picosecond region and afford bright, compact and reliable sources, particularly in the red and near-infrared region in an appropriate "window" for measuring tissue hemoglobin (13) and water signals (14).

A valuable piece of information for time-resolved spectroscopy and imaging has been the experimental finding that photon migration in the human head persists for over 10 ns, due to large scattering (μ_s') factors and small absorption (μ_a) factors. This finding mmediately revolutionized the system concept for such studies. Solid state lasers are unnecessary and laser diodes come into their own and flash tubes are not useful. At the same time, detectors of unsuitable time characteristics can now be considered useful for studies of most large objects, and expensive and inaccessible microchannel plate detectors are not essential. Furthermore, simplified designs using the phase modulation technique

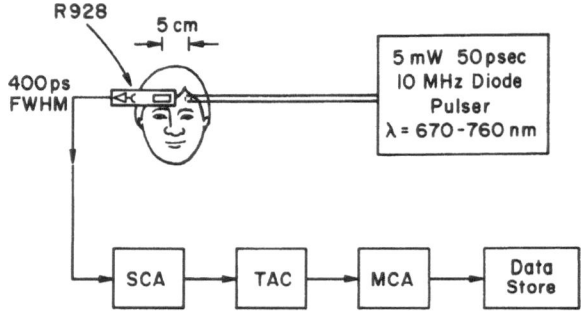

Fig. 1. Simplified TRS System.

(i.e. radio frequency modulation of the laser diode), heterodyne detectors and phase detectors have opened possibilities for new simplified, portable and economical equipment (see Fig. 1).

A number of questions previously unanswerable can now be addressed. The principal one is the determination of the optical path length of photon migration and scattering. For a pulse input response, the output envelope corresponds to the kinetics in Fig. 2, showing a rapid rise to T_{max} and an exponential decay thereafter. It was this exponential decay that attracted our attention. The decay rate is linearly related to the concentration of the absorber. Thus, each point on the decay curve corresponds to a solution of Beer's law, since the pathlength, L, and the attenuation of signal, I_0, are known, and with the existing extinction coefficient, ε, calculation of the concentration, c, is straightforward, $- 1/L \, \mathrm{Log}_e \, I/I_0 = \varepsilon c$. In cases where other absorbers contribute to the photon decay, a dual- or multi-wavelength method is employed using existing algorithms, since the pathlength is known.

The Design of Specific Equipment

The simplicity of pulse time studies for the human brain, where the input-output separation is 5 cm or more and is a distance for obtaining adequate deep photon migration to the cortical material. The light source consists of a 5-mW laser diode operating at 5-mW power, 10 MHz repetition rate, and a 50 ps output light pulse in the range of 670-820 nm. The light source is fiber-optic-coupled to the brain with a 1-3 mm diameter fiber at the surface of the forehead. The diode is connected either by the manufacturer's fiber coupling pigtail or by direct coupling to the laser diode itself, which in the Hamamatsu apparatus that we have employed is in a small probe with electrical couplings to its power supply.

The detector indicated here is the well-known R928 found to be useful in phase modu-lation fluorometry up to a few hundred MHz, and it has been tested here for pulse response and shown to be adequate for measuring the photon decay rate in human subject studies (see Fig. 2).

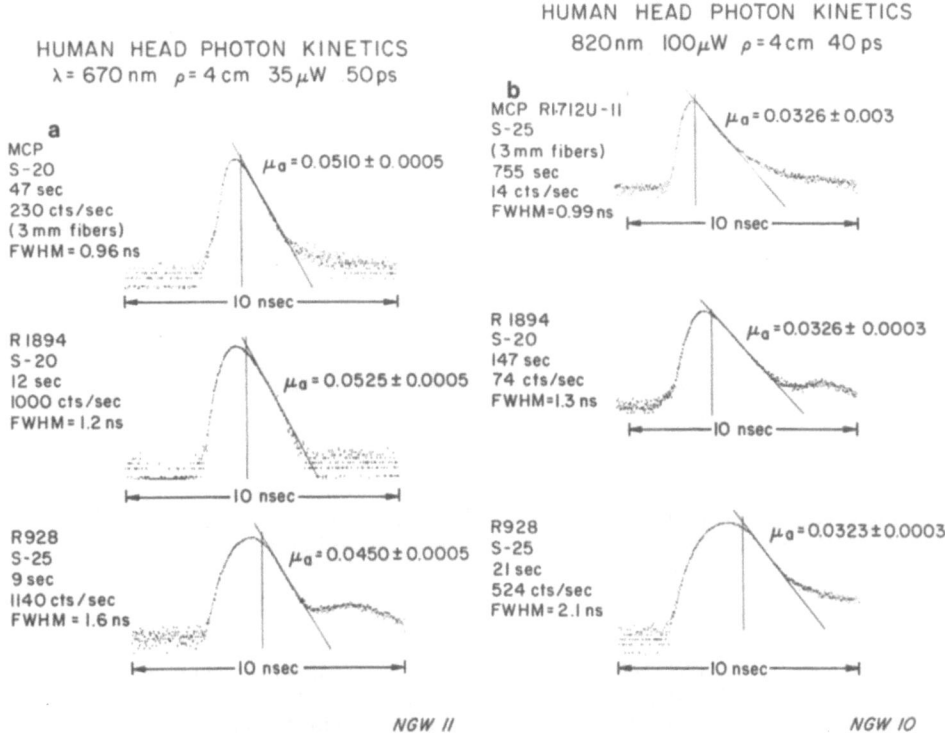

Fig. 2A & 2B. Time-resolved photon migration of the human forehead using various detectors and wavelengths; ρ = 4.0 cm.

Type 1894 is significantly faster but lacks the extended red response of the R928. As can be seen from Fig. 1, these detectors are coupled directly to the forehead to afford optimal gathering of scattered light and therefore optimal signal-to-noise ratio. In the case of R928, the area is approximately 100 mm^2, while in the case of the 1894 it is 20 mm^2. Similarly, when a microchannel plate tube is used, 3 mm fiber optic coupling can be employed in order to have appropriate spatial resolution for a 16-anode imager.

The apparatus consists of the usual single-photon counting equipment; single-channel analyzer, time-to-amplitude convertor, and computer registration of the counts so as to form the time distribution function of the signal. Counting intervals for clinical studies must be restricted to less than a minute, and 5-10 sec seems appropriate, since observations at multiple sites might be required where localization of changes of μ_a and μ_s' are expected.

Figure 2A indicates comparative response of these detectors in terms of the time required to accumulate 10^4 counts. Whilst little difference in performance of the three detectors is evinced at 670 nm, Fig. 2B illustrates the very superior extended red spectral sensitivity of the type R928. This consideration would lead us to improve circuits associated with the R928 to render it as accurate as the other cathodes in the red region by conventional band width expansion circuits.

The calculation of μ_s' of type R928 requires a more significant time resolution than can be obtained, unless rather significant μ_s' correction factors are used. It appears that the faster 1894 is more suitable and, of course where possible, the microchannel plate can be used.

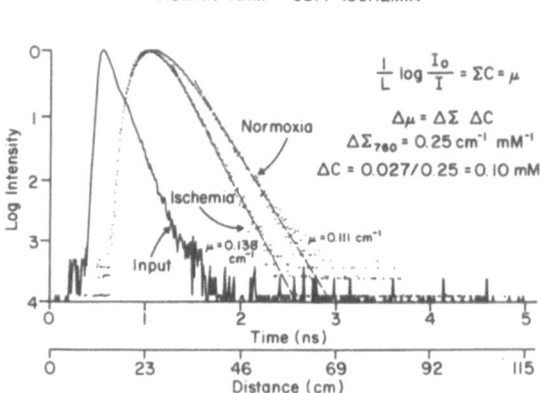

HUMAN ARM – CUFF ISCHEMIA

Fig. 3. Illustration of the effect of an increased absorption of deoxygenation of hemoglobin and myoglobin at 760 nm due to cuff ischmia of the human arm.

Concentration Changes

Direct reading of concentration changes has been previously demonstrated by us. For example, in Fig. 3 cuff ischemia of the arm caused patlength shortening at 760 nm, due to the increased absorbance of the deoxygenated form, with respect to the oxygenated forms of hemoglobin and myoglobin. The change in pathlength even at a single wavelength can be used to calculate the change of concentration (the decrease in pathlength times the extinction coefficient is the change in concentration). The change in concentration is 0.1 millimolar. This value repre-sents the amount of hemoglobin that is deoxygenated from the resting to the ischemic state, as observed by time-resolved spectroscopy. This is not the total hemoglobin/myoglobin signal but is that portion of the blood that is in the capillary bed plus the myoglobin signal. The latter is found elsewhere to represent 70% of the total signal, while the myoglobin represents 30%.

Scattering and Absorption Factors

Muscle contraction is studied in Fig.4. It can be seen that the scattering factor is essentially constant at a value of 7 cm^{-1} while the absorption factor rises from 0.095 to 0.15 cm^{-1}.

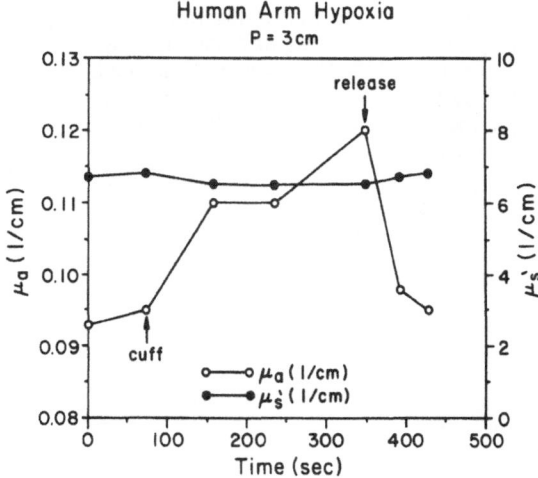

Fig. 4. The time course of μ_a and $\mu s'$ changes in cuff ischemia of a human arm of smaller diameter than in Fig. 3.

Localized Responses and Imaging

Figure 5. The top traces illustrate a photon migration pattern without a localized absorber, while the bottom traces indicate the migration pattern with a localized absorber present. It is apparent that the absorber is identified with this particular widely spaced input-output geometry (90°), and a relatively deep absorber by ablating the long pathways and allowing the short pathways to reach the detector.

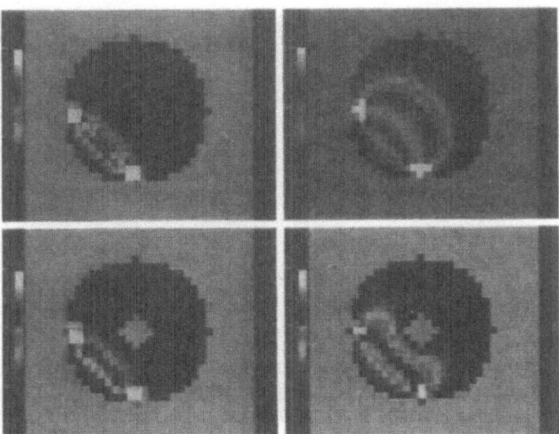

Fig. 5. Illustration of photon migration at shoft times (left) and longer times (right).

Figure 6 illustrates conversely that if the input-output space is close and the absorber deep (section A), the short path lengths predominate to the extent that in section A the migration pattern is not affected by the absorber, while at a larger separation, section B, the absorber is detected by the ablation of shorter path lengths. However, with further separation of the fibers, the migration of photon between the object is emphasized and longer paths will be observed.

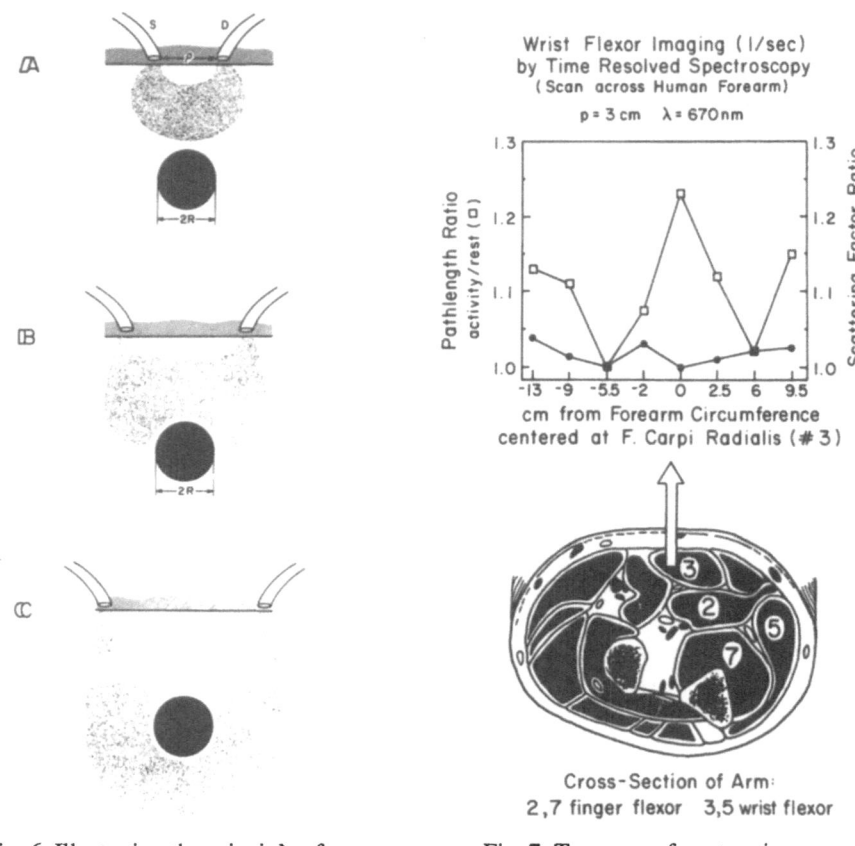

Fig. 6. Illustrating the principle of depth localization by variation of the separation of input-output fibers (ρ).

Fig. 7. Top-scan of contracting wrist flexor muscle (#3) of lower diagram. Points every 2-2.5 cm on circumference of arm. λ=670 nm.

As an experimental example, we illustrate in Fig. 7 the circumferential scan of exercise in wrist flexor muscle with a separation of input-output (ρ) of 3 cm. The circumferential scan of the arm traversing the wrist flexor muscles is initiated, and the pathlength and scattering factor are computed with respect to the resting state. When the absorber is close to the detector, the pathlengths are lengthened as in the model system and the scattering factor is constant.

In order to identify the wrist flexor muscle activated metabolism in the area we have scanned, we have obtained through the kindness of Dr. T. Brown and colleagues one of the first localized magnetic resonance images of the activation of phosphocreatine breakdown and the increase of inorganic phosphate, a volume of approximately 2 cm in diameter and 1 cm below the surface of the muscle (2 cm below) is clearly responding to the exercise. This is an important observation with respect to the wrist flexor exercise, since it verifies that the time-resolved phase modulation spectroscopy is capable of localizing metabolic events, the oxygenation changes of hemoglobin associated with the activation of metabolism by muscular contraction. We shall show below that similar activation of brain metabolism by cognitive processes that cause blood volume increases can be similarly identified.

Scattering Factors in Alzheimer's Disease

It was noted above that while considerable variation of the scattering factor was obtained in a normal control's scans across the forehead 8-18 cm[-1], in studies currently limited to a few Alzheimer's patients (in collaboration with A. Alavi), we find a very low scattering factor in time-resolved scans across the forehead in an Alzheimer's patient. In Fig. 8, we find uniformly low scattering in regions several centimeterss each side of the midline with a small but significant deoxygenation, i.e., shortening of the pathlength. We regard the ability to discriminate scattering and absorption to be of great advantage in the study of both hypoxia and ischemia, on the one hand, and neuronal disintegration, on the other hand.

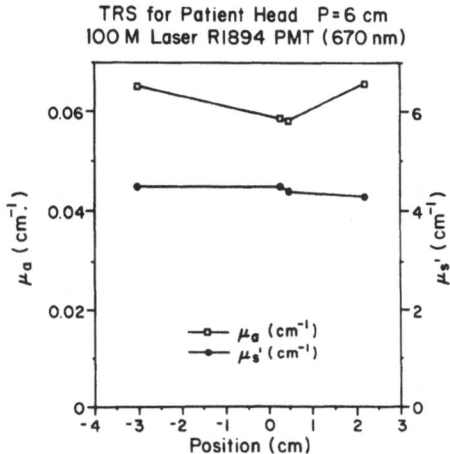

Fig. 8. Scan of the forehead of a patient with Alzheimer's disease; the zero point is the midline.

SUMMARY

The principles of photon migration in tissues, as exemplified by time and frequency domain studies, illustrates the applicability of these methods to quantitative spectrophotometry and to localization of absorbing and scattering objects in tissues. The fact that relatively simple equipment can be used and that resolution comparable to that of much more expensive and complicated methods can be obtained, makes this method most attractive for a convenient and effective spectroscopy and imaging system.

ACKNOWLEDGEMENTS

This work was supported in part by NIH Grants HL-44125, NS-26975 and Hamamatsu Photonics.

References

1. B. Chance, S. Nioka, J. Kent, K. McCully et al, Time Resolved Spectroscopy of Hemoglobin and Myoglobin Resting and Ischemic Muscle. Anal. Biochem. 174:698-707 (1988)
2. B. Chance, J.S. Leigh, H. Miyake, D.S Smith, et al Comparison of Time Resolved and Unresolved Measurements of Deoxyhemoglobin in Brain. Proc. Natl. Acad. Sci. USA 85:4971-4975 (1988)
3. E.M.Sevick, and B. Chance, Quantitation of Time- and Frequency-Resolved Optical Spectra for the Determination of Tissue Oxygenation. Anal. Biochem. 195:330-351 (1991)
4. J.G. Fujimoto, S. De Silversti, E.P. Ippen, R. Margolis,and A. Oseroff, Femtosecond Optical Ranging in Biological Sytstems. Opt. Lett., 11:150-152 (1986)
5. Hazeki, O. and Tamura, M. (1988) The Quantitative Analysis of Hemoglobin Oxygenation State of Rat Brain *in situ* as Monitored by Near-Infrared Spectrophometry. J. Appl. Physiol. 64:796-802
6. B. Chance, J. Haselgrove, J.S. Leigh, M. Patterson, and E. Sevick, Optical Ranging of Muscle and Brain. IEEE Transactions on Microwave Theory and Techniques (Symposium Issue of Spectrum - December 1991) In press (1991)
7. D.T. Delpy, M. Cope, P. van der Zee, S.R. Arridge, S Wray, and J.S. Wyatt, Estimation of Optical Pathlength through Tissue from Direct Time of Flight Measurement. Phys. Med. & Biol. 33:1433-1442 (1988)
8. D.T. Delpy, this volume (1991)
9. M.S. Patterson, B. Chance, and B.C. Wilson, Time Resolved Reflectance and Trans-mittance for the Noninvasive Measurement of Tissue Optical Properties. J. Appl. Optics 28:2331-2336 (1989)
10. M.S. Patterson, J.D.Moulton, B.C.Wilson, and B. Chance, Applications of Time-Resolved Light Scattering Measurements to Photodynamic Therapy Dosimetry. In: Photodynamic Therapy: Mechanisms II Proc. Soc. Photo Optical Instrum. Engr. 1203:62-75 (1990)
11. J. Fishkin, E. Gratton, M.J. vandeVen, and W.W. Mantulin, Diffusion of Intensity-Modulated Near-Infrared Light in Turbid Media. In: Proceedings of Time-Resolved Spectroscopy and Imaging of Tissues, SPIE (B. Chance, ed) SPIE, Bellingham, WA, 1431:122-135 (1991)
12. S.L. Jacques, and S.T. Flock, Effect of Boundary on Time-Resolved Reflectance: Measurements with a Prototype Endoscopic Catheter. In: Proceedings of Time-Resolved Spectroscopy and Imaging of Tissues, SPIE (B. Chance, ed) SPIE, Bellingham, WA, 1431:12-28 (1991)
13. B. Chance, M.T. Dait, C. Chang, T. Hamaoka, and F. Hagerman, Non-invasive Evaluation of Deoxygenation during Intense Exercise and Recovery of Oxygenation of Hemoglobin and Myoglobin in the Quadriceps Muscles of Elite Competitive Rowers. J. Appl. Physiol. 262:C000-C00 in press (1992)
14. W.L. Butler, Absorption of Light by Turbid Materials. J. Opt. Soc. Amer.52:292-299 (1962); see also P. Williams, and K. Norris, Near-Infrared Technology in the Agricul-tural and Food Industries. Am. Assoc. of Cereal Chemists, Inc, St. Paul, NM (1987)

EVALUATION OF THE ALGORITHM USED IN NEAR INFRARED

SPECTROPHOTOMETRY

Willy N.J.M. Colier, Biny E.M. Ringnalda, Jos A.M. Evers, and Berend Oeseburg

Department of Physiology, Faculty of Medical Sciences, University of Nijmegen Nijmegen, The Netherlands

INTRODUCTION

The major causes of death and disability in a neonatal intensive care unit (NICU) are cerebral ischemia and intra-ventricular hemorrhage. The monitoring of cerebral oxyge-nation and hemodynamics is of utmost importance for the care of the preterm infant. A small, inexpensive and continuous measuring bedside instrument is a substantial extension of the NICU. The technique of near infrared spectrophotometry (NIRS) might provide in this need.

This technique is based on the relative high transpa-rency of skin and skull for light in the near infrared region. This allows photon transmission through the brain. For the brain, it is assumed that changes in light absorption are caused by oxy- and deoxyhemoglobin (O_2Hb and HHb) and the cytochrome aa_3 oxidation level.

To monitor this three component system, a minimum number of three wavelengths is needed. The optical density OD can be written as:

$$\begin{pmatrix} OD(\lambda_1) \\ OD(\lambda_2) \\ OD(\lambda_3) \end{pmatrix} = \begin{pmatrix} \alpha_{O_2Hb}(\lambda_1) & \alpha_{HHb}(\lambda_1) & \alpha_{Cyt}(\lambda_1) \\ \alpha_{O_2Hb}(\lambda_2) & \alpha_{HHb}(\lambda_2) & \alpha_{Cyt}(\lambda_2) \\ \alpha_{O_2Hb}(\lambda_3) & \alpha_{HHb}(\lambda_3) & \alpha_{Cyt}(\lambda_3) \end{pmatrix} \cdot \begin{pmatrix} c_{O_2Hb} \\ c_{HHb} \\ c_{Cyt} \end{pmatrix} \cdot L + \begin{pmatrix} OD_R(\lambda_1) \\ OD_R(\lambda_2) \\ OD_R(\lambda_3) \end{pmatrix}$$

where L denotes the optical pathlength through the medium (cm), α the absorption coefficient ($mM^{-1}cm^{-1}$) and c the concen-tration (mM) of the absorbing components. OD_R represents the oxygen independent absorption losses caused by skull and tissue. The equation is based on Lambert-Beer's law.

If OD_R is assumed to remain constant during measure-ments, and by looking at changes of the absorption signal, OD_R will, by substraction, cancel out. Solving the set of linear equations will result in the algorithm for use in NIRS.

Oxygen Transport to Tissue XIV, Edited by W. Erdmann and D.F. Bruley, Plenum Press, New York, 1992

$$\Delta \vec{c}_{comp} = \vec{\alpha}^{-1} \cdot \Delta \vec{OD}(\lambda_n) \, / \, L$$

$\vec{\alpha}^{-1}$ represents the inverted matrix of the absorption coefficients for the different components at the wavelengths used in the system. Several authors have published the inverse absorption values used in their work (Cope et al., 1987, Reynolds et al., 1988, Livera et al., 1991). Extensive research into the absorption coefficients of hemoglobin and/or cytochrome has been done by e.g. Wray et al. (1988), Rea et al. (1985) and Zijlstra et al. (1991). However, the values of the inverse absorption matrix used by some groups in the field of NIRS show differences, although the same algorithm is used. In our opinion this makes a basic evaluation of the used coefficients necessary. In our studies we investigated the coefficients used by the group of Rolfe, University of Keele, Great Britain and by the group of Delpy, University College London, Great Britain. Furthermore we investigated the matrix of Radiometer, Copenhagen, supplied by their near infrared spectrophotometer.

The studies were performed in an *in vitro* system in which the saturation of the hemoglobin component could be altered. Upon same sets of absorption data, obtained in the *in vitro* system, calculations were performed to obtain the oxy- and deoxyhemoglobin concentration and the saturation, which can be derived by dividing the oxyhemoglobin concentration by the total hemoglobin concentration. Furthermore it is possible to calculate the cytochrome oxidation level. However, as in both hemoglobin solutions used in the experiments no cytochrome is present we should detect a constant zero level.

For quantitation of the changes in the hemoglobin concentration and cytochrome oxidation level the optical pathlength of the photons in the medium must be known (Wyatt et al., 1990).

MATERIALS AND METHODS

Hemoglobin solutions

Two types of solutions were used. The first was a, ready to use, human erythrocyte suspension, without leucocytes, obtained from the hospital's bloodbank.

The second type was a human stroma-free hemoglobin solution. The solution was obtained by shaking one part of an erythrocyte suspension with two parts of distilled water, then storing it in a refrigerator. After one hour the solution is frozen on a mixture of dry ice and acetone. After thawing, the mixture is spun and filtered to below 0.45 μm particle size to give a clear, stroma-free hemoglobin solution.

Experimental setup

The fluid of interest was pumped through a 5 mm cuvette and a membrane oxygenator. Saturation was changed by varying the gas supply, a humidified oxygen-nitrogen-carbon dioxide mixture. By means of dilution, using saline, the hemoglobin concentration was varied. Concentrations of methemoglobin, CO-hemoglobin, oxy- and deoxyhemoglobin were measured using an IL 482 CO-Oximeter. The absorption was measured with the Radiometer near infrared spectrophotometer, at wavelengths

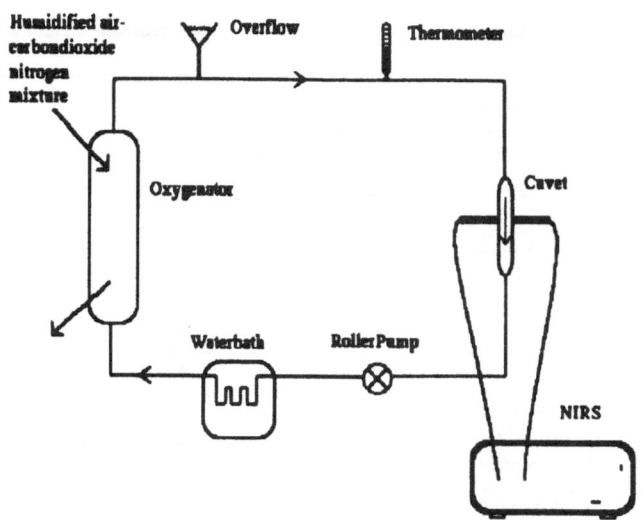

Fig.1. Experimental setup for the *in vitro* evaluation of the NIRS coefficients.

775, 805, 845 and 904 nm. Figure 1 represents a schematic view of the setup used.

Experiments

During an experiment the saturation of the solution was varied from about 3% to 98%. For faster oxygenation the erythrocyte suspension was kept at a temperature of 28°C. The stroma-free hemoglobin solution was kept at a temperature of 10°C, to avoid formation of methemoglobin.

The measured absorption data were corrected for losses due to divergence and reflection, using a blank sample.

RESULTS AND DISCUSSION

At first the algorithm was checked for its stability by applying a gradually changing saturation to the system. In figs. 2a and 2b it can be observed that the total hemoglobin concentration, especially in the case of the stroma-free solution, is not the totally flat curve it should be. The Rolfe and the Delpy coefficients show similar behaviour in the non-scattering case, which also holds for the cytochrome oxidation level. In the case of the Rolfe coefficients we can observe the expected constant zero cytochrome oxidation level. Almost the same result is obtained with the Delpy coefficients. In the case of the Radiometer coefficients an unreliable cytochrome oxidation signal level was observed.

The differences between the erythrocyte solution and the stroma-free solution are most probably caused by scattering effects. As long as these effects are limited to an offset in the signal this has no influence on the final results, as in NIRS only *changes* of the signal are calculated.

When the erythrocyte suspension and the stroma-free hemoglobin solution are diluted, to an extent of more than 50%, we get a linear relation between measured and calculated hemoglobin saturation (fig. 3). In a clinical situation this means that changes in cerebral blood volume can be detected very well. Although having a linear relationship, the concentration values do not match. This is partly due to the scat-

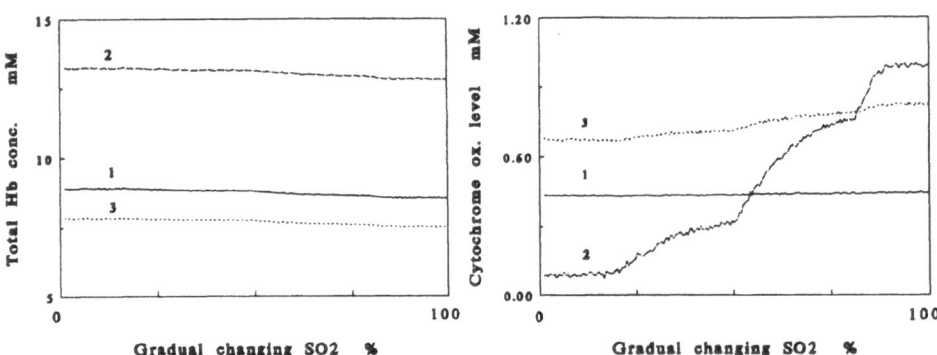

Fig 2a. The effect of a gradual changing saturation on the total hemoglobin concentration and the cytochrome oxidation level for the, scattering, erythrocyte suspension (1=Rolfe, 2=Radiometer and 3=Delpy).

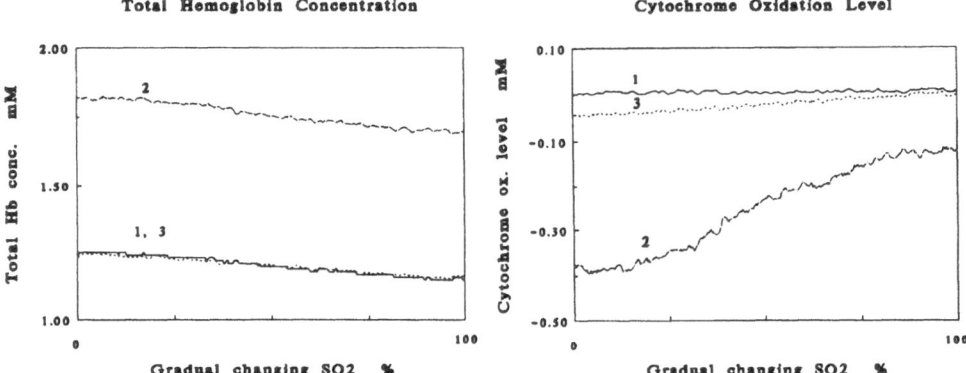

Fig. 2b. The same as fig. 2a, but now for the, non-scattering, stroma-free hemoglobin solution.

tering effect as we do not correct for the differences between optical and physical pathlength. Another disturbing effect is the system's optical losses, which can not be corrected as well as is possible for a (dual beam) spectrophotometer.

A linear relation is also found between calculated and measured saturation for the erythrocyte suspension as well as for the stroma-free hemoglobin solution (figs. 4a and 4b). Dilution of either of the solutions has no effect on the calculated saturation. Due to scattering the response of the erythrocyte suspension is reduced.

CONCLUSIONS

Only minor differences can be observed between the Rolfe and Delpy coefficients. Both are suitable for use in NIRS, this in contrast with the Radiometer coefficients.

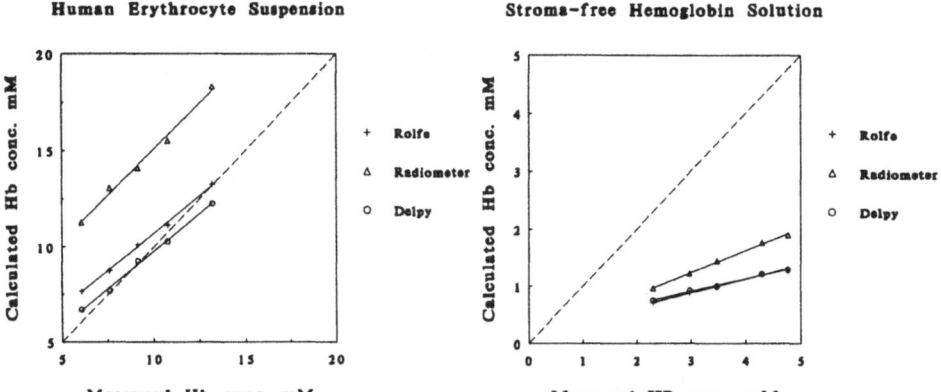

Fig. 3. The calculated hemoglobin concentration as a function of the measured hemoglobin concentration during dilution.

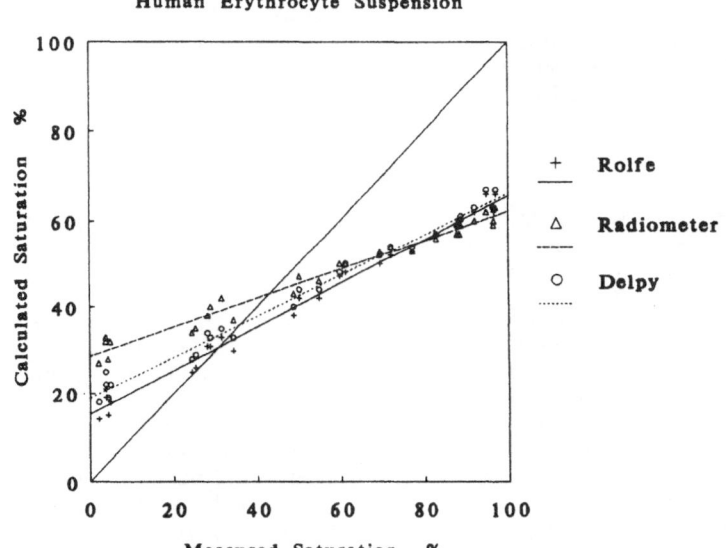

Fig. 4a. The calculated saturation as a function of the measured saturation (CO-Oximeter) for the erythrocyte suspension.

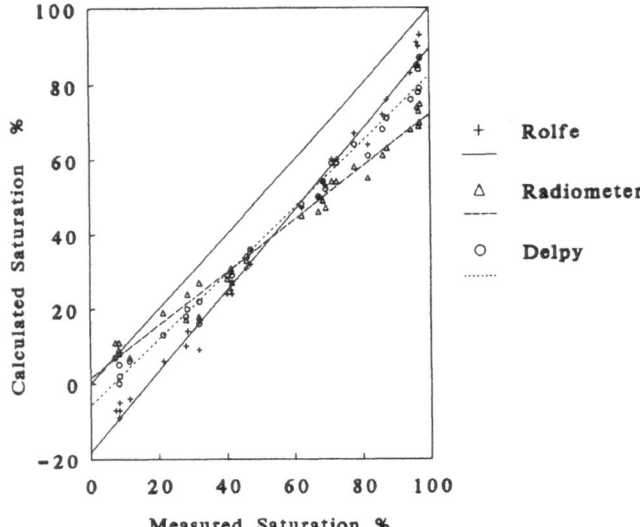

Fig. 4b. The calculated saturation as a function of th
e measured saturation for the stroma-free hemoglobin
solution.

In scattering media, incorporation of an optical pathlength factor in the algorithm, eventually analyzed per
wavelength, is essential for a good result.
    The cytochrome oxidation level is still very difficult
to derive and needs further investigations.

REFERENCES

Cope, M., Delpy, D.T., Reynolds, E.O.R., Wray, S., Wyatt, J.
    and van der Zee, P., 1987, Methods of quantitating
    cerebral near infrared spectroscopy data, Adv. Exp.
    Med. Biol., 222:183-189.
Livera, L.N., Spencer, A.S., Thorniley, M.S., Wickramasinghe,
    Y.A.B.D. and Rolfe P., 1991, Effects of hypoxaemia
    and bradycardia on neonatal cerebral haemodynamics,
    Arch. Dis. Child., 66:376-380.
Rea, P.A., Crowe, J., Wickramasinghe, Y. and Rolfe, P., 1985,
    Non-invasive optical methods for the study of cere-
    bral metabolism in the human newborn: a technique for
    the future?, J. Med. Eng. Technol., 9(4):160-166.
Reynolds, E.O.R., Wyatt, J.S., Azzopardi, D., Delpy, D.T.,
    Cady, E.B., Cope, M. and Wray, S., 1988, New non-
    invasive methods for assessing brain oxygenation and
    haemodynamics, Brit. Med. Bull., 44(4):1052-1075.
Wray, S., Cope, M., Delpy, D.T., Wyatt, J.S. and Reynolds,
    E.O.R., 1988, Characterization of the near infrared
    absorption spectra of cytochrome aa_3 and haemoglobin
    for the non-invasive monitoring of cerebral oxygena-
    tion, Biochim. Biophys. Acta., 933:184-192.

Wyatt, J.S., Cope, M., Delpy, D.T., van der Zee, P., Arridge, S.,Edwards, A.D. and Reynolds, E.O.R., 1990, Measurement of optical pathlength for cerebral near-infrared spectroscopy in newborn infants, _Dev. Neurosci._, 12:140-144.

Zijlstra, W.G., Buursma, A. and Meeuwsen-van der Roest, W.P., 1991, Absorption spectra of human fetal and adult oxyhemoglobin, de-oxyhemoglobin, carboxyhemoglobin and methemoglobin, _Clin. Chem._, 37(9):1633-1638.

THE LUNG, ENTRANCE PORT OF OXYGEN

INFORMATIVE IMAGING OF OXYGEN SUPPLY PARAMETERS IN CLINICAL

PRACTICE

A.W. Grogono, A.P.K. Verkaaik and W. Erdmann

Department of Anesthesiology, Erasmus University

Rotterdam, The Netherlands

Informative design principles are required to display logical and understandable computer-aided images of the numerous measurements that confront the practising clinician and the busy experimental scientist. Software directed feed-back mechanisms or control algorithms can be used to close the loop between transducer and effector in the therapeutic situation or experimental procedure (Erdmann, 1977; Erdmann, Prakash and Schepp, 1984).

The present scientific workplace consists of
- many instruments;
- many displays;
- complex machinery that can only be handled with sophisticated knowledge and technical experience;
- many simple threshold alarms;
- many scattered controls.

The scientist or clinical investigator is thus forced into the position to function as a complex integrator, processor and therapist, at the same time as being a mechanic or technician that adjusts, corrects and controls highly complex mechanical machinery (Verkaaik, Grogono and Erdmann, 1990).

Technical development has progressed in recent years to highly sophisticated and reliable low-cost systems where application of various microchips gathered in so-called dedicated computers are able to overtake a large amount of integration work of various parameters, automatic calculation of derived parameters, limit (alarm) signal processing and even feed-back control of the machinery to automatically adjust physiological parameters to the preset values.

The educated mind therefore today asks for a computer assisted workplace (Saunders and Jewett, 1983) to streamline, organize and help to concentrate the investigator on his main task: a physiologically correct performance of the investigation, immediate, logical and consequent reaction if parameter acquisition tends to go wrong.

A computer assisted work-place is placed around a computer as a "black box" that gathers all information and displays and memorizes the processed data according to the needs of the operator. The investigator who remains in control most often is a non-engineer and the handling of directions needed to activate respective software programs should be easy, logical and self-explanatory, thus the investigator does not need to know anything about the computer and computer programming (Cork, 1989).

The proposed role of the workstation thus is:
1. To provide a central and uniform format for the display of information.
2. To aid in providing control and acquisition of selected variables.
3. To aid in the detection of unexpected events.
4. To prepare a record of the procedure (measurements) as desired.

The general design of the display should allow for the human brain's limited capacity for integrating complex visual information. Thus, multiple parameters should be grouped logically and distinctly, parameters organized in block registrations of values belonging to each other, being dependent on each other or derived from each other.

GENERAL RULES FOR INFORMATIVE DATA DISPLAY

There are several important basic principles that should be followed to reach informative data display of multiple parameters and alarm signalisation without loss of information, together with a user-friendly operation mode of a system:

1. The human brain can maximally absorb simultaneously only three parameters (e.g. values, bars, curves, message information and users' instruction modes).
2. Multiple parameters should thus be grouped organizationally into information blocks of three. Different groups of information should be clearly distinct from each other (e.g. one in curves, one in bars, one in numbers and others in another form of visual display).
3. Alarm signals should not extinguish valuable on-line information and should always include for alertness asking simultaneous visual and audible signs (two senses alerted simultaneously).
   The visual signalisation should not be extinguishable unless the problem has been solved, the audible alarm should only be able to be switched off for very short intermittent periods when the problem has been resolved. Different alarm categories should be clearly distinguishable (visible and audible) whereby all tools of attention attraction should be applied (visual alerts always bright but different colours, continuous or flashing, information in letters versus easily understandable logical signs). The audible signalisation should use specific sound characteristics for the various fields of problems (low and high sounds, continuous versus interrupted signs, intermittent sequences of low and high, or short and long sound tracks), but should be limited to three different signals.
4. Operational information should clearly appear on the screen apart from the data acquisition, never more than three annotations at the same time. The setting up of operation should be clearly designed with step-by-step information after turning the system on. Optimal user comfort is reached when the users' manual is not at all necessary, even for the absolutely computer-naive operator.
5. All manual direction tools should be placed apart from the screen, which should be solely preserved for transmission of information (no distracting touch screen). The handling buttons (e.g. touch buttons) should be grouped according to operating modes clearly apart from each other, any change of an operational mode, data and preset values for further processing should only be activated by a secondary handling necessary pushing a separate OK-mode).
6. Readjustment information or control of standing mode should never be displayed at

cost of on-line information; however, longterm display, display of trend recording, generally can be missed for short periods.

In conclusion: The general informative display follows the rule of three. The operational keyboard should be designed accordingly.

Employing these principles, a new closed circuit system has been developed which facilitates the measurement of oxygen consumption, carbon dioxide production, respiratory quotient, and lung function mechanics during controlled or spontaneous ventilation. This physiologically flexible system (PhysioFlex®, Physio B.V., The Netherlands) offers flexibility of data acquisition as well as in the modes of display according to the needs of the individual.

The design of display is accordingly arranged:
A. Three blocks of curves.
    Block 1. On-line respiratory tidal volume and respiratory pressure.
    Block 2. On-line respiratory flow and CO_2 concentration.
    Block 3. Oxygen consumption and CO_2 production as trend recording.

B. Two blocks of bars:
    1. Oxygen, nitrogen and nitrous oxide concentrations.
    2. Concentration of inhalation anesthetics.

C. Two blocks of pertinent model information.

The operational keyboard is designed logically as well, again in three groups:
1. A group of buttons for activating screen information which is constrained so that no more than three items appear at the same time.
2. Buttons for selecting experimental or patient data, and for presetting limits; there must be an adjoining but separate key (OK-button) to confirm each selection as it is being made.
3. Buttons separated from the operational keyboard area for other items.

CONCLUDING REMARKS

The described informative design is an example of how the application of a high-tech computer system develops into an integrator, processor, imaging equipment of many parameters and a controller and automatic corrector of deviated parameters. Thus the computer aid can be made most valuable to daily animal experimentation and patient investigation/control. A most complicated mode of operation and most confusing, non-informative parameter display can be transformed to a valuable assistance following simple rules of information design.
The daily task of the investigator becomes:
- a system manager where the experimental model or patient represents a dynamic system;
- gathering and recording data from the system;
- interpreting the data to detect deviation;
- controlling the system to maintain a desired state.

The investigator develops from an overloaded integrator to a computer operator, and the computer helps with integration.

REFERENCES

1. R.C. Cork, Computers in relation to anesthesia, in: General Anesthesia, Fifth Edition, 320-327, J.F. Nunn, J.E. Utting, B.R. Brown (eds.), Butterworths, London (1989).
2. W. Erdmann, Closing the loop from sensor to therapy, International Congress Series 399:35-42, Excerpta Medica, Amsterdam-Oxford (1977).
3. W. Erdmann, O. Prakash and R. Schepp, Closing the loop from sensor to therapy, an update, International Congress Series 637:595-602, Excerpta Medica, Amsterdam-Oxford (1984).
4. R.J. Saunders and W.R. Jewett, System integration - the need in future anesthesia delivery systems, Med. Instrumentation 17:389-392 (1983).
5. A.P.K. Verkaaik, A.W. Grogono and W. Erdmann, Informative imaging of multiple parameters for the practice of the anesthesiologist, Acta Anaesth. Belg. 41:201-209 (1990).

DIFFUSION LIMITATION OF OXYGEN IN HETEROGENEOUS LUNG AND TISSUE MODELS

Johannes Piiper

Abteilung Physiologie
Max-Planck-Institut für experimentelle Medizin
Göttingen, Germany

In the analysis of gas exchange in lungs models have been developed to distinguish diffusion limitation from other mechanisms producing gas exchange inefficiency such as ventilation/perfusion ($\dot{V}_A/\dot{Q}$) inhomogeneity and shunt. It will be attempted to apply similar models to describe and explain gas exchange in tissues, in particular to establish the role played by diffusion limitation in O_2 supply to exercising muscle.

LUNGS

There exists always an O_2 equilibration deficit usually measured as alveolar-arterial P_{O_2} difference (AaD_{O_2}). The first approach is to attribute this AaD_{O_2} to diffusion limitation. However, by specific methods it has been possible to show that most part of this O_2 equilibration deficit is caused by unequal distribution of alveolar ventilation to perfusion ($\dot{V}_A/\dot{Q}$ inhomogeneity) and shunt (venous admixture) (reviewed e.g. by Piiper and Scheid, 1989). Only in hypoxia and in heavy exercise O_2 diffusion limitation proper is expected to occur in normal lungs. The O_2 diffusion limitation has usually been considered in a homogeneous model, after the effects of shunt and $\dot{V}_A/\dot{Q}$ inequalities were taken into account, by the ideal-alveolar air approach or its modifications (Riley and Cournand, 1951; Haab et al., 1964; Wagner et al., 1988). The diffusion conditions, mostly quantified in terms or diffusing capacity D (diffusion conductance), have been considered as uniformly distributed in lungs.

In modeling of diffusion limitation in lungs with $\dot{V}_A/\dot{Q}$ inhomogeneity the problem arose how to model the distribution of D. D was distributed proportionally to either $\dot{V}_A$ or $\dot{Q}$ (Chinet et al., 1971; Geiser et al., 1983; Hammond and Hempleman, 1987; Hempleman and Gray, 1988; Gray, 1991). In this report, a model is presented with primary unequal distribution of D to $\dot{Q}$, i.e. independent of $\dot{V}_A/\dot{Q}$ distribution, developed on lines established previously (Piiper, 1961). After introduction of the equilibration index $D/(\dot{Q}\beta)$ as a generally applicable index for diffusion limitation (Piiper and Scheid 1981, 1983), it is attempted to reconsider the effects of unequally distributed diffusion conditions in terms of this variable. Furthermore, the error in determination of D due to disregard of its inhomogeneous distribution will be considered.

Model

The model (Fig. 1) consists of a gas phase (alveolar space) in gas exchange contact with blood in two diffusion/perfusion units (x = 1 or 2) which receive equal blood flow, but the diffusing capacities are varied at constant overall diffusing capacity, D_{tot}. Calculations of O_2 transfer are performed for straight O_2 dissociation curves, i.e. for $\Delta C_{O_2}/\Delta P_{O_2} = \beta = $ const. For

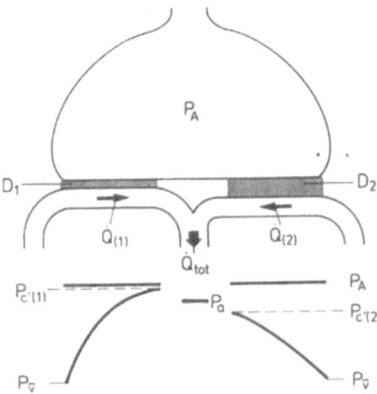

Fig. 1. Model for D/Q̇ heterogeneity comprising two diffusion-perfusion units with variable D/Q̇ ratio. Lower panel, gas/blood O_2 equilibration in both diffusion-perfusion units, and generation of arterial blood by mixing.

each unit (x) the O_2 gas/blood O_2 equilibration is given by the relationship (Piiper and Scheid, 1981),

$$\frac{P_A - P_{c'(x)}}{P_A - P_{\bar{v}}} = \exp [-D_{(x)}/(\dot{Q}_{(x)} \beta)] \qquad (1)$$

where P_A, $P_{c'}$ and $P_{\bar{v}}$ designate O_2 partial pressures in alveolar gas, in end-capillary and mixed venous blood.

Since $\dot{Q}_1 = \dot{Q}_2$, (mixed) arterial P_{O_2} is obtained as the arithmetic mean of $P_{c'(1)}$ and $P_{c'(2)}$. The apparent O_2 diffusing capacity D_{app} is calculated using modified eq. (1), disregarding the inhomogeneity:

$$D_{app} = (\dot{Q}_{tot} \cdot \beta) \cdot \ln \left[\frac{P_A - P_{\bar{v}}}{P_A - P_a} \right] \qquad (2)$$

D_{app} is compared to the true diffusing capacity, $D_{tot} = D_1 + D_2$.

Calculated results

With variation of the overall $D/(\dot{Q} \beta)$ from 5 to 0.5, corresponding to approximate values in normoxic rest to exercise in deep hypoxia (cf. Piiper and Scheid, 1981, 1983), the relative alveolar-arterial P_{O_2} difference, $(P_A - P_a)/(P_A - P_{\bar{v}})$, as well as the apparent/true diffusing capacity ratio, D_{app}/D_{tot}, were considered in function of the extent of D/Q̇ inhomogeneity, expressed as D_1/D_2. The following features were evident (Fig. 2).
(1) For the homogeneous model ($D_1/\dot{Q}_1 = D_2/\dot{Q}_2$) the ratio $(P_A - P_a)/(P_A - P_{\bar{v}})$ increases with decreasing $D/(\dot{Q} \beta)$, meaning increasing diffusion limitation.
(2) With increasing D/Q̇ inhomogeneity (displacement to the right on the abscissa),

320

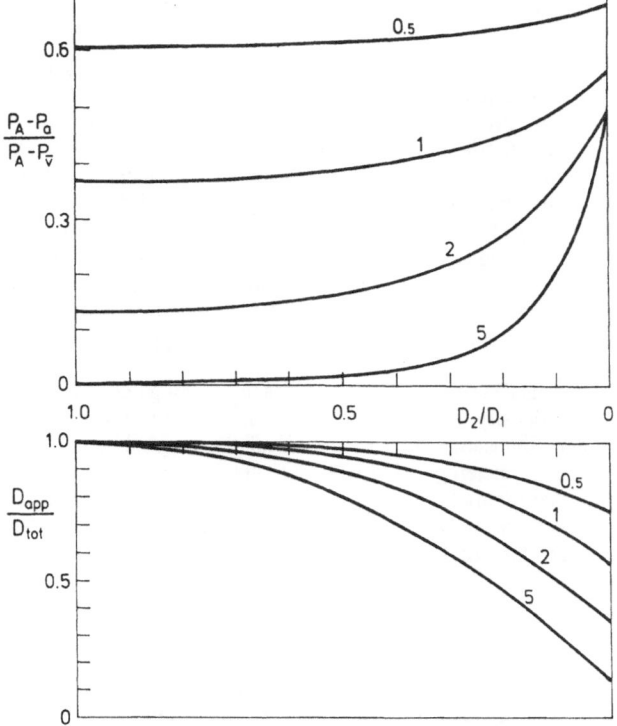

Fig. 2. Relative O_2 equilibration deficit, $(P_A - P_a)/(P_A - P_{\bar{v}})$, and apparent/true O_2 diffusing capacity ratio, D_{app}/D_{tot}, as function of the extent of $D/\dot{Q}$ inhomogeneity (abscissa) and the equilibration index, overall $D/(\dot{Q}\beta)$ (parameter).

$(P_A - P_a)/(P_A - P_{\bar{v}})$ increases, indicating increasing effect of inhomogeneity. Moreover, the effect is larger with higher $D/(\dot{Q}\beta)$.

(3) With increasing inhomogeneity, D_{app}/D_{tot} decreases, meaning underestimation of D when calculated using the homogeneous model. The effect is more pronounced with higher $D/(\dot{Q}\beta)$.

The calculated results show that in a lung model with $D/\dot{Q}$ inhomogeneity O_2 transfer efficacy is reduced compared to lung with the same D and $\dot{Q}$ but with equal distribution.

Mechanisms

Both D and $\dot{Q}$ are expected to show some dispersion and even more so the $D/\dot{Q}$ ratio. For example, a short capillary path would have a small D but a large $\dot{Q}$ and thus a much reduced $D/\dot{Q}$ ratio. In pulmonary diseases (edema, fibrosis) the tickness of the effective gas-blood barrier is expected to be inhomogeneously distributed.

The variance of $D/\dot{Q}$ may be also temporal due to pulsatile blood flow. This effect would be particularly enhanced in short capillary pathways or with large stroke volumes. In exercise, with high $\dot{Q}$ and high flow pulsatility, the O_2 transfer efficiency may be considerably reduced below the values expected on the basis of overall average D values. This factor may be involved in the large alveolar-arterial P_{O_2} differences observed to occur during extremely heavy exercise in normoxia (Wagner et al., 1988).

TISSUES

The limited O_2 extraction from blood in muscle during maximum O_2 uptake has been traditionally attributed to diffusion limitation, assuming a homogeneous diffusion/blood flow model (Stainsby and Otis, 1964; Hogan et al., 1988; Hogan et al., 1989; Roca et al., 1989). In analogy to lungs, unequal distribution of perfusion is expected to limit overall O_2 equilibration. This can easily be shown be model calculations. This is important since blood flow in resting and exercising muscle has been shown by many methods (e.g. inert gas washout, local xenon clearance, microsphere injection) to be markedly unequally distributed (reviewed by Piiper, 1990). Similar effects are produced by shunt, be it a true shunt flow or a functional shunt like arterio-venous diffusion shunt (Piiper, 1984). In many respects the effects of inhomogeneous blood flow and shunt resemble those of diffusion limitation. Unfortunately the methods developed to distinguish these mechanisms in lungs are not applicable in tissue.

To the unequal distribution of diffusing capacity in lungs corresponds in tissue unequal distribution of capillary density, reported to occur in many tissues (Egginton et al., 1988; Hoofd et al., 1985; Turek and Rakusan, 1981). Model simulations reveal some characteristic signs of diffusion and/or perfusion heterogeneity when O_2 uptake is O_2 delivery dependent: smaller maximum O_2 uptake than predicted on the basis of O_2 delivery and average capillary density; and a more gradual decrease of O_2 uptake with decreasing O_2 requirement or O_2 delivery (for more details, see Piiper, this volume).

CONCLUSIONS

In both lungs and tissue, unequal distribution of diffusing capacity (diffusion conductance) with respect to other gas transport variables (blood flow, O_2 consumption) leads to exchange inefficiency. In lungs, the arterialization of pulmonary effluent blood is decreased compared to homogeneous distribution of D to Q̇. In tissues, local anoxia develops at higher levels of overall O_2 delivery in the case of unequal distribution of diffusing distances as modeled by radii of Krogh cylinders. These effects are qualitatively comparable to effects of unequal blood flow distribution (in lungs, relative to ventilation; in tissue, relative to O_2 requirement). Thus quite generally, in inhomogeneous systems O_2 delivery is compromised.

REFERENCES

Chinet, A., Micheli, J.L., and Haab, P., 1971, Inhomogeneity effects on O_2 and CO pulmonary diffusing capacity estimates by steady-state methods. Theory, Respir. Physiol., 13: 1-22.
Egginton, S., Turek, Z., and Hoofd, L., 1988, Differing patterns of capillary distribution in fish and mammalian skeletal muscle, Respir. Physiol., 74: 383-396.
Geiser, J., Schibli, H., and Haab, P., 1983, Simultaneous O_2 and CO diffusing capacity estimates from assumed lognormal $\dot{V}_A$, $\dot{Q}$ and D_L distributions, Respir. Physiol., 52: 53-67.
Gray, A.T., 1991, The effects of D/Q̇ and D/$\dot{V}_A$ inequalities on pulmonary oxygen diffusing capacity estimates, Respir. Physiol., 84: 287-293.
Haab, P., Duc, G., Stucki, R., and Piiper, J., 1964, Les échanges gazeux en hypoxie et la capacité de diffusion pour l'oxygène chez le chien narcotisé, Helv. Physiol. Acta, 22: 203-227.
Hammond, M.D., and Hempleman, S.C., 1987, Oxygen diffusing capacity estimates derived from measured $\dot{V}_A/\dot{Q}$ distributions in man, Respir. Physiol., 69: 129-147.
Hempleman, S.C., and Gray, A.T., 1988, Estimating steady-state $D_{L_{O_2}}$ with nonlinear dissociation curves and $\dot{V}_A/\dot{Q}$ inequality, Respir. Physiol., 73: 279-288.
Hoofd, L., Turek, Z., Kubat, K., Ringnalda, B.E.M., and Kazda, S., 1985, Variability of intercapillary distance estimated on histological sections of rat heart, Adv. Exp. Med. Biol., 191: 239-247.
Hogan, M.C., Roca, J., Wagner, P.D., and West, J.B., 1988, Limitation of maximal O_2 uptake and performance by acute hypoxia in dog muscle in situ, J. Appl. Physiol., 65: 815-821.

Hogan, M.C., Roca, J., West, J.B., and Wagner, P.D., 1989, Dissociation of maximal O_2 uptake from O_2 delivery in canine gastrocnemius in situ, J. Appl. Physiol., 66: 1219-1226.

Piiper, J., 1961, Unequal distribution of pulmonary diffusing capacity and the alveolar-arterial P_{O_2} differences: theory, J. Appl. Physiol., 16: 493-498.

Piiper, J., 1984, Mechanisms of functional shunting in mammalian skeletal muscle, in: "Cardiovascular Shunts" (Alfred Benzon Symposium 21), P. Krogsgaard-Larsen, S. Brogger Christensen and H. Kofod, eds., Munksgaard, Copenhagen, pp. 467-485.

Piiper, J., 1990, Unequal distribution of blood flow in exercising muscle of the dog, Respir. Physiol., 80: 129-136.

Piiper, J., and Scheid, P., 1981, Model for capillary-alveolar equilibration with special reference to O_2 uptake in hypoxia, Respir. Physiol., 46: 193-208.

Piiper, J., and Scheid, P., 1983, Comparison of diffusion and perfusion limitations in alveolar gas exchange, Respir. Physiol., 51: 287-290.

Piiper, J., and Scheid, P., 1989, Gas exchange - theory, models and experimental data, in: "Comparative Pulmonary Physiology - Current Concepts, Vol. 39", S.C. Wood, ed., Marcel Dekker, Inc., New Yor, Basel, pp. 369-416.

Riley, R.L., and Cournand, A., 1951, Analysis of factors affecting partial pressures of oxygen and carbon dioxide in gas and blood of the lung: theory, J. Appl. Physiol., 4: 77-101.

Roca, J., Hogan, M.C., Story, D., Bebout, D.E., Haab, P., Gonzalez, R., Ueno, O., and Wagner, P.D., 1989, Evidence for tissue diffusion limitation of $\dot{V}_{O_2}$max in normal humans, J. Appl. Physiol., 67: 291-299.

Stainsby, W.N., and Otis, A.B., 1964, Blood flow, oxygen tension, oxygen uptake, and oxygen transport in skeletal muscle, Am. J. Physiol., 206: 858-866.

Turek, Z., and Rakusan, K., 1981, Lognormal distribution of intercapillary distance in normal and hypertrophic rat heart as estimated by the method of concentric circles: its effect on tissue oxygenation, Pflügers Arch., 391: 17-21.

Wagner, P.D., Gale, G.E., Moon, R.E., Torre-Bueno, J.R., Stolp, B.W., and Salzman, H.A., 1988, Pulmonary gas exchange in humans exercising at sea level and simulated altitude, J. Appl. Physiol., 61: 260-270.

GAS EXCHANGE IN THE LUNG, COMPUTER FEED BACK CONTROLLED

PHYSIOLOGICAL MATCHING OF ARTIFICIAL VENTILATION

A.P.K. Verkaaik, G. van Dijk, B. Westerkamp and
W. Erdmann
Department of Anesthesiology, Erasmus University Rotterdam, The
Netherlands

INTRODUCTION

Maintenance of adequate gas exchange in the critically ill can often only be achieved by artificial ventilation. Physiological negative pressure insufflation was primarily achieved by Sauerbruch's negative-pressure chamber and later developed to an artificial respirator generally called the "iron lung". However, for ventilation respectively insufflation the body of the patient needed to be enclosed in a chamber and access for nursing measures, positioning of the patient, cleaning e.g. was difficult. Furthermore, changes of parameters in ventilation, introduction of positive pressure gradients if needed and high frequent ventilation mode was not possible. Intermittent positive pressure ventilation via endotracheal tubes, as used generally today, proved to be more practical and efficient.

Considering this physiologically reversed mode of lung ventilation there exist several shortcomings: positive pressure ventilation, although guaranteeing a sufficient oxygen uptake has influence on pulmonary hemodynamics, lung mechanics and distribution of the ventilated volume. These parameters are of equal importance and cannot be separated from each other. The concept of modern artificial ventilation has thus to be summarized in the following way:

1. Achievement of optimal oxygenation of the blood (and CO_2 elimination) with the least increase of inspiratory oxygen concentration above the normal of 21%.
2. The least mechanical distention of lung tissue.
3. Avoidance of paradox bronchiolar ventilation and maldistribution of ventilated volume respectively biphasic distribution resulting in the "Pendelluft" phenomenon.
4. Limiting of mean intrapulmonary pressure to minimal values and keeping interference with pulmonary vascular resistance and cardiac output at a negligible level.

These guidelines, however, are often not achievable as dynamic ventilation modes are not possible with conventional fixed parameter adjustments. The mechanically engineered ventilators currently used are either pressure or volume controlled with adjustable steady inspiratory flows up to a certain pressure or tidal volume, a plateau with inflated alveolars kept open (oxygen uptake phase), expiration with adjustable delay modes (CO_2 elimination) and adjustable expiratory plateau pressures (positive end-expiratory pressure: PEEP). This is necessary to avoid alveolar collapse and/or fluid invasion, especially in case where the patient lies continuously on the back with consequent appearance of atelectasis and increase of pulmonary shunt. This mechanically achieved artificial ventilation mode, aimed to keep alveolar gas exchange as optimal as possible, however, did not allow to introduce physiologically dynamic inspiratory and expiratory flow patterns and several shortcomings up to date create major problems in long-term ventilation: both from the hemodynamic and lung mechanics viewpoint. The major problem is the appearance of unphysiological peak pressures during which period easily ventilated alveoli are over-extended partially combined with intermittent bronchiolar dilatation followed by floating back of partially deoxygenated air from the primarily over-extended areas into later opening ("Pendelluft"-phenomenon). Valves with changing resistance in the circuit are another problem, as are dry air and dry gases derived from compressed gas cylinders leading to mucociliar dysfunction. Increase of atelectasis and loss of lung compliance lead to increasing impairment of alveolar capillary oxygen transfer and ask for increase of inspiratory oxygen concentration and increase of mean intrapulmonary pressure with concomitant increase of pulmonary perfusion pressure (increase of pulmonary vascular resistance, appearance of increased shunt perfusion, increase of right heart workload and venous pooling). The final consequence is difficult weaning of the patient from the ventilator.

Aim of the presented work was to develop a system that permits application of any predesigned ventilatory mode to the patient. This to enable, e.g. the physician, lung physiologist, intensive care specialist to predetermine the most optimal artificial ventilation pattern specifically designed for the pathophysiology of each individual patient. To achieve this, the system must be capable to perform dynamic non-linear processes where necessary. Inclusion of automatic feed back control of respiratory gas concentrations according to preset values was also a prerequisite.

TECHNICAL AND FUNCTIONAL DESCRIPTION

A dedicated 40 Megabyte computer on the basis of a 16-bits microprocessor in VLSI-architecture was assembled and software controlled to initiate and control all wanted feed-back regulation processes (Figs.1-3).

To avoid any risk no moving parts were included into the central computer system (e.g. hard disks or tape streamers). The software is printed in Eprom and the memory is filed in a non-volatile random access memory (RAM). This guarantees safe and accurate control and regulation of all regulatory tasks assigned to the computer, whereby a critical mass is achieved that shows strong functional and synergistic effects. The patient is connected to a closed circuit through which the fully humidified air is rotated unidirectional with 70 l/min by a blower; this avoids any rebreathing and makes valves unnecessary. CO_2 is absorbed by soda lime in the expiratory part of the system and oxygen is fed into the inspiratory part via closed loop feed-back control of a paramagnetic oxygen sensor. A frequency valve injects 0.5 ml opening controlled by the central computer to match the preset inspiratory oxygen concentration. In this way oxygen inflow equals oxygen consumption by the patient (Erdmann et al, 1989).

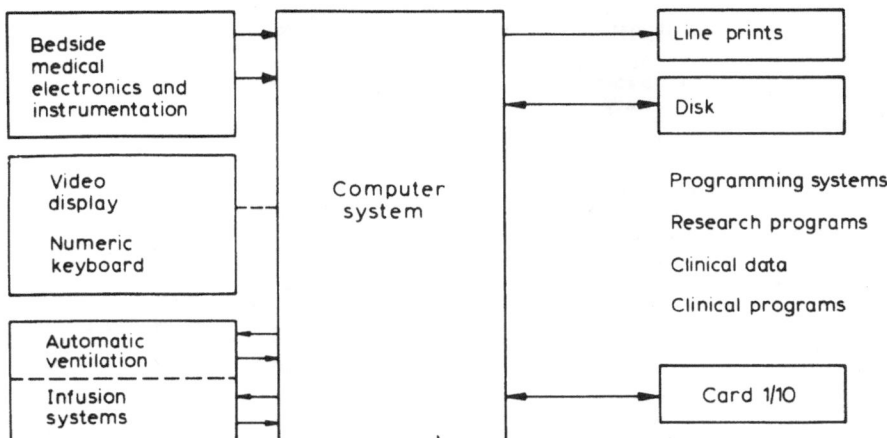

Fig.1.　Function of the automated care system: preparation of preoperative, intra-operative and postoperative orders; automatic measurement acquisition, display, storage and charting; measurement retrieval for review of trends, the cardiac output measurement, detection of impaired cardiac performance, blood loss trend analysis, blood gas measurement entry, acid base balance computation and assessment.

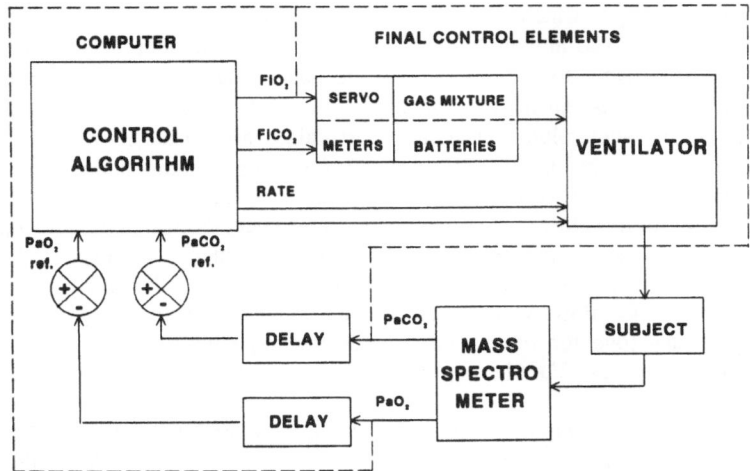

Fig.2.　Schematic diagram of automatic ventilation (continuous blood gas controlled ventilation - CBC ventilation), using the intensive care computer, IBM 1800, PDP 11/20. The blood gases are measured continuously via an arterial catheter with connection to a mass spectrometer, the respective values are fed into the control algorithm and compared to a desired PaO_2 reference or $PaCO_2$ reference, the control algorithm controls the servoventilator for tidal volume and rate (arterial pCO_2) or if desired, a gas mixing battery of a servo-motor at prefixed rate and tidal volume for $PaCO_2$ regulation. PaO_2 is always controlled by a servomotor driven gas mixing battery.

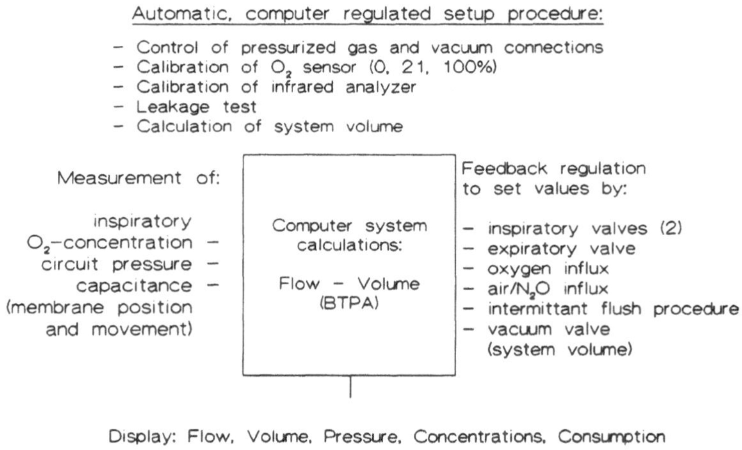

- Control of pressurized gas and vacuum connections
- Calibration of O_2 sensor (0, 21, 100%)
- Calibration of infrared analyzer
- Leakage test
- Calculation of system volume

Measurement of:

inspiratory
O_2-concentration –
circuit pressure –
capacitance –
(membrane position
and movement)

Computer system
calculations:

Flow – Volume
(BTPA)

Feedback regulation
to set values by:

- inspiratory valves (2)
- expiratory valve
- oxygen influx
- air/N_2O influx
- intermittant flush procedure
- vacuum valve
  (system volume)

Display: Flow, Volume, Pressure, Concentrations, Consumption

Fig.3.   Schematic description of data flow, derived calculated parameters, display and feed-back control functions of a computer feed-back controlled ventilator for physiological matching of artificial ventilation.

A flat membrane chamber (Fig.4) is integrated into the circuit with the circulating air. The rounded shape of the chamber avoids resistance increasing turbulances of the gas flow. A metal membrane separates the patient side with rotating gas stream from the external part through which the membranes are moved by either increasing or decreasing pressure. The moving bellow is a metal membrane hung up in a rolling seal that changes its distance to a metal plate in the upper chamber wall during ventilation movement. The displacement movement is measured by capacitance changes and registered in flow and volume. The measured values are corrected for pressure and compressible volume of the circuit to actual flow and volume administered to the patient, which are thus registered in BTPA (body temperature and pressure adjusted) values. The resulting values are fed into the microprocessor which in turn opens the inspiratory port of the external part of the ventilation chamber or the expiratory port, to achieve a ventilation pattern congruent with the predesigned ventilation curve characterized by predetermined flow, volume and pressure values. The inspiratory port has two inspiratory values (one for rough and one for fine regulation) connected to pressure reduced compressed air. For the expiratory port a one-valve system was found to be sufficient.

In summary, the computer controls feed back the interplay between the inspiratory pressure increasing and the expiratory pressure decreasing valves. Ventilation movements of the metal membranes are accordingly initiated to exactly redraw the ventilation pattern predesigned by the physician on the computer screen.

There are actually no limits to the ventilatory modes possible, as long as they are defined by the respective software. Any static or dynamic mode for ventilation can be introduced via the computer screen with the respective and defining parameters. The preset ventilation curve is precisely redrawn by the computer whereby volume as well as pressure parameters and any combination thereof can be transformed to a dynamic ventilation curve that is volume as well as pressure controlled. Peak pressures, redistribution of ventilated volume ("Pendelluft"), paradoxic bronchiolar ventilation e.g. are avoided.

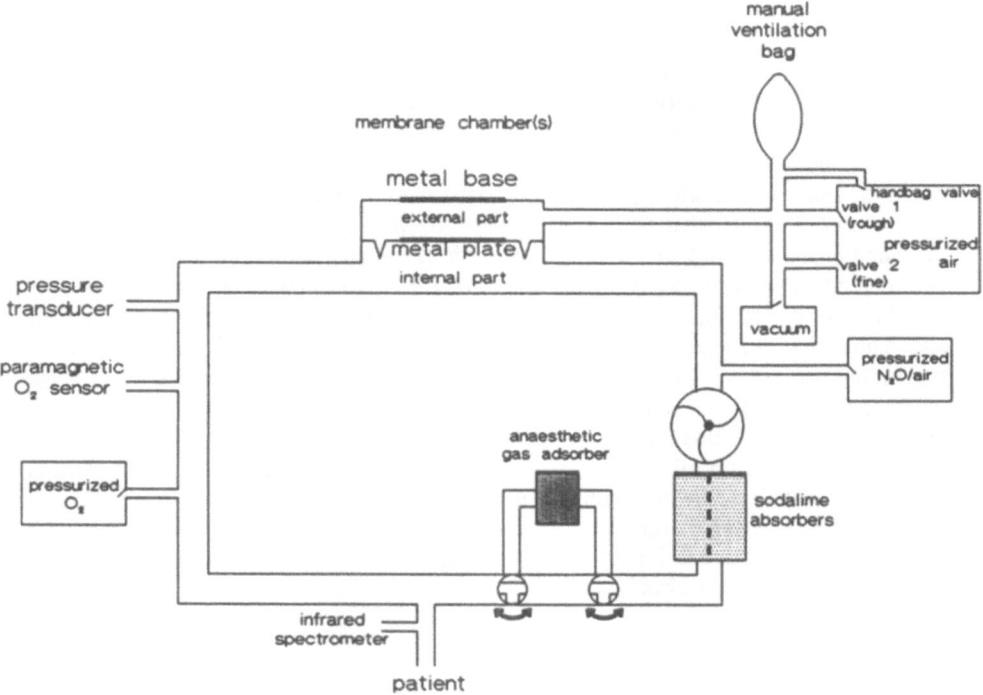

Fig.4.   High flow closed circuit system with gas-rotating blower.
Ventilation is performed by metal membranes fixed in a rolling seal separating the external part of the ventilation chamber from the patient side. Pressure changes (through injection of compressed air and reversively valve release) in the external chamber side move the membrane performing patient ventilation and change its distance to the metal base in the chamber wall; the distance between metal plate (membrane) and metal base is capacitively measured and calculated into flow and volume displacement. Inspiratory oxygen concentration and system volume is feed-back controlled as is the exact movement of the metal membrane.

APPLICATION

To avoid peak pressure during ventilation, the ventilator performs primarily a conventional insufflation according to the preset inspiratory flow and the preset tidal volume and determines the plateau pressure (after volume redistribution) necessary to expand the lungs with the wanted tidal volume. The inspiratory flow decreases when the necessary plateau pressure is approached but continues as long as the plateau pressure is not surpassed ($\pm$ 10% range), and only ceases when the preset tidal volume is reached. In case the preset inspiratory fraction of the respiratory cycle is too short to reach the preset tidal volume, an alarm signal appears. To correct for this, the I/E ratio has to be changed or the respiratory pressure limit (plateau pressure + 10%) has to be increased. During expiration the reverse happens: The preset expiratory flow determines the primary expiratory flow rate which dynamically approaches the set end expiratory pressure; if full expiration cannot be achieved in the allocated time for the expiration phase, an alarm signal alerts that primary expiratory flow has to be increased, or if feasible, the I/E ratio respectively the respiratory rate must be adjusted.

CONTROL VALUES

On-line measurement of expiratory CO_2 and oxygen saturation measurement via non-invasive pulse oximetry are the basic continuous control parameters supplemented if necessary by invasive blood gas analysis. In the critically ill hemodynamic measurements should also be available (pulmonary artery pressure or at least central venous pressure, and invasive arterial pressure). These values together with on-line oxygen consumption offer insight into a whole spectrum of oxygen supply parameters and whether matching or mismatching of supply/demand ratio is present. Instead of volume feed back control end expiratory CO_2 feed back control can be introduced as, for example, a possibility to control intracranial blood volume and intracranial pressure in neurosurgical and neuro-intensive care patients; this feed-back control mode uses tidal volume as regulation parameter. The described system has been tested experimentally in pigs during pressure and volume controlled ventilation as well as the combined pressure/volume controlled mode. Body plethymography was used to assess accurate measurement of the system as basis for all further feed-back control processes. More than 2000 patients have been treated meanwhile with full documentation through a 90 hours memorizing capacity of all measured and preset values. A specially designed easily used "teleflex" program permits data copying on any IBM-compatible PC-system cleaning the memory for new data storage (Verkaaik and Erdmann, 1990).

FUTURE ASPECTS

There is no limit for new software induced additional ventilatory modes or help functions (e.g. SIMV, CPAP, pressure or volume/inspiratory flow assisted ventilation). Using the described control values expert system development permits self-controlled assessment of the most optimal ventilation mode in critical disease state in respect to sufficient oxygen uptake with the lowest inspiratory oxygen concentration possible, and adequate CO_2 elimination, surveillance of interference with lung/cardiovascular hemodynamics and pulmonary compliance respectively lung mechanics.

Hardware and software adjustments permit acute stepwise changes in gas concentrations and thus uptake and clearance curve registration using infrared spectrometrically measurable gases (e.g. N_2O, halogenated carbohydrates). Non-invasive intermittent determination of pulmonary perfusion (cardiac output) and shunt perfusion comes within reach; this is due to two important facts:

1. the high rotation of system gas leading to continuous complete mixing without gas concentration differences within the system and,
2. full and complete removal of all exhaled halogenated carbohydrates in an active carbon adsorber.

REFERENCES

Erdmann, W., Veeger, A. I., and Verkaaik, A. P. K., 1989, Narkosebeatmungsgeräte: Gegenwart und Zukunft. In: Narkosebeatmung, low flow, minimal flow, geschlossenes System, J.P. Jantzen, P.P. Kleemann, eds., Schattauer, Stuttgart.

Verkaaik, A. P. K., Erdmann, W., 1990 Respiratory diagnostic possibilities during closed circuit anesthesia, Acta Anesth. Belg., 41:177-188.

NON-INVASIVE, ON-LINE MEASUREMENT OF OXYGEN CONSUMPTION DURING ANESTHESIA

Allan PK Verkaaik, Jan-Willem Kroon,
Helmie GM van den Broek, Wilhelm Erdmann,

Department of Anesthesiology
Erasmus University and University Hospital
Rotterdam, The Netherlands

SUMMARY

Oxygen consumpton is usually derived from values measured by a thermodilution catheter, i.e. an invasive procedure, with associated risks and calculation errors. Closed circuit ventilation provides a reliable, non-invasive means of access to this parameter of metabolism. Routine application of closed circuit ventilation requires overcoming many, mainly technical, difficulties (e.g. leakage problems, valve malfunctions, and calculations of gas uptake). We developed a computerized closed circuit anesthesia ventilator without these problems. With this system non-invasively measured oxygen uptake is continuously presented on-line. We discuss three representative patients presenting for surgical repair of an abdominal aneurysm. Besides actual changes in metabolism and depth of anesthesia, success or failure of the operation is visible in the pre- and direct post-clamping period. With resuscitative therapeutic interventions, increase in oxygen consumption gives valuable information under changing conditions. We conclude that closed circuit anesthesia is a safe and valuable method for measurement of oxygen consumption.

INTRODUCTION

On-line measurement of oxygen consumption (VO_2) is not yet established routine during anesthetic and surgical procedures, and in intensive care. Many reports, however, have stressed the need for measurement of this parameter [1-4].
When measurement of oxygen consumption is indicated, cardiac output is usually derived from values measured by a thermodilution catheter. By applying the Fick equation, $VO_2 = Q*(CaO_2-CvO_2)$, oxygen consumption is calculated.
It has been reported, however, that calculation of cardiac output derived by the thermodilution technique may have an error margin of 10% [5]. Also, integrated oxygen saturation

(SO_2) calculation programs in commercial blood gas analyzers
are of questionable use [6]. Calculation of additional parameters based on these values might increase a primary calculation error and significantly influence the final result (e.g.
VO_2).

General acceptance of Fick's formula, and its inherent shortcomings and complications [7], is probably due to the fact
that access to indicators of metabolism and cardiac output
remains difficult. Spirometry could give factual and reliable
insight into uptake of gases, and direct measurement of oxygen
consumption with this method is very reliable [8]. However,
use of a spirometer during anesthesia for surgery would be
cumbersome, if not impossible, because induction and maintenance of anesthesia and ventilation are themselves problematic.

On-line VO_2 measurement is possible with closed circuit ventilation [9]. This technique, however, is still not widely
accepted, mainly because the systems used for closed circuit
ventilation involves adjustments to existing ventilators originally designed as semi-closed systems. The resulting configuration leads to leakage problems, valves may stick in the
humid atmosphere and reliable measurement of metabolic and
anesthetic gas concentrations and uptake is difficult to realize [10].

Recently, we developed a reliable closed circuit anesthesia
ventilator which avoids all these problems [11]. The conventional bag-in-bottle system that has dominated ventilation technology so far was avoided by returning to the much older and
simpler high flow technique [12]. This was possible through
application of the power and flexibility of a modern digital
computer to overcome the problems that once led to its rejection. With this ventilator system accurate and direct on-line
measurement of VO_2 is possible [13].

PATIENTS AND METHODS

Closed Circuit Anesthesia Ventilator (Physioflex®, Physio BV,
Hoofddorp, The Netherlands) (Fig 1).
This closed circuit system is based on a spirometer normally
applied in lung function diagnostics. Valves are avoided by a
high unidirectional flow (70 L/min) using a blower. Instead of
adding fresh gases, the gases are made fresh in the inspiratory port. In fact, we reverted to the Ayre system [12] with
the aid of a spirometer; to integrate the ventilation mode
into the system the closed rebreathing spirometer is replaced
by a membrane ventilator. Changes in volume are controlled by
freely moveable air-tight membranes in membrane chambers. Up
to four separate membrane chambers are cumulatively in use
based on the principle to have the least possible compressible
volume in the system. The ventilatory membrane (a metal plate)
is placed on a rubber membrane and moves freely. Movements of
the membrane are capacitively measured, enabling continuous
registration of flow and volume. The carrier gas (nitrous
oxide or air) is feedback controlled, regulating the end-expiratory volume to the original setting by exact control of the
position of the membranes. Because of the high flow of circu-

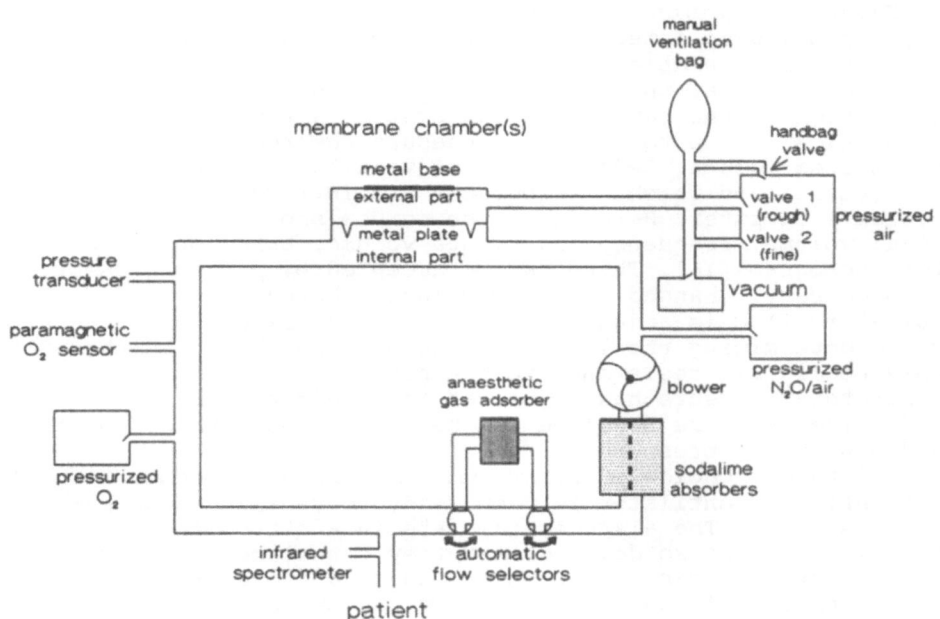

manual
ventilation
bag

membrane chamber(s)

handbag
valve

metal base

external part

valve 1
(rough)

pressurized
air

pressure
transducer

metal plate

internal part

valve 2
(fine)

vacuum

paramagnetic
O_2 sensor

pressurized
N_2O/air

anaesthetic
gas adsorber

blower

pressurized
O_2

sodalime
absorbers

infrared
spectrometer

automatic
flow selectors

patient

Fig 1. Schematic representation of the closed
circuit anesthesia ventilator.

lating gas (70 L/min) due to the blower system and free movement of membranes in their chambers, there is very low resistance in the system. During spontaneous respiration the patient breathes freely out of and into the high flow gas stream. Manual ventilation can be performed by a bag. Artificial ventilation is initiated by sophisticated interaction of two inspiratory valves (one for rough and one for fine regulation) connected to pressurized gas, and one expiratory valve connected to ambient air pressure or suction. This interaction is computer controlled. Development of adequate software is required to perform different modes of ventilation. CO_2 is continuously absorbed by soda lime. An infrared absorption spectrometer measures CO_2 and (if used) anesthetic gas concentrations, which are feedback controlled. Oxygen is analyzed by a paramagnetic oxygen analyzer and is also feedback controlled to the set value [13]. In the nitrous oxide mode, measurement of N_2O concentration is done by infrared absorption spectrometry, enabling to detect malfunction of the paramagnetic oxygen analyzer, accumulation of unwanted gases in the system, or leakage of gas at the patient side. To achieve this, before connection of the patient to the system, a test program is automatically run to check and measure the respective controls and system volume. Before start of ventilation, the patient's age (years) and weight (kilograms) are introduced into the system's computer. Hereafter, the screen shows the advised tidal volume, frequency and minute volume, based on the Radford nomogram [14]. These can be accepted by pressing OK or, if necessary, changed to other values. Thereafter, the inspiratory flow is adjusted, as is the inspiratory/expiratory ratio determining the length of the plateau phase. As there is only negligible resistance to the patient's expiration, end-expiratory pressure has to be set to determine the expiration mode. The pressure baseline can be set in the form of positive end-expiratory pressure.

Gas mixtures, choice of N_2O or air, and the O_2 concentration are set; the ventilator is then ready to ventilate the patient's lungs. The alarm systems are in a state of constant function and are so devised that they cannot be disregarded.

A flush program can be run on demand so as to decrease the concentration of unwanted gases. During two min (e.g. once hourly), a fresh gas flow of 2.5 L/min enters the circuit, while at the same time the expiratory valve opens in order to keep the volume unchanged.

The system enables continuous control of oxygen consumption, CO_2 production, physiologic parameters of the patient's lungs, and gas concentrations. During ventilation, pressure and flow curves are continuously presented. All these data are automatically saved on disk every minute. Airway resistance and lung compliance can be deduced.

Patients

The following examples (based on 3 representative patients) illustrate how direct measurement of oxygen consumption is presented during abdominal aortic bifurcation surgery. Anesthesia techniques are listed in Table I.

    Patient A is a 69-year-old man, weighing 97 kg, scheduled for repair of an abdominal aneurysm. Angiography showed a 5 cm dilatation of the aorta below the renal arteries, with extension of the dilatational process in the common iliac arteries and in the right internal iliac artery. Both external iliac and femoral arteries were normal. The aneurysm was an incidental finding, as the patient underwent nephrectomy because of an infected cyst 4 months previously. Otherwise, he was in good health, without concurrent medical problems and no physical complaints. Renal function was slightly disturbed; cardiovascular parameters were within

Table I. Anesthesia techniques used for abdominal aortic bifurcation surgery.

	Patient A	Patient B	Patient C
Gasmixture	O_2/air	O_2/air	O_2/air
Induction			
Hypnotic	Thiopental	Midazolam	Ketamine
Opiate	Fentanyl	Sufentanil	None
Muscle relaxant	Pancuronium	Pancuronium	Succinyl-choline
Maintenance			
Hypnotic	Isoflurane	Midazolam	Enflurane
Opiate	Fentanyl	Sufentanil	Sufentanil
Muscle relaxant	Pancuronium	Pancuronium	Pancuronium
Dopamine	Yes	No	Yes
Nitro-prusside	No	Yes	No

normal limits and lung function was relatively good: FEV_1/VC: 75 (83%). Coagulation prophylaxis with phenprocoumarol (Marcoumar) was started on admission. Preoperative values of Hb, urea and creatinin were 14.5 g%, 6.9 mmol/L and 105 mmol/L, respectively.

Patient B had a bodyweight of 70 kg. He was scheduled for repair of an abdominal atherosclerotic aorta bifurcation (5 cm), with angiographic evidence of involvement of one of the renal arteries, stenosis of both deep femoral arteries and obliteration of both superficial femoral arteries, with development of collaterals. His physical condition was inferior to that of patient A. A transient ischemic attack (basal artery) several years previously, with persistent insufficiency of blood flow to the brain stem, was first treated with aspirin, and later with phenprocoumarol (Marcoumar). Lung function showed a (moderate) obstruction: FEV_1/VC = 56 (76%). Medication: phenprocoumarol, digoxin 125 μg, aminophyl-

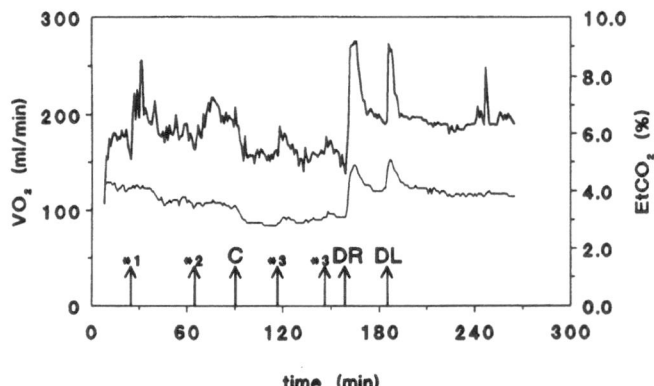

Fig 2. Recorded oxygen consumption (VO_2 = ▬) and end-tidal
CO_2 ($EtCO_2$ = ──) in patient A (97 kg) undergoing
elective aneurysm repair. After 20 min, at *1 FiO_2 was
changed from 0.30 to 0.40. Basal oxygen consumption was
175 ml/min. After clamping (C), this reduced to 160
ml/min. At *2, VO_2 increased due to increased release of
adrenaline caused by insufficient analgesia, which was
treated with fentanyl. At *3 dopamine was administered
to treat low blood pressure. Declamping of first the
right (DR), and later the left (DL) leg resulted in an
increase in VO_2, after a large overshoot to pay back an
oxygen debt. After the procedure VO_2 was higher than at
the beginning. Changes in end-tidal CO_2 are also
apparent, but less distinctive than the changes in VO_2.

line 600 mg/24 h. Preoperative values of Hb, urea and
creatinin were 13.5 g%, 10.1 mmol/L and 108 mmol/L, respectively.

Patient C was admitted to the emergency ward because of
a ruptured abdominal aneurysm. His weight was estimated at 80
kg, blood pressure on admission was 65/30 mmHg and no medical
history was available.

RESULTS

Patient A (Fig 2)
According to the Brody equation ($10*kg^{3/4}$) [15], resting VO_2
should be 320 ml/min. After induction of anesthesia and start
of ventilation, mean VO_2 before surgery was 175 ml/min. The
first major change in VO_2 was caused by an increase of FiO_2
from 0.30 to 0.40, as a result of a temporary influx of oxygen
into the circuit. After this correction, VO_2 returned to the

basal value. The increase in VO_2 due to an increase in endogenous adrenaline (probably caused by insufficient analgesia), was treated with intravenous fentanyl. Administration of exogenous catecholamines (in this case dopamine) is also reflected by a change in metabolism and, therefore, oxygen consumption. At C, the time the aorta was clamped, VO_2 decreased to 160 ml/min. At DR (declamping right leg) and at DL (declamping left leg) a substantial overshoot in oxygen con-

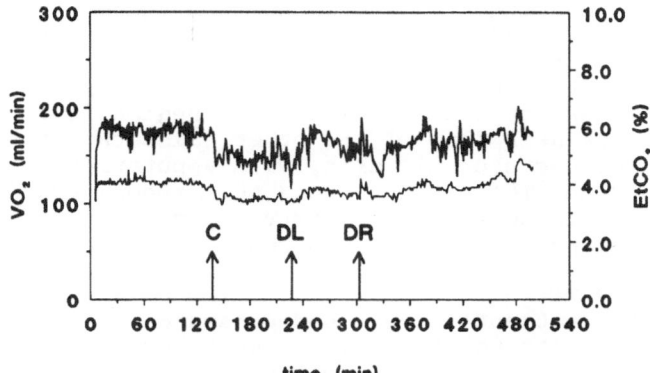

time (min)

Fig 3. Recording of VO_2 (▬▬) and $EtCO_2$ (▬▬) during aneurysm repair in patient B (70 kg; FiO_2: 0.40). Basal oxygen consumption was 175 ml/min. Clamping (C) resulted in a decrease in VO_2, that returned to normal after declamping of the left leg (DL). Declamping of the right leg (DR) failed to produce an increase in oxygen consumption. There is no overshoot to indicate repay of an existing oxygen debt. VO_2 after the procedure was the same as before. The operation was extended to a bilateral femoropopliteal bypass.

sumption occurred to 260 ml/min, lasting several minutes before returning to a point higher than the original VO_2. The increase in end-tidal CO_2 during these instances is probably influenced by the administration of sodium bicarbonate, given to the patient to decrease the sudden changes in acid-base balance. After the procedure, mean VO_2 had increased to 195 ml/min.
Four hours after intubation, the abdomen was closed and the patient transferred to the ICU for short-term ventilation. One day later the patient was returned to the ward and subsequently developed a light pneumonia which was treated with antibiotics. He left the hospital 10 days after the operation in good condition.

Patient B (Fig 3)

After induction and start of ventilation, mean VO_2 was 175 ml/min (Brody equation: 240 ml/min). As in patient A, VO_2 can be closely followed. At C, clamping of the aorta, VO_2 decreased to 145 ml/min. At declamping of the left leg (DL) a minor increase in VO_2, back to normal, was seen, but with no overshoot.

Declamping of the right leg (DR) failed to produce a change in oxygen consumption. Manual, and later ultrasound examination (Doppler), confirmed that there was no pulsatile blood flow to the patient's legs. The operation was finally extended to include a femoropopliteal bypass, left and right. After 11 hours of operation, the patient was transferred to the ICU. This extended operation failed to produce the desired result, namely improvement of circulation to the lower part of the body. It was necessary to exarticulate one of the hip joints several days later. Because of necrosis in the sacral field due to impairment of blood flow to that region it was later proposed that a skin flap of the other, also threatened, leg should replace the necrotic skin, with amputation of that leg at the same time. The patient found this recommendation unacceptable, he refused further surgery and died in hospital 13 days after the attempted aneurysm repair.

Patient C (Fig 4)

Because of the patient's low blood pressure, immediate therapy consisted of replacing the lost volume by crystalloids and colloids, while blood was crossmatched. A few minutes after admission, he entered the operation theater, was intubated and ventilation was started. Weighing 80 kg, the patient's mean VO_2, about 120 ml/min, was much lower than expected (270 ml/min, *vide supra*). The Hb concentration was 3.8 gr%. Fourty min after intubation, crossmatched blood was available and was immediately infused (BT) to increase oxygen delivery and, thus, VO_2. Indeed, oxygen consumption increased to 160 ml/min. Blood pressure was controlled by dopamine. Declamping resulted in an increase in mean VO_2 (210 ml/min), after a short overshoot to repay the debt that was accumulated during the anaerobic period in the lower parts of the body. The patient, in reasonably stable condition, was transferred to the ICU after 6 h of operation for further ventilation and therapy. During this period, his condition deteriorated. Blood pressure reduced in spite of volume loading guided by wedge pressure, vasoactive drugs and M.A.S.T. (military anti-shock trousers) therapy. He also developed anuria and ARDS. The following day, the patient was moribund and the decision was made to cease further treatment. Results from autopsy: 1. micronodular liver cirrhosis; 2. hemorrhagic pancreatitis; 3. anterior communal cerebral artery aneurysm (all three pre-existing); 4. extensive retroperitoneal bleeding. Direct cause of death was due to myocardial hypoxia. The patient was 69 years old.

DISCUSSION

The case of patient A (as compared to patient B), clearly demonstrates that repair was successful. The changes in VO_2

and, therefore, cardiovascular responses to pain, exogenous catecholamines, clamping and declamping, can be closely followed. The large overshoot of VO_2 after declamping of both legs in patient A demonstrates a large pay back of an existing oxygen debt, as a result of more tissue taking part in aerobic metabolism after a period of occlusion [16]. This phenomenon is clearly presented on-line as it happens, enabling both surgeon and anesthesiologist to continuously assess the clinical situation.

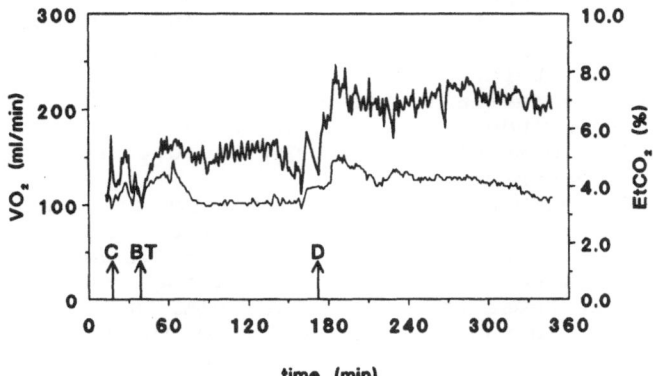

Fig 4. Oxygen consumption (VO_2 = ▬) and end-tidal CO_2 ($EtCO_2$ = —) during ruptured aneurysm repair in patient C (80 kg; FiO_2 from the start: 0.50). At clamping (C), VO_2 was not stable due to surgical manipulation. The relatively low VO_2 (120 ml/min) after start of ventilation was partly due to a low oxygen delivery, that improved after blood transfusion (BT). After declamping (D) and the resultant VO_2 overshoot the basal oxygen consumption increased to over 200 ml/min.

In patient B no such an overshoot was visible after declamping, so it was assumed that pay back of an existing oxygen debt did not take place. This was confirmed by absence of pulsatile blood flow to the legs when tested by routine methods. The changes in oxygen consumption are partially reflected in end-tidal CO_2. However, these variations are not as distinctive as the changes in VO_2 and, after declamping, may be caused by $NaHCO_3$, given sometimes to counterbalance the acid products of metabolism, entering the circulation during this period. This therapeutic intervention usually makes a diagnostic judgement less decisive. Further, these more gradual changes can also be influenced by ventilation. The low oxygen consumption compared to the Brody equation, after induction of anesthesia, is equivocal with the assumption that anesthesia itself lowers VO_2 [2,3].

The very low oxygen consumption in patient C after start of ventilation indicates that the patient had entered the dependent part of the VO_2/DO_2 relationship [16]. In a previous study, Bland suggested that surviving a medical disaster, such as major trauma or severe hemorrhage during surgery, is related to oxygen delivery [1]. The dependency of oxygen consumption to delivery in this case was probably due to anemia, as hemoglobin was only 3.8 g%. An important increase in VO_2 is seen immediately after blood transfusion. It is tempting to say that transfusion increased oxygen delivery above the critical threshold where oxygen consumption becomes dependent on oxygen delivery. After aortic clamping, VO_2 was not stable. An explanation could be that - in this emergency case - surgical manipulation and resultant changes in FRC led to volume changes in the closed circuit and, subsequently, in the amount of gases entering the system. The same instability can be seen just before declamping (D), usually a period when the occlusion is partially released and the vessel is immediately reoccluded if the anastomosis is not sufficiently stitched. After complete declamping, the overshoot in VO_2 evidenced a successful repair.

Changes in VO_2 are thought to be indicative of changes in metabolism, they may give insight to the actual depth of anesthesia and reflect changes in cardiac output. These changes are of vital importance for patients undergoing central vascular surgery, and/or those whose cardiovascular status will not tolerate large fluctuations. Usually, VO_2 is calculated using the Fick equation and values for cardiac output are derived with a thermodilution catheter. Arterial and venous oxygen content are calculated by measuring O_2 saturation, pO_2 and hemoglobin concentration in arterial and venous blood samples. It is generally accepted, however, that these measurements may have inherent and sometimes unpredictable errors, indicating that results of these calculations are not absolutely reliable. Conclusions drawn from these calculations may be acceptable only if it is realized that an error of unknown magnitude may exist. However, wrong conclusions may be drawn which might lead to unsafe or even dangerous therapeutic interventions [5,7]. Further, there may be a considerable lapse of time between measurement and actual calculation of VO_2, especially if the laboratory is not close to the operation theater or ICU. The resultant time delay after sampling may itself affect the outcome of calculations of pO_2 and SO_2. These shortcomings have, nevertheless, been generally accepted because no easy and reliable alternative has been available. Reliable on-line VO_2 measurement is now possible with closed circuit ventilation.

CONCLUSION

If anesthesiologists and surgeons are no longer obliged to catheterize the heart and pulmonary artery - with concomitant complications and risks of infection - to measure VO_2, this measurement may become more palatable. It is illustrated that non-invasive, on-line measurement of VO_2 with closed circuit

ventilation plays an indicative, if not decisive, role during
ventilation and anaesthesia.

REFERENCES

1.   Bland RD, Shoemaker WC. Common physiologic patterns in
     general surgical patients: Hemodynamic and oxygen trans-
     port changes during and after operation in patients with
     and without associated medical problems. Surg Clin North
     Am 1985; 65:793-809.
2.   Viale JP, Annat GJ, Tissot SM, Hoen JP, Butin EM, Bert-
     rand OJ, Motin JP. Mass spectrometric measurements of
     oxygen uptake during epidural analgesia combined with
     general anesthesia. Anesth Analg 1990; 70:589-593.
3.   Viale JP, Annat G, Bertrand OJ, Thouverez B, Hoen JP,
     Motin J. Continuous measurement of pulmonary gas ex-
     change during general anesthesia in man. Acta Anaes-
     thesiol Scand 1988; 32:691-697.
4.   Theye RA. Thiopental and oxygen consumption. Anesth
     Analg 1970; 49:69-72.
5.   Kaplan JA. Hemodynamic Monitoring. In: Kaplan JA, ed.
     Cardiac anesthesia. Vol I, 2nd ed. Orlando: Grune &
     Stratton, 1987:179-225.
6.   Breuer H-WM, Groeben H, Breuer J, Worth H. Oxygen satu-
     ration calculation procedures: a critical analysis of
     six equations for the determination of oxygen saturati-
     on. Intens Care Med 1989; 15:385-389.
7.   Todd TR. 2. Complications of monitoring systems. Cana J
     Surg 1988; 31:324-6.
8.   Severinghaus JW. The rate of uptake of nitrous oxide in
     man. J Clin Invest 1954; 33:1183-1189.
9.   Lowe H, Ernst E. The quantitative practise of anesthesi-
     a. Baltimore: Williams & Wilkins, 1981.
10. Erdmann W, Veeger AI, Verkaaik APK. Narkosebeatmungsge-
     räte: Gegenwart und Zukunft. In: Jantzen JPAH, Kleemann
     PP, eds. Narkosebeatmung (low flow, minimal flow, ge-
     schlossenes System). Stuttgart-New York: Schattauer,
     1989:5-17.
11. Verkaaik APK, Erdmann W. Respiratory diagnostic possibi-
     lities during closed circuit anesthesia. Acta Anaesthe-
     siol Belg 1990; 41:177-188.
12. Ayre P. The T-piece technique. Br J Anaesth 1956;
     28:520-524.
13. Versichelen L, Rolly G. Mass spectrometric evaluation of
     some low flow, closed circuit systems. Acta Anaesthesiol
     Belg 1990; 41:225-238.
14. Radford EP Jr, Ferris BG Jr, Kriete BC. Clinical use of
     nomogram to estimate proper ventilation during artifici-
     al respiration. New Eng J Med 1954; 251:877.
15. Brody S. Bioenergetics and growth. New York: Reinhold,
     1945.
16. Van der Zee H, Verkaaik APK. Cardiovascular implementa-
     tions of respiratory measurements. Acta Anaesthesiol
     Belg 1990; 41:168-175.

OXYGEN TRANSPORT THROUGH A MODEL LUNG SURFACTANT SURFACE LAYER:

INFLUENCE OF THE FILM COMPRESSION ON THE KINETICS

Erna Ladanyi[1], R.C.Ahuja[2], D.Möbius[2] and Karlheinz Stalder[1]

[1]Department of Occupational Health University of Göttingen, Windausweg 2, D-3400 Göttingen, F.R.G.
[2]Max-Planck-Institut für biophysikalische Chemie Am Faßberg, D-3400 Göttingen, F.R.G.

INTRODUCTION

Inhaled oxygen can reach the lung tissue only after having penetrated the so-called lung surfactant surface layer(LSSL) which lines the alveoles at the air/water interface.

We have a special interest in LSSL, the active part of the whole lung surfactant system. The main role of LSSL is to provide a stable surface tension which assures normal breathing by adapting the surface tension to the periodically changing surface area of the lung alveoli [Schürch, Bachofen and Goerke, 1990]. There is accumulating evidence which indicates that LSSL also plays an important role in oxygen transport processes.

An earlier indication concerning the interaction of oxygen with the LSSL was obtained ten years ago through the electrochemically gained observation, that bronchoalveolar lavages BAL of experimental animals contained more dissolved oxygen than the saline [Ladanyi, 1980]. The same happened to be true for BAL of healthy persons compared to BAL of pulmonary patients [Ladanyi and Stalder, 1989]. While looking for reasons of augmented oxygen storage in BAL compared to saline, we observed a difference in both the oxygen uptake and the release kinetics [Ladanyi and Stalder, 1990]. Comparing oxygen uptake and release in BAL itself, we found out that though the release was faster than the uptake, at equilibrium the uptake level was higher than the release level. Thus it seemed that the system never becomes totally free of oxygen.

In addition, it turned out that in the presence of BAL material, the half-wave potential of oxygen reduction at a mercury electrode was shifted towards more negative values [Ladanyi, 1980]. We concluded therefore that one or several surface active components of the BAL, adsorbed at the mercury electrode surface, represent a certain energetic barrier for the crossing oxygen. For screening the BAL components, electrochemical measurements were carried out on an interface layer transferred onto the surface of a hanging mercury drop electrode HMDE [Ladanyi 1989]. These experiments showed that the

spreading of surfactant lipids onto saline resulted in a strong
negative shift of oxygen reduction peak potential, whereas injection of
the hydrophilic surfactant specific protein Sp-A into the saline
underlying the lipid layer shifted the peak potential into the positive
direction. Thus it was inferred that the energetic barrier was mainly
determined by the phospholipids.

In the present work, we have attempted to mimic the *in vivo*
conditions of the oxygen transport through LSSL and to find out as to
what extent the oxygen permeation kinetics are influenced by the
compression state of the phospholipid film.

EXPERIMENTAL

The experiment consisted basically in allowing the oxygen from
the air to penetrate a phospholipid monolayer situated at an
air/deaerated saline interface and in measuring the oxygen

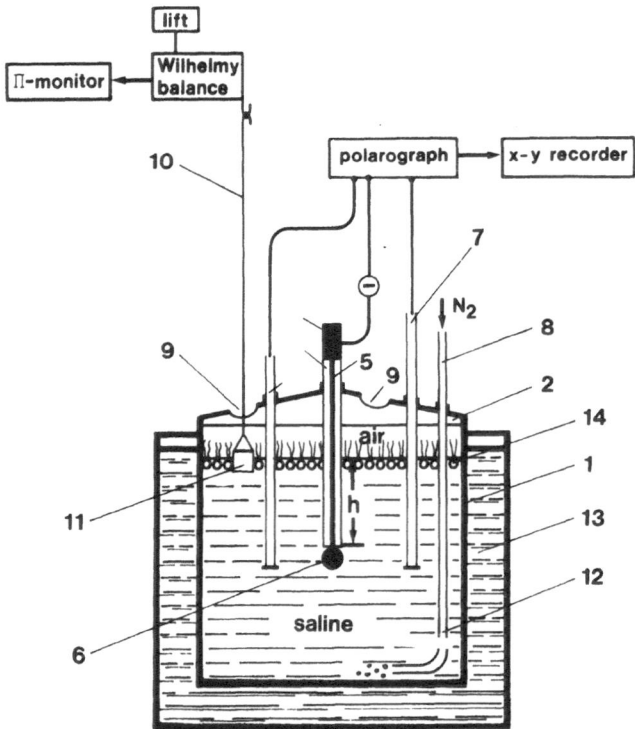

Fig. 1   Schematic representation of the experimental set-up for
the measurement of oxygen permeation kinetics through a
phospholipid layer.1.silanized vessel;2.lid;3.HMDE;4.Hg-
reservoir;5.capillary;6.Hg-drop;7.reference electrode;
8.auxiliary electrode;9.holes in the lid;10.wire holding
the Wilhelmy plate;11.Wilhelmy plate;12.deaeration tube;
13. thermostat.

concentration at a definite depth in the bulk of the hypophase. In order to mimic the breathing conditions, the measurements were carried out with two phospholipid layers of different surface concentrations such that the surface area/lipid molecule for the two layers differed by 30%. This difference corresponds to the maximal alveolar area change during the breathing process [Schürch, Bachofen and Goerke, 1990].

The measurement of the oxygen concentration in the bulk (cf.Fig.1) was carried out electrochemically, at 37^0C, during the first 25 minutes of oxygen contact with the interface. A Metrohm Polarograph equipped with a tri-electrode system and an x-y recorder was used to record the height of the second oxygen reduction wave at 1120 mV, measured as fast direct current DC_T. The distance of the hanging mercury drop to the interface was equal to 5 mm, with the drop lifetime being equal to 1 min. The measurements were carried out every 2 minutes.

The phospholipid films consisted of dipalmitoyl phosphatidylcholine DPPC, known to be the main component of the lung surfactant. A 2mM chloroform solution of DPPC was spread onto the air/saline interface. By spreading either 5µl or 7.5µl of the solution onto a 43cm^2 saline surface, the available area/DPPC molecule was equal to 0.72nm^2 or 0.48nm^2 respectively. Saline subphase was prepared from NaCl p.a. in doubly distilled deionized water.

The surface pressure at the interface was monitored using a Wilhelmy balance with a filter paper serving as the Wilhelmy plate.The dipping level of the filter paper was kept constant by using a vertical translation unit. The surface pressure corresponding to 0.72nm^2 and 0.48nm^2 area/molecule was equal to 28mN/m and 43mN/m respectively.

A typical experiment was carried out in the following way:The silanized electrochemical vessel(1) was filled with saline up to a constant level yielding an interfacial area equal to 43cm^2 and then it was firmly attached to an air tight lid (2) holding the 3 electrodes. The working electrode was a Hanging Mercury Drop Electrode HMDE (3) consisting of a mercury reservoir (4), the mercury-filled capillary (5) and the hanging mercury drop itself (6). As reference electrode served an Ag/AgCl electrode (7) and a Pt strip was used as auxiliary electrode (8). The lid has two holes (9) in contact with the air. Through one of these holes the wire (10) holding the Wilhelmy plate (11) was attached to the Wilhelmy balance. A deaeration tube (12) was tightly inserted into the lid. The electrochemical vessel was partially immersed in a thermostated water bath kept at 37^0C (13) and the saline was deaerated with pure nitrogen for 30 minutes. After stopping the deaeration, a slight overpressure of nitrogen was maintained over the saline in order to hinder the penetration of the airborne oxygen into the solution. A new drop was formed, and after 1 min the intensity of the DC_T current was recorded between 1000 and 1120 mV with an x-y recorder. This current value was considered as corresponding to the time 0. Then the phospholipid solution was carefully spread over the deaerated saline surface by inserting a micro syringe through the free hole of the lid. The surface flow of the nitrogen was now stopped, and the space between the lipid film and lid was evacuated three times using a Dräger sampler pump, connected for this purpose to the free hole of the lid. A chronometer was started simultaneously with the formation of a fresh mercury drop at the HMDE and the air was allowed to enter the two free holes. After one minute, the current intensity was recorded. At 2 minute intervals a new mercury drop was formed, and the current intensity was recorded again.

The oxygen permeation kinetics was measured 9 times for each DPPC concentration, starting always with a fresh, nitrogen-saturated saline. The data so obtained was analyzed using a linear regression algorithm.

RESULTS AND DISCUSSION

Fig. 2 shows the kinetic curves obtained with the linear regression analysis of 210 experimental data-pairs. As can be seen, the available area/DPPC molecule and thus the compression state of the film, has an evident influence on the oxygen permeation kinetics. The DPPC layer corresponding to 0.72nm^2/molecule (surface pressure 28 mN/m), is more permeable for oxygen than the layer having 30% less area/molecule (.i.e. 0.48nm^2/molecule and surface pressure = 43mN/m).In addition, the longer the observation time, the larger the difference in the oxygen concentration in the bulk.Thus the difference after 25 minutes is ca. 10%.

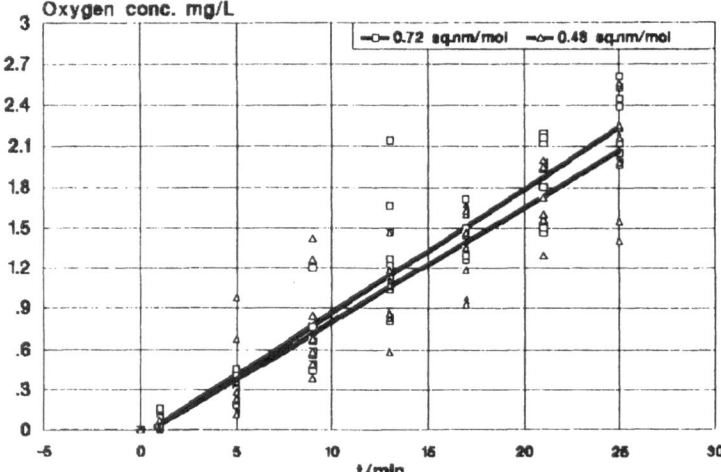

Fig. 2   Kinetics of oxygen permeation through phospholipid layers of two different compression states at the air/ deaerated saline interface.

These data suggest that the oxygen transport through the ELSSL has a different kinetics in the inspiration and expiration phase of breathing. Thus enlargement of the alveolar surface during the inspiration phase results in a two fold enhancement of oxygen transport to the tissue firstly due to the enlarged exchange surface of the lung and secondly as a result of faster oxygen penetration through the expanded ELSSL of each interface unit.

This is a new and rather important result. But it has to be regarded with criticism because of the imperfection of the used model experiment. We measured oxygen penetration into an oxygen-free hypophase, which is never the case in vivo; oxygen concentration was measured at a distance from the interface which is several orders of magnitudes higher than the thickness of the hypophase in the alveoli; observation time is too long compared to the time of the breathing cycle and, last but not least, the LSSL consists not only DPPC, but also several other phospholipids and four specific proteins. Each of these components might have some influence on the oxygen permeation.

Nevertheless, these are the first results concerning the influence of the LSSL compression state on oxygen permeation. Even if they do not give us absolute values, they show a certain trend. Thus tentatively, we may suggest that the oxygen transport to the alveolar wall might be additionally dependent on the compression state of the LSSL, which in turn is determined by the actual lung area during the breathing.

SUMMARY

The influence of the compression state of a model Lung Surfactant Surface Layer LSSL on the oxygen permeation kinetics was studied *in vitro* at 37^0C. In an attempt to mimic *in vivo* conditions, the oxygen from the air was allowed to cross a dipalmitoylphosphatidylcholine DPPC layer situated at an air/deaerated saline interface in an electrochemical vessel. The time dependent concentration change of the oxygen diffusing through this layer into the deaerated saline hypophase was measured electrochemically using a Hanging Mercury Drop Electrode HMDE, situated at a definite depth in the bulk of the saline. The surface pressure in the monolayer was monitored using a Wilhelmy balance. The oxygen permeability was measured through two differently compressed DPPC layers in which the area/phospholipid molecule differed by 30%. This is consistent with the difference in the alveolar area at the end-points of the compressed and expanded lung. The results, submitted to a linear regression analysis, showed that the DPPC film compression influences the oxygen permeation kinetics. The denser the lipid film, the slower the oxygen uptake by the deaerated hypophase. The results suggest that the LSSL might play an important role in the oxygen transport kinetics, the oxygen permeation through it being dependent on the actual lung area.

REFERENCES

Schürch,   S., Bachofen, H. and Goerke J.,1990 in: Basic Research on Lung Surfactant , von Wichert P. and Müller B. eds., Prog. Resp. Res. vol 25, pp 181-190, Karger, Basel

Ladanyi,   E., 1980, Polarographische Elektrosorptionsanalyse des oberflächenaktiven Systems der lunge (Lung surfactant), Dissertation, Technische Universität Clausthal

Ladanyi,   E., Miller, I., Popovitz-Biro, R., Marikovsky, J., von Wichert, P., Müller, B., Stalder, K., 1988, Molecular structure of the extracellular surface-layer of the human lung surfactant, in Progr.Resp.Res.,von Wichert P and Müller B. eds, Karger, Basel

Ladanyi,   E., 1989, in "Oxygen Transport to Tissue XI", Rakusan, Ed.,Plenum, New York Ladanyi, E., K. Stalder 1990, in "Oxygen Transport to Tissue XII", Piiper, Ed., Plenum, New York

Ladanyi,   E., K. Stalder, ISOTT 1990, Townsville, Australia

DIFFERENT SURFACTANT TREATMENT STRATEGIES FOR RESPIRATORY FAILURE INDUCED BY HYDROCHLORIC ACID ASPIRATION IN RATS

E.P. Eijking, D. Gommers, J. Kullander[1], E. de Buijzer, R.W. Beukenholdt and B. Lachmann

Dept. of Anesthesiology, Erasmus University, Rotterdam, The Netherlands and [1]Dept. of Anesthesiology, Halmstad Hospital, Halmstad, Sweden

INTRODUCTION

Massive aspiration of gastric contents is one of the most feared complications of general anesthesia and is known to be an important cause of the adult respiratory distress syndrome (ARDS) (for review, see Gibbs and Modell, 1990 and Fowler et al., 1983). Due to direct damage of the alveolar-capillary membrane, especially to the type I and II pneumocytes, protein-rich edema fluid enters the alveolar space (Awe et al., 1966; Greenfield et al., 1969; Grimbert et al., 1981; Kobayashi et al., 1990; Manny et al., 1986). It has been established that these plasma derived proteins are potent surfactant inhibitors (Ennema et al., 1988; Fuchimukai et al., 1987; Holm et al., 1988; Ikegami et al., 1984; Seeger et al., 1985). Besides this, it has been proposed that hydrochloric acid (HCl) directly damages the surfactant system (Greenfield et al., 1969; Winn et al., 1983). As the surfactant system seems to be involved in the pathophysiology of respiratory failure caused by HCl aspiration, a study was designed to investigate the effect of different treatment strategies with an exogenous surfactant preparation on lung function of rats suffering from respiratory failure after intratracheal HCl instillation.

MATERIALS AND METHODS

Exogenous surfactant

The surfactant used in these experiments is a freeze-dried natural surfactant isolated from bovine lungs in basically the same manner as previously described (Metcalfe et al., 1980). It consists of approximately 83% phospholipids, 1% hydrophobic proteins (SP-B and SP-C), the remainder being other lipids such as cholesterol, glyceride and free fatty acids. There is no SP-A in this surfactant preparation. In this study surfactant is suspended in saline.

Animal study

The studies were performed in 34 male adult Sprague-Dawley rats (body weight: 300-350 g). All rats were anesthetized with pentobarbital sodium (60 mg/kg, i.p.), tracheotomized and paralyzed with pancuronium bromide (0.5 mg/kg, i.m.). A catheter was inserted into the carotid artery. The rats were ventilated with a Servo Ventilator 900 C (Siemens-Elema, Solna, Sweden) at the following ventilator settings: pressure-controlled ventilation, $F_iO_2 = 1.0$, ventilation frequency = 30/min, peak airway pressure (P_{peak}) = 14 cmH$_2$O, positive end-expiratory pressure (PEEP) = 2 cmH$_2$O and inspiratory/expiratory ratio = 1:2. After reaching steady state (PaO$_2$ > 500 mmHg), all rats received 1.5 ml/kg hydrochloric acid (HCl) intratracheally (0.1 N; pH = 1.0) while lying on their right side, followed by a bolus of air (30 ml/kg). This was immediately followed by instillation of 1.5 ml/kg HCl, while lying on their left side, again followed by a bolus of air. Directly after instillation, P_{peak} was increased to 26 cmH$_2$O and PEEP to 6 cmH$_2$O in all rats; these ventilator settings were maintained throughout the whole observation period. After PaO$_2$ decreased <200 mmHg, the animals were divided into 5 groups: Group I (n=7) received no treatment (26/6); Group II (n=8) received surfactant intratracheally (Surf); Group III (n=7) was lavaged with saline followed by intratracheal surfactant instillation (Sal BAL + Surf); Groups IV and V (n=6 in both groups) were lavaged with saline or a diluted surfactant suspension, respectively, at two time points: the first lavage was after PaO$_2$ values decreased <200 mmHg, followed by a second lavage after 60 min (Sal BAL and Surf BAL, respectively). The amount of saline used in all bronchoalveolar lavages (BAL) was 30 ml/kg at 37°C; the amount of surfactant in the diluted surfactant suspensions was 100 mg/kg. The surfactant concentration used for intratracheal instillation (Groups II and III) was at a dose of 200 mg/kg. Blood samples for measurement of PaO$_2$ and PaCO$_2$ were taken from the carotid artery of each rat before intratracheal HCl instillation, at regular intervals post-instillation and 5, 30 and 60 min after treatment (ABL 330 Acid-Base Laboratory; Radiometer, Copenhagen, Denmark). Bicarbonate was administered intraperitoneally when required to correct for metabolic acidosis, as evidenced by low pH, low plasma bicarbonate level and low base excess. Also, saline was administered intra-arterially to correct for increase in hematocrit.

Statistical analysis

All data are expressed as mean ± standard deviation (SD). Statistical analysis of data was performed using the Mann-Whitney-U test for intergroup comparison or the Wilcoxon signed rank test for intragroup comparisons. Statistical significance was accepted at p≤0.05 (two-tailed).

RESULTS

Animal study

Figure 1 shows PaO$_2$ values for each group. Before instillation of HCl PaO$_2$ values are high and there is no significant intergroup difference. Also, there are no differences between pre-treatment PaO$_2$ values (t=0) of different groups. There is a significant increase in PaO$_2$ after treatment compared to pre-treatment in Groups III (Sal BAL + Surf) and V (Surf BAL). The only effect seen in Group II (Surf) is that gas exchange does not deteriorate any further after surfactant instillation; whereas in Groups I (26/6) and IV (Sal BAL) there is a further decrease in PaO$_2$ values. The PaO$_2$ values of both Groups I and IV do not differ significantly from each other. PaO$_2$ values of Group I are significantly lower compared to the remaining groups at t=5-30 min; at t=60 min the difference is significantly lower compared to Groups II and V. The difference between Group IV and the other groups is significant at t=5-60 min. There are no statistical differences between Groups III and V. Also, there are no significant differences between Groups II and III. The PaO_2 values of Group V are significantly higher compared to Group II.

Table 1 shows the $PaCO_2$ values for each group. There are no marked differences between all groups at any time, except those significant differences which are indicated.

Table 2 shows the PaO_2 and $PaCO_2$ values of Groups IV and V after the first and second lavage. At all time points PaO_2 values are significantly higher in Group V compared to Group IV, whereas there is no significant difference between $PaCO_2$ values of the two groups. After the second lavage there is an increase of PaO_2 in both groups compared to PaO_2 values after the first lavage, although this increase is not significant.

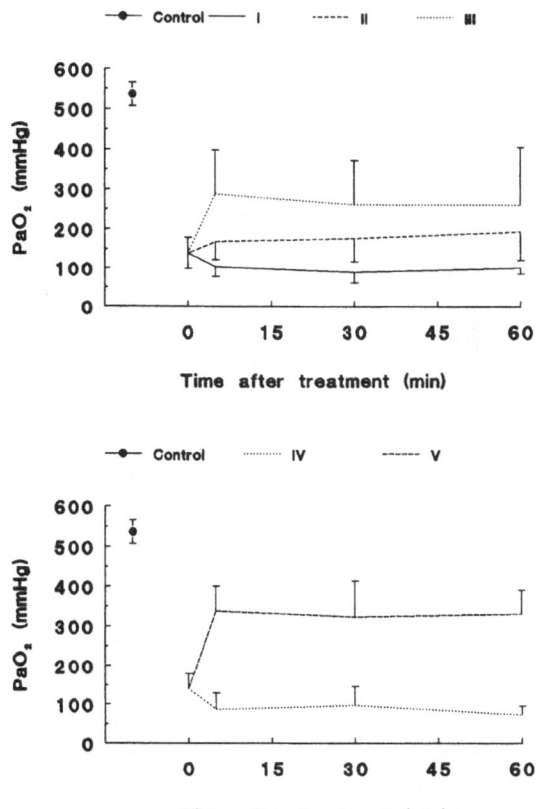

Figure 1. PaO_2 values in mmHg (mean ± SD) of the five treatment groups (for treatment schedule: see text); "control" indicates PaO_2 before HCl instillation; "t=0" indicates PaO_2 after HCl instillation, just before treatment.

DISCUSSION

The results in the present study demonstrate that ventilation with 100% oxygen, high peak airway pressure and PEEP was not able to maintain high PaO_2 levels in rats suffering from respiratory failure due to massive HCl aspiraton. The best way to treat these animals appears to be either to lavage the lungs with saline, directly followed by intratracheal instillation of a high dose of surfactant, or to lavage the lungs with a diluted surfactant suspension, as evidenced by an increase of PaO_2. Surfactant instillation without prior lavage did not restore lung function, although it appears that there is no further decrease of PaO_2.

It has been hypothetized that the mechanism of inhibition of surfactant by proteins is based on competition for space at the air-liquid interphase (Holm et al., 1985, 1987, 1988). Thus, the way to overcome the inhibition of surfactant by these proteins would

Table 1. $PaCO_2$ values (mmHg) of all groups.

Group	n	control	pre	5'	30'	60'
26/6 (I)	7	30.7 ± 3.19	58.1 ± 15.8	53.0[II] ± 11.6	62.4 ± 23.1	76.0[II] ± 13.6
Surf (II)	8	34.8 ± 4.58	67.5 ± 20.3	67.5 ± 13.9	68.6 ± 10.3	65.6[I] ± 8.81
Sal BAL + Surf (III)	7	32.6 ± 4.78	65.1 ± 15.8	65.9 ± 23.4	57.3 ± 15.2	57.4 ± 18.7
Sal BAL (IV-1st)	6	30.8 ± 3.48	54.7 ± 11.1	50.0[#1] ± 7.79	47.8[#2] ± 9.35	51.3[II/#2] ± 2.87
Surf BAL (V-1st)	6	33.5 ± 3.30	67.2 ± 17.3	63.7 ± 10.5	58.5 ± 8.46	57.5[I] ± 5.41

$PaCO_2$ values before HCl instillation ("control"), before treatment ("pre") and 5/30/60 min after treatment. All data are mean ± SD. [I] = one animal died.
[#1] $p < 0.05$ between IV-1st and both II and V-1st.
[#2] $p < 0.05$ between IV-1st and II.

be to get a relatively high surfactant concentration in relation to the amount of proteins. This favorable surfactant/inhibitor ratio could be accomplished in two ways. In the first place by treating the rats with a high dose of surfactant (without prior lavage), a high surfactant/inhibitor ratio could be achieved. The results from the present study do not demonstrate an increase in PaO_2 after instillation of surfactant at a dose of 200 mg/kg. Thus, according to this hypothesis, surfactant at a dose of 200 mg/kg was too low to neutralize all inhibitors present in the alveolar space after HCl aspiration. These results confirm other studies in which animals suffered from respiratory failure due to prolonged exposure to 100% oxygen or HCl aspiration; surfactant at the same (or less) concentration as used in the present study was not able to restore gas exchange in these lungs (Ennema et al., 1988; Kobayashi et al., 1990; Strohmaier et al., 1990). However, in another study in which also a large amount of plasma proteins was present in the alveolar space, surfactant instillation at very high doses (280-350 mg/kg) was able to restore gas exchange in guinea pigs suffering from severe respiratory failure after intravenous instillation of anti-lung serum (Lachmann et al., 1987). Thus, in that study a more favorable surfactant/inhibitor ratio was obtained by giving a large amount of surfactant.

In the second place a favorable surfactant/inhibitor ratio could be achieved by

Table 2. PaO_2 and $PaCO_2$ values (mmHg) after repeated BAL.

Group	n	control	pre	5'	30'	60'	5'	30'	60'
Sal BAL (IV-1st/2nd) PaO_2	6	536.3 ± 24.6	123.0 ± 32.4	86.3 ± 41.9	97.5 ± 48.1	72.7!! ± 23.8	106.3! ± 29.8	140.0 ± 67.0	140.5 ± 78.5
$PaCO_2$		30.8 ± 3.48	54.7 ± 11.1	50.0 ± 7.79	47.8 ± 9.35	51.3 ± 2.87	39.7 ± 10.5	41.0 ± 7.10	41.1 ± 9.02
Surf BAL (V-1st/2nd) PaO_2	6	533.8 ± 31.7	159.7 ± 27.8	338.2*1 ± 61.6	322.7 ± 90.7	331.0! ± 59.2	401.6 ± 41.1	421.2 ± 57.7	375.3! ± 99.3
$PaCO_2$		33.5 ± 3.30	67.2 ± 17.3	63.7*2 ± 10.5	58.5 ± 8.46	57.5 ± 5.41	56.4 ± 7.39	49.6 ± 5.46	53.3 ± 7.08

$PaCO_2$ and PaO_2 values before HCl instillation ("control"), before BAL ("pre"), 5/30/60 min after first BAL (IV/V-1st) and 5/30/60 min after second BAL (IV/V-2nd). All data are mean ± SD. ! = one animal died.

*1 $p < 0.05$ between pre-lavage and 5 min after lavage.

*2 $p < 0.05$ between Surf BAL and Sal BAL.

removing the inhibitors from the alveolar space by means of BAL. In the present study it has been demonstrated that PaO$_2$ values increased after surfactant treatment (200 mg/kg) after the lungs were lavaged. Similar findings were also reported by Kobayashi and colleagues (1990). In the present study, BAL with a diluted surfactant suspension (100 mg/kg) was also capable of restoring gas exchange. BAL with saline without additional surfactant treatment was not able to restore gas exchange. The difference between PaO$_2$ values of rats lavaged with a diluted surfactant suspension and rats lavaged with saline (not followed by surfactant treatment) can be explained as follows: after removal of a large amount of proteins in both groups, a small amount of surfactant remains in the lungs of the rats lavaged with a diluted surfactant suspension. This appears sufficient to establish surface active material at the air-liquid interphase, which keeps the lungs open. In rats lavaged with saline, followed by surfactant treatment, proteins are also removed. The reason that PaO$_2$ values are higher in the rats lavaged with a diluted surfactant suspension, compared to rats lavaged with saline followed by surfactant treatment, is unknown. However, it appears that for treatment of respiratory failure due to HCl aspiration only half of the amount of surfactant is needed when BAL is performed with the diluted surfactant suspension, compared to treatment with surfactant after BAL with saline. However, both treatment strategies result in a favorable surfactant/inhibitor ratio. PaO$_2$ values appeared to increase even more after the second lavage with the diluted surfactant suspension. There was also a slight increase after the second lavage with saline, although this was not significant (Table 2). Thus, although it seems that BAL with saline also removes inhibitors from the alveolar spaces, surfactant substitution is needed to restore lung function.

In conclusion, surfactant treatment can only succeed when there is a favorable surfactant/inhibitor ratio. It is demonstrated in this study that this can be achieved by BAL with saline, followed directly by surfactant instillation, or by BAL with a diluted surfactant suspension.

ACKNOWLEDGEMENT

This work was financially supported by The Dutch Foundation for Medical Research (SFMO). We thank Ari Kok for expert technical assistance.

REFERENCES

Awe, W.C., Fletcher, W.S. and Jacob S.W., 1966, The pathophysiology of aspiration pneumonitis, Surgery, 60: 232.

Ennema, J.J., Kobayashi, T., Robertson, B. and Curstedt, T., 1988, Inactivation of exogenous surfactant in experimental respiratory failure induced by hyperoxia, Acta Anaesthesiol. Scand., 32: 665.

Fowler, A.A., Hamman, R.F., Good, J.T., Benson, K.N., Baird, M., Eberle, D.J., Petty, T.L. and Hyers, T.M., 1983, Adult respiratory distress syndrome: risk with common predispositions, Ann. Intern. Med., 98: 593.

Fuchimukai, T., Fujiwara, T., Takahashi, A. and Enhorning, G., 1987, Artificial pulmonary surfactant inhibited by proteins, J. Appl. Physiol., 62: 429.

Gibbs, C.P. and Modell, J.H., 1990, Management of aspiration pneumonitis, in: "Anesthesia" (third edition), R.D. Miller, ed., Churchill Livingstone, New York, Edinburgh, London, Melbourne, 1293.

Greenfield, L.J., Singleton, R.P., McCaffree, D.R. and Coalson, J.J., 1969, Pulmonary effects of experimental graded aspiration of hydrochloric acid, Ann. Surg., 170: 74.

Grimbert, F.A., Parker, J.C. and Taylor, A.E., 1981, Increased pulmonary vascular permeability following acid aspiration, J. Appl. Physiol., 51: 335.

Holm, B.A., Notter, R.H. and Finkelstein, J.N., 1985, Surface property changes from interactions of albumin with natural lung surfactant and extracted lung lipids, Chem. Phys. Lipids., 38: 287.

Holm, B.A. and Notter, R.H., 1987, Effects of hemoglobin and cell membrane lipids on pulmonary surfactant activity, J. Appl. Physiol., 63: 1434.

Holm, B.A., Enhorning, G. and Notter, R.H., 1988, A biophysical mechanism by which plasma proteins inhibit lung surfactant activity, Chem. Phys. Lipids., 49: 49.

Ikegami, M., Jobe, A., Jacobs, H. and Lam, R., 1984, A protein from airways of premature lambs that inhibits surfactant function, J. Appl. Physiol., 57: 1134.

Kobayashi, T., Ganzuka, M., Tanigushi, J., Nitta, K. and Murakami, S., 1990, Lung lavage and surfactant replacement for hydrochloric acid aspiration in rabbits, Acta Anaesthesiol. Scand., 34: 216.

Lachmann, B., Hallman, M. and Bergmann, K.C., 1987, Respiratory failure following anti-lung serum: study on mechanisms associated with surfactant system damage, Exp. Lung Res., 12: 163.

Manny, J., Manny, N., Lelcuk, S., Alexander, F., Feingold H., Kobzik, L., Valeri, R., Shepro, D. and Hechtman, H.B., 1986, Pulmonary and systemic consequences of localized acid aspiration, Surg. Gynecol. Obstet., 162: 259.

Metcalfe, I.L., Enhorning, G. and Possmayer, F., 1980, Pulmonary surfactant-associated proteins: their role in the expression of surface activity, J. Appl. Physiol., 49: 34.

Seeger, W., Stöhr, G., Wolf, H.R.D. and Neuhof, H., 1985, Alteration of surfactant function due to protein leakage: special interaction with fibrin monomer, J. Appl. Physiol., 58: 326.

Strohmaier, W., Redl, H. and Schlag, G., 1990, Studies of the potential role of a semisynthetic surfactant preparation in an experimental aspiration trauma in rats, Exp. Lung Res., 16: 101.

Winn, R., Stothert, J., Nadir, B. and Hildebrandt, J., 1983, Lung Mechanics following aspiration of 0.1 N hydrochloric acid, J. Appl. Physiol., 55: 1051.

GAS EXCHANGE UNIFORMITY WITHIN INDIVIDUAL LUNG LOBES

Michael J.Emery, Mical E. Middaugh, Tim Tran,
Michael P. Hlastala

Department of Physiology and Biophysics
University of Washington
Seattle, WA 98195, USA

INTRODUCTION

Oxygen delivery to peripheral tissue is limited, in part, by the efficiency of gas exchange within the lungs. Pulmonary ventilation ($\dot{V}_A$) to pulmonary perfusion ($\dot{Q}$) matching ($\dot{V}_A/\dot{Q}$) in the normal lung is known to be heterogeneous (Hlastala and Robertson, 1978) (fig. 1). The degree of $\dot{V}_A/\dot{Q}$ mismatching in various lung regions has the major effect on the efficiency of blood oxygenation. Factors which increase or decrease the homogeneity of ventilation to perfusion matching have profound effects on arterial oxygen concentrations, and therefore tissue oxygenation.

The degree of $\dot{V}_A/\dot{Q}$ heterogeneity may be different in anatomically distinct lung regions. Mechanisms which cause gas exchange inhomogeneity are not expected to act uniformly throughout the lung. For instance, the gradient of intrapleural pressure within the thorax results in an inhomogeneous distribution of ventilation and perfusion between various lung regions (Milic-Emili, 1986). The shape of the thoracic space within which the lungs lie is not the same as excised lungs inflated to the same volume (Engel, 1986). Differential impingement of the chest wall, heart, and diaphragm on the lung surface may create regional differences in $\dot{V}_A$, $\dot{Q}$, and $\dot{V}_A/\dot{Q}$ ratios.

Several mixing mechanisms serve to homogenize the gas exchange process. Mixing mechanisms may also have larger or smaller effects in specific lung regions. Diffusive mixing of inspired gas increases ventilation homogeneity between nearby regions throughout the lung, but regions diffusively distant will not mix by this mechanism (Paiva and Engel, 1987). $\dot{Q}$ and $\dot{V}_A$ heterogeneity may be reduced by differing degrees in various lung regions due to cardiogenic oscillations. There is ample evidence to postulate regional heterogeneity of $\dot{V}_A$, $\dot{Q}$, and $\dot{V}_A/\dot{Q}$ matching because mixing mechanisms are not applied uniformly throughout the lung.

The total gas exchange of the lungs is a summation of the gas exchange of all the smaller pulmonary regions. Heterogeneity of $\dot{V}_A/\dot{Q}$ matching has been characterized for

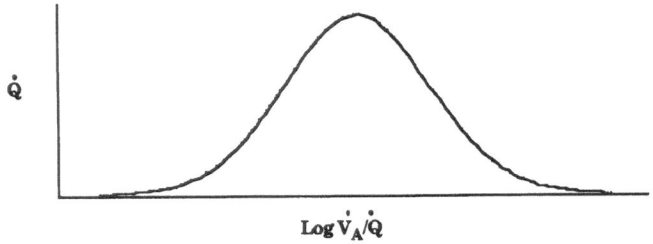

Figure 1. A gas exchange region can be characterized in terms of the amount of perfusion ($\dot{Q}$) going to regions with a log distribution of ventilation to perfusion ratios ($\dot{V}_A/\dot{Q}$).

normal and diseased whole lungs (Hlastala and Robertson, 1978) and for single lobe preparations (Ohlsson et al., 1989). To date there has been no attempt to quantify the contribution of pulmonary subunits to whole lung gas exchange heterogeneity.

Pulmonary lobes offer well defined regions, with their own ventilation and perfusion, to begin exploring the anatomical distribution of regional $\dot{V}_A/\dot{Q}$ matching. The lobes could be postulated to contribute to whole lung gas exchange heterogeneity in one of three general ways. Each lobe could have the same mean $\dot{V}_A/\dot{Q}$ ratio but differing degrees of $\dot{V}_A/\dot{Q}$ ratio inhomogeneity (fig. 2a). On the other hand, the pulmonary lobes

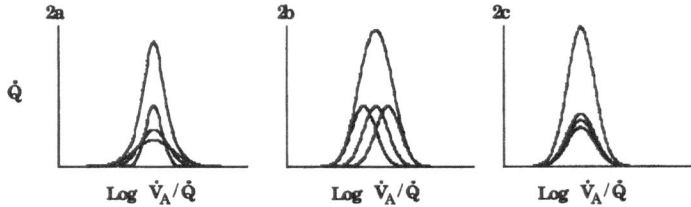

Figure 2. A hypothetical three lobe lung, and three possible ways in which lobar gas exchange properties can produce whole lung gas exchange heterogeneity (see text for details).

could have different mean ratios of $\dot{V}_A/\dot{Q}$, but the same degree of $\dot{V}_A/\dot{Q}$ ratio inhomogeneity (fig. 2b). The other possibility is that the lobes are uniform in their matching of ventilation to perfusion, and that the whole lung is simply a larger version of the single lobe (fig. 2c).

This study tests the hypothesis that the heterogeneity of the individual lobes of the lung is less than the heterogeneity of the whole lung.

METHODS

The multiple inert gas elimination technique (MIGET) was used in 13 dogs (20 to 30 kg) to determine the mean $\dot{V}_A/\dot{Q}$ ratio and the heterogeneity of $\dot{V}_A/\dot{Q}$ within specific lobes and the whole lung. MIGET characterizes the gas exchange properties of the lung by analyzing the pulmonary elimination and retention characteristics of six inert gases dissolved in saline and infused into the venous circulation.

Heterogeneity of gas exchange was calculated by the dispersion index (Ohlsson et al., 1989). The dispersion index is obtained from the sum of squares of the difference between actual retention and excretion of each inert gas, and the retention and excretion which would be expected in a perfectly homogeneous gas exchange region.

Induction and anesthesia in each animal was maintained with sodium pentobarbital. The animal was placed in the supine position and room air ventilation was then supported with a piston ventilator. Tidal volume was fixed at 15ml/kg and respiratory rate was adjusted to give an end tidal CO_2 of 30 to 35mmHg (monitored with a mass spectrometer). The lung was 'sighed' once every 10 minutes with twice the normal tidal volume and a one cycle breathhold. A femoral arterial catheter was advanced percutaneously into the thoracic aorta for sampling of mixed arterial blood oxygen and the amount of inert gas that was retained in the arterial blood. A femoral venous catheter was advanced to the thoracic inferior vena cava for fluid delivery. A Swan-Ganz catheter was advanced to the wedge position in the pulmonary venous bed for determination of cardiac output and the amount of inert gas and oxygen in blood entering the lungs. The whole lung excretion of inert gas was measured in the mixed expirate.

The thorax was opened wide down the sternal centerline and a positive end expiratory pressure (PEEP) of +5cmH$_2$O (+0.5kPa) was added to maintain a more normal transpulmonary pressure difference, and therefore a more normal lung volume. The lungs were visually inspected 1) for normal appearance, 2) to determine if all pulmonary veins from a particular lobe drained through a common lobar pulmonary vein and 3) to determine if all ventilation to a particular lobe took place through a common lobar bronchi. Only lobes with pulmonary venous and bronchial anatomy that allowed blood and gas sampling representing all circulation and ventilation from that single lobe were utilized. Manipulation of the pleural surface was kept to a minimum and the lungs were prevented from drying in the open chest by occasional misting with normal saline.

Catheters were placed in the pulmonary vein and lobar bronchi of chosen lobes. Pulmonary vein catheterization involved first placing a pursestring suture and loose tie of 5-0 silk in the exposed surface of a pulmonary vein. The vessel wall was punctured in the center of the sewn loop with a fine scalpel blade. The slightly flared tip of a length of polyethylene tubing (i.d.=0.76mm, o.d.=1.22mm) was then pushed through the opening and secured by tightening the pursestring tie. With practice, this procedure allowed placement of lobar vein sampling catheters without reduction of bloodflow or significant blood loss. The lobar bronchus was catheterized in basically the same manner as the pulmonary vein except that puncture of the bronchial wall was accomplished with needle electrocautery and 3-0 silk was used for the pursestring suture. A slightly flared length of polyethylene tubing (i.d.=1.75mm, o.d.=2.79mm) was also used as the gas sampling catheter. One to three pulmonary lobes were catheterized in each of the 13 animals utilized.

Inert gases were infused into the femoral vein at a constant rate. Steady state pulmonary retention and excretion of inert gases was assumed to be achieved in 30 minutes. The ventilator 'sigh' devise was disabled during sampling. Blood samples from the pulmonary artery and thoracic aorta were drawn slowly by withdrawl pumps over one minute. Simultaneously and manually, lobar blood outflow was continuously drawn and lobar expired gas was gathered in 0.2ml increments only during the mid-expiratory phase of ventilation. Lastly, a sample representing mixed expired gas from the whole lung was drawn from a mixing box.

The effect of a larger mean lung volume on lobar gas exchange characteristics was examined by repeating all measurements at a PEEP of +10cmH$_2$O (+1.0kPa) following a 30 minute equilibration period. The order of sampling was varied at random. The lung and lobes were sampled at least twice at each mean lung volume.

RESULTS

The left upper (LUL), middle (LML), and lower lobes (LLL) of the dog had the most accessible pulmonary veins and lobar bronchi. The results presented in this paper are from the left side lobes and the whole lung (WL) (table 1). Significance of differences in measured parameters between the whole lung and lobes, and between the various regions at PEEP = +5cmH$_2$O vs. +10cmH$_2$O, were tested by paired t-Tests (two-tailed).

The left side lobes revealed different mean $\dot{V}_A/\dot{Q}$ ratios at the smaller (+5cmH$_2$O PEEP), but not the larger (+10cmH$_2$O PEEP), mean lung volume. At +5cmH$_2$O PEEP, the LLL had a significantly lower mean $\dot{V}_A/\dot{Q}$ ratio than the LML (p<.006), LUL (p<.02), and WL (p<.04). The LML had significantly lower mean $\dot{V}_A/\dot{Q}$ than the WL (p<.0004), but not the LUL. The LUL had a higher mean $\dot{V}_A/\dot{Q}$ ratio than the WL (p<.005). At +10cmH$_2$O PEEP, all gas exchange regions sampled had mean $\dot{V}_A/\dot{Q}$ ratios that were not significantly different from one another.

The left upper (LUL), middle (LML), and lower lobes (LLL) were found to be similar in their degree of gas exchange heterogeneity at both the smaller and the larger mean lung volumes. At +5cmH$_2$O PEEP, the LUL and LML were found to be significantly less heterogeneous than the WL (WL vs. LUL, p<.02; WL vs. LML, p<.006). The heterogeneity of the LLL tended to be less than that of the WL at +5cmH$_2$O PEEP, but the difference was not statistically significant by the paired t-Test. The LUL, LML, and LLL were more homogeneous for gas exchange than the WL at +10cmH$_2$O (p<.04, .0001, .0001, respectively).

Gas exchange heterogeneity of the whole lung and all lobes sampled tended to increase at +10cmH$_2$O as compared to +5cmH$_2$O, but only in the WL and LUL was the increase statistically significant (p<.0002 and .008, respectively).

Lobar and WL oxygenation reflected the gas exchange properties of those regions. At +5cmH$_2$O PEEP the LLL effluent blood had a significantly lower P$_{O_2}$ than the LML (p<.02), and LUL (p<.05), but not the WL. The LML and LUL each had significantly higher P$_{O_2}$ than the WL at the smaller mean lung volume (p<.006 and p<.0002, respectively).

WL oxygenation of blood was not significantly altered at

Table 1. Whole lung and left lobar mean $\dot{V}_A/\dot{Q}$ ratios, heterogeneity of $\dot{V}_A/\dot{Q}$ ratios (dispersion index, see 'methods'), and blood oxygen levels, at PEEP = +5cmH$_2$O (**a**), and PEEP = +10cmH$_2$O (**b**). All results are expressed as the mean ± S.E.M.

a. PEEP = +5

	Whole lung (n = 13)	L.Upper (n = 8)	L.Middle (n = 6)	L.Lower (n = 10)
$\dot{V}_A/\dot{Q}$;				
mean	1.31 ± 0.18	1.60 ± 0.23	1.28 ± 0.14	1.19 ± 0.17
heter.	11.3 ± 1.1	7.56 ± 1.6	8.1 ± 1.5	8.3 ± 1.3
P$_{O2}$	107.6 ± 2.8	121.9 ± 2.3	123.5 ± 3.3	105.9 ± 4.7

b. PEEP = +10

	Whole lung (n = 11)	L.Upper (n = 6)	L.Middle (n = 4)	L.Lower (n = 9)
$\dot{V}_A/\dot{Q}$;				
mean	1.30 ± 0.21	1.31 ± 0.34	0.71 ± 0.09	1.34 ± 0.27
heter.	15.8 ± 1.1	11.9 ± 1.7	11.0 ± 0.8	8.5 ± 1.2
P$_{O2}$	106.6 ± 2.7	110.7 ± 4.6	100.2 ± 3.2	109.9 ± 3.6

+10cmH$_2$O as compared to the +5cmH$_2$O condition. LUL blood oxygenation was reduced at the larger lung volume ($p<.05$), as was LML ($p<.02$), but the LLL P$_{O2}$ was unaltered. In contrast to the pattern of lobar and WL oxygenation at the lower mean lung volume, LUL and LML P$_{O2}$ was not different from WL P$_{O2}$ at the higher mean lung volume, and the LLL P$_{O2}$ was larger than that for the WL.

CONCLUSIONS

Individual lobes in the supine dog are found to be similar in dispersion of $\dot{V}_A/\dot{Q}$, and the lobes are more homogeneous in their gas exchange function than the whole lung. This was true for all left side lobes at both mean lung volumes tested, with the exception of the LLL at PEEP = +5cmH$_2$O. The LLL tended to be more homogeneous than the WL at the smaller mean lung volume, but the difference was not statistically significant. Left side lobes with uniform $\dot{V}_A/\dot{Q}$ heterogeneity and different mean $\dot{V}_A/\dot{Q}$ ratios produce the more heterogeneous whole lung matching of ventilation and perfusion at PEEP = +5cmH$_2$O (as in fig. 2b). At PEEP = +10cmH$_2$O, left side lobes possess uniform $\dot{V}_A/\dot{Q}$ heterogeneity and similar mean $\dot{V}_A/\dot{Q}$ ratios to contribute to the whole lung gas exchange character (as in fig. 2c). The greater whole lung as compared to lobar $\dot{V}_A/\dot{Q}$ heterogeneity at the larger mean lung volume must result from the gas exchange character of lobes not sampled.

The cause of the observed unique lobar mean ventilation to perfusion matching at PEEP = +5cmH$_2$O is not known. In the supine dog, lobar gravitational gradients would not be expected to be different between lobes. Although the lobes may receive unique $\dot{V}_A$, $\dot{Q}$, and have a unique mean $\dot{V}_A/\dot{Q}$, mechanisms which can mix heterogeneously distributed ventilation and perfusion, such as diffusion and cardiogenic oscillations, produce fairly uniform intralobar gas exchange matching. The lobes demonstrated similar degrees of $\dot{V}_A/\dot{Q}$ heterogeneity at both mean lung volumes. Lobar differences in the mean matching of $\dot{V}_A/\dot{Q}$ were eliminated at the larger lung volume produced by PEEP = +10cmH$_2$O.

A functional consequence of each lobe possessing unique gas exchange properties, is unique lobar blood oxygenation. The various lobes can each contribute blood with different levels of oxygenation to the whole lung mixed arterial pooled blood. At the lower mean lung volume the LUL and LML mean $\dot{V}_A/\dot{Q}$ was higher, and gas exchange was more homogeneous, than that for the lung as a whole. This must have contributed to a LUL and LML P$_{O2}$ that was greater than that found for the whole lung. At the higher mean lung volume the lobar mean $\dot{V}_A/\dot{Q}$ decreased in the LUL and LML, and homogeneity of gas exchange also decreased. The result was a lower oxygenation of LUL blood. The LLL, on the other hand, did not reveal much change in it's gas exchange characteristics at the higher, as compared to the lower, mean lung volume. This was reflected in similar blood oxygenation of LLL blood at +5cmH$_2$O and +10cmH$_2$O PEEP.

In summary, the left side pulmonary lobar and the whole lung retention and excretion of inert gases and oxygenation of blood was explored. We found that lobes contribute to whole lung blood oxygenation in correspondence to their own unique gas exchange properties. Pulmonary lobes sampled were quite uniform in terms of gas exchange heterogeneity, and in general, more homogeneous than the whole lung. Each lobe can have it's own mean $\dot{V}_A/\dot{Q}$ ratio which, when summed for all the pulmonary lobes, produces the broader heterogeneity of gas exchange found for the lung as a whole.

REFERENCES

Engel, L.A., 1986, Dynamic distribution of gas flow, in: "Handbook of Physiology. The Respiratory System," P.T. Macklem and J. Mead ed., sect.3, vol.III, ch.32, p.575, Am. Physiol. Soc., Bethesda, Maryland.

Hlastala, M.P., Robertson, H.T., 1978, Inert gas elimination characteristics of the normal and abnormal lung, J. Appl. Physiol., 44(2):258.

Milic-Emili, J., 1986, Static distribution of lung volumes, in: "Handbook of Physiology. The Respiratory System," P.T. Macklem and J. Mead ed., sect.3, vol.III, ch.31, p.561, Am. Physiol. Soc., Bethesda, Maryland.

Ohlsson, J., Middaugh, M., Hlastala, M.P., 1989, Reduction of lung perfusion increases $\dot{V}_A/\dot{Q}$ heterogeneity, J. Appl. Physiol. 66(5):2423.

Paiva, M., Engel, L.A., 1987, Theoretical studies of gas mixing and ventilation distribution in the lung, Physiol. Rev. 67(3):750.

EFFECT OF ARTIFICIAL VENTILATION ON PULMONARY ANTIOXIDANT ENZYME ACTIVITIES IN A CONGENITAL DIAPHRAGMATIC HERNIA RAT MODEL

R. Tenbrinck[1], W. Sluiter[2], F. Silveri[2,4], A.P. Bos[1], E.C. Scheffers[1], A.T.J.I. Go[1], J.A.H. Bos[3], D. Tibboel[1] and B. Lachmann[3]

Depts of [1]Pediatric Surgery, Sophia Childrens Hospital; [2]Biochemistry I and [3]Anesthesiology, Erasmus University Rotterdam. [4]Clinica Reumatologica, Universita di Ancona, Jesi, Italy

INTRODUCTION

Treatment of infants with congenital diaphragmatic hernia (CDH) developing respiratory insufficiency within a few hours after birth remains unsatisfactory. The incidence of CDH is about 1:3000 newborns (Hazebroek et al, 1988), mortality for these high-risk infants ranges from 30%-60%. These infants require aggressive respiratory support, including high pressures and oxygen concentrations. Frequently the clinical course is complicated by pulmonary hypertension. Compared to premature infants, CDH survivors have a high incidence (40%) of bronchopulmonary dysplasia (BPD) (Molenaar et al, 1991; Redmond et al, 1987; O'Rourke et al, 1991). Because this disease occurs almost exclusively in premature infants who receive mechanical ventilation with increased inspiratory oxygen concentration, it was postulated (Northway et al, 1967; Crapo, 1986) that oxygen alone is toxic to the lung parenchyma.Other factors that may play a role in BPD include gestational age, barotrauma, infection, the presence of a persistent ductus arteriosus (PDA), pulmonary hypertension and reperfusion damage. It is difficult to separate the effect of oxygen from those of other factors that may influence the development of BPD. Therefore the need for a reliable animal model (preferably with CDH) to study the pathogenesis of BPD and investigate protective measurements has augmented. DeLuca described barotrauma in ventilated CDH lambs; there was no specific mention of oxygen toxicity or its defense mechanisms (DeLuca et al, 1987).

The defense mechanism against oxygen damage has been extensively described (Tanswell and Freeman, 1984; Gerdin et al, 1985; Frank and Sosenko, 1987a,b), but none of them deals with ventilated newborn animals. From the results of the cited O_2 toxicity studies has evolved the concept that baseline antioxidative enzyme activity (AOA) levels are of much less importance in determining resistance or susceptibility to O_2-induced lung damage than are the responses of the AOA to hyperoxic challenge. We are able to ventilate newborn rats with induced congenital diaphragmatic hernia (Tenbrinck et al, 1990) and investigated the AOA levels in these CDH newborns in comparison with the data from similar treated control animals.

Oxygen Transport to Tissue XIV, Edited by W. Erdmann and
D.F. Bruley, Plenum Press, New York, 1992

This study consists of two experiments: the first to detect whether the development of the baseline AOA levels in CDH rats differs from that of controls. We describe the developmental pattern of superoxide dismutase (CuZnSOD) (EC.1.15.1.1), catalase (EC.1.11.1.6) and glutathione peroxidase (GPX) (EC.1.11.1.9) in the CDH lung during late gestation from day 19 up to birth.

In the second experiment, animals were ventilated during 5 h with air or oxygen to establish different responses in AOA to hyperoxidative stress.

MATERIALS AND METHODS

CDH was induced in pregnant Sprague Dawley rats by means of the herbicide nitrofen (Tenbrinck et al, 1990). Two groups of animals were studied: C = controls and N = nitrofen. In the N group animals were obtained with CDH = NH and without CDH = NN. The distinction between NN and NH groups could only be made after autopsy. In the first experiment, imaginable differences in late gestational development of AOA profiles were studied; in a second experiment the changes in AOA during 5 h ventilation with either air or 100% oxygen were observed.

Animals

In the first experiment, the fetuses were obtained by hysterotomy (days 19, 20, 21 of gestation) or after spontaneous birth at day 22 of gestation. After determination of bodyweight the fetuses were killed by an intraperitoneal injection of pentobarbital (2 g/kg) (Tenbrinck et al, 1990).

The thoracic cavity was opened, the presence of a possible diaphragmatic defect and its size noted; after this the lungs were perfused with phosphate buffered saline (PBS, 0.07 mol/l, 4°C) via the pulmonary artery until they turned pale white. The perfused lungs were taken out, weighed, frozen in liquid nitrogen and stored at -70°C for biochemical assay.

For the second experiment adult animals from groups C and N were allowed to deliver the pups after a gestation period of approximately 22 days. The pups were anesthetized (pentobarbital 35 mg/kg every 4 h), relaxed (pancuroniumbromide 0.1 mg/kg/h) and intubated. The tubes were connected with a body box (Lachmann, 1981) and the pups were ventilated with either $FiO_2 = 0.21$ or $FiO_2 = 1.0$ for 5 h to establish a possible difference in AOA during ventilation between C, NN and NH groups. The ventilator settings of the Servo 900B (Siemens Elema, Sweden) were the same throughout the experiment in C, NN and NH groups: pressure controlled 17/2 cm H_2O; respiration rate 40/min; I:E ratio 1:2. The animals were treated as in the first experiment after they had completed their 5 h ventilation period.

Biochemical analysis

After weighing, each obtained lung was coded and homogenized separately so no pools were made. In the homogenate we determined the protein (Lowry et al, 1951) and DNA (Labarca et al, 1980) content expressed per mg wet lung weight. For AOA estimation the suspensions were centrifuged at 20,000 g for 30 min. The activity of the most prevailing SOD isoenzyme in the lung, the copper-zinc SOD (CuZnSOD) was assayed by the inhibition of xanthine xanthineoxidase catalyzed reduction of ferricytochrome-c at pH 10.2 in the presence of EDTA to chelate free copper; the unit of SOD activity was defined according to Fridovich. (Hayatdavoudi et al, 1981; Biemond et al, 1984). Catalase was measured as described earlier by Bergmeyer

(Bergmeyer, 1955). Glutathione peroxidase (GPX) activity was assayed according to Paglia (Paglia et al, 1967). The activities of SOD, catalase and GPX were expressed as units per mg lung DNA to eliminate lung weight differences.

Statistical analysis
The data are presented as mean with one standard deviation (SD). After rank transformation treatment effects were evaluated by analysis of variance (Conover, 1981).If a significant F-value was found, Bonferroni's correction method for multiple comparisons was used to identify differences among the groups (Glantz, 1987). A difference was considered statistically significant when the p-value was $< 5\%$. No further indication of p-values are made; however lower values were found.

RESULTS

Wet lung weights increased in C, NN and NH groups during gestation (Table 1), but the mean lung weight of NN and NH was significantly lower compared to C. The differences between NN and NH were also significant. No significant differences between body weights were found. The L/B ratio was also significantly lower in the NH group before birth, suggesting that lung hypoplasia develops before birth. Lung protein and DNA content expressed per mg wet lung remained virtually unchanged in all groups; this is shown by the protein/DNA ratio which did not alter over time and amounted to a mean value of 7 for all groups.
The CuZnSOD activity did not change significantly within the three groups between day 19 to birth (Fig 1A). Catalase activity showed a significant increase in activity between day 21 and birth (Fig 1B); however there was no significant difference between the groups. GPX activity (Fig 1C) increased in each group during gestation (time dependency in the three groups $P < 0.001$). The activity measured at birth for C, NN and NH was respectively 173%, 161% and 185% of the initial GPX activity at day 19 of gestation.

The effect of ventilation on AOA
In the NH group only 2 animals survived the 5 h ventilation period with $FiO_2 = 0.21$, the others died within a few hours after start of ventilation. Therefore it was decided to use data of animals immediately after birth as reference values rather than those who were not ventilated for 5 h, because none of the NH animals survived the 5 h without ventilation. There were no significant differences in lung weight, body weight, L/B ratio, DNA, protein and P/D ratio between the ventilated animals and their initial values immediately after birth. The existing differences between the groups remained the same.
Table 2 shows that in the C and NN groups the values for CuZnSOD, Catalase and GPX did not change significantly under the influence of ventilation, neither with air nor with 100% oxygen.
In the NH group CuZnSOD showed a tendency to decrease under both ventilatory conditions to about 85% of initial activity ($P = 0.18$ for NH 100% oxygen). This value was significant from the same value in the C and NN groups. Catalase activity in the NH rats remained at the initial level and was similar to values in the C and NN groups. Values for GPX activity decreased significantly to 78% and even 68% of initial values after ventilation with room air or 100% oxygen, respectively. Also compared to the C 100% value this means a significant decrease of 21%.

Table 1. Late gestational changes (day 19 until birth) in groups C, NN, NH in several lung parameters.

		LW mg	BW g	L/B ratio mg/g	DNA μg/mg LW	protein μg/mg LW	P/D ratio
C	19	43 (7)	1.51 (0.15)	28.2 (2.4)	10.4 (0.7)	71.5 (4.3)	6.9 (0.1)
	20	90 (5)	2.73 (0.3)	33.8 (1.7)	8.2 (1.4)	71.1 (4.7)	9.0 (1.8)
	21	122 (8)	3.90 (0.01)	33.5 (0.9)	9.6 (0.7)	62.8 (3.9)	6.6 (0.7)
	birth	136 (21)	5.43 (0.32)	28.1 (2.6)	10.3 (1.1)	64.1 (7.8)	6.5 (0.8)
NN	19	41 (4)	1.57 (0.14)	25.8 (1.6)	11.0 (0.5)	72.7 (4.3)	6.6 (0.1)
	20	66 (5) #	2.55 (0.45)	27.1 (1.8)	8.9 (1.4)	61.7 (6.2)	7.2 (0.8)
	21	137 (12)	4.49 (0.42)	30.8 (1.1)	7.5 (1.2)	59.8 (6.3)	8.3 (2.0)
	birth	109 (6) #	5.08 (0.07)	22.6 (1.2)	8.8 (0.5)	65.0 (6.8)	7.4 (0.8)
NH	19	35 (5) #	1.45 (0.14)	24.1 (1.8) #	11.2 (1.0)	76.9 (3.7)	6.7 (0.1)
	20	47 (18) #*	1.98 (0.73)	26.6 (2.7) #	9.7 (1.1)	65.2 (6.7)	6.7 (0.1)
	21	99 (17) #*	4.21 (0.72)	24.8 (5.1) #*	9.2 (0.8)	63.4 (3.5)	7.0 (0.9)
	birth	63 (10) #*	4.96 (0.44)	12.2 (2.1) #*	12.1 (1.6)	83.4 (11.3)	7.0 (0.1)

Abbreviations: values are expressed as mean $\pm$ (standard deviation); each parameter is assessed in 6 animals. LW lungweight; BW body weight; L/B ratio lung weight/ body weight; P/D ratio protein content/ DNA content. # significant ($p < 0.05$) from C; * significant ($p < 0.05$) from NN.

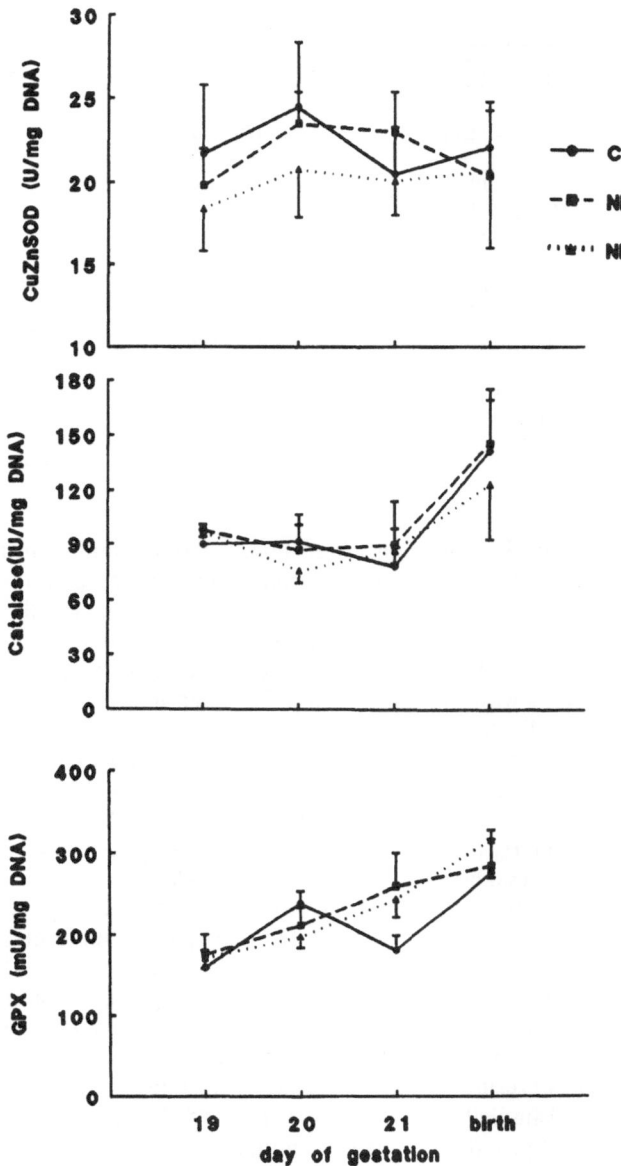

Figure 1. The development of AOA during gestation from day 19 until birth. The continuous lines represent the C; the striped line the NN and the dotted line the NH group. The CuZnSOD (a); catalase (b) and GPX (c) are shown. The values are expressed in activity per mg DNA and given as mean ± SD.

Table 2. The effect of ventilation with either air or 100% oxygen on SOD, catalase and GPX activities in the three groups (C, NN, NH).

		n	SOD (U/mg DNA)	Catalase (IU/mg DNA)	GPX (mU/mg DNA)
	birth	10	22.1 (2.2)	142 (27)	277 (40)
C	air	7	20.9 (3.5)	136 (23)	276 (19)
	100%	8	22.8 (4.5)	126 (14)	271 (41)
	birth	7	19.4 (4.4)	146 (30)	283 (44)
NN	air	6	24.5 (4.6)	117 (12)	264 (12)
	100%	9	24.6 (5.1)	144 (31)	255 (41)
	birth	7	20.7 (4.7)	124 (30)	315 (45)#
NH	air	2	17.5 (0.2)	119 (10)	246 (12)
	100%	6	17.9 (1.2)#	134 (40)	216 (24)*#

Values are expressed as mean $\pm$ (SD), * significant (p<5%) from birth value in the same group, # significant (p<5%) from controls.

DISCUSSION

CDH lungs are hypoplastic in humans (Areechon, 1963), as well as in the rat model (Tenbrinck, 1990) this is mainly based on morphological differences, however little is known about the biochemical compound of the hypoplastic lung. Frank reported that in the rat the chronology of development of AOA is very similar to that of the fetal surfactant system (Frank, 1987a). A different content of lecithin-sphingomyelin in amniotic fluid of fetuses with hypoplastic lungs has been found (Hisanaga, 1984).
CDH lungs of humans (Redmond, 1987) and fetal lambs (DeLuca et al, 1987) are very vulnerable to high inflation pressures. Nothing is known about their reaction to high inspiratory oxygen concentrations during ventilation. In healthy lungs the AOA levels are strongly correlated with the degree of protection that may be anticipated from O_2 radical induced lung injury. Increased enzyme activity has been consistently found in association with tolerance to hyperoxia, and reduced AOA usually leads to greater than normal susceptibility of the lung to high O_2 concentrations. This phenomenon was recently demonstrated in healthy premature rabbits compared with term rabbits (Frank and Sosenko, 1991). Literature search revealed no study in which the effect of ventilation on AOA in newborn CDH vs healthy (rat) lungs was described.
In the first experiment of this study we concluded that baseline AOA levels are almost the same throughout the three groups. We also found a similar developmental pattern in late gestation. These results are in accordance with those found by others (Frank and Sosenko, 1987a; Tanswell, 1984; Gerdin et al, 1985; Hayashibe et al, 1990). The lung weights in the NH group were lower than that in C and NN so the total amount of AOA is reduced. This indicates that the CDH lungs does not differ qualitatively from the C and NN lungs, but only quantitatively. This is also supported by the unchanged protein/DNA ratio in the groups.
In the second experiment we concluded that in the NH group the SOD and GPX activities tended to decrease during ventilation both with room air and oxygen. This supports our hypothesis that the CDH lung behaves like a premature lung and unlike C and NN lungs, would fail to mount a protective increase in AOA during ventilation with high FiO_2. This failure could be an explanation for the increased susceptibility to O_2-induced damage in CDH lungs.

Because the NH FiO$_2$ = 0.21 group consisted of only 2 surviving animals, care has to be taken with the interpretation of these results. Also in the larger NH FiO$_2$ = 1.0 group care has to be taken with simplification that the decreased SOD and GPX levels alone are the cause of the high incidence of BPD in human CDH survivors. On the basis of these experiments we could only speculate about the consequences of our results for the clinical practice. Of more importance is that this is the first applicable in vivo study that deals with the problem of oxygen toxicity in hypoplastic CDH lungs during ventilation.

REFERENCES

Areechon W, Reid L, 1963, Hypoplasia of lung with congenital diaphragmatic hernia, Br Med J, 1:230.

Bergmeyer HU, 1955, Zur messung von Katalase-aktivitaten, Biochem Z, 327: 255.

Biemond P, Swaak AJG, Koster JF, 1984, Protective factors against oxygen free radicals and hydogen peroxide in rheumatoid arthritis synovial fluid, Arthritis Rheum 27:760.

Conover WJ, Iman RL, 1981, Rank transformations as a bridge between parametric and nonparametric statistics, Amer Statist, 35: 124.

Crapo JD, 1986, Morphologic changes in pulmonary oxygen toxicity, Ann Rev Physiol,48:721.

DeLuca U, Cloutier R, Laberge JM, Fournier L and Guttman FM,1987, Pulmonary barotrauma in congenital diaphragmatic hernia: experimental study in lambs, J Pediatr Surg, 22: 311.

Frank L, Sosenko IRS, 1987a, Development of lung antioxidant enzyme system in late gestation: possible implications for the prematurely born infant, J Pediatr, 110:9.

Frank L, Sosenko IRS, 1987b, Prenatal development of lung antioxidant enzymes in four species,J Pediatr, 110:106.

Frank L, Sosenko IRS, 1991, Failure of premature rabbit to increase antioxidant enzymes during hyperoxic exposure: increased susceptibility to pulmonary oxygen toxicity compared with term rabbits, Pediatr Res, 29:292.

Gerdin E, Tyden O, Eriksson UJ, 1985, The development of antioxidant enzymatic defense in the perinatal rat lung: activities of SOD, GPX, and catalase, Pediatr Res, 19:687.

Glantz SA, 1987, Primer of biostatistics, 2nd ed. McGraw-Hill, New York.

Hayashibe H, Asayama K, Dobashi K, Kato K, 1990, Prenatal development of AOE in rat lung, kidney and heart: marked increase in immunoreactive superoxide dismutases, glutathione peroxidase and catalase in the kidney, Pediatr Res, 27: 472.

Hayatdavoudi G, O'Neill JJ, Barry BE, Freeman BA, Crapo JD, 1981, Pulmonary injury in rats following continious exposure to 60% O$_2$ for 7 days, J Appl Physiol, 51:1220.

Hazebroek FJW, Tibboel D, and Molenaar J, 1988, Congenital diaphragmatic hernia: the impact of preoperative stabilization. A prospective pilot study in 13 patients, J Pediatr Surg 23: 1139

Hisanaga S, Shimokawa H, 1984, Unexpectedly low lecithin/sphingomyelin ratio associated with fetal diaphragmatic hernia, Am J Obstet Gynecol 149:905.

Labarca C, Paigen K, 1980, A simple, rapid and sensitive DNA assay procedure, Anal Biochem, 102:344.

Lachmann B, Grossmann G, Freyse J, Robertson B, 1981, Lung thorax compliance in the artificially ventilated premature rabbit neonate in relation to variations in I:E ratio, Pediatr Res, 15:833.

Lowry OH, Rosebrough NJ, Farr AL, Randall RJ, 1951, Protein measurements with the folin phenol reagent, J Biol Chem, 193:265.

Molenaar JC, Bos AP and Tibboel D, 1991, Congenital diaphragmatic hernia, what defect?, J Pediatr Surg, 26: 248.

Northway WH, Rosan RC, Porter DY, 1967, Pulmonary disease following respiratory therapy of hyaline membrane disease, N Eng J Med, 267,357.

O'Rourke PP, Lillehei CW, Crone RK, Vacanti JP, 1991, The effect of ECMO on the survival of neonates with high-risk congenital diaphragmatic hernia: 45 cases from one institution, J Pediatr Surg,26:147.

Paglia DE, Valentine WN, 1967, Studies on the quantitative and qualitative characterization of erythrocyte glutathione peroxidase, J Lab Clin Med, 70: 158.

Redmond C, Heaton J, Calix J, Graves E, Farr G and Arensman R, 1987, A correlation of pulmonary hypoplasia, MAP and survival in congenital diaphragmatic hernia treated with ECMO, J Pediatr Surg, 22,1143.

Tanswell AK, Freeman BA, 1984, Pulmonary antioxidant enzyme maturation in the fetal and neonatal rat.I. Developmental profiles, Pediatr Res,18:584.

Tenbrinck R, Tibboel D, Gaillard JLJ, Kluth D, Lachmann B and Molenaar JC, 1990, Experimentally induced congenital diaphragmatic hernia in rats, J Pediatr Surg, 25: 426.

COMPARISON OF PRESSURE SUPPORT VENTILATION (PSV) AND INTERMITTENT MANDATORY VENTILATION (IMV) DURING WEANING IN PATIENTS WITH ACUTE RESPIRATORY FAILURE

F. Esen[1], T. Denkel[1], L. Telci[1], J. Kesecioglu[1,2], A.S. Tütüncü[1,2], K. Akpir[1], B. Lachmann[2]

[1]Department of Anesthesiology, University of Istanbul, Faculty of Medicine, Istanbul, Turkey
[2]Department of Anesthesiology, Erasmus University and Academic Hospital Dijkzigt, Rotterdam, The Netherlands

INTRODUCTION

Certain groups of mechanically ventilated patients with acute respiratory failure are difficult to wean from prolonged ventilation despite advanced respiratory support techniques. The common modes for providing support during weaning are intermittent mandatory ventilation (IMV), and spontaneous ventilatory trial (SVT) techniques; such as continuous positive airway pressure (CPAP) and T-piece. Weaning is often initiated with IMV, and a series of SVTs are introduced as the patient tolerates decreasing rates of IMV. However, there is some concern that this strategy of weaning may not be optimal for patient comfort, and muscle reconditioning [1].

Pressure support ventilation (PSV) is a new ventilatory mode which augments a spontaneous breath with a fixed amount of positive pressure. The level of positive pressure is set by the physician, and the patient has control over the respiratory frequency and inspiratory time. In many clinical studies, PSV appeared to be potentially useful during the weaning period by maintaining muscle activity with patient comfort [1,2].

Controversy still exists regarding the best approach to weaning, since there are not enough data to indicate the superiority of one technique over another. This study was designed to compare the effects of PSV and IMV on weaning of long-term artificially ventilated patients with acute respiratory failure (ARF) due to thorax trauma.

Oxygen Transport to Tissue XIV, Edited by W. Erdmann and
D.F. Bruley, Plenum Press, New York, 1992

METHODS

Fourty mechanically ventilated patients (male:21; female:19) with thorax injury were studied in two groups during the weaning period. Patients fulfilling the following criteria were included in the study on admission to the ICU:

1. Trauma Score (TS) > 5 and < 10.
2. PaO_2/FiO_2 ratio of greater than 150.
3. Total static lung compliance (TSLC) > 30 ml/cmH_2O
4. Glascow Coma Scale of 15.
5. Not undergoing abdominal or any other surgical procedure.

All patients received mechanical ventilatory support in a pressure controlled mode with a Servo 900C (Siemens-Elema). The decision to begin weaning a patient was based on the following criteria:

a) A spontaneous respiratory rate > 10/ min, < 36/ min when removed from mechanical ventilatory support.
b) Stable or improving chest radiographs.
c) PaO_2 > 60 mmHg with a FiO_2 < 0.4, $PaCO_2$ < 50 mmHg, PEEP < 5 cmH_2O.
d) Negative inspiratory effort < - 2 cmH_2O.
e) Hemodynamic stability as evidenced by a regular cardiac rhythm and mean arterial pressure (MAP) higher than 60 mmHg without any vasopressor.

The patients were then randomly assigned to a weaning mode of either PSV (n=20) or IMV (n=20). IMV mode was set according to clinical guidelines to supply minute volume (MV) of 120-150 ml/ kg, and a $PaCO_2$ of less than 50 mmHg. FiO_2 was set to provide a PaO_2 > 60 mmHg. The initial level of PSV was designed to result in a minute volume similar to the IMV group. A pressure of -2 cmH_2O in the inspiratory circuit triggered inspiratory flow to provide the level of inspiratory pressure. Again FiO_2 was set to provide PaO_2 > 60 mmHg.

During the weaning period, the following measurements were made at 2-hourly intervals.

1. Mandatory and spontaneous ventilatory rates (f) and tidal volumes (TV) measured by expiratory flows in the ventilatory circuit.
2. Peak airway pressure (Peak paw) measured in the inspiratory circuit of the ventilator.
3. Arterial O_2 saturation (SaO_2) determined from Hemoximeter OSM3 (Radiometer). Blood gas analyses were done by ABL300 (Radiometer, Copenhagen).
4. End-tidal CO_2 and O_2 consumption (VO_2) was measured by Datex-Multicap.
5. Arterial blood pressures were measured from a radial artery catheter using a Viggo-Spectramed transducer from a Horizon 1000 (Mennen Medical) monitor.

The accepted parameters during the weaning procedure were as follows:
1. PaO_2 > 60 mmHg.
2. Increase in $PaCO_2$ < 10 mmHg from the baseline value.
3. PH > 7.28 or PH < 7.55
4. Spontaneous respiration rate < 30/ min.
5. Heart rate (HR) < 140/ min.
6. MAP > 60 mmHg.

Patients were excluded from the protocol at any time during weaning when their parameters did not fulfil the given criteria. The patient was placed on the previous ventilatory support when significant deteriorations in SaO_2 or $PaCO_2$ were seen. All patients received endotracheal suctioning when necessary, and no sedatives or narcotics were administered during the weaning trial.

On the first day of weaning, the parameters were measured and calculated for mean ± SD. The patients were then treated with the same protocol until they could be disconnected from the ventilator. Before disconnection, a test was performed so that they should generate a maximal inspiratory effort.During the weaning period, patients were asked how well they tolerated the type of the ventilation.

The time from the beginning of weaning until its successful completion was accepted as weaning time (WT), and a successful wean was defined as 48 hours discontinuation of mechanical ventilation.

Data were compared between the two groups by Student's t-test. Significance was considered at $p < 0.05$.

RESULTS

Both groups of patients were of similar age and sex, and there was no significant difference of trauma scores between the two groups on admission. Characteristics of the PSV and IMV groups along with the mean baseline mechanical ventilation parameters are summarized in Table 1.

Table 1. Data for the patient groups on admission.

	PSV (n = 20)	IMV (n = 20)
Sex M/F	12 / 8	9 / 11
Age (years)	41.65 ± 17.1	43.7 ± 13.4
TS	7.69 ± 1.8	7.76 ± 1.4
PaO_2/FiO_2	183.8 ± 60.8	182.4 ± 61.4
TSLC (ml/cmH$_2$O)	39.4 ± 8.4	39.9 ± 10
SaO_2 (%)	97.4 ± 2.3	97.9 ± 3.1
$PaCO_2$ (mmHg)	36.2 ± 10.6	34.8 ± 8.4

TS = trauma score; TSLC = total static lung compliance;

Ventilatory parameters, gas exchange, blood pressure, heart rate and VO_2 values during the first day of weaning period in both groups are compared in Table 2. No significant differences were observed in the mean values of SaO_2, $ETCO_2$, arterial pressure and heart rate. Better gas exchange, and lower levels of VO_2 were seen in the PSV group, although statistically not significant.

Table 2. Ventilatory parameters, gas exchange, and hemodynamic parameters during PSV and IMV during the first day of weaning.

	PSV	IMV
MV (ml/kg/min)	124.8 ± 11.1	122.4 ± 10.9
frequency (/min)	15.1 ± 4.3[a]	23.2 ± 5.6
Peak paw (cmH₂O)	27.0 ± 4.1[a]	33.1 ± 6.1
SaO₂ (%)	97.4 ± 1.5	96.9 ± 1.7
PaCO₂ (mmHg)	34.3 ± 11.4	32.7 ± 9.3
MAP (mmHg)	66.9 ± 9.1	67.5 ± 10.5
HR (bpm)	88.8 ± 12.2	87.8 ± 13.4
VO₂ (ml/min)	199.8 ± 26.2	223.8 ± 29.1

[a] $p < 0.05$ between PSV and IMV
MV = minute ventilation; MAP = mean arterial pressure; HR = heart rate; VO₂ = oxygen consumption.

Applied peak airway pressure to provide adequate minute volume in the PSV group was ranged between 18 and 32 cmH₂O. It was noted that PSV was associated with a significantly lower peak airway pressure than IMV. Patients were observed to be clearly more comfortable in PSV mode.

Patients who generated a maximum inspiratory pressure of more than -20 cmH₂O with an adequate gas exchange of PaO₂ > 80 mmHg with FiO₂ < 0.35, were disconnected from the ventilator. A successful wean was accepted as 48 hours discontinuation of mechanical ventilatory support. Of the 40 patients, 32 were weaned successfully: 17 in the PSV group and 15 in the IMV group. Although total ventilation time (TVT) prior to weaning was similar in both groups, weaning time was observed to be significantly shorter in the PSV group than in the IMV group (Table 3).

Table 3. Comparison of weaning time (WT) and total ventilation time (TVT)

	PSV	IMV
WT (/day)	6.3 ± 3.1	9.9 ± 2.7[a]
TVT (/day)	19.1 ± 2.3	22.4 ± 3.1

[a] $p < 0.05$ between PSV and IMV

DISCUSSION

Pressure support ventilation is a recent form of ventilatory assistance, that can be used for mechanical ventilation and weaning. Although no study has yet demonstrated the superiority of this mode over other techniques of weaning, studies have shown a number of advantages [1-6].

The main findings of this study demonstrated that inspiratory pressure assist with PSV resulted in lower levels of peak airway pressure with a slower spontaneous ventilatory rate compared to IMV. Patients appeared to tolerate PSV better than IMV.

Recently, McIntyre and coworkers described the effects of various levels of PSV on airway pressure, gas exchange and patient comfort [1]. They found PSV to be a reasonable form of mechanical ventilatory support in patients with spontaneous ventilatory drives. It improved patient comfort and reduced the patient's ventilatory work. In the same study McIntyre found a significant positive correlation between the level of PSV and tidal volume, and a negative correlation with the frequency of spontaneous breathing. These findings are similar to ours, since we measured the same amount of minute volume at a decreased respiratory rate and peak airway pressure in PSV.

Kanak and coworkers described a decrease in oxygen consumption when the mode of ventilatory assistance was changed from IMV to PSV [7]. They noted that the patients' respiratory efforts decreased and that the respiratory rate was reduced during PSV. Fahey and coworkers also reported a significant reduction in the O_2 cost of breathing with PSV [8]. In their study, they found that PSV prevented diaphragmatic failure by diminishing the patient's work of breathing and total O_2 consumption. Brochard and coworkers showed that O_2 consumption decreased when PSV was added [2]. From their data, the work of breathing appeared to be the prime determinant of the O_2 cost of breathing. Extrapolating these results to ours, we noted a lower O_2 consumption in the PSV group. Although not significant, this difference was considerable when compared with IMV.

Prakash and Meij, in their study including 26 patients after aortocoronary bypass grafts, compared the cardiovascular effects of PSV with conventional ventilatory modes, and they found no negative effects of PSV [9]. Our results also showed no adverse effects on cardiovascular variables in patients of both groups.

Hurst and coworkers evaluated the effects of adding low levels of PSV in conjuction with intermittent mandatory ventilation [5]. They showed an increase in mean airway pressure, and improvement in oxygenation. McIntyre noted that PSV was associated with higher levels of mean airway pressure compared to IMV, however gas exchange was similar during both forms of ventilation [1]. Our results showed perfect efficacy of gas exchange in both PSV and IMV groups. We observed slightly and nonsignificantly better gas exchange with higher PaO_2 and lower $PaCO_2$ values in PSV mode.

The main goal in this study was to compare the effects of both modes on weaning time. In order to have separate data, we formed two groups by standardizing patients according to some determined criteria. Among the successfully weaned patients, we observed that weaning time was significantly shorter in patients weaned by PSV: Three patients in the PSV group needed to be connected in their first 24 hours after the disconnection. Five patients in the IMV group, showing fatigue at one time during their weaning procedure, needed to be put back to a control mode of ventilatory support.

We conclude that ARF patients can be weaned successfully by either PSV or IMV modes. However, being synchronized with the natural rhythm of breathing and maintaining patient comfort with muscle reconditioning, PSV seems to be more efficient than IMV since it significantly shortens weaning time in ARF patients on long-term ventilatory support.

REFERENCES

1. N. R. McIntyre, Respiratory function during pressure support ventilation, Chest. 5: 677-683 (1989).
2. L. Brochard, A. Harf, H. Lorino, and F. Lemaire, Inspiratory pressure support prevents diaphragmatic fatigue during weaning from mechanical ventilation, Am. Rev. Respir. Dis. 139:513-521 (1989).
3. F. Fiastro, M. Habib, and S. Quan, Pressure support compensation for inspiratory work due to endotracheal tubes and demand continuous positive airway pressure, Chest. 93:499-505 (1988).
4. J. P. Viale, G. J. Annat, Y. M. Bouffart et al, Oxygen cost of breathing in postoperative patients: pressure support ventilation vs positive airway pressure, Chest. 93:506-509 (1988).
5. J. M. Hurst, R. D. Branson, K. D. Davis, and R. R. Barette, Cardiopulmonary effects of pressure support ventilation, Arch. Surg. 124:1067-170 (1989).
6. L. Brochard, F. Pluskwa, and F. Lemaire, Improved efficacy of spontaneous breathing with inspiratory pressure support, Am. Rev. Respir. Dis. 136:411-415 (1987).
7. R. Kanak, P. J. Fahey, and C. Vanderwarf, Oxygen cost of breathing: changes dependent upon mode of mechanical ventilation, Chest. 87:126-127 (1985).
8. P. J. Fahey, C. Vanderwarf, and A. David, Comparison of oxygen cost of breathing during weaning with continuous positive airway pressure vs pressure support ventilation, Am. Rev. Respir. Dis. 131:A130 (1985).
9. O. Prakash, and S. Meij, Cardiopulmonary response to inspiratory pressure support during spontaneous ventilation versus conventional ventilation, Chest. 88:403-408 (1985).

ALTERATIONS OF THE RESPIRATORY MINUTE VOLUME CAUSE CHANGES IN MUSCLE

OXYGENATION OF RATS

Martina Günderoth

University of Tübingen, Medical Hospital I
7400 Tübingen, FRG*

INTRODUCTION

Various factors such as blood pressure, blood velocity, and muscle contraction, influence oxygenation of muscle tissue. One of the various functions of skeletal muscle is regulation of the blood supply to the organs. It is known that in case of a shock the blood flow of skeletal muscle is reduced prior to those of other important organs.

In rats too the regulation capacity of skeletal muscle is very important as, under normal conditions, 47 % of the heart time volume (HTV) is distributed to the skeletal muscle (Baker et al., 1979).

Only little is known about the regulation capacity of skeletal muscle oxygenation in rats under changing conditions. In a previous study it was shown (Günderoth, 1991) that short-lasting changes of heart rate (HR) and of cardiac output (CO) did not influence the muscle oxygenation of m. gracilis in healthy rats.

PURPOSE

This study was performed to find out whether respiration has an influence on muscle oxygenation in rats. Therefore, the respiratory minute volume (RMV) was changed in healthy and distressed (restricted lung function) rats.

MATERIAL AND METHODS

17 male Sprague-Dawley rats weighing 311.6 g (± 47.9 SD) were anesthetized with a combination of Ketanest (Parke Davies) and Dehydrobenz-peridol (Janssen) (100 mg and 25 mg respectively per kg body weight). Following a midline cervical incision, the trachea was intubated, and the animals were ventilated with room air using a constant volume pump (Ugo Basile, Italy) attached to the tracheal cannula.

---

*Current address: Eppendorf-Netheler-Hinz GmbH, 2000 Hamburg 65, FRG

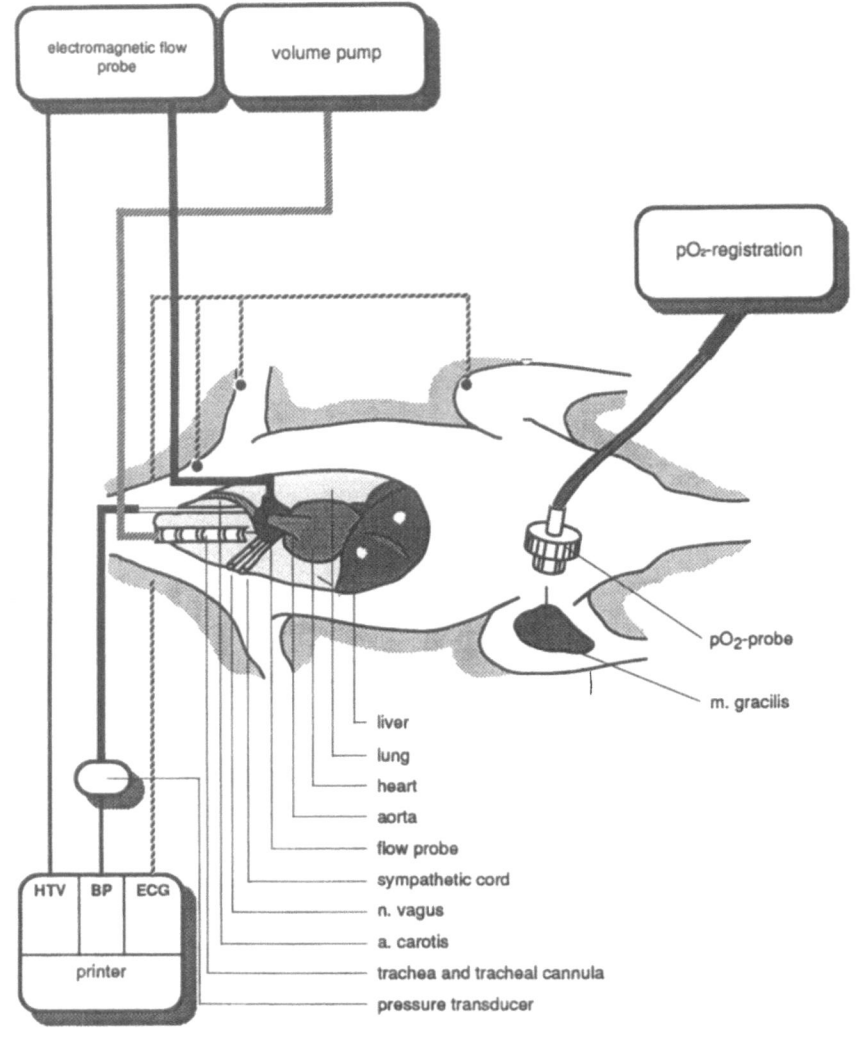

Fig. 1. Experimental design.

HTV	=	heart time volume
ECG	=	electrocardiogram
BP	=	blood pressure

The chest was opened by a midsternal incision. The left carotid artery was cannulated and the blood pressure was monitored with a pressure transducer (Statham P23 ID; Hellige, FRG) and recorded throughout the experiment on a polygraph (Graphtec Corp. Japan). An electromagnetic flow probe (Narco Bio Systems, USA) was then placed around the ascending aorta and blood flow was continuously recorded on the multichannel recorder. A bipolar electrocardiogram (Eindhoven) was continuously recorded on the polygraph and heart rate was determined from the interval between successive P waves. The mean arterial pressure (MAP) and the total peripheral resistance (TPR) were calculated (Fig.1).

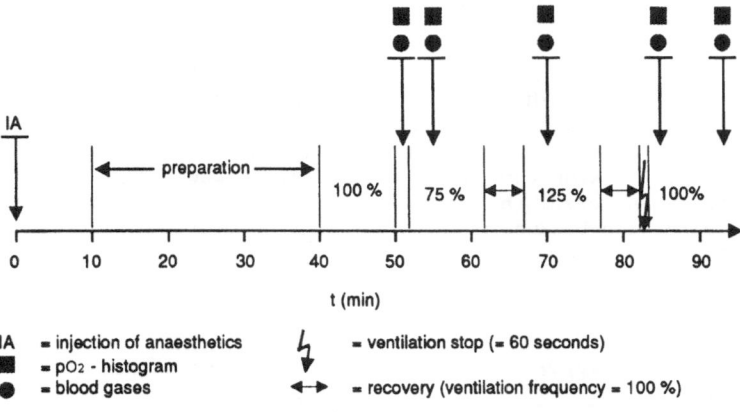

Fig. 2. Experimental schedule.

The tissue pO_2 was measured continuously on the right gracilis muscle with a multiwire surface electrode (Kessler and Lübbers, 1966) before and during the RMV changes. Additional pO_2 histograms were taken under normal breathing conditions, 3 minutes after the ventilation change, as well as directly and 10 minutes after the onset of the ventilation after the ventilation stop (Fig. 2).

In a previous study (unpublished) the normal ventilation frequency of anesthetized spontaneous breathing rats (Ketanest) without any further manipulation was found to be constant at $100/min^{-1}$. Depending on the body weight, the respiratory volume varied between 1.2 and 2.0 ml/breath (Table 1). Therefore, it seemed reasonable to adjust the respiratory volume individually while the RMV variations were induced by changing the ventilation frequency. Each change lasted 10 minutes. The animals were allowed to recover for 5 minutes after every RMV variation. At the end of every experiment the ventilation was stopped for 60 seconds. After the ventilation onset all parameters were recorded for at least 10 minutes.

Table 1.   Respiratory minute volume (RMV) of all animals.

Rat	1	2	3	4	5	6	7	8	9	10	MW	SD
Healthy rats												
BW	386	368	334	304	350	334	340	366	228	244	324	50
RV	2,0	2,0	1,8	1,8	2,0	1,8	1,8	2,0	1,2	1,2	1,8	0,3
75 %	159	152	138	125	144	138	140	151	94	101	134	22
100 %	212	202	184	167	193	184	187	201	125	134	179	29
125 %	265	253	230	209	241	230	234	252	157	168	224	36
Distressed rats												
BW	336	248	262	270	280	328	320				292	35
RV	1,8	1,2	1,5	1,5	1,8	1,8	1,8				1,6	0,2
75 %	136	100	106	109	113	133	130				118	14
100 %	181	134	142	146	151	177	173				158	19
125 %	227	167	177	182	189	221	216				197	24

BW = body weight (g)   MW = mean value
RV = respiratory volume (ml/breath)   SD = standard deviation
75 %, 100 %, 125 % = respiratory frequency

There were no significant differences in mean muscle pO_2 under normal breathing conditions in healthy and distressed rats although both the macrocirculatory parameters and the parameters of arterial blood were different. This again shows that the range of oxygen capacity in the resting muscles is wide enough to support distressed situations of the whole organism. However, the histogram form in distressed rats indicated a changed situation because more values in the lower pO_2 classes were obvious (Fig. 3).

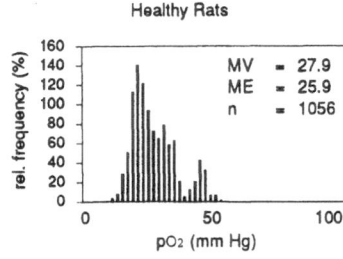 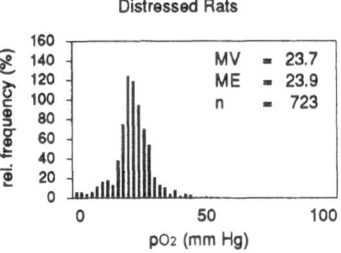

Fig. 3. pO_2 histograms under normal breathing conditions in healthy (n = 10) and distressed (n = 7) rats.

MV   =   mean pO_2 (mm Hg)
ME   =   median pO_2 (mm Hg)
n   =   number of pO_2 values

Healthy rats

Alterations of the RMV did not influence mean muscle pO_2 in healthy rats whereas the median pO_2 decreased during a reduced RMV. A significant fall of mean muscle pO_2 could be seen after the ventilation had stopped (Table 2, Fig. 4, Fig. 5). 10 minutes after the ventilation onset the mean muscle pO_2 had increased again but was still below the initial values. The recovery period was not finished as the increased HR and both the decreased HTV and MAP showed.

Table 2.   Mean and Median pO_2 values of healthy (n = 10) and distressed (n = 7) rats during changes of the respiratory minute volume.

RMV	Healthy rats			Distressed rats		
	MV	ME	n	MV	ME	n
75 %	27,0	20,0	1033	22,6	20,0	737
100 %	27,9	25,9	1056	23,7	23,9	723
125 %	25,5	24,7	1071	20,4	22,8	709
S 1	14,0	12,0	999	11,6	7,8	715
S 2	22,0	21,9	1020	14,6	13,7	615

100 % = normal ventilation frequency   MV = mean pO_2 (mm Hg)
75 % = reduced ventilation frequency   ME = median pO_2 (mm Hg)
125 % = increased ventilation frequency   n = number of pO_2 values
S 1   = 60 seconds of ventilation stop
S 2   = 10 minutes after ventilation onset

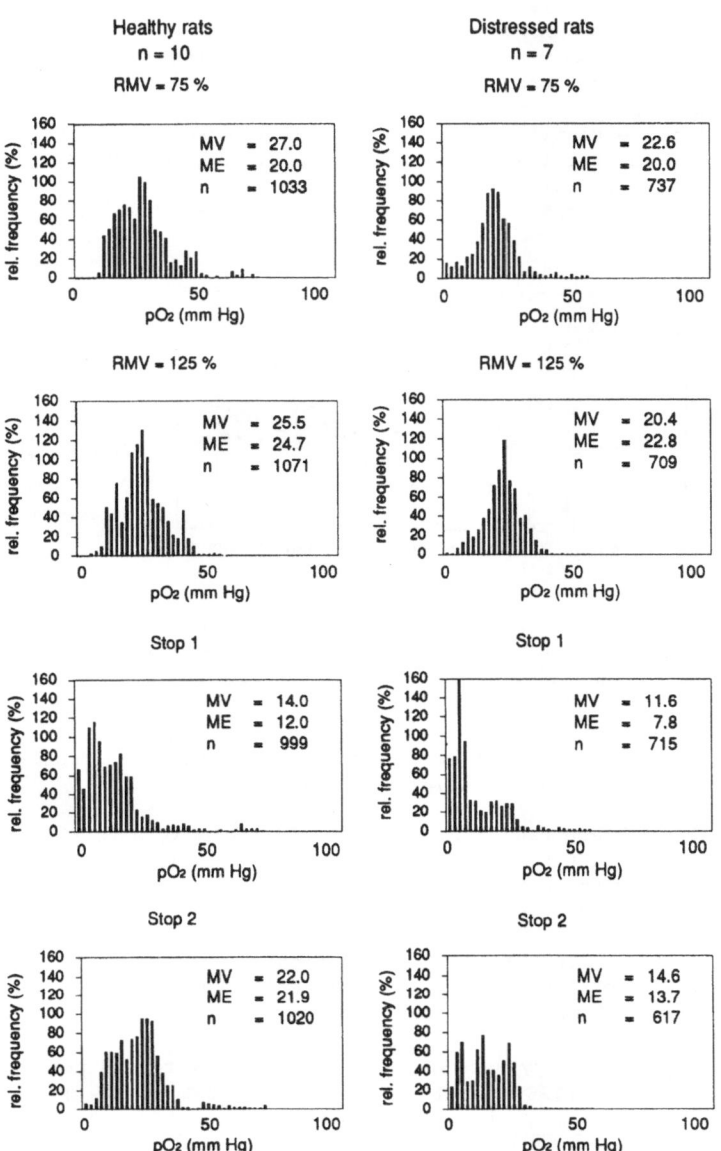

Fig. 4. pO₂ histograms during changed RMV in healthy and distressed
rats.

    MV   = mean pO₂ (mm Hg)
    ME   = median pO₂ (mm Hg)
    n   = number of pO₂ values
    RMV   = respiratory minute volume
    Stop 1 = 60 seconds of ventilation stop
    Stop 2 = 10 minutes after ventilation onset

Distressed rats

In contrast to healthy rats, a decreased RMV did not induce a fall of mean or median muscle pO_2 although an increase of pO_2 values in the lower pO_2 classes was obvious. During an increased RMV the mean muscle pO_2 remained almost unchanged and the histogram resembled that which was measured under normal breathing conditions. After the ventilation had stopped mean pO_2 decreased significantly (Table 2, Fig. 4, Fig. 5). 10 minutes after the ventilation onset, mean muscle pO_2 had slightly increased and was still significantly below the initial values. In comparison with the initial values, the HR seemed to be normalized whereas the HTV and the MAP were significantly lower (Fig. 6).

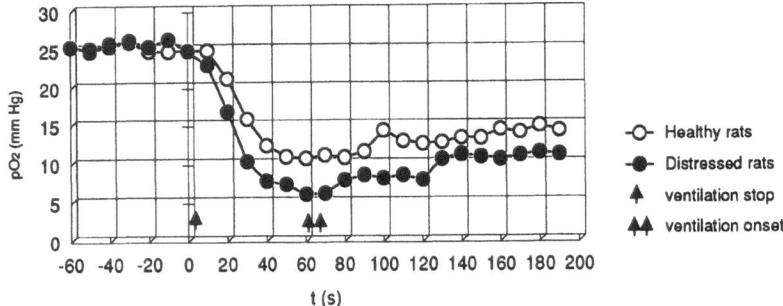

Fig. 5. Skeletal muscle pO_2 measured continuosly on m. gracilis in healthy (n = 10) and distressed (n = 7) rats before, during, and after a ventilation stop lasting 60 seconds.

Table 3. Mean values and standard deviation (±) of macrocirculatory parameters and arterial blood gas values.

HR   = heart rate (min^{-1}) O_2 Sat = O_2 saturation (%)
HTV = heart time volume (ml/min) $paCO_2$ = arterial pCO_2 (mm Hg)
RMV = respiratory minute volume paO_2 = arterial pO_2 (mm Hg)
MAP = mean arterial pressure (mm Hg)
S 1 = 60 s of ventilation stop
S 2 = 10 min. after ventilation onset
TPR = total peripheral resistance (mm Hg/ml/min)
75%,100%,125% = respiration frequency

RMV	Healthy rats					Distressed rats				
	75 %	100 %	125 %	S 1	S 2	75 %	100 %	125 %	S 1	S 2
HR	243	206	216	255	234	29	249	266	260	250
±	22	34	31	43	53	29	54	32	49	70
HTV	49	42	43	37	39	47	51	51	31	29
±	11	7	7	25	20	10	10	11	27	22
MAP	70	67	68	72	64	58	64	64	51	41
±	8	8	10	29	8	10	7	4	29	13
TPR	1,5	1,6	1,6	2,8	1,8	1,3	1,3	1,3	3,8	1,4
±	0,3	0,2	0,3	2,7	1,1	0,4	0,4	0,4	4,6	0,6
paO_2	113	127	129	96	91	83	89	94	62	68
±	24	19	27	35	29	42	17	24	12	9
$paCO_2$	24	19	23	27	26	37	30	34	45	39
±	6	4	5	7	11	7	6	3	15	16
O_2 Sat	97	98	98	94	93	87	93	94	78	83
±	2	1	1	5	5	6	3	4	10	7

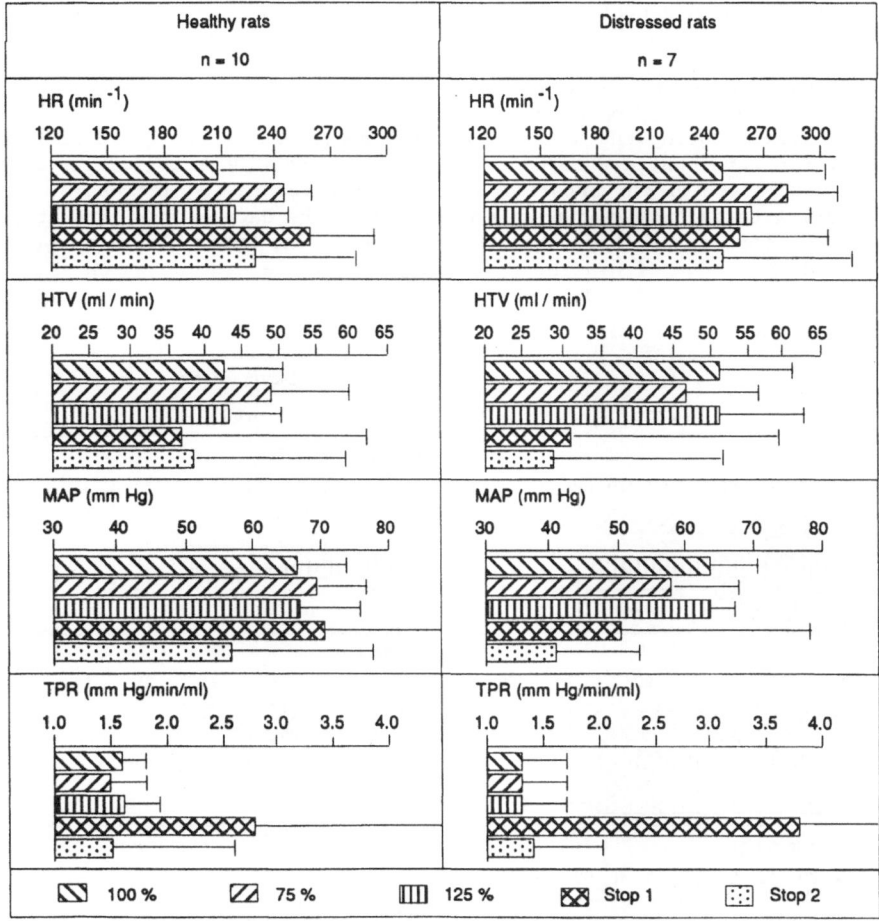

Fig. 6:   Mean values and standard deviation of hemodynamic parameters
          during changes of the respiratory minute volume (RMV) in
          healthy and distressed rats.
          100 %   = normal RMV
          75 %   = reduced RMV
          125 %   = increased RMV
          Stop 1 = 60 seconds of ventilation stop
          Stop 2 = 10 minutes after ventilation onset
          HR   = heart rate (min^{-1})
          HTV   = heart time volume (ml/min)
          MAP   = mean arterial pressure(mm Hg)
          TPR   = total peripheral resistance (mm Hg/min/ml)

Macrocirculatory parameters and blood gases

In healthy rats the HR, HTV, and MAP were lower compared with the data known from the literature (Baker et al. 1979). In this experiment measurements were obtained in open-chest rats. The opening of the chest seems to involve a decrease of HR, CO, and MAP. However, the regulation system of the macrocirculation was not impaired. The results show that macrocirculation was controlled on the lower level during RMV alterations (Table 3, Fig. 6).

In distressed rats the HR was low but higher than in healthy rats. The HTV seemed normal compared to the data of the literature whereas the MAP was lower. During the RMV changes, the reactions of the macrocirculatory system were contrary to those in healthy rats. The arterial pO_2 and the O_2 saturation of the arterial blood were decreased in distressed rats. The arterial pCO_2 was normal compared to the data of the literature but higher than in healthy rats. The changes during RMV alterations were comparable to those of healthy rats except the O_2 saturation which showed greater falls during a reduced RMV and after the ventilation stop (Table 3, Fig. 7).

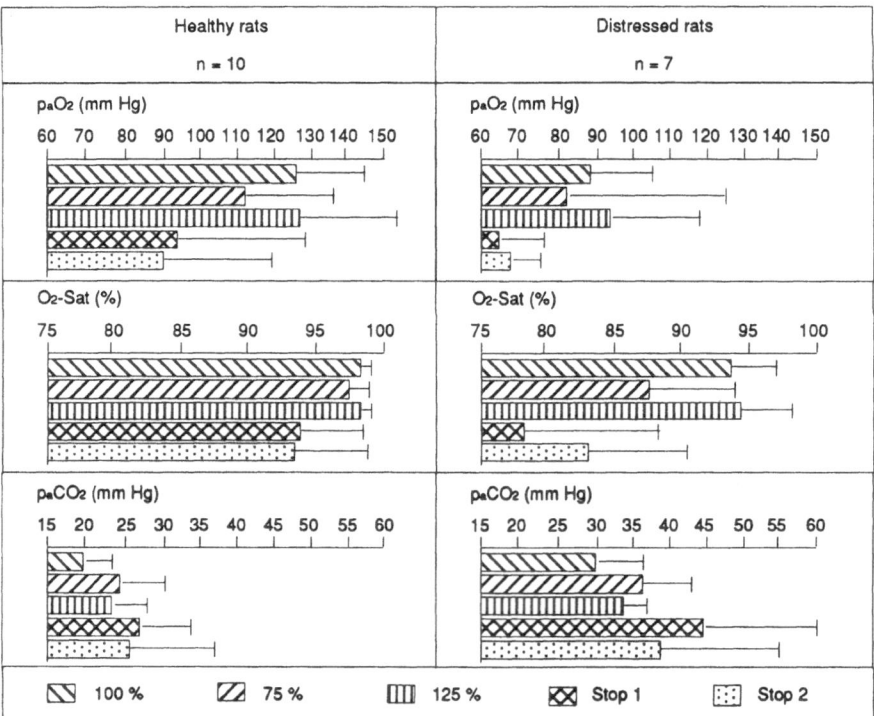

Fig. 7. Mean values and standard deviation of blood gas parameters during changes of the respiratory minute volume (RMV) in healthy and distressed rats.

100 %	= normal RMV
75 %	= reduced RMV
125 %	= increased RMV
Stop 1	= 60 seconds of ventilation stop
Stop 2	= 10 minutes after ventilation onset
p_aO_2	= arterial pO_2
p_aCO_2	= arterial pCO_2
O_2-Sat	= O_2 saturation of arterial blood

DISCUSSION

A wide range of oxygen regulation capacity in skeletal muscle was obvious as there was no significant difference between mean muscle pO2 of healthy and distressed anaesthetized rats. The only reference to a changed microcirculation in distressed rats is the filling of the lower pO2 classes between zero and 10 mm Hg.

The influence of RMV alterations on muscle oxygenation in healthy anaesthetized rats was very small. In distressed rats a reduced RMV induced an increase of pO2 values in the lower pO2 classes. A comparison of the changes of mean muscle pO2 during RMV alterations shows that the proportion in distressed and healthy rats is in the same range. Although the values of the blood gas parameters in distressed rats were contrary to those of healthy animals, the regulation capacity of muscle oxygenation was not modified.

A different behavior of muscle pO2 in healthy and distressed rats was shown 10 minutes after a ventilation stop lasting 1 minute. Healthy rats seem to recover faster than distressed rats.

Although the arterial blood gas parameters and the macrocirculatory regulation were significantly changed, the muscle oxygenation in distressed rats seemed to be regulated in the same way as in healthy rats. Therefore, an independent regulation mechanism in the microcirculation of skeletal muscle can be assumed. The aim of this mechanism is to maintain a sufficient oxygen supply to the tissue in normal and extremely changed situations. The results of this experiment do not give any information about the nature of this mechanism. Possibly the oxygen extracting rate was increased in distressed rats.

The efficacy of the mechanism was disturbed by a ventilation stop lasting 60 seconds. The oxygen values of the arterial blood decreased significantly. The oxygen delivery to the muscle tissue was no longer sufficient as the longer recreation phase in distressed rats shows.

SUMMARY

In the anesthetized, open-chest rat, alterations of the respiratory minute volume were induced by changes of the respiratory frequency. Mean tissue pO2 of skeletal muscle in healthy and distressed rats did not change significantly. 10 minutes after a ventilation stop lasting 60 seconds, healthy rats showed an approximately normalized pO2 histogram, whereas distressed rats did not seem to recover. The results indicate an independent regulation mechanism which maintains a sufficient oxygen supply to skeletal tissue of anesthetized rats. The efficacy of this mechanism seemed to be reduced in distressed rats after the ventilation stop.

REFERENCES

Kessler, M., Lübbers, D. W., 1966, Aufbau und Anwendungsmöglichkeiten
    verschiedener pO2-Elektroden, Plügers Arch. ges. Physiol., 291: 82.
Baker, H. J., Lindsey, J. L., Weisbroth, S. H., 1979, The laboratory
    rat, Vol. I & II, Academic Press, New York.
Günderoth, M., 1990, Tissue oxygenation of the skeletal muscle and of
    the heart during hemodynamic alterations in rats, Adv. Exp. Med.
    Biol., 277: 499.

OPTICAL MEASUREMENTS OF OXYGEN AND ELECTRICAL MEASUREMENTS OF OXYGEN CHEMORECEPTION IN THE CAT CAROTID BODY

William L. Rumsey[+], Sukhamay Lahiri, Rodrigo Iturriaga, Anil Mokashi, Daniel Spergel and David F. Wilson[*]

Department of Physiology and [*]Department of Biochemistry and Biophysics, University of Pennsylvania School of Medicine, Philadelphia, PA 19104

[+]Department of Radiopharmaceuticals, Bristol-Myers Squibb Pharmaceutical Research Institute, New Brunswick, NJ 08903

INTRODUCTION

Measurements of oxygen pressures within the carotid body have been of particular importance for elucidating the mechanism(s) of oxygen chemoreception. An optical technique based on the quenching of phosphorescence by oxygen that is capable of imaging oxygen within the microvasculature has been developed (Rumsey et al, 1988, Wilson et al, 1991). These optical measurements are non-invasive and in particular have the advantage of determining oxygen pressures throughout the entire tissue. We have previously applied this technology to the study of the isolated perfused/superfused cat carotid body and oxygen chemoreception (Rumsey et al, 1991), relying primarily on imaging of phosphorescence intensity. Although the latter is quite useful for following dynamic changes in oxygen pressures, calculation of oxygen pressure is only qualitative. In addition, the carotid body was perfused and superfused with cell-free media (Iturriaga et al, 1991). In the present report, we describe measurements of oxygen pressure determined from images in phosphorescence lifetimes (Wilson et al, 1991) in the cat carotid body using an *in vivo* preparation (Lahiri and DeLaney, 1975). These determinations were compared to those obtained previously with the in vitro preparation, thereby permitting comparison of the oxygenation of this organ in the presence and absence of blood, respectively. In both cases, chemosensory discharge was recorded in order to provide measurements of the expression of oxygen chemoreception.

METHODS AND MATERIALS

Cats (1.8-2.8 kg) of either sex were anesthetized with sodium pentobarbital (35 mg/kg, i.p., followed by 10 mg/hr i.v.). Femoral arteries were cannulated for determination of blood pressure and sampling of arterial blood gases. One femoral vein was cannulated for administration of drugs and the oxygen probe. Rectal temperature was monitored continuously with a thermistor and was maintained at 37.5 ± 0.5^0 C using a feedback system. The animal was paralyzed with gallamine triethodide (Flaxedil, 3 mg/kg/hr) and artificially ventilated via a cannula placed in the trachea within the inferior aspect of the neck. The inspired gas mixture was obtained from pure O_2, N_2, CO_2 and compressed air in order to achieve the desired levels of end tidal PO_2 and PCO_2. Tracheal O_2 and CO_2 were monitored continuously using a O_2, CO_2 analyzer (Beckman Instr). The carotid body was exposed by first reflecting the upper section of the trachea and esophagus anteriorly. The ganglioglomerular nerves were severed and a few chemosensory fibers from the carotid sinus nerve were isolated and placed on bipolar platinum electrodes. The carotid bifurcation including the carotid body was kept warm with paraffin oil. Carotid chemosensory discharges were selected with an electronic window discriminator and the frequency of discharge counted. All the signals from the measurements above were displayed on an electrostatic recorder (Gould ES 100).

The carotid bifurcation was prepared for *in vitro* perfusion as described previously (Iturriaga et al, 1991). Single pass perfusion and superfusion were established using hydrostatic pressures of 80 and 15 Torr, respectively. The perfusate and superfusate were a modified Tyrode solution, containing [in mM] Na^+ [154], Cl^- [123], K^+ [4.7], Ca^{2+} [2.2], Mg^{2+} [1.1], glutamate [42], glucose [5] and HEPES [5], pH 7.4. The perfusate was gassed with either room air (normoxia) or 100% O_2 (hyperoxia) and the superfusate was vigorously bubbled with 100% Ar. The temperature in the perfusion chamber was maintained at 35.5 ± 0.5^0 C. Superfusate flow was adjusted to assure that the tissue was bathed in the media. Paraffin oil was layered over the superfusate to a depth of 5 mm. The whole sinus nerve was placed on two platinum wire electrodes for recording of chemosensory discharge as used by Eyzaguirre and Koyano (1965).

Imaging of oxygen pressure by phosphorescence quenching has been described previously (Rumsey et al, 1988, 1991, Wilson et al, 1991). In some of these studies, Pd-coproporphyrin (3 uM, Porphyrin Products, Logan UT) was present in the perfusate throughout perfusion, remaining in the vascular space. The perfusate contained 1% bovine serum albumin (Fraction V, ICN Immunobiologicals, Lisle, IL) to bind the Pd-coproporphyrin. For experiments *in vivo*, Pd-meso-tetra(4-carboxyphenyl) porphine (50-90 mg, Porphyrin Products, Logan UT) was complexed with albumin and was infused via the right femoral vein. Addition of this oxygen probe to the circulation did not produce any measurable physiological responses, i.e., carotid sinus nerve discharge or blood pressure. Phosphorescence (> 645 nm) was detected with an intensified CCD camera (Xybion Electronics Systems, San Diego, CA). The image was magnified with a Wild Makroskope (Leitz, Germany) positioned about 5-7 cm above the carotid body. The tissue was illuminated (530-560 nm) with a flash lamp (xenon 45 watt) controlled by a 80836 microcomputer that also controlled the gating of the video intensifier. The decay of phosphorescence after the flash was detected by the video camera. The intensifier was turned on at 20, 40, 80, 160, 300, 600, and 2500 usec after the flash and leaving it on for 2500 usec. The integrated value and the pre-determined value of T_0 were then used to calculate oxygen concentration by the Stern-Volmer relationship. T_0 and k_q for Pd-coproporphyrin and Pd-meso-tetra(4-carboxyphenyl) porphine were determined ex vivo in buffered media with bovine serum albumin and found to be 865 usec and 575 usec, respectively, and 260 $Torr^{-1} \cdot sec^{-1}$ and 325 $Torr^{-1} \cdot sec^{-1}$, respectively, at 37^0 C and pH 7.22. The quenching constant and T_0 were found to be unaffected by pH when using Pd-coproporphyrin as the oxygen probe. When using Pd-meso-tetra(4-carboxyphenyl) porphine, phosphorescence lifetime at zero oxygen was found to be decreased by 13% at pH 6.8 whereas the quenching constant was increased by 9% (Wilson et al, 1991).

RESULTS

Figure 1A shows an image of phophorescence intensity arising from the microvasculature of the cat carotid body, in vivo. During this image acquisition, the cat was ventilated with room air. Although images of phosphorescence intensity were not used in this report to calculate oxygen pressures, they provide a view of the tissue distribution of the probe and thereby the vascular distribution within the organ. The two bright areas in the upper central section of the photograph indicate greater density of carotid body tissue. In Figure 1 B, the magnification was reduced from 32 X to 14 X in order to show the relationship of phosphorescence distribution to that arising from the carotid sinus (bright area in lower right of photograph). This image was obtained immediately after anoxia was induced by heart failure. Prior to imaging the tissue, the carotid body was identified under a dissection microscope and demarcated by placement of small pieces of black suture around its perimeter. A piece of this suture about 1 mm wide is seen as a black line in the lower left of the photograph.

In a separate experiment shown in Figure 2 A, the cat was also ventilated initially on room air. The carotid body is for the most part located within the center of the photograph. Arterial oxygen pressure was 97 Torr. Chemosensory discharge at this time was low, about 8.9 ± 0.3 imp/sec (averaged for 1 min) and mean arterial pressure was normal, 100 Torr. The distribution of oxygen was heterogeneous. Two areas in the upper and lower central sections of the map are lower (darker) in oxygen than the surrounding tissue. When the oxygen pressure was averaged in three regions of interest (40 X 40 pixels), the mean value was much less than that in the arterial blood, 40 vs 97 Torr, respectively. When the end tidal PO_2 was lowered ($PaO_2 = 33$ Torr), chemosensory discharge responded rapidly with an expected rise in activity, to

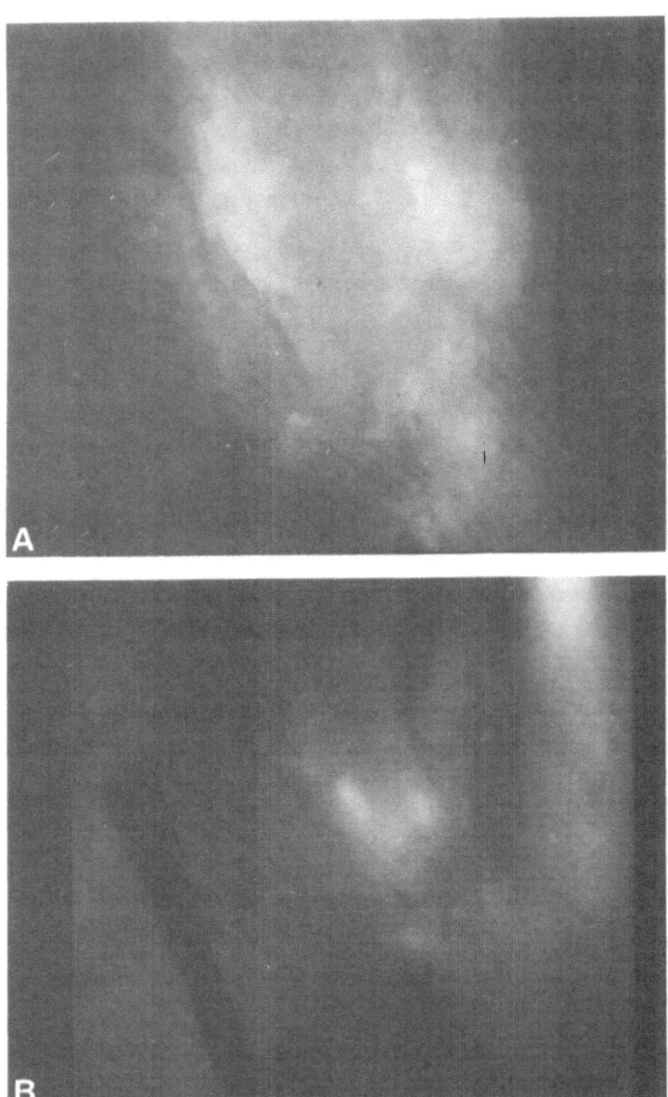

Figure 1 A and B. Phosphorescence intensity arising from the carotid body *in vivo* at high magnification (A, 32 X = 1.4 mm wide X 1.1 mm high) and low magnification (B, 14 X = 0.6 mm wide and 0.4 mm high). In B, the animal was in heart failure resulting in very low levels of oxygen. A portion of the carotid bifurcation can be seen on the right side of the photograph.

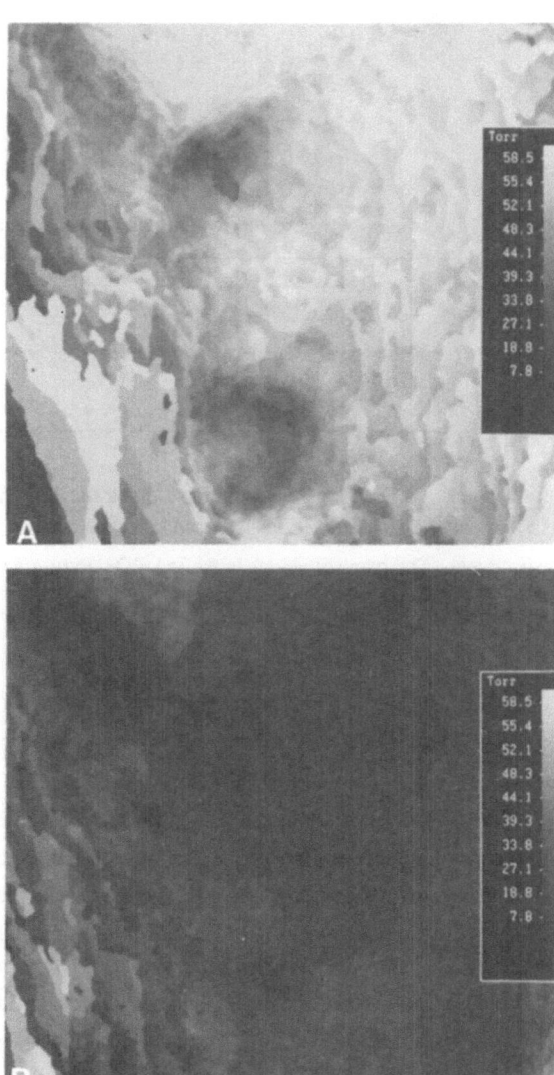

Figure 2 A and B. Oxygen pressure maps of the cat carotid body, *in vivo*. In (A), the cat was ventilated on room air. In (B), end tidal PO2 was lowered resulting in PaO2 of 33 Torr. In both cases the carotid body lies primarily in the central region of the photographs. Grey levels are calibrated in units of Torr shown on the scale on the right side of the photographs. Magnification was equivalent to 1.4 mm wide X 1.1 mm high.

16.2 $\pm$ 0.3 imp/sec. Associated with this change in chemosensory activity was an increase in phosphorescence intensity (data not shown) and a decrease in the oxygen pressure throughout the carotid body (Figure 1B). Using the same pixel areas as above to determine an average oxygen pressure in the carotid body, the mean carotid body PO_2 was 20 Torr, about a 50% decrease from the previous value obtained during ventilation with room air.

Table 1. Comparison of PO_2 from in vitro and in vivo preparations of the cat carotid body.

| | in vitro Preparation | | in vivo |
	Low PO$_2$ perfusate	High PO$_2$ perfusate	Preparation
Influent or Arterial PO$_2$	109 $\pm$ 5	259 $\pm$ 31	91 $\pm$ 8
Tissue PO$_2$	59 $\pm$ 13	96 $\pm$ 12	49 $\pm$ 3

Values (Torr) represent means $\pm$ S.E.M. for 5 *in vitro* preparations and 3 *in vivo* (arterial) experiments. Four and eight determinations were made for the low and high PO$_2$ perfusates, respectively. The range of values in the low PO$_2$ perfusate was 100 to 118 Torr and that in the high PO$_2$ perfusate was 155 to 395 Torr.

Comparison of values of oxygen pressures obtained in vivo with those resulting from perfusion of the carotid body in vitro with cell-free media (Rumsey et al, 1991) is shown in Table 1. During ventilation of three cats with room air, it was apparent that oxygen pressure within the capillaries of the carotid body in vivo was about 46% less than that in the arterial blood delivered to the organ, 49 $\pm$ 3 Torr vs 91 $\pm$ 8 Torr, respectively. Differences between carotid body PO$_2$ and influent PO$_2$ were also observed when the tissue was perfused in the isolated state with cell-free media. When the PO$_2$ of the perfusate was 109 $\pm$ 5 Torr, the oxygen pressure in the exchange vessels of the carotid body was 59 $\pm$ 13 Torr. Both of these values were 20% greater than their in vivo counterparts. Comparison of values of oxygen pressures obtained in vivo with those resulting from perfusion of the carotid body in vitro with cell-free media (Rumsey et al, 1991) is shown in Table 1. During ventilation of three cats with room air, it was apparent that oxygen pressure within the capillaries of the carotid body in vivo was about 46% less than that in the arterial blood delivered to the organ, 49 $\pm$ 3 Torr vs 91 $\pm$ 8 Torr, respectively. Differences between carotid body PO$_2$ and influent PO$_2$ were also observed when the tissue was perfused in the isolated state with cell-free media. When the PO$_2$ of the perfusate was 109 $\pm$ 5 Torr, the oxygen pressure in the exchange vessels of the carotid body was 59 ± 13 Torr. Both of these values were 20% greater than their in vivo counterparts. When the oxygen pressure of the perfusate was increased by more than two-fold, i.e., 291 ± 30 Torr, capillary oxygen pressure also increased. In the latter case, the oxygen pressure in the carotid body was 96 ± 12 Torr. This increase in oxygenation resulted in a lowering of the chemosensory discharge. Figure 3 shows the relationship between the oxygen pressures within the influent and the exchange vessels of the isolated perfused/superfused cat carotid body throughout a range of influent values.

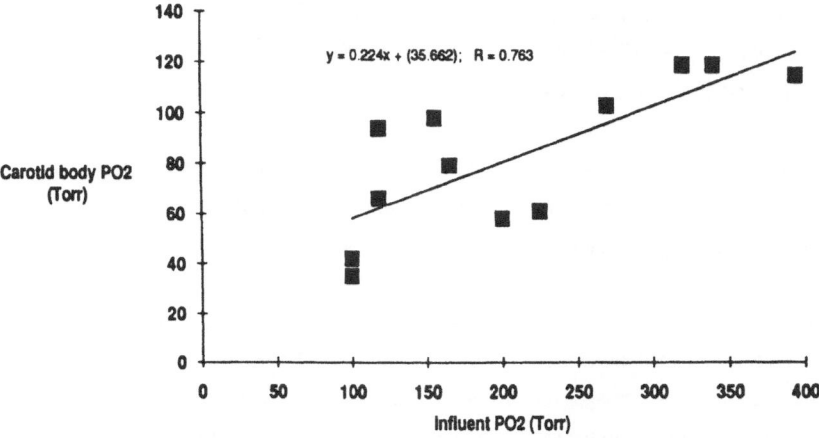

Figure 3. Relationship of perfusate PO_2 and carotid body PO_2 in the in vitro perfused/superfused cat carotid body.

It would appear that a linear correlation may exist between these two parameters albeit not a strong one ($R = 0.76$). It can be seen, however, that as the influent oxygen pressure increases, the level of oxygen in the carotid body also increases.

By interrupting the flow of perfusate to the carotid body in vitro, a marked stimulation of chemosensory discharge resulted regardless of the value of oxygen pressure in the perfusate media. The duration required to reach 50% of the maximal response to flow interruption, however, was prolonged when the media was equilibrated with the higher oxygen pressures.

DISCUSSION

The preliminary results provided in the present report show that it is possible to image the oxygen pressure within the exchange vessels of the intact blood perfused carotid body using measurements of phosphorescence lifetime. Optical measurements of oxygen may provide a useful methodology to elucidate the mechanism(s) responsible for sensing oxygen in the general circulation. Delineation of the mechanism(s) was not possible from the results of this brief report, however, the following findings were evident. 1) It was apparent that the oxygen pressure within the carotid body was not distributed evenly throughout the entire tissue, instead oxygen distribution showed a marked heterogeneity. In addition, the carotid body was not always found to be a compact unified organ rather it can be spread out in a somewhat diffuse manner. 2) Oxygen pressure in the exchange vessels of the carotid body is markedly less than that in the blood entering this tissue. When similar oxygen pressures were used in an in vitro perfused preparation, the level of oxygen in the carotid body was also less than that in the influent. In the latter case, the oxygen pressure in the carotid body was 20% greater than that in the in vivo preparation, however, the influent PO_2 was also 20% higher than that of the in vivo preparations.

3) By lowering or raising the oxygen pressure in either the blood or the saline-perfusate, the level of oxygen in the carotid body was increased or decreased, respectively. These changes were matched by appropriate manifestations of oxygen chemoreception, i.e., a decrease in carotid PO_2 resulted in a marked increase of chemosensory discharge.

The precise level of oxygen pressure in the carotid body during normoxic and other conditions has been a matter of controversy (for example see Acker 1989, or Fitzgerald and Lahiri, 1986). The range of values extends from as low as 10 Torr (Acker and Lubbers, 1977) to as high as 70 Torr (Nair et al, 1986, Whalen and Nair, 1983). The reasons for this disparity are not clear but the differences may be due at least in part to methodological difficulties associated with microelectrodes used in the previous investigations. It is known that the impalement of microelectrodes in tissue damages the surrounding area. On the other hand, from a mathematical analysis of oxygen pressure in the cat carotid body, Degner and Acker (1986) suggested that differences in the literature can be ascribed to variations in the length of the capillaries within the carotid bodies. Our results indicate that the level of oxygen pressure in the microcirculation of the cat carotid body in vivo during normoxic conditions is between 40 and 60 Torr. Whalen and Nair (1983) found that carotid body PO_2 was higher in saline-perfused than in blood-perfused cat carotid bodies. Our results obtained either from in vivo or in vitro preparations were similar.

Acknowledgments: This work was supported by a grants HL43413 and NS10939.

REFERENCES

Acker, H., 1989. PO_2 chemoreception in arterial chemoreceptors. Ann. Rev. Physiol. 51: 835-844.

Acker, H. and Lubbers, D.W., 1977, The kinetics of local tissue PO_2 decrease after perfusion stop within the carotid body of the cat in vivo and in vitro. Pfluegers Archiv. 369: 135-140.

Degner, F. and Acker, H., 1986, Mathematical analysis of tissue PO_2 distribution in the cat carotid body. Pfluegers Archiv 407: 305-311.

Eyzaguirre, C. and Koyano, H., 1965, Effects of hypoxia, hypercapnia, and pH on the chemoreceptor activity of the carotid body in vitro. J. Physiol. London 178: 385-409.

Fitzgerald, R.S. and Lahiri, S. 1986, Reflex responses to chemoreceptor stimulation. In: Handbook of physiology. The Respiratory System. Control of Breathing. Bethesda, MD: Am. Physiol. Soc., 313-362.

Iturriaga, R., Rumsey, W.L. Mokashi, A. Spergel, D. Wilson, D.F. and Lahiri, S., 1991. In vitro perfused-superfused cat carotid body for physiological and pharmacological studies. J. Appl. Physiol. 70:1393-1400.

Lahiri, S. and DeLaney, R.G., 1975. Stimulus interaction in the response of carotid body chemoreceptor single afferent fibers. Respir. Physiol. 24: 349-366.

Nair, P.K., Buerk, D.G., and Whalen, W.J., 1986, Cat carotid body oxygen metabolism and chemoreception described by a two-cytochrome model. Am. J. Physiol. 250 (Heart Circ. Physiol. 19): H202-H207.

Rumsey, W.L., Vanderkooi, J. and Wilson, D.F., 1988, Imaging of phosphorescence: a novel method for measuring oxygen distribution in perfused tissue. Science 241, 1649-1651.

Rumsey, W.L. Iturriaga, R., Spergel, D., Lahiri, S., and Wilson, D.F., 1991, Optical measurements of the dependence of chemoreception on oxygen pressure in the cat carotid body. Am. J. Physiol. 261 (Cell Physiol. 30): C000-C000.

Whalen, W.J. and Nair, P., 1983, Oxidative metabolism and tissue PO_2 of the carotid body. In: Physiology of the peripheral arterial chemoreceptors. Eds. Acker, H and O'Regan, R. Elsevier, Amsterdam, 117-132.

Wilson, D.F., Pastusko, A., DiGiacomo, J.E., Pawloski, M., Schneiderman, R., and Delivoria-Papadopoulos, M., 1991, Effect of hyperventilation on oxygenation of the brain cortex of newborn piglets. J. Appl. Physiol. 70 (6): 2691-2696.

DOSE-DEPENDENT IMPROVEMENT OF GAS EXCHANGE BY INTRATRACHEAL PERFLUBRON (PERFLUOROOCTYLBROMIDE) INSTILLATION IN ADULT ANIMALS WITH ACUTE RESPIRATORY FAILURE

A. S. Tütüncü[1,2], B. Lachmann[1], N. S. Faithfull[3], W. Erdmann[1]

[1]Dept. of Anesthesiology, Erasmus University Rotterdam, The Netherlands
[2]Dept. of Anesthesiology, Univ. of Istanbul, Faculty of Medicine, Istanbul, Turkey
[3]Alliance Pharmaceutical Corp., San Diego, CA, USA

INTRODUCTION

Liquid ventilation with oxygenated perfluorocarbons (PFC) has been shown to improve pulmonary gas exchange and lung mechanics in preterm animals with respiratory failure [1-4]. Histological studies also reveal that liquid ventilation is less harmful to the lung structures than conventional gas ventilation [3,5]. Moreover, the first human application of liquid ventilation has been reported with promising success in premature infants with severe respiratory distress in whom conventional therapies had failed [6].

This study was designed to investigate whether intratracheal instillation of PFC in combination with mechanical gas ventilation could also be effective to support pulmonary gas exchange in animals with acute respiratory failure. For this purpose, we applied intratracheal PFC in increasing doses to test the dose-response characteristics of this type of ventilatory support.

METHODS

Animal Preparation

Adult New Zealand rabbits weighing 2.8 ± 0.2 kg (n = 12) were anesthetized with 50 mg/kg of pentobarbital sodium iv and an endotracheal tube was inserted through a tracheotomy with its tip proximal to the carina. Ventilation with a Servo 900C ventilator (Siemens-Elema, Sweden) was initiated using pure oxygen and zero end-expiratory pressure with a constant tidal volume (TV) of 12 ml/kg, frequency (f) of 30/min and inspiratory/expiratory ratio (I/E) of 1:2. Anesthesia was maintained with additional doses of pentobarbital, as required, and pancuronium bromide was administered by a continuous infusion (0.1 mg/kg/h) for muscle paralysis. A solution of 5% Dextrose / 0.45% NaCl was administered continuously at a rate of 10 ml/h as a maintenance fluid. A femoral artery and femoral vein were cannulated for pressure recording and blood sampling. Blood gases, pH and Hb were determined by conventional methods (ABL-330 and Osm-2 Hemoximeter, Radiometer, Copenhagen, Denmark). Intravascular pressure monitoring was performed with a Statham P23XL transducer (Spectramed, USA) and all tracings, including ECG, were traced by a Sirecust 1280 monitor (Siemens, USA) and recorded by a Siredoc 220 recorder (Siemens, Germany).

Following the baseline observations, respiratory insufficiency was induced by repeated lung lavage [7,8] and confirmed by the presence of an arterial pO_2 below 100 mm Hg with the following ventilator settings: Volume controlled ventilation, $FiO_2 = 1.0$, TV = 12 ml/kg, f = 30/min, I/E = 1:2 and positive end-expiratory pressure (PEEP) = 6 cm H_2O.

Treatment:

The PFC tested in this experiment is perflubron (perfluorooctylebromide; Alliance Pharmaceutical Corp., San Diego, CA), a PFC which has a specific gravity of 1.918 g/cm^3 at 25°C, a surface tension of 18.1 dynes/cm, a vapour pressure of 3.6 mm Hg at 20°C and 10.5 mm Hg at 37°C, O_2 solubility of 53 ml/100 ml and CO_2 solubility of 210 ml/100 ml at 37°C and 1 atmosphere pressure.

After induction of respiratory insufficiency, animals were divided into two groups. Group 1 (n=6) was treated with perflubron; Group 2 (n=6) received saline and served as controls. Animals were ventilated for 15 minutes to achieve a steady state with the above-mentioned settings after instillation of each treatment dose, and arterial blood gases and lung mechanics were determined thereafter.

Statistical analysis:

The statistical significance of measurements between groups was assessed by the Wilcoxon signed rank test. A p value less than 0.05 was considered significant.

RESULTS

All animals tolerated perflubron administration without any observed adverse effects. PaO_2 values in the perflubron group increased significantly ($p < 0.05$) with each subsequent dose, in contrast to the saline group. PaO_2 in the perflubron group was no longer significantly different from the pre-lavage value after the dose of 12 ml/kg (Table 1).

Arterial pCO_2 decreased significantly after instillation of 3 ml/kg perflubron and remained stable throughout the experiment. In the saline group, $PaCO_2$ increased further.

Table 1. Blood gas values of the treatment groups (mean ± SD).

	PaO_2 mm Hg		$PaCO_2$ mm Hg	
	Perflubron	Saline	Perflubron	Saline
Pre-lavage	504 ± 40	487 ± 21	37 ± 3	34 ± 1
Post-lavage	75 ± 15	70 ± 7	49 ± 6	46 ± 7
3 ml/kg	160 ± 38*	51 ± 9*	43 ± 5*	50 ± 9
6 ml/kg	297 ± 54*	42 ± 9*	44 ± 5	55 ± 9*
12 ml/kg	398 ± 36*	39 ± 7*	45 ± 5	58 ± 8*

*$p < 0.05$ compared to pretreatment (post-lavage) value.

DISCUSSION

This report describes the efficacy of intratracheal instillation of perflubron in combination with conventional mechanical gas ventilation to improve pulmonary gas exchange in dose-dependent manner in animals with respiratory failure due to surfactant depletion.

Perfluorocarbons have been employed in liquid ventilation techniques due to their low surface tension and high capacity for dissolving respiratory gases. Liquid ventilation studies have demonstrated that PFC eliminates elevated surface tension at the air-liquid interface and improves ventilation-perfusion mismatch and, thus, supports pulmonary gas exchange in respiratory failure. Moreover, improvements in oxygenation and lung mechanics on reinstitution of conventional ventilation after liquid ventilation have been observed, and are speculated to be due to the increased alveolar stabilization by the residual PFC in the surfactant deficient lungs [1,2,9].

The results of this study demonstrate that pulmonary gas exchange can be improved even with small amounts of perflubron instillation (e.g. 3 ml/kg) in this new type of ventilatory support which does not require liquid ventilator or require extracorporeal oxygenation of the PFC. The treatment dose of 3 ml/kg is far below the functional residual capacity (FRC) volume. FRC has been measured to be 18.4 ± 2.7 ml/kg in healthy rabbits and 8.6 ± 1.0 ml/kg in lung-lavaged rabbits at identical conditions (an average body weight of 2.7 kg and with PEEP of 6 cm H_2O) [10]. From this data, it can be suggested that when administered into surfactant-depleted lungs, PFC spreads over the alveoli by the positive spreading coefficient and decreases surface tension at the air-liquid interface, and, thus, lung inflation can occur at low airway pressures and pulmonary gas exchange improves in a dose-dependent manner.

CONCLUSION

This study demonstrates that PFC liquids in combination with conventional mechanical gas ventilation could be an effective treatment modality to support animals with acute respiratory failure. Although based on short-term observation, the remarkable dose-dependent improvements in pulmonary gas exchange suggest the potential of this technique to offer simple, advantageous ventilatory support compared to liquid ventilation. However, further study is warranted to demonstrate the long-term effects of intratracheal PFC instillation on respiratory and cardiovascular parameters.

REFERENCES

1. T. H. Shaffer, D. Rubenstein, G. D. Moskowitz, and M. Delivoria-Papadopoulos, Gaseous exchange and acid-base balance in premature lambs during liquid ventilation since birth, Pediatr Res. 10:227-231 (1976).

2. T. H. Shaffer, C. A. Lowe, V. K. Bhutani, and P. R. Douglas, Liquid ventilation: effects on pulmonary function in meconium stained lambs, Pediatr Res. 18:47-52 (1983).

3. M. R. Wolfson, and T. H. Shaffer, Liquid ventilation during early development: theory, physiologic processes and application, J Dev Physiol. 13:1-12 (1990).

4. T. H. Shaffer, N. Tran, V. K. Bhutani, and E. M. Sivieri, Cardiopulmonary function in very preterm lambs during liquid ventilation, Pediatr Res. 17:680-684 (1983).

5. D. L. Forman, V. K. Bhutani, S. R. Hilfer, and T. H. Shaffer, A fine structure study of the liquid-ventilated newborn rabbit lung, Fed Proc. 43:647 (1984).

6. J. S. Greenspan, M.R. Wolfson, S. D. Rubenstein, and T. H. Shaffer, Liquid ventilation of human preterm neonates, J Pediatr. 117:106-111 (1990).

7. B. Lachmann, B. Robertson, and J. Vogel, In-vivo lung lavage as an experimental model of the respiratory distress syndrome, Acta Anaesthesiol Scand. 24:231-236 (1980).

8. B. Lachmann, B. Jonson, M. Lindroth, and B. Robertson, Modes of artificial ventilation in severe respiratory distress syndrome. Lung function and morphology in rabbits after wash-out of alveolar surfactant, Crit Care Med. 10:724-732 (1982).

9. T. H. Shaffer, P. R. Douglas, C. A. Lowe, and V. K. Bhutani, The effects of liquid ventilation on cardiopulmonary function in preterm lambs, Pediatr Res. 17:303-306 (1983).

10. D. Gommers, C. Vilstrup, J. A. H. Bos, A. Larsson, O. Werner, E. Hannappel, and B. Lachmann, Effects of exogenous surfactant therapy on lung function cannot be assessed by dynamic compliance, (submitted for publication).

GAS EXCHANGE AND LUNG MECHANICS DURING LONG-TERM MECHANICAL VENTILATION WITH INTRATRACHEAL PERFLUOROCARBON ADMINISTRATION IN RESPIRATORY DISTRESS SYNDROME

A. S. Tütüncü[1,2], K. Akpir[2], P. Mulder[3], N. S. Faithfull[4], W. Erdmann[1], B. Lachmann[1]

[1]Dept. of Anesthesiology, Erasmus University Rotterdam (EUR), The Netherlands
[2]Dept. of Anesthesiology, Univ. of Istanbul, Faculty of Medicine, Istanbul, Turkey
[3]Dept. of Epidemiology and Biostatistics, EUR, The Netherlands
[4]Dept. of Medical Research, Alliance Pharmaceutical Corp., San Diego, CA, USA

INTRODUCTION

Respiratory distress syndrome (RDS) is mainly characterized by hypoxemia and pronounced alveolar collapse, associated with high surface forces in the alveoli. Several methods of artificial ventilation have been introduced to support such lungs and maintain adequate oxygenation until recovery of lung function occurs [1-5]. Since perfluorocarbon (PFC) liquids were recognized as a useful media for pulmonary gas exchange, animal studies with PFC liquid ventilation have demonstrated that elimination of high surface forces in the PFC-filled lung can offer an alternative respiratory medium to improve gas exchange and lung expansion [6-8].

Recently we demonstrated that a simple method of PFC application, intratracheal PFC instillation in combination with conventional mechanical ventilation, could support pulmonary gas exchange in animals with acute respiratory failure [9]. With this technique, there was a dose-dependent improvement in oxygenation with well-preserved arterial CO_2 pressures. This study aimed to investigate the long-term efficacy of this dose-dependent improvement in pulmonary gas exchange in adult animals with acute respiratory failure over a 6-hour observation period.

METHODS

Animal Preparation

Adult New Zealand rabbits weighing 2.8 ± 0.2 kg (n=24) were anesthetized with pentobarbital sodium iv (50 mg/kg) via an auricular vein. The animals were positioned supine, tracheotomized and an endotracheal tube was inserted, after which mechanical ventilation with a Servo Ventilator 900C (Siemens-Elema, Sweden) was initiated with 100% oxygen, zero end-expiratory pressure, tidal volume (VT) of 12 ml/kg, frequency (f) of 30/min and inspiratory/expiratory (I/E) ratio of 1:2. An infusion of 5% dextrose / 0.45% NaCl solution was continuously administered (20 ml/h) via the auricular vein as a maintenance fluid. Anesthesia was maintained with continuous infusion of pentobarbital (10 mg/h) and fentanyl (2 μg/kg/min); pancuronium bromide was administered for muscle paralysis (0.1 mg/kg/h).

A femoral artery was cannulated with polyethylene catheter for arterial pressure monitoring and blood sampling. Arterial samples were analyzed for blood gases by conventional methods (ABL-330, Radiometer, Copenhagen, Denmark). End-tidal CO_2 was measured on-line at the proximal end of the endotracheal tube by CO_2 Analyzer 930 (Siemens, Sweden). Core temperature was maintained at $37 \pm 1\,°C$ by a heating blanket and monitored with an esophageal thermistor (Elektrolaboratoriet, Copenhagen, Denmark). Intravascular pressure was monitored by Statham P23XL transducer (Spectramed, USA), and tracings including ECG were traced by a Sirecust 1280 monitor (Siemens, USA) and recorded by a Siredoc 220 recorder (Siemens, Germany).

Treatment:
Acute respiratory failure was induced by lung lavage with 30 ml/kg warm saline [10] and repeated to achieve an arterial pO_2 below 100 mm Hg at the following ventilator settings: Volume controlled ventilation, $FiO_2 = 1$, VT = 12 ml/kg, f = 30/min, I/E = 1:2 and PEEP = 6 cm H_2O.

The PFC administered in this experiment was perflubron (perfluorooctylbromide; Alliance Pharmaceutical Corp., San Diego, CA), a PFC with a specific gravity of 1.918 g/cm^3 at 25 °C, a surface tension of 18.1 dynes/cm, O_2 solubility of 53 ml/100 ml and CO_2 solubility of 210 ml/100 ml at 37 °C and 1 atmosphere pressure.

Animals were randomly divided into 4 groups of 6 animals and each group was treated intratracheally with a different dose of perflubron: Group 1 = 3 ml/kg; Group 2 = 6 ml/kg; Group 3 = 9 ml/kg and Group 4 = 12 ml/kg. At instillation, animals were disconnected from the ventilator and perflubron (warmed to room temperature) was administered directly into the endotracheal tube; the ventilator was then immediately reconnected. The ventilator settings were kept constant as above and mechanical ventilation was maintained for 6 hours.

During the 6-hour observation period, arterial blood gases, ET CO_2 and hemodynamic parameters were recorded at 15 min, and at 30 min intervals, thereafter. After 6 hours animals were sacrificed with an overdose of pentobarbital.

Statistical analysis
Results were analyzed by analysis of variance (ANOVA) for repeated measurements, using a maximum likelihood technique (Program 5V of BMDP package). The statistical significance between all pairs of groups was also tested at seperate points in time by the Student-Newman-Keuls test. Measurements within groups were compared using the Wilcoxon signed-ranks test. A p value of less than 0.05 was accepted as statistically significant.

RESULTS

The measured and calculated variables were comparable in all groups before and after lung lavage. All animals in Groups 3 and 4 survived for the duration of the experiments. In Group 1, five animals developed pneumothorax after 3.5 h. In Group 2, three animals developed pneumothorax after 5 h.

Figure 1 depicts the response of arterial pO_2 over the course of the experiments to different doses of perflubron. Following treatment, arterial pO_2 was significantly less ($p < 0.05$) in Group 1 than in the other groups until 3.5 h; thereafter Groups 1 and 2 were less than Groups 3 and 4 at all time points. The mean values at 6 h remained significantly higher in Groups 3 and 4 (229.8 ± 84.1 mm Hg and 196.7 ± 129.5 mm Hg, respectively) compared to the pretreatment values.

Only in Groups 3 and 4 were arterial CO_2 values maintained below hypercarbic levels throughout the study; they were significantly different from Groups 1 and 2 after 4 h (Figure 2).

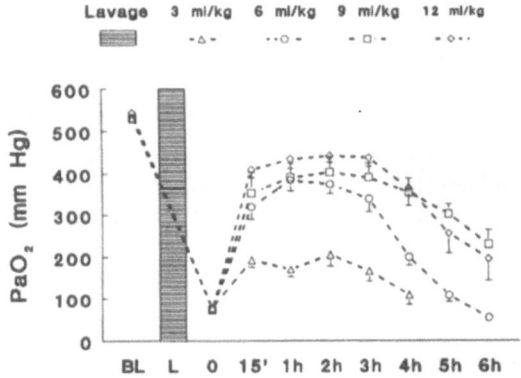

Figure 1. Following treatment, Groups 3 and 4 maintained significantly higher PaO_2 levels for 6 h compared to Groups 1 and 2. The data for Group 1 is not depicted after 4 h because of too few survivors. (BL= before lavage; L= lavage) (Mean±SEM)

After PFC instillation, airway pressures needed to inflate the lungs decreased significantly in all groups (Figure 3). Mean airway pressures were significantly lower in Groups 3 and 4 compared to Groups 1 and 2 from 3 h onwards.

The mean arterial pressure and heart rate of the treatment groups remained stable throughout the study (Table 1).

DISCUSSION

Since PFC liquids have been employed in liquid ventilation, studies have explored the feasibility of various models of liquid breathing applications for varying periods and

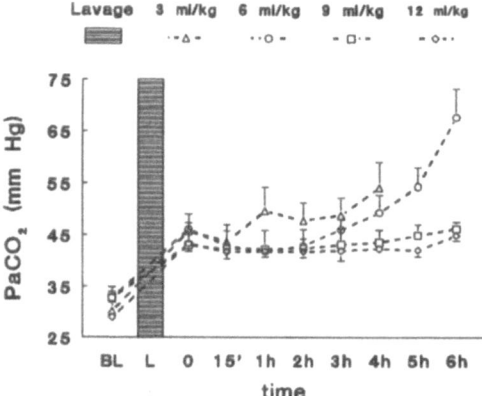

Figure 2. Following the lavage procedure (L), there was significant increase in $PaCO_2$ in all groups, but remained stable only in Groups 3 and 4 after treatment. The data for Group 1 is not depicted after 4 h because of too few survivors. (BL= before lavage) (mean±SEM)

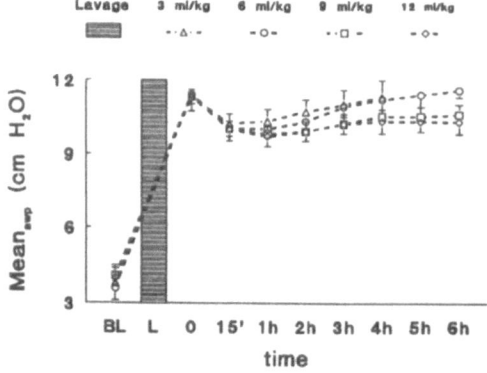

Figure 3. Treatment with intratracheal PFC decreased mean airway pressures significantly in all groups. However, they were still significantly' lower than the pretreatment values at 6 h only in Groups 3 and 4. The data for Group 1 is not depicted after 4 h because of too few survivors. (BL= before lavage, L= lavage) (Mean±SEM)

Table 1. Mean arterial pressure (MAP = mmHg), heart rate (HR = beats/min) and end-tidal CO_2 (ET CO_2 = %) values of treatment groups: 3 ml/kg (G1), 6 ml/kg (G2), 9 ml/kg (G3) and 12 ml/kg (G4) (mean ± SD).

		BL	AL	15 min	1 h	2 h	3 h	4 h	5 h	6 h
MAP:	G1	95 ± 9	95 ± 6	89 ± 10	87 ± 3	91 ± 4	90 ± 8	90 ± 14		
	G2	93 ± 8	89 ± 10	88 ± 10	86 ± 12	82 ± 5	84 ± 9	88 ± 10	84 ± 7	84 ± 3
	G3	93 ± 11	83 ± 12	92 ± 11	90 ± 14	89 ± 5	94 ± 9	95 ± 8	96 ± 9	97 ± 6
	G4	90 ± 17	97 ± 21	89 ± 15	90 ± 11	86 ± 8	88 ± 10	88 ± 10	86 ± 14	85 ± 16
HR:	G1	315 ± 27	308 ± 40	298 ± 37	280 ± 30	278 ± 31	268 ± 18	279 ± 13		
	G2	333 ± 36	328 ± 6	318 ± 15	300 ± 19	270 ± 16	278 ± 16	270 ± 21	258 ± 26	255 ± 30
	G3	315 ± 21	318 ± 11	313 ± 18	289 ± 13	285 ± 19	285 ± 13	280 ± 16	285 ± 23	288 ± 24
	G4	323 ± 13	310 ± 12	313 ± 15	298 ± 22	283 ± 31	288 ± 15	283 ± 15	288 ± 20	268 ± 22
ET CO_2:	G1	3.93 ± 0.5	4.70 ± 1.1	4.70 ± 1.0	4.98 ± 1.1	5.30 ± 1.1	5.33 ± 1.1	5.28 ± 1.0		
	G2	4.03 ± 0.2	4.77 ± 0.6	4.53 ± 0.4	4.42 ± 0.3	4.63 ± 0.4	4.87 ± 0.7	5.10 ± 1.0	5.15 ± 1.5	5.50 ± 2.0
	G3	3.57 ± 0.8	3.97 ± 0.6	4.02 ± 0.7	4.03 ± 0.5	4.23 ± 0.4	4.33 ± 0.5	4.53 ± 0.5	4.78 ± 0.6	4.80 ± 0.5
	G4	3.73 ± 0.4	4.45 ± 0.5	4.27 ± 0.4	4.18 ± 0.4	4.28 ± 0.4	4.30 ± 0.3	4.18 ± 0.3	4.22 ± 0.3	4.23 ± 0.4

Data for Group 1 is not depicted after 4 h because of too few survivors. (BL = before lavage, L = lavage)

their effects on pulmonary and cardiovascular parameters [11]. The literature indicate that animals can tolerate total PFC liquid ventilation for periods up to 8 hours and can be successfully recovered to gas breathing [12]. We have recently demonstrated that a simple low dose PFC application in combination with standard gas ventilation can effectively provide short term support of adult animals with RDS and improve oxygenation in a dose-dependent manner [9].

The present study provides further information on the dose-dependent character of this new supportive technique, namely the time-dependency of the beneficial effects achieved by different doses of perflubron.

The overall data demonstrated a clear difference between the four tested perflubron doses in their ability to improve lung function and to maintain these effects over the 6 h observation period without additional administration of the PFC. With the administration of low doses (3 ml/kg and 6 ml/kg), adequate oxygenation with well preserved arterial pCO_2 levels could be maintained using low inflation pressures for approximately 3h; this could be extended to 6 h using high doses of perflubron (9 ml/kg and 12 ml/kg). In other words, the time interval elapsed between PFC instillation and the impairment of pulmonary function correlated well with the dose administered.

When administered intratracheally, PFC liquids decrease elevated surface tension at the air-liquid interface in diseased lungs and thus decrease inflation pressures needed for lung inflation. However, an equal lowering of alveolar surface tension with the PFC liquid would not stabilize all different sized alveoli (Law of La Place). Nevertheless, the administration of greater volumes of liquid would fill more alveoli and keep them open and prevent the collapsing during expiration.

Unless evaporative losses are replaced, the duration of the favorable effects of PFC liquids depends, to a great extent, on the rate of evaporation of the liquid from the lungs. Our experimental data do indeed reveal a time-dependency of the observed beneficial effects of PFC administration and strongly suggest that this is due to evaporation of perflubron from the lung. This reduction of efficacy was more pronounced and occurred earlier following lower doses (Groups 1 and 2), as evidenced by the increased airway pressures and the impairment of pulmonary gas exchange. The overall data in Group 4 implies the optimal therapeutic PFC dose for long-term efficacy of mechanical ventilation in combination with intratracheal PFC administration.

CONCLUSION

Intratracheal PFC instillation in combination with mechanical gas ventilation provides adequate oxygenation, improves ventilation and keeps inflation pressures low without compromising hemodynamics in acute respiratory failure. Fuhrman and coworkers have recently demonstrated that adequate pulmonary gas exchange can be provided using the same technique in normal piglets [13]. All these favorable effects suggest the potential of intratracheal PFC treatment as an alternative means of achieving the goals of ventilatory support in humans with RDS, which is mainly to support lungs until the recovery from the primary disease. This study demonstrated that a PFC dose of 12 ml/kg is optimal for this type of ventilatory support for periods up to 6 h to achieve improvements in oxygenation and alveolar ventilation by eliminating the high surface forces along the alveolar lining-air interface. The very low vapor pressure makes perflubron the choice of PFC for this type of application with particular respect to evaporative losses and alveolar oxygen level.

REFERENCES

1. E. O. R. Reynolds, Effect of alterations in mechanical ventilator setting on pulmonary gas exchange in hyaline membrane disease, Arch Dis Child. 46:152-159 (1971).
2. B. Lachmann, E. Danzmann, B. Haendly, and B. Jonson, Ventilator settings and gas exchange in respiratory distress syndrome, In: O. Prakash (ed) Applied physiology in clinical respiratory care. Martinus Nijhoff Publishers , The Hague (1982).
3. U. Sjöstrand, High-frequency positive-pressure ventilation (HFPPV): a review, Crit Care Med. 8:345-364 (1980).
4. K. G. Hickling, S. J. Henderson, and R. Jackson, Low mortality associated with pressure limited ventilation with permissive hypercapnia in severe adult respiratory distress syndrome, Intensive Care Med. 16:372-377 (1990).
5. L. Gattinoni, A. Pesenti, M. L. Caspani, A. Pelizzola, D. Mascheroni, R. Marcolin, G. Iapichino, M. Langer, A. Agostoni, T. Kolobow, and G. Damia, Low frequency positive-pressure ventilation with extracorporeal CO_2 removal in severe acute respiratory failure, JAMA. 256:881-886 (1986).
6. T. H. Shaffer, D. Rubenstein, G. D. Moskowitz, and M. Delivoria-Papadopoulos, Gaseous exchange and acid-base balance in premature lambs during liquid ventilation since birth, Pediatr Res. 10:227-231 (1976).
7. R. Rufer, and H. L. Spitzer, Liquid ventilation in the respiratory distress syndrome, Chest 66 (Suppl):29S-30S (1974).
8. T. H. Shaffer, C. A. Lowe, W. K. Bhutani, and P. R. Douglas, Liquid ventilation: effects on pulmonary function in meconium stained lambs, Pediatr Res. 18:47-52 (1983).
9. A. S. Tütüncü, B. Lachmann, N. S. Faithfull, and W. Erdmann, Dose-dependent improvement of gas exchange by intratracheal perflubron (perfluorooctylbromide) instillation in adult animals with acute respiratory failure, Adv Exp Med Biol. (this volume).
10. B. Lachmann, B. Robertson, and J. Vogel, In-vivo lung lavage as an experimental model of the respiratory distress syndrome, Acta Anaesthesiol Scand. 24:231-236 (1980).
11. T. H. Shaffer, A brief review: liquid ventilation, Undersea Biomed Res. 14:169-178 (1987).
12. J. H. Modell, E. J. Newby, and B. C. Ruiz, Long-term survival of dogs after breathing oxygenated fluorocarbon liquid, Fed Proc. 29:1731-1736 (1970).
13. B. P. Fuhrman, P. R. Paczan, and M. DeFrancisis, Perfluorocarbon-associated gas exchange, Crit Care Med. 19:712-722 (1991).

PERFLUBRON (PERFLUOROOCTYLBROMIDE) INSTILLATION COMBINED WITH MECHANICAL VENTILATION: AN ALTERNATIVE TREATMENT OF ACUTE RESPIRATORY FAILURE IN ADULT ANIMALS

B. Lachmann[1], A. S. Tütüncü[1,2], J. A. H. Bos[1], N. S. Faithfull[3], W. Erdmann[1]

[1]Dept. of Anesthesiology, Erasmus University Rotterdam, The Netherlands
[2]Dept. of Anesthesiology, Univ. of Istanbul, Faculty of Medicine, Istanbul, Turkey
[3]Alliance Pharmaceutical Corp., San Diego, CA, USA

INTRODUCTION

Experimental studies have shown that adequate gas exchange can be accomplished during liquid ventilation with oxygenated perfluorocarbons (PFC) in animals with respiratory failure in which conventional mechanical ventilation could be insufficient to support life [1,2]. Moreover, a number of studies have reported improvements in lung functions after liquid ventilation which has been attributed to the residual PFC in the lungs [3,4].

In view of these accumulated data, we demonstrated recently that intratracheal administration of PFC in combination with conventional mechanical gas ventilation improves pulmonary gas exchange in a dose-dependent way in adult animals with respiratory failure due to lung lavage with saline [5]. This study aimed to investigate the effects of this new treatment modality with large doses of PFC on lung functions in long-term follow-up and to compare this with conventional mechanical ventilation.

METHODS

Adult New Zealand rabbits (n = 18) with a mean body weight of 2.8 kg (range 2.4-3.2 kg) were anesthetized with pentobarbital sodium iv (50 mg/kg). Tracheostomy was performed and an endotracheal tube was inserted with its tip proximal to the carina. Artificial ventilation with a Servo 900C ventilator (Siemens-Elema, Sweden) was initiated with 100% oxygen, zero end-expiratory pressure, tidal volume (TV) = 12 ml/kg, frequency (f) = 30/min, and inspiratory/expiratory time (I/E) = 1:2. An infusion of 5% dextrose / 0.45% NaCl solution was administered continuously as a maintenance fluid (10 ml/h). Anesthesia was maintained with an infusion of pentobarbital (10 mg/h) and fentanyl (2 μg/kg/h). Pancuronium bromide was administered for muscle paralysis (0.1 mg/kg/h).

A femoral artery and femoral vein were cannulated for arterial and central venous pressure monitoring and blood sampling. Arterial samples were analyzed for blood gases and pH using conventional methods (ABL-330, Radiometer, Copenhagen, Denmark). An indwelling catheter (Mikro-pO$_2$-Messkatheter, Licox, GMS, Kiel, Germany) was introduced into the other femoral artery for continuous oxygen-pressure monitoring

(Licox, GMS, Kiel, Germany). Intravascular pressure monitoring was performed using Statham P23XL transducers (Spectramed, USA) and all tracings including ECG were traced by a Sirecust 1280 monitor (Siemens, USA) and recorded by a Siredoc 220 recorder (Siemens, Germany).

Respiratory failure was induced by lung lavage with warm saline [6] to achieve an arterial pO_2 below 100 mmHg with the following ventilator settings: Volume controlled ventilation, $FiO_2 = 1$, TV = 12 ml/kg, f = 30/min, I/E = 1:2 and positive end-expiratory pressure (PEEP) = 6 cm H_2O. Animals were then divided randomly into 3 groups of 6 animals each. Group 1 was treated with 18 ml/kg of intratracheal perflubron; Group 2 was treated with 18 ml/kg of intratracheal saline and Group 3 received no treatment apart from continuous positive pressure ventilation (CPPV). Before and after treatments, the above-mentioned ventilator settings were kept constant for all three groups during the 3-hour observation period. Arterial blood gases and airway pressures were determined at 30 min intervals; meanwhile continuous arterial pO_2 monitoring enabled observation of changes in PaO_2.

Perflubron (perfluorooctylbromide; Alliance Pharmaceutical Corp., San Diego, CA) is a PFC which has a specific gravity of 1.918 g/cm^3 at 25°C, a surface tension of 18.1 dynes/cm, a vapor pressure of 3.6 mm Hg aMw20°C and 10.5 mm Hg at 37°C, O_2 solubility of 53 ml/100 ml and CO_2 solubility of 210 ml/100 ml at 37°C and 1 atmosphere pressure.

The statistical significance of the measured variables was assessed by the Mann-Whitney-U test for intergroup analysis. Intragroup comparisons were made by the Wilcoxon signed rank test. P values less than 0.05 were considered significant.

RESULTS

The PFC treatment was well tolerated in all animals. PaO_2 increased from 67 ± 11 to 424 ± 24 mm Hg at 30 min in the PFC group and remained stable for 3 hours. In the CPPV group, PaO_2 stayed around 60 mm Hg, while it further decreased in the saline group. $PaCO_2$ values were significantly lower in the PFC group compared to the other groups (Table 1).

Table 1. Blood gas values of the treatment groups (mean ± SD)

	PaO_2 (mm Hg)			$PaCO_2$ (mm Hg)		
	Perflubron	Saline	CPPV	Perflubron	Saline	CPPV
Pre-lavage	502 ± 20	494 ± 15	497 ± 22	36 ± 2	35 ± 2	34 ± 3
Post-lavage	67 ± 11	71 ± 11	70 ± 12	48 ± 3	47 ± 7	42 ± 4
1 h	422 ± 24[a]	38 ± 4	62 ± 18[b]	47 ± 6[b]	61 ± 5	49 ± 8[b]
2 h	418 ± 32[a]	40 ± 7	63 ± 29	47 ± 7[b]	58 ± 6	51 ± 8
3 h	405 ± 53[a]	39 ± 5	60 ± 19	48 ± 6[b]	62 ± 6	58 ± 13

[a] $p < 0.01$ compared to both the saline group and the CPPV group,
[b] $p < 0.05$ compared to the saline group by Mann-Whitney-U test.

Airway pressures decreased significantly in the PFC group and remained low throughout the observation period compared to the other groups (Table 2). It should be remembered that 6 cm H_2O of PEEP was applied post-lavage in all groups.

Table 2. Mean airway pressures of the treatment groups (mean ± SD)

	Mean Airway Pressure (cm H_2O)		
	Perflubron	Saline	CPPV
Pre-lavage	3.4 ± 0.3	3.6 ± 0.3	3.4 ± 0.2
Post-lavage	11.4 ± 0.4	11.5 ± 0.6	11.3 ± 0.4
1 h	10.3 ± 0.2[a]	12.3 ± 0.6	11.3 ± 0.4[b]
2 h	10.2 ± 0.3[a]	12.4 ± 0.6	11.7 ± 0.7
3 h	10.2 ± 0.3[a]	12.3 ± 0.6	11.8 ± 0.7

[a]$p < 0.01$ compared to the saline group and the CPPV group,
[b]$p < 0.01$ compared to the saline group by Mann-Whitney-U test.

DISCUSSION

We demonstrated recently that intratracheal PFC administration in combination with mechanical ventilation can improve pulmonary gas exchange in a dose-dependent manner in animals with acute respiratory failure [5]. The present results demonstrate that this new form of ventilatory support can provide adequate oxygen supply for several hours, and animals can tolerate even high volumes of intratracheal PFC instillation.

During total liquid ventilation, extracorporeally oxygenated PFC liquid is administered using a liquid ventilator after instillation of PFC in functional residual capacity (FRC) volume. We treated animals in this experiment with a PFC volume of 18 ml/kg, which corresponds to the FRC for healthy rabbits with identical average body weight [7]. When administered at FRC volumes, PFC eliminates the alveolar air-liquid interface and the tendency for alveoli to collapse at end-expiration. It has been demonstrated that the liquid-filled lung retains a collapse volume greater than the collapse volume of the same lung filled with air [8]. Thus, by preventing alveolar collapse during expiration, PFC increases effective lung volume to participate in alveolar gas exchange and, furthermore, keeps inflation pressures low compared to the gas ventilation of the diseased lungs due to its low surface tension. In addition, the extremely high capacity of PFC for dissolving O_2 and CO_2 maintains a respiratory medium for continuous gas exchange during both inspiration and expiration.

In the saline group, arterial blood gases deteriorated further, suggesting that the improvements in pulmonary gas exchange in the PFC group were due to the direct contribution of PFC itself.

In the CPPV group, pO_2 did not improve at identical ventilator settings and pCO_2 appeared to be higher during the observation period compared to the PFC group. Moreover, airway pressures increased, causing pneumothorax in one animal of the CPPV group. The present results related to the airway pressures indicate the potential of intratracheal PFC instillation combined with conventional mechanical ventilation in ensuring low inflation pressures in animals with acute respiratory failure.

CONCLUSION

This study suggests that, at identical ventilator settings, intratracheal PFC administration combined with mechanical ventilation is significantly more effective than conventional gas ventilation for improving pulmonary gas exchange at relatively low inflation pressures. This type of ventilatory support can be tolerated for several hours with a PFC volume approximately equal to the FRC volume of the animals, and appears to be a simple, alternative treatment of acute respiratory failure compared to liquid ventilation.

REFERENCES

1. T. H. Shaffer, D. Rubenstein, G. D. Moskowitz, and M. Delivoria-Papadopoulos, Gaseous exchange and acid-base balance in premature lambs during liquid ventilation since birth, Pediatr Res. 10:227-231 (1976).
2. T. H. Shaffer, N. Tran, V. K. Bhutani, and E. M. Sivieri, Cardiopulmonary function in very preterm lambs during liquid ventilation, Pediatr Res. 17:680-684 (1983).
3. T. H. Shaffer, C. A. Lowe, V. K. Bhutani, and P. R. Douglas, Liquid ventilation: effects on pulmonary function in meconium stained lambs, Pediatr Res. 18:47-52 (1983).
4. T. H. Shaffer, P. R. Douglas, C. A. Lowe, and V. K. Bhutani, The effects of liquid ventilation on cardiopulmonary function in preterm lambs, Pediatr Res. 17:303-306 (1983).
5. A. S. Tütüncü, B. Lachmann, N. S. Faithfull, and W. Erdmann, Dose dependent improvement of gas exchange by intratracheal perfluorooctylbromide (Perflubron) instillation in adult animals with acute respiratory failure, Adv. Exp. Med. Biol. (This volume).
6. B. Lachmann, B. Robertson, and J. Vogel, In-vivo lung lavage as an experimental model of the respiratory distress syndrome, Acta Anesthesiol Scand. 24:231-236 (1980).
7. D. Gommers, C. Vilstrup, J. A. H. Bos, A. Larsson, O. Werner, E. Hannappel, B. Lachmann, Effects of exogenous surfactant therapy on lung function can not be assessed by dynamic compliance, (submitted for publication).
8. J. A. Kylstra, and W. H. Schoenfisch, Alveolar surface tension in fluorocarbon-filled lungs, J Appl Physiol. 33:32-35 (1972).

CLINICAL USE OF OXYGEN STORES:

PRE-OXYGENATION AND APNEIC OXYGENATION

Rolf Zander and Fritz Mertzlufft

Institute of Physiology and Pathophysiology, Johannes Gutenberg-University Mainz, D-6500 Mainz, and Clinic of Anaesthesiology and Intensive Care Medicine, University of Homburg, D-6650 Homburg-Saar, Germany

INTRODUCTION

During states of respiratory arrest the human oxygen stores may be used therapeutically, regardless of the origin, i.e. either prior to the routinely induced apnea for endotracheal intubation or as an emergency measure in any other case of apnea. The present considerations focus on the clinical use of the oxygen stores available, applying

* pre-oxygenation
  and
* apneic oxygenation.

OXYGEN STORES

Under physiological conditions, the total O_2 store of the human body (65 kgBW) contains about

300 ml O_2	physically dissolved or bound to myoglobin,
800 ml O_2	bound to hemoglobin (750g Hb, 1.39 ml/g, 15% arterial blood with sO_2 = 100%, 85% venous blood with a mean sO_2 of 75%),
400 ml O_2	within the functional residual capacity (3000 ml x 0.135),
1500 ml O_2	TOTAL (for comp. see [Nunn, 1987]).

During respiratory arrest, this total O_2 store theoretically could guarantee sufficient O_2 supply for approx. 3 minutes. In this case the arterial sO_2 decreases to 50% (paO_2 25 mmHg), the mean venous sO_2 to 25% (i.e. an arterial plus venous sO_2 reduction of 50% = 500 ml O_2), simultaneously the alveolar pO_2 (pAO_2) from 100 to 25 mmHg (i.e. = 300 ml O_2).
Consequently, the normal O_2 consumption of 750 ml (3 min x 250 ml/min) is guaranteed by the above calculated 800 ml O_2.

PRE-OXYGENATION

Following optimal pre-oxygenation with total denitrogenation, the pAO_2 increases to 673 mmHg, corresponding to 88.6% O_2 (pB 760 mmHg, $pACO_2$ 40 mmHg, pAH_2O 47 mmHg). In this case the intrapulmonary O_2 store amounts to 88.6% O_2 within the FRC of 3000 ml and an intrapulmonary oxygen store of 2650 ml O_2 is achieved.

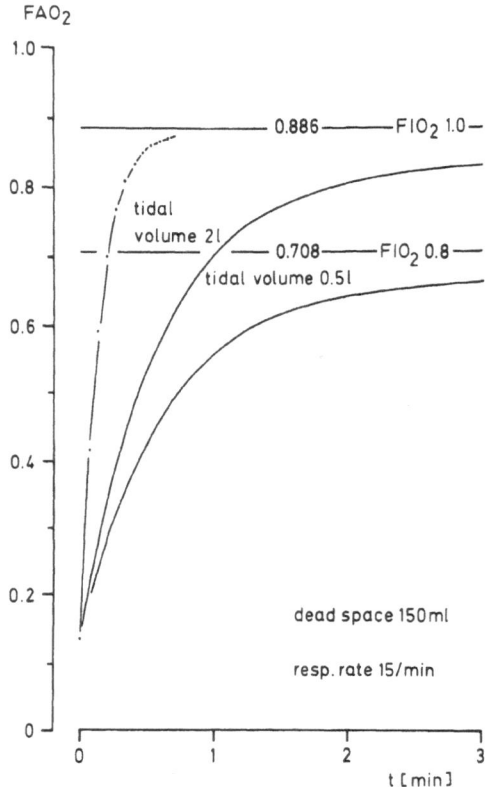

Fig. 1. Estimation of alveolar oxygen fraction (FAO_2) as a function of time during pre-oxygenation of a patient with pure O_2.
With normal ventilation (resp. rate 15/min, tidal volume 500 ml, dead space 150 ml) at FIO_2 1.0, the maximal FAO_2 of 0.886 is reached to 95% within about 3 min. When the inspired O_2 is diluted by ambient air (mask not tight, FIO_2 0.8), about 95% of the maximal FAO_2 of 0.708 are achieved within the same time. Hyperventilation with 10 breaths of 2 l O_2 shortens the time of pre-oxygenation to about 30 seconds.

Oxygen supply obviously is safely guaranteed for about 10 minutes among patients at rest (normal oxygen consumption of 250 ml/min), using this intrapulmonary O_2 store of 2650 ml alone.
However, pre-oxygenation per se is

- time consuming, and requires
- optimal systems for applying 100% O_2 (FIO_2 1.0), i.e. for elimination of alveolar nitrogen.

Under these conditions only, pre-oxygenation could be successfully performed within 3 min breathing spontaneously or, within only 1 min when increasing tidal volume. Corresponding estimations are shown in figure 1.

The plotted data are in good agreement with published proposals: For instance, Gold et al. (1981) propose pre-oxygenation by four deep breaths during 30 sec, whereas Braun et al. (1980) or Berthoud et al. (1983) recommend 3 min of pre-oxygenation during normoventilation and using gas-tight fits of face masks.

APNEIC OXYGENATION

In case of subsequent apnea all gases present (N_2, CO_2, O_2) will equilibrate between alveolar space and capillary blood. Within 10 min of apnea the following diffusion rates between alveolar space and capillary blood can be estimated:

1) Nitrogen

The N_2 diffusion from the blood into the alveolar space may be calculated using different data:

a) Under extreme assumptions, i.e. total elimination of N_2 from mixed venous blood (pN_2 626 mmHg: 760 mmHg - pO_2 40 mmHg - pCO_2 47 mmHg - pH_2O 47 mmHg), 5 l of blood (cardiac output) may release ca. 50 ml N_2/min (solubility = 0.012 ml/ml/atm).
Therefore, the maximal and unrealistic value amounts to 500 ml N_2 per 10 min of apnea.

b) Compared to the total N_2 pool of 800 ml (400 ml dissolved in each body water and body fat [Farhi, 1964]), within 10 min only 25% of the total pool are assumed to be eliminated, i.e. only 200 ml N_2.

c) Miles et al. (1956), in contrary, have determined a N_2 elimination rate by the lungs of 500 ml during 30 min following optimal pre-oxygenation. This corresponds to ca. 170 ml N_2 for a 10 min period of apnea.

d) However, according to Farhi (1964), the pN_2 in mixed-venous blood decreases to about 10% of the initial value, i.e. 60 mmHg, after 10 min of pre-oxygenation under optimal conditions. Thus, the maximum amount of N_2 diffusion would be only 50 ml per 10 minutes.

In conclusion, it can be predicted that, during a 10 min apnea not more than 200 ml of N_2 diffuse from blood into the alveolar space under the condition of an optimal pre-oxygenation.

2) Carbon dioxide

During apnea the pCO_2 of mixed-venous ($p\bar{v}CO_2$) blood comes into equilibrium with the alveolar pCO_2 ($pACO_2$) as well as with the arterial pCO_2 ($paCO_2$). Within the first minute, the $paCO_2$ increases by about 10 - 13 mmHg corresponding to the three following factors:

- equilibration of the $paCO_2$ (40 mmHg) by the $p\bar{v}CO_2$ (47 mmHg),
- increase of $paCO_2$ by about 3 mmHg above $p\bar{v}CO_2$ as a consequence of the Christiansen-Douglas-Haldane effect, i.e. O_2 uptake into the blood without CO_2 release [Mertzlufft et al., 1989], and
- increase of $paCO_2$ by about 3 mmHg resulting from CO_2 production.

In practice, an increase of about 3-4 mmHg/min from the 2nd to the 10th minute has been measured by various authors. The corresponding data of $paCO_2$ increase in humans during hyperoxic hypercapnia are listed in figure 2, the resulting increase for a 10 min apnea is shown in figure 3.

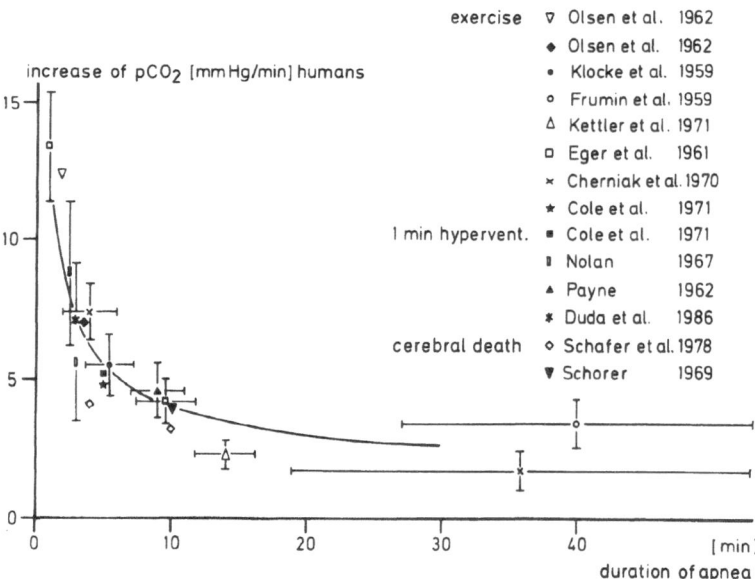

Fig. 2. Increase of arterial or alveolar pCO_2 (mmHg/min) during apnea in humans, i.e. hyperoxic hypercapnia, according to various authors.

The pCO_2 increase of 40 mmHg as shown in figure 3 for a 10 min period of apnea, corresponds to a 5.3% CO_2 increase within the alveolar space and, therefore, to only 160 ml CO_2 (FRC 3000 ml) entering the FRC.

The remaining 90% of the metabolically produced CO_2 are stored within the extracellular space of the body.

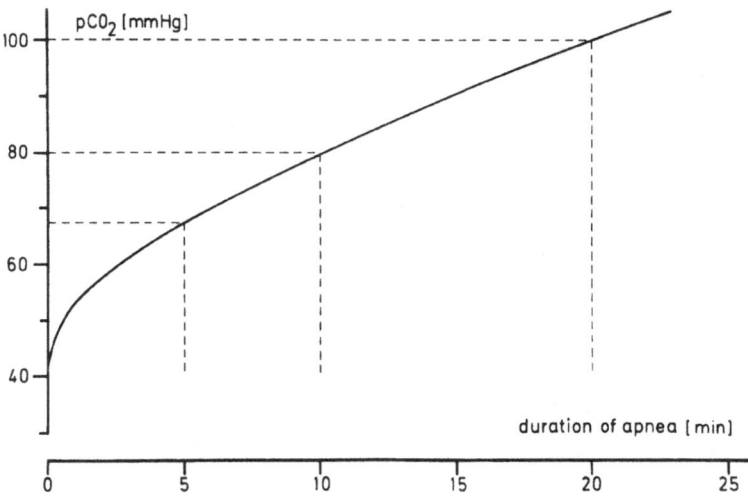

Fig. 3. Increase of the arterial pCO_2 in the human under "normal" conditions of hyperoxic hypercapnia during apnea.

Values of 160 - 170 mmHg of $paCO_2$ [Payne, 1962; Birt et al., 1965] up to 200 - 250 mmHg [Ellison et al., 1955; Frumin et al., 1959] have been measured in humans during hyperoxic hypercapnia without any ill effects.

3) Oxygen

The Oxygen consumption of 250 ml/min under normal conditions will be reduced to about 200 ml/min under anaesthesia. Thus, 2000 ml O_2 will have to diffuse from the alveolar space into the arterial blood. This is achieved by the steep alveolar-capillary pO_2 gradient (pAO_2 673 mmHg) due to optimal pre-oxygenation.

Patent airways provided, in the present case of a 10 min apnea the diffusion balance of all three gases in question causes a gas uptake by m a s s – m o v e m e n t as described by Volhard (1908) and Draper et al. (1944).

This mass-movement amounts to either

- 1640 ml during administration of pure oxygen (2000 ml
  O_2 - 160 ml CO_2 - 200 ml N_2) or to

- 1840 ml during the exposition to ambient air (2000 ml
  O_2 - 160 ml CO_2).

Ambient air

If a patient is exposed to ambient air instead of pure oxygen, the mass-movement will be increased to 1840 ml, as the N_2 of ambient air stops the N_2 diffusion from capillary blood into the alveolar space.

These 1840 ml of ambient air (21% O_2) entering the alveolar space by mass-movement, finally only contain 19.7% oxygen due to humidification (pH_2O of 47 mmHg) within the dead space.

Therefore, within 10 min of apnea only 2000 ml O_2 from the initial FRC store of 2650 ml O_2 are consumed, while 360 ml O_2 are replaced by mass-movement (1840 ml x 0.197).

The remaining 1010 ml O_2 (2650 ml O_2 - 2000 ml O_2 + 360 ml O_2) correspond to a FAO_2 of 0.336 or to a pAO_2 of 256 mmHg. Thus, the pAO_2 has been decreased from 673 mmHg to 256 mmHg, i.e. a decrease of 417 mmHg during the 10 min apnea or a decrease of about 40 mmHg per minute.

Pure oxygen

If the patient is connected to a source supplying pure oxygen, the mass-movement as compared to ambient air is only 1640 ml per 10 min but, containing 100% oxygen (corresponding to an O_2 concentration reduced to 93.8% after humidification).

Starting with a total FRC store of 2650 ml O_2 following pre-oxygenation, the patient consumes 2000 ml O_2 while 1540 ml O_2 are being replaced by 1640 ml mass-movement (1640 ml x 0.938, after humidification). Thus, 2190 ml O_2 remain within the FRC store (2650 ml O_2 store - 2000 ml O_2 consumption + 1540 ml O_2 replacement).

The remaining 2190 ml O_2 correspond to a pAO_2 of 555 mmHg. As the result, the pAO_2 has only slightly been decreased from 673 to 555 mmHg after a 10 min period of apnea.

The described mass-movement is the basic mechanism of the so-called apneic oxygenation, formerly termed diffusion respiration.

The possibility to achieve adaequate oxygenation with apneic oxygenation has been demonstrated in both animals and man, in man for up to 40 min of apnea. For details see e.g. Nunn (1987).

SUMMARY

Optimal pre-oxygenation prior to apnea plus administration of pure O_2 during apnea provide adaequate O_2 supply for patients with very high security, if required for at least 30 minutes.

REFERENCES

Berthoud, M., Read, D.H., Norman, J., 1983, Pre-oxygenation - how long?, Anaesthesia, 38:96.

Birt, C., Cole, P. V., 1965, Some physiological effects of closed circuit anaesthesia, Anaesthesia, 20:258.

Braun, U., Hudjetz, W., 1980, Dauer der Präoxygenation bei Patienten mit regelrechter und gestörter Lungenfunktion, Anaesthesist, 29:125.

Cherniak, N. S., Langobardo, G. S., 1970, Oxygen and carbon dioxide gas stores of the body, Physiol Rev, 50:196.

Cole, W. L., Stoelting, V. K., 1971, Blood gases during intubation following two types of oxygenation, Anesth Analg, 50:68.

Draper, W. B., Whitehead, R. W., 1944, Diffusion respiration in the dog anesthetized by pentothal sodium, Anesthesiology, 5:262.

Duda, D., Rudlof, B., El Gindi, M., Liessem-Sachse, R., Iversen, S., 1986, The influence of intubation apnea on the arterial gas state in patients undergoing coronary artery bypass surgery, Int Symp Anaesthesia for Cardiac patients, Abstracts, Munic

Eger, E. I., Severinghaus, J. W., 1961, The rate of rise of PA_{CO_2} in the apneic anesthetized patient, Anesthesiology, 22:419.

Ellison, R. G., Ellison, L. T., Hamilton, W. F., 1955, Analysis of respiratory acidosis during anaesthesia, Ann Surg, 141:375.

Farhi, L. E., 1964, Gas stores of the body, in: Handbook of Physiology, Respiration, Vol. I, Am. Physiol. Soc., Washington.

Frumin, M. J., Epstein, R. M., Cohen, G., 1959, Apneic oxygenation in man, Anesthesiology, 20:789.

Gold, M. I., Durate, I., Muravchick, S., 1981, Arterial oxygenation in conscious patients after 5 minutes and after 30 seconds of oxygen breathing, Anesth Analg, 60:313.

Kettler, D., Sonntag, H., 1971, Apnoische Oxygenation unter Verwendung von Trispuffer während Bronchographie, Anaesthesist, 20:94.

Klocke, F. J., Rahn, H., 1959, Breath holding after breathing of oxygen, J Appl Physiol, 14:689.

Mertzlufft, F. O., Brandt, L., Stanton-Hicks, M., Dick, W., 1989, Arterial and mixed venous blood gas status during apnea of intubation - proof of the Christiansen-Douglas-Haldane effect in vivo, Anesth Intens Care, 17:325.

Miles, G. G., Martin, N. T., Adriani, J., 1956, Factors influencing the elimination of nitrogen using semiclosed inhalers, Anesthesiology, 17:213.

Nolan, R. T., 1967, Pre-oxygenation and thiopentone-suxamemethonium induction, Brit J Anaesth, 39:794.

Nunn, J. F., 1987, Applied respiratory physiology with special reference to Anesthesia, Butterworths, London.

Olsen, C. R., Fanestil, D. D., Scholander, P. F., 1962, Some effect of apneic underwater diving on blood gases, lactate, and pressure in man, J Appl Physiol, 17:938.

Payne, J.P., 1962, Apnoeic oxygenation in anaesthetised man, Acta Anaesth Scandinav 6:129.

Schafer, J. A., Caronna, J. J., 1978, Duration of apnea needed to confirm brain death, Neurology, 28:661.

Schorer, R., 1969, Vermeidung von Hypoxämie und Acidose beim Atemstillstand, in, "Hypoxie", Anaesthesiology and Resuscitation 30, Springer, Berlin; pp 102 - 108.

Volhard, F., 1908, Über künstliche Atmung durch Ventilation der Trachea und eine einfache Vorrichtung zur rhythmischen künstlichen Atmung, Münch Med Wschr, 55:209.

A NEW DEVICE FOR THE OXYGENATION OF PATIENTS:

THE «NasOral-SYSTEM»

Fritz Mertzlufft and Rolf Zander

Clinic of Anaesthesiology and Intensive Care Medicine, University of Homburg, D-6650 Homburg-Saar, and Institute of Physiology and Pathophysiology, Johannes Gutenberg-University Mainz, D-6500 Mainz, Germany

INTRODUCTION

Outcome studies from the USA [Beecher and Todd, 1954], France [Tiret et al., 1986], the United Kingdom [Buck et al., 1987], and from Germany [Schulte-Sasse and Eberlein, 1990]) for example, report a mortality of 1 to 3 deaths per 10,000 anaesthesias, that solely attributable to anaesthesia.

Deteriorations of the respiratory system, combined with a lack of or inaedequate measures of prevention and/or surveillance, have been discussed as the major causes for these deaths [Cheyney, 1988; Larsen, 1989].

Apnea, either unanticipated or routinely induced for the induction of endotracheal intubation, still represents the most common challenge for outdoor practice and in emergency medicine, as well as for innerhospital settings.

This mainly depends on two facts. The duration of apnea is unpredictable and, in case of induced apnea, may be endangered by unexpected events due to unanticipated situations, and/or insufficient skills of the performer, i.e. generally the anaesthetist.
The combination of ongoing O_2 consumption (normal $\dot{Q}O_2$ ca. 250 ml/min, during anaesthesia ca. 200 ml/min or less) and increasing lack of adaequate O_2 supply during apnea the potential risk of hypoxia and hypercapnia both of which may cause devastating damage after only a few minutes. Even with control of the FIO_2 (standard procedure) and use of modern monitors (e.g. pulse oxymetry, capnometry, etc.), these damages can never be prevented.

However, apnea could be anticipated, i.e. routinely manageable and less hazardous, by renewing and improving knowledge on the oxygen stores of the human body, of which only the functional residual capacity (FRC) can be used therapeutically:

O$_2$ supply could be easily and safely guaranteed for up to 30 minutes of apnea (a maximum of clinically relevant security that could not be considered before), by combining the two relevant procedures:

1) optimal pre-oxygenation <u>prior</u> to apnea, and
2) maintenance of O$_2$ supply <u>during</u> apnea via apneic oxygenation with pure oxygen.

ACTUAL SITUATION

Practically, optimal pre-oxygenation at the onset of apnea and maintenance of O$_2$ supply during apnea involve a number of problems.

Sufficient oxygenation is either not achieved at all or only achieved inaedequately by conventional methods. Thus a need exists to reconsider the actual situation for performing both pre-oxygenation and apneic oxygenation.

Pre-oxygenation

Within the framework of what is referred to as pre-oxygenation, the filling of the FRC with only pure O$_2$ prior to therapeutic maneuvers (e.g. endotracheal intubation following induced or artificial apnea) is the goal of each clinician. Unfortunately, this goal is seldom achieved in clinical practice due to a variety of general and practical problems.

1) General problems

Failure to achieve adaequate pre-oxygenation may result in particular from four basic disadvantages which characterize all contemporary systems used (e.g. the variety of different masks, oropharyngeal or nasopharyngeal tubes, etc.) [Sandersen, 1972]:

■ The O$_2$ concentration never is 100% (due to system leakages or incomplete dead space washout),

■ re-inhalation of expired gases occurs (CO$_2$, N$_2$),

■ the humidification and warming of O$_2$ is insufficient,

■ the use of very high O$_2$ flows is necessary.

2) Practical problems

Practical problems concerning the performance of pre-oxygenation are mainly due to four major disadvantages:

■ All systems commonly used require complete washout of the apparatus and its functional dead space volume, which,

- in case of using half-closed circuit systems (e. g. Dräger circuits; Drägerwerk AG Lübeck, Germany), may represent a considerable volume of 6 up to 11 liters (depending on the kind of connected tubings plus the volume and number of the CO_2 absorbers) and which, therefore,
- need at least a 2 min washout time using the maximal oxygen flow of 15 liters per minute,

■ oxygenation of a patient by means of the commonly used masks requires the mask to be pressed closely fitting over both mouth and nose,

■ if only for about one breath ambient air should enter the system (a mishap that commonly occurs), some considerable time is needed to regain the former situation,

■ the procedure is time consuming and the performer has to handle the pre-oxygenation procedure very carefully with continuous control of the patient.

Even under the extreme, improbable assumption, that these four disadvantages are known and have been managed adaequately, it still remains very uncertain whether the patient will be securely oxygenated or not.

So far, however, none of the techniques available provides the two fundamental prerequisites needed for optimal pre-oxygenation:

1. enrichment of the intrapulmonary oxygen store (FRC) with pure oxygen only,

2. simultaneous and, possibly, complete elimination of other gases within the FRC, nitrogen (N_2) in particular.

Apneic oxygenation

Within the framework of what is referred to as apneic oxygenation, a patient at apnea could be sufficiently oxygenated if the consumed oxygen is substituted by an external oxygen influx to the alveolar space.

While pre-oxygenation is often handled more or less adaequately, apneic oxygenation is generally not performed at all, neither following pre-oxygenation nor during any other kind of oxygenation. The unavailability of a system to safely perform this type of oxygenation appears to be the major reason, furthermore the possibility of apneic oxygenation is merely unknown.

Therefore, a new oxygenation system is required. The goal of the present contribution is the introduction of such a new system that meets the requirements for both, optimal pre-oxygenation and apneic oxygenation.

The following three preconditions had to be met:

- supply of the FRC solely with pure O_2, without any need for dead space washout, plus concomittant and complete elimination of N_2 and CO_2,

- maintenance of O_2 diffusion from the alveolar space into the blood, and

- correct functioning, without any special technique or special skills of the user.

THE NasOral-SYSTEM

The developed oxygenation system was termed «NasOral-System» (cf. fig. 1) and tries to meet the described needs, i.e. feasibility for both adaequate pre-oxygenation and apneic oxygenation. It intends to eliminate the above described drawbacks of prior art apparatus.

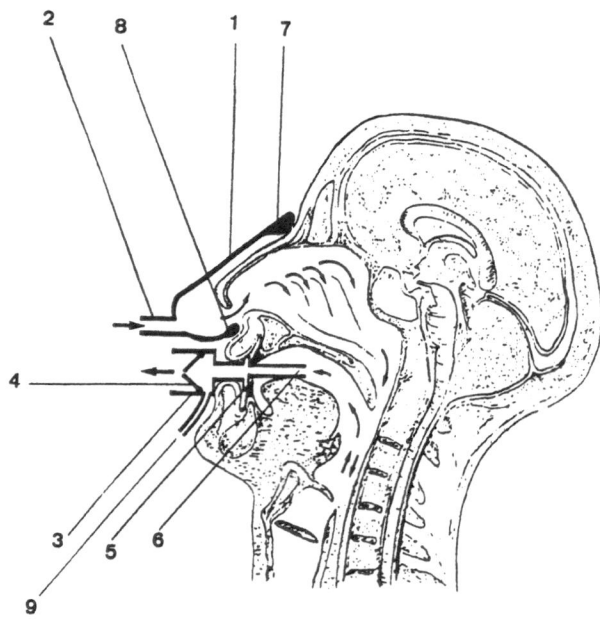

Fig. 1.   Embodiment of the NasOral-System diagrammatically and in vertical section:
- oxygen applicator   (1)
- oxygen tube   (2)
- one-way valve   (3)
- valve parts   (4)
- oral sealing plate   (5)
- tubular part of the oral sealing plate   (6)
- nasal sealing lips   (7, 8)
- gas outlet of the oral one-way valve   (9)

The new system supplies pure O_2 by means of an unidirectional flow (nas-oral) solely via the nose (nasal route) and comprises a one-way valve to be inserted in sealing-tight fashion into the mouth (oral), opening only during gas outflow, otherwise closing and allowing excess O_2 and expired gas to escape solely via the mouth (oral route).

The oxygen applicator of the system can be constructed in various ways. Expediently, however, a mask will be used as the oxygen applicator, the mask covering the nose in a substantially sealing-tight manner.

This **NasOral**-System guarantees

1) separation of inhalation and exhalation,
   ----
   i.e. inhalation of 100% O_2 via the "nasal" route and exhalation of excess O_2, N_2 and CO_2 via the "oral" route, i.e. a "nas-oral" system is achieved,

2) physiological gas flow,
   ----
   i.e. humidification and warming, retainment via the nasal route without increasing dead space and avoiding any re-inhalation,

3) maintenance of oxygen supply
   ----
   during apneic oxygenation by providing continuous external O_2 flux into the alveolar space.

It particularly ensures that, if breathing is present (e.g. during pre-oxygenation) only pure oxygen (warmed and humidified) can enter the respiratory passages and lungs via the nasal route, whereas all gases which should be eliminated (N_2, N_2O, CO_2) are completely (N_2, N_2O) or physiologically (CO_2) discharged via the oral route, since the one-way valve prevents the uncontrolled and undesired penetration of ambient air and thus large quantities of nitrogen.

Even during apnea, this system is capable of allowing the patient to draw in only pure oxygen (apneic oxygenation).
During apnea, the oral valve can be removed to allow for therapeutic or diagnostic measures, i.e. allowing performance of, for instance, endotracheal intubation in a smotth manner and independent of the skills of the performer.

Preliminary clinical trials have been performed to investigate how far pre-oxygenation prior to intubation procedures is improved and if apneic oxygenation could be a suitable measure for e.g., transportation of patients to the operation theatre (including the time for positioning) or to the recovery room.

With written informed consent, 10 patients (ASA classification II) scheduled for posterior cranial fossa surgery were examined (pre-medication 1 mg Lormetazepam the evening before and 2 hours prior to pre-oxygenation).

After pre-oxygenation by means of an optimal system during induction of anaesthesia and following endotracheal intubation, both the alveolar (AGM 1304, Brüel & Kjaer, Denmark) and the arterial O_2 partial pressures (STAT PROFILE 5, NOVA biomedical, Waltham, USA) were analyzed simultaneously:

1) The alveolar pO_2 (pAO_2) in fact was increased to 638 mmHg (at actual conditions: pB 734 mmHg, pH_2O 47 mmHg, actual pCO_2 and pN_2), i.e. a pAO_2 value that closely approaches the theoretical pAO_2 of 652 mmHg (pB 734 mmHg),

2) the arterial pO_2 (paO_2, radial artery) at these conditions increased to 618 $\pm$ 9 mmHg (preliminary values of 10 measured patients), i.e. a paO_2 value that was never achieved before and that, obviously, seems to contradict the theory of increasing $AaDO_2$ with increasing FIO_2 (e.g. [Falke, 1991]).

During these clinical settings another essential feature of the NasOral-System became evident:
It could be used uniquely, either for different patients or many times for the same patient, i.e. pre-oxygenation, onset and emergence from anaesthesia, post-operatively.

CONCLUSIONS

The described oxygenation system can be advantageously used for both pre-oxygenation and apneic oxygenation and will, therefore, be capable to reduce the overall risk of inaedequate oxygen supply in case of apnea.

It is an essential feature of this NasOral-System that the oxygen supply takes place solely via the nose (nasal inflow route), whereas oxygen overflow and exhalation pass through the mouth (oral outflow route) which is prevented from drawing in ambient air by the one-way valve, i.e. that gas flow is allowed following a "nas-oral" route only.

Applying this easy-to-use System, performance of both pre-oxygenation and apneic oxygenation is possible safely and immediately, in pre- and innerhospital settings as well as in outdoor practice, and regardless of the user's skills.

Further studies on pre-oxygenation and apneic oxygenation will have to prove, whether theoretical considerations and the described preliminary observations are reproducible in an extended patient population.

REFERENCES

Beecher, H. K., Todd, D. P., 1954, A study of the deaths associated with anesthesia and surgery based on a study of 599.548 anesthesias in ten institutions 1948-1952, inclusive, Ann Surg, 140:2.

Buck, N., Devlin, H. B., Lunn, J. N., 1987, The report of a confidential enquiry into perioperative deaths, Nuffield Provincial Hospitals Trust, London.

Cheyney, F. W., 1988, Anesthesia: Potential risks and causes of incidents, in: Safety and Cost Containment in Anesthesia, J. S. Gravenstein and G. Holzer, eds., Butterworths, Stoneham, MA.

Falke, K. J., 1991, Therapeutic thresholds for acute changes in arterial O_2 partial pressure, in: The Oxygen Status of Arterial Blood, R. Zander and F. Mertzlufft, eds., Karger, Basel; p. 64.

Larsen, R., 1989, Anaesthesie, Urban & Schwarzenberg, 3rd ed., München; p. 131 - 148.

Sandersen, R.G., 1972, The Cardiac Patient, W.B. Saunders, Philadelphia.

Schulte-Sasse, U., Eberlein, H. J., 1990, Patientensicherheit in der Anaesthesie, Anaesth Intensivmed, 31:97.

Tiret, L., Desmonts, J. M., Hatton, F., Vourch, G., 1986, Complications associated with anaesthesia - a prospective study in France, Can Anaesth Soc J, 33:336.

OXYGEN TRANSPORT IN BLOOD

HEMODILUTION AND OXYGEN TRANSPORT

A. Trouwborst, E.C.S.M. van Woerkens, R. Tenbrinck

Department of Anesthesiology, Erasmus University, Rotterdam
The Netherlands

INTRODUCTION

Several risks are associated with transfusion of homologous blood and blood products. The risk of alloimmunisation is well known. Transmission of viral infections from homologous blood transfusion (e.g. hepatitis, and especially AIDS) draw increasing the attention (Adler, 1984; Alter et al., 1972; Curran et al., 1984). In addition, it is suggested that transfusion of homologous blood promotes markedly increased tumor growth and growth of established metastases, and induces more postoperative infections, because blood transfusions exert a long-term (months) immunosuppressive action, increasing during the first week after transfusion (Schriemer et al., 1988; Waymack et al., 1989).

Transfusion with donor blood may be diminished by predeposited autologous blood (Toy et al., 1987), intraoperative autotransfusion with a cell-saver (Cutler, 1984) and hemodilution techniques. With hemodilution, fewer red cells are lost because of the non-linear decrease in packed cell volume during replacement of blood with plasma substitutes (Zetterström and Wiklund, 1986). Hemodilution can be achieved in several ways. First, by pre-operative withdrawal of blood or by peroperative blood loss and simultaneous infusion of plasma substitutes (normovolemic hemodilution),(Messmer, et al., 1986). Second, by rapid infusion of fluid without blood withdrawal (hypervolemic hemodilution), (Trouwborst, et al., 1990a; Trouwborst, et al., 1990b).

In general, oxygen flux into the tissue and finally into the cell depends on many factors as for example arterial oxygen content (Hb and Hb oxygen saturation), systemic hemodynamics (cardiac output), blood flow properties in the microcirculation, capillary density in the tissue, hematocrit values in the capillaries, Hb affinity for oxygen, release of oxygen from red cells, tissue affinity for oxygen, tissue diffusion coefficient and oxygen transport over the cell membrane. Regulatory mechanisms by hormones (e.g. catecholamines) or by local metabolites in relation to receptors in the macro- and microcirculation play an important role. Hemodilution might interfere with or change many of these factors and regulatory mechanisms.

HEMODILUTION AND SYSTEMIC HEMODYNAMICS

It has been postulated that acute isovolemic hemodilution induces a decrease in systemic vascular resistance (SVR) almost parallel to the decrease in blood viscosity, whereas cardiac output (CO) increases significantly without increase in myocardial contractility. Over a wide range of hematocrit (Hct) levels, the rise in CO compensates for the decreased oxygen transport capacity, thereby maintaining oxygen transport to the tissue (Messmer et al., 1973). The extent of the rise in CO, however, differs from one study to the other, even when studying the same proportional change in Hct. In anesthetized dogs rises varying between 40% and 125% are reported (Pavek and Carey, 1974; Cain, 1977; Cain and Chapler, 1978; Jan and Chien, 1977; Fan et al., 1980; Crystal et al., 1988), except in one study where the rise in CO did not exceed 10% (Geha, 1976). In this study, however, filling pressures (left atrial end diastolic pressure) did not change during so-called normovolemic hemodilution. Because of an increased venous return to the heart after a reduction of viscous resistance of blood by hemodilution, a significant increase in filling pressures of the heart during normovolemic hemodilution can be expected. Therefore, constant filling pressures in this study suggest the existence of hypovolemic instead of normovolemic hemodilution. In conscious dogs, comparing CO values of the different studies in relation to the proportional change in hematocrit, reveals that the CO response on hemodilution is less. Related to the degree of normovolemic hemodilution increases of CO of 75-120% were found (Glick, et al., 1964; Vatner, et al., 1972; von Restorff, et al., 1975a).

Surprisingly, comparing the studies in conscious dogs with the studies in anesthetized dogs, the increase in CO in all anesthetized animal studies is closely related to the increase in stroke volume (SV), while heart rate (HR) remains constant while, in contrast, in all conscious dog experiments an increase in CO is mainly attributable to an increase in HR. In conscious dogs, plotting the change in CO against the change in HR by regression analysis; a positive correlation with r = 0.85 was found (Pavek and Carey, 1974). The increase in SV, as observed during hemodilution in anesthetized species, has been attributed to several mechanisms; 1. increased venous return due to reduced whole blood viscosity with consequent increased filling pressures (Laks, et al., 1974). 2. facilitation of left ventricular emptying by reduced after load because of reduced viscosity and possible vasodilatation (Crystal, et al., 1988). 3. increased myocardial contractility due to activation of cardiac sympathetic nerves (Glick, et al., 1964).

Because dogs differ from humans in anatomy, distribution of coronary arteries and in sympathetic responses (Weaver et al., 1986), we studied cardiovascular responses, hemodynamics, oxygen transport to tissue during normoxic acute isovolemic hemodilution in sedated and in anesthetized pigs (Trouwborst, et al., 1990c; Trouwborst et al., 1990d; van Woerkens et al., 1991). Ample evidence exists to demonstrate that the pig is closely related to the human both anatomically and physiologically with respect to cardiovascular system, regional distribution of cardiac output, metabolism and maximum oxygen consumption (Swindle, 1984; Mc Krinan et al., 1986; Weaver et al., 1986). In the sedated pig we found an increase in CO of 40%, as in conscious dogs, mostly related to an increase in HR, while in the fully anesthetized pigs an increase in CO with 100% was observed, as in anesthetized dogs, mainly due to an increase in SV. In contrast, others found in anesthetized pigs an increase of 30% correlated with an increase in HR while SV remained unchanged (Nöldge, 1991). However, in this study the hemodilution was rather hypovolemic than normovolemic, because blood exchange was done keeping filling pressures of the heart unchanged, ignoring the normally increased filling pressures during normovolemic hemodilution.

Only a few studies report the effects of normovolemic hemodilution in anesthetized humans, while no report exists about acute hemodilution in conscious humans. In anesthetized humans, during normovolemic hemodilution (range Hct: 40-20), CO increases with 25-35%, as in anesthetized animals, mainly due to increased SV, while HR remains unchanged. (Laks, 1974; Coburg et al., 1976; Gisselsson et al., 1982; Vara-Thorbeck et al., 1985; Von Bormann, 1986; Mouren et al., 1989). Acute hypervolemic hemodilution in anesthetized humans induces an increase in CO, comparable with the increase in CO during normovolemic hemodilution and also without any change in HR (Trouwborst et al., 1990a).

In conclusion: 1. The increase in CO during hemodilution in anesthetized species is mainly due to increased SV, possibly due to reduced blood viscosity. 2. In consious or sedated animals, reduced viscosity is a less important determinant during hemodilution and the increase in CO is directly related to an increase in HR; 3. hemodilution can only be called iso- or normovolemic when filling pressures of the heart are increased, compared to baseline values obtained before hemodilution is induced.

HEMODILUTION AND SYSTEMIC OXYGENATION

Oxygen flux is the product of CO and arterial oxygen content. During hemodilution oxygen flux might remain unchanged or even increased, depending on the degree of increase in CO. Further decrease in Hct might be followed by a decreased oxygen flux, because the increase in CO as a compensatory factor is exhausted. Any further decrease in arterial oxygen content leads to an increased extraction ratio (ER = oxygen uptake divided by oxygen flux) until a certain maximum. Further reduction of Hct produces oxygen supply dependency of oxygen uptake (VO_2). In sedated pigs a significant drop in oxygen flux could be observed at a Hct of 15%, while oxygen supply dependency of the VO_2 started at a Hct of 10% (Trouwborst et. al., 1990d). The ER value at this point was 0.57, and the oxygen flux 350 $ml.m^{-2}.min^{-1}$ (= 15.0 $ml.kg^{-1}.min^{-1}$.). In anesthetized pigs a significant increase in ER could be observed at a Hct of 13%, while at the final step of hemodilution (Hct = 9.3%) VO_2 was well maintained (ER = 0.61), and the oxygen flux was decreased to 7.8 $ml.kg^{-1}.min^{-1}$ (van Woerkens et al., 1991). Cain found in anesthetized dogs during anemic hypoxia, when oxygen supply dependency of VO_2 started at a Hct of 10%, a critical whole body ER of 0.79, and a critical O_2 flux of 10 $ml.kg^{-1}.min^{-1}$ (Cain and Chapler, 1978; Cain, 1977). In another study in anesthetized dogs a critical systemic oxygen flux of 7.9 $ml.kg^{-1}.min^{-1}$ and a critical ER = 0.69 were established (Nelson et al., 1987).

In a fatal normothermic Jehovah Witness case we observed a decrease in oxygen flux when Hct dropped below 20%. Oxygen supply dependency of VO_2 started at a hemoglobin value of 4.0 gm% (Hct = 13%). The oxygen flux at this point was 184 ml.m^{-2} (= 4.9 ml.kg^{-1}.min^{-1}) and the ER 0.46. The patient died at a Hct value of 8%. During the same stages of hemodilution, the cardiovascular responses of this patient were similar in comparison to other anesthetized humans and this case might therefore be representative for anesthetized humans.

During hemodilution the ER might change not only due to changes in O_2 flux but also because of changes in VO_2. In some reports increases in VO_2 at the initial stages of stepwise induced normovolemic hemodilution are reported (Von Restorff et al., 1975b; Cain, 1977; van Woerkens et al., 1991). It has been suggested that the extra oxygen use by the heart alone with hemodilution could account for this increase. An increase of total circulating catecholamine level during anemic hypoxia has been discussed (Sylvester, et al., 1979) whereby the metabolic effect exerted by catecholamines is to increase whole-body resting O_2 (Cain, 1983). The critical point of hemodilution whereby any further decrease of Hct might induce oxygen supply dependency of VO_2 is of course also influenced by differences in VO_2. During full anesthesia VO_2 is lowered and therefore a lower Hct is better tolerated than during sedation alone or during consciousness. In sedated pigs we found a critical oxygen flux of 15.0 ml.kg^{-1}.min^{-1}, while in fully anesthetized pigs an oxygen flux of 7.5 ml.kg^{-1}.min^{-1} was still sufficient to maintain a constant VO_2. In patients the induction of anesthesia decreases VO_2 with 10%, while VO_2 the first hours after anesthesia is increased by 20% compared with preanesthetic values (Trouwborst et.al., 1990a). Therefore, in clinical practice, during surgery and anesthesia a more pronounced hemodilution is acceptable, than during the initial postoperative phase. Several factors might influence the increased oxygen demand postoperatively as for example shivering during recovery from anesthesia and increased catecholamine levels (Bay, et al, 1968).

In conclusion: The critical point of hemodilution with respect to systemic oxygenation differs in anesthetized and in conscious or sedated species because of differences in VO_2 through: 1. differences in basal metabolism and 2. possible differences in catecholamine levels.

HEMODILUTION AND REGIONAL HEMODYNAMICS AND OXYGENATION

In general, in the face of a reduction in O_2 flux, to meet the different metabolic demands of different organs, regional redistribution of blood away from lower extracting beds toward more critical tissues might occur. The effect of hemodilution on regional flows has been studied in several models. Most studies used electro-magnetic or Doppler flow probes and, in general, showed increases in flow to brain and heart (185-700% increase), out of proportion compared to the increase in CO, and with little or no change in flow to other organs (Geha, 1976; Race, et al., 1967; Von Restorff, et al., 1975b; Vatner, et al., 1972). Little or no change in flow to other organs was especially obvious in conscious animal models. Another study, using electromagnetic flow probes, but studying only the splanchnic area, showed that total hepatic blood flows and mesenteric arterial blood flow increased, but not compensating totally decreased oxygen content, inducing increased oxygen extractions of liver and small intestine (Nöldge, et al., 1991). However in this study only a small increase in CO is observed, because hemodilution was induced, thereby keeping filling pressures unchanged.

The microsphere technique for determining regional flows and oxygen flux has been used in four studies. In sedated rats and anesthetized dogs increased flows to heart (100-130%) and brain (26-140%) are reported, not related to increased CO (12-43%), while flow to the spleen decreased and flow to other organs measured did not change, or changed only slightly (Fan, et al., 1980; Woodson and Auerbach, 1982; Crystal et al., 1988). In fully anesthetized pigs with hemodilution (Dextran 40,50 g.l^{-1}, Isodex^R) flows increased out of proportion compared to increased CO (103%) to heart (420%) and brain (170%) while, in contrast to the other microsphere studies, flows to all other organs increased proportional to the increased CO, except for skin flow (70% increase) and skeletal muscle flow (70% increase) (van Woerkens, et al., 1991). In this study myocardial oxygen consumption increased (106%) as hematocrit was decreased. However, even at an Hct of 9%, left ventricular oxygen ER did not increase since the increase in myocardial blood flow produced an increase in myocardial oxygen flux sufficient to compensate the decreased arterial oxygen content and the increased myocardial oxygen demand. In conscious dogs myocardial VO$_2$ increased with 108% but was covered by more than seven-fold increase in coronary flow (Von Restorff, et al., 1975b).

Redistribution of CO might be explained by changes in circulating catecholamines and by enhanced sympathetic activity (Glick, et al, 1964). Carbon monoxide hypoxia, used as a model for anemic hypoxia, increases total catecholamine levels in anesthetized dogs (Sylvester, et al., 1979), while infusing drugs with ß-vasodilator activity during hypoxic hypoxia can cause a decreased ability to extract O$_2$ (Yonekawa, et al., 1981; Cain and Bredle, 1990). The use of α-adrenergic blockers was reported to lower the oxygen extraction rate in the hypoxic dog (Cain, and Chapler, 1980). It has been suggested that the main function of a vasoconstrictor response during hypoxia is to ensure oxygen delivery to the brain, and therefore generalized vasoconstriction should be caused everywhere except in the brain (Cain, 1983). This vasoconstriction might be overcome by local factors through the production of a vasoactive metabolite or through a direct effect of the oxygen tension on vascular tissue, especially in organs with high metabolic rate, as the heart (Cain, 1977). In our pig study, no increase of catecholamines during hemodilution has been found, maybe because of, in these experiments, high-dose fentanyl anesthesia, known for blocking stress responses and catecholamine release. This might explain why in this study, in contrast to the other microsphere studies, flow to most organs increased proportional to the increased CO (van Woerkens, et al., 1991).

In-vivo studies of the microcirculation show that in general the Hct in the capillaries (Hmic) is substantially lower than the systemic Hct. This is explained by several mechanisms. At first, in the microvessels red blood cells (RBC) travel faster than plasma, because RBC are located more in the centre of the vessel tube. The result is that blood adjoining to the vessel wall presents a "plasma layer". In general, when a solution floats through a tube, that part of the solution adjacent to the wall of the tube passes the tube more slowly due to the higher resistance. Therefore, plasma meets a higher resistance in floating through the capillaries. Because plasma needs more time to travel through the capillaries than RBC, at any moment of time more plasma than RBC will be present in the capillary resulting in a lower Hct. This phenomenon has been termed "plasma skimming" (Perkio and Keskinen, 1983). However plasma skimming alone is not sufficient to account for the observed reduction in Hmic, which is of the order of 30-50% from systemic Hct. Therefore mass balance considerations require that the low Hmic seen is also compensated by flow of blood through high Hct shunts. Such capillary vessels that can fulfil this function has been described and are termed "thoroughfare channels" (Lipowsky and Zweifach, 1974). During hemodilution until a systemic Hct of 20 is reached, different mechanisms, not yet totally clarified, tend to maintain constant RBC influx and therefore Hmic (Mirhashemi, et al., 1988). Another important factor to be considered during hemodilution is that normally RBC's lose oxygen during their travel through the arterioles (Dulling, and Beine, 1973). The loss of oxygen by the RBCs is a function of the transit time. Since hemodilution causes blood flow to increase in the microvasculature it will decrease the transit time of RBC and, therefore, blood will arrive at the capillaries with a greater amount of oxygen (Mirhashemi, et al., 1987). Finally, during extreme hemodilution, RBC influx into the capillary will decrease and the spacing between two cells in the capillary will increase. Until a Hmic of 20%, blood in the capillaries is a continuous and homogenous source of oxygen to the tissue (Federspiel and Sarelius, 1984). During extreme hemodilution, however, where spacing between red cells become dominant, RBC's act more as point sources of oxygen to the tissue and the tissue becomes increasingly sensitive to the passage of each source (Tsai, et al., 1990).

In conclusion: a. Regional redistribution of CO and regional oxygen flux might be regulated: 1) by enhanced sympathetic activity and might therefore be more pronounced in conscious or sedated animals than in fully anesthetized species and 2) by vasoactive metabolites, especially during extreme hemodilution and in organs with high metabolic rate. b. Until systemic Hct of 20% is reached, RBC's influx into capillaries remain constant, keeping Hct in capillaries (normally 30-50% lower than systemic Hct) more independent from changes in systemic Hct. c. during hemodilution RBC reaching capillaries carry more oxygen.

HEMODILUTION AND OXYHEMOGLOBIN DISSOCIATION CURVE

A shift to the left of the oxyhemoglobin dissociation curve (ODC) limits oxygen delivery when blood flow is limited (Malmberg, et al., 1979). Acute shift to the right of coronary sinus, venous and arterial blood have been reported during signs of hypoxia of the heart in humans (Colvard and Longmuir, 1973; Kostuk, et al., 1973; Shappell, et al., 1970). In two studies, chronically reduced Hb concentration is compensated for by improved oxygen unloading, afforded by the increased P_{50} (Edwards and Canon, 1972; Blumberg and Marti, 1972). In one study in dogs an acute change in ODC during acute hemodilution occurred but not before Hct dropped below 10% (Messmer, et al., 1973), while in anesthetized pigs at every step of hemodilution the ODC shifts to the right (Trouwborst et al., 1990b; Trouwborst et al., 1990e). A close relationship was found between the shift in ODC and the changed, for pH corrected, mixed venous PO_2 (Trouwborst et al., 1990e). This relationship is in agreement with the mentioned acute changes in P_{50} during signs of hypoxia of the heart. In a fatal Jehovah Witness case we found no change in P_{50} before Hct dropped below 10%. In the pig experiments a further direct significant linear relationship was observed between the shift to the right of ODC and the increase in O_2 ER also during oxygen supply dependency of VO_2, suggesting that the maximum of O_2 ER is also influenced by the position of the ODC. This might explain differences in the literature about the O_2 ER critical value because of possible differences during different experiments in the position of the ODC.

A new concept in monitoring systemic oxygenation that includes the effect of changes in ODC has been introduced (Trouwborst, et al., 1990e). Using the S_{35}

(saturation of hemoglobin at PO_2 = 35 mmHg), which changes with alterations in ODC, real arterial available oxygen content ($CavlO_2$) can be calculated being the maximum amount of oxygen that can be extracted from Hb in several organs before oxygen diffusion into tissue becomes compromised and VO_2 may decrease. The relation between VO_2 and $CavlO_2$ expressed by the extraction ratio of the arterial available oxygen content (ERav) should give a realistic indices of oxygen supply and VO_2, including the effect of changes in ODC. Comparing our pig experiments with the fatal Jehovah Witness case revealed differences in critical ER (0.57 vs 0.46) but not in critical ERav.

In conclusion: Changes in ODC might play a role in oxygen delivery during hemodilution and especially, as stated before (Rand, 1975), when oxygen reserves are minimal. The position of the ODC is an important determinant for the critical O_2 ER value during hemodilution, also when VO_2 becomes dependent on oxygen flux.

REFERENCES

Adler, S.P., 1984, Transfusion transmitted C.M.V. infections, Vox Sang., 46: 387.

Alter, H.J., Holland P.V., Purcell R.H., et al., 1972, Posttransfusion hepatitis after exclusion of commercial and hepatitis-B antigen positive donors, Ann. Intern. Med., 77: 691.

Bay, J., Nunn, J.F., and PrysRoberts, C., 1968, Factors influencing arterial PO_2 during recovery from anesthesia, Br. J. Anaesth., 40: 398.

Blumberg, A., and Marti, H.R., 1972, Adaptation to anemia by decreased oxygen affinity of hemoglobin in patients on dialysis, Kidn. Int., 1: 263.

Cain, S.M., 1977, Oxygen delivery and uptake in dogs during anemic and hypoxic hypoxia, J. Appl. Physiol., 42: 228.

Cain, S.M., and Chapler C.K., 1978, O_2 extraction by hind limb versus whole dog during anemic hypoxia, J. Appl. Physiol., 45: 966.

Cain, S.M., and Chapler, C.K., 1980, O_2 extraction of canine hind limb during αadrenergic blockade and hypoxic hypoxia, J. Appl. Physiol., 40: 630.

Cain, S.M., 1983, Peripheral oxygen uptake and delivery in health and disease, Clin. Chest Med., 4: 139.

Cain, S.M., and Bredle, D.L., 1990, Actions of a dopaminergic and ß$_2$ adrenergic agonist on O_2 extraction by canine skeletal muscle, Adv. Exp. Med. Biol., 277: 569.

Coburg, A.J., Husen, K., and Picklmayr, J., 1976, Kreislauf reaktionen bei hämodilution, Anaesthesist, 25: 150.

Colvard, M.L., and Longmuir, J.S., 1973, The effects of pacing on oxygen hemoglobin dissociation and oxygen carrying capacity in patients suspected of coronary artery disease, Am. Heart J., 85: 662.

Crystal, G.J., Rooney M.W., and Salem, M.R., 1988, Regional hemodynamics and oxygen supply during isovolemic hemodilution alone and in combination with adenosine induced controlled hypotension, Anesth. Analg., 67: 211.

Curran J.W., Lawrence D.N., Jaffe H., et al., 1984, Acquired immunodeficiency syndrome (AIDS) associated with transfusions, N. Eng. J. Med., 310: 69.

Cutler B.S., 1984, Avoidance of homologous transfusion in aortic operations: the role of autotransfusion, hemodilution and surgical technique, Surg., 95: 717.

Dulling, B.R., Beine R.M., 1973, Longitudinal gradients of periarteriolar oxygen tension, Circ. Res., 27: 669.

Edwards, M.J., and Canon B., 1972, Oxygen transport during erythropoietic response to moderate blood loss, New Eng. J. Med., 287: 115.

Fan, F.C., Chen R.Y.Z., Schuessler G.B., and Chien S., 1980, Effects of hematocrit variations on regional hemodynamics and oxygen transport in the dog, Am. J. Physiol., 238: H545.

Federspiel, W.J. And Sarelius I.H., 1984, An examination of the contribution of red cell spacing to the uniformity of oxygen flux at the capillary wall, Microvasc. Res., 27: 273.

Geha, A.S., 1976, Coronary and cardiovascular dynamics and oxygen availability during acute normovolemic anemia, Surgery, 80: 47.

Gisselson, L., Rosberg, B., and Ericsson, M., 1982, Myocardial blood flow, oxygen uptake and carbon dioxide release of the human heart during hemodilution, Act. Anaesthesiol. Scand., 26: 589.

Glick, G., Plauth, W.H., and Braunwald E., 1964, Role of the autonomic nervous system in the circulatory response to acutely induced anemia in unanesthetized dogs, J. Clin. Invest., 43: 2112.

Jan K.M., and Chien S., 1977, Effects of hematocrit variations on coronary hemodynamics and oxygen utilization, Am. J. Physiol, 233: H106.

Kostuk, W.J., Suwa, K., Berstein, E.P., and Sebel, B.E., 1973, Altered hemoglobin oxygen affinity in patients with acute myocardial infarction, Am. J. Cardiol., 31: 295.

Laks, H., Pilon, R.N., Klovekorn, W.P., et al., 1974, Acute hemodilution: its effect on hemodynamics and oxygen transport in anesthetized man, Ann. Surg., 180: 103.

Lipowsky, H.H., and Zweifach, B.W., 1974, Network analysis of the microcirculation of the cat mesentery, Microvasc. Res., 7: 73.

Malmberg, P.O., Hlastala, M.P., and Woodson, R.D., 1979, Effect of increased blood oxygen affinity on oxygen transport in hemorrhagic shock, J. Appl. Physiol., 47: 889.

Mc Krinan, M.D., White, B.D.G., and Bloor, C.M., 1986, Cardiovascular and metabolic responses to acute and chronic exercise in swine, in: "Swine in Biomedical Research", M.E. Tumbleson, ed., Plenum Press, New York and London.

Mirhashemi, S., Ertefai S., Messmer, K., and Intaglietta, M., 1987, Model analysis of the enhancement of tissue oxygenation by hemodilution due to increased microvascular flow velocity, Microvasc. Res., 34: 290.

Mirhashemi S., Breit, G.A., Chávez, R.H., Intaglietta, M., 1988, Effects of hemodilution on skin microcirculation, Am. J. Physiol., 254: H411.

Messmer, K., Görnandt, L., Jesch, E., Sinagowitz, E., Sunder Plassman, L., and Kessler, M., 1973, Oxygen transport and tissue oxygenation during hemodilution with dextran, Adv. Exp. Med. Biol., 37: 669.

Messmer K., Kreimeier U., and Intaglietta M., 1986, Present state of intentional hemodilution, Eur. Surg. Res., 18: 254.

Mouren, S., Baron, J.F., Hag. B., Arthaud, M., and Viars, P., 1989, Normovolemic hemodilution and lumbar epidural anesthesia, Anesth. Analg., 69: 174.

Nelson, D.P., King, C.E., Dodd, S.L., Schumacker, P.T., and Cain, S.M., 1987, Systemic and intestinal limits of O_2 extraction in the dog, J. Appl. Physiol., 63: 387.

Nöldge, G.F.E., Priebe, H.J., Bohle, W., Buttler, K.J., and Geiger, K., 1991, Effects of acute normovolemic hemodilution on splanchnic oxygenation and on hepatic histology and metabolism in anesthetized pigs, Anesthesiol., 74: 908.

Pavek, K., and Carey, J.S., 1974, Hemodynamics and oxygen availability during isovolemic hemodilution, Am. J. Physiol., 226: 1172.

Perkio, J., and Keskinen, R., 1983, Hematocrit reduction in bifurcations due to plasma skimming, Bull. Math. Biol., 45: 41.

Race, D., Dedichen, H., and Schenk, W.G., 1967, Regional blood flow during dextran induced normovolemic hemodilution in the dog, J. Thorac. Cardiovasc. Surg., 53: 578.

Rand, P.W., 1975, Is hemoglobin oxygen affinity relevant?, JAMA, 66: 5.

Schriemer P.A., Longnecker, D.E., Mints P.D., 1988, The possible immunosuppressive effects of perioperative blood transfusion in cancer patients, Anesthesiology, 68: 422.

Shappell, S.D., Murray, J.A., Masser, M.G., Willis, R.E. Torrance, J.D., and Lenfant, C.J.M., 1970, Acute change in hemoglobin affinity for oxygen during angina pectoris, New. Engl. J. Med., 282: 1219.

Swindle, M.M., 1984, Swine as replacement for dogs in the surgical teaching and research laboratory, Lab. Anim. Sci., 34: 383.

Sylvester J.T., Scharf, S.M., Gilbert, R.D., Fitzgerald, R.S., and Trystman, R.J., 1979, Hypoxic and CO hypoxia in dogs: hemodynamics, carotid reflexes, and cathecholamines, Am. J. Physiol., 236: H23.

Toy P.T., Strauss R.G., Stehling L.C., et al., 1987, Predeposited autologous blood for elective surgery. A national multicenter study, N. Eng. J. Med., 316: 517.

Trouwborst A., van Woerkens E.C.S.M., van Daele M., and Tenbrinck R., 1990a, Acute hypervolaemic haemodilution to avoid blood transfusion during major surgery, Lancet, 336: 1295.

Trouwborst, A., Hagenouw, R.R.P.M., Jeekel, J., and Ong, G.L., 1990b, Hypervolaemic haemodilution in an anaemic Jehovah's Witness, Br. J. Anaesth., 64: 646.

Trouwborst A., Tenbrinck R., Fennema M., Bucx M., v.d. Broek, W.G.M., and Trouwborst-Weber, B.K., 1990c, Cardiovascular responses, hemodynamics and oxygen transport to tissue during moderate isovolemic hemodilution in pigs, Adv. Exp. Med. Biol., 277: 873.

Trouwborst A., Tenbrinck R., and van Woerkens, E.C.S.M., 1990d, Blood gas analysis of mixed venous blood during normoxic acute isovolemic hemodilution in pigs, Anesth. Analg., 70: 523.

Trouwborst, A., Tenbrinck, R., and van Woerkens, E.C.S.M., 1990e, S_{35}: a new parameter in blood gas analysis for monitoring the systemic oxygenation, Scand. J. Clin. Lab. Invest., 50, S203: 135.

Tsai, A.G., Arfors, K.E., and Intaglietta, M., 1990, Analysis of oxygen transport to tissue during extreme hemodilution, Adv. Exp. Med. Biol., 277: 881.

Van Woerkens, E.C.S.M., Trouwborst, A., Duncker, D.J.G.M., Koning, M.M.G., Boomsma, F., and Verdouw, P.D., 1991, Catecholamines and regional hemodynamics during isovolemic hemodilution in anesthetized pigs, J. Appl. Physiol., in press.

VaraThorbeck, R., and GuerreroFernandez Marcote, J.A., 1985, Hemodynamic response of elderly patients undergoing major surgery under moderate normovolemic hemodilution, Eur. Surg. Res., 17: 372.

Vatner, S.F., Higgins, C.B., and Franklin D., 1972, Regional circulatory adjustments to moderate and severe chronic anemia in conscious dogs at rest and during exercise, Circ. Res., 30: 731.

Von Restorff, W., Höfling, B., Holtz, J., and Bassenge E., 1975a, Effect of increased blood fluidity through hemodilution on general circulation at rest and during exercise in dogs, Pflügers Arch., 357: 25.

Von Restorff, W., Höfling, B., Holtz, J., and Bassenge, E., 1975b, Effects of increased blood fluidity through hemodilution on coronary circulation at rest and during exercise in dogs, Pflügers Arch., 357: 15.

Von Borman, B., Weidler, B., Boldt, J., Jooss, D., Aigner, K., Peil, J., and Hempelman, G., 1986, Die akute normovolämische hämodilution bei großen operativen eingriffen, Chirurg, 57: 457.

Waymack J.P., Miskell Ph., and Gonce S., 1989, Alterations in host defence associated with inhalation anesthesia and blood transfusion, Anesth. Analg., 69: 163.

Weaver, M.E., Pantely, G.A., Bristow, J.D., and Ladley, H.D., 1986, A quantitive study of the anatomy and distribution of coronary arteries in swine in comparison with other animals and man, Cardiovasc. Res., 20: 907.

Woodson, R.D., and Auerbach, S., 1982, Effect of increased oxygen affinity and anemia on cardiac output and its distribution, J. Appl. Physiol., 53: 1299.

Yonekawa, H., Berk, J.C., Neuman, H.R., and Liu, C.C., 1981, Tissue hypoxia and increased physiological tissue shunt caused by beta-adrenergic stimulation, Eur. Surg. Res., 13: 325.

Zetterström H., and Wiklund L., 1986, A new normogram facilitating adequate hemodilution, Act. Anaesthesiol. Scand., 30: 300.

SECOND GENERATION FLUOROCARBONS

N. Simon Faithfull

Anesthesiologist and Vice President Medical Research, Alliance
Pharmaceutical Corp., 3040 Science Park Rd, San Diego, CA, USA

FIRST GENERATION PRODUCTS

Fluorocarbons, which should perhaps be more correctly termed
perfluorocarbons or perfluorochemicals (PFCs) are relatively simple
organic compounds in which all hydrogen atoms have been replaced by
fluorine. This statement is not entirely true as PFCs as a broad class of
chemicals also include compounds in which other halogen atoms are present.
PFCs are extensively used in industry because of their inertness, as
convective coolants, electrical insulating agents, and as non-flammable,
chemically stable, lubricants and polymers[1].

PFCs are metabolically inert due primarily to the very high bond
strength of the carbon to fluorine bond -116 kcal/mol[2]. One recent paper
has suggested that PFCs can be metabolized in biological systems [3]. However
the conditions under which the reactions were occurring do not exist in
vivo and hence are irrelevant to this discussion and in no way contradicts
or weakens the conclusions of the hundreds of experiments and clinical
trials which establish the innocuity and chemical inertness of PFCs in
vivo (Riess, 1991, personal communication). This conclusion is supported
by early work by Yokoyama et al[4] demonstrating that though excretion rates
affected the relative amounts of different PFCs present in the liver, no
new gas chromatograph peaks could be detected, indicating no metabolic
processing of the compounds. Additionally no increased fluoride
concentrations were seen in either animals[4] or man[5]. No evidence has been
advanced to show that fluorocarbons themselves, when chosen with due
regard to their vapor pressure and molecular weights, have any more
toxicity than polytetrafluoroethylene - teflon[6].

PFCs are almost completely insoluble in aqueous media and they must
first be emulsified before intravenous injection. This was first achieved
using albumin as the emulsifier[7]. Subsequently PFC emulsions were produced
using pluronic emulsifiers and these were able to support life in animals
with all blood replaced by the emulsion [8]. Following this the Green Cross
Corporation of Osaka, Japan has produced the two PFC emulsions Fluosol®
and Oxypherol (FC- 43).

Fluosol®, a dilute (20% w/v) emulsion of the two PFCs
Perfluorodecalin (FDC) and Perfluoro- tripropylamine (FTPA), entered
clinical trials many years ago and was licensed by the US Food and Drug

Agency (FDA) in 1989 for use during percutaneous transcoronary angioplasty. It has also undergone trials as a blood substitute and is currently in trials as a radio- and chemo-sensitiser in cancer therapy and as an adjuvant in thrombolysis following myocardial infarction; trials are imminent to study its efficacy in the treatment of sickle cell crisis. Fluosol is a 20% weight/volume (w/v) emulsion and its use may be limited by the large volumes that must be administered to achieve adequate oxygen delivery. Due to its limited stability, Fluosol® must be stored frozen and annex solutions must be added immediately before use. Although a number of companies have at one time or another had interest in the medical applications of PFCs, it is believed that only three companies are currently advanced in their development of second generation products.

SECOND GENERATION PRODUCTS

The Green Cross Corporation

The Green Cross Corporation of Osaka, Japan (the manufacturers of Fluosol®) are slowly continuing with development of perfluoro-N-methyldecahydroisoquinoline (FMIQ). This is produced as a 25% (w/v) preparation emulsified in egg yolk phospholipids [9]. The emulsion is neutral and is made isotonic by addition of NaCl. Stability at 5°C is quite good and the particle size only increased from about 0.13 micron to 0.16 micron over 18 months.

Oxygen transport characteristics are similar to those of Fluosol®, the emulsion transporting 5.5 to 6 ml of O2 per 100 ml of emulsion at 760 mm Hg. The whole body half life of FMIQ is said to be less than 10 days [10]. A certain amount of toxicology and pharmacology has been performed with FMIQ but it is understood that development of this emulsion is not being aggressively pursued at the present moment (Hildebrand, personal communication).

HemaGen/PFC

HemaGen of St Louis, Missouri, USA is producing an emulsion of perfluoro-decalin (FDC) emulsified in egg yolk phospholipids. This emulsion contains 40% volume /volume of FDC which comes to approximately 75% w/v. It also contains an oil of undisclosed composition. After storage at 25°C for 6 months the pH decreases from 7.53 to 7.13, though the particle size remains unchanged at 0.2 micron. Gross toxicity remains unchanged with an LD50 of about 50 ml per kg body weight. It is stated that clinical trials will start this year (1991).

Assuming a 75% w/v emulsion and an O2 solubility for FDC of 40.3 ml per 100ml at 760 mm Hg [11], this product should have an oxygen transport capability of approximately 18 ml per 100 ml of emulsion at 760 mm Hg. The whole body half life of FDC is 7 days; that for the oil (assuming it is a high vapor pressure PFC used to prevent Ostwald ripening [12]) will be considerably longer.

Alliance Pharmaceutical Corp.

Perflubron (perfluorooctyl bromide) is in the later stages of development by Alliance Pharmaceutical Corp. of San Diego, California, USA, as a second generation fluorocarbon emulsion. The presence of a single bromide atom at the end of the perfluorinated eight carbon chain confers radiopacity to the compound. This PFC is emulsified with egg yolk phospholipid and can be produced as highly concentrated emulsions [13]. Phospholipids have advantages as emulsifying agents as they have been shown not to induce complement activation [14] and have been in use for many years in parenteral fat emulsions. Stable emulsions of PFOB with concentrations as high as 125% weight/volume have been produced in large volumes [15].

Many preparations of Perflubron are in development, the "standard" emulsion is to be used in oxygen transporting applications as a 90% w/v emulsion. The solubility of oxygen in pure unemulsified Perflubron is 53 ml of O2 per 100 ml at 760 mm Hg; the 90% w/v emulsion, which is known as Oxygent™, will carry 26 ml O2 per 100 ml of emulsion.

Henry's law applies and the oxygen carried by PFCs is directly proportional to the oxygen tension. This is of course not so for hemoglobin, where oxygen uptake and release follows the familiar S-shaped curve. The oxygen content/PO2 relationships for blood, plasma, Fluosol, Oxygent, FMIQ and the HemaGen FDC product are shown for comparison in Fig 1.

At full saturation, 100 ml of normal blood contains approximately 20 ml of oxygen at a PO2 of about 100 mm Hg. The calculated value for Fluosol® at this tension comes to 0.65 ml of oxygen per 100 ml of emulsion. This amount was calculated from the oxygen solubility coefficients and densities of FDC and FTPA. This value does not agree with the 1.0 ml per 100 ml emulsion given by the manufacturers [16], but agrees much better with the measured value of 0.75 ml per 100 ml of emulsion as reported in the literature [17]. This latter value is used in the construction of the solubility line for Fluosol®.

In Fig.1, the vertical dotted line at a PO2 of 45 mm Hg represents mean tissue PO2. Blood, at a PaO2 of 100 mm Hg will deliver about 5 ml of blood while passing through the tissues at a normal cardiac output of 5 l/min. In view of the straight relationship between O2 content of the PFC emulsions and oxygen tension the PaO2 of Fluosol must be raised to about 710 mm Hg to obtain the same (normal) oxygen delivery to the tissues. The figures for the Green Cross FMIQ emulsion, the Hemogen FDC emulsion and the Alliance Perflubron emulsion are 705, 255 and 190 mm Hg.

The above figures assume that cardiac output remains constant. In view of the different viscosities of the various products, this will not be true in practice. As the viscosity for all the products is not known it is impossible to accurately compare them.; nevertheless the calculations taken in conjunction with Fig 1 do serve to indicate relative oxygen transport efficacy of the various emulsions.

In clinical practice, in an individual whose circulation contains both blood and PFCs, high oxygen extraction from the fluorochemical phase leads to far greater oxygen delivery than might have been expected from oxygen transport dynamics. In an early clinical case report, for instance, Fluosol® contributed 27 per cent to total oxygen delivery while at the same time contributing more than 50 per cent of the oxygen consumed [18]. A practical consequence of the administration of PFC emulsions is an increase in mixed venous oxygen tensions and a "sparing" of hemoglobin oxygen delivery. This concept is demonstrated in Fig 2 in which the percentage contribution of hemoglobin is shown for dogs hemodiluted to a hematocrit of 10%. In the control group of dogs 45-50% of oxygen consumption was delivered from hemoglobin, whereas in the dogs given 3 gm/Kg of Perflubron (3.3 ml/kg of 90% Perflubron emulsion) only 10-15 percent of consumption came from hemoglobin.

PFOB for use in emulsions is produced to a purity of 99.9%, in contrast to perfluorodecalin and perfluorotripropylamine, the two PFCs used in Fluosol®, which are of 98% and 96% purity respectively [11]. The major impurities are byproducts of the production process; some processes are much "cleaner than others. Depending on the production methods the major impurities will consist of other fluorocarbon products.

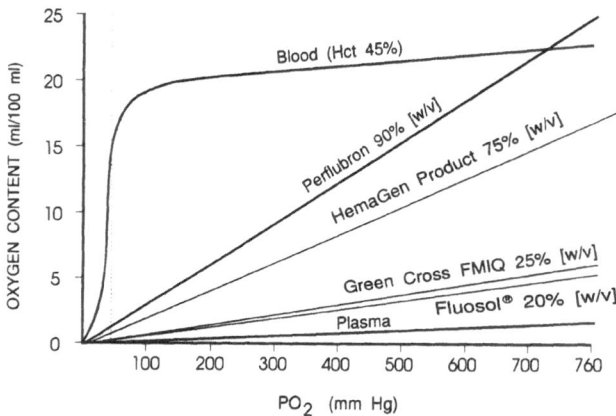

Figure 1. A comparison of the oxygen transport of normal blood with a hematocrit (Hct) of 45%, Plasma and various perfluorocarbon emulsions.

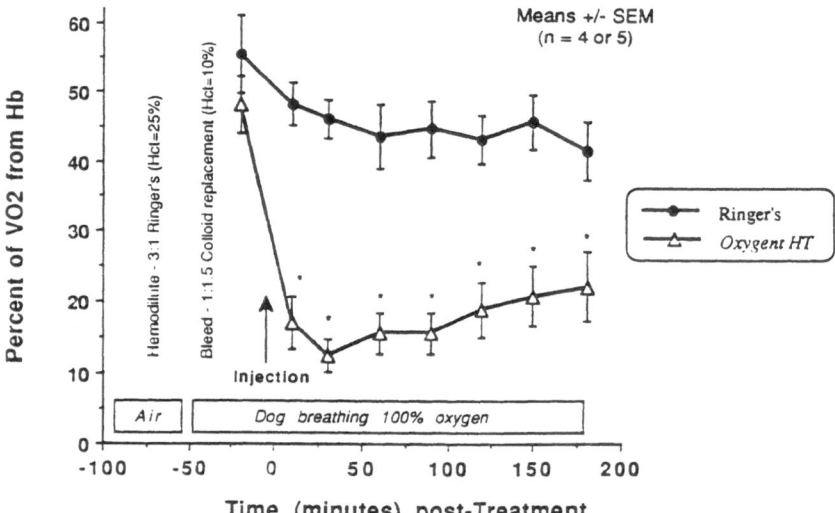

Figure 2. The percentage contribution of hemoglobin to oxygen consumption in dogs hemodiluted to an Hct of 10% and given 3 gm/kg perflubron (Oxygent™) or similar volumes of saline.

The primary pathway for elimination of PFCs is via the expired air and possibly to a very minor degree via transpiration through the skin [16, 19, 20, 21, 22]. Excretion in urine and feces is practically nil. In comparison to the PFCs in Fluosol® and FC-43, Perflubron has a very advantageous excretion rate combined with good emulsion stability due to the lipophilic character of its terminal bromine atom [6]. PFC particles circulating in the blood are taken up in the reticuloendothelial system (RES) and remain there depending on their vapor pressure (VP). The rate of excretion is directly proportional to the molecular weight and VP, the higher the VP the higher the excretion rate of a particular PFC. This RES uptake is a normal body mechanism for clearance of particulate materials and is seen after administration of fat emulsions, liposomes and PFC emulsions (Fluosol®, Oxypherol and others). The whole body half lives of FDC and FTPA, the two PFCs present in Fluosol® are 7 and 65 days respectively [4]; FDC is the primary PFC in the HemaGen emulsion. The half life of PFOB is 4 days.

APPLICATIONS OF SECOND GENERATION PRODUCTS

Though there are three main development products in the field of second generation PFC compounds, little is known of their preclinical development. The author of this article contacted representatives from both The Green Cross Corporation and HemaGen/PFC to ask for data to put in this paper; they were both unwilling to release information.

Radiological Contrast

Neat (unemulsified) PFOB has been used clinically for bronchography and alveolography [23] and is currently in clinical trial as a gastrointestinal contrast agent. Due to the low viscosity and high density of the agent gastrointestinal passage is rapid and lower bowel images can be obtained within 20-40 minutes.

Due to its lack of hydrogen, Perflubron gives no signal on hydrogen weighted Magnetic Resonance Imaging (MRI). It can also be used in its unemulsified state as a negative contrast agent in the gastrointestinal tract and can be used to distinguish bowel from other areas of interest in MRI scanning of the abdomen and pelvis [24, 25]. Full clinical trials have been performed and FDA approval is expected later this year (1991). Fluorine can also be detected on nuclear magnetic imaging though the signal strength obtained is not as strong as that from hydrogen. It is modulated by the presence of O2 (a paramagnetic molecule) and hence F-19 MRI imaging can be used to detect tissue oxygenation [26, 27] and to estimate tumor blood flow [28, 29].

As with all intravenously injected particulates, Perflubron emulsion particles are taken up in the reticuloendothelial system. It thus initially serves as a computer tomographic (CT) blood pool imaging agent and later as a liver and spleen contrast agent. It is thus an excellent agent for clinical identification of secondary malignant tumors in the liver [30, 31]. PFCs are selectively taken up by certain tumors outside the RES including cerebral neoplasms [32]. When PFOB is used high radiographic definition can be obtained. Osmotic gradients in the kidney can be identified on CT [33].

Owing to their compressibility, PFCs are good ultrasound contrast agents - PFOB is particularly good in this respect and development of special emulsions for this purpose is underway. These can be used to evaluate cardiac and vascular flow, detect tumors [34] and abscesses, assess myocardial infarction volume [35] and to assess renal function [36, 37]. Perflubron emulsions have been evaluated experimentally for lymphography [38]. However, when PFOB emulsions are injected subcutaneously or intramuscularly, the PFC particles are picked up in the lymphatics and transported to lymph nodes. These can then be imaged using CT [39]. Special emulsions are also being developed for this purpose.

Oxygen Transport

Perflubron emulsion at a concentration of 90% (w/v), and at relatively low doses, has been shown to be an effective deliverer of oxygen to tissues. In a dog gastrocnemius preparation, paced to produce maximum possible work, maximum oxygen consumption was increased by about 11% [40]. This observation is important because it demonstrates perflubron's ability to supplement oxygenation of a maximally stressed tissue.

In a dog model in which hemodilution was performed to an hematocrit of about 10 percent, 3.3 ml/kg of the same emulsion was able to initially supply just under 30% of oxygen consumption. This resulted in a 60% decrease in the amount of oxygen delivered from hemoglobin and the build up of a considerable safety "reserve" of oxygen carried in the red cell. Concomitantly mixed venous oxygen tensions increased significantly (Fig. 3), implying improved oxygenation at the tissue level.

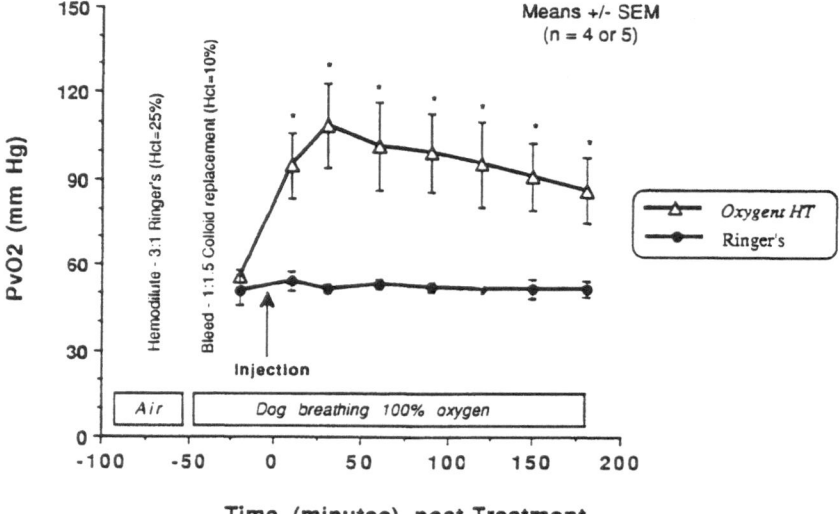

Figure 3. Changes in mixed venous oxygen tension (PvO2) in dogs hemodiluted to an Hct of 10% and given 3 gm/kg perflubron (Oxygent™) or similar volumes of saline.

In addition to the ability of Oxygent to transport and deliver oxygen to the tissues by reason of its high solubility for gases, there are indications that it may also, by providing increased oxygen content in the plasma phase, cause additional facilitation of oxygen diffusion to the tissues [41].

446

Salvage of Ischemic Tissues

The viscosity of blood varies with shear rate and increases very markedly under the low flow rates existing in the microcirculation. The viscosity of dilute PFC emulsions is almost independent of shear rate. This has been used as one explanation for the agents beneficial effects in supporting an ischemic microcirculation; this may not be as important as previously thought [42]. The small PFC particle size in emulsions is of considerable importance in ischemic hypoxia. The particles in Fluosol® are 0.1 micron in diameter [16] in PFOB they have a diameter of 0.25 micron. It used to be thought that increasing particle size correlated with increased toxicity and decrease in the LD50 [43]. This relationship does not hold good for PFOB emulsions and emulsion particles of more than one micron in diameter do not result in decreased LD50. Indeed experimental large particle ultrasound emulsions may have a higher LD50 than the "standard" emulsions.

PFCs may also improve ischemic hypoxia by their actions on leukocytes because fluorocarbon emulsions can inhibit neutrophil phagocytosis, chemotaxis and superoxide release [44, 45, 46]. This may explain the decreased inflammatory response and increased myocardial salvage observed following early post infarction hemodilution with PFCs [47] and their protection against the reperfusion syndrome [46, 48, 49]. Specially formulated Perflubron emulsions are showing promise as warm reperfusion agents following 12 hours storage of explanted hearts 4°C. This results in significantly better ventricular function than treatment with whole blood, blood cardioplegic solution, leukocyte depleted blood or Fluosol® and not significantly different from control, unstored hearts (Martin, personal communication)

Radiotherapy and Chemotherapy

A large amount of work has been performed into the use of concentrated perflubron emulsions as adjuncts to radiotherapy of tumors. Enhancement of tumor growth delay following single dose radiotherapy increased with increasing perflubron dose; further marked improvement was seen following daily administration of small doses of Perflubron [50].

Perflubron emulsions have been shown to increase the proportion of well oxygenated cells in tumors as evidenced by both tumor cell survival data and polarographic measurements of intratumoral PO2 [51]. Hyperbaric oxygenation has also been tested in combination with Perflubron emulsion and has been shown to cause further tumor sensitization to radiation. This can be achieved with no increase in radiation toxicity to bone marrow or skin, thus providing useful enhancement of therapeutic ratio [52]. Furthermore it was demonstrated that smaller doses of perflubron were necessary to produce maximal effects under HBO than under normobaric conditions.

Work has been performed to assess the effect of Perflubron emulsions in sensitizing the chemotherapeutic effect of a number of alkylating agents. The optimal dose varied with the particular drug, but the effects were maximal when the agents were prepared in the emulsion [53].

Liquid Breathing

Following the initial demonstration of the ability of small mammals to remain alive while breathing oxygenated PFCs [54] a lot of work has been performed to study the physiology of liquid breathing. Much of this work has dealt with premature animals [55,56] and has involved ventilation with PFCs oxygenated outside the body. Premature infants have been successfully ventilated using these techniques [57]. Recently it has been shown that gas ventilation after the lungs had been filled to functional residual capacity or less was quite feasible. These techniques could maintain normal gas exchange with virtually normal lung mechanics in normal animals [58]. Perflubron could also greatly increase oxygenation and reduce inflation pressures in animals with respiratory distress caused by surfactant washout [59,60]

The mechanism of action of PFC liquid breathing in premature infants is the reduction of the high alveolar/air surface tensions existing in the absence of surfactant. A secondry action probably involves splinting of alveoli be the PFC, thus preventing their collapse. It is anticipated that partial liquid breathing utilizing gas ventilators, in contrast to total liquid breathing that needs a liquid ventilator and oxygenator, will rapidly enter medical practice as a preatment for both adult and neonatal respiratory distress syndromes.

Perflubron is currently in clinical trials for radiology and should shortly enter clinical studies for use during perioperative hemodilution; its use as a radiosensitiser in cancer therapy or as a treatment for respiratory distress syndromes may not be far off.

REFERENCES

1. Slinn DSL, and Green SW, "Fluorocarbon fluids for use in the electronics industry.," Preparation, properties and industrial applications of organofluoride compounds., Ed. Banks RE (Chichester: Ellis Horwood, 1982) 45-82.
2. Sargent JW, and Sefel RJ, Properties of perfluorinated liquids, Fed Proceed 1970 29.5 (1970): 1699-1703.
3. MacNicol DD, and Robertson CD, New and unexpected reactivity of saturated fluorocarbons, Nature 332 (1988): 59-61.
4. Yokoyama K, Yamanouchi K, Ohyanagi H, and Mitsuno T, Fate of perfluorochemicals in animals after intravenous injection of hemodilution with their emulsions, Chem Pharm Bull 26.3 (1978): 956-966.
5. Sada T, Tohyama Y, Aizawa Y, and Murakami S, "Plasma Fluoride concentration and urinary fluoride excretion following Fluosol-DA administration in man," Oxygen Carrying Colloidal Blood Substitutes (Mainz, March 1981), Ed. H. B. a. K. S. R Frey (Munchen: Zuckschwerdt, 1982) 225-229.
6. Riess JG, Blood substitutes: where do we stand with the fluorocarbon approach?, Current Surgery 45.5 (1988): 365-370.
7. Sloviter HA, "Perfusion of the brain and other isolated organs with dispersed perfluoro compounds," Blood Substitutes and Plasma Expanders, (New York: Alan R Liss, 1978) 28-39.
8. Geyer RP, Monroe RG, and Taylor K, "Survival of rats totally perfused with a fluorocarbon-detergent preparation," Organ Perfusion and Preservation, Ed. JC Norman 1968) 85-96.
9. Mitsuno T, Ohyanagi H, Yokoyama K, and Suyama T, Recent studies on Perfluorochemical emulsion as an oxygen carrier in Japan, Biomater Artif Cells Artif Organs 16.1-3 (1988): 365-374.

10. Clark LC, Clark EW, Moore RE, Kinnett DG, and Inscho EI, "Room temperature-stable biocompatible fluorcarbon emulsions," Prog Clin Biol Res, (1982) 122: 169-180.

11. Riess JG, and Le Blanc M, Solubility and transport phenomena in perfluorocehmicals relevant to blood substitution and other biomedical applications, Pure & Appl Chem 54.12 (1982): 2383-2406.

12. Sharma SK, Davis SS, and Lowe KC, (Abstract) Novel perfluorochemical emulsions for biological uses, Biomater Artif Cells Artif Organs 15.2 (1987): 432.

13. Arlen C, Follana R, Le Blanc M, Long C, Long D, Riess JG, and Valla A, Formulation of highly concentrated fluorocarbon emulsions and assessment by near-total exchange perfusion of the conscious rat, Biomater Artif Cells Artif Org 16.1-3 (1988): 455-457.

14. Hammerschmidt DE, and Vercellotti GM, Limitation of complement activation by perfluorocarbon emulsions: Superiority of lecithin-emulsified preparations, Biomater Artif Cells Artif Org 16.1-3 (1988): 431-438.

15. Long DM, Long DC, Mattrey RF, Long RA, Burgan AR, Herrick WC, and Shellhamer DF, An overview of perfluoroctylbromide - Application as a synthetic oxcygen carrier and imaging agent for X-ray, ultrasound, and nuclear magnetic resonance, Biomater Artif Cells Artif Organs 16.1-3 (1988): 411-420.

16. Naito R, and Yokoyama K, Perfluorochemical Blood Substitutes Fluosol-43, Fluosol-DA, 20%, and 35% for preclinical studies as a candidate for erythrocyte substitution, Green Cross Corporation, 1978)

17. Zander R, and Makowski HV, "Life without hemoglobin?," Oxygen Carrying Colloidal Blood Substitutes (Mainz, March 1981), Ed. H. B. a. K. S. R Frey (Munchen: Zuckschwerdt, 1982) 133-141.

18. Tremper KK, Lapin R, Levine E, Friedman A, and Shomaker WC, Hemodynamic and oxygen transport effects of a perfluorochemical blood substitute, Fluosol-DA (20%), Crit Care Med 8.12 (1980): 738-741.

19. Biro GP, Blais P, and Rosen A, Perfluorocarbon Blood Substitutes, CRC Critical Reviews in Oncology/Hematology 6.4 (1987): 311-374.

20. Lowe KC, Perfluorocarbons as oxygen-transport fluids, Comp Biochem Physiol 87/a.4 (1987): 825-838.

21. Faithfull NS, Fluorocarbons, Anaesthesia 42 (1978): 234-242.

22. Pillai R, Bando K, Schueler S, Zebley M, Reitz BA, and Baumgartner WA, Leukocyte depletion results in excellent heart-lung function after 12 hours of storage, Ann Thorac Surg 50 (1990): 211-214.

23. Liu MS, and Long DM, Biological disposition of perfluoroctylbromide: Tracheal administration in alveolography and bronchography, Investigative Radiology 11 (1976): 479-485.

24. Mattrey RF, Trambert MA, Brown J, Bruneton JN, Young S, Kortman K, Wesby G, and Schooley G, Summary of phase III clinical trials of imagent GI (PFOB) as a negative oral MR contrast agent, (RSNA) Radiologic Society of North America, (Chicago, IL: 1990)

25. Rubin DL, Muller HH, Nino-Murcia M, Sidhu M, Christy V, and Young S, Intraluminal contrast enhancement and MR visualization of the bowel wall: Efficacy of PFOB, JMRI 1 (1991): 371-380.

26. Delpuech J-J, Hamza MA, and Serratrice G, Determination of oxygen by a nuclear magnetic resonance method, <u>Journal of Magnetic Resonance</u> 36 (1979): 173-179.

27. Reid RS, Koch CJ, Castro ME, Lunt JA, Treiber EO, Boisvert DJP, and Allen PS, The influence of oxygenation of the 19F spin-lattice relaxation rates of Fluosol-DA, <u>Phys Med Biol</u> 30.7 (1985): 677-686.

28. Authier B, Reactive hyperemia monitored on rat muscle using perflurorocarbons and 19F NMR, <u>Magnetic Resonance in Medicine</u> 8 (1988): 80-83.

29. Ceckler TL, Gibson SL, Hilf R, and Bryant R, In situ assessment of tumor vascularity using fluorine NMR imaging, <u>Magnetic Resonance in Medicine</u> 13 (1990): 416-433.

30. Bruneton JN, Falewee MN, Francois E, Cambon P, Philip C, Riess JG, Balu-Maestro C, and Rogopoulos A, Liver, spleen, and vessels: Preliminary clinical results of CT with perfluoroctylbromide, <u>Radiology</u> 170 (1989): 179-183.

31. Behan M, O'Connell D, Mattrey RF, and Carney DN, Perfluorooctylbromide (PFOB) as a contrast agent for sonography and computed tomography, <u>Invest Radiol</u> Submitted (1991):

32. Patronas N, Miller DI, and Girton M, Experimental comparison of EOE-13 and perfluoroctylbromide for the CT detection of hepatic metastases, <u>Invest Radiol</u> 19 (1984): 570-573.

33. Coley BD, Mattrey RF, Mitten RM, and Peterson T, The physiologic basis of the radiodense renal medulla after the administration of blood pool contrast agent PFOB, <u>Invest Radiol</u> 25 (1990): 1287-1293.

34. Mattrey RF, Scheible FW, Gosink BB, Leopold GR, Long DM, and Higgins CB, Perfluoroctylbromide: A liver/spleen-specific and tumor-imaging ultrasound contrast material, <u>Radiology</u> 145 (1982): 759-762.

35. Mattrey RF, and Andre MP, Ultrasonic enhancement of myocardial infarction with perfluorocarbon compounds in dogs, <u>Am J Cardiol</u> 54 (1984): 206-210.

36. Coley BD, Mattrey RF, Roberts A, and Keane S, Potential role of PFOB enhanced sonography of the kidney. II. Detection of partial infarction., <u>Kidney International</u> 39 (1991): 740-745.

37. Munzing D, Mattrey RF, Reznik VM, Mitten RM, and Peterson T, Potential role of PFOB enhanced sonography of the kidney. I. Detection of renal function and dacute tubular necrosis., <u>Kidney Inter</u> 39 (1991): 733-739.

38. Liebner EJ, Evaluation of perfluoroctylbromide for lymphography in the dog: Comparison with ethiodol, <u>Investigative Radiology</u> 12 (1977): 368-372.

39. Wolf GL, Long D, and Reiss J, (Abstract) Percutaneous lymphography with PFOB emulsions, <u>RSNA</u>, (Chicago: 1990)

40. Hogan MC, Willford D, Faithfull NS, and Wagner PD, Effect of low dose Oxygent added to blood on muscle VO2 max, <u>Adv Exp Med Biol</u>, 1991) In Press:

41. Faithfull NS, Perfluorocarbons as blood substitutes - Aspects of convective and diffusive oxygen delivery, <u>VIII World Congress of the International Society for Artificial Organs in conjunction with the IV International Symposium on Blood Substitutes</u>, Aug 19-22, 1991, (Montreal, Quebec: 1991) In Press:

42. Faithfull NS, Fluorocarbons - Current status and future applications, _Anaesthesia_ 42.3 (1987): 234-242.

43. Mitsuno T, Ohyanagi H, and Yokoyama K, Development of a perfluorochemical emulsion as a blood gas carrier, _Artif Organs_ 8.1 (1983): 25-33.

44. Virmani R, Fink LM, Gunter K, and English D, Effect of perfluorochemical blood substitutes on human neutrophil function, _Transfusion_ 24 (1984): 343-347.

45. Virmani R, Warren D, Rees R, Fink LM, and English D, Effects of perfluorochemical on phagocytic function of leukocytes, _Transfusion_ 23 (1983): 512-515.

46. Virmani R, Osmialowski AF, Kolodgie FD, and Forman MB, The effect of perfluorochemical Fluosol-DA (20%) on myocardial infarct healing in the rabbit, _Am J Cardio Path_ 3.1 (1990): 69-80.

47. Kolodgie FD, Dawson AK, Forman MB, and Viramani R, Effect of perfluorochemical (Fluosol-DA) on infarct morphology in dogs with permanent coronary artery occlusion, _Virchow Arch B_ 50 (1985): 119-134.

48. Forman MB, Bingham S, Kopeman HA, Wehr C, Sandler MP, Kolodgie F, Vaughn WK, Friesinger GC, and Virmani R, Reduction of infarct size with intracoronary perfluorochemical in a canine preparation of reperfusion, _Circulation_ 71.5 (1985): 1060-1068.

49. Parrish MD, Olson RD, Mushlin PS, Artman M, and Boucek RJ, Treatment of postischemic reperfusion cardiac injury with a perfluorochemical solution, _J Cardiovasc Pharmacol_ 6.1 (1984): 159-164.

50. Herman TS, and Teicher BA, Enhancement of radiation therapy by an experimental concentrated perfluorooctylbromide (PFOB) emulsion in the Lewis lung carcinoma, _VIII World Congress of the International Society for Artificial Organs in conjunction with the IV International Symposium on Blood Substitutes_, Aug 19-22, 1991 , (Montreal, Quebec: 1991) In Press:

51. Rockwell S, Kelley M, Irvin CG, Hughes CS, Porter E, Yabuki H, and Fischer JJ, Modulation of tumor oxygenation and radiosensitivity by a perfluorooctylbromide emulsion, _Radiother & Oncol_ In Press (1991):

52. Rockwell S, Irvin CG, Kelley M, Hughes CS, Yabuki H, Porter E, and Fischer JJ, Effects of Hyperbaric Oxygen and a Perfluorooctylbromide Emulsion on the Radiation Responses of Tumors and Normal Tissues in Rodents, _Int J Rad Oncol Biol and Phy_ In Press (1991):

53. Holden SA, Teicher BA, and Herman TS, Effect of a PFOB emulsion and carbogen breathing on the tumor cell survival of the FSaIIC fibrosarcoma after treatment with antitumor alkylating agents, _VIII World Congress of the International Society for Artificial Organs in conjunction with the IV International Synmposium on Blood Substitutes_, Aug 19-22, 1991, (Montreal, Quebec: 1991) In Press:

54. Clark LC, and Gollan F, Survival of mammals breathing organic liquids equilibrated with oxygen at atmospheric pressure, _Science_ 152 (1966): 1755-1756.

55. Shaffer TH, A brief review: liquid ventilation, _Undersea Biomed Res_ 14.2 (1987): 169-179.

56. Curtis SE, Fuhrman BP, and Howland DF, Airway and alveolar pressures during perfluorocarbon breathing in infant lambs, _J Appl Physiol_ 68.6 (1990): 2322-2328.

57. Greenspan JS, Wolfson MR, Rubenstein D, and Shaffer TH,
     Liquid ventilation of human preterm neonates, <u>J. Pediatr</u>
     117, (1990): 106-111.
58. Fuhrman BP, Paczan PR, DeFrancisis M, Perfluorcarbon-
     associated gas exchange, <u>Crit Care Med</u> 19 (1991): 712-
     722. 73-74.
59. Lachman B, Tutuncu AS, Bos JAH, and Faithfull NS,
     Intratracheal Perfluorooctylbromide (PFOB) in
     combination with Mechanical Ventilation, <u>Adv Exp Med</u>
     <u>Biol</u>, (1991) In Press:
60. Tutuncu AS, Lachman B, and Faithfull NS, I n t r a t r a c h e a l
     perfluorooctylbromide (PFOB) Instillation, <u>Adv Exp Med</u>
     <u>Biol</u>, (1991) In Press:

PROPERTIES OF CHEMICALLY CROSS-LINKED HEMOGLOBIN

SOLUTIONS DESIGNED AS TEMPORARY OXYGEN CARRIERS*

Peter E. Keipert

Department of Pharmacology, Alliance Pharmaceutical Corp.
3040 Science Park Road, San Diego, CA 92121

INTRODUCTION

Hemoglobin (Hb) is a ubiquitous oxygen-carrying molecule found throughout nature in almost all animal species. In humans, it exists as a 64,000 Dalton tetrameric protein comprised of two α-globin chains (141 amino acids each) and two β-globin chains (146 amino acids each), and is found within red blood cells at an extremely high concentration (32 g Hb/dl). Also present in red cells in equimolar amounts is 2,3-diphosphoglycerate (2,3-DPG), which interacts with the Hb as an allosteric effector to lower the affinity of Hb for oxygen (*Benesch & Benesch, 1969*).

When Hb is free in solution, affinity for oxygen is increased (due to the absence of 2,3-DPG), and the Hb tetramer dissociates spontaneously into 32 kDa dimers. Any free Hb resulting from the destruction of aging red cells (normally < 0.6 mg Hb/dL blood) is first picked up by the plasma protein haptoglobin, which can handle about 150-200 mg Hb/dL blood (*Lathem, 1959*). When plasma Hb levels exceed this binding capacity, glomerular filtration occurs, followed by reabsorption of some of the filtered Hb (up to 5 mg Hb/hr) in the proximal tubular cells of the kidney. Hemoglobinuria therefore only occurs when enough free Hb is present in the plasma to exceed these two processes (*Bunn et al., 1969*).

The controversy over Hb toxicity can be traced back to very early studies using crude red blood cell hemolysates for resuscitation. Intravascular coagulation following injection of hemolyzed blood was first reported over 120 years ago, causing some investigators at that time to conclude that the purpose for the red cell was to protect us from the toxic hemoglobin inside (*Naunya, 1868; Ponfick, 1875*). The vasoconstrictor effect of dissolved Hb was first demonstrated to be localized in the kidneys of dogs (*Mason & Mann, 1931*), and was later confirmed by Amberson, despite his attempts to remove the cell stroma by centrifugation (*Amberson et al., 1933*). Other studies, however, have not reported any obvious toxic side-effects (*Carrier et al., 1922*), and in the earliest clinical trial using human Hb in 1916, no permanent injurious side effects were observed (*Sellards & Minot, 1916*).

From later studies, it became clear that the stromal lipid contamination in earlier Hb preparations was the most likely cause of the acute toxicity observed following the infusion of these Hb solutions (*Hamilton et al., 1947; Amberson et al., 1949; Rabiner et al., 1967*). Many techniques such as controlled hypotonic lysis, high-speed centrifugation, microfiltration, crystallization, acidification, column chromatography, and organic solvent extractions

---

* Parts of this work were done at: (1) the Letterman Army Institute of Research while Dr. Peter E. Keipert held a LAIR Senior Research Associateship awarded by the U.S. National Research Council; and at (2) the College of Physicians &Surgeons of Columbia University while Dr. Peter E. Keipert held a MRC Fellowship awarded by the Medical Research Council of Canada.

have been used to varying degrees of success for preparing stroma-free hemoglobin (SFHb) solutions. To achieve the maximum purity, most current techniques for preparing SFHb include a combination of at least 2-3 of these methodologies.

Unmodified SFHb has two major shortcomings when used as an oxygen carrier and a blood substitute: (1) the intravascular retention time is too short (plasma $T_{1/2} < 3$ hours) due to rapid glomerular filtration of the 32 kDa Hb dimers (which results from the spontaneous dissociation of Hb tetramers); and (2) the affinity for oxygen is too high ($P_{50} = 10$ mmHg) compared to whole blood ($P_{50} = 27$ mmHg) due to the absence of 2,3-DPG. To prevent renal filtration of Hb, *Bunn et al. (1969)* cross-linked Hb using bis(N-maleimidomethyl)ether (BME). The ability to influence the oxygen affinity of Hb by chemical modification was the discovery of *Benesch et al. (1972)*, who were the first to demonstrate that a 2,3-DPG analogue, pyridoxal 5'-phosphate (PLP), could bind to deoxy Hb within the 2,3-DPG binding site and lower the affinity for oxygen effectively. Pyridoxylation of Hb remains one of the most widely used initial modification steps in the production of many polymerized or conjugated Hb derivatives. Over the past 20 years, significant efforts have focused on the development of different cross-linking reagents which will both prevent the spontaneous dimerization of Hb and reduce the oxygen affinity (*Winslow, 1989*).

A number of major commercial ventures are currently in the process of developing different chemically modified Hb products for future use as "blood substitutes". Northfield Labs, Chicago, IL, is polymerizing pyridoxylated-human Hb with glutaraldehyde; Biopure Corp., Boston, MA, is testing glutaraldehyde polymerized bovine Hb; while Hemosol Inc., Toronto, Ontario, is polymerizing human Hb using a novel polyaldehyde cross-linker derived from periodate-oxidized raffinose. Baxter Healthcare Corp., Chicago, IL, has been testing Hb cross-linked interdimerically with bis(3,5-dibromosalicyl) fumarate (DBBF); Enzon Inc., Plainfield, NJ, is conjugating bovine Hb to polyethylene glycol (PEG), and Ajinomoto Company in Japan is testing pyridoxylated-Hb conjugated to polyoxyethylene; while Somatogen Inc., Boulder, CO, has recently produced intramoleculary cross-linked human Hb in *E. coli* using recombinant techniques. Collaborative efforts between the Hb companies (to provide large scale quantities of pure Hb product) and academia (to evaluate these Hb solutions in appropriate animal models) will be required to assure the successful and timely development of a clinically useable product (*Winslow, 1991*).

In the past, numerous animal studies using massive exchange transfusion models have demonstrated the ability of Hb solutions to supply oxygen in the absence of adequate red cells (*Amberson et al., 1934; Moss et al., 1976; DeVenuto et al., 1977; Keipert & Chang, 1987*). Many of the commercial efforts have yielded very promising Hb derivatives, but to date, their main focus has been on production and manufacturing. When two different glutaraldehyde-polymerized Hb's (one from Northfield and one from Biopure) were tested in human clinical trials in the USA recently, some unexplained side-effects occurred that caused both trials to be halted (*Pool, 1990*). It is still not clear whether the side-effects which have been observed in animal studies and in human clinical trials have been due to a toxic contaminant or to the Hb itself (*Winslow, 1991*). Consequently, the clinical safety of chemically modified Hb solutions is still being questioned by the U.S. Food and Drug Administration (*CBER, 1991*), and significant research efforts are still needed to resolve the issue of Hb toxicity (*Moffat, 1991*).

This report will describe some of the physico-chemical characteristics of different chemically modified Hb derivatives, and will present data to show how these properties can affect the *in vivo* pharmacokinetics of plasma Hb clearance, tissue distribution, and renal clearance. Much of the data to be presented was collected in recent ADME (adsorption, distribution, metabolism, and excretion) studies in rats (*Keipert & Triner, 1989; Keipert et al., 1989a; 1990a; 1991a; 1991b; 1991c*), using a variety of radiolabeled Hb derivatives.

MATERIALS and METHODS

Intramolecular Cross-linking of Hemoglobin

Stroma-free Hb (SFHb) was prepared from washed red blood cells by osmotic lysis in hypotonic buffer. Techniques used to purify SFHb included centrifugation, ultrafiltration, and chromatography. In all cases, the SFHb was reacted under deoxygenated conditions (using nitrogen or argon gas) to stabilize Hb in the "T-state" (lower oxygen affinity). Pyridoxylated-Hb (PLPHb) was prepared according to *Benesch et al. (1972)*. The dialdehyde derivative, 2-nor-2-formylpyridoxal 5'-phosphate (NFPLP) was synthesized as described (*Benesch & Benesch, 1981*), and then reacted with Hb to yield NFPLPHb (covalently cross-linked between the β-chains). Bis-pyridoxal tetraphosphate [(bis-PL)P_4], synthesized according to *Benesch & Kwong (1988)*, was reacted with SFHb to yield (bis-PL)P_4Hb (cross-linked covalently between the β-chains). The Schiff's base formed during these reactions could then be reduced in a Tris buffer containing ^{3}H-labeled $NaBH_4$ to label the final cross-linked Hb derivatives (*Keipert et al. 1990b*).

Bis-(3,5-dibromosalicyl) fumarate (DBBF) was synthesized as described (*Chatterjee et al., 1984*), but using ^{14}C-fumaric acid, and then reacted with deoxy Hb (37°C, 2 hours). The resulting ^{14}C-labeled ααHb was pasteurized (75°C, 90 minutes), which both sterilizes the Hb by inactivating viruses (*Estep et al., 1989*), and purifies by precipitating the less stable noncross-linked Hb. This ^{14}C-ααHb was purified by centrifugation, filtration, and mixed-bed ion-exchange chromatography as described elsewhere (*Keipert et al., 1991b*).

Polymerization of Hemoglobin

Glutaraldehyde is the reagent most commonly used for polymerizing pyridoxylated-Hb (PLPHb, prepared as above). This reaction must be monitored either by chromatography, or by the changes in colloid osmotic pressure (*Keipert & Chang, 1984*), and requires that the Hb first be reacted with PLP or with NFPLP in order to achieve a reasonably low oxygen affinity in the final product. The glutaraldehyde polymerization of Hb is a rather nonspecific reaction, making batch to batch reproducibility difficult to achieve. The characterization of the final PLPpolyHb solution has been reported previously (*Keipert & Chang, 1984*). When bovine Hb is used as the starting material, a pyridoxylation step is not necessary, since the intrinsic P_{50} of bovine Hb (at physiological Cl^- concentrations) is significantly higher ($P_{50} >40$ mmHg) than unmodified human Hb (P_{50} <15 mmHg) under similar conditions (*Bucci et al., 1988*). The polymerized bovine Hb (B-polyHb) tested in these studies was kindly supplied by Bio-Vita Ltd., Guatemala (X-Polyheme).

A rather unique Hb cross-linking and polymerizing reagent has been developed recently by *Hsia (1989)*. This reagent was derived from the periodate-oxidation of raffinose, and did not require a pyridoxylation step to lower the O_2 affinity of the final product. This o-raffinose polymerized Hb (oRpolyHb) was specially prepared using ^{3}H-labeled raffinose, and was kindly provided by Hemosol Inc., Toronto, Ontario. The ^{3}H-oRpolyHb was fractionated into single peaks of ^{3}H-labeled components by size-exclusion preparative HPLC. Four major ^{3}H-labeled cross-linked fractions -- (I) MW=64 kDa; (II) MW=128 kDa; (III/IV) MW=192-256 kDa; and (V) MW=>576 kDa -- were then tested individually (in rats) in the presence of the nonlabeled oRpolyHb solution (*Keipert et al., 1990a; 1992a; 1992b, 1992c; Hsia et al. 1991*).

Prior to injection into rats, final Hb concentrations were adjusted by dilution in a balanced electrolyte solution to yield normal physiological colloid osmotic pressure (COP), measured with either a Weil (IL 186) or a Wescor (Model 4400) Oncometer. Endotoxin levels (EU/ml) were measured by the LAL gel-clot assay (Whittaker). The oxygen affinity (P_{50} value) was estimated from oxyHb dissociation curves, generated under standard conditions (37°C, pH 7.4, pCO_2=40 mmHg) using either a Hem-O-Scan (Aminco) or Hemox analyzer. Details for analytical methods have been described (*Keipert & Chang, 1984; Keipert et al., 1991b; 1992a*).

Animal Studies:

All Hb solutions were tested in male Sprague-Dawley rats (250-350 g) which had specially designed femoral arterial and venous catheters implanted chronically several days prior to treatment. The catheters were passed subcutaneously to the tail, exteriorized and protected by a special tail sheath assembly. The design of the catheters and the plastic tail sheath, described previously (*Keipert & Chang, 1987*), were both modified to assure patency for up to 21 days (*Keipert & Gonzales, 1991*). At least 3-4 days were allowed for recovery from surgery to ensure that body weight, hematocrits, eating and drinking behavior, and urine outputs were all normal. Treatment consisted of an isovolemic exchange transfusion (ET) using a precisely calibrated two-channel peristaltic pump. All rats were fully conscious during treatment and throughout the acute 10 hour post-ET monitoring phase, and were held in a customized minimal restraint cage (*Keipert & Chang, 1987; Keipert et al., 1991b*). Arterial blood pressure and heart rates were monitored continuously for 10 hours post-ET, via a Stathem pressure transducer (P-23XL) connected to a Gould chart recorder (Model 2200S).

Urine samples were collected, sterile filtered, and stored at 4°C. A continuous Ringer's-lactate infusion was administered intravenously (i.v.) at 3.0 ml/kg/hr to replace sampling volumes and maintain a normal state of hydration. Plasma and urine were collected and analyzed for Hb content by Drabkin's assay (at 540 nm), and for radioactivity levels by liquid scintillation counting (LSC). Some urine samples containing Hb were also analyzed by SDS-PAGE and by reversed-phase HPLC, as described elsewhere (*Keipert et al., 1991b*). To determine the distribution and metabolic fate of the different Hb derivatives, a total of 13 different tissues were harvested at time points ranging from 1 hr to 14 days post-ET, and were processed for scintillation counting by methods published elsewhere (*Keipert et al., 1989a; 1991c; Keipert & Triner, 1989*).

RESULTS and DISCUSSION

Physical and Chemical Characteristics

In order for acellular hemoglobin solutions to function as effective oxygen-carriers or "blood substitutes", they must possess certain basic properties. Ideally, any preparation of chemically modified Hb intended for ultimate use *in vivo* must possess adequate colloid osmotic properties to maintain blood pressure and intravascular blood volume. Also, the viscosity of these Hb preparations should not exceed that of whole blood, and the molecular size of the polymerized Hb molecules should be appropriate to assure prolonged retention within the vasculature. In addition, these Hb solutions must retain the ability to carry and deliver sufficient amounts of oxygen to the tissues.

Hemoglobin solutions generally possess advantageous rheological flow properties since these acellular solutions are more Newtonian in nature than blood (*Usami et al., 1971*). The total Hb concentration and the degree to which these solutions are polymerized, however, affects the viscosity of the final Hb preparation directly. A comparison of the viscosity of both unmodified and polymerized Hb solutions at concentrations of 4.0, 7.0, 10.0, and 14.0 g/dL is shown in **Figure 1** (left panel). The measured kinematic viscosities were normalized to that of isotonic (0.9%) saline, and are plotted as a relative viscosity. The range of viscosity values obtained for intramolecularly cross-linked Hb at 7.0 g/dL (not significantly different from 7.0 g/dL unmodified SFHb) and the polymerized Hb preparations at 14.0 g/dL fell in between the viscosities that were measured for plasma (1.19 cP) and whole blood (4.07 cP) (*Keipert, 1985; Keipert & Chang 1988*). Viscosity tended to increase with increasing Hb concentration, due presumably to a greater degree of protein-protein interaction at higher Hb concentrations (*Blank, 1984*), and this effect was clearly more pronounced for the heterogeneous polymerized Hb preparations which contained Hb macromolecules of differing size and shape.

The relationship between Hb concentration and the colloid osmotic pressure (COP) is dependent on the number of colloidal particles (or impermeant protein molecules) and is not linear, tending to deviate upwards to higher COP values at high protein concentrations due to an additional solute coefficient caused by increasing protein aggregation (*Blank, 1984*). As shown in the right panel of **Figure 1**, unmodified SFHb (which did not differ significantly from intramolecularly cross-linked hemoglobin) had a normal COP of 22 - 25 mmHg at a Hb concentration of only 7.0 - 7.5 g/dL (normal plasma COP = 25 mmHg), which corresponds to only $1/2$ the normal Hb concentration in blood. Consequently, one of the benefits of Hb polymerization was to lower the hyperoncotic nature (excessive COP) of highly concentrated (>8.0 g/dL) Hb solutions (*Keipert, 1985*). All of the polymerized Hb's had significantly lower COP, such that a normal COP range of 20 to 25 mmHg could be obtained at Hb concentrations of 14.0 - 15.0 g/dL (*Keipert & Chang, 1988*). These solutions contain the same concentration of Hb as found in blood and therefore possess a "normal" total oxygen <u>carrying</u> capacity relative to whole blood. The oxygen <u>delivering</u> capability of these polyHb solutions, however, depends to what extent the heme groups in each individual Hb polymer are still free to bind oxygen reversibly.

One of the most important properties of Hb that must be preserved during chemical cross-linking and/or polymerization procedures, is the ability to transport oxygen with low enough affinity to supply tissue demands adequately. When Hb is removed from the red cell, the oxyHb dissociation curve shifts to the left due to the loss of the allosteric effector 2,3-DPG and a higher pH (7.4 in the plasma versus 7.2 inside the red cell). This left shift represents Hb with a higher affinity for oxygen, which corresponds to a lower P_{50} (the pO_2 at which the Hb is still 50% saturated). As shown in **Figure 2**, the P_{50} of whole blood (27 mmHg) fell to about 12 mmHg for SFHb. Representative oxyHb dissociation curves have also been shown for a typical polymerized Hb preparation (PLPpolyHb, P_{50} = 18 mmHg) and for two Hb's cross-linked interdimerically: the diaspirin cross-linked $\alpha\alpha$Hb (P_{50} = 30 mmHg) and the very low affinity NFPLPHb (P_{50} = 47 mmHg). The (bis-PL)P_4Hb had a P_{50} = 30 mmHg with a similar dissociation curve to $\alpha\alpha$Hb, while B-polyHb and oRpolyHb had nonsigmoidal dissociation curves (similar to the PLPpolyHb curve) with P_{50} values of 21 and 28 mmHg respectively (curves not shown). Loss of normal heme-heme cooperativity resulting from polymerization of Hb was evident from the hyperbolic shape of the oxyHb dissociation curves at low pO_2 values. This is most likely due to the increased rigidity (steric hindrance) imposed by the multiple intermolecular covalent cross-links within these polyHb macromolecules.

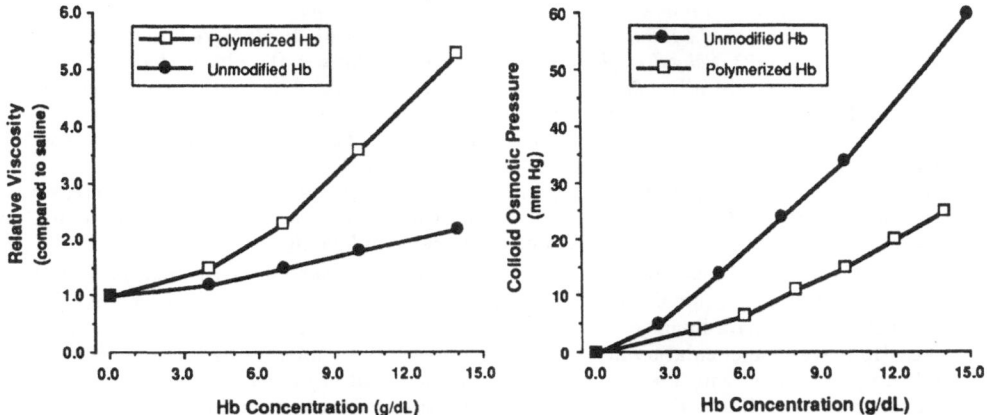

Fig. 1. **Left Panel:** The effect of polymerization on the viscosity of Hb solutions at different concentrations. The absolute viscosities (cP) were measured in a Canon Ubbelohde Viscometer at 37°C, and have been normalized to the viscosity of physiological saline. (Data from *Keipert & Chang, 1988*)
Right Panel: The effect of polymerization on the colloid osmotic pressure (COP) of Hb solutions, measured at different Hb concentrations. (Normal COP for plasma is about 20-25 mmHg.) (From *Keipert & Chang, 1988*)

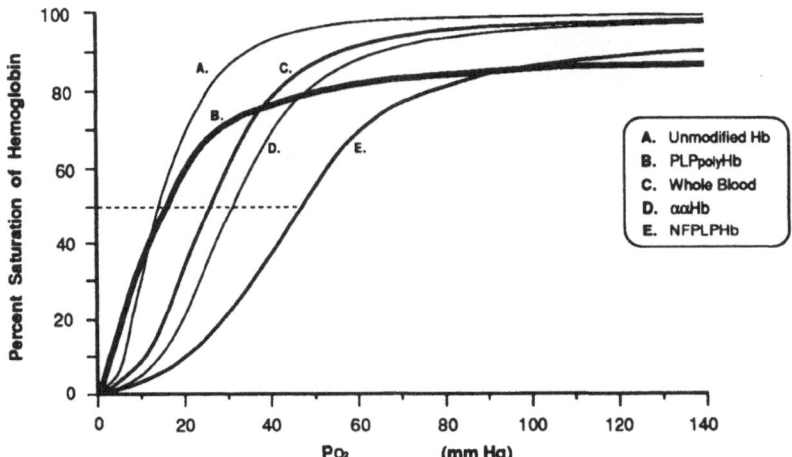

Fig. 2. Representative oxyHb dissociation curves measured *in vitro* at standard conditions (pH 7.4, pCO_2 = 40 mmHg, 37 °C). The dashed line indicates the P_{50} values, ie., the pO_2 at which the hemoglobin is still 50% saturated with oxygen. See Methods section for the description of the various Hb derivatives. The oxyHb dissociation curves for unmodified Hb (curve A) and whole blood (curve C) have been included for reference. (Data is from *Keipert, 1985; Keipert & Chang, 1988;* and *Keipert et al. 1991b*.)

Properties *In Vivo*: Intravascular Retention

One of the primary reasons for cross-linking or polymerizing Hb is to prolong the intravascular retention time of the Hb derivatives, since it has been shown clearly that unmodified Hb splits spontaneously into 32 kDa dimers which are then cleared rapidly from the circulation by renal glomerular filtration, and to a lesser extent by haptoglobin (*Lathem, 1959; Bunn et al., 1969; Rabiner et al. 1967; Moss et al. 1976; DeVenuto et al. 1977; Keipert & Chang, 1983; Sehgal et al. 1984*). As shown in **Figure 3**, the plasma half-life ($T^1/_2$) retention time of chemically modified Hb derivatives was found to be related linearly to the total dose of Hb administered. For the intramolecularly cross-linked Hb (MW = 64 kDa) solutions tested, the $T^1/_2$ values increased from about 2.5 hr up to 5.0 hr over the dose range injected in these studies. For polymerized Hb derivatives, however, the plasma $T^1/_2$ values were found to range from about 7 hr up to 28 hr, with substantially larger increases in plasma retention occurring at each incremental increase in dose (compared to similar increases in dose for the 64 kDa interdimerically cross-linked Hb's).

For the 64 kDa cross-linked Hb derivatives, a very close linear correlation existed between Hb plasma $T^1/_2$ and the total dose (slope=1.15; r^2=0.972). It is unclear whether this relationship will remain linear at much higher doses. The rate at which the 64 kDa cross-linked Hb leaves the circulation at much higher doses will be affected by the magnitude of fluid shifts associated with the movement of Hb (colloid) out of the circulation. In previous studies, however, a linear relationship was demonstrated for noncross-linked Hb's as well, although the effect (i.e., the slope) was much less. In this case, $T^1/_2$ values ranged from 1.0 hr to 2.5 hr for SFHb, and from 2.0 hr to 3.3 hr for PLPHb (*Keipert & Chang, 1987*). Thus, it is clear that a very large dose (about 7.0 g Hb/kg based on the data shown in **Figure 3**) would be required to achieve a plasma $T^1/_2$ of 10 hours with a 64 kDa cross-linked Hb derivative. For polymerized Hb, a similar dose (corresponding to 4.9 liters of a 10.0 g/dL polyHb solution), would yield a plasma $T^1/_2$ of 20 hours (slope = 2.33). As a result, the choice of whether to select a Hb that is cross-linked intramolecularly versus a polymerized Hb will, to some extent, be dictated by the intended clinical application - i.e., a short half-life intramolecularly cross-linked Hb may be more desirable for acute procedures like balloon angioplasty (PTCA), whereas a longer half-life polymerized Hb may be preferred for trauma resuscitation purposes.

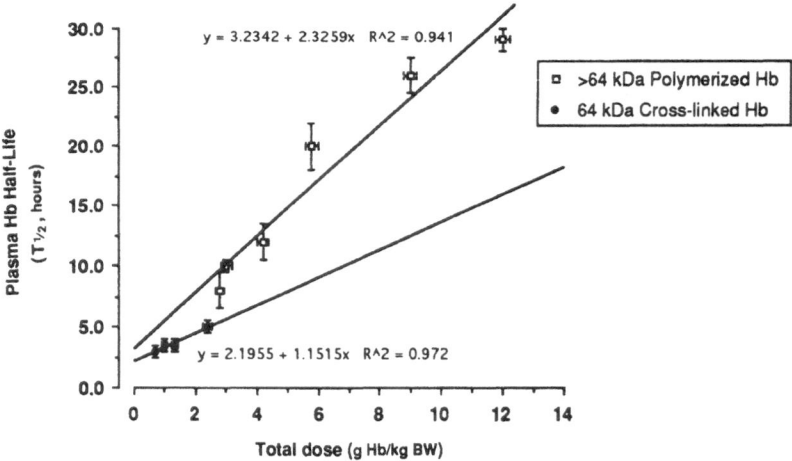

Fig. 3.   The effect of the total administered Hb dose on the intravascular retention time of both 64 kDa intramolecularly cross-linked Hb and the polymerized (>64 kDa) Hb derivatives. The plasma Hb half-life ($T^1/_2$) was measured either by the standard Drabkin's assay (540 nm), or by liquid scintillation counting (for radiolabeled Hb derivatives). Data shown are Means ± SEM (n=6 rats minimum per data point). Curves have been generated with data compiled from *Keipert & Chang, 1987;* and *Keipert et al. 1989b; 1990b; 1991b.*

Although the total dose of Hb administered had the greatest influence on the plasma retention of Hb, results from studies with individual ^{3}H-labeled components of o-raffinose polymerized Hb (oRpolyHb) have now clearly demonstrated that the molecular weight of the different polyHb molecules influenced their plasma clearance (*Keipert et al. 1990a; 1992a*). The half-lives of the four ^{3}H-labeled cross-linked fractions yielded $T^{1/2}$ values of 3.5, 9.0, 12.0, and 13.0 hours for fractions I - $(Hb)_1$ = 64 kDa, II - $(Hb)_2$ = 128 kDa, III/IV - $(Hb)_{3-4}$ = 192-256 kDa, and V - $(Hb)_{>9}$ = >576 kDa, respectively. In contrast, the noncross-linked component $(Hb)_{1/2}$ = 32 kDa, had a $T^{1/2}$ of only 0.6 hours. These studies have also shown that the relationship between the <u>molecular weight</u> of polyHb molecules and their respective plasma retention ($T^{1/2}$) did not remain linear at higher molecular weights (>320 kDa). As a result, there was no significant additional prolongation of $T^{1/2}$ once the MW's exceeded 300-400 kDa (*Keipert et al. 1992a*). Polymerization of Hb should therefore be controlled in such a way as to maximize the fraction of Hb macromolecules in the 192 to 320 kDa range, especially since *Feola et al. (1988)* have already demonstrated in rabbits that larger Hb polymers (>500 kDa) are potentially toxic because they can activate complement.

Properties *In Vivo*: Effects on the Kidneys

The most obvious benefit gained from the cross-linking and polymerization of Hb was a significant reduction in hemoglobinuria, i.e., the amount of Hb that filtered into the urine after crossing the kidney glomerulus. This effect has been demonstrated in many earlier studies (*Bunn et al. 1969; Keipert & Chang, 1983; Sehgal et al. 1984; Bleeker et al. 1989; Keipert et al. 1989b; Urbaitis et al. 1990*). Another consistent observation seen immediately upon completion of 50% exchange transfusions with different cross-linked Hb preparations was a transient, but significant, diuresis. Urine flow increased immediately post-ET, and then gradually returned to normal levels (0.75 ml/hr or 2.5 ml/kg/hr) by 10 to 24 hours. As shown in **Figure 4**, the degree of diuresis (shown as the peak urine flow rate seen during the first hour post-ET) correlated closely with the amount of free Hb appearing in the urine. The vast majority (>95%) of the urinary Hb content could be accounted for by the amount of noncross-linked Hb that was still present in the final Hb preparations, based on reversed-phase HPLC and SDS-PAGE analyses of these urine samples (*Keipert et al. 1992b*).

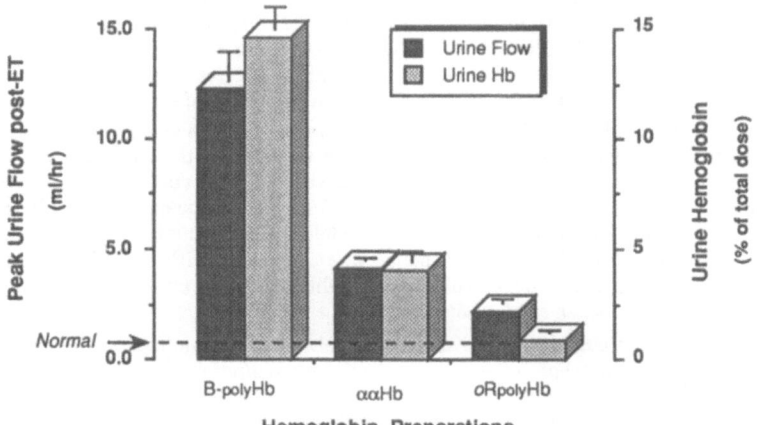

Fig. 4. Urine output (shown as the peak urine flow rate during the first hour post-ET), and the percent of the total Hb dose eliminated into urine (recovered during the first 10 hours post-ET), in rats treated with a 50% isovolemic exchange transfusion (ET) using either ααHb (n=24 rats), B-polyHb (n=6 rats), or oRpolyHb (n=40 rats). (Refer to the Methods section for a description of each Hb derivative.) Data shown are Means ± SEM. Normal urine output is indicated by the dashed line. (Data is from *Keipert et al. 1991b; 1992b*.)

It has not been possible previously, to study Hb clearance pharmacokinetics for the individual MW fractions comprising a polymerized Hb preparation. By performing the 50% exchange transfusions with radiolabeled Hb derivatives, however, it was possible to demonstrate that the amount of radioactivity getting into the kidney tissue was inversely proportional to the MW of the particular Hb species (*Keipert et al. 1991a; 1992a*). For the 64kDa intramolecularly cross-linked Hb preparations (^{3}H-NFPLPHb and ^{14}C-ααHb), the highest concentration of counts were always found in kidney tissue at all time points (ranging from 1 hour up to 14 days post ET). Once polymerized, the relative concentration of each ^{3}H-oRpolyHb component in kidney tissue decreased with increasing MW, which indicated that polymerization was successful at preventing renal filtration of hemoglobin. The opposite trend was apparent in liver tissue (which has the leakiest endothelium to macromolecules), where progressively greater amounts of radioactivity were found for the higher MW components, indicating that some aspects of the mechanisms for clearance of the 64 kDa Hb molecules were clearly different from those responsible for clearance of the larger Hb macromolecules (from 128 kDa to >576 kDa).

Properties *In Vivo*: Hemodynamic Effects

One of the hemodynamic effects of Hb solutions that has been observed and described repeatedly in the literature is vasoconstriction. When Hb solutions have been used *in vivo* to perform exchange transfusions, oxygen delivery could be maintained, but hemodynamic responses including an increase in mean arterial pressure and no elevation in cardiac output (despite significant hemodilution), were indicative of some degree of vasoconstiction (*Moss et al. 1976; Hauser et al. 1982*). This potential side-effect has been studied extensively by perfusing isolated hearts and kidneys with Hb solutions. To date, the differences in the degree of vasoconstriction observed (determined by increasing perfusion pressure under constant flow conditions) for various Hb preparations have been attributed primarily to the degree of purification and to a lesser extent polymerization (*Vogel et al. 1988*).

In our recent studies using a 50% isovolemic exchange transfusion model in fully conscious rats, it was found that both ααHb (a 64 kDa interdimerically cross-linked Hb; n = 30 rats) and the oRpolyHb (Hb cross-linked and polymerized with o-raffinose; n = 45 rats) caused essentially identical, statistically significant, elevations in mean arterial pressure (MAP). As shown in **Figure 5**, the peak change in MAP (which occurred at about 2 hours post-ET) was approximately 20% in both Hb-treated groups. Heart rates, on the other hand, tended to decrease transiently by about 20% in the Hb-treated rats. MAP remained elevated from about 0.5 to 6 hr post-ET (data not shown), while the heart rate returned to normal by about 4 hours (*Keipert et al. 1991b, 1992b*). The time courses of these hemodynamic effects were clearly not dependent on the absolute amount of Hb present in the plasma, since the plasma half-lives ($T^{1}/_{2}$) were quite different for these Hb preparations (i.e., 5.0 hours for the ααHb, versus 10.0 hours for oRpolyHb). The consistency of these hemodynamic effects was remarkable, and has since been observed in subsequent studies in both rabbits and in pigs, using both cross-linked and noncross-linked Hb solutions (*Macdonald, 1991*).

In contrast, a solution of human serum albumin (HSA, 6.0 g/dL in Ringer's-lactate) had the opposite hemodynamic effects when used to perform the identical 50% exchange transfusion in conscious animals (n = 4 rats). As seen in **Figure 5**, exchange transfusion with albumin solution caused a significant (p<0.0001) fall in mean arterial blood pressure (at about 2 hours post-ET) and a significant (p<0.0001) rise in the heart rate (which peaked earlier, at 0-1 hour post-ET). This response was consistent with what has been observed traditionally following acute normovolemic hemodilution with either crystalloid or colloid plasma expanders (*Messmer et al. 1986*).

Whether the decrease in heart rates represents a true Hb-induced bradycardia (due to a direct negative chronotropic effect as suggested by *Walter & Chang, 1988*), or simply a compensation mechanism in response to elevated blood pressure, secondary to increased peripheral resistance, has not yet been established. The isovolemic nature of these 50% exchange transfusions, and the fact that all Hb solutions were diluted to an appropriate Hb concentration that would be iso-oncotic with the plasma, would argue against any simple explanation of these hemodynamic side-effects based solely on fluid shifts within the body.

A more likely hypothesis might be formulated using the extensive body of literature that has appeared recently on the pharmacology of an endothelial-derived relaxing factor (EDRF) and its effects on isolated smooth muscle (*Lüscher et al. 1991*). At present, based on

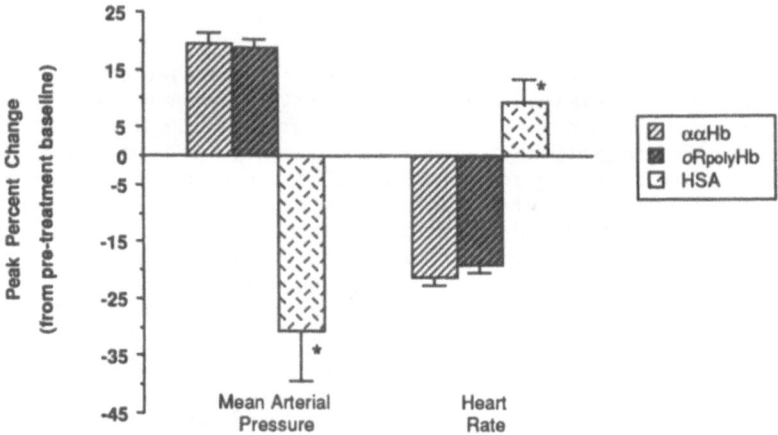

Fig. 5. Peak hemodynamic changes (expressed as a percent of the pre-treatment baseline values), observed in fully conscious chronically cannulated rats treated with a 50% isovolemic exchange transfusion (ET) using either ααHb (n = 30 rats) or oRpolyHb (n = 45 rats). The peak changes occurred at 2-2.5 hours post-ET, except for heart rate in the albumin (HSA) control group (n = 4) which peaked at about 0-1 hour after exchange transfusion with human serum albumin solution (6.0 g/dL HSA in Ringer's-lactate). The asterisk indicate a significant difference ($p<0.0001$) compared to the Hb-treated groups. Data shown are Means ± SEM.

pioneering studies by many groups including *Furchgott et al., Ignarro et al.,* and *Moncada et al.,* there is solid evidence that the identity of EDRF is most likely nitric oxide (NO), a simple gas which is constantly being released from vascular endothelial cells to cause local smooth muscle relaxation and thereby blood vessel dilatation (*Kolata, 1991*). Very dilute solutions of unmodified hemoglobin have often been used for *in vitro* studies as a pharmacological drug to inhibit EDRF. It therefore appears likely that the increased mean arterial pressure observed *in vivo* following the infusion of large doses of Hb solutions may result in part from an inhibition or binding of NO by the Hb. To what extent such a mechanism could account for the degree of peripheral vasoconstriction observed after the administration of Hb solutions remains to be determined. Clearly, evidence exists that hemoglobin solutions cause some degree of vasoconstriction and hypertension. This potential side-effect must be investigated carefully, and should be understood better before these Hb products can be used safely in clinical practice.

Summary and Conclusions

Current efforts in both academia and industry to develop chemically modified Hb solutions, are being driven by the desire to produce a universally administrable oxygen carrier for temporary red cell substitution. The clinical utility of such a so-called "blood substitute" has been recognized for over 100 years. Recently, however, a growing public awareness of the potential dangers associated with transfusion of virally contaminated homologous (donated) blood, has made the need even more acute. It is precisely because of the risks of transmitting viral hepatitis or HIV (AIDS), that companies are currently so interested in developing Hb-based oxygen carriers.

Clearly, solutions of cross-linked or polymerized Hb possess attractive physical and chemical properties. Their viscosity can be very low, they provide adequate colloid osmotic pressure to maintain blood volume, and they can be modified chemically in such a way that the oxygen affinity is lowered to the point where these solutions will be able to load and unload O_2 under physiological conditions and at ambient pO_2 levels. From extensive *in vivo* testing in different animal models, it is easy enough to become convinced that most Hb preparations can act as effective oxygen carriers. Moreover, intravascular retention ($T_{1/2}$) of these chemically modified Hb derivatives has now been shown to be both dose and size dependent, which means that more than one type of Hb modification could be designed in order to meet various clinical needs.

Despite significant efforts for many years, however, no such product has yet become available commercially. The reason lies in the fact that large scale production represents a formidable and unprecedented task. Hemoglobin, if removed from the red cell, must be chemically modified or genetically engineered in some way. It must then be purified completely of all contaminating red cell stromal lipid, endotoxin, and all other red cell constituents that may be toxic when released into the circulation. The entire production process must be reproducible, and must remain economically feasible when huge amounts (kilogram quantities) of Hb are being manufactured. In this regard, continuing support and collaborative efforts between the Hb industry and academic scientists will be vital to overcome the technological hurdles that still need to be resolved before Hb production capabilities will come close to being able to make appropriate amounts of such a product.

Finally, the nature of the "toxicity" of acellular Hb solutions needs elucidation. In an earlier clinical trial, acute renal dysfunction occurred after a low dose infusion (250 mg Hb/kg BW) of unmodified SFHb in normal, well hydrated human volunteers (*Savitsky et al. 1978*). In two more recent US clinical trials, different unexplained side-effects were noted and forced premature termination of both trials. The lack of obvious side-effects or toxicities in many animal studies to date indicates clearly that better "stressed" animal models need to be developed and tested before Hb solutions can be given safely to patients.

ACKNOWLEDGEMENTS

The author would like to gratefully acknowledge all aspects of technical support that has been provided by the following individuals for different parts of these studies: Dr. C. Clifford, M. Cross, C.L. Gomez, A. Gonzales, D. Green, S. Kwong, C. Lister, Dr. M. Marini, E. Moore, M. Verosky, R. Varma, and A. Zegna. The author would also like to express his appreciation to the following individuals for their different scientific input to these studies: Dr. R.E. Benesch, Dr. T.M.S. Chang, Dr. J.R. Hess, Dr. J.C. Hsia, Dr. V.W. Macdonald, Dr. L. Triner, and Dr. R.M. Winslow. The author would like to thank Dr. R.M. Winslow for his helpful comments and critical review of this manuscript, and G. Rosenberg for her diligent editorial assistance.

REFERENCES

Amberson, W.R., Mulder, A.G., Steggerda, F.R., Flexner, J., Pankratz, D.S., 1933, Mammalian life without red blood corpuscles. Science, 78(2014): 106-7.

Amberson, W.R., Flexner, J., Steggerda, F.R., Mulder, A.G., Tendler, M.J., Pankratz, D.S., Laug, E.P., 1934, On the use of Ringer-Locke solution containing hemoglobin as a substitute for normal blood in mammals. J. Cell. Comp. Physiol., 5: 359-82.

Amberson, W.R., Jennings, J.J., Rhode, C.M., 1949, Clinical experience with hemoglobin-saline solutions. J. Appl. Physiol., 1: 469-89.

Benesch, R., Benesch, R.E., 1969, Intracellular organic phosphates as regulators of oxygen release by hemoglobin. Nature, 221: 618.

Benesch, R.E., Benesch, R., Yung, S., Edalji, R., 1972, Affinity labeling of the polyphosphate binding site of hemoglobin. Biochemistry, 11: 3576-82.

Benesch, R.E., Benesch, R., 1981, Preparation and properties of hemoglobin modified with derivatives of pyridoxal. Methods in Enzymol., 76; 147-59.

Benesch, R.E., Kwong, S., 1988, Bis-pyridoxal polyphosphates: a new clas of specific intramolecular cross-linking agents for hemoglobin. Biochem. Biophys. Res. Acta, 156; 9-14.

Blank, M., 1984, Molecular association and viscosity of hemoglobin solutions. J. Theor. Biol., 108: 55-64.

Bleeker, W.K., van der Plas, J., Feitsma, R.I., Agterberg, J., Rigter, G., Vries-van Rosen, A., Pauwels, E.K.J., Bakker, J.C., 1989, *In vivo* distribution and elimination of hemoglobin modified by intra-molecular crosslinking with 2-nor-2-formylpyridoxal 5'-phosphate. J. Lab. Clin. Med., 113: 151-61.

Bucci, E., Fronticelli, C., Orth, C., Martorana, M., Aebischer, L., Angeloni, P., 1988, Bovine hemoglobin as a basis for artificial oxygen carriers. Biomat., Artif. Cells, Artif. Organs, 16; 197-204.

Bunn, H.F., Esham, W.T., Bull, R.W., 1969, The renal handling of hemoglobin. I. - Glomerular filtration. J. Exp. Med., 129: 909-24.

Bunn, H.F., Jandl, J.H., 1969, The renal handling of hemoglobin. II. - Catabolism. J. Exp. Med., 129: 925-34.

Carrier, E.B., Lee, F., Whipple, G.H., 1922, Determination of plasma and hemoglobin volumes after unit hemorrhages under controlled experimental conditions. Amer. J. Physiol., 61: 138-48.

CBER (Center for Biologics Evaluation and Research), U.S. Food and Drug Administration., 1991, Points to consider in the safety evaluation of hemoglobin-based oxygen carriers. Transfusion, 31: 369-371.

Chatterjee, R., Iwai, Y., Walder, R.Y., Walder, J.A., 1984, Structural features required for the reactivity and intracellular transport of bis(3,5-dibromosalicyl) fumarate and related anti-sickling compounds that modify hemoglobin-S at the 2,3-diphosphoglycerate binding site. J. Biol. Chem., 259: 14863-73.

DeVenuto, F., Moores, W.Y., Zegna, A.I., Zuck, T.F., 1977, Total and partial blood exchange in the rat with hemoglobin prepared by crystallization. Transfusion, 17: 555-62.

Estep, T.N., Bechtel, M.K., Miller, T.J., Bagdasarian, A., 1989, Virus inactivation of hemoglobin solutions by heat. In: "Blood Substitutes", Chang, T.M.S., Geyer, R.P., eds., Marcel Decker, New York: 129-34.

Feola, M., Simoni, J., Canizaro, P.C., Tran, R., Rauschbaum, G., Behal, F., 1988, Toxicity of polymerized hemoglobin solutions. Surg. Gynecol. Obstet., 166; 211-22.

Hamilton, P.B., Hiller, A., Van Slyke, D.D., 1947, Renal effects of hemoglobin infusions in dogs in hemorrhagic shock. J. Exp. Med., 86: 477-87.

Hauser, C.J., Kaufman, C., Frantz, R., Shippy, C., Schwartz, S., Shoemaker W.C., 1982, Use of crystalline hemoglobin as replacement of red blood cell mass. Arch. Surg., 117; 782-86.

Hsia, J.C., 1989, Pasteurizable, freeze-driable hemoglobin-based blood substitute. *United States Patent* No. 4,857,636.

Hsia, J.C., Song, D.L., Er, S.S., Wong, L.T.L., Keipert, P.E., Gomez, C.L., Gonzales, A., Hess, J.R., Macdonald, V.W., Winslow, R.M., 1991, Pharmacokinetic studies on a raffinose-polymerized human hemoglobin in the rat. Biomat., Artif. Cells, Immobil. Biotech., 19(2): 401.

Keipert, P.E., Chang, T.M.S., 1983, *In vivo* assessment of pyridoxylated crosslinked polyhemoglobin as an artificial red cell substitute in rats. Trans. Amer. Soc. Artif. Intern. Organs (ASAIO), 29; 329-33.

Keipert, P.E., Chang, T.M.S., 1984, Preparation and *in vitro* characteristics of a blood substitute based on pyridoxylated polyhemoglobin. Appl. Biochem. Biotech., 10; 133-41.

Keipert, P.E., 1985, Physiological Effects of Pyridoxylated Polyhemoglobin Solution as a Blood Substitute in Rats. (Doctoral Thesis), McGill University, Montreal.

Keipert, P.E., Chang, T.M.S., 1987, Effects of partial and total isovolemic exchange transfusion in fully conscious rats using pyridoxylated polyhemoglobin solution as a colloidal oxygen-delivering blood replacement fluid. Vox. Sang., 53: 7-14.

Keipert, P.E., Chang, T.M.S., 1988, Pyridoxylated-polyhemoglobin solution: A low viscosity oxygen-delivering blood replacement fluid with normal oncotic pressure and long-term storage feasiblity. Biomat., Artif. Cells, Artif. Organs, 16; 185-96.

Keipert, P.E., Triner, L., 1989, Catabolism and excretion of crosslinked hemoglobin. In: "The Red Cell - Seventh Ann Arbor Conference", Brewer, G.J., ed., Alan R. Liss, New York: 383-405.

Keipert, P.E., Verosky, M., Triner, L., 1989a, Plasma retention and metabolic fate of hemoglobin modified with an interdimeric covalent crosslink. ASAIO Transactions, 35: 153-9.

Keipert, P.E., Adeniran, A.J., Kwong, S., Benesch, R.E., 1989b, Functional properties of a new crosslinked hemoglobin designed for use as a red cell substitute. Transfusion, 29: 768-73.

Keipert, P.E., Gomez, C.L., Gonzales, A., Hess, J.R., Macdonald, V.W., Winslow, R.M., Hsia, J.C., Wong, L.T.L., Er, S.S., Song, D.L., 1990a, Distribution, metabolism, and excretion of a polymerized hemoglobin: Influence of molecular weight. Blood, 76: 402a.

Keipert, P.E., Bleeker, W.K., Bakker, J.C., 1990b, Radiolabeling of crosslinked hemoglobin derivatives for *in vivo* distribution and metabolism studies. FASEB J., 4(4): A1234.

Keipert, P.E., Gonzales, A., 1991, Chronic femoral cannulation in rats: A simplified catheter and externalization technique for improved long-term accessibility. J. Surg. Res., (in preparation).

Keipert, P.E., Gomez, C.L., Gonzales, A., Macdonald, V.W., Winslow, R.M., 1991a, The role of the kidneys in the metabolic elimination of chemically modified human hemoglobin from rats. Biomat., Artif. Cells, Immobil. Biotech., 19(2): 406.

Keipert, P.E., Gonzales, A., Gomez, C.L., Macdonald, V.W., Hess, J.R., Winslow, R.M., 1991b, Acute physiological and renal responses to major exchange transfusion using diaspirin cross-linked hemoglobin. J. Lab. Clin. Med., (submitted).

Keipert, P.E., Gomez, C.L., Gonzales, A., Macdonald, V.W., Hess, J.R., Winslow, R.M., 1991c, Organ distribution and long-term excretion of diaspirin cross-linked hemoglobin after major exchange transfusion. J. Lab. Clin. Med., (submitted).

Keipert, P.E., Gomez, C.L., Gonzales, A., Macdonald, V.W., Winslow, R.M., 1992a, The role of the kidneys in the excretion of chemically modified hemoglobins. Biomat., Artif. Cells, Immobil. Biotech., 20: (in press).

Keipert, P.E., Gonzales, A., Gomez, C.L., Macdonald, V.W., Hess, J.R., Winslow, R.M., Hsia, J.C., Wong, L.T.L., Er, S.S., Song, D.L., 1992b, o-Raffinose-polymerized hemoglobin. I - Influence of molecular weight on pharmacokinetic properties following 50% isovolemic exchange transfusion. Transfusion, (manuscript in preparation).

Keipert, P.E., Gomez, C.L., Gonzales, A., Macdonald, V.W., Hess, J.R., Winslow, R.M., Hsia, J.C., Wong, L.T.L., Er, S.S., Song, D.L., 1992c, o-Raffinose-polymerized hemoglobin. II - Influence of molecular weight on tissue distribution and metabolic elimination following 50% isovolemic exchange transfusion. Transfusion, (manuscript in preparation).

Kolata, G., 1991, Key signal of cells found to be a common gas. Science Times section, The New York Times, Tuesday, July 2, 1991; B5.

Lathem, W., 1959, The renal excretion of hemoglobin: regulatory mechanisms and the differential excretion of free and protein bound hemoglobin. J. Clin. Invest., 12: 493-98.

Lüscher, T.F., Richard, V., Tanner, F.C., 1991, Endothelium-derived vasoactive factors and their role in the coronary circulation. Trends Cardiovasc. Med., 1; 179-85.

Macdonald, V.W., 1991, Letterman Army Institute of Research, Presidio of San Francisco, CA (Personal Communication).

Mason, J.B., Mann, F.C., 1931, The effect of hemoglobin on the volume of the kidney. Amer. J. Physiol., 98: 181-5.

Messmer, K., Kreimeier, U., Intaglietta, M., 1986, Present state of intentional hemodilution. Eur. Surg. Res. 18; 254-63.

Moffat, A.S., 1991, Three li'l pigs and the hunt for blood substitutes. Science, 253: 32-4.

Moss, G.S., DeWoskin, R., Rosen, A.L., Levine, H., Palani, C.K., 1976, Transport of oxygen and carbon dioxide by hemoglobin-saline solution in the red cell free primate. Surg. Gynecol. Obstet., 142: 357-62.

Naunya, B., 1868, Beitrage zur Lehre vom Icterus. Arch. Anat. Physiol. Wiss. Med., 21:401-41.

Pool, R., 1990; Slow going for blood substitutes. Science, 250: 1655-56.

Ponfick, E., 1875, Experimentelle Beiträge auf Lehre von der Transfusion. Virchows. Arch., 62: 273-335.

Rabiner, S.F., Helbert, J.R., Lopas, H., Friedman, L.H., 1967, Evaluation of a stroma-free hemoglobin solution for use as a plasma expander. J. Exp. Med., 126: 1127-42.

Savitsky, J.P., Doczi, J., Black, J., Arnold, J.D., 1978, A clinical safety trial of stroma-free hemoglobin. Clin. Pharmacol. Therap., 23: 73-80.

Sehgal, L.R., Gould, S.A., Rosen, A.L., Sehgal, H.L., Moss, G.S., 1984, Renal clearance of polymerized pyridoxylated hemoglobin. Fed. Proc., 43(3): 607.

Sellards, A.W., Minot, G.R., 1916, Injection of hemoglobin in man and its relation to blood destruction, with special reference to the anemias. J. Med. Res., 34: 469-94.

Usami, S., Chien, S., 1971, Hemoglobin solution as a plasma expander: effects on blood viscosity. Proc. Soc. Exptl. Biol. Med., 136: 1232-36.

Urbaitis, B.K., Razynska, A., Corteza, Q., Fronticelli, C., Bucci E., 1990, Intravascular retention and renal handling of purified natural and intramolecularly cross-linked hemoglobins. J. Lab. Clin. Med., 117: 115-21.

Vogel, W.M., Lieberthal, W., Apstein, C.S., Levinsky, N., Valeri, C.R., 1988, Effects of stroma-free hemoglobin solutions on isolated perfused rabbit hearts and isolated perfused rat kidneys. Biomat., Artif. Cells, Artif. Organs, 16; 227-35.

Walter, S.V., Chang, T.M.S., 1988, Chronotropic effects of stroma-free hemoglobin and polyhemoglobin on cultured myocardiocytes derived from newborn rats. Biomat., Artif. Cells, Artif. Organs, 16; 701-3.

Winslow, R.M., 1989, Blood substitutes - Minireview. In: "The Red Cell - Seventh Ann Arbor Conference". Brewer, G.J., ed., Alan R. Liss, New York: 305-23.

Winslow, R.M., 1991, "Hemoglobin-Based Red Cell Substitutes". The Johns Hopkins University Press, Baltimore.

CURRENT ADDRESS & REPRINT REQUESTS

Peter E. Keipert, Ph.D., Director, Dept. of Pharmacology, Alliance Pharmaceutical Corp., 3040 Science Park Road, San Diego, CA 92121.   Tel. (619) 558-4300; FAX: (619) 558-4333.

ELABORATION OF FLUOROCARBON EMULSIONS WITH IMPROVED OXYGEN-CARRYING CAPABILITIES

Jean G. Riess and Marie-Pierre Krafft

Laboratoire de Chimie Moléculaire, Associé au CNRS, Université de Nice-Sophia Antipolis, Parc Valrose, 06034 Nice, France

INTRODUCTION

The notion one has of the fluorocarbon-based approach to injectable oxygen carriers is still too often associated with the earlier Fluosol®-type emulsions. The fact that Fluosol has been approved, although only for a limited number of applications, as an adjunct during percutaneous transluminal coronary angioplasty (PTCA), means that the intravascular administration of fluorocarbon emulsions is considered safe by the FDA, and that oxygen transport capacity has been recognized for this preparation, in spite of its low concentration and other limitations.[1] Very similar 20% w/v concentrated emulsions have been developed in the Soviet Union and in China under the names Ftorosan and Emulsions n° II, respectively.

Considerable progress has been achieved since these first-generation products were designed. The new emulsions presently under development are substantially more efficient and more stable. Both of these qualities have a decisive impact on the extent of the applications that can now be reached with fluorocarbon-based preparations.[2-4]

This paper will focus on the choice of ingredients, formulation and emulsion technology which have led to such improved emulsions. It will also emphasize the sensitivity of fluorocarbon emulsions characteristics to processing conditions and its consequences on the dependability and comparability of experimental data collected with such emulsions.

EMULSION FORMULATION AND ENGINEERING

Selection of the Key Ingredients, Fluorocarbons and Surfactants

The major criterion for selecting the fluorocarbon is imposed by the need for its being rapidly excreted from the body. For "true" fluorocarbons, this depends essentially on their molecular weight. As the excretion rate *vs* molecular weight is exponential, the range of molecular weights acceptable is narrow - 460 to 520 -, which leaves little latitude in the choice of the fluorocarbon.[5] Faster excretion is observed when the fluorocarbon has a lipophilic extremity, as is the case with perfluorooctylbromide.[6] Other important criteria are purity and industrial feasibility of the fluorocarbon. The presently available candidates which fulfil all these requirements best are perfluorooctylbromide (PFOB or perflubron, bis(F-butyl)ethene (F-44E) and F-decalin (FDC) (Table 1)).[1] Where the oxygen-dissolving capacity is concerned, there is some advantage of the linear over the cyclic products. Concerning stability, the amphiphatic PFOB molecule shows a definite advantage, and for a given molecular weight, linear compounds appear to be more advantageous than cyclic ones; thus F-44E gives stabler emulsions than FDC.

Table 1 . Fluorocarbons and surfactants used in first and second generation injectable fluorocarbon emulsions.

FDC	FTPA	PFOB	F-44E

$N(C_3F_7)_3$ $C_8F_{17}Br$ $C_4F_9CH=CHC_4F_9$

FMIQ	FTBA	FMCP	"FMA"

$N(C_4F_9)_3$ 2 CF_3

$HO(CH_2CH_2O)_n(CH-CH_2O)_p(CH_2CH_2O)_nH$
CH_3

a poloxamer, ex. Pluronic F-68

$RCOO-$
$RCOO-$ ... $O-P-O-NMe_3$

a glycerophosphocholine diester in EYP

The surfactant selected for the second-generation emulsions is egg-yolk phospholipids (EYP). With any of the above fluorocarbons, EYP's give more stable emulsions than Pluronic F-68, which is the primary surfactant in Fluosol. EYP's are, however, not particularly fluorophilic, a feature which one would obviously wish to have when the oil to be emulsified is a fluorocarbon. Considerable efforts are now devoted to developing more fluorophilic surfactants for future generations of emulsions.[7]

The selection of a satisfactory EYP is essential. The propensity of EYP to elicit host defense reactions to particles, triggering the arachidonic acid cascade, varies somewhat with the supplier.

Formulation

The main progress accomplished since Fluosol was developed is a considerable, three-to-five-fold increase in emulsion concentration, hence in oxygen-carrying capacity and efficacy. Fluosol contains 20% of fluorocarbon by weight, i.e. only ca 11% by volume. In other words, the useful oxygen-carrying compartment of the preparation represents only about a tenth of the injected volume, and can dissolve only about 6 vol.% of oxygen in pure oxygen under atmospheric pressure. The aqueous phase then has a dilution effect on the patient's blood, which in some cases may aggravate his anemic condition. For some reason, the development of Fluosol-DA 20%, and the abandonment of the Fluosol-DA 35% w/v project, made people think that higher concentrations would be difficult to reach, perhaps they believed the viscosity would be too high.

The demonstration that considerably more concentrated emulsions - 100% w/v or 52% v/v - could be prepared and were still well tolerated when injected i.v. in massive amounts, therefore came as a surprise.[8] The oxygen-carrying capacity was enormously increased and the excess diluent eliminated (fig. 1).

Since then an optimal balance between oxygen-carrying capacity and fluidity has been established with 90% concentrated emulsions containing 4% EYP in a phosphate buffer.[9] The lowered viscosity is especially desirable at the low shear rates that are relevant to capillary beds. The rheological profile is close to Newtonian (fig. 2). The physical characteristics of such emulsions also depend far less on processing parameters, they are easier to sterilize, and they display better resistance to mechanical stress.

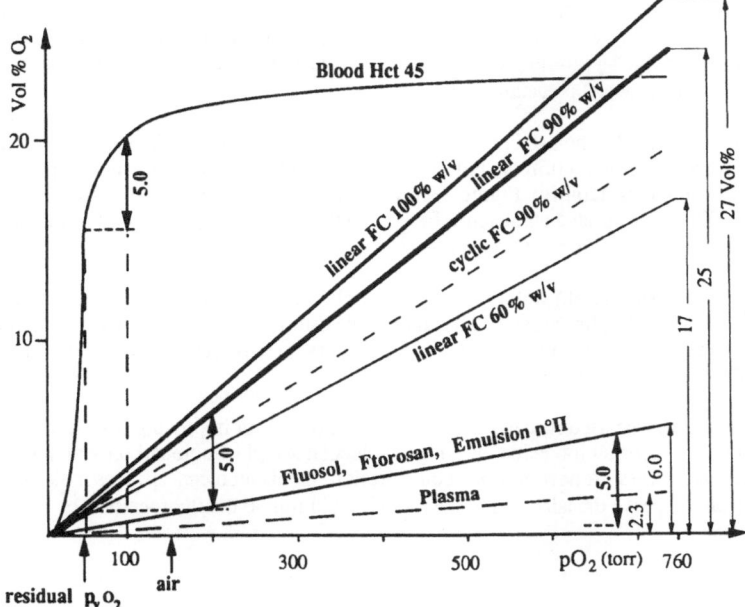

Fig 1. Volume of oxygen carried by blood or emulsions of linear (F-44E, PFOB) or cyclic (FDC, F-adamantane) fluorocarbons of various concentrations as a function of oxygen partial pressure.

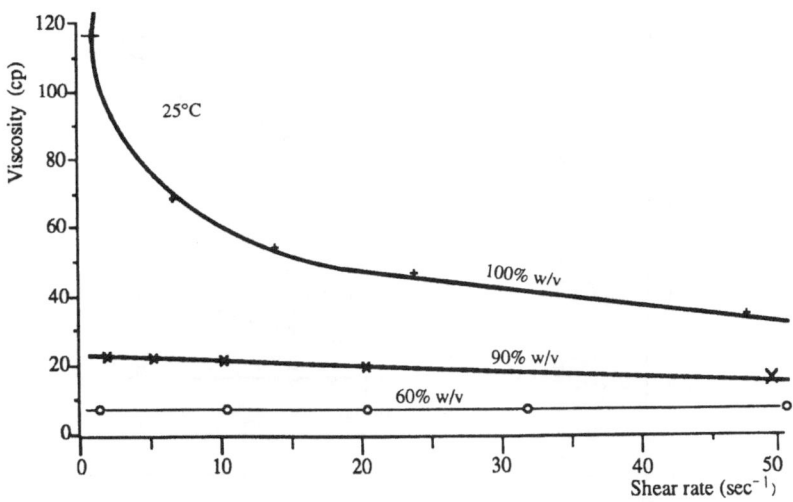

Fig 2. Viscosity at 25°C of variously concentrated PFOB/EYP emulsions as a function of shear rate

Another determinant improvement has been achieved where emulsion stability is concerned. Fluosol has to be shipped and stored in the frozen state; as a consequence, it comes as a frozen stem emulsion which must be thawed and mixed with two annex solutions prior to use. In contradistinction, the new concentrated emulsions are ready for use and their stability is quite remarkable, as illustrated by the successful close-to-total exchange perfusion of rats (75% survival at hematocrit 3-5%) with a 4-year-old preparation which had been stored unfrozen (and had crossed the Atlantic twice without special precautions).[10]

The amount of EYP to be used in the formulation is crucial. It must be sufficient to cover the small fluorocarbon droplets entirely, but an excess of EYP not only does not improve, but is detrimental to, emulsion stability. Figure 3 shows that the optimal amount of EYP is in the 3-6% range for 90-100% concentrated emulsions. Excess EYP results in the presence of fluorocarbon-free EYP vesicles, and in faster droplet size increase.[11]

The pH is adjusted at physiologic values, usually with a carbonate or, more effectively, a phosphate or amino acid buffer system. The appropriate osmolarity is obtained with sodium chloride or a polyol. Other typical ingredients include antioxidants such as tocopherol, and chelating agents such as EDTA to prevent the oxidation of the EYP.

Other agents may be used (usually in small 1-2% amounts) for improving emulsion stability, for example triglycerides or amino acids, or a high molecular weight fluorocarbon which hinders the Ostwald ripening process A new fluorocarbon emulsion stabilization concept uses molecular dowels to strengthen and improve the adhesion of the phospholipid film to the fluorocarbon.[12]

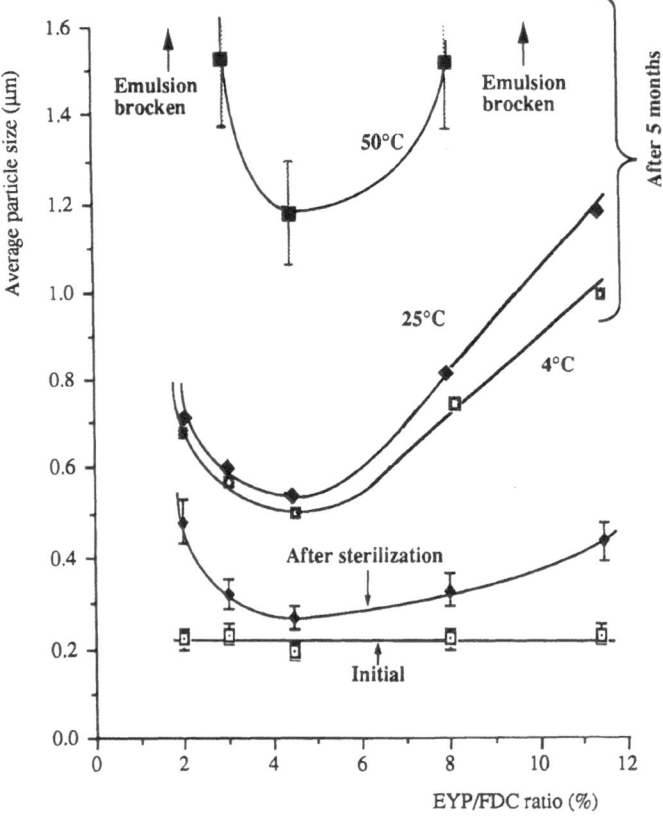

Fig 3. Influence of the EYP/FDC ratio on the stability of emulsions prepared by microfluidization (10 passes, 1000 bars, 30°C).

Processing

Processing includes the dispersion of the water-insoluble EYP in the aqueous phase, admixing of the fluorocarbon and preparation of a crude premix, emulsification, preferably by a high-pressure mechanical procedure, bottling and sterilization, inspection, labeling and packaging. Until the product is bottled, all these operations must be performed in a clean room under oxygen exclusion.

Since emulsions are, by nature, thermodynamically unstable systems, the preparation procedure used has a considerable influence on their characteristics and stability. Procedures applicable to obtaining fluorocarbon emulsions include sonication, high-pressure homogenization, microfluidization and hydroshear. Sonication is often used in Laboratories because it does not require any costly equipment and can be used for preparing small samples. However, as examplified in figure 4, then exist important differences in characteristics and stability between emulsions prepared by sonication and those prepared by microfluidization.[13] For every EYP/FC ratios, the microfluidized emulsions are consistently finer, more narrowly dispersed, resist heat sterilization better and display higher shelf stability than those obtained by sonication.The superiority of microfluidization over sonication becomes even more pronounced when lower EYP/FDC ratios are used. Moreover, sonication suffers from very poor reproducibility and requires an additional centrifugational step to remove large particles.

A set of parameters has to be defined and optimized for each step of the procedure . They include temperature, pressure, rate of addition, number of passes and, depending on the process, such less obvious variables such as shape of the vessel, size and position of the probe (sonication), configuration of the valve (high pressure homogenization), etc. The amount of energy pumped into the system during the process also has to be adjusted. Although additional passes through the microfluidizer result in smaller particles and narrower particle size distributions, it has been found that the resulting emulsions' shelf stability is lower (fig. 5), and its velocity higher. When industrial scale-up is considered, the number of passes has to be limited, and is usually in the 5-8 passes range.

The way the EYP is dispersed in the aqueous phase prior to addition of the fluorocarbon is also important. Electron microscopy studies correlated with stability studies show that when well-organized multilayer liposomes are formed, the incorporation of the fluorocarbon requires more energy and tends to give less stable emulsions than when the EYP is dispersed as coarse, imperfectly formed layers.[14]

Sterilization of injectable emulsions is achieved by heat. The required conditions (121°C, 15 min, 15 lb/in^2) represent a severe stability test for the emulsion. This is usually achieved in a rotatory autoclave and requires substantial specific know-how.[15]

Emulsion dependability

We wish to re-emphasize that the characteristics and behavior of an emulsion, including those which are relevant to biological tolerance such as particle size distribution, presence of unemulsified fluorocarbon, viscosity, pH, lysolechithin content, etc., depend strongly on formulation, processing conditions and history. The few examples discussed above illustrate the impact some variables of the processing parameters can have. Many more variables have been identified, the proper optimization of which is best achieved by computer-assisted planning and assessment[16]. The importance of such parameters as head space and head-space pressure in the bottles during sterilization is often overlooked in research laboratories. Shipping and storage conditions can also affect emulsion characteristics profoundly.

The preparation of reliable fluorocarbon emulsions requires current good manufacturing practices (CGMP) and extensive quality control, to assure the meeting of specifications, as well as batch-batch and vial-to-vial consistency; this supposes specific know-how and professional handling, almost impossible to achieve in non-specialist laboratoires that are not expressly equipped for such work.

It is therefore essential that before undertaking time-consuming and costly experimentation, a reliable source of properly controlled material be identified, and that the condition of the emulsion be checked. This will avoid discrepancies in data from one experiment to another and possible disappointment.

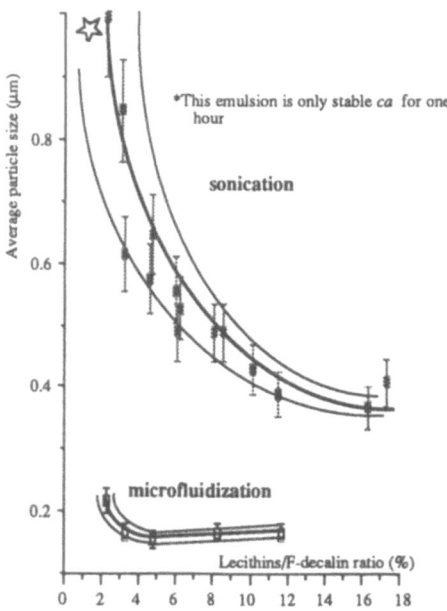

*This emulsion is only stable *ca* for one hour

sonication

microfluidization

Lecithins/F-decalin ratio (%)

Fig 4. Impact of the emulsification process - sonication vs microfluidization - for a range of EYP/FDC ratios.

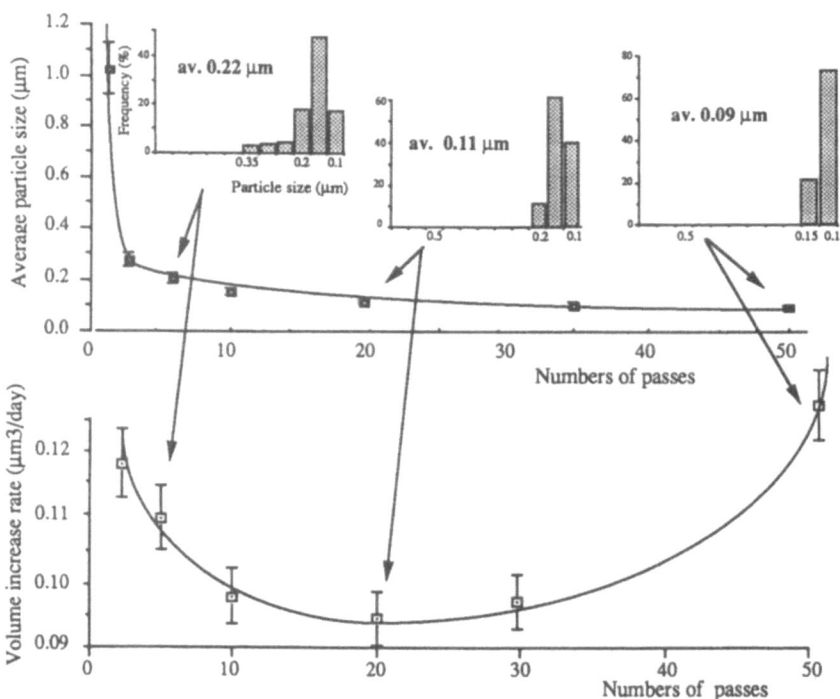

av. 0.22 μm

av. 0.11 μm

av. 0.09 μm

Numbers of passes

Numbers of passes

Fig 5. Influence of the number of passes (a) on the average particle size when measured immediatly after preparation and (b) on the average particle size increase upon aging in 90% w/v PFOB/4% w/v EYP emulsions prepared by microfluidization.

Table 2 collects information on the principal emulsions that have been the object of extended development efforts.

Table 2. Available and recently developed fluorocarbon emulsions.

Fluorocarbon[1]	Concentration v/v (w/v)	Surfactant[1]	Company	Trade Name	Observations	Availability	Ref
							17
FDC/FTPA 7:1	11% (20%)	Poloxamer EYP potassium oleate	Green Cross Corp.	Fluosol®	frozen stem emulsion + 2 annex solutions	commercial, approved for PTCA dec. 89	
FDC/FMCP	11% (20%)	Poloxamer EYP	Soviet Union	Ftorosan	high organ retention of FMCP	?	18
FDC/FTPA	11% (20%)	EYP Poloxamer	China	Emulsion n°II		?	19
FTBA	11% (20%)	Poloxamer	Green Cross Corp.	Oxypherol®	high organ retention,	commercial, for experimental work only	
FMIQ	13% (25%)	EYP K oleate	Green Cross Corp.			?	20
"FMA"	21% (40%)	EYP	Adamantech	Addox	low fluorocarbon definition	abandoned	21
F-44E	40% (78%)	EYP	DuPont	Therox B(40)®		on a collaborative research basis	22
FDC	40% (78%)	EYP	HemaGen		stabilized with triglycerides	?	23
PFOB	47% (90%)	EYP	Alliance Pharm. Corp.	Oxygent®		on a collaborative research basis	9

[1] see Table 1 for abreviations

PRESENT TRENDS

With respect to the first generation Fluosol-type preparations, the second generation of emulsions represents a definite breakthrough in terms of efficacy and stability. Little or no improvement has been made, however, where intravascular persistence is concerned.

Increasing intravascular persistence will certainly be one of the major objectives for future generations of emulsions. This requires that improved control over *in vivo* particle recognition be gained. Improving the "stealthiness" of the fluorocarbon droplets will be, in large part, determined by the external aspect or "tastiness" of the droplets, i.e. by the nature of the surfactant coating [4,24]. The surfactant system used should also allow increased mastery of emulsion characteristics, hence optimization of these characteristics for each one of the specific therapeutic indications considered·

REFERENCES

1   J. G. Riess, Fluorocarbon-based blood substitutes, Artif. Org. **14**:255 (1991).
2   J. G. Riess, Hemocompatible fluorocarbon emulsions, *in*: "Blood Compatible Materials and Devices" C. P. Sharma, M. Szycher, eds,, Technomics Publ. C°, Lancaster, Pa USA (1990), Chap.14, 237.
3   K. C. Lowe, Synthetic oxygen transport fluids based on perfluorochemicals: Applications in medicine and biology, Vox Sang. **60**:129 (1991).
4   J. G. Riess, Fluorocarbon-based *in vivo* oxygen transport and delivery systems, Vox Sang. **60**, xxx.
5   J. G. Riess, Reassessment of criteria for the selection of perfluorochemicals for second-generation blood substitutes: Analysis of structure/property relationships, Artif. Org. **8**:44 (1984).
6   R. . Mitten, A. R. Burgan, A. Hamblin, G. Yee, D. C. Long, D. M. Long and R. F. Mattrey, Dose related biodistribution & elimination of 100% PFOB emulsion, *in*: "Blood Substitutes", T. M. S. Chang, R. P. Geyer, eds, Marcel Dekker, New York (1989), p. 671.
7   J. G. Riess, C. Arlen, J. Greiner, M. Le Blanc, A. Manfredi, S. Pace and C. Varescon, Design, synthesis and evaluation of fluorocarbons and surfactants for in vivo applications. New perfluoroalkylated polyhydroxylated surfactants, *in*: "Blood Substitutes", T. M. S. Chang, R. P. Geyer, eds, Marcel Dekker, New York (1989), p. 421.
8   D. Long, D. M. Long, J. G. Riess, R. Follana, A. Burgan and R. Mattrey, Preparation and application of highly concentrated perfluoroctylbromide fluorocarbon emulsions, *in*: "Blood Substitutes", T. M. S. Chang, R. P. Geyer, eds, Marcel Dekker, New York (1989), p. 441.
9   J G. Riess, J. L. Dalfors, G. K. Hanna, D. H. Klein, M-P. Krafft, T. J. Pelura and E. G. Schutt, Development of highly fluid, concentrated and stable fluorocarbon emulsions for diagnosis and therapy, Proceed. Intl. Symp. Blood Substitutes,
10 R. Follana, D. Klein, M.-P. Krafft, D. M. Long, C. D. Long, J. G. Riess and A. Valla, Prolonged shelf life and biocompatibility of a concentrated injectable fluorocarbon emulsion, Proceed. Intl. Symp. Blood Substitutes,, Montreal, 1991.
11 M.-P. Krafft, J.-P. Rolland and J. G. Riess, Detrimental effect of excess lecithin on the stability of fluorocarbon/lecithin emulsions, J. Phys. Chem. **95**:xxxx (1991).
    12 J. G. Riess and M. Postel, Stability and stabilization of fluorocarbon emulsions, Proceed. Intl. Symp. Blood Substitutes, Montreal, 1991.
13 M.-P. Krafft, J. P. Rolland and J. G. Riess, unpublished results.
14 M.-P. Krafft, F. Giulieri and J. G. Riess, unpublished results.
15 J. L. Dalfors, Terminal sterilization of perfluorocarbon emulsions: Difficulties and possible solutions, Proceed. Intl. Symp. Blood Substitutes, Montreal, 1991.
16 G. K. Hanna, M. C. Ojeda and T. A. Sklenar, Application of computer-based experimental design to optimization of processing conditions for perfluorocarbon emulsion, Proceed. Intl. Symp. Blood Substitutes, Montreal, 1991.
17 R. Naito and K. Yokoyama, Perfluorochemical blood substitutes, FC-43 emulsion Fluosol-DA, 20% and 35%, Technical Information Ser. n°s **5** and **7** (The Green Cross Corp., Osaka, Japan (1978, 1981).
18 F. F. Beloyartsev, E. I. Mayevsky and B. I. Islamov, Ftorosan-oxygen carrying perfluorochemical plasma substitute, *Acad. Sci. USSR*, Pushchino (1983).
19 H.-S. Chen, Z.-H. Yang et al, Perfluorocarbon as blood substitute in clinical applications and in war casualties, in: "Blood Substitutes", T. M. S. Chang, R. P. Geyer, eds, Marcel Dekker, New York   (1989), p. 403.
20 H. Ohyanagi, Y. Saitoh, T. Mitsuno, M. Watanabe, K. Yamanouchi and K. Yokoyama, A new perfluorochemical emulsion: An overview, Intl. J. Artif. Org. **14**:199 (1990).
21 R. E. Moore, Physical properties of a new synthetic oxygen carrier, Biomat., Art. Cells, Art. Org. **16**:443 (1988).
22 Therox, Fluorochemical Oxygen Carriers, DuPont, presentation brochure
23 R. J. Kaufman, Fluorocarbon emulsions as blood substitutes, Workshop on Emulsions, Bergen, Norway, June 1991.
24 J. G. Riess, Fluorocarbon-based oxygen carriers: New orientations, Artif. Org. **14**:xxx (1991).

ACKNOWLEDGMENTS: We wish to thank ATTA and the Centre National de la Recherche Scientifique for their support.

THE RESPIRATORY POTENTIAL OF OXYGEN: A NEW QUANTITY TO CHARACTERIZE STATE, EFFECTS AND BIO-AVAILABILITY OF THE GAS IN ORGANISM

W.K.R. Barnikol

Institute for Physiology and Pathophysiology
Johannes Gutenberg University, Saarstraße 21
D-6500 Mainz, Germany

INTRODUCTION AND PROBLEM

A number of quantities are known which enable to characterize the state of oxygen in blood, for instance: the concentration, which means the physically solved mass per volume; or the content, which comprises the whole mass per volume irrespective of the molecular state; or capacity, which is the chemically bound mass of oxygen per volume and, relatively, the saturation, or the oxygen partial pressure. These various quantities may be divided into two types: the mass-related and the not-mass-related ones.

From the functional point of view a sufficient amount of oxygen has to pass from blood to the tissues with sufficient velocity. It is evident that neither mass-related nor not-mass-related quantities alone can fully characterize the bio-availability of oxygen: a high oxygen partial-pressure alone is not sufficient if there is not at the same time a sufficiently high binding capacity. On the other hand, a high binding capacity alone is not sufficient unless there is, at the same time, a sufficiently high oxygen partial pressure as the driving force for diffusion into the tissues. These considerations show that a new quantity which better characterizes the functional state of oxygen has to integrate both mass-related and not-mass-related quantities.

Proper adjustment of the above-mentioned simple oxygen quantities is an on-going problem in a number of fields: for instance, how to adjust the partial pressure of a 50% haemoglobin saturation and the oxygen carrying capacity of an artificial oxygen carrier in order to adequately supply the substituted organism with oxygen: or, how to choose parameters in intensive care ventilation (e.g. frequency, tidal volume, positive end-expiratory pressure, inspiratory oxygen fraction) in order to supply the patient with oxygen adequately or, how to judge the performance of the lung concerning oxygen uptake; or, how to understand the erythropoietin release: is it governed by a decrease of oxygen partial pressure or of oxygen content? The same question arises in connection with vascular reactions in case of oxygen deficit. Our special problem was to interpret the effect of oxygen shortage in anesthetized rats on ventilation[1].

The search for suitable quantities to characterize the status of oxygen is reflected in the literature[2,3]. Siggaard-Anderson et al. define a so-called O_2 status, which consists of several quantities[4]. Trouwborst et al. regard the oxygen utilization above a (critical) mixed venous oxygen partial pressure[5]. In general, this oxygen partial pressure plays an important role in the considerations of many authors. But the problems mentioned above cannot be tackled with the different quantities introduced by other authors, especially because of lack of a single integrating quantity. The latter would give an unequivocal directive for adjustment of the simple afore mentioned quantities. Therefore the so-called respiratory potential was introduced[6].

DEFINITION AND DESCRIPTION OF THE RESPIRATORY POTENTIAL OF OXYGEN

The respiratory potential of oxygen (RPO_2) is defined as follows:

$$(1) \quad RPO_2 \equiv \int_0^{MO_2} PO_2(MO_2) \, dMO_2$$

PO_2 = oxygen partial pressure, MO_2 = mass of oxygen in a blood volume (V_{BL}) regarded

Evidently the respiratory potential comprises partial pressure and mass of oxygen in a multiplicative manner. From this cardinal quantity two others are derived: first the density ($DRPO_2$) and second the flow ($FRPO_2$) of the respiratory potential.

$$(2) \quad DRPO_2 \equiv RPO_2 / V_{BL}$$

$$(3) \quad FRPO_2 \equiv DRPO_2 \cdot \dot{V}_{BL}$$

$\dot{V}_{BL}$ = blood flow

Substitution of RPO_2 in equation (2) by equation (1) gives

$$(4) \quad DRPO_2 = \int_0^{GO_2} PO_2 (GO_2) \, dGO_2$$

GO_2 = content of oxygen in blood

The density of the respiratory potential of oxygen is graphically visualized in figure 1 as an area (hatched).

The density of the respiratory potential of oxygen is explicitly

$$(5) \quad DRPO_2 = HFZ \, [Hb] \int_0^{SO_2 (PO_2)} PO_2 \, dSO_2 (PO_2) + \tfrac{1}{2} \, \alpha \, O_2 \, PO_2{}^2$$

HFZ = HÜFNER's number, $[Hb]$ = haemoglobin content of blood, $SO_2(PO_2)$ = oxygen binding curve of haemoglobin, $\alpha \, O_2$ = BUNSEN's solubility coefficient

The explicit formula shows that the density of the respiratory oxygen potential depends, besides on oxygen partial pressure, directly on HÜFNER's number, on haemoglobin content of blood, on the oxygen binding curve of haemoglobin and on

BUNSEN's solubility coefficient. It is indirectly influenced by all parameters of the oxygen binding curve like temperature, pH, carbon dioxide partial pressure and 2,3-diphosphoglycerate. This turns out the integrative character of the respiratory potential. With knowledge of all these parameters the density of the respiratory oxygen potential can be calculated.

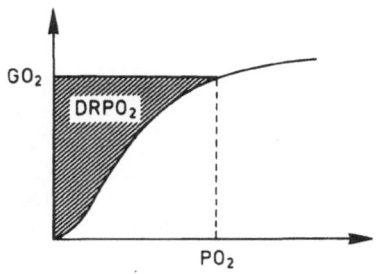

Fig. 1.   Graphical visualization of the density of the respiratory oxygen potential as the (hatched) area, symbols (see text).

APPLICATION OF THE DENSITY OF THE RESPIRATORY POTENTIAL OF OXYGEN ON THE VENTILATORY DRIVE IN ANESTHETIZED RATS INDUCED BY OXYGEN DEFICIT

As mentioned above our problem was to understand the effect of oxygen shortage on ventilation in anesthetized (Nembutal) and tracheotomized rats. Figure 2 show results of a typical experiment[A].

An isovolemic substitution of blood by plasma expander was performed stepwise (within five minutes) and at each step the oxygen fraction of inspiratory gas was decreased while oxygen partial pressure and oxygen content of arterial blood were measured which takes twenty five minutes moreover.

Figure 2 shows clearly that the ventilation of the anesthetized rat is neither un-equivocally dependent on oxygen partial pressure nor on oxygen content, but both quantities are effectors of ventilation.

Figure 2 also proves, that the hypoxic ventilatory drive may be stimulated in two ways, namely by decreasing oxygen partial pressure alone -a well known fact- or by only decreasing the oxygen content (1). According to the experimental design it is a matter of medium-term effects.

A. Data are part of the thesis of St. Guth

When decreasing only the inspiratory oxygen fraction, arterial oxygen partial pressure and oxygen content are functionally coupled, but by variation of inspiratory oxygen fraction and hematocrit it is possible to alter arterial oxygen content and oxygen partial pressure independently from each other. This point is further clarified by Figure 3; the diagrams are constructed from measured data by interpolation.

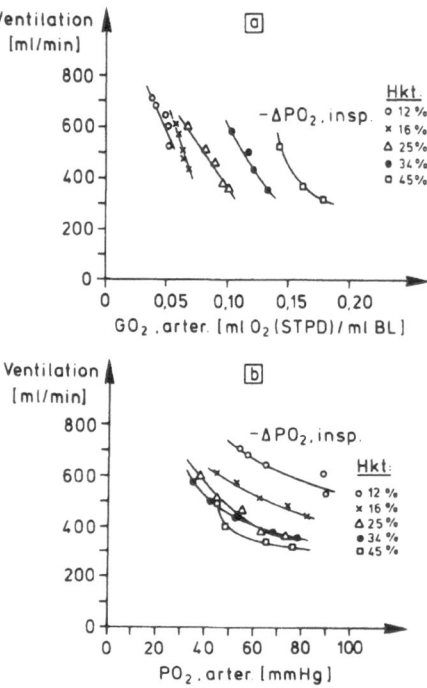

Fig. 2.   Ventilatory drive in anesthetized (Numbutal) and tracheotomized rat induced by hypoxia; Hk t= hematocrit, $GO_2 = O_2$ content, $PO_2 =$ oxygen partial pressure, isovolemic substitution of blood by a plasma expander combined with inspiratory hypoxia (-ΔPO_2, insp., experimental data), points of curves belong to equal hematocrit as indicated, ventilation dependent (a) on arterial oxygen content (b) on arterial oxygen partial pressure.

It is clear from Fig. 3a, that the ventilatory drive induced by decrease of the oxygen content is very weak above 0.10 ml O_2 (STPD) per ml blood, but it is strong below this assessed at different levels of ventilation. In contrast, as Figure 3b shows, the PO_2-answer is influenced decisively by oxygen content: unexpectedly, the PO_2-sensitivity is decreased with decreasing oxygen content of arterial blood. Figure 3 proves the ventilatory drive to be a typical two-effector-system.

The foregoing explanations demonstrate the complexity of the ventilatory answer to hypoxia. Applying the new quantity, based on the same data but by calculating the density of the arterial respiratory oxygen potential and plotting it against ventilation, results within experimental error, in an unequivocal function (Fig. 4). The application of the respiratory potential transforms the hypoxic ventilatory drive from a two-effector-system to a one-effector-system.

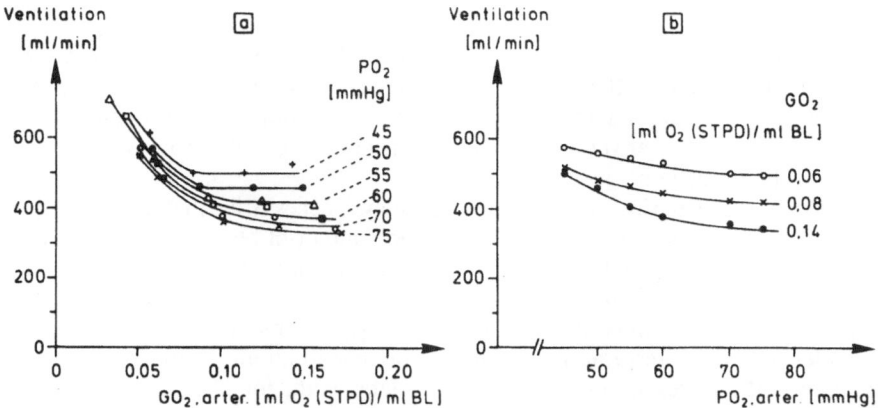

Fig. 3. Ventilatory drive in anesthetized rat induced by hypoxia (interpolated data), GO_2 = oxygen content, PO_2 = oxygen partial pressure: a) Ventilation dependent on arterial oxygen content at different oxygen partial pressures: b) Ventilation dependent on arterial oxygen partial pressure at different oxygen contents.

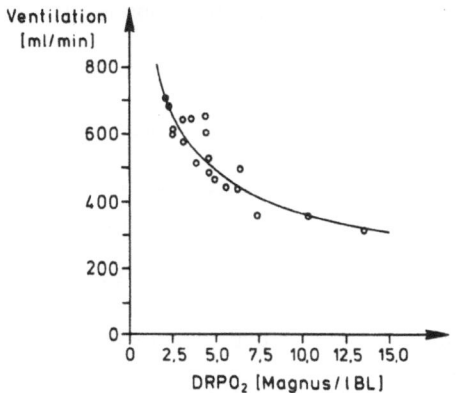

Fig. 4. Ventilation of an anesthetized (Nembutal) rat in dependence on the density of the arterial respiratory oxygen potential, 1 Magnus^B = 1 mmHg. L gas (STPD).

B. In honour of H.G. Magnus (born 1802, died 1870), who was the first to have determined exactly the oxygen and carbon dioxide content of arterial and venous blood from man, horse and cattle: "Über die im Blut enthaltenen Gase, Sauerstoff, Stickstoff und Kohlensäure" (Annalen der Physik und Chemie 40 (1837) 583-606)

DISCUSSION

In our experiment pH-value and carbon dioxide partial pressure were not stabilized in the animal: therefore the first increases and the second decreases during hypoxia. So both parameters could develop their damping effect on ventilation. An increased ventilatory aswer is expected when both pH and carbon dioxide partial pressure should have been stabilized.

From the experiments presented it follows necessarily that there must exist within organism a detecting system for oxygen content of blood, but it is not easy to understand how this systems should work, especially in a direct manner.

The unequivocal relationship between ventilation and the density of the arterial respiratory potential -in contrast, when using oxygen content and partial pressure as variables- prove the respiratory potential to be indeed an integrating quantity. Questionable is if the respiratory potential correlates with, for instance, the activity of special respiratory neurons.

As the definition of the respiratory potential shows, there is no restriction with regard to inhomogeneity of the system. Therefore the new quantity may be applied to blood and also to tissues. But only a micro inhomogeneity (on cellular level) is allowed, because otherwise the integral algorithm is no longer applicable.

To return to the problem mentioned in the beginning: the artificial oxygen carrier to develop should have a maximal density of the respiratory potential. Furthermore, the anesthesiologist should adjust his ventilation parameters in such a way to get a maximal arterial flow of the respiratory potential. Both, the release of erythropoietin and the vascular tone should be governed by the flow of the respiratory oxygen potential. The new quantity -the respiratory potential- may be applied also to carbon dioxide.

REFERENCES

1. St. Guth and W.K.R. Barnikol, Antrieb der Ventilation durch akute Anämie unter Luftatmung in der narkotisierten Ratte, Pfluegers Arch., 418 : R 112, (1991).
2. S.A. Gould, L.R. Sehgal, A.L. Rosen, H.L. Sehgal, H.D. Levine, C.L. Rice, G.S. Moss, Is a normal Hb or P_{50} more important? J. Surg. Res. 33: 189 (1982).
3. O. Müller-Plathe, "Computational" blood oxygen status: 50 (Suppl. 203): 129 (1990).
4. O. Siggaard-Andersen, I.H. Gothgen, P.D. Wimberley, N. Fogh-Andersen, The oxygen status of the arterial blood revised: relevant oxygen parameters for monitoring the arterial oxygen availability, Scand. J. Clin. Lab. Invest. 50 (Suppl. 203): 17 (1990).
5. A. Trouwborst, R. Tenbrinck, E.C.S.M. van Woerkens, S_{35}: A new parameter in blood gas analysis for monitoring the systemic oxygenation, Scand. J. Clin. Lab. Invest. 50 (Suppl. 203): 135 (1990).
6. W.K.R. Barnikol, Das Respiratorische Potential: Ein neuer quantitativer Begriff zur Charakterisierung des Zustandes der Atmungsgase in Körperflüssigkeiten, Pfluegers Arch. 418: R 112 (1991).

FACILITATION OF OXYGEN TRANSFER BY PERFLUBRON IN

HEMODILUTED DOGS

S.M. Cain, S. E. Curtis, and W. E. Bradley

Departments of Physiology and Biophysics and Pediatrics,
University of Alabama at Birmingham, Birmingham, Alabama
35294-0005, U.S.A.

INTRODUCTION

Intentional hemodilution is an alternative to avoid the potential problems associated with blood transfusions. Multipoint measurements of surface PO_2 in several organ systems have indicated that moderate hemodilution may actually increase tissue oxygenation (Messmer, 1973). Theoretical analyses, on the other hand, have indicated that hemodilution may hinder oxygen delivery to tissues because of the increased barrier to diffusion offered by an expanded plasma phase (Homer, 1981; Gutierriez, 1986). The explanation in simple terms is that oxygen can be removed faster from the plasma than it can be released from the red blood cell in the tissue capillary. This results from the very low solubility of oxygen in plasma and the resultant slow diffusion in the plasma phase. Homer et al. (1981) suggested that a faster release and better equilibrium between red cell and plasma would be achieved by increasing oxygen solubility in the plasma.

Faithfull and Cain (1988) followed that suggestion by comparing hemodilution with dextran to that with a commercial perfluorocarbon blood substitute. They tested the hypothesis by progressively hemorrhaging anesthetized dogs to lower total oxygen delivery and determining the critical delivery level at which oxygen demand could no longer be met. Their findings of a lower critical oxygen delivery and higher oxygen extraction at the critical point in the dogs diluted with perfluorocarbon seemed to support the hypothesis that increased plasma oxygen solubility would promote tissue oxygenation. A special circumstance in that study was the fact that the commercial perfluorocarbon emulsion, Fluosol, caused a severe hypotensive episode in the dogs. A similar formulation had been shown to cause significant microcirculatory disturbances (Faithfull et al.,1987). To make a valid comparison, Faithfull and Cain (1988) caused the same hypotensive response in both the dextran and perfluorocarbon diluted dogs with a small preliminary bolus of Fluosol and then allowed all animals to recover before beginning the experimental protocol. Our present study is in response to the question of whether a similar difference in critical oxygen delivery and extraction ratio would be seen if there had been no microcirculatory disturbance caused by the perfluorocarbon emulsion. Accordingly, we repeated the study of Faithfull and Cain but we used a brominated perfluorocarbon (perflubron) emulsion that had no observable effect on the cardiovascular system of anesthetized dogs.

Oxygen Transport to Tissue XIV, Edited by W. Erdmann and
D.F. Bruley, Plenum Press, New York, 1992

479

METHODS

A total of 16 mongrel dogs was used in two groups of 8 each. All were anesthetized (30 mg/kg pentobarbital sodium iv, supplemented as necessary), paralyzed (30 mg succinylcholine chloride im + 0.1 mg/min iv), and pump-ventilated with air to maintain $PaCO_2$ between 30 and 35 torr. Catheters were placed in carotid and pulmonary arteries and in the right femoral vein. Venous outflows from the left hindlimb muscles and from a segment of ileum were isolated as previously described (Stork et al., 1989). Both areas were autoperfused and innervated. Regional blood flows were measured each time arterial and venous blood samples were taken. Regional O_2 uptake was calculated from the results. Whole body O_2 uptake and cardiac output were calculated from measurements of expired gas and the O_2 content difference between arterial and mixed venous blood. Arterial and regional venous blood lactate concentrations were measured and lactate flux across muscle or gut was calculated as the product of arteriovenous difference and regional blood flow.

Each animal was isovolemically hemodiluted with dextran 70 to hematocrit of ~25%. Hematocrit was kept close to that value by the addition of donor red cells as needed. The last step of hemodilution was done with either 6 ml/kg of perflubron (n=8) or with the same volume of carrier emulsion without any perflubron (n=8) instead of dextran. The dose of perflubron was chosen to match the increase in plasma O_2 solubility that was achieved in the Fluosol study (Faithfull and Cain, 1988). Because perflubron was excreted with time, supplementary doses were given to keep the blood concentration near the initial level. When all measured variables were again steady, the experimental protocol was begun. After an initial set of measurements, the animals were progressively hemorrhaged in small steps of 5 ml/kg or less every 15 min with measurements made in the last 5 min of each step. In this manner, 8 to 10 simultaneous measurements of whole body and regional O_2 delivery and uptake and of arterial lactate and regional lactate flux were made. The critical O_2 delivery value was identified from the biphasic relationship Of O_2 uptake to O_2 delivery by the method of Samsel and Schumacker (1988). The critical O_2 extraction ratio was taken as the ratio Of O_2 uptake to delivery at the critical delivery point.

RESULTS

A clearly defined biphasic relationship between whole body O_2 uptake and delivery was seen in all cases so that there was no difficulty in selecting the critical values shown in Table 1. The hindlimb muscles never became limited even as the animal became moribund so that no critical values could be selected for that region. The gut segments in two of the animals in the control series were apparently supply dependent at the highest available O_2 delivery so only 6 of those values are included in Table 1.
There were no significant differences between the two groups in any of the comparisons.
In Figure 1, O_2 uptakes for the whole body and for the two regions are shown as functions of whole body O_2 delivery. For ease of viewing, the information has been binned for each 5 ml/kg/min of O_2 delivery. This figure illustrates the statement just made above; namely, that hindlimb muscle was able to maintain its O_2 uptake even at values of whole body O_2 delivery at which whole body and gut O_2 uptake had become supply dependent.

Table 1. Critical values of O_2 delivery and O_2 extraction ratio in whole body and in gut

	Control	Perflubron
Whole body DO_2	7.4±0.5	7.6±0.4
Gut DO_2	24.9±0.9	29.8±1.1
Whole body O_2ER	0.76±0.01	0.81±0.01
Gut O_2	0.83±0.01	0.79±0.01

Mean±S.E. DO_2 is toal O_2 delivery in ml/min per kg of body weight or organ weight. O_2ER is O_2 extraction ratio.

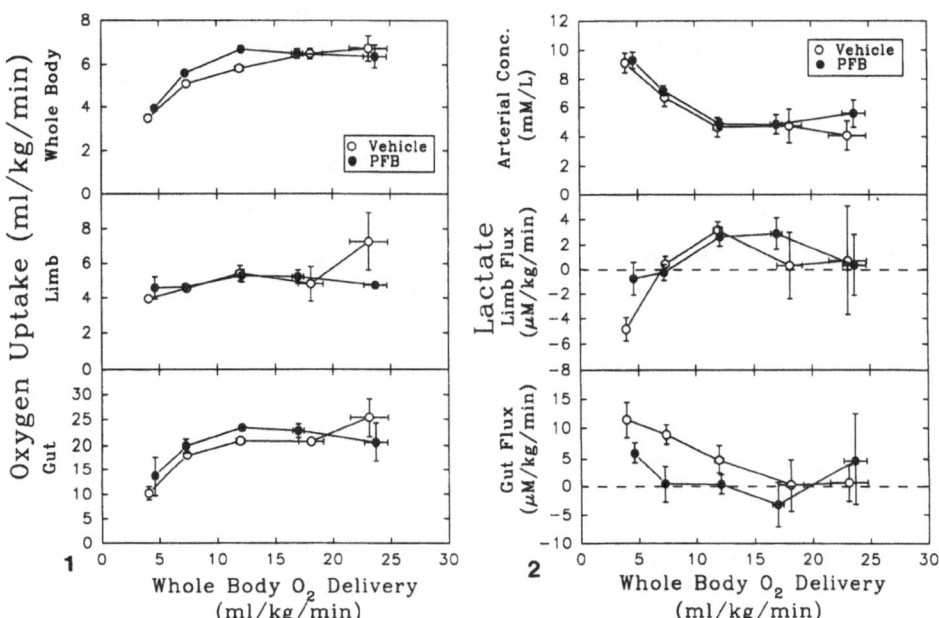

Figure 1. Mean values (± S.E.) for whole body and regional O_2 uptakes binned over intervals of 5 ml/kg/min of whole body total O_2 delivery.

Figure 2. Mean values (± S.E.) for arterial lactate concentration and regional lactate fluxes binned over intervals of 5 ml/kg/min of whole body O_2 delivery.

Figure 2 provides additional evidence that skeletal muscle was able to maintain tissue oxygenation in the face of progressive lowering of O_2 delivery. Arterial lactate concentration and the regional lactate flux rates are shown as a function of whole body O_2 delivery. A negative lactate flux represents uptake of lactate by the region and a positive value is efflux of lactate from the region. Again, the values have been binned for every 5 ml/kg/min of O_2 delivery. Hindlimb muscle can be seen to have increased its average uptake of lactate as the arterial concentration increased whereas lactate efflux began to increase in gut as O_2 delivery decreased below the critical value. There were no discernible differences between the two groups.

DISCUSSION

A common finding in anemia has been that venous PO_2 remains high even as O_2 uptake decreases because of inadequate supply (Cain, 1978; Heusser et al., 1989). This gives rise to the question: if an adequate driving pressure of PO_2 is available, why does the tissue not extract more O_2 to satisfy its demand? One reason for a high venous PO_2 is the fact that the full range of the dissociation curve is available so that even at high O_2 extraction rates, the venous PO_2 will still be higher than for an equivalent extraction in hypoxic hypoxia (Cain, 1983). Another reason is that organ systems with a high ratio of blood flow to O_2 uptake, such as in the kidney, would contribute a significant flow of high PO_2 blood to raise the mixed venous PO_2 (Cain and Chapler, 1978). Finally, the low solubility of O_2 in plasma allows the PO_2 to fall in a tissue capillary faster than it can be replaced from the red blood cell so that a disequilibrium is created in the tissue capillary (Homer, et al., 1981; Gutierrez, 1986). This is corrected after the blood leaves the capillary so that the venous PO_2 rises higher than the PO_2 in the tissue served by that capillary. It was this impediment to O_2 transfer from red cell to tissue that Homer et al. (1981) suggested might be improved by increasing the O_2 solubility in the plasma.

Perflubron has about five times the solubility for oxygen than does the same volume of Fluosol. Even though we only used 6 ml/kg of perflubron in these experiments, the increase in plasma solubility for O_2 should have been the same as in the earlier experiments of Faithfull and Cain (1988) who used 25 to 30 ml/kg of Fluosol to reach the same level of hemodilution. Since the animals were ventilated with air, there was an insignificant increase in actual dissolved O_2 in either case. The question then arises as to why we did not obtain any apparent facilitation of O_2 extraction as they did when the animal was challenged by progressive stepwise hemorrhage to lower O_2 delivery. We suggest that the answer lies in the fact that perflubron given intravenously did not elicit any peripheral vascular response in contrast to the profound hypotension seen when even small boluses of Fluosol were given to dogs (Faithfull and Cain, 1988).

When Faithfull et al. (1987) injected a bolus of a similar perfluorocarbon formulation, FC-43, into anesthetized dogs, they saw a similar severe hypotensive response as that noted with Fluosol. In this particular study, however, they showed that there was a persistent increase in perfusion heterogeneity even after the animals had recovered from the hypotension. There was also a blunting of microcirculatory reactivity to both hyperoxia and to hypoxia which they felt was attributable to the initial reaction to perfluorocarbon injection and which might have been the result of complement activation.

If the Fluosol caused a similar microcirculatory disturbance, then it would explain why Faithfull and Cain (1988) obtained a critical value cf O_2 extraction in their control group which more closely resembled that seen in animals treated with endotoxin. Endotoxin is known to activate complement and generally models the septic state quite closely (Cain and Curtis, 1991). Nelson et al. (1988) found a critical O_2 extraction ratio of 0.78 in their control group. That value was significantly different in the endotoxic group where critical O_2 extraction ratio was 0.54. Compare the latter value with the critical extraction ratio of 0.60 in the control group of Faithfull and Cain (1988) and the former value for the controls with 0.79 in their perfluorocarbon treated group. The two studies differed in that the animals of Nelson et al. (1988) were normocythemic whereas those of Faithfull and Cain were hemodiluted. Other studies have shown, however, that the same increase in O_2 extraction can be obtained in anemic hypoxia as in other forms of hypoxia (Cain, 1989). The implication of these findings is that the facilitation cf O_2 extraction that Faithfull and Cain attributed to the increased plasma solubility obtained with perfluorocarbon hemodilution was detectable only when there had been a marked disturbance in microcirculation.

Biro et al. (1991) also hemodiluted dogs with dextran but used stroma-free hemoglobin as an additive in one group to increase O_2 content of plasma. Instead of lowering whole body O_2 delivery, they lowered blood flow stepwise to the vascularly isolated hindlimb muscles of dogs and thus obtained critical values of O_2 delivery and extraction ratio. They were also unable to find any facilitation of O_2 extraction by increasing plasma O_2 but their values were not different from those found by others in normocythemic dog muscle (Bredle et al., 1989).

CONCLUSIONS

Our main conclusion is that a low plasma solubility for O_2 offers no measurable barrier to tissue O_2 extraction in hemodiluted dogs under normal conditions in which diffusion distances and perfusion heterogeneity are not abnormal. It remains to be seen whether treatment with perfluorocarbon to raise O_2 solubility in plasma will correct the extraction defect that has been observed in endotoxic dogs and septic patients in which such abnormalities are thought to be present.

ACKNOWLEDGEMENT

Funds were furnished by NIH Grant #HL 26926 and by Alliance Pharmaceutical Corp. who also furnished the perflubron emulsion and vehicle.

REFERENCES

Biro, G.P., Anderson, P.J., Curtis, S.E., and Cain, S.M., 1991, Stroma-free hemoglobin: Its presence does not improve oxygen supply to the resting hindlimb vascular bed of anesthetized dogs. Can.J. Physiol. Pharmacol. 69:1656-1662.

Bredle, D.L., Samsel, R.W., Schumacker, P.T., and Cain, S.M., 1989, Critical O_2 delivery to skeletal muscle at high and low PO_2 in endotoxemic dogs. J. Appl. Physiol. 66:2553-2558.

Cain, S.M., 1978, Oxygen delivery and uptake in dogs during anemic and hypoxic hypoxia. J. Appl. Physiol. 42:228-234.

Cain, S.M., 1983, Peripheral oxygen uptake in health and disease. Clin. in Chest Med. 4:139-148.

Cain, S.M. and Chapler, C.K., 1978, O_2 extraction by hind limb versus whole body during anemic hypoxia. J. Appl. Physiol. 45:966-970.

Cain, S.M. and Chapler, C.K., 1989, Circulatory adjustments to anemic hypoxia. Adv. Exp. Med. Biol. 227:103-115.

Cain, S.M. and Curtis, S.E., 1991, Experimental models of pathologic oxygen supply dependency. Crit. Care Med. 19:11 O .

Faithfull, N.S. and Cain, S.M., 1988, Critical levels Of O_2 extraction following hemodilution with dextran or Fluosol-DA. J. Crit. Care 3:14-18.

Faithfull, N.S., King, C.E., and Cain, S.M., 1987, Peripheral vascular responses to fluorocarbon administration. Microvasc. Res. 33:183-193.

Gutierrez, G., 1986, The rate of oxygen release and its effect on capillary PO_2 tension: a mathematical analysis. Resp. Physiol. 63:79-96.

Heusser, F., Fahey, J.T., and Lister, G., 1989, Effect of hemoglobin concentration on critical cardiac output and oxygen transport. Am. J. Physiol. 256:H527-H532.

Homer, L.D., Weathersby, P.K., and Kiesow, L.A., 1981, Oxygen gradients between red blood cells in the microcirculation. Microvasc. Res. 22:308-323.

Messmer, K., Sunder-Plassmann, L., Jesch, F., Gornandt, L., Sinagowitz, E., and Kessler, M., 1973, Res. Exp. Med. 159:152-166.

Nelson, D.P.,Samsel, R.W., Wood, L.D.H., and Schumacker, P.T., 1988, Pathological supply dependence of systemic and intestinal uptake during endotoxemia. J. Appl. Physiol. 64:2410-2419.

Samsel, R.W. and Schumacker, P.T., 1988, Determination of the critical O_2 delivery from experimental data: sensitivity to error. J. Appl. Physiol. 64:2410-2416.

Stork,R.L., Dodd, S.L., Chapler, C.K., and Cain, S.M., 1989, Regional hemodynamic responses to hypoxia and hypermetabolism in polycythemic dogs. J. Appl. Physiol. 67:96-102.

MONITORING OF INTRACAPILLARY HbO2 IN FOETAL SCALP DURING DELIVERY

J. Höper, M. Kessler, K. Frank, D. Tauschek, J. Zündorf,
N. Lang [*] , E. Mauch

Institut für Physiologie und Kardiologie der Universität Erlangen-Nürnberg,
Waldstraße 6, D-8520 Erlangen, [*] Frauenklinik der Universität
Erlangen-Nürnberg, Universitätstraße, D-8520 Erlangen

INTRODUCTION

During delivery the foetus passes periods with decreased blood supply caused by contractions
of the uterus. A healthy foetus is able to tolerate these periods of altered blood supply, while
under pathophysiological and pathological conditions more or less severe damage may be
caused by tissue hypoxia.

At present only two methods are available for monitoring the foetal situation:
cardiotocography (CTG) and foetal blood gas analysis. The CTG allows continuous
registration of the foetal heart rate. It changes due to maternal and foetal factors and thus can
be subject of misinterpretation. The number of wrong positive cardiotocograms is up to 20%.
The foetal blood gas analysis only allows measurements within certain time intervals and thus
does not allow continuous monitoring.

In a first study we investigated whether or not the redistribution of blood flow postulated by
Paulick (1987) can be detected by measurements of intracapillary haemoglobin oxygenation
and concentration in the scalp during delivery.

METHODS

The EMPHO (Erlangen micro-lightguide spectrophotometer, Frank et al. 1989) was used for
non invasive monitoring of oxygen supply to the foetal scalp during delivery. The instrument
enables the measurement of haemoglobin spectra in microvolumes and allows the
determination of intracapillary haemoglobin oxygenation and concentration and thus local
oxygen content. In this paper only the results of the determination of intracapillary
oxygenation will be demonstrated.

The micro-lightguide was integrated into the scalp electrode used for ECG measurements.
This allowed a safe adaptation to the skin surface.

RESULTS

Figure 1 shows the maximal, minimal and mean haemoglobin oxygenation measured during a normal delivery. During the last 2 1/2 hours a continuous decrease was observed. The lowest values approach 5%. These low values disappear within 10 min after birth.

In contrast, after complicated delivery (Fig.2) the number of low HbO_2 values decreased only after insufflation of oxygen (see also Fig. 6). The mean HbO_2 did not differ significantly between normal and complicated delivery.

In figures 3 and 4 the time dependent change in the integrated HbO_2 gradients after normal

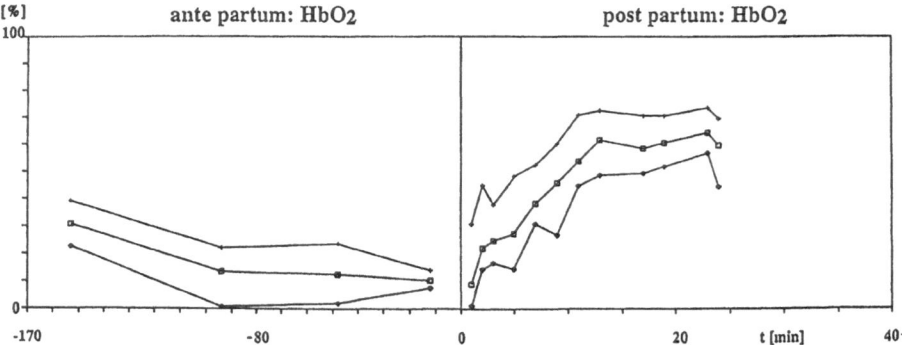

Fig. 1. Time dependent change in haemoglobin oxygenation in the scalp during a normal delivery. The maximal and minimal HbO_2 values as well as the mean value are shown. Note that the time scale changes with birth.

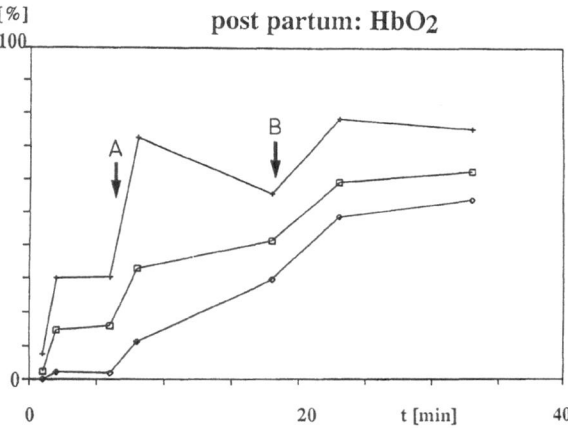

Fig. 2. Time dependent change in haemoglobin oxygenation in the early period after a complicated delivery. A: insufflation of pure oxygen B: mechanical ventilation

and complicated delivery are shown. It can be seen that in both cases a spontaneous increase in the highest HbO_2 values was observed, whereas the low values did not change after complicated delivery.

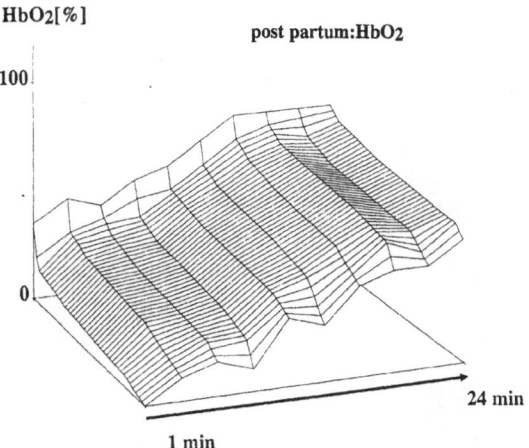

Fig. 3. Time dependent integrated HbO₂ gradients after birth (normal delivery).

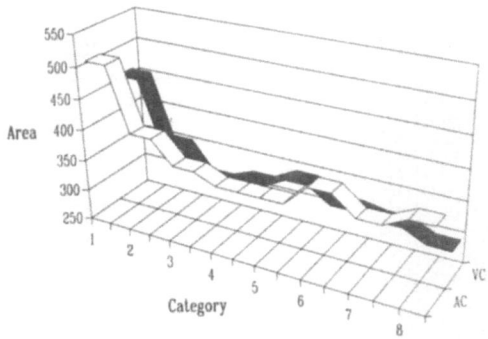

Fig. 4. Time dependent integrated HbO₂ gradients in the early post partal period after a complicated delivery. A: insufflation of pure oxygen, B: mechanical ventilation

Figures 5 and 6 show the percentage of HbO$_2$ values < 20% after normal and complicated delivery. The difference becomes more prominent: there is almost no change in the low values after complicated delivery until oxygen insufflation was started. Only then a rapid improvement was observed.

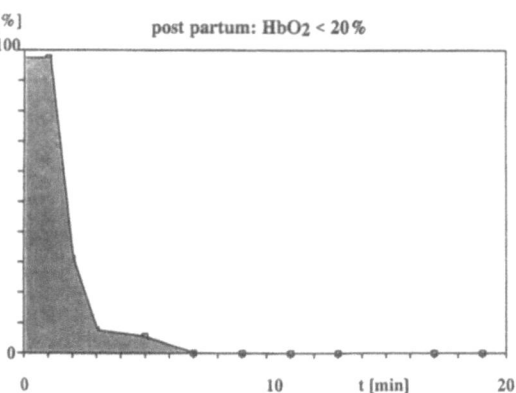

Fig. 5. Percentage of local HbO$_2$ lower 20% during the first 20 min after birth (normal delivery).

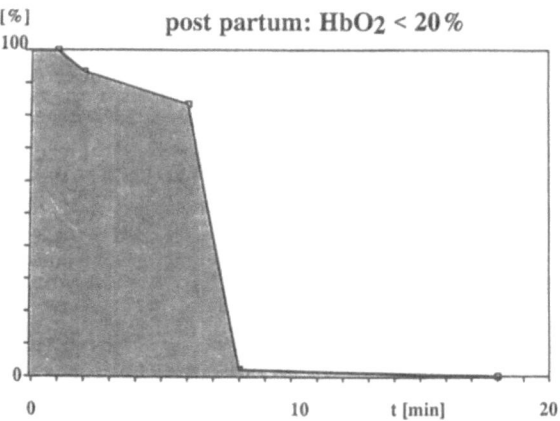

Fig. 6. Percentage of local HbO$_2$ lower 20% during the same period of time after a complicated delivery. Insufflation of oxygen was started after 6 min.

DISCUSSION

The results obtained during normal delivery clearly show that the O_2-supply to the foetal scalp declines during the last two hours ante partum. 12 min before delivery the lowest local HbO_2 values approach 5% saturation. This indicates that skin blood flow is low.

Already in 1960 Scholander suggested that foetal animals may react to asphyxia in a manner like that of diving seals. The diving reaction is characterized by a bradycardia and an increase in peripheral vascular resistance. Bradycardia has been evoked in foetal sheep (Bauer 1937) and human foetus by clamping the umbilical cord (Hon 1966) or by compression of the foetal head (Hon 1958). Franklin et al. (1963) demonstrated that during delivery the blood flow in the foetal abdominal aorta decreased.

From these facts it can be concluded that the changes measured in the foetal scalp during delivery reflect a physiological reaction.

Also during the early post partal period there are intracapillary HbO_2 values below 20% saturation which disappear within a few minutes after normal delivery. Elsner (1978) showed that aortic flow returned to normal values when breathing was well established. This occurs during the first minutes after cord parting.

In contrast to this normal course after complicated delivery the number of HbO_2 values below 20% saturation stays high although the mean value was not significantly different to normal delivery. Only after insufflation of pure oxygen was started the number of low HbO_2 values rapidly declines to normal values.

Summarizing the results it is evident that during delivery critically low peripheral HbO_2 values develop, indicating that the local oxygen reserve is almost exhausted. The most important question of the prognostic meaning of low oxygenation of the scalp during delivery must be analyzed by further investigations. In this context it will be of utmost importance to define the critical threshold of local oxygenation and local oxygen content, respectively. Furthermore the question about the meaning of certain numbers of critical values has to be answered.

REFERENCES

Bauer, D.J. (1937) The slowing of the heart rate produced by clamping the umbilical cord in the foetal sheep. J. Physiol. 90, 25P-27P

Elsner, R. (1978) Asphyxial survival: diving seals and fetal sheep. In: Fetal and Newborn Cardiovascular Physiology, ed. L.D. Longo, 399-411, Garland Publishing, New York.

Frank,K.H., Kessler, M., Appelbaum, K., Dümmler, W. (1989) The Erlangen micro-lightguide spectrophotometer EMPHO I. Phys.Med.Biol. 34, 1883-1900.

Franklin, D.L., Schlegel, W.L., Watson, N.W. (1963) Ultrasonic Doppler shift blood flow meter: Circuitry and practical applications. In: Biomedical Sciences Instrumentation, 390-415. Plenum Press, New York.

Hon, E.H. (1958) The electronic evaluation of the fetal heart rate. Prelim. Report. Am. J. Obstet. Gynecol. 75, 1215-1230.

Hon, E.H. (1966) The human fetal circulation in normal labor. In: The Heart and Circulation in the Newborn and Infant, ed. D.E. Cassels, 37-52. Grune & Stratton, New York.

Paulick, R., Kastendieck, E., Weth, B., Wernze, H. (1987) Metabolische, kardiovaskuläre und sympathoadrenale Reaktionen des Feten auf eine progrediente Hypoxie - Tierexperimentelle Untersuchungen. Z. Geburtsh. u. Perinat. 191, 130-139.

Scholander, P.F. (1960) Experimental studies on asphyxia in animals. In: Oxygen Supply to the Foetus, ed. J. Walker and A.C. Turnbull. Blackwell, Oxford.

O_2 TRANSPORT DURING EXERCISE AFTER CARDIAC TRANSPLANTATION

M. Meyer[1], P. Cerretelli[2], C. Cabrol[3] and J. Piiper[1]

[1]Department of Physiology, Max Planck Institute for Experimental Medicine, Göttingen, Germany ; [2]Département de Physiologie, Centre Médical Universitaire, Genève, Switzerland; [3]Département de Chirurgie Cardiovasculaire, Université de Paris VI, Hôpital de la Pitié-Salpétrière, Paris, France

INTRODUCTION

Over the past two decades cardiac transplantation has emerged as a viable treatment of terminal heart failure and actuarial survival statistics have approached 75% for the first 5 years after operation. Though substantial differences exist between the performance of the cardiac allograft and the normal heart, the patient´s quality of life is improved considerably and many cardiac allograft recipients have returned to a normal life (3, 7).

The most obvious consequence of cardiac transplantation is the increased resting heart rate due to withdrawal of parasympathetic flow and the delayed and more gradual pattern of response to dynamic exercise (3, 5, 6). Though the transplanted heart can increase its output considerably, maximal exercise capacity is reduced by 30 - 50% compared to normal individuals (1, 2, 3, 6, 7, 9). Since the contribution of heart rate to the increase of cardiac output at the onset of exercise is negligible, the altered time course of readjustment of cardiac output and the lower peak heart rates achieved have been considered as important factors limiting the physical performance of the allograft recipient (3, 7, 8).

The aim of the present study was to get a better understanding of the mechanisms whereby the transplanted heart increases its output in response to increasing metabolic demands and of the role of central and peripheral mechanisms limiting the individual´s exercise capacity after transplantation.

METHODS

The studies were performed in the Department of Cardiovascular Surgery of La Pitié-Salpétrière Hospital at Paris. Twenty-one short-term and long-term (over one year) recipients of an orthotopic cardiac transplant (HTX; mean age ± SD: 44.5 ± 8.1 yrs, range 28 - 59 yrs) were investigated 24.2 ± 30.0 mo (means ± SD, range: 1.3 - 137.1 mo) after operation in the course of their routine clinical evaluation. They received immunosuppressive treatment protocols (cyclosporine A, low quantities of oral prednisone) and were recreationally active. Ten normal sedentary untrained subjects (CTL) served as controls (mean age ± SD: 36.6 ± 9.8 yrs).

Continuous breath-by-breath monitoring of O_2 uptake ($\dot{V}O_2$) and CO_2 output ($\dot{V}CO_2$) was performed by a SensorMedics 4400tc analyzer. Beat-by-beat monitoring of stroke volume (SV) and cardiac output ($\dot{Q}$) was performed by impedance cardiography. Baseline thoracic impedance (Z_0), change of impedance (dZ/dt) and maximum of impedance derivative (dZ/dt_{max}) along with detection of systolic time intervals (pre-ejection period, left ventricular ejection time) and heart rate (HR) facilitated calculation of beat-by-beat stroke volume according to the formula of Kubicek et al. (4) and cardiac output ($\dot{Q} = SV \cdot HR$). Continuous recording of HR, SV and $\dot{Q}$ in a heart transplant recipient during exercise over 25 min is shown in Fig. 1.

Exercise testing was performed on an electromagnetically braked cycle ergometer with the subject in upright sitting position. The protocol consisted of 10 min rest followed by two 5 min 50 W bouts of square-wave exercise (*E1* and *E2*, respectively) separated by 5 min recovery (*R1*) with additional 5 min recovery following the 2nd bout of exercise (*R2*). Because the symptom-limited maximum power output ($\dot{V}O_{2max}$) of HTX generally ranges from about 50 - 70% of predicted $\dot{V}O_{2max}$ which corresponds to 75 - 100 W, respectively, a submaximal load of 50W was selected to enable all patients to comply with the imposed work load.

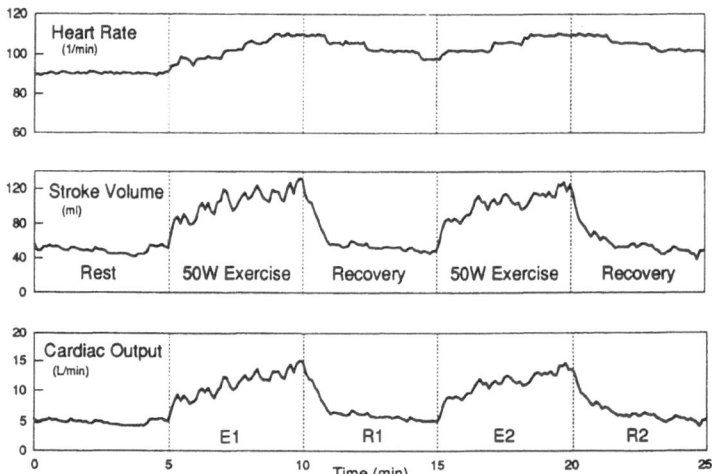

Figure 1. Continuous recording of heart rate (*upper panel*), stroke volume (*middle panel*) and cardiac output (*lower panel*) by impedance cardiography in a heart transplant recipient during two-stage 5 min 50 W exercise testing.

RESULTS

Metabolic and cardiovascular parameters during rest and submaximal square-wave 50 W exercise are compiled in Tab.1 for heart transplant recipients (HTX) and normal control subjects (CTL). O_2 uptake ($\dot{V}O_2$) and CO_2 output ($\dot{V}CO_2$) were essentially the same in both HTX and CTL. For the same submaximal work load (50 W) $\dot{V}O_2$ was similar in both groups indicating that, despite chronic immunosuppressive therapy (prednisone, cyclosporine A), the mechanical efficiency of exercise was unchanged in HTX.

TABLE 1. *Cardiopulmonary parameters in heart transplant recipients (HTX) and healthy control subjects (CTL) during rest and 50 W square-wave cycloergometric exercise.*

| | Rest | | Exercise (50 W) | |
	HTX	CTL	HTX	CTL
Age (yrs)	44.5 ± 8.1	36.6 ± 9.8		
Height (cm)	171 ± 6	177 ± 6		
Weight (kg)	69 ± 9	75 ± 13		
$\dot{V}O_2$ (L/min)	0.27 ± 0.05	0.28 ± 0.04	0.96 ± 0.1	0.95 ± 0.08
$\dot{V}CO_2$ (L/min)	0.24 ± 0.04	0.22 ± 0.03	1.02 ± 0.19	0.82 ± 0.06
HR (min-1)	102 ± 12	72 ± 9	120 ± 12	98 ± 16
BP s/d (Torr)	$145/105 \pm 15/10$	$120/85 \pm 20/10$	$175/95 \pm 20/15$	$150/80 \pm 25/15$
SV (ml)	71 ± 26	94 ± 16	130 ± 39	137 ± 50
$\dot{Q}$ (L/min)	7.2 ± 2.4	6.8 ± 1.5	15.6 ± 5.3	12.8 ± 3.6

$\dot{V}O_2 = O_2$ uptake, $\dot{V}CO_2 = CO_2$ output, HR = heart rate, BP = arterial blood pressure; s/d = systolic/diastolic, $\dot{S}V$ = stroke volume, $\dot{Q}$ = cardiac output. Mean values $\pm$ SD (HTX n = 21; CTL n = 10).

The high heart rate (HR) during rest (102 ± 12 min-1) and submaximal work load (120 ± 12 min-1) is characteristic for the chronically denervated heart. The heart rate of HTX increases with dynamic exercise but the pattern of response is different from that of CTL. At the onset of exercise HR rises more gradually and subsides during recovery from exercise (*cf*. Fig. 1). In the absence of autonomic control, the increase of HR is mainly mediated by increasing levels of circulating catecholamines though additional factors are expected to contribute to the persisting elevation of HR at the offset of exercise. At the end of the two sequential 50 W exercise tests HR approached similar levels in both HTX and CTL though the absolute values, about 120 min-1 in HTX and 100 min-1 in CTL, were different.

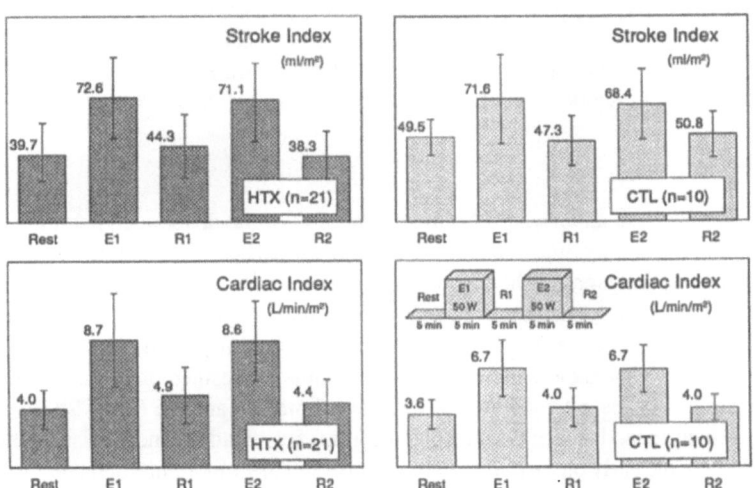

Figure 2. Stroke index and cardiac index in heart transplant recipients (HTX, *left*) and normal control subjects (CTL, *right*). E1, E2 and R1, R2 refer to individual stages of the experimental protocol (*insert, lower right*).

Cardiac output ($\dot{Q}$) was essentially the same between HTX and CTL at rest but stroke volume (SV) was less compared to CTL (*cf.* Tab. 1). The transplanted heart, in the absence of autonomic innervation, is able to increase its output in response to dynamic load. During 50 W exercise $\dot{Q}$, for the same level of $\dot{V}O_2$, in HTX exceeded that of CTL but SV was essentially similar in the two groups. The steady-state values of stroke index and cardiac index at the various stages of the experimental protocol are shown in Fig. 2. For repetitive testing with the same submaximal work load (50 W) both indices approached similar levels but the absolute value of cardiac index in HTX was greater than in normal controls. Thus, cardiac output of the transplanted heart, by virtue of compensatory mechanisms, may be similar (resting conditions) or slightly higher (submaximal dynamic exercise) than that of normally innervated hearts.

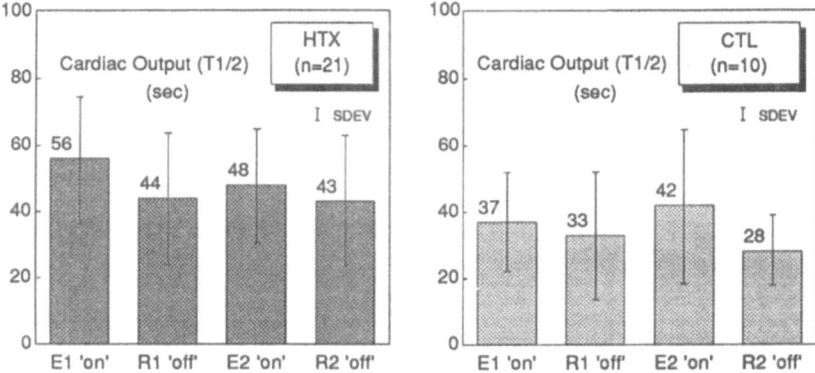

Figure 3. Dynamics of adjustment of cardiac output at the onset and offset of square-wave 50 W cycloergometric exercise in heart transplant recipients (HTX, *left*) and normal control subjects (CTL, *right*). The rest (work)-to work (rest) kinetics are quantified by half-time (t½, in sec) required for cardiac output to attain 50% of the steady-state or peak value following a rectangular increase of 50 W power.

The dynamics during exercise and recovery expressed by half-time (t½, sec) of the on- and off- responses of $\dot{Q}$ are summarized in Fig. 3. The results demonstrate that in HTX the kinetics of adjustment of $\dot{Q}$ is only moderately impaired and would be caused by delayed onset of venous return (*see below*). The kinetics for repetitive 50 W testing were essentially similar and were not different at higher work load (75 W) imposed on some of the HTX.

DISCUSSION

The mechanisms underlying the adjustment of $\dot{Q}$ of the denervated heart are different from those of the normally innervated heart. The innervated heart responds to exercise with an almost instantaneous increase of heart rate with little increase in stroke volume. Whereas in the normal heart the Frank-Starling mechanism plays a minor role for the adjustment of stroke volume, the increase of cardiac output early in exercise (before any change in heart rate) appears to be exclusively mediated by an increased stroke volume as a result of increased venous return from exercising muscles, *i.e.* mediated by augmented preload and the Frank-Starling mechanism. The unique dependence of the transplanted heart on preload as a mechanism of increasing cardiac output in the early phase of dynamic exercise and the importance of the Frank-Starling mechanism is well documented by the recording of Fig. 1. At higher work loads and/or prolonged exercise the chronotropic and inotropic effects of circulating catecholamines, augmented by hypersensitivity of the transplanted heart, cause a further increase of cardiac output by increasing heart rate, circumferential fiber shortening and ejection fraction. The present data suggest that the transplanted human heart retains a remarkable capacity to adjust to the metabolic demands of exercise. Thus, during submaximal exercise, the limiting role of $\dot{Q}$ for the physical performance of the allograft recipient does not appear to prevail over that in normal healthy subjects.

In the present study the adjustment of cardiac output of HTX has been analyzed in terms of steady-state responses to dynamic eyercise as well as during rest (work)-to-work (rest) transitions and compared to normal CTL. The dependence of the transplanted heart on preload and the Frank-Starling mechanism as a means of increasing cardiac output differs substantially from the normal heart with intact neural control. But the compensatory role of these mechanisms provides for almost normal kinetics of readjustment of cardiac output. Hence, O_2 transport to tissues would not be expected to be limited by the transplanted heart or gas exchange occuring at the lungs.

Unequivocally, the results from the present study have demonstrated that the cardiovascular adjustment in HTX is rapid and fully adequate during submaximal exercise and would explain the excellent degree of rehabilitation from terminal cardiac disease gained by most patients after cardiac transplantation. With the present information, $i.e.$ rapid cardiorespiratory kinetics, along with preliminary evidence for absence of an early massive formation of lactic acid in venous blood and normal kinetics of adjustment of peripheral blood flow, it may be concluded that central cardiorespiratory failure of O_2 transport to exercising skeletal muscles is unlikely to account for the impairment of maximum exercise capacity in HTX.

It is of interest to note that HTX (and kidney transplant recipients) undergoing cycloergometric exercise testing often complain of leg muscle fatigue and generally do not stop exercise prematurely because of symptom- or sign-limited factors. Generally, cardiac patients have experienced a long history of inactivity and physical deconditioning. A reduction of muscle mass possibly involving structural changes of skeletal muscle presents a common observation in HTX and, associated with myopathic effects of chronic immunosuppressive therapy, likely plays a major role in limiting maximum exercise performance. Studies in renal transplant recipients and the results of a 2-year exercise rehabilitation program ($cf.$ 3) have demonstrated that strengthening of peripheral muscles rather than improvements of central hemodynamics play an important role but may not yield full restoration of skeletal muscle function. It appears that pertinent evidence points to a peripheral skeletal muscle limitation that is independent of central and peripheral O_2 transport, and all allograft recipients may suffer from the side-effects of the immunosuppressive regime.

It is well recognized that other factors may ultimately contribute to the reduced exercise capacity of HTX. Systemic hypertension elicited or contributed to by immunosuppressive drug treatment is of substantial concern. Other factors include chronic anemia resulting from cyclosporine nephrotoxicity, chronic graft rejection and graft atheroslcerosis that may lead to impaired allograft function.

REFERENCES

1. Banner, N.R., Lloyd, M.H., Hamilton, R.D., Innes, J.A., Guz, A. and Yacoub, M.H., 1989, Cardiopulmonary response to dynamic exercise after heart and combined heart-lung transplantation. *Br. Heart J.* 61: 215-223.

2. Cerretelli, P., Grassi, B., Colombini, A., Carù, B. and Marconi, C., 1988, Gas exchange and metabolic transients in heart transplant recipients. *Respir. Physiol.* 74: 355-371.

3. Kavanagh, T., Yacoub, M.H., Mertens, D.J., Kennedy, J., Campbell, R.B. and Sawyer, P., 1988, Cardiorespiratory responses to exercise training after orthotopic cardiac transplantation. *Circulation* 77: 162-171.

4. Kubicek, W.G., Karnegis, J.M., Patterson, R.P., Witsoe, D.A. and Mattson, R.H., 1966, Development and evaluation of an impedance cardiac output system. *Aerosp. Med.* 37: 1208-1212.

5. Pflugfelder, P.W., McKenzie, F.N. and Kostuk, W.J., 1988, Hemodynamic profiles at rest and during supine exercise after orthotopic cardiac transplantation. *Am. J. Cardiol.* 61: 1328-1333.

6. Pope, S.E., Stinson, E.B., Daughters, G.T., Schroeder, J.S., Ingels, N.B. and Alderman, E.L., 1980, Exercise response of the denervated heart in long-term cardiac transplant recipients. *Am. J. Cardiol.* 46: 213-218.

7. Savin, W.M., Haskell, W.L., Schroeder, J.S. and Stinson, E.B., 1980, Cardiorespiratory responses of cardiac transplant patients to graded, symptom-limited exercise. *Circulation* 62: 55-60.

8. Schroeder, J.S., 1979, Hemodynamic performance of the human transplanted heart. *Transplant. Proc.* 11: 304-308.

9. Theodore, J., Morris, A.J., Burke, C.M., Glanville, A.R., Van Kessel, A., Baldwin, J.C., Stinson, E.B., Shumway, N.E. and Robin, E.D., 1987, Cardiopulmonary function at maximum tolerable constant work rate exercise following human heart-lung transplantation. *Chest* 92: 433-439.

EFFECT OF LOW DOSE OXYGENT™ ADDED TO BLOOD ON MUSCLE $\dot{V}O_2$MAX

M. C. Hogan[1], D. Willford[1], N. S. Faithfull[2] and P. D. Wagner[1]

[1]Department of Medicine, University of California San Diego
San Diego, La Jolla, California, U.S.A.
[2]Alliance Pharmaceutical Corp., San Diego, California, U.S.A.

Federspiel et al.[1] have suggested that the plasma space between red blood cells in capillaries does not contribute significantly to diffusive O_2 transport out of the muscle vasculature due to the low plasma solubility of O_2. To test this, we pump-perfused in situ isolated canine gastrocnemius muscle stimulated electrically to produce $\dot{V}O_2$max under two conditions [with blood flow 120 ml 100 gm^{-1} min^{-1}, [Hb] 8.7 gm dl^{-1} and $PaO_2 \sim 500$ torr in each case]: A) control, and B) with Oxygent (perflubron emulsion, Alliance) added at 6 gm perflubron/70 ml. Total plasma O_2 solubility was taken to be 0.003 ml 100 ml^{-1} $torr^{-1}$ in A) and computed to be 0.005 ml 100 ml^{-1} $torr^{-1}$ in B) from the increase in total $[O_2]$ in B) compared to A).

If Federspiel's hypothesis is correct, Oxygent administration should have led to a higher muscle O_2 diffusing capacity (DO_2) due to augmented plasma O_2 solubility, but this was not found. Instead, the significant increase in $\dot{V}O_2$max associated with Oxygent was explained purely by the increased convective delivery of O_2 into the muscle circulation (i.e., higher CaO_2) which, at the same blood flow rate as in control conditions, led to a higher mean capillary PO_2 (and hence higher $\dot{V}O_2$max for the same O_2 diffusing capacity). Our results: 1) do not support Federspiel's hypothesis of the importance of the plasma space between red cells for O_2 transport limitation, 2) suggest that the [Hb]-dependence of muscle O_2 diffusing capacity we have previously observed[2] is not explained by changes in inter-red cell spacing, and 3) show that even breathing 100% O_2, $\dot{V}O_2$max can be increased (under the present conditions) when O_2 supply is increased (in this case, by addition of Oxygent).

TABLE 1. RESULTS

	Condition	$\dot{V}O_2$	O_2 Delivery	Blood Flow	O_2 Extraction	PaO_2	CaO_2	PvO_2	$PCAP_{O2}$	DO_2
		ml min^{-1} 100 g^{-1}	ml min^{-1} 100 g^{-1}	ml min^{-1} 100 g^{-1}	%	torr	ml 100 ml^{-1}	torr	torr	O_2 diffusing capacity ml min^{-1} torr^{-1} 100 g^{-1}
A	CONTROL	9.3	16.1	119	57.0	498	13.65	31.6	56.6	.167
B	OXYGENT	10.3	17.7	120	58.0	552	14.82	33.1	65.5	.161
	P	.003	.001	NS	NS	.002	.001	NS	.001	NS

ACKNOWLEDGEMENTS

Supported by Alliance Pharmaceutical Corp. (R-05-07) and NIH grant HL 17731.

REFERENCES

1. W. J. Federspiel and, A. S. Popel, A theoretical analysis of the effect of the particulate nature of blood on oxygen release in capillaries, Microvasc. Res. 32:164-189 (1986).

2. M. C. Hogan, D. E. Bebout and, P. D. Wagner, Effect of hemoglobin concentration on maximal O_2 uptake in canine gastrocnemius muscle in situ, J. Appl. Physiol. 70(3):1105-1112 (1991).

FETAL OXYGENATION IN CHRONIC MATERNAL HYPOXIA;

WHAT'S CRITICAL?

Berend Oeseburg[1], Biny E.M. Ringnalda[1], Jane Crevels[2], Henk W. Jongsma[2], Paul Mannheimer[3], Jan Menssen[2] and Jan G. Nijhuis[2]

Perinatal Research Group, [1]Department of Physiology and of [2]Obstetrics & Gynecology. [3]Nellcor Incorporated Hayward CA.
University of Nijmegen, PO BOX 9101, 6500 HB Nijmegen, the Netherlands

INTRODUCTION

For the study in an animal model of the consequences of chronic maternal hypoxia on several physiological variables (e.g., fetal ECG variability, fetal breathing patterns), as potential clinical indicators for insufficient oxygen supply to the fetus, the continuous availability of a reliable signal on fetal oxygenation is necessary. This signal is needed as a feedback for the control of maternal inspiratory oxygen fraction (F^iO_2) in the model. Theoretically, fetal arterial oxygen saturation is the indicator of choice for the observation of oxygen availability, assuming hemoglobin concentration is normal as well as constant. The development [1] of pulse oximetry as a non-invasive optical technique seemed promising for the application in fetal studies [2], especially in chronically instrumentated fetal lambs. To evaluate the possibilities of this technique, we tested pulse oximetry transducers, both in transmission and reflection mode, against blood gas analysis before, during and after periods of maternal hypoxia in sheep. From the results of the blood gas measurements we tried to find objective criteria to determine the onset of fetal endanger, induced by maternal hypoxia.

MATERIALS AND METHODS

Operation was carried out on 7 pregnant sheep at 126-133 days of gestation age. The fetuses were partially exposed during caesarian section and instrumentated as follows. ECG-electrodes and electrocorticogram electrodes were sutured subcutaneously. A catheter for arterial sampling and arterial pressure monitoring was introduced in a brachial or carotid artery so far that the tip remained in a pre-ductal position. Standard transmission pulse oximetry sensors (Nellcor° Oxiband) were placed around a forelimb muscle using a stainless steel support to suppress motion artifacts. Experimental reflectance pulse oximetry sensors (Nellcor°) were sutured on skull or forelimb. After an intra-uterine catheter for the measurement of pressure the fetus was placed back and the uterus closed. All cables and canulas were tunnelled subcutaneously to the backside of the ewe and packed into a pocket. The ewe got a tracheal canula for the administration of inspiratory gas mixtures and after at least 3 days of recovery measurements were started. Continuous acquisition and processing of all fetal signals was performed on PC based systems. The pulse-oximeter signals were synchronized using the fetal ECG. Before starting to lower the oxygen supply to the ewe, baseline readings were obtained for at least 1 h. Step wise changes in maternal oxygen supply were induced by lowering the O_2 fraction of the gas supplied to the tracheal canula of the ewe (breathdown). A certain level of hypoxia was maintained until stable values on fetal arterial oxygen saturation were obtained. Measurements were continued, after restoring room air breathing, until pre-hypoxia values were found. Fetal arterial blood samples were drawn every 15 min and analyzed within 5 min using an IL 1312° for blood gases and IL 482° for oxygen saturation.

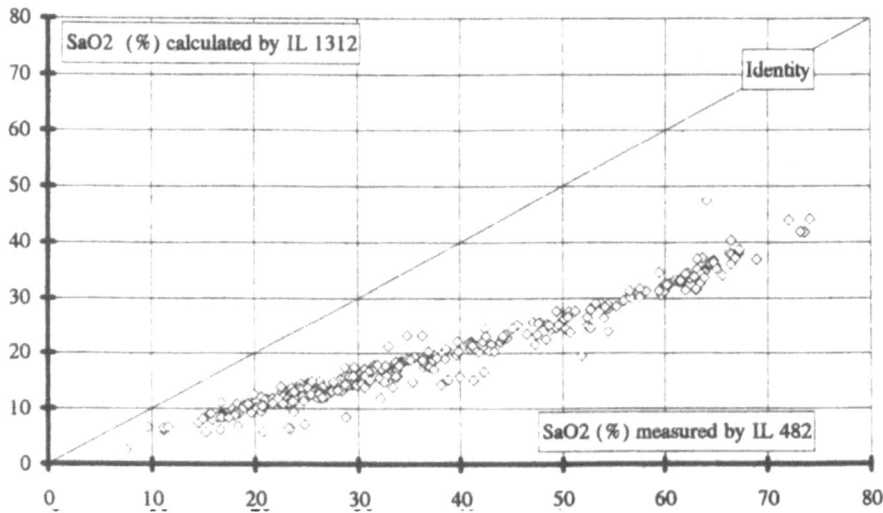

Figure 1. Relation between calculated and measured hemoglobin oxygen saturation in 237 fetal arterial samples.

A multi wavelength oximeter for the real measurement of hemoglobin oxygen saturation (SO_2) in stead of relying on calculated values from a blood gas machine is mandatory since the difference in hemoglobin oxygen affinity between human and sheep blood. Not only have fetal lambs a p(50) of 19 mm Hg in stead of 26.5 as assumed by the

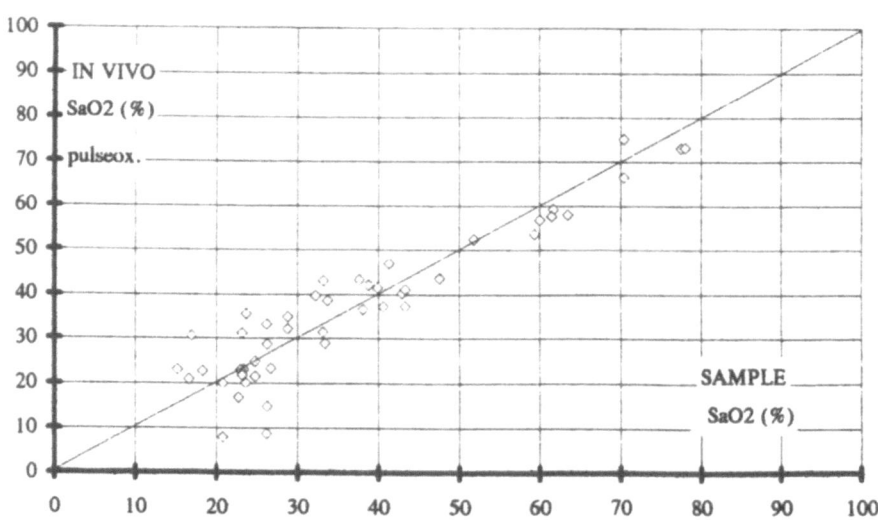

Figure 2. Calibration curve of a reflectance pulse-oximeter sensor after 8 days of implantation.

algorithm of the blood gas machine, but also the Bohr effect is smaller than in human blood [3]. Each animal was allowed to recover for at least 48 h before a new breathdown was started.

RESULTS

In 4 lambs a total of 12 successful breathdowns were carried out with stable fetal hypoxia periods of 15 to 300 min (figure 3 A and B).

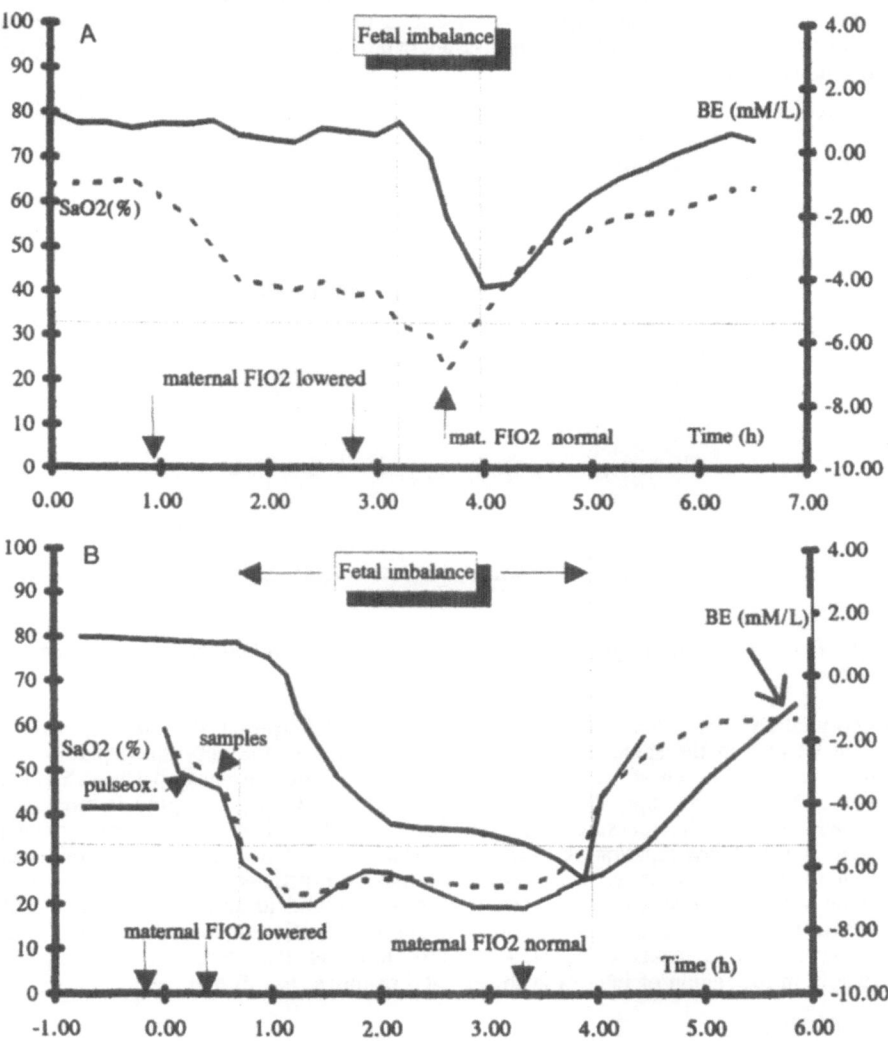

Figure 3 A and B. Short (A) and longer (B) term maternal hypoxia leading to transient fetal imbalance in 2 different sheep. Onset as well as offset of imbalance starts at fetal arterial oxygen saturation of about 30%.

A change in fetal base excess (BE) exceeding -1 mM/L was considered as a sign for the onset of fetal imbalance due to insufficient oxygen supply. In all experiments fetal arterial saturations at these points were at about 30% as shown in figure 3. This was found both at fast as well as step wise slow onset of hypoxemia. After restoring maternal normoxia fetal BE increased again at fetal arterial saturations of around 30% (figure 3). When reliable pulse oximetry signals were obtained the resultant oxygen saturation readings were consistent with the measured oxygen saturation values, as shown in the recording of figure 3B. However, as could be concluded from the scatter in the calibration line of figure 2, the pulse-oximeter recording in figure 3B shows clearly that so far the sample measuring procedure is still better than continuous pulse oximery, especially at the low saturation values.

CONCLUSIONS

The data obtained so far suggest that fetal compromise starts below a certain fetal arterial hemoglobin oxygen saturation value. This is indicated by the change in fetal arterial BE, caused by an increase in lactic acid concentration. The instantaneous decrease in lactic acid during recovery is an indication of the high capacity for lactic acid removal in fetal and placental tissues.

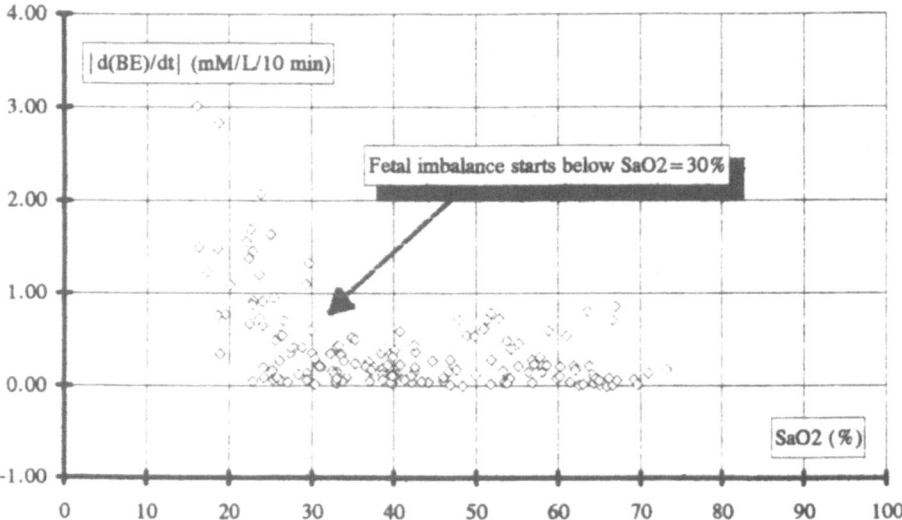

Figure 4. Relation between fetal arterial oxygen saturation and change in base excess.

No correlation was found between the change in fetal arterial BE and the arterial oxygen tension in the samples. This is presumably due to the very steep part of the oxygen dissociation curve as well as to the Bohr shift during imbalance hypoxia periods. These results suggest that the sufficiency of fetal oxygen supply is monitored more adequate by the measurement of oxygen saturation then by measuring oxygen tension. Further investigation is needed for elucidating these findings in diminished placental gas transfer as well as during uterine contractions. If a discrete critical absolute value for fetal arterial oxygen saturation is found in animal studies then still the clinical relevance needs to be established. The introduction of the non-invasive pulse oximetry technique seems a big step forward, however the scatter in these signals, especially in the region of interest at lower SaO_2 readings, is still high.

REFERENCES

1   Yoshiga I, Shimada Y and Tanaka K. Spectrophotometric monitoring of arterial oxygen saturation in the fingertip.
    Med. Biol. Eng. Comput. 1980; 18: 27-32
2   Wijkstra S, Schuiling G, Kwant G and Oeseburg B. Pulse-oximetry in fetal rats. in: Fetal and neonatal physiological measurements III.
    Eds: Gennser et al. Malmo 1989; 75-79
3   Kwant G, Oeseburg B, Zijlstra W.G. and Zwart A. Direct and indirect deter-mination of the CO_2 Bohr effect in human whole blood.
    J. of Physiol. 1985; 366: 56
4   Jongsma H.J, Crevels J, Menssen J.J.M, Arts T, Mulders L.G.M, and Nijhuis J.G. Application of transmission and reflectance pulse oximetry in fetal lambs. in: Fetal and neonatal physiological measurements IV. Eds: Lafeber et al. In press.

THE RELATION OF OXYGEN DELIVERY TO UTILIZATION DURING LIVER TRANSPLANTATION: IS THERE A CRITICAL VALUE?

Steltzer H, Hiesmayr M, Tuchy G, Zimpfer M

Department of Anesthesia and General Intensive Care, University of Vienna A - 1090 Vienna, Austria

Oxygen metabolism in man can be described by oxygen consumption (VO_2) and oxygen delivery (DO_2). Both in healthy individuals, irrespective of the technique of measurement, and in critically ill patients, if VO_2 is measured directly, VO_2 and DO_2 are not correlated (1). Hence the relationship between these two variables is of increasing interest and has been extensively investigated and reviewed. In the course of these studies, a number of clinical and experimental investigations have described a flow dependency of VO_2 below a critical value (2). In patients with severe critical illness, e.g. septicemia, ARDS and fulminant liver failure, this critical offset point with regards to flow dependency of VO_2, appears to be even higher. Consequently, there are various studies on these parameters resulting in different concepts of how to optimize DO_2 and VO_2. In liver transplant recipients it was assumed that inadequate oxygenation of some tissues occurred already with much higher levels of oxygen transport than normal (3). However, the relation between DO_2 and VO_2 in patients with liver failure has not been studied systematically. The goal of this study was to investigate the relationship of VO_2 and DO_2 during typical phases of liver transplantation and to determine whether a threshold value could be detected in this particular group of critically ill patients.

Methods: 50 consecutive patients undergoing liver transplantation without employment of an anhepatic veno-venous bypass technique have been evaluated. After approval of the Institutional Human Ethics Committee, we measured cardiac output (CO) using the thermodilution technique. Measurements of CO were performed in triplicate and were reported as means. Hemoglobin concentration and oxygen saturation as well as arterial (CaO_2) and mixed venous oxygen content (CvO_2) were measured directly (Hemoxymeter OSM 3, Radiometer Copenhagen). DO_2 (CO x CaO_2 x 10) and VO_2 (CO x CaO_2-CvO_2 x 10) were calculated by means of standard formulas and indexed on body surface area. A minimum of 2 measurements was done preanhepatically, anhepatically and neohepatically.

Statistical analysis: ANOVA with Tukey's method for multiple comparisons. Correlations between DO_2 and VO_2 were calculated by least-squares regression at each phase of operation and for each group of patients outcome. According to outcome (defined as leaving the hospital), patients were divided in survivors (n=40) and nonsurvivors (n=10, died after sepsis).

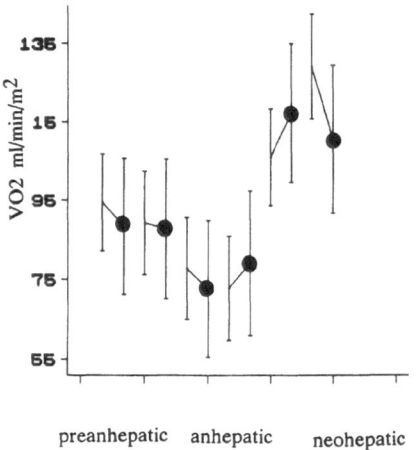

Fig. 1.   Oxygen consumption during the different phases transplantation in survivors and nonsurvivors (closed symbols)

Results: A similar pattern of VO_2 and DO_2 occurred in survivors and nonsurvivors during each distinct perioperative stage and significant anhepatic decreases ($p < 0.001$) and neohepatic increases ($p < 0.001$) in DO_2 and VO_2 was found in both groups (figure 1 and 2). Correlations between DO_2 and VO_2 were: preanhepatic: r=0.35 (fig 3), anhepatic: r=0.40, neohepatic: r=0.53 (fig 4). The slopes were comparable during all stages, but there was a significant decrease in the intercept during the anhepatic stage from 450 ± 59 to 138 ± 39 ml/min/m2. The intercept returned to the preanhepatic values in the postanhepatic stage.

Discussion: Recently, a number of clinical studies have shown that in patients at risk to develop multiple organ failure or in those with fulminant liver failure, VO_2 may be flow dependent over a wide range of cardiac output (4). We found that during all stages of liver transplantation there is some correlation between DO_2 and VO_2, but these correlations seem to be independent of the respective DO_2 level. Moreover, above DO_2 of 400 ml/min/m2 an enhanced scatter of the two variables was detected (fig 3 + 4). Therefore, no "critical value" can be defined according to our data. Normally, DO_2 equals approximately 4 times the amount of oxygen actually consumed (3). However, in our liver graft recipients DO_2 was 4 times VO_2 even during the reduced blood flow of the anhepatic stage and it was 6-7 times VO_2 during all other phases. The discrepancy

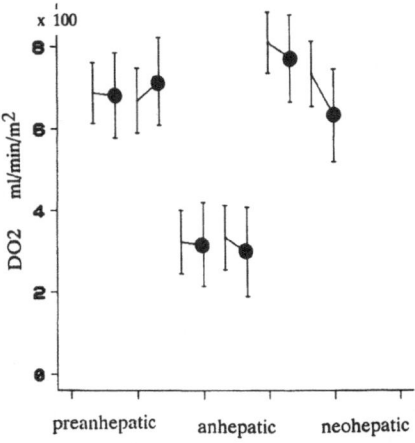

Fig. 2.   Oxygen delivery during different phases of liver transplantation in survivors and nonsurvivors (closed symbols)

between our findings and the current concepts of delivery limited VO_2 may be the result of studying a different group of patients. The lack of a threshold value cannot be explained by differences in the statistical analysis and the calculations of DO_2 and VO_2. In order to keep our results comparable to those of other investigators we were employing the same controversial method of simple regression analysis to present a DO_2/VO_2 relationship (fig 3,4). However, the analysis was adjusted to the factors outcome and stage of operation, respectively. The neohepatic increase of the coefficient from 0.35 (preanhepatic) to 0.53 (neohepatic) points at a greater impact of problems associated with mathematical coupling (5) at higher levels of cardiac output. On the other hand, we speculated about the dependency of DO_2 and cardiac output from VO_2, as a result of increasing metabolism under conditions of good neohepatic function of the new liver. In addition, an abnormal relationship between DO_2 and VO_2 could be related to higher neohepatic lactate levels (6) and repayment of anhepatic oxygen debt (7).

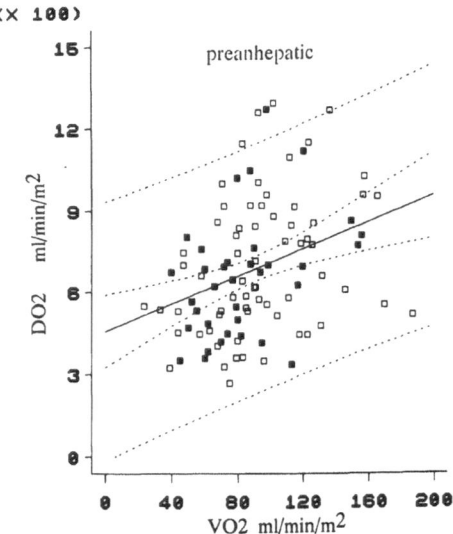

Fig. 3.   Preanhepatic correlation between oxygen transport (DO₂) and consumption
(VO₂) in all patients (C=0.35)

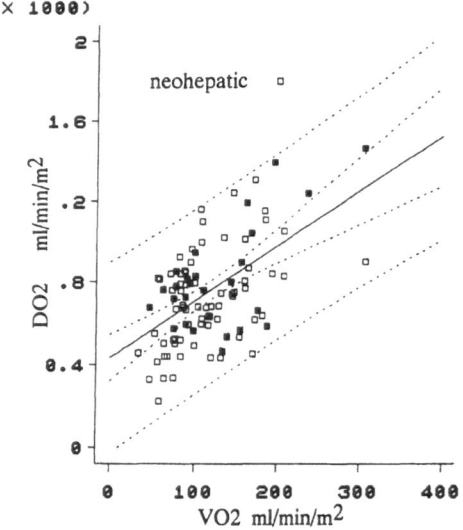

Fig. 4.   Neohepatic correlation between oxygen delivery (DO₂) and consumption (VO₂)
in all patients (C=0.53).

Conclusion: Our results emphasize the independency of oxygen consumption from oxygen delivery during various periods of changing cardiac function, metabolic needs and body temperature in this group of patients. Therefore, in liver graft recipients it seems to be difficult to detect either a critical value or to recommend optimum levels for DO_2. Thus, if a critical value is uncertain, it remains unclear how much an increase of DO_2 should be achieved and which price has to be paid.

Literature

1. G. Annat, J. P. Viale, C. Percival, M. Froment and J.Motin, Oxygen delivery and uptake in the adult respiratory distress syndrome, Am Rev Resp Dis 133:999 (1986)
2. P.T. Schumacker and S.M. Cain, The concept of a critical oxygen delivery, Intensive Care Med 13:223 (1987)
3. J.V. Snyder, Postoperative evolution of extrahepatic organ function, Transplantation Proc 21:3508 (1989)
4. D.Bihari, M. Smithies, A. Gimson and J. Tinker, The effects of vasodilation with prostacyclin on oxygen delivery and uptake in critically ill patients, New Engl J Med 317:397 (1987)
5. L.F. Moreno, H.H. Stratton, J.C Newell and P.J. Feustel, Mathematical coupling of data: correction of a common error for linear calculations, J Appl Physiol 60:335 (1986)
6. J.J. Fath, N.L. Ascher and F.N. Konstantinides, Metabolism during hepatic transplantation: indicators of allograft function, Surgery 94:664 (1984)
7. H. Steltzer, M.Hiesmayr, G.Huemer and M.Zimpfer, Oxygen debt and oxygen deficit as determinants of liver allograft failure, Anesthesiology 71:A 183 (1989)

COMPARATIVE STUDY OF THE ACCURACY OF TWO FIBEROPTIC MIXED VENOUS SATURATION CATHETERS (SPECTRACATH® VS OPTICATH®) DURING ACUTE CHANGES IN HEMATOCRIT AND CARDIAC OUTPUT IN HUMANS

ECSM van Woerkens, A Trouwborst, L Snel, A. van Dorp van Vliet, R Tenbrinck

Department of Anesthesiology, Erasmus University
Rotterdam, The Netherlands

INTRODUCTION

The mixed venous oxygen saturation of hemoglobin (SvO_2) reflects, under many circumstances, the state of tissue oxygenation. Changes in cardiac output (CO), arterial oxygen content, and oxygen uptake by the tissue influence this parameter (Miller, 1982; Trouwborst et al., 1990a). Therefore continuous fiber optic monitoring of SvO_2 has been developed. Due to scattering of the erythrocyte wall itself, however, sudden changes in hematocrit (Hct) might influence the reliability of SvO_2 values, measured by such fiber optic systems (Martin et al., 1973). Two manufacturers claim that their system overcome this problem, one because of using three wavelengths (Opticath®) and the other using two wavelengths via one fiber but receiving backscattered light by two other fibers (Spectracath®). The manufacturers claim that with their systems the reliability of SvO_2 measurements are not affected by sudden changes in Hct and that any change in Hct will not entail the need to update or recalibrate the device. In this study, during acutely induced hypervolemic hemodilution (Trouwborst et al., 1990b) followed by surgical blood loss and with no recalibration of the systems during the study period, the in vivo SvO_2 values measured with the fiber optic devices were compared with the in vitro SvO_2 values obtained with a multiwave length spectrophotometer.

PATIENTS AND METHODS

One of the studied catheters (Spectracath®, Viggo-Spectramed, Oxnard, USA) has three fibers terminated in the plane of the tip of a balloon-tipped, 7.5 F thermodilution catheter. A light emitting fiber directs infrared red (IR 805 nm) light and red light (R: 660 nm) into the blood. A second fiber (the near fiber) is terminated adjacent to the emitting fiber and receives backscattered light from red blood cells.
A third fiber (the far fiber) is terminated two fiber diameters (500 microns) from the source fiber and also receives backscattered light. The ratio of the IR and R signals from the near fiber is designed as X = IR near/R near. X was found to be

highly dependent on saturation, but also dependent on Hct. However the ratio R = IR near fiber/IR far fiber directly relates to Hct but is independent of saturation because 805 nm is an isobestic wavelength. Using incorporated algorithms based on these observations should give reliable SvO_2 values independent from Hct changes in the range 20% to 50%. The other studied catheter (Opticath®, Oximetrix, Mt View; CA, USA) has two fibers terminated at the tip of a balloon-tipped, 7.5 F thermodilution catheter. Light emitting diodes generate alternating pulses of three different wavelengths (between 600 and 1000 nm), 244 times per second via a light emitting fiber. A second fiber receives backscattered light from red blood cells and conducts this light to a photodetector. The oxygen saturation of hemoglobin (Hb) is derived by a computer from the relative intensities corresponding to three different wavelengths. The study comprised measurements (Opticath®: n = 52; Spectracath®: n = 54) in 12 consecutive Jehovah's Witness patients scheduled for major surgery. Randomly divided between the patients, either a Spectracath® or Opticath® SvO_2 catheter was inserted via the internal jugular vein into the pulmonary artery. Before insertion, according to the manufacturer's specifications, the SvO_2 catheters were calibrated in vitro and no recalibration was performed during the entire study period. The design of the study was the same as in a previously reported study (van Woerkens et al., 1991).

In brief: values of Hct, blood gases, hemodynamics and oxygenation were obtained before and after induction of anesthesia, after each step of hypervolemic hemodilution, after every 500 ml of blood loss, at the end of surgery, 20 min, 2 h and 4 h postoperatively. The in vivo SvO_2 values of the devices were compared with the in vitro SvO_2 values obtained with a multiwave length Spectrophotometer (OSM_3, Radiometer, Copenhagen).

Statistical analysis was performed by the paired students T-test and the Wilcoxon signed rank test. Furthermore the correlation values, coefficients of determination and regression lines were determined (Saxena, 1985). These relations were also tested by means of a Fisher's Z transformation (correlation coefficient) and by means of the test of the regression slope (Shavelson, 1988). The accepted probability for a statistical difference between means was $P < 0.05$.

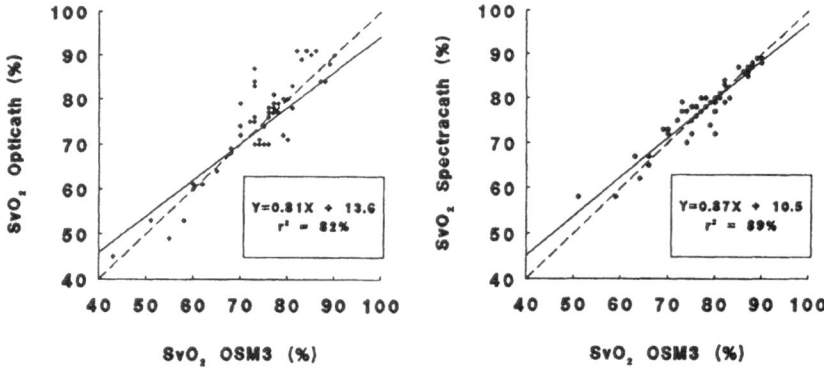

Figure 1. Scatterplot of all fiberoptic mixed venous hemoglobin saturation (SvO_2) data of both catheters- Opticath® left and Spectracath® right- against the in vitro reference spectrophotometer values (OSM3). The dotted line represents the line of identity (x=y) and the solid line represents the lineair regression lines with their coefficient of determination (r^2).

Table 1. Selected data on systemic hemodynamics and oxygenation before and after induction of anesthesia; after each step of hyper volemic hemodilution (H1, H2, H3); at the end of surgery (ES); 20 min, 2 and 4 hours postoperatively (PO).

	Ht (O) %	Ht (S) %	CO (O) l/min	CO (S) l/min	PWP (O) mmHg	PWP (S) mmHg	PaO$_2$ (O) mmHg	PaO$_2$ (S) mmHg	Opt®-OSM3 %	Spec®-OSM3 %
pre induction	38.3 4.1	36 3.0	5.1 0.8	5.2 0.7	6.8 3.1	6.4 2.3	89 5	92 13	1.33 2.87	-0.83 3.60
post induction	36.5 4.2	35 3.4	4.3+ 0.9	4.2+ 1.0	5.8 3.3	5.8 2.9	179+ 30	183+ 31	0.50 5.11	-0.67 1.20
H1	29.3+* 3.1	29.5+* 3.0	5.5* 1.3	5.8* 1.3	10+* 3.1	16+* 3.0	165+ 45	182+ 42	0.33 3.90	-0.67 1.21
H2	25.7+* 2.5	27.5+* 2.4	5.7* 1.4	6.2* 1.5	15.3+* 1.8	21.8+* 3.0	149+ 49	187+ 28	-0.67 4.29	-0.67 0.82
H3	24+* 2.2	26+* 1.5	6.1* 1.7	6.2* 1.4	19.3+* 1.9	26.4+* 2.7	149+ 44	186+ 28	-0.67 5.31	-0.83 1.16
ES	22.3+* 6.7	24.3+* 3.1	5.4* 1.5	6.3* 1.7	18.2+* 2.1	19+* 7.0	143+ 44	139+ 35	-0.17 5.33	1.17 2.28
20 min PO	29.8+* 2.6	30+* 3.2	6.3* 2.8	6.5* 1.1	13+* 5.6	7.4 2.3	98* 45	111+ 44	3.67 5.30	1.33 3.37
2h PO	30.2+* 3.7	30 3.3	6.7* 2.3	5.6* 0.5	8.3 3.3	4.8 2.2	122+ 55	122+ 45	1.00 5.10	0.67 4.31
4h PO	30.2+* 2.5	29.5+ 3	7.3* 2.8	6.0* 0.5	7.4 3.2	3.2 0.4	79* 6	113+ 33	2.67 4.21	1.00 3.39

Legend: (O) concerns the Opticath®, (S) concerns the Spectracath® Ht = hematocrit; CO = cardiac output; PWP = pulmonary wedge pressure; PaO$_2$ = arterial oxygen content. The last colums represent the difference in mixed venous saturation between catheter minus OSM3. + = P< 5% compared to preanesthetic values; * = P < 5% compared to post anesthetic values; No differences with P< 5% between Spectracath and Opticath values at the same level were observed.

RESULTS

The study of each patient varied between 9-12 h depending on the duration of the surgical procedure. Directly after the in vitro calibration of the SvO_2 catheters gives differences in SvO_2 values compared to the OSM-3 SvO_2 of 1.33 ± 2.87% for the Opticath® and -0.83 ± 3.60% for the Spectracath® (Table 1). At the end of the study period the catheter SvO_2 values differed from the OSM-3 SvO_2 by 2.67 ± 4.21 for the Opticath® and by 1.00 ± 3.39 for the Spectracath®. Throughout the study period at the same levels of the study, no significant difference between Opticath® and Spectracath® SvO_2 values could be observed, while for both catheters the difference between SvO_2 catheter value and the Spectrophotometer SvO_2 value at the beginning of the study period was not significantly different from the differences in SvO_2 values at the end of the experimental procedure. Plotting Hct against the difference between in vivo SvO_2 and in vitro SvO_2 of both systems showed that both catheter systems gave SvO_2 values with an accuracy independent from changes in Hct (Figs.2a and c). Changes in CO slightly influenced the accuracy of the Opticath® (r^2 = 9.2%) but not the accuracy of the Spectracath® (r^2 = 3.2%) (Figs.2b and d). Plotting in vivo determination of SvO_2 by the fiberoptic system against the in vitro reference value of all data points obtained during the entire study gives a correlation coefficient r^2 = 91% (Y = 0.81 X + 13.6) for the Opticath® and r^2 = 94% (Y = 0.87 X + 10.5) for the Spectracath®.
Between -5% and +5% of SvO_2 catheter minus OSM_3 SvO_2 values are 42 (81%) of the Opticath® and, 51 (94%) of the Spectracath® of all measured pairs during the whole study period. Other relevant data selected during the study are presented in Table 1.

DISCUSSION

The value of continuous fiberoptical determination of SvO_2 depends on how accurately in vivo SvO_2 approximates the reference spectrophotometric measured saturations under different physiologic conditions. Specific obstacles as vessel wall artifact and confounding effects of varying Hct and secondary effects including blood flow, erythrocyte shape might influence the saturation measurement (Landsman, et al., 1978). The Hct of the blood floating along the tip of the catheter might be influenced by flow properties (linear flow vs turbulent flow), and by systemic Hct value. Because many situations during surgery and intensive care lead to sudden changes in Hct, devices must produce reliable continuous in vivo SvO_2 measurements independent from changes in Hct. Furthermore, SvO_2 itself is an indicator of the critical point of hemodilution and continuous registration of in vivo SvO_2, if accurate, might be used as an extra parameter during hemorrhage to assess the oxygen transport capacity of the blood (Trouwborst et al., 1990a). During stepwise-induced isovolemic hemodilution a gradual decline in SvO_2 was observed while the oxygen extraction ratio (ER) increased. A direct and strong correlation was found between SvO_2 and ER confirming that the SvO_2 during hemodilution reflects the overall balance between VO_2 and oxygen delivery (Trouwborst, et al., 1990; Räsänen, 1990; Trouwborst and van Woerkens, 1991).
In the present study in humans the reliability of the Spectracath® and Opticath® continuous SvO_2 devices was tested during sudden changes in Hct, cardiac output, and pulmonary wedge pressure (PWP). The results of the study, demonstrate that both systems are able to be calibrated in vitro prior to insertion.

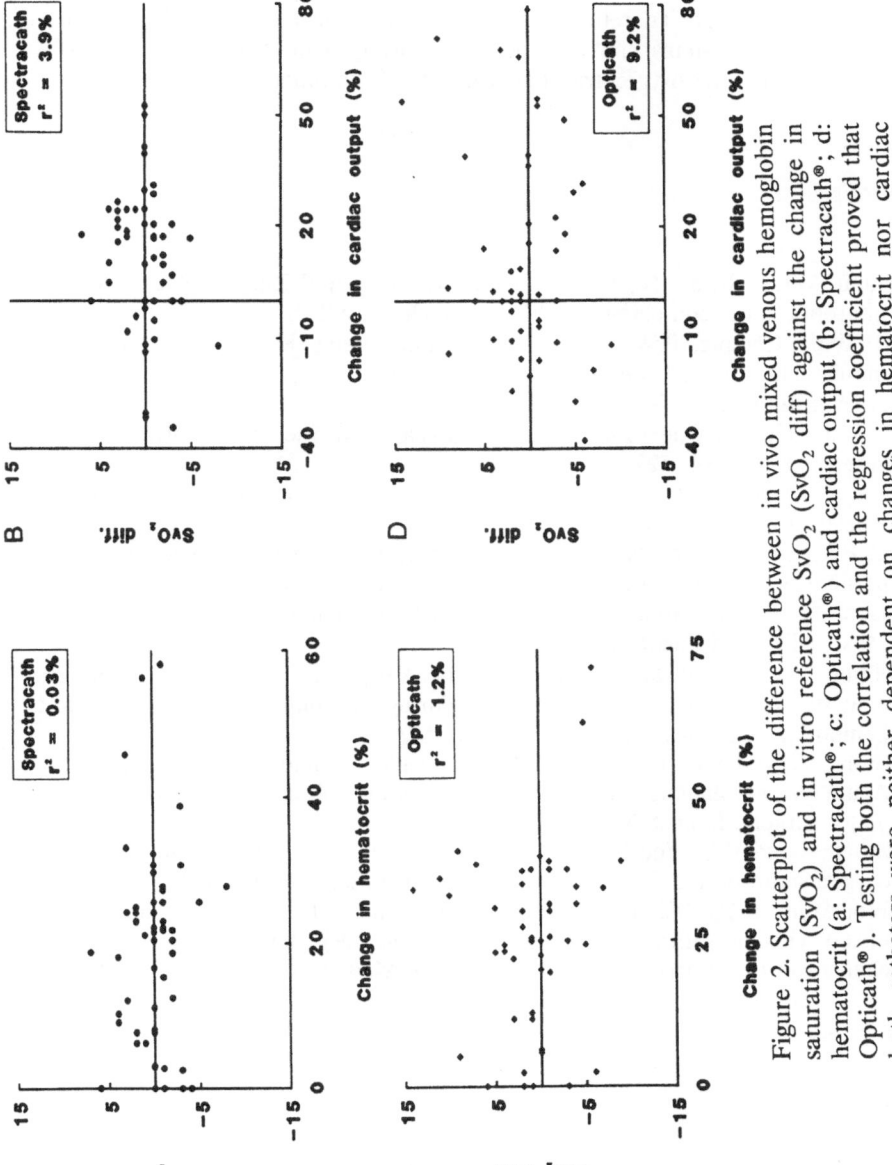

Change in hematocrit (%)

Figure 2. Scatterplot of the difference between in vivo mixed venous hemoglobin saturation (SvO$_2$) and in vitro reference SvO$_2$ (SvO$_2$ diff) against the change in hematocrit (a: Spectracath®; c: Opticath®) and cardiac output (b: Spectracath®; d: Opticath®). Testing both the correlation and the regression coefficient proved that both catheters were neither dependent on changes in hematocrit nor cardiac output in the measured range.

Furthermore, the reliability of both devices was not affected by changes in Hct or PWP. Cardiac output changes had a slight influence on the reliability of the Opticath® but not on the accuracy of the Spectracath®. No drift was observed in either system during the course of the study. The Spectracath® was slightly, but not statistically significant, more accurate than the Opticath® system. It is concluded that both devices reflect measured SvO_2 over a wide range of time, also during abrupt changes in Hct. Therefore the continuous displayed in vivo SvO_2 values derived with both systems can also be used during acute hemorrhage as extra parameter to assess the oxygen transport capacity of the blood.

REFERENCES

Landsman, M.L.J., Knop, N., Kwant, G., Mook, G.A., and Zijlstra, W.G., 1978, A fiberoptic reflection oximeter, Pflügers Arch., 373: 273.

Martin, W.E., Cheung, P.W., Johnson, C.C., and Wong, K.C., 1973, Continuous monitoring of mixed venous oxygen saturation in man, Anesth. Analg., 52: 784.

Miller, M.J., 1982, Tissue oxygenation in clinical medicine: an historical review, Anesth. Analg., 61: 527.

Räsänen, J., 1990, Mixed venous oximetry may detect critical oxygen delivery, Anesth. Analg., 71: 567.

Saxena, P.R., 1985, An interactive computer programme for data management and parametric and non-parametric statistical analysis, Br. J. Pharmacol, 86: S818.

Shavelson, R.J., 1988, Statistical reasoning for the behavioral sciences, Boston Mass: Allyn and Bacon Inc.

Trouwborst, A., Tenbrinck, R., and van Woerkens, E.C.S.M., 1990a, Blood gas analysis of mixed venous blood during normoxic acute isovolemic hemodilution in pigs, Anesth. Analg., 70: 523.

Trouwborst, A, van Woerkens, E.C.S.M., van Daele, M, and Tenbrinck, R, 1990b, Acute hypervolaemic haemodilution: avoiding blood transfusions during major surgery. Lancet, 336: 1295.

Trouwborst,A and Van Woerkens, ECSM, 1991, Extraction ratio and mixed venous oxygen saturation: a false relationship; in response, Anesth. Analg., 72:716.

Van Woerkens, E.C.S.M., Trouwborst, A., Tenbrinck, R., 1991, Accuracy of a mixed venous saturation catheter (Spectracath®) during acutely induced changes in hematocrit in humans, Crit. Care Med, 19:1025.

ERYTHROPOIETIN INDUCTION BY HYPOXIA

A COMPARISON OF IN VITRO AND IN VIVO EXPERIMENTS [*]

H. Pagel[*], A. Engel[*] and W. Jelkmann[#]

Institute of Physiology[*]
Medical University of Lübeck[#]
University of Bonn[#], Germany

INTRODUCTION

The limited lifespan of erythrocytes renders their continuous replacement necessary. Furthermore, red cell production increases following blood losses and under conditions of an increased oxygen demand of the organism. There is a clear relationship between the degree of anemia or tissue hypoxia and the rate of erythropoiesis. The most important regulator of the proliferation of erythroid committed progenitors, the maturation of erythroblasts and the release of reticulocytes into the blood is the glycoproteid hormone erythropoietin [Epo] Jelkmann, 1986; Krantz, 1991).

Epo is produced primarily by tubular (Maxwell et al., 1990) or peri-tubular (Koury et al., 1988) cells in the cortex of the kidney. The oxygen sensing mechanism in the control of the synthesis of Epo is still not completely understood. It is unknown, as yet, whether extrarenal sites are involved in the oxygen sensing and its transduction into the renal synthesis of Epo (Pagel et al., 1988; Scholz et al., 1991).

In order to get more information about the feedback mechanism triggering the renal Epo synthesis, investigations were carried out in the isolated hypoxemic perfused rat kidney and in rats after a reduction of the blood oxygen availability. The level of Epo production in vitro and in vivo was compared quantitatively.

METHODS

Isolated Perfused Rat Kidneys

Right kidneys from adult male Sprague-Dawley rats (354 ± 12 g) were perfused in a recirculation system as described in detail elsewhere (Pagel et al., 1991). The kidneys were perfused at a constant pressure of 100 mmHg and 37 $^{\circ}$C for 3 h. The perfusion medium (0.2 l) was a substrate enriched Krebs-Henseleit solution supplemented with 60 g/l bovine serum albumin and 40% freshly drawn and 5-times washed human erythrocytes. To ensure a high stability of the isolated organ, the perfusion medium was regenerated by

---

[*] Dedicated to Prof. Ch. Weiss, Lübeck, in his 65th year

Table 1. Synopsis of the Experimental Conditions

	in vitro	in vivo
Experimental period (h)	3	3
Hydrostatic pressure (mmHg)	100	122 ± 4
Arterial pO_2 (mmHg)	35	42
Hematocrit (%)	40	46 ± 2
O_2 supply (μmol/min/g kidney)	40 ± 3	35

dialysis against 5 l of a protein- and cell-free medium equilibrated with 5% O_2, 5% CO_2 and 90% N_2. Aliquots of the perfusion media were taken for the determination of Epo concentration, which was carried out by a recombinant human Epo based radioimmunoassay as described recently (Fandrey et al., 1990, Pagel et al., 1991).

The O_2 saturation of the perfusion medium and the renal O_2 supply were calculated from continuous measurements of the perfusion flow rate and the constants and equations described by Siggaard-Andersen (1974).

Exposure of Rats to Hypoxia

Adult male Sprague-Dawley rats (383 ± 8 g) were exposed to hypoxia at 0.42 atm in a hypobaric chamber (corresponding to 7,000 m simulated altitude) for 3 h. Thereafter, a blood sample was taken for the determination of the hematocrit and the plasma concentration of Epo.

Mean arterial blood pressure was measured in 5 restrained conscious animals with a sphygmomanometer equipped with a latex tail cuff (Harvard Indirect Rat Tail Blood Pressure Monitor, Harvard Apparatus, Edenbridge, England). The renal O_2 supply was calculated according to the renal blood flow data measured by Pagel et al. (1988) and the equations of Siggaard-Andersen (1974). The arterial pO_2 was estimated to be 42 mmHg at 0.42 atm.

Table 1 presents the experimental conditions of both groups.

RESULTS

Table 2 summarizes the results of the measurements of the production of Epo in the isolated perfused kidney and in hypoxemic rats.

The data indicate that Epo production started to increase after 1 h of hypoxemic perfusion of the isolated kidney. Similarly, in vivo studies have shown that the plasma Epo level increases in humans and rodents 1 h after the exposure to hypoxic hypoxia (Eckardt et al., 1989a, 1989b). In the following, therefore, Epo formation rates were calculated for the period of continuous increase of the hormone (between 60 and 180 min).

Table 2. Comparison of Epo-Production in vitro and in vivo

Time (min)	30	60	90	120	150	180
Epo in vitro (mU/g kidney; n=5)	10 ± 2	13 ± 2	53 ± 4	138 ± 5	149 ± 15	240 ± 15
Epo in vivo (mU/ml plasma; n=5)	--	--	--	--	--	370 ± 21

In the isolated kidney Epo formation was 240 mU/g kidney. Corrected for the urinary Epo clearance rate this value corresponds to an Epo formation rate of 130 mU/h/g kidney (see Appendix).

Plasma Epo amounted to 370 mU/ml in the intact rat after exposure to hypoxic hypoxia. Taking into account the plasma volume and the hepatic Epo clearance rate, this concentration value corresponds to a production rate of about 2800 mU/h/g kidney (see Appendix).

The results of these calculations indicate that the rate of Epo formation was in vivo more than 20-fold higher than in the isolated kidney.

DISCUSSION

In this study, the rate of Epo formation was determined in the isolated hypoxemic perfused rat kidney in comparison to that in hypoxemic rats. The results indicate that the isolated kidney responds to a lowering of its oxygen supply with a moderate increase of the production of Epo. Epo synthesis was much more stimulated in intact rats exposed to hypoxic hypoxia.

Note that exocrine function of the isolated perfused kidney was well preserved during the 3-h perfusion period, as judged from measurements of parameters of kidney function like the glomerular filtration or the sodium reabsorption rate (cf. Pagel et al., 1991). In addition, it could be demonstrated that the isolated perfused kidney is able to produce considerably more Epo than in the present study. Epo production rate was more than 2-fold higher, if the pO_2 of the perfusion medium was reduced to 20 mmHg (Pagel et al., 1991).

The observations described herein agree with previous studies from our laboratory. If the renal blood flow and, thus, the oxygen supply to the kidney alone is greatly reduced, an only moderate elevation of plasma Epo occurs. In contrast, if the systemic oxygen supply is lowered, the increase in Epo concentration is by two decades greater (Pagel et al., 1988, 1989).

Clinically, renal artery stenosis leads to an only slight elevation of the Epo concentration in plasma (Grützmacher & Schoeppe, 1989). Polycythemia is seen very rarely, and if very moderately in these patients (Tarazi et al., 1966).

There is some evidence indicating that the hypothalamic-hypophyseal system plays an important role in the regulation of the renal production of Epo. Epo activity evoked by hypoxia is significantly lower in hypophysectomized animals than in sham-operated controls (Halvorsen et al., 1968; Peschle et al., 1978). Electric stimulation of distinct hypothalamic regions leads to an increased Epo plasma concentration (Mirand et al., 1964; Halvorsen, 1966).

In conclusion, the exclusiveness of a renal oxygen sensing mechanism controlling the production of Epo must be questioned. In view of the strong Epo response following a reduction of the whole body oxygen offer, the possibility should be considered that the renal synthesis of Epo is partly under the control of extrarenal oxygen sensitive mechanisms.

Comparison of Epo-Production in vitro and in vivo

Calculations

in vitro	in vivo

<u>in vitro</u>

240 mU/(2h*g kw)

baseline value: 10±2 mU/g kw

230 mU/(2h*g kw)

Cl_R: 0.25±0.07 mU/(min*g kw)
= 30 mU/(2h *g kw)
(n=10; Pagel, unpublished data)

260 mU/(2h*g kw)

130 mU/(h*g kw)

<u>in vivo</u>

370 mU/(ml*2h*2kidneys)

normal value: 20±6 mU/ml
(Pagel et al., 1988)

350 mU/(ml*2h*2kidneys)

175 mU/(ml*2h*1kidney)

1 kidney: 1.50±0.05 g

117 mU/(ml*2h*g kw)

V_p: 40 ml/kg bw
(Baker et al., 1979)
383 g bw = 15.3 ml plasma

1787 mU/(2h*g kw)

V_z: 1.26±0.04   (n=20)
(after Cotes et al., 1989)

2252 mU/(2h*g kw)

Cl_R: neglectable
(Emmanouel et al., 1984)

Cl_H: t(1/2) = 1.5 h
(Spivak & Hogans, 1989)
Assumption: lin.pharmacokinet.,
i.e. 1.grade reaction according
to: $C(0)=C(t)/e^{-kt}$;   k=ln2/1.5h

5675 mU/(2h*g kw)

2837 mU/(h*g kw)

130 mU/(h*g kw)

Abbrevations

kw - kidney weight;   bw - body weight;

Cl_R - renal (urinary) clearance; Cl_H - extrarenal (hepatic) clearance;

V_p - plasma volume; V_z - Epo distribution space related to plasma volume;

t(1/2) - elimination half-life;   C(0) - virtual concentration at t(0);

C(t) - actual concentration at the point of time t.

REFERENCES

Baker, H.J., Lindsey, J.R., Weisbroth, S.H. (1979) Selected normative data. In: The Laboratory Rat, vol. 1, H.J.Baker, J.R.Lindsey, S.H.Weisbroth (eds.) Academic Press, New York, p. 411.

Cotes, P.M., Pippard, M.J., Reid, C.D.L., Winearls, C.G., Oliver, D.O., Royston, J.P. (1989) Characterization of the anaemia of chronic renal failure and the mode of its correction by a preparation of human erythropoietin. Quart.J.Med.70: 113-137.

Eckardt, K.U., Kurtz, A., Bauer Ch. (1989a) Determinants of erythropoietin formation. Proc.IUPS 17: 449 (abstr.).

Eckardt, K.U., Boutellier, U., Kurtz, A., Schopen, M., Koller, E.A., Bauer, Ch. (1989b) Rate of erythropoietin formation in humans in response to acute hypobaric hypoxia. J.Appl.Physiol.66: 1785-1788.

Emmanouel, D.S., Goldwasser, E., Katz, A.I. (1984) Metabolism of pure human erythropoietin in the rat. Am.J.Physiol.247: F168-F176.

Fandrey, J., Seydel, F.P., Siegers, C.P., Jelkmann, W. (1990) Role of cytochrome P-450 in the control of the production of erythropoietin. Life Sci.47: 127-134.

Grützmacher, P., Schoeppe, W. (1989) Renal artery stenosis and renal polyglobulia. In: Erythropoietin, W.Jelkmann, A.J.Gross (eds.), Springer, Berlin, pp. 111-121.

Halvorsen, S. (1966) The central nervous system in regulation of erythropoiesis. Acta Haematol.35: 65-79.

Halvorsen, S., Roh, B.L., Fisher, J.W. (1968) Erythropoietin production in nephrectomized and hypophysectomized animals. Am.J.Physiol.215: 349-352.

Jelkmann, W. (1986) Renal erythropoietin: properties and production. Rev.Physiol.Biochem.Pharmacol.104: 139-215.

Koury, S.T., Bondurant, M.C., Koury, M.J. (1988) Localization of erythropoietin synthesizing cells in murine kidneys by in situ hybridization. Blood 71: 524-527.

Krantz, S.B. (1991) Erythropoietin. Blood 77: 419-434.

Maxwell, A.P., Lappin, T.R.J., Johnston, C.F., Bridges, J.M., McGeown, M.G. (1990) Erythropoietin production in kidney tubular cells. Brit.J.Hematol.74: 535-539.

Mirand, E.A., Grace, J.T., Johnston, G.S., Murphy, G.P. (1964) Effects of hypothalamic stimulation on the erythropoietic response in the rhesus monkey. Nature 204: 1163-1165.

Pagel, H., Jelkmann, W., Weiss, Ch. (1988) A comparison of the effects of renal artery constriction and anemia on the production of erythropietin. Pflügers Arch.413: 62-66.

Pagel, H., Jelkmann, W., Weiss, Ch. (1989) Oxygen supply to the kidneys and the production of erythropoietin. Respir.Physiol.77: 111-118.

Pagel, H., Jelkmann, W., Weiss, Ch. (1991) Isolated serum-free perfused rat kidneys release immunoreactive erythropietin in response to hypoxia. Endocrinology 128: 2633-2638.

Peschle, C., Rappaport, I.A., Magli, M.C., Marone, G., Lettieri, F., Cillo, C., Gordon, A.S. (1978) Role of the hypophysis in erythropoietin production during hypoxia. Blood 51: 1117-1124.

Scholz, H., Schurek, H.J., Eckardt, K.U., Kurtz, A., Bauer, Ch. (1991) Oxygen-dependent erythropoietin production by the isolated perfused kidney. Pflügers Arch.418: 228-233.

Siggaard-Andersen, O. (1974) The Acid-Base Status of the Blood, ed. 4, Munksgaard, Copenhagen.

Spivak, J.L., Hogans, B.B. (1989) The in vivo metabolism of recombinant human erythropoietin in the rat. Blood 73: 90-99.

Tarazi, R.C., Frohlich, E.D., Dustan, H.P., Gifford, R.W., Page, I.H. (1966) Hypertension and high hematocrit. Am.J.Cardiol.18: 855-858.

ALTERED CONCENTRATIONS OF ALDOSTERONE IN NEONATAL CALVES

DURING CHRONIC HYPOXIA AND THE SUBSEQUENT RECOVERY PERIOD

Howard Tyler and Harold Ramsey

Department of Animal Science
North Carolina State University
Raleigh, North Carolina USA 27695

INTRODUCTION

Significant changes occur in both blood composition and plasma volume of the calf within the first few hour after parturition (McEwan *et al.*, 1968). These changes require sensitive and reliable regulatory mechanisms to maintain electrolyte balance and prevent fluid dyshomeostasis. Although the time required for maturation of the renal system in calves is much shorter than in most experimental animals (Dalton, 1968), significant electrolyte imbalances are a common clinical finding in neonatal calves.

Aldosterone influences sodium reabsorption and potassium excretion in the distal renal tubule, and is therefore of critical importance in maintaining fluid and electrolyte homeostasis. Recent reports of reduced aldosteronogenesis in response to decreases in oxygen tension within the physiological range (Raff *et al.*, Am. J. Physiol. 1989; Brickner *et al.*, Proc. Endocr. Soc. 1990) suggest that significant imbalances may arise in calves remaining hypoxic postnatally. Hypoxia may occur spontaneously in newborn animals maintaining a patent ductus arteriosus or experiencing respiratory distress due to prematurity or disease.

The shift from placental to pulmonary respiration is accompanied by dramatic changes in oxygen availability to the newborn. This sudden increase in Po_2 occuring in all mammals, as well as the high incidence of spontaneous dysfunctions in the oxygen delivery system of the newborn, emphasizes the importance of understanding the relationship between oxygen and aldosteronogenesis during this period. In a preliminary investigation, plasma samples from six hypoxic and six normoxic neonatal calves were assayed to determine if aldosterone concentrations were affected by hypoxia in the newborn.

METHODS

Twelve Holstein calves were obtained at birth and assigned to one of two treatment groups; hypoxic (HYP) or normoxic (NORM). For hypoxic calves, inspired gas mixtures contained 10.5% oxygen between 0 and 24 h and 21% oxygen between 24 and 72 h. For normoxic calves, inspired air contained 21% oxygen throughout the experiment, 0 to 72 h. Hypoxia was induced experimentally by reducing oxygen content of inspired air with nitrogen.

Blood samples were obtained every 6 h from birth through the second day, and every 12 h through the third day postpartum. Blood was sampled via puncture from the jugular vein into sterile evacuated tubes containing potassium oxalate and sodium fluoride (Becton Dickinson, Rutherford, NJ). Plasma was separated by centrifugation at 1286 x g for 15 min and stored at -20 C for later analysis.

Samples were analyzed by radioimmunoassay to determine concentrations of aldosterone. The radioimmunoassay procedure utilized commercially available antibody-coated tubes, although the adult human standards provided in the kit were replaced with standards prepared in neonatal calf serum. The assay was then validated using these standards. Under these conditions, the assay proved to be precise, specific, and exhibited linearity throughout the physiological range. No significant cross-reactivity was noted with other relevant steroids. Inter- and intra-assay coefficients of variation were 5.16% and 3.38%. Sensitivity of the assay was 3 ng/dl.

Arterial samples were obtained via puncture of the carotid artery or the ventral coccygeal artery. Samples were store on ice in sealed, heparinized syringes and analysis was completed within 30 min of collection. Whole blood was analyzed on an Instrumentation Laboratories 1302 Blood Gas System (Instrumentation Laboratory, Lexington, MA) for the measurement of Po_2 and Pco_2.

Prior to analysis, data were divided into 4 periods; birth, day 1, day 2, and day 3. The data then were sorted by period and analyzed statistically using the General Linear Models Procedure of SAS (SAS Institute, Cary, NC). Differences between means were evaluated by the method of least squares means using ANOVA. Level of significance was placed at $p < .05$ and results are reported accordingly.

RESULTS

The goal of the hypoxic treatment was to maintain arterial Po_2 near prenatal levels. Published values for arterial Po_2 in the fetal calf vary from 19.5 mm Hg (Gahlenbeck et al., 1968) to 29.5 mm Hg (Reeves et al., 1972) depending on sampling site and use of anaesthesia. Previous results from this laboratory using unanaesthetized fetal calves (Tyler, 1991) confirm these findings. Therefore, based on values obtained in this experiment (mean Po_2 = 24 mm Hg), our treatment for inducing hypoxia was successful. Arterial Po_2 and Pco_2 were maintained near 26 and 45 mm Hg, respectively, for the first 24 h postnatally in hypoxic calves, thus inducing a eucapnic hypoxic condition. Oxygen tension and Pco_2 in normoxic calves averaged 73 and 46 mm Hg, respectively, during the same period.

Concentrations of aldosterone at birth were not significantly different between treatment groups and averaged 351 ± 88 ng/dl. Plasma aldosterone concentrations increased during hypoxia (Figure 1), with means during day 1 of 850 and 1430 ng/dl for normoxic and hypoxic calves, respectively ($p < .01$). By 24 h, concentrations of aldosterone had decreased to levels similar to those observed in normoxic calves (64 and 87 ng/dl in normoxic and hypoxic calves, respectively). However, following cessation of hypoxia at 24 h, arterial oxygen tension rose to near 80 mm Hg and aldosterone levels increased. Concentrations of aldosterone in post-hypoxic calves during day 2 averaged 770 ng/dl vs 188 ng/dl in normoxic calves ($p < .01$). Aldosterone concentrations continued to increase through day 3 in post-hypoxic calves to 1012 ng/dl, whereas levels in normoxic calves decreased to 133 ng/dl ($p < .01$).

DISCUSSION

Mature ewes exposed to hypoxic conditions exhibit a decreased urinary aldosterone excretion (Curran-Everett et al., 1988). Cultured bovine adrenocortical cells respond to decreases in Po_2 with a reversible decrease in the synthesis of aldosterone (Raff et al., 1989). In contrast to in vitro findings in adult bovine adrenocortical cells, newborn calves in the present study apparently increased aldosterone synthesis in response to extended periods of hypoxia. There may be a compensatory mechanism preventing potentially dangerous decreases in aldosterone in vivo.

By 24 h, aldosterone concentrations had returned to levels seen in normoxic calves, suggesting responsiveness of the system and a return to homeostasis following the apparent previous overcompensation induced by hypoxia. The dramatic increases in aldosterone seen in post-hypoxic calves suggests that the increased oxygen availability to adrenocortical cells at this time may have induced an increase in aldosteronogenesis. The prolonged nature of this increase in aldosterone concentrations could well be the most important finding of this study. Long-term deviations in aldosterone concentrations and the resulting alterations in sodium and potassium reabsorption in kidney tubules may damage the immature renal system and make maintenance of electrolyte balance difficult.

In conclusion, given the unavoidable fluctuations in Po_2 occurring in the perinatal period, regulation of aldosteronogenesis may differ between the neonate and the adult. Oxygen availability does apparently alter rates of aldosteronogenesis, and compensatory mechanisms are only partially effective in maintaining homeostasis. The magnitude and the duration of the response seen in these calves emphasizes the importance of understanding this relationship. The physiological consequences of this response may be significant and warrant further investigation.

REFERENCES

Brickner, R.C., B. Jankowski, and H. Raff. 1990. Potassium-stimulated aldosteronogenesis is oxygen sensitive. Proc. Endocr. Soc. 72nd Annual Meeting, Atlanta, page 275.

Curran-Everett, D.C., J.R. Claybaugh, K. Miki, S.K. Hong, and J.A. Krasney. 1988. Hormonal and electrolyte responses of conscious sheep to 96 h of hypoxia. Am. J. Physiol. 255:R274-R283.

Dalton, R.G. 1968. Renal function in neonatal calves: Inulin, thiosulphate, and paraaminohippuric acid clearance. Br. Vet. J. 124:498-502.

Gahlenbbeck, H., H. Frerking, A.M. Rathsschlag-Schaefer, and H. Bartels. 1968. Oxygen and carbon dioxide exchange across the cow placenta during the second part of pregnancy. Resp. Physiol. 4:119-126.

McEwan, A.D., E.W. Fisher and I.E. Selman. 1968. The effect of colostrum on the volume and composition of the plasma of calves. Res. Vet. Sci. 9:284-286.

Raff, H., D.L. Ball, and T.L. Goodfriend. 1989. Low oxygen selectively inhibits aldosterone secretion from bovine adrenocortical cells in vitro. Am. J. Physiol. 256:E640-E644.

Reeves, J.T., F.S Daoud, and M. Gentry. 1972. Growth of the fetal calf and its arterial pressure, blood gases, and hematologic data. J. Appl. Physiol. 32:240-245.

Tyler, H.D. 1991. Regulation of small intestinal development during the perinatal period in calves and piglets. PhD Thesis. North Carolina State University, Raleigh.

OXYGENATION OF THE HEART, THE MOVING MOTOR

MYOCARDIAL OXYGEN SUPPLY UNDER CRITICAL CONDITIONS, THE EFFECTS OF

HEMODILUTION AND FLUOROCARBONS

M. Fennema, W. Erdmann and N.S. Faithfull[*]

Department of Anesthesiology, Erasmus University, Rotterdam, The Netherlands and [*]Department of Medical Research, Alliance Pharmaceutical Corp., San Diego, U.S.A.

INTRODUCTION

Acute myocardial ischemia continues to be a major cause of morbidity and mortality in spite of considerable advances in the prevention and treatment of this condition. Ischemia is characterized by an imbalance between oxygen supply and demand which in turn is usually caused by insufficient coronary blood flow (CBF). In the healthy heart oxygen supply is usually easily maintained, even under extreme conditions when myocardial oxygen demand may increase dramatically. This might not be the case in patients with atherosclerotic obstructive coronary disease. One of the major problems is that the heart is an exclusively aerobic organ whose main source of energy is oxidative phosphorylation; an oxygen debt is not tolerated. Currently, much effort is directed towards investigation of methods aiming to produce an increased salvage rate of the ischemic myocardium. Many deaths still occur in the immediate post-infarction period. These are usually the result of arrhythmias which occur within 2 hours of the beginning of the ischemic period (Corr and Sobel, 1979) and are the result of electrophysiological disturbances caused by hypoxia and the accumulation of metabolites as a consequence of the decrease in perfusion. A better understanding of the factors affecting myocardial oxygen supply and demand will enable us to influence these factors to prevent hypoxia, improve oxygenation in ischemic areas and reduce damage caused by infarction.

This article reviews knowledge on coronary pathophysiology with emphasis on the effects of hemodilution, specifically with fluorocarbons. This analysis is supplemented by intramyocardial oxygen tension (PmO_2) measurements (Fennema, 1988). Most studies of oxygenation that have been performed in the beating heart have been carried out in dogs (Moss, 1968; Reves et al., 1978; Rude et al., 1984; Schuchhardt, 1985; Crystal et al., 1988; Parsons et al., 1990). This species is known to have a number of microscopic collateral vessels, in contrast to humans or pigs that have few such structures (Berne and Rubio, 1979). Further evidence exists demonstrating that the pig resembles the human with respect to the cardiovascular system, regional distribution of cardiac output (CO), metabolism and maximum oxygen consumption (Swindle, 1984; MacKrinan et al., 1986; Weaver et al., 1986). For this reason results presented here from our department have been supplied from experiments performed in the anesthetized pig. Anesthesia can influence myocardial oxygenation (van Daal et al., 1989; Stowe, 1991) and also influence the effects of hemodilution on oxygenation (Trouwborst, 1992). Therefore great care must be taken in extrapolation of the results discussed below to humans (anesthetized or not) with a normal or compromised heart, during rest, exercise, and/or stress.

MYOCARDIAL OXYGENATION (see figure 1)

According to Braunwald (1971), activity of the non-contracting heart (basal metabolism) represents about 20% of the myocardial oxygen consumption (MVO_2). Even at rest, the heart has one of the highest rates of oxygen consumption: 8 ml/min.100 g compared to an average oxygen consumption rate of the entire organism of 0.4 ml/min.100 g. During exercise the MVO_2 can increase to 40-50 ml/min.100 g. The supply of oxygen to the myocardium is, as in all tissues, determined by blood flow and the oxygen content of blood. Under normal conditions,

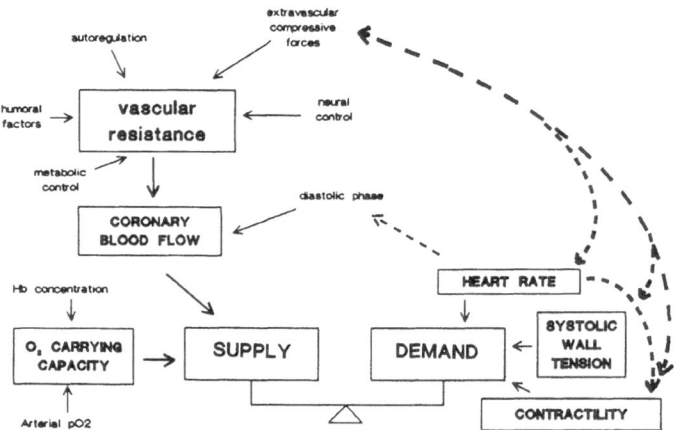

Fig. 1. Schematic drawing of the factors influencing myocardial oxygen supply and demand. Note that factors can influence both supply and demand.

myocardial oxygen extraction is already high ($\pm$ 70%) so that large increases in MVO_2 can only occur when CBF (normally about 80 ml/min.100 g) also increases. When an increase in CBF is restricted, it is only possible to meet limited increases in oxygen demand by a small increase of oxygen extraction to $\pm$ 85% (Weber and Janicki, 1979). Therefore, normally myocardial oxygen consumption is determined by CBF.

Coronary blood flow (CBF)

Blood flow in the coronary vessels is determined by the pressure gradient across the vascular bed and the resistance. Resting resistances are high and 35-40% of the total resistance is due to extravascular forces (Schremmer and Dhainaut, 1990). During systole perfusion pressures are not enough to overcome the extravascular resistance. Gregg (1963) has shown that there might even be a small back flow in the coronary arteries at the beginning of systole. This is especially evident in the subendocardium of the left ventricle as the extravascular compressive forces are higher than in the subepicardium. Effective perfusion occurs during diastole when the pressure gradient is determined by the coronary pressure minus the extravascular pressure, the coronary venous pressure, or the diastolic zero-flow pressure, whichever gradient is smallest (Feigl, 1983). Any factor that shortens the diastolic period, such as tachycardia, will impair blood supply to the subendocardium. Because of greater wall stress and increased myocyte shortening compared to the subepicardium, the subendocardium also has a higher MVO_2. Even though the subendocardium has about a 20% higher blood flow than the subepicardium (Hoffmann, 1987), this combination still makes the subendocardium susceptible to ischemia. It has been shown that, because of the complicated interrelationships it is, in a specific area of myocardial tissue, difficult to predict the point of maximum oxygenation during the cardiac cycle (Faithfull et al., 1984). To meet oxygen demands, coronary vessels must dilate to accommodate a four to sixfold increase in flow. This regulative mechanism is complicated and can be analyzed by splitting up the process into four major factors: 1) autoregulation; 2) metabolic control; 3) humoral control; and 4) neural control.

Autoregulation

Autoregulation is the intrinsic tendency of an organ to maintain constant blood flow despite changes in arterial perfusion pressure, while organ metabolism remains constant. This is difficult to prove in the heart because aortic blood pressure not only determines coronary perfusion pressure, but also represents the afterload for the left ventricle. Systolic pressure generated by the left ventricle is an important determinant of MVO_2. Therefore a change in coronary blood flow secondary to a change in aortic blood pressure will include the effects of the modified myocardial metabolism and may not be ascribed to autoregulation. Berne (1964) doubted the importance of myocardial autoregulation although Shaw and coworkers (1962) had already proven that there is a plateau in coronary flow between 60 and 160 mm Hg. Of course, in vivo MVO_2 is changing constantly so that the level of the plateau is also changing - in other words autoregulation occurs in the setting of an integrated response (Schremmer and Dhainaut, 1990). There are three hypotheses to account for autoregulation: 1) myogenic; 2) metabolic; and 3) tissue pressure.

The myogenic hypothesis states that the stretch imposed by an increase in arterial pressure stimulates arteriolar smooth muscle to vasoconstrict and limit increase in flow. This might involve stretch-activated calcium channels (Lansman et al., 1987).

The metabolic hypothesis proposes that the coronary blood flow is kept constant because the vascular smooth muscle tonus is controlled by tissue levels of metabolites or metabolic substrates. If the concentration of these substances (discussed in the next section) decreases because of an increase in perfusion pressure and blood flow, vascular smooth muscles constrict and coronary blood flow is limited until the concentration of these substrates is re-established. This phenomenon must not be confused with local metabolic control, discussed below, which is based on changes in metabolism.

The third hypothesis of autoregulation, tissue pressure, is based on the observation that an increase in perfusion pressure can result in an increase in tissue fluid volume due to ultra-filtration. This could cause increased interstitial pressure, which would tend to collapse small veins, thereby limiting blood flow. This mechanism is probably not so important in the heart, but more so in the kidney due to its stiff capsule so that an increase in volume would more readily result in an increase in pressure.

Whatever the mechanism of autoregulation, its importance becomes apparent in coronary artery disease. The perfusion pressure distal to an atherosclerotic plaque may fall below the limits of autoregulation ($\pm$ 60 mm Hg), rendering the distal vasculature bed maximally dilated and dependent on the perfusion pressure.

Metabolic control

Changes in metabolism are closely related to CBF, and a change in metabolism is immediately followed by a change in flow - within one cardiac cycle (Marcus 1988). To make things complicated, it is possible that a change in CBF is followed by a change in metabolism (Gregg's phenomenon). Different authors have suggested many substances that might control blood flow. Whatever this substance might be, it must be constantly produced by the myocardium, released into the interstitium and cleared by the blood flow. This metabolite might act as a potent vasodilator. When metabolism exceeds a previously established steady-state, the production of this metabolite increases, inducing coronary vasodilation. This, in turn, facilitates metabolite clearance, and the steady state concentration is re-established. It is also possible that the metabolite does not work directly, but through an intermediary. Probably more than one substance is involved in this control mechanism.

Oxygen itself would seem to be a logical vaso-active substance. It is possible that an increase in oxygen concentration causes an increase in coronary smooth muscle contraction, causing vasoconstriction and a decrease in flow; the opposite is most certainly true (Berne and Rubio, 1969). It is possible that oxygen does not act directly, but through messenger prostaglandins and endothelial-derived relaxing factor (EDRF) because this effect is more intense in the presence of an intact endothelium. In neural tissue hypercapnia and/or acidosis also causes vasodilation (Fennema et al., 1989) especially in the precapillary sphincters. Whether this is an important mechanism in the heart, or concurrent to other metabolic changes, is unclear (Feigl, 1983). Potassium is the major intracellular cation. During exercise there is an efflux of potassium from the skeletal muscular cells to the extracellular space, causing local dilation of the arterioles. It is possible that this also takes place in coronary vessels (Feigl, 1983). It is still debatable whether these, or other metabolites, play an important role because after brief coronary occlusion, hyperemic vasodilation persists long after the concentrations of these substances has returned to normal (Faithfull et al., 1984; Walfridsson et al., 1985).

Probably, the most important metabolite effective in controlling local coronary flow is adenosine (Ardehali and Ports, 1990). During metabolism, energy is produced by converting ATP into AMP. AMP is, in turn, converted into adenosine which is released by the myocardial cell. Normally this is cleared by blood flow and converted into hypoxanthine via inosine. With increased metabolic rates adenosine accumulates, induces vasodilation, causing an increase in blood flow, and the steady-state concentration of adenosine is restored. As mentioned before, not one, but many agents are probably involved in metabolic coronary blood flow control, and there may be a cascade of messengers involved, with interacting effects.

Humoral control

Certain agents, such as angiotensin II, serotonin, thromboxane, prostacyclin and bradykinin are involved in coronary flow. Coronary endothelial cells seem to have receptors for these agents. Binding to a specific receptor stimulates release of substances with smooth muscle relaxation properties (endothelial-derived relaxing factor, EDRF) or constrictor properties (Peach et al., 1985). The effect of EDRF is mediated through a rise in the intracellular concentration of cyclic guanosine monophosphate (cGMP), causing vasculature smooth muscle relaxation and a decrease in coronary resistance. Different parts of the coronary system (large arteries, microvasculature, collaterals) may respond differently to the above-mentioned agents. Certain diseases, such as hypertension, diabetes and atherosclerosis, can damage the endothelium so that the effects of these humeral substances may be changed.

Neural control

In the larger vessels of the epicardium, α-adrenergic stimulation induces coronary constriction, while ß-adrenergic and parasympathetic stimulation induces coronary vasodilation. In the epicardial arterioles, α-receptor stimulation causes vasodilation. This stimulation has no effect on coronary collateral vessels as they lack α-adrenergic receptors. Regional responses can be different in the presence of coronary disease. Normally acetylcholine causes a release of EDRF and therefore vasodilation, but it has been shown that when the endothelium is damaged, acetylcholine causes vasoconstriction (Blomberg et al., 1990). It is possible that in these circumstances "false neurotransmitters" may also play a role (Cohen, 1985).

In addition to the effects of the autonomic nervous system, the central nervous system also has an effect on coronary resistance (Bonham et al., 1987). Stimulation of certain foci in the brainstem due to, for instance, cerebral bleeding or edema, can cause coronary vasoconstriction.

Through this mechanism, emotions probably have a direct effect on coronary resistance, parallel to the indirect effect due to the stress response with the production of catacholamines.

Myocardial Oxygen demand

As mentioned before, basal metabolism (energy required for maintaining organelle systems, cost of electric depolarization and the activation of the contractile process) consumes only a fraction of the total energy requirements of the heart. The three most important factors determining myocardial metabolism are: 1) heart rate (HR), 2) contractility, and 3) systolic wall tension. These factors are not independent of each other and they can be influenced by other factors in a synergistic or antagonistic manner. Heart rate is felt to be the most important determinant of oxygen consumption. Fortunately HR can reasonably easily be regulated by therapeutic methods. Contractility is a function of the shortening capabilities of muscle fibers. This is determined by the load resisting shortening (afterload), the length of the fiber during shortening, and the contractile state of the fiber. Contractility is influenced by the autonomic nervous system, HR, blood calcium level, temperature, etc. Systolic wall tension is proportional to ventricular pressure, ventricular radius, and inversely proportional to ventricular wall thickness (Law of Laplace). Preload influences ventricular radius, whereas afterload dictates the magnitude of systolic pressure generation. Systolic wall tension decreases with increasing ventricular wall thickness (a compensatory mechanism in chronic hypertension).

The important aspect one must realize is that many factors influence myocardial oxygen supply and demand. Some factors influence both. Changes taking place in the organism will have different effects on the heart, and therefore a complicated effect on myocardial oxygen supply and demand, resulting in a change in myocardial oxygenation. These effects can also be different in different parts of the myocardium, and are in particular relevant to the risk of ischemia in the subendocardium.

CORONARY ARTERY DISEASE

As mentioned in the introduction, most critical heart conditions are caused by insufficient CBF. In the normal heart, CBF can meet oxygen requirements even under extreme conditions (Brazier et al., 1974). Myocardial oxygenation becomes critical in certain conditions when the coronary arteries are affected by atherosclerosis. Other diseases affecting myocardial oxygenation (for instance cardiomyopathies, valve diseases, or sepsis) are beyond the scope of this article.

Progressive coronary artery disease, especially in its earlier stages, has its greatest effect on coronary perfusion pressure. The pressure drop across an atherosclerotic plaque is proportional to the fourth power of the radius, the length of the stenosis, and the magnitude of the flow. The reduction in perfusion pressure is compensated by vasodilation distal to the stenosis until a critical threshold level is reached and flow becomes perfusion pressure dependent. In later stages, when the endothelium is affected, other factors regulating coronary flow can play a role (EDRF).

If in a post-stenotic area ischemia occurs, myocytes will die and infarction will occur. It is possible for tissue around the ischemic area (the border area) to redistribute limited amounts of blood to the ischemic area by regulation methods described above (Reves et al., 1978). It is even possible, probably due to a decrease in oxygen demand resulting from cell death in the ischemic area, coupled with the oxygen "stolen" from the border area, that oxygen pressures in the ischemic area return to normal. Also, as we have shown (Faithfull et al., 1986a), the oxygen partial pressures in the border area of the myocardium can decrease because of a combination of this "steal effect" and an increase in oxygen demand due to inefficient contractions because the infarcted area hampers the bordering area (see figure 2). Walfridsson et al. (1985) doubted this steal effect, but their measurements took place in the epicardium. Probably this effect is more pronounced in the subendocardium, especially considering that this area is already more susceptible to ischemia. Cells in an infarcted area are irreversibly destroyed so that treatment will be of no avail. Acute treatment will only be useful in salvaging the compromised border area and preventing lethal arrhythmias. Long-term treatment will be useful in preventing further ischemic attacks.

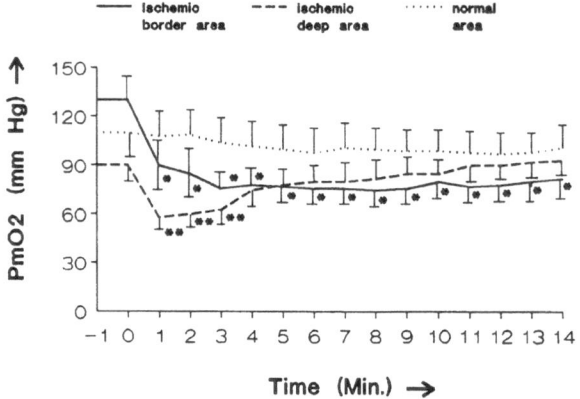

Fig. 2. Changes in PmO_2 in three regions of the porcine heart after occlusion of the left anterior descending artery (LAD) at time zero. Significance from pre ischemia: ** = $p < 0.01$, * = $p < 0.001$.

TREATMENT OF CORONARY DISEASE

Behavioral modifications and the correction of risk factors (hypertension, cholesterol, etc.) in patients with coronary artery disease are beyond the scope of this article. Revascularization with the aid of arterial or venous grafts, balloon dilatation, laser, etc. are a separate subject and are usually applied to obtain long-term results. Pharmacological therapies are divided into three groups of drugs: 1) nitrates, 2) ß-adrenergic receptor blockers, and 3) calcium channel blockers (see figure 3).

Nitrates

Nitrates relax venous and coronary arterial smooth muscles. Coronary arterial vasodilation increases supply of oxygen to the myocardium, but might also increase the above-mentioned steal effect as well as increasing shunt perfusion due to an increased collateral blood flow (Cohen et al., 1973). Probably the most important effect is that venous vasodilation decreases the return of blood to the heart, reducing preload, ventricular size and systolic wall tension with a net result of a decrease in oxygen consumption. Probably due to a decrease in wall tension, nitrates also cause a redistribution of blood flow to the endocardium (Bache et al., 1975).

Beta-blockers

ß-adrenergic receptor blockers inhibit the effects of circulating and neurally released catecholamines. This causes a decrease in HR and contractility, thereby reducing myocardial oxygen consumption, especially during periods of stress when the production of catecholamines is stimulated. A slower HR also has the extra benefit that the diastolic phase is prolonged, which has a favorable effect on subendocardial blood flow.

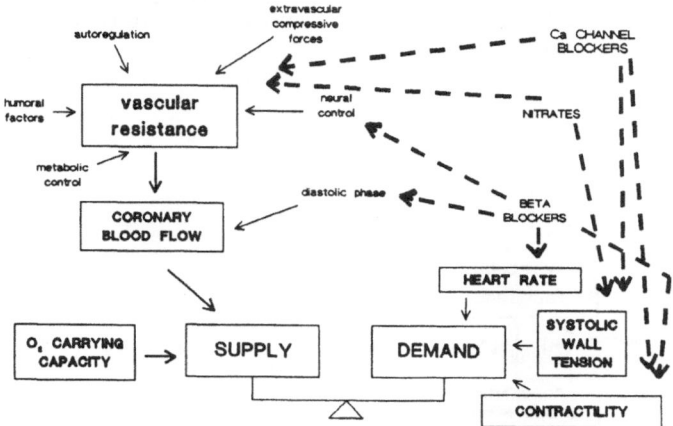

Fig. 3. The effects of medication on myocardial oxygenation.

Calcium channel blockers

These inhibit the entry of calcium into cardiac and vascular smooth muscle cells through voltage sensitive channels. All calcium channel blocking agents cause vasodilation, negative inotropy, chronotropy and dromotropy, but to different degrees. The net effect is a favorable influence on the balance of myocardial oxygen supply and demand.

HEMODILUTION (see figure 4)

Hemodilution is a technique which is often applied during anesthesia, or in intensive care situations. This treatment is used successfully in conditions where blood transfusion is undesirable or impossible, as shown by work done in our department (Trouwborst et al., 1990a; Trouwborst et al., 1990b). The effects of hemodilution in organisms with a normal heart are quite well documented, but the effects of hemodilution when the heart is compromised by coronary disease must also be taken into consideration. Hemodilution has an effect on systemic oxygen supply and, of course, an effect on myocardial oxygen demand and supply. First the effects of hemodilution in the healthy heart will be discussed.

Hemodilution in the healthy heart

Hemodilution decreases blood viscosity which causes a decrease in systemic vascular resistance (SVR) and an increase in CO. Because of the lower hemoglobin concentration, oxygen content also falls, but this is compensated by the increase in CO so that oxygen supply is maintained over a wide range of hematocrits (Messmer et al., 1973). Also, this group (Trouwborst et al., 1989; Trouwborst et al., 1990c) have shown that hemodilution can cause an acute shift of the oxyhemoglobin dissociation curve to the right, improving availability of oxygen to the tissues. In experiments with conscious animals, the increase in CO is attributed to an increase in HR, while in anesthetized animals, CO is closely related to stroke volume (SV) (Trouwborst, 1992). This increase in SV can be attributed to an increase in preload and a decrease in afterload due to the drop in viscosity (Laks et al., 1974). Also the autonomic nervous system can be involved (Glick et al, 1964), although this mechanism might be suppressed during anesthesia (Van Woerkens et al., 1991). As mentioned above, HR is the most important factor determining MVO$_2$, so that MVO$_2$ will not rise significantly during anesthesia and hemodilution (Gisselsson et al., 1982). This might not be the case in the intensive care unit, where tachycardia is often seen.

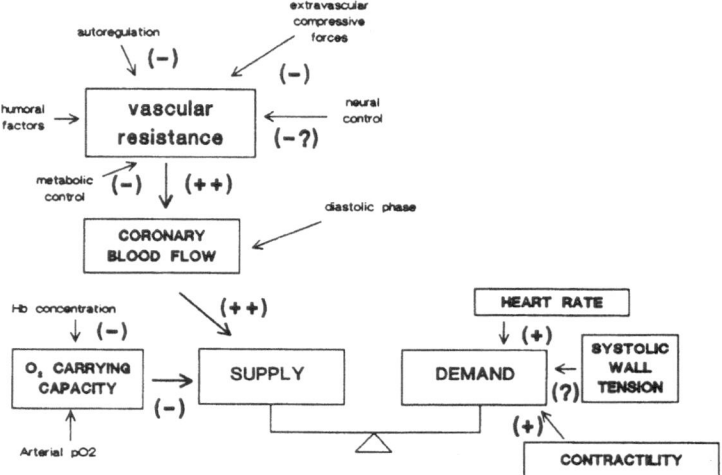

Fig. 4. The effects of hemodilution on myocardial oxygenation.

Because arterial pressure remains constant during hemodilution (Geha, 1976), coronary perfusion pressure also remains constant. Therefore the increases seen in CBF are due to a decreased coronary vascular resistance, partly due to a decrease in viscosity, and partly due to metabolic vasodilator mechanisms (Crystal and Salem, 1991). Van Woerkens et al. (1992a) have shown that hemodilution can cause a four-fold increase in CBF in the left ventricle, which is more than enough to compensate for the increase in MVO$_2$. Interestingly, the increase in flow is more pronounced in the subepicardium than in the subendocardium, so that in extreme conditions, when the limits of vasodilatory capacity are meet, the subendocardial area could still become ischemic (Brazier et al., 1974). It is also interesting to note that Vogel et al. (1989) reported that although coronary flow increases by 400%, myocardial microflow, measured with hydrogen clearance, only increases slightly during hemodilution. They postulated that this might be due to preferential perfusion of "high-flow channels".

There are also other mechanisms involved in maintaining oxygenation in the microcirculation during hemodilution, at least until the hematocrit falls below 20%. In normal conditions, the hematocrit of the microcirculation is already 30 to 40% lower than systemic hematocrit. A partial explanation is that plasma skimming occurs, a phenomenon whereby plasma passes slower through small capillaries than do erythrocytes. This means that there are relatively fewer erythrocytes in the capillaries than in the larger vessels at a certain point in

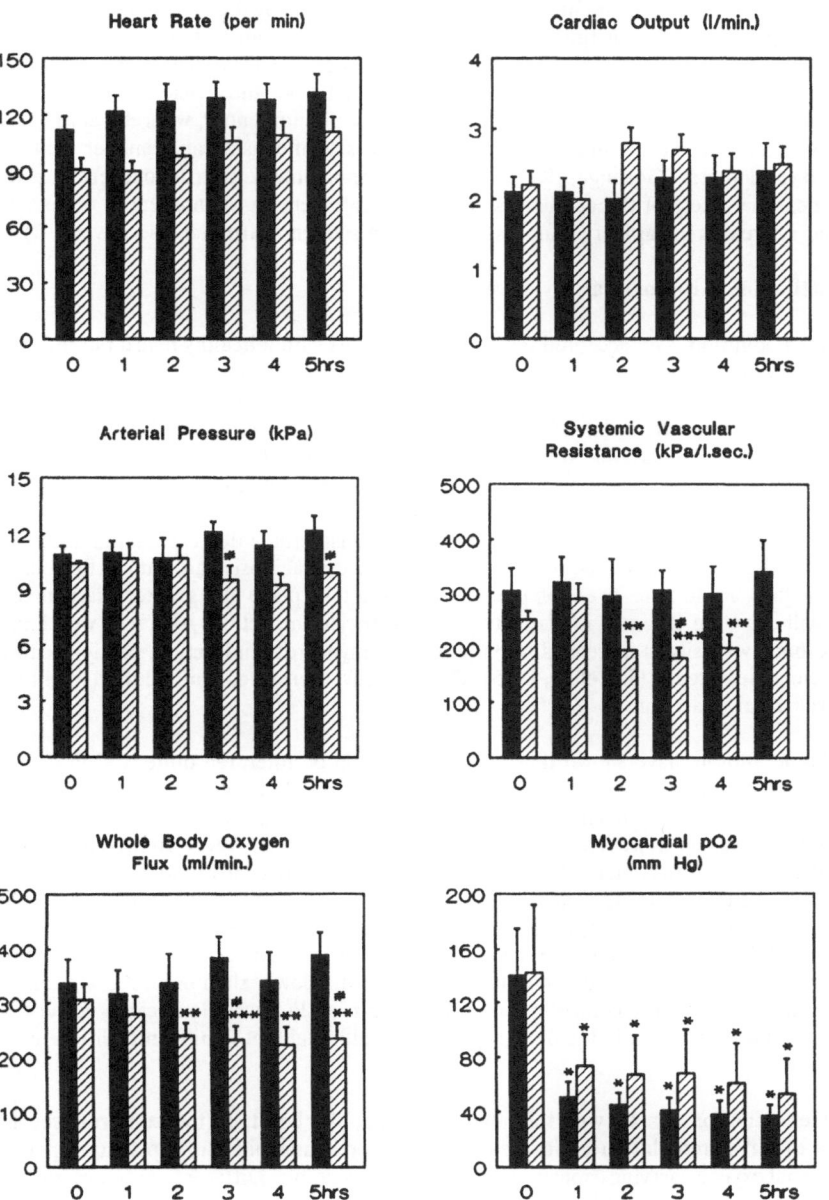

Fig. 5. Changes in hemodynamics and oxygenation after LAD occlusion in control pigs (solid bars) and pigs hemodiluted with Dextran-40 between 1 and 2 hours (striped bars). Myocardial oxygen measurements took place in the area with the greatest drop in PO_2 after the first hour of infarction. Significance from pre-occlusion: * = $p < 0.05$, ** = $p < 0.01$, *** = $p < 0.001$; significant difference from the control group: # = $p < 0.05$.

time, so that the hematocrit is lower. "Thoroughfare channels" are channels where blood with a high hematocrit flows to compensate for plasma skimming. Furthermore, other mechanisms not yet understood tend to maintain a constant erythrocyte influx during hemodilution, and therefore a constant capillary hematocrit. Release of oxygen by the erythrocyte in the arterioles is a function of the transit time. Since hemodilution causes an increase in blood flow in the arterioles, transit time will decrease so that erythrocytes will arrive at the capillaries with relatively more oxygen. For a review of these phenomena, see the article by Trouwborst (1992). Another important effect is the "Fahraeus-Lindquist phenomenon", whereby erythrocytes can flex to such an extent that they can pass through vessels of considerably smaller diameter than there own diameter (Goslinga, 1982). This phenomenon only holds true to a certain extent and, at critical diameters of the capillaries, the so-called "inversion phenomenon" occurs, whereby further decreases in capillary diameter results in a sudden sharp increase in viscosity.

Hemodilution in coronary artery disease

Hemodilution is an accepted technique in patients with a healthy heart, but little is known about the effects in patients with coronary artery disease. During acute hemodilution, the heart is the only organ with an increase in oxygen consumption, so that it is the limiting factor; this becomes even more critical in heart disease. One must bear in mind that the increase in cardiac output necessary to maintain total tissue oxygenation might be impaired during hemodilution because of a deterioration in ventricular function due to ischemia.

Cohn et al. (1974) have shown that there is a considerable decrease in pH in the ischemic area. According to Schmid-Schönbein et al. (1973) this acidosis can reduce flexibility of the erythrocytes, so that the inversion phenomenon will take place at a greater vascular diameter. This will cause an increase in viscosity and erythrocyte impaction. Also it is well known that whole body viscosity is increased in patients with myocardial infarction, partly because of an increase in hematocrit (Jan et al., 1975). Theoretically, therefore, hemodilution might be useful in myocardial infarction.

Some authors have shown that hemodilution can be tolerated quite well in the ischemic heart, even to a hematocrit of 6% (Yoshikawa et al., 1973; Tucker et al., 1980). Others have even postulated an improvement of ischemic myocardial injury due to hemodilution, but this was only the case in short-term coronary occlusion (Cohn et al., 1975). In this case vasodilation is probably already maximal in the post-stenotic area, so that improvement in the ischemic area would be due to an increased flow of the collateral circulation (Stam and de Jong, 1977). This group (Faithfull et al., 1986b) performed experiments in pigs in which one hour after heart infarction 20 ml/kg of hemodilution was carried out with 5% dextran 40 (see figure 5). Five hours after infarction whole body oxygen flux was significantly lower than before infarction, and significantly lower than the control group. Intramyocardial PO_2 was significantly lower than pre-infarction measurements in the most hypoxic area, although not significantly different from the control group.

Hemodilution can be well tolerated in the healthy heart. In the compromised heart, it seems that if hemodilution is deemed necessary, or if unavoidable, a plasma expander with substantial oxygen carrying capacities would be more suitable. Different oxygen-carrying plasma substitutes are now available, such as highly polymerized human hemoglobin (Barnikol and Burkhard, 1990) but only fluorocarbons will be discussed in this article.

FLUOROCARBONS

Fluorocarbons, also known as perfluorocarbons or perfluorochemicals (PFC's), are almost completely inert chemical substances with a high solubility for oxygen. This property was dramatically demonstrated by Clark and Gollan (1966) in mice surviving total immersion in these liquids. As PFC's are almost completely insoluble in water (or blood), they must be emulsified before use. The product used in our experiments was Fluosol-DA 20% (Green Cross Corporation, Osaka, Japan). A number of other PFC's are now undergoing trials, and new PFC-containing preparations are being developed, which might avoid some of the disadvantages mentioned below (Faithfull, 1992).

Fluosol-DA 20% (FDA20) can carry about 0.75 ml/100 ml.100 mm Hg (Grote et al., 1985). This capacity is low in comparison to the content in normal human blood under similar conditions. Hence, in practice, it is necessary to breathe as high a percentage of oxygen as possible. At an arterial PO_2 of 550 mm Hg, being about the maximal achievable with an FiO_2 of 1.0, FDA20 contains about 4.1 ml O_2 per 100 ml emulsion. Normal blood contains about five times that amount of oxygen but, because of the linear relationship between PO_2 and oxygen content in PFC's, the delivery of oxygen to tissue by FDA20 is not much lower than that from blood. PFC's can also release oxygen in a very low PO_2 environment, whereas the release of oxygen by hemoglobin is limited (Trouwborst et al., 1989). There is evidence that a combination of blood and FDA20 will result in a higher oxygen supply than would be expected from normal oxygen transport dynamics (Faithfull and Cain, 1987).

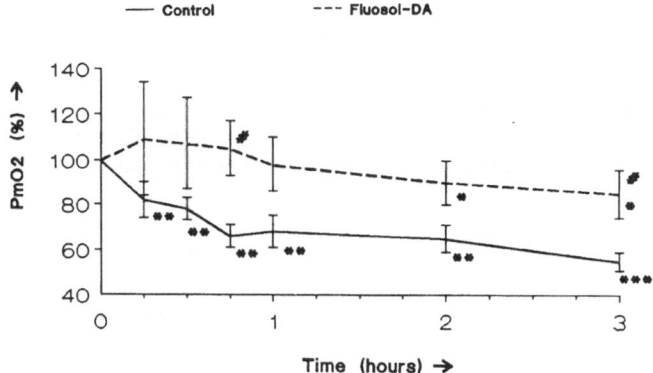

Fig. 6. Changes in whole body oxygen flux following LAD occlusion. Hemodilution was performed between 1 and 2 hours after occlusion. For significance see figure 5.

FDA20 has a very low viscosity, about 0.55 times that of blood. Furthermore, whereas the viscosity of blood increases markedly under the low-flow shear rates existing in the microcirculation, the viscosity of FDA20 is almost independent of shear rate. FDA20 is only about 0.34 times as viscous as blood under these conditions (Olson et al., 1983). Biro (1981) has suggested that, in conditions such as myocardial infarction, reduction in viscosity by hemodilution may improve perfusion through available collaterals. Harrison et al. (1985) have shown that there is in fact an increase in viscosity from moderate to extreme hemodilution with FDA20, and that the improvement seen in myocardial PO_2, even in these extreme circumstances, must be explained in another way. Vogel et al. (1989) have shown that after FDA20 is given, there is a redistribution of oxygen supply in favor of the poorly oxygenated sites.

The size of the particles of emulsified PFC's in FDA20 is very small. The mean diameter is 0.118 μm and more than 90% of the particles are less than 0.2 μm in diameter (Naita and Yokoyoma, 1978). It is possible that a combination of low viscosity and small particle emulsion size confers a unique ability for penetrating deeply into collateral capillary pathways and for bypassing stiffened erythrocytes. It is also possible that reoxygenation of these impacted red cells occurs with subsequent further improvement of penetration into the ischemic area. In this way the vicious circle of stagnant acidosis can be broken. PFC's can also affect oxygenation in other ways, and can be used for the treatment of many conditions. This has been extensively reviewed by Faithfull (1987).

In dogs, FDA20 has been shown to cause hypotension. Ohyanagi and Mitsuno (1975) considered this to be a result of serotonin and histamine release, while Suyama et al. (1979) concluded that the hypotension was due to direct and indirect effects on α-adrenergic receptors. This has been opposed by Faithfull and Cain (1988). These hemodynamic changes do not seem to occur with the newer lecithin emulsifications (Hammerschmidt and Vercellotti, 1988).

We have shown that hemodynamic changes due to FDA20 are different in the pig (Faithfull et al., 1989). These changes are probably similar to those seen in humans (Tremper et al., 1984). As seen with other plasma expanders, PFC emulsions cause an increase in CO. The drop in systemic vascular resistance is also similar. The most important effect is a transient, but significant, rise in pulmonary artery pressure. This does not always occur, but when it does, the result is a decrease in arterial blood pressure and sometimes CO. The alternative pathway in complement activation is probably involved (Vercellotti et al., 1982).

Hemodilution with PFC's after myocardial infarction

As postulated above, hemodilution with PFC's might be a useful treatment after myocardial infarction. Kessler et al. (1983) have reported near normal epicardial PO_2 histograms in dogs breathing 30% oxygen at a hematocrit of only 8% and a "fluorocrit" of 16%. Biro (1983), also working in dogs, showed a marginal decrease in infarcted muscle mass after giving PFC's, while

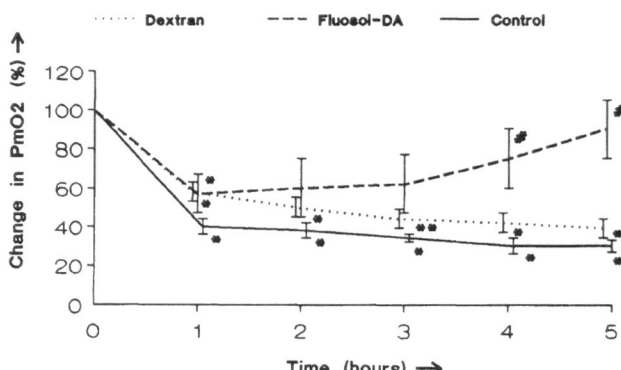

Fig. 7. Percentage change in myocardial PO_2 in the most hypoxic area after LAD occlusion. Hemodilution was performed between 1 and 2 hours after occlusion. For significance see figure 5.

a marked improvement in blood flow to the ischemic area was observed. Nunn et al. (1983) found significantly less necrosis in dogs when comparing FDA20 to control groups. Schaer et al. (1990) have shown that this improvement could be due to a reduction in reperfusion injury.

For reasons mentioned above, our work was performed in anesthetized pigs. To enhance oxygen carrying capacities, experiments were performed while the pigs were ventilated with 100% oxygen. An infarction was induced by occluding the distal third of the left anterior descending coronary artery (LAD). No significant or clinically relevant hemodynamic changes occurred, although an infarct in the myocardium was visible (Faithfull et al., 1986b). After hemodilution with FDA20, CO was significantly higher, and SVR and arterial pressure significantly lower than the control group. However, after 5 hours these values were no longer significantly different to pre-LAD occlusion values. These results are similar to those shown in figure 5 where dextran was used. Even though there was no difference in CO between the FDA20 group and the control group after 5 hours of infarction, oxygen flux was still maintained (329 ± 24.7 ml/min resp. 389 ± 42.3 ml/min.), this in contrast to the effects of dextran (235 ± 28.1 ml/min.), as mentioned above (see figure 6). The oxygen extraction coefficient in the FDA20 group was significantly lower than that of the control group at the end of hemodilution. After hemodilution with FDA20, intramyocardial PO_2 measurements of the most hypoxic area showed an improvement, so that after 3 hours of hemodilution, there is no significant difference from pre-occlusion values, while PO_2 measurements in the control group are still significantly decreased (see figure 7).

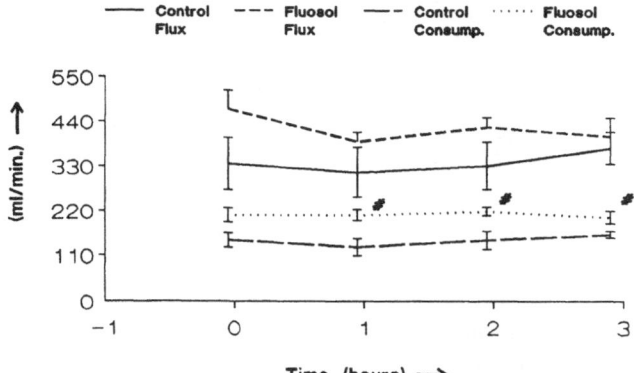

Fig. 8. Changes in whole body oxygen flux and consumption following LAD occlusion in pigs pre-treated with Fluosol compared to controls. Significant difference from controls: # = p<0.05.

As mentioned above, the most hypoxic area is probably the border area, so that this is the area that is salvaged by treatment with PFC's. It seems that a certain amount of time is necessary before PO_2 measurements start to improve. This is probably because it takes time before the flexibility of the erythrocyte membrane has returned to normal. Pathological examination showed areas of ischemic change in both the FDA20 and control groups. No significant difference in the degree of ischemia could be seen between the two groups. A reason for this could be that the period of time between infarction and death was relatively short (5 hours) for pathological examination. Rice et al. (1990) have shown, in rabbits, that 24 hours after infarction and treatment with PFC's, infarction size is smaller than in control groups, but that this is dose-dependent with a maximal reduction in infarction size at 30 ml/kg hemodilution. Virmani et al. (1990) showed similar results in rabbits that were sacrificed 1 to 14 days after infarction.

Hemodilution with PFC's before myocardial infarction

As mentioned above, hemodilution is an accepted technique during anesthesia, but it is limited in coronary disease. Therefore it would be interesting to combine this technique with PFC's. We performed experiments in pigs, not only to show the improvement in oxygenation with PFC's, but to show that they can also decrease myocardial damage during infarction (Faithfull et al, 1988a). As would be expected, after hemodilution with FDA20 (20 ml/kg), CO was significantly higher, and SVR significantly lower than in the control group. Whole body oxygen flux and consumption were higher in the FDA20 group, and this did not significantly change in the course of time (see figure 8). As in our other experiments, LAD occlusion caused no significant changes in hemodynamics. In work described above, we showed that the ischemic border area was the most susceptible to hypoxia, with significant decrease in PO_2 after LAD occlusion. In pigs pre-treated with FDA20, there was no significant change in the intra-myocardial PO_2 of the ischemic border area after LAD occlusion (Faithfull et al., 1988b). When

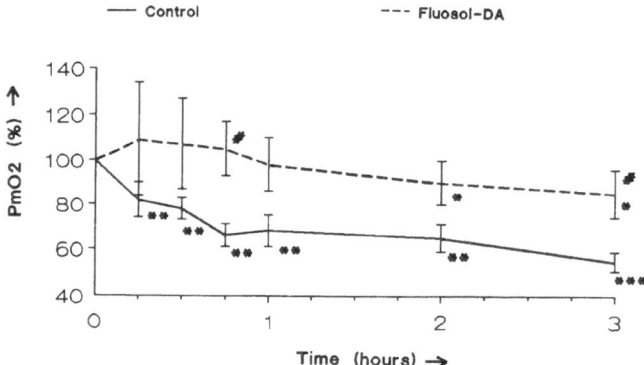

Fig. 9. Percentage change of the average myocardial PO_2 in different areas following LAD occlusion in pigs pre-treated with Fluosol compared to controls. For significance see figure 5.

looking at measurements from areas showing the greatest decrease in PO_2 over the first hour of LAD occlusion, there was already a significant drop in PO_2 after 1 minute. In the FDA20 treated group, this took 45 minutes. When looking at all PO_2 measurements of the heart, including ischemic and non-ischemic areas, it takes 15 minutes for a significant drop in mean intramyocardial PO_2 (see figure 9). This is not seen until 2 hours after LAD occlusion in the FDA20 pre-treated group, indeed during the first 15 minutes mean PmO_2 actually rose (Faithfull et al., 1988a).

This work shows that, although prior hemodilution with FDA20 cannot completely prevent decreases in PmO_2, the changes can be reduced substantially and delayed in onset. In clinical situations both of these effects would be beneficial.

CONCLUSIONS

Hemodilution is an accepted technique during anesthesia, as it can often avoid blood transfusion. Although theoretically useful in heart disease, experimental work has shown that care must be taken, especially when ischemia already exists. Fluorocarbons, as plasma expanders, have the advantage of improving oxygenation so that it might be useful in coronary artery disease. Studies have shown that it might be helpful after myocardial infarction, but it is probably more potent in preventing or delaying ischemia. In clinical situations, this therapeutic gain would probably be adequate to allow myocardial revascularization procedures to be instituted. In the near future it might be possible to use fluorocarbons instead of other plasma expanders, especially in coronary compromised patients and/or patients with a contra-indication for receiving blood. The use of second generation fluorocarbon emulsions and clinical research with the aid of intensive monitoring (Fennema et al., 1991, Van Woerkens et al., 1992b) will help hasten this process.

SUMMARY

This article reviews the factors influencing myocardial oxygen supply and demand. The regulative mechanisms in coronary blood flow, especially in critical conditions, are explained. Myocardial oxygenation in coronary artery disease is discussed with special reference to pharmacological intervention. An extensive evaluation of the effects of hemodilution on both the healthy and diseased heart is presented. Effects of hemodilution with fluorocarbons for the treatment or prevention of myocardial ischemia are shown with the aid of intramyocardial oxygen partial pressure measurements.

REFERENCES

Ardehali, A., and Ports, T. A., 1990, Myocardial oxygen supply and demand, *Chest*, 98:699.

Bache, R. J., Ball, R. M., Cobb, F. R., Rermbert, J. C., and Greenfield, J. C., Jr., 1975, Effects of nitroglycerin on transmural myocardial blood flow in the unanesthetized dog, *J. Clin. Invest.*, 55:1219.

Barnikol, W. K. R., and Burkhard, O., 1990, Low viscosity of densely and highly polymerized human hemoglobin in aqueous solution - the problem of stability, *Adv. Exp. Med. Biol.*, 248:335.

Berne, R. M., 1964, Regulation of coronary blood flow, *Phys. Rev.*, 44:1.

Berne, R. M., and Rubio, R., 1969, Acute coronary occlusion: Early changes that induce coronary dilation and the development of collateral circulation, *Am. J. Cardiol.*, 24:766.

Berne, R. M., and Rubio, R., 1979, Coronary circulation, *In*: "Handbook of Physiology", R. M. Berne, N. Sperelakis and S. R. Geriger, eds, American Physiology Society, Bethesda, Maryland.

Biro, G. P., 1981, Effects of hemodilution with dextran, stroma-free hemoglobin solution and Fluosol-DA on experimental myocardial ischemia in the dog, *Bibliotheca Haematologica*, 47:54.

Biro, G. P., 1983, Fluorocarbon and dextran hemodilution in myocardial ischemia, *Can. J. Surg.*, 26:163.

Blomberg, S., Emanuelsson, H., Kvist, H., Lamm, C., Pontèn, J., Waagstein, F., and Ricksten. S. E., 1990, Effects of thoracic epidural anesthesia on coronary arteries and arterioles in patients with coronary artery disease, *Anesthesiology*, 73:840.

Bonham, A. C., Gutterman, D. D., Arthur, J. M., Marcus, M. L., Gebhart, G. F., and Brody, M. J., 1987, Neurogenic regulation of coronary blood flow: evidence for central nervous system pathway, *Circ. Res.*, 61:1142.

Braunwald, E., 1971, Control of myocardial consumption: Physiological and clinical considerations, *Am. J. Cardiol.*, 27:416

Brazier, J., Cooper, N., Maloney, J. V., and Buckberg, G., 1974, The adequacy of myocardial oxygen delivery in acute normovolemic anemia, *Surgery*, 75:508.

Clark, L. C., and Gollan, F., 1966, Survival of mammals breathing organic liquids equilibrated with oxygen at atmospheric pressure, *Science*, 152:1755.

Cohen, M. V., Downey, J. M., Sonnenblick, E. H., and Kirk, E. S., 1973, The effects of nitroglycerin on coronary collaterals and myocardial contractility, *J. Clin. Invest.*, 52:2836.

Cohen, R. A., 1985, Platelet-induced neurogenic coronary arterial contractions due to the accumulation of the false neurotransmitter 5-hydroxytryptamine, *J. Clin. Invest.*, 75:286.

Cohn, L. H., Fujiwara, Y., and Collins, J. J. Jr, 1974, Mapping of ischemic myocardium by surface Ph determination, *J. Surg. Res.*, 16:210.

Cohn, L. H., Lamberti, J. J., Florian, A., Moses, R., Vandewater, S., Kirk, E., and Collins, J. J. Jr, 1975, Effects of hemodilution on acute myocardial ischemia, *J. Surg. Res.*, 18:523.

Corr, P. B., and Sobel, B. E., 1979, The importance of metabolites in the genesis of ventricular dysrhythmia induced by ischemia, *Modern Concepts Cardiovasc. Dis.*, 48:43.

Crystal, G. J., Rooney, M. W., and Salem M. R., 1988, Regional hemodynamics and oxygen supply during isovolemic hemodilution alone and in combination with adenosine-induced controlled hypotension, *Anesth. Analg.*, 67:211.

Crystal, G. J., and Salem, M. R., 1991, Myocardial and systemic hemodynamics during isovolemic hemodilution alone and combined with nitroprusside-induced controlled hypotension, *Anesth. Analg.*, 72:227.

Faithfull, N. S., Erdmann, W., and Fennema, M., 1984, Oxygen supply to the myocardium, *Adv. Exp. Med. Biol.*, 180:411.

Faithfull, N. S., Fennema, M., Erdmann, W., Dhasmana, M., and Eilers, G., 1986a, The effects of acute ischemia on intramyocardial oxygen tensions, *Adv. Exp. Med. Biol.*, 200:339.

Faithfull, N. S., Erdmann, W., Fennema, M., and Kok, A., 1986b, Effects of haemodilution with fluorocarbons or dextran on oxygen tension in the acutely ischaemic myocardium, *Br. J. Anaesth.*, 58:1031.

Faithfull, N. S., 1987, Fluorocarbons: Current status and future applications, *Anaesthesia*, 42:234.

Faithfull, N. S., and Cain, S. M., 1987, Critical oxygen delivery levels during shock following normoxic and hyperoxic haemodilution with fluorocarbons or dextran, *Adv. Exp. Med. Biol.*, 215:79.

Faithfull, N. S., and Cain, S. M., 1988, Cardiorespiratory consequences of fluorocarbon reactions in dogs, *Biomater. Artif. Cells Artif. Organs*, 16(1-3):463.

Faithfull, N. S., Fennema, M., and Erdmann, W., 1988a, Protection against myocardial ischaemia by prior haemodilution with fluorocarbon emulsions, *Br. J. Anaesth.*, 60:773.

Faithfull, N. S., Fennema, M., and Erdmann, W., Myocardial oxygen tensions during ischaemia in fluorocarbon diluted pigs, 1988b, *Adv. Exp. Med. Biol.*, 222:473.

Faithfull, N. S., Fennema, M., and Erdmann, W., 1989, Haemodilution and myocardial ischaemia - studies with fluorocarbons and dextran in pigs, *In*: "Innovations in Physiological Anaesthesia and Monitoring", R. Droh, R. Spingte, eds., Springer-Verlag, Heidelberg, New York, London, Paris, Tokyo, Hong Kong, 151.

Faithfull, N. S., 1992, The new generation of perfluorocarbons, *Adv. Exp. Med Biol.*, this volume.

Feigl, E. O., 1983, Coronary Physiology, *Phys. Rev.*, 63:1.

Fennema, M., 1988, Tissue oxygen tensions under physiological and pathological conditions, *Ph.D. Thesis*, Erasmus University, Rotterdam.

Fennema, M., Wessel, J. N., Faithfull, N. S., and Erdmann, W., 1989, Tissue oxygen tension in the cerebral cortex of the rabbit, *Adv. Exp. Med. Biol.*, 248:451.

Fennema, M., van Krugten, R. J., de Boer, H. J., Prakash, O., and Erdmann, W., 1991, Continuous intra-arterial PO_2 monitoring during thoracic surgery, *Adv. Exp. Med. Biol.*, in press.

Geha, A. S., 1976, Coronary and cardiovascular dynamics and oxygen availability during acute normovolemic anemia, *Surgery*, 80:47.

Gisselsson, L., Rosberg, B., and Ericsson, M., 1982, Myocardial blood flow, oxygen uptake and carbon dioxide release of the human heart during hemodilution, *Acta Anaesth. Scand.*, 26:589.

Glick, G., Plauth, W. H., and Braunwald, E., 1964, Role of the autonomic system in the circulatory response to acutely induced anemia in unanesthetized dogs, *J. Clin. Invest.*, 43:2112.

Goslinga, H., 1982, The viscosity of blood. An experimental study into the effects of alteration in blood viscosity during shock. *Ph.D. Thesis*, University of Utrecht, Utrecht.

Gregg, D. E., 1963, Physiology of the coronary circulation. The George E. Brown memorial lecture, *Circulation*, 27:1128.

Grote, K., Steuer, R., Müller, R., Söntgerath, C., and Zimmer, K., 1985, O_2 and CO_2 solubility of the fluorocarbon emulsion Fluosol-DA 20% and O_2 and CO_2 dissociation curves of blood-Fluosol-DA 20% mixtures, *Adv. Exp. Med. Biol.*, 191:453.

Hammerschmidt, D. E., and Vercellotti, G. M., 1988, Limitations of complement activation by perfluorocarbon emulsions: superiority of lecithin-emulsified preperations, *Biomater. Artif. Cells Artif. Organs*, 16(1-3):431.

Harrison, D. K., Günther, H., Vogel, H., Ellerman, R., Brunner, M., Höper, J., and Kessler, M., 1985, Oxygen supply and microcirculation of the beating heart after haemodilution with Fluosol-DA 20%, *Adv. Exp. Med. Biol.*, 191:445.

Hoffmann, J. I. E., 1987, Transmural myocardial perfusion, *Prog. Cardiovasc. Dis.*, 29:429.

Jan, K. M., Chien, S., and Bigger, J. T. jr, 1975, Observations on blood viscosity changes after acute myocardial infarction, *Circulation*, 51:1079.

Kessler, M., Vogel, H., Günther, H., Harrison, D. K., and Hoper, J., 1983, Local oxygen supply of the myocardium after extreme hemodilution with Fluosol-DA, *Prog. Clin. Biol. Res.*, 122:237.

Laks, H., Pilon, R. N., Klovekorn, W. P., Anderson, W., MacCallum, J. R., and O'Connor, N. E., 1974, Acute hemodilution: its effects on hemodynamics and oxygen transport in anesthetized man, *Ann. Surg.*, 180:103.

Lansman, J. B., Hallman, T. J., and Rink, T. J., 1987, Single stretch-activated ion channels in vascular endothelial cells as mechano-transducers?, *Nature*, 325:811.

MacKrinan, M. D., White, B. D. G., and Bloor, C. M., 1986, Cardiovascular and metabolic responses to acute and chronic exercise in swine, *In*: "Swine in Biomedical Research", M. E. Tumbleson, ed., Plenum Press, New York and London.

Marcus, M. L., 1988, The regulation of myocardial perfusion in health and disease, *Hosp. Prac.*, 23:203.

Messmer, K., Görnandt, L., Jesch, X., Sinagowitz, E., Sunder-Plassman, L., and Kessler, M., 1973, Oxygen transport and tissue oxygenation during hemodilution with dextran, *Adv. Exp. Med. Biol.*, 37:669.

Moss, A. J., 1968, Intramyocardial oxygen tension, *Cardiovasc. Res.*, 3:14.

Naito, R., and Yokoyama, K., 1978, Perfluorochemical blood substitutes, *In*: "Green Cross Corporation Technical Information Service", No. 5, Osaka, Japan.

Nunn, G. R., Dance, G., Peters, J., and Cohn, L., 1983, Effects of fluorocarbon exchange transfusion on myocardial infarction size in dogs, *Am. J. Cardiol.* 52:203.

Ohyanagi, H., and Mitsuno, T., 1975, Biophysiological effects of perfluorochemicals as artificial blood, *In*: "Proceedings of the 10th International Congress of Nutrition: Symposium on PFC Artificial Blood", Igakushobo, Osaka, 21.

Olson, R. D., Parrish, M. D., Mushlin, P. S., and Broucek, R. J. jr, 1983, Perfluorocarbon artificial blood: superior rheologic properties, *Pediatric Research*, 17: 120.

Parsons, W. J., Bembert, J. C., Bauman, R. P., Greenfield, J. C. jr., and Piantadosi, C. A., 1990, Dynamic mechanisms of cardiac oxygenation during brief ischemia and reperfusion, *Am. J. Physiol.*, 259(5):1477.

Peach, M. J., Loerb. A. L., Singer, H. A., and Saye, J. A., 1985, Endothelium-derived vascular relaxing factor, *Hypertension*, 7:94.

Reves, L. G., Erdmann, W., Mardis, M., Karp, R. B., King, M., and Lell, W. A., 1978, Evidence for existence of myocardial steal, *Adv. Exp. Med. Biol.*, 94:755.

Rice, H. E., Virman R., Hart, C. L., Kolodgie, F. D., and Farb, A., 1990, Dose-dependent reduction of myocardial infarct size with the perfluorochemical Fluosol-DA, *Am. Heart J.*, *120:1039*.

Rude, R. E., Bush, L. R., and Tilton, G. D., 1984, Effects of fluorocarbons with and without oxygen supplementation on cardiac hemodynamics and energetics, *Am. J. Cardiol.*, 54:880.

Schaer, G. L., Karas, S. P., Santoian, E. C., Gold, C., Visner, M. S., and Virman, R., 1990, Reduction in reperfusion injury by blood-free reperfusion after experimental myocardial infarction, *J. Am. Coll. Cardiol.*, 15:1385.

Schmid-Schönbein, H., Weiss, J., and Ludwig, H. A., 1973, A simple method for measuring red cell deformability in models of the microcirculation, *Blut*, 16:369.

Schremmer, B., and Dhainaut, J. F., 1990, Regulation of myocardial oxygen delivery, *Intensive Care Med.*, 16:S157.

Schuchhardt, S., 1985, Myocardial oxygen pressure: Mirror of oxygen supply, *Adv. Exp. Med. Biol.*, 191:21.

Shaw, R. F., Mosher, P., Ross, J., Joseph, J. I., and Lee, A. S. J., 1962, Physiological principles of coronary perfusion, *J. Thorac. Cardiovasc. Surg.*, 44:608.

Stam, H. M., and de Jong, J. W., 1977, Sephadex-induced reduction of coronary flow in the isolated rat heart: a model for ischemic heart disease, *J. Molec. Cell Cardiol.*, 9:633.

Stowe, D. F., Marijic, J., Bosnjak, Z. J., and Kampine, J. P., 1991, Direct comparative effects of halothane, enflurane, and isoflurane on oxygen supply and demand in isolated hearts, *Anesthesiology*, 74:1087.

Suyama, T., Watanabe, M., Hanada, S., Yano, K., Yokoyama, K., and Naito, R., 1979, Pharmacological analysis of the mode of transient hypotensive action of Fluosol-DA found in the dog, *In*: "Proceedings of the 4th International Symposium on PFC Blood substitutes", Exerpta Medica, Amsterdam, 257.

Swindle, M. M., 1984, Swine as replacement for dogs in the surgical teaching and research laboratory, *Lab. Anim. Sci.*, 34:383.

Tremper, K. K., Vercellotti, G. M., and Hammerschmidt, D. E., 1984, Hemodynamic profile of adverse clinical reactions to Fluosol-DA 20%, *Crit. Care Med.*, 12:428.

Trouwborst, A., v.d. Broek, W. G. M., Tenbrinck, R., Groenland, T. H. N., Bucx, M., and Faithfull, N.S., 1989, Alterations in oxyhemoglobin dissociation curve during normoxic acute normovolemic hemodilution, *Adv. Exp. Med. Biol.*, 248:419.

Trouwborst, A., van Woerkens, E. C. S. M., van Daele, M., and Tenbrinck, R., 1990a, Acute hypervolaemic haemodilution to avoid blood transfusions during major surgery, *Lancet*, 336:1295.

Trouwborst, A., Hagenouw, R. R. P. M., Jeekel, J., and Ong, G. L., 1990b, Hypervolaemic haemodilution in an anaemic Jehovah's Witness, *Br. J. Anaesth.*, 64:646.

Trouwborst, A., Tenbrinck, R., Fennema, M., Bucx, M., v.d. Broek, W. G. M., and Trouwborst-Weber, B. K., 1990c, Cardiovascular responses, hemodynamics and oxygen transport to tissue during moderate isovolemic hemodilution in pigs, *Adv. Exp. Med. Biol.*, 277:873.

Trouwborst, A., 1992, Hemodilution and oxygen transport, *Adv. Exp. Med. Biol.*, this volume.

Tucker, W. Y., Bean, J., Vandewater, S., and Cohn, L. H., 1980, The effect of hemodilution on experimental myocardial infarct size, *Eur. Surg. Res.*, 12:1.

Van Daal, G. J., Lachmann, B., Schairer, W., Tenbrinck, R., van Woerkens, L.J., Verdouw, P., and Erdmann, E., 1989, The influence of different anesthetics on the oxygen delivery to and consumption of the heart, *Adv. Exp. Med. Biol.*, 248:527.

Van Woerkens, E. C. S. M., Trouwborst, A., Duncker, D. J. G. M., Koning, M. M. G., Boomsma, F., and Verdouw, P. D., 1991, Catecholamines and regional hemodynamics during isovolemic hemodilution in anesthetized pigs, *J. Appl. Physiol.*, in press.

Van Woerkens, E. C. S. M., Trouwborst, A., Duncker, D. J. G. M., and Verdouw, P. D., 1992a, Regional cardiac hemodynamics and oxygenation during isovolemic hemodilution in anesthetized pigs, *Adv. Exp. Med. Biol.*, this volume.

Van Woerkens, E. C. S. M., Trouwborst, A., Tenbrinck, R., Trouwborst-Weber, B. K., and Snel, L. A., 1992b, Comparative study of two fibre optic mixed venous saturation catheters (Spectracath[R] vs Opticath[R]) during acute changes in hematocrit and cardiac output in humans, *Adv. Exp. Med. Biol.*, this volume.

Vercellotti, G. M., Hammerschmidt, D. E., Craddock, P. R., and Jacobs, H. S., 1982, Activation of plasma compliment by perfluorocarbon artificial blood: probable mechanism of adverse pulmonary reactions in treated patients and rationale for corticosteroid prophylaxis, *Blood*, 59:1299.

Virmani, R., Osmialowski, A. F., Kolodgie, F. D., and Forman, M. B., 1990, The effect of perfluorochemical Fluosol-DA (20%) on myocardial infarct healing in the rabbit, *Am. J. Cardiovasc. Pathol.*, 3:69.

Vogel, H., Günther, H., Harrison, D. K., Anderer, W., Kessler, M., and Peter, K., 1989, Hemodilution and myocardial oxygen supply. the influence of Fluosol-DA, *Adv. Exp. Med. Biol.*, 248:653.

Walfridsson, H., Odman, S., and Lund, N., 1985, Myocardial oxygen pressure across the lateral border zone after acute coronary occlusion in the pig heart, *Adv. Exp. Med. Biol.*, 191:203.

Weaver, M. E., Pantely, G. A., Bristow, J. D., and Ladley, H. D., 1986, A quantitive study of the anatomy and distribution of coronary arteries in swine in comparison with other animals and man, *Cardiovasc. Res.*, 20:907.

Weber, K. T., and Janicki, J. S., 1979, The metabolic demand and oxygen supply of the heart: Physiological and clinical considerations, *Am. J. Cardiol.*, 44:722.

Yoshikawa, H., Powell, W. R. jr, Bland, J. H. L., and Lowenstein, E., 1973, Effect of acute anemia on experimental myocardial ischemia, *Am. J. Cardiol.*, 32:670.

REGIONAL CARDIAC HEMODYNAMICS AND OXYGENATION DURING

ISOVOLEMIC HEMODILUTION IN ANESTHETIZED PIGS

E.C.S.M. van Woerkens, A. Trouwborst, D.J.G.M. Duncker[*],
and P.D. Verdouw[*]

Departments of Anesthesiology and Cardiology[*]
Erasmus University Rotterdam
The Netherlands

INTRODUCTION

Hemodilution causes a drop in hematocrit, thereby lowering the oxygen carrying capacity of blood, necessitating an increase in flow or an augmented oxygen extraction by the tissues to meet their oxygen demands. Under normal conditions the myocardial oxygen extraction is already high and the capacity to increase is limited. In the heart the decrease in arterial oxygen content during hemodilution therefore is mainly compensated by an increase in coronary flow. This increase in flow can be achieved by both reduction of the blood viscosity and coronary vasodilatation. The hemodynamic changes during hemodilution may increase myocardial oxygen consumption and, therefore, an even greater increase in coronary flow might be needed.

In order to establish whether the subendocardium, the most vulnerable of the myocardial layers, is still adequately perfused during isovolemic hemodilution we evaluated the effects of stepwise normovolemic hemodilution on regional myocardial blood flows and oxygen fluxes, and myocardial oxygen consumption in anesthetized pigs. The pig was selected because the responses of pigs under several conditions of stress are similar to those of humans.

MATERIAL AND METHODS

Twelve pigs (28 ± 1 kg) were sedated with midazolam (0.3 mg/kg i.m.) and ketamine (10 mg/kg i.m.) and subsequently anesthetized with thiopental (5 mg/kg i.v.). Anesthesia was maintained with fentanyl (12.5 μg/kg bolus followed by 12.5

μg/kg/h) and midazolam (0.45 mg/kg bolus followed by 0.45 mg/kg/h), while muscle relaxation was achieved with pancuronium bromide (0.1 mg/kg bolus followed by 0.3 mg/kg/h).

Catheters were positioned in the aortic arch, pulmonary artery, descending aorta and left ventricle. A midsternal thoracotomy was performed and an electromagnetic flow probe placed around the ascending aorta. The left and right atria and the great cardiac vein were cannulated.

Systemic hemodynamics, blood gases (OSM2 Radiometer, Copenhagen) and hematocrit were measured at baseline and after each step of hemodilution. In control animals measurements were made at corresponding time points. Regional blood flows of the heart were measured using the radioactive microsphere technique (Saxena and Verdouw, 1985).

Systemic and myocardial oxygen fluxes and consumptions were calculated.

In six pigs hemodilution was induced by exchanging blood with iso-oncotic dextran 40, 50 g/l in 0.9% saline (Isodex ^{R}). Stepwise hemodilution was induced by two steps of 10 ml/kg followed by two steps of 15 ml/kg. Six other animals served as controls. All data are presented as arithmetic means $\pm$ SE of mean.

RESULTS

Hemodilution produced a doubling of the cardiac output, primarily due to an increase in stroke volume with only a moderate increase in heart rate. The mean arterial pressure was maintained until the last step of hemodilution because of systemic vasodilation (table 1) (van Woerkens et al., in press). Mixed venous oxygen tension and saturation decreased slightly, resulting in an increase in systemic oxygen extraction ratio. Total body oxygen consumption was well maintained (table 2).

Total myocardial blood flow increased 420%. This increase was not homogeneously distributed. Blood flow to right and left atria increased by up to 120 $\pm$ 20% and 390 $\pm$ 70%, respectively, while oxygen fluxes in both atria were maintained. Blood flow to right and left ventricle increased by up to 510 $\pm$ 55% and 440 $\pm$ 60%, respectively, resulting in an increase in oxygen fluxes to both ventricles.

Despite the increase in left ventricular myocardial oxygen consumption, coronary venous saturation after the last step of hemodilution was slightly increased, indicating a decreased left ventricular oxygen extraction ratio.

The increase in left ventricular blood flow was more pronounced in subepicardial layers than in subendocardial layers. Therefore, the subendocardial-subepicardial blood flow ratio decreased from 1.22 $\pm$ 0.03 to 1.06 $\pm$ 0.07 (P<0.05) (table 3).

DISCUSSION

Hemodilution produces a reduction in the oxygen carrying capacity of the blood. In this study this is compensated for mainly by an increase in cardiac output but, at the last step of hemodiltion, also by an increased systemic oxygen extraction ratio.

In order to meet the metabolic demands of different organs a redistribution of the cardiac output can occur. Hemodilution and its effects on myocardial blood flow and oxygen delivery has been studied in anesthetized dogs (Crystal et al., 1988; Brazier et al., 1974; Geha, 1976; Crystal and Salem, 1991; Kettler et al., 1976).The decrease in arterial oxygen content is counteracted by an increase in cardiac output mainly due to an increase in stroke volume.

Table 1 Systemic and pulmonary hemodynamic effects of hemodilution in anesthetized pigs

Exchange		0		0		0		0		0	
Blood volume		0		10		20		35		50	
Hct	C	26.7 ± 0.9		27.2 ± 1.2		26.2 ± 0.9		25.3 ± 0.9		26.2 ± 0.7	
	H	27.5 ± 0.7		20.3 ± 0.4	*	16.7 ± 0.2	*	12.2 ± 0.5	*	9.3 ± 0.3	*
CO	C	2.83 ± 0.14		2.85 ± 0.11		2.87 ± 0.14		2.98 ± 0.13		2.92 ± 0.12	
	H	2.67 ± 0.23		3.75 ± 0.26	*	4.26 ± 0.21	*	5.02 ± 0.17	*	5.42 ± 0.18	*
HR	C	142 ± 7		146 ± 7		152 ± 10		152 ± 10		153 ± 14	
	H	113 ± 7	+	118 ± 7		130 ± 3		134 ± 5		140 ± 4	
SV	C	19.9 ± 0.4		19.4 ± 0.7		19.2 ± 1.4		18.8 ± 2.1		19.8 ± 2.0	
	H	22.8 ± 2.9		29.9 ± 2.9	*	30.5 ± 2.5	*	34.3 ± 2.7	*	34.9 ± 2.2	*
LVEDP	C	4.5 ± 0.5		4.4 ± 0.6		4.3 ± 0.5		4.8 ± 0.5		4.8 ± 0.6	
	H	4.5 ± 0.7		5.8 ± 0.8	*	5.7 ± 0.8	*	7.3 ± 0.8	*	6.8 ± 0.7	*
LVdP/dt$_{max}$	C	4290 ± 270		4200 ± 200		4420 ± 270		4460 ± 390		4410 ± 570	
	H	3370 ± 220	+	3680 ± 290	*	3980 ± 220	*	4000 ± 310	*	4050 ± 480	
MAP	C	108 ± 3		108 ± 3		111 ± 3		114 ± 4		113 ± 7	
	H	109 ± 8		110 ± 6		112 ± 7		103 ± 6		90 ± 5	*
SVR	C	37.9 ± 1.8		36.7 ± 1.3		38.5 ± 1.8		37.1 ± 2.5		38.4 ± 2.5	
	H	39.0 ± 4.9		32.5 ± 3.3	*	27.7 ± 2.4	*	21.7 ± 1.3	*	17.6 ± 1.1	*
PAP	C	18.0 ± 1.6		18.1 ± 1.8		18.7 ± 1.6		19.8 ± 1.3		20.9 ± 1.7	
	H	19.5 ± 1.3		21.8 ± 1.9		22.9 ± 1.5		24.2 ± 2.0		24.5 ± 2.0	
PVR	C	5.0 ± 0.8		4.8 ± 0.9		5.0 ± 0.7		5.1 ± 0.6		5.6 ± 0.8	
	H	6.8 ± 1.1		5.1 ± 0.7		4.7 ± 0.5	*	3.9 ± 0.5	*	3.6 ± 0.4	*

Blood volume exchange in ml/kg; C = control (n=6); H = isovolemic hemodilution (n=6); Hct = hematocrit (%); CO = cardiac output (l/min); HR = heart rate (beats/min); SV = stroke volume (ml); LVEDP = left ventricular end-diastolic pressure (mm Hg); LVdP/dt$_{max}$ = maximum rate of rise of left ventricular pressure (mm Hg/sec); MAP = mean arterial blood pressure (mm Hg); SVR = systemic vascular resistance (mm Hg/l/min); PAP = pulmonary arterial blood pressure (mm Hg); PVR = pulmonary vascular resistance (mm Hg/l/min); Data are mean ± SEM; * change from baseline significantly different ($P < 0.05$) from the change in the control animals. + significant difference in baseline values ($P < 0.05$). From: van Woerkens et al., in press.

Table 2 Systemic and left ventricular oxygenation during hemodilution in anesthetized pigs

Hct	C	26.7 ± 0.9	27.2 ± 1.2 *	26.2 ± 0.9	25.3 ± 0.9	26.2 ± 0.7
	H	27.5 ± 0.7	20.3 ± 0.4	16.7 ± 0.2 *	12.2 ± 0.5 *	9.3 ± 0.3 *
P_aO_2	C	139 ± 16	138 ± 17	141 ± 15	139 ± 17	135 ± 22
	H	166 ± 6	174 ± 11	164 ± 3	164 ± 9	158 ± 8
S_aO_2	C	97 ± 0.3	97 ± 1.1	97 ± 0.7	96 ± 1.2	96 ± 1.6
	H	97 ± 0.3	97 ± 0.3	96 ± 0.3	95 ± 0.6	94 ± 0.5
P_vO_2	C	36 ± 4	35 ± 3	34 ± 4	35 ± 5	35 ± 6
	H	38 ± 3	39 ± 4	38 ± 3	35 ± 3	31 ± 3 *
S_vO_2	C	56 ± 7	55 ± 8	53 ± 8	52 ± 1	52 ± 13
	H	57 ± 6	57 ± 7	52 ± 5	47 ± 6 *	37 ± 7 *
$V_{syst}O_2$	C	141 ± 10	150 ± 11	155 ± 11	163 ± 13	163 ± 16
	H	128 ± 8	135 ± 9	144 ± 8	147 ± 8	141 ± 5
ER_{syst}	C	0.40 ± 0.03	0.42 ± 0.03	0.45 ± 0.03	0.46 ± 0.04	0.46 ± 0.05
	H	0.41 ± 0.03	0.42 ± 0.03	0.47 ± 0.03	0.54 ± 0.03 *	0.61 ± 0.02 *
$P_{cv}O_2$	C	21 ± 1	20 ± 1	21 ± 1	20 ± 1	20 ± 1
	H	24 ± 1	24 ± 1	25 ± 1	28 ± 1	30 ± 2
$S_{cv}O_2$	C	23 ± 2	20 ± 1	23 ± 1	18 ± 2	21 ± 1
	H	21 ± 2	22 ± 1	23 ± 1	26 ± 3	28 ± 2
$V_{lv}O_2$	C	14.8 ± 0.8	12.9 ± 2.5 *	15.1 ± 0.8	11.2 ± 3.4 *	15.9 ± 1.7
	H	11.7 ± 0.7 +	14.4 ± 0.7 *	17.8 ± 0.7 *	20.3 ± 0.6 *	24.1 ± 0.5 *
ER_{lv}	C	0.75 ± 0.02	0.78 ± 0.01	0.76 ± 0.01	0.82 ± 0.02	0.77 ± 0.01
	H	0.77 ± 0.02	0.77 ± 0.01	0.75 ± 0.02	0.72 ± 0.02 *	0.70 ± 0.02 *

C= control (n=6); H= isovolemic hemodilution (n=6); Hct= hematocrit (%); P_aO_2 = arterial oxygen tension (mm Hg); S_aO_2 = arterial oxygen saturation (%); P_vO_2 = mixed venous oxygen tension (mm Hg); S_vO_2 = mixed venous oxygen saturation; $V_{syst}O_2$ = systemic oxygen consumption (ml/min); ER_{syst} = systemic extraction ratio; $P_{cv}O_2$ = coronary venous oxygen tension (mm Hg); $S_{cv}O_2$ = coronary venous oxygen saturation; $V_{lv}O_2$ = left ventricular oxygen consumption (ml/min/100g); ER_{lv} = left ventricular extraction ratio; Data are mean ± SEM; * change from baseline significantly different (P < 0.05) from the change in the control animals. From: van Woerkens et al., in press.

Table 3. Regional myocardial blood flows (ml/min/100g) and oxygen fluxes (ml/min/100g).

Regional myocardial blood flows

Region						
RA	C	163 ± 25	148 ± 23	153 ± 21	142 ± 29	142 ± 27
	H	157 ± 29	197 ± 34 *	234 ± 45 *	289 ± 51 *	330 ± 37 *
LA	C	82 ± 6	81 ± 9	105 ± 12	100 ± 19	85 ± 12
	H	75 ± 18	106 ± 19 *	154 ± 20 *	216 ± 35 *	322 ± 34 *
RV	C	124 ± 8	119 ± 11	127 ± 11	130 ± 13	132 ± 18
	H	104 ± 6	169 ± 11 *	271 ± 16 *	438 ± 54 *	631 ± 74 *
LV	C	170 ± 12	171 ± 15	175 ± 14	185 ± 22	180 ± 25
	H	130 ± 5 *	221 ± 11 *	341 ± 33 *	505 ± 57 *	690 ± 72 *
LV endo	C	192 ± 13	185 ± 19	194 ± 17	201 ± 26	207 ± 29
	H	145 ± 6 *	254 ± 21 *	390 ± 44 *	567 ± 79 *	717 ± 94 *
LV epi	C	154 ± 1	158 ± 14	160 ± 13	174 ± 20	161 ± 23
	H	119 ± 4 *	201 ± 8 *	314 ± 30 *	472 ± 49 *	670 ± 67 *

Regional myocardial oxygen fluxes

Region						
RA	C	18.8 ± 2.3	17.2 ± 2.0	17.6 ± 1.9	15.3 ± 2.8	16.3 ± 2.6
	H	19.3 ± 3.5	18.1 ± 3.0	17.3 ± 3.0	15.7 ± 2.1	14.0 ± 1.2
LA	C	9.5 ± 0.6	9.5 ± 0.8	12.1 ± 1.2	10.5 ± 1.6	9.8 ± 1.1
	H	9.3 ± 2.2	9.8 ± 1.8	11.4 ± 1.3	11.9 ± 1.6	13.0 ± 1.1
RV	C	14.5 ± 0.7	13.9 ± 0.8	14.8 ± 1.1	13.9 ± 1.1	15.1 ± 1.6
	H	12.7 ± 0.8	15.5 ± 1.1 *	20.1 ± 1.0 *	24.0 ± 2.0 *	26.6 ± 1.9 *
LV	C	19.8 ± 0.8	20.0 ± 1.3	20.2 ± 1.1	19.8 ± 1.8	20.7 ± 2.3
	H	15.9 ± 0.6 *	20.2 ± 1.1	25.3 ± 2.2 *	27.8 ± 2.5 *	29.4 ± 2.6 *
LV endo	C	22.4 ± 1.1	21.7 ± 1.8	22.5 ± 1.6	21.5 ± 2.2	23.8 ± 2.6
	H	17.7 ± 0.7 *	23.2 ± 1.8	28.9 ± 2.9 *	31.2 ± 3.6 *	30.6 ± 3.6 *
LV epi	C	17.9 ± 0.8	18.4 ± 1.1	18.5 ± 1.1	18.6 ± 1.6	18.5 ± 2.0
	H	14.6 ± 0.5 *	18.4 ± 0.8 *	23.3 ± 1.9 *	26.0 ± 2.0 *	28.4 ± 2.3 *
endo/epi	C	1.25 ± 0.05	1.17 ± 0.06	1.21 ± 0.05	1.15 ± 0.03	1.28 ± 0.02
	H	1.22 ± 0.03	1.25 ± 0.06	1.23 ± 0.03	1.18 ± 0.05	1.06 ± 0.07

C = control (n=6); H = isovolemic hemodilution (n=6); RA = right atrium; LA = left atrium; RV = right ventricle; LV = left ventricle; LV endo = left ventricular endocardial region; LV epi = left ventricular epicardial region; endo/epi = endocardial epicardial flow ratio; Data are mean ± SEM; * change from baseline significantly different (P < 0.05) from the change in the control animals. + significant difference in baseline values (P < 0.05). From van Woerkens et al., in press.

In the study by Geha (1976) no increase in cardiac output was observed during moderate hemodilution. However filling pressures did not change either, probably indicating hypovolemia (Trouwborst, in press). Left ventricular oxygen consumption was maintained primarily by an increase in left ventricular blood flow, but an additional increase in left ventricular oxygen extraction ratio was observed. Due to the increase in coronary flow the vasodilatory reserve capacity was significantly reduced. In most studies the increased myocardial blood flow was sufficient to maintain regional myocardial oxygen supply (Crystal et al., 1988; Crystal and Salem, 1991; Kettler et al., 1976).

Brazier et al. (1974) estimated the myocardial oxygen consumption by the tension time index, which did not change significantly during moderate and severe hemodilution. The myocardial oxygen demand to supply ratio, however, fell with moderate hemodilution and decreased further during severe hemodilution. The endocardial/epicardial flow ratio decreased as the hemoglobin level fell below 5 g% and the augmented cardiac output could not be sustained.

Reduction of hematocrit to half the baseline value did not change coronary sinus oxygen tension in one study (Crystal and Salem, 1991), while in another study coronary venous oxygen saturation even increased during hemodilution, despite the slight increase in myocardial oxygen consumption (Kettler et al., 1976).

Stepwise hemodilution in anesthetized baboons produced an increase in left ventricular flow which was able to maintain both left ventricular oxygen delivery and left ventricular oxygen consumption until hematocrit levels of 6%. No increase in left ventricular oxygen consumption was observed. At a hematocrit of 10% maximal coronary vasodilatation had occurred and a further decrement in oxygen supply led to myocardial anaerobic metabolism as shown by left ventricular lactate production (Wilkerson et al., 1988).

Moderate decreases in hematocrit levels in anesthetized humans (hematocrit reduced from 38% to 28%) also produces an increase in myocardial blood flow with a slight, but not significant increase in myocardial oxygen consumption. Coronary sinus oxygen tension did not change. The increase in cardiac output was associated with an increase in external cardiac work solely due to an increase in stroke volume. This can be obtained without a substantial rise in need of energy (Gisselson et al., 1982).

In conscious dogs the elevated cardiac output during hemodilution is mainly due to an increase in heart rate (von Restorff et al., 1975a, 1975b). Myocardial oxygen consumption increases, necessitating an additional increase in myocardial blood flow in these studies in conscious dogs. During exercise myocardial oxygen consumption increases even further. The increased myocardial oxygen consumption was met both by an elevated coronary blood flow and an increased oxygen extraction ratio, accompanied by a decrease in coronary sinus oxygen saturation. The peak reactive hyperemic flow increased under hemodilution, while at rest dilatory capacity was left even at a hematocrit of 10-15%. At moderate levels of exercise the dilatory capacity was exhausted at a hematocrit of 20-25% (von Restorff et al, 1975b). Reduction of hematocrit to 13% produced an increase in left ventricular blood flow in conscious dogs which is mainly distributed to the epicardium, indicating that the coronary reserve in subendocardium is almost completely exhausted (von Restorff et al., 1975b).

In the present study an increase in myocardial blood flow of 420% with a doubling of the cardiac output mainly due to an increase in stroke volume was observed. In contrast to the studies discussed, in this study hemodilution resulted in a doubling of left ventricular oxygen consumption. However the increase in myocardial blood flow produced a rise in left ventricular oxygen flux sufficient to meet the elevated left ventricular oxygen consumption.

The increase in left ventricular blood flow exceeded the increase in cardiac output. Since the subendocardial region is subjected to the greatest intramyocardial forces of the myocardium, this region is perfused mainly during diastole. Therefore, it has to have a lower vascular resistance in order to receive the same amount of blood flow as the subepicardial region. The vasodilatory reserve capacity in the subendocardial layers is therefore lower and maximal vasodilatation occurs. During severe hemodilution a redistribution of the coronary flow away from the subendocardial layers was observed in the present study. When maximal vasodilatation and maximal oxygen extraction have occurred, a further increase in oxygen demand will lead to myocardial ischemia and the heart will not be able to sustain the increased cardiac output. This is more likely to happen when myocardial oxygen consumption is increased (e.g. hypertension, valvular heart disease) or when myocardial oxygen delivery is further reduced, Myocardial oxygen delivery can be reduced by reduction of the arterial oxygen content (hypoxia), a shortening of the diastolic filling period (tachycardia) or a lowered coronary driving pressure. Maximal coronary vasodilatation is compromised when coronary artery disease is present.

In conclusion hemodilution in anesthetized animals produces an increase in myocardial blood flow sufficient to meet the myocardial oxygen consumption. However, at low levels of hematocrit the endocardial/epicardial flow ratio is decreased and the vasodilatory reserve capacity of the coronary arteries is exhausted in the subendocardium. This limits the degree of hemodilution which can be reached. Additional caution is necessary if factors elevating the myocardial oxygen consumption or further reducing the (maximal) myocardial oxygen delivery are present.

REFERENCES

Brazier, J., Cooper, N., Maloney, J. V., and Buckberg, G., 1974, The adequacy of myocardial oxygen delivery in acute normovolemic anemia. *Surgery* 75:508-516.

Crystal, G. J., Rooney, M. W., and Salem, M. R., 1988, Regional hemodynamics and oxygen supply during isovolemic hemodilution alone and in combination with adenosine-induced hypotension. *Anesth. Analg.* 67:211-218.

Crystal, G. J., and Salem, M. R., 1991, Myocardial and systemic hemodynamics during isovolemic hemodilution alone and combined with nitroprusside-induced hypotension. *Anesth. Analg.* 72:227-237.

Geha, A. S., 1976, Coronary and cardiovascular dynamics and oxygen availability during acute normovolemic anemia. *Surgery* 80:47-53.

Gisselsson, L., Rosberg, B., and Ericsson, M., 1982, Myocardial blood flow, oxygen uptake and carbon dioxide release of the human heart during hemodilution. *Acta Anesth. Scand.* 26:589-591.

Kettler, D., Hellberg, K., Klaess, G., Kontokollias, J. S., Loos, W., and de Vivie, R., 1976, Hämodynamik, Sauerstoffbedarf und Sauerstoffversorgung des Herzens unter isovolämischer Hämodilution. *Anaesthesist* 25:131-136.

Restorff, W. von, Höffling, B., Holtz, J., and Bassenge, E., 1975a, Effect of increased blood fluidity through hemodilution on coronary circulation at rest and during exercise in dogs. *Pflügers Arch.* 357:15-24.

Restorff, W. von, Holtz, J., Bard, P., and Bassenge, E., 1975b, Transmural distribution of myocardial blood flow under normovolemic hemodilution. *Pflügers Arch.* 355:R9.

Saxena, P. R., and Verdouw, P. D., 1985, 5-Carboxamide tryptamine, a compound with high affinity for 5-HT binding sites, dilates arterioles and constricts arteriovenous anastomoses. *Br. J. Pharmacol.* 84:533-544.

Trouwborst, A. Oxygen consumption and hemodilution. *Adv. Exp. Med. Biol.* this volume.

Wilkerson, D. K., Rosen, A. L., Sehgal, L. R., Gould, A. S., Sehgal H. L., and Moss, G. S., 1988, Limits of cardiac compensation in anemic baboons. *Surgery* 103:665-670.

Woerkens, E.C.S.M. van, Trouwborst, A., Duncker, D.J.G.M., Koning, M.M.G., Boomsma, F., and Verdouw, P.D. Catecholamines and regional hemodynamics during isovolemic hemodilution in anesthetized pigs. *J. Appl. Physiol.* in press.

GRADIENTS OF CAPILLARIZATION IN THE SUBENDOCARDIUM

OF RAT HEART SEPTUM

S. Batra, P. Veprek, and K. Rakusan

Department of Physiology
Faculty of Medicine, University of Ottawa
Ottawa, Ontario, Canada K1H 8M5

INTRODUCTION

Coronary capillaries and cardiac myocytes constitute the principal components of the heart. The importance of the relationship between these two components is underscored by the critical role of capillaries in supplying oxygen to the working myocytes. The geometrical conditions for the diffusion of oxygen are influenced not only by the capillary density, but also by the spatial distribution of capillaries within the tissue. In the case of the subendocardium, it has long been suggested that oxygen may diffuse directly to the innermost layers of the subendocardium from the ventricular cavity, thus providing a direct, nonvascular source of oxygen to the subendocardium (Hort, 1968). Despite this longstanding idea, we are not aware of any previous studies that have considered the extent of oxygen supply from the ventricular cavity to the subendocardium. If this pathway for oxygen is appreciable, we hypothesized that this may influence the geometry of the local microvascular bed, the primary means of oxygen supply to subendocardium.

To test this hypothesis, we have considered the extent of oxygen transport from the ventricular cavity, i.e. the depth in the subendocardium that ventricular oxygen would compromise local capillary geometry. To further this question, we hypothesized that there may be differences in the oxygen transport between the left and right ventricles, as these cavities contain blood of different oxygen saturation. To identify capillaries we have used a histochemical technique, first developed by Lojda (1979), and previously used in our lab (Batra et al., 1989), that serves to distinguish arteriolar capillaries (AC, high PO_2) and venular capillaries (VC, low PO_2) on the basis of color.

The present study takes advantage of this differential staining sensitivity to analyze capillary geometry in the rat heart septum. From transverse tissue sections, capillary supply regions were determined by the method of capillary domains. All capillary domain data were classified according to the histochemical type of each capillary, distance from endocardial surface, for both left and right ventricular septum. This study presents for the first time, quantitative data regarding the spatial distribution of capillaries as a function distance from the endocardial surface in the rat heart septum.

Oxygen Transport to Tissue XIV, Edited by W. Erdmann and
D.F. Bruley, Plenum Press, New York, 1992

METHODS

Six male Sprague-Dawley rats (Body Weight = 285 ± 10 g) were used in this study. From each animal, following anaesthesia with pentobarbital sodium, hearts were rapidly excised, weighed, dissected and frozen in liquid nitrogen. From each heart, a block was dissected perpendicular to the principle axis (base-apex) to render subendocardial capillaries on cross section (Fig. 1). Cryostat sections were taken (16 μm) from each block and prepared for differential staining as described previously (Batra et al. 1989). Briefly, the histochemical technique serves to localize Alkaline Phosphatase (AP) and Dipeptidyl Peptidase IV (DPP IV) in the capillary endothelium. AP activity is specific to arteriolar capillary (AC) regions, as stains blue in color. DPP IV activity has been localized to venular capillary (VC) regions, and stains red in color. In this manner, capillaries were represented as a bivariate point pattern in the tissue plane, and all data were derived from the subendocardium as a function of capillary staining response (AC and VC regions).

From scale drawings made from subendocardial regions, in direct proximity to the ventricular cavity, direct distances from capillary to cavity were measured (Fig. 2). In addition, capillary profiles were digitized and entered into our software program for capillary domains (Hoofd et al. 1985). This program serves to divide the tissue plane into polygonal regions of tissue, each polygon enclosing one capillary (Fig. 2). The area subtended by these polygons, the capillary domain, is closer to the enclosed capillary than any other.

A 3-way analysis of variance (ANOVA) test was employed to consider the possible main effects of distance from cavity, right versus left ventricle, and capillary histochemical type, on capillary domain area. Post-hoc analysis was performed by the Scheffe method where appropriate.

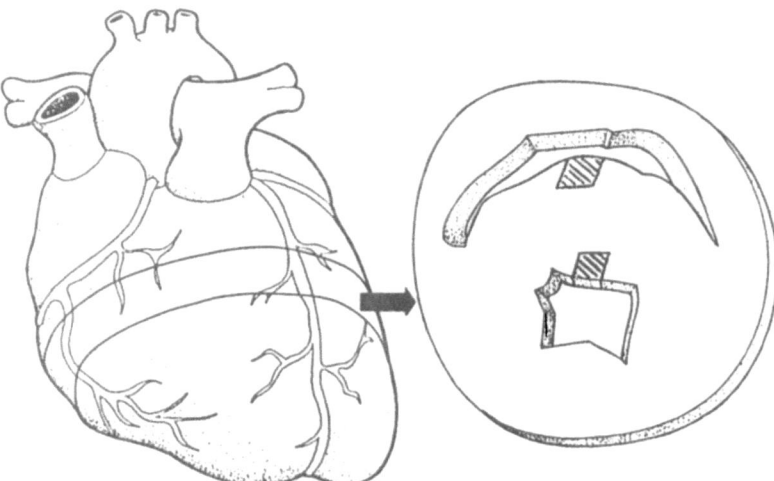

Figure 1. Representation of entire heart (left panel) with broken lines delineating the block of tissue used for sectioning. The right panel shows the heart septum, with shaded areas representing the region from which capillary domains were measured.

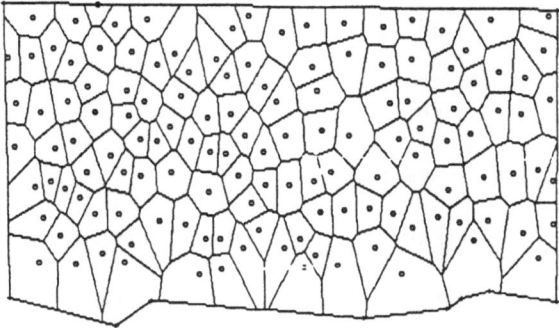

Figure 2. Illustration of measurements of distance from capillary to endocardial surface (upper panel). Illustration of capillary domain areas delineated at different depths from the endocardial surface (lower panel).

RESULTS

Capillary domain areas were significantly larger for capillaries situated closer to the ventricular cavity. With increasing distance from the endocardial surface, domain area was progressively reduced, and beyond approx. 40 μm from the cavity, were similar to deeper subendocardial regions. Regression lines of domain area versus distance from ventricular cavity are shown in Figure 3.

Distances from capillary to cavity were subsequently divided into 8 categories, representing 20 μm intervals. As shown in Figure 4 and Table 1, domain areas were largest in category 1 (0-20μm). There were no differences in domain area between all other categories. In all cases, AC domain areas were larger than VC domain areas (Table 1).

There were no overall differences in these data between the right and left ventricles. The gradient in capillary domain area, however, was greater in the right ventricular septum, with values ranging from 510 ± 21 μm (category 1) to 310 ± 34 μm (category 8). In the left ventricle, the gradient was much less pronounced, with values ranging from 441 ± 26 (category 1) to 340 ± 61 (category 8). There was a significant two-factor interaction (P<0.05) between category and ventricle. In the right ventricle, domain area showed progressive and stepwise decreases from category 1 to 8. In the left ventricle, domain area displayed a bimodal pattern with high values observed in categories 1 and 6.

TABLE 1

Capillary Domain Areas in
Rat Subendocardium

Category	AC Domain Area (μm^2)	VC Domain Area (μm^2)
1	503 ± 25†	456 ± 22†
2	401 ± 19	356 ± 18
3	366 ± 10	314 ± 12
4	351 ± 11	327 ± 12
5	351 ± 13	346 ± 10
6	386 ± 14	322 ± 11
7	357 ± 24	316 ± 19
8	375 ± 13	298 ± 34
Pooled	**384 ± 06**	**345 ± 06**[*]

All values are mean ± SEM. data from left and right ventricular septum have been combined. AC = arteriolar capillary; VC = venular capillary. * denotes significant difference from corresponding AC value at $P<0.01$. † denotes significant difference from all other categories at $P<0.01$.

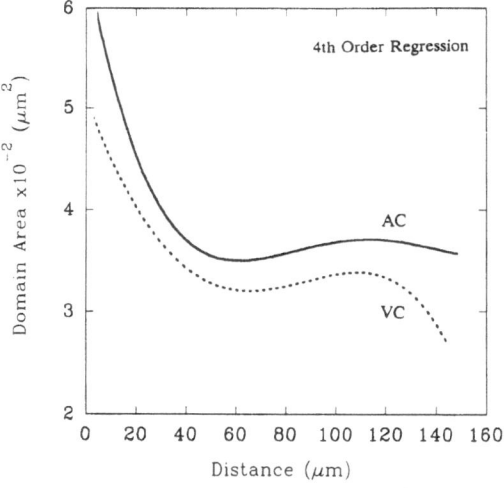

Figure 3. Regression lines of arteriolar (AC, solid line) and venular (VC, broken line) capillary domain area versus distance from endocardial surface.

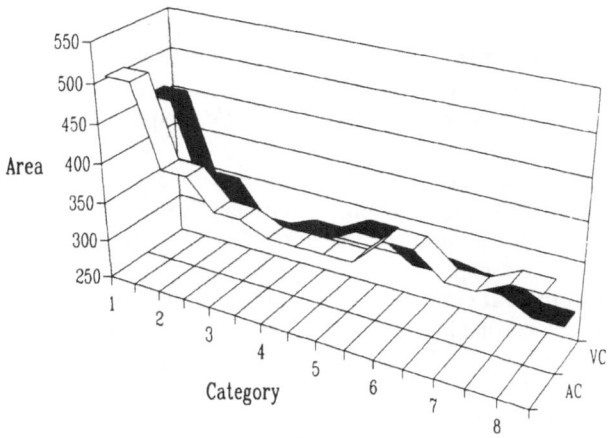

Figure 4. Plot of arteriolar (AC) and venular (VC) domain areas (μm^2) as a function of distance from endocardial surface as defined by 20 μm categories.

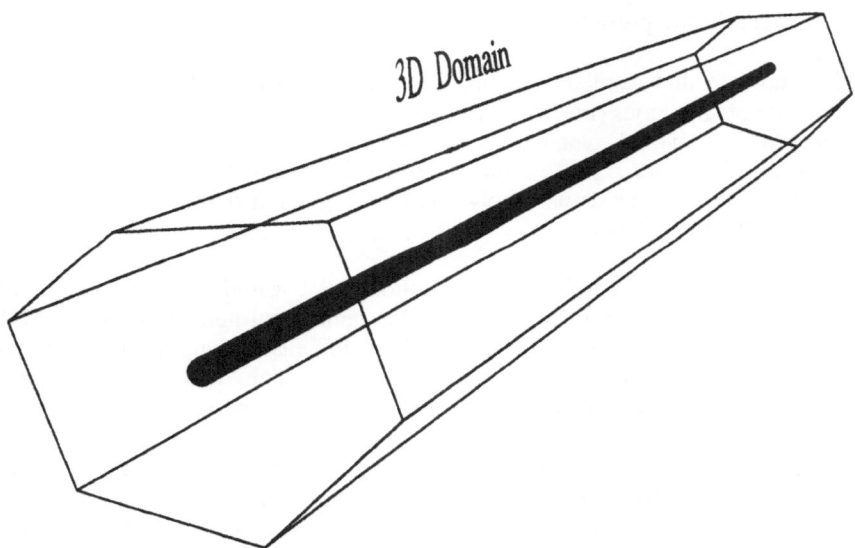

Figure 5. Three dimensional schema of capillary domains, illustrating the reduction in domain area along the capillary length from arteriole to venule.

DISCUSSION

The preeminent finding in this study was that capillary supply in the initial myocyte layers of the subendocardial septum was attenuated by a potential nonvascular source of oxygen from the ventricular cavity. The reduced capillarization, as displayed by an increase in capillary domain area, produced an effect to a depth of approx. 40 μm. Assuming myocyte cross sectional areas of 327 and 241 μm^2 for LV subendocardium and RV, respectively (Campbell *et al.* 1989), this effect would correspond to 3 and 4 myocyte layers (LV and RV, respectively).

Although there were no overall differences in capillary domain area between the right and left ventricles, the gradient of domain area was greater on the right ventricular septum as compared to the left side. This was a surprising result for which we have no direct explanation. Given the presence of blood with higher PO_2 and O_2 content in the left ventricle as compared to the right side, one would expect that the first subendocardial layers of the left ventricular septum to be less dependent upon oxygen supply from capillaries, and therefore to have larger capillary domain areas than the right side. If anything, the opposite appears to be true. The results of this study would be tenable if there was higher oxygen consumption in the deepest layers of the left ventricle, leading to increased capillarization as compared to the right ventricle. Unfortunately there are no data available concerning local oxygen consumption on this level in the heart muscle. Weiss and coworkers (1978) measured oxygen consumption of the canine septum divided into three layers (left, middle and right). They found no regional differences when these larger layers were compared. We are not sure if there is a gradient in oxygen consumption of the first several layers in the subendocardium; and further, if there are differences in this gradient between the left and right ventricles. Such a relationship may help to explain the data observed in the present study.

It is notable that capillary domain areas decreased from the arteriolar to the venular side of capillaries (Fig. 5). This decrease would provide for shorter diffusion distances for oxygen on the venular side of capillaries, where PO_2 values are generally considered to be lower. These data would call for a modification of the classical Krogh tissue cylinder; the tissue supply volume being better described as a truncated cone rather than a cylinder.

Of interest to note was that the proportion of VC capillaries was considerably higher than we have previously observed in the midmyocardium (Batra *et al.* 1991). This may reflect either regional differences in endothelial enzyme profiles; or more likely, a different architecture of arterio-venous capillary paths, with a greater number/density of collecting venules are found in the subendocardium.

To summarize, the direct, nonvascular source of oxygen from the ventricular cavity plays an important role in affecting the capillary supply geometry in the subendocardial regions close to the endocardial surface. The depth of this affect is limited to approximately the first three or four myocyte layers; is not different between the right and left ventricular septum; and is preserved in both arteriolar and venular capillary regions. This morphometric approach may also be valuable for studies of capillary geometry at various depths from the epicardial surface, to assist in the interpretation of PO_2 data obtained from microelectrodes.

ACKNOWLEDGMENTS

This work was supported by the Medical Research Council of Canada. P. Veprek is a visiting scholar from Kardiocentrum in Prague, Czechoslovakia.

REFERENCES

Batra, S., Rakusan, K., and Kuo, C. (1989). Spatial distribution of coronary capillaries: A-V segment staggering. *In*: "Oxygen Transport to Tissue-XI" (K. Rakusan, G.P. Biro, T.K. Goldstick, Z. Turek, Eds.), pp. 241-247. Plenum Press, New York and London.

Batra, S., Rakusan, K., and Campbell, S.E. (1991). Geometry of capillary networks in hypertrophied rat heart. *Microvasc. Res.* **41**, 29-40.

Campbell, S.E., Rakusan, K., and Gerdes, A.M. (1989). Change in cardiac myocyte distribution in aortic-constricted neonatal rats. *Basic Res. Cardiol.*, **84**, 247-258.

Hoofd, L., Turek, Z., Kubat, K., Ringnalda, B.E.M., and Kazda, S. (1985). Variability of intercapillary distance estimated on histological sections of rat heart. *In*: "Oxygen Transport to Tissue VII" (F. Kreuzer, S.M. Cain, T.K. Goldstick, Eds.) pp. 239-247, Plenum Press, New York and London.

Hort, W. (1968). Capillarisation of the myocardium under normal and pathological conditions. *In:* "Oxygen Transport in Blood and Tissue" (D.-W. Lubbers, U.C. Luft, G. Thews, E. Witzleb, Eds.) pp. 150-158, Georg Thieme Verlag, Stuttgart.

Lojda, Z. (1979). Studies on Dipeptidyl(Amino)Peptidase IV (Glycyl-Proline Naphthylamidase). *Histochem.* **59**, 153-166.

Weiss, H.R., Neubauer, J.A., Lipp, J.A., and Sinha, A.K. (1978). Quantitative determination of regional oxygen consumption in the dog heart. *Circ. Res.* **42**, 394-401.

OXYGEN PRESSURE HISTOGRAMS CALCULATED IN A BLOCK OF RAT HEART TISSUE

Louis Hoofd and Zdenek Turek

Department of Physiology, University of Nijmegen

Geert Groote Plein Noord 21, 6525 EZ Nijmegen, The Netherlands

INTRODUCTION

Measurements of oxygen pressure in tissue are usually presented in the form of oxygen histograms. In such histograms, measurements are grouped into classes of equal width and their frequency represented. An alternative view is presentation in cumulative histograms, adding the frequencies of all lower classes. Particularly for muscle tissue, including rat heart, very differing histograms were described, the lowest values being from cryophotometric measurements of myoglobin saturation (Honig and Gayeski, 1987).

In the past years we developed a method to calculate oxygen partial pressures in a 2-dimensional field ("tissue slab") for arbitrary capillary distribution (Hoofd et al., 1989) as well as coupling of capillary pressures for an array of slabs (Hoofd et al., 1990). The array of slabs forms a tissue cylinder or block, for circular or rectangular fields respectively. Here, we present calculations of oxygen histograms in a tissue block of working rat heart tissue, in order to investigate the influence of different sets of input data on tissue oxygenation.

METHODS

The multicapillary model was applied to calculate oxygen pressure, pO_2, in a tissue block. This model is 3-dimensional, piling a number of tissue slabs to form a tissue block. In each slab, pO_2 is calculated assuming 2-dimensional diffusion starting from the capillaries. For the first slab, initial values have to be assigned to each capillary of, e.g., capillary flow, hematocrit and O_2 content. For the following slabs, capillary pO_2 is calculated by subtracting the amount consumed by the surrounding tissue.

Calculation of tissue pO_2 in each slab at location (x,y) is by the multi-source equation incorporating Mb-facilitated diffusion (Hoofd et al., 1989):

$$pO_2 + p_F S = C_p - \frac{M}{4\mathcal{P}} \left\{ \sum_{n=1}^{N} \frac{A_n}{\pi} \ln \left[\frac{(x-x_n)^2 + (y-y_n)^2}{r_{cn}^2} \right] - \Phi(x,y) \right\}$$

where p_F is facilitation pressure (Hoofd et al., 1990), S is myoglobin saturation, C_p is a constant, M is oxygen consumption, $\mathcal{P}$ is oxygen permeability (product of diffusion coefficient D and solubility a), A_n is the oxygen supply area of the nth capillary located at (x_n,y_n) and with radius r_{cn}, and Φ is a function depending on geometry (in fact: O_2

consumption topography). For the rectangular case here, this "field function" is (Hoofd et al., 1990; Hoofd, in press):

$$\Phi(x,y) = x^2 + y^2 + \frac{1}{\pi}\sum_{k=1}^{4}\left[\Delta x_k \Delta y_k \ln(\Delta x_k^2 + \Delta y_k^2) + (\Delta x_k^2 - \Delta y_k^2)\arctan(\frac{\Delta y_k}{\Delta x_k})\right]$$

where $\Delta x_k = x - x_k$, $\Delta y_k = y - y_k$ and (x_k, y_k) are the coordinates of the corner points of the rectangular field. In these two equations, all constants are known except C_p and the supply areas of the N capillaries, A_n. These are derived from the boundary conditions that for each capillary the average capillary rim pO_2 is given and that the sum of all supply areas is equal to the total rectangle area (Hoofd et al., 1989).

Coupling to the next slab is done by subtracting the amount of O_2 delivered, for each capillary:

$$\Delta cO_{2,n} = -\frac{M}{F_n}(A_n - \pi r_{cn}^2)\Delta z$$

where F_n is the blood flow of the nth capillary, Δz is the thickness of the slab and $cO_{2,n}$ is its total oxygen content; mainly hemoglobin-bound. The capillary area πr_{cn}^2 must be subtracted from the area A_n because the O_2 "delivered" there does not leave the capillary.

Capillary rim pO_2 will be lower than erythrocytic pO_2 because the latter has to deliver its oxygen up to this location. Being mathematically quite complex, this problem was circumvented by assuming a difference of 2 kPa (15 mm Hg) for an average capillary, and a proportionally larger or smaller value for capillaries delivering a larger or smaller amount respectively. This pressure drop is called *Extraction Pressure* and is denoted by EP (Hoofd et al., 1990). It substitutes for the whole complex of intracapillary delivery phenomena: hemoglobin-O_2 reaction and diffusion and erythrocyte, plasma and capillary-wall diffusion. The value of 2 kPa for a capillary, m, with an average supply area A_m was estimated based on a simplified model of erythrocytic O_2 release (Hoofd, in press). Then, for the other capillaries:

$$EP_n = EP_m \frac{A_n - \pi r_{cn}^2}{A_m - \pi r_{cm}^2}$$

Capillaries were modelled as straight tubes perpendicular to the slabs. In principle, capillary locations could be different in each slab but this would violate the assumption of 2-dimensional diffusion *within* each slab and there is no straightforward way to estimate the consequences on pO_2. With straight capillaries, the net effect of diffusion in the third dimension is negligibly small; for a detailed description of the mathematical treatment see our earlier publications (Hoofd et al., 1989; Hoofd et al., 1990; Hoofd, in press).

In rat heart, capillary spacing is distributed lognormally (Turek et al., 1986). Here, the capillary locations were read in from a typical photomicrograph of normal rat heart tissue slice (Turek et al., 1986), i.e., with a capillary distribution resembling the overall distribution (Fig. 1; left panel).

For the first slab, values of capillary pO_2's had to be chosen. An arteriolar pO_2 of 13.3 kPa (100 mm Hg) was assigned to some of the capillaries, but it is unlikely that all of them would start with this high pressure in the same slab. Therefore, the other capillaries were assigned pO_2's of the same pO_2 distribution that the other "arteriolar" capillaries reached at 250 μm, half their length. This mimics staggering, in a simple way. The field was divided into 6 equal parts, as shown in Fig. 1; in the three parts indicated **A** the high pO_2 was assigned and in the parts indicated **V** the "half-way" pO_2's.

The block was divided vertically into 50 slabs so that each slab was $5\,\mu m$ thick; this was tested earlier as a sufficiently small vertical step (Hoofd et al., 1990). In each slab, pO_2 was calculated in a grid with the same spacing, $5\,\mu m$. Some portions were left out from the histogram:

- The boundary conditions allow O_2 to diffuse into and out of the slab at the borders, assuming tissue with other capillaries present there; only the net overall exchange is zero. Capillaries just outside the border might alter the local pO_2; therefore, a border region of $10\,\mu m$ was left out of consideration (short-dashed lines in Fig. 1);

- Also the top $10\,\mu m$ layer was left out, for two reasons. Firstly, the arterial pO_2 might be unrealistic for capillaries due to precapillary O_2 loss. Secondly, the top layer may have some O_2 loss due to diffusion into the third dimension (there is no influx from a layer above it);

- A region of $3\,\mu m$ around each capillary was left out (examples shown dotted in Fig. 1), in order to avoid close-to-the-capillary effects (see DISCUSSION below).

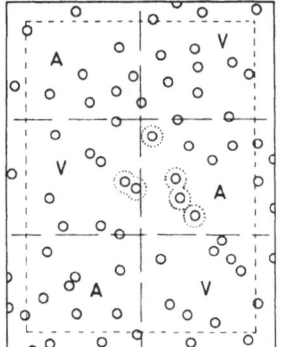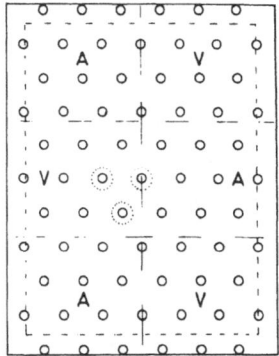

Fig. 1. Capillary distribution in the rat heart (left panel) and with regular spacing (right panel). Circles: capillaries, with high (zones A) or medium (zones V) initial pO_2. Border (short-dashed lines) and close-to-capillary (dotted circles) regions not included in the histogram.

This leaves about 30 000 points for the histogram.

Data, for working rat heart at 37°C, were the same as in earlier publications; unless stated different. Slab dimensions were $177.5\,\mu m \times 140.4\,\mu m$, covering 71 capillaries and selected to adequately represent mean and distribution of capillary distances of the whole tissue (Turek et al., 1986). Capillary radius was $2.4\,\mu m$; all capillaries had the same flow and hematocrit. $M/\mathcal{P}$ was $0.0253\ kPa/\mu m^2$; myoglobin data were $p_F = 2\ kPa$ and $p_{50} = 0.707\ kPa$; $\alpha F/\mathcal{P}$ was $10.56\,\mu m$; blood data were $c_{Hm}/\alpha = 844.4$ kPa (c_{Hm} is heme concentration) and $p_{50} = 4.93\ kPa$, Hill-$n = 2.7$ (Hoofd et al., 1990). As mentioned above, slab thickness was $5\,\mu m$, block thickness $250\,\mu m$ (50 slabs) and EP was 2 kPa.

RESULTS

The pO_2 histogram calculated for the above set of data is referred to as the standard situation and represented by the open bars and dotted lines in each panel of the

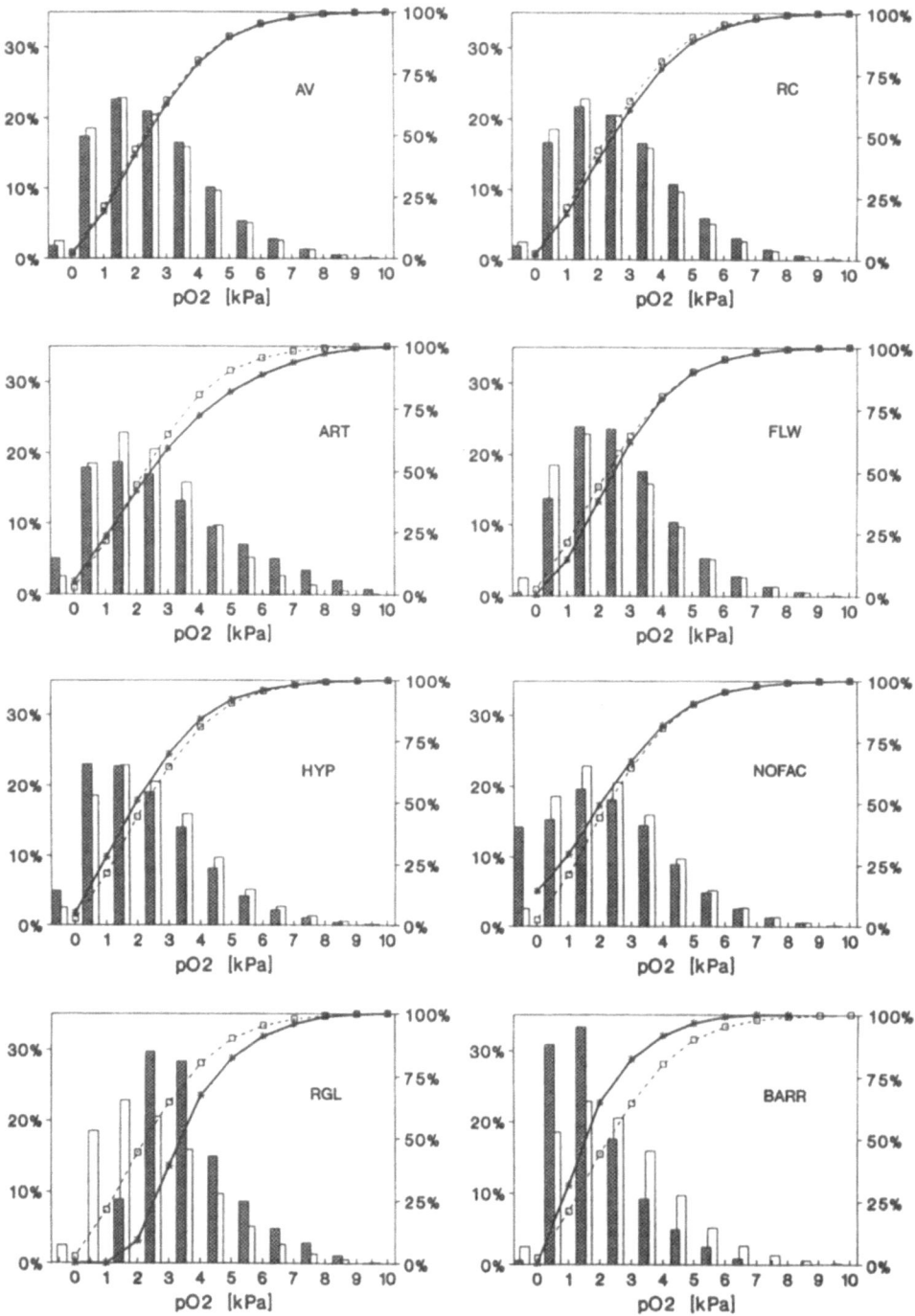

Fig. 2. pO$_2$ histograms (bars; left axis) and cumulative histograms (lines; right axis) of standard situation (open bars, dashed lines) compared with different other possibilities (see text).

histograms of Fig. 2. A small percentage of tissue had calculated pO_2 below zero, which will be referred to as "anoxic". Median pO_2 was around 2 kPa, 64% was below 3 kPa and 90% below 5 kPa, whereas the mean end-capillary pO_2 was 4.56 kPa. So, tissue pO_2 is calculated to be considerably lower than venous pO_2.

Several other possibilities for the input data, referred to as "tissue factors", were considered and shown in the different panels of Fig. 2 by the dark bars and the solid lines. The first case, top left panel, denoted **AV** in the figure, is accounting for different spacing of arteriolar and venular capillary ends. According to the data of Batra et al. (1991) for normal rat heart, the **A** regions above were distended by 2.3% and the **V** regions shrunken by 2.3%. As seen in the **AV** panel this hardly affected the histograms. Arteriolar and venular capillary ends may have different radii; the **RC** panel was calculated for $r_c = 2.5 \mu m$ for the A-region and $r_c = 2.3 \mu m$ for the V-region capillaries. When all the capillaries started at an arterial pO_2 of 13.3 kPa (and now travelled 500 μm) the resulting histogram was as shown in panel **ART**; note that the histograms extends to higher pO_2 values but that the low part is hardly affected. The next panel, **FLW**, compared with a situation in where capillary flow was matched to the oxygen supply, i.e., capillary flow was taken linearly proportional to the end-capillary O_2 supply area. So, the capillaries that supply more O_2 will get more blood resulting in almost uniform end-capillary pO_2. As expected, this removed very low pO_2 values from the histogram but the overall effect was remarkably few. The next panel, **HYP**, was calculated for 5% hypertrophy, i.e., all distances were 1.05 times larger; the value 5% was chosen as comparable to the **AV** case. The histogram was shifted a few tenths of a kPa to the right. Abolishing Mb-facilitated diffusion (**NOFAC**; $p_F = 0$) had a dramatic effect on the amount of anoxic regions but hardly affected the histogram at high pO_2 values.

The last two panels of Fig. 2 apply to a situation where not the actual capillary locations of the photomicrograph were used but instead the capillaries were distributed regularly, keeping the same average capillary distance. The resulting pattern is represented in the right panel of Fig. 1 above; the number of capillaries is the same, 71. As seen from the panel **RGL**, the resulting histogram was completely different, without any low-pO_2 values and much more peaked, around 3 kPa; the corresponding cumulative histogram being much steeper. This absence of heterogeneity in capillary spacing even would make the myoglobin redundant. It regained its functionality for an additional O_2 transport resistance; panel **BARR** shows a situation where EP was doubled to 4 kPa, or, in other words, an additional barrier of 2 kPa added around the capillaries. Then, low pO_2's prevailed, without anoxic regions, and a steep cumulative histogram largely below 3 kPa.

DISCUSSION

Most of the tissue factors investigated here had remarkably few influence on the calculated pO_2 histograms. Some were not unexpected; a/v differences and hypertrophy of 5% (**AV** and **HYP** respectively) in fact are small. For instance, for hypertrophied rat heart (Batra et al., 1991) the histogram alterations turned out to be much larger as shown in a companion paper (Turek et al, this issue). The same holds for the **RC** case. Absence of staggering (**ART**) only leads to few lower values of pO_2, as best seen in the small increase in anoxic regions. Mb indeed is functional only at the low-pO_2 side (Hoofd et al., 1987; Kreuzer and Hoofd, 1987) as seen from **NOFAC**. Heterogeneity in capillary spacing, however, is a dominating factor in the calculation of pO_2 histograms; the regular-pattern case (**RGL**) is completely different from the standard case. This leads to the conclusion that it is imperative to consider at least this type of heterogeneity. Single Krogh cylinder calculations come very close to the regular case thus also do not apply to heterogeneous tissue. By the same argument this holds true for models with only a few capillaries. Similar conclusions were already drawn from the multi-Krogh or multicylindrical models (e.g., Turek et al., 1986; 1991; this issue). These models assumed non-varying O_2 supply area along the capillary, whereas the model here allows that area to vary as needed since A_n is recalculated for each length step Δz (Hoofd et al., 1990).

The model is not capable to handle anoxic regions correctly. It will assume O_2 consumption even where pO_2 is below zero. Instead, less O_2 will flow towards these regions so that some will have no consumption, at $pO_2 = 0$, and draw no O_2; consequently, the O_2 will be available at other locations, uplifted to $pO_2 > 0$. In most cases covered here this hardly will affect the histograms, since the percentage of anoxic regions was small. Especially the **NOFAC** case, however, would come out different, the anoxic area being lower than the 14% indicated here and the percentage between 0 and 1 kPa correspondingly higher.

It is remarkable that matching capillary flow does so little (**FLW**). One would think of the cases handled here as two limiting ones, the standard case having the same flow for each capillary as compared to a situation in where flow is matched to O_2 need (almost, but not exactly, equal to what is called "same specific flow"). Obviously, the continuing readjustment of O_2 supply areas A_n in the standard case already accounts for some redistribution of capillary O_2 extraction. Although other schemes of flow distribution might be imaginable it looks questionable whether these could alter the pO_2 histograms notably.

None of the histograms matches the very-low-pO_2 measurements of Honig and Gayeski (1987), with all or almost all values below 1 kPa (7.5 mm Hg). These authors did not include measurements close to the capillaries which was accounted for here by leaving out the 3 μm zone around them, avoiding accidentally high values. Even though, our histograms are much smoother; even the regular-spacing cases (**RGL**, **BARR**) are not that steep. So, for such histograms other tissue factors must be considered (Turek et al., 1991).

REFERENCES

Batra, S., Rakusan, K., and Campbell, S. E., 1991, Geometry of capillary networks in hypertrophied rat heart, Microvasc. Res., 41:29-40.

Honig, C. R., and Gayeski, T. E. J., 1987, Comparison of intracellular PO_2 and conditions for blood-tissue O_2 transport in heart and working skeletal muscle, in: "Oxygen Transport to Tissue-IX," Adv. Exper. Med. Biol., Vol. 215, I. A. Silver and A. Silver, eds., Plenum Press, New York and London, pp. 309-321.

Hoofd, L., in press, Updating the Krogh model - assumptions and extensions, in: "Modelling of oxygen transport from environment to cell," S. Egginton and H. F. Ross, eds., Cambridge University Press, Cambridge, U.K.

Hoofd, L., Olders, J., and Turek, Z., 1990, Oxygen pressures calculated in a tissue volume with parallel capillaries, in: "Oxygen Transport to Tissue-XII," Adv. Exper. Med. Biol., Vol. 277, J. Piiper, T. K. Goldstick, and M. Meyer, eds., Plenum Press, New York and London, pp. 21-29.

Hoofd, L., Turek, Z., and Olders, J., 1989, Calculation of oxygen pressures and fluxes in a flat plane perpendicular to any capillary distribution, in: "Oxygen Transport to Tissue-XI," Adv. Exper. Med. Biol., Vol. 248, K. Rakusan, G. P. Biro, T. K. Goldstick, and Z. Turek, eds., Plenum Press, New York and London, pp. 187-196.

Hoofd, L., Turek, Z., and Rakusan, K., 1987, Diffusion pathways in oxygen supply of cardiac muscle, in: "Oxygen Transport to Tissue-IX," Adv. Exper. Med. Biol., Vol. 215, I. A. Silver and A. Silver, eds., Plenum Press, New York and London, pp. 171-177.

Kreuzer, F., and Hoofd, L., 1987, Facilitated diffusion of oxygen and carbon dioxide, in: "Handbook of Physiology: The Respiratory System: Gas Exchange," Section 3, Vol. IV, Chapter 6, L. E. Fahri and S. M. Tenney, eds., American Physiological Society, Bethesda, Maryland, pp. 89-111.

Turek, Z., Hoofd, L., Batra, S., and Rakusan, K., this issue, The effect of realistic geometry of capillary networks on tissue pO_2 in hypertrophied rat heart.

Turek, Z., Hoofd, L., and Rakusan, K., 1986, Myocardial capillaries and tissue oxygenation, Can. J. Cardiol., 2:98-103.

Turek, Z., Rakusan, K., Olders, J., Hoofd, L., and Kreuzer, F., 1991, Computed myocardial Po_2 histograms: effects of various geometrical and functional conditions, J. Appl. Physiol., 70:1845-1853.

THE EFFECT OF REALISTIC GEOMETRY OF CAPILLARY NETWORKS ON TISSUE PO_2

IN HYPERTROPHIED RAT HEART

Zdenek Turek[+], Louis Hoofd[+], Sanjay Batra[*], and Karel Rakusan[*]

Departments of Physiology, [+]Faculty of Medicine, Catholic University, Nijmegen, and [*]School of Medicine, University of Ottawa, Ottawa, Ontario, Canada

INTRODUCTION

A multicylindrical Kroghian model was developed at our Departments. The distribution of the radii of the cylinders was logarithmic-normal and was defined by the mean or median value and the logarithmic standard deviation (i.e., SD of log-transformed variates, $\sigma log_{10}x$, usually denoted by the abbreviation log SD) serving as the heterogeneity index. The mean value and log SD were obtained from capillary spacing on histological cross-sections, using the method of capillary domains (Hoofd et al., 1985). The model allowed calculation of PO_2 histograms in a block of tissue and was specifically designed to depict the effect of the geometric heterogeneity in capillary spacing on tissue oxygenation in normal and hypertrophied heart (Rakusan et al., 1984; Turek et al., 1986). The model was later expanded by including also the facilitation of O_2 by myoglobin, an additional resistance between blood and tissue (capillary barrier), and PO_2-dependent O_2 consumption. It was applied to skeletal (Turek et al., 1989) and cardiac muscle (Turek et al., 1991).

The model is based upon idealized capillary geometry, assuming straight unbranched capillary pattern with the same capillary distribution at the arteriolar and the venular side (Krogh cylinder). It is, however, possible to sample these distances utilizing the histochemical method for distinguishing arteriolar and venular capillary regions by color as proposed by Lojda (1979). This method has been applied to the study of capillary geometry in the rat heart by Batra et al. (1989). The data clearly established stepwise decreases in intercapillary distance with corresponding decreases in heterogeneity in capillary spacing, particularly in hypertrophied heart (Batra et al., 1991).

We have adapted our model so that the effect of the realistic capillary geometry of the arteriolar and venular regions of capillaries on tissue oxygenation could be assessed. In the present communication the first results are shown.

METHODS

Experimental data on the intercapillary distance and heterogeneity in capillary spacing were taken from Batra et al. (1991). Other input data were the same as used by Turek et al. (1991).

The model consisting of a set of tissue cylinders has been described in detail elsewhere (Turek et al., 1989; Turek et al., 1991). In the present study, we have incorporated data regarding intercapillary distance as sampled at 25% intervals along

Oxygen Transport to Tissue XIV, Edited by W. Erdmann and
D.F. Bruley, Plenum Press, New York, 1992

the capillary path from arteriole to venule. Half the intercapillary distance was used for the radius of tissue cylinder in each interval using data from Table 2 of Batra et al. (1991). It was assumed that the first and second interval have log SD of the arteriolar end and that the third and fourth interval have log SD of the venular end of the capillary (Table 1 from the same source). We assumed, that all capillaries begin at the same arteriolar level, i.e., there is no staggering. In the first interval the computation started from arterial blood gases (P_{O_2} = 100 mm Hg, P_{CO_2} = 40 mm Hg, pH = 7.4). The values of P_{O_2}, P_{CO_2} and pH calculated in the capillaries at the end of the interval were further applied as input data of P_{O_2}, P_{CO_2} and pH at the beginning of the following interval. This procedure was repeated for the remaining intervals. P_{O_2} histograms were calculated for each interval separately and, finally, a combined histogram was constructed for the whole tissue block. This model arrangement we call the truncated cone model. For comparison we also calculated P_{O_2} histograms by the same way, in four intervals, but using in each interval the same value of the radius of the cylinder and of log SD, obtained after averaging the data of Table 1 and 2 of Batra et al. (1991). This we call here the cylinder model.

The model with more intervals consisting of sets of tissue cylinders requires that P_{O_2}, P_{CO_2} and pH values in all capillaries at the beginning of each interval are identical. This is only possible when always the same specific flow (flow per volume of tissue) and zero-order O_2 consumption are assumed and when no appreciable percentage of anoxic tissue is present. The reason for the last requirement is that anoxic tissue does not consume oxygen and therefore in a large cylinder with an anoxic periphery, P_{O_2} at the end of the capillary will be larger than in a small cylinder without any anoxic zone but with the same specific flow. With zero order O_2 consumption usually a small percentage of anoxic tissue occurs. In order to test whether this percentage of anoxic tissue might distort the results, we compared the capillary P_{O_2} at the end of each interval calculated by the computer program with that obtained by calculation according to Fick principle. With a large percentage of anoxic tissue these values are expected to be different. In our case, they were always almost the same. We have checked that with the cylinder model with four intervals the combined P_{O_2} histograms and venous blood gas values were almost identical to those derived with our previously presented model, that calculated P_{O_2} histograms in one move from the arteriolar to the venular end.

The requirement that P_{O_2} in all capillaries at the beginning of each interval must be the same makes it impossible to use a large value for the capillary barrier. We have demonstrated before that a combination of large capillary barrier with zero-order O_2 consumption results in a large percentage of anoxic tissue. A combination of a large capillary barrier with P_{O_2}-dependent O_2 consumption, that can result in uniform and low P_{O_2} within the whole tissue, cannot be used here, as in that case we obtain heterogeneous end-capillary P_{O_2}, P_{CO_2} and pH at the end in the intervals and thus cannot apply single values as input data for the following interval, as required by the model in its present form. In the present communication we show results obtained with a moderate capillary barrier corresponding to ΔP_{O_2} in average-sized tissue cylinder of 10 mm Hg.

RESULTS

Fig. 1 depicts the comparison between four intervals truncated cone (upper panel) and cylinder (lower panel) calculations. In the truncated cone model, low P_{O_2} values (< 10 mm Hg) occur in all intervals, whereas in the cylinder model low P_{O_2} values can be seen mainly towards the venular end of the capillary.

Fig. 2 depicts a comparison between combined four intervals cylinder and truncated cone model. In the cylinder model the histograms are wider, percentage of both very low as well as very high P_{O_2} values is higher than in the truncated cone model. The P_{O_2} histograms calculated in the truncated cone model are more homogeneous, showing a more distinct peak around about 30 mm Hg, with moderate capillary barrier.

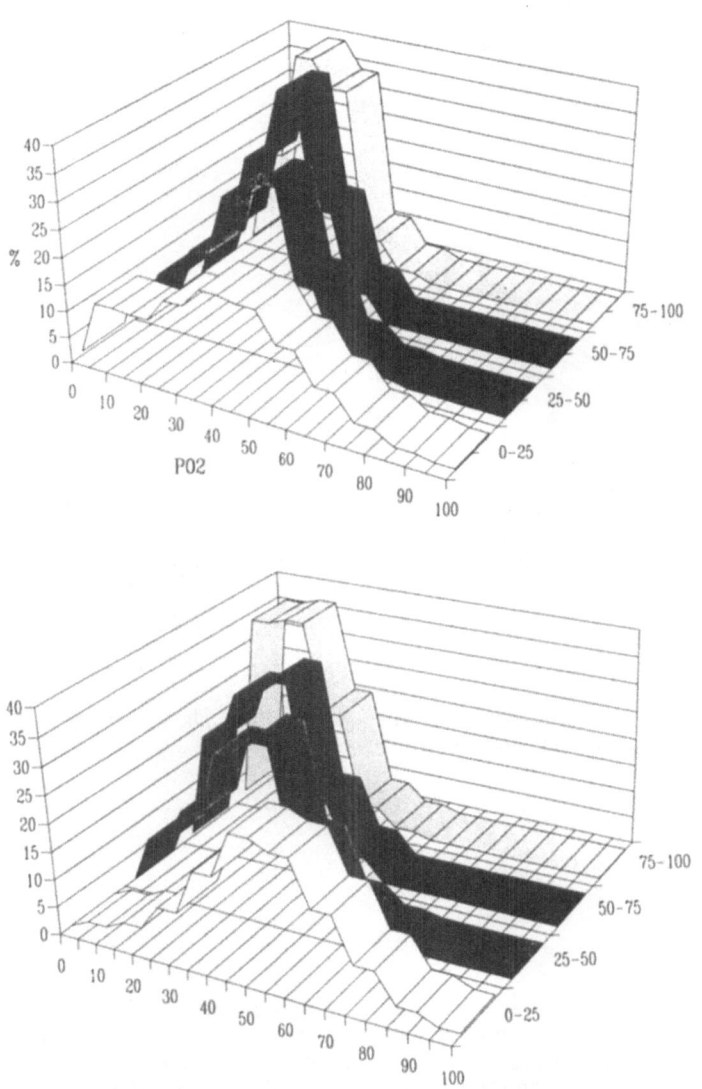

Fig. 1. Comparison between P_{O_2} histograms of four intervals truncated
cone (upper panel) and cylinder (lower panel) calculations.

The model also computes the percentage of anoxic tissue ($P_{O_2} < 0.005$ mm Hg). This is shown in Fig. 3 for cylinder and truncated cone model. Note that whereas in the cylinder model the percentage of anoxic tissue increases very distinctly along the length of the capillary, being almost zero in the first interval and approaching about 1.5 % in the last interval, in the truncated cone model the highest percentage occurs in the first interval and then decreases when moving to deeper intervals. When all intervals are combined, the percentage of anoxic tissue is about $1.7 \times$ higher in the cylinder model than in the truncated cone model.

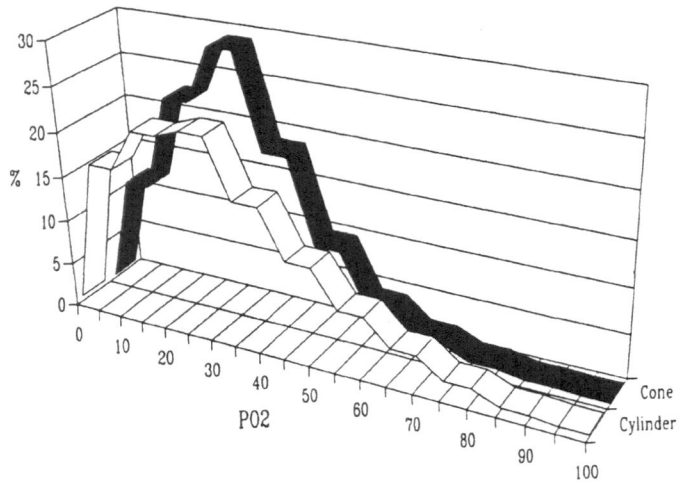

Fig. 2. P_{O_2} histograms for combined four intervals truncated cone and four intervals cylinder model.

DISCUSSION

In this communication we show only results calculated for hypertrophied rat hearts. The observation that both the intercapillary distance as well as the heterogeneity in capillary spacing decrease towards the venular (as against the arteriolar) end of capillaries is true also in control animals (Batra et al., 1991). We calculated P_{O_2} histograms also in control animals and the trend of differences between the cylinder model and the truncated cone model was the same as in rats with pressure-induced hypertrophy. However, the differences were less pronounced and even difficult to depict when using graphical methods (see also Hoofd et al., this issue). Therefore, we show here only the results derived on hypertrophied hearts.

Present results were computed for the realistic heterogeneous pattern of microvascular bed. However, for reasons described in Methods, an otherwise rather homogeneous situation had to be assumed. In particular an identical capillary specific blood flow and an identical O_2 consumption (as long as $P_{O_2} > 0$), as implicit for zero-order

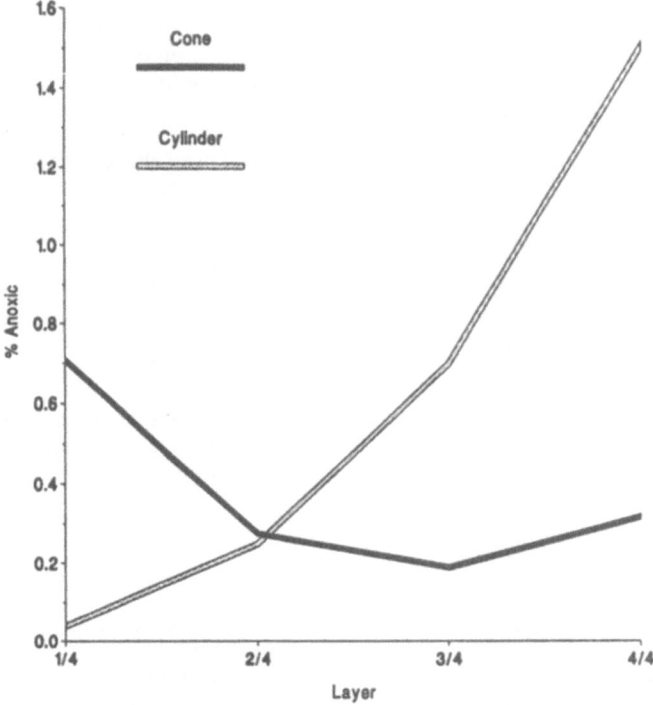

Fig. 3. Percentage of anoxic tissue ($P_{O_2} < 0.005$ mm Hg) within four intervals of truncated cone and cylinder model. Note that in the
truncated cone model the percentage of anoxic tissue in the first
interval is higher than in the last one.

O_2 consumption, as well as the absence of capillary staggering are probably unrealistic in the working heart. It may be interesting that with the multicapillary model a
rather small effect was demonstrated of staggering concerning the percentage of low
tissue P_{O_2} (Hoofd et al., this issue). Previously we have shown that the heterogeneity
in capillary spacing had a much more pronounced effect on the heterogeneity of calculated tissue P_{O_2} when also capillary flow was heterogeneous (Rakusan and Turek,
1985; Turek et al., 1991) and we expect it to happen also here. Unfortunately, we do
not see, at this moment, how these model limitations could be overcome in the model
consisting of a set of Kroghian tissue cylinders.

When the traditional methods of capillary visualization were used it was impossible to distinguish on cross-sections whether capillaries were cut closer to the arteriolar or to the venular end. Hence, when in our previous work histological sections
of the myocardium were analyzed using our routine method of capillary domains, we
obtained values of the radius of tissue cylinder and of log SD that were somewhere
between the arteriolar and venular value. In our previous studies we used these values
as input data for our model thus assuming that capillaries were parallel and straight
vessels that did not bifurcate at any point, i.e., we used the cylinder model. The present results show that when the more realistic values of the truncated cone model are
used instead, the calculated P_{O_2} histograms become less heterogeneous and the percentage of P_{O_2} lower then 10 mm Hg and the percentage of anoxic tissue decreased.
This agrees with the intuitive feeling that a shorter intercapillary distance and a less
pronounced heterogeneity would be more effective when the capillary P_{O_2} is already
low to start with. However, the difference between the cylinder and the truncated
cone model is rather small, particularly in control animals, and this remains so even

when a moderate capillary barrier is incorporated. This suggests that, at least within the limits of present assumptions, our conclusions previously obtained with the cylinder model probably are valid also for the truncated cone model. In this respect we are mainly concerned about the importance of heterogeneity in capillary spacing on tissue P_{O_2} (Turek et al., 1989; Turek et al., 1991).

In this communication we were not able to analyze the effect of the truncated cone versus the cylinder model in a situation with a large capillary barrier and P_{O_2}-dependent O_2 consumption, for reasons explained in Methods. We have shown previously that when the combination of large capillary barrier and P_{O_2}-dependent O_2 consumption was applied in order to obtain calculated histograms similar to experimental results of Honig and Gayeski (1987), the effect of both the capillary spatial and flow heterogeneity on the form of histograms was negligible (Turek et al., 1991). We do not expect that using the truncated cone instead of cylinder model would change this conclusion.

Our study suggests that the shorter intercapillary distance and smaller heterogeneity in capillary spacing at the venular versus arteriolar end of capillaries leads to a more homogeneous tissue P_{O_2} distribution and decreases the chance that anoxic regions occur.

REFERENCES

Batra, S., Rakusan, K., and Campbell, S. E., 1991, Geometry of capillary networks in hypertrophied rat heart, Microvasc. Res., 41:29-40.

Batra, S., Rakusan, K., and Kuo, C., 1989, Spatial distribution of coronary capillaries: A-V segment staggering, in: "Oxygen Transport to Tissue-XI," Adv. Exper. Med. Biol., Vol. 248, K. Rakusan, G. P. Biro, T. K. Goldstick, and Z. Turek, eds., Plenum Press, New York and London, pp. 241-247.

Honig, C., and Gayeski, T. E. J., 1987, Comparison of intracellular P_{O_2} and conditions for blood-tissue transport in heart and working red skeletal muscle, in: "Oxygen Transport to Tissue-IX," Adv. Exper. Med. Biol., Vol. 215, I. A. Silver, and A. Silver, eds., Plenum Press, New York and London, pp. 309-321.

Hoofd, L., and Turek, Z., this issue, Oxygen pressure histograms calculated in a block of rat heart tissue.

Hoofd, L., Turek, Z., Kubat, K., Ringnalda, B. E. M., and Kazda, S., 1985, Variability of intercapillary distance estimated on histological sections of rat heart, in: "Oxygen Transport to Tissue-VII," Adv. Exper. Med. Biol., Vol. 191, F. Kreuzer, S. M. Cain, Z. Turek, and T. K. Goldstick, eds., Plenum Press, New York and London, pp. 239-247.

Lojda, Z., 1979, Studies on dipeptidyl(amino)peptidase IV (glycyl-proline naphtylamidase), Histochemistry, 59:153-166.

Rakusan, K., Hoofd, L., and Turek, Z., 1984, The effect of cell size and capillary spacing on myocardial oxygen supply, in: "Oxygen Transport to Tissue-VI," Adv. Exper. Med. Biol., Vol. 180, D. Bruley, H. I. Bicher and D. Reneau, eds., Plenum Press, New York and London, pp. 463-475.

Rakusan, K., and Turek, Z., 1985, The effect of heterogeneity of capillary spacing and O_2 consumption - blood flow mismatching on myocardial oxygenation, in: "Oxygen Transport to Tissue-VII," F. Kreuzer, S. M. Cain, Z. Turek, and T. K. Goldstick, eds., Plenum Press, New York and London, pp. 257-261.

Turek, Z., Hoofd, L., and Rakusan, K., 1986, Myocardial capillaries and tissue oxygenation, Can. J. Cardiol., 2:98-103.

Turek, Z., Olders, J., Hoofd, L., Egginton, S., Kreuzer, F., and Rakusan, K., 1989, P_{O_2} histograms in various models of tissue oxygenation in skeletal muscle, in: "Oxygen Transport to Tissue-XI," Adv. Exper. Med. Biol., Vol. 248, K. Rakusan, G. P. Biro, T. K. Goldstick, and Z. Turek, eds., Plenum Press, New York and London, pp. 227-237.

Turek, Z., Rakusan, K., Olders, J., Hoofd, L., and Kreuzer, F., 1991, Computed myocardial P_{O_2} histograms: effects of various geometrical and functional conditions, J. Appl. Physiol., 70:1845-1853.

CONTRACTILE DYSFUNCTION OF "REPERFUSED" NEONATAL RAT HEART CELLS: A MODEL FOR STUDYING "MYOCARDIAL STUNNING" AT THE CELLULAR LEVEL?

Peter Boekstegers, Alexander Pfeifer, Wolfgang Peter, Karl Werdan

Department of Internal Medicine I, Klinikum GroBhadern
Univ. of Munich, FRG

INTRODUCTION

Myocardial stunning defined by prolonged contractile dysfunction of reperfused hearts after ischemia (Braunwald et al. 1982) was demonstrated both in vivo and in isolated heart models such as the Langendorff preparation. Evidence exists that different mechanisms (e.g. oxygen radicals, "no reflow" phenomenon, calcium overload) and different cells (endothelial cells, leukocytes, cardiomyocytes) (Bolli 1990, Marban 1991, Lefer et al. 1991, Bagchi et al. 1990) are involved in the development of myocardial stunning. However, the contribution of each to the finally observed contractile dysfunction of the cardiomyocyte is not clear (Bolli 1990). Intact heart, in which different cell types are present, do not allow differentiation between cell mediated injury of the cardiomyocyte and the susceptibility of the cardiomyocyte itself to myocardial stunning.

Isolated cardiomyocytes exposed to anoxia and reoxygenation have been used to simulate ischemia and reperfusion at the cellular level (Barry et al. 1980, Piper et al. 1984, Stern et al. 1985, Musters et al. 1991). After reoxygenation "oxygen and calcium paradox" could be demonstrated in adult heart cells (Stern et al. 1985, Rodrigo 1990). However, dynamic changes of contractility similar to myocardial stunning of whole hearts are difficult to study in resting adult heart cells (Bond 1991). Therefore, spontaneously beating neonatal rat heart cells might be a more suitable model to investigate contractile dysfunction of the cardiomyocyte exposed to "ischemia" and "reperfusion".

MATERIALS AND METHODS

Cell culture

The preparation and culture of cardiac muscle and non-muscle cells from neonatal rats have been described in detail previously (Werdan 1989). In brief the procedures are: preparation of hearts from 1 to 3-day-old rats under sterile conditions, disaggregation of heart tissue at 37 °C with trypsin-collagenase-salt solution (Ca^{2+}- and

Mg^{2+}-free), and seeding of the cells (1-2x10^5/cm^2) in plastic flasks in CMRL 1415 ATM medium supplemented with 10% fetal calf serum, 10% horse serum and 0.02 mg/ml gentamycin. After 24 h of culture, serum containing medium was replaced by a serum free culture medium (CMRL 1415 ATM medium containing 2 μM insulin, 0.1 μM dexamethasone, 0.4 μM iron saturated transferrin and 0.4 μM bovine serum albumin) supplemented 10mM HEPES, pH 7.4. The cells were cultured for 48 h in the presence of the indicated additions with daily medium changes.

Simulation of ischemia and reperfusion

In order to simulate ischemia at the cellular level an in vitro cell chamber suitable for a multidish 6-well plate (Primaria, Becton Dickinson, USA) was developed (figure 1). Previous experiments had shown that anoxic PO$_2$ values (defined by PO$_2$ < 1.0 mmHg) in the culture medium above the cell layers were difficult to obtain by solely gassing the atmosphere with nitrogen despite a chamber volume of about 192 cm^3 and a high gas flow of about 3-4 l/min (figure 2A). Therefore, water saturated (37 °C) purified nitrogen (Oxysorb®, Messer Griesheim, FRG) was directly introduced into the culture medium by continuously bubbling each well separately in addition to gassing of the chamber atmosphere.

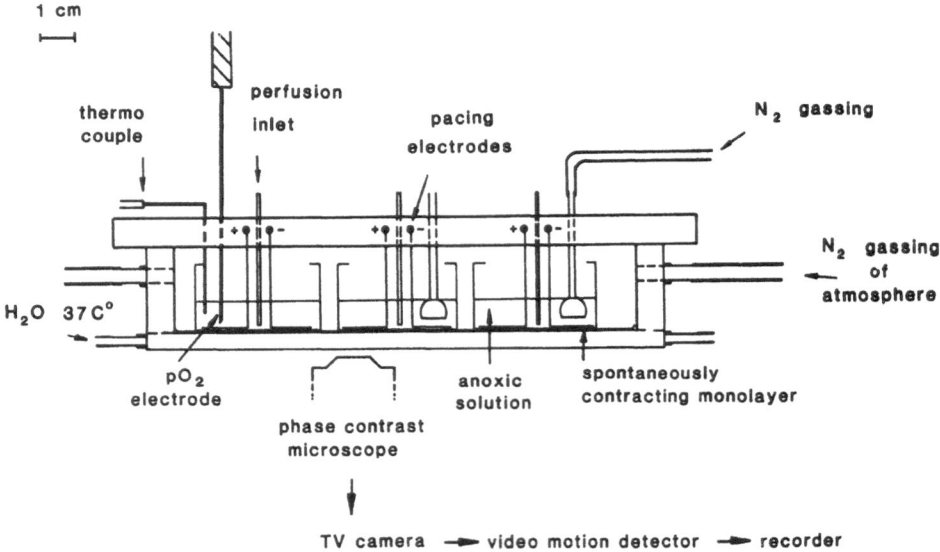

Fig.1. Hypoxia and anoxia cell chamber.

Thus, anoxia could be rapidly achieved (figure 2B). Furthermore, anoxic perfusion (up to 2ml/min) using different solutions (preequilibrated with nitrogen) was possible due to the entrance of the perfusion inlet into the gassing stream of each well (fig.1). As a consequence of the continuous bubbling with nitrogen there was a small but constant movement of the fluid (3 ml) above the cell layers. PO$_2$ and temperature were equal in different heights (0.05-10mm) above the cells and at different randomly chosen points within the well if the solution was bubbled. However, without directly bubbling the fluid, the PO$_2$ within the fluid was dependent on the distance from the surface of the fluid, from the cell layer and from the PO$_2$ in the atmosphere. The temperature within the cell chamber was kept between 36-37 °C by a continuous stream of prewarmed gas and by heated water circulating through the space below the cell layer.

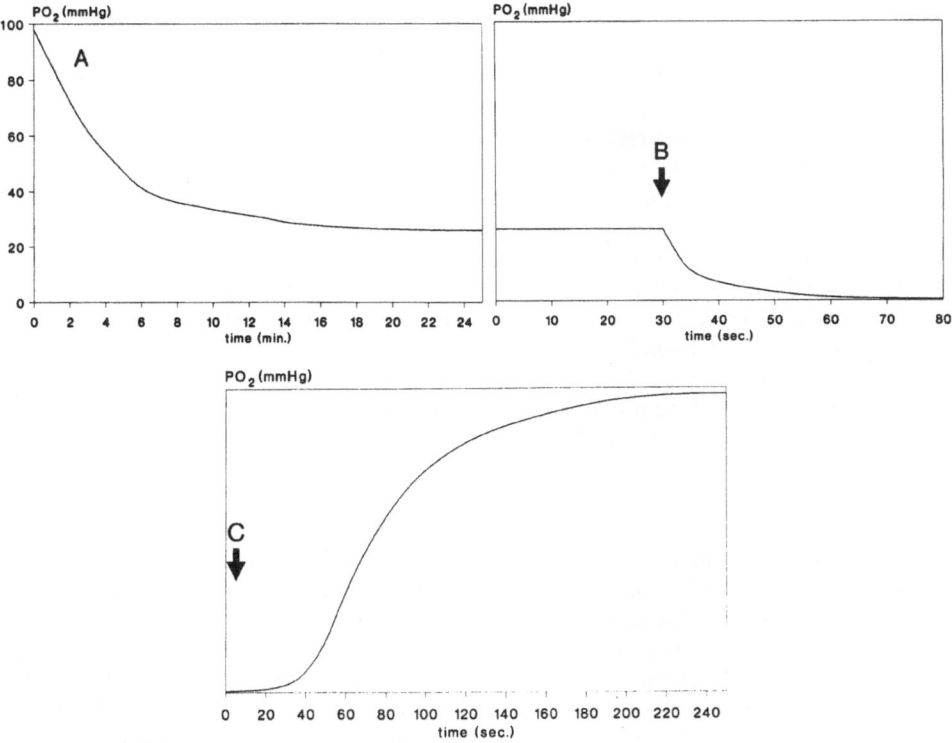

Fig.2. PO_2 recording within supernatant of cultured heart cells exposed to A) start of N_2 gassing (4l/min) of the cell chamber atmosphere, B) start of N_2 bubbling of the supernatant C) start of O_2 bubbling of the supernatant.

In addition to anoxia for simulating ischemia the serum free culture medium was replaced by an unbuffered salt solution (1.3 mM $CaCl_2$, 3.5 mM KCL, 0.97 mM $MgSO_4$, 0.36 mM NaH_2PO_4, 1.4 mM Na_2HPO_4, pH 7.4). The volume of fluid above the cell layers was restricted to 3 ml during "ischemia". After "ischemia", oxygen and serum free culture medium was reintroduced within 60 sec into each well (figure 2C) for simulating reperfusion.

PO_2 and temperature

PO_2 within the fluid above the cell layers was measured using polarographic electrodes. The kinetics of gas exchange of the cell chamber were determined by fast responding needle electrodes (t90 < 500ms, drift < 2%/h) (Fleckenstein 1982). The tip of the electrode could be moved stepwise (0.1 mm) forward or backward by a micromanipulator (PO_2 Histograph®, Eppendorf, FRG). Long term measurements of PO_2 below 2.0 mmHg were performed using more stable electrodes (drift < 0.5%/lh, MT-1-AC, Eschweiler, FRG). In each experiment PO_2 and temperature (thermocouple, GMS, FRG) within the chamber were continuously measured.

Beating frequency and contractility

The experimental setup for determination of frequency and contractility of spontaneously beating neonatal rat heart cells has been described in detail elsewhere (Werdan 1989). In brief: the beating cell in the multidish 6-well plate is observed through an inverted phase-contrast microscope (Leitz Diavert, Wetzlar, FRG) in a thermostated chamber (37 °C) at x300 magnification. It is recorded with a television camera (Grundig FA 70B, FRG). The image is displayed on a monitor (Grundig BG 23T, FRG) and simultaneously transduced to a video recorder (Sony Umatic VO-5630, Japan). Contraction of the cells during beating causes changes in light intensity around the cell wall, the dark cell body contrasting with the light surroundings. These light intensity changes are recorded by a photocell (Siemens BPY 61, FRG). The output signal of the photocell is amplified and processed by a differentiating preamplifier, whereupon it is electronically filtered and monitored on a multichannel recorder. This systems permits continuous measurement of beating frequency and amplitude (y) as well as velocity (dy/dt) of cellular wall motion (contraction and relaxation velocity).

Due to the frequency dependence of contraction velocity and relaxation (Werdan 1989), measurements of contractility were performed under pacing using a stimulator (10-100V, pulse duration 5-10 ms, frequency 60-150/min, Grass SD, USA). During and up to 3 min after stop of pacing, cells were perfused at a constant rate of 2ml/min.

Energy metabolism

Each of the six wells of a plate were extracted with 400 μl perchloric acid and cooled on ice for 10 min. 300 μl of the supernatant were removed, neutralized with 2 M KOH and centrifuged. The remaining monolayer was lysed in 1 ml of 0.1 M NaOH for determination of cell protein according to Lowry et al. For measurement of energy rich phosphates and NAD, 100 μl of the neutralized supernatant were injected into the HPLC system (Gilson 712, Gilson Medical Electronics, USA). ATP, ADP, AMP, CP, GTP and NAD were determined by HPLC using a Hypersil ODS column (Pannosch, Austria) as described in detail elsewhere (Furst et al. 1991). In brief: buffers were prepared with bidistilled water and degassed by sonification. Buffer A consisted of 0.1 M NaH_2PO_4 and 2g/1 tetrabutylammoniumhydrogensulfate, adjusted to pH 5.5 with NaOH. Buffer B was a 75/25 (v/v) mixture of buffer A and acetonitrile. The elution profile was started with 5 min isocratic elution at 100% buffer A followed by a 20 min linear gradient up to 70% buffer B. After 3 min constant solvent delivery with the 30:70 (v/v) mixture of buffer A and buffer B and a 2 min gradient back to 100% buffer A, the column was flushed with buffer A for 20 min.

Statistics

Unless otherwise indicated data are expressed as means +/- SEM. Data were compared using the Student's test for unpaired or paired observations.

RESULTS

Beating frequency and contractility

All experiments on beating frequency were performed with spontaneously beating cells without pacing. A typical registration of a neonatal rat heart cell during "ischemia" and "reperfusion" is shown in figure 3. During "ischemia" beating frequency decreased progressively and arrhythmias occurred. Reintroduction of oxygen restored spontaneous beating within seconds. Figure 4 summarizes the experiments on beating frequency. Complete recovery of frequency to preischemic values lasted 60-90 min after "reperfusion".

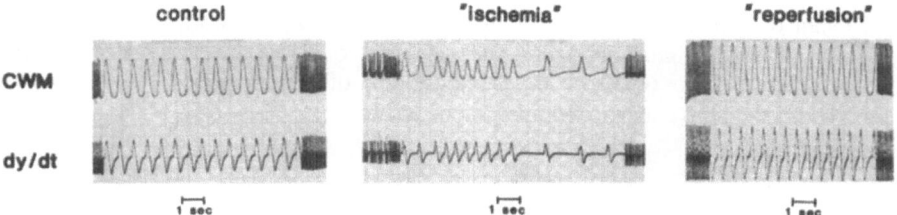

Fig. 3. Recording of cellular wall motion (CWM) and velocity (dy/dt) of cellular wall
motion (contraction and relaxation velocity) of a neonatal rat heart cell exposed
to "ischemia" (60 min) and "reperfusion" (15 min).

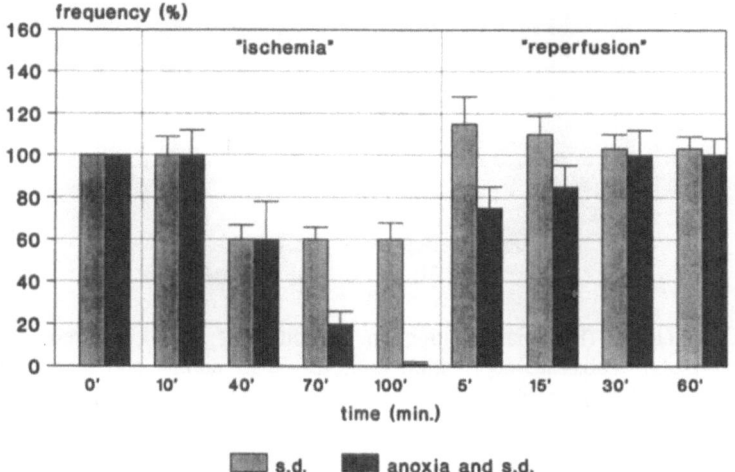

Fig. 4. Beating frequency of cultured heart cells exposed to substrate deprivation (s.d.,
n=3) or anoxia and s.d. ("ischemia", n=6) (n= number of experiments) (for
details s. methods).

Contractility of cells was estimated by maximum velocity (Vmax) of cell wall motion under pacing. Vmax progressively declined during "ischemia" (figure 3,5). In contrast substrate free solution without anoxia (s.d.) did not affect Vmax. "Reperfusion" resulted in different responses in individual cells of the monolayers despite a homogeneous decline of Vmax during "ischemia" in all cells (figure 6). About 38% of the cells showed hypercontractility defined by exceeding 110% of the preischemic value of Vmax after 15 min of "reperfusion". 28% of the cells had a delayed recovery reaching less than 75% of the preischemic value after 15 min of "reperfusion" and had still significantly lower Vmax (66%, $p < 0,01$) after 60 min of "reperfusion". The other cells lay in between the above mentioned two groups (figure 6).

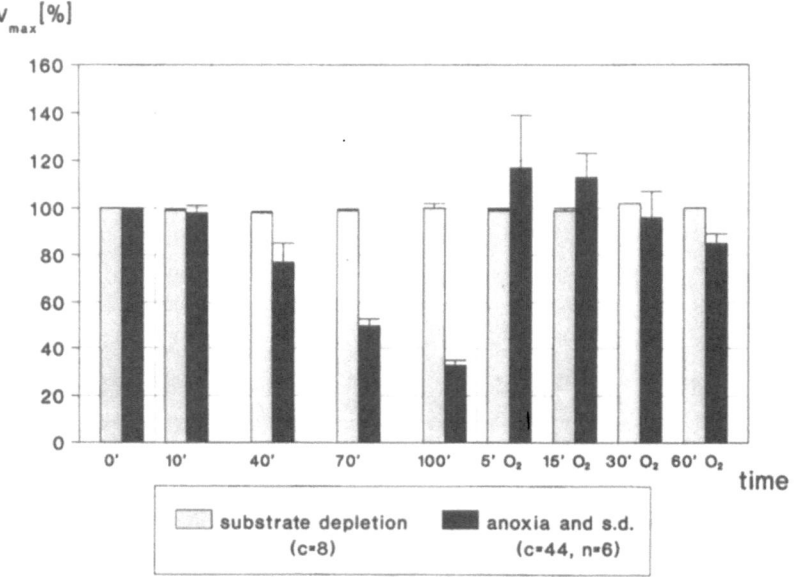

Fig. 5. Contractility (maximal contraction velocity $= V_{max}$) of cultured heart cells exposed to substrate deprivation (s.d., c=8) or anoxia and s.d. ("ischemia", c=44) and "reperfusion", c=number of cells (for details s. methods).

Energy metabolism

Representative chromatograms which were obtained from monolayers (s.methods) before anoxia (A), after 90 min of anoxia (B) are shown in figure 7. Mean values of ATP, ADP, AMP, CP, GTP and NAD substantially decreased after 90 min of "ischemia" ($p < 0.001$) whereas there was no significant difference in the concentration of energy metabolites after 60 min of "reperfusion" compared to preischemic values (figure 8). Calculated energy charge ((ATP+ADP/2)/ (ATP+ADP+AMP)) decreased from 0.76 to 0.64 (p<0.01) during "ischemia" and was restored to 0.77 after 60 min of "reperfusion".

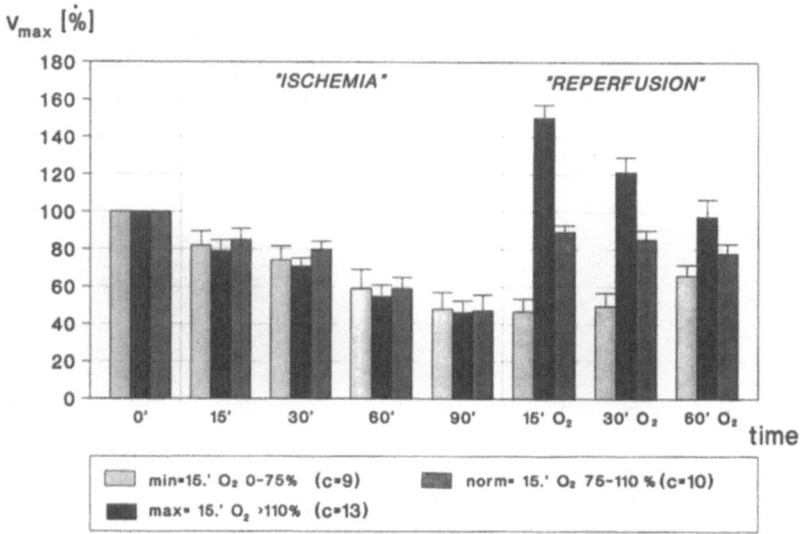

Fig.6. Different response of contractility (maximal contraction velocity=Vmax) in cultured heart cells during "reperfusion", c=number of cells (for details s. methods).

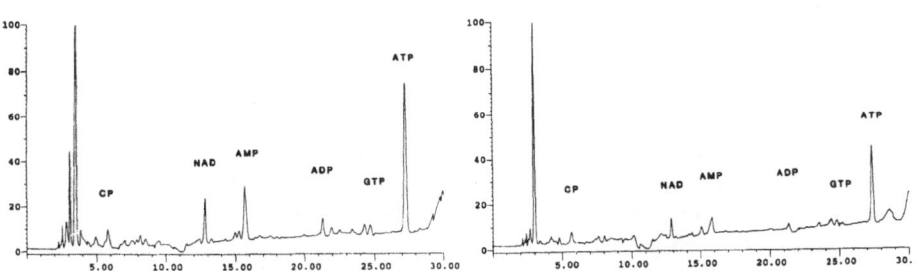

Fig.7. Representative chromatograms for determination of energy rich phosphates and NAD. Left Panel: control before "ischemia". Right panel: 90 min "ischemia".

DISCUSSION

"Ischemia" in cultured heart cells

Previous studies on spontaneously beating isolated heart cells exposed to hypoxia or anoxia have shown that loss of contractility and beating activity mainly depended on the degree of oxygen deprivation (Karsten et al. 1973, Barry et al.1984). Therefore, we aimed to develop a model which meets the condition of reproducible oxygen deprivation (anoxia defined by $PO_2 < 1mmHg$) while allowing rapid fluid and gas exchange. As

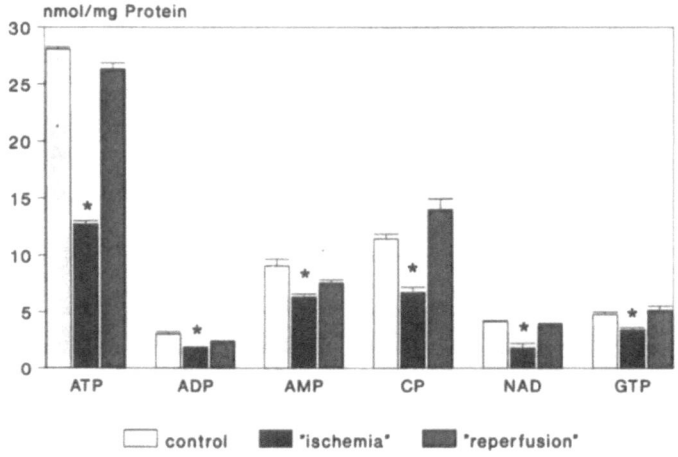

Fig. 8.: High energy phosphates and NAD in cultured heart cells before "ischemia", after 90 min of "ischemia" and after 60 min of "reperfusion" (n=6, * = p<O. 01).

shown by PO_2 measurements (s. methods, fig. 2) this was achieved by directly introducing nitrogen into the supernatant of the monolayers. No effects on beating frequency or contractility of neonatal rat heart cells were observed as long as ambient air was bubbled through the supernatant instead of nitrogen (up to 120 min, results not shown).

If cultured neonatal rat heart cells were exposed to anoxia and substrate deprivation ("ischemia") beating frequency and contractility decreased in a highly reproducible manner (fig. 4,5,6). Using salt solution without glucose and the volume of the supernatant restricted to 3ml, cells were exposed to anoxia, extracellular substrate deprivation and volume restriction, which is very similar to conditions of "ischemia" in whole hearts (Vemuri 1985). Though loss of beating activity mostly occurred after 60 min of "ischemia", pacing of the cells was possible and contractility was not completely abolished (fig.5,6). This finding of persisting weak contractility despite severe anoxia is in agreement with results from embryonic chick heart cells (Barry 1984) and was explained by ATP-production from glycogen stores as long as glycolysis was not inhibited (Barry 1984). In our model 90 min of "ischemia" resulted in a 40-60%

reduction in adenosinetriphosphate (ATP), creatinephosphat (CP) and nictonamide adenine dinucleotide (NAD) (fig.8) which is comparable to energy depletion of isolated canine heart cells after 60-90 min of ischemia (Hohl et al. 1991) and comparable to energy depletion after 10 min of complete ischemia in whole hearts in vivo (Belle 1986).

"Reperfusion" in cultured heart cells

"Reperfusion" elicited a different recovery of contractility in individual cells of the monolayer (fig.6) despite contractility of all cells was similarly affected during "ischemia". Since energy metabolism was measured in extracts from the whole monolayer, it cannot be excluded that energy depletion might have been different in individual cells during "ischemia". However, homogenous decrease of contractility during "ischemia" suggested that ATP production and availability was affected quite similarly in all cells during "ischemia". In contrast to in vivo models, extracellular conditions were similar for all cells in our model at the onset of "reperfusion".
Heterogeneity of contractile recovery after "reperfusion" in our model agrees well with heterogeneity in cellular response of other in vitro models using isolated heart cells (Hayashi et al. 1990, Hohl et al. 1991, Li et al. 1989, Bond et al. 1991). Distinct hypercontractility in about one third of cells (fig.6) supports the assumption that rise in intracellular Ca^{2+} after "ischemia" was heterogeneously distributed among the cells of the monolayer as shown in adult heart cells by Hayashi et al. Further studies using inotropic agents are required to find out whether Ca^{2+}-sensitivity of ischemic or reperfused cells is heterogenously affected, particularly with regard to those cells with delayed recovery of contractility after "reperfusion" (fig.6). Though 60 min after "reperfusion" energy metabolism was similar to "preischemic" values (fig.8), contractility was less than 80% of the "preischemic" value in more than 50% of the cells (fig.6). Again energy metabolism of individual cells cannot be estimated from our measurements. However, with regard to several studies of energy metabolism in stunned myocardium using different models (Piper 1984, Bolli 1990, Zucchi 1990, Bond 1991), energy depletion or delayed resynthesis are unlikely to explain the observed contractile dysfunction of our cells.

In conclusion, neonatal rat hearts cells exposed to "ischemia and reperfusion" seem to be a suitable model for studying postischemic contractile dysfunction (stunning) at the cellular level. However, heterogeneity in the degree of postischemic contractile dysfunction of individual cells has to be considered if mechanisms responsible for myocardial stunning are aimed to be elucidated in cultured cardiomyocytes.

REFERENCES

Acosta D, Puckett M, 1977, Ischemic myocardial injury in cultured heart cells: Preliminary observations on morphology and beating activity, In Vitro 13:818-823.
Bagchi D, Das DK, Engelman RM, Prasad MR, Subramamnian R, 1990, Polymorphonuclear leukocytes as potential source of free radicals in the ischaemic-reperfused myocardium, European Heart J 11:800-813.
Barry WH, Pober J, Marsh JD, Franlel StR, Smith ThW, 1980, Effects of graded hypoxia on contraction of cultured chick embryo ventricular cells, Am J Physiol 239:H651-657.
Bolli R, 1990, Mechanism of myocardial "stunning", Circulation 82:723-738.
Bond JM, Herman B, Lemasters JJ, 1991, Recovery of cultured rat neonatal myocytes from hypercontracture after chemical hypoxia, Res Comm Chem Pathol Pharmacol 71:195-208

Braunwald E, 1982, The stunned myocardium: Prolonged, postischemic ventricular dysfunction, Circulation 66:1146-1149.

Fleckenstein W, 1982, In vivo measurements Of PO_2 histograms using a hypodermic needle electrode system, Pflügers Arch 392: R209.

Furst W, Hallström S, Schlag G, 1991, Simultaneous determination of myocardial nucleosides, purine bases and creatine phosphate by high performance liquid chromatography, J Chromatogr (in press).

Hayashi H, Miyata H, Kobayashi A, Yamazaki N, 1990, Heterogeneity in cellular response and intracellular distribution of Ca2+ concentration during and after metabolic inhibition, Cardiovascular Research 24:605-608.

Hohl CM, Altschuld RA, 1991, Response of isolated adult canine cardiac myocytes to prolonged hypoxia and reoxygenation, Am J Physiol 260:C383-391.

Karsten U, Kössler A, Janiszewski E, Wollenberger A, 1973, Influence of variations in pericellular oxygen tension on individual cell growth, muscle characteristic proteins, and lactate dehydrogenase isoenzyme pattern in cultures of beating rat heart cells, In Vitro 9:139-146.

Lefer AM, Tsao PS, Lefer DJ, MA XL, 1991, Role of endothelial dysfunction in the pathogenesis of reperfusion injury after myocardial ischemia, FASEB J 5:2029-2034.

Li Q, Hohl CM, Altschuld RA, Stokes BT, 1989, Energy depletion-repletion and calcium transients in single cardiomyocytes. Am J Physiol 257:C427-434.

Lowry OH, Rosebrough NJ, Farr AL, Randall RJ, 1951, Protein measurement with the Folin phenol reagent, J Biol Chem 193:265

Marban, E, 1991, Myocardial stunning and hibernation: The physiology behind the colloquialisms, Circulation 83:618-688.

Musters RJP, Post JA, Verkleij AJ, 1991, The isolated neonatal rat-cardiomyocyte used in an in vitro model for "ischemia". I. A morphological study, Biochimica et Biophysica Acta 1091:270-277.

Piper HM, Schwartz P, Spahr R, HUtter JF, Spieckermann PG, 1984, Absence of reoxygenation damage in isolated heart cells after anoxic injury, Pflügers Arch 401:71-76.

Rodrigo GC, Chapman RA, 1991, The calcium paradox in isolated guinea-pig ventricular myocytes: effects of membrane potential and intracellular sodium, J Physiology 434:627-645.

Stern MD, Chien AM, Capogrossi MC, Pelto DJ, Lakatta EG, 1985, Direct observation of the "oxygen paradox" in single rat ventricular myocytes, Circ Res 56:899-903.

Van Belle H, Wynants J, Xhonneux R, Flameng W, 1986, Changes in creatine phosphate, inorganic phosphate, and the purine pattern in dog hearts with time of coronary artery occlusion and effect thereon of mioflazine, a nucleoside transport inhibitor. Cardiovascular Research 20:658-664

Werdan K, Erdmann E, 1989, Preparation and culture of embryonic and neonatal heart muscle cells: modification of transport activity, In: Methods in Enzymology 173:634-662.

Zucchi R, Limbruno U, Di Vincenzo A, Mariani M, Ronca G, 1990, Adenine nucleotide depletion and contractile dysfunction in the "stunned" myocardium, Cardiovascular Research 24:440-446.

MONITORING OF REDOX-STATE OF RESPIRATORY ENZYMES AND MYOGLOBIN OXYGENATION IN THE WORKING RAT HEART IN NORMOXIA AND OXYGEN DEFICIENCY

Zündorf J., Tauschek D., Frank K., Ito K[1]., Nioka S[2]., Kessler M., Chance B[2]

Institut für Physiologie und Kardiologie der Universität Erlangen-Nürnberg, Waldstr. 6, D-8520 Erlangen
[1] Biophysics Division Research Institute of Applied Electricity, Hokkaido University, Sapporo 060, Japan
[2] Department of Biochemistry and Biophysics, University of Pennsilvania, Philadelphia, PA, USA

SUMMARY

The cellular oxygen supply in the isolated, hemoglobin-free perfused, working rat heart can be determined by measurements of myoglobin oxygenation. However, for a precise analysis of mitochondrial hypoxia and anoxia ($pO2 < 0.01$ Torr) redox-state of respiratory enzymes must be known.

By use of the EMPHO (Frank et al. 1989) it is possible to perform a high speed spectrometry within very small tissue volumes.

Because of the characteristic absorption spectra of oxygenated and deoxygenated myoglobin and of the oxidized and reduced cytochrome aa3 within the wavelength interval from 500 to 630 nm it is possible to isolate these two pigments from the remission spectra and to determine the oxygenation state of myoglobin and the redox-state of cytochrom aa3.

METHODS

The experiments were performed on isolated hearts of male wistar rats using the model of the isolated, hemoglobin-free perfused, working rat heart. It was a modified version of the model described by Turek and Olders (Olders et al. 1990).

The animals were anesthetized with ether whereafter 0.5 ml heparine (5000 i.U./ml) were injected into the vena femoralis. After five minutes the thorax was opened and the heart was preparated for the perfusion.

The perfusion medium we used was a modified Tyrode's solution which contained (mmol/l): NaCl (130), KCl (5.6), CaCl$_2$ (2.16) MgCl$_2$ (0.56), NaHCO$_3$ (25.0) NaH$_2$PO$_4$ (1.2) glucose (11.1) and pyruvate (5.0). As colloidosmotic substance we added 60 g/l hydroxyethylstark (40.000 Dalton).

Two gas mixtures were used to equilibrate the perfusion medium. It was 95 % O$_2$ + 5 % CO$_2$ and 95 % N$_2$ + 5 % CO$_2$. To decrease the oxygenation of the medium stepwise, these two gas mixtures were used in combination.

The perfusion medium coming from a thermostated oxygenator (25°C) was fed into the left atrium through an atrial canula. The preload pressure was installed hydrostatically between 12 and 17 cmH$_2$O.

Through a canula placed in the aorta the perfusion medium left the heart by cardiac work. To absorp pressure peaks and to establish diastolic aortic pressure a chamber with a bubble of air was placed directly behind the canula of the aorta. The pressure amplitude could be regulated by the volume of the air bubble within this chamber.

GLOBAL PARAMETERS

Aortic pressure and preload pressure were monitored and documented continuously. The inflow PO$_2$ was measured by use of a PO$_2$-electrode (Clark-Type) placed in the line coming from the oxygenator. Coronary flow was measured intermittandly by weighing the perfusion medium which was collected through a canula placed in one of the two pulmonal arteries. A flow probe (Spectramed SP2202 Statham) was installed in the aortic line to determine the aortic flow. After the experiments the heart performance was calculated from the mean aortic pressure multiplied by the aortic flow.

LOCAL PARAMETERS

Tissue remission spectra were measured with the Erlangen Micro-lightguide spectrophotometer (EMPHO I) (Frank et al. 1989). The tip of the lightguide was placed on the surface of the left ventricle.

At each step of the graded hypoxia 200 to 350 spectra were measured within the wavelength range of 500-630 nm. Data were recorded by use of a computer based (Olivetti 80286) recording system developed by Appelbaum (Appelbaum 1987).

DATA EVALUATION

The evaluation system for the remission spectra developed by Dümmler (Dümmler 1988) performes a mathematical colorimetry by mixture of two known absorption spectra. It is based upon Kubelka and Munk's theory and fits the spectrum measured by use of two reference spectra (Frank et al. 1989).

EXPERIMENTAL PROTOCOL

The experiments were started with normoxic perfusion (PaO$_2$ >600 Torr). During this initial phase preload and afterload were adjusted. The measurements of the local parameters started at the end of this phase when the heart was adapted and warmed up to 25°C. The data gained from the first measurements were used as control for high oxygen supply conditions.

During the experiments the inflow PO$_2$ was decreased stepwise from PaO$_2$ >600 to PaO$_2$ <50 Torr.

At each step we performed measurements of the global parameters such as the left atrial pressure, the aortic pressure, the aortic flow, the coronary flow, the inflow PO$_2$ and the coronary venous PO$_2$.

RESULTS

Three typical groups of tissue remission spectra are depicted in the figures 1-3. These spectra were measured under three different oxygen supply conditions during graded hypoxia. The spectra of fig. 1 were measured during the initial phase with high oxygenated perfusion medium (PaO$_2$ = 635 Torr). Two absorption maxima of oxygenated myoglobin are detectable. At a wavelength of 605 nm where reduced cytochrome aa$_3$ shows a pronounced peak no sign of reduction of cytochrome oxidase was found.

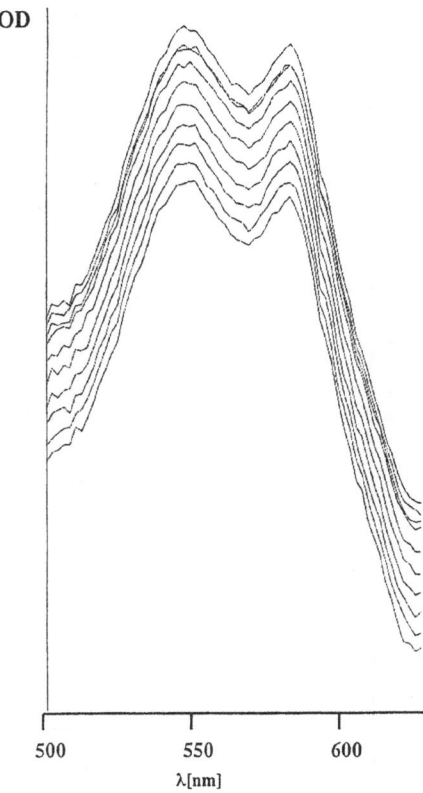

Fig. 1. The spectra were received from the left ventricle during the control phase using high oxygen tension of the perfusion medium (PaO$_2$=635). The absorption peaks of oxygenated myoglobin are easily detectable. There is no peak at the wavelength of 605 nm which shows that cytochrome aa$_3$ is completely oxidized.

The spectra depicted in fig. 2 consist mainly of the absorption spectrum of partly deoxygenated myoglobin.A peak at the wavelength 584 nm shows, that the myoglobin is not completely deoxygenated and indeed the quantitative evaluation results in an oyxgenation value of 6.54 %. A small shoulder at the wavelength 605 nm reveals, that the cytochrom aa3 shows a beginning reduction. Under these conditions the PaO$_2$ was 175 Torr.

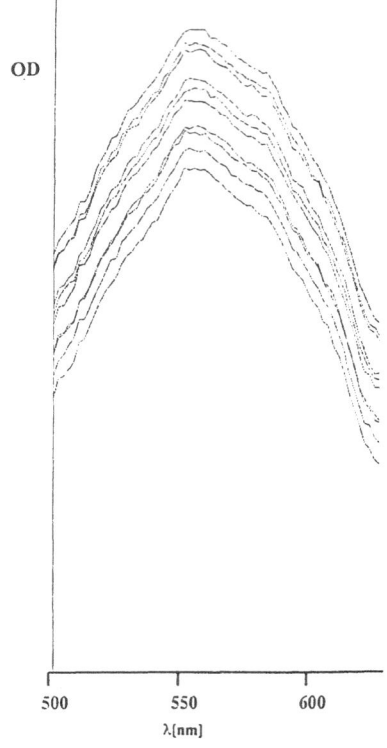

Fig. 2. The spectra were received from the left ventricle during a critical step of graded hypoxia (PaO$_2$=175). The absorption peaks of partly oxygenated myoglobin are detectable. There is a slight shoulder at the wavelength of 605 nm which shows that cytochrome aa3 is partly reduced.

The Spectra shown in fig. 3 were measured during the perfusion with 35 Torr PaO$_2$. The myoglobin is completely deoxygenated and the peak of reduced cytochrome aa3 at the wavelength 605 nm is easily detectable. Despite of this sizeable reduction of Cytochrome aa3 (30 %) which characterizes a partial tissue anoxia, the myocardial performance has not completely ceased but shows a residual value of 14.1 %.

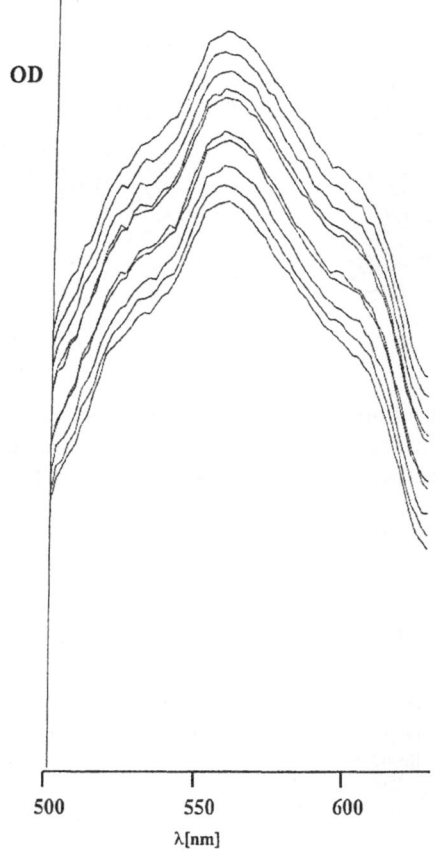

Fig. 3. The spectra were received from the left ventricle during perfusion with an oxygen tension of 35 Torr. In this situation myoglobin is completely deoxygenated and at least a large part of cytochrome aa3 is reduced.

In figure 4 the integrated MbO₂ gradients (Kessler et al. this volume) of myoglobin oxygenation during different steps of graded hypoxia are depicted. From back to front the arterial PO₂ decreased (see table 1). Myoglobin oxygenation starts to fall when oxygen delivery falls below 9.34 mlO₂/100g/min. First zero oxygenation values are detectable when oxygen delivery falls below 1.69 mlO₂/100g/min. In this situation the heart performance reaches 24.5% of control.

The integrated gradients of the redox-state of cytochrome aa₃ during graded hypoxia are

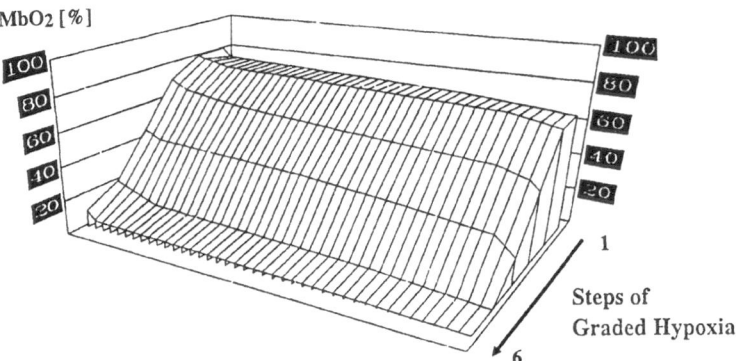

Fig. 4. Integrated gradients of myoglobin oxygenation measured during six different steps of graded hypoxia. Oxygen tension of the perfusion medium was decreased from step 1 to step 6. The numerical values belonging to this figure are listed in table 1.

Tab. 1. Numerical values underlying fig. 4. Additionally the global parameters of the experiment are listed.

Step	PaO₂ [Torr]	O₂-Delivery [mlO₂/100 g/min]	Perform-ance [% of control]	Mean MbO₂ [%]	Min. MbO₂ [%]	Max. MbO₂ [%]	n
1	634	17.15	100	71.08	67.03	79.51	280
2	398	9.34	104.6	72.23	67.19	78.43	284
3	271	5.50	72.6	55.63	42.69	69.97	315
4	214	3.91	52.4	33.65	18.92	42.65	279
5	175	1.69	24.5	6.54	0	15.09	209
6	35	0.26	14.1	1.45	0	9.16	210

shown in fig. 5. The results presented in this figure are based upon the same spectra as plotted in figure 4. Table 2 shows the corresponding numerical values of figure 5. Reduction of cytochrome aa₃ can be detected when oxygen delivery falls below 2 mlO₂/100g/min. At the first four steps of graded hypoxia no reduction of cytochrome aa₃ can be observed. Within the fifth step when cytochrome aa3 becomes partly reduced, the first zero values of myoglobin oxygenation can be measured.

As can bee seen, first values of reduced cytochrome aa3 were recorded when heart performance fell to 24.5 % of control. This result indicates that regulation of performance is

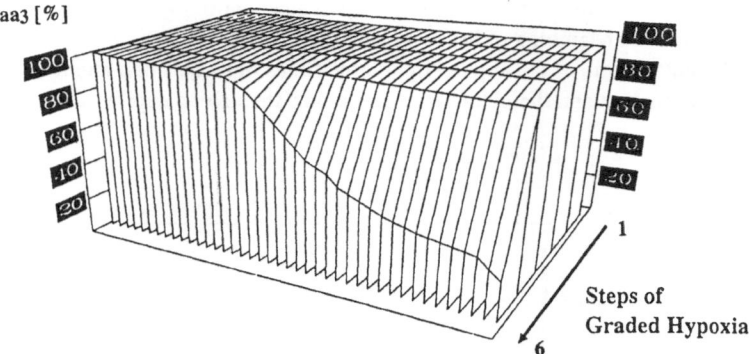

aa3 [%]

Steps of
Graded Hypoxia

Fig. 5. Integrated gradients of the redox-state of cytochrome aa3 measured during six different steps of graded hypoxia. Oxygen tension of the perfusion medium was decreased from step 1 to step 6. The numerical values belonging to this figure are listed in table 2.

Tab. 2. Numerical values underlying fig. 5. Additionally the global parameters of the experiment are listed.

Step	PaO2 [Torr]	O2- Delivery [mlO2/100 g/min]	Performance [% of control]	Mean aa3 [%]	Min. aa3 [%]	Max. aa3 [%]	n
1	634	17.15	100	100	100	100	280
2	398	9.34	104.6	100	100	100	284
3	271	5.50	72.6	100	100	100	315
4	214	3.91	52.4	100	100	100	279
5	175	1.69	24.5	99.93	94.10	100	209
6	35	0.26	14.1	70.10	20.85	100	210

not based upon cellular anoxia. The fact that tissue anoxia appears only at this low value of heart performance gives evidence that the previous decrease might be interpreted as a downregulation of heart function.

DISCUSSION

By use of the EMPHO I it is possible to measure tissue spectra from the isolated, hemoglobin-free perfused, working rat heart with a very high spatial and time resolution.

The spectra can be analyzed in such a way that we can get a valuable information about local myoglobin oxygenation and redox-state of cytochrome aa3.

For an accurate and correct estimation of physiological relationship between these two parameters, two prerequisites are necessary:

1. In order to prevent disturbances in capillary flow an adequate perfusion of all capillaries must be attained. Furthermore it is of utmost importance that blood coagulation of the animal is inhibited before the beginning of heart explantation.

2. The catchment volume of the applied lightguide fibres must be small enough to resolve the myoglobin oxygenation and the cytochrome aa3 redox-state along capillaries. A sensor with a

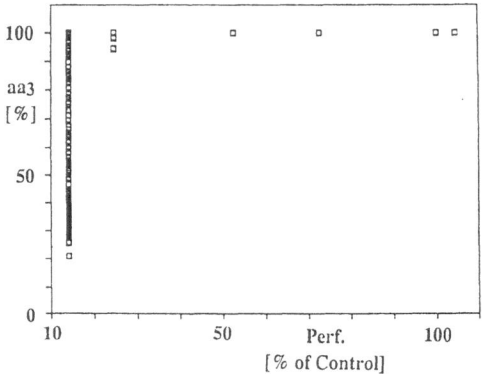

Fig. 6. Relationship between redox-state of cytochrome aa3 and heart performance. The first reduction of cytochrome aa3 can be observed when the heart performance falls below 30 % of control. (n=1577)

too large cachment volume will lead to integrated signals thus mixing tissue values originating from well and badly perfused myocardium. Such an integration of spatially distributed tissue signals will result in an apparent early reduction of respiratory enzymes.

In a perfusion model with homogeneous capillary perfusion myocardial performance can be decreased to less than 50 % without apparence of tissue anoxia. These important results give evidence that a sensor system may exist in the isolated heart able to induce a downregulation of myocardial function when tissue oxygenation falls short off critical pO_2 values of 3 mmHg. As shown in figure 6 a reduction of cytochrome aa3 starts only at myocardial performance below 25 % of control.

How can we explain these results?

Chance (1989) and Lübbers (Starlinger et al. 1973) found that the critical pO_2 of isolated mitochondria at which oxidative phosphorylation starts to decrease lies around 0.01 mmHg. Gayeski (1987) could show that critical pO_2 values for oxidative phosphorylation in myocytes are also very deep (0.5 mmHg).

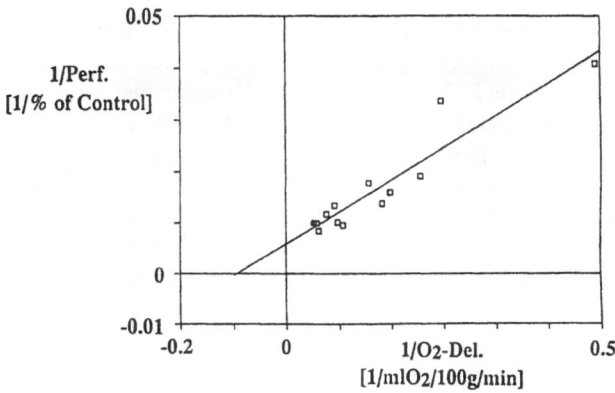

Fig. 7. Relationship between 1/Oxygen delivery and 1/performance. The Km-value of the linear relationship between the two parameters is 0.473 mmolO$_2$/100g/min.

Based on these results we can conclude in our context that by downregulation of myocardial function oxidative phosphorylation can be maintained down to extremely low tissue oxygenation values.

Due to this protective mechanism anaerobic glycolysis only starts when almost all oxygen molecules have been used up. Practically seen this means that myocytes can survive for long time despite a very low capillary flow when their oxygen demand is decreased by hypokinesia to 20 % of control.

However, when anaerobic glycolysis starts because of beginning cellular anoxia the oxidative energy formation decreases from 38 to 2 ATP molecules and thus the chances of survival fall off drastically.

In our experiments the relationship between oxygen delivery and heart performance follows a Michaelis-Menton kinetic with a Km-value of 0.473 mmolO$_2$/100g/min. (fig. 7) indicating that the mechanism of hypokinesia could be induced by an oxygen dependent enzyme which shows typical characteristics.

REFERENCES

Appelbaum K. (1987) Entwicklung eines rechnergestützten Aufnahme- und Auswertesystems für photometrische Meßdaten. Studienarbeit, Tech. Fak. d. Friedrich-Alexander Universität, D-Erlangen -Nürnberg.

Chance B. (1989) Metabolic Heterogeneities in Rapidly Metabolizing Tissues. J. Appl. Card. 4:207-221.

Dümmler W. (1988) Bestimmung von Hämoglobin-Oxygenierung und relativer Hämoglobin- Konzentration in biologischen Systemen durch Auswertung von Remissionsspektren mit Hilfe der Kubelka-Munk-Theorie. Dissertation, Nat. Fak. d. Friedrich- Alexander Universität, D-Erlangen -Nürnberg.

Frank K. H., Kessler M., Appelbaum K., Dümmler W. (1989) The Erlangen Micro-Lightguide Spectrophotometer EMPHO I. Phys. Med. Biol. 34:1883-1900.

Gayeski T.E.J., Connett R.J., Honig C.R. (1987) Minimum Intracellular Po$_2$ for Maximum Cytochrome Turnover in Red Muscle In Situ. Am. J. Physiol. 252:H906-915.

Kessler M., Höper J., Spatial Distribution of Oxygen Supply Units in Heart and Skeletal Muscle and their Regulatory Significance. Adv. Exp. Med. Biol. (this Volume).

Olders J., Boumans T., Evers J., Turek Z. (1990) An Experimental Set-up for the Blood Perfused Working Isolated Rat Heart. Adv. Exp. Med. Biol. 277:151-160.

Olders J., Turek Z., Evers J., Hoofd L., Oeseburg B., Kreuzer F. (1990) Comparison of Tyrode and Blood Perfused Working Rat Hearts. Adv. Exp. Med. Biol. 277:403-413.

Starlinger H., Lübbers D.W. (1973) Polarographic Measurements of the Oxygen Pressure Performed Simultaneously with Optical Measurements of the Redox State of Respiratory Chain in Suspensions of Mitochondria under Steady-State Conditions at Low Oxygen Tensions. Pflügers Arch. 341:15-22.

SPATIAL DISTRIBUTION OF OXYGEN SUPPLY UNITS IN HEART AND SKELETAL MUSCLE AND THEIR REGULATORY SIGNIFICANCE

Kessler, M., Höper, J.

Institut für Physiologie und Kardiologie der Universität
Erlangen-Nürnberg, Waldstraße 6, D-8520 Erlangen

Introduction

Investigations of local pO_2 in skeletal and heart muscle revealed unexpected results. In both organs local pO_2 measurements demonstrated that tissue oxygenation is structured in such a way that 10 - 15 characteristic O_2 supply units do exist (see figures 1 and 2). Each individual supply unit shows a rather homogeneous distribution of the spatial pO_2 profiles (Kessler et al. 1984). In skeletal muscle the size of the individual unit is

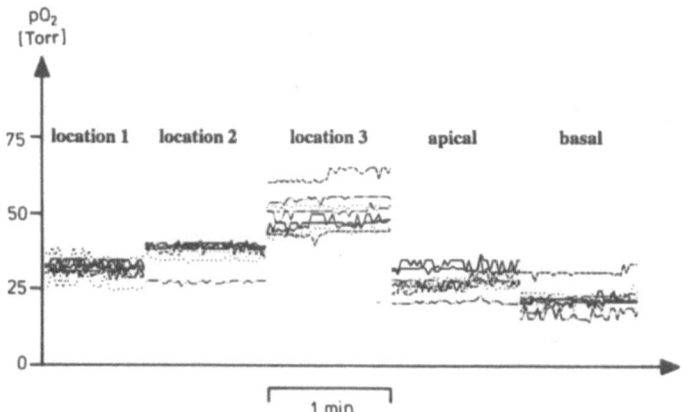

Figure 1. Continuous recordings of local pO_2 in adjacent areas. The measurements were performed in the beating dog heart. Note that there are very distinct differences between the areas, whereas the pO_2 values are very similar within each individual field.

approximately 400 - 600 μm in length with a diameter of approximately 150 - 200 μm. In such functional structures with relatively homogeneous patterns of capillary blood flow large pO_2 gradients between the individual supply units are not induced. This prevents loss of oxygen by diffusion shunt.

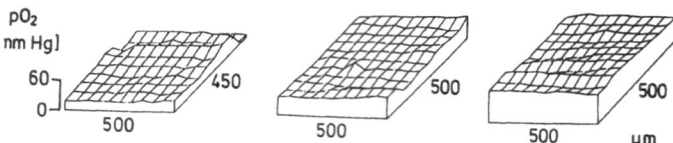

Figure 2. Spatial distribution of local pO$_2$ values measured at the surface of a resting skeletal muscle of a rat. The different networks were measured in adjacent areas. The distance between these areas was approximately 300 to 500 μm (Beier, 1987).

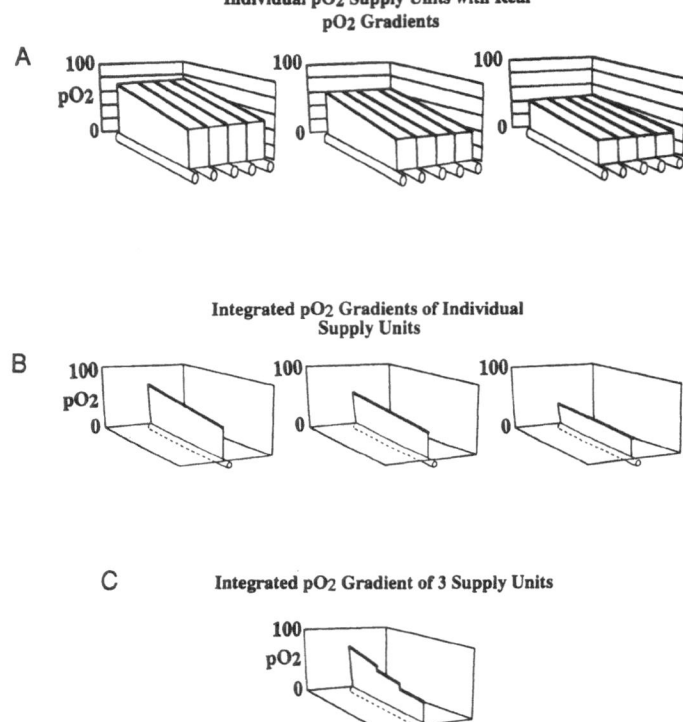

Figure 3. Schematic drawing of 3 individual supply units. Upper part: 3 units with real pO$_2$ gradients are shown. Middle part: Statistical measurements result in the integrated gradients of the individual supply units. Lower part: Statistical measurements in the 3 supply units result in only one integrated gradient.

Results

When pO_2 measurements are performed in the beating dog heart in areas of 3 X 3 mm by use of multiwire Pt-electrodes functional structures characteristic for pO_2 supply units become visible. This is shown in figure 1 where very different pO_2 patterns are found at 5 different locations. Within each individual field of measurement (300 X 300 μm) relatively small pO_2 gradients are found. This is indicative for 5 separate pO_2 supply units.

When the pO_2 network is analyzed by systematic movements of the pO_2 electrode in steps of 50 μm both, different supply units as well as small pO_2 gradients within the individual unit become apparent (Fig. 2).

When identical measurements of intracapillary HbO_2 were performed by moving micro-lightguides in steps of 50 μm similar structures were found. In figures 3 and 4 typical HbO_2 and pO_2 supply units representing the early warning and the high flow units are depicted (A). The HbO_2 and pO_2 unit plotted between the two units with lowest and highest oxygenation values lies somewhere in between the extremes and thus characterizes the mean tissue oxygenation.

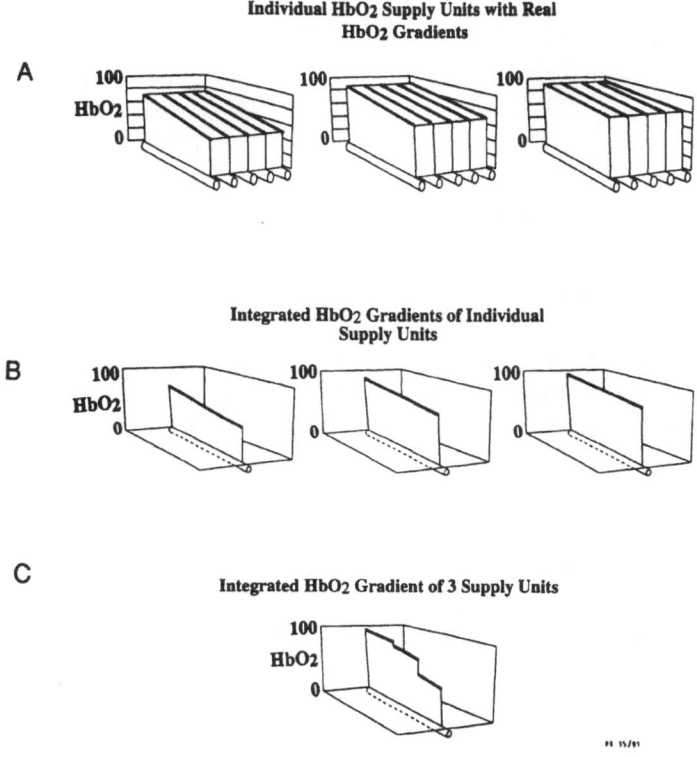

Figure 4: Schematic drawing of 3 individual supply units. In this figure the intracapillary HbO_2 gradient is shown. By use of a precise scanning technique real intracapillary gradients can be determined (A). Statistical measurements result in the determination of integrated gradients of the individual supply units (B). Mixing the statistical values of 3 supply units results in only one integrated gradient (C).

When systematic monitoring of tissue oxygenation is performed four principle types of local measurements can be applied:

1. Recording of the real gradients in individual HbO₂ and pO₂ supply units by moving single sensors in small steps or by application of light fibre or electrode arrays (A).
2. Recording of integrated HbO₂ and pO₂ gradients within an individual supply unit by statistical measurements at approximately 300 different locations (B).
3. Recording of integrated HbO₂ and pO₂ gradients of 3 or more supply units within a representative regional area(C).
4. Interregional Recording of tissue oxygenation in a larger number of regional areas.

Typical examples of interregional monitoring in the beating heart are depicted in figures 5 and 6.

Fig. 5 shows an interregional field of 12 integrated HbO₂ gradients which were recorded in the beating human myocardium during open heart surgery under control conditions.

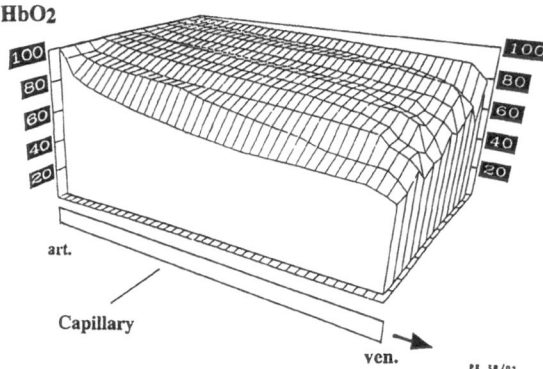

Figure 5. Interregional field of integrated HbO₂ gradients. Each curve represents the integrated gradient of a representative number of individual supply units. The measurements were performed in the beating human heart during surgery.

An interregional field of 24 integrated pO₂ gradients measured in the beating dog heart are plotted in figure 6.

Both figures document the regional and interregional heterogeneity of tissue oxygenation.

Discussion

The normal physiological function of an organ depends on a sufficient O₂-supply to all its cells. Among other factors this is determined by capillary flow. Investigations of capillary blood flow revealed a heterogeneous distribution pattern. The exact analysis of this flow pattern shows that it is primarily induced by the heterogeneous perfusion rate through terminal arterioles supplying blood to a certain number of capillaries. Thus different individual capillary supply units exist within tissue. The blood flow to these individual supply units is regulated in such a way that under physiological conditions all cells are sufficiently supplied with oxygen. Under conditions of increased O_2-uptake or hypoxic hypoxemia typical changes of these patterns can be observed.

For the interpretation of these results it is necessary to understand the meaning of local measurements. In figures 3 and 4 the principle is explained. Using a precise scanning technique it is possible to measure the true pO_2-gradients in tissue caused by single capillaries. The same yields for spectrophotemetric measurements of HbO_2-gradients along a capillary.

Statistical measurements with such a technique will result in the comprehension of an integrated gradient of three or more supply units (see fig.3).

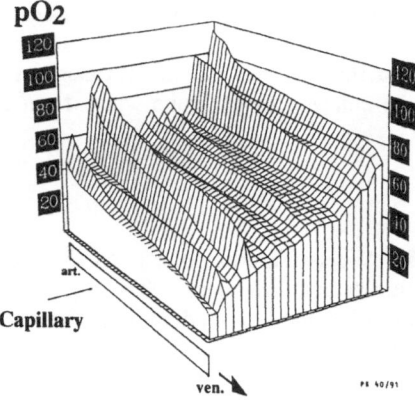

Figure 6.   Interregional field of integrated pO_2 gradients. The measurements were performed in the beating dog heart. As explained in figure 4 each curve represents the integrated pO_2 gradient of a representative number of individual supply units.

If a series of measurements is performed in adjacent areas an interregional field of integrated pO_2 or HbO_2 gradients can be constructed (Zündorf 1992).

The results obtained in individual supply units indicate that the distribution of local O_2 supply within an individual supply unit is regulated in such a way, that the gradients along the different capillaries in each unit are very similar. The highest values at the arterial end of the capillaries range from 100% to 70% HbO_2, the lowest values at the venous end from 80% to 50% HbO_2. Supply units with high flow channels show high arterial and venous saturation and have a low vascular resistance.

An increase in local oxygen consumption or a hypoxic hypoxemia leads to a situation in which the intracellular pO_2 falls below a critical threshold in units with low HbO_2 and pO_2 at the venous end of capillaries. Its absolute value characterizes the pO_2 level below which regulatory responses are induced. From our data we can conclude that these units serve as "early warning" supply units influencing the distribution of blood flow between adjacent terminal arterioles. Such a regulation requires different prerequisites:

1. The cells must have O_2-dependent "sensors", so called signal oxidases.

2. Cellular signal chains must exist between the cells at the venous capillary end and the smooth muscle of terminal arterioles.

Earlier investigations have shown that MAO-B (Höper and Kessler 1981) is the responsible enzyme of a signal oxidase used by the cell as oxygen sensor. The signal generated seems to be transferred in antegrade direction via the capillary endothelium to the arterioles. Measurements of the change in the pattern of capillary flow indicate that in high-flow channels a vasoconstriction occurs whereas the arterioles supplying the normal-flow units dilate. Because its unlikely that different signals or transmitters are generated the two types of arterioles must have different receptors.
The principle of regulation by signal oxidases under conditions of critical supply is such that the capillary flow through high flow supply units is decreased at the benefit of the low flow supply units.
These considerations lead to the concept that the autonomous local regulation depends upon the interaction between adjacent supply units which form a cooperative group.

References

Beier, I. Die Verteilung des Sauerstoffpartialdruckes an der Oberfläche des Musculus gracilis der Ratte. Dissertation, Erlangen, 1987.

Höper, J., Kessler, M. pO_2 and sodium dependent mechanism regulating liver blood flow. In: Oxygen Transport to Tissue. Ed. A.G.B. Kovach, E. Dora, M. Kessler, I.A. Silver. Advances in Physiological Science, vol. 25, Akademiai Kiado-Pergamon, Budapest-London, 1981, 163-164.

Kessler, M., Höper, J., Harrison, D.K., Skolasinska, K., Klövekorn, W.P., Sebening, F., Volkholz, H.J., Beier, I., Kernbach, C., Rettig, V., Richter, H. Tissue O_2 supply under normal and pathological conditions. In: Oxygen Transport to Tissue-V. Ed. D.W. Lübbers, H. Acker, E. Leniger-Follert, T.K. Goldstick, Plenum 1984, 69-80.

Zündorf, J. Intraoperative Untersuchungen zur intrakapillären Sauerstoffversorgung am Myokard des Koronarpatienten und experimentelle Untersuchungen zur zellulären Sauerstoffversorgung am isoliert arbeitenden Rattenherzen. Dissertation, Medizinische Fakultät Erlangen, 1992.

EFFECTS OF CPPV, PC-IRV, AND LFPPV-ECCO2R ON RIGHT VENTRICULAR FUNCTIONS IN PIGS WITH ARDS

L. Telci[1], J. Kesecioglu[1,2], F. Esen[1], T. Denkel[1], A. S. Tütüncü[1,2], K. Akpir[1], B. Lachmann[2]

[1]Department of Anesthesiology, University of Istanbul, Faculty of Medicine, Istanbul, Turkey
[2]Department of Anesthesiology, Erasmus University and Academic Hospital Dijkzigt, Rotterdam, The Netherlands

INTRODUCTION

Adult respiratory distress syndrome (ARDS), which is characterized by pronounced alveolar collapse and edema with ensuing hypoxia, is frequently associated with pulmonary hypertension. The sensitivity of the right ventricle to an acute increase in pulmonary arterial pressure is particularly relevant in critical states associated with ARDS [1].

Several new modes of ventilatory support in the treatment of ARDS have been shown to improve oxygenation in experimental and clinical studies, and their effects on cardiac functions have also been evaluated. However, the effects of mechanical ventilation on right ventricular function are quite complex since they can involve both a decrease in preload and an increase in afterload [2,3].

It generally was believed that increased intrathoracic pressure generated by mechanical ventilation hinders venous return and thus reduces preload [4]. However, recent observations have suggested alternative explanations, such as decreased compliance of the ventricles with the application of positive pressure and positive end-expiratory pressure (PEEP) [4,5]. In this experimental study, we aimed to evaluate the effects of different modes of ventilation on right ventricular function in pigs with ARDS. These modes included continuous positive pressure ventilation (CPPV), pressure controlled inverse ratio ventilation (PC-IRV), and low frequency positive pressure ventilation with extracorporeal CO_2 removal (LFPPV-ECCO2R).

RESULTS

The mean values of respiratory parameters, PaO_2, $PaCO_2$, mean airway pressure (Mean Paw), and peak airway pressure (Peak Paw) measured during the administration of each mode along with the baseline measurements are listed in Table 1. As shown in the data, the best gas exchange was seen to be associated with LFPPV-ECCO2R. In the inverse ratio ventilation, with slightly lower peak airway pressure, PaO_2 reached higher levels compared to CPPV. However, there were no statistically significant differences between M3, and M4 concerning PaO_2.

Hemodynamic variables during each mode trial are compared in Table 2, and the changes in RVEF and end-diastolic volume (EDV) are depicted in Figure 1. As shown in these data, there was a significant decrease in RVEF with an increase in mean pulmonary arterial pressure (MPAP) after the induction of ARDS. In M3 setting, high values of PEEP were accompanied by a reduction in CO, MAP, and EDV. However, there was no significant change in RVEF and heart rate (HR) when PEEP was increased from 2 to 10 cmH_2O.

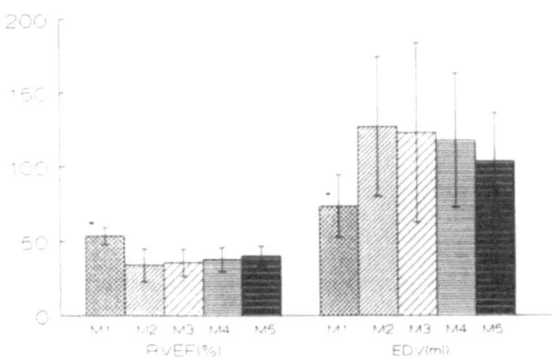

Figure 1. Right ventricle ejection fraction (RVEF), end diastolic volume (EDV) values during each mode trial. (* = p< 0.05)

In the pressure controlled mode with 80% inspiratory time, significant increase in mean Paw did not cause the expected reduction in CO compared with M3. No further increase in MPAP was observed. A slight decrease in EDV with no significant change in RVEF was noted.

During LFPPV-ECCO$_2$R, an average PEEP of 19.3 ± 6.1 cmH_2O was applied. Compared to the modes M3 and M4, no significsnt changes were observed in MAP and CO. There was a slight but nonsignificant reduction in EDV and increase in RVEF.

METHODS

Eleven male pigs 42.36 ± 5.76 kg (range, 35-50kg) were used in the study. Pigs were premedicated with midazolam (0.5 mg/kg) intramuscularly, and anesthesia was induced with thiopentone (2-4 mg/kg) through a 20 G cannula placed into an ear vein. Tracheostomy was performed and a portex tube with an internal diameter of 7 mm was inserted. Lungs were ventilated with a Servo 900C (Siemens-Elema) thereafter. Anesthesia was maintained by infusion of midazolam (0.2 mg/kg/min) and fentanyl (2 μgr/kg/min). Pancuronium bromide (0.08 mg/kg/min) was given for muscle relaxation.

A triple lumen catheter (Abbott Critical Systems, 7F) and a modified pulmonary arterial catheter (Swan Ganz catheter 93A-431H, 7.5F, Edwards Lab.) were inserted to the right jugular and femoral veins, respectively. Two cannula (Cook, 20F) were inserted to the left internal jugular and femoral veins for the administration of extracorporeal circulation. Femoral artery was cannulated for blood sampling, and invasive blood pressure monitoring. An 18F Foley catheter was placed into the bladder by systostomy.

All pressure monitoring was done by Viggo Spectramed transducers on a Horizon 2000 (Mennen Medical). Cardiac output (CO), and right ventricle ejection fraction (RVEF) were measured by thermodilution technique using a Baxter REF1 computer (Edwards Lab.). Arterial and mixed venous blood gases were determined by ABL300 (Radiometer, Copenhagen). Centrifugal pump (Biomedicus), and two membrane lungs with a total surface area of 7 m^2 (Sci-med) were used for extracorporeal CO_2 removal.

A wash-out technique was used to establish a model of ARDS which is reported to be rather stable, so that different modes can be evaluated in a single animal [6]. After all the measurements were completed and recorded as control values, lung lavage was perfMwmed with 2 l of warm saline solution to induce ARDS. PaO_2 below 100 mmHg was accepted as ARDS.

The modes of ventilation were applied in the following protocol:
Control mode (M1) : IPPV with PEEP of 2 cmH_2O, tidal volume (Vt) of 10 ml/kg, frequency (f) of 12/min, I:E ratio of 1:2.
M2   : Same as M1 after lung lavage.
M3   : CPPV with PEEP of 10 cmH_2O, Vt of 10 ml/kg , f of 12/min, I:E ratio of 1:2.
M4   : PC-IRV with PEEP of 4 cmH_2O, f of 12/min, I:E ratio of 4:1.
M5   : LFPPV with a f of 5/min, PEEP of 15-25 cmH_2O, $ECCO_2R$ with a pump speed of 20% of CO.

FiO_2 was 1.0 in all modes. The pigs were treated for 4 hours with each mode, and the measurements were made thereafter. ARDS was reconfirmed by switching back to M2 before the next mode was applied in a random order.

All data are presented as mean ± SD and values were analyzed statistically by using the Wilcoxon pairs test. $p < 0.05$ was accepted as statistically significant.

DISCUSSION

Artificial ventilation by the application of positive airway pressure and the addition of supplemental inspired O_2 represents the mainstays of support in ARDS.

Table 1. Respiratory parameters before lung lavage (M1) and during each mode.

	M1	M2	M3	M4	M5
PaO_2 (mmHg)	427.1 ± 104.3	72.0 ± 25.4^a	210 ± 93.1	252.7 ± 58.2	290.8 ± 79.2^c
$PaCO_2$ (mmHg)	30.3 ± 7.1	47.1 ± 14.7^b	42.1 ± 10.7	36.2 ± 8.2	29.3 ± 7.4
Peak paw (cmH_2O)	17.7 ± 4.6	31.1 ± 9.2	37.4 ± 8.7^d	32.3 ± 7.4	31.4 ± 6.4
Mean paw (cmH_2O)	5.9 ± 2.0	9.2 ± 3.7	16.7 ± 3.8	25.4 ± 7.0^c	19.2 ± 3.9

a = significantly different from M1, M3, M4 and M5
b = significantly different from M4 and M5
c = significantly different from M3 and M5
d = significantly different from M1, M2, M4, and M5
e = significantly different from M2, M3, and M4

Table 2. Hemodynamic parameters before lung lavage (M1) and during each mode.

	M1	M2	M3	M4	M5
CO (l/min)	5.5 ± 1.6^a	5.0 ± 1.6	4.7 ± 1.9	4.6 ± 1.5	4.9 ± 1.4
MAP (mmHg)	142.0 ± 15.6	133.5 ± 21.6^b	125.4 ± 19.1	116.9 ± 21.5	121.0 ± 15.4
MPAP (mmHg)	25.8 ± 8.4	35.0 ± 8.1	37.1 ± 10.0	35.3 ± 9.1	34.8 ± 8.1
PCWP (mmHg)	11.3 ± 4.2	10.5 ± 5.4^b	13.8 ± 4.2	15.4 ± 5.8	13.7 ± 3.8

a = significant difference between M1, M3, M4 and M5 ($p < 0.05$)
b = significant difference between M3, M4 and M5 ($p < 0.05$)
c = significant difference between M2, M3, M4 and M5 ($p < 0.05$)

However, positive pressure ventilation results in cardiovascular interactions that may impede blood flow, thus decreasing O_2 delivery to the tissue. The effects of mechanical ventilation on the left ventricle has long been considered, nevertheless right ventricular function during the application of positive pressure ventilation has been a subject of growing interest in recent years [7].

The thermodilution technique has become routine in the evaluation of right ventricular function at the bedside of the critically ill [8]. The reliability of this method of measuring RVEF and EDV has been proved by several experimental and clinical studies. In this study we also used a thermodilution technique to evaluate the effects of different ventilatory support modes on right ventricle function.

As mentioned earlier, ARDS is frequently associated with pulmonary hypertension, thus increasing right ventricle afterload. Several studies have shown a correlation between the mean pulmonary arterial pressure, and the RVEF; suggesting that RVEF is largely dependent on right ventricle afterload. Parallel to that, we observed a significant reduction in RVEF after the induction of ARDS.

By the administration of 10 cmH$_2$O of PEEP in M3, no change in RVEF was observed. The decrease in EDV could be due to a reduced ventricular filling induced by restricting venous return to the right heart. That right ventricle function is not impaired with high levels of PEEP was confirmed by other clinical studies [9,10]. Martin and coworkers showed no impairment of RVEF with high levels of PEEP [2].

No previous work has described changes in right ventricular performance during PC-IRV and LFPPV-ECCO$_2$R. Most studies reported little influence of IRV on cardiac functions when the I:E ratio is less than 4:1 [11,12]. However, Cole and colleagues demonstrated a decrease in cardiac output and O_2 transport [13]. In our study expected improvement in oxygenation was achieved with higher mean airway pressures in IRV with an I/E ratio of 4:1. Insignificant decrease in systemic blood pressure and CO was noted with no negative changes in RVEF. Although pulmonary blood pressure and vascular resistance increased, the decrease noted in EDV was accepted as the predominating effect of the reduction in preload caused by the increase in mean airway pressure.

LFPPV-ECCO$_2$R is a non-conventional technique for ventilatory support, representing a new approach to the management of ARDS. Predictions regarding the potential utility of this technique are encouraging, since many clinical and experimental studies support the progressive improvement in gas exchange and the decrease in the mortality rate of ARDS [14]. In our previous experimental work, we have shown the positive effects of this non-conventional technique over others [15]. The results of this study support the data of our previous work, since better gas exchange was observed with no cardiocirculatory depression. Mean airway pressure, approximating the applied PEEP level, showed no impairment of right ventricle function.

According to the results, we conclude that continuous positive pressure ventilation with high levels of PEEP, IRV with 80% inspiratory time, and LFPPV-ECCO$_2$R technique can be used in ARDS to improve oxygenation without expecting any adverse effects on right ventricular performance and cardiocirculatory stability.

REFERENCES

1. W. J. Sibbald, A. A. Driedger, D. G. Cunningham, and H. Cheung, Right and left ventricular performance in acute hypoxemic respiratory failure, Crit. Care Med. 14: 852-857 (1986).
2. C. Martin, P. Saux, J. Albarase, J. J. Bonneni, and F. Gouin, Right ventricular function during positive end-expiratory pressure: Thermodilution evaluation and clinical application, Chest . 92:99-1004 (1987).
3. D. S. Schulman, J. W. Brondi, R. A. Matthay, P. G. Barash, B. L. Zancet, and R. Soufer, Effect of positive end-expiratory pressure on right ventricular performance: Importance of baseline right ventricular function, Am. J. Med. 84:57-67 (1987).
4. J. Manny, G. Grindlinger, A. A. Mathe, and H. B. Hechtman, Positive end-expiratory pressure, lung stretch and decreased myocardial contractility, Surgery. 84:127-133 (1978).
5. S. R. Powers, and R. F. Dutton, Correlation of positive end-expiratory pressure with cardiovascular performance, Crit. Care Med. 3:64-68 (1975).
6. B. Lachmann, B. Robertson, and J. Vogel, In-vivo lung lavage as an experimental model of the respiratory distress syndrome, Acta Anaesth. Scand. 24:231 (1980).
7. J. L. Vincent, and A. Lenaers, Right heart function and its evaluation, Perspectives In Critical Care. Volume 2, 1:141-154 (1989).
8. J. L. Vincent, Measurement of ventricular ejection fraction in the acutely ill, Br. J. Anaesth. 60:1135-1155 (1988).
9. W. P. Santamore, A. A. Boye, and J. L. Heckman, Right and left ventricular pressure-volume response to positive end-expiratory pressure, Am. J. Physiol. 246:114-119 (1984).
10. J. J. Marini, B. H. Culver, and J. Butler, Effect of positive end-expiratory pressure on canine ventricular function curves, J. Appl.Physiol. 51:1367-1374 (1981).
11. E. Abraham, and G. Yoshihara, Cardiorespiratory effects of pressure controlled inverse ratio ventilation in severe respiratory failure, Chest . 96:1356-1359 (1989).
12. L. Gattinoni, R. Marcolin, M. L. Caspani et al, Constant mean airway pressure with different patterns of positive pressure breathing during the adult respiratory distress syndrome, Clin. Physiol. 21:275-279 (1985).
13. A. G. H. Cole, S. F. Wellerf, and M. K. Sykes, Inverse ratio ventilation compared with PEEP in adult respiratory failure, Intensive Care Med. 10:227-232 (1984).
14. L. Gattinoni, A. Pesenti, R. Marcolin et al, Extracorporeal support in acute respiratory failure. Intensive Care World. 5:42 (1988) .
15. J. Kesecioglu, K. Akpir, L. Telci, F. Esen, and T. Denkel, Effects of LFPPV-ECCO$_2$R, CPPV, and PC-IRV on respiratory and cardiocirculatory parameters in ARDS, Intensive Care Med. 16:S41 (1990).

OXYGEN SUPPLY OF THE TISSUE

FACTORS THAT DETERMINE THE OXYGEN SUPPLY OF THE CELL AND THEIR POSSIBLE DISRUPTION

W. Erdmann, M. Fennema and R. van Kesteren

Dept. of Anesthesiology, Erasmus University, Rotterdam
The Netherlands

INTRODUCTION

Oxygen transport to tissue and cell occurs in three major steps:
1. oxygen uptake in the lung;
2. oxygen transport in blood;
3. diffusion of oxygen from the capillaries, through the tissue and into the cell.

When pulmonary gas exchange, cardiac output and oxygen transport capacity (oxygen flux) are within normal range, oxygen supply to the tissue is furthermore dependent on the following microphysiological parameters:
1. distribution/perfusion ratio of the capillary meshwork;
2. oxygen consumption of the cell;
3. oxygen diffusion parameters from capillaries to tissue, through the tissue and across the cell membrane into the cell.

Normally, capillary perfusion and oxygen consumption are in balance and oxygen supply to the cell is continuously autoregulated to its needs. This autoregulation occurs through several mechanisms:
1. redistribution of capillary perfusion with decreasing and increasing shunt;
2. change of perfusion pressure due to precapillary vasodilation respectively constriction and/or blood pressure changes;
3. probably, changes of diffusion resistance in the tissue and across the cell membrane;
4. changes in viscosity concomitant with changes in oxygen transport capacity (erythrocytes);
5. change in oxygen extraction ratio.

However, many pathophysiological conditions can severely interfere with this balance (e.g. in sepsis, sickle cell crisis, metabolic acidosis and alkalosis) independent from disturbances of oxygen uptake and transport. This can result in an imbalance of oxygen consumption and oxygen delivery which then leads to a decrease in oxygen consumption, the only parameter in clinical practise that can be measured on-line.

Excluding decrease of oxygen flux to the tissue, the following parameters might be impaired:

1. distribution/perfusion ratio of the capillary meshwork;
2. perfusion of capillaries (e.g. compression by edema, occlusion by stiff erythrocytes);
3. oxygen diffusion across the capillary membrane, through the tissue and across the cell membrane;
4. oxygen binding to hemoglobin.

All the above-mentioned factors lead to impairment of cellular oxygen supply or, in other words, to disbalance of supply/demand ratio of the cell. The available free tissue oxygen disappears in as short a time as 2-8 seconds in the brain, the most sensitive organ for damage due to failure of adequate oxygen supply. After 12 seconds, deep coma occurs and the clinically measurable electrocorticogram activity falls to zero after 30-40 seconds. After four minutes, enough time has already passed to produce histologically demonstrable necrosis of the brain parenchyma, especially in the cortex. As far as we know at the moment, the human brain can not survive total ischemia of more than 10 minutes. It has, however, been shown in cats that the brain can, in some cases, be revived even after a total circulatory arrest of more than 30 minutes when oxygen supply to tissue is fully re-established.

Oxygen supply to the tissue and its relationship to cell metabolism and cell function has always been of major interest to medical scientists, physiologists, biochemists, pharmacologists, morphologists, pathophysiologists and bioengineers. Experimental studies have been performed in recent years on microphysiological assessment of cellular PO_2 and cell function. These studies were performed with microelectrodes working according to the polarographic principle, "Clark-type" microelectrodes. The electrodes were arranged as single electrodes (Fig. 1) or as multi-electrode systems (Fig. 2) with seven simultaneously measuring electrode tips hexagonally equally distributed in a tissue area, thus simultaneously recording from different locations (supply conditions) in the intercapillary meshwork (Erdmann et al., 1973, Fennema and Erdmann, 1985).

For registration of spontaneous pacemaker activity (neurogenic cell function) a series resistor is introduced into the electronic circuit (Fig. 3). The resulting voltage division between the variable PO_2 dependent resistance of the electrode surface and the fixed series resistor over which action potentials are recorded leads to changes of the polarisation voltage available across the noble metal/medium interphase (Helmholtz double layer). This problem is avoided by introduction of an electronic feed-back circuit (Fig. 4) maintaining the polarisation voltage constant at a set value (Kunke, Erdmann and Metzger, 1972).

Acute hypoxic hypoxia induced by 100% nitrogen respiration (gerbil) for 1 minute leads to an acute decrease of brain cortex tissue (Fig. 5). PO_2 drops in less than 5 seconds to nearly zero, followed by a breakdown of spontaneous pacemaker activity (as a sign of cellular function) with a delay of up to 50 seconds. Reoxygenation is followed by a hyperoxic phase and an immediate recovery of action potential rate with a slight overshoot preceding recovery of oxygenation to prehypoxic values.

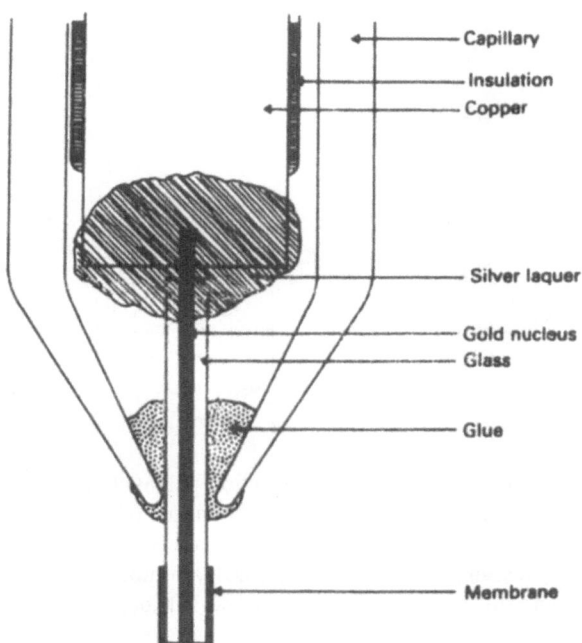

Fig. 1.   Oxygen microlectrode.

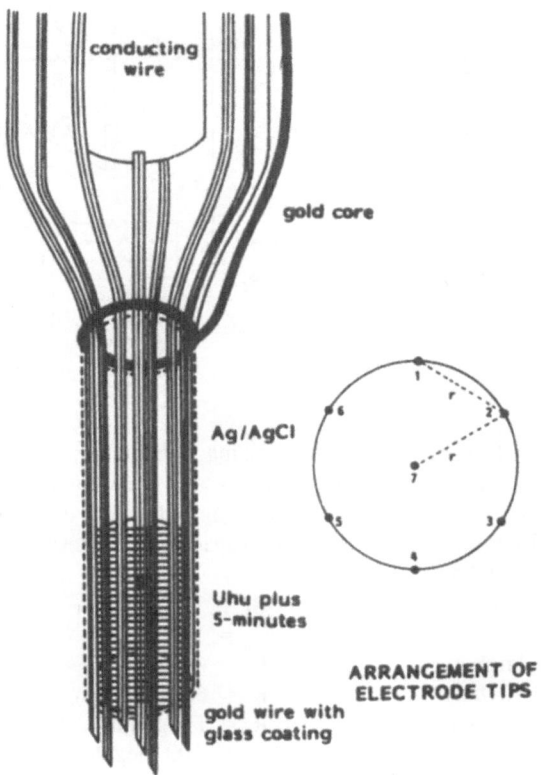

Fig. 2.   Diagram of multi-microelectrodes.

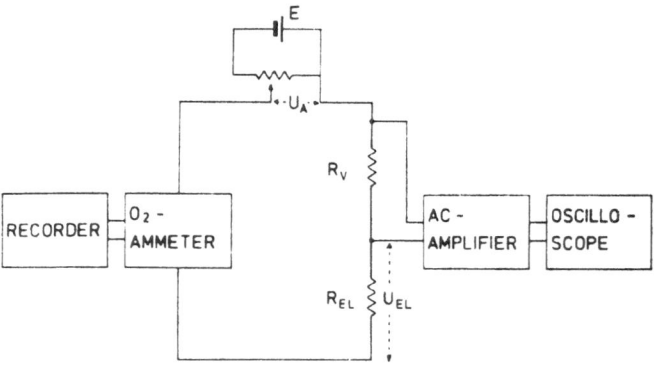

Fig. 3.   Electronic circuit for simultaneous action potential and PO_2 measurements. Polarization voltage U_{EL} is generated by means of a battery and a potentiometer. Series resistor R_V for AP measurement is connected in series to electrode resistance R_{EL}. R_{EL} is defined as quotient between polarization voltage and diffusion limiting current and PO_2 dependent.

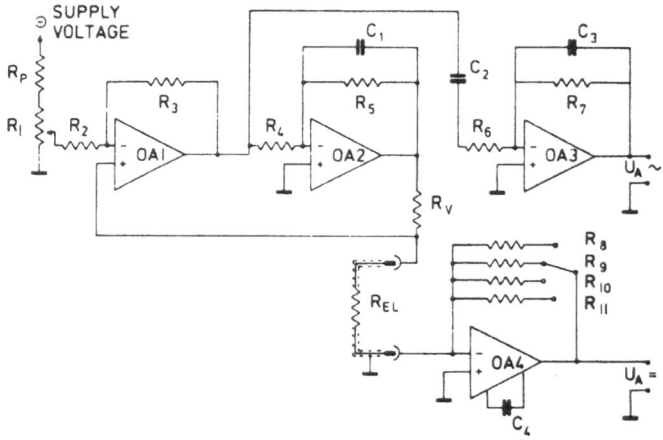

Fig. 4.   Electronic feedback circuit maintaining polarization voltage constant at a desired value. Polarization voltage is measured by operational amplifier OA1. OA2: inverting amplifier; OA3: amplifier for AC-amplification; OA4: O_2-amplifier. List of resistance and capacitance values:

R_P = 4.7 kilohm	R_6 = 10 kilohm	C_1 = 100 nf
R_1 = 500 ohm	R_7 = 1.2 megohm	C_2 = 220 nf
R_2 = 10 kilohm	R_8 = 10 megohm	C_3 = 8 pf
R_3 = 50 kilohm	R_9 = 80 megohm	C_4 = 1 nf
R_4 = 10 kilohm	R_{10} = 680 megohm	
R_5 = 1 megohm	R_{11} = 1 gigohm	
	R_V = 100 megohm	

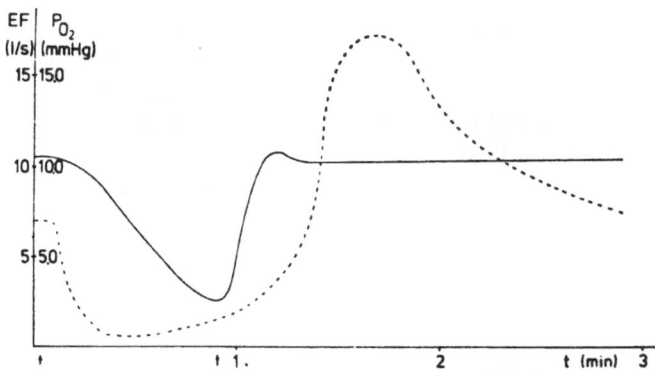

Fig. 5.   Simultaneous registration of local PO_2 (continuous line) and AP rate (dotted line) in response to a 1-min nitrogen respiration. The measurements were performed in occipital rat cortex by means of a gold microelectrode. The figure shows a delayed reaction of AP-rate and overshoot-undershoot phenomena in the time course of PO_2 and AP-rate. Changing points of inspiratory gas are marked by arrows. AP-rate was obtained by counting the number of spikes for intervals of 10 sec. A mean curve was plotted through single values of AP-rate.

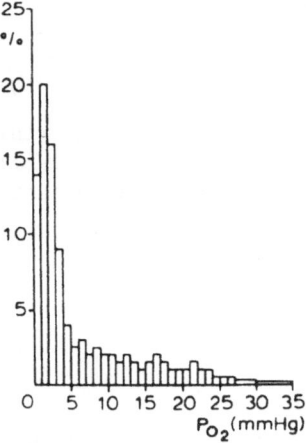

Fig. 6a.   Frequency distribution of PO_2 values in the rat brain cortex. More than 50% of the tissue has PO_2 values below 5 mm Hg.

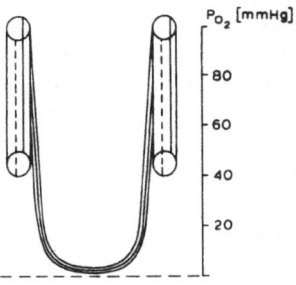

Fig. 6b. PO_2 distribution in brain tissue transferred from our results into the Krogh capillary cylinder model: there are steep gradients at the side of the capillary and nearly level gradients around 3-5 mm Hg PO_2 in the middle of the intercapillary meshwork.

Active neurons are formed in well supplied areas with tissue PO$_2$ values of 30 mm Hg or more, as well as areas where cells have to survive with tissue PO$_2$ values of around 5 mm Hg, or maybe even less (Fig 6a, 6b). The question now arises whether under conditions of impaired O$_2$ supply the worst supplied areas are the first to fail in function or not (Erdmann, 1976).

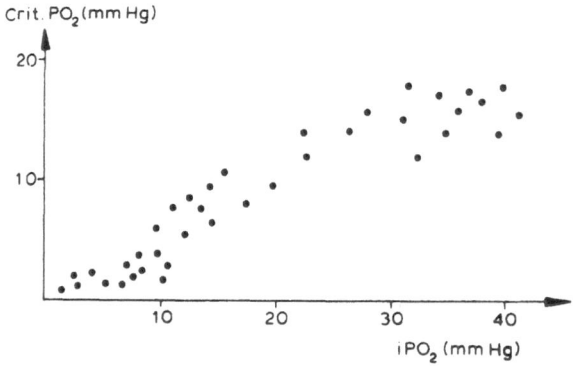

Fig. 7.   Critical PO$_2$ values for neuronal function: cells usually functioning in a high PO$_2$ environment cease cell function at 15-20 mm Hg; other cells can stand PO$_2$ values of 2 mm Hg, remaining fully and normally active (environmental adaption of neuronal cells).

Graded hypoxia showed that the tissue critical PO$_2$ for active pacemaker function of neurons has a broad range between 1 and 20 mm Hg PO$_2$ which is closely related to the normal supply conditions of a cell (Fig 7). Thus, there exists something like environmental adaptation of cells. The same studies revealed that relative drop of tissue PO$_2$ during hypoxic hypoxia was much less in the already low tissue PO$_2$ areas (lethal corners?) than in the primary well supplied areas, and cell function in the basically well supplied areas was forced to cease much earlier than in the low-PO$_2$ adjusted areas. This, however, was only the case when hypoxia significantly increased tissue perfusion (measured with H$_2$ clearance). On the other hand, decrease of hematocrit (Ht) is fully autoregulated down to 20%, and followed by a steep decrease of tissue PO$_2$ below a Ht of 20% (Fig 8); however, the tissue PO$_2$ decrease due to failing oxygen transport capacity of the blood is percentage wise much more prominent in the originally badly supplied tissue areas, the same has been found in low perfusion state (Erdmann, 1976).

The question now arises as to how environmental adaptation of the spontaneously active neuronal cells to the O$_2$-supply conditions is achieved. This becomes increasingly interesting as from clinical conditions we know that children with congenital cardiovas-

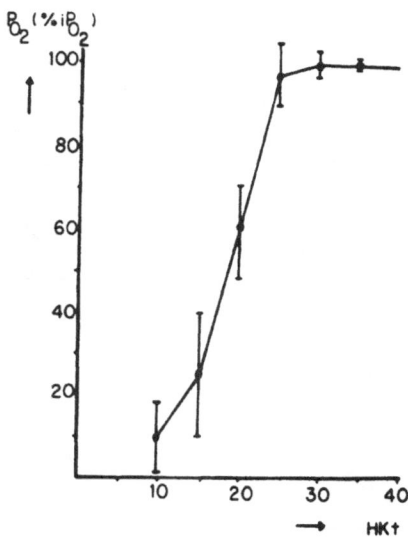

Fig. 8.   Decrease of hematocrit and tissue PO$_2$ autoregulation.

Conclusion 1: the neurons are highly sensitive to oxygen supply failure. Active neurons are equally distributed over the intercapillary tissue and present in well supplied areas as well as badly supplied areas. They show environmental adaptation and tissue PO$_2$ values show a broad range between 5 to more than 40 mm Hg. The critical PO$_2$ below which active pacemaker function ceases is not a fixed value but shows a range from < 1 mm Hg to nearly 20 mm Hg PO$_2$ and is closely related to the basic PO$_2$ level with which a cell has to get around with (environmental adaptation). Hypoxic hypoxia is more dangerous for normally well supplied areas as compared to the less fortunate areas if tissue autoregulation (reactive increase of perfusion) is intact. Hypoxic effects due to decrease in oxygen transport capacity is more vulnerable to the under normal conditions marginally supplied areas as is hypoxia due to vasoconstriction respectively decrease of capillary perfusion in response to decrease of the perfusion pressure gradient (ischemic hypoxia).

cular abnormalities can function intelligence wise quite adequately with arterial PO$_2$ values that would lead to deep coma and subsequently brain death in the normal human. The same stands for the elderly with, for instance, slowly developing lung emphysema and a gradual decrease of arterial oxygenation. Behavioral studies in rabbits exposed to slowly and repeatedly decreasing hypobaric pressures (comparable to high altitude living) have indicated that secondary adaptation to hypoxia is possible with less loss of neuronal capability as compared to the acute exposure. To gain some insight into the overall aspect of this question investigators had to change to single cell neuron studies. A quite feasible object are the pacemaker neurons that can be easily prepared free out of the abdominal ganglion of the Californian sea snail (aplysia californensis). The cells are fixed onto a plate slightly covered by sea water in a gas flow through chamber with an opening to introduce microelectrodes under microscopic view (Fig. 9). Moving the electrodes from outside the cell through the cell membrane into the oxygen metabolizing interior showed a steep gradient of the PO$_2$ (diffusion barrier) across the cell membrane with no measurable intracellular profile (Fig 10), and no pacemaker activity being registered at oxygen concentrations of the fresh gas flow of just 10% (Chen, Erdmann and Halsey, 1978).

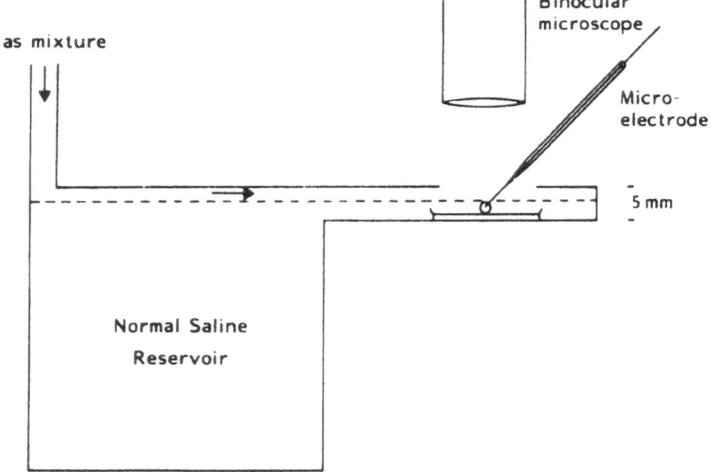

Fig. 9.   Measurement chamber with saline reservoir. The oxygen concentration of
            the constant flow gas mixture (oxygen-nitrogen) can be feed back controlled
            via extracellular PO_2 measurement with a second electrode controlling the
            oxygen flow regulating electronic valve. Extracellular PO_2-electrode and
            both reference electrodes, extra- as well as separate intra-cellular, are not
            indicated in this schematic drawing of the experimental set up.

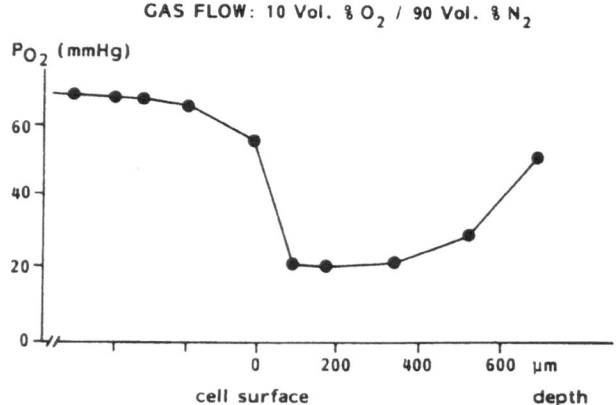

Fig. 10.   Oxygen partial pressure profile of a pacemaker neuron of aplysia califor-
            nensis. After a steep decrease of the oxygen partial pressure during penetra-
            tion through the cell membrane and the immediate inner layer of the neuro-
            genic membrane, intracellular PO_2 (IPO_2) shows no further significant
            changes inside the cell.

This is, or should have been, predictable as the pacemaker neuron receives its oxygen by diffusion (and convection) from the surrounding endolymph with normally a PO_2 of 15-20 mm Hg. Thus, further studies were performed under, for aplysia neurons, normal environmental PO_2 conditions. They could be conducted with spontaneous pacemaker activity present. A steep gradient of PO_2 across the cell membrane, however, was present with an intracellular PO_2 of 5 mmHg which was quite persistent in the various neurons examined. Simultaneous measurement of extracellular PO_2 and intracellular PO_2 together with the spontaneously discharging spike activity revealed that intracellular PO_2 was not proportionally changing to extracelluar PO_2. Decrease of extracellular PO_2 is immediately "counteracted" after a short-term slight decrease of intracellular PO_2. Once a critical extracelluar PO_2 value of 7-8 mm Hg is reached, further decrease of extracelluar PO_2 causes a decrease of intracellular PO_2, as a consequence this is immediately followed by deterioration of the pacemaker pattern. On the other hand, extracellular hyperoxia is

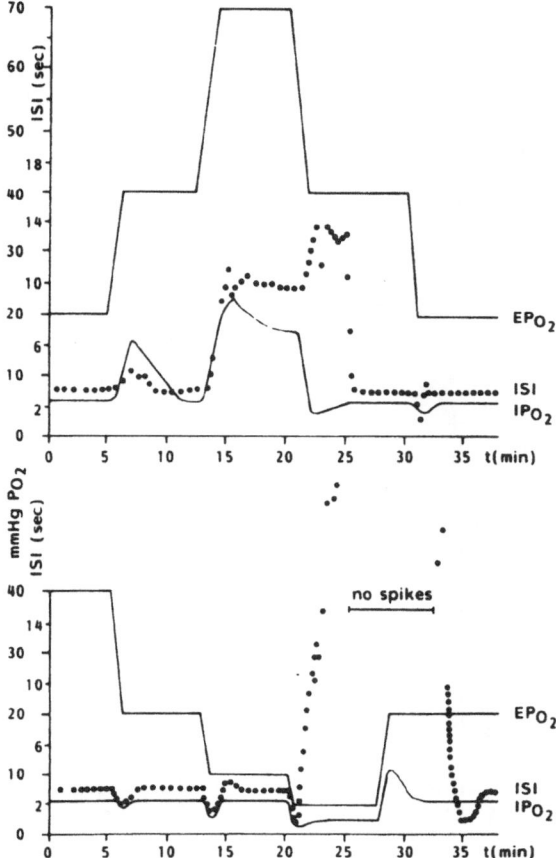

Fig. 11.   Responses of IPO$_2$ (intracellular PO$_2$) to changes of EPO$_2$ (extracellular PO$_2$) and effects on bioelectronic function of aplysia giant neurons (R6, spontaneously discharging spikes). The intermittent spike interval (ISI) is registered. ISI changes are closely related and very sensitive to IPO$_2$ changes, but not to changes of EPO$_2$. Regarding only the intracellular PO$_2$ there is no hypoxic-hyperoxic tolerance of cellular neurogenic function to be seen.

compensated by an increase of the partial pressure gradient across the cell membrane up to values of 30-40 mm Hg PO_2. Higher values result in an increase of intracellular PO_2 answered by numerous intermittent spike bursts and later disappearance of spontaneous activity (Fig. 11).

Intracellular PO_2 plotted against extracelluar PO_2 showed fairly constant intracellular PO_2 values of 5-6 mm Hg with extracelluar PO_2 values ranging from 10-50 mm Hg; value ranges also present in mammalian brain tissue (Fig. 12).

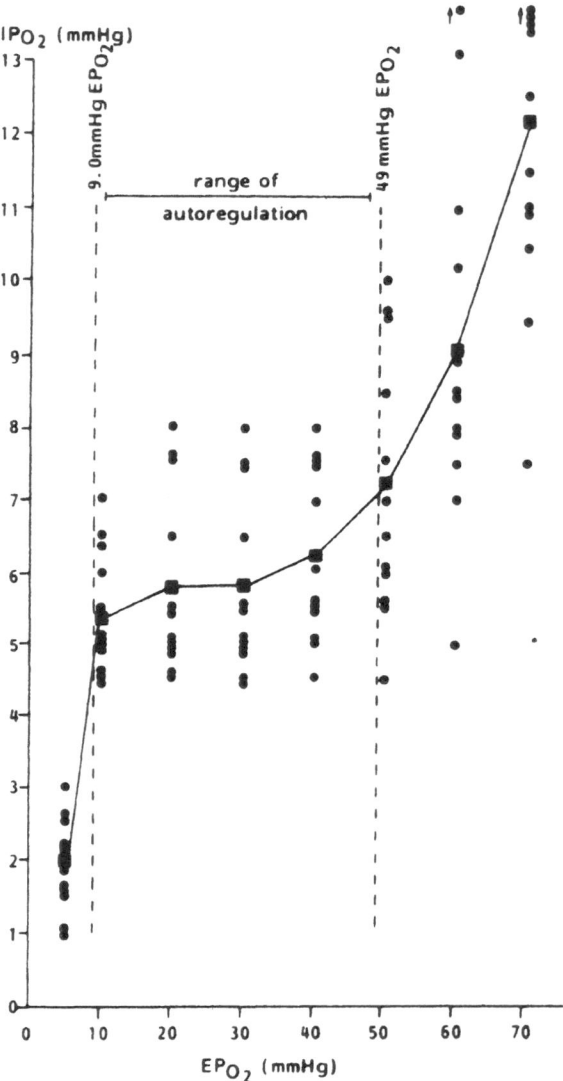

Fig. 12.   Intracellular PO_2 (IPO_2) in R6 of the abdominal ganglion of aplysia californensis plotted against extracellular PO_2 (EPO_2). Between 10 mm Hg and 50 mm Hg EPO_2 the IPO_2 is kept fairly constant (autoregulative changes of the diffusion resistance of the cell membrane?).

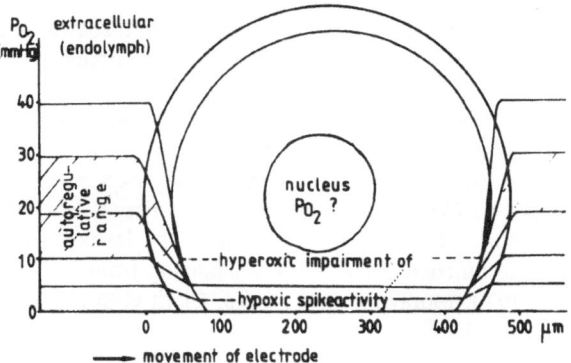

Fig. 13.   Oxygen gradient through the cell wall of a neuron in the abdominal ganglion of aplysia californensis. These cells are pacemaker cells and have a diameter of 500 microns. The cell membrane changes the diffusion resistance for oxygen by autoregulation and keeps the intracellular PO_2 constant over a broad range of extracellular PO_2 values. Slight reaction of intracellular PO_2 beyond this range, however, result in severe changes of the spike pattern.

Conclusion 2: there exists some sort of an autoregulative adaptation of diffusion resistance for oxygen across the cell membrane aimed at keeping the intracellular PO_2 fairly constant in a broad range of environmental supply conditions. Description of the mechanisms involved can so far only be hypothetical (morphological, biochemical or kinetical); even the possibility of changing facilitated oxygen diffusion/transport that has been discussed years ago might be involved.

In any case, the assumption of static conditions of the various diffusion barriers in the oxygen cascade from air to the cell cannot be fully maintained. The findings described above draw more and more attention to diffusion as a possible major factor influencing, defining and autoregulating the cellular oxygen supply/demand ratio in a continuously changing environment fully dependent on oxygen flux to tissue in general (Fig. 13).

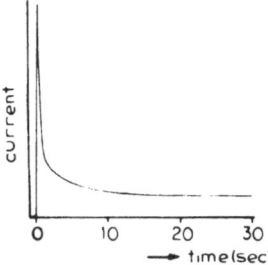

Fig. 14.   Time course of the diffusion limited current after closing of the electric circuit for PO_2-measurement.

Methodologically a start had been made by Krell from Thews' department pointing out the possibility to use the time course of the diffusion limited current of bare noble metal electrodes (Fig. 14) to directly determine the oxygen diffusion coefficient in tissue, a method which has become very valuable after the exact biomathematical description in recent years (Roh, Goldstick and Linsenmeier, 1990). The changing time course is directly related to the diffusion coefficient and consequently experiments have revealed that these oxygen diffusion coefficients vary from tissue to tissue. For the brain cortex the values range from 1.7 to 1.9 x 10^{-5} cm^2/sec while in the less oxygen consuming white matter the oxygen diffusion coefficients are lower ranging from 1.5 to 1.6 x 10^{-5} cm^2/sec (Erdmann and Krell, 1976, Morawetz et al., 1978, Clark et al., 1978).

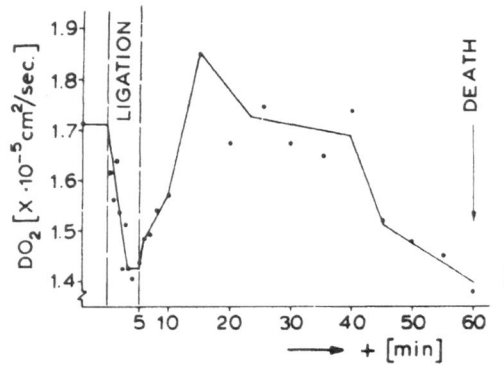

Fig. 15.   Oxygen diffusion (DO$_2$) in the gerbil brain during reperfusion after a 5 minute carotid ligation period. A primary increase of the oxygen diffusion coefficient is followed by a constant decrease until death occurs.

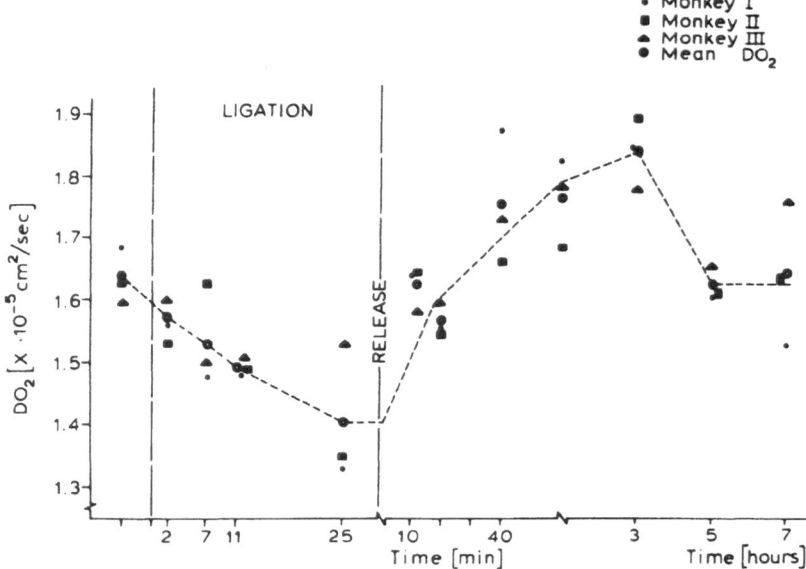

Fig. 16.   Tissue diffusion after middle cerebral artery occlusion in the monkey brain (maccaca iris). This partial occlusion of blood supply permits full recovery within short time.

Interventions such as short-term anoxia (ischemic anoxia) decrease the oxygen diffusion coefficient decisively by up to 25%. After reoxygenation oxygen diffusion coefficients increase primarily but secondarily decrease to subnormal values and after short-term global ischemia (gerbil brain) death might follow (Fig. 15). After local ischemic anoxia (monkeys) recovery to normal coefficient values (Fig. 16) was observed (Nemoto et al., 1977, Morawetz et al., 1978). Two questions arise from these results:

1. Why do oxygen diffusion characteristics change primarily?
2. What is the reason for a secondary decrease while tissue oxygenation has been fully re-established?

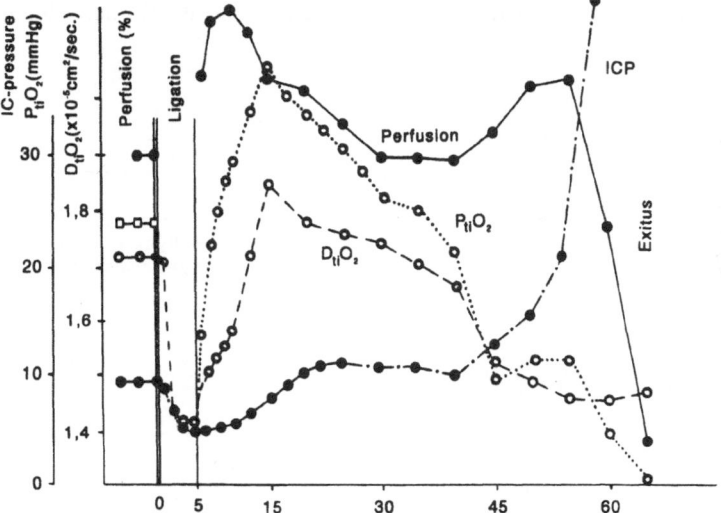

Fig. 17. Postischemic changes of microphysiologic parameters determining oxygen supply - without treatment. $D_{ti}O_2$ = tissue diffusion coefficient, $P_{ti}O_2$ = tissue oxygen partial pressure, ICP = intracranial pressure.

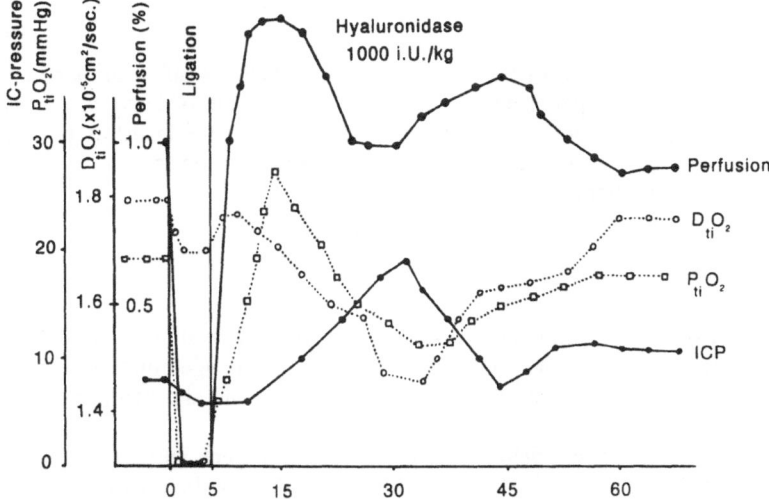

Fig. 18. Postischemic changes of microphysiologic parameters determining cellular oxygen supply - treatment with hyaluronidase administration at point of deterioration of $D_{ti}O_2$ (tissue oxygen diffusion coefficient).

Experiments were therefore performed simultaneously recording tissue perfusion, tissue PO_2 and intracranial pressure. As shown (Fig. 17) intracranial pressure increases secondarily concomitant with decrease of tissue perfusion, however, primarily counter-regulated to above normal values by blood pressure increase. The oxygen diffusion coefficient shows little correlation and decreases constantly already during the phase of normal perfusion and intracranial pressure values. In a second series of experiments an attempt was made to interfere pharmacologically with the glucosamino oxidase (hyaluronidase). Glucosamino oxidases (in the brain ß-glucoronidase) regulate the transfer from depolymerized mucopolysaccharide to polymerized mucopolysaccharide which form the supporting structure (network) of the intercellular space and might be as thin membranes the main diffusion barriers. The experiments have shown (Fig. 18) that administration of

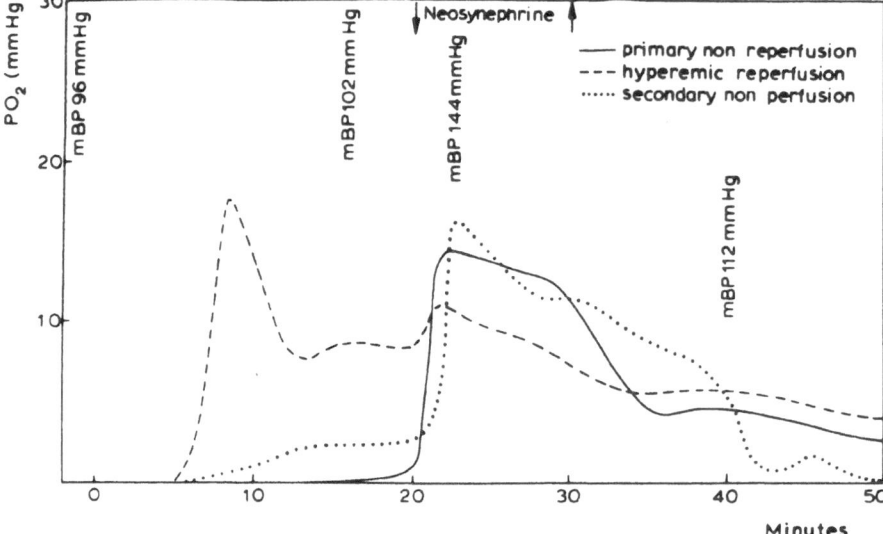

Fig. 19.   Reoxygenation of the microarea after ischemic insults: primary non-perfusion and hyperperfusion may exist just a couple of hundred microns away from each other. Increase of perfusion pressure leads to reperfusion of primarily non-reperfused areas. Secondary non-reperfusion is a more general appearance.

hyaluronidase (the only commercially available glucosamino oxidase) reverses the process of intracranial pressure increase and at the same time the oxygen diffusion coefficient is recovering with concomitant increase of tissue PO_2 (Erdmann and Faithfull, 1982).

Conclusion 3: Oxygen diffusion is a changing and very vulnerable factor in the oxygen supply chain of the cell. It seems to be autoregulated, but under certain conditions may deteriorate. There are substances present that have a regulatory factor probably on the polymerization-depolymerization of mucopolysaccharide as the main substance forming the interstitial structure and thus the oxygen diffusion barrier.

A remaining question left open is that of the changes of tissue perfusion in the postischemic period. Experiments in gerbil brains show that after release of occlusion (ischemic anoxia), primary hyperperfusion appears concomitant with areas totally lacking reperfusion (non-reperfusion). Increase of perfusion pressure leads to reperfusion of the primarily non-perfused areas. Secondarily, non-perfusion might appear again but as a more general appearance (Fig. 19). Primary non-reperfusion is explained generally as being an effect of hypoxic and acidotic stiffening of erythrocytes as it is present in much more serious form in patients with sickle cell crisis. Secondary non-reperfusion is a more general phenomenon and is explained by development of brain edema which as a conclusion of these experiments is probably partially due to secondary tissue hypoxia following the increase of diffusion resistance to oxygen (see Fig. 15). Not knowing about the effects of glucosamino oxidases, artificial decrease of cellular oxygen consumption by pharmacological intervention has been advocated. Clinical experiences give the feeling that the outcome is better, however this has never been proven statistically. On the other hand, this might have been due to a wrong choice as barbiturates were used in these studies in order to reduce cellular oxygen metabolism; gamma-hydroxy butyric acid might have given better results (Fig. 20).

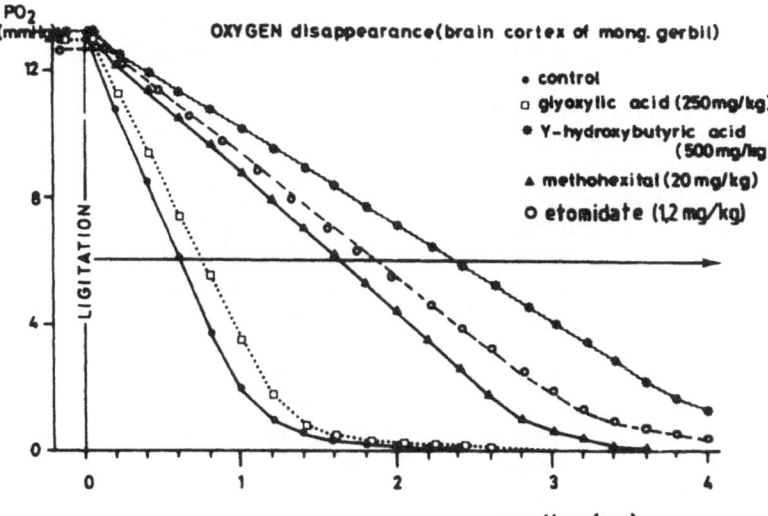

Fig. 20. Registration of oxygen disappearance as a determinant of oxygen consumption ratio of the tissue (method of Reneau and Halsey, 1978). The more rapid oxygen disappears after occlusion of the arteria carotis (gerbil experiments), the higher consumption is. Control studies (no treatment) show 90% of the oxygen present disappear in 1 second. After administration of glyoxylic acid, methohexital, etomidate and gamma-hydroxybutyric acid oxygen consumption is decreased with very little effect with 90% of present oxygen consumed in 1.2 seconds (glyoxylic acid) to large effects with 90% of oxygen reserve lasting for up to 4 seconds. Gamma-hydroxybutyric acid and etomidate have the largest oxygen consumption reducing effects.

In summary, the factors involved in the oxygen supply of the micro-area and its possible disruption are:
1) capillary perfusion disturbed by viscosity changes, the consistency of the erythrocytes and perfusion pressure;

2) passive (or active) oxygen diffusion in erythrocytes, the blood medium, capillary wall, interstitium and the cellular membrane. Changes of diffusion resistance might have a much larger impact on certain supply failure than so far assumed; and

3) oxygen consumption of the cells.

ISOTT works actively in these fields and the reported data for this overview were, if not original, gleaned from own contributions to ISOTT Proceedings from recent years. The feed-back of an open question in clinical practice to the bio-engineer to model and predict a way to go, the testing of the model and derived hypotheses by experimentally working scientists and the clinical application and testing of found insights and knowledge in clinical practice closes the circle of streamlined research in such a group where experts in the field for all necessary areas of research are present.

REFERENCES

Chen, C., Erdmann W., and Halsey, J. H., 1978, The sensitivity of aplysia giant neurons to changes in extracellular and intracellular PO_2, Adv. Exp. Med. Biol., 94:691-696.

Clark, D. K., Erdmann, W., Halsey, J. H., and Strong, E, 1978, Oxygen diffusion, conductivity and solubility coefficients in the microarea of the brain, Adv. Exp. Med. Biol. 94:697-704.

Erdmann, W., Kunke, S., and Krell, W., 1973, Tissue-PO_2 and cell function - an experimental study with multi-microelectroeds in the rat brain. In: Oxygen supply, Kessler et al. eds., Urban & Schwarzenbergn, München, Berlin, Wien.

Erdmann, W., 1976, Oxygen and the microarea as related to anesthesia, Excerpta Medica International Congress Series, 399:483-487.

Erdmann, W., and Krell, W., 1976, Measurement of diffusion parameters with noble metal electrodes, Adv. Exp. Med. Biol. 75:225-228.

Erdmann, W., and Faithfull, N. S., 1982, The disturbance of cellular oxygen supply in the post hypoxic period, pp 183-190, in: Protection of Tissues against Hypoxia, A. Wauquier et al., eds., Elsevier Biomedical Press, Amsterdam.

Fennema, M., and Erdmann, W., 1985, On line determination of PO_2, oxygen diffusion, tissue perfusion and action potentials with a 10 micron tip electrode system, Sensor 5:821-817.

Kunke, S., Erdmann, W., and Metzger, H., 1972, A new method for simultaneous PO_2 and action potential measurement in microareas of tissue, J. Appl. Physiol. 32:436-438.

Morawetz, R., Strong, E., Clark, D. K., and Erdmann, W., 1978, Effects of ischemia on the oxygen diffusion coefficients in the brain cortex, Adv. Exp. Med. Biol. 94:629-632.

Nemoto, E.M., Erdmann, W., Strong, E., Rao, G.R., and Moosy, J., 1977, Regional brain PO_2 after global ischemia in monkeys: evidence for regional differences in critical perfusion pressures, Stroke 8:558-564.

Reneau, D.D., and Halsey, J.H., Jr., 1978, Interpretation of oxygen disappearance rates in brain cortex following total ischaemia, Adv. Exp. Med. Biol. 94:189-198.

Roh, H.-D., Goldstick, T.K., and Linsenmeier, R.A., 1990, Spatial variation of the local tissue oxygen diffusion coefficient measured in situ in the cat retina and cornea, Adv. Exp. Med. Biol. 277:127-136.

OXYGEN SUPPLY BY PERFUSION AND DIFFUSION IN HETEROGENEOUS TISSUE MODELS

Johannes Piiper

Abteilung Physiologie
Max-Planck-Institut für experimentelle Medizin
Göttingen, Germany

INTRODUCTION

Ample evidence for tissue perfusion inhomogeneity has been provided in resting and stimulated skeletal muscle preparations by microsphere embolization (Cerretelli et al., 1984; Iversen and Nicolaysen, 1989, 1990; Iversen et al., 1989; Marconi et al., 1988; Pendergast et al., 1985; Piiper et al., 1985), by inert gas washout (Grønlund et al., 1989; Piiper and Meyer, 1984), by local radioactive xenon clearance (Cerretelli et al., 1984), by observation of the variance of transit times, of the local tissue P_{O_2} heterogeneity and by other methods (cf. Piiper, 1985; Piiper, 1990). There is also morphometrical evidence for the irregularity of the capillary array in the myocardium (Turek and Rakusan, 1981; Hoofd et al., 1985; Rakusan et al., 1984) and in skeletal muscle (Egginton et al., 1988).

It is of interest to estimate the possible effects of both diffusion and perfusion heterogeneity on O_2 supply, particularly with increased O_2 requirement. In this report results of model calculations on both kinds of heterogeneity are presented. It will be shown that their effects on O_2 supply are qualitatively similar.

MODELING

O_2 supply limitation by perfusion or diffusion

O_2 supply may be quantified as the flux of O_2 provided by arterial blood inflow, $\dot{Q} \cdot C_a$ ($\dot{Q}$, blood flow; C_a, arterial O_2 content) or, in the case of diffusive transport, as $D \cdot \Delta P$ (D, diffusive conductance or diffusing capacity; ΔP, O_2 partial pressure difference effective for blood/tissue diffusion). In the individual compartments of heterogeneous models (or in the single compartment of homogeneous models) the following characteristics of the conditions for O_2 supply by perfusion or diffusion should be discerned.

a) Sufficient O_2 supply: O_2 supply > O_2 requirement; O_2 uptake = O_2 requirement.
b) Critical O_2 supply: O_2 supply = O_2 requirement; O_2 uptake = O_2 requirement.
c) Insufficient O_2 supply : O_2 supply < O_2 requirement; O_2 uptake = O_2 supply.

Perfusion-limited O_2 supply

The model is composed of three compartments, each with equal tissue volume and specific O_2 requirement, but with perfusion in the relation 9:3:1. Diffusion limitation is assumed

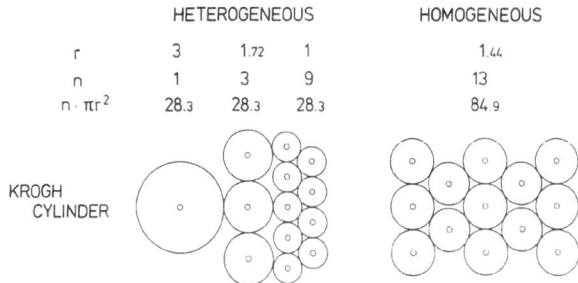

	HETEROGENEOUS			HOMOGENEOUS
r	3	1.72	1	1.44
n	1	3	9	13
$n \cdot \pi r^2$	28.3	28.3	28.3	84.9

KROGH CYLINDER

Fig. 1. The heterogeneous and the homogeneous diffusion limitation models for capillary/tissue O_2 transfer. The models are composed of Krogh cylinder disks. r, outer radius of disks; n, number of disks; $n \cdot \pi r^2$, total surface area; all in relative units. The inner (capillary) radius is 0.172 (in the same relative units).

to be absent, i.e. O_2 can be extracted from blood down to $P_{O_2} = 0$. Computations are based on Fick principle and flow-proportional mixing of effluent blood leaving the compartments (Piiper and Haab, 1991). The O_2 uptake/O_2 requirement ratio at increasing O_2 requirement is compared to that of a homogeneous model.

Diffusion-limited O_2 supply

Cross-sectional elements of Krogh cylinders in form of disks with a central hole (capillary lumen) are used as models (Fig. 1). Unequal distribution of diffusion conditions is represented by three sizes of such disks, with radii in relation 3: 3:1, but of equal cross-sectional area, i.e. the numbers of disks of each size are in relation 1:3:9. The homogeneous reference model is a disk with equal average cross-sectional area. The capillary radius was assumed as 10% of the radius of the medium-size disks.

The calculations of the critical O_2 supply radius (where P_{O_2} becomes 0) were performed using modifications of the Krogh-Erlang diffusion equation (Krogh, 1919). Specific O_2 consumption was assumed to be equal to specific O_2 requirement at $P_{O_2} > 0$, and zero at $P_{O_2} = 0$. The ratio O_2 uptake/O_2 requirement was obtained as the fraction of the cross-sectional area supplied with oxygen. Details of the calculations have been described elsewhere (Piiper and Scheid, 1991).

RESULTS OF CALCULATIONS

Perfusion-limited O_2 supply (Fig. 2, bottom)

In the homogeneous model (broken line), O_2 uptake can match the O_2 requirement up to a critical value (fat arrow) and falls thereafter in line with O_2 delivery. In the heterogeneous model (solid line) the critical O_2 requirement value is smallest for the least perfused compartment, whereby, upon increasing O_2 requirement, the first drop of the O_2 uptake/O_2 requirement for the whole system is produced (arrow 3). The critical condition for the medium-perfused compartment is marked by a further accelerated drop (arrow 2) and that for the best perfused compartment, by a further slight kink of the O_2 uptake/O_2 requirement ratio (arrow 1). Further increase of the O_2 requirement has the same O_2 uptake/O_2 requirement ratio as the

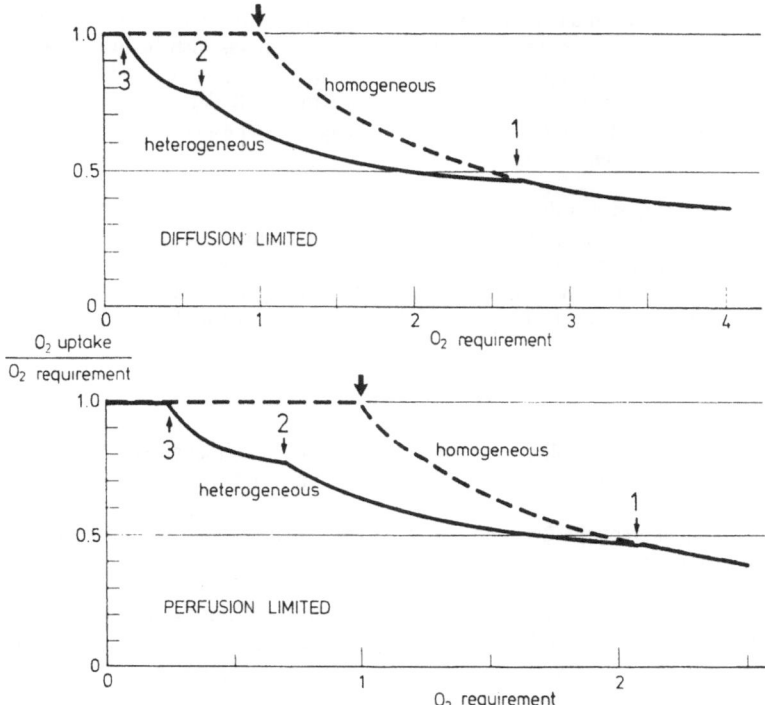

Fig. 2. The O_2 uptake/O_2 requirement ratio with increasing O_2 requirement in heterogeneous and homogeneous models. Top, diffusion-limited models. Bottom, perfusion-limited models. Arrows at kinks mark onset of O_2 supply limitation in a compartment. For explanations, see section Results of Calculations.

homogeneous model, because in these conditions O_2 uptake (O_2 supply) is below the O_2 requirement in all compartments of the heterogeneous system as well as in the single compartment of the homogeneous system.

Diffusion-limited O_2 supply (Fig. 2, top)

The relationship between the O_2 uptake/O_2 requirement is similar to that for the perfusion-limited case. The critical condition is here due to P_{O_2} dropping to zero at the outer radius of the Krogh cylinder disk of certain radius. The three kinks of the falling O_2 uptake/O_2 requirement curve for the heterogeneous model (solid line) mark the critical condition in the different size disks (arrows 1, 2 and 3), whereas in the homogeneous system (broken line) only one kink is seen (fat arrow).

DISCUSSION

Choice of models

The three-compartment model for perfusion heterogeneity represents a blood flow distribution in order-of-magnitude agreement with perfusion heterogeneity found by various methods in skeletal muscle (cf. Piiper, 1990). The diffusion inhomogeneity model was shaped after that of the perfusion model. But the number of compartments as well as the existence of discrete compartments is a simplifying assumption. A continuous distribution is more probable and is in agreement with microsphere data. Evidently a model comprising both aspects, perfusion and diffusion limitation, could have been used, e.g. a Krogh cylinder of finite length. But in the context of this model study it was preferable to investigate both aspects, perfusion and diffusion heterogeneity, in separation.

It has been argued that the Krogh model may be unsuitable for the analysis of O_2 supply to skeletal muscle because of tortuosity of capillaries (Groom et al., 1984; Piiper and Scheid, 1986). In such conditions, the solid cylinder (Hill) model may represent a better choice. Comparative calculations of the effects of diffusion heterogeneity for the solid cylinder and Krogh cylinder models revealed some quantitative differences, but no basically different patterns (Piiper and Scheid, 1991).

Simplifying assumptions

A large number of simplifying assumptions were made, although known to be in conflict with the real situation. In the following, some factors are listed.

(1) All (diffusive) interactions between the cylinders and between afferent and efferent blood vessels were disregarded.

(2) Diffusion and reaction kinetics limitation to O_2 transfer from intraerythrocyte oxyhemoglobin through red cell interior, red cell membrane and plasma was not specifically included in the model.

(3) The O_2 diffusion facilitating effect of myoglobin was not considered.

(4) The sharp delimitation of O_2 consumption equal to O_2 requirement at $P_{O_2} > 0$ from zero O_2 consumption at $P_{O_2} = 0$ certainly is not realistic.

Blood flow and hematocrit

The decisive parameter determining O_2 supply in perfusion-limited O_2 transport systems is the O_2 availability, equal to blood flow multiplied by the arterial O_2 content. At given O_2 saturation, this latter is close to proportional to blood hemoglobin concentration and hematocrit. Variations of hematocrit may importantly modify the effects of unequal perfusion distribution on O_2 transport. At the microcirculatory level, large spatial and temporal heterogeneity of hematocrit has been reported (e.g. Desjardins and Duling, 1987).

Conclusion: effects of perfusion vs. diffusion heterogeneity

The similarity of the patterns of the effects of perfusion and diffusion heterogeneity on O_2 supply are not surprising, since both processes, diffusion and perfusion, determine the availability of O_2 for consumption in tissue. The common feature of both heterogeneities is that upon increase of O_2 requirement the critical condition is reached at a lower O_2 requirement than in the homogeneous models. The steps and notches marking trespassing of the critical condition in individual compartments will flatten or disappear when more compartments or a continuous distribution function is assumed. Thus only the early and flat decrease of the O_2 uptake/O_2 requirement ratio with increasing O_2 requirement is characteristic of inhomogeneity, be it due to diffusion or perfusion limitation.

REFERENCES

Cerretelli, P., Marconi, C., Pendergast, P., Meyer, M., Heisler, N., and Piiper, J., 1984, Blood flow in exercising muscle by xenon clearance and by microsphere trapping, J. Appl. Physiol., 56: 24-30.

Desjardins, C., and Duling, B.R., 1987, Microvessel hematocrit: measurement and implications for capillary oxygen transport, Am. J. Physiol., 252: H494-H503.

Egginton, S., Turek, Z., and Hoofd, L., 1988, Differing patterns of capillary distribution in fish and mammalian skeletal muscle, Respir. Physiol., 74: 383-396.

Grønlund, J., Malvin, G.M., Hlastala, M.P., 1989, Estimation of blood flow distribution in skeletal muscle from inert gas washout, J. Appl. Physiol., 66: 1942-1955.

Groom, A.C., Ellis, C.G., and Potter, R.F., 1984, Microvascular architecture and red cell perfusion in skeletal muscle, in: "Progress in Applied Microcirculation, Vol. 5", F. Hammersen and K. Messmer, eds., Karger, Basel, pp. 64-83.

Hoofd, L., Turek, Z., Kubat, K., Ringnalda, B.E.M., and Kazda, S., 1985, Variability of intercapillary distance estimated on histological sections of rat heart, Adv. Exp. Med. Biol., 191: 239-247.

Iversen, P.O., and Nicolysen, G., 1989, Heterogeneous blood flow distribution within single skeletal muscles of the rabbit: role of vasomotion, sympathetic nerve activity and effect of vasodilation, Acta Physiol. Scand., 137: 125-133.

Iversen, P.O., and Nicolaysen, G., 1990, The distribution of blood flow and glucose uptake within single skeletal muscles in the awake rabbit, Acta Physiol. Scand., 140: 373-381.

Iversen, P.O., Standa, M., and Nicolaysen, G., 1989, Marked regional heterogeneity in blood flow within skeletal muscle at rest and during exercise hyperemia in the rabbit, Acta Physiol. Scand., 136: 17-28.

Krogh, A., 1919, The number of distribution of capillaries in muscles with calculations of the oxygen pressure head necessary for supplying the tissue, J. Physiol., 52: 409-415.

Marconi, C., Heisler, N., Meyer, M., Weitz, H., Pendergast, D.R., Cerretelli, P., and Piiper, J., 1988, Blood flow distribution and its temporal variability in stimulated dog gastrocnemius muscle, Respir. Physiol., 74: 1-14.

Pendergast, D.R., Krasney, J.A., Ellis, A., McDonald, B., Marconi, C., and Cerretelli, P., 1985, Cardiac output and muscle blood flow in exercising dogs, Respir. Physiol., 61: 317-326.

Piiper, J., 1985, Mechanisms of functional shunting in mammalian skeletal muscle, in: "Cardiovascular Shunts" (Alfred Benzon Symposium 21), K. Johansen and W.W. Burggren, eds., Munksgaard, Copenhagen, pp. 467-485.

Piiper, J., 1990, Unequal distribution of blood flow in exercising muscle of the dog, Respir. Physiol., 80: 129-136.

Piiper, J., and Haab, P., 1991, Oxygen supply and uptake in tissue models with unequal distribution of blood flow and shunt, Respir. Physiol., 84: 261-271.

Piiper, J., and Meyer, M., 1984, Diffusion-perfusion relationship in skeletal muscle: models and experimental evidence from inert gas washout, in: "Oxygen Transport to Tissue V" (Adv. Exp. Med. Biol. 169), D.W. Lübbers, H. Acker, E. Lehniger-Follert and T.K. Goldstick, eds., Plenum Press, New York, pp. 457-466.

Piiper, J., Pendergast, D.R., Marconi, C., Meyer, M., Heisler, H., and Cerretelli, P., 1985, Blood flow distribution in dog gastrocnemius muscle at rest and during stimulation, J. Appl. Physiol., 64: 241-251.

Piiper, J., and Scheid, P., 1986, Cross-sectional P_{O_2} distribution in Krogh cylinder and solid cylinder models, Respir. Physiol., 64: 241-251.

Piiper, J., and Scheid, P., 1991, Diffusion limitation of O_2 supply to tissue in homogeneous and heterogeneous models, Respir. Physiol., 85: 127-136.

Rakusan, K., Hoofd, L., and Turek, Z., 1984, The effect of cell size and capillary spacing on myocardial oxygen supply, Adv. Exp. Med. Biol., 180: 463-475.

Turek, Z., and Rakusan, K., 1981, Lognormal distribution of intercapillary distance in normal and hypertrophic rat heart as estimated by the method of concentric circles: its effect on tissue oxygenation, Pflügers Arch., 391: 17-21.

IN SITU DETERMINATION OF CONVECTION AND DIFFUSION

PROFILES IN HETEROGENEOUS MEDIA

Nathan A. Busch and Martin L. Yarmush
Department of Chemical and Biochemical Engineering
Rutgers University
Piscataway, New Jersey

INTRODUCTION

Biologic tissue in general, represent a complex media composed of, on a microscopic level, parenchymal tissue and the vasculature that is involved with transport of substances to and from the parenchymal tissue. In a variety of situations, it is important for one to know the concentration of a particular species (e.g. O_2) within the tissue with high precision and resolution. For example if one had the ability to easily measure the oxygen concentration in a specific region of brain tissue, then one might be able to predict and, consequently prevent certain catastrophic events such as stroke.

Traditionally for O_2 measurements in brain, ultra- microelectrodes with tip diameter on the order of $1\mu m$ are inserted and measurements taken using polarographic techniques. The spatial resolution is of the order of $10\mu m$, and is only a point measurement. Moreover, these techniques are necessarily invasive and destructive to the living organism. Recently non - invasive techniques such as Magnetic Resonance Imaging (MRI) tomography have been developed and applied to provide information primarily concerning anatomic structures and vasculature flow patterns in tissue ([9], [10], [16], [20], [21]).

Although these techniques are noninvasive, they suffer from being of poor spatial resolution, very expensive and quite time intensive. Thus new high resolution methodology which is capable of accurately and noninvasively measuring specific solute concentrations in tissue is needed. This report describes a mathematical approach towards utilizing tomographic data in a more efficient manner towards this end.

The hypothesis of this work is that the transport of oxygen and tissue metabolic activity may be directly monitored in a heterogeneous system by employing tomographic technology. Using this mathematical treatment individual species concentration in heterogeneous media can be produced based on measured tomographic information. As descibed below, the described method depends upon the linear additivity of the concentration dependent transilluminating radiation extinction.

BACKGROUND: Applied Tomography

Computed tomography is the recovery of a function of three variables from its integrals over planes, or the recovery of a two dimensional function from its line integrals. The integrals of the functions are called Radon Transforms. In two dimensions, the Radon transform is the line integral:

$$R(G) = \int_L G(x,y)ds = \breve{G}(p,\xi). \tag{1}$$

The observed attenuated signal, or Radon Transform, is a function of both the angle of transmitted signal with reference to a fixed laboratory co-ordinate system and to the position of the pencil, (bundle of lines), of attenuated signal with respect to the center of the sample. The objective of tomography is then to obtain a reconstructed image from the collection of Radon Transforms. With current techniques if one wished to determine species concentrations and /or specific relevant transport parameters from tomographic data, one would first obtain the image and then, from the image, determine the relevant parameters. The mathematical treatment described below simplifies the process by transforming the collected data directly without having to go through the image construction step. Given that our ultimate goal is to measure O_2) concentration in tissue, it is worthwhile noting that, to date, there is no available straightforward tomographic technique for doing so. MRI tomography with $^{17}O_2$ might be considered but it would suffer from the drawbacks of low $^{17}O_2$ abundance and low sensitivity. Alternatively, one might envision that near infrared (NIR) techniques [14] may succeed. Regardless of which modality is eventually to be used, the treatment described below should provide a clear path to determining desired solute concentration profiles.

THEORY

Concisely the problem to be addressed is as follows: Determine the spatially dependent oxyhaemoglobin and deoxyhaemoglobin concentration in a heterogeneous media. Consider an attenuated radiation which has a wavelength dependent extinction coefficient $\epsilon_{OH,\lambda}$ for oxyhaemoglobin (here we only consider HbO_2) and $\epsilon_{H,\lambda}$ for deoxyhaemoglobin. According to the Beer-Lambert law (neglecting scattering), the intensity transilluminating a tissue section of thickness L is simply

$$I_\lambda = I_{0,\lambda} \exp\left(-\epsilon\,[\mathrm{C}]\,L\right). \tag{2}$$

where [C] represents the oxy/deoxy haemoglobin concentration, and ϵ is the respective extinction coefficient. $I_{0,\lambda}$ is the incident radiation intensity for wavelength λ. For the case of time resolved spectroscopy, $L = ct/n$, (c is the velocity of light, n is the refractive index, and t is the photon time of flight through the tissue sample) then

$$I_\lambda = I_{0,\lambda} \exp\left(-k\,[\mathrm{C}]\,t\right). \tag{3}$$

where $k = \epsilon c/n$.

Presume that the attenuation of the radiation may be linearised as $I_\lambda(s) = I_{0,\lambda}\left(-\epsilon\,[\mathrm{C}]\,l\right)$ and orientate the radiation input / detector system such that they are both co-incident with the transmitted beam, (cf. Figure{1}). Then the measured beam will represent the net intensity decrease along the line of the transilluminating radiation.

When a collection of transilluminating beams have been recorded with the orientation (p, ϕ), (cf. Figure{2}) then the measured intensity is equivalent to the radon transform of the attenuation due to the presence of an absorbing species.

Let the concentration of oxy / deoxy haemoglobin be, at least, piecewise continuous with compact support on the unit disc ($r \in [0, 1]$) and be represented by the

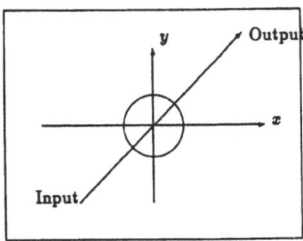

Figure 1. Basic radiation input / output arrangement.

function $c_{OH}(x, y)$ for oxyhaemoglobin and $c_H(x, y)$ for deoxyhaemoglobin. Let the absorbance at wavelength, λ, be linearly dependent upon concentration as $I_{0,\lambda}\left(\epsilon_{OH,\lambda}c_{OH}(x, y)\right.$ $\left.+\epsilon_{H,\lambda}c_H(x, y)\right)$. Then the intensity of the transilluminating radiation at coordinate (p, ϕ) is easily given by the radon transform

$$
\begin{aligned}
\check{g}_\lambda(p, \phi) &= \check{I}_\lambda(p, \phi) \\
&= I_{0,\lambda} \int_s \left\{ \epsilon_{OH,\lambda}c_{OH}(x, y) + \epsilon_{H,\lambda}c_H(x, y) \right\} ds,
\end{aligned} \tag{4}
$$

where ds is a differential distance along the line defined by (p, ϕ). Recognise that $x = p\cos(\phi) - s\sin(\phi)$, $y = p\sin(\phi) + s\cos(\phi)$ then write

$$
\begin{aligned}
\check{g}_\lambda(p, \phi) = I_{0,\lambda} \int_s &\left\{ \epsilon_{OH,\lambda}c_{OH}(p\cos(\phi) - s\sin(\phi), p\sin(\phi) + s\cos(\phi)) \right. \\
&\left. + \epsilon_{H,\lambda}c_H(p\cos(\phi) - s\sin(\phi), p\sin(\phi) + s\cos(\phi)) \right\} ds.
\end{aligned} \tag{5}
$$

The stated problem is reduced to determining $c_{OH}(x, y)$ and $c_H(x, y)$ given a measured $\check{g}_\lambda(p, \phi)$.

Following Deans [7] let $f(x, y)$ (here f(x,y) represents either $c_{OH}(x, y)$ or $c_H(x, y)$) be expanded as an orthonormal series as

$$
f(\underline{x}) = \sum_{\ell=0}^{\infty} \sum_{s=0}^{\infty} \left(\sum_{m=1}^{N(n,l)} A_{\ell,m,s} Z_{\ell+2s}^{\ell}(r) S_{\ell,m}(\underline{\omega}) \right). \tag{6}
$$

where n is the dimension of space, $Z^l_{l+2s}(r)$ is the Zernike polynomial defined in terms of the Jacobi polynomial $(P^{0,|l|}_s(2r^2-1))[1]$ as

$$Z^l_{l+2s}(r) = r^{|l|} P^{0,|l|}_s(2r^2-1), \tag{7}$$

$$S_{l,m}(\omega) = \begin{vmatrix} \cos(l\phi) & \text{if } m=1 \\ \sin(l\phi) & \text{if } m=2 \end{vmatrix}, \tag{8}$$

and

$$\begin{aligned} A_{l,m,s} &= \frac{2(|l|+2s+1)}{\pi} \int_0^{2\pi} \int_0^1 f(r\cos(\theta), r\sin(\theta)) \\ &\quad \cdot Z^l_{l+2s}(r) S_{l,m}(\omega) r\,dr\,d\theta \end{aligned} \tag{9}$$

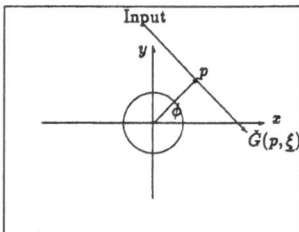

Figure 2: Basic radiation input / output arrangement.

is the two dimensional fourier coefficient of $f(x,y)$. It is easily demonstrated that the radon transform of $f(x,y)$ as represented in Equation{6} is

$$\check{f}(p,\phi) = \sum_{l=0}^{\infty} \sum_{s=0}^{\infty} \left(\frac{2A_{l,s}\sqrt{1-p^2}}{|l|+2s+1} \right) U_{|l|+2s}(p) \tag{10}$$

where for convenience $A_{l,s}$ is defined as

$$A_{l,s} = \sum_{m=1}^{N(n,l)} A_{l,m,s} S_{l,m}(\omega). \tag{11}$$

Equation{6} and Equation{10} represent a radon transform pair. Since the radon transformation (Equation{5}) is a linear operation, then

$$\check{g}_\lambda(p,\phi) = I_{0,\lambda} \left\{ \epsilon_{\text{OH},\lambda} \check{c}_{\text{OH}}(p,\phi) + \epsilon_{\text{H},\lambda} \check{c}_{\text{H}}(p,\phi) \right\}. \tag{12}$$

To be concise then

$$\check{g}_\lambda(p,\phi) = I_{0,\lambda} \sum_{l=0}^{\infty} \sum_{s=0}^{\infty} \left(\frac{2A_{l,s}\sqrt{1-p^2}}{|l|+2s+1} U_{|l|+2s}(p) \left\{ \epsilon_{\text{OH},\lambda} A^{\text{OH}}_{l,s} + \epsilon_{\text{H},\lambda} A^{\text{H}}_{l,s} \right\} \right). \tag{13}$$

Expand $\breve{g}_\lambda(p,\phi)$ as in Equation{10} and equate coefficients of equal (ℓ,s) and obtain that

$$A_{\ell,s}^\lambda = \left\{ \epsilon_{OH,\lambda} A_{\ell,s}^{OH} + \epsilon_{H,\lambda} A_{\ell,s}^{H} \right\}, \tag{14}$$

where $A_{\ell,s}^\lambda$ is a coefficient of $\breve{g}_\lambda(p,\phi)$ in circular chebyshev - fourier space (cf. Equation{11}).

It is clear that the determination of the coefficients $A_{\ell,s}^{OH}$ and $A_{\ell,s}^{H}$ from Equation{14} is ill-posed. However, if a second wavelength is used then under the proper conditions the two coefficients may be uniquely determined. Noting the change in notation $\lambda \to \lambda_1, \lambda_2$ write

$$\begin{bmatrix} \epsilon_{OH,\lambda_1} & \epsilon_{H,\lambda_1} \\ \epsilon_{OH,\lambda_2} & \epsilon_{H,\lambda_2} \end{bmatrix} \begin{bmatrix} A_{\ell,s}^{OH} \\ A_{\ell,s}^{OH} \end{bmatrix} = \begin{bmatrix} A_{\ell,s}^{\lambda_1} \\ A_{\ell,s}^{\lambda_2} \end{bmatrix} \tag{15}$$

or in shorter notation (cf. Appendix{})

$$\underline{\epsilon} \underline{A}_{\ell,s} = \underline{A}_{\ell,s}^\lambda. \tag{16}$$

With Equation{16} the generalisation to multiple species is obvious. The solution $(\underline{A}_{\ell,s})$ is uniquely determined provided that $\underline{\epsilon}$ is not singular. With the coefficient vectors $\{\underline{A}_{\ell,s}\}$ known, and combining Equation{11} and Equation{6} the spatial distribution of each species may be directly computed (say for OH) as

$$\begin{aligned} c_{OH}(x,y) &= c_{OH}(r\cos(\theta), r\sin(\theta)) \\ &= \sum_{\ell=0}^{\infty} \sum_{s=0}^{\infty} A_{\ell,s}(\underline{\omega}) Z_{\ell+2s}^\ell(r). \end{aligned} \tag{17}$$

With Equation{17} the problem of obtaining spatially distributed concentration profiles from transilluminating radiation is completely determined. It is important to note that the current method does not depend upon first obtaining the section image as typically done in analysing metabolic activity in tissue. Rather this method works directly with experimental data.

COMPUTATIONAL ASPECTS

Central to the success of the current method is the efficacy of obtaining the coefficients $\{A_{\ell,s}^\lambda\}$ $\forall \ell, s$ in Equation{10} for the measured intensity of transilluminated radiation (the measured radon transform). Write Equation{10} as

$$\begin{aligned} \breve{f}(p,\phi) &= \sum_{\ell=0}^{\infty} \sum_{s=0}^{\infty} \frac{2\sqrt{1-p^2}}{|\ell|+2s+1} U_{|\ell|+2s}(p) \\ &\cdot \left\{ A_{\ell,1,s} S_{\ell,1}(\underline{\omega}) + A_{\ell,2,s} S_{\ell,2}(\underline{\omega}) \right\} \end{aligned} \tag{18}$$

and require an even expansion in ϕ, then the coefficients $\{A_{\ell,2,s}\} = 0 \forall \ell, s$. Then rewrite Equation{18} as

$$\frac{\breve{f}(p,\phi)}{\sqrt{1-p^2}} = \sum_{\ell=0}^{\infty} \sum_{s=0}^{\infty} A_{\ell,1,s}^* U_{|\ell|+2s}(p) \cos(\ell\phi), \tag{19}$$

with

$$A_{\ell,1,s}^* = \frac{2A_{\ell,1,s}}{|\ell|+2s+1}. \tag{20}$$

633

The coefficients of Equation{19} may be obtained using standard Gram- Schmidt orthonormalisation procedures. Kudo and Saito [15] have argued that Equation{10} satisfies the Helgason - Ludwig consistency condition ([17], [28], [18], [19], [31], and [29]) and that the coefficients $\{A_{\ell,s}^{\lambda}\}$ $\forall \ell, s$ are efficiently computed, whereas the moments ([13], [15]) are difficult and expensive to compute.

CONCLUSION

The current mathematical methodology, for deconvolution of multiple frequency tomographic to obtain spatially resolved concentration profiles of several species, is both simple and straightforward in nature. It relies on the linear additivity of the concentration dependent transilluminated radiation extinction coefficients in the media. The measured data, or radon transform, is simply represented by series in circular fourier-chebyshev space and decomposed into contributing species concentrations based upon known extinction coefficients of each relevant species in the media.

Whilst the methodology is not entirely new in content, it is an important contribution as the tomographic image need not be reconstructed before deconvolution into constituent species concentration profiles. This step is important as errors introduced in the radon transform inversion, to image, do not enter into the analysis of species concentrations.

ACKNOWLEDGMENT

This research was supported in part by NIH Training grant T32 GM08339. Conference travel was supported in part by NSF Travel Award BCS 901 8556.

APPENDIX

Definition of Extinction Matrix

For clarity the matricies used in Equation{16} are defined below:

$$\underline{\epsilon} = \left[\begin{array}{cc} \epsilon_{OH,\lambda_1} & \epsilon_{H,\lambda_1} \\ \epsilon_{OH,\lambda_2} & \epsilon_{H,\lambda_2} \end{array} \right], \tag{21}$$

$$\underline{A}_{\ell,s} = \left[\begin{array}{c} A_{\ell,s}^{OH} \\ A_{\ell,s}^{H} \end{array} \right] \tag{22}$$

$$\underline{A}_{\ell,s}^{\lambda} = \left[\begin{array}{c} A_{\ell,s}^{\lambda_1} \\ A_{\ell,s}^{\lambda_2} \end{array} \right] \tag{23}$$

REFERENCES

[1] Abramowitz, M., I. A. Stegun, *Handbook of Mathematical Functions*, Dover Publications, Inc., New York, NY, (1972).

[2] Ahn, C. B., S. Y. Lee, O. Nalcioglu, and Z. H. Cho, "An Improved Nuclear Magnetic Resonance Diffusion Coefficient Imaging Method Using an Optimized Pulse Sequence," *Mecial Physics*, Volume 13, 789-796, (1986).

[3] Allers, A., and F. Santosa, "Stability and Resolution analysis of a Linearized Problem in Electrical Impedance Tomography," *Inverse Problems*, Volume 7, 515-533, (1991).

[4] Carr, H. Y., and E. N. Purcell, "Effects of Diffusion on Free Precession in Nuclear Magnetic Resonance Experiments," *Physics Review*, Volume 94 630-632, (1954).

[5] Chance, B., J. S. Leigh, H. Miyake, D. S. Smith, S. Nioka, R. Greenfeld, M. Finander, K. Kaufmann, W. Levy, M. Young, P. Cohen, H. Yoshioka, and R. Boretsky, "Comparison of Time-Resolved and -Unresolved Measurements of Deoxyhaemoglobin in Brain," *Proceedings of the National Academy of Science, USA.*, 4971-4975, (1988).

[6] Cho, Z. H., C. B. Ahn, S. C. Juh, H. K. Lee, R. E. Jacobs, S. Lee, J. H. Yi, J. M. Jo, "Nuclear Magnetic Resonance Microscopy with $4\mu m$ Resolution: Theoretical Study and Experimental Results," *Medical Physics*, Volume 15, 815-824, (1988).

[7] Deans, S. R., *The Radon Transform and Some of Its Applications*, J. Wiley and Sons, New York, (1983).

[8] Diederich, C. J., S. Clegg, and R. B. Roemer, "A Spherical Source Model for the Thermal Pulse Decay Method of Measuring Blood Perfusion: A Sensitivity Analysis," *Journal of Biomedical Engineering*, Volume 111, 55-61, (1989).

[9] Guardo, R. C. Boulay, B. Murray, and M. Bertrand, "An Experimental Study in Electrical Impedance Tomography Using Backprojection Reconstruction," *IEEE Transactions on Biomedical Engineering*, Volume 38 617-627, (1991).

[10] Guo, Q., G. Kashmar, and O. Nalcioglu, "NMR Angiography With Enhanced Quasi-Half Echo Scanning," *Magnetic Resonance Imaging*, Volume 9, 129-139, (1991).

[11] Hammer, B. E., C. A. Heath, S. D. Mirer, G. Belfort, "Quantitative Flow Measurements in Bioreactors by Nuclear Magnetic Resonance Imaging," *Bio/Technology*, Volume 8, 327-330, (1990).

[12] Harding, G., J. Kosanetsky, and U. Neitzel, "X-ray Diffraction Computed Tomography," *Medical Physics*, Volume 14, 515-525, (1987).

[13] Howard, J., "Tomography and Reliable Information," *Journal of the Optical Society of America*, Volume 5, 999-1014, (1988).

[14] Katon, J. E., A. M. Somme, and P. L. Lang, "Infrared Microspectroscopy," *Applied Spectroscopy Reviews*, Volume 25, 171-211, (1990).

[15] Kudo, H., and T. Saito, "Sinogram Recovery With the Method of Convex Projections for Limited Data Reconstruction in Computed Tomography," *Journal of the Optical Society of America,* Volume 8, 1148- 1160, (1991).

[16] Lanzer, P. W. McKibbin, D. Bohning, B. Thorn, G. Gross, G. Cranney, N. Nanda, and G. Pohost, "Aortoiliac Imaging by Projective Phase Sensitive MR Angiography: Effects of Triggering and Timing of Data Acquisition on Imaging Quality," *Magnetic Resonance Imaging,* Volume 8 107-116, (1990).

[17] Lewitt, R. M., and R. H. T. Bates, "Image Reconstruction from Projections III: Projection Completion Methods (theory)," *Optik,* Volume 50, 189-204, (1978).

[18] Louis, A. K., "Picture Reconstruction from Projections in Restricted Range," *Mathematical Methods in Applied Science,* Volume 2, 209- 220, (1980).

[19] Natterer, F., *The Mathematics of Computerized Tomography,* Wiley, New York, 158-179, (1986).

[20] Nishimura, D. G., A. Macovski, and J. M. Pauly, "Magnetic Resonance Angiography," *IEEE Transactions on Medical Imaging,* Volume 5, 140-151, (1986).

[21] Nishimura, D. G., A. Macovski, and J. I. Jackson, "Magnetic Resonance Angiography by Selective Inversion Recovery Using a Compact Gradient Echo Sequence," *Magnetic Resonance in Medicine,* Volume 8, 96- 103, (1988).

[22] Pan, J. W., J. R. Hamm, D. L. Rothman, and R. G. Shulman, "Intracellualar pH in Human Skeletal Muscle by ^{1}H NMR," *Proceedings of the National Academy of Sciences,* Volume 85, 7836-7839, (1988).

[23] Pangrle, B. J., E. G. Walsh, S. Moore, and D. DiBiasio, "Investigation of Fluid Flow Patterns in a Hollow Fiber Module Using Magnetic resonance Velocity Imaging," *Biotechnology Techniques,* Volume 3, Number 1, 67-72, (1989).

[24] Pangrle, B. J., D. DiBiasio, E. G. Walsh, and S. Moore, "Magnetic Resonance Imaging: A Novel Technique for Fluid Flow Determinations in Ceramic Bioreactors," *Proceedings, First International Conference on Inorganic Membranes,* 187-192, (1989).

[25] Pangrle, B. J., *Experimental and Theoretical Study of Fluid Flow in Porous Tube Systems Using Magnetic Resonance Imaging and 2-D Finite Element Methods,* Ph. D. Dissertation, Worcester Polytechnic Institute, Boston, MA, (1989).

[26] Park, J. K., and H. N. Chang, "Flow Distribution in the Fiber Lumen Side of a Hollow-Fiber Module," *AIChE. J.,* Volume 32, Number 12, 1937-1947, (1986).

[27] Patterson, M. S., B. Chance, and B. C. Wilson, "Time Resolved Reflectance and Transmittance for the Non-Invasive Measurement of Tissue Optical Properties," *Applied Optics,* Volume 28, Number 12, 2331-2336, (1989).

[28] Peres, A., "Tomographic Reconstruction from Limited Angular Data," *Journal of Computational Assisted Tomography,* Volume 3, 800-803, (1979).

[29] Prince, J. S., and A. S. Willsky, "Constrained Sinogram Restoration for Limited - Angle Tomography," *Optics Engineering*, Volume 29, 535-544, (1990).

[30] Roberts, D. Aaron, "Analysis of Vessel Absorption Profiles in Retinal Oximetry," *Medical Physics*, Volume 14, Number 1, 124-130, (1987).

[31] Saito, T., and H. Kudo, "An Image Reconstruction from Limited View Angle Projection Data," in *Proceedings of the International Conference on Acoustics, Speech, and Signal Processing*, (Institute of Electrical and Electronics Engineers), New York, 1187-1190, (1987).

[32] Siderits, R. H., C. J. Evans, and Joel S. Welling, "A Three Dimensional Reconstruction of Metastatic Adenocarcinoma in Lymph Node," *BioComputing*, Volume 8, Number 7, 670-672, (1990).

[33] Ståhlberg, F., B. Nordell, A. Ericsson, T. Greitz, B. Perrson, and G. Sperber, "Quantitative Study of Flow Dependance in NMR Images at Low Flow Velocities," *Journal of Computed Assisted Tomography*, Volume 10, 1006-1010, (1986).

[34] Takesawa, S., M. Tarasawa, M. Sakagami, T. Kobayshi, H. Hidai, and K. Sakai, "Nondestrive Evaluation by X-ray Tomography of Dialysate Flow Patterns in Capillary Dialysers," *ASIAO*, Volume 34, 794-798, (1988).

TISSUE ALTERATIONS BY THE PENETRATION OF A pO$_2$ SENSING NEEDLE PROBE

Klaus Wagner[1], Wolfgang Bossen[1], Uda Schramm[2]

[1]Dept. of Physiology (Head: Prof. Dr. Ch. Weiss),
[2]Dept. of Anatomy (Head: Prof. Dr. W. Kühnel),
Medical University of Lübeck, Lübeck, Germany

Introduction

Oxygen partial pressure measurements in living tissue with needle probes inevitably lead to tissue alterations. The recent introduction of microprobe based pO$_2$ measuring methods for clinical applications (Weiss and Fleckenstein, 1986) has invigorated research addressing the influence of anaesthetic gases on liver oxygen partial pressure distribution. Thus far, histological changes due to the penetration of this pO$_2$ probe have been studied in skeletal muscle by Schramm et al. (1990).

Therefore, the aim of our investigation was to assess the extent and the degree of histological changes in liver tissue of rats caused by the insertion of a pO$_2$ sensing needle probe.

Material and Methods

7 male Wistar rats (250-800 g) were used. After anaesthesia with pentobarbital (60 mg/kg i. p.), a midline laparotomy was performed and the 4 liver lobes were exposed. The right and the left lobe were fixed to thin cork plates with histacryl glue. The pO$_2$ needle probes (350 μm diameter) were inserted into these two liver lobes (Figure 1, schematic drawing of the needle in situ). There was no bleeding from the insertion point. Immediately afterwards the lobes were either immersed in formalin (4%) or in Bouin's solution (picrin acid : formalin : acetic acid, 15:5:1 ppv) for embedding in paraffin, or they were immersed in glutaraldehyde (2,5%) for embedding in araldite. After fixation of the tissue for 48 hours the needle probes were removed, the lobes dissected in small blocks and routinely prepared for embedding in paraffin or araldite. The paraffin blocks were cut in 5 μm thick serial sections and the sections stained for light microscopy with Masson's trichrome, azan, haematoxylin eosin (HE), PAS, toluidin blue, cresyl violet, alcian blue at pH 0.5, pH 2.5 and pH 4.5, berlin blue, and scarlet red

(sudan IV). Araldite blocks were cut in semi thin sections (0.5 μm) and stained with methylene blue - azur II according to von Richardson (1960). Altogether 11 tracks of needle probes were studied.

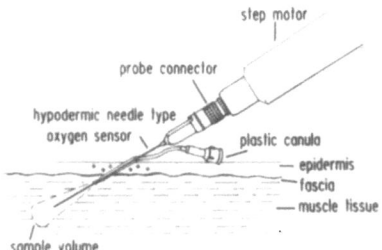

Figure 1. Schematic drawing of the needle probe in tissue

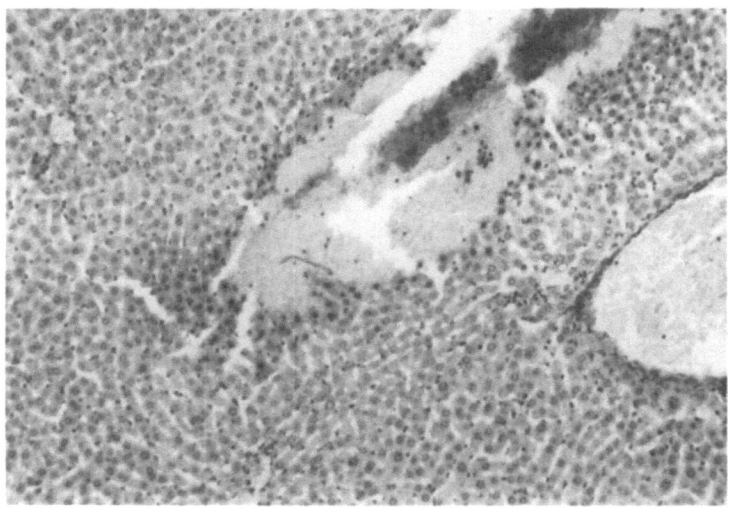

Figure 2. Longitudinal section of the probe canal. Displacement of hepatocyte columns, which is more marked in the region of the probe shaft. Cells along the probe canal with pyknotic nuclei. Cresyl violet staining, magn. x 125.

Results

The needle probes passed straight through the liver tissue. They were not deviated by connective tissue structures. The tips of the needle probes were found to lie mostly within liver cell columns or liver sinusoids (Figure 2).

In nine tracks no bleeding into the probe canal occurred and no ruptured arteries, arterioles or veins were found in the course of the probe track. In two tracks a bleeding into the probe canal was found. No hematoma in the vicinity of the probe canal was observed.

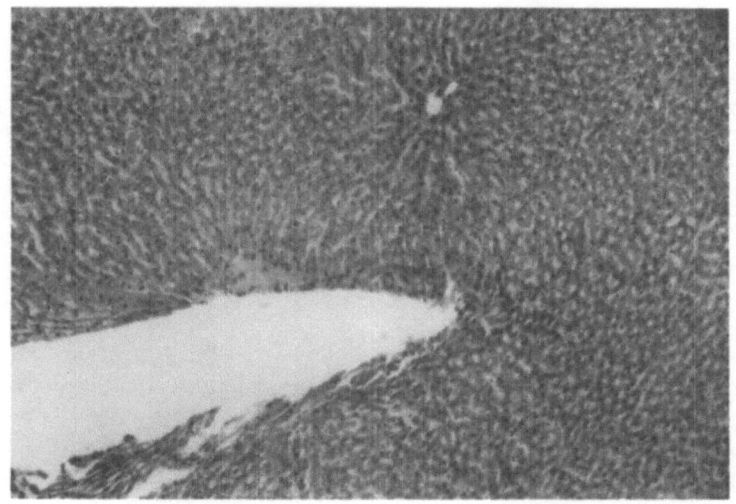

Figure 3. Longitudinal section of the probe canal on the left and a vein. Hepatocytes of zone 1 with small, dark cells and few pyknotic nuclei. Abraded cells and bleeding into the probe canal. PAS, magn. x 200.

The inner surface of the probe canal was rough. There were groups of cells protruding into the lumen of the canal and there were areas of missing tissue (Figure 2, 3, and 4). Within the probe tracks abraded single cells and abraded cell groups were seen (Figure 4 and 6). The abraded cells and some of the hepatocytes adjacent to the probe canal had pyknotic nuclei. This was not found in more distant lying hepatocytes (Schäfer and Höper, 1978).

In figures 2 and 3 a tissue layer of more darkly stained cells lying directly adjacent to the probe track contrasted to the staining of the surrounding cells. This difference in staining was found with azan, Masson's trichrome, PAS, cresyl violet, and toluidin blue. We interpret this as the result of a pH-shift to lower values within the hepatocytes.

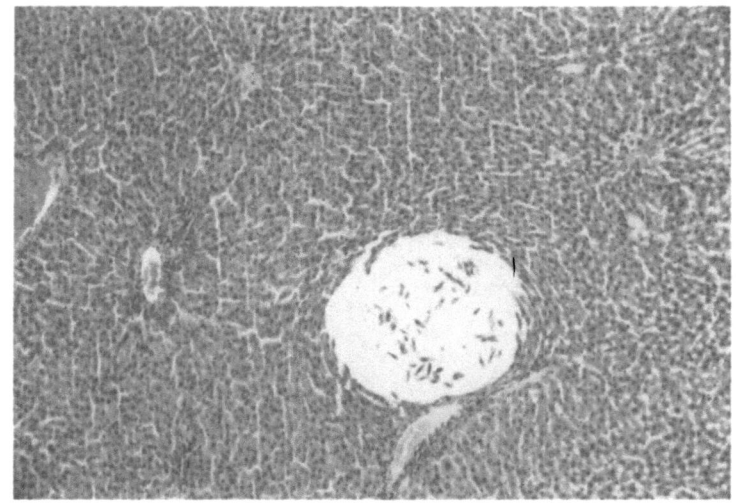

Figure 4. Cross section through the shaft area of a probe canal. Compression of hepatocyte columns and sinusoids leading to a concentric orientation of the adjacent tissue around the probe. Abraded hepatocytes within the probe canal. N.B.: Cells of zone 1 show no different staining properties compared to normal tissue. HE, magn. x 125.

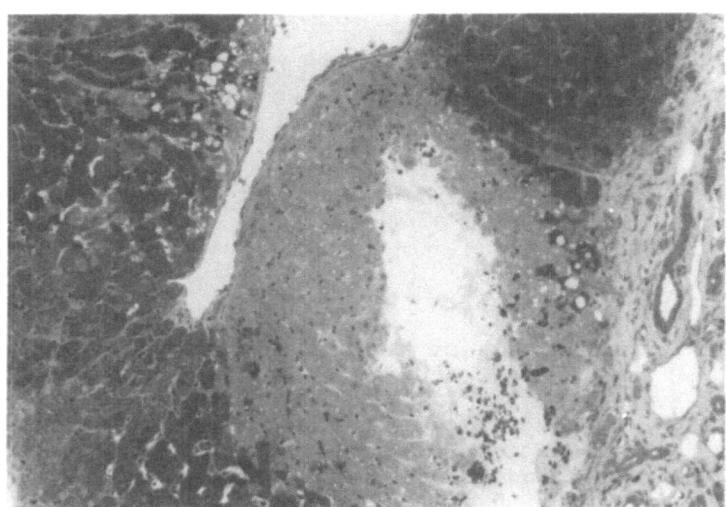

Figure 5. Semithin section of the tip of a probe canal. Zone 1 with differently staining, grey cells lining the probe canal, some with pyknotic nuclei. Cell margins hardly visible. Compressed vein above the tip of the probe canal (top left). Transition from zone 1 to zone 2 with intracellular vacuoles. von Richardson staining, magn. x 200.

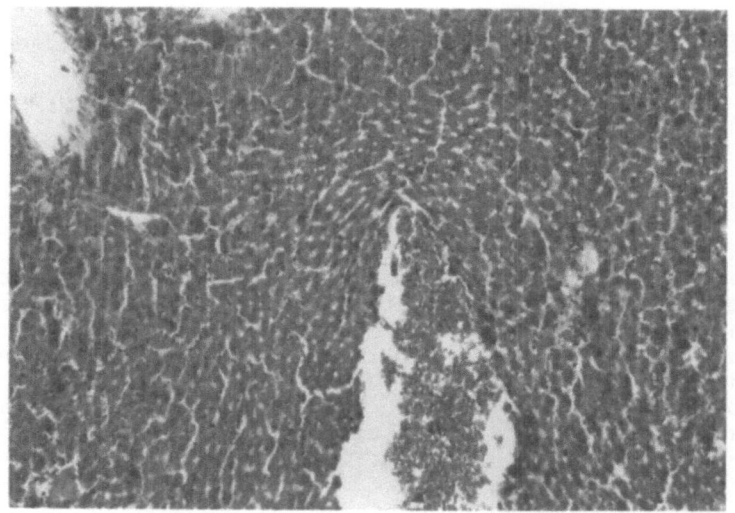

Figure 6. Oblique section through the tip of the probe canal. Displacement of hepatocyte columns and compression of the sinusoids. Trichrome staining according to Masson/Goldner, magn. x 200.

For the methylene blue - azur II staining according to von Richardson (1960) the coloring of the lining cells and the cells far from the canal was inverse to the above described stains (Figure 5): normal cells showed a dark coloured cyto-plasm and slightly lighter nuclei, whereas the cells along the probe track had a grey appearance and light nuclei.

In figure 5 part of the irregularly lined tip of a needle probe track and part of a vein is seen. Due to the tissue displacement by the probe tip a vein in the centre of the preparation is severely compressed. The tissue in between the probe canal and the vein contrasted with a clear border towards the normally stained cells on the right side of the slide. Just above the compressed vein a group of cells with large and small vacuoles was seen, here a gradual transition between light and dark stained hepatocytes occured. This picture suggests that the probe compressed the vein and damaged the liver tissue above the vessel. As expected, the histological changes were more pronounced in areas lying in close proximity to the probe canal than in more distant areas. Hepatocyte columns near the probe canal were compressed and the sinusoids collapsed. Occasionally, the columns were arranged in a shell-like fashion in front of the tip (Figure 6). In these areas hepatocytes frequently showed the different staining properties described above. We call this area the zone 1, it extended 20 - 100 μm from the probe shaft and in front of the probe tip. Zone 1 was followed by zone 2. The tissue of zone 2 stained normally for all stains used but the tissue architecture was still altered. The columns of the hepatocytes did not show the normal orientation but converged towards the needle probe track (Figure 2). In addition, the size of the liver sinusoids within zone 2 was altered. While the diameter of the sinusoids adjacent to zone 1 appeared slightly reduced, the diameter of more distant sinusoids appeared slightly enlarged. Zone 2 extended 100 - 250 μm from the probe shaft and 100 - 350 μm in front of the probe tip.

Discussion

Whereas in rat skeletal muscle the needle probe mostly advanced along the sheath of connective tissue surrounding the muscle fibres (Schramm et al, 1990), in liver tissue it frequently penetrated into the parenchyma.
With regard to the extension of up to 350 μm of zone 2 the steplength of 700 μm for the stepwise forward movement of the pO_2 probes it can be assumed, that with each step the probe tip reaches unalterd 'virgin' tissue.

References

von Richardson, K.C., Jarret, L., Finke, E.H., 1960, Embedding in epoxy resins for ultrathin sectioning in electron microscopy. Stain Techn, 35: 313.

Schäfer, D. and Höper J., 1978, Alterations in rat liver cells and tissue caused by needle electrodes, in: "Ion and enzyme electrodes in biology and medicine", M. Kesser, J. Clark and D. W. Lübbers, ed., Urban und Schwarzenberg, München.

Weiss, CH. and Fleckenstein W., 1986, Local tissue pO_2 measured with 'thick' needle probes, in: "Funktions-analyse biologischer Systeme 15", J. Grote and G Thews, ed., Steiner, Stuttgart.

Schramm, U., Fleckenstein, W., Weber, C., 1990, Morphological assessment of skeletal muscular injury caused by pO_2 measurements with hypodermic needle probes, in: "Clinical oxygen pressure measurement II", A. M. Ehrly, W. Fleckenstein, J. Hauss and R. Huch, ed., Blackwell Ueberreuter Wissenschaft, Berlin.

A NEW PROGRAM TO EVALUATE DATA OF PO$_2$ HISTOGRAPH ON

PERSONAL COMPUTERS UNDER MS-DOS

K.F. Wagner and B.G. Steppan

Department of Physiology, Medical University of Lübeck and
Technical Highschool Lübeck, Germany

INTRODUCTION

In clinical situations such as vascular occlusion and for basic research the oxygen pressure distribution in the diseased tissue may be of great importance. Tissue pO$_2$ distribution can be measured with a commercially available pO$_2$ histograph (Eppendorf-Netheler-Hinz GmbH, Hamburg, Germany). Since the instrument was developed in the early 1980s, the operating system uses a CP/M compatible format that was common at that time, but is incompatible with today's MS-DOS PCs.

Therefore a program was developed to read and evaluate the pO$_2$ data of the pO$_2$ histograph on an IBM compatible personal computer.

METHODS

pO$_2$ data measured with the pO$_2$ histograph are stored on a 3.5 inch floppy disk. The physical and the logical structure of the pO$_2$ histograph data format is incompatible with the data format of MS-DOS. Therefore a new block device driver was developed and integrated into MS-DOS. This block device driver simulated the MS-DOS format structures for the pO$_2$ histograph floppy disks thus allowing the pO$_2$ histograph disks to be read.

The program is written in turbo pascal and assembler, it requires a Hercules, EGS or VGA graphics card. The different functions of the program were programmed in separate units that were linked to constitute the complete program. This modular architecture enables the programmer to update independently the different existing functions and add new functions without having to rewrite the whole program.

THE PROGRAM

Since one of the main functions of the program is the graphic presentation of pO$_2$

Oxygen Transport to Tissue XIV, Edited by W. Erdmann and
D.F. Bruley, Plenum Press, New York, 1992

histograms, a graphic surface was chosen to circumvent the constant switching between text mode and graphic mode. A help screen is available (Figure 1).

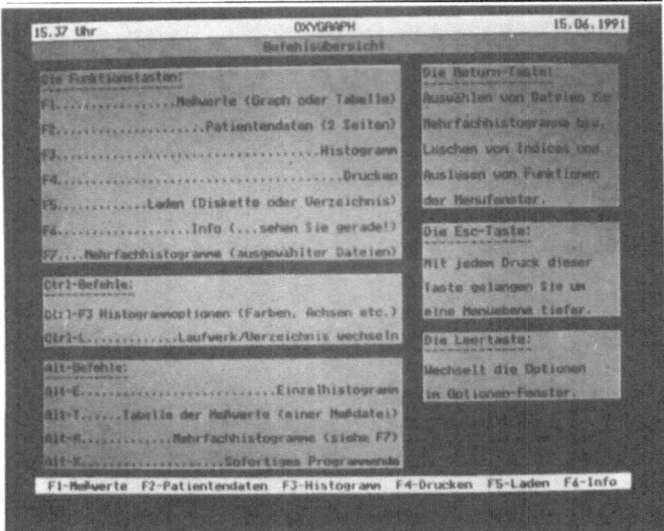

Figure 1.   The help screen.

The execution of the different program functions is steered through function keys. The nomination of the function keys of the pO$_2$ histograph. Figure 2 shows the disk content screen.

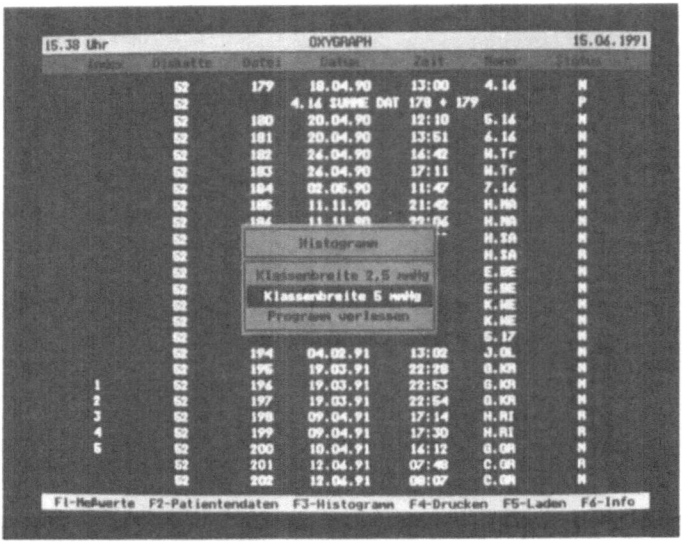

Figure 2.   The disk content screen with pop-up menu.

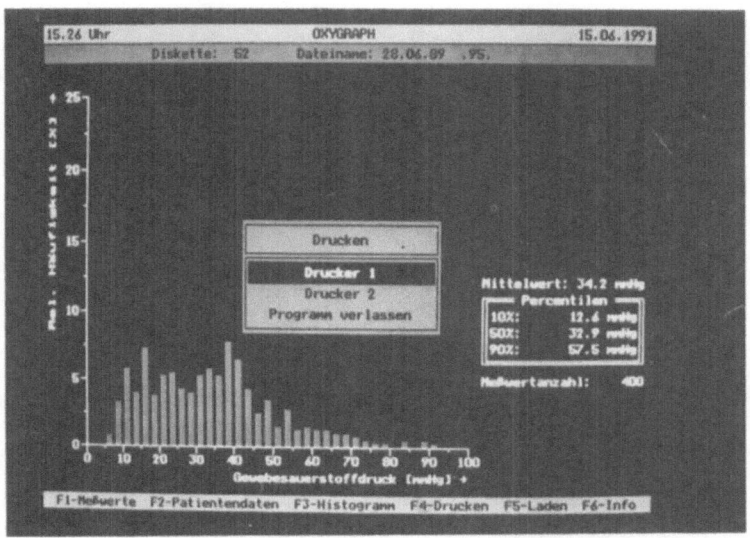

Figure 3.   Single histogram with a width of the pO₂ classes of 2.4 mm Hg.

Histograms are selected with the cursor. Pushing a function key results in the presentation of a pop-up window allowing the selection of options for the chosen function (Figure 2). The pO₂ data can be presented as graphs, tables and histograms. The width of the classes (Figure 3) and the scale of the ordinate can be modified. The colour and the lining of the histogram bars can be selected (Figure 4) and mean, median and 10% and 90% percentiles of the histogram are shown.

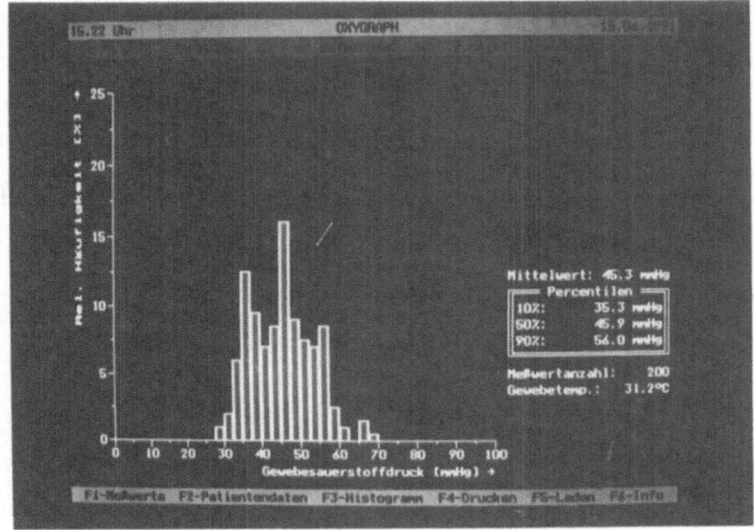

Figure 4.   Single histogram with lining of the bars.

Particularly important for sequential investigations, a 3-D organization of up to 5 selected histograms in a single graph is possible (Figure 5). For multiple histogram presentations the mean pO_2 value of the individual histograms is indicated by an arrow. The user is alerted if an overflow of the scale of the relative frequency of pO_2 values occurs.

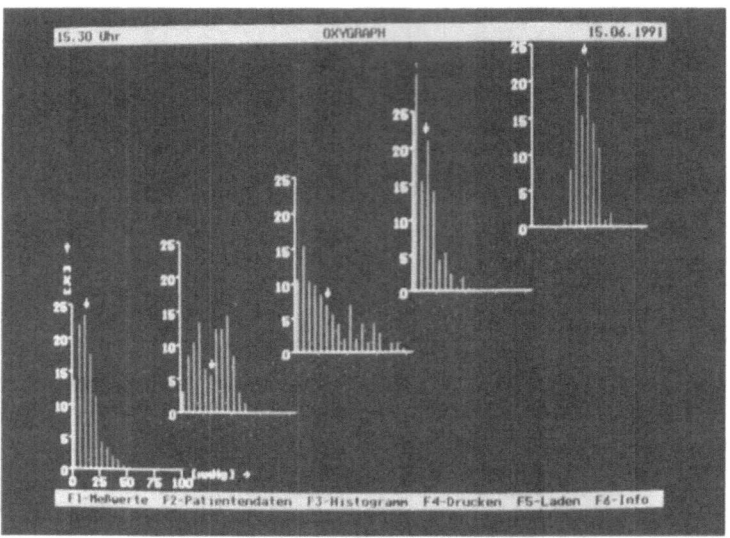

Figure 5.   3-D arrangement of 5 histograms.

High quality printout of single or multiple histograms on standard 24 needle printers are part of the program features. The new block device driver permits copying of the pO_2 histograph files to the hard disk of the personal computer. This markedly shortens file access time and allows pO_2 histograph data back-up with standard back-up programs.

The newly developed program facilitates the evaluation of pO_2 data and improves data exchange and access.

Acknowledgements: The program was written by B. Steppan as part of his Ph.D. thesis.

A PLACENTAL PERFUSION pO$_2$ LOGGER

[1]D. J. Maguire, [1]R. Blums, [1]R. Morgan, [2]J. Collie
and G. R. Cannell[2]

[1]Division of Science and Technology, Griffith University
Nathan, Queensland, Australia
[2]Conjoint Internal Medicine Laboratory, Royal Brisbane
Hospital, Herston, Queensland, Australia

INTRODUCTION

It has been estimated that half of the oxygen is depleted from maternal blood during its circulation through the placenta and much of this oxygen is used by the tissues of the placenta. Obviously respiratory demands (involving the electron transport pathway) will account for a large proportion of the oxygen consumed in this tissue. This can be calculated from consideration of lactate production and glucose utilization rates. Oxygen consumption rate is therefore an important parameter to be considered when monitoring tissue viability during *in vitro* placental perfusions. Manual methods and automated methods for measuring oxygen partial pressure in discrete samples have been available in routine Clinical Chemistry Laboratories for many years and in that environment have provided a valuable tool in the process of caring for critically ill patients. In terms of monitoring perfusate oxygen levels such an approach is tedious and subject to errors. Continuous monitoring systems using various electrode technologies with outputs linked to chart recorders have also been available for a considerable period of time and have been valuable research tools for studying oxygen consumption by homogeneous, discrete cultures of many different cell types. Tissue prepared for moderately long perfusions (of the order of six hours) represent a considerable investment of time and energy. Such preparations warrant the development of purpose-built logging devices which record and display real-time data for the tissue under examination. An ideal monitoring device will display current oxygen levels, report trends in oxygen levels and calculate and display oxygen consumption history. Such devices have been feasible since the advent of microprocessors and are available commercially in a generalized form. The technology described in the present work is a step towards the on-line collection and display of data from *in vitro* human placental perfusions.

INSTRUMENT DESIGN

In order to monitor the perfusion system in use in our laboratory (Fig 1), up to six galvanic oxygen electrodes (Lazar DO-166FT) are used. These were connected to a signal conditioning and switching board, input/output (I/O) card and analog to digital conversion (A/D) on board an IBM personal computer. Electrode output required amplification involving pre-amp (Fig 2) and post-amp stages and for efficiency was

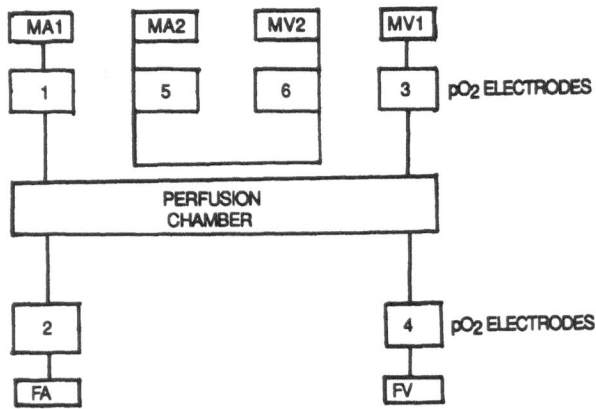

Fig. 1. Placental perfusion circuit, showing positioning of the six pO$_2$ electrodes. MA; maternal artery , MV; maternal vein perfusion lines in first (1) or second (2) circuit, FA; fetal artery, FV; fetal vein.

sampled by switching (Fig. 3) at the latter stage using a 4067 16-1 multiplexer controlled by a 4 bit address from the I/O card. This was a CMOS IC with operating rails from supply voltage (upper) to ground voltage (lower) not exceeding 15V. Analog to digital conversion involved a non-negative voltage range. A block diagram of the completed circuit is shown (Fig. 4)

Software, written in C, was designed in a modular format for ease of writing, editing and future expansion. A total of six units were used, each for a specific set of instructions addressed by the main program (Fig. 5)

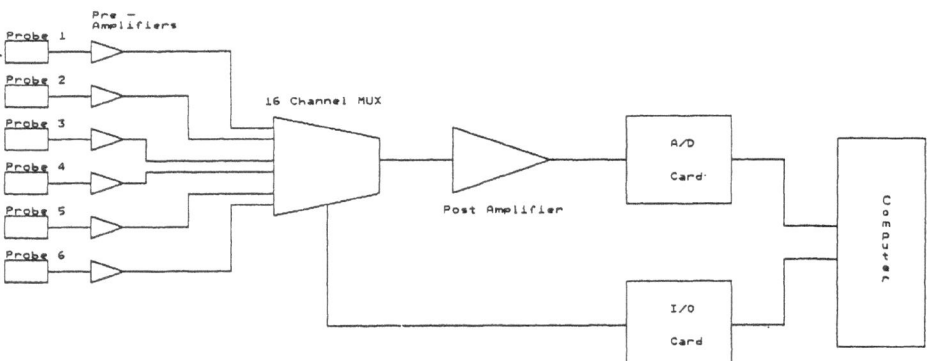

Fig. 2. Circuit diagram of a single pre-amplification stage.

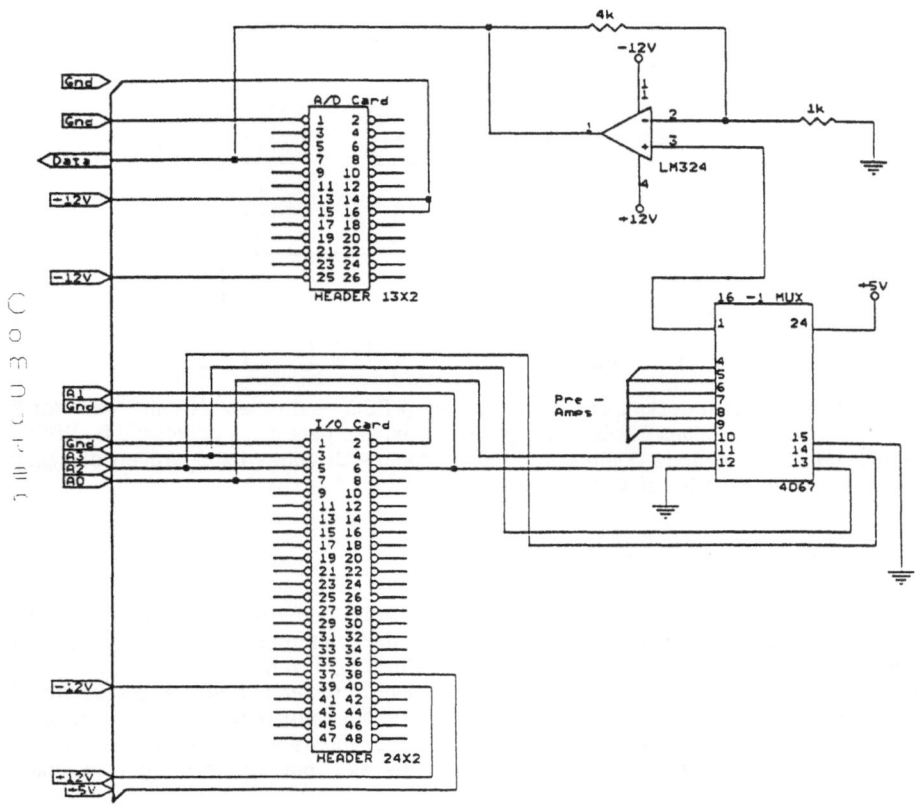

Fig. 3. Circuit diagram of switching and post-amplification stage.

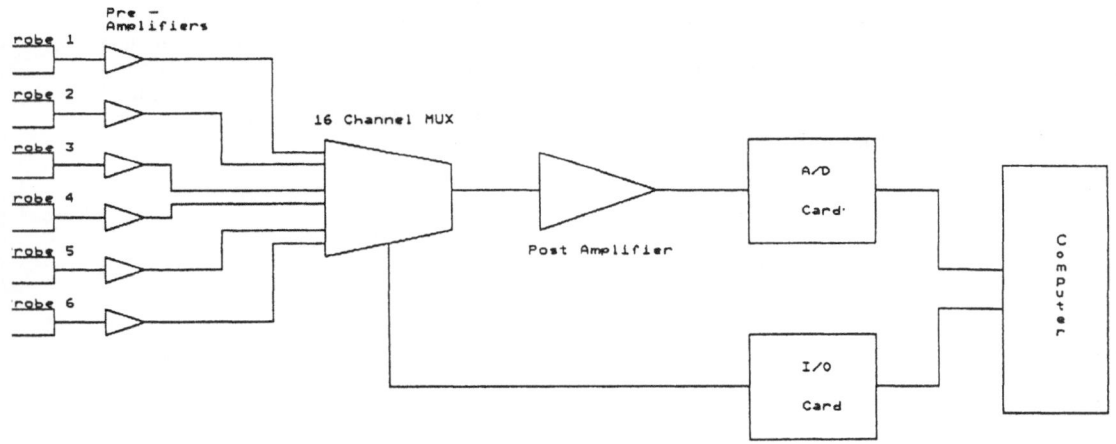

Fig. 4. Block diagram of complete circuit.

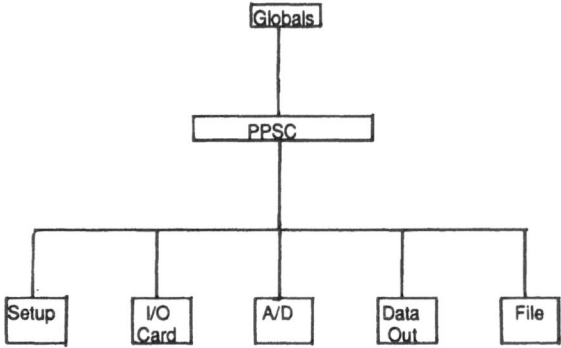

Fig. 5. Main program design. The main program (PPSC) addresses six units, each for a specific set of instructions (Globals; defines variables and data types addressed by PPSC, Setup; data for electrode control, I/O, input/putput, A/D; codes for operation of digital to analog card, File; handling of file input/output)

INSTRUMENT PERFORMANCE

Total power consumption from the computer including cards was less than 10 watts. Thermal drift was shown to be negligible. Instrument performance was tested using a pseudo-perfusion system in which perfusion solutions at various fixed oxygen levels were delivered to the electrodes. After standardizing on deaerated and oxygen-saturated solutions, intermediate values logged by the instrument were within acceptable ranges. The only difficulties experienced were with the electrodes and involved leakage under pressure and occasional bubbles collecting in the electrode chamber after switching perfusion circuits.

CONCLUSIONS

It is vital to be able to obtain information about oxygen concentrations during experiments on whole animals, organs and isolated tissues. The value of this information is greatly enhanced if it can be manipulated during those experiments into more readily interpretable parameters such as oxygen consumption rates. The low cost instrument described in this work will enable the collection of such data from an *in vitro* human placental perfusion.

A METHOD FOR SIMULTANEOUS RECORDING OF TISSUE P_{O_2} AND EP

IN THE BRAIN CORTEX OF A TEST ANIMAL WITH A SINGLE ELECTRODE

Herman Vermariën, Koen van Rossem, Rifat Turgut Altan and Karin Decuyper

Laboratory of Physiology and Physiopathology, University of Brussels VUB

Laarbeeklaan 103, B-1090 Brussels, Belgium

ABSTRACT

The time course of local variables as tissue P_{O_2} and evoked potential (EP) in brain cortical tissue of a test animal during a sudden experimental intervention (such as a photochemically induced focal infarction) can be evaluated more accurately if these variables are continuously measured at the very same site. Moreover, application of a single electrode also has practical advantages (simplified implantation procedure, reduced risk of affecting tissue).

A measuring apparatus prototype has been developed for using with chronically implanted membrane covered electrodes, constructed for polarographic P_{O_2} monitoring. Simultaneous and continuous measurement of both P_{O_2} and EP is based on their separated frequency spectra : 0 to 0.5 Hz for P_{O_2} and 5 to 1 000 Hz for EP. As required by the polarographic principle, the system is designed with a very low input impedance in the first range; in the second frequency interval input impedance is made high satisfying conditions for biopotential recording. The measuring performance of the system is quantified by the current (P_{O_2}) and the voltage (EP) transfer functions.

Two rabbits each of them having 3 electrodes implanted have been used to adjust the system's design parameters in vivo. Errors due to the process of simultaneous recording appeared extremely small. The applicability of the principle described is not restricted to the experimental conditions envisaged. Nevertheless, as the results depend on the interaction between electrode and amplifier impedances, the latter is to be adjusted if electrodes are connected having properties too different from the group handled in this study.

INTRODUCTION

The evolution of local physiological variables during a suddenly induced experimental event can be evaluated more precisely if these variables are continuously recorded at the same site in tissue. In the application envisaged oxygen tension (P_{O_2}) is studied together with somesthetic evoked potential (EP) in the brain cortical tissue of a rabbit before, during and following photochemical induction of a focal infarction (van Rossem et al., 1990 (b)). For this purpose a measuring system has been designed which allows combined recording using a single electrode, originally intended for polarographic P_{O_2} derivation. The electrode (van Rossem et al., 1990 (a)) is made with a platinum wire ($\varnothing$ 0.1 mm), insulated except for it's tip (length 1 mm) which is covered with a homogenous membrane of cellulose acetate (0.02 mm thickness). In vitro calibration tests showed an almost perfect linear relation between P_{O_2} and electrode current (polarization voltage V_p = -0.6 V). These electrodes are chronically implanted, 1 mm in cortical tissue; a disk-shaped Ag/AgCl reference electrode is attached to the animal's ear. Measurements are performed with an accurate current-to-voltage convertor providing a stable polarization voltage (V_p = -0.6 V). EP is derived using Ag electrodes ($\varnothing$ 0.3 mm) placed on the cortical surface and a needle electrode as a reference inserted in the skin above the animal's nose. Stimulation of the forepaw, amplification, filtering (3.2 to 800 Hz) and averaging are performed with the aid of a conventional apparatus (Medelec).

Oxygen Transport to Tissue XIV, Edited by W. Erdmann and
D.F. Bruley, Plenum Press, New York, 1992

DEFINITIONS, PRINCIPLES AND PHYSICAL MODELING

According to literature on polarographic P_{O_2} electrodes and biopotential electrodes (Newman, 1978; Mylrea, 1988) recording principles for both applications can be summarized as presented in figure 1a : a measuring electrode and a reference electrode (Ag/AgCl) are connected to a measuring system via a polarization voltage source (V_p) (the latter being redundant in the case of biopotential recording). The system receives an input current I_i and observes an input voltage V_i; it's input impedance Z_i determines a relation between both variables.

$$Z_i = V_i/I_i \qquad [1]$$

Although P_{O_2} polarograms show non-linear relations between electrode current and voltage, linearity can be assumed for small alterations (e.g. ± 0.1 V) around V_p (- 0.6 V) and linear system theory can be applied. The interface between the measuring electrode and the tissue can be modeled by the electrode impedance Z_e with a current source I_{O_2} parallel to it (representing current generation as a result of oxygen reduction). Biopotentials (V_b) generated by the excitable tissue between the measuring and the reference electrode can be represented by a voltage source in series (Figure 1b). If the reference electrode is large and stable, it's contribution to the model can be disregarded.

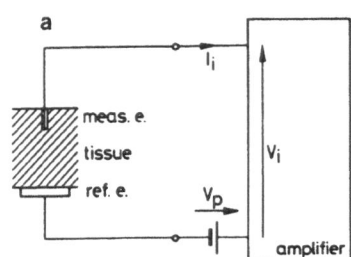 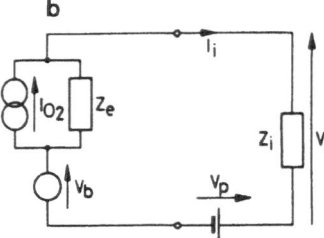

Figure 1. a) A measuring electrode and a reference electrode connected to an amplifier via a polarization source (V_p) (left). b) A physical model for the electrode-amplifier situation (right).

P_{O_2} measurement requires a $Z_i = 0$ (short circuit). In that case,

$$V_i = 0 \quad \text{and} \quad I_i = I_{O_2} + (V_b - V_p)/Z_e \qquad [2]$$

Biopotential measurement requires a $Z_i = \infty$ (open circuit). In that case,

$$I_i = 0 \quad \text{and} \quad V_i = Z_e I_{O_2} + V_b - V_p \qquad [3]$$

P_{O_2} information can be found in a constant term and in low frequencies (maximum frequency f_1). Biopotential information extends to higher frequencies; if EP is envisaged the higher frequencies are the only important ones (minimum frequency f_2). A "middle" frequency f_m can be defined as

$$f_m = \sqrt{f_1 f_2} \qquad [4]$$

In relation [3] the constant term V_p is redundant and a possible influence of $Z_e I_{O_2}$ is eliminated by the EP averaging process. In relation [2] $-V_p/Z_e$ represents the residual current of the electrode ($P_{O_2} = 0$) and V_b/Z_e is small in the frequency range envisaged. As will be shown later, experimental results support these approximations. In both cases "cross-talk" terms can be disregarded.

Combined recording (I_i and V_i) can be accomplished meaningfully by designing the measuring amplifier with a specific input impedance which is essentially low in the P_{O_2} frequency range and high in the EP range. If Z_i is neither zero nor infinite, then

$$I_i = I_{O_2} Z_e/(Z_i + Z_e) + (V_b - V_p)/(Z_i + Z_e)$$
$$V_i = I_{O_2} Z_i Z_e/(Z_i + Z_e) + (V_b - V_p)Z_i/(Z_i + Z_e) \qquad [5]$$

If, for frequencies (f) below f_1, $Z_i \ll Z_e$, then relations [5] approximate [2]; if, for $f > f_2$, $Z_i \gg Z_e$, then relations [5] approximate [3].

Based on these relations measuring transfer functions can be defined : a current transfer function $T_{\Pi} = I_i/I_{O_2}$ ($V_b = 0$) and a voltage transfer function $T_{EE} = V_i/V_b$ ($I_{O_2} = 0$). As such,

$$T_{\Pi} = Z_e/(Z_i + Z_e) \,, T_{EE} = Z_i/(Z_i + Z_e) \text{ and } T_{\Pi} = 1 - T_{EE} \qquad [6]$$

Ideally, the amplitude transfer should equal 1 (0 dB) and the phase transfer should equal 0, for both transfer functions in their correspondent frequency ranges. Errors may originate from deviating amplitude or phase transfers or/and from deviating bandwidths.

For further analysis the electrode impedance has to be discussed. The most simple model describing the electrode behaviour is a resistance R_e in parallel with a capacitance C_e (Figure 2a). In accordance to literature (MacDonald, 1977) the results will show that this model is oversimplified. Although inaccurate, the model agrees with the findings that the electrode impedance is high for low frequencies (R_e) and that it diminishes towards higher frequencies ($1/(2\pi f C_e)$). This property is advantageous for solving the problem of combined measurement.

The most simple input impedance satisfying the requirements established previously is an inductance L_i (with an unavoidable resistance R_i in series) (Figure 2b). The input impedance is then low for low frequencies (R_i) and increases towards higher frequencies ($2\pi f L_i$). The electrode-amplifier model represents a damped resonant system with a resonant frequency f_r determined by C_e and L_i, equal for both T_{Π} and T_{EE} ($f_r = 1/(2\pi\sqrt{L_i C_e})$).

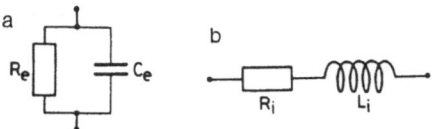

Figure 2. a) Electrode impedance model : a resistance R_e in parallel with a capacitance C_e (left). b) Amplifier input impedance model : a resistance R_i in series with an inductance L_i (right).

MATERIALS AND METHODS

The system design for combined measurement of P_{O_2} and EP is based on results obtained in vivo in 2 rabbits, each of them having 3 P_{O_2} electrodes implanted (van Rossem et al., 1990 (b)).

Firstly, spectral limit frequencies have to be determined for P_{O_2} (f_1) and for EP (f_2). P_{O_2} is measured with a properly designed current-to-voltage convertor FYSPpO$_2$1 (external or internal (-0.6V)V_p, $Z_i = 0$, conversion factor 10^7 VA^{-1}, 4 identical channels). Spectral analysis is performed with a spectrum analyser HP3582A in the 0 - 1 Hz range, $V_p = -0.6$ V and $V_p = 0$ V. EP is obtained with a conventional 2-channel apparatus (Medelec) and f_2 is determined by comparing a conventionally filtered EP (3.2 Hz; channel 1) with other high-pass filtered versions of the same EP (channel 2; 0.8, 1.6, 8 and 16 Hz, -3 dB frequencies; average of 100 stimuli to the forepaw).

Electrode impedance is approximated in 2 ways. Firstly, polarograms are obtained with the aid of the current-to-voltage convertor FYSPpO$_2$1, a waveform generator providing an external polarization voltage (a triangular alteration (0.1 V, 0.0016 Hz superposed on the normal constant -0.6 V) and a XY-recorder (registering I_{O_2} as a function of V_p). Electrode resistance (R_e) at the very low frequencies is derived from the slope of the polarogram. Secondly, Z_e is obtained by spectral analysis in a range from 0.3 to 1 000 Hz. Referring to figure 1a, a test voltage source V_t (white noise signal) is connected in series with the reference electrode and the resulting current is measured with the FYSPpO$_2$1. The spectrum analyser HP3582A then calculates V_t/I_i, representing Z_e provided the test source sufficiently exceeds the physiological sources. Nevertheless, V_t should be smaller than 0.1 V to avoid non-linear effects.

A prototype for combined recording FYSPpO$_2$2 has been built, allowing 3 modes of operation: P_{O_2} alone, EP alone and P_{O_2} and EP simultaneously available. When electrode impedances are known one can estimate the inductance value required. The inductance is realized with a gyrator circuit applying only resistors, capacitors and active components. For obtaining T_{EE}, a voltage test source V_t is connected in series with the reference electrode and the sprectrum analyser calculates V_i/V_t (same remark as for Z_e); for obtaining T_{Π}, a current test source I_t is connected parallel to the measuring electrode and the reference electrode and the spectrum analyser calculates I_i/I_t (same remark as for Z_e).

RESULTS

P_{O_2} and EP limit frequencies

Figure 3a shows typical spectral density curves of I_i $(nA/\sqrt{Hz})$ with $V_p = -0.6$ V (I_{O_2}) and $V_p = 0$ V. The latter represents the physiological noise level and corresponds with the term V_b/Z_e in eq. [2]. The I_{O_2} spectral density more or less shows a plateau (0 - 0.1 Hz) and then gradually diminishes towards higher frequencies. To obtain a quantitative estimate for f_1 an average density is calculated in the 0 - 0.1 Hz range for the 6 electrodes and f_1 is then defined as the frequency at which the current density is fallen to 10 % of the average value : current density (average $\pm$ S.D.) in 0 - 0.1 Hz amounts 3.39 $\pm$ 0.12 $nA/\sqrt{Hz}$; $f_1 = 0.31 \pm 0.06$ Hz (minimum 0.19 Hz, maximum 0.37 Hz). 0.5 Hz is then established as limit frequency for I_{O_2}.

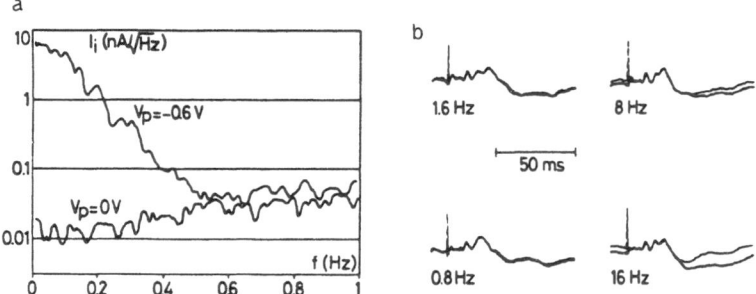

Figure 3. a) Typical current density curves $(nA/\sqrt{Hz})$, $V_p = -0.6$ V and $V_p = 0$ V, as a function of frequency (Hz) (left). b) Evoked potentials as processed by different high-pass filters ($f_{-3dB} = 0.8$, 1.6, 8 and 16 Hz) with 3.2 Hz as a reference (right). The peak in the signal is the stimulation artefact.

For EP no meaningful difference is observed between filter 3.2 Hz and respectively 1.6 and 0.8 Hz (Figure 3b); between 3.2 Hz and 8 Hz a difference in shape appears on the slow wave, which is increased when 3.2 Hz is compared with 16 Hz. Based on these observations the minimum frequency for EP is estimated at 5 Hz. According to [4] $f_m = 1.6$ Hz.

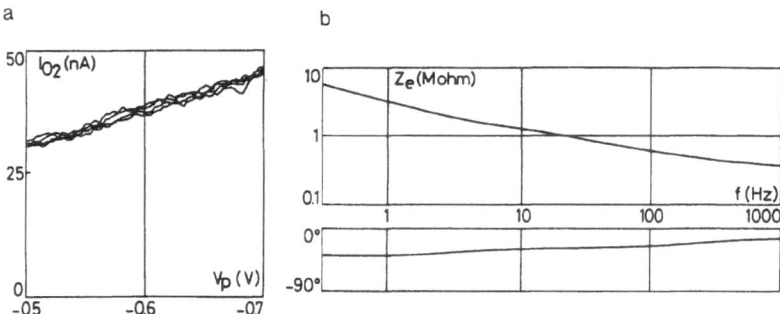

Figure 4. a) A typical polarogram (electrode K4E3) (left). Alterations are due to the normal fluctuations of P_{O_2} level in vivo. b) Amplitude and phase of electrode impedance (electrode K4E3) (right).

Electrode impedance

Polarography allows to estimate R_e at the very low frequencies ($R_e(0)$) : 23.8 $\pm$ 9.9 MΩ. Figure 4a shows a typical polarogram and figure 4b Z_e (amplitude and phase) as a function of frequency. At 1 Hz the equivalent resistance value $R_e(1)$ amounts 3.15 MΩ ($\pm$ 0.64 MΩ) and the capacitance value $C_e(1)$ 49.3 nF ($\pm$ 12.9 nF). The slope on the amplitude curve is smaller than -20 dB/decade and the phase delay less than -90° as one could expect from the model (Figure 2).

Current and voltage transfer functions

Transfer functions were firstly recorded for an inductance of 33 kH. Amplitude and phase errors were sufficiently small in the pass-bands but bandwidths had to be adjusted. Based on these in vivo results the second and final estimate for L_i is 100 kH. For reasons of lowering electronic noise level a relatively high series resistor R_i is used in the gyrator circuit (180 kΩ). Figures 5a and 5b show typical results for T_{II} (P_{O_2} alone; P_{O_2} + EP) and for T_{EE} (EP alone; EP + P_{O_2}). Both transfer functions in combined mode show a small overshoot corresponding to damped resonance. Beyond these effects curves level with each other in the significant bandwidths. Pass-bands are indicated by a -3 dB frequency or a 45° frequency. Mean values and standard deviations are given in table 1.

Table 1. Average values ($\overline{X}$) and standard deviations (S.D.) of T_{II} and T_{EE} parameters (n = 6).

| | T_{II} | | T_{EE} | | | | | |
	f_{-3dB} (Hz)	$f_{45°}$ (Hz)	f_{-3dB} (Hz)	$f_{45°}$ (Hz)	α_{-3dB} (-)	$\alpha_{45°}$ (-)	f_{m-3dB} (Hz)	$f_{m45°}$ (Hz)
$\overline{X}$	2.39	1.34	1.25	2.15	1.95	1.61	1.72	1.70
S.D.	0.19	0.10	0.20	0.17	0.32	0.07	0.16	0.12

Limit frequencies are -3 dB frequency (f_{-3dB}) and 45° frequency ($f_{45°}$). α_{-3dB} and $\alpha_{45°}$ represent ratios between correspondent frequencies. f_{m-3dB} and $f_{m45°}$ are middle frequencies defined according to eq. [4].

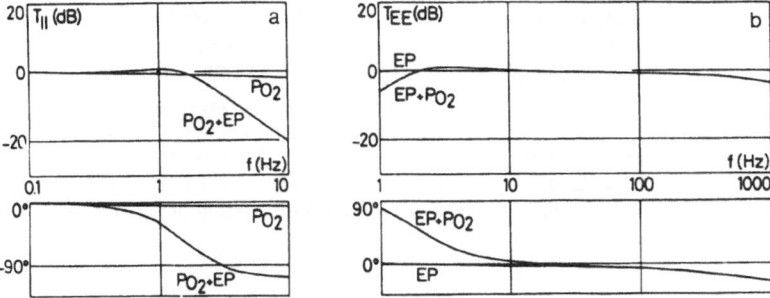

Figure 5. a) Current transfer functions (amplitude and phase) in modes P_{O_2} and P_{O_2} + EP (electrode K4E3) (left). b) Voltage transfer functions (amplitude and phase) in modes EP and EP + P_{O_2} (K4E3) (right).

Recordings

Figure 6a shows a typical recording of I_{O_2} (mode P_{O_2} alone and combined P_{O_2} + EP); the switching-over from one mode to another causes only a short transient effect. Figure 6b shows EP's obtained in the mode EP alone and EP + P_{O_2}.

DISCUSSION AND CONCLUSION

The middle frequencies calculated on the limit frequencies of the measuring characteristics according to both criteria (-3 dB and 45°) are respectively 1.72 and 1.70 Hz with small standard deviations (Table 1). Considering the signal middle frequency previously determined (1.6 Hz) a small enlargement of the inductance could provide excellent matching, but this seems unnecessary as the values have been determined on a small population of electrodes and test animals. More-over, all limit frequencies are contained within the accepted range (0.5 to 5 Hz) : the lowest limit frequency for T_{II} is 1.20 Hz (45°), the highest limit frequency for T_{EE} is 2.36 Hz (45°).

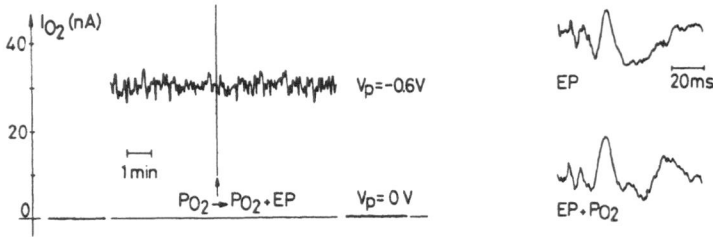

Figure 6. a) Current recording in modes P_{O_2} and P_{O_2} + EP ($V_p = 0$, $V_p = -0.6$ V) (left). b) EP in modes EP and EP + P_{O_2} (right).

The series resistor R_i can cause an error with respect to P_{O_2} measurement : it has to be small compared to $R_e(0)$: only 1 % of $\overline{R_e(0)}$ has been allowed. If designed otherwise, a large R_i would decrease sensitivity; nevertheless, the latter problem can be met via the calibration process.

The measuring device has been designed and adjusted for a specific application mentioned before. Spectral analysis has only been performed on normal animals; nevertheless large differences in results are not expected. Following infarct induction P_{O_2} pattern changes (van Rossem et al., 1990 (b)) : large slower waves and/or disappearance of variations are observed; such will not affect the estimate for f_1. Infarction gives rise to cell necrosis and new tissue generation; it is not expected that these alterations can alter significantly the electrode properties. Nevertheless additional tests are advised to prove this assumption.

The applicability of the measuring device is not restricted to the experimental situation cited. It could be used for other electrodes, but in that case the input properties are to be adjusted in order to obtain an optimal separation between both signals. E.g., if an electrode is used with a larger sensitive surface, C_e will be higher and R_e lower : in that case a smaller L_i and R_i can be used, respectively satisfying middle frequency requirements and P_{O_2} error limitation.

To check the measuring quality in vivo during routine procedures, sofisticated test apparatus is unnecessary. With the aid of a waveform generator one can easily determine the limit frequency (45°) of T_{EE}. If $\alpha_{45°}$ is known for the specific group of electrodes used one can estimate the limit frequency of T_{II} without the need for a current source. Further, with a fixed test waveform provided in the apparatus, one could easily adjust the inductance according to the limit frequency requirements.

ACKNOWLEDGEMENTS

This investigation is supported by grant 3.0023.91 of the Belgian Fund for Medical Scientific Research and by the OZR VUB.

REFERENCES

van Rossem, K., Vermariën, H., and Bourgain, R., Construction, calibration and evaluation of pO_2 electrodes for chronical implantation in the rabbit brain cortex, *in* : "Adv Exp Med and Biol (ISOTT 90)", Plenum Press, New York, in press 1990.

van Rossem, K., Vermariën, H., Decuyper, K., and Bourgain, R., Photothrombosis in rabbit brain cortex : follow up by continuous pO_2 measurement, *in* : "Adv in Exp Med Biol (ISOTT 90)", Plenum Press, New York, in press 1990.

Neuman, M. R., 1978, Biopotential electrodes, *in* : "Medical instrumentation", J. G. Webster, ed., Houghton Mifflin Company, Boston, 215 - 272.

Mylrea, K. C., 1988, Oxygen sensors, *in* : "Encyclopedia of medical devices and instrumentation", J. G. Webster, ed., John Wiley & Sons, New York, 2169 - 2174.

MacDonald, D. D., 1977, "Transient techniques in electrochemistry", Plenum Press, London.

COMPUTER SIMULATION OF ERYTHROCYTE TRANSIT

IN THE CEREBROCORTICAL CAPILLARY NETWORK

Antal G. Hudetz

Department of Physiology
Medical College of Wisconsin
Milwaukee, WI

INTRODUCTION

The fundamental question as to how erythrocyte transit time and flow path length ensure adequate oxygen extraction from blood during compromised blood supply or increased cerebral metabolic rate is still unanswered. A longer mean transit time would allow more complete deoxygenation of the arterial blood, however, it would attenuate the diffusion gradient of oxygen tension in the tissue. The importance of optimal erythrocyte flow path length for oxygen transport may be reflected in the characteristic tortuous pattern of cerebral capillaries (Wiederhold et al, 1976), which is different in various cytoarchitectural layers of the cerebral cortex (Duvernoy et al, 1981). Tortuosity may play a role in the prolongation of transit time to compensate for the high velocity of erythrocytes in cerebrocortical capillaries.

The distributions of erythrocyte transit time and flow path length in the anastomosing cerebral capillary network may have a complex relationship which is not clearly understood. In general, microvascular transit time is assumed to increase with the length of the flow path and/or with the lowering of flow velocity. Thus, cerebral transit time increases in hypotension (Ferrari et al, 1989), and secondary increases in transit time in partial cerebral ischemia may reflect increases in flow path length due to microvascular obstruction (Little et al, 1981). In contrast, theoretical models predict a decrease in transit time with the progressive elimination of alternative flow pathways (Hudetz and Werin, 1986). Cerebrocortical microflow pathways may be redistributed also during increased metabolic demand (Lübbers and Leniger-Follert, 1978), which would influence the distribution of transit times within the network.

In order to understand and to mathematically model cerebral microflow distribution and oxygen transport, one would like to know the distributions of both erythrocyte transit time and flow path length (Groebe, 1988). While cerebrocortical transit times can be estimated using the indicator dilution technique, functional flow path lengths cannot be assessed using this method. An alternative approach to estimate microvascular path lengths and transit times is the computer simulation of hemodynamics in reconstructed capillary networks.

The objective of this study was to determine the quantitative

Oxygen Transport to Tissue XIV, Edited by W. Erdmann and
D.F. Bruley, Plenum Press, New York, 1992

relationship between erythrocyte transit time and flow path length in the cerebrocortical microcirculation using a probabilistic computer simulation of erythrocyte transit in a true three-dimensional microvascular network of the rat cerebral cortex reconstructed from in vivo images of vascular casts of this circulation. The non-Newtonian characteristics of blood and the phase separation of blood flow at bifurcations were accounted for in the simulation. Another novel feature of this simulation was the iterative variation of boundary conditions to optimize vascular wall shear stress in the network. As it will be shown, the simulation suggests a nonlinear relationship between erythrocyte path length and transit time in the reconstructed network.

METHODS

Network reconstruction

A cerebrocortical microvascular network was reconstructed from a vascular corrosion cast of the rat brain. The cast was prepared as described previously (Hudetz et al, 1989) by bilateral intracarotid injection of Batson's No. 17 casting solution (Polysciences, Inc., Warrington, PA) following perfusion with 0.9% NaCl solution containing 10 IU/ml heparin and 10^{-7}g/ml papaverine and subsequently with 4% buffered formalin for 10 minutes at a pressure of 100 cmH$_2$O. Thick sections (0.5 to 1 mm) of the cortex were cut and placed in 20% NaOH for 12 to 24 hours followed by 5% NaClO for 1 to 3 weeks. The cast was then dehydrated and embedded in epoxy.

The geometry and connection pattern of vessels as represented by the corrosion cast were reconstructed using a modified version of the Bioquant IV image analysis system (R & M Biometrics, Nashville, TN). Briefly, the microscopic video image of the vascular cast was displayed on the computer screen and the three dimensional course of vessels was traced using a superimposed cursor and graphics overlay. The z-dimension was assessed by optically sectioning the traced vessel segment by adjusting the elevation of the microscope stage. Vessel diameters were measured using the same system at several locations along each segment. Vascular branch points, arterial and venous terminals, and vessels which were cut at the section surfaces were typified and labeled using the software and stored in the data base. Further details of the tracing and reconstruction system have been described (Hudetz et al, 1989).

Computer simulation

Pressure-flow distribution. A first approximation of flow distribution in the network was calculated assuming constant blood viscosity. The hydraulic resistance R_{ij} of each vessel segment between branch points i and j was estimated from the measured values of vessel diameter and length based on Poiseuille flow. Equations for flow were written for each vascular segment, and at each branch point the flow balance was written as a second equation. The equations were combined to yield

$$\sum_{j=1}^{n_i} (P_i - P_j)/R_{ij} + S_i = 0 \qquad (i \neq j)$$

where n_i is the number of segments connected to node i (i=1,n), P_i and P_j are the intravascular pressures at the two ends of a vessel segment, and S_i is an unknown flow source which is different from zero at the terminal nodes (arterial, venous or cut vessels, see below). The resulting system of linear equations was solved by matrix inversion using fixed pressure boundary conditions at the terminal nodes. The intravascular pressure of 30mmHg was assumed for arteries and 20mmHg for veins.

An iterative optimization scheme was used to account for unknown boundary conditions at the broken end of 52 capillaries which were cut by sectioning. Briefly, a first approximation of flow distribution was obtained assuming an intravascular pressure of 25mmHg at the cut end of these vessels. The cut end pressures were then varied at random by ±0.5 mmHg increments within the range of 20 to 30mmHg with the constraint of minimizing the variation in wall shear stress in the network. We called this procedure "hemodynamic optimization" (Hudetz and Kiani, 1989). Wall shear stress (s) in each vessel was calculated as

$$s = (P_i - P_j) r / 2L$$

where r is the vessel radius, and L is the vessel length.

Next, erythrocyte flow was calculated in the network. The partitioning of red blood cells at bifurcations was approximated by the following formula (Klitzman and Johnson, 1982):

$$[(1 - f)/f] = [(1-q)/q]^B$$

where $q = Q_i/Q$ and $f = F_i/F$ are fractional blood flow and red cell flux in branch i (i=1,2) with respect to the parent vessel. We demonstrated earlier (Hudetz, 1990) that small variations in the parameter B result in significantly different statistical distributions of vascular hematocrit. In the present work, the value B=1.15 (Klitzman and Johnson, 1982) was used.

In order to account for non-Newtonian flow, a semi-empirical mathematical model of apparent relative viscosity as a function of vessel diameter and vessel discharge hematocrit was developed earlier (Kiani and Hudetz, 1991) in the following form:

$$\mu_{app} = \mu_p[1 - (1 - \mu_p/\mu_c)(1 - 2w/D)^4]^{-1}[1 - (D_m/D)^4]^{-1}$$

$$\mu_c = \exp(0.48 + 2.35H_d)$$

$$w = 2.03 - 2H_d$$

where the following notations have been used:
μ_p = viscosity of plasma (=1.7cp)
μ_c = core viscosity in a large tube
w = width of marginal layer
D = vessel diameter
D_m = minimum vessel diameter that the erythrocyte can traverse (=2.7μm)
H_d = discharge hematocrit

Using the formulae above, the distribution red cell flux was calculated in an iterative manner. From vessel diameter and hematocrit, apparent viscosity and vascular resistance were updated, and the pressure distribution was recalculated by solving the system of linear equations. Red cell flux was then calculated using the bifurcation formula. This procedure was repeated several times until convergence in red cell flow was reached (maximum of 100 iterations). All calculations were performed on the Apollo 10000 computer.

Path length and transit time. The transit of erythrocytes through the network was simulated using the Monte Carlo method as described earlier (Hudetz et al, 1989). Briefly, the probabilistic pathway of cells traveling from one of the two arterioles to one of the two veins was simulated using a random number generator. At each bifurcation, the probability of cell entry into a daughter vessel was selected to be proportional to the calculated fractional cell flux in that vessel. Path length and transit time for each cell passage were calculated as the sums of the passed segment lengths or segment transit times, respectively. Segment transit time for vessels larger than $8\mu m$ was calculated assuming a near-parabolic velocity profile, $v(r_c)$, in the following form:

$$v(r_c) = 2v_a[1 - (r_c/r)^2]$$

where v_a is the average cell velocity and r_c is the radial location of the cell satifying $r_c=(r-D_m/2)(1-x^2)$ where x is a random variable, $0<x<1$.

RESULTS

Network geometry

The reconstructed cerebral capillary network contained 172 vascular segments, 116 of which were internal, and 52 external (cut) segments (Figure 1). The network included two terminal arteries and two small veins. Vascular diameters ranged between 3.6 and $20.4\mu m$ (lognormal distribution, mean$\pm$SD = 6.8$\pm$2.2). Segment lengths varied between 9.8 and 302.4 μm (lognormal distribution, mean$\pm$SD = 74.4$\pm$62.9). Frequency distributions of vessel diameter and segment length are displayed in Figure 2.

Fig. 1. Two dimensional projection of the reconstructed cerebral capillary network. A=artery, V=vein. Only the trunks of cut segments are shown, although the course of these vessels was traced all the way to point of transection.

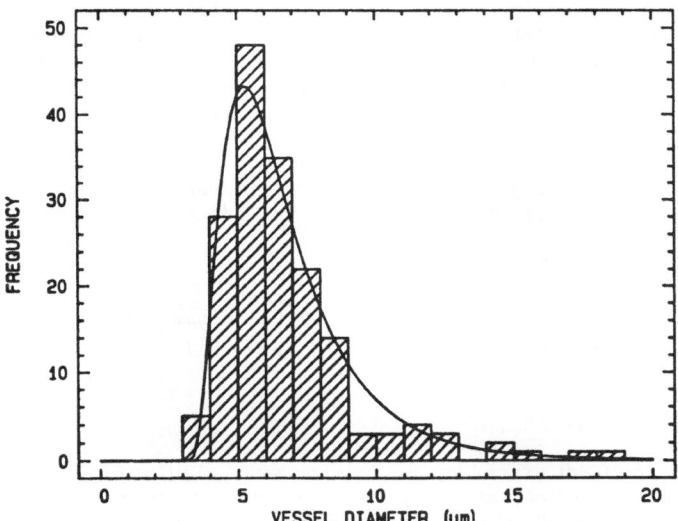

Fig. 2. Frequency distributions of vessel diameter (top) and segment length
(bottom) from the network shown in Figure 1. The continuous curves
indicate the best fit of the lognormal distribution to data.

Table 1. Computer simulated path lengths and associated mean transit times of erythrocytes in the cerebrocortical microvascular network.

Path length(μm)	594	701	815	909	1185	1234	1350	1417
Transit time(s)	1.98	1.84	1.99	3.37	4.4	15.3	12.6	10.2
Relative frequency(%)	21.3	17.4	37.6	20.7	0.9	0.4	0.8	0.5

Pressure-flow distribution

Intravascular pressure, wall shear stress, and red cell flux in each vessel were calculated with and without hemodynamic optimization, that is, the iterative adjustment of cut end pressures. Ten simulation runs were performed with optimization, each time using a different random number seed. Data from these simulations were lumped and compiled into frequency distributions. The frequency distributions of wall shear stress before and after hemodynamic optimization are displayed in Figure 3. As shown there, optimization resulted in more equalized wall shear stress in the network. Simultaneously, the distribution of intravascular pressures was widened and transformed into a normal distribution (Figure 4). The optimization did not influence the statistical distribution of red blood cell flux. The final microvascular (discharge) hematocrit ranged between 0.1 and 0.4 (Figure 5).

When the calculated red cell flows in vessels of similar diameter (within 2μm) were averaged, a power law relationship between cell flow (F_c) and vessel diameter (D) was found in the following form:

$$F_c = K \, D^n$$

where K is a constant and n is the so called "diameter exponent" whose value was 2.87 from the best fit (r=0.99, p<0.001). The regression is displayed in Figure 6.

Path length and transit time

A total of 10000 erythrocyte transits were simulated in this study. The computer simulation predicted eight different flow pathways of erythrocytes through the network. Flow path lengths together with mean transit times along these paths are listed in Table 1. Four of the paths were used much more frequently (had higher cell flow) than the remaining ones.

The frequency distribution of transit times is shown in Figure 7. As seen there, most transit times were distributed around the two major peaks of 1.91 and 3.16 seconds. The variation in transit time around each peak was due to the intravascular velocity profile. The coefficient of variation around any of the two major transit time peaks was about 10 percent.

The predicted flow path lengths and transit times did not seem to correlate linearly with each other. Instead, there was a significant correlation (r=0.91) between the major path lengths and the logarithms of the corresponding mean transit times (Figure 8). Thus, erythrocyte transit time increased exponentially as a function of path length within the network.

DISCUSSION

The objective of this study was to test whether functional flow path lengths can be predicted from transit times in the cerebrocortical microcirculation. The quantitative relationship between microvascular path

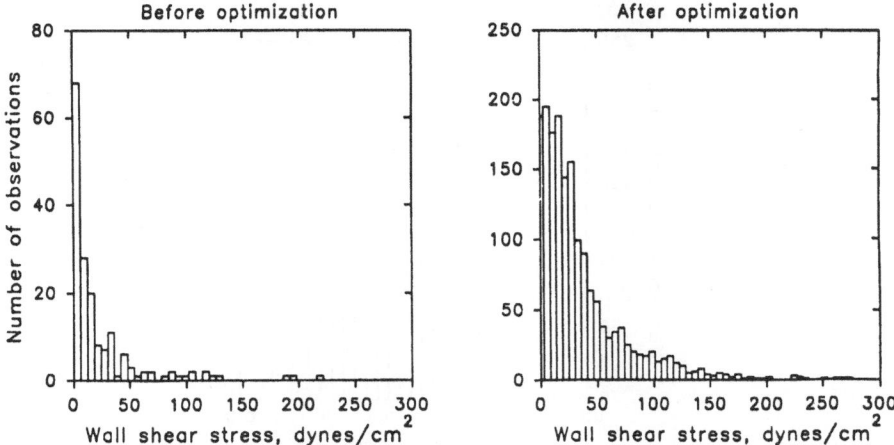

Fig. 3. Frequency distributions of calculated wall shear stress before
(left) and after (right) hemodynamic optimization (explanation in
text). The total number of calculated values (observations) is ten
times higher after than before optimization because the
optimization program was run ten times.

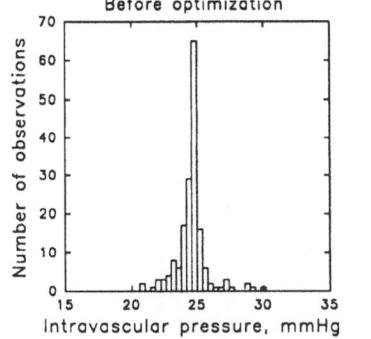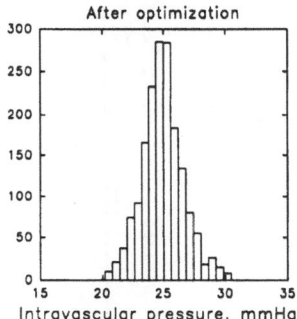

Fig. 4. Distribution of capillary pressures before (left) and after (right)
hemodynamic optimization. Note how the distinct peak at 25mmHg
associated with the cut vessels disappears after optimization.

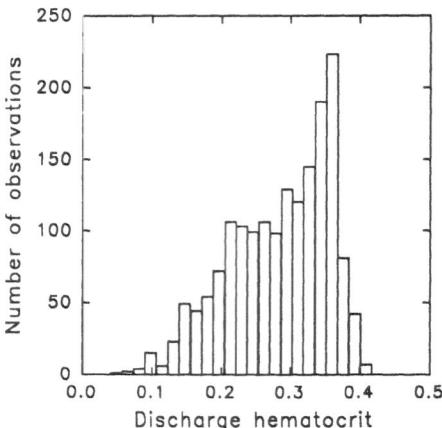

Fig. 5. Frequency distribution of calculated discharge hematocrit in the cerebrocortical microvascular network.

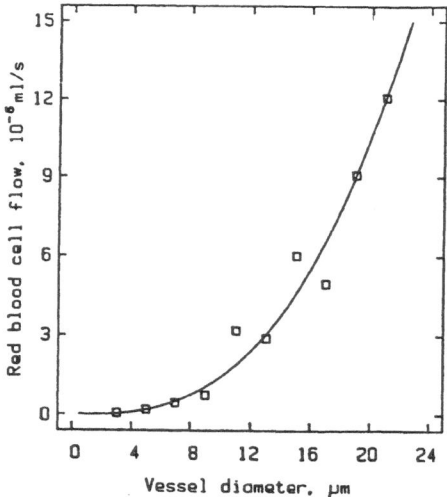

Fig. 6. Predicted power law relationship between red cell flow and vessel diameter in the cerebrocortical microvascular network.

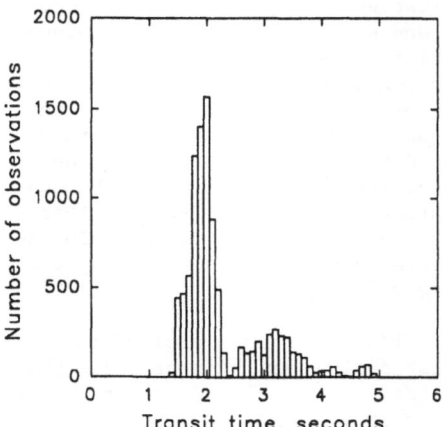

Fig. 7. Frequency distribution of erythrocyte transit times as predicted by
Monte Carlo simulation.

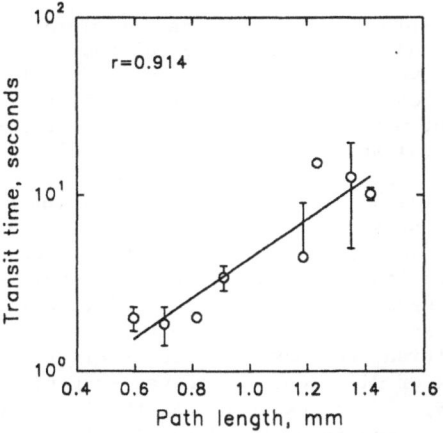

Fig. 8. Exponential relationship between computer simulated erythrocyte
transit time and flow path length in the cerebrocortical
microvascular network.

length and transit time has not been evaluated before because of the obstacles to measure functional flow path length in the complex, three-dimensional cerebral microvascular bed. To overcome this limitation, cerebrocortical microvascular path length and transit time of erythrocytes were estimated by computer simulation in this work.

Most of the applied methods have been developed and described before (Hudetz et al, 1989; Hudetz, 1990). A novel feature in the present simulation was the treatment of unknown boundary conditions by hemodynamic optimization (Hudetz and Kiani, 1989). This technique has not been validated, which remains the subject of future work. We used this approach in the present simulation study because we found previously (Hudetz and Kiani, 1991) that network adaptation to mean wall shear stress resulted in a decrease in total power dissipation without rarefaction of the network. These results suggested that a better approximation of overall network hemodynamics may be obtained by optimizing the boundary conditions to achieve minimum variation if wall shear stress, than using an arbitrary intravascular pressure for all cut end vessels. In the present study, the normalization of intravascular pressure as a result of hemodynamic optimization supported this hypothesis.

Another result of this work that supported the role of wall shear stress in microvascular design was the near-cubic power law relationship of flow and vessel diameter. A cubic flow-diameter relationship which complies with the constant shear stress hypothesis was originally suggested by Murray (1926) based on optimization principles. The validity of this relationship was supported by physiological data in the cremaster muscle microcirculation (Mayrovitz and Roy, 1983) as well as in cat pial arteries of 20 to 200 μm diameter (Kobari et al, 1984). The present results extend this finding to cerebrocortical microvessels having diameter of 3.6 to 20.4 μm. The obtained red cell flow-diameter relationship was not the consequence of hemodynamic optimization (which itself was based on the shear stress hypothesis) because a similar relationship was already present before the optimization.

Microvascular erythrocyte transit time was predicted to vary mostly between 1.5 and 4 seconds. Since these calculated values depend on the selected arterial and venous pressures in the model, they only represent an approximate physiological condition. Nevertheless, the obtained transit times are in agreement with those measured in the cerebral cortex of the cat (Tomita et al, 1983) and the dog (Ferrari et al, 1989).

This computer simulation predicted relatively few erythrocyte flow pathways compared to the several anatomical paths contained by the richly anastomosing network used in this study. The predicted main pathways may be analogous to the so called thoroughfare channels whose existence in the cerebral cortex was suggested by Hasegawa et al (1967). Other microvascular paths may contain "plasma" vessels with no or very low cell flow (Yamakawa et al, 1987). In addition, the physiological redistribution of capillary flow between pathways of different length was suggested by Lübbers and Leniger-Follert (1978). It is possible that the low number of major flow pathways in our network was in part the result of cut vessels. We believe, however, that cut vessels had little effect on the predicted relationship between path length and transit time.

The present model predicted an exponential increase in transit time with functional flow path length. This result is similar to that obtained by Groebe (1988) in a parallel capillary model of skeletal muscle microcirculation, in which red cell flow was inversely proportional to flow path length. The same assumption may explain the predicted extra long transit times in the anastomosing cerebrovascular network. Thus, long

668

transit times associated with long paths may be further prolonged by a low erythrocyte flow velocity due to higher flow resistance of these paths, provided that systematic variations in vascular diameter on resistance does not obscure the effect of path length. If our findings were true in vivo, then transit time distributions assessed under various physiological conditions would not strictly reflect the distributions of flow path length. However, accepting the predictions of the present model, flow path length distributions could be estimated by the logarithmic transformation of measured transit time distributions.

The present data should be interpreted with attention to that the computer simulation was based on the anatomical representation of a dilated microvascular bed. The potential effects of precapillary vascular tone and local variations in microvessel diameter (Kiani et al, 1991) yet have to be evaluated. The results should also be confirmed in other, preferably larger, cerebrocortical microvascular networks. The influence of boundary conditions on the simulation results would be less significant in a larger network. From physiological point of view, the consequences of the predicted nonlinear relationship between transit time and path length with respect to the transport of oxygen to cerebral tissue should be evaluated.

SUMMARY

The relationship between erythrocyte flow path length and transit time was studied using a probabilistic computer simulation of erythrocyte transit in morphometrically reconstructed cerebrocortical microvascular network of the rat. The results suggest (1) a near-cubic relationship between erythrocyte flow and vessel diameter, (2) preferential pathways of erythrocyte flow, (3) an exponential relationship between erythrocyte transit time and flow path length.

ACKNOWLEDGEMENT

The author expresses his thanks to Dr. Richard J. Roman for reviewing the manuscript. This work was supported by the NSF grant BCS-9001425 and the NIH grant HL-29578.

REFERENCES

Duvernoy, H. M., Delon, S., and Vannson, J. L., 1981, Cortical blood vessels of the human brain, Brain Res. Bull., 7:519.
Ferrari, M., Wilson, D. A., Hanley, D. F., and Traystman, R. J., 1989, Near infrared determined cerebral transit time and oxy- and deoxyhemoglobin relationships during hemorrhagic hypotension in the dog, Adv. Exp. Med. Biol., 248:55.
Groebe, K., 1988, Coupling of hemodynamics to diffusional oxygen mass transport, Adv. Exp. Med. Biol., 222:3.
Hasegawa, T., Ravens, J. R., and Toole, J. F., 1967, Precapillary arteriovenosus anastomoses. "Thoroughfare channel" in the brain. Arch. Neurol., 16:217.
Hudetz, A. G., and Kiani, M. F., 1989, Hemodynamic optimization of computer generated capillary networks, Int. J. Microcirc., 8(Suppl. 1):S16.
Hudetz, A. G., and Kiani, M. F., 1990, Dependence of cerebral capillary hematocrit on red cell flow separation at bifurcations: A computer simulation study. Adv. Exp. Med. Biol., 277:31.
Hudetz, A. G., and Kiani, M. F., 1991, The role of wall shear stress in microvascular network adaptation. Adv. Exp. Med. Biol., (in press).

Hudetz, A.G., Spaulding, J.G. and Kiani, M.F., 1989, Computer simulation of cerebral microhemodynamics. Adv. Exp. Med. Biol., 248:293.

Hudetz, A. G., and Werin, S., 1986, Percolation and transit in microvascular networks, Adv. Exp. Med. Biol., 200:79.

Kiani, M. F., Cokelet, G. R., and Sarelius, I. H., 1991, Effect of diameter variability along the capillary segment on apparent pressure drop. Proc. Fifth World Congr. Microcirculat., Louisiville, KY, p. 50.

Kiani, M. F., and Hudetz, A. G., 1991, Mathematical model of apparent blood viscosity as a function of vessel diameter and discharge hematocrit. Biorheology, 28:65.

Klitzman, B., and Johnson, P. C., 1982, Capillary network geometry and red cell distribution in hamster cremaster muscle. Am. J. Physiol., 242:H211.

Kobari, M., Gotoh, F., Fukuuchi, Y., Tanaka, K., Suzuki, N., and Uematsu, D., 1984, Blood flow velocity in the pial arteries of cats, with particular reference to the vessel diameter. J. Cereb. Blood Flow Metabol., 4:110.

Little, J. R., Cook, A., Cook, S. A., and MacIntyre, W. J., 1981, Microcirculatory obstruction in focal cerebral ischemia: albumin and erythrocyte transit, Stroke: 12:218.

Lübbers, D. W., and Leniger-Follert, E., 1978, Capillary flow in the brain cortex during changes in oxygen supply and state of activation, In: "Ciba. Fnd. Symp." 56:21.

Mayrovitz, H. N., and Roy, J., 1983, Microvascular blood flow: evidence indicating a cubic dependence on arteriolar diameter, Am. J. Physiol., 245:H1031.

Murray, C. D., 1926, The physiological principle of minimum work. I. The vascular system and the cost of blood volume, Proc. Natl. Acad. Sci. U.S.A., 12:207.

Tomita, M., Gotoh, F., Amano, T., Tanahashi, N., Kobari, M., Shinohara, T., and Mihara, B., 1983, Transfer function through regional cerebral cortex evaluated by photoelectric method. Am. J. Physiol., 245:H385.

Wiederhold, K.-H., Bielser, W., Schultz, U., Jr., Veteau, M.-J., and Hunziker, O., 1976, Three dimensional reconstruction of brain capillaries from frozen serial sections, Microvasc. Res., 11:175.

Yamakawa, T., Yamaguchi, S., Niimi, H., and Sugiyama, I., 1987, White blood cell plugging and blood flow maldistribution in the capillary network of cat cerebral cortex in acute hemorrhagic hypotension: An intravital microscopic study. Circulat. Shock, 22:323.

INSTRUMENTATION AND TECHNOLOGY FOR MULTIPARAMETRIC MAPPING OF

INTRAPARENCHYMAL CIRCULATION IN THE BRAIN CORTEX

Andras Eke

Experimental Research Department and 2nd Institute of Physiology
Semmelweis University of Medicine, Budapest, Hungary 1082 and Departments
of Neurology and Pathology, University of Alabama, Birmingham AL 35294, U.S.A.

INTRODUCTION

The intraparenchymal distribution pattern of erythrocytes and plasma should certainly interact with the oxygen supply and energy balance in tissue microareas. To enhance our knowledge of how this interaction takes place in the brain cortex, a non-destructive videophotometric method has been developed. The current state of this method has evolved through a number of intermediate steps [1, 2, 3, 4, 5] representing an interplay between the actual level of understanding of how microcirculation can be assessed by intraparenchymal indicator-dilution and the level of instrumentation and technology available to do quantitative spatial imaging as the means of detecting intraparenchymal indicator dilution from capillary beds of the intact brain cortex. The latter has always been a limiting factor in achieving the ultimate goal of obtaining high resolution maps of local erythrocyte and plasma volumes, mean transit times, volume flows, local tube and discharge hematocrit, blood volume and flow in a repeatable manner. The purpose of this paper is to present the current state of instrumentation and technology that has made multiparametric assessment of cerebrocortical circulation for these parameters possible at a spatial resolution of 10000 sites/10 mm^2.

METHODS AND RESULTS

A closed cranial window-technique and in situ quantitative television densitometry were used to record and process the television images of an epiilluminated area of the brain cortex during bolus perfusion by erythrocyte and plasma indicators (isosmotic saline and Evans blue solutions respectively) for maps of the aforementioned microcirculatory parameters according to a published protocol.[5] The schematic representation of the quantitative videoreflectometric instrumentation and experimental arrangement is shown in Fig. 1.

Erythrocyte and Plasma Indicators

Intraparenchymal indicator dilutions for erythrocytes and plasma can be nondestructively detected by epiillumination television reflectometry when erythrocyte and plasma indicators (transparent and dense to light) are being monitored in their transit through a 100x100 square array of cylindrical volumes of tissue each of which measures 0.0003 - 0.007 cubic mm and are located superficially in the brain cortex in terms of their changing the tissue's reflectance in proportion to their local, intracapillary concentration.[1, 2, 5]

The erythrocyte-indicator, saline, is a diluent solution for erythrocytes and stays in the

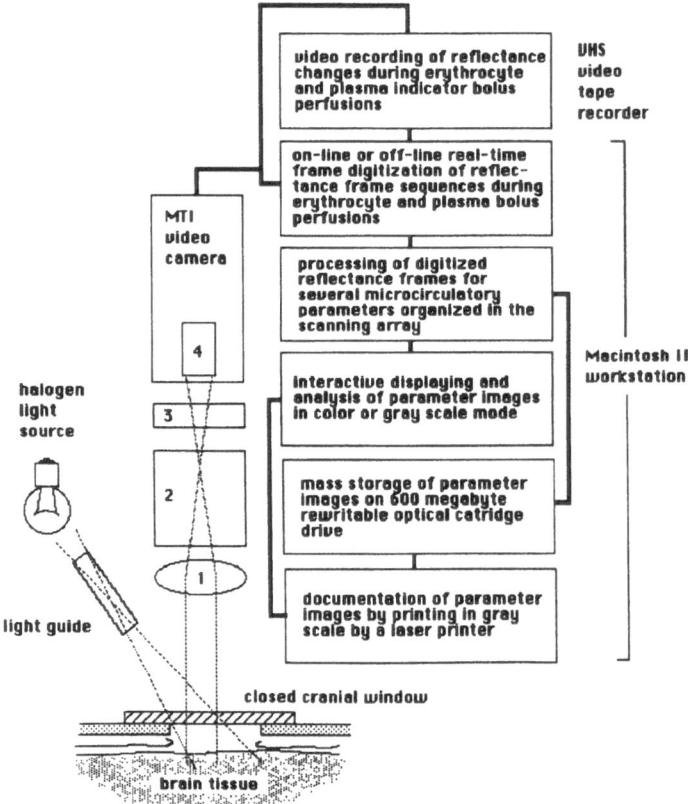

Fig. 1. Block diagram and schematic representation of the videoreflectometric setup and experiemental arrangement for multiparametric mapping of microregional circulation in the brain cortex.

1 = Macro lens assembly
2 = Extension tube.
3 = Interference filter (at 589 nm).
4 = Newicon pickup tube of a Dage MTI 67 instrumentional video camera

intracerebral vascular compartment. In essence the endogenous red blood cells act as indicators themselves. Care should be taken to make the indicators isosmotic to plasma. The reflection coefficient of the blood-brain barrier for the components of the indicator solutions has to be as close as possible to 1.0. Dyes bound to plasma proteins certainly perform better as plasma-indicators than carbon black solution does, since the latter tend to sediment within the microvessels and adhere to the endothelium when flow is low and stagnant or when the endothelium is most likely damaged.

Closed Cranial Window Technique for Epiillumination Reflectometry of the Brain Cortex

The area of the brain cortex to be imaged for the microcirculatory parameters is exposed and epiilluminated through a cranial window with white, cold light (Figs. 1 and 2). Cooled halogen lightsource can provide sufficiently high intensity illumination. The actually needed intensity is determined by the sensitivity of the video camera at settings with optimal signal to noise ratio. Epiillumination is to be carried out so that the angle between the incident and detection light paths be as small as possible. Annular fiber optics have proven excellent in achieving this geometry and capable of providing an even incident intensity profile across the field (Fig. 3).

In the visible range of wavelengths of light (i.e. 589 nm), it is necessary to implant an optical quality glass window in the skull in order to carry out reflectometry of the brain cortex. This cranial window should not only be of high optical quality (high transmittance, no distortion), but should also provide a leak-tight seal for the cerebrospinal fluid compartment of the brain. This is essential in restoring physiologic conditions in cerebral hemodynamics. In an open skull preparation, the brain surface may change its position relative to the focal plane of the video camera due to unbalanced fluctuations in cerebral blood volume. Consequently, this would result in artificial fluctuations in the reflectance signal, too.

In species of larger skull size, it is easier to come up with a suitable design for the cranial window (cat, rabbit, monkey). In rat, a special design had to be worked out that meets not only the aforementioned criteria but allows for an easy, flexible, yet very reliable positioning of the brain surface relative to the focal plane of the video camera (Figs. 2 and 3). The most characteristic feature of this design is that the video camera remains stationery, while the animal's body and head are positioned in space so that the brain's surface can be precisely brought into the focus of the video camera.

Videoreflectometry of the Brain Cortex

The television camera as an inherently scanning device is ideally suited for the task of scanning the illuminated area of the brain cortex for a square array of backscattered light intensities (reflectance) during the transit of the indicators through microregional tissue elements (voxels defined by the size of the scanning raster elements and the depth of detection in the tissue). It is done at 589 nm, one of the isosbestic points of hemoglobin. The Dage MTI Model 67 professional camera shown in Fig. 1 meets the standards for high precision spatial intensity data acquisition. It has a Newicon pickup tube with adequate response time and linear response in its output to reflected (back scattered) light intensity. Reflectance images during indicator-perfusions are stored on a VHS cassette tape recorder. High quality tape and recording mechanism are prerequisites to prevent image degradation. A proper timing signal needs to be stored on the recording tape along

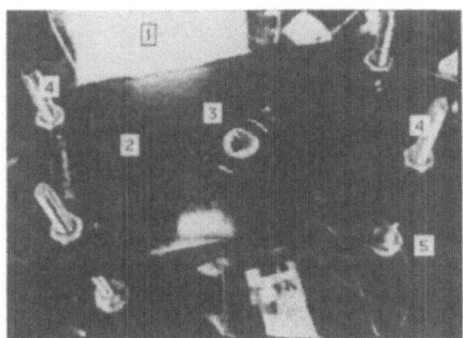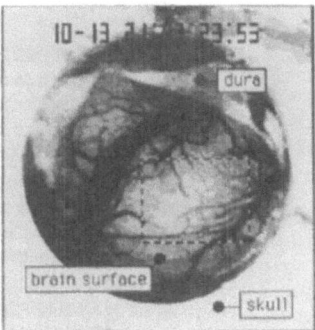

Fig. 2. Head holder assembly for high precision video reflectometry of the brain cortex in the rat. A close-up view of the cranial window's area indicated by item-3 is shown on the right. The region of interest can be freely selected and is marked by dashed lines.

1 = Anesthetized experimental animal (rat).
2 = Platform to support the animal's head and to keep the cranial window's area (3) in focus.
3 = Cranial window's area, also shown enlarged on the right.
4 = Threaded rods to adjust the relative position of the platform (2) and the supporting base plate (5).
5 = Base plate to support the animal's body (1) and the head holder platform (2).

Also see Fig. 3. for a side view of the head holder assembly!

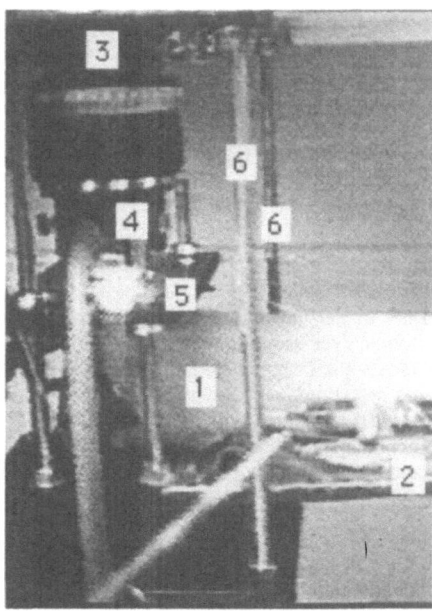

Fig. 3. Side view of the rat head holder assembly for epiillumination reflectometry.

1 = Anesthetized experimental animal (rat).
2 = Base plate to support the animal's body (1) and the head holder platform (5).
3 = Front plate of the video camera's body supporting the macro lens assembly, and the threaded rods holding the base plate (2) and the head holder platform (5).
4 = Annular fiber optic centered around the optical axis and attached to the front of the macro lens assembly. It provides epiillumination of even intensity across the field (shown as a bright spot under the lens).
5 = Platform to support the animal's head and to keep the cranial window's area in focus of the epiillumination and that of the camera's lens assembly.
6 = Threaded rods to adjust the relative position of the platform (5) and the supporting base plate (2) by firmly attaching it to the front plate of the television camera .

Note, that this system provides flexible yet firm means for a matching alignment of the camera's focal plane and the brain surface.

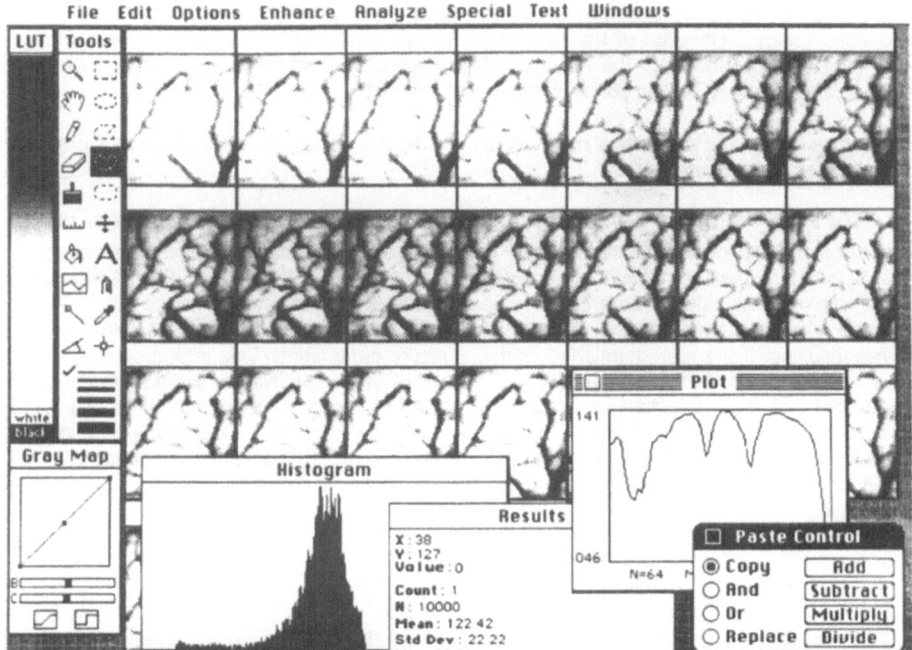

Fig. 4. Digitized sequence of cerebrocortical reflectance images recorded onto vide tape during bolus perfusion of the brain cortex by a plasma indicator (Evan's blue dye) as shown on the monitor of a Macintosh II computer system.

The images are being displayed in gray scale representation by program "Image" (a public domain software by Dr. Wayne Rasband at the NIH) on the monitor of a Macintosh II computer. Darker values in the images indicate increasingly lower light intensities during the transit of the indicator bolus through the epiilluminated cerebrocortical vasculature. Note the sophisticated tools (labeled by Tools, Histogram, Results, Plot, Paste Control) that are available for necessary manipulation of the reflectance records in preparation for their multiparametric evaluation protocol as well as for a complex analysis of the final microcirculatory parameter images resulted from this evaluation protocol.

with the video signals to provide a timebase for image digitization with reference to the time of the indicator injection.

Real-Time Video Image Digitization to Create a Digital Movie of Microregional Indicator-Transit Detected by Videoreflectometry

The magnitude of the video signal intensity at any given pixel location corresponds to the reflectance intensity from the tissue voxel imaged by that pixel. Its conversion into a digital representation of the reflectance intensity has long been a formidable task for an 8-bit microcomputer system.[2, 5, 7] In recent years, however, conversion of incoming video signals to a stack of frames of video data at video rates of 50 - 60 Hz (in real-time) has become available on 16-32 bit microcomputer systems. The system presented in Fig. 1 utilizes 32 bit Macintosh II computer technology. The Macintosh II workstation performs reflectance image digitizing, densitometric and methodological calculations, image reconstruction and displaying under computer control (Figs. 4, 5, and 6). The Macintosh computer is ideally suited for implementing the complex tasks of the multiparametric imaging method, which requires extreme computational power and flexible imaging capabilities. These tasks begin with videoreflectometric data preparation, quality control and digitization implemented by the *Input Software Module* based on a public domain software called "Image" created by Dr. Wayne Rasband (NIH, Bethesda, Maryland, U.S.A.). Real-time video image digitization of the reflectance record results in a stack of reflectance images, essentially a digital movie of the microregional transit of the indicators (Fig. 4) from which the microregional indicator-dilution curves can be reconstructed.

Reconstruction and Processing of an Array of Microregional Indicator-Dilution Curves for Microcirculatory Parameter Maps

The *Data Processing Software Module* developed by the author takes care of the data processing and methodological calculations[5] utilizing stacks of reflectance images generated by the Input Software Module from the video records of microregional indicator-transit. First, it reconstructs a microregional indicator-dilution curve for each pixel of the reflectance image stack. Then, it performs a number of methodological calculations on each of these microregional indicator-dilution curves resulting in the aforementioned microcirculatory parameters. This is a substantial task in terms of the amount of data to be processed. It can only be done efficiently under software control, which does not require human assistance for computing maps of local erythrocyte and plasma volumes, mean transit times, flows, tube and discharge hematocrits, blood volume and flow from the reflectance data set. The result can be hundreds of microcirculatory parameter images (Figs. 5 and 6) automatically databased for further numerical or statistical analyses.

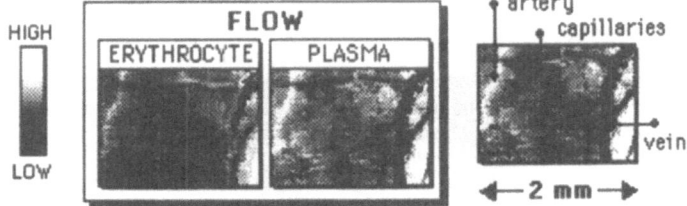

Fig. 5. Examples of high resolution maps of microregional erythrocyte and plasma flows that can be obtained in a repetitive manner by the multiparametric videoreflectometric method.

Note the high spatial definition of these microcirculatory parameter maps. Typically, each map is assembled from the value of 10000 individual measurements arranged in the scanning array of the videocamera. These parameter maps can be visually analyzed either in gray or color coded mode. In the presented case, the microregional flow values were displayed according to an intensity (gray) scale shown on the left. Values corresponding to measuring sites overlapping pial arteries and veins stand out from the map allowing for a detailed topographical analysis of the microflow distribution within the parenchymal (capillary) areas in relation to sites of their supply and drainage from and to the pial vascular network.

Visualization and Interactive Analyses of Microcirculatory Parameter Maps

Microcirculatory parameter image data can be handled with ease and in a very complex manner using the graphic tools of the Macintosh II computer with 8 megabyte of core memory and a 600 megabyte storage capacity of a rewritable optical cartridge drive (Figs. 1 and 4). Program "Image" can be very effectively used to implement the tasks of this final segment of the methodological protocol, the *Output Software Module*. plotting profiles across regions of interest, generating histograms, differential maps, etc. (See Fig. 4).

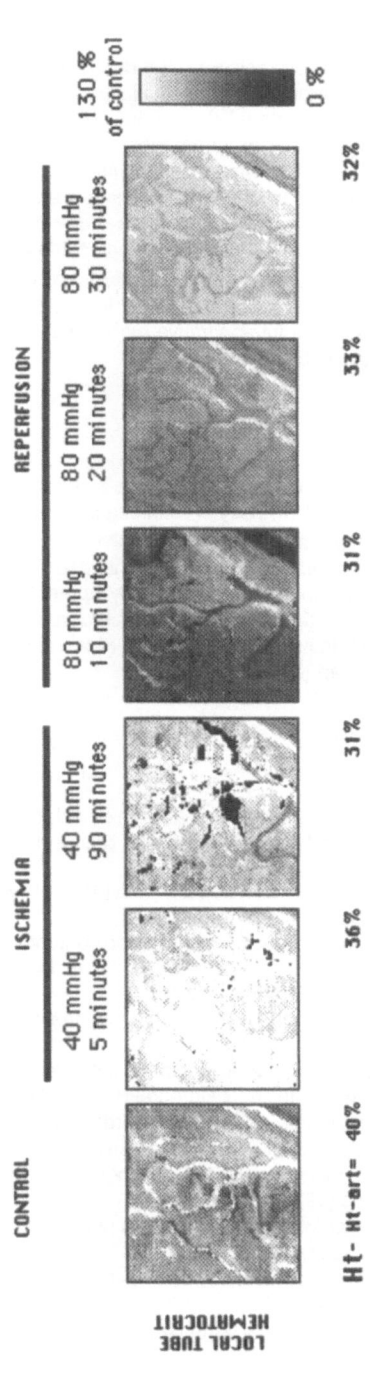

Fig. 6. Local (microregional) tube hematocrit maps obtained within a cerebrocortical area of 3 x 3 mm in an anesthetized cat in control condition and at the given time intervals during global cerebral ischemia induced by central arterial hemorrhage followed by a period of reperfusion.

Arterial hemorrhage was carried out by letting the animal's blood via a catheter introduced into the abdominal aorta into a pressurized reservoir under computer control so that the mean arterial blood pressure could be lowered to and maintained at 40 mmHg for a period of 90 minutes sufficient to produce global cerebral ischemia. The maps are being composed of 100 x 100, that is 10000 individual, simultaneous measurements of local tube hematocrit. Lighter areas indicate a rise in hematocrit. Solid black areas on the second and third maps from the left indicate microareas where valid values of local hematocrit could not be obtained due to severely stagnant flow conditions (stasis).

CONCLUSIONS

The method of multiparametric imaging of cerebrocortical microcirculation has been utilized in a number of studies involving different species (rat, rabbit, cat and monkey) and experimental models. [6, 7, 8, 9, 10, 11] It is capable of monitoring rapid transients in cerebral microhemodynamics such as taking place during the onset of an epileptic seizure.[1, 6] It does not only allow for separate measurements of erythrocyte and plasma parameters and local hematocrit but can provide a pictorial presentation of their intraparenchymal spatial distribution. Microcirculatory parameter distributions along two out of the three dimensions of intracerebral tissue space are now made available by this method in a repetitive and nondestructive manner within the scale of a few square millimeters. Since the imaging raster overlaps a part of the pial network (macrocirculation) and the underlying parenchymal areas (microcirculation) a complex analysis of the relative contribution of extra- and intraparenchymal factors in the observed intraparenchymal distribution patterns can be carried out (Fig. 6).[11] The current level of sophistication and productivity proved necessary to make the method of multiparametric imaging of cerebrocortical microcirculation a powerful tool of research in the lab and potentially a useful tool of intraoperative monitoring and even diagnosis in the operating theater under conditions of human brain surgery.

ACKNOWLEDGMENTS

This work has been jointly supported from personal funds of the author, by grants-in-aid of the Departments of Neurology and Pathology of the University of Alabama at Birmingham, Birmingham, U.S.A., the Experimental Research Department and 2nd Institute of Physiology of the Semmelweis University of Medicine, Budapest, Hungary and in part by OTKA Grant 2040. The cooperation of Dr. Wayne Rasband at the NIH, Bethesda, Maryland, U.S.A. is greatly appreciated in incorporating features of the Input and Output Software Modules in his "Image" program. The author can be reached for further technical details and with inquiries on installation of the method's hardware and software on the following telephone/fax number: (36-1) 134-3162.

REFERENCES

1.   A. Eke, Gy. Hutiray, and A. G. B. Kovach, Induced hemodilution detected by reflectometry for measuring microregional blood flow and blood volume in cat brain cortex, Am. J. Physiol. 236(5): H759-H768 (1979).
2.   A. Eke, Reflectometric mapping of microregional blood flow and blood volume in the brain cortex, J. Cereb. Blood Flow Metabol. 2:41-53 (1982).
3.   A. Eke, Integrated microvessel diameter and microregional blood content as determined by cerebrocortical video reflectometry, in: "The Cerebral Veins," L. M. Auer, and F. Loew, eds., Springer Verlag, Vienna, New York (1983).
4.   A. Eke, Repetitive mapping of tissue hematocrit over microareas of the brain cortex, Intl. J. Microcirc. Clin. Exp. 3(3/4):548 (1984).
5.   A. Eke, Reflectometric imaging of local tissue hematocrit in the cat brain cortex, in: "Cerebral Hyperemia and Ischemia: From the Standpoint of Cerebral Blood Volume," M. Tomita, T. Sawada, N. Naritomi and W. D. Heiss, eds., Elsevier, Amsterdam (1988).
6.   A. Eke, Heterogeneity of cerebrocortical microflow in epileptic seizure, in: "Cerebral Blood Flow, Metabolism and Epilepsy," M. Baldy-Moulinier, D. H. Ingvar, and B. S. Meldrum, eds., John Libbey Eurotext, London, Paris (1983).
7.   A. Eke, J. H. Halsey, Distribution of cerebrocortical microflow in normo- and hypertensive rats, Adv. Exp. Med. Biol. 180:203-210 (1984).
8.   A. Eke, J. H. Halsey, Distribution of cerebrocortical microflow in normo- and hypertensive rats: studies in ischemia, Adv. Exp. Med. Biol. 191:107-209 (1985).
9.   A. Eke, Redistribution of cerebrocortical microflow during increased neuronal activity, Adv. Exp. Med. Biol. 191:101-105 (1985).
10. A. Eke, Imaging of red blood cells and plasma dispersion in the brain cortex, Adv. Exp. Med. Biol. 215:21-27 (1987).
11. A. Eke, Hematocrit changes in the extra- and intraparenchymal circulation of the feline brain cortex in the course of global cerebral ischemia, Adv. Exp. Med. Biol. 248:439-449 (1988).

THE BRAIN

CLASSIFICATION OF OXYGEN TRANSPORT
TO TISSUE WITH NEURAL NETWORKS

Wolfgang Babel, Norbert Hetterich, Thomas Müller

Diehl GmbH
Fischbachstrasse 16
D 8505 Roethenbach, Germany

INTRODUCTION

When spectrometric techniques are applied for tissue monitoring in patients [1] a high sampling rate of spectra can improve the accuracy and reliability of the measurements by increasing the signal to noise ratio. Furthermore a high time resolution for recording of rapid tissue kinetics can be attained. A considerable amount of data must be computed for real time evaluation.

In practice this means that very powerful signal processors in combination with adequate algorithms must be available. When tissue measurements are performed by use of the EMPHO II 100 spectra per second, each consisting of 256 samples, have to be classified and visualized on a PC monitor in real time. This special classification is performed by a neural network, which has proven to be a powerful approach in the field of pattern recognition.

The paper will describe the characteristics of neural Networks, their design due to the system requirements and the implementation aspects.

PATTERN RECOGNITION IN GENERAL

Pattern recognition is the transformation of feature space $\underline{f}$ into the decision space $\underline{d}$

$$t : f \rightarrow \underline{d} \quad \underline{f} \in IR^N, \underline{d} \in IR^K$$

where N is the number of features in the feature space $\underline{f}$ and K is the number of classes in the decision space $\underline{d}$. The main task in pattern recognition is to find the optimum transformation between the feature space and the decision space. One of the best methods to solve this task is the classification via a neural network.

It should be mentioned, that the availability of a representative data base is an important precondition for the proper design.

NEURAL NETWORK DESIGN

The first step in designing a pattern recognition system is to analyze the system requirements. The major requirements for the EMPHO II are

- o Optimized man-machine interface so that nurses can handle the equipment
- o High system reliability
- o Realtime processing and visualization of the tissue spectra
- o High confidence level of pattern recognition (representative data base)
- o High resolution in classification (1%)
- o Data processing and visualization on a PC 486

All these requirements will influence the expenditure of hardware and software.

For the EMPHO II a resolution of 1% for the classification of oxygen was required by the physician. That resolution should be available for the heart data as well as for the skin data. Figure 1 shows typical spectra for heart and skin data.

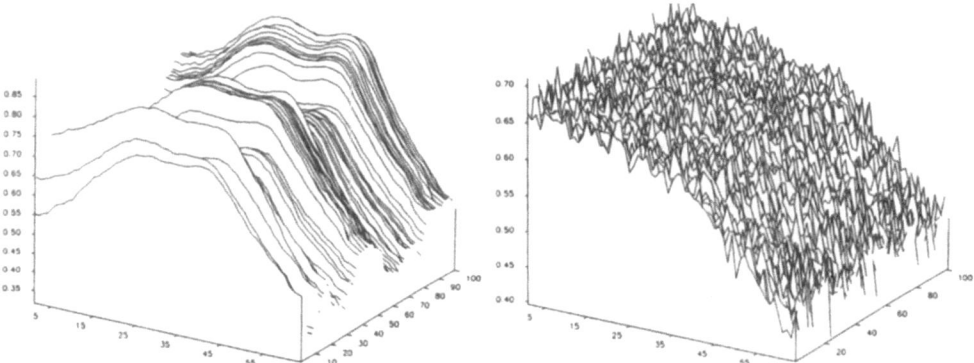

Figure la. heart tissue spectra   Figure 1b. skin tissue spectra

The next step is to determine the minimum number of features which permits to separate the predefined classes. Therefore the experience of the physician is required as well as the knowledge of the computer scientist.

For designing a network which is able to classify both the heart data as well as the skin data 64 spectral values were given to the input, where every input line is the mean value of 4 consecutive spectral lines. In other words the features in this application will consist of all available spectral information. Exactly this behavior is one of the most important reasons for the use of neural backpropagation networks in relation with supervised learning: Time or frequency information processing without a selection of special signal features.

DESIGN OF NETWORK AND SUPERVISED LEARNING

Due to the required number of classes and the selected features the network topology can be determined. The following numbers of neurons are chosen:

Input layer   : 64 neurons
Hidden layer   : 15 neurons
Output layer   : 7 neurons (binary coded output)

This network was trained by supervised learning via the backpropagation algorithm for the heart data as well as for the skin data. The necessary training data sets had been preclassified by physicians [2]. The backpropagation algorithm or general delta rule works due to the steepest descent in the complex error domain [3]. An adequate error function is given by the equation:

$$E = 1/2 \ \sum_p \sum_j \ (t_j - o_j)**2$$

p: number of input vectors
j: number of classes
t: desired value for the j-th class

This error has to be minimized during the training phase until a predefined confidence level is reached. The change of weights from the output layer to the hidden layer is

$$W_{ij}(T+1) = W_{ij}(T) + \Delta\delta_i o_j$$

$$\delta = o_i(1-o_i)(t_i-o_i)$$

The change for the weights from hidden layer to the input layer

$$W_{jn}(T+1) = W_{jn}(T) + \Delta\delta_n o_n$$

$$\delta = o_j(1-o_j) \ \sum \ \delta_i W_{ij}$$

where

i: i-th neuron in the hidden layer
j: j-th neuron in the output layer
n: n-th neuron in the input layer

The flow chart in figure 2 shows the structure of the full learning procedure.

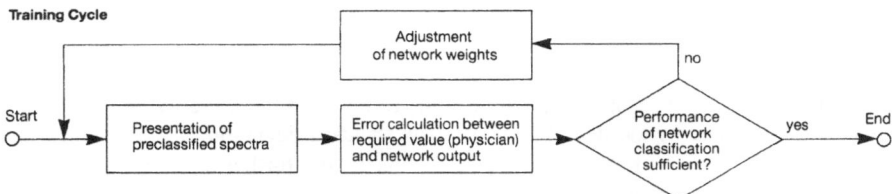

Figure 2. learning procedure of neural network

RESULTS OF TRAINING AND TESTING FOR UNLEARNED HEART AND SKIN DATA

At first, the neural network is trained to discriminate human heart spectra with a maximum resolution of 1%. Two training data sets were recorded during a bypass operation - the first one before and the second one after the bypass surgery. For optimizing the network's weights we used the features from the data set "before bypass".

When the iteration process has converged to a predefined error, the untrained data set ("after bypass") can be used to evaluate the network's robustness and its ability to generalize. The neural network is properly trained, when nearly the same performance for both the trained and the untrained data can be achieved.

A good means for visualizing the performance of a classifier is the confusion matrix in which the target class of a feature vector is plotted versus the output class given by the classifier. Thus the brightness of the matrix element (i,k) is a measure of the conditional probability that the classifier will decide class k provided a feature vector of class i is given to the classifier's input.

Figure 3 shows the confusion matrix for the oxygen classification of the trained heart data ("before bypass") for 100 classes with a resolution of 1% per matrix element.

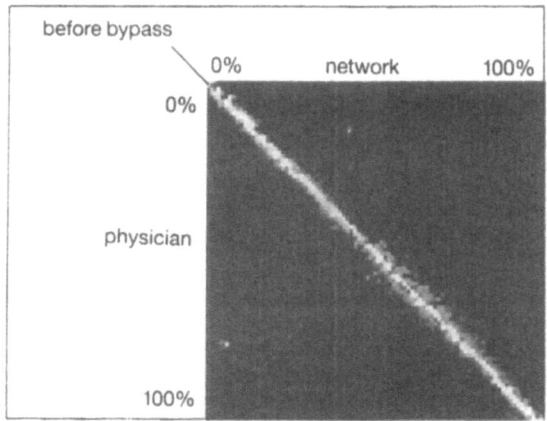

Figure 3. confusion matrix for trained heart data

As we can see there is a very good accordance between the target class and the class which was calculated by the neural network. The standard deviation of the diagonal has a value of about 1%. This deviation can be reduced by further training but then the network's ability to generalize will also get worse. Therefore the computer scientist always has to make a compromise between the performance of the class training data set and the classifier's robustness. For the untrained test data ("before bypass") the confusion matrix is displayed in Figure 4.

Because of the successful bypass surgery there are no spectra with a very low oxygenation. The rest of the pixels is strongly concentrated in the diagonal of the matrix which means a very good robustness of the designed classifier.

For optimizing the network's weights for the skin data the full 64 point spectrum was given to the neural network's input.

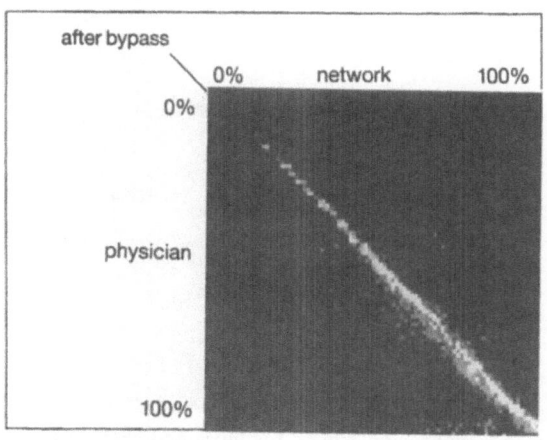

Figure 4. confusion matrix for untrained heart data

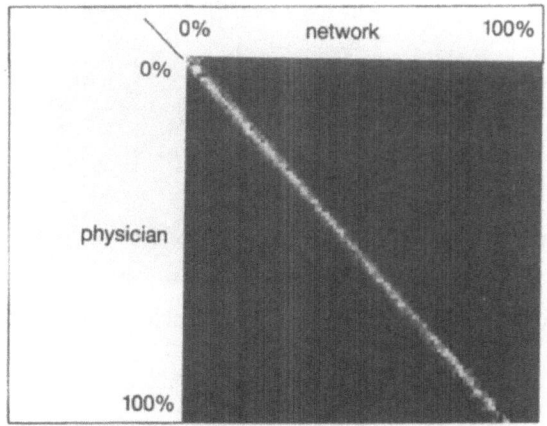

Figure 5. confusion matrix for trained skin data

In Figure 5 we see the confusion matrix of patient #1/data set#1 after training. Despite the high noise level in the skin data a standard deviation of less than 1% can be achieved by using the 64 point skin spectra. The confusion matrices for patient#1/data set#2 and patient#2/data set#1 are shown in Figure 6 and Figure 7, respectively. Again the pixels

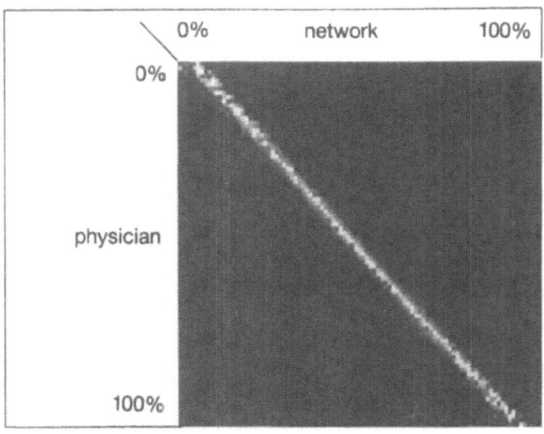

Figure 6. confusion matrix for untrained skin data

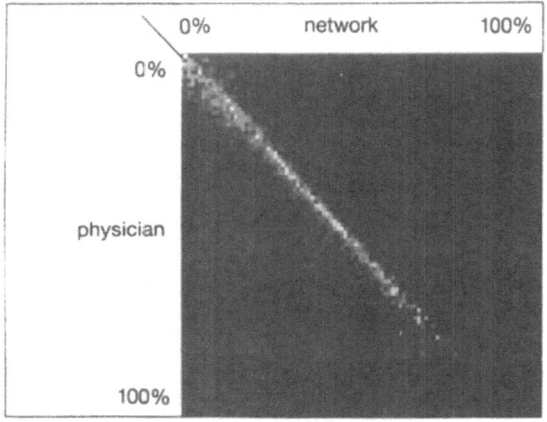

Figure 7. confusion matrix for untrained skin data (patient 2)

are strongly concentrated to the diagonal of the matrices. It should be noted, that in Figure 7 the deviations in the right lower part are due to the fact that there was only one spectrum available for each of the high oxygenation classes.

HARDWARE- SOFTWARE REALIZATION OF A NEURAL NETWORK

The designed network has to be implemented for realtime processing. It has to be considered whether a pure software implementation is possible or a special hardware configuration is necessary. For the EMPHO II a 1% percent resolution is required.
If the network topology is given by N neurons in the input layer, I neurons in the hidden layer and J neurons in the output layer, the following mathematical operations have to be processed:

number of weights	$: N*I + I*J + I + J$
number of additions	$: N*I + I*J + 2*I + J + N$
number of multiplications	$: N*I + I*J + I + J$
number of sigmoids	$: I + J$

For the EMPHO II application an IBM compatible TMS 320C30 signal processor board (cycle time 62.5 nsec) was used for the software implementation of the neural network. For the chosen network topology of 64 input, 15 hidden and 7 output neurons about 830 μsec are necessary for processing one spectrum. Therefore about 1200 spectra per second can be evaluated which means a very high performance.

SUMMARY

In the current stage of the development of the EMPHO II it was shown that the transport of oxygen to tissue can be classified for all applications in real time by using neural networks. The resolution of the oxygen classification is 1%. The implementation of the neural network is possible on a commercial signalprocessor TMS 320C30 board which is compatible with an IBM PC. The implementation of the neural network is done fully in software and no special neurocomputer is necessary. The results are very promising so that the neural network design goals will be established in the future. In the future further efforts will be made in order to increase the data base and to train the network in more detail.

LITERATURE

1. M. Brunner, R. Ellermann, K.H. Frank, and M. Kessler, Measurements and Processing of Intracapillary Hemoglobin Spectra by Using a Micro-Lightguide Spectrophotometer in Connection with a Microcomputer in: Advances in Experimental Medicine and Biology, Volume 191 Plenum Press (1985)
2. W. Duemmler, Bestimmung von Haemoglobin-Oxygenierung und relativer Haemoglobin-Konzentration in biologischen Systemen durch Auswertung von Remissionsspektren mit Hilfe der Kubelka-Munk-Theorie. Dissertation Institut für Physiologie und Kardiologie, Universitaet Erlangen-Nuernberg (1988)
3. D.E. Rumelhart and J.L. McClelland, Parallel Distributed Processing, Vol.I: Foundations, The MIT Press, Cambridge, Massachusetts, (1987)

OXYGENATION OF THE CORTEX OF THE BRAIN OF CATS DURING OCCLUSION OF THE MIDDLE CEREBRAL ARTERY AND REPERFUSION.

D.F. Wilson[1], S. Gomi[2], A. Pastuszko[1] and J.H. Greenberg[2]

Department of Biochemistry and Biophysics[1] and Department of Neurology[2]
Medical School, University of Pennsylvania
Philadelphia, PA 19104.

INTRODUCTION

Cerebral function is lost within seconds of deprivation of oxygen and within 2-3 minutes anoxic depolarization begins (see Silver, 1977, 1978). This depolarization is characterized by massive loss of K^+ and uptake of Cl^- into cells (Van Harrevald, 1971). If depolarizing conditions continue, there is progressive accumulation of intracellular Ca^{2+} and degradation of essential cellular components. The extent to which the loss of cellular and brain function can be reversed depends on the time and degree of oxygen deprivation as well as the conditions during and after reoxygenation. Attempts to define the biochemical and physiological parameters responsible for the irreversible damage have met with limited success (for review see Welsh et al, 1982; Raichle, 1983). Analytical limits in the quantitation of oxygen have been particularly vexing. Oxygen deprivation is a relative term and it is important to know the extent to which the residual flow and collateral circulation are providing oxygen to the tissue. Some measurements have been made by inserting small oxygen electrodes into the tissue (Silver, 1977; 1978; Nair et al, 1987; Fennema et al, 1989) or by multielectrode surface electrodes (Leniger-Follert, 1977; Grote et al, 1984). The former are limited to measuring at single points in the tissue and cause significant mechanical disruption of the tissue during insertion.

A quantitative, noninvasive optical method has recently been developed using the oxygen dependent quenching of phosphorescence (see for examples, Vanderkooi et al, 1987; Wilson et al, 1988). More recently it has been demonstrated that the phosphorescent oxygen probe can be injected into the blood and the phosphorescence of tissue imaged using a video camera. When the video camera is equipped with a intensifier which can be gated (turned on or off) in less than a microsecond, it is possible to collect the data needed to generate quantitative maps of the phosphorescence lifetimes and of oxygen pressure in the tissue (see for example, Rumsey et al, 1988; Wilson et al, 1991; this proceedings; Shonat et al, 1992). The spatial resolution of these maps of oxygen pressure is limited only by the optics of the system, allowing examination of the microheterogeneity of the oxygen distribution at the level of few microns. In the present paper, we report the use of this method to follow the oxygen pressure in the veins and capillary beds of the cortex in response to occlusion of the middle cerebral artery and reperfusion.

MATERIALS AND METHODS

Preparation of animals for video imaging of the effect of MCA occlusion on cortical oxygen pressure.

The observations of phosphorescence were made through a cranial window (see Harbig et al, 1976; Dora, 1984; Dora et al, 1986) Adult cats were anesthetized, the head placed in a steriotaxic unit, the scalp removed from the left parietal region and the skull exposed. A hole approximately 1.5 cm in diameter was made in the skull over the parietal hemisphere and the dura carefully reflected. A cranial window was placed in the hole, sealed with bone wax and cemented into place with dental acrylic. The surface of the brain was superfused with artificial cerebral spinal fluid. Observations were made using a Wild Macrozoom microscope with an epifluorescence attachment.

The Pd complex of meso tetra-(4-carboxyphenyl) porphine was injected into the blood stream and the camera was focused on the surface of the brain prior to injection of the Pd-porphyrin. The optical filters for measuring phosphorescence (excitation through an interference filter with a center wavelength of 537 nm and a bandwidth at half height of 45 nm; emission through a 630 nm cutoff filter) were put in place and the room darkened. Under these conditions, there was no detectable phosphorescence before the dye was injected. When Pd-porphyrin was injected, the arteriolar system brightened first, followed by the capillary bed and then the venous system. A few minutes (15 to 30) was allowed to establish a steady state of normoxia and then the middle cerebral artery was occluded.

The illuminating light for the epifluorescence attachment was a EG&G 45 watt xenon flashlamp with the timing of the flashes and the gating of the camera controlled by a microcomputer. The number of flashes averaged for each delay time was 8 and the delay times after the flash were 20 usec, 40 usec, 80 usec, 160 usec, 300 usec, 600 usec, and 2,500 usec. The gate width in all cases was 2,500 usec.

The images of phosphorescence are digitized as 512 x 480 pixel arrays of data. These data arrays were filtered and used to calculate a phosphorescence lifetime for each pixel of the image set. The result was two dimensional maps of the distribution of phosphorescence lifetimes in the observed area of the cortex. The lifetime maps were then converted to oxygen pressure maps using the Stern-Volmer relationship (see Vanderkooi et al, 1988; Wilson et al, 1991).

RESULTS

The phosphorescence imaging system was used to follow the oxygen pressure through a control period, 60 minutes of occlusion of the middle cerebral artery and then 4 hours after release of the occlusion. The calculations of phosphorescence lifetimes were carried out using a best fit to a linearized single exponential decay. The probe used gives a strictly single exponential decay of phosphorescence when it is in an environment with a uniform oxygen pressure. In general, the correlation coefficients are greater than 0.95, indicating that for most regions of the phosphorescence lifetime maps the phosphorescence decays closely approximate a single exponential.

The phosphorescence lifetimes can be converted to oxygen pressure at each pixel of the map using the Stern-Volmer relationship (see Vanderkooi et al, 1988; Wilson et al, 1991; this proceedings), providing a two dimensional map of the oxygen pressure. The quenching constant (k_Q) and lifetime at zero oxygen pressure were determined *in vitro*. Such calibrations apply to the probe *in vivo* provided the measurements are made for the appropriate conditions i.e. with the probe in the presence of excess albumin and at the pH and temperature of the blood in the animal (see Wilson et al, 1991).

A set of six oxygen pressure maps of an area 4.7 mm high and 3.6 mm wide are presented in Figure 1. These illustrate the data obtained from the control period (A), and 2.5 minute (B), 10 minutes (C), 60 minutes (D), 90 minutes (E) and 3 hours (F) after release of the occlusion. No map is shown for the period of occlusion because the calculated oxygen pressures were less than 2 Torr. The oxygen pressure scale appear as grey wedges on the left side of the maps. Each scale is linear in oxygen pressure from 0-72 Torr (A), 0-60 Torr (B), 0-72 Torr (C), 0-18 Torr (D), 0-18 Torr (E) and 0-18 Torr (F). The large changes in oxygen pressure which occurred during the course of the experiment are readily apparent.

In early reperfusion, the tissue oxygenation is generally extremely heterogeneous with some regions remaining near zero oxygen while others have already achieved normal oxygen levels. This heterogeneity is observed in all animals in the early phase of reperfusion but became less apparent by 10 to 15 minutes reperfusion. After 1 hour of reperfusion the oxygen pressure in most of the area of observation fell to well below the control values. The oxygen pressure map obtained after three hours reperfusion shows that the cortex was generally hypoxic (average less than 10 Torr) but that some small regions had particularly low oxygen pressures. These small areas develop as part of the general decrease in oxygen pressure which generally developed by about 1 hour after release of the occlusion.

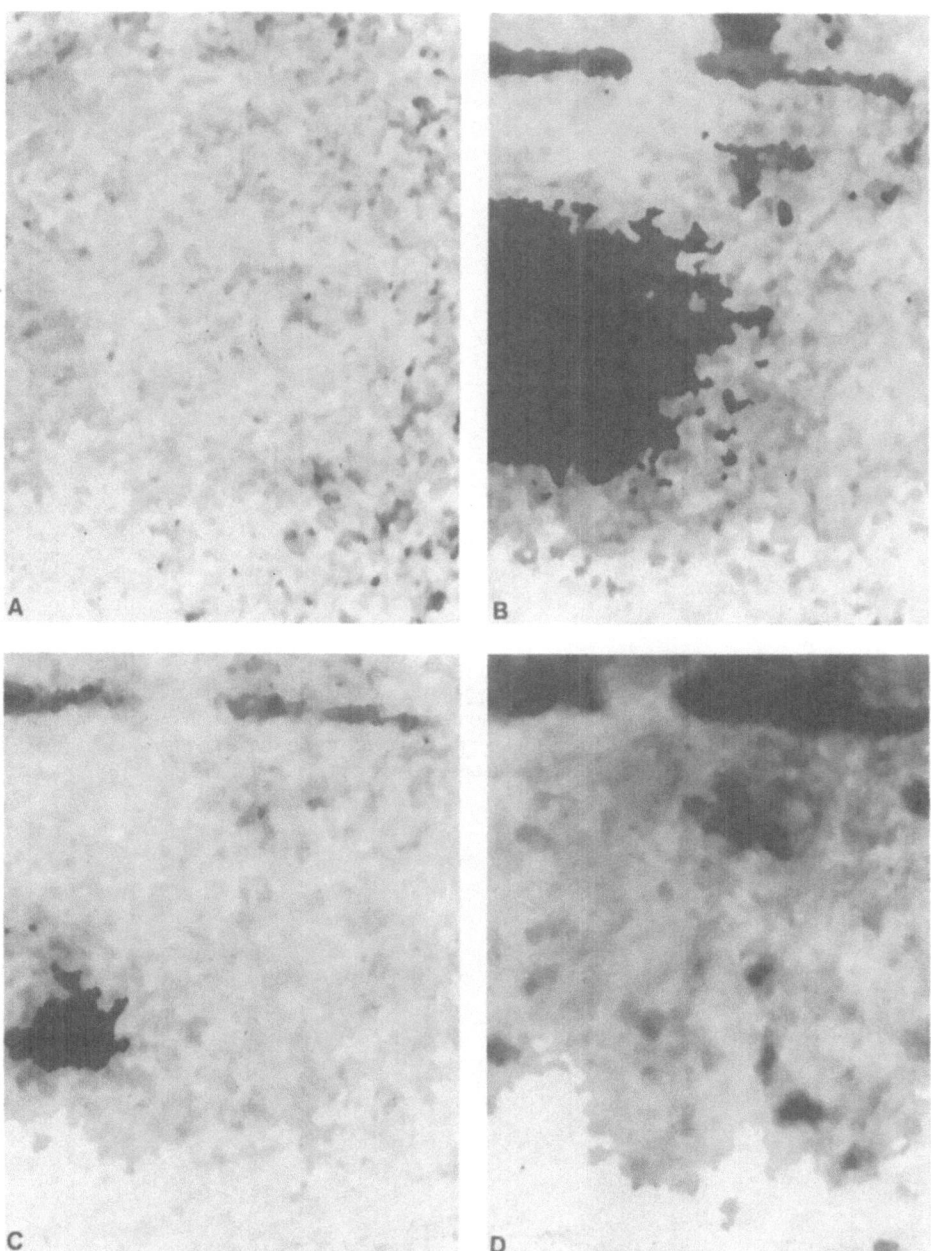

Figure 1. (part 1) Two dimensional maps of oxygen pressure in the surface of the cortex. The observed area was 4.7 mm high and 3.6 mm wide. The computer program then calculated the oxygen pressure for each pixel position of the images array. The phosphorescence decay constants were converted to oxygen pressures using the Stern-Volmer relationship and a quenching constant of 325 $Torr^{-1}sec^{-1}$ and a lifetime in the absence of oxygen of 600 usec, the values for this probe at pH 7.2-7.4 and 38°. The time points are a control prior to occlusion of the MCA (A), 2.5 minutes (B), 10 minutes (C), and 60 minutes after release of the occlusion (D). The grey wedges (black to white) on the left are linear representations of oxygen pressure from 0-72 Torr (A), 0-60 Torr (B), 0-72 Torr (C), and 0-18 Torr (D).

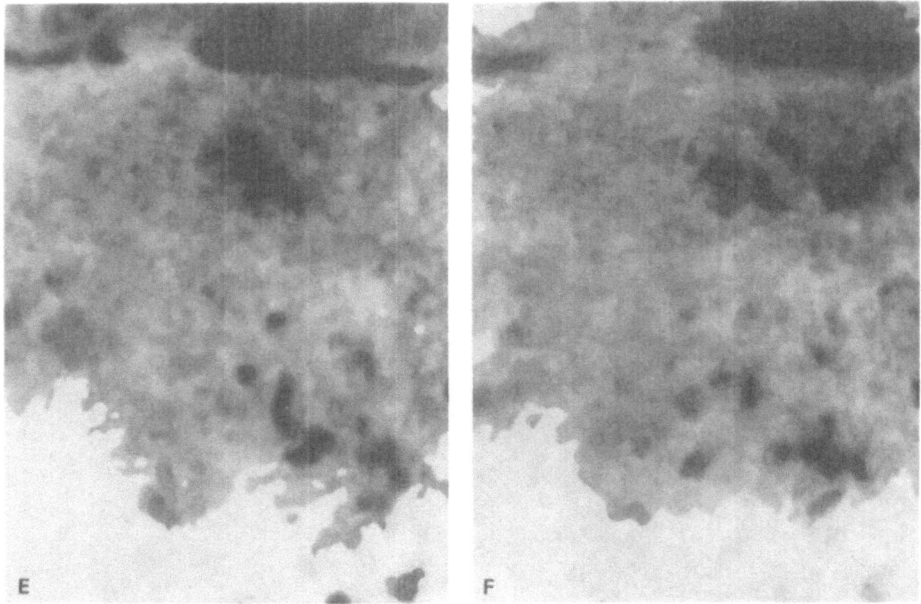

Figure 1. (part 2) Two dimensional maps of oxygen pressure in the surface of the cortex. Sets of images were taken as described in Materials and Methods. The oxygen pressure maps are for 90 minutes (E) and three hours (F) after release of the occlusion. The grey wedges (black to white) on the left are linear representations of oxygen pressure from 0 Torr to 18 Torr.

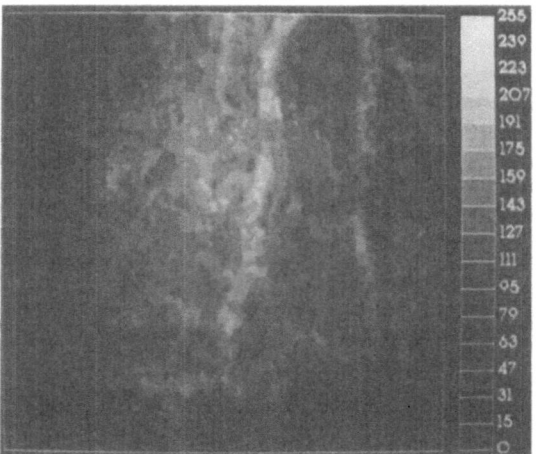

Figure 2. The initial intensity of phosphorescence as obtained by extrapolation of the best fit to single exponential to time zero under control conditions. The initial intensity of phosphorescence is directly related to the concentration of blood in the tissue at the oxygen pressure near that of the determined lifetime. These phosphorescence intensity maps are determined in the fitting process and allow visual identification of some morphological features such as blood vessels.

The vessels seen in the maps of initial phosphorescence under most conditions (see for example Figure 2) do not appear in the phosphorescence lifetime or oxygen pressure maps because the oxygen pressure in the vessels (veins) is not substantially different from that in the underlying capillary beds. The vessels "disappear" when the phosphorescence lifetime maps and oxygen pressure maps are calculated, because lifetime and oxygen pressure are independent of phosphorescence intensity *per se*.

DISCUSSION

The oxygen pressure maps of regions of 5 mm diameter and greater show substantial variability in oxygen pressure from one region to another in the cortex of the brain under normal conditions. The variability is generally in the range of a few Torr, about a mean value of 30-40 Torr. The oxygen pressures increase and decrease in each region relative to the mean value, suggesting control of the vascular resistance is a dynamic process set to operate within a range of oxygen pressures. In some animals there were intermittent areas of marked hypoxia of 50 to 200 um in diameter which appeared and disappeared over a period of minutes, suggesting vascular spasms at the level of small arteriols, possibly induced by the surgical procedure.

Tissue oxygen pressures rapidly rise to above normal values after release of the MCA occlusion, consistent with the onset of a period of reactive hyperemia. By about 1 hour after release, however, the oxygen pressures again fall to below normal. The extent of the decrease below normal appears to be related to the degree of irreversible damage which has occurred but this remains to be quantitated. It has been reported that there is a delayed period of local hypoperfusion (Ames et al, 1968; Ginsberg et al, 1980), but the contribution of this to postischemic damage has been questioned (see for example Levy et al, 1975A,B; 1979; Welsh, 1982) since it could be attributed to a period of suppressed metabolic rate. Our data show a delayed period of tissue hypoxia, indicating impaired delivery of oxygen. If the rate of oxygen metabolism was suppressed, it was to a lesser extent than the suppression of oxygen delivery or the tissue would not have been hypoxic. The oxygen pressures in the cortex shown in Figure 1D are well below normal and would be expected to compromise the ability of neurons to maintain their integrity during the period of delayed hypoxia (see Wilson et al, 1979; 1982; 1988; Robiolio et al, 1988; for review see Erecinska and Wilson, 1982).

The vascular damage giving rise to the delayed period of hypoxia appears to involve primarily the microvasculature. Closure of a major vessel (arteriol or larger) produces a readily identifiable pattern in the two dimensional maps of oxygen pressure in the cortex. These are characterized by well defined regions of marked hypoxia in surrounding regions of normoxia. The delayed period of hypoxia, however, is characterized by onset of a diffuse hypoxia associated with capillary bed regions i.e. regions without observable vessels. From the presented data, it is reasonable to conclude that damage to the vascular system plays a major role in the pathology of cerebral ischemia/reperfusion injury.

Acknowledgements: This research was supported by grant NS-10939 from the National Institutes of Health. The authors are indebted to Dr. Martin Reivich for his continuing support and encouragement.

References

Ames, A. III, Wright, R.L., Kowada, M., Thurston, J.M., Majno, G. (1968) Cerebral ischemia. II. The no reflow phenomenon. Amer. J. Path. 52: 437-453.

Dora, E. (1984) A simple cranial window technique for optical monitoring of cerebrocortical microcirculation and NAD/NADH redox state. Effect of mitochondrial electron transport inhibitors and anoxic anoxia. J. Neurochem. 42: 101-108.

Dora, E., Tanaka, K., Greenberg, J.H., Gonatas, N.H. and Reivich, M. (1986) Kinetics of microcirculatory, NAD/NADH, and electrocorticographic changes in cat brain cortex during ischemia and recirculation. Ann. Neurol. 19: 536-544.

Erecinska, M. and Wilson, D.F. (1982) Regulation of cellular energy metabolism. J. Memb. Biol. 70: 1-14.

Fennema, M., Wessel, J.N, Faithful, N.S., and Erdmann, W. (1989) Tissue oxygen tension in the cerebral cortex of the rabbit. Adv. Exptl. Med. Biol. 248: 451-460.

Ginsberg, M.D., Welsh, F.A., and Budd, W.W. (1980) Deleterious effect of glucose pretreatment on recovery from diffuse cerebral ischemia in the cat. I. Local cerebral blood flow and glucose utilization. Stroke II: 347-354.

Grote, J., Zimmer, K., and Schubert, R. (1984) Tissue oxygenation in normal and edemitous brain cortex during arterial hypocapnia. Adv. Exptl. Med. Biol. 180: 179-184.

Harbig, K., Chance, B., Kovach, A.G.B., and Reivich, M. (1976) In Vivo measurements of pyridine nucleotide fluorescence from cat brain cortex. J. Appl. Physiol. 41: 480-488.

Leniger-Follert, E. (1977) Direct determination of local oxygen consumption of the brain cortex *in vivo*. Pflugers arch. 372: 175-179.

Levy, D.E., Brierley, J.B., and Plum, F. (1975A) Ischemic brain damage in the gerbil in the absence of "no reflow". J. Neurol. Neurosurg. Psychiatry 38: 1197-1205.

Levy, D.E., Brierley, J.B., Silverman, D.G., and Plum, F. (1975B) Brief hyopoxia-ischemia initially damages cerebral neurons. Arch. Neurol. 32: 450-456.

Levy, D.E., Van Uitert, R.L., and Pike, C.L. (1979) Delayed postischemic hypoperfusion: A potentially damaging consequence of stroke. Neurology (Minneap.) 29: 1245-1252.

Nair, P.K., Buerk, D.G., and Halsey, J.H. (1987) Comparisons of oxygen metabolism and tissue PO_2 in cortex and hippocampus of gerbil brain. Stroke 18: 616-622.

Raichle, M.E. (1983) The pathophysiology of brain ischemia. Ann. Neurol. 13: 2-10.

Robiolio, M., Rumsey, W.L., and Wilson, D.F. (1989) Oxygen diffusion and mitochondrial respiration in neuroblastoma cells. Amer. J. Physiol. 256: C1207-C1213.

Rumsey, W.L., Vanderkooi, J.M., and Wilson, D.F. (1989) Imaging of phosphorescence: a novel method for measuring oxygen distribution in perfused tissue. Science, Wash. DC 241: 1649-1651.

Shonat, R.D., Wilson, D.F., Riva, C.E., and Pawlowski, M. (1992) Oxygen distribution in the retina and choroidal vessels of the cat as measured by a new phosphorescence imaging method. Applied Optics in press.

Silver, I.A. (1977) Changes in PO_2 and ion fluxes in cerebral hypoxia-ischemia. Adv. Exptl. Med. Biol. 78: 299-312.

Silver, I.A. (1978) Cellular microenvironment in relation to local blood flow in cerebral vascular smooth muscle and its control. Elsevier, New York, p. 49-61. (Ciba Found. Symp. 56).

Vanderkooi, J.M., Maniara, G, Green, T.J., and Wilson, D.F. (1987) An optical method for measurement of dioxygen concentration based on quenching of phosphorescence. J. Biol. Chem. 262: 5476-5482.

Van Harreveld, A. (1971) "The extracellular space in the central nervous system" in: Structure and Function of Nervous Tissue. (Bourne, G.E. ed.) Academic Press, London, pp. 447-511.

Welsh, F.A., O'Connor, M.J., Marcy, V.R., Spatacco, A.J., and Johns, R.L. (1982) Factors limiting regeneration of ATP following temporary ischemia in cat brain. Stroke 13: 234-242.

Wilson, D.F., Erecinska, M., Drown, C., and Silver, I.A. (1979) The oxygen dependence of cellular energy metabolism. Arch. Biochem. Biophys. 195: 485-493.

Wilson, D.F. and Erecinska, M. (1982) Effect of oxygen concentration on cellular metabolism. Chest 88S: 229S-232S.

Wilson, D.F., Pastuszko, A., DiGiacomo, J.E., Pawlowski, M., Schneiderman, R., Delivoria-Papadopoulos, M. (1991) Effect of hyperventilation on oxygenation of the brain cortex of newborn piglets. J. Appl. Physiol. 70(6): 2691-2696.

Wilson, D.F., Rumsey, W.L., Green, T.J., and Vanderkooi, J.M. (1988) The oxygen dependence of mitochondrial oxidative phosphorylation measured by a new optical method for measuring oxygen concentration. J. Biol. Chem. 263: 2712-2718.

ACTIVE AND BASAL WHOLE BRAIN BLOOD FLOW, OXYGEN AND GLUCOSE

METABOLISM IN MONKEYS

Edwin M. Nemoto, Liping Yao, Howard Yonas, and
Joseph Darby

Departments of Anesthesiology/CCM and
Neurological Surgery, University of Pittsburgh
Pittsburgh, PA   15261

INTRODUCTION

By definition, active cerebral metabolic rate for oxygen
(ACMRO2) is the oxygen consumption associated with the
generation of spontaneous electrical activity.[1] The balance,
or basal CMRO2, is presumably, the oxygen consumed to maintain
neuronal viability through ion pumping and reparative and
synthetic metabolic pathways.[2] This concept evolved out of
interest in the mechanism of action of anesthetics and the
hypothesis that the effect of anesthetics on cerebral
metabolism is secondary to a direct effect on cerebral
function.[3]
Michenfelder[1] clearly showed that the effect of
thiopental at least, produced an effect on CMRO2 that appeared
to be directly related to a direct effect of thiopental on
cerebral function where CMRO2 was reduced only as far as
cerebral function was suppressed. Doses of thiopental
exceeding that required to suppress all cerebral electrical
activity did not further depress CMRO2. This property of
thiopental in suppressing only active CMRO2 is not necessarily
true for all general anesthetics since halothane at higher
than clinical concentrations continues to decrease CMRO2 with
increasing dose even after all brain electrical activity has
ceased.[4] The basis for this difference between thiopental and
halothane is unknown. Nevertheless, the ability of thiopental
to reduce CMRO2 only as far as that required for spontaneous
brain electrical activity can be used to measure the
proportion of ACMRO2 and BCMRO2.
Previous measurements of ACMRO2 and BCMRO2 were made in
anesthetized animals which might be expected to introduce an
error into a measurement based upon an awake brain. Thus, our
objective was to determine the partitioning of active and
basal CMRO2, CMRG and CBF in fully conscious monkeys.

Oxygen Transport to Tissue XIV, Edited by W. Erdmann and
D.F. Bruley, Plenum Press, New York, 1992

METHODS AND MATERIALS

Restraint-chair acclimated Rhesus monkeys of either sex weighing between 4.5 and 7 kg body weight were studied in the following protocol approved by the institutional animal care and use committee. Monkeys fasted overnight were anesthetized with ketamine 10 mg/kg, IM. Peripheral venous catheters and femoral artery catheters were inserted using aseptic surgical techniques. Physiological saline was infused at 5 ml/kg/hr. Platinum microelectrodes (50 um tip diam) and catheters were inserted into the torcula to monitor cerebral H_2 clearance to measure global cerebral blood flow (CBF) and to sample cerebral venous blood. Silastic catheters were inserted into the nasopharynx to infuse H_2 gas into the inspired air at 1-5% and for sampling and monitoring of end-tidal CO_2. Subperiosteal needle electrodes were inserted biparietal to monitor cortical EEG. The monkeys were allowed 2 hr to recover from ketamine anesthesia in the restraint chair while monitoring rectal temperature, end-tidal CO_2, and arterial blood pressure.

After 2 hr recovery, three sequential measurements of CBF and cerebral metabolic rate for oxgen and glucose were made with 10 to 15 min periods of H_2 equilibration and clearance during which time arterial and cerebral venous samples were obtained for measurements of CMRO2, and cerebral metabolic rate for glucose (CMRG). Global CBF was calculated from the monoexponential H_2 clearance curves using the T1/2 method, ignoring the first 40 secs of the clearance curve to eliminate the influence of the arterial input.

Following the 3 sets of CBF and CMR measurements in the fully conscious monkeys, thiopental was infused IV to induce anesthesia, the monkeys were intubated, and mechanically ventilated on air with respiration adjusted to maintain end-tidal CO_2 similar to that observed in the awake portion of the study. Thiopental infusion was then continued to progessively reduce EEG activity until an isoelectric EEG was obtained. Titrated IV infusion of norepinephrine was used to maintain mean arterial blood pressure (MABP) at about 100 mmHg. EEG was kept isoelectric by titrated infusion of thiopental while a second set of 3 measurements of CBF and CMR were made. CMRO2 compartments were calculated as follows: Conscious CMRO2 = total CMRO2 (TCMRO2) and isoelectric EEG CMRO2 = BCMRO2. Active CMRO2 (ACMRO2) = TCMRO2-BCMRO2. The oxygen-glucose index (OGI) = (CMRO2/CMRG) X 6. The same derivations apply for other cerebral variables, namely, CBF and CMRG; active CBF and CMRG and basal CBF and CMRG.

Statistical analyses were by one and two-way analysis of variance for within and between group comparisons. Posthoc analyses were by Student Newman Keuls test with a maximum significant p value of 0.5. All values presented are mean $\pm$ SD.

Because there were no significant differences between cerebral variables measured in sets 1 through 3 within the groups of unanesthetized and thiopental anesthetized monkeys, the three measurements were combined and averaged (Table 2).

RESULTS

Physiological variables were unchanged during thiopental anesthesia compared to the unanesthetized state with MABP at about 100 mmHg, $PaCO_2$ at about 35 mmHg, PaO_2 between 80 and 90 mmHg and pHa at about 7.38 (Table 1).

Basal and active CBF distribution was 60% and 40%, respectively, whereas CMRO2 represented 50% of total CMRO2 in both compartments (Table 2). On the other hand, CMRG in the active compartment was opposite to the CBF distribution with 60% in the active compartment and 40% in the basal compartment reflected by a significantly higher OGI.

These results suggest a relative over perfusion of the basal compartment. They are substantiated by a higher ($p < 0.05$) ratio of BCBF/BCMRO2 (13.97 ± 2.46, X ± SD) compared to ACBF/CMRO2 (10.37 ± 3.96). The ratio of TCBF/TCMRO2 was 11.8 ± 2.70.

DISCUSSION

Surprisingly, the active and basal CMRO2 reported by Michenfelder[2] in 1.0 to 1.5% halothane anesthetized dogs of 45% active and 55% basal compares very well with our values of 47% active and 53% basal CMRO2 distribution in unanesthetized Rhesus monkeys. Surprisingly, because 1.0 to 1.5% halothane should cause a 25% decrease in CMRO2, presumably active CMRO2. Nevertheless, the results show that previous estimates of active and basal CMRO2 are well-correlated with the active/basal CMRO2 distribution observed in the unanesthetized Rhesus monkey and that approximately 50% of the oxygen consumed by the brain is used to support spontaneous EEG activity and 50% for the maintenance of brain viability.

The results clearly show a relative over perfusion of the basal compartment compared to the active compartment as indicated by the %TCBF and the ratio of BCBF/BCMRO2 and at the same time a higher glucose utilization in the active compared to the basal compartment. These results would suggest that the active metabolic compartment is working at a closer tolerance to oxygen supply and demand compared to the basal compartment while the active compartment consumes a greater amount of glucose than the basal compartment. The significance of this observation is uncertain. However, the ability of the brain to respond to oxygen and substrate demands may depend upon the closeness of the tolerance to borderline hypoxia or ischemia. Therefore, the relatively lower perfusion and greater metabolic rate for glucose in the active compartment may provide this compartment with a greater sensitivity to oxygen or substrate deficiency.

The importance of the distribution of CBF, CMRO2 and CMRG in the active and basal compartments relates to the responses of the brain to ischemia and hypoxia. We recently provided evidence that recirculation of the brain following complete global brain ischemia in rats indicated that the distribution of the active and basal brain oxygen compartments is drastically altered postischemia.[5] With the great emphasis recently placed on attempting to normalize the postischemic metabolic state for fear of cerebral over–activation and its exacerbation of the severity of the injury sustained, our findings suggest that all of hypermetabolic activity may not be detrimental to the brain if it represents an attempt to restore cellular ionic gradients postischemia.

Table 1. Physiological variables in 6 unanesthetized and 6 thiopental anesthetized Rhesus monkeys

Measurement No.		MABP	PaCO$_2$ (mmHg)	PaO$_2$	pHa
			Unanesthetized		
1	X	108	33	84	7.38
	SD	11	9	4	0.04
2	X	107	33	85	7.38
	SD	17	5	6	0.03
3	X	104	35	82	7.38
	SD	17	2	6	0.04
			Thiopental Anesthetized		
1	X	102	37	87	7.38
	SD	18	9	13	0.08
2	X	93	36	91	7.38
	SD	19	3	10	0.05
3	X	94	35	91	7.40
	SD	18	3	8	0.03

Table 2. Distribution of CMRO2 and CMRG in normal Rhesus monkeys.

Variable		TOTAL	BASAL	ACTIVE
CBF	X	76.1	47.9[*]	27.4[*+]
	SD	20.5	13.9	9.7
%TCBF	X	100	60.0	40.0
	SD	0	5.0	5.0
CMRO2	X	5.95	3.10[*]	2.85[*]
	SD	0.54	0.51	0.78
%TCMRO2	X	100	53.0	47.0
	SD	0	11.0	11.0
CMRG	X	8.09	3.13[*]	5.32[*]
	SD	2.78	0.77	3.54
%TCMRG	X	100	41.0	59.5
	SD	0	19.5	20.3
OGI	X	0.992	1.266	0.874[*]
	SD	0.098	0.252	0.151

CBF and CMRO2 expressed in ml/100g/min. CMRG = mg/100g/min.
* = p <0.05 compared to total. + = P <.05 compared to basal.

ACKNOWLEDGMENTS

This research was supported in part by the Public Health Service, research grant no. 5 R01 HL27208-09 and the departments of Anesthesiology and Critical Care Medicine and Neurological Surgery, University of Pittsburgh School of Medicine.

REFERENCES

1.   J. D. Michenfelder, The interdependency of cerebral functional and metabolic effects following massive doses of thiopental in the dog, Anesthesiology 41:231 (1974).

2.   J. Astrup, P. M. Sorensen, H. R. Sorensen, Oxygen and glucose consumption related to Na^+-K^+ transport in canine brain, Stroke 12:726 (1981).

3.   S. S. Kety, Discussion. Pharmacol Rev 17:230 (1965).

4.   J. D. Michenfelder, R. A. Theye, In vivo toxic effects of halothane on canine cerebral metabolic pathways, Am. J. Physiol.229:1050 (1975).

5.   E. M. Nemoto, J. A. Melick, P. M. Winter, Active and basal cerebrometabolic rate for oxygen (CMRO2) after complete global brain ischemia in rats, in: Oxygen transport to tissue X.M. Mochizuki, C. R. Honig, T. Koyama, T. K. Goldstick, D.F. Bruley, eds., Plenum Publishing Corp., New York, 1988.

THE REGIONAL CEREBRAL BLOOD FLOW RESPONSE TO CORTICAL MICROELECTRODE INSERTION IS NEUTROPHIL DEPENDENT

Mark W. Uhl, Patrick M. Kochanek, Joanne K. Schiding, John A. Melick, and Edwin M. Nemoto

Departments of Anesthesiology/Critical Care Medicine and Pediatrics, University of Pittsburgh, Pittsburgh PA 15213, USA

INTRODUCTION

Tissue blood flow measurements using the H_2-clearance technique with implanted polarographic microelectrodes was described by Aukland et al. in 1964 (1). The H_2-clearance technique has important advantages over other well-established methods such as radioactively labeled microspheres or ^{14}C-iodoantipyrine quantitative autoradiography. Its primary benefits are the capacity for repeated regional cerebral blood flow (rCBF) measurements in the same animal and the avoidance of radioactive tracer material. However, CBF values obtained by the H_2-clearance technique frequently differ from those obtained by radioactive tracer methods. Rat CBF values measured by iodoantipyrine (2) were significantly higher than measurements obtained by H_2-clearance (3). Chronic implantation of electrodes leads to higher CBF values (4). These concerns were addressed recently by Tomida et al. who reported low rCBF values measured by H_2-clearance shortly after electrode insertion in gerbils (5). CBF progressively increased by 24 hours, at which time the H_2-clearance and the iodoantipyrine measurements correlated well. "Recovery" of rCBF was attributed to gradual resolution of spreading depression after the local trauma of electrode insertion (5).

Tomida et al. stated that the minimal histological tissue damage caused by electrode insertion ruled out contribution of local tissue injury to the observed CBF effect (5). Others, however, have observed greater tissue injury after electrode placement. Persson noted that neutrophils accumulate in the injured area after cortical microelectrode stab wounds (6). In other tissues neutrophil accumulation after trauma is accompanied by increased blood flow or hyperemia. Hyperemia is a well-described component of the classic acute inflammatory response to tissue injury (7).

This study was designed to test the hypothesis that "recovery" of rCBF after microelectrode insertion is due in part to neutrophil-mediated hyperemia during the acute inflammatory response to trauma. To test the contribution of neutrophils, we performed CBF studies after microelectrode placement in normal and neutrophil-depleted rats. Our results suggest that neutrophils and the acute inflammatory response are in part responsible for the increase in rCBF after the trauma of electrode placement.

MATERIALS AND METHODS

This protocol was approved by the Animal Care and Use Committee of the University of Pittsburgh. Male Wistar rats (350-450 g) were anesthetized in plastic containers insufflated with 4% halothane in oxygen, then endotracheally intubated and mechanically ventilated. Surgical anesthesia was maintained with 1% halothane, 66% nitrous oxide and balance oxygen. Benzathine penicillin G (100,00 units) and gentamicin (10 mg/kg) were administered IM. Using aseptic technique PE-50 femoral venous and arterial catheters were inserted. Pancuronium bromide (0.1 mg/kg/hr) was administered IV for sustained muscle relaxation. Mean arterial pressure was continuously monitored. Core temperature was maintained at $37.0 \pm 0.5°C$ with a heated water blanket.

The dorsal calvarium was exposed by a midline incision and the rats were fixed in a stereotaxic device (David Kopf, Tujanga, CA). A platinum microelectrode (25 μm tip) was inserted 1 mm into the right parietal cerebral cortex through a small burr hole. The dura remained intact except at the electrode insertion site which was sealed with agar gel to prevent diffusion of H_2 gas. After completion of surgical procedures the inspired halothane concentration was reduced to 0.4% with 66% nitrous oxide and balance oxygen. Arterial blood samples were obtained at specified time points for hematologic and arterial blood gas measurements. Care was taken to maintain the blood gases within the normal physiological range. Complete blood counts were determined by Coulter counter and by manual differential counts of WBCs and platelets.

Electrodes were tested prior to use, and only those showing an immediate response in a H_2-saturated solution were used. Polarization voltage was +250 mV with the output signal monitored on a chemical microsensor (Diamond Electrotech, Ann Arbor, MI) and chart recorder.

Three hourly CBF measurements were taken beginning 30 min after microelectrode insertion. H_2 gas (5-10%) was introduced into the ventilatory gas circuit, and the electrode signal was monitored until a stable saturation baseline was obtained. After desaturation, rCBF as ml/min/100 g brain was calculated from the clearance curves by the $t_{1/2}$ method (8). The first 15 seconds of each curve was ignored because in our preparation arterial H_2 concentration does not reach $\leq 5\%$ of baseline until 15 s after discontinuation of H_2.

Neutropenia was induced in 13 rats by a single injection of vinblastine sulfate (0.5 mg/kg IV) via the femoral vein 5 days before the CBF studies.

The rCBF response to microelectrode insertion was determined in 49 normal and in 13 neutropenic rats. Physiologic parameters were monitored in all rats. Hematologic parameters were measured in 16 normal rats and 13 neutropenic rats.

Data are presented as mean $\pm$ standard error. Statistical comparisons within groups were made by repeated measures ANOVA followed by Student-Neuman-Keuls multiple comparison test. Between group comparisons were made by t-test with correction for multiple comparisons where appropriate. A value of $p < 0.05$ was considered statistically significant.

RESULTS

Mean arterial pressure (MAP), temperature, p_aCO_2, pH, and base excess did not differ significantly between normal and neutrophil-depleted groups at any time point (Table 1). p_aO_2, however, was significantly lower (although still within the normal physiologic range of 100-150 mm Hg) after 0.5 h in the normal group (Table 1).

The neutropenic group had a 98% reduction in absolute neutrophil count (ANC) ($p < 0.0001$), a 78% reduction in total WBC count ($p < 0.0001$), and a small (8%) but significantly significant decrease in hematocrit (Hct) ($p < 0.001$) (Table 2). Platelet count was not different between groups.

Regional CBF as measured by H_2-clearance increased progressively during the 3.5 h monitoring period only in the normal rats (* $p < 0.05$) comparing the 2.5 h and 3.5 h CBF values with the initial 0.5 h value (Figure 1). Regional CBF failed to increase in the neutrophil-depleted group and was significantly less than values in the normal rats at 3.5 h (** $p < 0.05$ vs normal) (Figure 1).

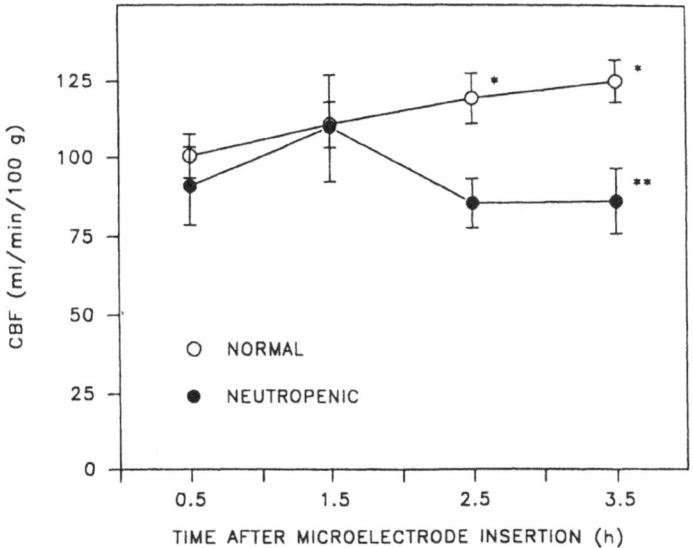

Figure 1. H_2-clearance rCBF in normal and neutropenic rats over time after microelectrode placement. (* $p < 0.05$ vs initial rCBF at 0.5 h; ** $p < 0.05$ vs normal).

DISCUSSION

After H_2-clearance microelectrode placement in the normal rat brain we observed an increase in CBF over time comparable to that described by others (5). Our data from neutrophil-depleted rats, however, suggests that the "recovery" of rCBF after cortical microelectrode insertion in part represents neutrophil-mediated hyperemia in damaged tissue. In the absence of inflammation, damaged tissue would have a lower than normal rCBF. Neutrophil-depleted rats showed no increase in rCBF over time and had a lower rCBF than reported in normal rat cortex at 3.5 h after electrode insertion (Figure 1).

It is unlikely that the statistically significant but physiologically minor effect of vinblastine treatment on Hct and p_aO_2 contributed to the effect of neutrophil depletion on rCBF observed in this study.

The postulated role of neutrophils may differ importantly in ischemic cerebral injury and traumatic injury. In cerebral ischemia, neutrophils accumulate in areas of low blood flow (9), and further reduce flow by adhesion and aggregation in the microcirculation. Neutrophil depletion in this setting increases CBF over ischemic levels. In contrast, in the acute inflammatory response to trauma, neutrophils may contribute to the marked hyperemia occurring in the first 24 h (10). Our data support the occurrence of an acute inflammatory response to the trauma of cortical microelectrode insertion.

Table 1. Physiologic Parameters

| | Time after microelectrode insertion | | | |
	0.5 h	1.5 h	2.5 h	3.5 h
MAP (torr)				
Normal	136 (2)	137 (2)	136 (2)	138 (2)
Neutropenic	130 (4)	132 (3)	136 (3)	140 (3)
Temp (C)				
Normal	37.2 (0.1)	37.4 (0.1)	37.5 (0.1)	37.5 (0.1)
Neutropenic	37.7 (0.2)	37.7 (0.1)	37.5 (0.1)	37.3 (0.1)
PaCO2 (torr)				
Normal	35.8 (0.6)	36.5 (0.6)	36.5 (0.6)	37.5 (0.6)
Neutropenic	37.2 (1.2)	36.3 (1.3)	35.6 (1.0)	37.9 (1.3)
pH				
Normal	7.38 (0.01)	7.38 (0.01)	7.38 (0.01)	7.38 (0.01)
Neutropenic	7.38 (0.01)	7.38 (0.01)	7.39 (0.01)	7.37 (0.01)
PaO2 (torr)				
Normal	121 (3)	118 (2)	114 (2)	112 (3)
Neutropenic	132 (6)	131 (3) *	134 (3) *	132 (2) *
Base Excess (mmol/L)				
Normal	-2.9 (0.3)	-2.4 (0.3)	-2.7 (0.3)	-1.7 (0.3)
Neutropenic	-2.3 (0.5)	-2.5 (0.5)	-2.3 (0.5)	-2.3 (0.7)

Data expressed as mean (SEM)

* $p < 0.05$ vs Normal

Table 2. Hematologic Parameters

	Normal	(n)	Neutropenic	(n)
Hct (%)	39.6 (0.6)	16	36.6 (1.0) *	11
ANC (#/mm3)	7189 (738)	16	164 (68) **	13
WBC (#/mm3)	9420 (870)	16	2250 (410) **	13
Platelet (# x 1000/mm3)	805 (44)	5	780 (76)	3

Data expressed as mean (SEM)

* $p < 0.05$ vs Normal

* $p < 0.001$ vs Normal

ACKNOWLEDGMENT

Supported, in part, by the Brain Trauma Foundation and the American Heart Association, Pennsylvania Affiliate.

REFERENCES

1.   K. Aukland, B. F. Bower, and R. W. Berliner, Measurement of local blood flow with hydrogen gas, Circ Res 14:164 (1964).
2.   O. Sakurada, C. Kennedy, J. Jehle, J. D. Brown, G. L. Carbin, and L. Sokoloff, Measurement of local cerebral blood flow with iodo[^{14}C]antipyrine, Am J Physiol 234:H59 (1978).
3.   O. U. Scremin, A. A. Rovere, A. C. Raynald, and A. Giardini, Cholinergic control of blood flow in the cerebral cortex of the rat, Stroke 4:2332 (1973).
4.   J. L. Haining, D. Turner, and R. M. Pantall, Measurement of local cerebral blood flow in the unanesthetized rat using a hydrogen clearance method, Circ Res 23:313 (1968).
5.   S. Tomida, H. G. Wagner, I. Klatzo, and T. S. Nowak, Effect of acute electrode placement on regional CBF in the gerbil: a comparison of blood flow measured by hydrogen clearance, [^{3}H]nicotine, and [^{14}C]iodoantipyrine techniques, J Cereb Blood Flow Metab 9:79 (1989).
6.   L. Persson, Cellular reactions to small cerebral stab wounds in the rat frontal lobe, Virchows Arch B Cell Path 22:21 (1976).
7.   A. C. Issekutz, Vascular responses during acute neutrophilic inflammation, Lab Invest 45:435 (1981).
8.   W. Young, H_2 clearance measurement of blood flow: a review of technique and polarographic principles, Stroke 11:552 (1980).
9.   J. M. Hallenbeck, A. J. Dutka, T. Tanishima, P. M. Kochanek, K. K. Kumaroo, C. B. Thompson, T. P. Obrenovitch, and T. J. Contreras, Polymorphonuclear leukocyte accumulation in brain regions with low blood flow during the early postischemic period, Stroke 17: 246 (1986).
10.   R. J. Schoettle, P. M. Kochanek, M. J. Magargee, M. W. Uhl, and E. M. Nemoto, Early polymorphonuclear leukocyte accumulation correlates with the development of posttraumatic cerebral edema in rats, J Neurotrauma 7:207 (1990).

CEREBRAL BLOOD FLOW AND BRAIN MITOCHONDRIAL REDOX STATE

RESPONSES TO VARIOUS PERTURBATIONS IN GERBILS

A. Mayevsky

Department of Life Sciences
Bar-Ilan University
Ramat-Gan 52900, Israel

INTRODUCTION

The interrelation between cerebral blood flow and metabolism is an important factor in understanding brain functions under normal and pathological situations. Since few of the situations, such as spreading depression and epilepsy, are transient and are not in steady state it became necessary to monitor the various parameters under observation in real time and in a continuous mode. Until recently CBF monitoring was limited to a single time point per animal using mapping technique or to a 10-15 minutes interval between single measurements. The development of the Laser Doppler flowmeter (LDF) opened up the possibility of monitoring CBF in a real time continuous mode (Stern et al 1977). As shown by a few groups, the CBF monitored by the LDF technique was significantly correlated to the flow measured by other quantitative techniques (Haberl et al 1989; Dirnagl et al 1989). Since CBF is only one of the important parameters to be measured we incorporated a laser Doppler probe into the multiprobe assembly developed by us a few years ago (Friedli et al 1982; Mayevsky 1983). Since all the probes used in the MPA are surface probes we also monitored the CBF from the surface of the brain.

In order to compare the responses obtained under various brain perturbations we used the Mongolian gerbil allowing an easy way to induce partial or near complete ischemia (Mayevsky & Breuer 1990; Mayevsky 1990). Using this model we were able to expose the same animal to all three types of brain perturbations, namely systemic hypoxia, ischemia and spreading depression (SD). In order to monitor the flow-metabolism coupling and correlations under those conditions we have developed a combined fiber optic probe for the simultaneous measurement of the relative CBF and mitochondrial redox state (NADH fluorescence). Incorporation of this metaflow probe (metabolism-flow) into the multi-probe assembly (Mayevsky 1990) provide us with information regarding the metabolic, ionic and electrical activities. In the present study we monitored the extracellular levels of K^+ and Ca^{2+} using ion selective minielectrodes as well as the DC steady potential and the ECoG. Preliminary results were described recently (Mayevsky et al 1991).

METHODS

Multiprobe assembly. This approach was developed by us several years ago (Friedli et al 1982; Mayevsky 1983,1990) and has been used successfully in rats and gerbils. The multiprobe assembly (MPA) used in the present study has been modified (Fig. 1) and contained the new optical probe named metaflow probe (MFP). The optical fibers of the two instruments, namely, the fluorometer and the Laser Doppler flowmeter, are mixed in the common part located on the surface of the brain. The Laser Doppler flowmeter used in the present study was made by TSI Inc. (Laserflo BPM) as described recently (Borgos 1988), but similar results were obtained by using the instrument made by Perimed Inc. (Piscataway N.J.) being used in our laboratory during the last year.

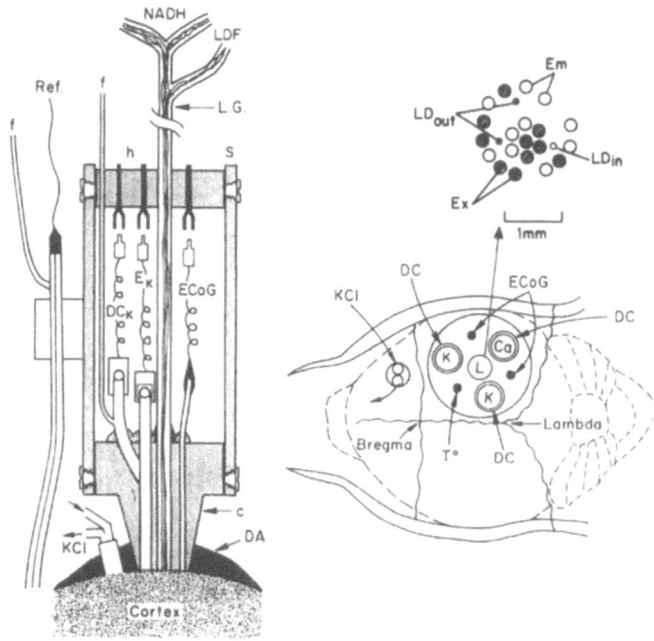

Figure 1. Schematic presentation of the experimental setup used.

LG -	Light guide monitoring the NADH redox state and relative CBF (LDF)
ECoG -	Electro Corticography electrodes
K, Ca, -	Minielectrodes for extracellular Potassium and Calcium monitoring

To -	Temperature thermistor probe	**DC** -	DC steady potential electrodes
DA -	Dental acrylic cement	**Ref** -	Reference electrode
KCl -	push pull cannula for KCl application (spreading depression).		
Ex, Em -	Excitation and Emission fibers of the fluorometer		
LD$_{in}$, LD$_{out}$ -	Fibers connected to the Laser Doppler flowmeter		
C -	Plexiglass probe holder	**h** -	connectors holder
s -	Plexiglass sleeve	**f** -	feeling tube of reference an X electrode

As shown in Fig. 1 in the present study we used 2 potassium electrodes and one calcium electrode. All details regarding the construction of the metaflow probe as well as the preparation of the minielectrodes and the MPA were published very recently (Mayevsky et al 1991).

Ten adult male Mongolian gerbils (60-80 gr) were used in the present study. The animal was anesthetized by Equithesin (each ml contains: pentobarbital 9.72 mg, chloral hydrate 42.51 mg, magnesium sulfate 21.25 mg, propylene glycol 44.34% w/v, alcohol 11.5% and water) injected IP (0.3 ml/100 gr). Body temperature was maintained at 36.5 ± 0.5°C and Equithesin was added during the monitoring period (0.03 ml per gerbil every 30 minutes).

The two common carotid arteries were isolated and a 4-0 silk suture was located around it for later lifting for occlusion by an aneurysm clip. The skull was exposed and a 5.5 mm hole was drilled in the parietal bone area. After gentle removal of the dura mater the MPA was located on the exposed cortex using a micromanipulator and cemented to the skull by dental acrylic. A push-pull cannula was located 3 mm anterior to the MPA in order to induce spreading depression (KCl application).

RESULTS

Each gerbil was exposed to all three perturbations tested, namely hypoxia, ischemia and spreading depression. Fig. 2 shows typical responses to unilateral (Roccl) and bilateral (Loccl) carotid artery occlusion. The two step reduction in blood supply was detected by the decrease in CBF (LDF) and increase in NADH levels (CF). During the period of ischemia, accumulation of K^+ in the extracellular space was recorded (K_1^+, K_2^+) but the DC steady potential and the Ca^{2+} levels remained unchanged during the occlusion period.

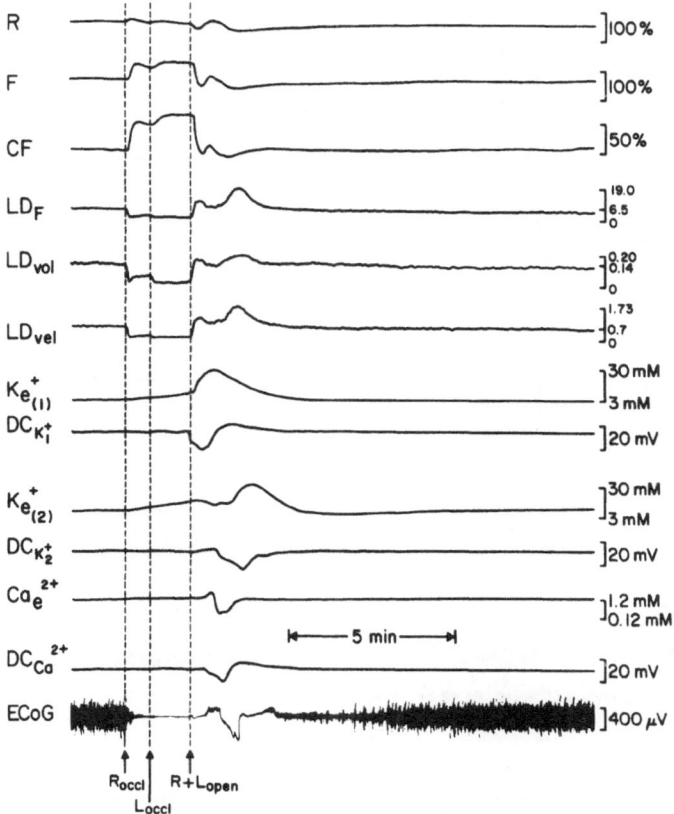

Figure 2. The effects of unilateral (Roccl) and bilateral (Loccl) carotid occlusion on the metabolic, hemodynamic, ionic and electrical activities in the gerbil brain.
R   366 nm reflectance   F -   450 nm fluorescence **CF** - NADH corrected fluorescence
LD$_F$, LD$_{vol}$, LD$_{vel}$ -   Laser Doppler flow, volume and velocity
$K^+_{e(1)}$, $K^+_{e(2)}$, Ca^{2+}_e - Extracellular potassium (two electrodes) and calcium electrodes
DC$_K^+{}_1$, DC$_K^+{}_2$, DC$_{Ca}2+$ - DC steady potential measured concentric to the three electrodes
ECoG - Electrocorticogram

The ECoG reached the isoelectric level very soon after the second occlusion. During the reopening of the two carotid arteries a fast reperfusion was recorded together with the oxidation of NADH. A spontaneous wave of SD (Spreading Depression) was developed during the recovery phase, characterized by the large increase in K^+_e and a decrease in Ca^{2+}_e together with a negative shift in the DC steady potential. During the recovery from the SD wave a large increase in CBF (300%) was recorded accompanied by an oxidation wave of the NADH (decrease CF).

The effects of graded hypoxia are shown in Fig. 3. Due to the decrease in oxygen delivery, detected by the increase in the intramitochondrial NADH levels, an increase in CBF was recorded (LDF). At the same time the reflectance trace (R) showed a decrease direction and is correlated to the increase in blood volume. The ionic responses to the decrease in energy availability, was similar to those recorded under ischemia (Fig. 2).

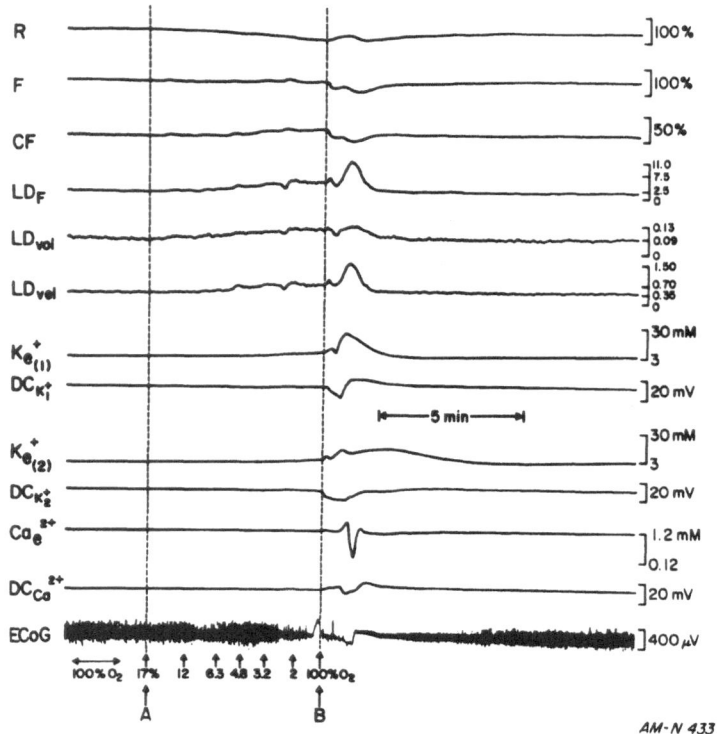

Figure 3. Hemodynamic metabolic, ionic and electrical responses to graded hypoxia in the Mongolian gerbil. At point A the gerbil was exposed to 17% O_2 and recovery started at point B (100% O_2). All abbreviations are as in Figure 2.

The ECoG was affected mainly when the hypoxic level was more severe. During the recovery period a typical response to SD was recorded also. The effects of repetitive cycles of spreading depression (SD), induced by 1M KCl, on the hemodynamic metabolic and ionic activities are shown in Fig. 4. The application site of KCl is seen in the scheme shown inside the figure. The front of the propagating SD wave affected the various electrodes at slightly different timing. As seen (Line A), the sequence was that the Ca^{2+} and $K^+_{(1)}$ were the first site to be affected followed by the light guide area and the $K^+_{(2)}$ was the last one to respond. Lines B and C show that the peak of the NADH oxidation precede the peak of the maximal CBF recorded (LDF). Since the 1M KCl was left in the washing cannula, repetitive cycles were recorded. Due to the depolarization wave developed as seen by the negative shift in the DC potential and the increase in the extracellular K^+, the energy metabolism was activated. Intramitochondrial

NADH became oxidized (decrease in CF signal) and a large increase in CBF was recorded in order to compensate for the extra O_2 needed (Leao 1944a,b; Bures et al 1984; Lauritzen 1987).

In order to study the correlation between mitochondrial activity and the hemodynamic responses under various conditions the analog signals recorded were analyzed quantitatively. The resting level of the signal before the perturbation was considered as a 100% value and the change during the perturbation was calculated in percent change relative to the resting level.

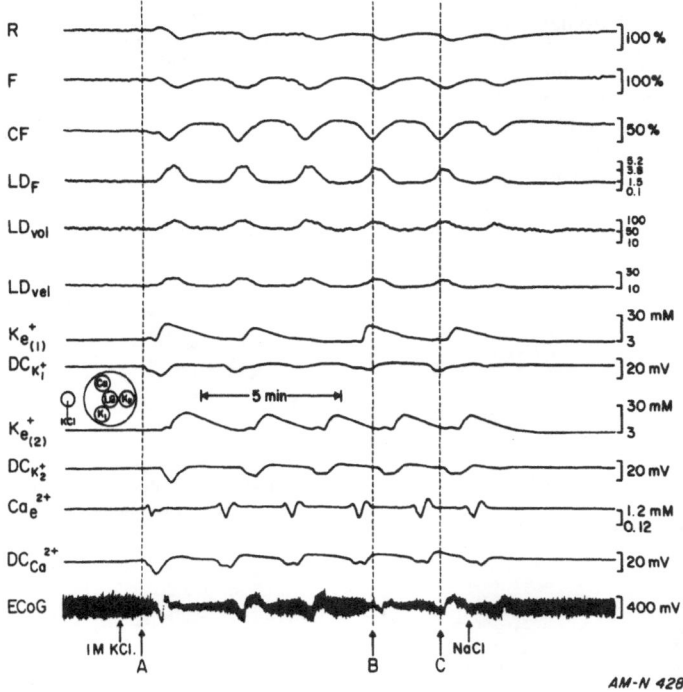

Figure 4. Metabolic, hemodynamic, ionic and electrical responses to cortical spreading depression induced by epidural application of KCl solution. All abbreviations are as in Figure 2.

Figure 5 shows the relationship between the CBF (measured by LDF) and mitochondrial redox state (NADH-CF) as well as between the reflectance and blood volume measured under ischemia, hypoxia and spreading depression in the gerbil brain. Since various gerbils have a different anatomy of the anterior part of the circle of Willis the level of ischemia created (part A), cover the range from 10% to 100% ischemia. Also, during the graded hypoxia a large increase in CBF up to 400% was noted simultaneously with the increase in NADH. During SD, the NADH became oxidized while the flow was increased by up to 400%. The correlation shown in part B presents the changes in blood volume measured by the Laser Doppler flowmeter versus the changes in the intensity of the reflected light (366 nm) measured by the NADH fluorometer. The correlations between the 3 different parameters measured by the LDF are shown in Fig. 6. A very clear positive correlation can be seen between the two pairs of parameters. The calculations of the correlation coefficients (R) reveal that most of the values were statistically significant.

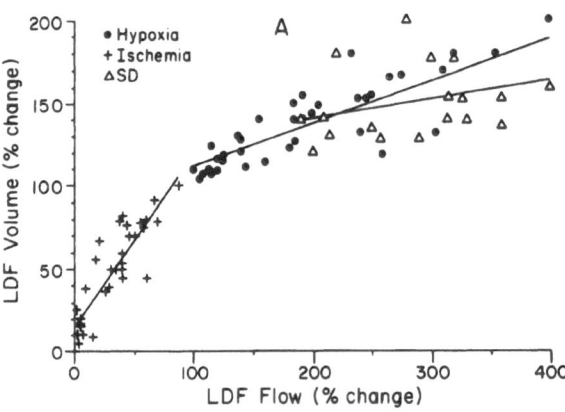

Figure 5. ʿCorrelations between parameters measured by the Laser Doppler flowmeter (flow and volume) and those monitored by the fluorometer reflectometer (NADH fluorescence and reflected light). The gerbils were exposed to ischemia, hypoxia and spreading depression.

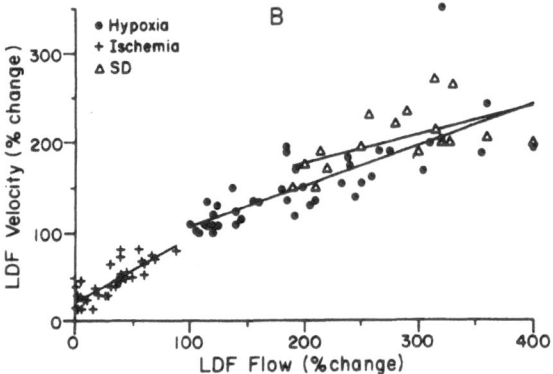

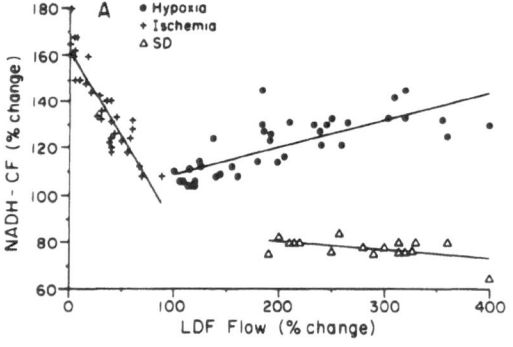

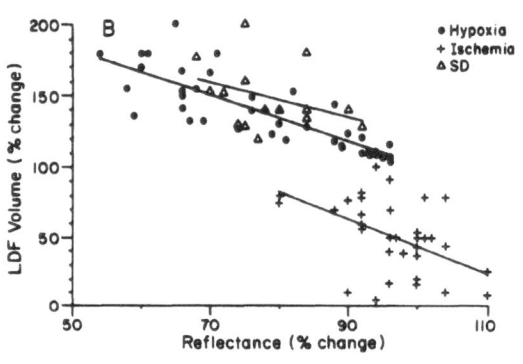

Figure 6. The relationship between all three parameters measured by the laser Doppler flowmeter.

In Table 1 we are presenting two sets of R values. In the upper right side, the values of each individual treatment, and in the lower left the pooled values of all 92 pairs of numbers. Under spreading depression most of the R values were not statistically significant except NADH and CBF as well as CBF velocity. The interrelation between energy supply and the ability of the brain to recover from spreading depression is shown in Fig. 7. Line A shows the initial effect on $K^+_{(2)}$ electrode followed by the other sensors. Due to the occlusion (Roccl) a small decrease in CBF was recorded (LDF) simultaneously with an increase in NADH redox state (CF). The propagation of the SD wave was blocked by the ischemic episode and recovery processes started only after the reopening of the occluded artery(R_{open}).

Table 1. The correlation coefficient values calculated between the various metabolic and hemodynamic parameters monitored during hypoxia, ischemia and spreading depression.

	Flow	Volume	Velocity	Reflectance	Corrected Fluorescence
Flow		0.667 0.753 0.095 (N.S)	0.592 0.676 0.319 (p<0.02)	0.742 0.158 (p<0.02) 0.176 (N.S)	0.592 0.808 0.241 (p<0.05)
Volume	0.799		0.490 0.561 0.005 (N.S)	0.684 0.259 (p<0.002) 0.156 (N.S)	0.653 0.682 0.057 (N.S)
Velocity	0.862	0.762		0.598 0.148 (p<0.05) 0.019 (N.S)	0.606 0.665 0.010 (N.S)
Reflectance	0.706	0.678	0.621		0.822 0.091 (N.S) 0.100 (N.S)
Corrected Fluorescence	0.307	0.339	0.328	0.067 (p<0.01)	

The calculations were done on 41 pairs of hypoxia (upper number), 34 pairs of ischemia (middle number) and 17 pairs of spreading depression (lower number).
The values appearing as single numbers per box were calculated by using all 92 pairs collected in the three kinds of perturbations. The level of significance was p<0.0001 for all correlations except in cases that it was lower and are presented in the parentheses.

In one of the gerbils monitored, an abnormal response to SD was recorded and is shown in Fig. 8. Due to unclear reason the delivery of O_2 under the SD wave was not sufficient and the NADH became mainly reduced (increase CF) concomitantly with the initial decrease in CBF (Line A). In the second cycle shown (Line B) the response did improve in terms of O_2 supply. This kind of response was found in rats exposed to SD under ischemic conditions (Mayevsky et al 1982).

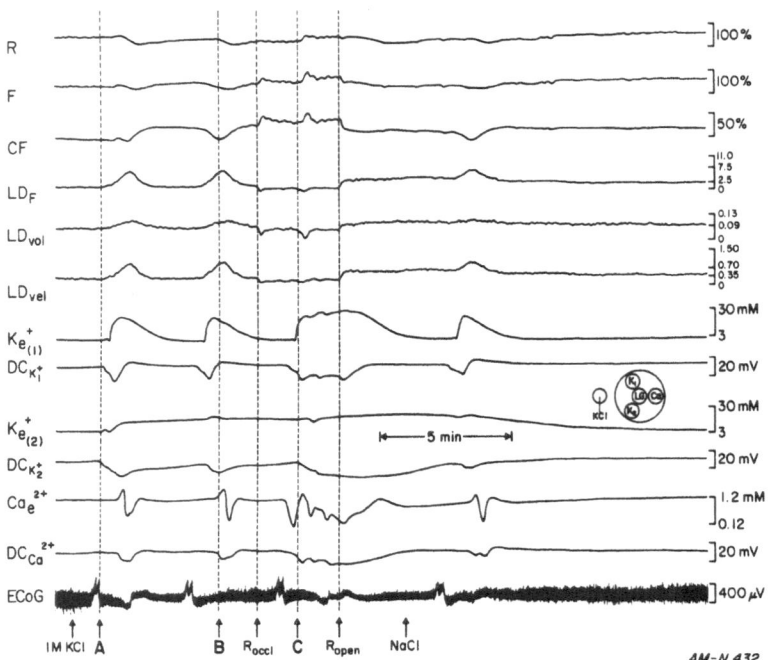

Figure 7. The effects of partial ischemia induced by unilateral carotid artery occlusion (Roccl) on the responses to spreading depression. All abbreviations are as in Figure 2.

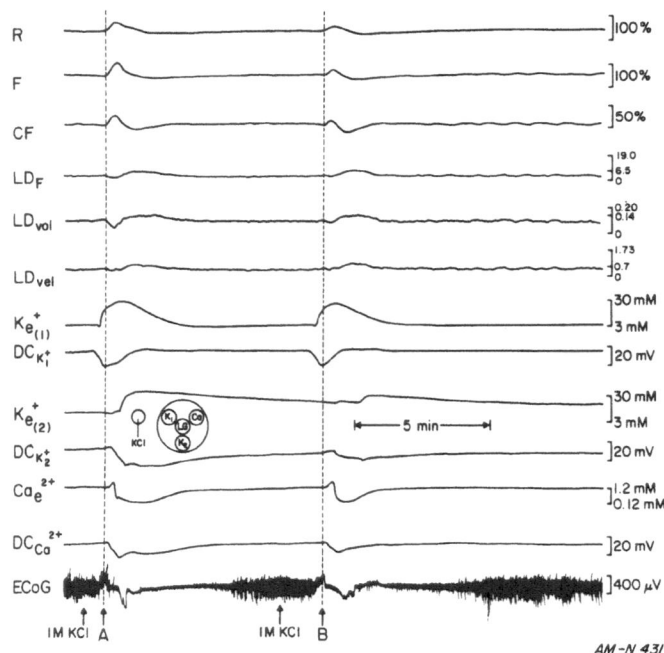

Figure 8. A non typical response to spreading depression waves measured in one of the gerbils. All abbreviations are as in Figure 2.

Discussion

The main purpose of the present work was to study in detail the coupling between CBF and mitochondrial activity under various metabolic perturbation. Since many of the pathological situations are transient and not in steady state it is important to monitor the responses in a real-time and continuous mode. The surface fluorometry technique developed by Chance and collaborators (Chance et al 1962,1973; Mayevsky, 1984) to monitor the NADH redox state could provide information in real-time. During all the years it was necessary to correlate the NADH monitoring to another in vivo real-time technique in order to verify the real functional state of the brain in terms of O_2 delivery or CBF. The development of the Laser Doppler flowmetry (Stern 1977) opened up this possibility for the first time and the present study is providing preliminary data.

In order to correlate the two parameters, namely, the NADH redox state and CBF it was necessary to integrate the probes so that both of them will "sample" approximately the same tissue volume. This was achieved by developing the MFP (Metaflow probe) consisting of two sets of optical fibers connecting the surface of the brain to the NADH fluorometer and to the Laser Doppler flowmeter. As seen in Fig. 1 the three fibers of the LDF are mixed between the NADH fibers and since the light wavelength of the LDF is longer (>600 nm) it will penetrate better than the NADH fibers.

The three kinds of perturbation used, namely, ischemia, hypoxia and spreading depression are covering a wide range of clinical pathological situations, thus the information regarding the correlation between flow and NADH redox state is very significant. As seen in Figure 5A, the relationship of CBF and NADH redox state was different under each of the 3 perturbations used. Under ischemia the decrease in CBF is expected by definition (decrease in blood supply) and indeed the CF signal increased according to the decrease in CBF. The response to hypoxia was characterized by an increase in CBF (due to autoregulatory response) and the NADH also increased. The response to spreading depression was completely different. The SD wave led to an increase in ion pumping activity which stimulate the energy consumption (Erecinska and Silver 1989). The CBF, under those conditions is stimulated in order to compensate for the extra-oxygen consumed (Mayevsky & Weiss 1991) as detected by the oxidation of the NADH (decrease of CF signal). As shown in Fig. 5B the changes in the reflectance trace were correlated to the blood volume changes measured by the LDF. The correlation was very clear under hypoxia and not so clear under SD.

According to the manufacturer of the LDF (TSI Inc.), the volume parameter represent only the volume of the flowing fraction of the blood and not the total volume, therefore the meaning of it is not clear as yet. As seen in Fig. 6 all three parameters measured by the LDF, namely, the flow, volume and velocity were correlated very well between each other. Recently, we started using the LDF produced by Perimed Inc. and the information obtained regarding the volume is different. In the Perimed instrument the total backscattering light is used to evaluate the changes in blood volume and therefore it may be more adequate for comparison with the reflectance changes.

Conclusion

The results presented in this paper suggest that the combination of the two optical techniques, namely, the NADH redox state and CBF, created a new powerful tool. It has been shown that the metaflow probe-MFP provide information which is more meaningful as compared to each one of the individual parameter. Also, the criticism about the lack of absolute calibration of the two techniques is diminished by using the MFP and on the other hand the information is continuous and in real-time mode which cannot be achieved using the quantitated techniques.

Acknowledgements

This work was supported by NIH grant NS-22881, the Dr. Jaime Lusinchi Research Center in Applied Life Sciences, the HSRC at the Dept. of Life Sciences, Bar-Ilan University and by a grant from the research committee at Bar-Ilan University, Ramat-Gan, Israel.

References

Borgos, J.A., TSI's LDV blood flowmeter in: Laser Doppler blood flowmetry. Edited by Shepherd, A.P., Oberg, P.A. Kluwer Academic Publishers pp. 73-91 (1988).

Bures, J., and Buresova, O., 1984, Susceptibility to spreading depression and anoxia: regional differences and drug control. In: Mechanisms of cerebral Hypoxia and Stroke. ed. G. Somjen Plenum Press N.Y. pp. 253-267.

Chance, B., Cohon, P., Jöbsis, F.F., and Schoener, B., 1962, Intracellular oxidation-reduction states *in vivo*. Science 137:499-508.

Chance, B., Oshino, N., Sugano, T., and Mayevsky, A., 1973, Basic principles of tissue oxygen determination from mitochondrial signals. Internat. Symposium on Oxygen Transport to tissue. In: Adv. Exp. Med. Biol. Vo. 37A, Plenum Pub. Corp., N.Y. 239-244.

Dirnagl, U., Kaplan, B., Jacewicz, M., and Pulsinelli, W., 1989, Continuous measurement of cerebral cortical blood flow by Laser Doppler Flowmetery with a rat stroke model. *J. CBF and Metabol.* 9:589-596.

Erecinska M., and Silver, I.A., 1989, ATP and brain function. *J. CBF and Metabol.* 9:2-19.

Friedli, C.M., Sclarsky, D.L., and Mayevsky, A., 1982, A new multiprobe assembly for surface monitoring of ionic, metabolic and electrical activities in the awake brain. *Am. J. Physiol.* 243:R642-R469.

Haberl, R.L., Heizer, M.L., Marmarou, A., Ellis, E.F., 1989, Laser doppler assessment of brain microcirculation: Effect of systemic alterations. *Am. J. Physiol.* 256:H1247-1254.

Lauritzen, M., 1987, Cerebral blood flow in migraine and cortical spreading depression. *Acta Neurol. Scand.* 76 (Suppl 113):1-14.

Leao, A.A.P., 1944a, Spreading depression of activity in cerebral cortex. *J. Neurophysiol.* 7:359-390.

Leao, A.A.P., 1944b, Pial circulation and spreading depression of activity in the cerebral cortex. *J. Neurophysiol.* 7:391-396.

Mayevsky, A., 1983, Multiparameter monitoring of the awake brain under hyperbaric oxygenation *J. Appl. Physiol.* 54:740-748.

Mayevsky, A., 1984, Brain NADH redox state monitored *in vivo* by fiber optic surface fluorometry. *Brain Res. Rev.* 7:49-68.

Mayevsky, A., 1990, Level of ischemia and brain functions in the Mongolian gerbil *in vivo*. *Brain Res.*, 524:1-9.

Mayevsky, A., and Breuer, Z., The Mongolian geruil as a model for cerebral ischemia, In: "Cerebral Ischemia and Cerebral Resuscitation", Schurr, A., and Rigor, B.M. Eds., CRC Press pp. 27-46 (1990).

Mayevsky, A., and Weiss, H.R., 1991 Cerebral blood flow and oxygen consumption in cortical spreading depression. *J. CBF and Metabol.* (In press).

Mayevsky, A., Flamm, E.S., Pennie, W., and Chance, B., 1991 A fiber optic based multiprobe system for intraoperative monitoring of brain functions. *SPIE Proc.* Vol 1431 pp 303-313.

Mayevsky, A., Zarchin, N., and Friedli, C.M., 1982, Factors affecting the oxygen balance in the awake cerebral cortex exposed to spreading depression. *Brain Res.* 236:93-105.

Stern, M.D., Lappe, D.L., Bowen, P.D., Chimosky, J.E., Holoway, G.A., Keiser, H.R., and Bowen, P.D., 1977, Continuous measurement of tissue blood flow by Laser-Doppler spectroscopy. *Am. J. Physiol.* 256:H1247-1254.

LOCAL TISSUE P_{O_2} DURING AND AFTER

FOCAL BRAIN CORTICAL INFARCTION IN RABBITS

Koen van Rossem, Herman Vermariën, Karin Decuyper, Jos Van Reempts[*], Murielle Laureys[+] and René Bourgain

Laboratory of Physiology and Physiopathology, University of Brussels VUB
Laarbeeklaan 103, B-1090 Brussels, Belgium

* Department of Morphology, Janssen Research Foundation
Turnhoutseweg 30, B-2340 Beerse, Belgium

+ Department of Clinical Chemistry, University Hospital VUB
Laarbeeklaan 101, B-1090 Brussels, Belgium

INTRODUCTION

Photochemical induction of thrombosis is an elegant technique (Watson et al., 1985) which is increasingly applied in order to produce reproducible focal cerebral infarctions. In this stroke model focal illumination of the brain cortex following intravenous injection of Rose Bengal, induces vascular thrombosis in the illuminated area, due to local production of singlet oxygen which causes endothelial damage. To date, attention has mainly been focussed on the evaluation of morphologic changes, alterations in local cerebral blood flow and metabolism, and functional deficits at certain time periods following photothrombotic infarction, induced by illumination of the brain through the intact skull in anesthetized rats.

We have developed a technique which enables continuous polarographic monitoring of local tissue P_{O_2} in and near the infarction zone, during and after direct illumination of a well demarcated cortical area, avoiding light scattering in the skull (van Rossem et al., 1990 (b)). Infarctions are induced in awake rabbits, steering clear of possible inconvenient effects of anesthesia including changes of blood gas levels and alteration of vascular responsiveness. The aim of the present study was to evaluate the time course of local P_{O_2} and it's reproducibility in this model. Moreover, in order to improve the interpretability of the results, we investigated morphologic changes related to different stages in the evolution of P_{O_2}.

MATERIALS AND METHODS

P_{O_2} monitoring

Measurements were performed polarographically with platinum electrodes having a cylindrical measuring tip (length 1 mm, $\varnothing$ 100 μm) covered with a homogeneous cellulose acetate membrane (thickness 20 μm $\pm$ 2.5 μm) (van Rossem et al., 1990 (a)). They were fixed into polymetylmeta-crylate (PMMA) frames (6 x 8 mm) containing a central shaft in which an optic fiber ($\varnothing$ 3 mm) can be mounted above a perfectly delineated transparent area (inner diameter 3.17 mm). In each frame two electrodes respectively were fixed in the center and 1 mm in front of the illumination area, the measuring tip extending perpendicularly beneath the bottom plate. A polarizing voltage of -600 mV was applied with respect to a Ag/AgCl electrode which in vivo was fixed to the animal's ear. Calibration was performed in Ringer solution at 38 °C. Since our electrodes show only moderate long-term stability (van Rossem et al., 1990 (a)) we can only mention absolute values of electrode current (i_{O_2}). However, because of their permanent linear behaviour and good short-term stability, relative changes of i_{O_2} within one recording can directly be related to changes in P_{O_2}.

Animal preparation and treatment

Twenty fed male Dutch rabbits HSD/POC weighing 1.8 - 2.8 kg were used for the experiments. All experiments were performed according to the locally established ethical rules.

Electrode implantation was performed as described earlier (van Rossem et al., 1990 (b)). In short, rabbits were anesthetized (Hypnorm® 0.5 ml/kg i.m.) and fixed in a stereotaxic apparatus. After exposure of the skull a hole was drilled according to the contours of the PMMA frame and the dura mater was removed. The center of the hole was located 1 mm behind and 4 mm lateral (left hemisphere) to the bregma. With the aid of a micromanipulator the frame was then placed onto the cortical surface so that electrode tips were inserted into the cortex completely. Finally, the frame was fixed to the skull applying cyanoacrylate gel and dental resin.

Ten days after implantation a cortical infarction was induced photochemically. Rose Bengal (7.5 mg/ml in 9 g/l NaCl, subjected to 0.45 μm filtration) was injected intravenously (10 mg/kg) over a 2 min interval via the marginal ear vein. An optic fiber was then mounted into the shaft of the implanted frame and the brain cortex was illuminated with cold green light (spectral width 450 - 600 nm, intensity 110 mW/cm²) during 20 min. Thirty min before infarction, blood samples for determination of blood gasses, pH, platelet count and plasma glucose level were taken from the central artery of the ear, and rectal temperature was measured.

Six animals were treated according to the following schedule: P_{O_2} was monitored continuously from 30 min before until 4 hours after infarction. Furthermore, measurements lasting 30 min, were performed 24 h, 48 h, 5 d, 9 d, 14 d, 21 d and 28 d following infarction. The other rabbits were divided into 7 pairs and were monitored as mentioned above. They were sacrificed respectively 1 h, 4 h, 24 h, 48 h, 5 d, 9 d, and 14 d following onset of illumination. No anesthesia was applied during infarction and P_{O_2} measurements. All rabbits looked comfortable during the experiments.

Immediately after the final measurements the rabbits were anesthetized with pentobarbital (30 - 60 mg/kg). Horseradish peroxidase (Sigma HRP type II, 12 500 U/kg) was then slowly injected intravenously and five minutes later perfusion procedure was started. Animals were transcardially perfused with Karnovsky's fixative, containing 2 % paraformaldehyde and 2.5 % glutaraldehyde, delivered at a pressure of 16 kPa for 5 min. Perfusion was always preceded by a 30 - 40 s rinse with Haemacel®. Following decapitation the brain was kept in situ in Karnovsky's fixative during 4 - 24 h. The skull - with the fixed electrodes - was removed carefully and the cerebrum was taken from the cranial vault and transported in Karnovsky's fixative at 4 °C to the neuropathology lab. Horizontal Vibratome sections were cut alternately at 100 μm and 200 μm. All 100 μm sections were routinely stained with azure A - eosin B at pH 4.5 and reacted with 3 - 3' diaminobenzidine (DAB) for demonstration of HRP. Intermediate 200-μm sections were postfixed in OsO₄, routinely embedded in Epon and cut at 2 μm for detailed microscopy.

Statistical analysis

Differences between $\bar{I}_{O_2}$ before and respectively 1 h, 2 h, 3 h and 4 h following onset of irradiation were evaluated applying the t-test for paired data.

RESULTS

Physiological variables, obtained half an hour before infarction are summarized in table 1. Electrode sensitivity in vitro measured 6.67 ± 1.68 nA/kPa (mean ± S.D.).

Table 1 Fysiological variables obtained half an hour before photochemical induction of cortical infarction.

P_{a,O_2} (kPa)	11.4	± 0.7
P_{a,CO_2} (kPa)	5.2	± 0.5
pH	7.44	± 0.03
plasma glucose (g/l)	1.75	± 0.31
platelets (10^3 mm^{-3})	481	± 121
temperature (°C)	38.6	± 0.6

Values represent means ± S.D. (n = 20).

Table 2 $\bar{I}_{O_2}$ (nA) before, during and after illumination in the absence of rose bengal (control), and following i.v. injections of rose bengal (infarction).

	Control			Infarction				
	before illumination (n = 20)	during illumination (n = 20)	following illumination (n = 20)	0 h (n = 14)[*]	1 h (n = 14)[*]	2 h (n = 14)[*]	3 h (n = 14)[*]	4 h (n = 14)[*]
A	21.6 ± 5.8	22.1 ± 6.9	22.0 ± 6.6	21.5 ± 6.1[i]	3.5 ± 2.0[r]	3.3 ± 1.9	3.2 ± 1.9	3.2 ± 2.0
B	26.7 ± 6.3	26.4 ± 6.2	26.4 ± 5.9	31.3 ± 6.0	36.6 ± 7.2[a]	31.2 ± 9.6	28.8 ± 8.6	27.7 ± 7.7[b]

Values represent means ± S.D. A: illuminated area ; B: 1 mm rostral to the illuminated area. [*]: In 4 animals recording following infarction lasted less than 4 h; in 2 others, for unknown reasons, $\bar{I}_{O_2}$ was abnormally high and instable so that only changes in fluctuations could be evaluated. [i]: Initial current I_i. [r]: Residual current I_r. [a]: Significantly higher than I_i (p < 0.016). [b]: Significantly lower than I_i (p < 0.023).

In vivo recordings obtained before infarction showed normal fluctuations of i_{O_2} (frequency 7 - 11 /min) around a mean value $\bar{I}_{O_2}$ which remained stable in time (max. deviation 10 %). Illumination in the absence of rose bengal did not alter $\bar{I}_{O_2}$ (table 2) nor fluctuations significantly.

Time course of P_{O_2} during and after infarction

In all animals the time course of $\bar{I}_{O_2}$ showed a comparable pattern (fig. 1).

In the illuminated area $\bar{I}_{O_2}$ decreased significantly ($\bar{I}_{O_2} \leq 80$ % of initial current I_i) after a latency period (LP$_1$) of 1.7 ± 1.6 min (mean ± S.D.). LP$_1$ was followed by a decline interval (DI) of 8.9 ± 9.2 min in which $\bar{I}_{O_2}$ decreased gradually until residual current I_r (table 2), assumed to correspond with zero P_{O_2}, persisted (flat period : FP). I_r was always reached within illumination time, but two animals showed a temporary partial recovery of i_{O_2}. Figure 2 shows the mean time course of relative P_{O_2} in the illuminated area following onset of illumination. Eight animals were followed long enough in order to observe gradual restoration of both $\bar{I}_{O_2}$ and fluctuations following FP. FP amounted between 5 and 9 days in six of them, between 2 and 5 days in one animal, and between 9 and 13 days in the other one. Two (six animals) or three weeks (two animals) following infarction both $\bar{I}_{O_2}$ and fluctuations were normalized.

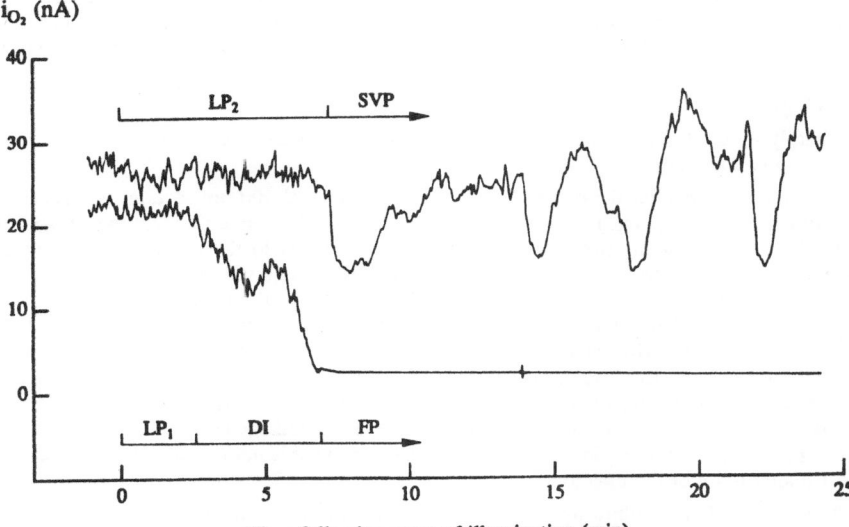

Figure 1 Recordings of i_{O_2} before and during photochemically induced infarction, in (lower curve) and 1 mm rostral to (upper curve) the illuminated area. Parameters describing the time course of $\bar{I}_{O_2}$ are indicated : LP$_1$ and LP$_2$: latency periods; DI : decline interval; FP : flat period, SVP : slow variation period.

One mm rostral to the illuminated area, the latency period (LP$_2$) (5.5 ± 2.4 min) lasted until the abrupt onset of a slow variation period (SVP) of 46.3 ± 12.3 min during which normal fluctuations were replaced by low frequent (1.5 - 0.17 / min) variations with large amplitudes (fig. 1). The minima of these slow variations only exceptionally approached zero level, and if so, i_{O_2} always returned to higher levels within a few seconds. The maxima reached values between 80 % and 200 % of I$_i$. At the end of SVP, slow variations disappeared and fluctuations returned completely normal or showed only slightly increased amplitude and/or slightly decreased frequency. Immediately following SVP, $\bar{i}_{O_2}$ was significantly higher compared to I$_i$ (p < 0.016) (table 2). Four hours after onset of illumination $\bar{i}_{O_2}$ appeared significantly lower than I$_i$ (p < 0.023). It should be mentioned that this general tendency did not apply to all subjects.

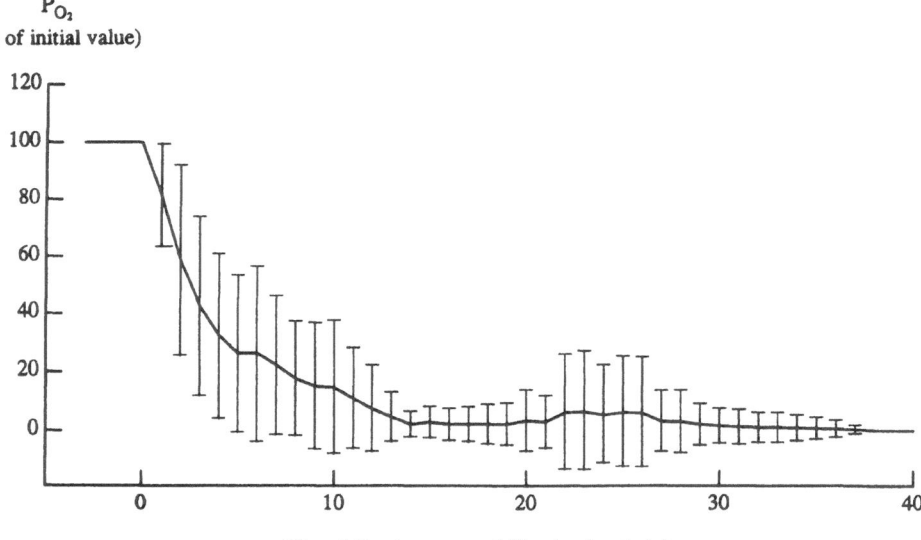

Figure 2 Mean time course of relative P$_{O_2}$ in the illuminated area following onset of illumination (n = 20). The curve connects mean values calculated per minute following onset of illumination. Bars indicate standard deviations. Large S.D. between 20 and 30 min are due to partial recovery of i_{O_2} in only two animals.

Morphology

Sharply delineated lesions extending to at least 1 mm beneath the end of the measuring tip were found in all animals.

One and 4 hours after the onset of illumination, horizontal 100-μm sections revealed circular lesions (diameter 3.3 - 3.9 mm) characterized by general vascular congestion and focal hemorrhages (fig. 3). Tissue surrounding the electrode located 1 mm rostral to the illuminated area remained unaltered. The distance between this electrode and the lesion edge varied between 0.5 mm and 0.7 mm. The border of the lesions showed a dark HRP rim (± 50 μm) which was indicative for blood brain barrier leakage. This rim was most pronounced in tissues which were reacted with DAB as soon as possible after perfusion fixation. Two-μm sections showed platelet thrombus formation, perivascular edema and focal hemorrhages over the entire lesion. A sharp transition was noticed from normal tissue to a markedly edematous border zone (status spongiosus) in which most neurons were irreversibly damaged as could be derived from severe shrinkage and nuclear pyknosis. In the center of the lesion edema was less pronounced and a considerable number of neurons appeared still viable 1 h after onset of illumination.

Twenty-four hours following infarction neuronal tissue within the lesion was completely necrotic. Disappearance of peroxidase staining in vessels gradually progressed from the center towards the periphery and an increasing number of apparently perfusable vessels was noticed from day 2. An expanding rim of macrophages and neutrophylic granulocytes was observed in the border zone. Macrophages were also found diffusely spread over the lesion.

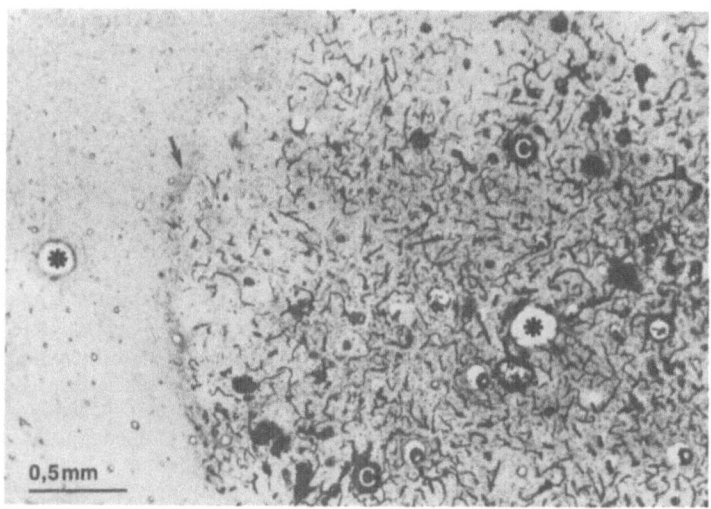

Figure 3 Azure-eosin-stained 100-μm horizontal section through the cerebral cortex of a rabbit sacrificed 4 hours after photochemical induction of thrombotic infarction. Electrode shafts are cut perpendicularly (asterisks). The circular lesion is sharply delineated and is characterized by darkly DAB-stained residual erythrocytes in congested blood vessels (C). At the border of the lesion a small dark rim (arrow) is indicative for blood brain barrier leakage. Tissue surrounding the electrode located 1 mm rostral to the illuminated area appears normal.

From day 9, neovascularization was noticed only in the peripheral zone of the lesion. Shrinkage of the infarcted region was clearly visible from two weeks following irradiation. At four weeks a small lesion (diameter 2.3 - 2.9 mm) containing amorphous granular material and macrophages was still present.

DISCUSSION

In this study we applied a modified photochemical stroke model, allowing for continuous monitoring of local mean cortical P_{O_2} during and after standardized induction of a focal cortical thrombotic infarction in awake rabbits. The time course of local P_{O_2} inside as well as 1 mm near the illuminated area showed a reproducible pattern. Parameters (latency period, decline interval) describing the gradual decrease of P_{O_2} to assumed zero level in the infarcted area, showed a considerable variation. To the best of our knowledge, no studies applying measuring techniques with corresponding time resolution have been performed to date. Yet, our observations are in agreement with those from a study (Dietrich et al., 1986 (a)) in which local cerebral blood flow was found to be extremely low in the central ischemic zone 30 min and 5 days following irradiation, showing partial recovery at day 15. Interestingly, while P_{O_2} still remained zero, morphologic evidence for congestion of large blood vessels in the center of the lesion was lost already at 24 h following irradiation.

The slow and large variations of P_{O_2} occurring 1 mm rostral to the illuminated area are probably related to changes in local cerebral blood flow (LCBF), possibly induced by varying imbalances between counteracting vasoactive substances such as thromboxane A_2 and serotonin being released during thrombosis, and prostacyclin and EDRF. Such marked changes of either P_{a,O_2} or local cortical oxygen consumption in normally breathing awake animals at rest seem to be most unlikely. During SVP, P_{O_2} reached values far below as well as far above the initial value. Autoradiographic studies cannot reveal such dynamic processes and could only state non-significant differences in LCBF in the cortex adjacent to the zone of infarction 30 min after irradiation. In the border zone, P_{O_2} at 4 h following infarction was significantly lower compared to initial value. This may, from a theoretical point of view, support the documented increase of glucose utilization at that time (Dietrich et al., 1986 (b)), in absence of a significantly increased LCBF at the periphery of the lesion (Dietrich et al., 1986 (a)).

Sharply delineated circular lesions showing classical signs of photochemically induced infarcts (Watson et al., 1985; Van Reempts et al., 1987; Dietrich et al., 1987) were demonstrated, their diameter only slightly exceeding the one of the illuminated area. Interestingly, in the center of the illuminated area, a considerable number of neurons appeared still viable 1 hour following infarction, although P_{O_2} at that moment was already at assumed zero level during at least half an hour. In contrast, at the border of the lesions almost all neurons were irreversibly damaged. Within this region ischemia may be either incomplete or accompanied by a certain oxygen supply via diffusion from the better perfused adjacent area. Such conditions may lead to more severe neuronal damage compared to complete ischemia (Siesjö, 1981).

Another important observation is the markedly sharp transition at the border of the lesion, from apparently normal viable tissue to the area demonstrating irreversible damage to almost all neurons. Differences in LCBF, P_{O_2}, arachidonate metabolism (Dorman et al., 1990) or formation of either oxygen or lipid radicals (Siesjö, 1981) might be very critical in this area. P_{O_2} monitoring at this location could possibly reveal additional important information.

ACKNOWLEDGEMENTS

We gratefully acknowledge the technical assistance of Jean Jacqueloot, Willy Smets, and Mies Van de Ven.

This work is supported by grant No 3.0023.91 of the Belgian Fund for Medical Scientific Research and by the OZR VUB.

REFERENCES

Dietrich, W. D., Ginsberg, M. D., Busto, R., and Watson, B. D., 1986, Photochemically induced cortical infarction in the rat. 1. Time course of hemodynamic consequences, *J. of Cerebral Blood Flow and Metabolism*, 6 : 184 - 194.

Dietrich, W. D., Ginsberg, M. D., Busto, R., and Watson, B. D., 1986, Photochemically induced cortical infarction in the rat. 2. Acute and subacute alterations in local glucose utilization, *J. of Cerebral Blood Flow and Metabolism*, 6 : 195 - 202.

Dietrich, W. D., Watson, B. D., Busto, R., Ginsberg, M. D., and Bethea, J. R., 1987, Photochemically induced cerebral infarction. I. Early microvascular alterations, *Acta Neuropathol.*, 72 : 315 - 325.

Dorman, R. V., Damron, D. S., and Hamm T. F. R., 1990, Description and manipulation of ischemia-induced alterations of cerebral arachidonic acid metabolism, *in* : "Lipid mediators in ischemic brain damage and experimental epilepsy", Baran, N. G., ed., Karger, 36 - 66.

Siesjö, B. K., 1981, Cell damage in the brain : a speculative synthesis, *J. of Cerebral Blood Flow and Metabolism*, 1 : 155 - 185.

Van Reempts, J., Van Deuren, B., Van de Ven, M., Cornelissen, F., and Borgers, M., 1987, Flunarizine reduces cerebral infarct size after photochemically induced thrombosis in spontaneously hypertensive rats, *Stroke*, 18 : 1113 - 1119.

van Rossem, K., Vermariën, H., and Bourgain, R., Construction, calibration and evaluation of pO_2 electrodes for chronic implantation in the rabbit brain cortex, *In* : "Adv. Exp. Med. Biol. (ISOTT 1990)", in press 1990.

van Rossem, K., Vermariën, H., Decuyper, K., and Bourgain, R., Photothrombosis in rabbit brain cortex : follow up by continuous pO_2 measurement, *In* : "Adv. Exp. Med. Biol. (ISOTT 1990)", in press 1990.

Watson, B. D., Dietrich, W. D., Busto, R., Mitchell, S., Wachtel, B. S., and Ginsberg, M. D., 1985, Induction of reproducible brain infarction by photochemically initiated thrombosis, *Ann. Neurol.*, 17 : 497 - 504.

BRAIN SURFACE pO$_2$ AND rCBF IN RABBITS WITH A FOCAL CEREBRAL LESION AND PULMONARY HYPOXIA UNDER FENTANYL-, ISOFLURANE- OR THIOPENTAL-ANESTHESIA

Berger[1] S, Murr[2] R, Schürer[3] L, Enzenbach[2] R,
Peter[2] K and Baethmann[1] A

Institute for Surgical Research[1], Institute for
Anesthesiology[2] and Dept. for Neurosurgery[3], LMU
München, Klin. Großhadern, D-8000 Munich 70, FRG

INTRODUCTION

Patients with severe head injury frequently suffer from general hypoxia in the early posttraumatic phase. General hypoxia must be considered a major cause of secondary brain damage. The anesthetic methods used for diagnostic or therapeutical procedures in these patients thus may have an influence on the formation of secondary damage by modifying blood flow and metabolism of the brain. As these mechanisms are still incompletely understood, we analysed the influence of different anesthetic agents on regional cerebral blood flow, brain surface pO$_2$ and edema formation in an experimental model of head injury combined with systemic hypoxia.

METHODS

In three groups of 6 rabbits each, anesthesia was induced by Thiopental and then maintained by either Fentanyl (1.0/0.5 μg/kg b.w. x min), Isoflurane (2.1 vol% = 1 MAC) or Thiopental (32.5 mg/kg b.w. x h). Animals of the Isoflurane group were infused with Angiotensin II (0.13 μg/kg b.w. x min) to stabilize the mean arterial pressure (MAP). After tracheotomy and relaxation with pancuronium bromide the animals were artificially ventilated. Catheters were implanted into the aorta and caval vein for administration of fluid and drugs, blood sampling and pressure monitoring. Mean arterial pressure, central venous pressure, blood gases and body temperature were monitored continuously or at intervals. A skull trephination was made then and the dura mater was opened for exposure of the left cerebral hemisphere in an open cranial window under oil. The brain surface pO_2 was determined by an 8-channel surface electrode (Kessler and Lübbers, 1966) with a weightless suspension system that allowed remote controlled rotations of the electrode for acquisition of pO_2-histograms. The H_2-clearance was employed for regional cerebral blood flow (rCBF) measurement with platinum needle electrodes (75 μm ø, 1.2 mm length) implanted into the cerebral cortex. Both parameters, pO_2 and blood flow, were obtained at 6 and 14 mm from the lesion. After a control period of 90 min a focal lesion (Klatzo) was induced at the brain surface by freezing for 1 min with a probe of 6 mm diameter (-70 °C). 7 min later pulmonary hypoventilation was instituted to decrease the arterial pO_2 to 35 mm Hg for 20 min. All parameters were followed for 6 hrs after trauma. At termination of the experiment the brain was removed for assessment of the specific gravity as a measure of brain edema. Specific gravity was measured in regional samples of cortical tissue by a linear Percoll® density gradient (Tengvar et al., 1982).

RESULTS

The systemic parameters such as temperature, central venous pressure or blood gases were not significantly different between groups. The average MAP was 77 mm Hg with Fentanyl, 70 mm Hg with Isoflurane and 78 mm Hg with Thiopental anesthesia after trauma.
Table 1 shows the temporal course of the median of the brain surface pO_2-histograms, the frequency of hypoxic pO_2 values (0-5 mm Hg) and the rCBF close to the lesion. The first histogram was obtained under control conditions before cerebral trauma with Fentanyl, Isoflurane or Thiopental. In the second set of measurements the pO_2 histogram contour was shifted to the left due to systemic hypoxia. The following measurements represent the pO_2-distribution at two, four and six hours after trauma. The regional cerebral blood flow was also measured during control and later on. With Fentanyl, rCBF remained largely unchanged,

TABLE 1. Medians of pO$_2$-histograms, frequency of hypoxic pO$_2$-values (0-5 mm Hg) and rCBF close to the lesion. The rCBF is given in ml/100g/min. 2, 4, 6 h = hrs after lesion.

Time	FENTANYL			ISOFLURANE			THIOPENTAL		
	Median pO$_2$ mm Hg	hypoxic values %	rCBF	Median pO$_2$ mm Hg	hypoxic values %	rCBF	Median pO$_2$ mm Hg	hypoxic values %	rCBF
Control	21.9	12.4	46.4	33.5	3.1	54.4	30.1	0.8	58.4
Hypoxia	10.1	29.3	--	9.1	32.3	--	13.6	13.7	--
2 h	27.2	6.0	49.6	35.6	2.3	59.4	29.3	1.9	47.8
4 h	34.1	2.9	44.8	27.5	5.72	61.1	29.4	4.2	44.6
6 h	15.0	29.6	47.5	32.3	2.8	59.9	35.6	9.8	44.2

TABLE 2. Medians of pO$_2$-histograms, frequency of hypoxic pO$_2$-values (0-5 mm Hg) and rCBF distant from the lesion. The rCBF is given in ml/100g/min. 2, 4, 6 h = hrs after lesion.

Time	FENTANYL			ISOFLURANE			THIOPENTAL		
	Median pO$_2$ mm Hg	hypoxic values %	rCBF	Median pO$_2$ mm Hg	hypoxic values %	rCBF	Median pO$_2$ mm Hg	hypoxic values %	rCBF
Control	22.2	10.7	48.9	31.3	8.0	60.1	29.1	6.3	65.5
Hypoxia	7.9	29.8	--	6.2	45.1	--	7.0	39.3	--
2 h	27.9	10.0	45.8	36.5	1.0	66.4	38.2	2.1	63.1
4 h	25.4	4.9	47.3	39.1	0.7	69.4	39.1	4.4	60.4
6 h	23.5	9.9	50.1	46.0	0.1	75.2	46.4	1.9	57.7

TABLE 3. Specific gravity of cortical grey matter in g/cm^3. Values are given as mean +/- standard error of mean. n is the number of samples removed from 6 brains at the respective location. (Slice 1 = frontal, slice 8 = caudal)

SLICE	FENTANYL		ISOFLURANE		THIOPENTAL	
	LEFT	RIGHT	LEFT	RIGHT	LEFT	RIGHT
1	1.038685 ±0.000665 n = 11	1.039286 ±0.000544 n = 11	1.039384 ±0.000312 n = 14	1.039083 ±0.000268 n = 14	1.038068 ±0.000863 n = 11	1.038056 ±0.000686 n = 11
2	1.039466 ±0.000493 n = 18	1.040190 ±0.000360 n = 18	1.038773 ±0.000653 n = 19	1.040072 ±0.000297 n = 18	1.038770 ±0.000506 n = 18	1.038564 ±0.000667 n = 18
3	1.037474 ±0.000870 n = 22	1.039599 ±0.000311 n = 23	1.038868 ±0.000664 n = 26	1.039969 ±0.000305 n = 26	1.037383 ±0.000443 n = 25	1.038228 ±0.000462 n = 26
4	1.037141 ±0.000774 n = 28	1.038903 ±0.000302 n = 31	1.038565 ±0.000467 n = 27	1.039939 ±0.000221 n =30	1.036618 ±0.000586 n = 29	1.037435 ±0.000356 n = 30
5	1.038443 ±0.000596 n = 27	1.040341 ±0.000305 n = 31	1.038279 ±0.000653 n = 25	1.039547 ±0.000345 n =30	1.037093 ±0.000497 n = 26	1.037781 ±0.000325 n = 31
6	1.037934 ±0.000511 n = 31	1.039378 ±0.000235 n = 37	1.038075 ±0.000450 n = 33	1.039114 ±0.000233 n = 35	1.036835 ±0.000479 n = 34	1.037193 ±0.000343 n = 41
7	1.037226 ±0.000355 n = 29	1.038929 ±0.000242 n = 30	1.037162 ±0.000533 n = 22	1.038592 ±0.000274 n = 23	1.035299 ±0.000534 n = 35	1.037049 ±0.000435 n = 32
8	1.035395 ±0.000513 n = 19	1.038518 ±0.000295 n = 22	1.037220 ±0.000480 n = 19	1.038261 ±0.000420 n = 18	1.035993 ±0.000358 n = 25	1.036779 ±0.000521 n = 23

while it slightly increased with Isoflurane and decreased with Thiopental. Six hours after lesion the pO_2-histogram was markedly left-shifted under Fentanyl (median = 15.0 mm Hg, 29.6% of values in the range from 0 to 5 mm Hg) while it shifted to the right with Isoflurane or Thiopental.

Distant from the lesion (table 2) the same tendency became obvious: a left-shift of the pO_2-histogram (median = 23.5 mm Hg) with an increased frequency of hypoxic values was found with Fentanyl. A shift to higher pO_2-medians developed especially under Thiopental (median = 46.4 mm Hg) although rCBF somewhat decreased in this group. No major changes of rCBF were found with Fentanyl while a slight increase was seen under Isoflurane. In table 3 the spatial distribution of the tissue water content in the brain is given as specific gravity. The brains were cut in 8 slices each and samples were taken from the cortex of the left and right hemisphere. A decrease in specific gravity indicates a higher water content that indicates brain edema. The brain water content was significantly increased close to the lesion and at distant areas in animals with Thiopental as compared to Isoflurane or Fentanyl.

DISCUSSION

Contrary to former experiments (Murr et al., 1989) on the above anesthetic agents -however without hypoxia- CBF was only minimally affected by trauma although it was somewhat decreased when Thiopental was used. Nevertheless Thiopental and Isoflurane were associated with an increase of the brain surface pO_2 at 6 hrs after trauma while the pO_2 was decreasing with Fentanyl. On the other hand, perifocal brain edema was most pronounced in animals with Thiopental. Thus, although the cerebral oxygen-supply appeared to be well preserved with Thiopental, formation of traumatic brain edema was enhanced. In conclusion, vasogenic brain edema appeared to develop independently from the effects of different anesthetic methods studied on rCBF and metabolism.

ACKNOWLEDGEMENT

We gratefully acknowledge the technical assistance of U. Goerke and E. Schütz.
Supported by DFG Grant Ba 452/6

REFERENCES

Kessler M, Lübbers DW (1966) Aufbau und Anwendungsmöglichkeiten verschiedener pO_2-Elektroden. Pflügers Archiv 291:82

Klatzo I, Piraux A, Laskowski EJ (1958) The relationship between edema, blood-brain-barrier and tissue elements in a local brain injury. J Neuropathol Exp Neurol 17:548-564

Murr R, Berger S, Schürer L, Baethmann A (1989) Regional cerebral blood flow and tissue pO_2 after focal trauma: effects of isoflurane and fentanyl. In: Hammersen F, Messmer K, (Eds.) Cerebral Microcirculation. Progr Appl Microcirc, Vol. 16, Karger, Basel, pp. 61-70

Tengvar C, Forssen M, Hultström D, Olsson Y, Pertoft H, Petterson A (1982) Measurement of edema in the nervous system. Acta Neuropathol (Berl) 57:143-150

TREATMENT OF HEMORRHAGIC HYPOTENSION WITH HYPERTONIC SALINE/DEXTRAN: EFFECTS ON BRAIN SURFACE OXYGEN TENSION IN EXPERIMENTALLY TRAUMATIZED BRAIN

C. Dautermann[*], L. Schürer, R.Härtl, F. Röhrich, A. Baethmann, K. Messmer

Institute of Surgical Research and Department of Neurosurgery[*], Ludwig-Maximilians-University, Marchioninistr.15, 8000 München 70, Germany

INTRODUCTION

In cases of severe hemorrhagic shock small volumes of hypertonic-hyperoncotic solutions (HHS) are efficient in restoring cardiovascular function immediately (1-3). While the macrocirculatory changes following infusion of HHS after hemorrhagic shock are well understood (4-6), little is known about the effects of this treatment on the central nervous system in particular in the presence of cerebral injury comprising deleted autoregulation and brain edema (7). We have, therefore, investigated the changes in oxygen supply of the brain in the presence of a cryogenic lesion after hemorrhagic shock and infusion of 7.2 % NaCl/10 % Dextran 60.

METHODS

New Zealand rabbits weighing 3.22 ± 0.19 kg were anesthetized with thiopental (total dose 25 - 35 mg/kg b.w.), tracheotomized, paralyzed (pancuronium bromide 0,4 mg/kg b.w.) and mechanically ventilated (f_iO_2 = 0,3). Ventilation was adjusted to a p_aCO_2 of approx. 35 mmHg and anesthesia was maintained with α-chloralose (50 mg/kg) given over a period of 30 min. Another 25 mg/kg b.w. of α-chloralose were applied after 6 hours to continue anesthesia. Catheters were inserted into the abdominal aorta and vena cava inferior for monitoring of arterial (MAP) and venous blood pressure (CVP), sampling of blood, and infusion of fluid and drugs. Via a left lateral thoracotomy an electromagnetic flow probe was placed around the pulmonary artery for continuous measurement of cardiac output (CO). 4 ml/kg/h of isotonic electrolyte solution was continuously infused for fluid replacement, relaxation was continued with pancuronium bromide (0.3 mg/kg/h).
After fixation of the animals head in a stereotactic holder a cranial window was made to expose the left cerebral hemisphere.

At the edge of the window a wall was formed with dental cement and the resulting pool was filled with paraffin oil to protect the brain surface from drying, temperature loss and pressure derangements. The dura mater was incised microsurgically and reflected.

The regional oxygen surface tension of the exposed brain was measured over the occipital part of the exposed cortex using a multiwire electrode according to KESSLER et al. (8). To avoid compression of the brain parenchyma by the weight of the pO_2-electrode as well as to compensate for movements of the brain the electrode was fixed to a phonoplayer-like arm which allowed a weightless contact with the brain surface . Tissue pO_2-histograms were generated by a remote control device which induced rotation of the electrode in 24 single steps of about 10°. pO_2-values of about 150-180 points (8 single electrodes, 24 rotations) were accumulated for acquisiton of each pO_2-histogram and stored in a computer after A/D-conversion together with calibration data.

EXPERIMENTAL PROTOCOL

The animals were randomly assigned to four groups of 6 animals each. After the surgical preparation baseline values of MAP, CO, hematocrit (hct) and brain surface oxygen tension were obtained during a control phase of 40 min.

In *group I* (sham operated group) measurements were continued for another 180 min. In *group II* (treatment with HHS) 4 ml/kg b.w. of 7.2 % NaCl/10 % Dextran 60 (Schiwa GmbH, Glandorf, Germany) were infused through an ear vein over a period of 2 min at 90 min after start of measurements. In *group III* (hemorrhagic shock and treatment with HHS) hemorrhagic shock was induced at 60 min after start of measurements by bleeding to a MABP of 40 mmHg. MAP was kept constant at this level for 30 min by withdrawal or reinfusion of blood. At the end of the hemorrhagic phase HHS was administered in the same manner as in group II.

In *group IV* (trauma, hemorrhagic shock and treatment HHS) a focal cerebral lesion according to KLATZO et al. (9) was made in the occipital cortex by freezing the tissue to -68° C for 1 min immediately after baseline measurements. Hemorrhagic shock was induced 20 min later in the same way as described above and HHS was given after a shock period of 30 min.

p_aO_2-histograms were obtained after trauma, during hemorrhagic shock and in 30 min-intervals up to 120 min after administration of HHS. At the end of the experiment animals were sacrifized by rapid injection of 10 ml of saturated KCL-solution.

STATISTICS

All data are mean values ± standard error of the mean (SEM). p_aO_2-data were analyzed by calculation of the median value for each histogram and were ranked in groups of 10 mmHg.

RESULTS

In group I <u>cardiac output</u> remained constant over the entire observation period. Infusion of HHS (group II) led to a steep increase of CO which did not reach baseline values until

Table 1. Changes of Cardiac output in the four experimental
groups

CARDIAC OUTPUT ml/kg*min^{-1}	Sham	HHS	Hemorrhagic shock and HHS	Trauma, shock and HHS
control	94 ± 9	83 ± 11	84 ± 6	84 ± 7
"hemorrhagic shock"	91 ± 10	81 ± 11	29 ± 5	26 ± 6
1 min after HHS	91 ± 8	137 ± 13	85 ± 13	88 ± 7
60 min after HHS	95 ± 7	93 ± 14	80 ± 6	76 ± 9
90 min after HHS	93 ± 8	93 ± 12	70 ± 7	75 ± 9
120 min after HHS	93 ± 6	89 ± 12	69 ± 5	72 ± 8

Table 2. Changes of mean arterial pressure in the four
experimental groups

MEAN ARTERIAL PRESSURE mmHG	Sham	HHS	Hemorrhagic shock and HHS	Trauma, shock and HHS
control	89 ± 7	80 ± 4	83 ± 4	88 ± 4
"hemorrhagic shock"	87 ± 5	80 ± 5	40 ± 0	41 ± 1
1 min after HHS	87 ± 5	86 ± 3	72 ± 3	67 ± 2
60 min after HHS	92 ± 6	82 ± 4	55 ± 5	58 ± 4
90 min after HHS	88 ± 5	82 ± 3	53 ± 4	60 ± 5
120 min after HHS	89 ± 6	79 ± 4	53 ± 5	58 ± 5

Table 3. Changes of hematocrit in the four experimental groups

HEMATOCRIT %	Sham	HHS	Hemorrhagic shock and HHS	Trauma, shock and HHS
control	35 ± 1	37 ± 1	37 ± 1	38 ± 1
"hemorrhagic shock"	34 ± 4	37 ± 1	28 ± 1	28 ± 1
1 min after HHS	35 ± 1	27 ± 2	18 ± 2	16 ± 1
120 min after HHS	35 ± 1	33 ± 1	21 ± 2	20 ± 1

Table 4. Median p_aO_2-values in the four experimental groups.

Median of p_aO_2-values mmHg	Sham	HHS	Hemorrhagic shock and HHS	Trauma, shock and HHS
control	23	20	29	25
"hemorrhagic shock"	23	21	25	20
60 min after HHS	22	27	25	20
120 min after HHS	27	29	28	25

the end of the experiment. During hemorrhagic shock (group III and IV) CO fell drastically but rose again following infusion of HHS. Values were slightly below baseline 2 h after treatment. Trauma (group IV) had no influence on CO.

Mean arterial pressure did not change during the entire experimental course in group I. It was not altered by infusion of HHS (group II). In group III and IV resuscitation from hypovolemia with HHS rapidly increased MAP to values only sightly below baseline level. MAP decreased very slowly during the observation period.

Hematocrit remained stable in group, but infusion of HHS led to a significant decrease (group II). While hematocrit was already lowered by hemorrhage in group III and IV infusion of HHS induced a further decrease. A slight increase was observed at the end of the experiment.

For determination of brain surface oxygen tension of the four experimental groups single values of 6 experimental animals per group were summarized in one histogram (up to 2100 single measurements per histogram). Median values obtained during the steady state phase ranged from 21 to 25, the shape of the histograms was slightly skewed leftward. In sham operated animals (group I) a discrete rightshift was observed at the end of the measurements, with an increased number of high p_aO_2-values and a decrease in low values (< 5 mmHg). In group II the median increased after infusion of HHS and hypoxic values were reduced. This effect was even more pro-nounced 120 min after infusion. Hemorrhagic hypotension in group III reduced the median value while the number of hypoxic p_aO_2-values increased. This resulted in a pronounced leftshift of the histogram. Infusion of HHS did not alter the median value at once but resulted in a broadening of the histogram towards high values and in a reduction of hypoxic values. 2 hours later an improvement of tissue oxygenation is expressed by a higher median and a very low rate of hypoxic values. In group IV the combination of cerebral trauma and hemorrhagic hypotension resulted in a marked decrease of the median and an increased number of hypoxic p_aO_2-values. As in group III the improvement of tissue oxygenation after infusion of HHS was most pronounced 2 hours after start of infusion.

DISCUSSION

The effects of 7,2 % NaCl/10 % Dextran 60 on brain surface oxygen tension were investigated in a standardized model of hemorrhagic shock following focal cerebral lesion. The effects on the cardiovascular system, which were found by other investigators (1-3) could be confirmed. The persistent high CO-values as well as the markedly lowered hematocrit in group II are probably due to the dose of HHS, which was given. In contrast to other authors who replaced about 10 % of the shed blood volume by HHS, the amount of HHS given in this study was about 20 % of the shed blood volume.

The multiwire surface electrode used in this study is able to demonstrate the whole range of p_aO_2-values present in the cortex surface, our findings correspond very well with the findings of other investigators (10-13).

No adverse effect of the treatment with HHS on brain surface oxygen tension was found under normal conditions inspite of the decrease of the hematocrit. The rightshift of p_aO_2-histogram and the decrease of hypoxic and anoxic tissue p_aO_2-values indicate that HHS improves brain surface tissue-p_aO_2 in normotension as well as after resuscitation from hemorrhagic shock.

It could also be demonstrated that infusion of hypertonic-hyperoncotic solutions in animals with focal cerebral injury and hemorrhagic shock results in an improvement of the previously reduced oxygen tension in the brain surface tissue.

ACKNOWLEDGMENT

Supported by Deutsche Forschungsgemeinschaft, Ba 452/6-7

REFERENCES

1. Smith G.J., Kramer G.C., Perron P., Nakayama S.I., Gunther R.A., and Holcroft J.W., 1985, A comparison of severel hypertonic solutions for resuscitation of bled sheep, J. Surg. Res., 39: 517-528.
2. Kramer G.C., Perron P.R., Lindsey D.C., Ho H.S., Gunther R.A., Boyle W.A. and Holcroft J.W., 1986, Small volume resuscitation with hypertonic saline dextran solution Surgery 100: 239-247.
3. Kreimeier U. and Messmer K., 1988, Small volume resuscitation with hypertonic-hypverosmotic solutions – state of the art review, in Fluid Resuscitation, W. Kox, J. Gamble, Eds., Ballière's Clinical Anaesthesiology: 545-577
4. Hannon J.P., Wade C.E., Bossone C.A., Hunt M.M., Coppes R.I. and Loveday J.A., 1987, Effects of 7,5 % NaCl/6 % Dextran on oxygen transport and demand in conscious swine subjected to hemorrhagic hypotension, The Physiologist, 30:229
5. Mazzoni M.C., Borgstroem P., Arfors K.E. and Intaglietta M., 1988, Dynamic fluid distribution in hyperosmotic resuscitation of hypovolemic hemorrhage, Am. J. Physiol, 255: h629-h637
6. Rocha e Silva M., Negraes G.A., Soares A.M., Pontieri V., and Loppnow L., 1986, Hypertonic resuscitation from severe hemorrhagic shock: Patterns of regional circulation. Circ. Shock 19: 165-175.
7. Gunnar W.P., Merlotti G.J., Jonasson O., Barrett J., 1986, Resuscitation from hemorrhagic shock. Alterations of the intracranial pressure after normal saline, 3 % saline and dextran 40, Ann. Surg., 204(6):686-692
8. Kessler M., Höper J. und Krumme B.A., 1976, Monitoring of tissue perfusion and cellular function. Anesthesiology 45: 184-197

9. Klatzo I., Piraux A., Laskowski E.J., 1958, The relationship between edema, blood brain barrier and tissue elements in a local brain injury, J. Neuropathol. Exp. Neurol. 17: 548- 564

10. Lübbers D.W., 1972, Physiology der Gehirndurchblutung in Der Gehirnkreislauf. Physiologie, Pathologie, Klinik. Gänshirt H., ed., Georg Thieme Stuttgard, 214-260

11. Maekawa T., McDowall D.G. and Okuda Y., 1979, Brain surface oxygen tension and cerebral cortical blood flow during hemorrhagic and drug induced hypotension in the cat, Anesthesiol., 51: 313-320

12. Kozniewska E., Weller L., Höper J., Harrison D.K. and Kessler M., 1987, Cerebrocortical microcirculation in different stages of hypoxic hypoxia, J. Cereb. Blood Flow Metab., 7: 464-470

13. Ivanov K.P., Kislanov Y.Y. and Samoilov M.O., 1979, Microcirculation and transport of oxygen to neurons of the brain, Microvasc. Res., 18: 434-446

MONITORING OF CORTICAL INTRACAPILLARY HEMOGLOBIN OXYGENATION

IN PATIENTS DURING BRAIN SURGERY - FIRST RESULTS

Tauschek, D., J. Höper, M.R. Gaab[*], M. Kessler

Institut für Physiologie und Kardiologie der Universität Erlangen-Nürnberg,
Waldstr. 6, D-8520 Erlangen
[*]Neurochirurgische Klinik der Medizinischen Hochschule Hannover,
Konstanty-Gutschow-Str. 8, D-3000 Hannover 61

INTRODUCTION

As early as 1907 Goldmann [4] made the observation that the vascular density of normal tissue surrounding a tumor is increased. Since that time many investigations were conducted to study tumor growth and neovascularisation as well as oxygen supply to tumor [8, 9].

In a first study we investigated the oxygen supply to human brain tissue surrounding different tumors. The results of this study may shine new light on the process occuring in the vicinity of neoplastic tissue.

PATIENTS AND METHODS

The measurements were performed in four patients with different brain tumors (oligo-dendroglioma, adenoma of the hypophysis, metastasis of unknown origin and a meningeoma). In this paper we will concentrate on the results obtained in a patient with an oligodendroglioma.

At the time of surgical intervention the patient was 37 years old. The tumor was located in the left temporal lobe a few millimeters below the cortex surface. The diameter was approx. 3 cm. Intraoperative histology confirmed the diagnosis oligodendroglioma without anaplasia.

The patient was anesthezised by neurolept anaesthesia and hyperventilated (endexpiratory pCO_2=31.5 Torr)

After trepanation and opening the dura we recorded integrated gradients of intracapillary hemoglobin oxygenation and spatial distribution of intracapillary hemoglobin concentration above normal brain cortex, tumor border zone and above the tumor. Due to the fact that the tumor was lying beneath brain cortex, we could not determine the exact distance between the site of measurement and the tumor.

For irradiation of light from a xenon high pressure lamp into the tissue a 250 μm micro-light-guide was used. The remitted light was collected by 6 fibers of 70 μm diameter each [2]. According to experiments performed in suspensions of scattering particles the catchment volume of the applied lightguide system used in our brain measurements reaches to a depth of approximately 250 µm [3].

The diameters of the lightguide fibers enabled to gather signals of microvolumes and thus to resolve the spatial heterogeneity of intracapillary hemoglobin oxygenation and concentration. The fibers used for the measurements were built into a holder (10 mm diameter and 100 mm length) in order to allow a better handling and to shield off light of operating theatre lamps. The fiber bundle was slowly moved over the cortex surface (area approx. 1 cm^2). Due to the high recording rate of 70 spectra per second it was possible to gather local values of hemoglobin oxygenation and concentration without movement artefacts. At each measuring field spectra from 500 different points were recorded. By computer evaluation the values of hemoglobin oxygenation and hemoglobin concentration were determined [1].

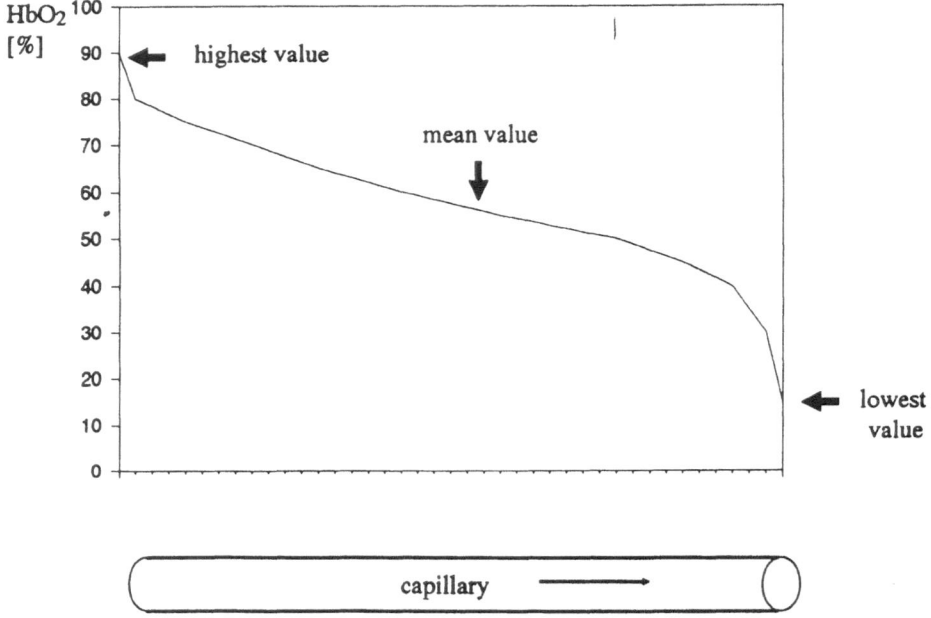

Fig. 1. Integrated gradient of hemoglobin oxygenation of the macroscopically normal control area.

RESULTS

The measurements in the brain of the patient were performed in such a manner that the neurosurgeon recorded spectra in the macroscopically normal brain cortex, above the tumor border zone and above the brain tumor. The net measuring time for 4000 spectra in 8 measuring sites was 60 seconds.

For routine evaluation the hemoglobin oxygenation values are plotted as cumulative curves which corresponds to the integrated gradients of intracapillary hemoglobin oxygenation [5].

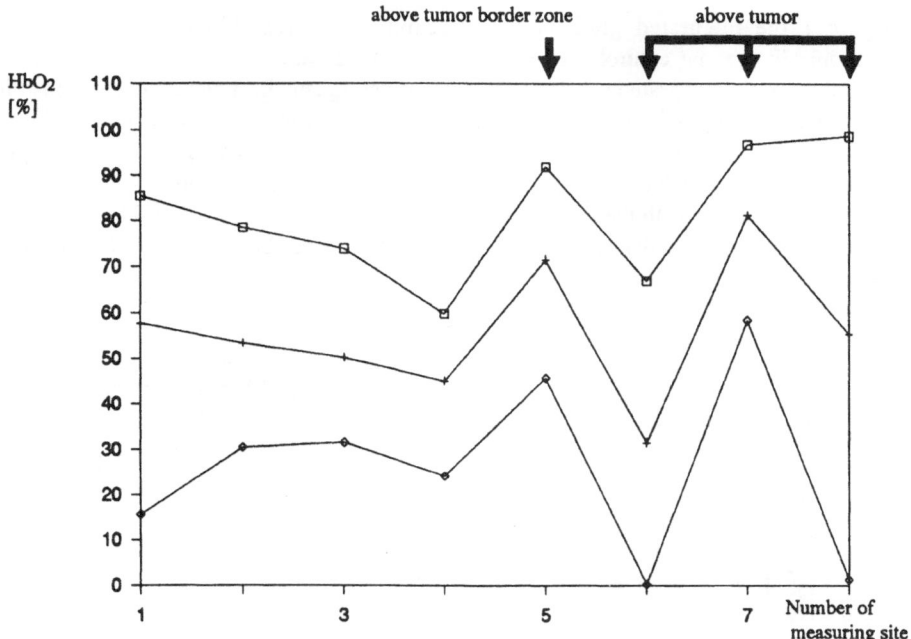

Fig. 2. Lowest, mean and highest values of hemoglobin oxygenation (HbO₂), from left to right side decreasing distance to tumor.

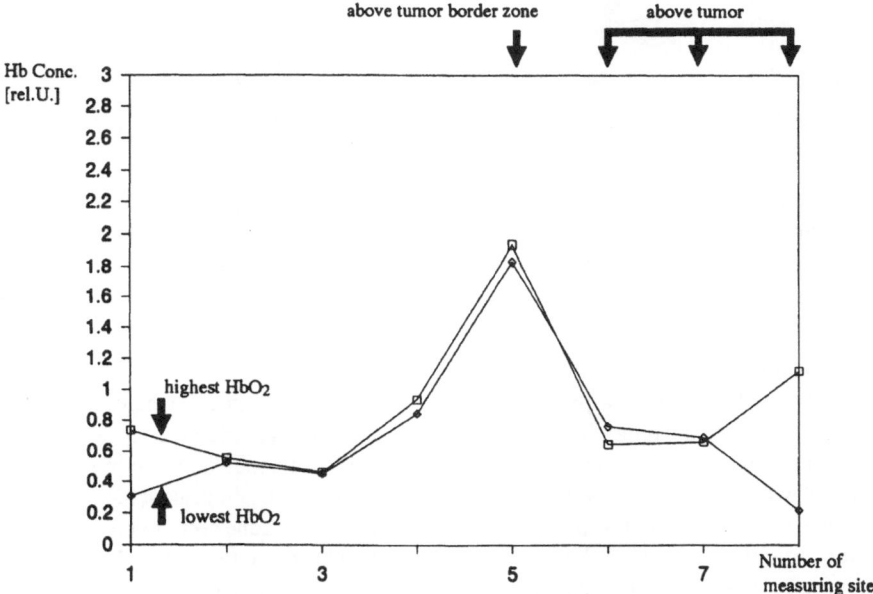

Fig. 3. Values of hemoglobin concentration (Hb Conc.) that are found at the same points as the highest hemoglobin oxygenation (marked "highest HbO2") and the lowest hemoglobin oxygenation (marked "lowest HbO2"); from left to right side decreasing distance to tumor.

In figure 1 the integrated gradient of intracapillary hemoglobin oxygenation of the macroscopically normal control area is plotted. Figure 2 shows the highest, mean and the lowest values of the integrated gradients of each measuring site. In figure 3 the corresponding values of local hemoglobin concentration are depicted. Multiplying corresponding values of hemoglobin oxygenation and concentration gives the relative local intracapillary oxygen content. The resulting values are shown in figure 4. Measuring site number 1 corresponds to the control. The holder with the light fibers was moved from the control site (No. 1 in figures 2-4) towards the tumor border zone (No. 5) and on to the zones above the tumor (No. 6-8).

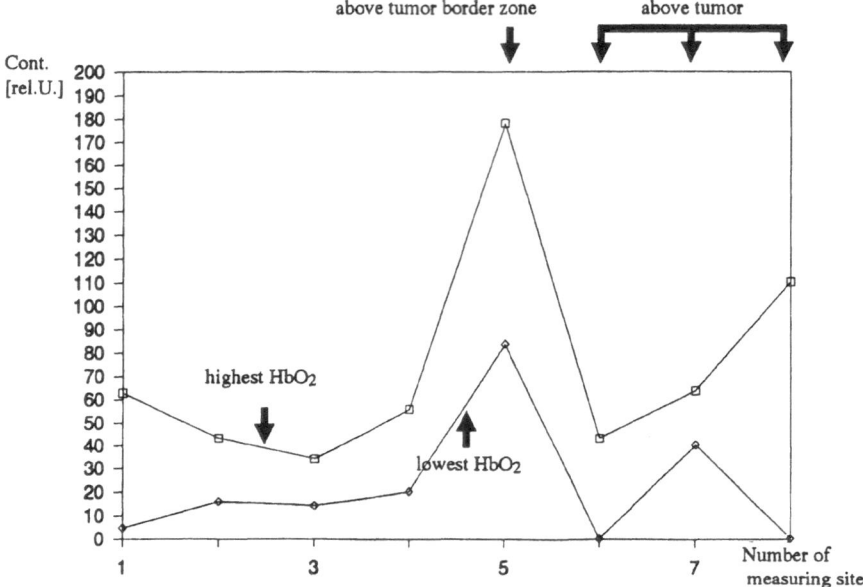

Fig. 4. Values of local oxygen content (Cont.) that are found at the same points as the highest hemoglobin oxygenation (marked "highest HbO2") and the lowest hemoglobin oxygenation (marked "lowest HbO2"); from left to right side decreasing distance to tumor.

Three interesting observations can be made :
1) In the tumor border zone a peaking of oxygenation values appears, whereas towards the outer zone as well as the tumor itself a drop in hemoglobin oxygenation was observed.
2) The hemoglobin concentration values between the control zone and the tumor zone show low values with the exception of the tumor border zone, where again peak values can be seen. Those peak values found in this patient above the tumor border zone seem to be in accordance with the investigations of Goldman [4] and many others. In the areas 2-7 only small differences between the highest and lowest hemoglobin concentration values were observed.
3) Also the oxygen content values show a peaking in the measuring site above the tumor border zone. In this area the difference between highest and lowest oxygen content values is larger than in the adjacent areas 4 and 6. This corresponds to a steep integrated gradient.
Similar patterns of HbO2-distribution where obtained in the other patients. However, the distribution of intracapillary hemoglobin concentrations was more heterogeneous.

DISCUSSION

In general surprisingly low HbO_2 values were observed in all areas measured. In this context, one must consider that the patient was hyperventilated (endexpiratory pCO_2=31.5 mm Hg). Thorensen et al [10] report that reduction of endexpiratory pCO_2 from 37.5 to 30 mm Hg lead to a reduction of cerebral blood flow by 40%.

As mentioned above Goldman [4] described an increasing vascularisation in the tissue surrounding a tumor. This may correspond to the area above the tumor border zone (No. 5). Here the HbO_2 values as well as the hemoglobin concentration values are high compared to the neighboring areas. Not only neovascularisation but also increased local perfusion due to vasodilation could be responsible for these high values. However, in both cases the tumor seems to influence its vicinity over a certain distance. For this effect tumor mediators could be of importance: Suddith [6] reported that in vitro neural tumor cell lines produce an endothelial proliferative factor.

For the further discussion some assumptions must be made.

The local hemoglobin oxygenation must show the highest values at the arteriolar beginning of capillaries, the lowest at the the venous end, reflecting the gradient of hemoglobin oxygenation along capillaries [5]. The steepness of this gradient depends on local oxygen supply and on oxygen consumption. The calculated local oxygen content can be considered to be a measure for oxygen supply.

For areas 2-7 we can draw the following conclusions: The hemoglobin concentration values that are found at the very same locations as the highest and lowest hemoglobin oxygenation values are almost identical. This indicates that these values correspond to the arteriolar and venous end of the same capillary supply unit [5]. Therefore the difference in oxygen content found in each measuring site may reflect the oxygen uptake rate. Above the tumor border zone a large difference in oxygen content compared to adjacent areas was found. Here the oxygen uptake rate may be higher than in the neighboring areas. Because the profiles of hemoglobin oxygenation, concentration and oxygen content are related to the distance to the tumor, an influence of the neoplastic tissue on these parameters as well as on oxygen uptake rate may be assumed.

REFERENCES

[1]   Dümmler, W.: Bestimmumg von Hämoglobin-Oxygenierung und relativer Hämoglobin-Konzentration in biologischen Systemen durch Auswertung von Remissionsspektren mit Hilfe der Kubelka-Munk-Theorie. 1988, Diss., Friedrich Alexander Universtität, Erlangen-Nürnberg.

[2]   Frank, K.H., M. Kessler, K. Appelbaum, W. Dümmler: The Erlangen micro-lightguide spectrophotometer EMPHO I. Phys. Med. Biol., 34, 18883-1900 (1989).

[3]   Frank, K.H., M. Kessler, J. Wiesner, A. Wokaun: Analysis of multiple multipole scattering by time-resolved spectroscopy and angular dependent spectrometry. Same volume.

[4]   Goldman, E: The growth of malignant disease in man and the lower animals with special reference to the vascular system. Proc. R. Soc. Med., 1, 1-13 (1907)

[5]   Kessler, M., J. Höper: Spatial distribution of oxygen supply units in heart and skeletal muscle and their regulatory significance. Same volume

[6]   Suddith, R.L., P.J. Kelly, H.T. Hutchinson, E.A. Murray, B. Haber: In vitro demonstration of an endothelial proliferative factor produced by neural cell lines. Science, 190, 682-684 (1975)

[7] Thorensen, M., L. Walloe: Changes in cerebral blood flow during hyperventilation and CO_2 breathing. J. Physiol., 298, 53-54 (1979)

[8] Vaupel, P.: Hypoxia in neoplastic tissue. Microvasc. Res., 13, 399-408 (1977)

[9] Vaupel, P., R. Manz, W. Müller-Klieser, W.A. Grunewald: Intracapillary HbO_2 saturation in malignant tumors during hypoxia and hyperoxia. Microvasc. Res., 17, 181-191 (1979)

MUSCLE

INERT GAS WASHOUT MEASUREMENT OF MUSCLE BLOOD FLOW

DISTRIBUTION - ROLES OF HYPOXIA AND DIFFUSION LIMITATION

Michael P. Hlastala, Gary M. Malvin,
Christopher Quartararo and Jørgen Grønlund

University of Washington
Seattle, WA 98195, USA

INTRODUCTION

The delivery of oxygen to peripheral tissue depends on local perfusion, hemoglobin storage characteristics and the properties of the tissue. The local tissue property that has major impact on O_2 delivery dynamics is the matching of blood flow to oxygen demand ($\dot{Q}/\dot{V}O_2$). Factors which play a role include perfusion to volume heterogeneity, oxygen uptake to volume heterogeneity, diffusion limitation, diffusional shunt and true shunt (Piiper et al, 1984; Piiper and Haab, 1991). To date, no method has been developed to simultaneously quantify all of these factors, hence, differentiation of all of the factors has been problematic. In general, it is thought that diffusion-dependent mechanisms play little role in limiting tissue inert gas exchange (Wagner, 1989). However, Hills (1967) argues that diffusion-dependent processes may be important because diffusion coefficients in tissue may have been overestimated.

This study uses a continuous muscle washout method, previously developed in our laboratory (Grønlund et al, 1989), to quantitate the role of possible diffusion dependent processes in tissue gas exchange. We measured the washout of two indicator inert gases with differing molecular weights to separate the role of diffusion from perfusion heterogeneity. To assess whether diffusion dependent processes are affected by arterial PO_2, washouts were performed both during normoxia and during hypoxia.

METHODS

Mongrel dogs [n = 6; Wt = 26.7 ± 3.2 kg (Mean ± SD)] of both sexes were used. Anesthesia was induced with thiopental sodium (30 mg/kg) followed by a combination of α-chloralose (30 mg/kg) and urethan (100 mg/kg). Additional α-chloralose and urethan were given during the experiment to maintain anesthesia. Animals were mechanically ventilated with a tidal volume of 15 ml/kg and a frequency sufficient to maintain end-expiratory P_{CO_2} (monitored with a Perkin-Elmer MGA 1100 mass spectrometer) between 30 and 35 torr. Animals were heparinized initially with 5000 units followed by 1000 units every 30 minutes.

The gracilis muscle was isolated except for the two arteries and two veins that were accessible from the medial aspect of the muscle. The arteries branched from the femoral artery, and the veins drained into the femoral vein. No other blood vessels supplying the muscle were seen. The muscle was left in place, and ties were made at both the insertion and origin. The nerve supplying the muscle was left intact. Both veins were cannulated and connected to a single cannula with a T connector. A plexiglass box was placed over the muscle, and warm humidified air from a humidifier was blown through the box. The temperature at the muscle surface was recorded with a thermistor probe and maintained at 37 ± 0.5 °C. After 20 min, blood flow through the venous cannula was measured with a graduated cylinder and a stop watch. The box was removed from the muscle, and the two arteries were cannulated and connected with a T connector to a single cannula. The muscle was perfused via the arterial cannula by a perfusion pump (Masterflex Standard Servodyne), which withdrew blood from the carotid artery. Pump flow was adjusted to match the venous outflow measured before the arteries were cannulated. Flow through the venous cannula was measured again and found to be equal to pump flow in every experiment. Venous pressure was measured with a pressure tranducer connected to the venous cannula and set to 5 cmH_2O by adjusting the height of the venous cannula tip. The venous effluent from the muscle was collected with a beaker and then returned every 3 min to the jugular vein via an intravenous drip. Blood samples were taken from the arterial and venous lines of the muscle before inert gas washin for analysis of blood P_{O_2}, P_{CO_2}, pH (Instrument Laboratories 1302), and O_2 content (Lex-O_2-Con, Lexington Intruments). Muscle O_2 uptake was calculated from blood flow and the arteriovenous O_2 content difference.

A blood gas catheter was connected to a Balzers quadrupole mass spectrometer (model QMG 511) for continuous recording of the partial pressure of acetylene and krypton in the venous effluent (Grønlund et al, 1989). The membrane-covered tip of the blood gas catheter was positioned in the venous catheter 20 cm from the muscle. The blood gas catheter and half the length of the arterial tubing were included inside the box for temperature control. The output from the mass spectrometer was recorded and corrected by the catheter response time.

After 20 min of stable perfusion, a 5% dextrose solution, equilibrated with acetylene and krypton, was infused with a Harvard syringe pump into the perfusion line proximal to the perfusion pump. The infusion rate was adjusted to 5% of the muscle blood flow. After P_{ac} and P_{kr} in the venous effluent had both reached a plateau (60-90 min), inert gas infusion was stopped (causing a step decrease in arterial P_{ac} and P_{kr}) and the decrease in venous P_{ac} and P_{kr} was recorded. At the end of the experiment, the muscle was cut from the animal and weighed.

The model used to analyze the experimental washout curves consists of 50 parallel muscle compartments, each with one capillary and a tissue volume. The following assumptions are made: no gas exchange between compartments, homogeneously mixed compartments, end-capillary inert gas partial pressure (P_{IG}) equals tissue P_{IG}, no diffusional shunting directly between arterial and venous blood, and no time-dependent variation in compartmental blood flow. This transformation assumes 50 parallel compartments, non-negativity of blood flow in each compartment and smoothness in the distribution (similarity between adjacent compartments). The theoretical development of the analysis method can be found in Grønlund et al (1989).

RESULTS

Table 1 shows mean arterial and venous PO_2, PCO_2, pH, Hct and muscle temperature.

TABLE 1. Blood Gas Values

	P_aO_2	P_aCO_2	pH_a	P_vO_2	P_vCO_2	pH_v	Hct	Temp
Normoxia	92.4	31.8	7.37	54.7	37.3	7.33	45.5	38.2
	±10.1	±3.0	±.05	±5.8	±5.2	±.06	±3.7	±1.0
Hypoxia	33.5	31.2	7.29	26.6	39.1	7.24	42.4	38.3
	±4.4	±4.6	±.20	±6.5	±6.8	±.20	±2.4	±0.7

Average experimental acetylene and krypton washout curves are shown for the six dogs during normoxia in Figure 1. The acetylene washout is significantly faster than krypton washout in all the animals. The difference between the two was less during hypoxia. The hypoxia washout curves for both Ac and Kr were more rapid than the normoxia curves.

At the initial stage of washout, there is a rapid and significant drop in venous inert gas partial pressure. This is due to incomplete equilibration of the muscle during the washin phase (about 60 min) and/or the presence of a diffusional shunt. The relative lack of equilibration was quantitated from the rapid initial drop in venous P_{IG} at the beginning of washout by fitting a mono- exponential function to the fourth through the fifteenth data point. Use of these points eliminated the initial transient points and the later points affected by slower time constant regions. Average values for the intercept value (relative tissue desaturation at the end of washin are: Normoxia (I_{Ac} = 0.684 ± 0.059, I_{Kr} = 0.753 ± 0.076) and Hypoxia (I_{Ac} = 0.628 ± 0.070, I_{Kr} = 0.691 ± 0.062).

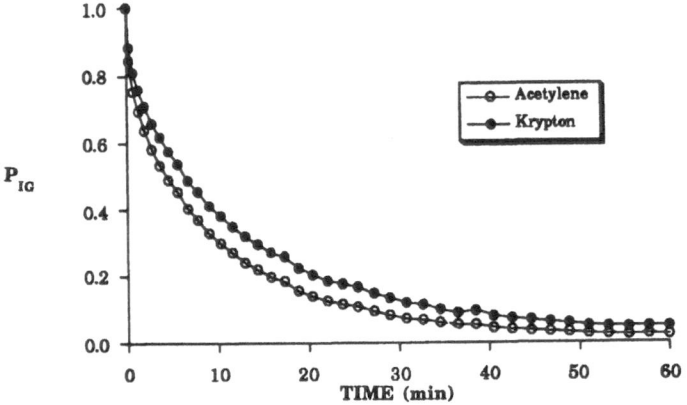

Figure 1. Average inert gas washout curves for Ac and Kr.

The washout curves were transformed into distributions of perfusion to volume using the algorithm previously developed in our laboratory for determining perfusion distribution in skeletal muscle (Grønlund et al, 1989). Inert gas washout in the resting skeletal muscle reveals a heterogeneous perfusion distribution. Figure 2 shows the distributions determined separately from the Ac and Kr washouts during normoxia. There are two distinct peaks determined with both gases. The right peak is a more rapid washout, higher perfusion/volume ratio region. The appearance of this fast washout region may, in part, be due to the diffusional shunt and the apparent rapid initial washout. The large central peak demonstrates that the majority of the perfusion is relatively uniformly distributed with an average $\dot{Q}$/Vol ratio of 10^{-1} or 0.10.

The relative dispersion or the amount of heterogeneity can be expressed as a log standard deviation (σ_Q). This term gives an indication of the width of the perfusion distribution. For the muscle shown in Figure 2, the dispersion for Kr is similar to that for Ac. This finding was uniform throughout all the data. The heterogeneity of the Kr-derived distributions were not different from the heterogeneity of the Ac-derived distributions.

DISCUSSION

The difference between the P_{IG} values for acetylene and krypton during washout was significant both during normoxia

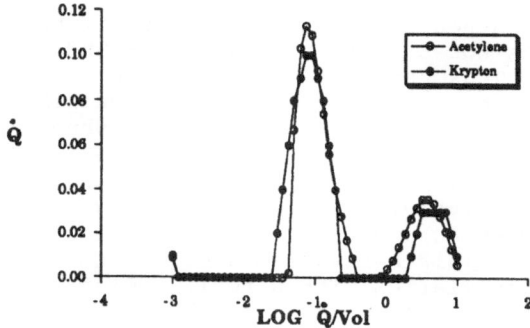

Figure 2. Perfusion to volume distribution determined by acetylene and by krypton for one animal in normoxia.

and hypoxia. The acetylene washout was in all cases faster than the krypton washout implying a mechanism at work which favors the lower molecular weight gas. Such behavior could be explained either by a diffusion limitation within the muscle tissue retarding Kr relative to Ac or by a diffusional shunt allowing more back diffusion by Ac from the venous vessels to the arterial vessels decreasing the effluent P_{Ac} relative to P_{Kr}. The more rapid washout of Ac compared to Kr cannot be used to separate the two diffusion-related mechanisms.

Washin of inert gas was continued until the venous partial pressures were apparently constant. However, the washin time was limited and the muscle may not have been completely equilibrated at the end of the washin. This is suggested by the rapid initial drop in venous P_{IG} during washout. This relative lack of equilibration is quantitated by the intercept values listed above. The fact that the intercepts are different for acetylene versus krypton reveals a diffusion-dependent tissue equilibration deficit. I_{Ac} and I_{Kr} averaged 0.656 and 0.722, respectively, for all runs (normoxia and hypoxia). The relative disequilibration of the tissue during washin is less for Krypton, the heavier molecular weight and slower diffusing gas. Such a molecular weight-dependent deficit is consistent with the presence of a diffusional shunt. During washin of the gases, the more rapidly diffusing gas, Ac, diffuses through the shunt reducing the amount delivered to the muscle. If the molecular weight dependent difference occurred by virtue of a diffusion limitation, then one would expect a lesser equilibration for Kr and a lower I_{Kr} than I_{Ac}.

Hypoxia resulted in a more rapid washout of both the Ac and Kr, although the increase was significant at only the early phase of washout. The heterogeneity of the $\dot{Q}$/Vol distribution was not different between Ac- and Kr-derived distributions. Thus the difference between normoxia and hypoxia appears to be in the relative disequilibrium with washin and/or the magnitude of diffusional shunt. In fact the difference between I_{Ac} and I_{Kr} was similar with hypoxia (0.628 and 0.691, respectively) compared to normoxia (0.684 and 0.753, respectively). It is interesting that the difference between Ac and Kr decreased in magnitude with the hypoxic washout. This behavior may be due to a recruitment of additional capillaries within small regions causing a decrease in the diffusion distances within tissue.

In summary, the simultaneous washout of two gases of differing molecular weight from muscle reveals a significant molecular weight dependence of gas exchange. The poorer equilibration of acetylene compared to krypton during washout and the more rapid washout of acetylene suggest a role of a diffusional shunt rather than diffusion limitation as the dominant diffusion-related factor in tissue gas exchange.

REFERENCES

Grønlund, J, Malvin, GM, and Hlastala, MP, 1989, Estimation of blood flow distribution in skeletal muscle from inert gas washout, J Appl Physiol 66:1942-1955.

Hills, BA, 1967, Diffusion versus blood perfusion in limiting the rate of uptake of inert non-polar gases by skeletal rabbit muscle, Clin Sci Lond 33:67-87.

Piiper, J, Meyer, M, and Scheid, P, 1984, Dual role of diffusion in tissue gas exchange: blood-tissue equilibration and diffusion shunt, Respir Physiol 56:131-144.

Piiper, J and Haab, P, 1991, Oxygen supply and uptake in tissue models with unequal distribution of blood flow and shunt. Respir Physiol 84:261-271.

Wagner, PD, 1989, Peripheral inert-gas exchange, In: Handbook of Physiology, Vol 4, Respiration Physiology, Ed., Farhi, LE and Tenney, M, American Physiological Society, Bethesda, MD, pp. 257-281.

EFFECTS OF ENDOTOXIN ON CANINE SKELETAL MUSCLE OXYGEN

DELIVERY-UPTAKE RELATIONS DURING PROGRESSIVE HYPOXIC HYPOXIA

Scott E. Curtis, W.E. Bradley, and Stephen M. Cain

Departments of Pediatrics and Physiology and
Biophysics, University of Alabama at Birmingham
Birmingham Alabama, 35294

INTRODUCTION

Septic patients, who may also have the adult respiratory distress syndrome (ARDS) often show apparent O_2 supply dependency with elevated arterial lactate levels even at normal to high rates of O_2 delivery (DO_2) and appear unable to increase O_2 extraction appropriately (Gilbert et al. 1986; Astiz et al. 1987). Cain (1986) has suggested that multiple events in the microcirculation may be responsible for this apparent defect in O_2 extraction. These include microembolization and loss of endothelial functions and integrity. If the capillary membrane becomes leaky, then pericapillary edema may ensue and offer a diffusional block to the movement of O_2 into tissue. The combination of microembolization and loss of endothelial functions that include microregional vascular control can lead to coexistence of underperfused and overperfused areas in a peripheral tissue, such as skeletal muscle. The outcome of these conditions would be a reduced ability of the tissues to extract O_2 from arterial blood and an increased impediment to O_2 diffusion.

In an earlier study, our laboratory showed that defective O_2 extraction was present in skeletal muscle of dogs given a bolus of endotoxin (Bredle et al. 1989). In that study, critical DO_2 was raised and critical O_2 extraction ratio (O_2ERc) was lowered when muscle blood flow was progressively reduced in comparison to control animals. Increased arterial PO_2 did not benefit O_2 extraction in that study. Samsel et al. (1988a) saw no significant impairment in muscle O_2 extraction in endotoxic dogs when they lowered blood flow in the whole animal by progressive hemorrhage. These two studies indicated that muscle was relatively resistant to the effect of endotoxin on tissue oxygenation and raised some doubt that diffusion impairment was the root cause. We hypothesized that if impaired O_2 diffusion was responsible for the O_2 extraction problem, then O_2 extraction should be much worse if DO_2 was decreased by progressive hypoxia rather than by progressive ischemia. A clearer demonstration of that possibility was the object of the present studies.

Oxygen Transport to Tissue XIV, Edited by W. Erdmann and
D.F. Bruley, Plenum Press, New York, 1992

MATERIALS AND METHODS

12 adult dogs in 2 groups of 6 each were anesthetized with pentobarbital, intubated, and pump ventilated with room air. Carotid and pulmonary arterial catheters were placed. We then surgically isolated the arterial inflow and venous outflow of the left hindlimb as previously described (Bredle et al. 1991). The right femoral artery was cannulated to supply flow to a roller-pump membrane oxygenator circuit, which provided blood flow to the isolated left femoral artery and hindlimb. A gas flow mixer delivered O_2, N_2, and CO_2 to the oxygenator as needed to obtain the desired limb PaO_2. The animal was given 1000 U/kg of IV heparin and then was paralysed with succinylcholine. Systemic VO_2 was determined from analysis of exhaled gas. DO_2 and VO_2 of the limb were measured ten minutes after each change in DO_2. Then, simultaneous hindlimb and whole body arterial and venous blood samples were obtained for measurement of blood gas tensions, pH, and co-oximetry. Using the Fick principle, cardiac output (Qt) and hindlimb VO_2 were calculated. Systemic and hindlimb vascular resistance were calculated as pressure divided by indexed flow and reported as $PRU \cdot kg$. DO_2 was calculated as the product of CaO_2 and flow. Regression lines were fitted to the delivery independent and delivery dependent portions of the delivery-uptake curve using a least-squares method (Samsel et al. 1988b). The intercept of these two lines defined the critical DO_2 (DO_2c); that is, the DO_2 at which VO_2 began to fall with further declines in DO_2. Critical O_2ER (O_2ERc) was taken as the ratio of VO_2 to DO_2 at DO_2c. Red blood cells and dextran were administered as needed throughout the study to keep Qt at or above baseline and to maintain Hct. $NaHCO_3$ was given as needed to keep arterial bicarbonate levels greater than 20 meq/dl.

Control group (CONT): After surgery was complete and all monitored values were stable, hindlimb blood flow was measured and whole body arterial, mixed venous, and hindlimb venous blood samples were obtained. Mean arterial pressure (MAP), heart rate, and temperature were also recorded. These measurements were repeated every 15 min. For the first 3 measurements, the hindlimb was perfused with normoxic blood at estimated flows of 120, 90, and 60 $ml \cdot min^{-1} kg^{-1}$. Further decreases in DO_2 were done at 15 min intervals by changing the oxygenator sweep gas to produce hindlimb PaO_2's of 60, 40, 35, 30, 25, 20, and 15 Torr, with the flow kept at 60 $ml \cdot min^{-1} kg^{-1}$.

Endotoxin group (ENDO): After the preparation was complete and all measured variables were stable, 2 mg per kg of endotoxin (Diffco *E. coli* LPS) dissolved in 60 ml of saline was infused over 60 min. Data collection began in this group immediately following the end of this infusion. The progressive hypoxia protocol and all other maneuvers in this group were identical to controls.

RESULTS

MAP was > 110 torr in CONT and > 80 torr in ENDO throughout the study. Qt did not differ between groups and was always > 100 $ml \cdot min^{-1} \cdot kg^{-1}$. To maintain Qt, CONT required 20 ± 3 ml/kg of dextran plus RBC compared to 71 ± 19 ml/kg in ENDO (p<0.05). Hct, blood gas tensions, and pH remained normal in both groups. Systemic DO_2 varied from about 12-26 $ml \cdot min^{-1} \cdot kg^{-1}$ with no significant differences between groups. Systemic VO_2 correlated poorly with DO_2 in CONT (r= 0.36), but had a positive slope of 0.25 and r=0.62 in ENDO (Fig 1).

In each study hindlimb DO_2 began near 20-25 $ml \cdot min^{-1} kg^{-1}$ and was decreased over two hours to about 2 $ml \cdot min^{-1} \cdot kg^{-1}$. Individual DO_2-VO_2 curves for the 12 dogs were constructed, from which critical values of DO_2 and O_2ER were determined. Group means are presented in Table 1. Fig. 2, which pools all the hindlimb DO_2-VO_2 data, highlights several important differences between the 2 groups. First, VO_2 at DO_2's above critical were significantly higher in ENDO than in CONT. Second, the relation of VO_2 to DO_2 in ENDO was less clearly biphasic than in CONT, with the suggestion of supply dependence over the entire range of DO_2. Third, the slope of the regression line in the supply dependent zone does not appear different between groups. Indeed, mean O_2ERc was 66+4% in CONT and 72+8% in ENDO (p=N.S.). Despite a similar ability to extract delivered oxygen, the elevated resting VO_2 caused VO_2 to decline at a higher DO_2 in ENDO compared to CONT.

In CONT, hindlimb vascular resistance varied directly with DO_2 (Fig 3). The curve was displaced downward in ENDO, indicating lower resistance at all flows. The ENDO curve was also flattened, suggesting a loss of vascular regulation. All 6 CONT dogs had vigorous reactive hyperemia in response to 30 sec of femoral artery occlusion at the end of the study, but only 1 of 6 ENDO dogs did.

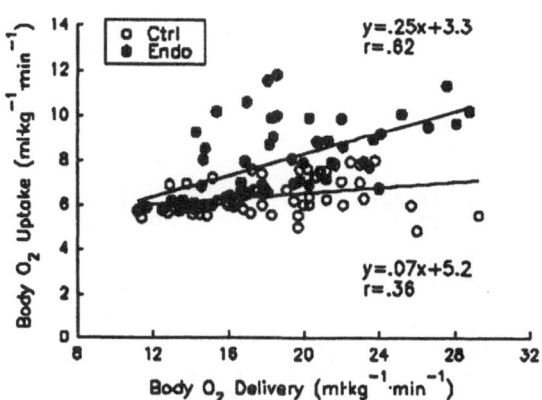

Figure 1

TABLE 1. Mean DO_2-VO_2 data

Parameter	CONT	ENDO
Body VO_2[a]	6.4+0.1	7.9+0.2*
Limb VO_2 above DO_2c[a]	4.4+0.1	8.0+0.5*
Limb O_2ERc[b]	66+4	72+8
Limb DO_2c[a]	6.9+0.6	11.4+1.7*

[a] in $ml \cdot min^{-1} kg^{-1}$ [b] in %
Data are mean+SEM
* p<.05 vs. CONT by ANOVA

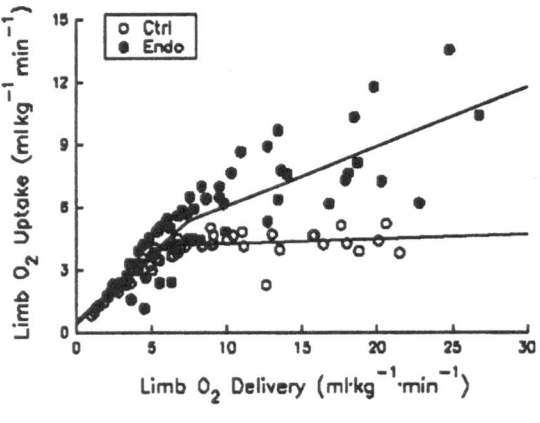

Figure 2

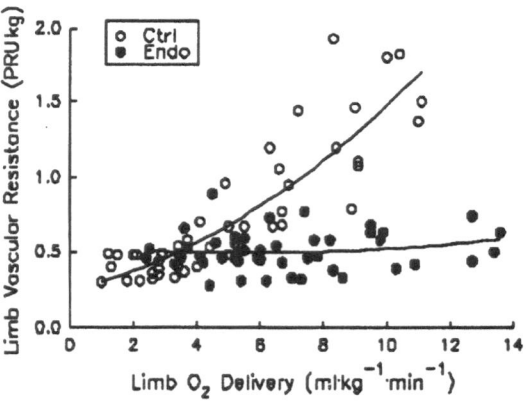

Figure 3

DISCUSSION

    Patients with hyperdynamic sepsis have low systemic vascular resistance (SVR) and normal to high cardiac output, two features that were well reproduced with our endotoxin model. Qt did not differ between ENDO and CONT, but SVR was 30% lower in ENDO. Hindlimb vascular resistance was also markedly decreased in ENDO, and did not vary with changes in DO_2. Considering that skeletal muscle receives about 40% of resting blood flow in the dog, abnormal dilatation of this vascular bed may have profound effects upon systemic blood pressure and blood flow to other organs. The loss of vascular reactivity in 5/6 ENDO dogs suggests that in the hindlimb, vessels were already maximally dilated, or that vascular tone was no longer coupled to blood flow and local metabolic factors. Both hyper- and hypo- vascular reactivity have been described in different animal models of sepsis, and multiple mediators have been postulated to play a role in this abnormal vasomotion, including bradykinin, histamine, prostaglandin E, prostacyclin, and others (Messmer et al. 1989).

    Loss of vascular tone will also significantly affect the ability of a tissue to direct blood flow to where it is most needed. Beer and Yonce (1972) demonstrated a high correlation (r>0.8) between capillary perfusion density and oxygen extraction in a study that varied flow to isolated gastrocnemius muscle. Somewhat later, Granger and Shepherd (1979) advanced the theory that cellular metabolites, such as adenosine, act upon upstream arteriolar sphincters to increase flow and direct it precisely toward cells with unmet oxygen demand. This requires a backdrop of high resting vascular tone to be effective, so that cells with less metabolic need are not overperfused. This notion is supported by the work of Cain (1978) in which dogs pretreated with the α-blocker phenoxybenzamine had significantly decreased O_2ER when exposed to whole body hypoxia. Indiscriminate vasodilation in sepsis could easily disrupt the precise matching between oxygen delivery and consumption. However, abnormal vasomotion is not the only possible cause of shunting within a tissue. Endotoxin can activate both the coagulation and complement cascades, leading to plugging of capillaries by platelets and leukocytes, respectively (Messmer et al. 1989). Such plugging would decrease blood flow to downstream cells, while shunting the blood to other cells that are overperfused relative to need.

    Microembolization appears to adversely affect O_2ER. In a canine muscle preparation, Landau et al. (1982) injected 15μm microspheres and caused critical DO_2 to increase from 2.85 to 4.00 ml·min^{-1}·100^{-1}g tissue. Also, reactive hyperemia was not seen in any tissues after embolization, suggesting that capillaries could no longer be recruited as needed. Cain et al. (1988) also demonstrated decreased O_2ER in embolized dog hindlimbs subjected to hypoxic hypoxia. Extreme perfusion heterogeneity caused by abnormal vasomotor control or microembolization can greatly increase diffusion distance to hypoxic areas. Even raising PO_2 in the perfusing blood may not overcome the increased diffusional barrier. Bredle et al. (1989) tested this hypothesis by challenging endotoxin-treated dog hindlimbs with progressive ischemia at PaO_2's of 60 and then 200 torr. Hindlimb critical O_2ER was significantly reduced in the endotoxin group but was not affected by the level of PaO_2. This supported the notion of severe intramuscular shunting, and spoke against the alternative possibility that a simple diffusional defect was primarily responsible for the decreased O_2ERc in sepsis. We further hypothesized that progressive hypoxic hypoxia would have no different effect than ischemic hypoxia for similar DO_2's. Indeed, hindlimb O_2ERc of endotoxin-treated dogs in the present study, $72\pm8\%$, was not significantly different from what was seen in either of the two progressive ischemia, endotoxin studies previously performed in this laboratory (Bredle et al. 1989, 1991).

A surprising finding in this study was the relatively low O_2ER seen in non-endotoxin control dogs, compared to normals (Bredle et al. 1991) undergoing progressive ischemia ($66\pm4\%$ vs. $81\pm5\%$, $p<0.05$). This would suggest that hypoxic hypoxia itself can cause an inefficiency in oxygen use by tissues. However, we noted that the methods of this study differed from that of the previous ischemia study in that a membrane oxygenator was not present in the perfusion circuit of the prior study. We considered the possibility that the oxygenator may have interfered with microvascular control by activation of inflammatory mediators. We therefore repeated the progressive ischemia studies with a membrane oxygenator in line, with the sweep gas set to maintain similar normoxic gas tensions. We found no difference in critical O_2ER between the ischemia groups with or without the membrane oxygenator (manuscript in progress). Hypoxic hypoxia may interfere with oxygen extraction relative to ischemic hypoxia by the known vasodilator effects of low PO_2. The recent work of Pohl and Busse (1989) showed that intraluminal hypoxia (PO_2 24 ± 8 torr) in isolated rabbit aorta or femoral artery stimulated the release of endothelium-derived relaxant factor and caused vascular dilatation. Any factor that reduces vascular tone may potentially interfere with the ability of the tissue to match O_2 supply to O_2 demand by local metabolic vasodilators.

One final aspect of this study that warrants discussion are the VO_2 data. A common feature of sepsis/ARDS is hypermetabolism, with systemic VO_2 increased up to 50% or more above normal (Cerra 1987). It is not known how, why, or where this excess oxygen is being consumed. This increase in VO_2 was replicated by our model: systemic VO_2 in ENDO was increased 23% over CONT ($p<0.05$). Also, the systemic DO_2-VO_2 relation had a positive slope of .25 and an r value of .62, suggestive of pathologic supply dependency. Assuming that skeletal muscle makes up 40 to 50% of body mass in dogs, the increase in systemic VO_2 may be entirely explained by the approximately 82% increase in muscle VO_2. It should be remembered that continuous paralysis was maintained in all dogs with succinylcholine and that core temperature did not differ between the two groups. Thus, how the muscle burned this oxygen is an interesting question. We believe that these data are the first to suggest a location for the excess O_2 consumption of sepsis.

CONCLUSIONS

Six dogs were treated with a 2mg/kg endotoxin infusion and volume resuscitated. This produced a state of normal cardiac output with reduced systemic and hindlimb vascular resistance and a significant increase in oxygen consumption. Hindlimb critical O_2ER during hypoxic hypoxia was not different than previous studies in which critical O_2ER was determined using ischemic hypoxia, lending further support to the hypothesis that a diffusion block to oxygen is not the primary cause of the sepsis-induced decrease in O_2ERc. A surprising finding was that limb muscles in control dogs that were subjected to hypoxic hypoxia and who did not receive endotoxin demonstrated abnormally low O_2ERc compared to historic controls, suggesting that hypoxic hypoxia itself causes an inefficiency in O_2 extraction relative to ischemic hypoxia.

ACKNOWLEDGEMENT

The work was supported by NHLBI Grant #HL 26927 to S.M. Cain. Dr. Curtis is supported in part by an American Lung Association Edward Livingston Trudeau Research Scholar Award.

REFERENCES

Astiz, M.E., Rackow, E.C, Falk, J.L, et al., 1987, Oxygen delivery and consumption in patients with hyperdynamic septic shock. *Crit. Care Med.*, 15:26.

Beer, G. and Yonce, L.R., 1972, Blood flow, oxygen uptake, and capillary filtration in resting skeletal muscle. *Am. J. Physiol.*, 223:492.

Bredle, D.L, Samsel, R.W., Schumacker, P.T., and Cain, S.M., 1989, Critical O2 delivery to skeletal muscle at high and low PO_2 in endotoxemic dogs. *J. Appl. Physiol.*, 66:2533.

Bredle, D.L., and Cain, S.M, 1991, Systemic and muscle O_2 uptake/delivery after dopexamine infusion in endotoxic dogs. *Crit. Care Med.*, 19:198.

Cain, S.M., 1978, Effects of time and vasoconstrictor tone on O_2 extraction during hypoxic hypoxia. *J. Appl. Physiol.*, 45:219.

Cain, S.M., 1986, Assessment of tissue oxygenation. *Crit. Care Clin.*, 2:537.

Cain, S.M., King, C.E., and Chapler, C.K., 1988, Effects of time and microembolization on O_2 extraction by dog hindlimb in hypoxia. *J. Crit. Care*, 3:89.

Cerra, F.B., 1987, Hypermetabolism, organ failure, and metabolic support. *Surgery*, 101:1.

Gilbert, E.M., Haupt, M.T., Mandanas, R.Y., et al., 1986, The effect of fluid loading, blood transfusion, and catecholamine infusion on oxygen delivery and consumption in patients with sepsis. *Am. Rev. Respir. Dis.*, 134:873.

Granger, H.J., and Shepherd, A.P., 1979, Dynamics and control of the microcirculation. *Adv. Biomed. Eng.*, 7:1.

Landau, S.E., Alexander, R.S., Powers, S.R., et al., 1982, Tissue oxygen exchange and reactive hyperemia following microembolization. *J. Surg. Res.*, 32:38.

Messmer, K., Kreimeier, U., and Hammersen, F., Changes in the microcirculation in sepsis and septic shock, in: "Sepsis," K. Reinhart and K. Eyrich, eds., Springer-Verlag, Berlin (1989).

Pohl U., and Busse, R., 1989, Hypoxia stimulates release of endothelium-derived relaxant factor. *Amer. J. Physiol.* 256:H1595.

Samsel, R.W., Nelson, D.P., Sanders, W.M., et al., 1988a, Effect of endotoxin on systemic and skeletal muscle O2 extraction. *J. Appl. Physiol.*, 65:1377.

Samsel, R. and Schumacker, P.T., 1988b, Determination of the critical O2 delivery from experimental data: sensitivity to error. *J. Appl. Physiol.*, 64:2074.

SKELETAL MUSCLE CAPILLARY FLOW AND OXYGENATION IN HYPOXIC HYPOXIA:

EFFECT OF A 5-HT$_2$ RECEPTOR ANTAGONIST

U Gustafsson[1], DH Lewis[1] and P Thorborg[2]

[1]Clinical Research Center, University Hospital, S-58185 Linköping, Sweden and [2]Department of Anesthesiology, University of Rochester Medical Center, Rochester, NY 14642, USA

INTRODUCTION

Serotonin (5-HT) seems to be essential to mammal life, however, its various actions have hither to been difficult to assess due to their complexity; further more, it is not until the last few years that serotonin receptors have been typed and subtyped. Following this recent development of selective and pure serotonin receptor antagonists, it has been shown that serotonin is involved in the modulation of tissue blood flow and oxygenation in response to various stimuli. Hyperoxemia is known to cause arteriolar vasoconstriction, the disturbed tissue oxygenation and capillary flow in skeletal muscle can be normalized by ketanserin or ritanserin, 5-HT$_2$ receptor antagonists (1,2). The purpose of the present study was to assess skeletal muscle capillary flow and oxygenation during hypoxic hypoxemia with particular reference to the action of 5-HT$_2$ receptors.

MATERIALS AND METHODS

Experiments were carried out on Swedish land-race rabbits (n=14) 2.5-3.0 kg. Anesthesia was induced by i.v. ketamine (25 mg/kg) and xylazine (2 mg/kg) through a butterfly cannula in the posterior marginal ear vein. Anesthesia was maintained stable by a continuous venous infusion of ketamine and xylazine. After tracheotomy, normocapnic ventilation was achieved using a ventilator and assessed by repeated blood gas analyses. An arterial cannula was introduced in the left carotid for blood pressure monitoring and for blood gas sampling. The left vastus medialis muscle was exposed and freed from its fascia. An eight channel electrode was carefully placed on the muscle surface in a lucite holder for measurements of capillary flow (1,3,4) (n=7) or tissue pO$_2$ (n=7) (5). Only muscle surface devoid of visible blood vessels was accepted. Capillary flow measurements were performed using the local hydrogen clearance technique and flow values were calculated from clearance curves (see 1,4 for further details). Catchment zones for each electrode is a half sphere with a 25 micron radius (when measuring oxygen pressure) or 35 microns (when measuring hydrogen pressure). When placed on predominantly white muscle intercapillary distances are such that it is most probable that each electrode only measures from one capillary when measuring oxygen pressure; in the case of hydrogen pressure in the few cases when washout curves had more than one component (thus registering flow from more than one capillary), only the fastest component was calculated.

A laser-Doppler probe was placed on the contralateral muscle surface for regional blood flow measurements (PF1d, Perimed, Sweden). It is based on the fraction of back-scattered light from moving red blood cells (6). From the continuous strip recordings of laser-Doppler flow signals, averages were calculated from the intervals preceding measurements of tissue oxygenation or capillary flow.

After preparations, animals were allowed to stabilize for 30 min before measurements during normoxemia (A) (FiO_2 0.21), followed by hypoxemia (B) (FiO_2 0.10) and finally sustained hypoxemia during which ritanserin (C) (0.035 mg/kg) was injected i.v. Ritanserin is a pure and selective $5-HT_2$ receptor antagonist, the chosen dosage is in the low dose range. From previous studies we know that ritanserin effects on the microcirculation are evident after 30 min and it remains effective (T1/2>40h) more than 40h in the rabbit. No systemic effects are seen after ritanserin injection. Duration of each state was 2 h. During administration of hydrogen gas boluses (FiH_2 0.30), gas with FiO_2 0.15 was used to maintain paO_2 constant. Calibrated gas mixtures 10.0 % and 15.0 % were delivered by AGA Specialty Gases Department, Lidingö, Sweden.

For statistical analysis the Wilcoxon signed rank sum test was used and differences were considered as statistically significant at the $p<0.05$ level. Values are given as mean ± SD.

All experiments were approved by the local Ethics Committee for animal research.

RESULTS

Arterial blood gas results are given in the Table for normoxemia (A), hypoxemia (B) and hypoxemia with ritanserin (C).

Table 1. Results from analyses of arterial blood gases. Gas pressure values are given in kPa, BE in mmol/L.

	paO_2	$paCO_2$	pH	BE
(A)	12.5 ± 1.9	4.2 ± 0.2	7.5 ± 0.03	1.7 ± 1.7
(B)	3.9 ± 0.5	4.3 ± 0.2	7.4 ± 0.06	-2.4 ± 3.8
(C)	4.1 ± 0.4	4.4 ± 0.4	7.2 ± 0.1	-11.7 ± 6.5

Mean arterial blood pressure (MAP) was 74 ± 9 torr during (A), 55 ± 11 torr during (B) and 50 ± 10 torr during (C), respectively.

The mean *laser-Doppler flowmetry* results, set to 100 % during normoxemia were 100 ± 38 % during (A), 77 ± 31 % during (B) ($p<0.001$) and 80 ± 36 % during (C) (n.s.).

Mean capillary flows, set at 100 % at normoxemia, were 100 ± 82 %, 73 ± 85 % ($p<0.01$) and 117 ± 76 % ($p<0.001$), respectively. These mean values were based on 362 (A), 372 (B) and 382 (C) values, respectively. In the distributions of capillary flow values (see Figure), the number of flow values in the lowest class (0-1 ml/100g/min) increased from 9.4 % in normoxemia to 35.4 % in hypoxemia but decreased after ritanserin to 4.9 %.

Mean values for *skeletal muscle pO_2* were set to 100 % during normoxemia. Results for states (A), (B) and (C) were 100 ± 37 %, 78 ± 44 % (p<0.001) and 104 ± 61 % (p<0.001), respectively. These mean values were calculated from 1119 (A), 1118 (B) and 1118 (C) values, respectively. The percentage of low ptO₂ values (below 5 torr) increased from zero in normoxemia (A) to less than 3 % in hypoxemia (B) to zero percent after ritanserin (C) in hypoxemia.

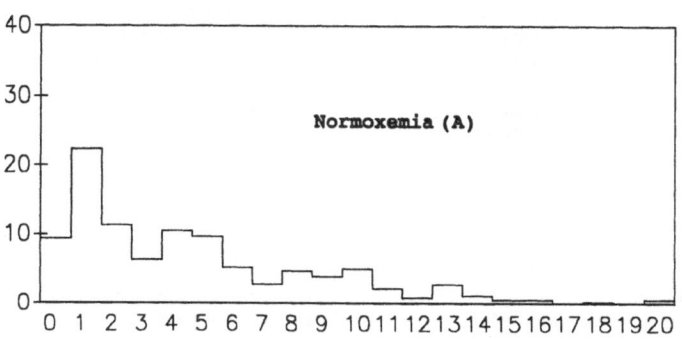

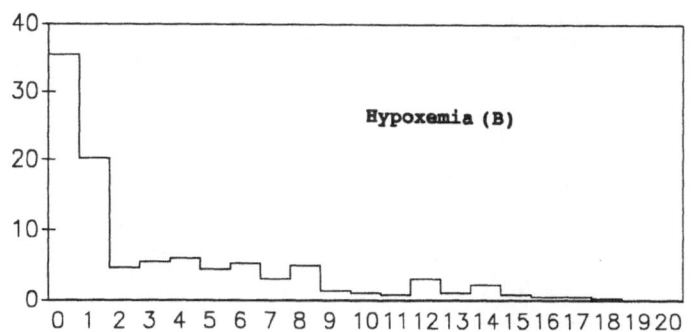

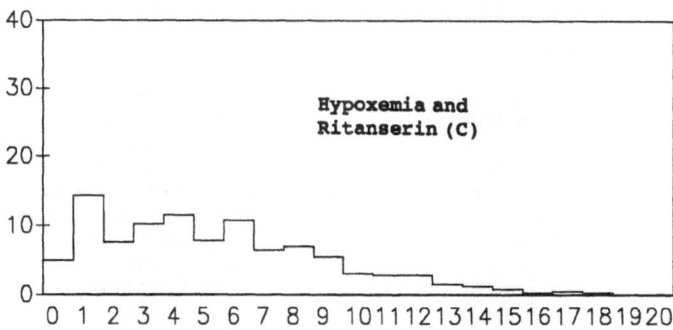

Figure 1. Capillary flow distributions during (A), (B), and (C). Ordinate: percentage of flow values.
Abscissa: flow values class in ml/100g/min.

DISCUSSION

Previous measurements of tissue pO$_2$ and capillary flow using the same type of electrodes under hypoxic hypoxemia (FiO$_2$ 0.10) (7-10) have consistently shown increased flow in larger conduit vessels but varying degrees of decreased mean capillary flow and mean tissue pO$_2$. In the dog model (7-10) MAP increased during hypoxemia in contrast to the decrease seen in this rabbit model. Decreased capillary flow has also been observed iv vivo using video microscopy techniques (11).

In the study by Harrison et al (9) there were less than 5 % of skeletal muscle pO$_2$ values in the lowest class, 0-5 torr, during hypoxemia. In the present study less than 3 % of the ptO$_2$ values were in the same class during hypoxemia, which is comparable to the above-mentioned study. In the present study, however, the decrease in capillary flow during hypoxemia was more pronounced than that noted by Harrison et al (9). The decrease was of the order of 27 % during hypoxemia. This could be due to the fact that in the present study there was a pronounced decrease in blood pressure, not observed by Harrison et al (9). In addition, in the present study, regional blood flow, as measured with laser-Doppler flowmetry, decreased 23 %, paralleling the decrease in capillary flow. In general then, the changes seen during hypoxemia in the present study were comparable to the results reported in other studies. It should also be pointed out that muscle relaxants were not used in the present study. Pancuronium is known to produce slight ganglionic blockade, which may affect tissue blood flow.

The results of the present study also raise the question of the effect of hypoxemia on the distribution of capillary flow within the capillary bed, with possible changes in the degree of shunting of capillary flow. Such a redistribution is a physiological regulatory mechanism in which anatomico-physical determinants (according to the Hagen-Poiseuille law) (12) conceivably are responsible for flow redistribution. For the present study, this would mean that the shift from normoxic distribution to that during hypoxemia would entail a redistribution of capillary flow from nutritive capillaries to flow through shorter and/or larger diameter capillaries, acting as functional shunt vessels. Such an interpretation has been proposed by other research groups (13). In the present study the increase in the number of capillary flow values in the range between 0-1 ml/100g/min from 9.4 % to 35.4 % is in agreement with this interpretation (see Figure).

The main interest with this study was the (somewhat unexpected) results obtained after ritanserin was administered during continuous hypoxemia. There was a parallel increase in both capillary flow and tissue pO$_2$ without significant increase in regional flow. This is interpreted as decreased shunting of capillary flow to nutritive flow, i.e., a redistribution of capillary flow within the muscle. This is also substantiated by the number of capillary flow values in the lowest class (0-1 ml/100g/min) that decreased from 35.4 % to 4.9 % after ritanserin (Figure). The development of base excess with time (Table) shows that after 2 h BE had developed from 1.7 to -2.4 and after another 2 h after ritanserin decreased to -11.7, more than the time dependent factor could account for. In our opinion, it should be interpreted as an increased washout of acid metabolites due to improved nutritive flow.

It is, however, not known why a 5-HT$_2$ receptor antagonist has these effects on skeletal muscle capillary flow redistribution during hypoxemia. Focus in research has moved from systemic control (14) to local control during hypoxemia. Hypoxemia is well known to induce vasodilation (15) by direct action of the hypoxic blood on the blood vessel wall (16). It is now known (for review see 17) that endothelial cells in response to various stimuli release vasoactive agents (autacoids) that effect vascular tone, namely the vasodilators prostacyclin (PGI_2) and nitric oxide (formerly called EDRF or endothelium-derived relaxant factor) and the vasoconstrictor endothelin. Hypoxemia (as opposed to true anoxia) induces a calcium dependent release of PGI_2 and nitric oxide. Serotonin has marked effects on vascular reactivity (for review see 18) mainly due to activation of $5-HT_2$ receptors in peripheral blood vessels but also by enhancing the effect of other vasoconstrictors. It also triggers the release of vasodilators as nitric oxide and pGI_2 from endothelial cells. The composite effect will depend on the balance between vasoconstrictor and vasodilator influences which additionally may vary in different types of organs and vessel sizes. Antagonism of $5-HT_2$ receptors may push the balance towards vasodilation. The results from this study do, however, not favor any simple interpretation of vasodilation effects due to elimination of $5-HT_2$ receptor effects in the skeletal muscle blood vessels. Instead it appears that selective antagonism of $5-HT_2$ receptors improves nutritive flow selectively without increasing shunt flow in spite of a relatively low MAP.

To further expand knowledge in this field of flow regulation, one would need to locate $5-HT_2$ receptors in the microcirculation. One could speculate that they may appear in increased concentration in endarterioli before departure of larger sized or shorter capillaries that may act as functional shunt vessels. That would explain some effects seen in this and previous studies (1,2).

From these results it appears that $5-HT_2$ receptor antagonists may become an interesting clinical tool in hypoxic hypoxemia in that it improves nutritive blood flow in skeletal muscle.

ACKNOWLEDGEMENTS

This project was in part sponsored by AGA Medical Research (021/89), by the Östergötland County Council (90/112), by the Swedish Medical Research Council (02142), Lions Research Foundation and by Linköping University, Sweden. Ritanserin was kindly provided by Janssen Pharmaceuticals, Beerse, Belgium.

REFERENCES

1. Thorborg P, Gustafsson U, Sjöberg F, Harrison DK, Lewis DH (1990), Effect of hyperoxemia and ritanserin on skeletal muscle microflow. J Appl Physiol 68(4):1494-1500

2. Thorborg P, Lund N (1989), Serotonin as a modulator of skeletal muscle oxygenation: effects of ketanserin and ritanserin on oxygen pressure distributions. Int J Microcirc: Clin Exp 8:191-203

3. Harrison DK, Kessler M (1989), A multiwire hydrogen electrode for in vivo use. Phys Med Biol 34(10):1397-1412

4. Harrison DK, Kessler M (1989), Local hydrogen clearance as a method for the measurement of capillary blood flow. Phys Med Biol 34(10):1413-1428

5. Kessler M, Höper J, Krumme BA (1976), Monitoring of tissue perfusion and cellular function. Anesthesiology 45(2): 184-197

6. Nilsson GE, Tenland T, Öberg Å (1980), A new instrument for

continuous measurement of tissue blood flow by light beating spectroscopy. IEEE Trans Biomed Eng 27:12-19

7. Harrison DK, Knauf SK, Vogel H, Günther H, Kessler M (1985), Redistribution of microcirculation in skeletal muscle during hypoxaemia. Adv Exp Med Biol 191:387-397

8. Harrison DK, Birkenhake S, Knauf S, Hagen N, Beier I, Kessler M (1988), The role of high flow capillary channels in the local oxygen supply to skeletal muscle. Adv Exp Med Biol 222: 623-630

9. Harrison DK, Birkenhake S, Hagen N, Knauf S, Kessler M (1989), Regulation of capillary blood flow: a new concept. Adv Exp Med Biol 248:583-589

10. Harrison DK, Kessler M, Knauf SK (1990), Regulation of capillary blood flow and oxygen supply in skeletal muscle in dogs during hypoxaemia. J Physiol 420:431-446

11. Bertuglia S, Colantuoni A, Coppini G, Intaglietta M (1991), Hypoxia- or hyperoxia-induced changes in arteriolar vasomotion in skeletal muscle microcirculation. Am J Physiol 260 (Heart Circ. Physiol. 29):H362-H372

12. Potter RF, Groom AC (1983), Capillary diameter and geometry in cardiac and skeletal muscle studied by means of corrosion casts. Microvascular Research 25:68-84

13. Gutierrez G, Lund N, Acero AL, Marini C (1989), Relationship of venous PO2 to muscle PO2 during hypoxemia. J Appl Physiol 67(3):1093-1099

14. Chalmers JP, Korner PI, White SW (1966), The control of the circulation in skeletal muscle during arterial hypoxia in the rabbit. J Physiol 184:698-716

15. Jackson WF, Duling BR (1983), The oxygen sensitivity of hamster cheek pouch arterioles. In vitro and in vivo studies. Circ Res 53:515-525

16. Pohl U, Busse R, Kessler M (1982), Vascular resistance and tissue PO2 in skeletal muscle during perfusion with hypoxic blood. In Cardiovascular system dynamics: models and measurements, pp 521-530. Eds T Kenner, R Busse and H Hingofer-Szalkay. Plenum Press, New York-London

17. Pohl U (1990), Endothelial cells as part of a vascular oxygen-sensing system: Hypoxia-induced release of autacoids. In Experientia 46,pp 1175-1179. Birkhäuser Verlag, Basel

18. Van Nueten JM, Janssen WJ, Vanhoutte PM (1985), Serotonin and vascular reactivity. Pharmacological Research Communications 17(7):585-608

DISTRIBUTION PATTERN OF CAPILLARY AND VENULAR RED BLOOD CELL VELOCITY

FOLLOWING ISCHEMIA-REPERFUSION IN STRIATED MUSCLE[*]

Michael D. Menger[1], Gernot Feifel[2], and Konrad Messmer[1]

[1]Institute for Surgical Research, University of Munich, FRG and
[2]Department of General Surgery, University of Saarland, Homburg/
Saar, FRG

INTRODUCTION

Microvascular red blood cell (RBC) velocity in striated muscle fluc-
tuates with time and varies from capillary to capillary (Ellis et al.,
1990). The heterogeneity of microvascular RBC-velocity is known to be
enhanced in pathologic conditions, such as ischemia-reperfusion (Tyml and
Budreau, 1988; Menger et al., 1988). Previous studies have shown that microvascular ischemia-reperfusion injury is located primarily in nutritive
capillaries and postcapillary venules (Menger et al., 1989; Lehr et al.,
1991; Granger et al., 1989). The aim of the present study was to analyze the
distribution pattern of capillary and venular RBC-velocity following ischemia-reperfusion to striated muscle. Using the hamster dorsal skinfold
preparation intravital fluorescence microscopy provides a direct access to
the striated muscle microcirculation, allowing repeated observations over a
prolonged period of time in the awake animal, including quantitative analysis of different microhemodynamic parameters (Endrich et al., 1980).

MATERIAL AND METHODS

Model

Our studies were carried out in Syrian golden hamsters, equipped with
a dorsal skinfold chamber. In 10 animals (age: 6-8 weeks, 60-80 g body
weight) a dorsal skinfold chamber and a permanent venous catheter were
implanted. The chamber and implantation procedure have been described
previously by Endrich et al. (1980). Briefly: Under Nembutal anesthesia
(50mg/kg body weight; Abbott, North Chicago, USA) the animals were fitted
with two symmetrical titanium frames, positioned on the dorsal skinfold,
sandwiching the extended double layer of skin. One layer was completely
removed in a circular area of approximately 15mm in diameter, and the
remaining layer, containing striated muscle and subcutaneous tissue was
covered with a removable cover slip, incorporated in one of the titanium

---

[*]supported by the Deutsche Forschungsgemeinschaft Me 900/1-1, and Me 900/1-2

frames. A permanent catheter was placed into the jugular vein for intra-venous injection of fluorescent compounds.

Experimental protocol

Allowing a recovery period of 48 hours from anesthesia and surgical trauma, the microcirculation of the striated muscle was analyzed prior to and 15 minutes, 1, 2, 4, and 24 hours after 4 hours of tourniquet-induced ischemia by means of intravital fluorescence microscopy. Microhemodynamic analysis included determination of red blood cell velocity in nutritive capillaries (n=10-15/per animal) and postcapillary venules (n=9-12/per animal) using the dual slit technique (Intaglietta et al., 1975).

Tourniquet-ischemia

Tourniquet-ischemia was induced by an 'O'-shaped silicon stamp, fixed to the skinfold chamber. Under transillumination an external pressure of 40-50mmHg was applied to the stamp by means of an adjustable screw clamp, occluding all vessels feeding and draining the outer part of the observation window without mechanical alteration of the tissue under investigation (Menger et al., 1989).

Intravital fluorescence microscopy

For in vivo microscopic observations, the awake animals were immobi-lized in a plexi-glass tube, and the skinfold preparation was attached to the microscope stage. Repeated scanning of individual microvessels was enabled by use of a computer-controlled microscope desk. After intravenous injection of 5% fluorescein isothiocyanate (FITC)- labeled dextran 150.000 (Sigma Chemical Co., St. Louis, MO, USA) in vivo microscopy was performed using a modified Leitz Orthoplan microscope with a 75W, XBO, xenon lamp, attached to a Ploemo-Pak illuminator with an I_2 blue filter block (Leitz, Wetzlar, FRG) for epi-illumination. The changes in the preparation were recorded by means of a low light level CCD (coupled charge device)- camera (COHU FK 6990, Prospective Measurements Inc., San Diego, CA, USA) and transferred to a video system for off-line evaluation.

Statistics

Although a large number of microvessels were analyzed from each in-dividual animal, statistics were performed on corresponding mean values. Data are given as mean ± standard deviation. Wilcoxon-test accompanied by Bonferroni probabilities was used in order to detect significant differen-ces. Differences were considered significant at p<0.05 level.

RESULTS

Mean capillary RBC-velocity was found 0.24 ± 0.03 mm/s under baseline conditions. Furthermore, the distribution pattern of single values showed a normal histogram (Gauss' distribution) with only minor skewness to the left (Figure 1). Postischemic reperfusion revealed a significant (0.11 ± 0.06 mm/s; p<0.01) decrease of capillary RBC-velocity during the early reper-fusion period without recovery after 24 hours of reperfusion (0.12 ± 0.06

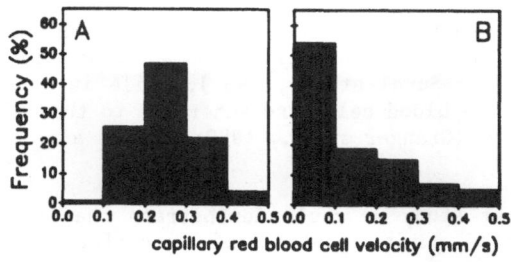

Figure 1. Distribution pattern of capillary RBC-velocity (n=113) prior to (A) and 24 hours after 4 hours of tourniquet-induced ischemia (B) in striated muscle of Syrian golden hamsters.

mm/s). Analysis of distribution pattern within the histogram of single values showed a marked shift to the left accompanied by a pronounced increase of skewness as compared to baseline histogram (Figure 1).

Mean RBC-velocity in postcapillary venules was found 0.46 ± 0.21 mm/s under baseline conditions. The distribution pattern of single values showed a wider range with less pronounced skewness as compared to capillary RBC-velocity (Figure 2). During postischemic reperfusion there was a slight, but statistically not significant decrease of mean venular RBC-velocity to 0.29 ± 0.16 mm/s after 24 hours of reperfusion. However, although this difference was not found significant, distribution pattern of single values demonstrated a marked reduction in velocity as indicated by the left shift of the histogram with distinct skewness to the left (Figure 2).

CONCLUSION

Prolonged ischemia of striated muscle is known to cause microvascular damage during postischemic reperfusion (Korthuis et al., 1985; Messina,

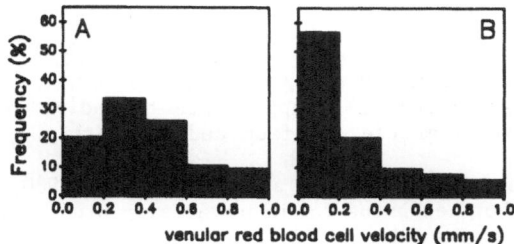

Figure 2. Distribution pattern of venular RBC-velocity (n=104) prior to (A) and 24 hours after 4 hours of tourniquet-induced ischemia (B) in striated muscle of Syrian golden hamsters.

1990; Lehr et al., 1991). Previous studies have demonstrated that "no-reflow" is one of the characteristic phenomena within the capillary bed (Menger et al., 1988; Suval et al., 1987), while in postcapillary venules accumulation of white blood cells and adherence to the endothelial wall was found most prominent (Granger et al., 1989; Lehr et al., 1991; Nolte et al., 1991).

In the present study we could demonstrate that RBC-velocity is significantly decreased during postischemic reperfusion in capillaries of striated muscle, while in postcapillary venules no significant changes were observed. However, analysis of distribution pattern of single values revealed in both capillaries and venules a marked shift of the histograms to the left with pronounced skewness as mean values approached zero, indicating impairment of flow properties and flow conditions of blood as cause for perfusion heterogeneity during postischemic reperfusion.

The alteration of microvascular RBC-velocity during ischemia-reperfusion might be caused by an increase of microvascular hematocrit due to a shift of ions and water from plasma into endothelial cells since the deficit of energy disables for active separation of cations across the cell membrane (Lewis, 1984). In addition, decrease of tissue pH during postischemic reperfusion (Messina, 1990) results in an impairment of red cell deformability. Both of these mechanisms may contribute to the impairment of flow properties and flow conditions of the blood during ischemia-reperfusion in striated muscle.

SUMMARY

Ischemia-reperfusion induced changes in distribution pattern of RBC-velocity of nutritive capillaries and postcapillary venules was analyzed in striated muscle using intravital fluorescence microscopy in awake hamsters. Four hours of ischemia to striated muscle contained in the dorsal skinfold preparation (n=10) revealed a significant (p<0.01) decrease of capillary, but only a slight decrease of mean venular RBC-velocity during the entire reperfusion period. However, analysis of distribution pattern of single values revealed a shift of the histograms to the left, including pronounced skewness as mean values approached zero. This characteristic phenomenon may be due to an increase of microvascular hematocrit and/or decrease in deformability of red blood cells during ischemia-reperfusion. Beside other factors, the impairment of flow properties and flow conditions of the blood may contribute to the development of microvascular reperfusion injury.

REFERENCES

Ellis, C.G., Wrigley, S.M., Potter, R.F., and Groom, A.C., 1990, Temporal distributions of red cell supply rate to individual capillaries of resting skeletal muscle, in frog and rat, Int. J. Microcir.: Clin. Exp., 9:67.
Endrich, B., Asaishi, K., Götz, A., and Messmer, K., 1980, Technical report - A new chamber technique for microvascular studies in unanesthetized hamsters, Res. Exp. Med., 177:125.
Granger, D.N., Benoit, J.N., Suzuki, M., and Grisham, M.B., 1989, Leukocyte adherence to venular endothelium during ischemia-reperfusion, Am. J. Physiol., 257:G683.
Intaglietta, M., Silverman, N.R., and Tompkins, W.R., 1975, Capillary flow velocity measurements in vivo and in situ by television method, Microvasc. Res., 10:165.

Korthuis, R.J., Granger, D.N., Townsley, M.I., and Taylor, A.E., 1985, The
    role of oxygen-derived free radicals in ischemia-induced increases in
    canine skeletal muscle vascular permeability, Circ. Res., 57:599.
Lehr, H.A., Guhlmann, A., Nolte, D., Keppler, D., and Messmer, K., 1991,
    Leukotrienes as mediators in ischemia-reperfusion injury in a micro-
    circulation model in the hamster, J. Clin. Invest., 87:2036.
Lewis, D.H., 1984, The response of the microvasculature in skeletal muscle
    to hemorrhage, trauma; and ischemia, Prog. Appl. Microcirc., 5:127.
Menger, M.D., Sack, F.U., Barker, J.H., Feifel, G., and Messmer, K., 1988,
    Quantitative analysis of microcirculatory disorders after prolonged
    ischemia in skeletal muscle: Therapeutic effects of prophylactic
    isovolemic hemodilution, Res. Exp. Med., 188:151.
Menger, M.D., Hammersen, F., Barker, J., Feifel, G., and Messmer, K., 1989,
    Ischemia and reperfusion in skeletal muscle: Experiments with tourni-
    quet-ischemia in the awake Syrian golden hamster. Prog. Appl. Micro-
    circ., 13:93.
Messina, L.M., 1990, In vivo assessment of acute microvascular injury after
    reperfusion of ischemic tibialis anterior muscle of the hamster, J.
    Surg. Res., 48:615.
Nolte, D., Lehr, H.A., and Messmer, K., 1991, Adenosine inhibits postische
    mic leukocyte-endothelium interaction in postcapillary venules of the
    hamster, Am. J. Physiol., in press.
Suval, W.D., Durán, W.N., Boric, M.P., Hobson, R.W., II., Berendsen, P.B.,
    and Ritter, A.B., 1987, Miocrovascular transport and endothelial cell
    alterations preceding skeletal muscle damage in ischemia and reper-
    fusion injury, Am. J. Surg., 154:211.
Tyml, K., and Budreau, C.H., 1988, Heterogeneity of microvascular response
    to ischemia in skeletal muscle, Int. J. Microcirc.: Clin. Exp., 7:205.

OXYGEN CONSUMPTION OF HUMAN SKELETAL MUSCLE BY NEAR INFRARED SPECTROSCOPY DURING TOURNIQUET-INDUCED ISCHEMIA IN MAXIMAL VOLUNTARY CONTRACTION

R.A. De Blasi[1], M. Cope[2], M.Ferrari[3,4]

[1]Ist. di Anestesiologia e Rianimazione, I Univ. Policlinico Umberto I, Roma, ITALY; [2]Dept. of Medical Physics, Univ. College London, UK; [3]Dip. di Scienze e Tecnologie Biomediche, Univ. dell'Aquila, ITALY; [4]Lab. di Biologia Cellulare, Ist. Superiore di Sanita', Roma, Italy

INTRODUCTION

Among tissues, skeletal muscle shows the highest variability in energy turnover from the resting state through to maximal activity so that muscle is particularly suitable for investigation of metabolic regulation and oxygen consumption (VO_2)(Wittenberg and Wittenberg, 1989). Biochemical assessment of energetic turnover can be obtained "in vitro" by muscle biopsy (Hultman et al., 1990) and "in vivo" by ^{31}P-NMR spectroscopy (Chance et al., 1986). Different experimental animal models have been employed to evaluate the relationship existing between the phosphorylation rate of energetic substrates and VO_2. If VO_2 is not limited by O_2 availability, all models show a correlation between VO_2 and the phosphorylation state of the adenine nucleotides and/or creatinine (Mahler, 1985). Changes in respiratory chain enzyme redox states, independent of phosphorylation state, play a very minor role in regulating VO_2 in red muscle (Connett and Honig, 1989).

A non invasive measurement of muscle oxygenation can be obtained by fiber optic near infrared (700-1100 nm) spectroscopy (NIRS). NIRS has been developed experimentally and clinically to non-invasively monitor brain and muscle hemoglobin/myoglobin (Hb/Mb) oxygenation and cytochrome $a-a_3$ redox state (Tamura et al., 1989). Recently NIRS has been employed to measure the rate of muscular oxygen utilization on patients with mitochondrial myopathy (Sobolewski et al., 1990) and heart failure (Wilson et al., 1989). These studies provide only a qualitative analysis of Hb/Mb oxygen saturation. The evaluation of saturation changes has been conventionally made considering full saturation during 100% O_2 breathing and complete desaturation after 10 min ischemia (Hampson and Piantadosi, 1988). New techniques of time and frequency-resolved spectroscopy allow the measurement of the distribution of optical pathlengths in order to quantify the absorption changes of NIRS in tissues (Chance, 1991).

The aim of the present study was to evaluate the saturation

changes and VO$_2$ of human skeletal muscle at rest and subjected to maximal increase of energetic request. Muscle VO$_2$ can be measured by inducing an abrupt flow limitation and evaluating the Hb/Mb desaturation rate (Cheatle et al., 1990). Combining spectral information obtained by a fast scanning spectrometer with pathlength data measured with time resolved spectroscopy it is possible to quantify oxy and deoxy Hb/Mb concentration changes in the monitored area.

MATERIALS AND METHODS

Six healthy subjects were recruited from the laboratory. Informed consent was obtained from each subject. Spectral measurements were made using a fast scanning spectrophotometer (400-1100 nm) (mod. 6500, NIRSystems, Silver Spring, MD). The procedure for spectral analysis was described recently (Ferrari et al., 1989). Measurements were performed on the proximal forearm brachio-radial muscle. Two optic fibers (200 cm long and 0.5 cm active diameter) were applied 3-3.5 cm apart with a black rubber support so that a stable fiber geometry was achieved. NIRSystems software was utilized to automatically collect a scan every 5 sec. Each subject was submitted to two consecutive experiments.

After a stabilization period of 10 min, an abrupt flow interruption was achieved by a pneumatic cuff loosely wrapped around the arm. Arterial occlusion was obtained by inflating the cuff to a pressure of 240-260 mmHg. In the first protocol (**A**) two isometric maximal voluntary contractions (MVC) of 15 sec duration were executed 15 and 60 sec respectively after the beginning of ischemia. The cuff was released 145 sec after the occlusion started. In the second protocol (**B**) the cuff occlusion was maintained for 7 min in a resting muscle. The cuff was then released and a 3 min recovery phase followed.

Spectra were analyzed according to a modified Lambert-Beer law in order to obtain quantification of Hb/Mb changes during the experimental procedures. Difference spectra (ΔA) of the muscle tissue were calculated relative to the pre-ischemic period. These were converted into muscle absorption coefficient ($\Delta \mu_a$) using $\Delta \mu_a = \Delta A/(Bd)$ where d was the physical separation of the optodes on skin surface and B was 3.59, a factor which took into account the effective optical pathength in muscle tissue (van der Zee et al., 1991). Changes in muscle absorption coefficient were assumed to result only from changes in the concentration of oxy-Hb/Mb and deoxy-Hb/Mb. The results were expressed as micromoles per liter of tissue (μM/L).

No difference in the absorption spectra of Hb and Mb in the near infrared region have been reported "in vitro" (Sassaroli and Rousseau, 1987). In this paper [Hb] and [HbO$_2$] represent the combined concentrations of deoxy-Hb/Mb and oxy-Hb/Mb respectively.

The tissue absorption coefficient spectra ($\Delta \mu_a$) were split into Δ[Hb] and Δ[HbO$_2$] using multilinear regression analysis (Cope et al., 1988) of the Hb and HbO$_2$ spectra (Wray et al., 1988). This statistical analysis led to values for the standard error of the Δ[Hb] and Δ[HbO$_2$] regression coefficients through the variance-covariance matrix and the sum of the squares of the residual errors at each wavelength. The regression analysis was performed over the wavelength region 750 to 900 nm with data points at 2 nm intervals.

VO$_2$ was measured by calculating the rate of change of the conversion of oxy to deoxyhemoglobin in the isolated muscle (i.e. 0.5 d {Δ[HbO$_2$]-Δ[Hb]} /dt) and taking into account the molecular ratio between haemoglobin and oxygen. In protocol **A** VO$_2$ was calculated during MVC, and in protocol **B** for the first 260 sec of ischemia when the desaturation process was linear. It was assumed that in that period changes in saturation were mainly due to haemoglobin. Note that the sum of [Hb] and [HbO2] should represent changes in hemoglobin content as Mb content should not change during these experiments.

RESULTS

Figure 1 (left panel) shows a typical set of results of protocol **A**. After an initial ischemic period at 15 sec the addition of MVC provoked a faster desaturation rate that slowed when the MVC was interrupted. The second MVC, performed after 30 sec, did not give rise to any further desaturation and a plateau was maintained until 95 sec.

Cuff release was accompanied by a rapid recovery of hemoglobin content and a slower recovery of saturation. The

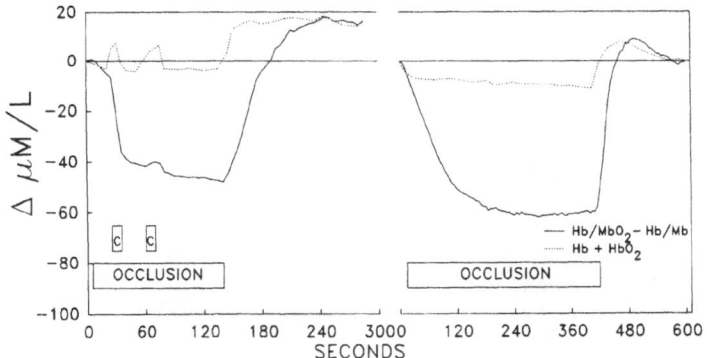

Fig. 1. Typical desaturation pattern during ischemia with (left panel) and without MVC (C)(right panel). The first MVC provoked a faster desaturation rate than that occurring in the occlusion without MVC.

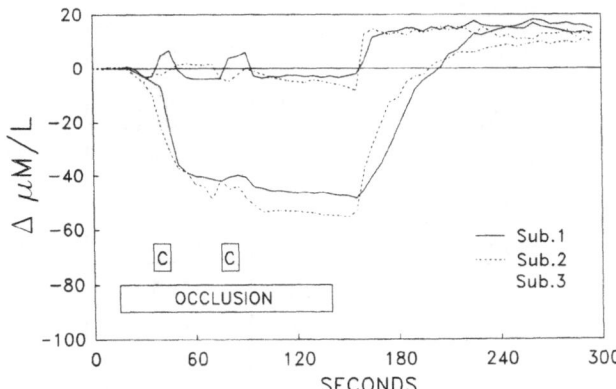

Fig. 2. Effects of MVC during ischemia on 3 different subjects. Upper tracing: (Hb+HbO$_2$); lower tracing: (Hb/MbO$_2$-Hb/Mb).

occlusion without MVC provoked a slower desaturation rate but surprisingly the total desaturation was larger than that observed in the first protocol (Figure 1, right panel).

This effect is clearly shown in Figure 2 which displays protocol **A** performed on the subject of Figure 1 and two other volunteers. The mean ± standard error of VO_2 for the 6 volunteers during procedure **A** and **B** are reported in Figure 3 (upper and lower panel respectively).

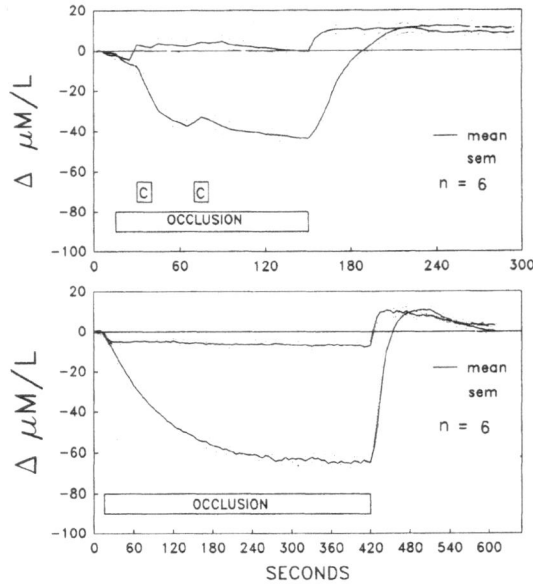

Fig. 3. Means ± sem for 6 volunteers during procedures A and B (see text) are reported on upper and lower panel respectively. Upper tracing: $(Hb+HbO_2)$; lower tracing: $(Hb/MbO_2-Hb/Mb)$.

Table 1. Forearm muscle oxygen consumption during a constant tourniquet compression in resting condition and with MVC (means±sem; n=6).

N_o	Resting condition		MVC	
	VO_2 $\mu M/min/$ $100gT$	HbO_2-Hb $\mu M/L$	VO_2 $\mu M/min/$ $100gT$	HbO_2-Hb $\mu M/L$
1	3.07	61.05	19.94	45.27
2	4.15	60.71	23.55	42.45
3	4.15	60.61	12.36	41.00
4	3.34	68.08	21.11	41.87
5	6.77	37.89	18.14	35.89
6	7.58	66.76	11.91	32.75
Mean ± sem	4.96 0.76	65.04 1.91	17.06 1.88	42.86 2.54

During constant tourniquet pressure in resting state muscle VO_2 was 4.96 ± 0.76 $\mu M/100g/min$ (mean±sem)(Table 1). In the six subjects in which the MVC was performed, oxygen consumption increased by 344 ± 38 % of the resting value during the first MVC. Conversely no change of VO_2 was found during the second MVC. Maximum deoxygenation in the working and resting conditions are reported on the same table.

DISCUSSION

In resting skeletal muscle the heterogeneity in the Hb content and capillary flow involves a non uniform O_2 and substrate distribution that could limit the oxidative metabolism of a large number of cells (Duling and Damon, 1987). Consequently a limitation in muscular oxygen uptake would result despite an organ non-limiting availability of O_2 and substrates.

NIRS reflects the equilibrium between oxygen supply and consumption. In this study VO_2 evaluation was performed during the first 60 sec of the experimental procedures when Mb remained stable i.e. almost completely oxygenated due to its "in vivo" p50 of 5 mmHg (Gayeski and Honig, 1991). In addition Mb concentration is only 25% of total muscle Hb plus Mb concentration (Wang et al., 1990). The source of O_2, either HbO_2 or MbO_2, does not effect the VO_2 calculation, however dissolved oxygen in the cells will introduce a systemic underreading in VO_2 of up to 10%.

Our finding of a mean VO_2 of 4.96 $\mu M/100g/min$ (range 3.07–7.58) in skeletal muscle at rest is lower than that reported in animal studies on leg utilizing flow data and the difference between the arterial and venous oxygen content (Duran and Renkin, 1974). The results obtained by Cheatle et al. (1990) on human leg by NIRS show values ranging from 4.46 to 24.55 $\mu M/100g/min$. The difference could be explained by the different muscle groups studied. The broad variability of data found here was related to differences in VO_2 among subjects. Multiple measurements performed on the same subject showed smaller deviations.

Studies on muscle energy regulation report two different types of respiratory control at rest and at work. At rest, muscle respiration undergoes extrinsic control by O_2 and substrate availability to cells (Chinet and Mejsnar, 1989). The observation of a relatively stable deoxygenation level 4 min after the occlusion can be explained by a rapid activation of anaerobic glycolysis leading to negligible oxygen consumption. In working muscle, phosphate energetics and/or calcium modulated enzymes activities within the mitochondria may account for VO_2 control, while the respiratory chain enzymes play a very small role (Connett and Honig, 1989) and the availability of O_2 (Gayeski et al., 1985) and substrates are not limiting factors (Wolfe et al., 1987). This could explain our finding of a significantly smaller degree of maximal deoxygenation in working muscle with respect to the resting state.

The presence of respiratory suppression with O_2 supply to the cells is also supported by the occurrence that in working skeletal muscle O_2 gradients from sarcolemma to cell interior are very shallow because of myoglobin-facilitated O_2 diffusion (Gayeski and Honig, 1986).

In conclusion:
1) fast scanning spectroscopy and path length data allow the measurement of oxygen consumption of skeletal muscle at rest and working conditions,
2) O_2 extraction is inhibited despite the presence of a significant concentration of O_2 both in working and resting muscle,
3) in ischemic muscle the maximal deoxygenation in the working condition is less than that observed at rest,
4) in working muscle VO_2 is not due to the reduced O_2 availability.

AKNOWLEDGMENTS

This research was supported in part by CNR contribution 90.01475.04 and 91.0025304. This research has been performed in the framework of the "Centro di ricerca interuniversitario: studio dei meccanismi molecolari coinvolti nel danno tissutale da ipossia e iperossia e di molecole che modificano tali lesioni". R.A. De Blasi partecipation at the meeting was supported by Lepetit.

REFERENCES

Chance, B., Leigh, J.S., Kent, J, McCully, K., Nioka, S., Clark,B.J., Maris, J.M., and Graham T. (1986) Multiple controls of oxidative metabolism in living tissues as studied by phosphorus magnetic resonance. Proc. Natl. Acad. Sci. USA 83: 9458-9462.

Chance, B. (1991) In Time Resolved Spectroscopy and Imaging of Tissue, Britton Chance, Editor, Proc. SPIE 1431 .

Cheatle, T.R., Potter, L.A., Cope, M., Delpy, D.T., Coleridge Smith, P.D., and Scurr, J.H. (1990) Near infrared spectroscopy: a new technique for metabolic assessment in peripheral vascular disease. Br. J. Surg. 77: 1416.

Chinet, A., and Mejsnar, J. (1989) Is resting muscle oxygen uptake controlled by oxygen availability to cells? J. Appl. Physiol. 66: 253-260.

Connett, R.J., and Honig, C.R. (1989) Regulation of VO_2 in red muscle: do current biochemical hypotheses fit in vivo data? Am. J. Physiol. 256: R898-R906.

Cope, M., Delpy, D.T., Reynolds, E.O.R., Wray, S., Wyatt, J. and van der Zee, P. (1988) Methods of quantitating cerebral near infrared spectroscopy data. Adv. Exp. Med. Biol. 222: 183-189.

Duling, B.R., and Damon, D.H. (1987) An examination of the measurement of flow heterogeneity in striated muscle. Circ. Res. 60: 1-13.

Duran, W.N.,and Renkin, E.M. (1974) Oxygen consumption and blood flow in resting mammalian skeletal muscle. Am. J. Physiol. 226: 173-177.

Ferrari, M., Wilson, D. A., Hanley, D.F., Hartman, J.F., Traystman, R.J., Rogers, M.C. (1989) Non invasive determination of cerebral venous hemoglobin saturation in the dog by derivative near infrared spectroscopy. Am. J. Physiol. 256: H1493-H1499.

Fisher, M., Guyot, A., Sobolewsky, E., Chance, B., Patti, L., Peterson, P.L. (1990) The evaluation of treatment strategies in mitochondrial myopathy with near infrared reflectance spectroscopy. Neurology 40 (S1): 296.

Gayeski, T.E.J., Connett, R.J., and Honig, C.R. (1985) Minimum intracellular PO_2 for maximum cytochrome turnover in red muscle in situ. Am. J. Physiol. 252: H906-H915.

Gayeski, T.E.J., and Honig C.R. (1986) O_2 gradients from sarcolemma to cell interior in red muscle at maximal VO_2. Am. J. Physiol. 251: H789-H799.

Gayeski, T. E., and Honig, C.R. (1991) Intracellular PO_2 in individual cardiac myocytes in dogs, cats, rabbits, ferrets, and rats. Am. J. Physiol. 260: H522-H531.

Hultman, E., Greenhaff, P.L., Ren, J.M., and Söderlund K. (1990) Energy metabolism and fatigue during intense muscle contraction. In: Biochemistry of exercise. Ed. Sudgen P.L.

Mahler, M. (1985) First-order kinetics of muscle oxygen consumption and an equivalent proportionality between VO_2 and phosphoryl-creatine level. Implications for the control of respiration. J. Gen. Physiol. 86: 135-165.

Sassaroli, M. and Rousseau, D. (1987) Time dependence of near-infrared spectra of photodissociated hemoglobin and myoglobin. Biochemistry 26: 3092-3098.

Sobolewski, E., Guyot, A., Fisher, M., Chance, B., Peterson, P.L. (1990) Near infrared reflectance spectroscopy (NIRS) of mitochondrial myopathy (MM). Neurology 40 (S1): 645.

Tamura, M., Hazeki, O., Nioka, S., and Chance, B. (1989) In vivo study of tissue oxygen metabolism using optical and nuclear magnetic resonace spectroscopies. Annu. Rev. Physiol. 51: 813-834.

van der Zee P, Cope, M., Arridge, S.R., Essenpreis, M., Potter, L.A., Edwards, D., Wyatt, J.S., McCormick, D.C., Roth, S.C., Reynolds, E.O.R., Delpy, D.T. (1991) Experimentally measured pathlength for the adult head, calf, and forearm and the head of the newborn infants as a function of interoptode spacing. Adv. Exp. Med. Biol. (in press).

Wang, D.J., Wang, Z., Noyszewski, E., Nioka, S., Hirao, K., Cheng-Du, T., Chance, B. (1990) Correlation of optical and ^{1}HNMR of Hb and Mb deoxygenation in canine gastrocnemius. Soc. Mag. Reson. Med. Ninth Annual Meeting, New York 1: 175.

Wilson J.R, Mancini, D.M., McCully, K., Ferraro, N., Lanoce, V., Chance, B. (1989) Non invasive detection of skeletal muscle underperfusion with near-infrared spectroscopy in patients with heart failure. Circulation 80: 1668-1674.

Wittenberg, B.A. and Wittenberg, J.B. (1989) Transport of oxygen in muscle. Ann. Rev. Physiol. 51: 857-78.

Wolfe, B.R., Graham, T.E., and Barklay, J.K. (1987) Hyperoxia, mitochondrial redox state, and lactate metabolism of in situ canine muscle. Am. J. Physiol. 253: C263-C268.

Wray, S., Cope, M., Delpy, D.T., Wyatt, J.S., Reynolds, E.O.R. (1988) Characterization of the near infrared absorption spectra of cytochrome aa_3 and haemoglobin for the non invasive monitoring of cerebral oxygenation. Biochim. Biophys. Acta. 933: 184-192.

TISSUE OXYGENATION MEASUREMENT: A DIRECTLY APPLIED CLARK-TYPE ELECTRODE

IN MUSCLE TISSUE

S. O. P. Hofer[*], A. J. Kleij van der[**], K. E. Bos[***]

Departments of Experimental Surgery[*], Hyperbaric Medicine[**]
and Plastic & Reconstructive Surgery[***], Academic Medical
Center, Meibergdreef 9, 1105 AZ, Amsterdam, The Netherlands

INTRODUCTION

Adequate assessment of tissue oxygen tension has been proven a
reliable indicator of tissue blood perfusion[1]. Monitoring of tissue oxygen
tension therefore offers a useful method in the clinical management of
patients[2]. Several devices for measurement of tissue oxygenation are
available, however no single one is universally accepted. A method for
measuring oxygen tension (PO_2) inside tissue is tissue tonometry. This
method uses a tonometer consisting of an implanted silastic tube through
which an anoxic fluid comes in equilibrium with the surrounding tissue. The
tonometer also can be used to insert PO_2 electrodes[1]. It has its limita-

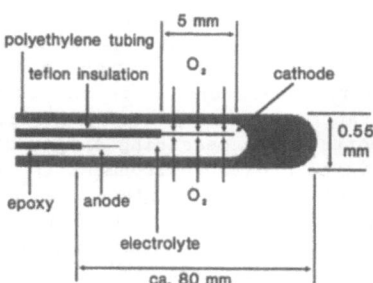

Figure 1. Cross-section of intravascular PO_2 sensor (Continucath).

tions, it is difficult to use routinely, time consuming, sensitive to
movement, and requires frequent recalibration. We evaluated a commercially
available intravascular oxygen sensor which was inserted into microsurgically revascularized muscle transplants. This oxygen sensor overcomes
numerous difficulties faced with tissue tonometry.

Oxygen Transport to Tissue XIV, Edited by W. Erdmann and
D.F. Bruley, Plenum Press, New York, 1992

MATERIALS AND METHODS

A commercially available sterile intravascular oxygen tension mea-surement device (Continucath 1000, Biomedical Sensors, Shiley, High Wycombe, U.K.) was applied to assess tissue oxygenation. This PO_2 sensor is a Clark-type oxygen electrode, mounted in a flexible polyethylene catheter (OD 0.55 mm.) (Fig.1). The surface of the catheter has a covalent heparin bonded coating with resultant low susceptibility to fibrin deposition, a prerequisite for long term stability. The PO_2 sensor is designed for introduction into the arterial system through a 18 or 20 gauge catheter. The sensor is connected to a portable A/C battery charged monitor, which applies a polarization voltage over cathode and anode. The current produced is proportional to the oxygen tension in the tissue, which is displayed in mm Hg or kPa. The monitor has a digital serial RS232 output. The sensor is calibrated in its saline-filled covering sheath so that tissue oxygen tension can be measured in absolute values (mm Hg or kPa). Partial oxygen pressure is calculated using the formula:

$$PO_2 = 20.93 \ [\ (PB - PH_2O)/100 \ ].$$

The fractional concentration of oxygen in the air is 20.93, PB is the barometric pressure and PH_2O is the saturated water vapor pressure, which varies at different temperatures. Because the sensor has a temperature coefficient of 4% per $^{\circ}C$, a temperature compensation function between 20$^{\circ}C$ and 42$^{\circ}C$ is built into the monitor, which allows correction for environmen-tal temperature changes. Two values of polarization potential allow routine PO_2 monitoring and monitoring PO_2 in the presence of N_2O. For this function the "N_2O-button" has to be pressed. In addition, the monitor is equipped with high and low alarm limits, audible as well as visible.

IN VITRO CHARACTERISTICS

The PO_2 sensor characteristics were tested in a custom made calibra-tion unit, which allowed easy calibration at different oxygen partial pressures and optimal dispersion of the calibration gas bubbles. The top of the unit is sealed with a mercury thermometer. Temperature was kept stable at 38$^{\circ}C$ using a heated water pump. We tested 4 Continucath PO_2 sensors, 2 new and 2 used in patients. In the Continucath, the cathode is polarized with -800 mV versus the anode. At this potential the PO_2 sensor has a sensitivity of around 6 nA/mm Hg and an oxygen consumption of 1.55×10^{-14} mol O_2/sec x mm Hg. The residual current is the current in the absence of electro-reducible substances at a given polarization voltage[3]. A low residual current is important for measuring oxygen at low tensions[3]. In all sensors the zero-point in an anoxic environment (N_2 100%) was 0 to 1 mm Hg. Temperature alterations influence the surface area of the electrode and the diffusion coefficient of substances in solution[4]. Temperature sensitivity of the PO_2 sensor was determined by recording the calibration line at different temperatures. This resulted in a temperature coefficient of approximately 4% per $^{\circ}C$. This is in agreement with other authors[5,6]. Linearity of the PO_2 sensors was tested after calibration in 20.93% oxyge-n(=air) by using a calibration cell filled with NaCl 0.9%, which was equilibrated with O_2 0%/N_2 100%. Subsequent step changes of oxygen and nitrogen were performed until the level of O_2 100%/N_2 0% was reached. From 0% to 20% oxygen step changes were 1% and from 20% to 100% oxygen step changes were 10% (Fig. 2). Response time of the PO_2 sensors was measured by using three calibration cells filled with NaCl 0.9% equilibrated with 100% nitrogen, 5% oxygen, and 20.93% oxygen(=air). In this fashion $T_{50\%}$ and $T_{90\%}$ were assessed, $T_{X\%}$ being the time after which X% of the final value is reached after application of a step change of PO_2. The 90% response time to a step change in oxygen tension was 60 to 80 seconds. The 50% response time

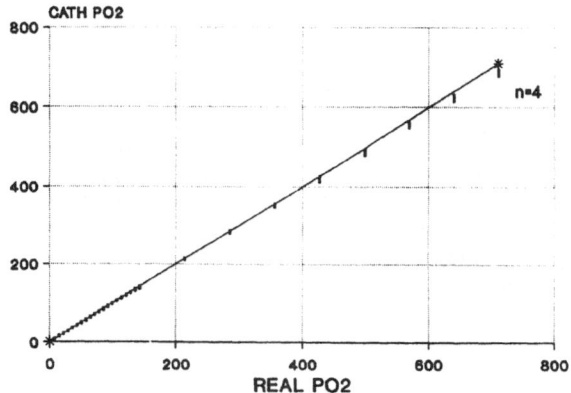

Figure 2. Liniarity test of the Continucath PO$_2$ sensor.

to a step change in oxygen tension was 20 to 30 seconds. New and used PO$_2$ sensors showed no significant differences. The sensors proved to be very stable. A drift of -7% to 0.5% during 96 hours was observed using a calibration cell at room temperature equilibrated with O$_2$ 20.93%.

CLINICAL CHARACTERISTICS

In 14 patients, the PO$_2$ sensor was inserted into the muscular part of microsurgically revascularized myocutaneous transplants (latissimus dorsi muscle). Failure of these transplants has disastrous consequences. Early detection of hypoxic/anoxic conditions caused by arterial or venous obstruction due to e.g. thrombosis, edema, kinking or vascular spasm, is vital for a successful reintervention. The PO$_2$ sensor was inserted into the muscle with an 18 gauge hollow needle. The intramuscular PO$_2$ was monitored continuously for a minimum of 36 hours postoperatively. The PO$_2$ values were compared to clinical parameters, e.g. color of the muscle, hemorrhage and capillary refill. PO$_2$ values in these transplants were between 15 and 35 mm Hg. PO$_2$ values between 10 and 15 mm Hg were considered to be critical for the survival of the transplant. PO$_2$ values below 10 mm Hg had to be reoperated in all cases. Transplants appearing with an impaired circulation c.q. oxygenation were submitted to an oxygen challenge test. With this test FiO$_2$

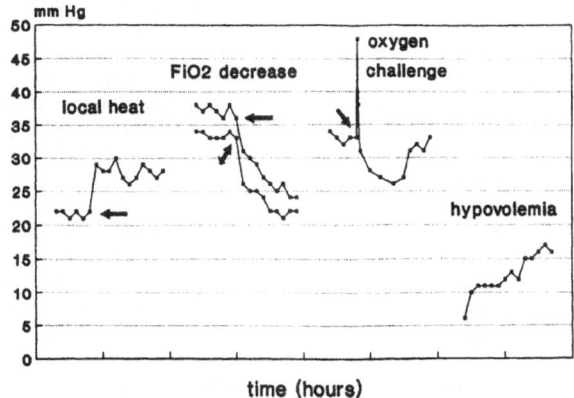

Figure 3. Response of PO$_2$ sensor to alterations in temperature, FiO$_2$, blood pressure.

100% is given to distinguish respiratory from circulatory failure. If tissue PO_2 rises with FiO_2 100% a respiratory problem was most likely to be present. When tissue PO_2 did not rise with FiO_2 100% circulatory failure had to be considered. Stable postoperative readings matched with good clinical parameters. The PO_2 sensor showed responses within 5 minutes to various alterations in muscle temperature, FiO_2 and hemodynamic parameters (Fig. 3). In three cases, the PO_2 sensor detected early circulatory failure in transplants which required reintervention. Two transplants showed arterial compromise (thrombosis and arterial spasm), the third showed venous outflow obstruction. In all cases clinical parameters lagged behind tissue PO_2 measurements.

DISCUSSION

The clinical management of monitoring the circulatory/oxygenation status of postoperative patients is mainly based on indirect parameters, such as blood pressure, cardiac output, urinary output, skin temperature and capillary refill. Tissue perfusion may be inadequate whereas blood pressure and urinary output are within normal limits[2]. It is desirable to monitor tissue perfusion directly. Tissue PO_2 has been shown a reliable parameter of tissue perfusion. It is efficient in assessing perfusion and it detects volume changes quickly[7]. It is attractive to measure tissue oxygen tension in subcutaneous[8] and skeletal muscle[9] tissue because its perfusion is the first to be sacrificed during volume loss. Tissue oxygen tension has also proved valuable in the monitoring of microsurgically revascularized myocutaneous transplants[10,11]. In these transplants quick reintervention in case of circulatory failure increases the chances of tissue survival. Despite the obvious advantages, assessment of tissue oxygen tension has remained difficult in clinical practice. Numerous techniques have been assessed, however none is without obvious disadvantages or advantages when used clinically. Among the methods that have found recognition in clinical settings are indirect measurements such as transcutaneous PO_2 measurements, Silastic tonometry and direct tissue PO_2 measurement using polarographical, potentiometrical or photometrical methods. In transcutaneous PO_2 measurements a heated electrode is attached to the skin. Heating the skin alters local perfusion, skin lipid structure, and the oxyhemoglobin dissociation curve. This change in physiology has to be considered when interpreting the results of transcutaneous PO_2 measurements. However, transcutaneous PO_2 measurements have shown low oxygen values in problem wounds[12] and distal to vascular obstructions[13], probably because the heat does not influence blood flow. Subcutaneous and skeletal muscle PO_2 measurements using polarographic oxygen electrodes have been proven a useful monitor of tissue PO_2. Tissue PO_2 has been shown to increase when local hyperthermia was applied[11,14]. Hypovolemia has been shown to result in a low tissue PO_2[7,11]. In patients with a low tissue PO_2 the incidence of wound infection increased significantly[2]. In patients that had received microsurgically revascularized tissue transfers to the lower extremity, sitting up in bed caused a decrease in tissue PO_2 probably due to impaired venous return[11].

The oxygen-permeable Silastic tonometer containing hypoxic saline protects early poisoning of the PO_2 electrode by tissue protein deposition which may affect calibration, and so the final results. Saline may leak and intermittent flushing of the catheter is necessary. The equipment must be observed carefully while in use, during which time the patient must remain immobile. Although in theory the method seems very simple, its practical application is very time consuming, requires experienced personnel and patience on behalf of the patient. The new single entry sensor (Continucath) also is an invasive technique. In our experience placement is less complicated than that of the Silastic tonometer. Once the sensor is in place no further manipulations are needed. The oxygen measurement part of the Continucath consists the first 5 cm of the catheter. The value measured is an integrated mean value of the surrounding tissue. The sensor is easily calibrated to absolute values. Calibration is performed sterile in the sensors' covering sheath with an auto-calibration procedure where only the external temperature has to be keyed into the monitor. The sensor does not suffer from unacceptable drift-problems like the Silastic tonometer method which has to be recalibrated regularly. The sensor shows a quick response to PO_2 changes. Within the values of interest for tissue PO_2 monitoring the sensor is perfectly linear. It seems this sensor is large enough to be stable yet small enough to keep the error of its own oxygen consumption insignificant low. Polyethylene and Silastic are both non-reactive materials. No problems of clinical infection or mechanical breakdown have been seen with the Continucath, which has been approved for continuous use in the vascular system. The main purpose of testing this sensor was the lack of an easy reliable method to detect early circulatory failure in muscle transplants. The sensor is commercially available (±US$140), which guarantees uniformity and availability of the product. A disadvantage may be that the sensor, for clinical application, is a single use probe which can not be resterilized. Clinically the sensor proved to be a useful method. Arterial as well as venous compromise were readily detected by a decrease in tissue PO_2. This indicates that tissue oxygen tension is a useful tool in the monitoring of transplants, especially in those where no proper clinical assessment is possible due to their positioning.

CONCLUSION

The presented oxygen monitoring sensor offers a promising method for assessment of tissue oxygenation in case of circulatory impairment. The method is comparable to the standard Silastic tonometry, but in our opinion has proven superior in reliability as well as ease of use. It is a stable, direct continuous monitor of tissue oxygen tension which has shown to provide accurate absolute PO_2 values over several days. More clinical use may result in a broader application of this PO_2 sensor in different fields of surgery.

REFERENCES

1. F. Gottrup, R. Firmin, N. Chang, et al., Continuous direct tissue oxygen tension measurement by a new method using an implantable silastic tonometer and oxygen polarography, Am. J. Surg. 146: 399 (1983).
2. K. Jonsson, J. A. Jensen, W. H. Goodson III, et al., Assessment of perfusion in postoperative patients using tissue oxygen measurements, Br. J. Surg. 74: 263 (1987).
3. P. W. Davies, The oxygen cathode, in: "Physical techniques in biological research. Vol IV, Special methods," W. L. Nastuk, ed., Academic Press, New York, London (1962).
4. I. A. Silver, Polarography and its biological applications, Phys. Med. Biol. 12: 285 (1967).
5. W. J. Whalen, J. Riley, P. Nair, A microelectrode for measuring intracellular PO_2, J. Appl. Physiol. 23: 798 (1967).
6. A. J. Kleij van der, in: "Skeletal muscle PO_2 in shock," Thesis, Nijmegen, The Netherlands (1984).
7. F. Gottrup, S. Gellett, L. Kirkegaard, et al., Effect of hemorrhage and resuscitation on subcutaneous, conjunctival, and transcutaneous oxygen tension in relation to hemodynamic variables, Crit. Care. Med. 17: 904 (1989).
8. T. K. Hunt, B.H. Zederfeldt, T. K. Goldstick et al., Tissue oxygen

tensions during controlled hemorrhage, Surg. Forum 18: 3 (1967).

9. A. J. Kleij van der, J. Koning de, J. Beerthuizen, et al., H. P. Kimmich, Early detection of hemorrhagic hypovolemia by muscle oxygen pressure assessment: Preliminary report, Surgery 93: 518 (1983).

10. J. L. Mahoney, F. R. Lista, Variations in flap blood flow and tissue PO_2: a new technique for monitoring flap viability, Ann. Plast. Surg. 20: 43 (1988).

11. S. O. P. Hofer, E. J. F. Timmenga, R. Christiano, et al., An intravascular oxygen tension monitoring device used in myocutaneous transplants (submitted, 1991).

12. P. J. Sheffield, Tissue oxygen measurements, in: "Problem wounds: The role of oxygen," J. C. Davis, T. K. Hunt, eds., Elsevier, New York, (1988).

13. T. R. S. Harward, J. Volny, F. Goldbranson, et al., Oxygen inhalation-induced transcutaneous PO_2 changes as a predictor of amputation level, J. Vasc. Surg. 2: 220 (1985).

14. J. M. Rabkin, T. K. Hunt, Local heat increases blood flow and oxygen tensions in wounds, Arch. Surg. 122: 221 (1987).

OTHER TISSUES

OXYGEN TENSION AND BLOOD FLOW IN THE RETINA OF NORMAL AND DIABETIC RATS

Stephen Cringle, Dao-Yi Yu, Valerie Alder, and Er-Ning Su

Lions Eye Institute and Department of Surgery

University of Western Australia, Nedlands, Western Australia 6009
Australia

INTRODUCTION

In man and most mammals the retina is nourished by two highly specialised vascular systems, the choroidal and the retinal. Despite the high metabolic demands of the retina the extent of the retinal vasculature is constrained by the requirement of minimal disruption of the light path. In contrast, the choroidal circulation, behind the retina, is highly vascularised and has an unusually high flow rate. The avascular layers of retina between these two circulations are dependent on diffusion of metabolites over considerable distances. It is perhaps then not surprising that systemic diseases which affect the microvasculature, such as diabetes, often produce sight threatening complications in the eye. The role of hypoxia and ischaemia in such cases is widely accepted but there have been few direct measurements reported. The rat offers a very convenient model in which to study the early stages of diabetes, and the sequence of histological changes are well documented (Robison *et al.*, 1990). We have developed microelectrode based techniques for the direct measurement of oxygen tension and blood flow in the eyes of normal and diabetic rats (Alder *et al.*, 1990; Cringle *et al.*, 1990; Yu *et al.*, 1991). This paper is essentially a review of our preliminary findings in what we believe will become a valuable preparation for the study of retinal oxygen supply in the presence of systemic vascular disease.

MATERIALS AND METHODS

Our computer controlled system for microelectrode placement and data acquisition was described at last years ISOTT (Cringle *et al.*, 1990). Microsurgical techniques are used to place the microelectrode in the eye and direct visualisation of the electrode tip and retinal vasculature is accomplished by a combination of a plano-concave contact lens and an operating microscope. The high quality stereoscopic view obtained allows judgement of electrode contact with the retinal surface to better than 10 μm. The oxygen tension in the vicinity of retinal arteries, veins and intervascular regions was mapped as a function of distance from the retina. Stepwise penetrations into the retina were also performed and intraretinal oxygen profiles recorded.

We have extended this system to utilise the hydrogen clearance technique to measure localised blood flow using the same type of microelectrodes. Hydrogen is delivered as a bolus of hydrogen saturated saline into the blood flowing into the eye by retrograde injection through the lingual artery. The microelectrode records the arrival of the hydrogen and the subsequent exponential decay of the response is used to calculate tissue blood flow (Kety, 1951). The unusual vascular distribution in the retina and the use of non-equilibrium delivery of hydrogen means that some of the theoretical requirements for clearance measurements could not be met. Initial studies therefore, concentrated on establishing the validity of the technique (Yu *et al.*, 1989, 1991). The flow determination was found to be independent of both the volume and rate of hydrogenated saline injected. The standard injection parameters chosen were 100 μl at a rate of 100 μl/s, which produced a clearly visible bolus and a reliably recorded response without compromising the fluid balance of the animal.

Oxygen Transport to Tissue XIV, Edited by W. Erdmann and
D.F. Bruley, Plenum Press, New York, 1992

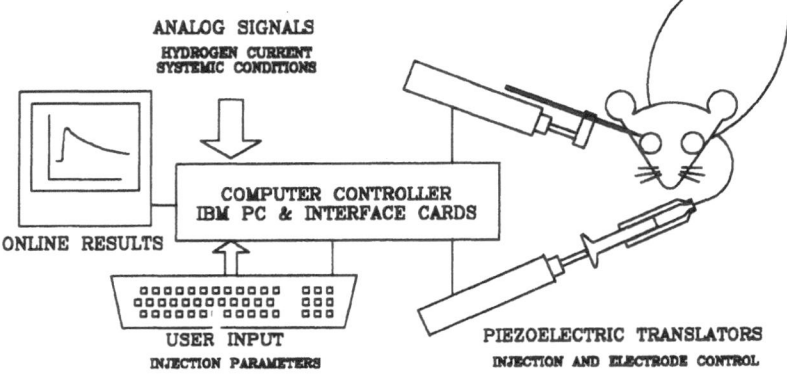

ANALOG SIGNALS
HYDROGEN CURRENT
SYSTEMIC CONDITIONS

COMPUTER CONTROLLER
IBM PC & INTERFACE CARDS

ONLINE RESULTS

USER INPUT
INJECTION PARAMETERS

PIEZOELECTRIC TRANSLATORS
INJECTION AND ELECTRODE CONTROL

Figure 1.   Schematic of the computer control of electrode movements, hydrogen delivery, and data acquisition. The same system was used for both oxygen and hydrogen measurements.

The reproducibility of the clearance measurement with this system is excellent. The bolus injection technique also overcomes the limitations of lung delivered hydrogen in tissues with a high flow rate (Metzger, 1988). The ability to make repeated measurements in the same animal (unlike microsphere techniques) is a significant advantage and offers the potential to study regional and temporal distribution of blood flow in a single experiment. In the present study, hydrogen clearance curves were measured with the electrode on the surface of the retina and also during stepwise penetrations of the retina and choroid. The hydrogen tension distribution through the retina as a function of time after bolus injection was mapped by combining the clearance curves at various retinal depths into a 3D plot. This analysis provides a useful insight into the hydrogen delivery and clearance mechanisms.

These oxygen tension and blood flow measurements were performed in both normal and in streptozotocin (STZ) induced diabetic rats after 5-6 weeks of hyperglycaemia. Details of the animal model and maintenance of systemic conditions have been reported by Alder et al., (1991). Great care was taken to match blood gases and blood pressure in the two groups.

RESULTS

The rat has a retinal and vitreal oxygen tension distribution (Cringle et al., 1990) that is similar to that seen in the cat (Alder et al., 1983; Linsenmeier 1986). The key features are the presence of steep gradients in the vicinity of retinal arteries and the existence of a minimum oxygen tension within the retina. The most striking difference in the STZ animals was the relative absence of oxygen gradients near retinal arteries. Figure 2 shows the mean and standard deviations for the ratio of arterial to vitreal oxygen tensions in the two groups. The vitreal oxygen tension was not significantly different in the two groups.

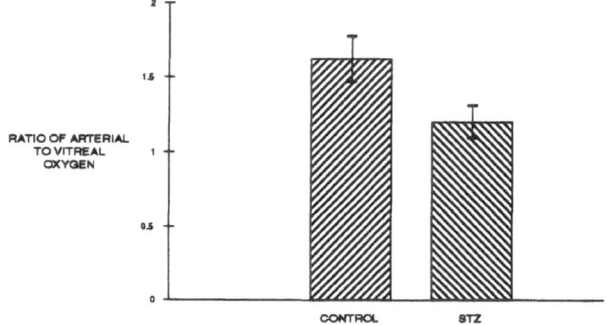

RATIO OF ARTERIAL
TO VITREAL
OXYGEN

CONTROL   STZ

Figure 2.   Mean and standard deviation of the ratio of the oxygen tension adjacent to a retinal artery compared to the central vitreous in control and STZ rats after 5-6 weeks hyperglycaemia.

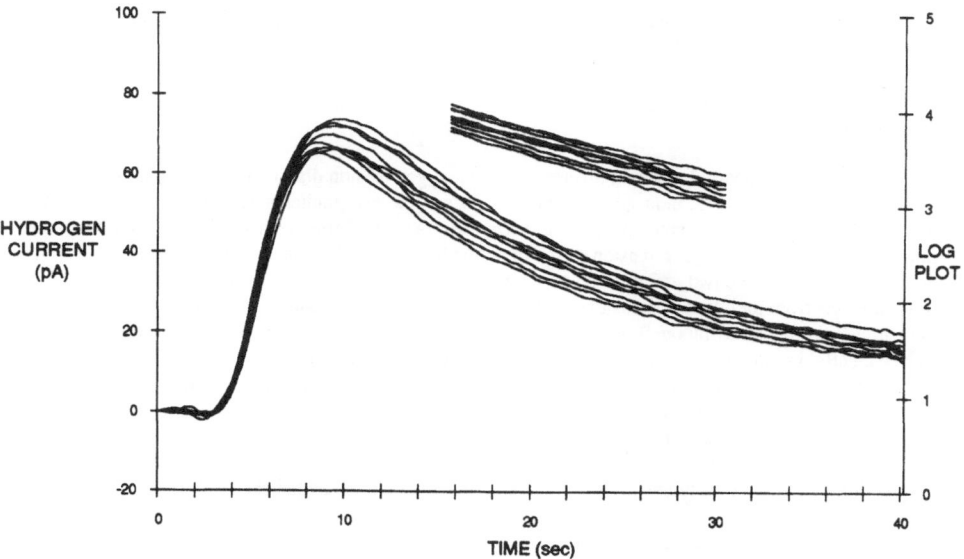

Figure 3.   Demonstration of the reproducibility of the hydrogen clearance responses from a single
retinal location and successive injections of hydrogen saturated saline. The flow rate is
calculated from the slope of the truncated semi-log plots shown above the clearance curves.

Intraretinal oxygen distribution was relatively unaffected and maintained the same characteristic
shape seen in control animals (Cringle *et al.*, 1990).

A set of hydrogen clearance responses from a control animal at a single location are shown in
Figure 3, together with the semi-log plot of the clearance phase, the slope of which is used to calculate
the flow rate (Yu *et al.*, 1991). The monoexponential nature of the clearance curves greatly simplifies the
interpretation of such clearance measurements. More complex clearance curves were obtained adjacent
to retinal arteries, or as the electrode penetrated the retina and became more influenced by the choroidal
circulation.

The combined retinal blood flow data from the control and STZ groups is shown in Figure 4.
The mean flow rate in the STZ animals was 476 ± 167 ml/min/100g (SD) which is significantly higher
($p<0.01$) than in control animals 330 ± 53 ml/min/100g. The regional heterogeneity of blood flow was
far more pronounced in the STZ group, and this was supported by visible fundus changes in which the
fundus appeared more "pinkish" and the visibility of dilated arterioles and venules indicated a
redistribution of blood flow.

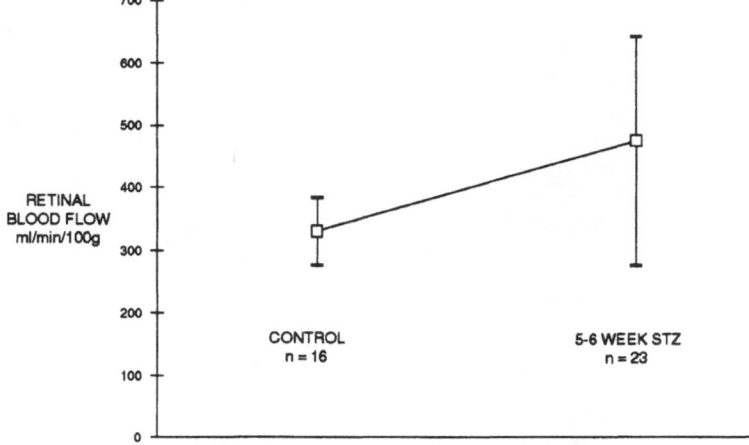

Figure 4.   Comparison of the retinal blood flow measurements in the control and STZ animals. The
mean value $\pm$ SD is shown.

Trypsin digest preparations of whole mounted retinae showed no gross histological changes to the retinal vasculature at this 5-6 week stage of diabetes.

DISCUSSION

The results clearly indicate a disruption to the oxygen tension distribution and retinal blood flow after only 5-6 weeks of STZ induced diabetes. The loss of oxygen gradients near retinal arteries can be explained either by the development of a barrier to oxygen diffusion across the vessel wall, or alternatively by a lower oxygen tension of the blood within the vessel. Since systemic blood gases were carefully matched in the two groups, the second explanation would necessitate a considerable increase in oxygen losses from the blood in transit from the heart to the site of measurement in the retina. The 40% increase in mean retinal blood flow may be a compensatory mechanism and has been demonstrated independently in microsphere measurements in the same animal model (Tilton *et al.*, 1989). The comparable vitreal, and intraretinal oxygen tension measurements in the two groups suggests that there is not a general hypoxia in the retina at this 5-6 week stage in the STZ rats. The complex relationship between oxygen supply and consumption is difficult to unravel in a multi-layered structure like the retina but the form of the intraretinal oxygen profile indicates that the oxygen supply of the inner retina is still derived from the retinal vasculature, whilst the outer retina is supplied by the choroid.

The hydrogen clearance curves at various retinal depths are difficult to interpret unambiguously. We have not yet attempted to extract quantitative flow information from the biphasic clearance curves that are found in the outer retina and choroid, although meaningful data may well be contained in such measurements. Conclusions may be drawn, however, from the nature of the 3D plots of hydrogen tension as a function of time and retinal depth (Fig 5 left). It is clear that the choroid is acting as the dominant source of hydrogen, which rapidly diffuses throughout the retina. During the clearance phase measured at the retinal surface the hydrogen is distributed much more uniformly throughout the retina than one may have expected, and a pseudo equilibrium condition exists (Fig 5 right). A further interesting observation is that the peak hydrogen tension at the retinal surface is attained several seconds after the visible transit of the saline bolus through the retinal circulation. It appears that relatively little hydrogen is contributed by the retinal circulation, indicating that losses in transit to the capillary bed are significant for the highly diffusible hydrogen molecules.

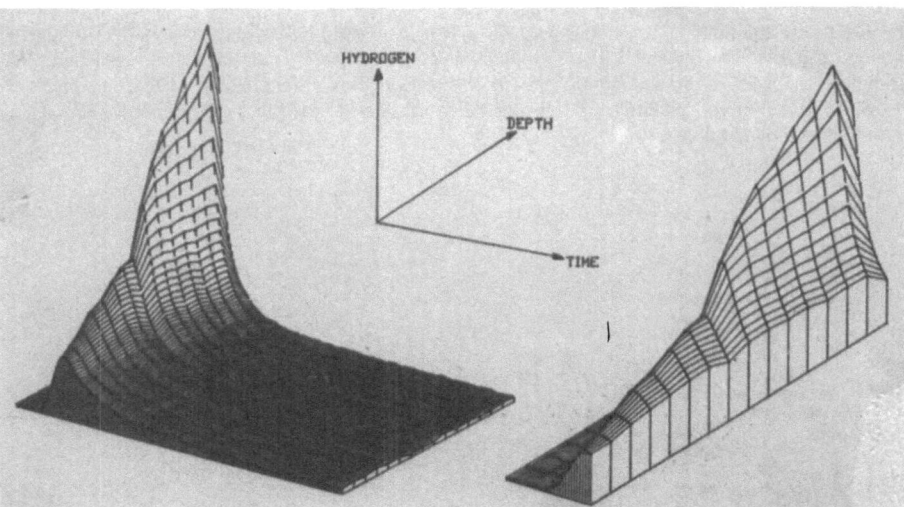

Figure 5.   Hydrogen clearance curves at increasing retinal depth (25 μm steps) are shown combined into a single 3D figure (left). Sampling is performed for a total of 40 seconds and the injection starts at 2 seconds. The same data is shown sectioned at the beginning of the clearance phase from which the flow rate is calculated (right).

Multi-compartmental clearance measurements are frequently encountered in other organs. This preparation in which the electrode position with respect to vascular elements is ascertained by direct visualisation and in which the hydrogen delivery is precisely controlled may also prove useful in studying the applicability of hydrogen clearance techniques in non-uniformly perfused tissue.

SUMMARY

The use of microelectrode techniques for studying oxygen distribution and blood flow in the eye of a physiologically well maintained rat provides a very convenient model in which to study oxygen supply to the retina. The availability of rat models of vascular disease such as diabetes and hypertension, and the existence of several models of retinal degeneration, make studies of oxygen supply in the rat eye of particular relevance. The experiments reported in this paper demonstrate changes in oxygen distribution and blood flow very early in STZ induced diabetes. Thus, we have established a preparation in which the role of changes in oxygen supply can be correlated with the pathological events that are apparent later in the disease.

ACKNOWLEDGEMENTS

The technical assistance of Mr Michael Brown, Wilf Brewster, Luis Aravena, and Paul van Saarloos is gratefully acknowledged.

This work was supported by grants from the Juvenile Diabetes Foundation International, the National Health and Medical Research Council of Australia, and the Juvenile Diabetes Foundation of Australia.

REFERENCES

Alder, V. A., Yu, D-Y., Cringle, S. J., and Su, E-N., 1990, Changes in vitreal oxygen tension distribution in the streptozotocin diabetic rat, Diabetologia, 34:469.

Alder, V. A., Cringle, S. J., and Constable, I. J., 1983, The retinal oxygen profile in cats, Inv. Ophthalmol. Vis. Sci., 24:30.

Cringle, S. J., Yu, D-Y, and Alder, V. A ., 1990, Intraretinal and intravitreal oxygen distribution in the rat. Adv. Exp. Med. Biol., Oxygen Transport to Tissue XIII.

Kety, S. S., 1951, The theory and applications of the exchange of inert gas at the lungs and tissues, Pharmacol. Rev., 3: 1.

Linsenmeier, R. A., 1986, Effects of light and darkness on oxygen distribution and consumption in the cat retina, J. Gen. Physiol., 88:521.

Metzger, H. P., 1988, The hydrogen gas clearance method for liver blood flow examinations: Inhalation or local application of hydrogen ?, Adv. Exp. Med. Biol., 248:141.

Robison, W. G., Tillis, T. N., Laver, N., and Kinoshita, J. H., 1990, Diabetes related histopathologies of rat retina prevented with an aldose reductase inhibitor, Exp Eye Res., 50: 355.

Tilton, R. G., Chang, K., Pugliese, G., Eades, D. M., Province, M. A., Sherman, W. R., Kilo, C., and Williamson, J. R., 1989, Prevention of hemodynamic and vascular albumin filtration changes in diabetic rats by aldose reductase inhibitors, Diabetes, 38: 1258.

Yu, D. Y., Alder, V. A., and Cringle, S. J., 1989, The validity of hydrogen clearance measurements of retinal blood flow, Exp. Eye Res., 50:533.

Yu, D. Y., Alder, V. A., and Cringle, S. J., 1991, Measurement of blood flow in the rat eye by hydrogen clearance, Am. J. Physiol. : Heart Circ. Physiol., Vol. 261 (30).

ARTERIOLAR SPASM AND ISCHEMIA IN THE OCULAR FUNDUS OF NaCl-

LOADED SALT SENSITIVE DAHL RATS: VASCULAR PROTECTION BY LONG-

TERM TREATMENT WITH THE CALCIUM ANTAGONIST NITRENDIPINE

F. Thimm, M. Frey, K. Spitzmüller, W. Hofgärtner, G. Fleckenstein-Grün

Study group for Calcium Antagonism, Physiological Institute, University of Freiburg, Federal Republic of Germany

INTRODUCTION

In vascular smooth muscle cells of arteries and arterioles a transmembrane supply of calcium ions is necessary for active tension development (Grün and Fleckenstein, 1972). Thus an excessive influx of calcium ions into the smooth muscle cells is responsible for phasic and tonic hyperactivity of the arterial and arteriolar vasculature, culminating in vasospasm that impairs tissue oxygen supply. Furthermore, calcium-overloaded arteries exhibit sclerosis of their walls with periarteritis-nodosa-like changes in their structure. Good models for the study of this pathogenic situation are spontaneously hypertensive Okamoto rats (SHRs) and NaCl-fed hypertensive Dahl-S rats. Here, the most spectacular alterations occur in the retinal arterioles which undergo both spasms and morphologic changes. A noninvasive ophthalmoscopic method for the examination of the arteriolar diameter and shape in the ocular fundus was introduced by Takahashi in 1972. It was the aim of this study to demonstrate the retinal arteriolar spasms in hypertensive rats (particularly NaCl-loaded salt-sensitive Dahl rats) leading to underperfusion and hypoxia of the retina, and to show the possibilities of vascular protection with the use of the calcium antagonist nitrendipine.

Experimental protocol

Animal experiments were performed (a) in NaCl-loaded salt resistant Dahl rats (Dahl-R), (b) in salt-sensitive Dahl rats (Dahl-S) receiving 8% NaCl in the diet, beginning at the age of six weeks, and (c) in salt-loaded Dahl-S rats additionally treated with nitrendipine (50-300 mg/kg body weight p.o.). Salt resistant Dahl-R rats served as controls. In all animal groups measurement of systolic blood pressure was performed by tail cuff method.

Ophthalmoscopic examination

In order to document arteriolar retinal width in the ocular fundus of rats we used a KOWA RC 2 camera (Japan) without anesthesia specially developed for small animals. For dilation of the pupilles Mydriaticum Roche was given.
Relative arteriolar width of retinal vessels was determined as ratio of arteriolar diameter over corresponding venous diameter (venous diameter = 1, unchanged during hypertension until death).

RESULTS

Development of hypertension - Antihypertensive effects of the calcium antagonist nitrendipine

The systolic blood pressure of salt-sensitive Dahl-S rats, when they received a diet containing 8% NaCl, rose, within two months, from 130 to 230 mmHg (Fig. 1). However, with oral doses of nitrendipine the effect of the NaCl-rich diet on blood pressure elevation was totally neutralized. Moreover, under the same NaCl-rich regimen, the blood pressure level of nitrendipine treated Dahl-S rats fell even below that of Dahl-R rats which served as controls.

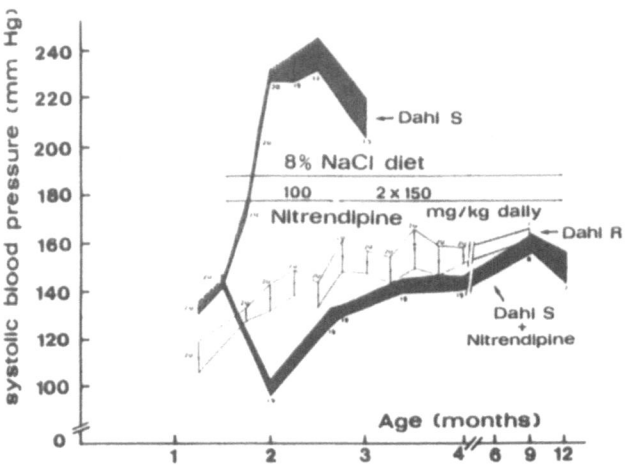

Fig. 1.
Normalization of the systolic arterial blood pressure of NaCl-loaded Dahl rats by chronic oral application of nitrendipine.

Systemic arteriolar spasms, microaneurysms and retinal hypoxia in Dahl-S rats assessed by ophthalmoscopic examination of the ocular fundus - Prevention by nitrendipine

The generalized arteriolar spasm that underlies the development of hypertension manifests itself most dramatically in the arterioles of the ocular fundus of salt-loaded Dahl-S rats. In non-hypertensive salt-resistant Dahl-R rats, the veins are typically about twice as large as the arterioles (Fig 2 a). This ratio changes in the NaCl-loaded Dahl-S rats when hypertension develops (Fig. 2 b). With severe hypertension, parts of the arterioles may even escape ophthalmoscopic examination because they become invisible due to severe spasm and ischemia. Most obvious was the appearance of multiple arteriolar aneurysms. Long-term treatment with nitrendipine was highly efficient in inhibiting arteriolar spasms and aneurysm-like arteriolar irregularities. Consequently, the insufficient perfusion of retinal tissue disappeared. The usually pale ocular fundus of untreated Dahl-S rats did not occur in nitrendipine-treated animals. Table 1 shows the retinal arteriolar constriction in untreated Dahl-S rats and the vasodilating effect in Dahl-S chronically treated with nitrendipine.

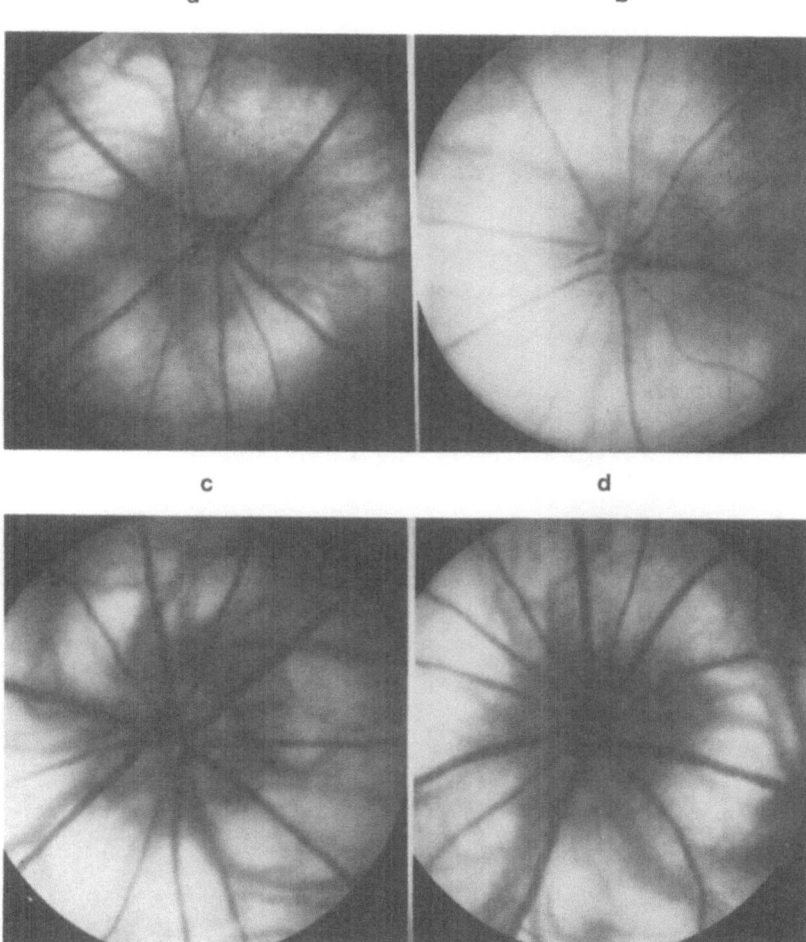

Fig. 2.
Ocular fundus of Dahl rats. (a) Normal ocular fundus of a nonhypertensive salt resistant Dahl-R rat after 2.5 months salt-rich diet. Note, that the veins are twice as large as the arterioles. (b) Development of arteriolar spasms, of small aneurysm-like arteriolar protuberances and a pale, ischemic ocular fundus of a Dahl-S rat two months after feeding a NaCl-rich diet. (c) Neutralization of arteriolar spasm and structural protection in a Dahl-S rat by 2.5 months administration of nitrendipine additionally to the salt-rich diet. (d) Preservation of arteriolar integrity in a Dahl-S rat during a chronic treatment with nitrendipine in addition to the salt-rich regimen until an age of 12 months.

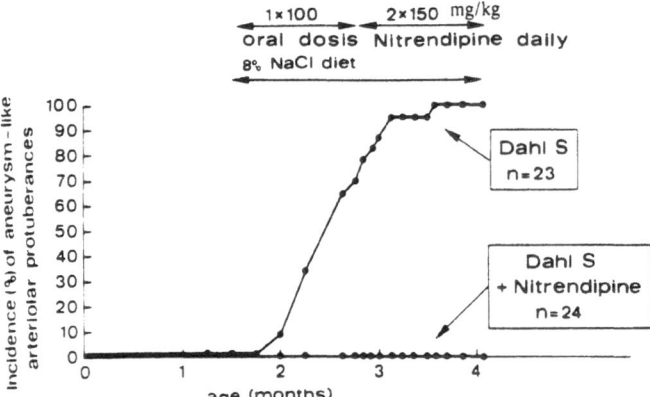

Fig. 3. Prevention of aneursym-like arteriolar protuberances in the ocular fundus of salt-loaded Dahl-S rats by long-term treatment with nitrendipine over 2.5 months. Without nitrendipine, all retinal arterioles (100%) exhibit structural damage, whereas nitrendipine therapy provided full protection.

Table 1.   Mean relative width of retinal arterioles $\pm$ SEM

Dahl-R	Dahl-S	Dahl-S + Nitrendipine
0.563	0.480	0.588
$\pm$0.017	$\pm$0.013	$\pm$0.012
n=20	n=20	n=20

Prolongation of survival rate of NaCl-fed Dahl-S rats by long-term treatment with the calcium antagonist nitrendipine

The best indicator for the validity of an antihypertensive drug is <u>not</u> the absolute fall in blood pressure, but the improvement of life quality and the prolongation of life expectancy. Figure 4 shows the fourfold increase in survival rate of NaCl-loaded salt-sensitive Dahl rats by chronic administration of nitrendipine. Without nitrendipine treatment only 20% of Dahl-S rats survived the NaCl-rich diet for 2.5 months. Under the influence of nitrendipine, however, all Dahl-S rats were still alive after 8.5 months; 37.5% even tolerated the salt-rich regimen for 10.5 months. Accordingly, long-term treatment with nitrendipine over 10.5 months was also highly effective in inhibiting the formation of microaneurysms. Thus the preservation of normal arteriolar width and retinal blood supply remained normal (Fig. 2 d).

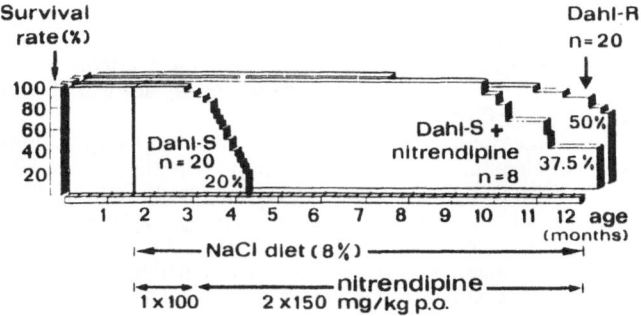

Fig. 4.
Prolongation of life expectancy of NaCl loaded salt-sensitive Dahl-S rats by long-term treatment with nitrendipine.

DISCUSSION

The oxygen supply to retinal tissue is related to the caliber of arteries and arterioles. In Dahl-S rats a suitable model exists in which the alteration of retinal arterioles influences the blood perfusion via abnormal transmembrane calcium influx into arteriolar smooth muscle cells. The alteration of retinal vessels manifests itself in a functional aspect e.g. vasopasm, and in a structural aspect e.g. aneurysm-like arteriolar protuberances. This transition from functional to structural abnormalities proceeded rapidly in NaCl-fed salt-sensitive Dahl rats, in which aneurysm-like arteriolar irregularities began to develop almost simultaneously with the increase in systolic blood pressure.
The enhanced transmembrane calcium influx into smooth muscle cells of Dahl-S rats evokes arteriolar spasms which lead consequently to arterial hypertension. 'NaCl-loaded

Dahl-S rats are particularly well suited for such studies, because in contrast to hypertensive Okamoto rats (SHRs), they develop calcium-overload, hypertension, retinal spasms, and microaneurysms with a minimal delay (2 months versus 16-20 months). But in principle, the pathogenic roots of calcium-overload-induced hypertension and vascular damage seem to be, in both types of rats, i.e. Dahl-S rats and Okamoto rats, practically identical. Therefore, it is also not surprising that the antihypertensive and vasoprotective influence of nitrendipine is not restricted to this particular calcium antagonist. In reality, all calcium antagonists, particularly the derivatives of the prototypical agent nifedipine, possess the same anticalcinotic membrane action on vascular smooth muscle cells with all beneficial consequences.

REFERENCES

Fleckenstein, A., 1983, "Calcium antagonism in heart and smooth muscle - experimental facts and therapeutic prospects". John Wiley and Sons, New York.

Fleckenstein, A., Fleckenstein-Grün, G., Frey, M., Zorn, J., 1987, Future directions in the use of calcium antagonists. Am. J. Cardiol. 59: 177B-187B.

Fleckenstein, A., Frey, M., Zorn, J., Fleckenstein-Grün, G., 1990, Calcium - a neglected key factor in hypertension and arteriosclerosis, in: "Hypertension: Pathophysiology, Diagnosis and Management", J.H. Laragh and B. M. Brenner, eds., Raven Press Ltd., New York, 471-509.

Garthoff, B., Kazda, S., 1981, Calcium antagonist nifedipine normalizes high blood pressure and prevents mortality in salt-loaded DS substrain of Dahl rats. Eur. J. Pharmacol. 74: 111-112.

Grün, G., Fleckenstein, A., 1972, Die elektro-mechanische Entkoppelung der glatten Gefäßmuskulatur als Grundprinzip der Coronardilatation durch 4-(2'-Nitrophenyl)-2,6-dimethyl-1,4-dihydropyridin-3,5-dicarbonsäure-dimethylester (Bay a 1040, Nifedipin). Arzneimittelforschung 22: 334-344.

Luckhaus, G., Garthoff. B., Kazda, S., 1986, Prävention und Therapie der kochsalzinduzierten hypertensiven Arteriopathie bei Ratten durch Nitrendipin, in: "Calcium-Antagonisten in der Hochdrucktherapie", A. Distler, ed., Schattauer, Stuttgart-New York.

Rapp, J. P., 1982, Dahl salt-susceptible and salt-resistant rats. A review. Hypertension, 4:753-763.

INFLUENCE OF CRYOPRESERVATION ON VIABILITY AND

NUTRITIONAL MICROCIRCULATION OF ISLETS OF LANGERHANS[*]

Michael D. Menger[1], Johanna Pattenier[2], Beate Wolf[2], Sabine Jäger[2], and Gernot Feifel[2]

[1]Institute for Surgical Research, University of Munich, FRG and [2]Department of General Surgery, University of Saarland, Homburg/Saar, FRG

INTRODUCTION

Cryopreservation of isolated islets of Langerhans would be a necessary procedure for realizing clinical pancreatic islet grafting for treatment of diabetes mellitus (Rajotte et al., 1981; Ricordi et al., 1988). However, little is known about the viability of isolated islets after the cryopreservation procedure, particularly in regard to their potential for revascularization after transplantation.

Therefore, the aim of the study was to analyze viability of isolated islets of Langerhans following different time periods of cryopreservation, and the nutritive microcirculation developed after free transplantation. Using the model of the dorsal skinfold chamber of the Syrian golden hamster, the microcirculation can be observed by means of intravital fluorescence microscopy (Menger et al., 1989; Menger et al., 1990).

MATERIAL AND METHODS

Pancreatic islets were isolated from 12 weeks old Syrian golden hamsters (120g body weight) using a modified collagenase digestion technique (Serva, Heidelberg, FRG) according to the method described previously by Lacy and Kostianovsky (1967). After desintegration of the pancreatic tissue the islets were hand-picked in order to guarantee almost exocrine free islets for cryopreservation (Jäger et al., 1990).

Subsequently after isolation, the islets were placed for 24 hours in tissue culture at 37°C and 5% CO_2 in air. Following culture procedure, islets were frozen in portions of 100 islets each using the technique originally described by Rajotte et al. (1981). Therefore the islets were incubated in 0.2 ml RPMI 1640 (Gibco, Eggenstein, FRG) and 0.1 ml 2M dimethyl sulfoxide (Me_2SO). After 5min incubation at 25°C 0.1ml 2M Me_2SO was added. The islets were incubated for further 25min at 25°C, and after addition of 0.4ml 3M Me_2SO incubation was continued further for 15 minutes at 0°C. Then, the

[*]supported in part by the Deutsche Forschungsgemeinschaft Me 900/1-1

freezing procedure was performed automatically by computer-assistance with 0.25°C/min to a temperature of -40°C. Finally, the islets were stored in liquid nitrogen for 1 and 10 weeks, respectively. After these time periods, islets were thawed at 250°C/min using a 37°C water bath, and placed into 37°C culture for another 24 hours.

Viability of the islets was assessed from 1 (group 1) and 10 weeks (group 2) cryopreserved islets, as well as from freshly prepared cells (group 3, controls). For each test 600 islets were analyzed from each group. Trypane blue test (Wiegand, 1966; Serva, Heidelberg, FRG)) was used in order to select non-viable cells; in addition, viable cells were analyzed by staining with neutral red (Sigma Chemical Co., St. Louis, MO, USA). Neutral red is known to stain secretory granules of vital endocrine cells dark red (Bensley, 1911). Finally, dithizone (Sigma Chemical Co., St. Louis, MO, USA) was used, staining vital islets of Langerhans pink-red, due to Zn-binding (Latif et al., 1988).

For analysis of the microcirculation, cryopreserved as well as freshly isolated islets (diameter: ~150μm, n = 8-10) were implanted into skinfold chambers of syngeneic Syrian golden hamsters (6-8 weeks old, body weight: 60-80g), which allow for repeated observations of the microcirculation by means of intravital fluorescence microscopy. The technique used has been described previously in detail (Menger et al., 1989; Menger et al., 1990; Menger et al., 1991a). The analysis included the determination of the total size of the microvascular network of 1 and 10 weeks cryopreserved, as well as non-preserved islets on days 6, 10 and 14 after implantation. Contrast enhancement for the in vivo fluorescence microscopy procedure was achieved by intravenous injection of 0.1ml 5% Fluorescein-Isothyocyanate (FITC)-Dextran (Sigma Chemical Co., St. Louis, Mo, USA). Microscopic images were recorded using a low light level CCD (coupled charge device)- camera (FK 6990, Cohu Prospective Measurements, San Diego, CA, USA), stored on video-tape, and evaluated off-line using a computer assisted image analysis system (Zeintl et al., 1986).

Values are given as mean and standard deviation. Data were tested for normal distribution (BMDP Statistical Software Inc.; Los Angeles, CA, USA) and either an analysis of variance and unpaired Student's t-test (normal distribution) or Kruskal-Wallis analysis and Mann-Whitney-U-test (non-normal distribution) were performed in order to test significant differences between the groups. For comparison within the groups paired Student's t-test including Bonferroni probabilities (normal distribution) or Wilcoxon signed rank test (non-normal distribution) was used. Differences were considered significant at a $p < 0.05$ level.

RESULTS

After the isolation procedure pancreatic hamster islets show translu-cent consistence, sometimes contaminated with exocrine tissue particles. Neither 24 hours 37°C tissue culture, nor 1 or 10 weeks cryopreservation influenced this characteristic. However, after cryopreservation exocrine tissue particles could not be detected. Islet count was 94 ± 3 % after 1 week and 93 ± 4 % after 10 weeks cryopreservation.

Trypane blue staining revealed 2% non-viable islets (analyzed from n=600) after 1 week cryopreservation and 1% non-viable islets (from n=600) after 10 weeks cryopreservation. Secretory granules of both, 1 and 10 weeks preserved islets were stained positive by neutral red. Dithizone staining showed 94 ± 4 % (from n=600) viable pancreatic islets after 1 week cryopreservation and 96 ± 3 % (from n=600) after 10 weeks cryopreservation. Differences between the groups were not found significant.

In all three groups analysis of the microvascular network of the islets revealed a significant increase (p<0.01) in total diameter of the microvascular network from day 6 to day 10 after implantation into dorsal skinfold chambers (table 1). Revascularization was completed after 10 days, and no further changes within the microvasculature were observed on day 14 after implantation. Comparison between 1 and 10 weeks cryopreserved islets and controls did not reveal statistically significant differences (table 1).

Table 1. Total diameter of the microvascular network (µm) of pancreatic islets implanted into the dorsal skinfold chamber of Syrian golden hamsters during revascularization (day 6, 10), and after completion of revascularization (day 14). Prior to transplantation, islets have been cryopreserved for 1 week and 10 weeks, respectively, or have been transplanted directly after the isolation procedure (controls). Mean ± SD, n = six animals per group (8 to 10 islets each animal), paired Student's t-test, Bonferroni-correction, **p<0.01 vs. day 10 and 14 (comparison within each group); no significant differences between the groups.

| | time after transplantation | | |
group	day 6	day 10	day 14
1-wk cryopreservation	371.6 ± 29.9**	420.8 ± 55.2	428.0 ± 19.4
10-wks cryopreservation	358.3 ± 13.4**	406.5 ± 25.2	408.7 ± 8.0
controls	358.5 ± 28.7**	411.7 ± 53.9	411.8 ± 40.0

CONCLUSION

Cryopreservation may be a necessary step in the managment of pancreatic islet transplantation (Kneteman et al., 1989). Viability of the islets and a regular potential for revascularization after transplantation are essential prerequisites for the development of a microcirculation sufficient for the nutritional blood supply.

In the present study we could demonstrate that 1 and 10 weeks cryopreservation did not alter their viability, in contrast, the preservation procedure had the potential to increase purification of the endocrine cells. In addition, the islets reveal a process of revascularization similar as compared to non-preserved islets. This results are in agreement with histological findings from Kneteman and coworkers (1989), demonstrating viable revascularized islets after cluster transplantation beneath the renal subcapsular space, presenting with normal morphology and vasculature.

In addition, recent studies from our laboratory could demonstrate that cryopreserved pancreatic islets do not only present with normal microvascular morphology, but also with microhemodynamics, i.e. functional capillary density, capillary red blood cell velocity, capillary diameters and capillary blood flow, similar as compared to non-preserved islets (Menger et al., 1991b).

SUMMARY

Isolated pancreatic islets of Syrian golden hamsters were cryopreserved for 1 and 10 weeks, respectively. Following thawing islet mass was found more than 90%. Analysis of viability of these islets by dithizone staining revealed 94 $\pm$ 4 % (1 week cryopreservation) and 96 $\pm$ 3 % (10 weeks cryopreservation) positively stained islets. After implantation into dorsal skinfold chambers of syngeneic animals, intravital fluorescence microscopy (5% FITC-dextran, i.v.) showed a regular microvascular network 10 days after implantation. The total diameter of the microvascular network was found similar in cryopreserved islets (day 10: 420.8 $\pm$ 55.2 µm (1 week preservation) and 406.5 $\pm$ 25.2 (10 weeks preservation)) as compared to non-preserved controls (day 10: 411.7 $\pm$ 53.9 µm). Therefore we conclude that cryopreservation of hamster islet isografts for 1 and 10 weeks, respectively, does not alter viability as well as the potential for revascularization. Cryopreservation seems to be an adequate technique for long time storage prior to free transplantation.

REFERENCES

Bensley, R.R., 1911, Studies on the pancreas of the guinea pig, Am. J. Anat., 12:298.
Jäger, S., Menger, M.D., Göhde, W., and Feifel, G., 1990, A specific fluorescent dye for ex situ staining of vital islets of Langerhans: Neutral red, Eur. Surg. Res., 22:8.
Kneteman, N.M., Alderson, D., Scharp, D.W., and Lacy, P.E., 1989, Long-term cryogenic storage of purified adult human islets of Langerhans, Diabetes, 38:386.
Lacy, P.E., and Kostianovsky, M., 1967, Method for the isolation of intact islets of Langerhans from the rat pancreas, Diabetes, 16:35.
Latif, Z.A., Noel, J., and Alejandro, R., 1988, A simple method of staining fresh and cultured islets, Transplantation, 45:827.
Menger, M.D., Jäger, S., Walter, P., Feifel, G., Hammersen, F., and Messmer, K., 1989, Angiogenesis and hemodynamics of microvasculature of transplanted islets of Langerhans, Diabetes, 38/I:199.
Menger, M.D., Jäger, S., Walter, P., Hammersen, F., and Messmer, K., 1990, A novel technique for studies on the microvasculature of transplanted islets of Langerhans in vivo, Int. J. Microcirc.: Clin. Exp., 9:103.
Menger, M.D., Wolf, B., Höbel, R., Schorlemmer, H.-U., and Messmer, K., 1991a, Microvascular phenomena during pancreatic islet graft rejection, Langenbecks Arch., 376:214.
Menger, M.D., Pattenier, J., Wolf, B., Jäger, S., Feifel, G., and Messmer, K., 1991b, Cryopreservation of islets of Langerhans does not affect angiogenesis and revascularization after free transplantation, Eur. Surg. Res., submitted.
Rajotte, R.V., Sharp, D.W., Downing, R., Preston, R., Molnar, G.D., Ballinger, W.F., and Greider, M.N., 1981, Pancreatic islet banking: the transplantation of frozen-thawed rat islets transported between centers, Cryobiology, 18:357.
Ricordi, C., Kneteman, N.M., Scharp, D.W., and Lacy, P.E., 1988, Transplantation of cryopreserved human pancreatic islets into diabetic nude mice, World J. Surg., 12:861.

Wiegand, D., 1966, Vergleichende Untersuchungen über Färbemethoden, die eine Unterscheidung zwischen lebenden und toten Säugetierzellen ermögli- chen, <u>Arch. Exp. Veterinärmed.</u>, 21:693.

Zeintl, H., Tompkins, W.R., Messmer, K., and Intaglietta, M., 1986, Static and dynamic microcirculatory video image analysis applied to clinical investigations, <u>Prog. Appl. Microcirc.</u>, 11:1.

O$_2$ SUPPLY: CLINICAL PROBLEMS

ROLE OF ARACHIDONIC ACID METABOLITES IN PULMONARY OXYGEN TOXICITY

J. Klein, A. Trouwborst, W. Erdmann

Department of Anesthesiology
Erasmus University, Rotterdam
The Netherlands

Arachidonic acid metabolites have biologic properties that can mimic the pulmonary changes produced by hyperoxic exposure. They have potent vasoactive, bronchoactive, and chemoattractant properties, and can increase vascular permeability; all of these are features of hyperoxic lung injury.

Mounting evidence suggests that reactive oxygen metabolites can initiate the release and metabolism of arachidonic acid (1). Increases in levels of cyclooxygenase as well as lipoxygenase pathway products in plasma and bronchoalveolar lavage (BAL) fluid have been associated with hyperoxic lung injury (2), but the administration of a cycloooxygenase inhibitor to block the synthesis of prostaglandins does not result in a decrease but rather in an increase of hyperoxic lung injury (2). The early prostaglandin increases which have been documented, therefore, may rather reflect an overall increase in arachidonic acid metabolism, with the increase in lipoxygenase pathway products being at least as important or, perhaps, having a more primary role in mediating the hyperoxic lung injury. Reduced mortality, inhibition of PMN infux and a reduction in the increase of BAL leukotriene B$_4$ levels in a rat hyperoxia model after treatment with the lipoxygenase inhibitor AA861 (3), and attenuation of rat lung injury induced by hydrogen peroxide with the use of various leukotriene antagonists and inhibitors (1,4) has been reported. A primary etiologic role for lipoxygenase pathway products would provide an explanation for the seemingly contradictory results of the studies in which the use of a cyclooxygenase inhibitor resulted in a exacerbation of prostaglandin associated lung injury (5). Blockade of just the cyclooxygenase pathway probably results in shunting of arachidonic acid metabolism to the lipoxygenase pathway (Fig. 1).

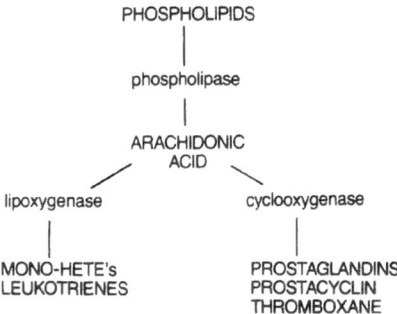

Fig. 1.   Arachidonate pathway (reprinted with permission
           from the International Anesthesia Research
           Society from "Normorbaric pulmonary oxygen toxi-
           city", by Klein, J., Anesth Analg 1990;70:195-
           207).

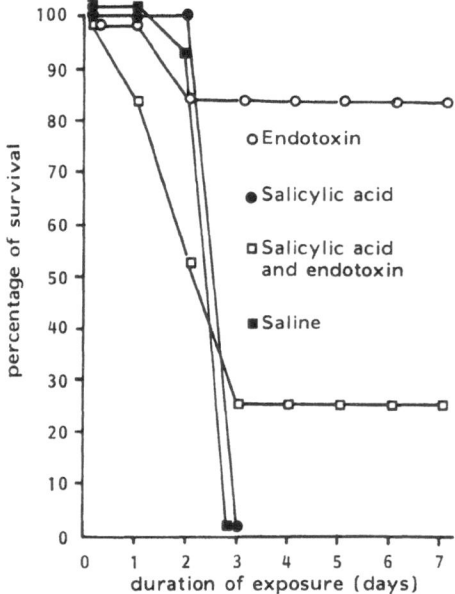

Fig. 2.   Survival of rats of exposed to over 95% oxygen
           for 7 days. The survival curve for rats pre-
           treated with endotoxin is significantly (p <
           0.01) to the right of the survival curve for
           saline-treated rats. This prolonged survival was
           partly reversed by concurrent treatment with L-
           ASA (p < 0.05) (reprinted with permission from
           the International Anesthesia Research Society
           from "Endotoxin protection against oxygen toxi-
           city and its reversal by acetylsalicylic acid,
           by Klein, J., et al., Crit Care Med 1986;14 (1):
           31-33).

This shunting would result in increased production of lipoxygenase products and, as a concequence, in increased lung injury. Understanding of this seemingly paradoxical effect of cyclooxygenase inhibitors and the role of lipoxygenase pathway products as mediators of hyperoxic lung injury awaits studies in which measurements of both prostaglandins and leukotrienes can be performed and the effects of selective inhibitors can be determined.

There could be another mechanism by which eicosanoids play an important role in pulmonary oxygen toxicity. To this point, the most effective pharmacologic agent described for increasing O_2 tolerance in rats is bacterial endotoxin. The protection provided by endotoxin is species-specific (rats and lambs develop O_2 tolerance, but mice and hamsters do not; primates have not been tested) (6,7,8). The mechanism of endotoxin protection against hyperoxic injury is not known. The improved tolerance has been associated with increases in lung SOD and other antioxidant enzymes during hyperoxic exposure (9,10). However, we found that the protective effect of endotoxin is blocked by the cyclooxygenase inhibitor lysine acetyl salicylate, the soluble form of aspirin (5) (Fig. 2).

Endotoxin treatment stimulates the production of at least three potent cytokines: tumor necrosis factor/cachectin, interleukin 1, and interferon. All three factors have been implicated as playing an important rol in endotoxin's protective action; pretreatment of rats with either tumor necrosis factor/cachectin and interleukin 1 (11), interferon inducers (12), or simply serum of endotoxin protected rats (13) decreases lung injury and mortality in hyperoxia. The early mechanisms which mediate cytokine induced tolerance are unclear. Several groups of investigators have suggested that either IL-1 (14), or TNF (15) can induce manganese SOD Mrna and protein in a variety of mammalian and human cells. Within the lung, Mn-SOD is a relatively small component of total SOD activity-representing, at most, 10-15% of that total. In addition, the lung contains 43 cell types. It appears that many of these are not the primary targets of injury by hyperoxia. Thus, early changes induced in only a fraction of lung's total SOD (i.e. Mn-SOD) may be important and sufficient to protect. This has been suggested in studies of exogenous enzyme augmentation (16,17). Additional early adaptive changes in lung surfactant, circulating antioxidants, or other defence mechanisms may also be induced by IL-1 and/or TNF.

Recently it has been reported that the cyclooxygenase inhibitor lysine acetyl salicylate also decreases tolerance of rats to lethal hyperoxic lung injury conferred by IL-1 and TNF (18). Although the mechanism whereby this cyclooxygenase inhibitor blocks endotoxin and IL-1 + TNF-mediated protection is unknown, it is possible that cyclooxygenase products are involved in inducing pulmonary antioxidants.

REFERENCES

1. Farrukh, I.S., Michael, J.R., Peters, S.P., et al., 1988,
    The role of cyclooxygenase and lipoxygenase media-

tores in oxydant-induced lung injury, Am.Rev.Respir. Dis., 137: 1343.

2. Klein, J., Zijlstra, F.J., Vincent, J.E., van Strik, R., Tak, C.J.A.M., and van Schalkwijk, W.P., 1989, Cellular and eicosanoid composition of broncho-alveolar lavage fluid in endotoxin protection against pulmonary oxygen toxicity, Crit. Care. Med., 17: 247.

3. Taniguchi, H., Taki, F., Takagi, K., Satake, T., Sugiyama, S., and Ozawa, T., 1986, The role of leukotriene B_4 in the genesis of oxygen toxicity of the lung, Am. Rev. Respir. Dis., 133: 805.

4. Burghuber, O., Strife, R., Zirrolli, J., et al., 1986, Leukotriene inhibitors attenuate rat lung injury induced by hydrogen peroxyde. Am. Rev. Respir. Dis., 133: 805.

5. Klein, J., Trouwborst, A., and Salt, P.J., 1986, Endotoxin protection against pulmonary oxygen toxicity and its reversal by acetylsalicylic acid, Crit. Care Med., 14: 32.

6. Frank L, Neriishi K. Endotoxin treatment protects vitamin E-deficient rats from pulmonary oxygen toxicity. Am J Physiol 1984;247(3 Part 2):R520-6.

7. Frank L, Roberts RJ. Endotoxin protection against oxygen-induced acute and chronic lung injury. J Appl Physiol 1979;47:577-81.

8. Hazinski TA, Kennedy KA, France ML, Hansen TN. Pulmonary O_2 toxicity in lambs: physiological and biochemical effects of endotoxin infusion. J Appl Physiol 1988; 65:1579-85.

9. Frank L. Endotoxin reverses the decreased tolerance of rats to greater than 95% O_2 after preexposure to lower O_2. J Appl Physiol 1981;51:577-83.

10. Frank L, Summerville J, Massaro D. Protection from oxygen toxicity with endotoxin. Role of endogenous anti-oxidant enzymes of the lung. J Clin Invest 1980; 65:1104-10.

11. White CW, Ghezzi P, Dinarello CA, Caldwell SA, McMurtry IF, Repine JE. Recombinant tumor necrosis factor/cachectin and interleukin 1 pretreatment decreases lung oxidized glutathione accumulation, lung injury, and mortality in rats exposed to hyperoxia. J Clin Invest 1987;79:1868-73.

12. Kikkawa Y, Yano S, Skoza L. Protective effect of interferon inducers against hyperoxic pulmonary damage. Lab Invest 1984;50:62-71.

13. Berg JT, Smith RM. Protection against hyperoxia by serum from endotoxin treated rats: absence of superoxide dysmutase induction. Proc Soc Exp Biol Med 1988; 187:117-22.

14. Masuda A, Longo DL, Kobayashi Y, Apella E, Oppenheim JJ, Matsushima K. Induction of mitochondrial manganese superoxide dismutase by interleukin 1. FASEB J 1988; 2: 3087-91.

15. Wong GHW, Goeddel DV. Induction of manganous superoxide dismutase by tumor necrosis factor: possible protective mechanism. Science 1988; 242:941-4.

16. Turrens JF, Crapo JD, Freeman BA. Protection against oxygen toxicity by intravenous injection of liposome

encapsulated catalase and superoxide dismutase. J Clin Invest 1984; 73:879-85.

17. Freeman BA, Young SL, Crapo JD. Liposome-mediated augmentation of superoxide dismutase in endothelial cells prevents oxygen injury. J Biol Chem 1985; 258: 12534-42.

18. White CW, Ghezzi P. Protection against pulmonary oxygen toxicity by interleukin-1 and tumor necrosis factor: Role of antioxidant enzymes and effect of cyclo-oxygenase inhibitors. Biotherapy 1989; 1:361-7.

THE RELATIONSHIPS BETWEEN OXYGEN DELIVERY AND CONSUMPTION AND CONTINUOUS MIXED VENOUS OXIMETRY ARE PREDICTIVE PARAMETERS IN SEPTIC SHOCK

F. Giunta, L.S. Brandi, T. Mazzanti, M. Oleggini, G. Tulli,
A.M.R. Cuttano.
Cattedra di Anestesiologia e Rianimazione Università degli Studi di
Pisa, Italy

INTRODUCTION

Life is essentially a process of energy that consumes oxygen and produces carbon dioxide. The survival of all mammalian cells depends on a continuous supply of oxygen. To ensure this survival, oxygen is transported from the environment to the cells of the body via a complicated delivery system that is dependent on three organ systems: lungs (ventilation), blood (oxygen content) and cardiovascular system (cardiac output). Living organisms maintain their identity and integrity only by a continuing process of consumption of energy. Under normal conditions, this energy is captured by a mechanism that results in the formation of high energy posphate bonds, mostly as ATP. When the energy demand of all cells increases, there are adjustments in pulmonary oxygen exchange, cardiac output, oxygen binding of hemoglobin and capillary resistance that facilitate cellular availability of oxygen. Moreover, when one component of this system fails, adjustments in the remaining components satisfy systemic oxygen requirements until oxygen delivery falls below a critical level.

From a physiological point of view, systemic oxygen delivery (DO_2) is normally matched to the metabolic requirements for oxygen (i.e. oxygen consumption VO_2). Under basal conditions, systemic oxygen consumption amounts to approximately one fourth of oxygen delivery, yielding an oxygen extraction ratio of 25-30%. When oxygen delivery decreases because of a decrease in either cardiac output, hemoglobin or arterial oxygen saturation, oxygen consumption remains constant and equal to oxygen demand, due to an increase in the extraction ratio [1,2]. This condition is taken as evidence of tissue well-being, or oxygen-supply independency (Fig.1).

Once oxygen extraction has been maximized, further decreases in oxygen delivery are accompanied by a parallel decrease in oxygen consumption, so-called physiological oxygen supply dependency, and an oxygen debt is established. This condition is described by the slope of the regression curve depicted by the continuous line in Figure 1. In other words, when oxygen delivery is reduced, oxygen consumption remains constant and independent of oxygen delivery until delivery reaches a critical level. Continued decrease in oxygen delivery below this level, will be accompanied by a corresponding decrease in oxygen consumption, which has now become supply dependent. At this critical point, the organism's final way of compensation is partial reliance on anaerobic metabolism, which results in lactic acidosis [3,4]. Recent clinical studies in critically ill patients with acute respiratory distress syndrome [5,6], sepsis and septic shock [7,8], acute liver failure [9], acute heart failure [10], demonstrated that these conditions exibit a pattern of oxygen utilization that differs from normal patterns.

Clinical tools to increase oxygen delivery in these critically ill patients are accompanied by a corresponding increase of oxygen consumption even at supranormal oxygen delivery values. This condition is currently termed pathologic supply dependency (Fig.1). For patients exhibiting supply dependency, therapeutic interventions

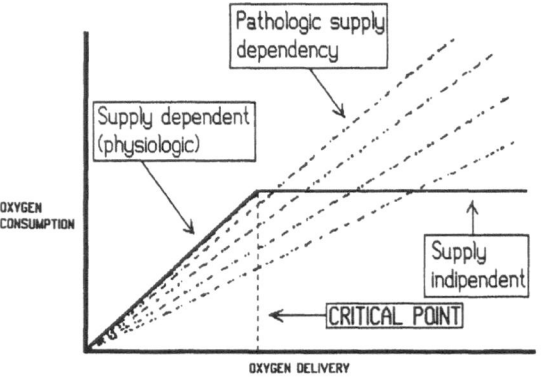

Fig. 1

(fluid, vasoactive drugs) designed to increase oxygen delivery might be expected to restore systemic oxygen balance, resolve lactic acidosis, and minimize the extent of risk of systemic organ failure. Some clinical studies have demonstrated that when oxygen delivery is limited such therapeutic interventions (volume, vasoactive drugs) improve oxygen delivery with a corresponding increase in oxygen consumption and resolution or improvement of blood lactate concentration [7,8]. Much confusion has been caused due to the assumption that all oxygen consumption that varies in proportion to oxygen delivery describes the condition of oxygen supply dependency. On the contrary, oxygen consumption and delivery usually vary together, because oxygen demand rather than oxygen supply is normally the independent variable. Thus, oxygen consumption of most mammals, varies several-fold during a normal day. This increased oxygen consumption generally is not supported by increased oxygen extraction ratio, but rather by increased oxygen delivery, that increases to satisfy the increased oxygen demand, as occurs in exercise. Considering the oxygen consumption/oxygen delivery relationship of normal subjects, one might erroneously conclude that in these subjects an oxygen-supply dependency relation exists, and that they are all in shock state!

The results of some previously published studies investigating oxygen supply dependency, might be explained by increases in basal oxygen requirements, either spontaneous or caused by the therapeutic intervention causing oxygen delivery to increase, rather than the opposite.

Since this is extremely difficult to test directly, we prospectively examined the spontaneous variations in oxygen consumption and delivery in critically ill postsurgical patients, in whom we performed no direct experimental interventions. The hypothesis was that if oxygen consumption and delivery were dependent under conditions where no intervention was applied, then the most likely explanation was an underlying change in oxygen demand.

To verify our hypothesis we studied two groups of similar patients. In the first group of 12 postoperative critically ill patients we measured oxygen consumption and delivery using the thermodilution technique while in the second group of 26 patients, to avoid mathematic coupling of data, we measured oxygen consumption using an independent method (indirect calorimetry).

PATIENTS AND METHODS

Studies were done in the morning with the patients lying supine, completely relaxed and awake in an air-conditioned environment. All patients were monitored with a 7 French, flow-direct thermosistor-tipped catheter (except the septic patients in whom an Optical Catheter P7110 Abbott was used for continuous monitoring of SvO_2) advanced into the pulmonary artery through the right internal jugular vein, and an indwelling radial or femoral arterial catheter. Before the study, the corrected position of the pulmonary arterial catheter was checked by chest X-ray. Systemic arterial, pulmonary artery, central venous pressures, and heart rate were monitored continuously. All pressure values were recorded using a strain gauge transducer (Medex Medical Eng, UK) leveled to the midchest position, zeroed to atmosphere and calibrated to a known mercury standard. Cardiac output ($L*min^{-1}$) was measured using the thermodilution technique and a cardiac computer (Critical Care System 3300, Abbott, Chicago, Ill). Injections of 10 ml of a room temperature (21-24°C) solution of 5% dextrose in water were used. Injection times were always less than 4 sec, thus eliminating possible effects of varying injection rates on calculations. The morphology of thermodilution curves was monitored. The reported cardiac output was the average of three serial measurements obtained within 2-3 minutes, providing that the inter-measurement variance was < 10%. If variance was > 10% two additional measurements were made and high and low values rejected. Arterial and mixed venous blood samples were obtained simultaneously from the catheter in the radial or femoral artery and from the distal port of the pulmonary arterial catheter.

Blood gases were measured immediately by an automated blood gas laboratory (Instrumentation Laboratory System IL 1312) that was calibrated before the study. Hemoglobin concentration and oxyhemoglobin saturation were measured directly using a CO-Oximeter (Instrumentation Laboratory IL 282). In the first group of patients cardiac output determinations and blood gases were obtained (in triplicate) every 30 minutes spaced over 210 minutes. In the second group of patients the same measurements were obtained over a 60- minute period at 20 minute intervals.

Indirect calorimetry estimated oxygen consumption (VO_2), carbon dioxide production (VCO_2) and respiratory quotient (RQ). A computerized, continuous open-circuit system was used to measure gas exchange through a 40-liter plexiglas canopy. Air flow and oxygen carbon dioxide concentrations in the inspired and expired air were measured by means of a Metabolic Measurement Cart Horizon apparatus (Sensormedics Corporation, Anaheim, U.S.A.) which uses, respectively; a flow meter, a polarographic oxygen sensor, and an infrared carbon dioxide analyzer. The software of this instrument implements routines for frequent calibration, which protects against shifts in the sensitivity of the gas analyzer during prolonged tests.

Forty-five minutes before the study the transparent plastic ventilated hood was placed over the patient's head and made air-tight around the neck. Gas exchange measurements were taken during a 60-minute period after the subjects had adapted to the hood and stabilized their breathing pattern (steady state condition, stable VO_2, VCO_2 and RQ; deviation was within the error of VO_2, VCO_2 and RQ measurement). Every effort was made to minimize potential errors in data measurements. Gas and volume calibration were performed before each test. With the metabolic monitor VO_2, VCO_2, and RQ were measured continuously and every 3 minutes the average values of measured VO_2, VCO_2, and RQ were printed. Starting at the onset of cardiac output measurement, 12 consecutive measurements of gas exchange were averaged. Energy expenditure derived from indirect calorimetry was calculated according to the following equation :

$$EE = 3.91 \, VO_2 + 1.10 \, VCO_2 - 3.34N.$$

Urinary non-protein nitrogen excretion was estimated (N=14.4g*die^{-1}).
The study was statistically determined with the regression analysis and Student's test t.
In the first group of patients, using the thermodilution technique, a total of 87 measurements of oxygen consumption and delivery were obtained (Fig.2).

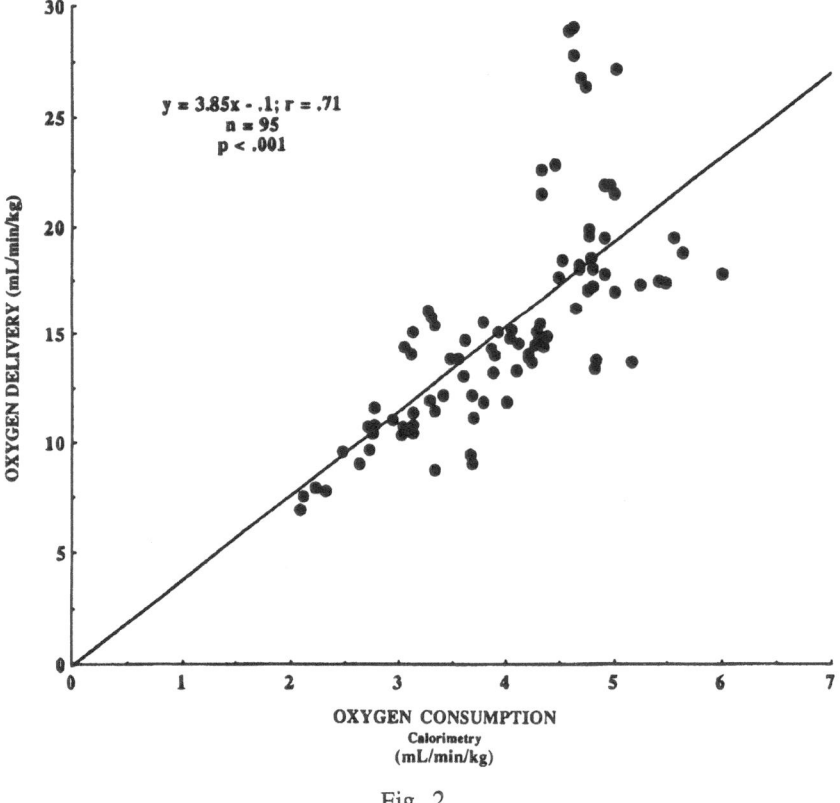

Fig. 2

When considering all data, a significant correlation was found between oxygen consumption and oxygen delivery (Fig.2).
In the second group of patients, using indirect calorimetry, a significant correlation was also found between oxygen consumption and oxygen delivery (Fig.3).

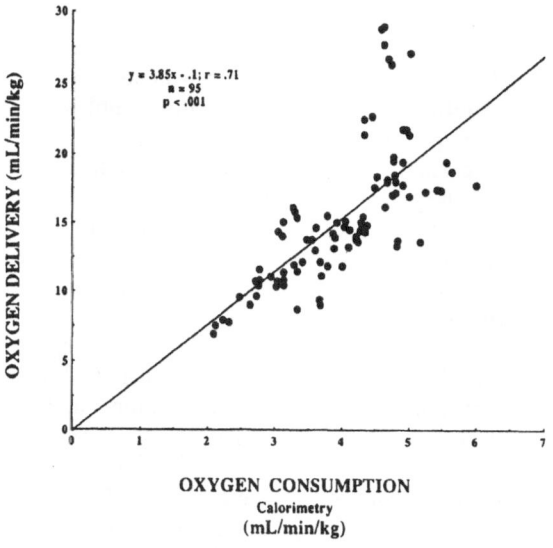

Fig. 3

Since no experimental interventions were performed in these two groups of patients to modify oxygen delivery, a logical hypothesis is that changes in oxygen consumption and oxygen delivery may be explained by spontaneous changes in oxygen demand. We suggest that these observed changes probably reflect primary changes in oxygen demand and, therefore, in energy metabolism. In fact, when we plotted energy expenditure measured by indirect calorimetry vs oxygen delivery measured by thermodilution, we demonstrated a significant linear correlation (Fig.4).

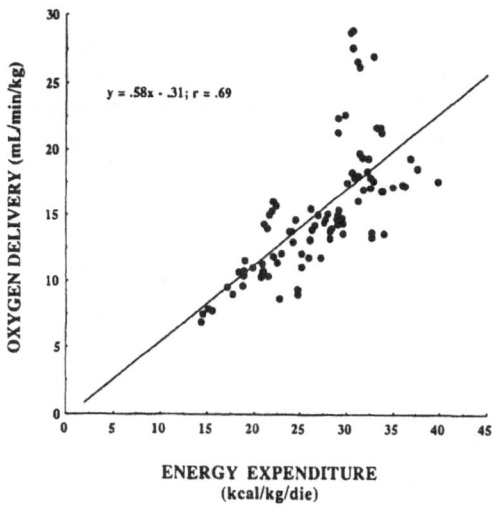

Fig. 4

Therefore great care has been taken on the performance and interpretation of studies examining the putative pathologic oxygen supply dependency. Our study gave these results: a) validation of VO_2/DO_2 by the Fick method vs indirect calorimetry, b) VO_2/DO_2 is a very good clinical parameter when studied in the same patient during follow-up, or when studied in different patients having the same clinical features.

The SvO$_2$ is another controversial parameter with regard to critically ill patients, because there are some differences regarding conditions, utilization of information and patient population. The importance of this parameter is to inform on the quantity of supply adapted to the demand of oxygen. SvO$_2$ is the easiest method for continuous monitoring of changes in the whole body O$_2$ balance.

The problem lies in the fact that SvO$_2$ reflects only global and not single organ oxygen balance and when major changes occur we do not know immediately if that they are due to disturbance of the oxygen supply or oxygen demand. In septic shock "if there is close coupling between oxygen consumption and oxygen delivery, mixed venous oxygen saturation would not be expected to change even though there are significant changes in the patient's clinical condition" (L.D.Nelson, 1991). Therefore the most important point in the treatment of septic shock according to the opinion of various authors is represented by the flow-dependency of VO$_2$. The VO$_2$ becomes flow--dependent when DO$_2$ is below critical values. It has been determined that the critical DO$_2$ value is 9 ml/min/kg in hypoxic or anemic animals [1]; 8 ml/min/kg in patients under anesthesia in open-heart surgery (2); 21 ml/min/kg in patients with ARDS (5); 15 ml/min/kg in septic shock [6]; 14-16 ml/min/kg for lactic acidosis [3]. The conclusions of Tuchschmidt [11] about septic shock are: ExO$_2$ rises, VO$_2$ decrease, lactic acidosis spreads out when DO$_2$ is under the critical limit. This means that DO$_2$ under that limit is inadequate.

We assume that the oxygen dynamic parameters as VO$_2$, DO$_2$, ExO$_2$, CaO$_2$, CvO$_2$, SaO$_2$ and SvO$_2$ have a characteristic slope in septic shock evolution. It has been observed that in early stages of septic shock there is enhanced flow and oxygen consumption while in terminal stages the flow, oxygen consumption, arterial-venous oxygen difference fail, and SvO$_2$ increases.

A study of critical septic shock was performed to evaluate this hypothesis and to find a correlation between the venous mixed saturation (SvO$_2$) (a continuous direct parameter) and the flow-dependency of oxygen consumption (VO$_2$) measured by thermodilution in patients were the oxygen availability was below the critical point. Fifteen patients with critical septic shock were studied during treatment in the Intensive Care Unit (ICU), from November 1989 to July 1990. Sepsis was diagnosed by repeated blood cultures, during abdominal or pulmonary infections. The hemodynamic data presented a

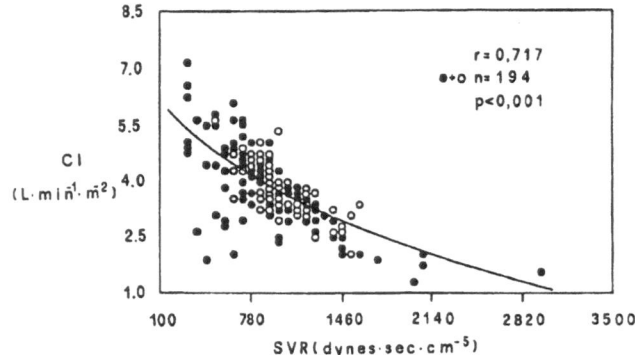

Fig. 5

typical feature of septic shock (Fig.5); in all the patients the oxygen availability (DO_2) was always equal to or < 15 ml/min/kg.

This study analysed 194 hemodynamic and hemogasanalytic measurements obtained from 15 patients during the first 4 days of observation in the ICU: 55 measurements were from 4 patients who survived (S group) and 139 were from the patients which did not survive (nS group). The typical septic shock scenario is evident by the SVR/CI relationship (Fig.5). The relationship between the averages of all the first measurements obtained from the S and nS groups shows a sensible difference only for the VO_2 ($p < .001$) and CaO_2 ($p < .05$) that are greater in the patients with a good. follow-up (Table 1).

Table 1

	SURVIVED	NOT SURVIVED	p
CO ($L.min^{-1}$)	$6,73 \pm 1,4$	$6,69 \pm 1,73$	NS
CI ($L.min^{-1}.m^{-2}$)	$3,83 \pm 0,7$	$3,83 \pm 1.03$	NS
SVR ($dynes.sec.cm^{-5}$)	1002 ± 246	926 ± 372	NS
$Da-\bar{v}O_2$(ml%)	$3,79 \pm 0,7$	$3,47 \pm 1,5$	NS
ExO_2(%)	$27,1 \pm 5,3$	$25,2 \pm 9,2$	NS
$S\bar{v}O_2$(%)	$69,8 \pm 4,5$	$71,8 \pm 9,8$	NS
CaO_2(ml%)	$14,3 \pm 1,8$	$13,7 \pm 2,3$	0,05
$C\bar{v}O_2$(ml%)	$10,6 \pm 1,7$	$10,2 \pm 2,0$	NS
$\dot{D}O_2$($ml.min^{-1}.Kg^{-1}$)	$13,7 \pm 3,6$	$13,2 \pm 3,5$	NS
$\dot{V}O_2$($ml.min^{-1}.Kg^{-1}$)	$3,67 \pm 0,9$	$3,10 \pm 0,6$	0,001

There is a good correlation between $Da-vO_2$ and SvO_2 in all the measurements and in each group alone: in all the patients presenting critical septic shock, high SvO_2 values corresponded to low $Da-vO_2$ values with a linear relationship that was independent from the follow-up (Fig.6).

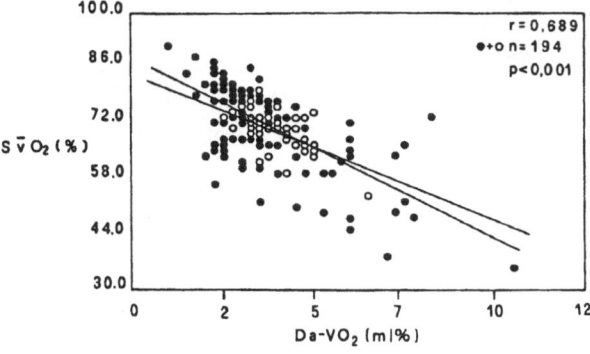

Fig. 6

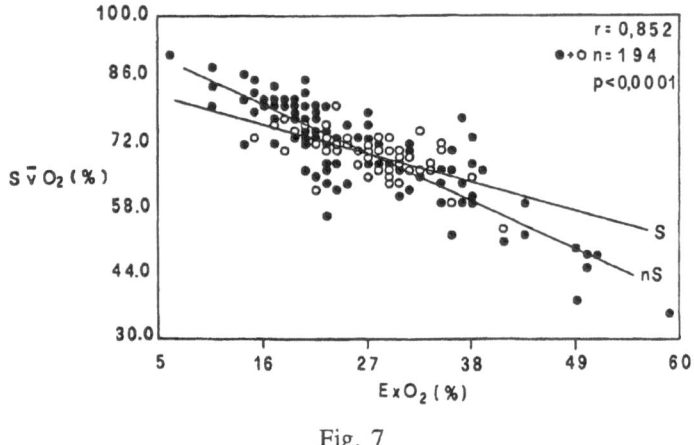

Fig. 7

The SvO_2 had a good correlation with the ExO_2 (Fig.7).

The SvO_2 showed no correlation with DO_2 in the S group, while this correlation was good (r=0.699) in the nS group (the increment of SvO_2 followed the DO_2, as if the oxygen was not extracted from the peripheral tissues) (Fig.8).

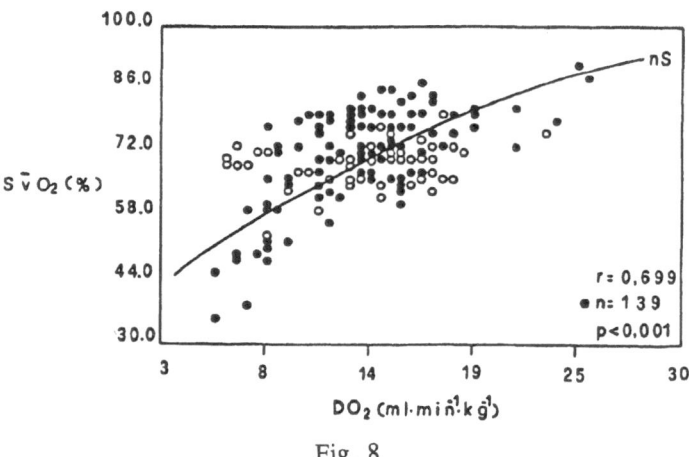

Fig. 8

The SvO_2 was correlated to the VO_2 in a special manner: high SvO_2 corresponded to low VO_2 (r=0.477) in the nS group, while in the S group this was not observed (Fig.9), because it was very difficult for the peripheral tissue to extract oxygen.

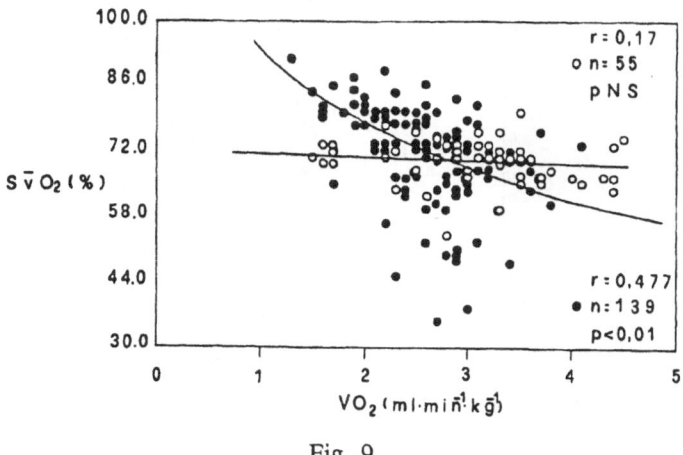

Fig. 9

Therefore, a condition with high DO_2 values and low ExO_2 ratio is possible, especially, in patients with poor prognosis (Fig.10).

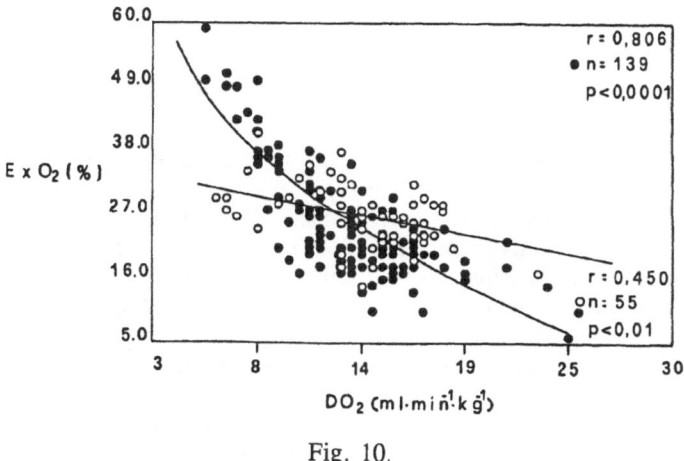

Fig. 10.

This ExO_2/DO_2 is confirmed by the VO_2/DO_2 relation (Fig.11).

Our experience confirms the conclusions of Tuchshmidt about DO_2 and ExO_2 only in the S group (r=0.728), while in the nS group (r=0.064) we observed loss of

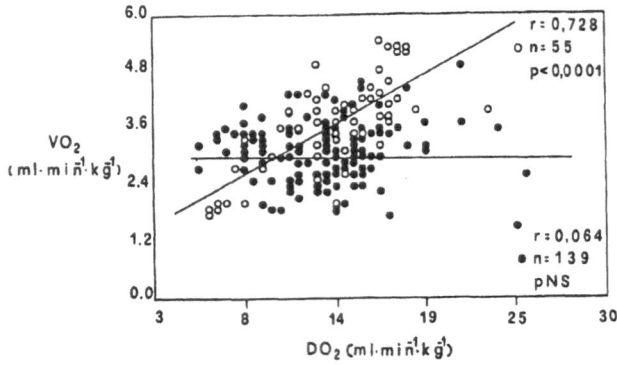

Fig. 11

flow-dependency because the VO_2 remained constant in the presence of a varying DO_2. Even if it is not widely accepted that there is a critical point for aerobic metabolism, it is possible to confirm a good correlation between DO_2 and VO_2 in critical septic shock with $DO_2 < 15$ ml/min/kg.

The continuous monitoring of SvO_2 can help to indicate the moment of a pathologic change of ExO_2. Therefore we consider the SvO_2 a very good indicator for hemodynamic, hemogasanalytic, and metabolic control.

REFERENCES

1. S.M. Cain, Oxygen delivery and uptake in dogs during anemic and hypoxic hypoxia, J. Appl. Physiol. 42:228 (1977).

2. K. Shibutani, T. Kamatsu, K. Kubal, V. Sanchala, V. Kumar, D.V. Bizzarri, Critical level of oxygen delivery in anesthetized man, Crit. Care Med. 11:640 (1983).

3. D.H. Simmons, A.P. Alpas, D.P. Tashkin, A. Coulson, Hyperlactatemia due to arterial hypoxemie or reduced cardiac output, or both, J. Appl. Physiol. 195:H100 (1978).

4. S.M. Cain, Appearance of excess lactate in anesthetized dogs during anemic and hypoxic hypoxia, Am. J. Physiol. 209:604 (1965).

5. S.J. Danek, J.P. Lynch, J.G. Weg, D.R. Dantzker, The dependence of oxygen uptake and oxygen delivery in the adult respiratory distress syndrome, Am. Rev. Resp. Dis. 122:387 (1980).

6. Z. Mohsenifar, P. Goldbach, D.P. Tashkin, D.J. Campisi, Relationship between O_2 delivery and O_2 consumption in the adult respiratory distress syndrome, Chest 84:267 (1983).

7. B.S. Kaufman, E.C. Reckow, J.L. Falk, The relationship between oxygen delivery and consumption during fluid resuscitation of hypovolemic and septic shock, <u>Chest</u> 85:336 (1984).

8. M.E. Astiz, E.C. Rackow, J.L. Falk, B.S. Kaufman, M.H. Weil, Oxygen delivery and consumption in patients with hyperdynamic septic shock, <u>Crit. Care Med.</u> 15:26 (1987).

9. D.J. Bihari, A.E. Giusan, R. Williams, Cardiovascular, pulmunary and renal complication of fulminant hepatic failure, <u>Semin. Liver. Dis.</u> 6:119 (1986).

10. Z. Mohsenifar, D. Amin, A.C. Jasper, P.K. Shah, S.K. Koerner, Dependence of oxygen consumption and oxygen delivery in patients with chronic congestive heart failure, Chest 92:447 (1987).

11. J. Tuchschmidt, J. Fried, R. Swinney, O.P. Sharma, Early hemodynamic correlates of survival in patients with septic shock, <u>Crit. Care Med.</u> 17:719 (1983).

OXYGEN DELIVERY AND POSTOPERATIVE MORTALITY

Hoyte van der Zee MD and Robert C. Evans RN

Department of Anesthesiology
Albany Medical Center, Albany, New York

INTRODUCTION

In several studies the levels of Oxygen Consumption (VO_2) and Oxygen Delivery (DO_2) and their relationship have recently been identified as the most accurate prognosticators for outcome of the critically ill patient (2,11,13). Many other hematological and hemodynamic variables have been found to be lacking reliable predictability (12), whereas the availability of the VO_2 and DO_2 measurements have aided importantly in recognizing prodromal signs of life-threatening conditions. It has to be stressed that early recognition of an impending fatal course is of paramount importance to effectively treat the physiologic derangement that may lead to the patient's death.

PURPOSE

The purpose of the presented, retrospective study is to report the findings with respect to physiological profiles, including VO_2 and DO_2, of critically ill surgical patients. In order to give consistent and meaningful interpretations of VO_2 and DO_2, we tested the hypothesis that:

(1)   the critical level of DO_2 of 8.2 ml/kg/min as indicated by Shibutani is a valid critical level of minimally adequate O_2 delivery to the tissues, and

(2)   a subcritical DO_2 may be tolerated for a limited time due to compensatory mechanisms at the cellular level, but that there exists a magnitude of oxygen debt that cannot be compensated for and leads to death.

We further hypothesized that if the above postulates were true:

(3) a significant difference would exist between post-operative Survivors and Non-survivors with respect to values above and/or below this assumed critical DO_2 of 8.2 ml/kg/min, termed Positive Net DO_2 and Negative Net DO_2 or Oxygen Debt, respectively. If a significant difference was found to exist between these two groups, then the Oxygen Debt was compared to the percentage incidence of the mortality for patients with that magnitude of Oxygen Debt, and, in so doing, to arrive at a maximal level of Oxygen Debt that could be tolerated with impunity.

METHODS

The critical level of a DO_2, defined as that level of delivery below which the VO_2 becomes DO_2-dependent and falls short of the required O_2 demand, was taken from the work of Shibutani (10). This investigator conducted a study in which VO_2 was measured at different levels of DO_2. He observed a critical level of Oxygen Delivery in anesthetized man at 8.2 ml/min per kg. Oxygen Consumption remained relatively fixed at oxygen delivery values well above the level of 8.2 ml/min per kg. However, the VO_2 decreased proportionally with DO_2 below this critical value, indicating that tissue oxygen deprivation may be occurring. With the result of this study as a guide we have examined, in a retrospective manner, whether the critical level of DO_2 of 8.2 ml/kg/min had validity as a basis on which to prognosticate the outcome of the critically ill surgical patient. In these patients the whole of the oxygen transport system was analyzed for all of its components, which were calculated as a total over the whole observation period (24 hrs) as well as on an hourly basis. Oxygen Delivery was plotted as a function of time. The graphic representation of the function of the DO_2 vs time describes a curve. The area under the curve is equal to the actual amount of oxygen in milliliters transported by the blood to the tissues during the observed time period. By utilizing Shibutani's value of 8.2 ml/min per kg as a zero reference line, DO_2 measurements then can take on positive or negative values. When DO_2 values are greater than 8.2 ml/min per kg, the area under the curve equals the amount of oxygen delivered in excess of the critical level of oxygen delivery. When DO_2 values are less than 8.2 ml/min per kg, the area under the curve equals the amount of Oxygen Debt in milliliters incurred. Thus, the oxygen debt is described by a negative value and the excess oxygen delivery is described by a positive value. The sum of the areas will provide a value equivalent to the patient's Net Oxygen Delivery status. If the latter is negative, then the patient is in Negative Net Oxygen Delivery (Oxygen Debt). If the value is positive or zero, the patient is meeting metabolic oxygen demands, or is in a Positive Net Oxygen Delivery status.

The terms used above are described in a graphic manner in Fig.1

Patient Population, Methods and Statistics

The medical records of 500 Albany Medical Center ASA Class IV and Class V patients, who had placement of a thermodilution pulmonary artery floatation catheter and serial measurements of Cardiac Output (CO), Arterial blood gas (ABG) values and Hemoglobin (Hgb) were collected throughout the intra-operative and immediate post-operative period, totalling 24 hours.

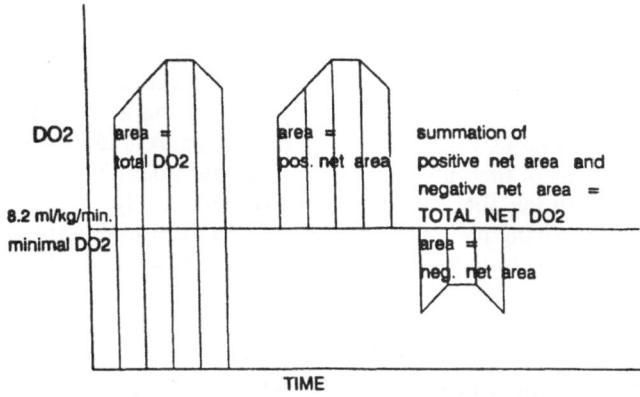

TOTAL DO2 minus MINIMAL DO2 = NET DO2 (neg. or pos.)
NEG. NET DO2 plus POS. NET DO2 = TOTAL NET DO2
AREA / MIN. / WEIGHT = AVERAGE VO2 / ML / KG / MIN.

FIGURE 1

Patients with the following conditions were excluded from the study for various reasons (3, 4,6,13). They were: 1) cyanide poisoning; 2) septic shock, indicated by a temperature greater than 37°C, and a documented source of infection by local bacterial or blood culture; and 3) patients with adult respiratory distress syndrome (8) and a history of chronic obstructive pulmonary disease as well as patients diagnosed as having pulmonary abnormalities, e.g. decreased functional residual capacity, decreased dynamic and static compliance, ventilation perfusion mismatch, increased pulmonary vascular resistance, and alveolar surfactant changes.

Of the 500 records, 125 Survivors and 25 Nonsurvivors were found to meet the study criteria. These groups were found to match one another in age, sex, weight and severity of their surgical condition and ASA classification. DO_2 was calculated using the standard equation:

$$(DO_2 = [(Hgb)(\%O_2\ Sat)(1.34\ ml/dl) + (0.0031)(PO_2)][CO]).$$

As stated above, the value of 8.2 ml/kg/min was taken as the zero reference point for calculation of areas. Areas were calculated employing the trapezoid rule. Areas below the zero reference were negative and those above were positive. The Total Positive Net Oxygen Delivery and the Total Negative Net Oxygen Delivery were then summed. This value is the Total Net Oxygen Delivery. The total oxygen delivered and the total net oxygen delivered were then divided by the number of minutes in the twenty-four hour period to yield an Average Oxygen Delivery and an Average Net Oxygen Delivery. The Average Net Oxygen Delivery was then compared to the minimal oxygen delivery and described as a percentage of the minimal oxygen delivery. All available DO_2's calculable within the 24 hour time period were used to calculate area. The total amount of oxygen delivered was calculated for each patient during the twenty- four hours of the combined intra-operative and post-operative period. Table 1 sums up the variables utilized in comparing Survivors and Nonsurvivors.

In addition, in order to provide a composite picture of the Survivor and Nonsurvivor groups the Total Net Oxygen Delivery and its components: Cardiac Output (CO), Hemoglobin (Hgb), Arterial Oxygen Saturation (SaO$_2$), and Partial Pressure of Oxygen (PaO$_2$), were calculated on a hourly basis. This procedure, we reasoned, would provide us with information on the nature and time of change in oxygen delivery status of the patients during their inter-operative and post-operative period. All the above parameters were individually compared between surviving and nonsurviving patients using a Student's t-test to determine if significant differences existed between these two groups, with $p < 0.05$ considered to be significant.

Finally, the frequency distribution of the Negative Net Oxygen Delivery was used to determine the percent mortality for that level of oxygen debt.

RESULTS

For the total population the significance level (t value) of the variables, such as Total Oxygen Delivery and its derivatives, by taking into account Shibutani's criterion of 8.2 ml O$_2$/kg /min, are listed below. However, their actual values are not given, since these are considered less meaningful because of the lack of time reference points during the total observation period.

TABLE 1

		(t value)
Total Oxygen Delivery	not significant	(0.065)
Average Total Oxygen Delivery	significant	(2.282)
Positive Net Oxygen Delivery	not significant	(1.247)
Negative Net Oxygen Delivery	significant	(3.403)
Total Net Oxygen Delivery	not significant	(1.777)
Average Net Oxygen Delivery	not significant	(1.680)
Percent Oxygen Delivered	not significant	(0.539)

The plotted frequency distribution of Negative Net Oxygen Delivery showed that when this value was above 300,000 ml, the mortality was 100 percent. Below a value of 250,000 ml of Negative Net Oxygen Delivery, mortality was consistently 16 percent of the average (Fig. 2).

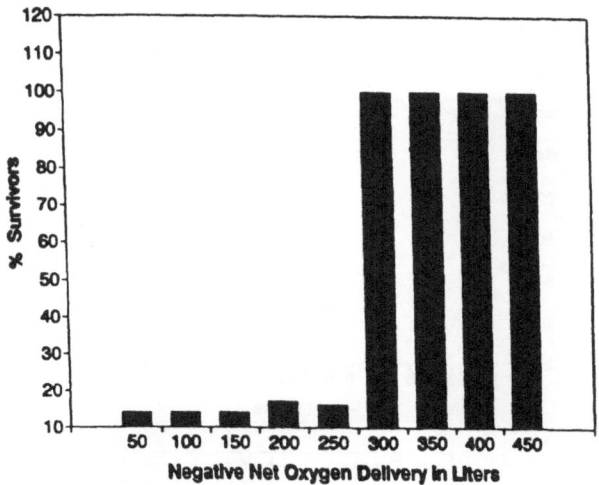

FIGURE 2. The plot of the cumulative frequencies vs percentage survivors shows a watershed occurring at 250 L O_2, of Negative Net Oxygen Delivery, higher values carry an almost 100% mortality.

Time reference points, calculated as hourly means and their level of significance, are listed below. Hourly means of the components of the oxygen delivery system i.e., CO, HgG, PaO_2, SaO_2, Total Oxygen Delivery and Total Net Oxygen Delivery. For the last three variables, data is represented in Figures 3, 4 and 5. Data of hourly means of CO, HgG and PaO_2 are not shown.

TABLE 2

		(t value)
Total Oxygen Delivery	significant	(2.79)
Total Net Oxygen Delivery	significant	(2.11)
Cardiac Output	not significant	(1.69)
Hemoglobin	not significant	(0.65)
SaO_2	significant	(6.51)
PaO_2	significant	(2.36)

The differences in hourly means of Total Oxygen Delivery and Total Net Oxygen Delivery between survivors and Nonsurvivors became significant at the three to four hour period and at the fourteen to twenty-four hour period. The average length of surgery was 7.1 hours. Surprisingly, in the 2-4 hour period (mid-surgery), the cardiac output was significantly lower in the Survivors rather than in the Nonsurvivors, but in the later (post-operative period) this situation was reversed. The SaO_2 in the Survivors was consistently higher throughout the intra- and post-operative period. No significant differences were found for Hgb values, except for the first 2 hours of surgery.

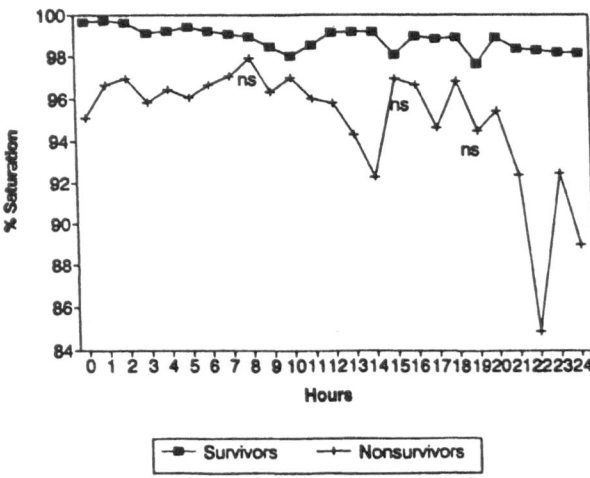

FIGURE 3. The hourly means of SaO₂S indicated significant lower values for the Nonsurvivors throughout the 24 hr. observation period (significance of data not denoted, data points of no significance denoted with ns).

It is noteworthy, that comparing Survivors and Nonsurvivors in the post-operative period, the hourly means attained a level of significance for SaO₂ with a t value = 6.51, for PaO₂ with a t value = 2.36, and for CO with a t value = 1.69, with the CO only coming into play in the last stages of the 24 hr period (Tables 1 and 2).

DISCUSSION

Studies (2,11,12,13) have found significant differences between the oxygen transport values of surviving and nonsurviving surgical patients and that the most critical time period for patient survival was found to exist in the immediate 24 hour postoperative period. Because oxygen cannot be stored at the mitochondrial level, normal aerobic cellular metabolism is dependent on a continuous adequate oxygen delivery to the tissues (1,5,9,15). The differences found between the DO_2 and VO_2 values in survivors' and nonsurvivors' values have therefore been attributed to the inability of nonsurvivors to increase their oxygen delivery in an effort to compensate for an alleged intra-operatively accrued oxygen debt. In these studies, post-operative hyperdynamic state was often observed and interpreted as a homeostatic response to increase oxygen delivery (2,3,12,14). The hyperdynamic response was found to be consistent over a wide range of illnesses and surgical procedures investigated. The authors (12,13) noted an inverse correlation between DO_2 and patient mortality; lower average in oxygen deliveries were correlated with higher percentages of patient mortality. They concluded that the performance of the oxygen transport system was an indicator of the body's ability to supply the necessary oxygen when increased cellular demand existed.

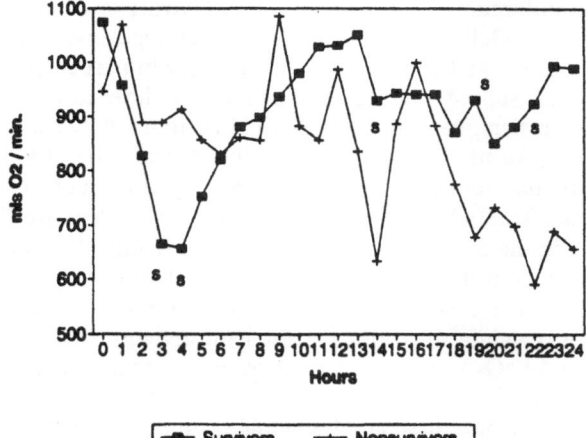

FIGURE 4. The hourly means of the Total Do$_2$ indicated significant differences at only a few of the observations points; being lower for the Survivors intra-operatively, but higher at some points post-operatively.

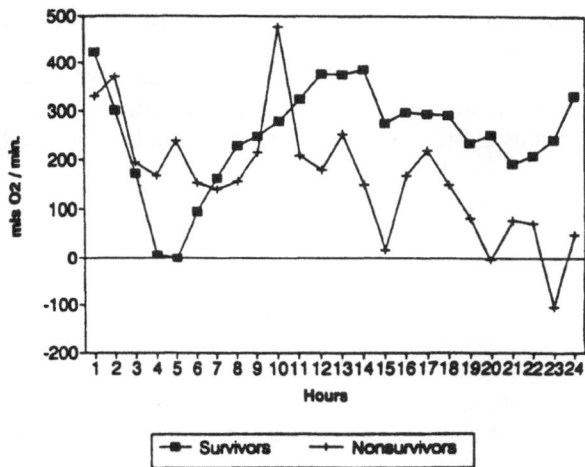

FIGURE 5. The hourly means of the Total Net DO$_2$ are consistently and significantly lower in the post-operative period for the Nonsurvivors. Intraoperatively Survivors have even some lower values.

The presented study shows similar and dissimilar results. However, the primary goal of this retrospective study was to find support for Shibutani's postulate of a minimally adequate oxygen delivery level of 8.2 ml/kg/min. The fact that in this study a significant difference was found between Survivors and Nonsurvivors in Average Total Oxygen Delivery, Negative Net Oxygen Delivery, hourly means of Total Oxygen Delivery and Total Net Oxygen Delivery in particular throughout the post-operative course, supports the concept of a critical level at about this magnitude. It is less likely that a significant difference would be found between survivors and nonsurvivors in negative net oxygen delivery if the critical level of minimal oxygen delivery was greatly different than Shibutani's level. It is of interest to note that intra-operatively the Total Oxygen Delivery and Total Net Negative Delivery were significantly lower in the Survivors than in the Nonsurvivors. This was mainly on account of a lower CO in the 2nd, 3rd and 4th hour of surgery; post-operatively. However, throughout the post-operative period Total Oxygen Delivery and Total Net Oxygen Delivery was significantly lower in the Nonsurvivors. This was mainly on account of SaO_2 and PaO_2 values, the SaO_2's having a greater level of significance than the PaO_2's (Table 2). There was no difference in Hgb content between Survivors and Nonsurvivors. It should be emphasized here that patients with diagnosed pulmonary abnormalities were excluded beforehand from the study. This leads one to conclude that changes in the study population's oxygen delivery status as a whole was more related to hypoxemia than to anemia or decreased cardiac output. The cause of the hypoxemia remained undetected, but may have been due to atelectasis, ventilation perfusion mismatch, interstitial and/or alveolar edema in the post operative period. It may be concluded from this study that increasing Negative Net Oxygen Delivery or Oxygen Debt is associated with increased mortality as is shown in Fig 2. This study placed the maximum level of oxygen debt that can be tolerated in a twenty-four hour period at 250,000 ml. Below this level mortality decreases with decreasing oxygen debt, averaging 16 percent. Above this level mortality was almost 100 percent. Future research should investigate prospectively whether increases in negative net oxygen delivery are indeed associated with increased mixed venous oxygen extraction and increases in lactic acid levels (7) and other indicators (15) of anaerobic metabolism.

REFERENCES

1.   Baue, A.E. Mitochondrial Function in Shock. The Cell in Shock. The Upjohn Co., 1974, 11-15.

2.   Bland, R.D. Common physiologic patterns in general surgical patients. Surgical Clinics of North America. Vol. 65, No. 4, August 1985, 973-809.

3.   Cain S.M. Oxygen delivery and uptake in dogs during anemic and hypoxic hypoxia. Journal of Applied Physiology. Vol. 42, No. 2, 1977, 228-234.

4.   Cain, S.M. Supply dependency of oxygen uptake in ARDS: Myth or reality. The American Journal of the Medical Sciences. Vol. 288, No. 3, Oct. 1984, 119-124.

5.   Connett, R.J. Defining hypoxia: a systems view of VO_2, glycolysis, energetics, and intracellular PO_2. Journal of Applied Physiology. Vol. 68, No. 3, 1990, 833-842.

6. Edwards, J.D. Use of survivors' cardiorespiratory values as therapeutic goals in septic shock. Critical Care Medicine. Vol. 17, No. 11, Nov. 1989, 1098-1103.

7. Groeneveld, A.B.J. Relation of arterial blood lactate to oxygen delivery and hemodynamic variables in human shock states. Circulatory Shock Vol. 22, 1987, 35-53.

8. Mohsenifar, Z. Relationship between O_2 delivery and O_2 consumption in the adult respiratory distress syndrome. Chest Vol. 84, 1983, 267-271.

9. Schumer, W. "The Effects of Shock on the Energy Pathways of the Cell. The Cell in Shock. The Upjohn Co., 1974, 8-10.

10. Shibutani, K. Critical level of oxygen delivery in anesthetized man. Critical Care Medicine Vol. 11, No. 8, August 1983, 640-643.

11. Shoemaker, W.C. Pathophysiology, monitoring, outcome prediction, and therapy of shock states. Critical Care Clinics Vol. 3, No. 2, April 1987, 307-356.

12. Shoemaker, W.C. Physiologic patterns in surviving and nonsurviving shock patients. Archives of Surgery Vol. 106, May 1973, 630-636.

13. Vermeij, C.G. Oxygen delivery and oxygen uptake in postoperative and septic patients. Chest Vol. 98, No.2, August, 1990, 415-420.

14. Waxman, K. Physiologic responses to massive intraoperative hemorrhage. Archives of Surgery Vol. 117, April 1982, 470-475.

15. Yang, S.C. Oxygen delivery and consumption and P50 in patients with acute MI. Circulation Vol. 73, No. 6, June 1986, 1183-1185.

IS OXYGEN CONSUMPTION MEASUREMENT DURING ANESTHESIA FOR LIVER TRANSPLANTATION VALUABLE FOR A RAPID ASSESSMENT OF ADEQUATE FUNCTION OF THE GRAFT?

THN Groenland, CGOT Bouman, A Trouwborst, W Erdmann

Department of Anesthesiology, Erasmus University Rotterdam, The Netherlands

INTRODUCTION

Several metabolic tests have been used to estimate the initial function of the transplanted liver. Most of them are clearance tests of exogenous or endogenous substances. Hyperkalemia and hypoglycemia during the immediate postoperative period are signs for poor liver function, while decreases in lactate and amino-acids levels, indicate good transplant function (Shapiro et al., 1986). Unfortunately, measurements of these latter parameters are only available postoperatively. Therefore intraoperatively, other available parameters must be relied on. A persistent reperfusion hyperglycemia peroperatively seems to be an early prognostic sign for impaired graft function (Mallett et al., 1989).

An increase in oxygen consumption (VO_2) after reperfusion might be an early indication of immediate well functioning of the new liver (Svensson et al., 1987; Lübbe et al., 1988). The aim of this study was to investigate oxygen transport variables, and to assess these changes and their impact on the clinical outcome.

PATIENTS AND METHODS

Ten consecutive patients (range 26-53 yr) with liver failure, undergoing auxiliary heterotopic liver transplantation (HLT), were included in the study. The mean operative risk score according to Pugh et al. for patients with liver disease was more than 12 and placed the patients in a high-risk category (Pugh et al., 1973). Three patients (range 18-31 yr), undergoing HLT (n = 1) or orthotopic liver transplantation (OLT, n = 2) followed by a worse clinical outcome, were reviewed. These patients were placed in the same high-risk category as the other ten patients. Anesthesia consisted of sufentanil, midazolam and pancuronium. Ventilation was maintained with a F_iO_2 of 0.4 in nitrogen, and the ventilator setting was adjusted to achieve an end-expiratory CO_2 which corresponded to a $PaCO_2$ of 35-40 mm Hg. A 7.5F Swan Ganz thermodilution catheter (Edwards Laboratories, Santa Anna, U.S.A.) was floated into the pulmonary artery through the right jugular vein. Hemodynamic measurements, including heart rate (HR),

Oxygen Transport to Tissue XIV, Edited by W. Erdmann and
D.F. Bruley, Plenum Press, New York, 1992

mean arterial pressure (MAP), mean pulmonary arterial pressure (MPAP), right atrial pressure (RAP), pulmonary capillary wedge pressure (PCWP), cardiac index (CI), hemoglobin and blood gas determinations were recorded throughout the procedure. All measurements were taken before clamping of the recipient inferior caval vein (ICV) and portal vein (PV), after clamping of ICV and PV, after circulation of the graft and at the end of the procedure. Based on these measurements, systemic and pulmonary vascular resistance (SVR, PVR) and, ac-cording to the Fick equation, oxygen availability (O_2Av/m^2), oxygen consumption (VO_2/m^2) and oxygen extraction ratio (O_2ER) were calculated for each of these exactly timed measurement procedures. Hemoglobin level was kept between 8.8 mg/dl and 10.4 mg/dl, the pH was kept within the normal range (7.35-7.45) and metabolic acidosis was treated with bicarbonate. All data are presented as means $\pm$ SD. Obtained data were analyzed using non-parametric statistical methods. Values obtained prior to cross clamping were used as controls. Statistical significance of differences between mean values was determined using the Sign test for paired samples. A P value below 0.05 was considered to be statistically significant.

RESULTS

Hemodynamic variables of both groups are presented in Tables I and II. Decreases in CI, MAP, PCWP and RAP occurred with (partial) clamping of ICV and PV (for each variable different but ranging between 2.7-31.8%), whereas SVR and PVR increased. After unclamping, and consequent recirculation of the graft, CI, PCWP and RAP returned to their preclamping values, but a further slight decrease in MAP was noted (up to 12% compared to initial value). SVR decreased to baseline levels. Increase in PVR occurred after recirculation compared to preclamping values. At the end of the procedure all hemo-dynamic variables had returned to baseline values.

Oxygen transport variables of the first group (n = 10) are presented in Table I. O_2Av and VO_2 decreased after partial clamping of ICV and PV with 11.9% and 13.8% respectively returning to preclamping values after unclamping. A further increase in O_2Av (27.1%) and VO_2 (20.7%) was noted at the end of the operation. During the procedure no changes in O_2ER oc-curred. Body temperatures decreased during the operation to 35.4 $\pm$ 0.4°C prior to partial clamping of ICV and PV. There was a further decrease of 1.1 $\pm$ 0.5°C after insertion of the cold liver that had been stored at 0-4°C and after flushing of the liver with one liter of 0.9% saline solution at 20°C before insertion.

Oxygen transport variables of the second group (n = 3) are shown in Table II. In this group O_2Av and VO_2 also decreased after clamping of ICV and PV with 23.7% and 14.2% respectively, with O_2Av returning to and VO_2 surpassing preclamping value (11.9%) after unclamping. At the end of the procedure VO_2 had returned to preclamping value. No changes in O_2ER occurred. Changes of body temperature in this group were comparable with the first group.

DISCUSSION

Patients undergoing HLT or OLT are generally in poor physical condition, as reflected by the operative risk score (Pugh et al., 1973; Cuervas-Mons et al., 1986).

Therefore hemodynamic changes during anesthesia have to be minimized as far as possible. Changes in hemodynamics were acceptable, physiological status remained stable, and were comparable with hemodynamics in patients undergoing HLT, as described in our preliminary report (Groenland et al., 1988). Hemodynamic changes resulted from surgical manoeuvres, due to clamping ICV, PV and recirculation of the donor liver, and at the end of the procedure all hemodynamic variables returned to baseline levels.

After hepatectomy, based on normal physiologic metabolism of the liver, a 20-25% decrease in VO_2 is expected, since splanchnic oxygen consumption represents an average of 20-25% of whole body consumption from which more than 90% is due to the high oxygen metabolism of the liver (Van Thiel et al., 1986). In patients with a normal liver function, following hepatectomy (e.g. due to liver tumor), a 20% reduction in VO_2 has been reported (Svensson et al., 1987; Lübbe et al., 1988). Hepatectomy in patients with an impaired liver function results in a lesser reduction in VO_2 (Lübbe et al., 1988). In this study in the first group (all patients undergoing HLT, without hepatectomy), a 14.2% reduction in VO_2 occurred after partial clamping of ICV and PV. This corresponds to the results found by Lübbe et al. during hepatectomy. Thus, although only slight hemodynamic changes occur during partial clamping of ICV and PV, flow to the already diseased liver is probably diminished, resulting in a decrease in VO_2. In the second group (1 patient undergoing HLT, 2 patients undergoing OLT with hepatectomy) an identical reduction in VO_2 was noted, confirming Lübbe's results. A contributing factor to VO_2 reduction might be the decrease of body temperature by a maximum of 0.6°C. Only a minor proportion ($< 5\%$), of this small decrease, can be held responsible for the reduction in VO_2 by 14.2% (Bigelow et al., 1950; Harper et al., 1961; Hegnauer et al., 1954; Michenfelder et al., 1968; Penrod, 1949).

During the procedure, changes in O_2Av occurred, mostly due to changes in cardiac output. In spite of these changes, O_2Av was still sufficient, because of an unchanged O_2ER, which remained within the normal range.

An increase in VO_2 after reperfusion might be an early indication of immediate well functioning of the new liver (Svensson et al., 1987; Lübbe et al., 1988). The initial and following values of VO_2 in the (small) second group were higher, which could not be explained. In the first group VO_2 returned to preclamping levels immediately after reperfusion and towards the end of the procedure VO_2 had increased significantly to 20.5% above preclamping levels. In the second group, VO_2 increased shortly after recirculation of the graft surpassing the preclamping values and returned to preclamping levels towards the end of the operation. The increase of VO_2 after recirculation is probably due to an oxygen debt. Oxygen debt was clearly observed in two of the three patients under-going OLT with veno-venous bypass and clamping of ICV and PV. Splanchnic circulation was decreased due to venous congestion. In the first group ICV and PV were only partially clamped, without a great effect on splanchnic circulation. An oxygen debt in this group was less likely. As no differences between first and last measurements of body temperature were found, changes in VO_2 could not be related to temperature dependency.

Differences do exist between the two groups: in the first group an increase in VO_2 was observed towards the end of the procedure and all the transplanted livers were functioning well; in the second group, there was no increase in VO_2 as compared to preclamping values and in this group primary non-function of the transplanted livers was present.

Thus, an increase in VO_2 at the end of the operation, which surpasses the preclamping level, could be an early indication of adequate function of the new liver. However, more patients need to be included in the study for a definite conclusion.

Table I. Hemodynamic and oxygen transport variables during HLT.

Variable	Before clamping of ICV and PV	After clamping of ICV and PV	After recirculation of graft	End of operation
HR	87.6 ± 9.8	83.3 ± 9.2	80.6 ± 11.6*	90.0 ± 18.4
MAP	75.4 ± 8.6	73.4 ± 7.2	69.0 ± 6.1	72.5 ± 6.3
CI	4.74 ± 0.99	4.15 ± 0.89*	4.52 ± 1.17	5.35 ± 0.75
MPAP	20.3 ± 6.7	18.9 ± 6.3	23.6 ± 6.5	21.6 ± 7.4
PCWP	13.2 ± 1.9	11.7 ± 2.9	14.7 ± 4.1	12.6 ± 2.3
RAP	11.0 ± 3.5	9.0 ± 3.5*	12.0 ± 3.1	10.8 ± 3.6
SVR	578 ± 177	706 ± 221*	580 ± 175	513 ± 126
PVR	56.8 ± 36.7	72.3 ± 34.3	88.5 ± 46.5*	71.4 ± 39.5
O_2Av/m^2	576 ± 107	507 ± 92	569 ± 135	732 ± 125
VO_2/m^2	87.3 ± 15.2	74.9 ± 16.9*	84.0 ± 21.0	105.2 ± 25.7*
O_2ER	15.6 ± 4.0	15.5 ± 4.7	15.1 ± 3.0	14.5 ± 3.4
Temperature	35.4 ± 0.4	34.8 ± 0.3	34.3 ± 0.5	35.6 ± 0.4

HR (bpm), MAP (mmHg), CI ($l/min/m^2$), MPAP (mmHg), PCWP (mmHg), RAP (mmHg), SVR (dynes x sec/cm^5), PVR (dynes x sec/cm^5), O_2Av/m^2 (ml/min), VO_2/m^2 (ml/min), O_2ER (%), Temp (°C). For abbrevations see text. Values are means ± SD, n=10, * statistically significantly different from preclamping values.

Table II. Hemodynamic and oxygen transport variables during HLT, OLT and primary non-function.

Variable	Before clamping of ICV and PV		After clamping of ICV and PV		After recirculation of graft		End of operation	
HR	88	± 16.8	82.3	± 17.2	80.3	± 15.3	93.3	± 17.2
MAP	91.3	± 16.2	87.3	± 6.4	80.3	± 15	89	± 7.5
CI	6	± 1.2	4.4	± 1.1	5.9	± 1.4	5.8	± 1
MPAP	24	± 3.6	17.3	± 5.5	22.7	± 7.8	19.3	± 3.1
PCWP	15.7	± 1.2	10	± 3.1	14.7	± 8.4	12	± 2
RAP	11.7	± 1.2	9.0	± 4	11.7	± 7.1	11	± 3
SVR	558	± 92	1092	± 453	587	± 202	669	± 198
PVR	59.7	± 20.6	73.3	± 34.5	69	± 27.2	61	± 4.4
O_2Av/m^2	784	± 250	598	± 270	749	± 116	823	± 236
VO_2/m^2	103.3	± 26.7	88.6	± 37.6	115.6	± 19.2	107.3	± 17.2
O_2ER	13.5	± 2.4	14.9	± 0.9	15.4	± 1.1	13.8	± 2.2
Temperature	35.6	± 0.2	35.1	± 0.3	34.7	± 0.3	35.7	± 0.2

HR (bpm), MAP (mmHg), CI (l/min/m^2), MPAP (mmHg), PCWP (mmHg), RAP (mmHg), SVR (dynes x sec/cm^5), PVR (dynes x sec/cm^5), O_2Av/m^2 (ml/min), VO_2/m^2 (ml/min), O_2ER (%), Temp (°C). For abbreviations see text. Values are means ± SD, n=3.

REFERENCES

Bigelow, W.G., Lindsay, W.K., Harrison, R.C., Gordon, R.A., and Greenwood, W.F., 1950, Oxygen transport and utilization in dogs at low body temperatures, Am. J. Physiol., 160: 125.

Cuervas-Mons, V., Millan, I., Gavaler, J.S., Starzl, T.E., and van Thiel, D.H., 1986, Prognostic value of preoperatively obtained clinical and laboratory data in predicting survival following orthotopic liver trans-plantation, Hepatology, 6: 922.

Groenland, T.H.N., Visser, L., Terpstra, O.T., Terpstra, J.L., Reuvers, C.B., Baumgartner, D., and Schalm, S.W., 1988, Stable hemodynamics during heterotopic auxiliary partial liver transplantation for end-stage liver cirrhosis, Transplantation Proceedings, 20 (suppl 1): 538.

Harper, A.M., Bain, W.H., Glass, H.I., Glover, M.M., and Mackey, W.A., 1961, Temperature difference in organs and tissues with observations on total oxygen uptake in profound hypothermia, Surg. Gynecol. Obstet., 112: 519.

Hegnauer, A.H., and D'Amato, H.E., 1954, Oxygen consumption and cardiac output in the hypothermic dog, Am. J. Physiol., 178: 138.

Lübbe, N., Bornscheuer, A., Grosse, H., Ringe, B., Gubernatis, G., and Seitz, W., 1988, Veränderungen des intraoperativen Gesamtsauerstoffverbrauchs bei Lebertransplantationen, Anaesthesist, 37: 211.

Mallett, S.V., Kang, Y.G., Freeman, J.A., Aggarwal, S., Gasior, T., and Fortunato, F.L., 1989, Prognostic significance of reperfusion hyperglycemia during liver transplantation, Anesth. Analg., 68: 182.

Michenfelder, J.D., and Theye, R.A., 1968, Hypothermia: Effect on canine brain and whole-body metabolism, Anesthesiology, 29: 1107.

Penrod, K.E., 1949, Oxygen consumption and cooling rates in immersion hypothermia in the dog, Am. J. Physiol., 157: 436.

Pugh, R.N.H., Murray-Lyon, I.M., Dawson, J.L., Pietroni, M.C., and Williams, R., 1973, Transection of the oesophagus for bleeding oesophageal varices, Br. J. Surg., 60: 646.

Shapiro, M.J., Wood, R.P., Shaw, B.W., and Grenvik, A., 1986, Postoperative care of liver transplantation patients, in: Hepatic transplantation - anesthetic and perioperative management, Winter, P.M., Kang, Y.G., eds., p 177, Praeger, New York, Westport, London.

Svensson, K.L., Sonander, H.G., Henriksson, B.A., and Stenquist, O., 1987, Whole body oxygen consumption during liver transplantation, Transplantation Proceedings, 19 (suppl 3): 56.

Van Thiel, D.H., Tarter, R., and Stone, B.G., 1986, Pathophysiology of liver disease, in: Hepatic transplantation - anesthetic and perioperative management, Winter, P.M., Kang, Y.G., eds., p 19, Praeger, New York, Westport, London.

ASSESSMENT OF CEREBRAL OXYGENATION AND HEMODYNAMICS BY NEAR INFRARED

SPECTROPHOTOMETRY DURING INDUCTION OF ECMO: PRELIMINARY RESULTS

K.D. Liem, J.C.W. Hopman, L.A.A. Kollée and B. Oeseburg, on behalf of the ECMO Research group

Faculty of Medical Sciences,
University of Nijmegen, The Netherlands

INTRODUCTION

Extra Corporeal Membrane Oxygenation (ECMO) has been proven to improve survival in newborn infants with severe respiratory failure, who do not respond to adequate conventional treatment[1,2]. However, induction of ECMO is not without risk. A high incidence of cerebrovascular injury has been described in infants treated with ECMO[3,4,5]. The etiology of cerebrovascular injury associated with ECMO might be multifactorial[4].

During the initial phase of ECMO, significant changes in cerebral circulation might occur due to ligation of the right common carotid artery and internal jugular vein, institution of cardiopulmonary bypass with decreased arterial pulsatility and rapid changes in gas exchange. These changes could be a potential risk for cerebrovascular injury. However, the influences of these changes on cerebral oxygenation and hemodynamic are not well known. Near infrared spectrophotometry (NIRS) offers the possibility of non-invasive and continuous cotside investigation of cerebral oxygenation in newborn infants[6,7]. Using Near Infrared Spectrophotometry (NIRS) we investigated the alteration of cerebral oxygenation and hemodynamics in relation to changes in physiological parameters during induction of ECMO in newborn infants.

MATERIAL AND METHODS

Five newborn infants (3 male and 2 female, gestational age 35 - 40 weeks, birth weight 2100 - 3470 grams) who required ECMO treatment for various reasons, were studied. Veno-arterial ECMO was instituted lege artis using cannulae in the right common carotid artery and the internal jugular vein. During bypass, blood was drained by gravity from right atrium into a reservoir and pumped into a membrane oxygenator by a roller pump. The oxygenated blood was returned to the body through a heat exchanger. After starting bypass ECMO flow rate was increased gradually in several minutes till a rate (ranging from 120 - 220 ml/kg/h) to maintain an appropriate SaO_2 (95 - 100%) as measured by pulse oximetry.

NIRS was used to measure the changes in the concentration of oxyhemoglobin (O Hb), deoxyhemoglobin (HHb), total hemoglobin (tHb) and oxidized cytochrome aa₃ (ox. aa₃) in cerebral tissue. The basis of this method is the spectrophotometric measurement of changes in the absorption properties of haemoglobin and cytochrome aa₃ in the near infrared region[8]. The NIRS equipment used was developed by the Department of Biomedical Engineering and Medical Physics, University of Keele (Prof. P. Rolfe) and made by Radiometer (Copenhagen, Denmark). The details have been described previously[7]. Light at 4 wavelengths (904, 845, 805 and 775 nm) is transmitted through the skull by fibre optic bundles. The transmitting and receiving optrodes were fixed to the skull as previously described[9] and placed at an angle of approximately 90°, one at the anterior fontanel, the other at the parieto-temporal region of the skull. The distance between the optrodes was measured with calliper and multiplied with 4.4 to obtain the optical pathlength[10]. Using the algorithms previously described in the literature[7], the concentration changes of O_2Hb, HHb, tHb and ox. aa₃ were calculated. Using the obtained optical pathlength and a value of 1.05 for brain specific gravity the concentration changes were expressed as μmol/100 g.

The following parameters were also continuously recorded: heart rate, mean arterial blood pressure (MABP), transcutaneous (tc) pCO_2 and pO_2 (Transend, Sensormedics) and SaO_2 by pulse oximetry (Nellcor N 200).

Measurements were started 30 min before cannulation and continued till 30 min after starting ECMO. The mean values over 20 sec periods at 10, 20 and 30 min after starting ECMO were calculated as relative changes in relation to the value 1 min before starting ECMO.

RESULTS

After ligation of the right common carotid artery and the internal jugular vein no changes in O_2Hb, HHb, tHb and ox.aa₃ concentration have been observed.

Immediately after starting ECMO, O_2Hb and ox.aa₃ concentration increased in all infants. The HHb concentration decreased (with exception of infant 3). The tHb concentration increased, despite decreased hematocrit due to ECMO hemodilution. The registration of the other parameters showed the following results: increase in SaO_2 and $tcpO_2$, decrease in $tcpCO_2$ (with exception of infant 5), more or less constant heart rate and increase in MABP (fig. 1). Blood gas analyses showed also improvement of PaO_2 and decrease in $PaCO_2$ (with exception of infant 5) (table 1).

Table 1. PaO_2 (kPa), $PaCO_2$ (kPa) and hemoglobin level (mmol/l) before and during ECMO in each infant.

infant	before ECMO			during ECMO		
	PaO_2	$PaCO_2$	Hb	PaO_2	$PaCO_2$	Hb
1	3.2	7.6	6.9	38.0	4.2	5.8
2	4.8	13.9	9.1	6.7	6.4	7.1
3	6.3	4.5	9.0	34.0	3.8	7.1
4	2.5	7.1	7.1	6.0	5.5	5.1
5	4.5	6.1	9.1	16.0	6.5	7.9

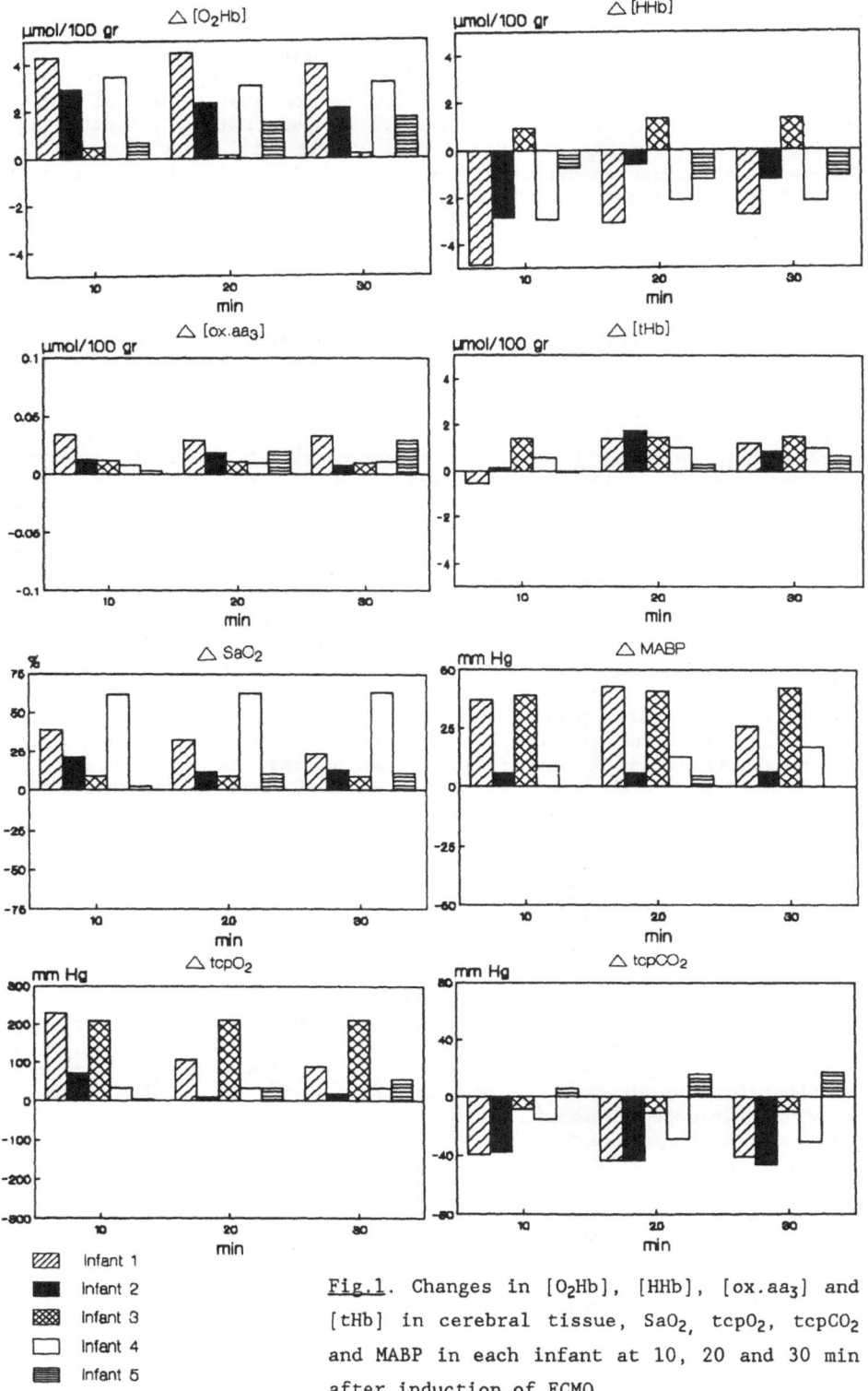

<u>Fig.1</u>. Changes in [O$_2$Hb], [HHb], [ox.aa$_3$] and [tHb] in cerebral tissue, SaO$_2$, tcpO$_2$, tcpCO$_2$ and MABP in each infant at 10, 20 and 30 min after induction of ECMO.

DISCUSSION

The goals of ECMO are to provide appropriate oxygenation and CO_2 removal and to minimize pulmonary barotrauma and O_2 toxicity. The typical patient is hypoxemic for hours before institution of ECMO. Induction of ECMO improves the arterial oxygenation as reflected by increasing SaO_2, $tcpO_2$ and PaO_2. This result will also improve O_2 delivery to tissue. As expected, the cerebral O_2 availability increases as indicated by increased O_2Hb and decreased HHb concentration. Finally an improvement of intracellular O_2 uptake in cerebral tissue occurs as indicated by increased ox. aa_3 concentration.

After induction of ECMO we observed an increase in tHb concentration in cerebral tissue. Since there is a decrease in hemoglobin concentration due to ECMO hemodilution, the increased tHb concentration must be the result of increased cerebral blood volume (CBV). A high correlation between cerebral blood flow (CBF) and CBV has been shown in animal experiment[11]. The increased CBV indicates increased CBF. An increased CBF has also been observed during non-pulsatile cardiopulmonary bypass in healthy normoxic cats[12]. On the other hand, in healthy normoxic newborn lambs initiation of ECMO does not alter CBF[13].Using Doppler-ultrasound an increase in mean cerebral blood flow velocity in the anterior cerebral artery has been shown during ECMO in newborn infants[14]. This reflects an increased CBF, since there is a significant correlation between the mean cerebral blood flow velocity and the mean CBF measured by [133]xenon clearance[15]. However, there is no clearcut explanation for this increasing CBF after induction of ECMO. Changes in arterial bloodgases are not responsible for this increased CBF; PaO_2 increased and $PaCO_2$ decreased, normally resulting in a decrease of CBF. Ligation of the right internal jugular vein did not alter tHb concentration, which means that the cerebral blood volume was also unchanged. This procedure does not result in cerebral venous congestion as suggested earlier[5]. The following mechanisms, whether or not in combination with each other, might be operative. Firstly, increased CBF might result from reactive hyperperfusion due to the prolonged hypoxia before ECMO. After hypoxemia in newborn piglets a significant hyperperfusion in all brain regions was found even 30 min after the restoring of normoxia and still present in the cerebrum and cerebellum after 60 min[16]. This might explain, why ECMO does not alter CBF in normoxic newborn lambs[13]. Secondly, we observed an increase in MABP after induction of ECMO. If the increased CBF is the result of increased blood pressure, then there must be a diminished autoregulation. Impaired autoregulation has been described in distressed newborn infant[17].
Thirdly, ECMO results in decreased arterial pulsatility. Using Doppler-ultrasound a pressure-passive blood flow velocity in middle cerebral artery indicating loss of autoregulation has been described during hypothermic non-pulsatile cardiopulmonary bypass in adults without previous prolonged hypoxia[18]. It has been shown that normothermic non-pulsatile cardiopulmonary bypass resulted in increased CBF, unrelated to metabolic needs, indicating a "luxury perfusion syndrome"[19]. Since ECMO results in reduced pulsatility, this might lead to decreased tone in cerebral resistance vessels resulting in increased CBF[13]. Fourthly, hemodilution will result in decreasing arterial O_2 content, being compensated by an increased CBF[20]. However, this mechanism does not seem very likely, since the decreased O_2 content is already compensated for by the significantly increased SaO_2.

Whatever the reason for increased CBF during the induction of ECMO, it might be harmful for cerebral tissue. Increased CBF has been mentioned as a risk factor in the etiology of cerebral bleeding in high risk newborns[21], especially in the presence of endothelial injury of

844

cerebral capillary vessels due to prolonged hypoxia. This increased CBF, in combination with heparinization, might contribute to the high incidence of cerebral hemorrhage during ECMO, which occurs in about 30% of ECMO treated infants[3,4]. Further study is necessary to determine the causes of increased CBF and its exact role in the etiology of cerebral hemorrhage during ECMO, before preventive measures could be suggested.

CONCLUSION

NIRS proved to be a valuable tool for studying cerebral oxygenation and hemodynamics. From these results it became clear that after starting ECMO the cerebral O_2 availability increases due to improvement of arterial oxygenation, finally resulting in improvement of intracellular O_2 uptake in cerebral tissue. There is an increase in CBV, which indicates increased CBF. This might be the result of reactive hyperemia or loss of autoregulation due to prolonged hypoxia before ECMO, or decreased arterial pulsatility due to cardiopulmonary bypass, or compensation for hemodilution caused by ECMO. The increased CBF might contribute to the high incidence of cerebral hemorrhage during ECMO.

ACKNOWLEDGEMENT

We gratefully acknowledge the cooperation of the coworkers of the departments of Anesthesiology, Central Animal Laboratory, Extracorporeal Circulation, Pediatrics, Pediatric Surgery, Child Neurology, Physiology and Thoracic Surgery who participate in the ECMO programme, directed by Prof. C. Festen.

REFERENCES

1. E. Stork, Extracorporeal membrane oxygenation in the newborn and beyond, Clin. Perinatol. 15:815 (1988).
2. P. P. O'Rourke, R. K. Crone, J. P. Vacanti, J. H. Ware, C. W. Lillehei, R. B. Parad and M. F. Epstein, Extracorporeal membrane oxygenation and conventional medical therapy in neonates with persistent pulmonary hypertension of the newborn: a prospective randomized study, Pediatrics 84:957 (1989).
3. R. E. Cilley, J. B. Zwischenberger, A. F. Andrews, R. A. Bowerman, D. W. Roloff and R. H. Bartlett, Intracranial hemorrhage during extracorporeal membrane oxygenation in neonates, Pediatrics 78:699 (1986).
4. G. A. Taylor, B. L. Short and C. R. Fitz, Imaging of cerebrovascular injury in infants treated with extracorporeal membrane oxygenation, J. Pediatr. 114:635 (1989).
5. A. Matamoros, J. C. Anderson, J. McConnell and D. L. Bolam, Neurosonographic findings in infants treated by extracorporeal membrane oxygenation (ECMO), J. Child Neurol. 4 (suppl):S52 (1989).
6. J. S. Wyatt, M. Cope, D. T. Delpy, S. Wray and E. O. R. Reynolds, Quantification of cerebral oxygenation and haemodynamics in sick newborn infants by near infrared spectrophotometry, Lancet ii:1063 (1986).
7. M. S. Thorniley, L. N. Livera, Y. A. B. D. Wickramasinghe, S. A. Spencer and P. Rolfe, The non-invasive monitoring of cerebral tissue oxygenation, Adv. Exp. Med Biol. 277:323 (1990).
8. F.F. Jobsis, Non-invasive, infrared monitoring of cerebral and myocardial oxygen sufficiency and circulatory parameters, Science 98:1264 (1977).

9. K. D. Liem, B. Oeseburg, J. C. W. Hopman, Method for the fixation of optrodes in near infrared spectrophotometry, Med. Biol. Eng. Comput. (1991) (accepted).

10. J. S. Wyatt, M. Cope, D. T. Delpy, P. van der Zee, S. Arridge, A. D. Edwards and E. O. R. Reynolds, Measurement of optical pathlength for cerebral near infrared spectroscopy in newborn infants, Dev. Neurosci 12:140 (1990).

11. J. Risberg, D. Ancri and D. H. Ingvar, Correlation between cerebral blood volume and cerebral blood flow in the cat, Exp. Brain Res. 8:321 (1969).

12. G. G. Santillan, J. M. Chemnitius and R. J. Bing, The effect of cardiopulmonary bypass on cerebral blood flow, Brain Res. 345:1 (1985).

13. B. L. Short, L. K. Walker, C. A. Gleason, M. D. Jones and R. J. Traytsman, Effect of extracorporeal membrane oxygenation on cerebral blood flow and cerebral oxygen metabolism in newborn sheep, Pediatr. Res. 28:50 (1990).

14. G. A. Taylor, B. L. Short, P. Glass and R. Ichord, Cerebral hemodynamics in infants undergoing extracorporeal membrane oxygenation: further observations, Radiology 168:163 (1988).

15. G. Greisen, K. Johansen, P. H. Ellison, P. S. Fredriksen, J. Mali and B. Friis-Hansen, Cerebral blood flow in the newborn infant: comparison of Doppler ultrasound and [133]xenon clearance, J. Pediatr. 104:411 (1984).

16. J. P. Odden, T. Stiris, T. W. R. Hansen, and D. Bratlid, Cerebral blood flow during experimental hypoxaemia and ischaemia in the newborn piglet, Acta Paediatr. Scand. 360 (suppl.):13 (1989).

17. C. L. Lou, N. A. Lassen and B. Friis-Hansen, Impaired autoregulation of cerebral blood flow in the distressed newborn infant, J. Pediatr. 94:118 (1979).

18. T. Lundar, K. F. Lindegaard, T. Froysaker, R. Aaslid, A. Grip and H. Nornes, Dissociation between cerebral autoregulation and carbon dioxide reactivity during nonpulsatile cardiopulmonary bypass, Ann. Thorac. Surg. 40:582 (1985).

19. K. Andersen, J. Waaben, B. Husum, B. Voldby, A. Bodker, A. J. Hansen and A. Gjedde, Nonpulsatile cardiopulmonary bypass disrupts the flow-metabolism couple in the brain, J. Thorac. Cardiovasc. Surg. 90:570 (1985).

20. M. D. Jones, R. J. Traytsman, M. A. Simmons and R. A. Molteni, Effect of changes in arterial O_2 content on cerebral blood flow in the lamb, Am. J. Physiol. 240:H209 (1981).

21. B. Friis-Hansen, Perinatal brain injury and cerebral blood flow in newborn infants, Acta Paediatr. Scand. 74:323 (1985).

OXYGEN UPTAKE AND STATIC LUNG COMPLIANCE DURING AUTOMATIC

VENTILATION

Y.A. Ruetsch[1], C.P. Naumann[1], A.P.K. Verkaaik[2],
W. Erdmann[2] and G.A. Zäch[1]

Departments of Anesthesiology. [1]Swiss Paraplegic Center, Nottwil,
Switzerland and [2]Erasmus University, Rotterdam, The Netherlands

INTRODUCTION

On-line registration of oxygen uptake has proven to be of value for determination of any impairmant of tissue oxygen supply when oxygen delivery has dropped to critical values (Van der Zee and Verkaaik, 1990). Disregulations of metabolism, e.g. in malignant hyperthermia, are seen in a pre-crisis phase (increase of oxygen consumption and of CO_2 production), and therapy can be started extremely early and before a disastrous condition has developped (Ellis, 1990; Verkaaik and Erdmann, 1990).

Oxygen uptake is extremely dependent on proper lung function. An additional easily registered value is on-line monitoring of static lung compliance. It guarantees an adequate control of lung function and early detectable deteriorations (e.g. atelectasis).

METHODS

An automatic feed-back controlled totally closed circuit system (PhysioFlex®) (Erdmann et al, 1989; Verkaaik and Erdmann, 1990) has been developed for quantitative application of anesthesia and ventilation. Four membrane chambers are introduced into the circuit through which the system gas is moved unidirectionally around by a blower with a flow of 70 liter/minute. The number of ventilation chambers integrated into the ventilation process is automatically chosen according to the preset tidal volume, thus keeping the compressible gas volume of the system as low as possible. Ventilation movements of the metal membranes in the chambers are registered capacitively and recorded in flow and displaced volume (ventilation). Pressure is controlled in the chambers and static lung compliance is also easily derived (static lung compliance = tidal volume/pressure difference where pressure difference = plateau pressure - peep pressure). The inspiratory oxygen concen-tration is computer feed-back controlled to set values, whereby the necessary oxygen supply (via electronic frequency valves) corresponds to oxygen uptake of the patient. To avoid wrong determination of oxygen uptake the system performs a leakage test prior to application with a leakage of 7 mL total gas leakage/minute not to be exceeded. Expiratory CO_2 is on-line registered whereafter absorbed via an in-line soda-lime absorber canister.

Oxygen Transport to Tissue XIV, Edited by W. Erdmann and
D.F. Bruley, Plenum Press, New York, 1992

The ventilation mode is designed on the screen before the patient is connected. The membrane movements are directed in such a way that the predesigned ventilation mode is meticulously copied accordingly.

A central computer based on a 16 Bite (40 megaBite memory capacity) VLSI architecture microprocessor serves for all feed-back controlled regulatory tasks of the system, data management, screen display and data storage. To avoid any risk no moving parts are included into this central computer system (e.g. hard disks or tape streamers). The software is printed in Eprom, the memory is filed in a non-volatile RAM (random access memory). This guarantees safe and accurate control and regulation of all regulatory tasks assigned to the computer, whereby a critical mass is achieved that shows strong functional synergistic effects.

The above-mentioned computer runs a self-calibration program and controls all measurements, calculations and feedback control processes including the accuracy of the various sensors and leakage-free functioning of the gas circuit and ventilatory system.

The system has been employed in routine anesthesia and in long-term intensive care ventilation. In 80 seriously ill patients, oxygen uptake has been related to the oxygen extraction ratio and static lung compliance changes were plotted against oxygen uptake, tidal volume and ventilatory pressures applied. Different modes of ventilation were employed as well as different anesthetics during anesthesia.

RESULTS

The individual patient shows a huge variety of physiological and pathophysiological changes during controlled ventilation and a broad spectrum of parameter interactions. The diagnostic possibilities of integrated description of the various phenomenons seen can only be described in a single case history of the various reaction patterns for the individual patient.

Table 1

Description of the evaluated parameters.

Upper parameter block:
VO_2	oxygen uptake	mL/min
Compl stat	static lung compliance	mL/cmH$_2$O

Lower parameter block:
Compl stat	static lung compliance	mL/cmH$_2$O
TV	tidal volume	mL
Delta press	plateau - peep pressure	cmH$_2$O

In the following four typical examples of VO_2, static lung compliance, ventilatory pressure support (plateau pressure minus positive end-expiratory pressure: PEEP) and tidal volume are reported separately for each patient and the interactions between these parameters are described.

The static lung compliance was derived on-line by dividing the measured tidal volume through delta press.

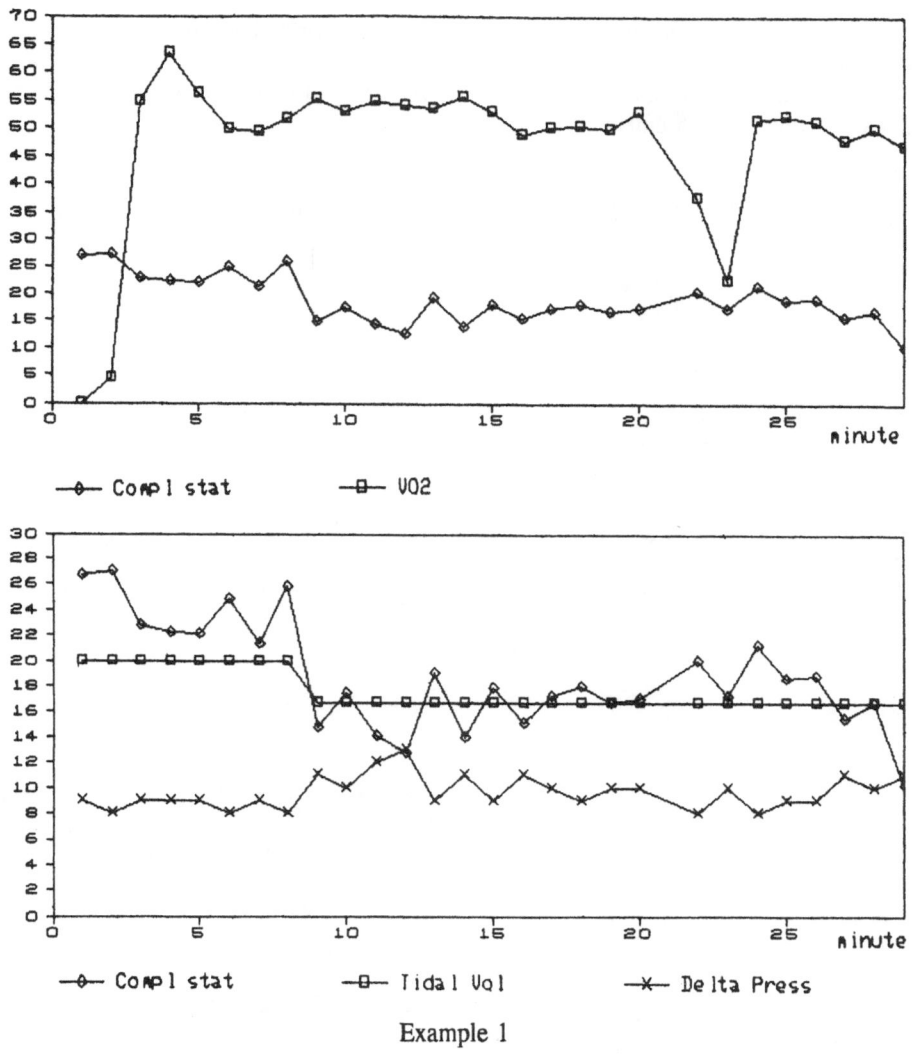

Example 1

A 55-year-old female with body weight of 45 kg undergoing spinal fusion L2-L3 after fracture. Balanced anesthesia with N_2O/O_2, ethrane 0.8% and addition of fentanyl. The anesthesia is performed using the PhysioFlex® closed circuit anesthesia ventilator. Relaxation is achieved with atracurium.

Upper parameter block: Static lung compliance (Compl stat, mL/cmH$_2$O) and oxygen consumption (VO$_2$, mL/min, measured value x 0.5) during the first 30 minutes of anesthesia. Oxygen uptake starts with 130 mL/min and decreases within 5 minutes to 100 mL/min under the influence of deepening anesthesia. After 20 minutes, the patient has an acute arterial bleeding with severe blood pressure drop that could be restored by volume replacement in 3 minutes. The patient has dropped in acute oxygen delivery (supply) dependency demonstrated by acute decrease of total oxygen consumption with the event. The static lung compliance (tidal volume divided by plateau pressure minus PEEP) decreases from a mean value of nearly 30 mL/cmH$_2$O in the first 10 minutes to 15 during the further course of anesthesia. The above mentioned bleeding event did not affect the static lung compliance.

Lower parameter block: Tidal volume (TV, mL, measured value x 0.1), pressure difference (delta press = plateau pressure - PEEP measured in cmH₂O). The tidal volume is maintained at 400 mL, lung compliance decreases after 10 min while delta press increases from 10 to 25 cmH₂O after 15 min.

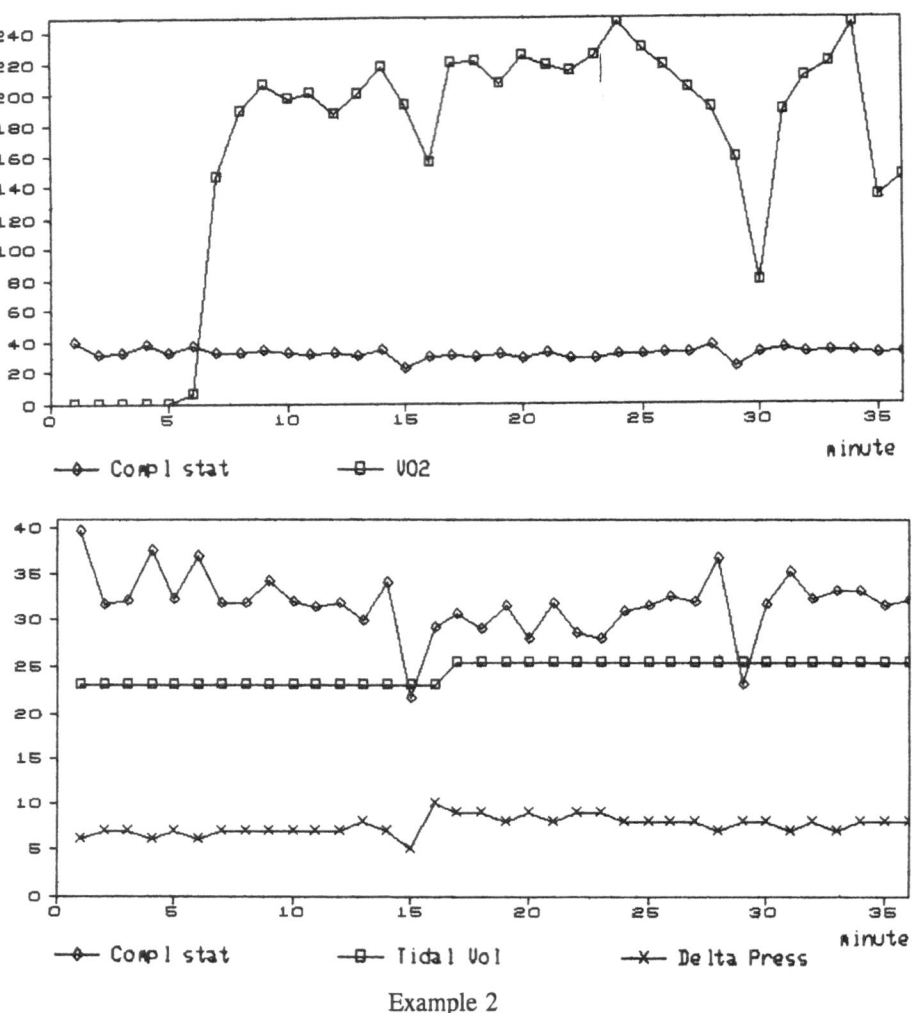

Example 2

A 60-year-old male, body weight 80 kg presenting for plastic surgery. The patient is bought under anesthesia and is maintained with N_2O/O_2 and ethrane 0.7% using the Physioflex® closed circuit anesthesia ventilator. Full relaxation is achieved with atracurium.

Upper parameter block: Static lung compliance (Compl stat, mL/ cmH₂O) and oxygen consumption (VO_2, mL/min) during the first 35 minutes of anesthesia. Static-lung compliance drops from 40 to 30 mL/cmH₂O in the first 15 minutes without further change. Oxygen consumption remains about 220 mL/min. The patient showed two acute bradycardic episodes (at 16 and 31 min) with concomitant severe blood pressure drop (again supply dependency), immediate treatment with methylatropine recovered the situation. Oxygen consumption drop was preceded by a drop in lung compliance which recovered rapidly and before the decrease in oxygen consumption was obvious.

Lower parameter block: Static lung compliance (Compl stat, mL/cmH$_2$O), tidal volume (TV, mL, measured value x 0.05), pressure difference delta press = plateau pressure - PEEP, cmH$_2$O). The first compliance drop (see upper parameter block) goes parallel with a drop in ventilation pressure. Compliance is not changed with increase of tidal volume from 460 to 520 mL.

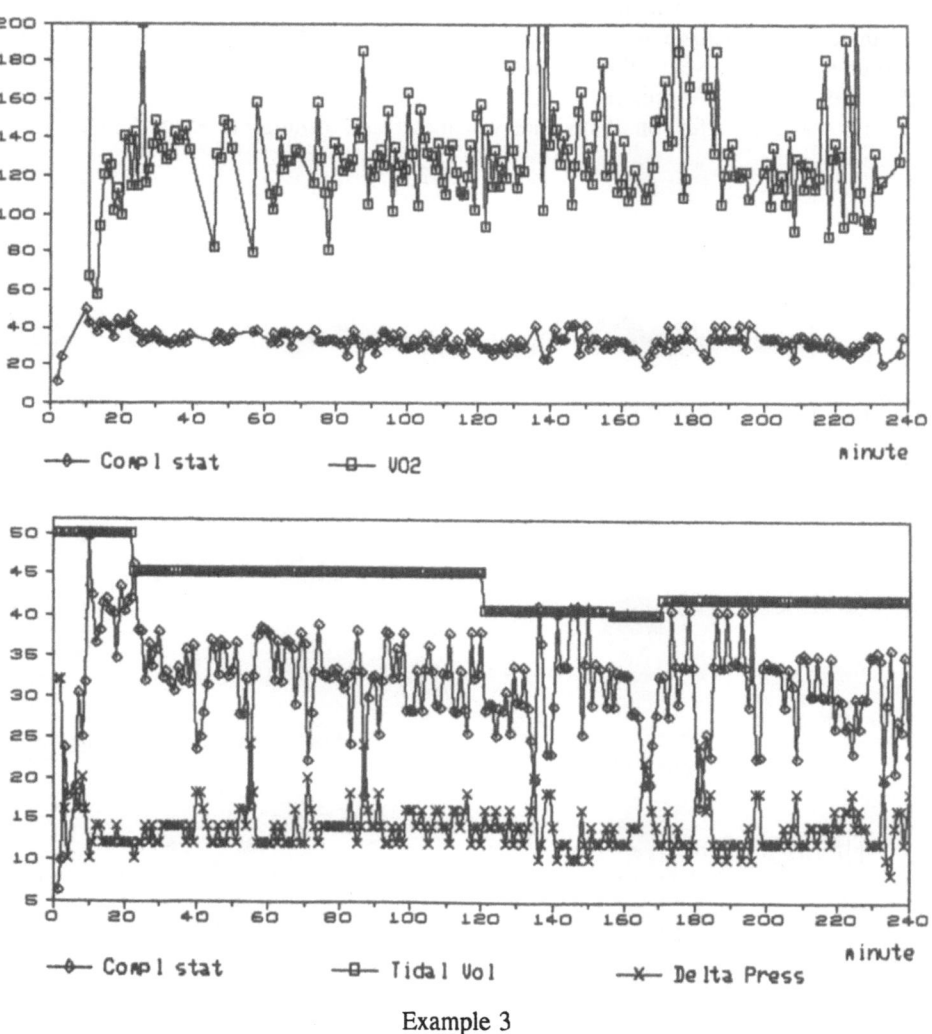

Example 3

A 23-year-old male, body weight 56 kg. Patient presents for major spinal column correction, because of postraumatic tetraplegia. Anesthesia is performed using the Physioflex° closed loop anesthesia ventilator, isoflurane 0.5% is used with N$_2$O/O$_2$ as carrier gas and increments of small doses of fentanyl. Relaxation is performed with pancuronium bromide.

Upper parameter block: Static lung compliance (compl stat, mL/cmH$_2$O) and oxygen consumption (VO$_2$, mL/min). Static lung compliance decreases from 45 to 35 mL/cmH$_2$O during the first 40 minutes and thereafter remains stable with short-lasting acute changes around 35 ml/cm H$_2$O. Oxygen consumption remains around 130 mL/min throughout the whole procedure of 230 minutes with short-lasting changes concomitant with those of static lung compliance.

Lower parameter block: Static lung compliance (Compl stat, mL/cmH$_2$O), pressure difference (delta press = plateau pressure - PEEP, cmH$_2$O), tidal volume (TV, mL, measured value x 0.1). Tidal volume had to be reduced from 500 to 440 mL throughout the procedure to keep the end tidal pCO$_2$ between 30 and 35 mmHg. The above mentioned fluctuations of the compliance are more obvious due to the difference in scaling. Fluctuations in ventilation pressure needed to achieve the tidal volume (volume controlled ventilation) are also present but much less marked than the changes in compliance. An intermittent change in tidal volume to 420 mL (minutes 120 to 170) has a slight effect on compliance which increases from 30 to 35 mL/cmH$_2$O.

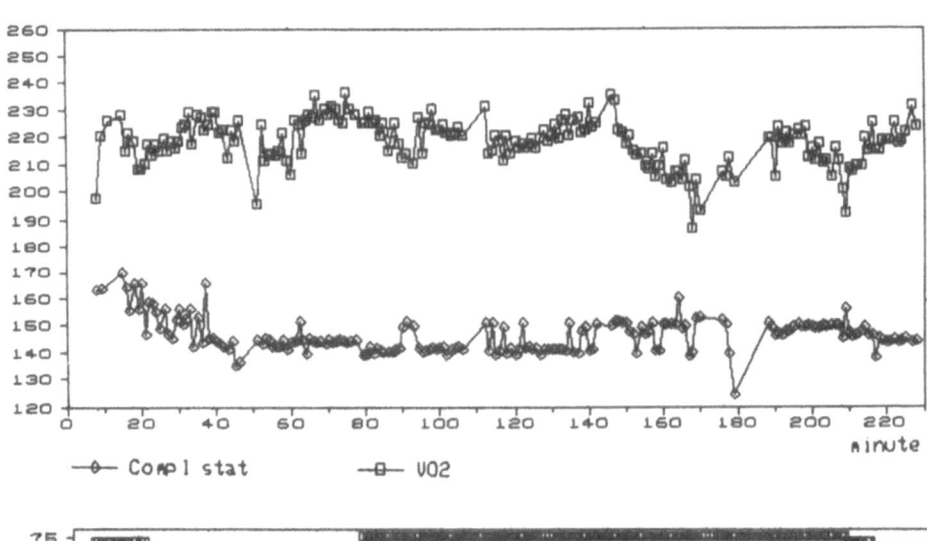

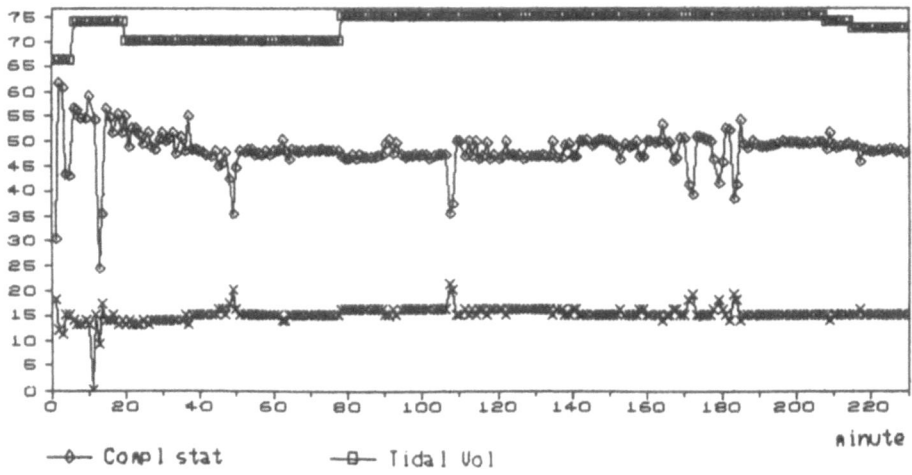

Example 4

A 35-year-old male, body weight 80 kg. Paraplegic patient presenting for spinal cord revision. Anesthesia is performed using PhysioFlex® closed loop anesthesia ventilator and N_2O/O_2 for basic analgesia supplemented by fentanyl. Sleep is achieved by continuous infusion of diprivan, no volatile anesthetic is used. The patient remained under full relaxation throughout the whole anesthesia by intermittent doses of atracurium.

Upper parameter block: Static lung compliance (Compl stat, mL/cmH_2O, measured value x 3), oxygen uptake (VO_2, mL/min). Static lung compliance decreases from 55 mL/cmH_2O during the start of ventilation to 48 in the first 50 minutes and thereafter remains unchanged throughout the whole procedure. The reason for a dip at 180 min was immediately resolved being due to loss of relaxation. Oxygen consumption ranged around 220 mL/min slightly decreasing to 210 after 160 minutes.

Lower parameter block: Static lung compliance (Compl stat, mL/cmH_2O), tidal volume (TV, mL, measured value x 0.1), pressure difference (delta press = plateau pressure - PEEP, cmH_2O). The 80 kg patient needed a tidal volume of 750 mL. The decrease of static lung compliance did not correspond with either changes in tidal volume or ventilation pressure which remained stable at 15 cmH_2O throughout the procedure, with just a few peaks when relaxation became less (at 160 to 180 minutes).

DISCUSSION

Example 1

Acute and severe arterial bleeding at minute 20 brings the patient into a rapidly developing oxygen supply dependency condition diagnosed by an acute decrease of oxygen uptake. This bleeding (hypovolemia) induced a tissue supply failure, but does not affect compliance and related pulmonary ventilation parameters.

Example 2

Parasympathicotonic decrease of heart rate and cardiac output may lead to oxygen supply dependency seen in an acute decrease of oxygen consumption of the patient. This parasympaticotonically induced tissue oxygen supply failure is preceeded by a pulmonary reactive drop of compliance and ventilatory support pressure needed to maintain the preset tidal volume.

Example 3

Under normal eventless conditions pulmonary compliance might show significant fluctuations with concomitant changes of needed pressure support to achieve preset tidal volumes in volume controlled ventilation. Changes of tidal volume do not markedly influence compliance and related pulmonary ventilatory pressures.

Example 4

At 160 to 180 minutes a slight drop of oxygen uptake is seen concomitant with instability and drop of compliance simultaneous with regulatory increases of ventilatory pressure support. The changes were due to loss of relaxation and administration of an additional dose of atracurium resolved the problem.

CONCLUSION

The previously unavailable continuous control of the two parameters, oxygen uptake and static lung compliance, proved to decisively enlarge the possibilities of early detection of tissue oxygen supply impairment as well as pulmonary dysfunction, impairment of gas exchange and inadequate anesthesia respectively relaxation.

Immediate therapeutic actions, medication, changes of the ventilatory mode, blood volume replacement and/or cardiovascular therapy can be employed in such a way under continuous control that re-establishment of normal physiology can be overseen and achieved before further deterioration occurs.

SUMMARY

Oxygen uptake is an important parameter to control proper tissue oxygen delivery. Oxygen uptake is dependent on adequate lung function and easily disturbed by changes in lung compliance and related parameters such as the tidal volume controlling pressure support.

Simultaneous on-line registration of oxygen uptake and lung compliance together with ventilatory pressures applied to achieve preset tidal volumes has been made possible using the computer feedback controlled closed circuit ventilatory system Physio-Flex® (Physio Co, Hoofdorp, The Netherlands). The system guarantees for leakage free functioning (maximal leakage 7 mL gas loss/minute) and, therefore, patient oxygen consumption measurements with an accuracy of more than 95%. A specially developed membrane ventilation mode registers on-line flow and displaced volume automatically corrected for temperature, pressure and compressible volume.

The current investigation has shown:

Decrease oxygen consumption versus oxygen delivery supply dependency may be induced by reflectory decreases of heart rate and cardiac output; in this case a reactive pulmonary parameter change is preceeding the event in form of a drop in compliance and corresponding changes in ventilatory pressures necessary to maintain the preset tidal volume.

In contrast, decrease of oxygen uptake following changes in cardiac output due to acute hypovolemia has no effect on pulmonary function parameters. This can be diagnostically used as moderate changes of tidal volume also have no significant influence on pulmonary parameters. However, changes due to reduction in depth of anesthesia and relaxation have some influence and need to be excluded.

REFERENCES

Ellis, F. R., 1990, Editorial II, Predicting malignant hyperthermia, Brit J Anaesth, 64: 411-412.

Erdmann, W., Veeger A. I. and Verkaaik A. P. K., 1989, Narkosegeräte: Gegenwart und Zukunft, In: Narkosebeatmung (low flow, minimal flow, geschlossenes System), Eds Jantzen J.-P.A.H. and Kleemann P.P., Schattauer, Stuttgart, New York, pp 5-17.

Van der Zee, H. and Verkaaik A. P. K., 1990, Cardiovascular implementations of respiratory measurements, Acta Anaesth Belg, 41:167-175.

Verkaaik, A. P. K. and Erdmann W., 1990, Respiratory diagnostic possibilities during closed circuit anesthesia, Acta Anaesth Belg, 41: 177-188.

OXYGEN UPTAKE / SUPPLY DEPENDENCY IN HUMAN SEPSIS: DOES IT INCREASE THE RISK OF MULTISYSTEM ORGAN FAILURE?

F. Esen[1], L. Telci[1], K. Akpir[1], J. Kesecioglu[1,2], T. Denkel[1], K. Pembeci[1]

[1]Department of Anesthesiology, University of Istanbul, Faculty of Medicine Istanbul, Turkey
[2]Department of Anesthesiology, Erasmus University and Academic Hospital Dijkzigt, Rotterdam, The Netherlands

INTRODUCTION

Multisystem organ failure (MSOF) is a major cause of death during critical illness, especially following sepsis. Although the pathophysiology of MSOF is poorly understood, the abnormal relationship between O_2 delivery (DO_2) and consumption (VO_2) is thought to be the potential cause of MSOF by causing cellular hypoxia and dysfunction [1].

Tissue O_2 uptake depends both on metabolic activity and O_2 delivery. VO_2 becomes limited by O_2 supply when DO_2 is reduced to a critical level. This relationship has been observed in a variety of critical illnesses, including sepsis. Clinical studies of patients in sepsis have shown O_2 uptake/supply dependency [2-7]. In these studies DO_2 was increased using crystalloid or colloid infusions, red blood cells transfusion, and dobutamine. Vincent and coworkers used a short-term dobutamine infusion to demonstrate the O_2 uptake/supply dependency in a group of critically ill patients with heart failure or sepsis [7]. Bihari and coworkers, using a fixed dose of prostocyclin, showed that the phenomenon of O_2 uptake/supply dependency was associated with increased mortality in acute respiratory failure [8].

To date, there has been very little clinical research to confirm whether the early recognition of this relationship may have important clinical consequences, such as MSOF. In this prospective study, we attempted to evaluate if O_2 uptake/supply dependency in the early course of sepsis increases the risk of MSOF.

METHODS

We examined 52 septic patients aged between 13-68 years (47.59 ± 19.16) meeting the three following criteria:
    1. Positive blood cultures and/or documented source of infection.
    2. Leukocytosis ($> 10\,000/\ mm^3$) or leukopenia ($< 6000\ /\ mm^3$).
    3. Core temp. above 38 °C or below 36.5 °C.

Oxygen Transport to Tissue XIV, Edited by W. Erdmann and
D.F. Bruley, Plenum Press, New York, 1992

In all patients, hypovolemia had been previously corrected by fluids. Patients with anemia (Hb level < 9g/dl) and severe persisting hypoxemia (PaO_2/FiO_2 < 150 mmHg) were not included in the study. All patients were artificially ventilated with a Servo 900C (Siemens-Elema), arterial and pulmonary arterial catheterization were performed for hemodynamic monitoring.

After measuring the baseline hemodynamic profile, each patient received dobutamine infusion starting with 5 μgr/kg/min, and increasing to 10 and 15 μgr/kg/min at 30 min intervals while mean arterial pressure (MAP), mean pulmonary arterial pressure (MPAP), pulmonary capillary wedge pressure (PCWP), cardiac output (CO), VO_2, DO_2 blood gases, and serum lactate were determined at each dose administration. No change in the therapy was allowed prior to the baseline measurements and during the 30 min dobutamine infusions. None of the patients were receiving vasopressors or inotropes.

All pressures were monitored with a Viggo-Spectramed transducer on a Horizon 2000 (Mennen Medical). CO was measured by the thermodilution technique (Hempro 1-Spectramed), and the cold bolus of injection was started at the end of the inspiratory phase. Five measurements were averaged to obtain each CO value. Arterial and mixed venous blood gas analyses were determined by ABL 300 (Radiometer, Copenhagen); hemoglobin concentration and saturation were measured with Hemoximeter OSM3 (Radiometer). Arterial lactate concentration was determined by TDX Abbott.

Oxygen consumption (VO_2 ml/min) was obtained by direct measurement of expired gas (Datex, Multicap). Oxygen delivery (DO_2 ml/min) was calculated by the simplified formula:
$$DO_2 \ = \ CO \ x \ Hb \ x \ SaO_2 \ X \ 13.9$$

After the dobutamine test, all patients were evaluated prospectively for the development of organ system failure throughout their stay in the ICU. Each organ failure was defined according to the criteria given by Friedman et al. [9].

The measured values obtained before and during dobutamine administration were compared using Student's t-test, and statistical significance was accepted as $p < 0.05$.

RESULTS

Each dose of dobutamine infusion was well tolerated in all patients. Principal hemodynamic parameters measured before and after every dose administration are presented in Table 1.

A slight increase in arterial pressure, and a decrease in pulmonary arterial pressure were observed with the increasing dose of dobutamine. CO markedly increased as expected. No significant differences were noted in blood gases and hemoglobin saturation.

O_2 delivery dramatically increased in all patients, with a substantial increase in mixed venous O_2 saturation. Mean values of O_2 consumption showed a parallel increase

Table 1. Hemodynamic parameters before and after 30 min infusion of each dose dobutamine.

	Baseline	5 μgr/kg/min	10 μgr/kg/min	15 μgr/kg/min
MAP (mmHg)	73.4 ± 6.2	73.7 ± 5.6	75.5 ± 4.8	76.1 ± 4.6
MPAP (mmHg)	31.4 ± 3.9	30.8 ± 3.8	29.8 ± 3.5	29.3 ± 3.8
PCWP (mmHg)	19.5 ± 2.1	18.7 ± 1.5	17.8 ± 1.5	16.6 ± 1.8
HR (bpm)	94 ± 18	100.7 ± 14.3	104.4 ± 15.6	105 ± 15.2
CO (l/min)	6.3 ± 1.9	7.6 ± 1.7[a]	8.7 ± 1.8[b]	10.0 ± 2.2[b]
DO_2 (ml/min)	621.4 ± 112.3	826.1 ± 211.6[a]	949.7 ± 235.0[b]	1088.0 ± 250.0[b]
VO_2 (ml/min)	182.1 ± 42.3	205.6 ± 60.2[a]	245.7 ± 75.4[b]	307.4 ± 112.0[b]
Lactate (mEq/l)	4.3 ± 1.2	3.5 ± 1.8[a]	3.4 ± 1.8	3.3 ± 1.7
SaO_2 (%)	96.2 ± 2.1	96.5 ± 1.6	96.8 ± 1.5	96.9 ± 1.8
SvO_2 (%)	63.1 ± 4.7	65.6 ± 4.1	66.8 ± 3.8	69.6 ± 4.5

[a] $p < 0.05$ from baseline value
[b] $p < 0.05$ from the previous dose administration
MAP = Mean arterial pressure, MPAP = Mean pulmonary arterial pressure, PCWP = Pulmonary capillary wedge pressure, HR = Heart rate, CO = Cardiac output, DO_2 = Oxygen delivery, VO_2 = Oxygen consumption.

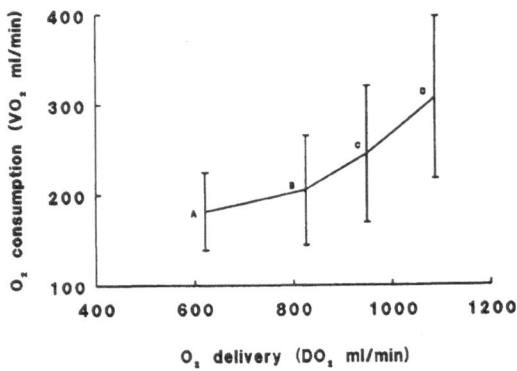

Figure 1. Relationship between mean values of DO_2 and VO_2. A = baseline; B = 5 μgr/kg/min, C = 10 μgr/kg/min, D = 15 μgr/kg/min dobutamine infusion.

with each dose of dobutamine administration. Individual data demonstrated a consistent increase in VO_2 associated with an increase in DO_2. Relationship between mean values of DO_2 and VO_2 is shown in Figure 1.

Lactate levels were higher than normal before dobutamine infusion, mean value being 4.1 ± 1.9 mEq/l, and decreased to 3.5 ± 1.8 mEq/l after 5 μgr/kg/min dobutamine infusion. However, the values did not alter after 10 and 15 μgr/kg/min dobutamine administration.

Adult respiratory distress syndrome (ARDS) defined as hypoxemia (PaO_2 < 60 mmHg with a FiO_2 > 0/4) with diffuse pulmonary infiltrates in the chest radiograph occurred in 41 patients throughout their stay in the ICU. Renal failure occurred in 35 patients and 38 patients (73%) died of at least 3 organ system failures, mostly respiratory, renal and liver failure. Development of organ system failures are summarized in Table 2.

Table 2. Clinical data for the development of organ failure in 52 patients

Organ systems	no of patients
Cardiovascular failure	8 (15.8%)
Respiratory failure	41 (78.8%)
Renal failure	35 (67.3%)
Hemotological problems	36 (69.2%)
Neurologic disorders	5 (9.6%)
Liver failure	34 (65.3%)

DISCUSSION

Occult cellular hypoxia has been shown to develop in sepsis and cause MSOF. Clinical studies of septic patients have demonstrated that VO_2 was dependent on DO_2 [2-10]. Evidence to support this pathologic relationship is associated with the development of lactic acidosis which is thought to reflect the presence of tissue hypoxia. Some of the literature suggested that lactic acidosis identifies delivery dependence of VO_2 in human sepsis [3,4,11], but other studies showed that acute decrease in O_2 delivery has not always been associated with lactic acidosis, since O_2 extraction capabilities can vary with the type of disease [12]. On the other hand, lactic acidosis can also be associated with disease processes other than circulatory failure, such as insufficiency of lactate clearance in liver failure. Thus, other than the interpretation of the lactic acid level, the pathologic relationship of VO_2 and DO_2 was thought to be the real evidence of tissue hypoxia in sepsis.

Rapid increase in O_2 delivery was used as experimental models of O_2 uptake/supply dependency. In this study, we increased O_2 delivery by using an inoptropic agent dobutamine. Although an adrenergic agent is said to increase cellular O_2 consumption by activating adenosine monophosphate, studies observed no change in VO_2 with doses of dobutamine up to 15 μgr/kg/min [7]. In their study, Vincent and coworkers reported data indicating that dobutamine can be safely administered to document the phenomenon of O_2 uptake/supply dependency. However, not many clinical studies exist, where dobutamine was used in septic states. Shoemaker and colleagues used dobutamine in septic and postoperative patients, and they observed a significant increase in VO_2 independent of the initial CO values [3]. Vincent and coworkers observed that the increase in VO_2 by increasing DO_2 with dobutamine was only significant in the presence of lactic acidosis. They claimed that lactic acidosis was seen when O_2 uptake/supply dependency was present. In this study all patients showed lactate levels higher than normal prior to dobutamine infusion. Therefore, evaluation of the VO_2 and DO_2 in states of normal lactate concentrations was not possible. However, Annat and colleagues did not observe O_2 uptake/supply dependency in ARDS patients with normal lactate levels [11]. This data was supported by Vincent's observation. In accordance with Annat and Vincent, our results showed that each septic patient individually had a significant increase in O_2 consumption associated with an increase in O_2 delivery by dobutamine infusion. Although some decrease in lactate concentration with the increase in DO_2 by 5 μgr/kg/min dobutamine was seen, values did not alter with doses of 10 and 15 μgr/kg/min. This data supported the fact of increased O_2 demand during sepsis.

The determination of O_2 consumption in finding out the O_2 uptake/supply dependency is still controversial. Numerous studies showed that mathematic coupling of shared variables CO and CaO_2, causes an artifactual correlation between calculated VO_2 and DO_2 [13,14]. Therefore in this study VO_2 was measured directly from a modified respiratory gas analyser. The reliability of this system has been demonstrated in patients undergoing artificial ventilation. Additionally, this system has the advantage of giving a continuous measurement of VO_2 in contrast to the Fick method.

Present observations have suggested that sepsis is the major cause of MSOF and it induces a pathologic dependence of VO_2 and DO_2. Bihari and coworkers, in their study with ARDS patients, demonstrated that O_2 uptake/supply dependency was seen in non-survivors, but not in survivors [8]. Retrospective studies have found greater DO_2 and VO_2 in survivors than in non-survivors early in the course of ARDS and sepsis [15]. However, these reports are not sufficiently clear to evaluate the relationship between dependent states and subsequent MSOF. The question "should DO_2 be increased to supra levels in sepsis in order to prevent MSOF?" is still not clearly answered.

The results of this study are of clinical importance as this uptake/supply dependency is closely associated with the development of MSOF and death. Therefore this phenomenon appears to be an alarm to correct the underlying O_2 deficit to the tissues. In conclusion, these data necessates further studies to investigate whether the pathologic relationship between O_2 uptake and delivery can be a prognostic indicator of MSOF in sepsis. Morover, prevention of the development of organ system failure by increasing O_2 delivery to supra normal levels deserves further clinical investigation.

REFERENCES

1. J. A. Russel, O_2 delivery and consumption in adult respiratory distress syndrome and sepsis , Update in Intensive Care and Emergency Medicine. (ed. by JL Vincent), Berlin, Heidelberg, Springer-Verlag; 175-181 (1991).
2. B. S. Kaufman, E. C. Backow, and J. L. Falkj, The relationship between O_2 delivery and consumption during fluid resuscitation of hypovolemic and septic shock, Chest. 85:336-340 (1984).
3. M. T. Haupt, E. M. Bilbert, and R. W. Carlson, Fluid loading increases O_2 consumption in septic patients with lactic acidosis, Am. Rev. Respir. Dis. 131:912-916 (1985).
4. E. M. Gilbert, M. T. Haupt, R. Y. Mandanas, and A. J. Huaringen, The effect of fluid loading, blood transfusion, and catecholamine infusion on O_2 delivery and consumption in patients with sepsis, Am. Rev. Respir. Dis. 134:873-878 (1986).
5. Y. G. Wolf, S. Coter, A. Perel et al, Dependence of O_2 consumption on cardiac output in sepsis, Crit. Care Med. 15:198-203 (1987).
6. M. Astiz, E. Rackow, J. Falk et al, O_2 delivery and consumption in patients with hyperdynamic septic shock, Crit. Care Med. 15:26-29 (1987).
7. J. L. Vincent, A. Roman, D. de Baker, and R. J. Kahn, O_2 uptake/supply dependency: Effects of short term dobutamine infusion, Am. Rev. Respir. Dis. 141:2-7 (1990).
8. D. Bihari, M. Smithies, A. Gimson, and J. Tinker, The effects of vasodilatation with prostacyclin on O2 delivery and uptake in critically ill patients, New Engl. J. Med. 317:397-403 (1987).
9. B. C. Friedman, J. F. Williams, and W. A. Knaus, Multisystem organ failure: Outcome and clinical implications, The American Society of Anesthesiologists. 83-94 (1989).
10. S. J. Danek, J. P. Lynch, J. G. Weg, and D. R. Dantzker, The dependence of O_2 uptake on O_2 delivery in the adult respiratory distress syndrome, Am. Rev. Respir. Dis. 122:387-395 (1980).
11. G. Annat, J. P. Viale, C. Percival, M. Froment, and J. Motin, O_2 delivery and uptake in the adult respiratory distress syndrome: lack of relationship when measured independently in patients with normal blood lactate concentrations, Am. Rev. Respir. Dis. 133:999-1001 (1986).
12. W. C. Shoemaker, P. L. Appel, and H. B. Kram, Hemodynamic and oxygen transport effects of dobutamine in critically ill general surgical patients, Crit. Care Med. 14:1032-1037 (1986).
13. C. G. Vermeij, B. W. A. Feenstra, and H. A. Bruining, O_2 delivery and O_2 uptake in postop and septic patients Chest. 98:415-420 (1990).

14. H. H. Stratton, R. J. Feustel, and J. C. Newell, Regression of calculated variables in the presence of shared measurement error, J. Appl. Physiol. 62:2083-2093 (1987).
15. J. A. Russel, J. J. Ronco, D. Lockhat, A. Belzberg, M. Kiess, and P. M. Dodek, Oxygen delivery and consumption and ventricular preload are greater in survivors than in nonsurvivors of the adult respiratory distress syndrome, Am. Rev. Respir. Dis. 141:659-665 (1990).

INTRA-ANESTHETIC ON-LINE MONITORING OF OXYGEN CONSUMPTION USING

A CLOSED CIRCUIT SYSTEM

H. Wauer*, M. Schädlich*, A.P.K. Verkaaik** and W. Erdmann**

* Department of Anesthesiology and Intensive Care, Charité, Humboldt University, Berlin, Germany and ** Erasmus University Rotterdam, The Netherlands

INTRODUCTION

There are three main reasons to advocate the use of a closed circuit system in anesthesia and intensive care:

1. Economical: cost reduction by decrease of gas consumption (oxygen, nitrous oxide and volatile anesthetics);
2. Environmental: reduction of anesthetic gas losses into the atmosphere and direct work place environment;
3. Monitoring: Continuous on-line monitoring of oxygen consumption and CO_2 production, thus oxygen metabolism.

The necessity to decrease anesthetic gas consumption because of financial problems (nitrous oxide was too expensive) stimulated Jackson to experiment with a closed circuit system as early as the beginning of this century [1]. Today we know that a decrease of anesthetic gas consumption is not only a question of economy but is urgently needed to reduce environmental pollution.

A few thousand million years ago no oxygen was yet present in our atmosphere. In the surface water, oxygen appeared by several biochemical processes as a waste product. Excess oxygen escaped into the atmosphere where in the upper layers it became present as ozone and blocked solar radiation. Anesthetic gases, such as nitrous oxide, inhibit ozone formation and serve to destroy the existing ozone layer.

If a totally closed circuit system is introduced the most decisive feed-back control to be introduced is on-line adjustment of inspiratory oxygen concentration. As the basic principle of a closed circuit system is that inflow into the system equals uptake by the patient, oxygen supplementation needed equals oxygen consumption by the patient. Having achieved a functional feed-back control of oxygen inflow the closed circuit enables on-line registration of oxygen consumption of the patient. This decisive parameter continuously informs about cellular oxygen demand/supply ratio and changes of cellular oxygen consumption (e.g. sepsis, temperature increase and increase of metabolic rate). Introducing capnography and thus getting insight into CO_2 production accomplishes the goal for on-line monitoring of oxygen metabolism [2]. In the present study, a microprocessor controlled system was introduced into patient care.

Oxygen Transport to Tissue XIV, Edited by W. Erdmann and
D.F. Bruley, Plenum Press, New York, 1992

METHODS

The heart of the developed ventilator (PhysioFlex®, Hoofddorp, The Netherlands) is a 16-bit computer. Usually a computer produces data, display and stores data which are later available to be taken out and filed, furthermore short-term trend recordings are made available and the integrated data are on-line transferred to a data management system. In addition the computer can be used to compare measured and integrated parameters to preset upper and lower parameter limits which, when surpassed, trigger off alarm signals. These signals are specific for the respective parameters in failure and are defined by visual and acoustic modes of alarm. Special calculated programs, integrated into the computer, enable

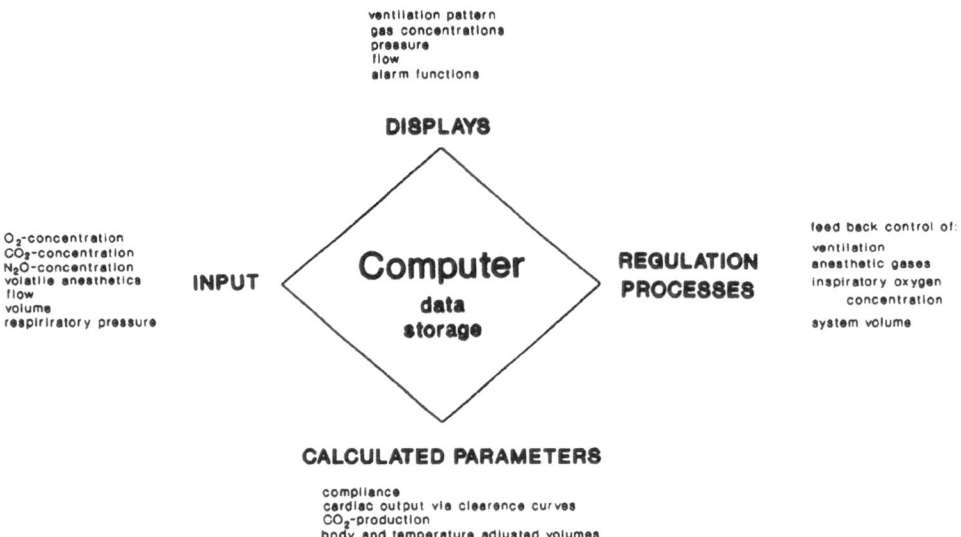

Fig. 1: Basic principle of computer controlled closed circuit ventilation.

automatic calculation of desired parameters and either display them or just store them as required (e.g. compliance, resistance, CO_2 production, anesthetic gas consumption). (Fig. 1).

The aim of the current study was to assign a third task to the computer: the feed-back control of measured parameters to preset values by automatic adjustment of the electronically regulated hardware (e.g. fresh and anesthetic gas dosaging, inspiratory/expiratory valves and ventilatory modes).

The basic principle is an open T-piece as described by Ayre in 1956 [3]. A blower guarantees the unidirectional flow of 70 l per minute preventing rebreathing of the patient. Ventilation is achieved by four membrane chambers dividing the system in two parts: the external circuit from which ventilation is performed and the internal, anesthesia circuit. The membranes separate pressurized air from anesthetic gas and serve for registration of volume displacement in the system by capacitively measuring movements of the metal membrane with displacements sensors. The computer controls the membrane movement, and therefore the ventilation pattern, corresponding to the set ventilation mode (inspiratory and expiratory flow, tidal volume, frequency).

Three feed-back regulation systems are integrated into the ventilator:
1. For feed-back control of oxygen concentration a sensor in the inspiratory part sends the measured values to the computer. If the oxygen concentration is lower than the set value a frequency valve is activated, the rate being dependent on the difference between set and measured value. The valve is connected to a stable oxygen pressure source achieved by a pressure reducing valve. Per beat the frequency valve injects 0.5 ml of oxygen into the circuit. The number of beats per minute multiplied by 0.5 is the total amount of oxygen needed to keep the oxygen concentration constant and corresponds to the amount of oxygen disappeared from the system into the patient.
2. Inhalation anesthetics are injected into the system via a step motor driven syringe. The speed of injection corresponds to the difference between end expiratory (alveolar) volatile anesthetic concentration and the set end expiratory concentration. Thus with a large difference a large volume will be injected and the inspiratory concentration becomes high and guarantees a speedy saturation of the patient. With increasing end expiratory values and decreasing differences to the set value, the injection rate decreases with a corresponding decrease of the inspiratory anesthestic concentration until end expiratory and inspiratory anesthetic concentration are the same (the patient is fully equilibrated to the pre-set end expiratory anesthestic concentration).
3. For N_2O another mechanism was chosen. As inspiratory oxygen concentration is maintained constant, volume loss from the system is only due to N_2O uptake. End expiratoryly the volume loss is measured and N_2O is injected (again by means of a frequency valve as described before for oxygen) until the system volume is the original. To fully comply with the goal to totally avoid environmental pollution, an active coal absorber for volatile anesthetics is connected to the circulatory flow to adsorb excess gases at the moment the set value is reduced or brought to zero (end of anesthesia).

About 200 patients aged between 8 and 81 years were ventilated with this new computer-controlled closed circuit system PhysioFlex®.

RESULTS

Besides an enormous reduction of anesthetic gas consumption to below 10% of the consumption in conventional semi-closed systems, the on-line parameter of oxygen consumption has proven to be a highly valuable additional control parameter of the physiological status of the patient. We have seen that changes in oxygen uptake were related to special clinical situations. The absolute amount of oxygen depends on various individual parameters. Increased oxygen uptake may be due to reduced or inadequate sedation, less relaxation or can be a first symptom of malignant hyperthermia. Decreased oxygen uptake show imbalances of demand/supply ratio which may originate from pulmonary or circulatory disturbances and cell failure (Figs. 2,3).

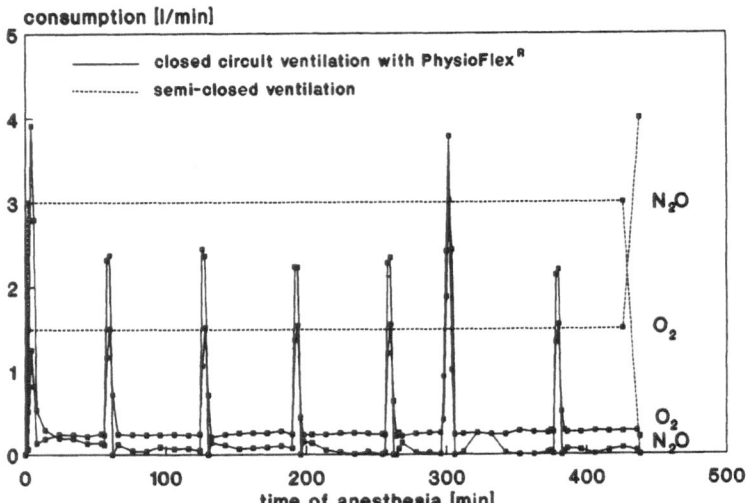

consumption [l/min]

Fig. 2.   Comparison of oxygen and nitrous oxide consumption during semi-closed ventilation (N$_2$O 3 l/min: O$_2$ 1.5 l/min) and ventilation with the closed circuit system (PhysioFlex®), male patient, 47 years, 78 kg body weight. The consumption per hour with the semi-closed system is about 90 l/h O$_2$ and 180 l/h N$_2$O; with the PhysioFlex® 18 l/h O$_2$ and 10.6 l/h N$_2$O. The peaks in the PhysioFlex® curves were due to flushing of the system for elimination of non-wanted gases such as e.g. nitrogen, methane, acetone.

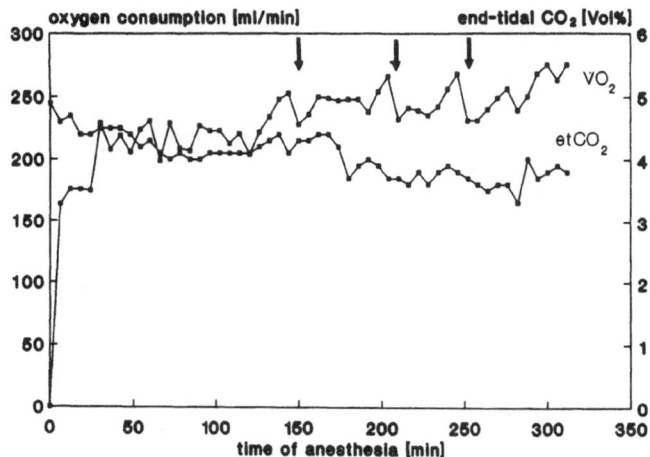

Fig.3.   Oxygen uptake and end-tidal CO_2 during anesthesia ventilation. The patient is a 28 year old woman (84 kg body weight) undergoing a gynecological operation. The patient received pancuronium bromide, fentanyl (0.4 mg) and dehydrobenzperidol (15 mg). She was given 0.1 mg fentanyl and 2 mg pancuronium bromide at marked moments. Note the increased oxygen consumption before the doses were given.

CONCLUSION

The introduction of the first totally closed circuit system into clinical routine has shown:
1.   easy operation with the computer overtaking a great part of the anesthetists daily work as integrator and processor. The anesthetist gets his hands free and has more time for primary patient care;
2.   cost reduction for anesthetic gases can be as high as 90%;
3.   most important is the acquisition of a continuous oxygen consumption measurement giving insight into the state of oxygen metabolism respectively demand/supply ratio of the patient.

REFERENCES

1.   P. E. Jackson, A new method for the production of general analgesia and anesthesia with a description of the apparatus used, J. Lab. Clin. Med., 1: 1-12 (1915).
2.   A. P. K. Verkaaik, J. Rupreht, and W. Erdmann, Computer-kontrolliertes geschlossenes Kreissystem zur Narkose- und Intensivtherapiebeatmung und seine Möglichkeiten der Patientenüberwachung, Anesthesiologie & Reanimation, 16:208-219 (1991).
3.   P. Ayre, The T-piece technique, Br. J. Anaesth. 28: 520-524 (1956).

PO₂-PROFILES IN HUMAN MUSCLE TISSUE AS INDICATOR OF

THERAPEUTICAL EFFECT IN SEPTIC SHOCK PATIENTS

C.P. Naumann[1], Y.A. Ruetsch[1], W. Fleckenstein[1], M.Fennema[2],
W. Erdmann[2] and G.A. Zäch[1]

Departments of Anesthesiology. [1]Swiss Paraplegic Center, Nottwil,
Switzerland and [2]Erasmus University, Rotterdam, The Netherlands

INTRODUCTION

Septic shock is characterized by an impaired cell metabolism. In the early, hyper-dynamic phase of septic shock, the cells are prevented from properly utilizing oxygen, glucose, etc. instead of an adequate supply. Maldistribution of blood flow and increased anatomic arterio-venous shunting increase this tendency.

During the course of septic shock, the hyperdynamic phase tends to convert very rapidly and repeatedly into a hypodynamic state with inadequate oxygen supply of the tissue, due to a relative hypovolemia and/or noxious substances from endogenous or exogenous origin.

Symptomatic therapy usually includes, besides fluid replacement, the administration of vasoactive drugs such as dopamine, dobutamine, epinephrine, nitroglycerine, etc. with the aim to establish an adequate cardiac output, adequate perfusion pressures and an adequate oxygen delivery.

Dopamine, besides its well-known beneficial effects on hemodynamic variables, can improve blood flow through skeletal muscle in animals as in humans[1,2,3,4]. It was also shown to increase tissue oxygen partial pressure (tissue pO_2) in healthy human volunteers and, to a lesser extent, in intensive care patients. However, an increase of muscular blood flow does not per se mean an increase in tissue pO_2. In case of maldistribution, which has to be expected in shock, e.g. due to sepsis, it may even lead to a fall of mean tissue pO_2[5,6,7].

Therefore, in the present investigation, the effects of a continuous intravenous infusion of dopamine on muscular pO_2 in septic patients was studied, as well as predictability of these effects by calculating oxygen delivery and other parameters, derived from hemodynamic measurements.

METHODS

Twenty patients with an established septic shock and multi-organ failure were included in the study. All patients were on Controlled Mechanical Ventilation (CMV) with an inspired oxygen fraction (FiO_2) between 0.4 and 0.6 and a positive end-expiratory pressure (PEEP) between 10 and 20 cmH$_2$O. Arterial CO$_2$ partial pressure (PaCO$_2$) was kept between 35 and 45 mmHg, arterial O$_2$ partial pressure (PaO$_2$) between 75 and 100 mmHg, and arterial O$_2$ saturation (SaO$_2$) between 95 and 100%. Sedation was obtained with morphine and midazolam. Criteria for inclusion were a minimal mean arterial pressure (MAP) of 60 mmHg and a cardiac index at or above the predicted normal value. All patients had an arterial catheter, inserted for blood pressure monitoring and for intermittent arterial blood gas sampling. A Swan-Ganz catheter was placed for mixed venous blood sampling and for hemodynamic controls: cardiac output (CO), central venous pressure (CVP), pulmonary arterial pressure (PAP), pulmonary capillary wedge pressure (PCWP). Derived parameters, calculated from these measurements included cardiac index (CI), pulmonary vascular resistance (PVR), systemic vascular resistance (SVR), arterial and venous oxygen content (CaO$_2$, CvO$_2$), arterio-venous oxygen content difference (avDO$_2$) and oxygen delivery (DO$_2$).

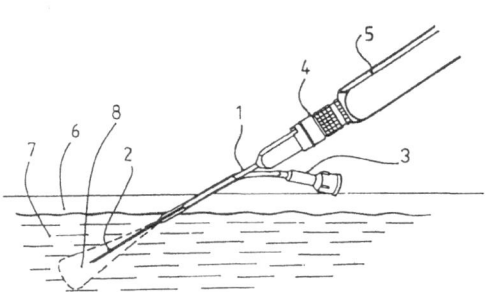

Fig. 1. Schematic drawing of pO$_2$ probe and its placement into the muscle tissue: the probe is inserted through a plastic catheter (Abbokath') as a guide into the muscle. (1): Upper thicker part of the probe, thickness 0.8 mm. (2): Flexible thin part of the probe, thickness 0.3 mm. (3): Luer fitting upper part of the Abbokath'. (4): Waterproof connection. (5): Preamplifier. (6): Skin, fat and fascia. (7): Muscle. (8): Tissue area that is measured during one measuring cycle: 200 pO$_2$ values are registered along the penetration pathway of the electrode tip, a profile is recorded and plotted in the form of a histogram (figure 2).

For the measurements of muscular tissue pO$_2$, automatically stepwise driven pO$_2$ electrodes were inserted transcutaneously in the venter vastus lateralis of musculus quadriceps femoris (figure 1). Signal processing and the mechanical movement of the probe in the tissue were carried out by the KIMOC 250 microcomputer system (Mfg. Gesellschaft für medizinische Sondentechnik, Germany). By moving the electrode forward into the tissue, a pO$_2$ profile was recorded (figure 2a). A pO$_2$ histogram was plotted from the 200 individual measurements gained in a measurement period of 6 minutes (figure 2b).

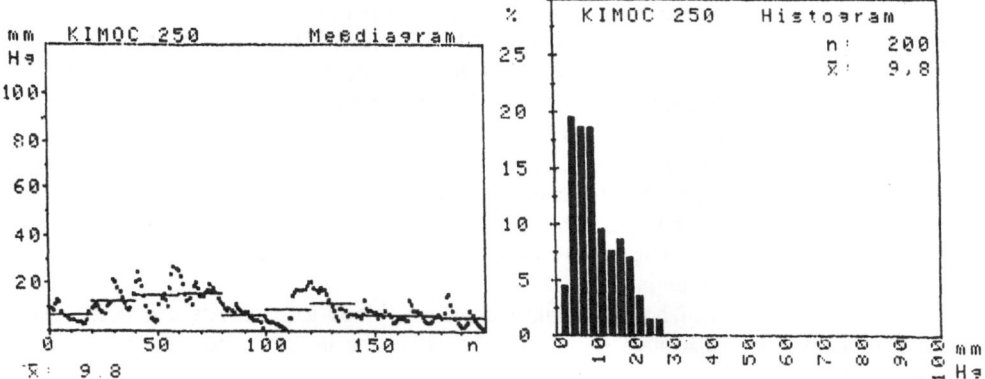

Fig. 2. Example of a pO$_2$ profile obtained during one measurement cycle (200 pO$_2$ values recorded in 6 minutes) (figure 2a, left) and the corresponding pO$_2$ histogram giving information over the number of measurements obtained (n), the frequency (%) of each measured pO$_2$ value group (mmHg), the configuration of the histogram and the mean tissue pO$_2$ (X) (figure 2b, right).

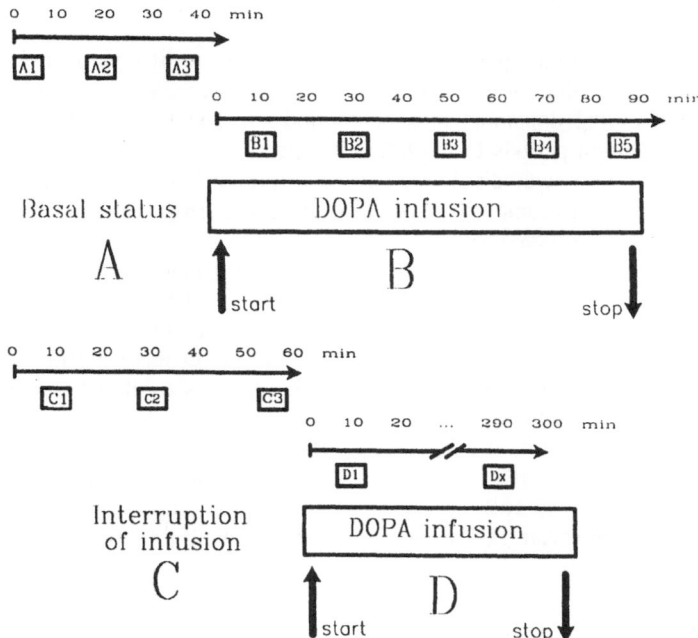

Fig.3. Measurement procedure: the measurements were obtained during 4 successive periods. (A): basal status before treatment, 3 pO$_2$ profiles (A1 to A3) recorded over a period of 40 minutes. (B): infusion of dopamine (5 µg/kg body weight per minute), 5 profiles (B1 to B5) registered during a period of 90 min. (C): period of 60 min without dopamine administration, 3 profiles (C1 to C3) were performed. (D): continuous infusion of dopamine (same dosage as period B), 5 hours or more period with records every 20 min. For each recording of a pO$_2$ profile the corresponding arterial and mixed venous blood gases were taken and oxygen delivery to the tissue was calculated (see text).

Figure 3 shows the measuring procedure with the periods A, B, C and D. During period A, the basal status was recorded at A1, A2 and A3, before start of an intravenous infusion of dopamine. During period B, the stepwise changes to a new steady state were followed by 5 measurement cycles (B1 to B5) spread over 90 minutes of continuous intravenous infusion of dopamine at 5 μg/kg BW/minute (BW body weight). At the beginning of the period C, the infusion of dopamine was stopped and stepwise changes of parameters to a new steady state were followed with 3 measurements (C1 to C3) within the following 60 minutes. The infusion of dopamine was restarted in period D and the same measurement procedure during period B was repeated for up to 5 hours, applied for each 20 minute interval (D1 to Dx).

All measurements of hemodynamic parameters, blood gas analysis and tissue pO_2 measurements were performed simultaneously within the intervals indicated by the arrows in figure 3.

RESULTS

Tissue pO_2

Mean tissue pO_2 of all 20 patients was 20.6 mmHg in the pre-treatment period (A) respectively intermittent treatment pause (period C, 19.9 mmHg) and significantly increased during dopamine infusion (period B to 25.7 respectively D to 25.1 mmHg) (figure 4).

Mean muscular tissue pO_2 showed moderate to severe hypoxia in the pre-treatment period (A) in more than 50% of the patients and a tendency in these patients towards normalization during dopamine infusion (period B). The findings of period A and B were reproducible in periods C and D in all patients.

There was a very wide distribution of values (Table 1) and a marked shift to the left (moderate to severe hypoxia) in 11 of 20 histograms during period A, but only in 8 of 20 histograms during period B.

The changes of muscular pO_2 histograms of the 5 patients with severe pre-treatment hypoxia are shown in figure 5. The moderate increase of mean pO_2 from period A to period B (9.6 respectively 12.5 mmHg) after start of dopamine infusion was reproducibly seen in the changes occurring from period C to D. In these 5 patients with severe tissue hypoxia, the basal status histograms during period A exhibited an extreme shift to the left as shown in figure 5.

As shown in figure 6, the percentage of individual pO_2 values below 5 mmHg in these histograms was extremely high (37%) before dopamine infusion (period A) and decreased significantly to 16% during dopamine infusion (period B). After stop of dopamine administration severe shift to the left reappeared with time (C2 and C3: 23% resp. 32.5% below 5 mmHg).

Mean tissue pO_2 and oxygen delivery

The changes of mean tissue pO_2 and O_2 delivery from period A (before dopamine) to period B (during dopamine infusion) in 20 septic shock patients are plotted against each other in figure 7. Mean tissue pO_2 increased by 5.07 mmHg during dopamine infusion concomitant with an increase of O_2 delivery ratio by 6.33%. However, only 14 of the 20 patients showed the same tendency for both parameters: out of these 14 patients 11 showed both an increase of mean tissue pO_2 and O_2 delivery while in 3 patients both parameters decreased (figure 7).

In the other 6 patients an inverse correlation was found: in 4 of them the mean tissue pO_2 increased despite decreased O_2 delivery, 2 patients showed a decrease of mean tissue pO_2 despite an increase in O_2 delivery.

The correlation coefficient r squared was 0.34 and thus not sufficiently significant to manifest a dependency of these two parameters on each other. Changes of one of these parameters cannot necessarily predict concomitant changes of the other.

Table 1. Number of septic shock patients with a mean tissue pO_2 falling into the range of normoxic, very mild, mild, moderate and severe tissue hypoxia before treatment (period A).

degree of hypoxia	pO_2 (mmHg)	number of patients
none	> 35.0	3
very mild	27.5 - 34.9	2
mild	20.0 - 27.4	4
moderate	12.5 - 19.9	6
severe	5.0 - 12.4	5

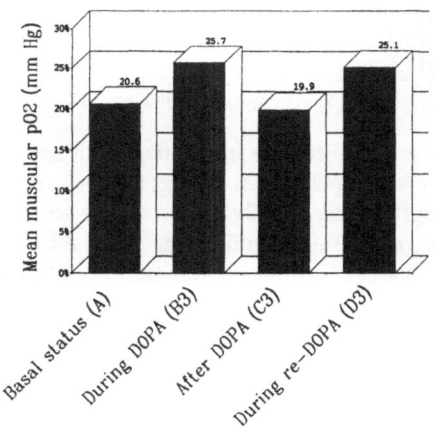

Fig. 4. Mean muscle tissue pO_2 of all 20 patients for periods A, B, C and D (see legend figure 2).

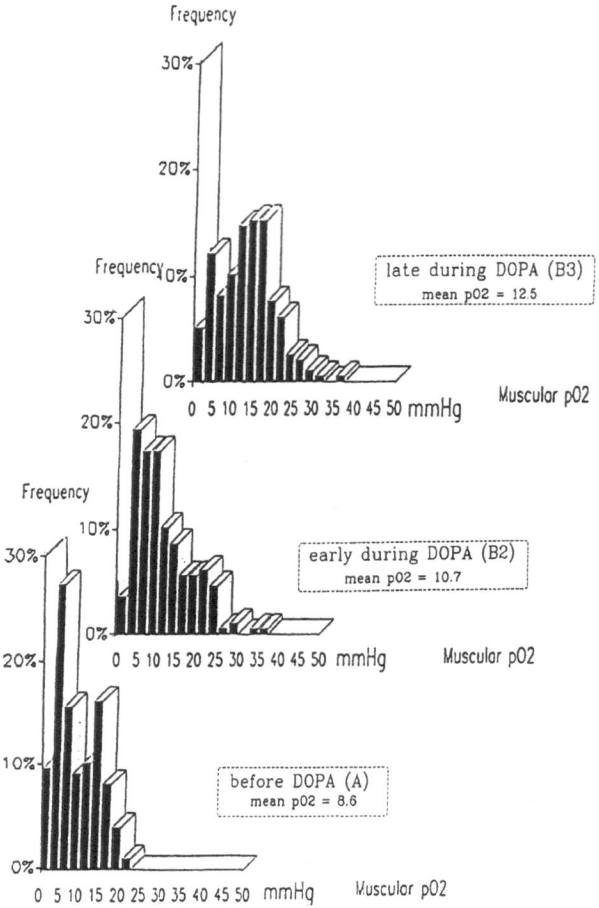

Fig. 5. Typical shift of a pO$_2$ histogram in a patient with severe muscle tissue hypoxia before treatment (mean pO$_2$ = 8.6 mmHg, bottom). With dopamine infusion the tissue pO$_2$ histogram very rapidly shows a shift to the right (B2) with an increase of mean pO$_2$ to 10.7 mmHg, the right shift becomes more dominant during the later phase of treatment (B3) and mean tissue pO$_2$ increases to 12.5 mmHg.

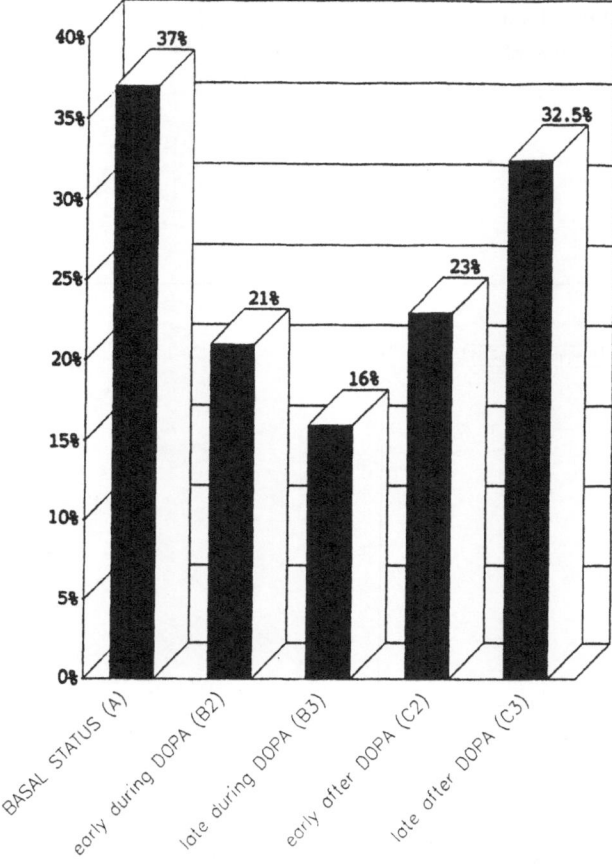

pO2 values < 5 mm Hg (%)

Fig. 6. Percentage of muscle tissue pO_2 values below 5 mmHg (calculated from all 20 patients) in the pre-treatment period (A), during dopamine infusion (B) and in the treatment pause (C). A gradual percentage-wise reduction of tissue pO_2 values below 5 mmHg (tissue areas with severe hypoxia) is seen from 37% to 21% (early phase of period B) and to 16% (later phase of period B). The percentage of tissue areas with severe hypoxia immediately increases again when dopamine infusion is stopped.

DISCUSSION

1. Mean tissue pO_2 before dopamine infusion exhibited severe to mild tissue hypoxia in 15 of 20 patients without signs of a hypodynamic septic state. This is in contrast to the expected results.

2. Only 13 of 20 patients showed an increase of mean tissue pO_2 during dopamine infusion. The amount of increase was extremely variable. This may be related to maldistribution and/or shunting of blood flow Dopamine administration to patients in septic shock may or may not increase mean tissue pO_2, independently of whether a hyperdynamic or a hypodynamic state has been suggested by hemodynamic measurements.

3. An increase in O_2 delivery does not necessarily mean a bettered tissue oxygena-

tion for septic shock patients. Oxygen delivery is a poor indicator in septic shock.

4. In questionable situations during septic shock direct tissue oxygen profiles measurements may help to improve the patient's condition and might be one guide for pharmacological therapy.

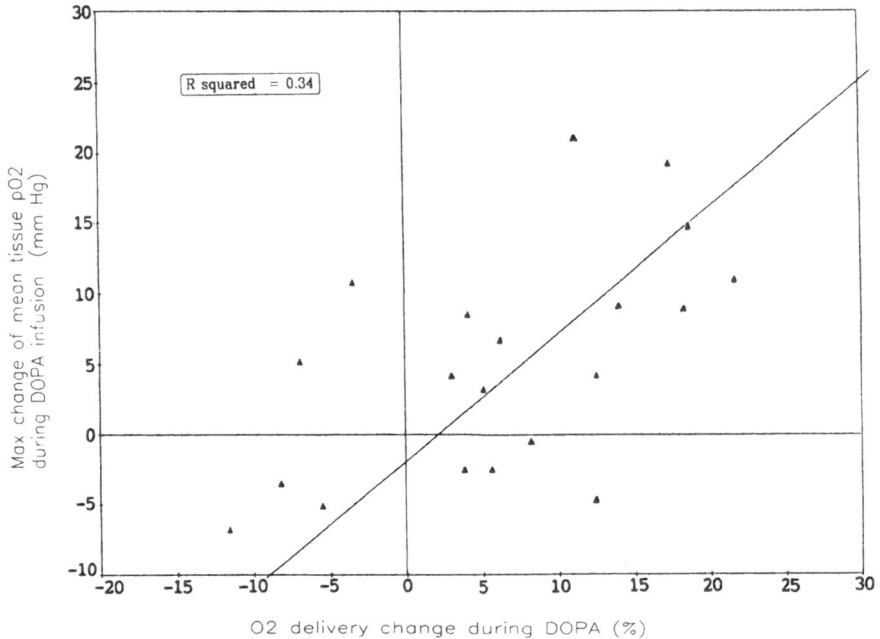

Fig. 7. Maximal changes of mean tissue pO_2 for each of the 20 patients (as percent change of baseline value plotted against the respective oxygen delivery values). Although in some patients there appears to be some correlation, in others this is not the case and in some even a negative correlation is seen (discussion, see the text).

CONCLUSION

Improvement of cardiovascular and blood gas parameters following pharmacological treatment of septic shock syndrome does not per se indicate that tissue oxygen supply is really improved. It is advised to keep this point in mind and in questionable situations introduce direct tissue oxygen profile measurement.

SUMMARY

Automatically stepwise driven pO_2 electrodes were transcutaneously inserted into muscle tissue of severely ill septic shock patients. The pO_2 profile was plotted from 200 individual measurements registered during 5 minutes and a histogram plotted for documentation. Arterial and venous blood gases, cardiac output, systemic and pulmonary vascular resistance were measured continuously on-line. In septicemia multiple drug schemes are suggested all intending to increase oxygen supply to the tissue and to improve oxygen demand/supply mismatch. So far the attending physician is bound to conclude and continue respectively change the treatment scheme according to the above described macrophysiological parameters. Perfusion distribution and local inhomogenities of tissue oxygen supply remain undetected. In the described study pretreatment pO_2 profiles in musculus quadriceps femoris were obtained and measurements repeated in intervals of 10 minutes after start of pharmacological treatment.

The changes of pO_2 profiles of 20 patients, monitored in such a way over days and weeks, were carefully correlated to the described cardiocirculatory parameters and blood gas analyses. Dopamine was used to improve cardiac function and tissue oxygen supply as well. The investigations show that resulting changes of cardiovascular and blood gas parameters do not always indicate that tissue oxygen supply has really improved. On the other hand there was never an improvement in tissue oxygen supply when no changes of the other parameters had occurred. It is advised to add as a further diagnostic parameter tissue pO_2 measurements to get insight if improvement in cardiac and pulmonary function really has the intended effect of improvement of tissue oxygen supply.

REFERENCES

1. J. L. Willens and M. G. Bogeart, Dopamine-induced neurogenic vasodilatation in isolated perfused muscle preparation of the dog, Naunyn-Schmiedberg Arch Pharmacology, 286: 413-428 (1975).
2. C. Bell and A. Stubbs, Localisation of vasodilator dopamine receptors in the canine hindlimbs, Br J Pharmacology, 64: 253-257 (1978).
3. O. E. Brodde, F. J. Meyer, W. Schemuth and J. Freistühler, Demonstration of specific vascular dopamine receptors mediating vasodilatation of the isolated perfused muscle preparation of the dog, Naunyn-Schmiedberg Arch Pharmacology, 316: 24-30 (1981).
4. L. K. Jackson, B. M. Key and S. M. Cain, Total hindlimb O_2 uptake and blood flow in hypoxic dogs given dopamine, Critical Care Med, 10: 327-331 (1982).
5. W. Schröder and W. Rathscheck, Investigation of the influence of acetylcholine on the distribution of capillary flow in the skeletal muscle of the guinea pig by recording of the pO_2 in the muscle tissue, Pflügers Arch, 345: 335-346 (1973).
6. J. Hauss, K. Schönleben, H. U. Spiegel and H. Bünte, Therapie-Kontrolle in der Intensivbehandlung durch kontinuerliche Gewebe-pO_2-Messung, in: Messung des Gewebessauerstoffdruckes bei Patienten, G. Witzstock, ed., A.M. Ehrly, Baden-Baden, New York (1981).
7. N. Lund, Skeletal muscle surface oxygen pressure fields in normal volunteers and in critically ill patients, in: Messungen des Gewebesauerstoffdruckes bei Patienten, G. Witzstock, ed., A.M. Ehrly, Baden-Baden, New York (1981).

PREOPERATIVE GASTROCNEMIUS MUSCLE PO$_2$ AS PREDICTOR OF HEALING

AFTER BELOW KNEE AMPUTATION

Wagner K.F., Noah E.M.*, Perner R., Busse F.-W.*,
Bruch H.-P.*
Department of Physiology, Clinic of Surgery*
Medical University of Lübeck, Lübeck, Germany

Introduction

In patients with occlusive peripheral arterial disease a limb amputation frequently has to be performed to avoid further health hazards. As a prerequisite for successful rehabilitation it is desirable to perform a below knee amputation. However, the routinely employed preoperative diagnostic procedures only give limited predictive information as to whether wound healing will occur after a below knee amputation. Since tissue oxygenation is an important factor influencing wound healing we hypothesized, that lower leg muscle oxygenation might have a predictive value for primary wound healing.
Therefore, the aim of this study was to compare preoperative lower leg muscle oxygenation of patients where wound healing after a below knee amputation occurred with those of patients where wound healing after a below knee amputation failed.

Methods

The study comprised 14 patients (9 men, 5 women), their age ranged from 53 - 90 years (mean age 72.1 years). Lower leg muscle oxygenation was measured with a hypodermic needle type polarographic oxygen sensor with a tip diameter of 250 μm (Weiss and Fleckenstein, 1986). A computer controlled pO$_2$ histograph (Eppendorf - Netheler - Hinz GmbH, Hamburg, Germany) was used for calibration, steering of the oxygen probe, and data acquisition.
Since oxygen is delivered through distinct vessels to skeletal muscle, the oxygenation within a muscle must be inhomogeneous. Therefore, the measurement of a single oxygen partial pressure value does not give a representative information about the oxygenation of the tissue. Representative information of the oxygenation of a tissue is given by the oxygen partial pressure distribution within the tissue. It can only be attained by measuring as many as possible randomly distributed single pO$_2$ values within the tissue. For practical purposes it is sufficient to measure 100 -200 randomly distributed single pO$_2$ values.

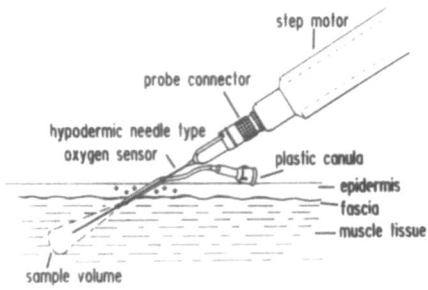

Figure 1. Schematic drawing of the oxygen sensor in muscle.

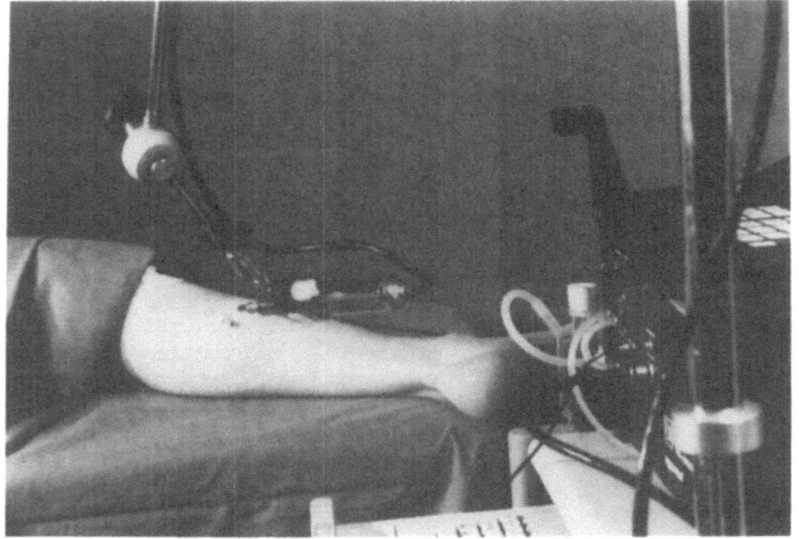

Figure 2. Picture of an oxygen partial pressure measurement in the gastrocnemius muscle of a patient.

These 100 -200 single pO_2 values are graphically presented as pO_2 histograms, where on the abscissa are plotted the oxygen pressure classes and on the ordinate the relative frequency of pO_2 values for each class. Several pO_2 histograms can be combined to a 'pooling' pO_2 histogram.

For the tissue pO_2 measurement a gage 20 plastic cannula is inserted through the skin to a depth of 0.5 - 1.0 cm into the muscle after intracutaneous local anaesthesia (Figure 1). Through a tangential slit in the cannula the oxygen sensing needle probe is positioned into the muscle tissue (Figure 2). Each single pO_2 value is measured in the stand still period 1 second after a fast forward step of 700 μm of the probe. This step length is large enough to push the probe with each step into unaltered tissue (Schramm et al., 1990).

Between 20 and 40 single pO_2 values were measured in sequence. Thereafter the probe was withdrawn, repositioned with a different penetration angle and the pO_2 measurement repeated until 200 single local pO_2 values were obtained. At the end of each measuring period muscle temperature was measured with a thermocouple through the plastic cannula left in situ.

Since the operative technique of a below knee amputation uses the gastrocnemius muscle to cover the bone stump, muscle pO_2 was measured in the gastrocnemius muscle. The measuring point lay in the medial portion at the level of the maximal calf circumference. During the measurements the patients were in supine position at neutral ambient temperature. In all but one patient 200 single pO_2 values were obtained. In addition, for every tissue pO_2 measurement arterial blood gases and the systolic/diastolic blood pressure (method of Riva-Rocci) were recorded. Doppler occlusion pressures of the brachial, anterior tibial and dorsalis pedis artery were routinely measured preoperatively. The amputation was performed within 24 hours after the tissue pO_2 measurement.

Statistical analysis was performed with the Kolgomorov-Smirnov test with the level of significance set at $p < 0.05$. All results are expressed plus/minus standard error.

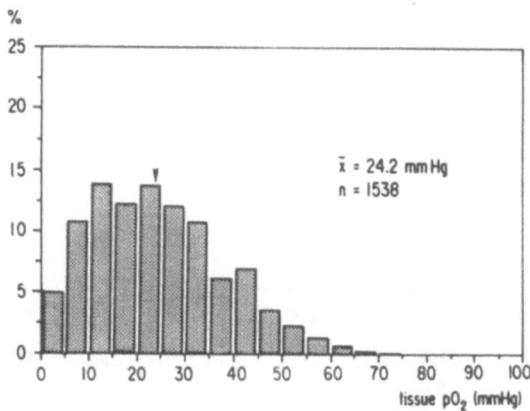

Figure 3. pO_2 'pooling' histogram of the muscle oxygenation in patients where wound healing occurred after below knee amputation.

Results

Healing of the wound after the below knee amputation was found in 8 patients (healer group), whereas in 6 patients no healing occurred (non-healer group).

The individual histograms of the patients of the healer group and those of the non-healer group were combined to pooling histograms. They are shown in figures 3 and 4.

The healer group had a significantly higher mean tissue pO_2 of 24.2 ± 2.7 mm Hg compared to 14.7 ± 0.9 mm Hg ($p < 0.02$) of the non-healer group. The pooling histogram of the healer group (Figure 3) had a nearly normal shape of the histogram curve, 10 % of the pO_2 values read above 43.3 mm Hg and the highest pO_2 values lay just below 70 mm Hg. 15.7 % of all values lay below 10 mm Hg.

These results contrasted markedly with the pO_2 data of the non-healer group (Figure 4). The curve of this pooling histogram was left shifted with 35.5 % of all pO_2 values lying below 10 mm Hg. Only 10 % of the pO_2 values lay above 27.8 mm Hg and the highest measured pO_2 values lay at 53.6 mm Hg.

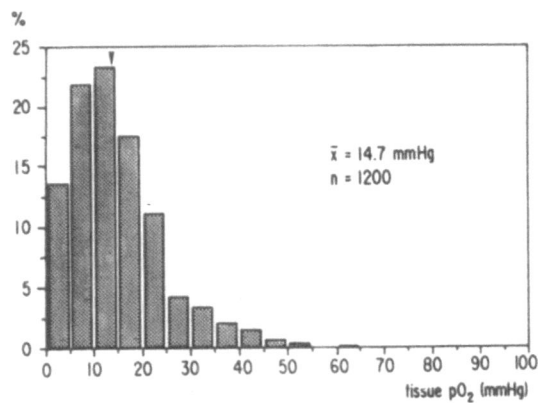

Figure 4. pO_2 'pooling' histogram of the muscle oxygenation in patients with failed healing after below knee amputation.

Mean gastrocnemius muscle temperature (Table 1) in the healer group was 34.2 ± 0.5 °C compared to 32.5 ± 0.5 °C in the non healer group.

The doppler occlusion pressure index of the tibialis anterior artery/brachial artery and the dorsalis pedis artery/brachial artery (Table 1) were - as expected - below normal value for the healer and the non-healer group. Surprisingly, the indices of the non-healer group with 0.74 ± 0.21 and 0.80 ± 0.26, respectively were higher than the corresponding indices of the healer group with 0.65 ± 0.07 and 0.62 ± 0.08, respectively.

Table 1. Data of the muscle oxygenation, muscle temperature and doppler indices.

	Healer (n = 8)	Non-Healer (n = 6)
Mean age (yrs., range)	74,4 (60-90)	66,0 (53-79)
Mean Muscle pO_2 mm Hg ± s.e.	24,2 ± 2,7	14,7 ± 0,9
Median musc. pO_2 mm Hg ± s.e.	23,1 ± 3,1	13,1 ± 1,0
Percentage value 0 - 10 mm Hg	15,7 %	35,5 %
mean muscle temp. °C ± s.e.	34,2 ± 0,5	32,5 ± 0,5
Doppler index Tib. ant. art./ Brachial art.	0,65 ± 0,07	0,74 ± 0,21
Doppler index Dors. ped. art./ Brachial art.	0,62 ± 0,08	0,80 ± 0,26

Discussion

We found, that patients of the non-healer group had a significantly lower mean gastrocnemius muscle pO_2 and pO_2 values between 0 and 10 mm Hg were more than twice as frequent compared to the healer group.

Comparing the box and whisker plots (Figure 5) of the individual mean muscle pO_2 values it can be seen that scattering was low for the mean muscle tissue pO_2 in the non healer group compared to the healer group and that overlap of the mean muscle pO_2 between both groups was small.
As expected, mean gastrocnemius muscle temperature tended to be lower in the non-healer group.

The mean doppler occlusion pressure indices were decreased in both groups. Surprisingly, they were higher in the non-healer group than in the healer group, possibly due to the media sclerosis in two non-healer patients.

These data indicate that patients with failed wound healing after below knee amputation tended to have a lower gastrocnemius muscle oxygenation than patients where healing of the amputation stump occurred. Further studies are needed to assess the sensitivity and specificity of muscle pO_2 measurement as predictor of wound healing.

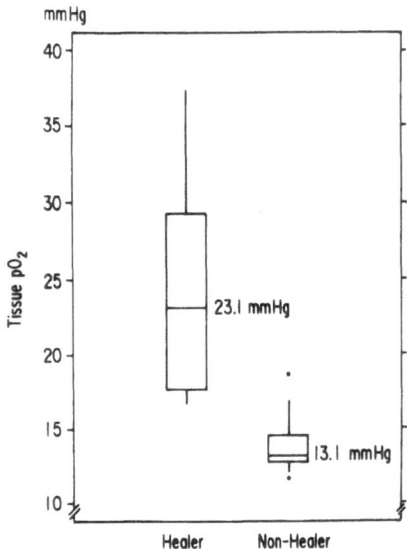

Figure 5. Box and whisker plot of
the individual mean pO$_2$ values of
the healer and the non-healer groups.

References

Schramm, U., Fleckenstein, W., Weber, C., 1990, Morphological
    assessment of skeletal muscular injury caused by pO$_2$
    measurements with hypodermic needle probes, in:
    "Clinical oxygen pressure measurement II", A. M. Ehrly,
    W. Fleckenstein, J. Hauss and R. Huch, ed., Blackwell
    Ueberreuter Wissenschaft, Berlin.
Weiss, CH. and Fleckenstein W., 1986, Local tissue pO$_2$
    measured with 'thick' needle probes, in:
    "Funktionsanalyse biologischer Systeme 15", J. Grote
    and G Thews, ed., Steiner, Stuttgart.

CHANGES OF TISSUE PO$_2$ IN THE LOWER LEG MUSCLES AFTER

VASCULAR SURGERY

[1]K. Wagner, [2]U. Krüger, [2]R. Schäfer, [3]M. Albrecht,
[4]G. Hohlbach

[1]Dept. of Physiology, [2]Dept. of Anaesthesiology,
 Medical University of Lübeck,
[3]Dept. of Anaesthesiology, Clinics of Mannheim,
[4]Clinic of Surgery, University Bochum, Germany

Introduction

The clinical benefit of vascular surgery in patients with occlusive peripheral arterial disease is obvious and well established. A marked improvement of muscle tissue oxygenation was found intraoperatively (1) and on the first postoperative day after vascular reconstruction (2). However, little is known about tissue oxygenation during the 2nd - 14th postoperative days. It was tacitly assumed, that the intraoperatively and immediately postoperatively measured increase of muscle pO$_2$ would persist.

The aim of the present study was to investigate the sequence of changes of oxygen partial pressure in lower leg skeletal muscle prior to and during vascular surgery and within the first 14 days following the operation in patients with severe peripheral arterial disease.

Methods

The study comprised 13 patients between 47-69 years. In each of three female and ten male patients five muscle tissue pO$_2$ measurements according to the protocol given in table 1 were performed.
Since patients with occlusive peripheral disease have a high incidence of complications during inhalation anaesthesia, a combination of inhalation anaesthesia (with a reduced concentration of the anaesthetic gas (enflurane)) and a peridural analgesia was chosen. Lower leg muscle oxygenation was measured with a hypodermic needle type polarographic oxygen sensor with a tip diameter of 350 μm (3). A computer controlled pO$_2$ histograph (Eppendorf - Netheler - Hinz GmbH, Hamburg, Germany) was used for calibration, the mechanical steering of the oxygen probe, and data acquisition.

Since oxygen is delivered through distinct vessels to skeletal muscle, the oxygenation within a muscle must be

Table 1. Study protocol, GA denotes general anaesthesia.

1. Measurement	Preoperative
2. Measurement	Intraoperative after induction of GA
3. Measurement	Intraoperative 10-30 min after reconstruction
4. Measurement	2. postoperative day
5. Measurement	14. postoperative day

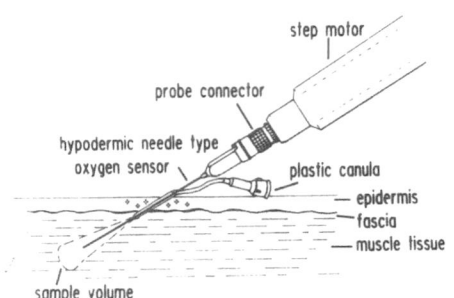

Figure 1. Schematic drawing of an oxygen partial pressure measurement in muscle.

inhomogenious. Therefore, the measurement of a single oxygen partial pressure value does not give representative information about the oxygenation of the tissue. Representative information on the state of oxygenation of a tissue is given by the oxygen partial pressure distribution within the tissue. It can only be attained by measuring as many as possible randomly distributed single pO_2 values within the tissue. For practical purposes it is sufficient to measure 50 - 200 randomly distributed single pO_2 values. These 50 - 200 single pO_2 values are graphically presented as pO_2 histograms, where on the abscissa are plotted the oxygen pressure classes and on the ordinate the relative frequency of pO_2 values for each class. Several pO_2 histograms can be combined to a 'pooling' pO_2 histogram.

For the tissue pO_2 measurement a gage 20 plastic cannula was inserted through the skin to a depth of 0.5 - 1.0 cm into the muscle after intracutaneous local anaesthesia (Figure 1). Through a tangential slit in the cannula the oxygen sensing needle probe was positioned into the muscle tissue. Each single pO_2 value was measured in the standstill period 1 second after a fast forward step of 700 μm of the probe. This step length was large enough to push the probe with each step into unaltered tissue (4).

Between 20 and 40 single pO_2 values were measured in sequence. Thereafter the probe was withdrawn, repositioned with a different penetration angle and the pO_2 measurement continued until 200 single local pO_2 values were obtained. At the end of each measuring period muscle temperature was determined with a thermocouple through the plastic cannula left in situ.

Tissue pO_2 was measured in the anterior tibial or the gastrocnemius muscle, depending on whether the anterior tibial or the posterior tibial artery were patent. pO_2 measurements from these two muscles in patients with pelvic or femoral occlusions were not statistically different (5) and, therefore, the results obtained from pO_2 measurements in both muscles were included in our analysis.

Every patient had at least 1 hr bed rest prior to measurements.

For every tissue pO_2 measurement the following parameters were measured: Mean arterial blood pressure (For pre- and intraoperative measurements in the radial artery with a disposable statham system, for postoperative measurements using

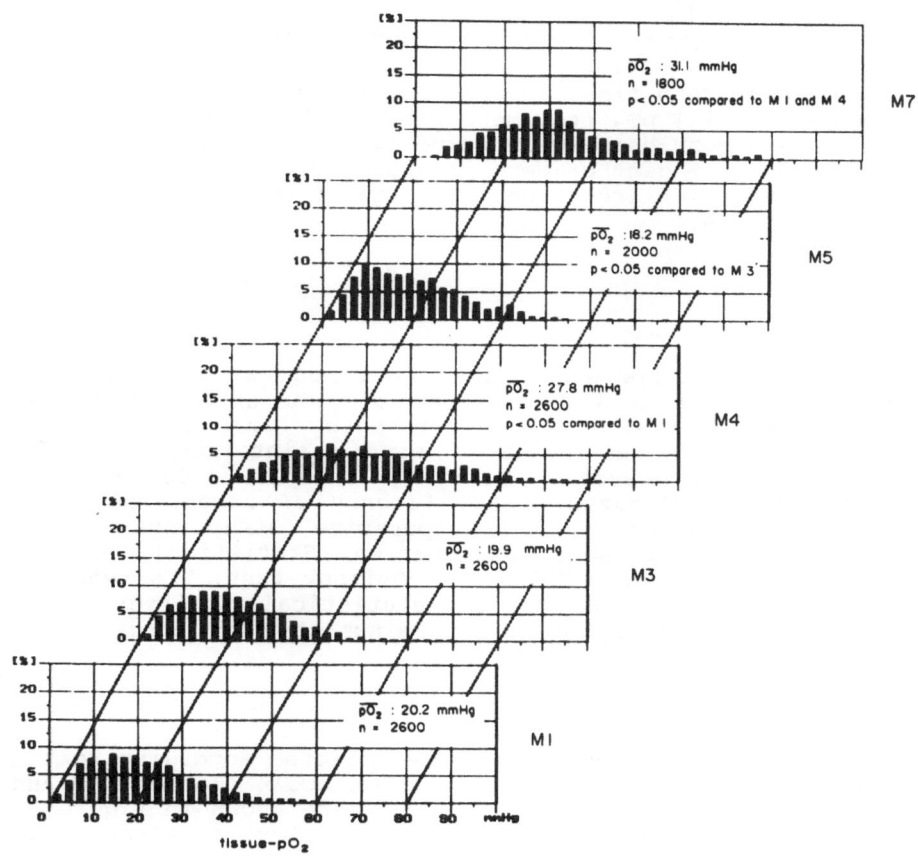

Figure 2. pO_2 'pooling' histograms of the measurements according to the study protocol. Numbers on the abscissa denote the measurements.

the method of Riva-Rocci and calculation of the mean arterial pressure from the systolic and diastolic blood pressure), heart rate, central venous pressure (not for postoperative tissue pO_2 measurements), arterial and venous blood gases (pO_2, pCO_2, pH, standard bicarbonate, base excess, O_2 saturation of hemoglobin, (Eschweiler System 2000, Eschweiler & Co, Kiel, FRG)), hemoglobin concentration (Corning 2500 Co-oxymeter, Corning, Medfield, USA), hematocrit, serum Na^+, K^+ and Ca^{++} (Eschweiler System 2000), body core temperature.

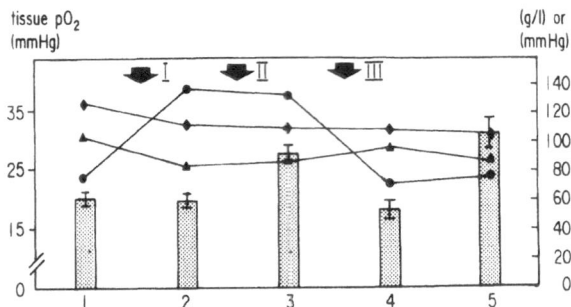

Figure 3. Mean tissue pO_2 (bars) ± standard error, mean arterial pO_2 (●), mean arterial blood pressure (▲), and mean hemoglobin value (♦) for measurements 1 to 5. I. Begin of general anaesthesia. II. Vascular desobliteration. III. End of general anaesthesia.

Muscle tissue temperature was measured with a Philips thermocouple with a tip diameter of 260 μm as described above. Interstitial fluid pressure on the second postoperative day was measured with a interacompartmental pressure monitor system (Stryker Surgical Cooperation, monitor 295-1, Kalamazoo, USA). Patency of reconstructed or anastomosed vessels for postoperative measurements was investigated by ultrasonic examination. Statistical analysis was performed using the Kolgomorov-Smirnov test. The means of the groups were chosen to be statistically different at p<0.05. Results are expressed plus/minus standard error.

Results

The individual histograms were selected according to the measuring periods of the study protocol (Table 1) and combined to 'pooling' histograms. They are shown in figure 2. The 'pooling' histogram of the preoperative tissue pO_2 measurements (Figure 2, measurement 1 (M1)) was markedly shifted to lower oxygen values. The mean tissue pO_2 lay at 20.2 ± 1.3 mm Hg and 20.2 % of the pO_2 values lay below 10 mm Hg.

General anaesthesia with artificial ventilation and peridural analgesia (Figure 2, M2) essentially did not change muscle tissue oxygenation. This is noteworthy, since the mean arterial pO_2 had increased substantially from 78.3 ± 2.6 mm Hg during M1 to 137.3 ± 9.4 mm Hg during M2 (Figure 3). But the shape of the 'pooling' histogram M2 was nearly identical to the shape of the 'pooling' histogram of the preoperative pO_2 measurements (Figure 2, M1) and the mean tissue pO_2 was even minimally decreased to 19.9 ± 1.2 mm Hg compared to the preoperative value.

Under essentially unchanged conditions for hemoglobin value, mean arterial blood pressure and arterial pO_2 the muscle tissue pO_2 was measured 10 - 30 minutes after vascular reconstruction (Figure 2, M3). Mean tissue pO_2 had risen significantly to 27.8 ± 1.4 mm Hg ($p<0.05$). There was a marked broadening of the histogram and pO_2 values above 40 mm Hg became abundant. The intraoperatively measured increase of tissue oxygenation did not persist in the postoperative period. On the 2nd postoperative day mean tissue pO_2 had decreased significantly ($p<0.05$) to 18.2 ± 1.4 mm Hg (Figure 2, M4), which was even lower than the preoperative median value of 20.2 ± 1.3 mm Hg. But on the 14th postoperative day (Figure 2, M5) median tissue pO_2 had increased significantly ($p<0.05$) to 31.1 ± 2.5 mm Hg, thus lying in the normal range.

Discussion

Preoperative lower leg muscle oxygenation in the patients studied was already markedly reduced. The pooling histogram was leftshifted compared to the physiological lower leg skeletal muscle histogram (6). Preoperative mean tissue pO_2 measured 20.2 ± 1.3 mm Hg and lay well within those reported in the literature for patients with peripheral arterial disease (5, 7).

The second tissue pO_2 measurement was performed after induction of a general anaesthesia. General anaesthesia led to a decrease of the mean arterial blood pressure from 102 to 84 mm Hg. This decrease was probably caused by two factors. First enflurane - like all inhalation anaesthetics - has a negative inotropic effect. The second probable cause was the peridural analgesia leading to a sympathetic nerve blockade followed by a vasodilation and hypotension. To partially counteract the decrease in mean arterial blood pressure a solution of 0.9 % NaCl was infused. This hemodilution led to a decrease of the hemoglobin value (Figure 3). The artificial ventilation with an increased inspiratory oxygen fraction led to a significant large increase of the arterial pO_2 from 78.3 ± 2.6 to 137.3 ± 9.4 mm Hg. But despite this nearly two-fold increase in arterial pO_2 the mean muscle pO_2 did not increase. On the contrary, a small further decrease of muscle oxygenation was found. The increase in arterial pO_2 resulted in a broadening of the histogram with more higher values. It could be argued that arterial pO_2 was increased but oxygen delivery actually was decreased. Although we did not measure blood flow to the leg a rough calculation showed that oxygen delivery was maintained at the preoperative level.

The intraoperative sympathetic nerve blockade did not lead to an increase in muscle oxygenation. Our data confirm the results of Singbartl et al.(8), who did not find a increase of skeletal muscle oxygenation after three days of continuous

peridural anaesthesia in patients with severe occlusive peripheral arterial disease.
Ten to thirty minutes after vascular reconstruction the tissue oxygenation had increased significantly to a mean value of 27.8 ± 1.4 mm Hg (p < 0.05 compared to the preoperative mean tissue pO_2). Our results confirm the intraoperative rise of the oxygenation of previously ischemic muscles immediately after vascular reconstruction that was first documented by Sunder-Plassmann et al (1) using a muscle surface oxygen probe. They found an increment from 26.0 mm Hg–37.0 mm Hg in 4 patients with a femoropopliteal bypass graft.
This increased level of muscle oxygenation was not maintained postoperatively but there was a significant decrease of mean muscle oxygenation to 18.2 ± 1.6 mm Hg on the second postoperative day rendering mean pO_2 values below the already pathologically low preoperative values. But on the 14th postoperative day muscle oxygenation (mean pO_2 of 31.1 ± 2.5 mm Hg) lay within the physiological range.

Our finding of a significant transient postoperative decrease of tissue oxygenation led us to measure tissue pO_2 in the early postoperative period at shorter time intervals.

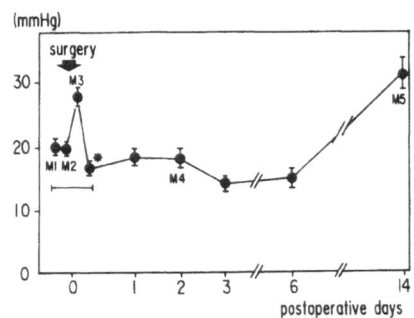

Fig. 4.   Mean tissue pO_2 ± standard error from all measurements, |———| denotes day of operation,
* measurement 6 hr postoperatively

In 3 of the 13 patients additional tissue pO_2 measurements were carried out at 6 hours postoperatively and on the 1st, 3rd and 6th postoperative day. The data from these additional measurements were combined with the data measured according to the protocol of table 1. They are shown in figure 4. Already 6 hours postoperatively mean lower leg muscle oxygenation was significantly decreased to 16.7 ± 1.2 mm Hg. On the first postoperative day a small increase of mean tissue pO_2 to 18.4 ± 1.4 mm was found. On the second postoperative day mean tissue pO_2 was 18.2 ± 1.6 mm Hg and on the third and sixth postoperative day mean tissue pO_2 was decreased again to 14.1 ± 1.3 and 14.9 ± 1.7 mm Hg, respectively.

Thus after a significant intraoperative rise of muscle oxygenation following revascularization muscle oxygenation was significantly reduced below preoperative values 6 hr postoperatively and stayed low until the sixth postoperative day. 14 days after the operation muscle oxygenation had again reached physiological values.

Krawzak et al (7) measured lower leg skeletal muscle pO_2 two and six weeks after vascular surgery. They found a longer time course until muscle oxygenation had reached physiological values. After two weeks mean muscle pO_2 was still decreased at 22.1 mm Hg and reached normal values after six weeks.

An early postoperative vessel or graft occlusion as plausible explanation for a low tissue pO_2 could be excluded for all patients by ultrasonic examination. Some patients develop postoperative edema at the site of the operation as a reaction to intraoperative manipulations or the bypass graft material. An increased interstitial fluid pressure could lead to compression of capillaries and thus reduce oxygen supply to the muscle. On examination 2 patients had minor pitting edema on the second postoperative day, but measuring interstitial fluid pressure of these two patients and 3 others showed a mean value of 5.2 ± 1.9 mm Hg (normal value < 10 mm Hg (9)). A comparison of the mean tissue pO_2 and the mean arterial pO_2 of the intraoperative measurements 10 - 30 minutes after vascular reconstruction and of the measurements on the postoperative day (figure 3) could suggest a correlation of the two parameters. However, the rise of tissue pO_2 between measurements 2 and 3 and measurements 4 and 5, with the arterial pO_2 remaining constant, does not support this explanation.

A more likely explanation seems to be a damage of the skeletal muscle microcirculation leading to a transiently reduced tissue pO_2. In a recent investigation Persson et al (10) found after routine vascular reconstruction an increase of lipid peroxidation in muscle biopsies after reperfusion and Neglen et al (11) showed an increased venous creatine content 30 minutes after recirculation. Increased lipid peroxidation and increased leakage of intracellular enzymes are part of the tissue damage in ischemia/reperfusion injury (12). It may lead to perfusion stop of capillaries and, ultimately, to tissue necrosis (13). Obviously, non perfusion of capillaries reduces oxygen supply to the tissue. With the cellular oxygen demand remaining constant the imbalance of oxygen supply and oxygen demand would lead to a decrease of tissue oxygenation.

Therefore, we assume, that the transient early postoperative decrease of muscle oxygenation is the result of an ischemia/reperfusion injury to the skeletal muscle.

References:

1. L. Sunder-Plassmann, M. Kessler, K. Messmer, D.W. Lübbers, "Quantitative assessment of microvascular integrity by tissue oxymetry in patients". 1st ed., Academic Press, New York. 276 (1981).

2. J. Hauss, K. Schönleben , H. Spiegel, "Therapiekontrolle durch Überwachung des Gewebe-pO_2". Huber, Bern (1982).
3. C. Weiss, W. Fleckenstein, Local tissue pO_2 measured with 'thick' needle probes, in: "Funktionsanalyse biologischer Systeme 15", J. Grote, G. Thews eds., Steiner, Stuttgart, 155-66 (1986).
4. U. Schramm, W. Fleckenstein, C. Weber, Morphological assessment of skeletal muscular injury caused by pO_2 measurements with hypodermic needle probes, in: "Clinical oxygen pressure measurement II", A. M. Ehrly, W. Fleckenstein, J. Hauss and R. Huch, ed., Blackwell Ueberreuter Wissenschaft, Berlin (1990).
5. A. Ehrly, W. Schroeder, Oxygen pressure values in the ischemic muscle tissue of patients with chronic occlusive arterial disease, in: "Advances in experimental medicine and biology", I. Silver, M. Erecinska, H. Bicher, eds., Vol 94, Plenum Press, New York, 401-5 (1978).
6. K. Kunze, "Das Sauerstoffdruckfeld im normalen und pathologisch veränderten Muskel", Spinger, Berlin (1969).
7. H. Krawzak, R. Heinrich, H. Strosche, Development of muscular tissue pO_2 after vascular reconstructive surgery, in: "Advances in experimental medicine and biology", K. Rakusan, G. Biro, T. Goldstick, Z. Turek, eds., Vol 248, Plenum Press, New York, 713-8 (1989).
8. G. Singbartl, R. Stögbauer, M. Gölzenleuchter, G. Metzger, 1990, Influence of lumbar sympathetic nerve blockade on tissue pO_2 of the anterior tibial muscle in patients with peripheral arterial occlusive disease, in:" Clinical oxygen pressure measurement II", A. Ehrly, W. Fleckenstein, J. Hauss, R. Huch eds., Blackwell Ueberreuter, Berlin.
9. V. Echtermeyer, "Das Kompartmen-Syndrom", Springer, Berlin (1985).
10. N. H. Persson, D. Bergqvist, G. Fex, S. L. Marklund, B. Nilsson, R. Takolander, Lipid peroxidation and activity of antioxidant enzymes in muscle of the lower leg before and after arterial reconstruction. Eur J Vasc Surg, 3(5):399 (1989).
11. P. Nelgen, C. M. Japs, B. Eklof, Plasma metabolic disturbance and reperfusion injury following partial limb ischemia in man. Eur J Vasc Surg, 3(2):165 (1989).
12. K. Harris, P. M. Walker, D.A.G. Mickle et al., Metabolic response of skeletal muscle to ischemia. Am J Physiol, 250:H213-H220.(1986).
13. A. Ames III, R. Wright, M. Kowada, J. M. Thurston, G. Majno, Cerebral ischemia. II. The no-reflow phenomenon. Am J Pathol, 52:437 (1968).

COMPARISON OF DIFFERENT MODES OF ARTIFICIAL VENTILATION WITH EXTRACORPOREAL CO₂ ELIMINATION ON GAS EXCHANGE IN AN ANIMAL MODEL OF ACUTE RESPIRATORY FAILURE

J. Kesecioglu[1,2], L. Telci[2], T. Denkel[2], A. S. Tütüncü[1,2], F. Esen[2], K. Akpir[2], B. Lachmann[1]

[1]Department of Anesthesiology, Erasmus University and Academic Hospital Dijkzigt, Rotterdam, The Netherlands
[2]Department of Anesthesiology, University of Istanbul, Faculty of Medicine, Istanbul, Turkey

INTRODUCTION

Where optimal oxygenation is the predominant aim as a life support for patients with adult respiratory distress syndrome (ARDS), special care must be taken to avoid untoward side effects of mechanical ventilation while promoting gas exchange. High tidal volumes (Vt) and high peak inspiratory pressures (PIP) have been shown to cause progressive lung injury[1] and studies in the last decade have concentrated on designing a method of oxygenation without damaging the lung tissue. Pressure controlled inverse ratio ventilation (PC-IRV) has been reported to avoid high PIP and therefore avoid barotrauma and overdistension[2,3]. Low frequency positive pressure ventilation with extracorporeal CO₂ removal (LFPPV-ECCO₂R) is another method designed to provide rest to the lungs and even intrapulmonary distribution of gases, avoiding pressure related complications of mechanical ventilation[4].

The main aim of this study was to compare the effects of volume controlled ventilation (VCV) with positive end-expiratory pressure (PEEP), PC-IRV and LFPPV-ECCO$_2$R on oxygenation in an animal model of ARDS.

METHODS

Ten male pigs, 54.3 ± 3.4 kg (range, 50-60 kg) were used in this study. Anesthesia was induced with ketamine (10 mg.kg^{-1}) intramuscularly and maintained with midazolam (0.2 mg.kg^{-1}.h^{-1}) and fentanyl (2 μg.kg^{-1}.h^{-1}). Pancuronium bromide (0.08 mg.kg^{-1}.h^{-1}) was given for muscle relaxation. Tracheostomy was performed after ketamine administration and a portex tube (ID 7 mm) was inserted into the trachea. Lungs were ventilated with a Servo 900C ventilator (Siemens-Elema) thereafter. A three lumen catheter (Abbott Critical Care Systems) and a Swan Ganz 7F thermodilution catheter (Spectramed Model SP-5537-H) were inserted into the right and left internal jugular veins for fluid replacement and hemodynamic monitoring. Two canula (Cook 20 French) were inserted into the right and left femoral veins for administration of extracorporeal circulation. A catheter (ID 1 mm) was placed through the tracheostomy tube, advanced to the level of the carina and connected to a source of O$_2$ (1-2 l.min^{-1}) for the LFPPV-ECCO$_2$R model. Femoral artery was cannulated for blood sampling and invasive blood pressure monitoring (P23XL Spectramed-Statham). Cardiac output (CO) and hemodynamic parameters were calculated with a Horizon 2000 (Mennen Medical). Blood gas analyses were done by ABL 300 (Radiometer, Copenhagen). Centrifugal pump (Biomedicus) and membrane lungs (Scimed) were used for ECCO$_2$R. The membrane lungs were each ventilated with 10 l of humidified O$_2$. The following modes were used in the study:

Control mode (CM) 1: VCV with PEEP of 2 cmH$_2$O, Vt of 10 ml.kg^{-1}, frequency (f) of 12 min^{-1}, I:E ratio of 1:2 and FiO$_2$ of 1.

Lung lavage was performed with 2 l of warm saline solution to induce ARDS. PaO$_2$ below 100 mmHg was accepted as ARDS.

CM 2 : After lung lavage. Same as CM 1.
M 1   : VCV with measured best PEEP, Vt of 10 ml.kg^{-1}, f 12 min^{-1} and I:E ratio of 1:2.
M 2   : PC-IRV with PEEP (5 cm H$_2$O), f 12 min^{-1} and I:E ratio of 2:1.

M 3 : LFPPV with PEEP (15-25 cm H_2O), Vt ofMw ml.kg^{-1} and f 5 min^{-1}; ECCO$_2$R with a pump speed of 20% of CO and membrane lungs with a surface area of 9 m^2

FiO$_2$ was 1 in all modes. The pigs were treated for 7 hours with each mode and arterial PO$_2$ measurement was made at 1, 3, 5 and 7 h. The rest of the parameters were measured and recorded at the 7th h. ARDS was reconfirmed when switching from one mode to the other by CM 2 and lung lavage was repeated when PaO$_2$ > 100 mmHg.

Data were compared among and between the groups by the Mann-Whitney-U test and presented as mean ± SD. Significance was considered at p≤0.05.

RESULTS

The PaO$_2$ value of CM 1 was 471.2±66.9 mmHg. This was significantly reduced to 76.6±14.1 mmHg with CM 2 after the lavage (p < 0.001). The PaO$_2$ values obtained with the trial modes are shown in Figure 1. Highest PaO$_2$ was obtained by M 3 throughout the trial. M 1 and M 2 did not prove to be different in providing oxygenation. Signi-ficantly higher PaO$_2$ values were achieved with each mode when compared to CM 2 (p< 0.001).

The measured best PEEP for M 1 was 14.27±4.71 mm Hg and PEEP applied in M 3 was 18.46±5.38. Data concerning other respiratory and hemodynamic parameters are summarized at Table 1.

DISCUSSION

LFPPV-ECCO$_2$R introduced by Gattinoni and colleagues[5] is based on the rationale to reduce ventilation of the healthy zone of the lungs and add an artificial lung to the patient to eliminate CO$_2$. In the meantime O$_2$ is delivered to the lungs by apneic diffusion. This gas exchange can only be achieved by a mode of ventilation which keeps the lung open during the whole breathing cycle.

The present study shows that LFPPV-ECCO$_2$R mode provided the best oxygenation and CO$_2$ elimination. Moreover, ventilation being limited to a small tidal

Table 1 Parameters measured during the administration of trial modes

	CM1	CM2	M1	M2	M3
$PaCO_2$ (mmHg)	40.7 ± 8.2[a]	55.5 ± 10.8[b]	42.7 ± 6.2	39.2 ± 6.5	28.6 ± 3.6[c]
PIP (cmH$_2$O)	22.6 ± 5.4[d]	29.4 ± 5.5[e]	34.2 ± 5.3	32.9 ± 4.9	27.3 ± 5.2[c]
MAP (cmH$_2$O)	6.7 ± 1.6[f]	9.6 ± 2.9[g]	15.6 ± 3.9[h]	19.9 ± 5.0	21.2 ± 6.0
CO (L/min)	5.21 ± 1.24	4.98 ± 2.23	4.88 ± 1.16	4.57 ± 1.67	4.61 ± 1.95
MPAP (mmHg)	23.9 ± 5.9[i]	29.7 ± 4.3	30.1 ± 3.7	31.7 ± 4.9	33.6 ± 6.3
Q_s/Q_t (%)	18.1 ± 12.3	45.4 ± 16.2[j]	27.9 ± 12.1	24.0 ± 10.7	18.8 ± 12.9

PIP = Peak inspiratory pressure; MAP = mean airway pressure; CO = cardiac output; MPAP = mean pulmonary artery pressure; Q_s/Q_t = alveolar shunt fraction.

a = significantly different from CM2 and M3 ($p < 0.001$); b = significantly different from M1, M2 and M3 ($p < 0.001$); c = significantly different from M1 and M2 ($p < 0.05$); d = significantly different from CM2, M1, M2 and M3 ($p < 0.001$); e = significantly different from M1 and M2 ($p < 0.05$); f = significantly different from CM2, M1, M2 and M3 ($p < 0.001$); g = significantly different from M1, M2 and M3 ($p < 0.001$); h = significantly different from M2 and M3 ($p < 0.01$); i = significantly different from M3 ($p < 0.05$); j = significantly different from CM1, M1, M2 and M3 ($p < 0.05$).

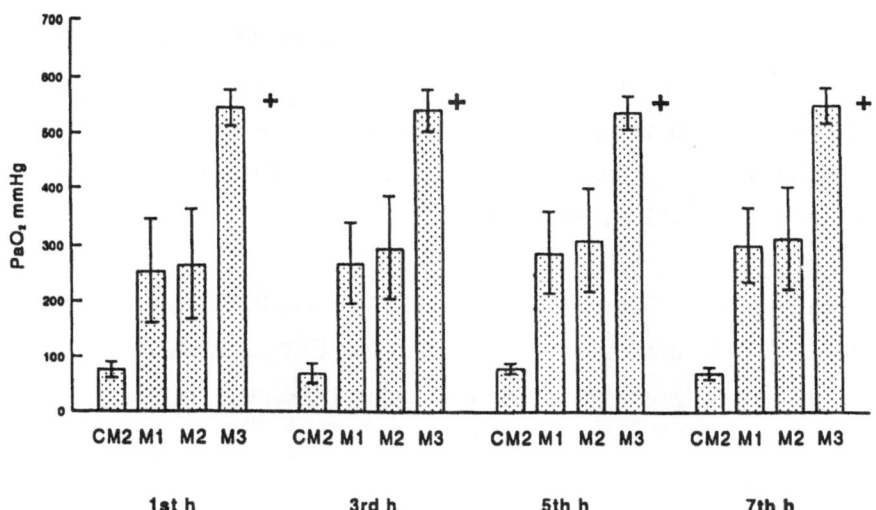

Figure 1. PaO$_2$ data of the trial modes. + = significantly different from M 1 and M 2 (p < 0.01); no significant change of PaO$_2$ was found within each mode during the trial.

volume and frequency, PIP obtained were much lower than the other modes. This remarkably high PaO_2 can be attributed to the two-fold effect of LFPPV-$ECCO_2$R, where oxygenation is achieved by apneic diffusion of the evenly distributed gases in the well recruited alveoli and by ventilation of the membrane lungs with O_2.

On the other hand, in spite of prolonged inspirium, PC-IRV mode with I:E ratio of 2:1 failed to provide significant increases in PaO_2 under these experimental conditions suggesting the necessity of further increasing the inspiratory time in severe respiratory failure to reach a higher end expiratory pressure which would prevent alveolar collapse during the expiratory phase.

Our findings were similar with VCV. Although best PEEP was measured and patients thus ventilated, low MAP was inadequate for homogeneous opening of the alveoli. This was clearly seen by lower PaO_2 obtained. One important difference of this mode from PC-IRV was the lack of possibility of increasing MAP without damaging the alveoli due to increased levels of PIP.

MAP was higher in the PC-IRV and LFPPV-$ECCO_2$R models obviously due to the prolonged inspirium time in the former and high PEEP levels applied in the latter. But, their effect on CO were not considerable in this ARDS model of otherwise healthy animals.

In conclusion, PC-IRV provided adequate oxygenation with safe airway pressures avoiding potential danger of barotrauma, higher I:E ratios being needed for adequate recruitment. LFPPV-$ECCO_2$R seemed to be an ideal mode in providing a high PaO_2 and alveolar opening within safe limits of airway pressure changes. VCV with measured best PEEP failed to provide ideal ventilatory parameters.

REFERENCES

1. K. G. Hickling, Ventilatory management of ARDS: can it affect the outcome? Intensive Care Med. 16:219-226 (1990).
2. B. Lachmann, B. Haendly, H. Schulz, and B. Jonson, Improved arterial oxygenation, CO_2 elimination, compliance and decreased barotrauma following changes of volume generated PEEP ventilation with inspiratory/expiratory (I/E) ratio of 1/2

to pressure generated ventilation with I/E ratio of 4:1 in patients with severe adult respiratory distress syndrome (ARDS), Intensive Care Med. 6:64 (1980).

3. B. Lachmann, E. Danzmann, B. Haendly, and B. Jonson, Ventilator settings and gas exchange in respiratory distress syndrome, in: O. Prakash (ed). Applied physiology in clinical respiratory care. Martinus Nijhoff Publishers, The Hauge (1982).

4. L. Gattinoni, A. Pesenti, M. L. Caspani, A. Pelizzola, D. Mascheroni, R. Marcolin, G. Iapichino, M. Langer, A. Agostoni, T. Kolobow, D. G. Melrose, and G. Damia, The role of total static lung compliance in the management of severe ARDS unresponsive to conventional treatment, Intensive Care Med. 10:121-126 (1984).

5. L. Gattinoni, T. Kolobow, T. Tomlinson, G. Iapichino, M. Samaja, D. White, and J. Pierce, Low-frequency positive pressure ventilation with extracorporeal carbon dioxide removal (LFPPV-ECCO$_2$R): an experimental study, Anesth Analg. 57:470-477 (1978).

EVALUATION OF OXYGENATION WITH DIFFERENT MODES OF VENTILATION

IN PATIENTS WITH ADULT RESPIRATORY DISTRESS SYNDROME

J. Kesecioglu[1,2], L. Telci[2], F. Esen[2], T. Denkel[2], K. Akpir[2], A. S. Tütüncü[1,2], W. Erdmann[1], B. Lachmann[1]

[1]Department of Anesthesiology, Erasmus University and Academic Hospital Dijkzigt, Rotterdam, The Netherlands.
[2]Department of Anesthesiology, University of Istanbul, Faculty of Medicine, Istanbul, Turkey

INTRODUCTION

Positive end-expiratory pressure (PEEP) was suggested to be the therapy of the adult respiratory distress syndrome (ARDS) by Asbaugh and colleagues[1]. However recent articles clearly indicate that any type of ventilatory management is merely a supportive measure to provide adequate gas exchange and has little effect in treatment of the underlying pathology[2-4]. In spite of this knowledge, new mechanical ventilatory approaches have been continuously introduced in the last two decades, aiming to provide adequate oxygenation and to avoid damage to the lungs which could be caused by the ventilation mode itself (for review see B. Lachmann et al.[5]).

There is sufficient evidence in the literature to suggest the adverse effects of high peak inspiratory pressures (PIP) and tidal volumes (Vt) which can cause barotrauma or progressive lung injury[6-8]. Hickling and coworkers[9] have recently shown decreased mortality in ARDS by avoidance of large Vt and limiting PIP. Lee and colleagues[10] have also demonstrated the safe use of low tidal volumes in a selected population of patients. However, this type of ventilation with permissive hypercapnia, can prove harmful to some patients with, for example, high intracranial pressure, ischemic heart disease or hypertension.

On the other hand Lachmann and coworkers[5,11] have reported higher oxygenation and lower PIP with pressure controlled inverse ratio ventilation (PC-IRV). Several case reports and studies are published reporting the benefits of PC-IRV in patients with ARDS[12-14]. In this study, volume controlled ventilation (VCV) with PEEP was compared with PC-IRV with three different I:E ratios in patients with ARDS, and their effects on oxygenation, and cardiac output (CO) were investigated.

METHODS

Study group consisted of 22 patients (male: 13; female: 9), with polytrauma, admitted to the intensive care unit with respiratory failure. Patients were included in the study when the following criteria were fulfilled:
1) Being in a diseased state known to cause ARDS,
2) Diffuse infiltrates on the chest x-ray,
3) PaO_2/FiO_2 ratio < 100 at the control mode,
4) Exclusion of diseases as heart failure, atelectasis or chronic lung diseases,
5) Decreased total static lung compliance (TSLC).

All patients had continuous monitoring of arterial blood pressure with P 23 XL (Spectramed Statham) transducer. A Swan Ganz 7F thermodilution catheter was inserted in each patient. CO was measured and hemodynamic parameters were calculated with a Hemopro 1 (Spectramed) computer. Arterial and mixed venous blood gases were measured with ABL 300 (Radiometer, Copenhagen). TSLC and best PEEP were measured with Compli 80 System (Kontron).

All patients were sedated with appropriate doses of benzodiazepines and paralyzed for the duration of the trial with nondepolarizing muscle relaxants (pancuronium bromide or vecuronium bromide). Narcotics were used to provide analgesia.

The patients were ventilated with a Servo 900C (Siemens Elema). PIP and mean airway pressures (MAP) displayed by the ventilator were recorded. The peak pressures at PC-IRV modes were adjusted to get a physiologic arterial CO_2 tension. The randomly applied modes of ventilation were as follows:
Control mode: VCV with PEEP of 4 cmH$_2$O, Vt of 10-15 ml.kg-1, I:E ratio of 1:2, frequency (f) of 12 min^{-1} and FiO$_2$ of 1.
Mode (M) 1 : VCV with measured best PEEP, Vt of 10-15 ml.kg^{-1}, f of 12 min^{-1} and I:E ratio of 1:2.
M 2   : PC-IRV with PEEP (4 cm H$_2$O), f of 12 min^{-1} and I:E ratio of 2:1.
M 3   : PC-IRV with PEEP (4 cm H$_2$O), f of 12 min^{-1} and I:E ratio of 3:1.
M 4   : PC-IRV with PEEP (4 cm H$_2$O), f of 12 min^{-1} and I:E ratio of 4:1.

Patients were ventilated with the control mode before the beginning of the trial and prior to conversion from one mode to the other for 30 min. Blood gas measurements were performed thereafter and patients showing recovery were excluded from the study (PaO_2 > 100 mm Hg). The rest of the modes were applied for 2 h each, FiO$_2$ being 1 during the last 30 min, prior to the measurements. All patients were given 5-10 μg. kg^{-1}.min^{-1} dopamine during the trial.

Data were compared between the groups by Student's t-test and presented as mean ± SD. Significance was considered at $p \leq 0.05$.

RESULTS

The average age of the study group was 33.15 ± 16.02 years (range 9-63 years) and the average body weight was 68.35 ± 16.80 kg. Average PaO_2/FiO_2 ratio and TSLC measured at the beginning of the study were 73.80 ± 9.61 and 26.78 ± 2.46 ml.cm H$_2$O^{-1}, respectively.

PaO$_2$ values obtained with each mode is shown in Figure 1. Increased I:E ratios resulted in significant increases in PaO$_2$ values. PaCO$_2$ values were M 1: 40.29±9.78; M 2: 37.76±6.89; M 3: 36.04±7.20; and M 4: 38.03±8.59 mmHg. There were no statistically significant differences between the modes concerning these data. The measured best PEEP for VCV mode was 13.23 ± 3.09 cmH$_2$O.

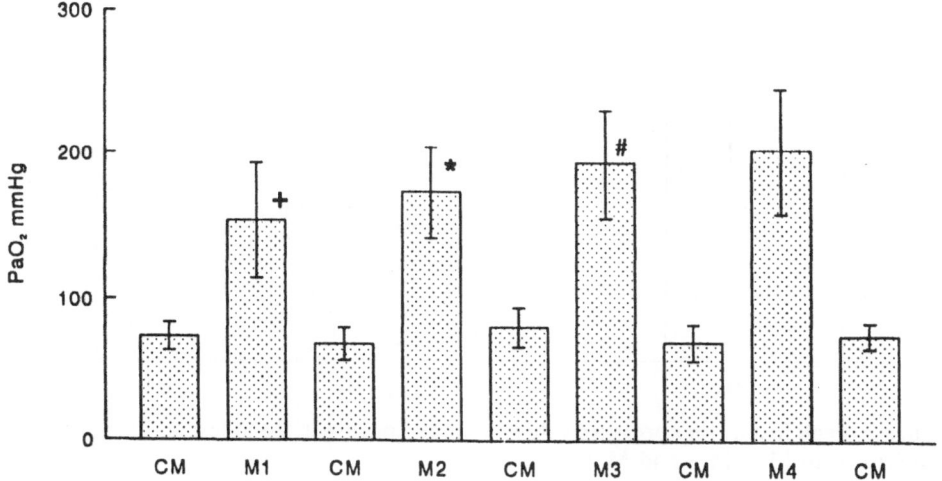

Figure 1. PaO$_2$ data of the control and trial modes. + = significantly different from M 2 ($p < 0.05$), M 3 and M 4 ($p < 0.001$); * = significantly different from M 3 ($p < 0.05$) and M 4 ($p < 0.01$); # = significantly different from M 4 ($p < 0.05$). No significant differences were found between the control groups.

Figure 2 summarizes airway pressure changes that occurred when switching one mode to the other. Remarkable decrease of PIP was observed with PC-IRV modes. An increase of MAP was seen with the inverted I:E ratios.

Data concerning the hemodynamic parameters are summarized in Table 1. A slight decrease in mean arterial pressure was observed with increased I:E ventilation ratios.

DISCUSSION

The aim of ventilatory therapy in ARDS is to provide adequate gas exchange and to obtain recruitment in functioning alveolar units, while avoiding complications as barotrauma, progressive lung injury and depression of hemodynamic functions. These can be achieved by keeping PIP to the minimum acceptable level and providing optimal CO and vascular pressures to enable adequate oxygen transport to the tissue.

The results of this study show that patients in the PC-IRV ventilation modes had higher PaO$_2$. The oxygenation was improving progressively with prolonged inspiratory time. Furthermore, close PaO$_2$ values obtained by CM, applied between modes ensured

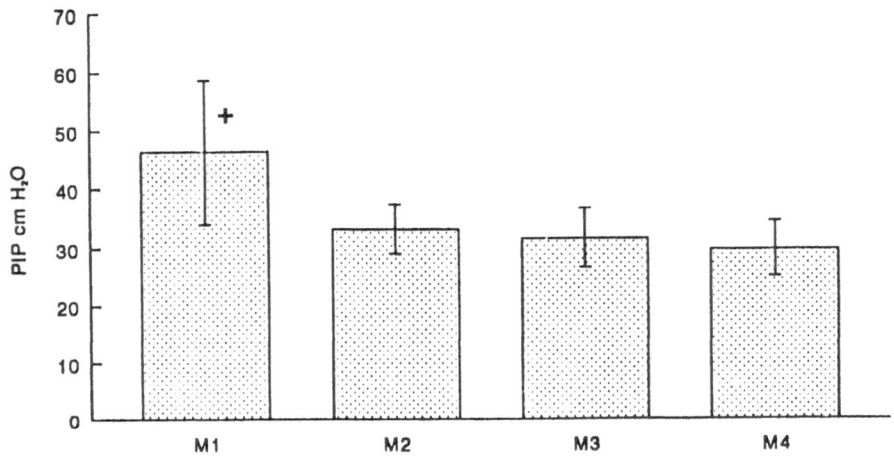

Figure 2. Peak inspiratory pressure values obtained during trial modes: + = significantly different from M 2, M 3 and M 4 (p<0.001).

Table 1. Parameters measured during the administration of trial modes of ventilation

	M1	M2	M3	M4
n	22	22	22	22
CO (L/min)	9.50 ± 2.96	9.27 ± 2.69	11.28 ± 3.30[a]	9.35 ± 2.63
MAP (mmHg)	83.39 ± 22.93	81.61 ± 16.66	82.37 ± 17.26	73.78 ± 12.87[b]

n = number of patients; CO = cardiac output; MAP = mean arterial pressure.
a = significantly different from M1, M4 (p<0.05) and M2 (p<0.01)
b = significantly different from M1, M2 and M3 (p<0.05)

us of the continuation of the diseased state, eliminating any possibility of recovery during the trial period. All modes provided adequate ventilation resulting in sufficient CO_2 removal.

Increased functional residual capacity (FRC) is suggested to be the mechanism by which PEEP exerts its beneficial effects[15]. The same mechanism is suggested by Tharratt[16] and coworkers to cause an increase in PaO_2 values with PC-IRV. In this study, during the application of the PC-IRV modes, while the oxygenation was improved impressively, PIP were remarkably lower. These findings are in accordance with Lachmann[8] and coworkers who have shown in rabbits that VCV with PEEP could produce optimal oxygenation only at dangerous PIP. In the same study, oxygenation was achieved by PC-IRV with much lower PIP. The morphological examination within the same work showed bronchiolar epithelial necrosis and desquamation, when PIP was 40 cmH_2O. Recently, in a review article, Hickling[17] also stressed the importance of high PIP as a cause of barotrauma and pulmonary interstitial emphysema. This, in addition to the animal experiments, explains the importance of low PIP during PC-IRV for the prevention of iatrogenic lung damage.

One special point of interest is that this study was not designed to investigate the long-term effects of various ventilation models nor did it aim to compare mortality or morbidity rates between them. It covers a study period of 10 h at the beginning stage of the disease to determine the immediate advantages of one mode over the other. The results obtained, certainly stress the importance of the use of IRV, but the limited study period prevents further elaboration over progressive lung injury or weaning of the patient from the ventilator.

REFERENCES

1. D. G. Asbaugh, T. L. Petty, D. B. Bigelow, and T. M. Harris, Continuous positive-pressure breathing (CPPB) in adult respiratory distress syndrome, J Thorac Cardiovasc Surg. 57:31-41 (1969).
2. J. A. Weigelt, Current concepts in the management of the adult respiratory distress syndrome, World J Surg. 11:161-166 (1987).
3. D. J. Shale, The adult respiratory distress syndrome- 20 years on, Thorax. 42:641-645 (1987).
4. R. W. Bolin, and D. J. Pierson, Ventilatory management of acute lung injury, Crit Care Clin. 2:585-599 (1986).
5. B. Lachmann, E. Danzmann, B. Haendly, and B. Jonson, Ventilator settings and gas exchange in respiratory distress syndrome, in: O. Prakash (ed). Applied physiology in clinical respiratory care. Martinus Nijhoff Publishers, The Hauge (1982).
6. L. J. Greenfield, P. A. Ebert, and D. W. Benson, Effect of positive pressure ventilation on surface tension properties of lung extracts, Anesthesiology. 25:312-316 (1964).
7. E. O. R. Reynolds, and A. Taghizadeh, Improved prognosis of infants mechanically ventilated for hyaline membrane disease, Arch Dis Child. 49:505-515 (1974).
8. B. Lachmann, B. Jonson, M. Lindroth, and B. Robertson, Modes of artificial ventilation in severe respiratory distress syndrome. Lung function and

morphology in rabbits after washout of alveolar surfactant, <u>Crit Care Med.</u> 10:724-732 (1984).

9. K. G. Hickling, S. J. Henderson, and R. Jackson, Low mortality associated with pressure limited ventilation with permissive hypercapnia in severe adult respiratory distress syndrome, <u>Intensive Care Med.</u> 16:372-377 (1990).

10. P. C. Lee, C. M. Helsmoortel, S. M. Cohn, and M. P. Fink, Are low tidal volumes safe? <u>Chest.</u> 97:425-429 (1990).

11. B. Lachmann, B. Haendly, H. Schultz, and B. Jonson, Improved oxygenation, CO_2 elimination, compliance and decreased barotrauma following changes of volume-generated PEEP ventilation with inspiratory/expiratory (I/E) ratio of 1:2 to pressure-generated ventilation with I/E ratio of 4:1 in patients with severe adult respiratory distress syndrome (ARDS), <u>Intensive Care Med.</u> 6:64 (1980).

12. S. J. Boros, S. V. Matalon, R. Ewald, A. S. Leonard, and C. E. Hunt, The effect of independent variations in inspiratory-expiratory ratio and end-expiratory pressure during mechanical ventilation in hyaline membrane disease: the significance of mean airway pressure, <u>J Pediatr.</u> 91:794-798 (1977).

13. A. G. Cole, S. F. Weller, and M. K. Sykes, Inverse ratio ventilation compared with PEEP in adult respiratory failure, <u>Intensive Care Med.</u> 10:227-232 (1984).

14. L. Gattinoni, A. Pesenti, M. L. Caspani, A. Pelizzola, D. Mascheroni, R. Marcolin, G. Iapichino, M. Langer, A. Agostoni, T. Kolobow, and G. Damia, Low frequency positive-pressure ventilation with extracorporeal CO_2 removal in severe acute respiratory failure, <u>JAMA.</u> 256:881-886 (1986).

15. R. R. Kirby, J. B. Downs, J. M. Civetta, J. H. Modell, F. J. Dannemiller, E. F. Klein et al, High level positive end-expiratory pressure (PEEP) in acute respiratory insufficiency, <u>Chest.</u> 67:156-163 (1975).

16. R. S. Tharratt, R. B. Allen, and T. E. Albertson, Pressure controlled inverse ratio ventilation in severe adult respiratory failure, <u>Chest.</u> 94:755-762 (1988).

17. K. G. Hickling, Ventilatory management of ARDS: can it affect the outcome? <u>Intensive Care Med.</u> 16:219-226 (1990).

GROUP PHOTOGRAPH

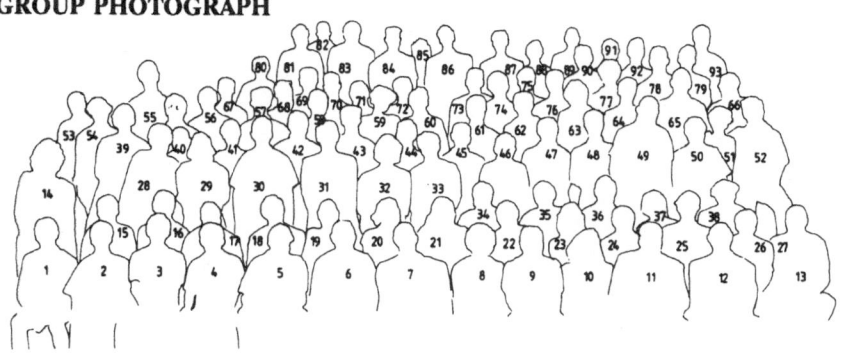

1. DUNPHY E.	33. MATEESCU, G.D.	64. SERBINOVA, E.
2. HORI, H.	34. LIEM, K.D.	65. ZEE, van der H.T.
3. KANAIZUMI, H.	35. INCE, C.	66. HÖPER, J.
4. NAKAGAWA, Y.	36. RAKUSAN, K.	67. BE BLASI, R.A.
5. ZÜNDORF, J.	37. LADANYI, E.	68. DAÜTERMANN, C.
6. FRANK, F.	38. PIIPER, J.	69. GOETZ, A.
7. KESSLER, M.	39. ELWELL, C.E.	70. BERGER, S.
8. CAIN, H.A.	40. VALERIANO, E.	71. WILSON, D.F.
9. CAIN, S.M.	41. ESEN, F.	72. VOTER, W.A.
10. HAAS, D.	42. LACHMANN, B.	73. WAUER, H.
11. KIBBELAAR, D.	43. MINCHINTON, A.I.	74. GROEBE, K.F.
12. ERDMANN, W.	44. TERRIS, D.J.	75. SCHOEMEIJER, P.
14. DE MAN, V.	45. HERIQUEZ, R.	76. TYLER, H.
15. SCHEUREGGER, W.	46. KANHAI, I.	77. VAN ROSSEM, K.
16. OESEBURG, B.	47. GROGONO, A.W.	78. CRINGLE, S.
17. HAGENOUW, R.	48. BUSCH, N.A.	79. MENGER, M.
18. TROUWBORST, A.	49. JÖBSIS van der VILIET, F.	80. HOOFD, L.J.
19. LUTZ, J.	50. LAMANNA, J.C.	81. ROG, H.
20. LONGMUIR, S.	51. GOLDSTICK, M.	82. HART, N.
21. LONGMUIR, I.S.	52. GOLDSTICK, T.K.	83. PINEDO, F.
22. MERTZLUFFT, F.	53. RIESS, J.	84. BIERVLIET, J.
23. KAYAR, S.R.	54. THIMM, F.	85. LEITO, N.
24. VAUPEL, P.W.	55. EIJKING, E.P.	86. PEPER, J.A.K.
25. FAITHFULL, S.N.	56. RANDOLPH, K.	87. KLEIJ, van der A.J.
26. DELPY, D.T.	57. TÜTÜNCÜ, A.	88. BLOEMENDAAL, S.G.
27. NAUMANN, C.P.	58. KEIPERT, P.E.	89. ANDERSON, P.J.
28. WERLEMAN, G.	59. SHINA, J.	90. HAYES, T.J.
29. VISSER, L.	60. BRULEY, D.F.	91. EMERY, M.J.
30. HUDETZ, A.G.	61. RUMSEY, W.	92. TAUSCHEK. D.
31. EKE, A.	62. MAGUIRE, D.	93. HLASTALA, M.P.
32. GÜNDEROTH, M.	63. ZANDER, R.	94. NEMOTO, E.M.

907

AUTHOR INDEX

Afuwape, S.A. 183
Ahuja, R.C. 343
Akpir, K. 371, 401, 599, 855, 893, 901
Albrecht, M. 885
Alder, V. 787
Altan, R.T. 653
Ashruf, J.F. 267

Babel, W. 681
Baethmann, A. 723, 731
Bakker, D.J. 95, 115, 121, 125
Barnikol, W.K.R. 473
Batra, S. 553, 567
Berger, S. 723
Beukenholt, R.W. 349
Biervliet, J.D. 115, 121
Blums, R. 649
Boekstegers, P. 573
Bos, A.P. 363
Bos, J.A.H. 363, 409
Bos, K.E. 779
Bossen, W. 639
Bouman, C.G.O.T. 835
Bourgain, R. 717
Boyer, S. 221
Bradley, W.E. 479, 751
Brandi, L.S. 813
Broek, H.G.M. van der 331
Brown, D. 221
Brown, J.M. 153, 177
Bruch, H.P. 879
Bruining H.A. 267, 277
Bruley, D.F. 3, 183
Buijzer, E. de 349
Busch, N.A. 629
Busse, F.W. 879

Cabrol, C. 491

Cain, S.M. 35, 479, 751
Cannell, G.R. 649
Cerretelli, P. 491
Chance, B. 297, 583
Chang, K. 221
Clarkson, R.B. 221
Colier, W.N.J.M. 305
Collie, J. 649
Cope, M. 235, 771
Coremans, J.M.C.C. 267, 277
Crevels J. 499
Cringle, S. 787
Curtis, S.E. 479, 751
Cuttano, A.M.R. 813

Daal, G.J. van 47
Darby, J. 695
Dautermann, C. 731
DeBakey, M.E. 247
De Blasi, R.A. 771
Decuyper, K. 653, 717
Delpy, D.T. 235
Denkel, T. 371, 599, 855, 893, 901
Dijk, G. van 325
Dorp van Vliet, A. van 509
Duncker, D.J.G.M. 545
Dunphy, E.P. 153

Edwards, A.D. 235
Eijking, E.P. 349
Eke, A. 671
Elwell, C.E. 235
Emery, M.J. 357
Engel, A. 515
Enzenbach, R. 723
Erdmann, W. 7, 315, 325, 331, 397, 401, 409, 527, 607, 807, 835, 847, 863, 869, 901

Esen, F. 371, 599, 855, 893, 901
Evans, R.C. 825
Evers, J.A.M. 305

Faithfull, N.S. 55, 397, 401, 409, 441, 497, 527
Feifel, G. 765, 799
Fennema, M. 527, 607, 869
Ferrari, M. 771
Fleckenstein, W. 869
Fleckenstein-Grün, G. 793
Frank, K. 203, 261, 485, 583
Frey, M. 793

Gaab, M.R. 737
Gast, P. 221
Giunta, F. 813
Glockner, J.F. 221, 229
Go, A.T.J.I. 363
Gomi, S. 689
Gommers, D. 47, 349
Greenberg, J.H. 689
Groebe, K. 21
Groenland, T.H.N. 835
Grogono, A.W. 315
Gronlund, J. 745
Grundmann, A. 213
Günderoth, M. 377
Gustafsson, U. 759

Härtl, R. 731
Henny, Ch.P. 125
Hetterich, N. 681
Hiesmayr, M. 503
Hlastala, M.P. 357, 745
Hoeckel, M. 139
Hofer, S.O.P. 779
Hofgärtner, W. 793
Hogan, M.C. 497
Hohlbach, G. 885
Hoofd, L. 561, 567
Hopman, J.C.W. 841
Höper, J. 203, 485, 593, 737
Hori, H. 255
Hu, H. 221
Hudetz, A.G. 659

Ince, C. 267, 277
Ito, K. 583
Iturriaga, R. 387

Jäger, S. 799

Jansen, H.M. 115, 121
Jelkmann, W. 515
Jongsma, H.W. 499

Kang, K.A. 183
Keipert, P.E. 453
Kesecioglu, J. 371, 599, 855, 893, 901
Kessler, M. 203, 261, 485, 583, 593, 737
Kesteren, R. van 607
Kleij, A.J. van der 115, 121, 125, 607, 779
Klein, J. 807
Kochanek, P.M. 701
Kollée, L.A.A. 841
Krafft, M.P. 465
Kroon, J.W. 331
Krüger, U. 885
Kullander, J. 349

Lachmann, B. 47, 349, 363, 371, 397, 401, 409, 599, 893, 901
Ladanyi, E. 343
Lahiri, S. 387
LaManna, J.C. 107
Lang, N. 485
Laureys, M. 717
Lee, J. 169
Lewis, D.H. 759
Liem, K.D. 841
Liu, K.J. 221
Lubbers, M. 125
Lübbers, D.W. 213
Lutz, J. 131

Maguire, D.J. 649
Malvin, G.M. 745
Mannheimer, P. 499
Maris, M. 297
Mauch, E. 485
Mayevsky, A. 707
Mazzanti, T. 813
Melick, J.A. 701
Menger, M.D. 765, 799
Menssen, J. 499
Mertzlufft, F. 413, 421
Messmer, K. 731, 765
Meyer, M. 491
Middaugh, M.E. 357
Minchinton, A.I. 153, 177
Möbius, D. 343
Mokashi, A. 387

Morgan, R. 649
Moussavi, M. 221
Mulder, P. 401
Müller, T. 681
Murr, R. 723

Nakagawa, Y. 255
Naumann, C.P. 847, 869
Nemoto, E.M. 695, 701
Nijhuis, J.G. 499
Niijima, T. 255
Nilges, M. 221
Nioka, S. 297, 583
Noah, E.M. 879
Noon, G.P. 247
Norby, S.W. 221
Nose, Y. 247

Oeseburg, B. 305, 499, 841
Ojima, H. 255
Okunieff, P. 161, 169
Oleggini, M. 813

Pagel, H. 515
Pastuszko, A. 689
Pattenier, J. 799
Pembeci, K. 855
Peper, J.A.K. 115, 121
Perner, R. 879
Peter, K. 723
Peter, W. 573
Pfeifer, A. 573
Pierik, E.G.J.M. 267
Piiper, J. 319, 491, 623
Purucker, E. 131

Quartararo, C. 745

Rakusan, K. 553, 567
Ramsey, H. 521
Reempts, J. van 717
Reynolds, E.O.R. 235
Riess, J.G. 465
Ringnalda, B.E.M. 305, 499
Röhrich, F. 731
Roos, C.M. 115, 121
Rossem, K. van 653, 717
Ruetsch, Y. 847
Ruetsch, Y.A. 869
Rumsey, W.L. 387

Sanderse, E.A. 267

Schaefer, C. 161
Schädlich, M. 863
Schäfer, R. 885
Scheffers, E.C. 363
Schiding, J.K. 701
Schlenger, K. 139
Schramm, U. 639
Schürer, L. 723, 731
Sevick, E. 297
Silveri, F. 363
Sluiter, W. 363
Smirnov, A. 221
Snel, L. 509
Spergel, D. 387
Spitzmüller, K. 793
Stalder, K. 343
Steltzer, H. 503
Steppan, B.G. 645
Su, E.N. 787
Swartz, H.M. 221, 229

Takatani, S. 247
Tauschek, D. 203, 485, 583, 737
Telci, L. 371, 599, 855, 893, 901
Tenbrinck, R. 363, 431, 509
Terada, H. 255
Terris, D.J. 153
Thews, G. 21
Thimm, F. 793
Thorborg, P. 759
Tibboel, D. 363
Tran, T. 357
Trouwborst, A. 431, 509, 545, 807, 835
Tüchy, G. 503
Tulli, G. 813
Turek, Z. 561, 567
Tütüncü, A.S. 371, 391, 401, 409, 599,
  893, 901
Tyler, H. 521

Uhl, M.W. 701
Vahidi, N. 221
Vaupel, P. 139, 161, 169
Veprek, P. 553
Verdouw, P.D. 545
Verkaaik, A.P.K. 315, 325, 331, 847,
  863
Vermariën, H. 653, 717

Wagner, P.D. 75, 497
Wagner, K.F. 639, 645, 879, 885
Walczak, T. 221

Wang, N.G. 297
Wauer, H. 863
Werdan, K. 573
Westerkamp, B. 325
Wiesner, J. 261
Willford, D. 497
Wilson, D.F. 195, 387, 689
Woerkens, E.C.S.M. van 431, 509, 545
Wokaun, A. 261
Wolf, B. 799
Wu, M. 221
Wyatt, J.S. 235

Yao, L. 695
Yarmush, M.L. 629
Yonas, H. 695
Yu, D.Y. 787

Zäch, G.A. 847, 869
Zander, R. 413, 421
Zee, H. van der 825
Zimpfer, M. 503
Zündorf, J. 203, 485, 583

SUBJECT INDEX

Adult respiratory
    distress syndrome (ARDS), 35, 47
  effects of ventilation 401-406
    during weaning 371-376
  endotoxin induced, 36, 37, 42
  broncholavage, 47, 350
  hydrochloric acid induced 349-354
  oxygenation by ventilation 901-906
  polymorphonuclear neutrophiles, 47
  surfactant treatment 349-354
    perfluorocarbon
        instillation 401-406, 409-412
Aldosterone (chronic hypoxia), 521-524
Alveolar proteinosis, 118, 121
Alveolar-arterial PO_2 difference, 319
  high altitude, 76, 77
  during exercise, 78
Alzheimer's disease
  NIRS scattering, 303
Aplysia californensis, 616
Arachidonic avoid
    pulmonary toxicity, 807
Artificial ventilation, 325-330
  versus ECMO, 893-899
ATP
  tumor, 162-166
Ayre system, 332

Benzotriazine N-oxide, 177
  effects of hypoxia, 179
Bioengineering, 3
  bioprocess engineering, 3, 5, 6
Blood
  high altitude, 87
Blood flow
  cerebral, 671, 695-699, 701, 707
  Langerhans islets, 799

neutrophil dependency (brain), 701-705
  retina, 789
  tumor, 140, 141
Brain
  blood flow, 235-243, 695-699, 701, 707, 723
    mapping, 671-679
  intracapillary HbO_2, 737-741
  ischemia
    artery occlusion, 689-693, 717
    reperfusion, 690
  oxygenation, 841-846
  oxygen tension, 611, 612
    (partial pressure) 691, 692, 723, 731
    local tissue PO_2, 617-722
  oxygen transport to, 681-687
  rat, 611
Brody equation, 336

Calcium antagonists, 793
Calf
  aldosterone and hypoxia, 522
Capillaries
  cerebral network,
    erythrocyte transit, 659-669
    pressure flow distribution, 664
  red blood cell velocity, 765
  fetal, HbO_2, 485-489
  HbO_2, 595, 596, 737-741
  oxygen exchange, 30, 31
    in hypoxic hypoxia, 759
  skeletal muscle, 759
  subendocardial, 553-558
Carbon dioxide retention
    elimination in ARDS, 372
  apneic, 416
  end tidal, 336, 337
  high altitude, 79, 80

Cardiac transplantation
  O$_2$ transport during exercise, 491-495
Cardiac output
  hemodilution, 432, 433
  after transplantation, 492
Carotid body
  oxygen chemoreception, 387-394
  phosphorescence quenching
        (oxygen), 388
Cat
  brain, 689-693
Catalase, 364
Computer simulation
  oxygenation during
        hyperthermia, 183-191
Cyanine dyes, 255-260
Cytochrome P-450
  oxygenation, 281, 307, 308

Delpy coefficient, 307
Diaphragmatic hernia, congenital
  artificial ventilation, 363-369
  antioxydant enzyme activity, 363-369
Dichloroacetate (DCA), 41, 42
Diffusion
  limitation, 319-322
Diffusion coefficient
  perfusion equilibrium, 319
  inhomogeneity, 320
Dipalmitoyl phosphatidylcholine
        (DPPC), 345
Dog
  hemodilution with PFC's, 479-483
Dopamine, 338, 871, 876

Electrode
  polarographic, 584, 779
ECMO, 841
  comparison to ventilation, 893-896
Electron paramagnetic resonance
        (EPR), 221-228
EMPHO, 262, 584, 687
Endotoxin
  O$_2$ delivery/uptake ratio, 751-757
Enzymes, respiratory, 583-590
Erythrocyte
  O$_2$, 30, 31
  saturation, 309
  transit computer simulation, 659-669
  velocity, 765-768
Erythropoetin, 515-518

Fetus,
  oxygenation
    maternal hypoxia, 499
  scalp
    intracapillary HbO$_2$, 485-489
Fick's law, 22, 23
  equation, 331
Fluorcarbon, see Perfluorocarbon
Fluorescence, microscope, 766
  Video fluorimetry, 271, 283
Fluori-Phosphorimetry, 267-274
Fluosol, 59, 444
  lung glutathione, 131-135
Fusinite, 229

Gerbil
  brain blood flow, 707-715
Glucose, metabolism, 695
Glutathione peroxidase
  lung, 131-135, 364

Haber-Weiss reaction, 15
Heart
  arterial disease, 531-533
    hemodilation, 536, 545-551
  blood flow, 528-531
  capillary network, 567-572
  contractility, 576
  exercise, 81, 491-495
  high altitude, 81-87
  function,
    effects of ventilation, 600
    extracorporal CO$_2$ removal, 602
  hemodilution, 533
  ischemia, infarction, 580
    hemodilution-PFC's, 537-540
  myoglobin oxygenation, 583-590
  neonatal reperfused, 573-581
    myocardial stunning, 573
  oxygenation, 545-551
    myocardial, 534, 575
  oxygen deficiency, 583-590
    after transplantation, 493, 494
    hemodilution, 546
  oxygen supply
    critical hemodilution, PFC's,
        527-541
  oxygen tension
    histogram, 561-566, 567-572
    by EPR, 224
  pd-porphyrin phosphorescence, 273
  perfused, 573-581

redox state of resp. enzymes, 583-590
subendocardial capillarization,
  553-559
Hemodilution
  oxygen dissociation curve, 436
  oxygen extraction, 434
  oxygen supply heart, 527-541
  oxygen transport, 431-437
    systemic hemodynamics, 432-436
    perfluorocarbon replacement, 446
Hemoglobin
  concentration, 309
    high altitude, 88
  deoxygenated, 300
  high altitude, 87
  human, 306
  oxygenation saturation tumors, 143
  polymerization, 58, 455
  rat, fetal, 499
  saturation-ARDS, 374
    fetal, 500
  solution
    chemically cross linked, 453-462
    hemodynamic effects, 460
    medical application, 58
  stroma-free, 306
    O_2 dissociation curve, 56
    saturation, 310
Hill's equation, 26
HT_2 receptor antagonist, 759
Hüfner's number, 474
Hyaluronidase, 619
Hyperoxia
  hyperbaric, 95
  medical application, 97-103
  lung lavage, 115-120
  lung glutathione, 131-135
Hyperthermia oxygenation, 183-191
Hypoxia
  adaption, 107-113
  heart function, 82
  high altitude, 75
  hypoxic hypoxia
    O_2 supply/demand ratio, 753
  NADH-fluorescence, 271, 273
  pd-porphyrin phosphorescence,
    272, 273
  regional blood flow, 110

Ischemia
  cerebral, oxygen diffusion, 619
  reperfusion, 620, 765

hemodilution-PFC's, 537-540
salvage from, 447
skeletal muscle, 771

Kidney
  hemoglobin solution, 459
  oxygenation, 227
Krogh cylinder, 624

Lactate
  L/P ratio, 41
Lamb
  hypoxia fetus, 501
Langerhans islets, 799
Latex beads, 263
Light absorbance
  scattering, 112
Liver transplantation
  oxygen delivery, 503-508
  oxygen consumption, 835-840
Lung
  antioxydant enzyme, 363-369
  compliance on-line, 848
  gas exchange, 325-330, 357, 397-400,
    401-407
  glutathione, 131-135
  lavage, 115, 403, 404, 411
    hyperbaric, 115
    tissue oxygenation, 121-123
  instillation of PFC's, 397-400,
    401-407, 409-412
  oxygen toxicity, 807-810
  surfactant
    model, 343-347, 349
    surface layer, 343
  ventilation-perfusion
    inequality (heterogeneity), 80, 361
    ratio, 358
Metabolism
  tumor, 162-166
Microcirculation
  Langerhans islets, 799
Mie's theory, 264
Mitochondria
  NADH fluorimetry, 267-274
  spectro(photo)metry, 262-265
Mitrochondrial oxygen
  redox state (brain), 707-715
Monkeys' brain, 695-699
Mucopolysaccharides (tissue), 620
Muscle
  blood flow

inert gas washout, 745
high altitude (review) 89, 90
oxygen extraction-consumption
  effect of oxygent, 497
tissue oxygenation, 779-784, 869-877,
  879-884
  effects of ventilation, 377-385
    spatial distribution, 593-599
    after leg surgery, 858-892
  high altitude, 89
  hyperbaric oxygenation, 125-128
  hypoxia and diffusion limitation,
    745-750
Myocardium, see heart
Myoglobin
  deoxygenated, 300

NADH
  fluorimetry, 267-274, 277-291
NADH/NAD redox state, 282
Nasoral system, 421-426
NIR spectroscopy, 235-243, 771
  during ECMO, 841-846
Nitrendipine, 793

Oncotic pressure
  hemoglobin solutions, 57
Oxidative phosphorylation, 8, 9
Oximetry, mixed venous, 813-822
Oxygenation
  apneic, 413-418, 421, 423
    brain (occlusion), 689-693
    skeletal muscle, 761
  ARDS ventilation, 901-906
  hemodilution, 433, 434
Oxygen bioavailibility, 473-478
Oxygen consumption, 35
  ARDS, 372
  human skeletal muscle, 771
  hemodilution, 433
  liver transplantation, 504, 506, 835
  on-line measurement, 331-341, 621,
    847, 863
Oxygen delivery, 35, 813-822, 825-832
  effect of dopexamine, 42
  heart, 591
  high altitude, 87
  hyperbaric, 99
  liver transplantation utilization,
    503-507
  perfluorocarbons, 481
  sepsis, 856

Oxygen diffusion, 9, 21, 618
  barriers, 13
  capacity, 29
  conductivity, 28, 29, 30
  convection, 9, 629-634
  diffuse transport, 24, 30
  facilitated, 28
  flux, 21
  profiles, 629-634
Oxygen dissociation curve, 25, 319
  hemodilution, 436
  stroma-free hemoglobin, 56, 57, 457
Oxygen fraction
  alveolar, 114
Oxygen electrodes
  polarographic, 214, 609, 653
  tissue alterations, 639-644
Oxygen expenditure, 818
Oxygen extraction, 35, 322, 512, 819,
  821
  defect, 36
  critical, 33
  hemodilution, 433
  PFC's, 481
Oxygen free radicals, 13, 14, 15
  protective factors, 16
Oxygen genesis, 7
Oxygen hemoglobin, 25
  equilibrium curve
    deficit, 319
    myoglobin, 27
    unmodified, 56, 57
Oxygen metabolism (brain), 695-699
Oxygen molecular structure, 8
Oxygen partial pressure
  critical, 512
  cortical surface PO_2, 723, 731
  focal brain infarction, 717-722,
    723-729
  muscle, 799
  retina, 787
Oxygen photosynthesis, 7
Oxygen profile (histogram), 29,
  380-382, 611, 869
  carotid body, 329
    heart, 561-566, 567-572
    PC program, 645-648
  muscle human, 869-877
    predictor healing, 879-884
    after leg surgery, 885
Oxygen release
  respiratory potential, 473-478

Oxygen supply dependency, 35
  pathological, 36
    high altitude, 75
Oxygen supply, 315-317
  autoregulation, 613, 616
  by perfusion and diffusion, 623-627
  cell, factors determining, 507-522
  demand ratio, 36, 517, 753, 821
  heart hemodilution-PFC's, 527-541
  heart, spatial distribution, 593-598
  informative display, 317
  limitations, 623-626
  measurement, on-line, 316
  septic shock, 813-822
Oxygent, 497
Oxygen toxicity pulmonary, 15, 807, 810
Oxygen transport, 9, 30
  artificial carriers, 55
  capacity, 9
    perfluorocarbons, 465-471
  equation, 23, 24, 25
  exercise, 491-495
  carriers, 10
  in birds, 12
  in fish, 11
  in insects, 10
  hemodilution, 431-437
  human, 12
    high altitude, 76
  lung, 343-347
  oxygen flow cascade, 13, 14
    hyperbaric, 99
  perfluorocarbons, 446
Oxygen transport to tissue (classification), 681-687
Oxygen uptake, 35
  high altitude, 83
  supply dependency, 855
    sepsis, 855-860
Oxyhemoglobin saturation
  fiberoptic catheters, 509-519
Oxypherol, 59

P_{50}, 57
PEEP, 326
Pendelluft, 325
Pd-Porphyrin phosphorescence, 267-274
Perflubron, 60
  facilitation of O_2 transfer, 479-483
Perfluorocarbons, 55, 59
  and surfactants, 465

liquid breathing, 447
lung glutathione, 131-135
  instillation in ARDS, 397-400, 401-406, 409-412
medical application, 61, 62
metabolism, 61
oxygen supply
  heart, 527-541
oxygen transport, 60, 465-471
review, 59-62
second generation (review), 441-448
Perfluorochemical emulsions, 444
  viscosity, 467
Phosphorescence quenching, 195-200, 388
Phthalocyanine lithium, 229, 230
Physioflex, 332
Pig
  coronary occlusion, 535
  hemodilution
    regional hemodynamics + oxygenation, 545-551
  perfusion PO_2 logger, 649-652
Pluronic, 59
Positron technique
  blood flow, 169-175
Pulmonary, see Lung
Pulmonary shunt
  high altitude, 78, 79
Pulmonary hypertension, 79
Pyruvate, 41
  dehydrogenase, 41, 42

Rabbit
  brain PO_2, 723, 731
Rat
  ARDS, 349
  chronic hypoxia, 107
  diaphragmatic hernia
    congenital model, 363-369
    antioxydant enzymes, 363-369
  heart
    capillary network, 567-572
    oxygen deficiency, 583-590
    oxygen histogram, 561-566
  neonatal heart
    reperfused 573-581
  retina oxygenation and blood flow, 787
    arteriolar spasm + ischemia, 793-798
  pulmonary capillary
    oxygen respiratory potential, 475

skeletal muscle oxygenation, 377-385
subendocardial capillarization,
    553-558
ventilatory drive of oxygen, 475-477
Red blood cell, see erythrocyte
Reflectance
spectrophotometry, 247-252
Respiratory quotient
failure
    hydrochloric acid, 349-354
    surfactant treatment, 349
Retina
rat
    oxygenation and blood flow, 787
    ischemic spasm, 793
Rolfe coefficient, 307

Sepsis
oxygen uptake/supply dependency,
    855-861
Septic shock, hyperdynamic
muscle PO_2 histogram, 869-877
predictive parameters, 813-822
Skeletal muscle
capillary flow, 759-763
endotoxin
    DO_2/VO_2 ratio, 753
hypoxic hypoxia, 751-757, 759
ischemia reperfusion, 765-769
Tourniquet ischemia, 771
Spectrometry
angular dependent, 261-265
continuous wave, 261
Spectrophotometry
multi-wavelength, 297-304
reflectance, 671-677
    multiple scattering, 204-207,
        261-265
    NIR, 261-265
    review, 203-212

Spectroscopy (NIR)
frequency-time resolved, 297-304
optical, 388
time resolved, 261-265
Superoxide dismutase, 364
Surfactant
deficiency, 48
inhibitor ratio, 352
lung lavage model, 49, 344, 350
oxygen toxicity, 50
replacement, 49
surface layer, 343
tracheal installation, 51

Transcutaneous PO_2
measurement technique, 213-218
Tumor
blood flow, 139-148, 169-175
oxygenation, 139-148, 161-167
tissue oxygen tension, 144-147,
    153-158

Ventilation
computer controlled, 325-330
gas exchange, 401-406
    versus ECMO, 893-899
    in ARDS, 901-906
Ventilation distribution
perfusion inequality, 80
inhomogeneity, 319

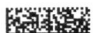